Lecture Notes in Artificial Intelligence   12117

Subseries of Lecture Notes in Computer Science

More information about this series at http://www.springer.com/series/1244

Denis Helic · Gerhard Leitner ·
Martin Stettinger · Alexander Felfernig ·
Zbigniew W. Raś (Eds.)

Foundations
of Intelligent Systems

25th International Symposium, ISMIS 2020
Graz, Austria, September 23–25, 2020
Proceedings

 Springer

Editors
Denis Helic
Graz University of Technology
Graz, Austria

Gerhard Leitner
University of Klagenfurt
Klagenfurt, Austria

Martin Stettinger
Graz University of Technology
Graz, Austria

Alexander Felfernig
Graz University of Technology
Graz, Austria

Zbigniew W. Raś
University of North Carolina at Charlotte
Charlotte, NC, USA

ISSN 0302-9743   ISSN 1611-3349   (electronic)
Lecture Notes in Artificial Intelligence
ISBN 978-3-030-59490-9   ISBN 978-3-030-59491-6   (eBook)
https://doi.org/10.1007/978-3-030-59491-6

LNCS Sublibrary: SL7 – Artificial Intelligence

This Springer imprint is published by the registered company Springer Nature Switzerland AG
The registered company address is: Gewerbestrasse 11, 6330 Cham, Switzerland

Preface

This volume contains the papers selected for presentation at the 25th International Symposium on Methodologies for Intelligent Systems (ISMIS 2020), which was held in Graz, Austria, 2020. The symposium was organized by the Department of Software Technology at the Technical University of Graz, Austria. ISMIS is a conference series that started in 1986. Held twice every three years, it provides an international forum for exchanging scientific, research, and technological achievements in building intelligent systems. In particular, major areas selected for ISMIS 2020 included: Explainable AI (XAI), machine learning, deep learning, data mining, recommender systems, constraint based systems, autonomous systems, applications (configuration, Internet of Things, financial services, e-health, etc.), intelligent user interfaces, user modeling, human computation, socially-aware systems, digital libraries, intelligent agents, information retrieval, natural language processing, knowledge integration and visualization, knowledge representation, soft computing, and web and text mining. This year, the following sessions were organized: special sessions for invited talks, for best paper nominees and best student paper nominees, thematic sessions on natural language processing, deep learning and embeddings, digital signal processing, modeling and reasoning, machine learning applications, and finally a section containing short paper presentations on diverse related topics.

We received 79 submissions that were carefully reviewed by three or more Program Committee members or external reviewers. After a rigorous reviewing process, 35 regular papers, 8 short papers, and 3 invited papers were accepted for presentation at the conference and publication in the ISMIS 2020 proceedings volume. An additional selection of papers was assigned to a specific industrial track. It is truly a pleasure to thank all the people who helped this volume come into being and made ISMIS 2020 a successful and exciting event. In particular, we would like to express our appreciation for the work of the ISMIS 2020 Program Committee members and external reviewers who helped assure the high standard of accepted papers. We would like to thank all authors of ISMIS 2020, without whose high-quality contributions it would not have been possible to organize the conference. We are grateful to all the organizers of and contributors to a successful preparation and implementation of ISMIS 2020.

The invited talks for ISMIS 2020 were: "Complementing Behavioural Modeling with Cognitive Modeling for Better Recommendations," given by Marko Tkalcic, University of Primorska in Koper, Slovenia; "Fairness is not a number: Methodological implications of the politics of fairness-aware systems," given by Robin Burke, University of Colorado Boulder, USA; and "Brevity*," given by Robert West, EPFL Lausanne, Switzerland. We wish to express our thanks to all the invited speakers for accepting our invitation to give plenary talks. We are thankful to the people at Springer (Alfred Hofmann, Anna Kramer, and Aliaksandr Birukou) for supporting the ISMIS

2020. We believe that the proceedings of ISMIS 2020 will become a valuable source of reference for your ongoing and future research activities.

July 2020

Denis Helic
Gerhard Leitner
Martin Stettinger
Alexander Felfernig

Organization

General Chair

Alexander Felfernig   Graz University of Technology, Austria

Program Co-chairs

Denis Helic   Graz University of Technology, Austria
Gerhard Leitner   University of Klagenfurt, Austria
Martin Stettinger   Graz University of Technology, Austria

Steering Committee Chair

Zbigniew Ras   UNC Charlotte, USA

Steering Committee Members

Troels Andreasen   Roskilde University, Denmark
Annalisa Appice   Università degli Studi di Bari, Italy
Jaime Carbonell   CMU, USA
Michelangelo Ceci   Università degli Studi di Bari, Italy
Henning Christiansen   Roskilde University, Denmark
Juan Carlos Cubero   University of Granada, Spain
Floriana Esposito   Università degli Studi di Bari, Italy
Alexander Felfernig   Graz University of Technology, Austria
Mohand-Said Hacid   Université Claude Bernard Lyon 1, France
Nathalie Japkowicz   American University, USA
Marzena Kryszkiewicz   Warsaw University of Technology, Poland
Jiming Liu   Hong Kong Baptist University, Hong Kong
Jerzy Pawel Nowacki   Polish-Japanese Academy of IT, Poland
George A. Papadopoulos   University of Cyprus, Cyprus
Olivier Pivert   Université de Rennes 1, France
Zbigniew Ras   UNC Charlotte, USA, and Polish-Japanese Academy of IT, Poland
Henryk Rybinski   Warsaw University of Technology, Poland
Andrzej Skowron   Polish Academy of Sciences, Poland
Dominik Slezak   University of Warsaw, Poland

Programm Committee

Esra Akbas   Oklahoma State University, USA
Marharyta Aleksandrova   University of Luxembourg, Luxembourg

Aijun An	University of York, UK
Troels Andreasen	Roskilde University, Denmark
Annalisa Appice	Università degli Studi di Bari, Italy
Martin Atzmueller	Tilburg University, The Netherlands
Arunkumar Bagavathi	Oklahoma State University, USA
Ladjel Bellatreche	University of Poitiers, France
Robert Bembenik	Warsaw University of Technology, Poland
Petr Berka	University of Economics, Prague, Czech Republic
Maria Bielikova	Slovak University of Technology in Bratislava, Slovakia
Gloria Bordogna	CNR, Italy
Jose Borges	University of Porto, Portugal
François Bry	Ludwig Maximilian University of Munich, Germany
Jerzy Błaszczyński	Poznań University of Technology, Poland
Michelangelo Ceci	Università degli Studi di Bari, Italy
Jianhua Chen	Louisiana State University, USA
Silvia Chiusano	Politecnico di Torino, Italy
Roberto Corizzo	Università degli Studi di Bari, Italy
Alfredo Cuzzocrea	ICAR-CNR, University of Calabria, Italy
Marcilio De Souto	LIFO, University of Orleans, France
Luigi Di Caro	University of Torino, Italy
Stephan Doerfel	Micromata, Germany
Peter Dolog	Aalborg University, Denmark
Dejing Dou	University of Oregon, USA
Saso Dzeroski	Jožef Stefan Institute, Slovenia
Christoph F. Eick	University of Houston, USA
Tapio Elomaa	Tampere University of Technology, Finland
Andreas Falkner	Siemens AG, Austria
Nicola Fanizzi	Università degli studi di Bari "Aldo Moro", Italy
Stefano Ferilli	Università degli Studi di Bari, Italy
Gerhard Friedrich	University of Klagenfurt, Austria
Naoki Fukuta	Shizuoka University, Japan
Maria Ganzha	Warsaw University of Technology, Poland
Paolo Garza	Politecnico di Torino, Italy
Martin Gebser	University of Klagenfurt, Austria
Bernhard Geiger	Know-Center GmbH, Austria
Michael Granitzer	University of Passau, Germany
Jacek Grekow	Bialystok University of Technology, Poland
Mohand-Said Hacid	Université Claude Bernard Lyon 1, France
Hakim Hacid	Zayed University, UAE
Allel Hadjali	LIAS, ENSMA, France
Mirsad Hadzikadic	UNC Charlotte, USA
Ayman Hajja	College of Charleston, USA
Alois Haselboeck	Siemens AG, Austria
Shoji Hirano	Shimane University, Japan
Jaakko Hollmén	Aalto University, Finland

Andreas Holzinger	Medical University and Graz University of Technology, Austria
Andreas Hotho	University of Wüerzburg, Germany
Lothar Hotz	University of Hamburg, Germany
Dietmar Jannach	University of Klagenfurt, Austria
Adam Jatowt	Kyoto University, Japan
Roman Kern	Know-Center GmbH, Austria
Matthias Klusch	DFKI, Germany
Dragi Kocev	Jožef Stefan Institute, Slovenia
Roxane Koitz	Graz University of Technology, Austria
Bozena Kostek	Gdańsk University of Technology, Poland
Mieczysław Kłopotek	Polish Academy of Sciences, Poland
Dominique Laurent	Université de Cergy-Pontoise, France
Marie-Jeanne Lesot	LIP6, UPMC, France
Rory Lewis	University of Colorado at Colorado Springs, USA
Elisabeth Lex	Graz University of Technology, Austria
Antoni Ligeza	AGH University of Science and Technology, Poland
Yang Liu	Hong Kong Baptist University, Hong Kong
Jiming Liu	Hong Kong Baptist University, Hong Kong
Corrado Loglisci	Università degli Studi di Bari, Italy
Henrique Lopes Cardoso	University of Porto, Portugal
Donato Malerba	Università degli Studi di Bari, Italy
Giuseppe Manco	ICAR-CNR, Italy
Yannis Manolopoulos	Open University of Cyprus, Cyprus
Małgorzata Marciniak	Polish Academy of Science, Poland
Mamoun Mardini	University of Florida, USA
Elio Masciari	University of Naples, Italy
Paola Mello	University of Bologna, Italy
João Mendes-Moreira	University of Porto, Portugal
Luis Moreira-Matias	NEC Laboratories Europe, Germany
Mikolaj Morzy	Poznań University of Technology, Poland
Agnieszka Mykowiecka	IPI PAN, Poland
Tomi Männistö	University of Helsinki, Finland
Mirco Nanni	ISTI-CNR, Italy
Amedeo Napoli	LORIA, CNRS, Inria, Université de Lorraine, France
Pance Panov	Jožef Stefan Institute, Slovenia
Jan Paralic	Technical University Kosice, Slovakia
Ruggero G. Pensa	University of Torino, Italy
Jean-Marc Petit	Université de Lyon, INSA Lyon, France
Ingo Pill	Graz University of Technology, Austria
Luca Piovesan	DISIT, Università del Piemonte Orientale, Italy
Olivier Pivert	IRISA-ENSSAT, France
Lubos Popelinsky	Masaryk University, Czech Republic
Jan Rauch	University of Economics, Prague, Czech Republic
Marek Reformat	University of Alberta, Canada
Henryk Rybiński	Warsaw University of Technology, Poland

Hiroshi Sakai	Kyushu Institute of Technology, Japan
Tiago Santos	Graz University of Technology, Austria
Christoph Schommer	University of Luxembourg, Luxembourg
Marian Scuturici	LIRIS, INSA Lyon, France
Nazha Selmaoui-Folcher	University of New Caledonia, France
Giovanni Semeraro	Università degli Studi di Bari, Italy
Samira Shaikh	UNC Charlotte, USA
Dominik Slezak	University of Warsaw, Poland
Urszula Stanczyk	Silesian University of Technology, Poland
Jerzy Stefanowski	Poznań University of Technology, Poland
Marcin Sydow	PJIIT and ICS PAS, Poland
Katarzyna Tarnowska	San Jose State University, USA
Herna Viktor	University of Ottawa, Canada
Simon Walk	Graz University of Technology, Austria
Alicja Wieczorkowska	Polish-Japanese Academy of Information Technology, Poland
David Wilson	UNC Charlotte, USA
Yiyu Yao	University of Regina, Canada
Jure Zabkar	University of Ljubljana, Slovenia
Slawomir Zadrozny	Polish Academy of Sciences, Poland
Wlodek Zadrozny	UNC Charlotte, USA
Bernard Zenko	Jožef Stefan Institute, Slovenia
Beata Zielosko	University of Silesia
Arkaitz Zubiaga	Queen Mary University of London, UK

Additional Reviewers

Max Toller	Michelangelo Ceci
Henryk Rybiński	Giovanni Semeraro
Allel Hadjali	Michael Granitzer
Giuseppe Manco	Simon Walk
Aijun An	

Publicity Chair

Müslüm Atas	Graz University of Technology, Austria

Publication Chair

Trang Tran	Graz University of Technology, Austria

Web and Local Committee

Jörg Baumann	Graz University of Technology, Austria
Elisabeth Orthofer	Graz University of Technology, Austria
Petra Schindler	Graz University of Technology, Austria

Invited Talks

Complementing Behavioural Modeling with Cognitive Modeling for Better Recommendations

Marko Tkalčič🆔

University of Primorska, Faculty of Mathematics, Natural Sciences and Information Technologies, Glagoljaška 8, SI-6000 Koper, Slovenia
marko.tkalcic@famnit.upr.si

Abstract. Recommender systems are systems that help users in decision-making situations where there is an abundance of choices. We can find them in our everyday lives, for example in online shops. State-of-the-art research in recommender systems has shown the benefits of behavioural modeling. Behavioural modeling means that we use past ratings, purchases, clicks etc. to model the user preferences. However, behavioural modeling is not able to capture certain aspects of the user preferences. In this talk I will show how the usage of complementary research in cognitive models, such as personality and emotions, can benefit recommender systems.

Keywords: Recommender systems · Behavioural modeling · Cognitive modeling

Fairness Is Not a Number: Methodological Implications of the Politics of Fairness-Aware Systems

Robin Burke(iD)

Department of Information Science, University of Colorado, Boulder
robin.burke@colorado.edu

Abstract. This invited talk will explore the methodological challenges for intelligent systems development posed by questions of algorithmic fairness. In particular, we will discuss the ambiguity of fairness as a concept and the gap between simple formalizations of fairness questions and the challenges of real applications, where there are a multiplicity of stakeholders who may perceive fairness in different ways.

Keywords: Fairness · Machine learning

1 Fairness in Machine Learning and Recommendation

The problem of bias and fairness in algorithmic systems generally and in machine learning systems in particular is a critical issue for our increasingly data-centric world. However, the emergence of this topic has thrust computer scientists into unfamiliar territory, especially with regard to system development. A substantial body of research on fairness in machine learning, especially in classification settings, has emerged in the past ten years, including formalizing definitions of fairness [1, 5, 7, 10] and offering algorithmic techniques to mitigate unfairness ~ [8, 11–13]. However, as noted in [4], there is little published research that investigates the complexity of real-world practices around fairness and reports findings from practical implementations. As a consequence, results from fairness-aware machine learning research often lack relevance for developers of real systems, and methodologies are lacking.

Fairness may be thought of as a property of particular system outcomes, like other system characteristics, such as reliability, and like these properties has to designed into systems, accounted for in testing, and monitored in deployment. Fairness is also a product of a complex set of social norms and interactions. Fair outcomes will be achieved and sustained throughout a system's lifecycle, if and only if, the process or methodology by which the system is developed and maintained itself incorporates the

This work has been supported in part by the National Science Foundation under award no. IIS-1911025.

multiple viewpoints of different stakeholders and provides a mechanism for arbitrating between them.

Economists recognize four rubrics under which fairness can be defined [9]:

1. Fairness as exogenous right: The definition of and necessity for fairness is imposed from outside, by legal requirement, for example.
2. Fairness as reward: Fairness is served by providing extra benefits from a system for those who contribute more to it. For example, a salary bonus in recognition of excellent work.
3. Fairness as compensation: Providing a benefit to those otherwise disadvantaged. A handicap in golf, for example, helps players of different abilities compete on a level basis.
4. Fairness as fitness. A subtle efficiency-based condition that states that a fair distribution is one in which resources go to those best able to utilize them. Thus, in dividing an estate, the inheritance of a piano might be fairly given to the most musically-inclined family member.

It is worth noting that the definitions do not all agree with each other or even point in the same direction in particular cases, particularly #2 and #3.

Fairness in machine learning is most often framed in terms of rubric #1. If government regulations require non-discriminatory treatment, organizations must comply. Such a stance has the promotes a legalistic approach in which organizations treat algorithmic fairness as just another compliance concern along with environmental regulations or financial disclosures. Note that such regulations themselves are inevitably the product of societal embrace of one or more of the other fairness rubrics, as applied to a particular issue.

A legalistic approach to fairness sometimes results in a "head-in-the-sand" approach. An organization may avoid evaluating its systems for discriminatory outcomes, because such a finding would incur liability for remedying the problem. Costs for building and maintaining fairness properties in systems can thus be deferred or avoided by neglecting to test for them, leaving it to external parties (usually those facing discrimination) to identify unfair impacts and seek redress under the law. Such an approach may be practically appealing, but it is not ethical for system designers who can reasonably anticipate such harms not to seek to mitigate them in advance: note the assertion in the ACM's code of ethics that computing professionals should strive to "minimize negative consequences of computing."

In addition to leading organizations to avoid proactive approaches, the focus on legalistic, exogenous aspects of fairness tends to reduce the issue to a matter of meeting some specific court-established legal test. This hides something crucial about fairness: namely, its contested and inevitably political nature. Regulations around discrimination or other fairness concerns do not arise from the sage wisdom of legal scholars; they arise from multi-vocal contestation in a political sphere where claims to various kinds of rights are asserted, argued and eventually codified (or not) through regulatory action.

Thus, as designers and developers of intelligent systems, it is incumbent on us to be proactive in the incorporation of fairness into system design. In doing so, we will further need to recognize that our view of fairness will necessarily shift from a focus on fixed constraints imposed from *without*, to an understanding of fairness as a nuanced, multiply-interpreted, and sometimes-contested construct derived from *within* a real-world organizational context, where different stakeholders understand fairness in different ways. Such a point of view is consistent with scholarship in sociology [3], organizational justice [2], welfare economics [9], and public administration [6].

As examples of this perspective in action, we can consider cases of systems operating in organizational contexts where fairness derives from organizational mission. In these cases, the legalistic framing is less salient and the contested nature of fairness is more evident. In this talk, we will look particularly at recommendation in two such contexts: the non-profit microlending site Kiva.org and a (planned) open news recommender site. We will examine the different claims to fairness that arise and how the design and implementation of intelligent systems in these areas can be open to the on-going dialog between them.

References

1. Chouldechova, A.: Fair prediction with disparate impact: a study of bias in recidivism prediction instruments. Big data **5**(2), 153–163 (2017)
2. Colquitt, J.A.: On the dimensionality of organizational justice: a construct validation of a measure. J. Appl. Psychol. **86**(3), 386 (2001)
3. Cook, K.S., Hegtvedt, K.A.: Distributive justice, equity, and equality. Ann. Rev. Sociol. **9**(1), 217–241 (1983)
4. Cramer, H., Holstein, K., Vaughan, J.W., Daumé III, H., Dudík, M., Wallach, H., Reddy, S., Garcia-Gathright, J.: Challenges of incorporating algorithmic fairness into industry practice (2019)
5. Dwork, C., Hardt, M., Pitassi, T., Reingold, O., Zemel, R.: Fairness through awareness. In: Proceedings of the 3rd Innovations in Theoretical Computer Science Conference. pp. 214–226. ACM (2012)
6. Frederickson, H.G.: Social Equity and Public Administration: Origins, Developments, and Applications: Origins, Developments, and Applications. Routledge (2015)
7. Hardt, M., Price, E., Srebro, N., et al.: Equality of opportunity in supervised learning. In: Advances in Neural Information Processing Systems. pp. 3315–3323 (2016)
8. Kamiran, F., Calders, T., Pechenizkiy, M.: Discrimination aware decision tree learning. In: 2010 IEEE 10th International Conference on Data Mining (ICDM), pp. 869–874. IEEE (2010)
9. Moulin, H.: Fair Division and Collective Welfare. MIT press (2004)
10. Narayanan, A.: Translation tutorial: 21 fairness definitions and their politics. In: Proceedings of the Conference on Fairness Accountability and Transparency, New York, USA (2018)
11. Pedreshi, D., Ruggieri, S., Turini, F.: Discrimination-aware data mining. In: Proceedings of the 14th ACM SIGKDD International Conference on Knowledge Discovery and Data Mining, pp. 560–568. ACM (2008)
12. Zemel, R., Wu, Y., Swersky, K., Pitassi, T., Dwork, C.: Learning fair representations. In: Proceedings of the 30th International Conference on Machine Learning (ICML-13), pp. 325–333 (2013)
13. Zhang, L., Wu, X. Anti-discrimination learning: a causal modeling-based framework. Int. J. Data Sci. Anal. **4**, 1–16 (2017). https://doi.org/10.1007/s41060-017-0058-x

Brevity

Kristina Gligorić[1] and Ashton Anderson[2] and Robert West[1]

[1] EPFL, Lausanne, Switzerland
{robert.west, kristina.gligoric}@epfl.ch
[2] University of Toronto, Toronto, Ontario, Canada
ashton@cs.toronto.edu

Abstract. In online communities, where billions of people strive to propagate their messages, understanding how wording affects success is of primary importance. In this work, we are interested in one particularly salient aspect of wording: brevity. What is the causal effect of brevity on message success? What are the linguistic traits of brevity? When is brevity beneficial, and when is it not? Whereas most prior work has studied the effect of wording on style and success in observational setups, we conduct a controlled experiment, in which crowd workers shorten social media posts to prescribed target lengths and other crowd workers subsequently rate the original and shortened versions. This allows us to isolate the causal effect of brevity on the success of a message. We find that concise messages are on average more successful than the original messages up to a length reduction of 30–40%. The optimal reduction is on average between 10% and 20%. The observed effect is robust across different subpopulations of raters and is the strongest for raters who visit social media on a daily basis. Finally, we discover unique linguistic and content traits of brevity and correlate them with the measured probability of success in order to distinguish effective from ineffective shortening strategies. Overall, our findings are important for developing a better understanding of the effect of brevity on the success of messages in online social media.

Keywords: Causal effects · Experimental methods · Linguistic style · Twitter · Crowdsourcing

1   Introduction

How to convey a message most successfully is an age-old question. In online communities, where billions of people consume and strive to propagate their messages, understanding how wording affects success is of primary importance. In this work, we are interested in one particularly salient aspect of wording: brevity, or conciseness. What is the causal effect of brevity on message success? What are the linguistic traits of brevity? When is brevity beneficial? To establish a causal link between brevity and success, one would need to compare posts that convey the exact same semantic

---

This is an abridged version of a longer paper with a longer title [4].

Fig. 1. Schematic diagram of the experimental design. The experiment consists of two parts, designed to replicate the *production* and *consumption* of textual content in online social media. The goal of the content production part (Tasks 1–4) is, for a given set of input tweets, to output shortened versions, having validated that the meaning is preserved. In the content consumption part of the experiment (Task 5), we show participants pairs of tweets, one treated short tweet, and the control long original tweet and ask which one will get more retweets. The outputs of the setup are binary votes (several per pair), based on which we compute the probabilities of success for each tweet version.

information and differ only in the number of characters used to express the fixed semantic content in a specific lexical and syntactic surface form.

Observational studies have striven to approximate this ideal goal by carefully controlling for confounding factors [2, 3, 6]. However, prior research arrived at contradictory conclusions and has not been able to guarantee that the semantic content of compared posts is identical, or that length is not confounded with other factors such as the inherent attractiveness of a message.

2   Experimental Setup

In order to overcome the aforementioned methodological hurdle inherent in observational designs and to more closely approximate the ideal of comparing two messages— one long, one short—expressing the exact same semantic content, we adopt an experimental approach instead, depicted schematically in Fig. 1.

Our experimental strategy consists of two steps: first, in the content production phase, we extract shortened versions of original tweets, and second, in the content consumption phase, we measure the quality of shortened versions compared with the unshortened version.

By ensuring (via additional checks) that length is the only difference between the unshortened tweets and the tweets shortened to prespecified lengths, we can attribute differences in quality to be causally related to brevity. Building on the fact that regular crowd workers can reduce the length of text by up to 70% without cutting any major content [1] and can accurately estimate which of two messages in a pair (for a fixed topic and user) will be shared more frequently [6], we deployed our experiment on Amazon Mechanical Turk.

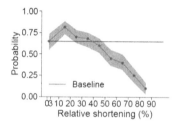

Fig. 2. The effect of conciseness as a function of the level of reduction, with 95% bootstrapped confidence interval.

3 Results

To the best of our knowledge, we are the first to study the effect of brevity in online social media in an experimental fashion. Applying our experimental framework, we collect short versions at 9 brevity levels for 60 original tweets, judged against the original tweets in a total of 27,000 binary votes. Based on this dataset, we address our research questions. We find that concise versions are on average more successful than the original messages up to a length reduction of 45%, while the optimal reduction is on average between 10% and 20% (resulting length 211–215 characters, down from the original 250) (Fig. 1). The observed effect is robust across different subpopulations of participants. Studying linguistic traits of brevity, we find that the shortening process disproportionally preserves verbs and negations—parts of speech that carry essential information—in contrast to, e.g., articles and adverbs. The shortening also preserves affect and subjective perceptions (quantified via the LIWC dictionary), and the effect is strongest for negative emotions. Addressing the question of when brevity is beneficial and when it is not, we find initial evidence that it is effective to omit certain function words and to insert commas and full stops, presumably as it increases readability by structuring or splitting long sentences. Ineffective editing strategies include deleting hashtags as well as question and exclamation marks, which have the potential to elicit discussion and reactions.

4 Discussion

In this work, our goals are threefold: to measure the effects of length constraints on tweet quality, to determine the linguistic traits of brevity, and to find when brevity is beneficial. To address these goals, we designed a large experiment that measured the quality of the shortened versions compared with the originals.

In brief, there are significant benefits of brevity. We observe that tweets can be successfully reduced up to 45% of their original length with no reduction in quality. The optimal range of shortening is consistently between 10% and 20% of the original length.

Practically, our findings are important for developing a better understanding of the effect of wording on the success of messages in online social media.

Methodologically, our experimental design highlights the power of a novel data collection paradigm, where crowd workers supply research data in the form of slightly edited text [1, 5, 7].

References

1. Bernstein, M.S., Little, G., Miller, R.C., Hartmann, B., Ackerman, M.S., Karger, D.R., Crowell, D., Panovich, K.: Soylent: a word processor with a crowd inside. Commun. ACM, **58**(8), 85–94 (2015)
2. Bramoulle, Y., Ductor. L.: Title length. J. Econ. Behav. Organ. **150**, 311–324 (2018)
3. Gligorić, K., Anderson, A., West, R.: How constraints affect content: the case of Twitter's switch from 140 to 280 characters. In: Proceedings of the International Conference on Web and Social Media (ICWSM) (2018)
4. Gligorić, K., Anderson, A., West, R.: Causal effects of brevity on style and success in social media. In: Proceedings of the ACM Conference on Computer-Supported Cooperative Work and Social (CSCW) (2019)
5. Ribeiro, M.H., Gligorić, K., West, R.: Message distortion in information cascades. In: Proceedings of the World Wide Web Conference (WWW) (2019)
6. Tan, C., Lee, L., Pang, B.: The effect of wording on message propagation: Topic- and author-controlled natural experiments on Twitter. In: Proceedings of the Annual Meeting of the Association for Computational Linguistics (ACL) (2014)
7. West, R., Horvitz, E.: Reverse-engineering satire, or paper on computational humor accepted despite making serious advances. In: Proceedings of the AAAI Conference on Artificial Intelligence (AAAI) (2019)

Contents

Deep Learning and Embeddings

Digital Signal Processing

Modelling and Reasoning

Machine Learning Applications

Short Papers

Invited Talk

Complementing Behavioural Modeling with Cognitive Modeling for Better Recommendations

Marko Tkalčič(✉) (iD)

Faculty of Mathematics, Natural Sciences and Information Technologies,
University of Primorska, Glagoljaška 8, 6000 Koper, Slovenia
marko.tkalcic@famnit.upr.si

Abstract. Recommender systems are systems that help users in decision-making situations where there is an abundance of choices. We can find them in our everyday lives, for example in online shops. State-of-the-art research in recommender systems has shown the benefits of behavioural modeling. Behavioural modeling means that we use past ratings, purchases, clicks etc. to model the user preferences. However, behavioural modeling is not able to capture certain aspects of the user preferences. In this talk I will show how the usage of complementary research in cognitive models, such as personality and emotions, can benefit recommender systems.

Keywords: Recommender systems · Behavioural modeling · Cognitive modeling

1 Introduction

Recommender Systems (RS) help people make better choices when there is an abundance of choice. Early approaches were limited in finding, for each user, such items that would maximize the user's utility [1]. Examples of this kind of recommender systems can be found in online services, for example in Amazon or Netflix. Today, the RS research community is addressing more diverse topics than plain utility prediction, such as multiple stakeholders (company's profit vs. user satisfaction), trust and transparency, preference elicitation, diversity, group recommendations and new algorithms (e.g. Deep Learning).

In the vast majority of cases, the RS community is addressing the issues above by collecting user behaviour data (e.g. clicks, purchases, viewing time etc.) and using machine learning algorithms to perform the necessary predictions. I will refer to the usage of behavioural data to do recommendations as *behavioural modeling*.

I argue that it's important to understand not only the user behaviour, but also the reasoning that a user does before performing a certain action. I will refer to the modeling of the user's cognitive processes as *cognitive modeling*. I will demonstrate the need for complementing behavioural modeling with cognitive modeling with three stories.

© Springer Nature Switzerland AG 2020
D. Helic et al. (Eds.): ISMIS 2020, LNAI 12117, pp. 3–8, 2020.
https://doi.org/10.1007/978-3-030-59491-6_1

1.1   The Netflix Story

Netflix is known for its heavy use of recommendations in their user interface. Neil Hunt from Netflix, in his keynote speech at the ACM RecSys 2014 conference, explained how they approached the recommendations[1] [10]. He stated that Netflix is optimizing for viewing hours. However, not all viewing hours are equal and their system did not distinguish between customer value and addiction.

1.2   The Nature and Nurture Story

This is an old dilemma, which debates how much of the human mind is built-in, and how much of it is constructed by experience. Throughout history there have been several manifestations of this dilemma. A prominent one has been the debate between Skinner and Chomsky on language acquisition. Skinner believed that children learned language starting as a tabula rasa through reinforcement, while Chomsky claimed that children have innate capabilities to learn language, the so-called Language Acquisition Device (LAD) [2]. Today, a similar debate is going on in machine learning between connectionists, who model mental phenomena with neural nets trained with behavioural data, and symbolists, who model them with ingrained concepts. The debate has been documented heavily online[2]. In this debate, Gary Marcus is arguing to merge both behaviour-based modeling and cognitive modeling: *To get computers to think like humans, we need a new A.I. paradigm, one that places top down and bottom up knowledge on equal footing. Bottom-up knowledge is the kind of raw information we get directly from our senses, like patterns of light falling on our retina. Top-down knowledge comprises cognitive models of the world and how it works*[3].

1.3   The "La vita e' bella" Story

The third story revolves around the film *Life is Beautiful* (La vita e' bella). The film narrates a sad story that takes place during the holocaust. It is also full of hilarious scenes, such as the one in which the main character, played by Roberto Benigni, translates the instructions of a German officer. In a movie recommender system, if a user gives this film a high rating, the system doesn't really know if the user liked it because of the excellent drama or because of the hilarious jokes. Again, we see that observing only the behaviour (i.e. the high rating) we may miss important information about the user, which can lead to bad recommendations.

Based on these three stories we can conclude that understanding the cognitive processes of the user is important to interpret correctly the behaviour. It is hard, especially because the data is not readily available.

---

[1] https://youtu.be/lYcDR8z-rRY?t=4727.

[2] https://twitter.com/ylecun/status/1201233472989356032.

[3] Marcus, Gary, Artificial Intelligence Is Stuck. Here's How to Move It Forward. New York Times, July 29, 2017.

2   Cognitive Modeling in Recommender Systems

There are many models that describe various cognitive processes. In one such model of decision making, two important factors that influence decision-making are emotions and personality [13]. Other models of the human mind also include emotions and personality, for example the model of artificial general intelligence [8].

2.1   Personality and Emotions

Personality is a set of traits that are supposed to be relatively stable across long periods of time. The most widely used model of personality is the five-factor model (FFM), which is composed of the following factors: openness, conscientiousness, extraversion, agreeableness, and neuroticism [14]. Personality can be acquired either using extensive questionnaires, such as the Big-Five Inventory [14] or using algorithms that predict personality from social media usage [7,11,16,18].

Emotions, on the other hand, change very rapidly with time. We usually model them either with discrete labels (e.g. happy, sad, angry etc. [3] or with dimensions, such as valence, arousal and dominance [17]. There are off-the-shelf solutions that use various sensors (e.g. cameras, physiological sensors etc.) to infer the current emotion of a person[4].

2.2   Personality in Recommender Systems

Personality has been used in a variety of ways to improve recommendations. For example, personality has been shown to be important for mood regulation. In an experiment by Ferwerda et al., the authors induced various moods in people using film music. Among other things, they observed that when people are sad they don't always regulate their mood towards being happy. People with different personalities had different needs in such situations, which has clear implications for personality-mood-based music recommendations [6]. Personality has also shown to useful in calculating user similarities for neighbourhood-based RS. Both, Tkalcic et al. and Hu et al., managed to solve the new user problem by introducing personality in the calculation of user similarity [9,20]. Matrix factorization-based RS were also augmented successfully with personality information [4,5]. Finally, personality has been shown to be useful also for adjusting the diversity of recommended items [25].

2.3   Emotions in Recommender Systems

Emotions have also been used in different ways to improve recommendations. Odic et al., as well as Zheng et al., have shown in a series of experiments that

---

[4] https://www.affectiva.com/.

emotions are useful as contextual variables [15,26]. In an early adoption of emotions, Tkalcic et al. showed that the induced emotions of images are effective features for doing content-based recommendations [19,23]. Emotions are also useful as feedback, which can be used without disrupting the user interaction with the system. Vodlan et al. used the emotion of hesitation to infer whether the list of recommended items is good [24]. Emotions were also used in a pairwise preference acquisition system to replace the user's explicit rating [21,22]. Finally, emotions should be taken in account also in group recommendation settings. In group dynamics, user preferences are influenced by other group members through emotional contagion. The impact of emotional contagion has been demonstrated also in virtual groups, such as the experiment conducted on Facebook users [12].

3 Conclusion

Recommender systems algorithms take advantage of past behaviour of users to make recommendations. However, this behaviouristic approach does not provide explanations for the user behaviour, which can lead to bad recommendations. In this contribution I surveyed how cognitive modeling, in particular the usage of personality and emotions, can be used to make better recommendations.

References

1. Adomavicius, G., Tuzhilin, A.: Toward the next generation of recommender systems: A survey of the state-of-the-art and possible extensions. IEEE Trans. Knowl. Data Eng. **17**(6), 734–749 (2005). https://doi.org/10.1109/TKDE.2005.99, http://ieeexplore.ieee.org/lpdocs/epic03/wrapper.htm?arnumber=1423975ieeexplore.ieee.org/xpls/abs_all.jsp?arnumber=1423975
2. Chomsky, N.: A review of BF skinner's verbal behavior. Readings Philos. Psychol. **1**, 48–63 (1980)
3. Ekman, P.: Basic emotions. In: Dalglesish, T., Power, M.J. (eds.) Handbook of Cognition and Emotion, vol. 1992, pp. 45–60. John Wiley & Sons Ltd., Chichester (2005). https://doi.org/10.1002/0470013494.ch3
4. Elahi, M., Braunhofer, M., Ricci, F., Tkalcic, M.: Personality-based active learning for collaborative filtering recommender systems. In: Baldoni, M., Baroglio, C., Boella, G., Micalizio, R. (eds.) AI*IA 2013. LNCS (LNAI), vol. 8249, pp. 360–371. Springer, Cham (2013). https://doi.org/10.1007/978-3-319-03524-6_31
5. Fernández-Tobías, I., Braunhofer, M., Elahi, M., Ricci, F., Cantador, I.: Alleviating the new user problem in collaborative filtering by exploiting personality information. User Model. User-Adapt. Inter. **26**(2), 1–35 (2016). https://doi.org/10.1007/s11257-016-9172-z
6. Ferwerda, B., Schedl, M., Tkalcic, M.: Personality & emotional states : understanding users music listening needs. In: Cristea, A., Masthoff, J., Said, A., Tintarev, N. (eds.) UMAP 2015 Extended Proceedings (2015). http://ceur-ws.org/Vol-1388/
7. Ferwerda, B., Tkalcic, M.: Predicting users' personality from instagram pictures. In: Proceedings of the 26th Conference on User Modeling, Adaptation and Personalization - UMAP 2018, pp. 157–161. ACM Press, New York (2018). https://doi.org/10.1145/3209219.3209248, http://dl.acm.org/citation.cfm?doid=3209219.3209248

8. Goertzel, B., Iklé, M., Wigmore, J.: The Architecture of Human-Like General Intelligence, pp. 123–144. Atlantis Press, Paris (2012). https://doi.org/10.2991/978-94-91216-62-6_8
9. Hu, R., Pu, P.: Using personality information in collaborative filtering for new users. In: Proceedings of the 2nd ACM RecSys 2010 Workshop on Recommender Systems and the Social Web, pp. 17–24 (2010). http://www.dcs.warwick.ac.uk/~ssanand/RSWeb_files/Proceedings_RSWEB-10.pdf#page=23
10. Hunt, N.: Quantifying the value of better recommendations (2014)
11. Kosinski, M., Stillwell, D., Graepel, T.: Private traits and attributes are predictable from digital records of human behavior. Proc. Natl. Acad. Sci. United States Am. **110**(15), 5802–5805 (2013). https://doi.org/10.1073/pnas.1218772110
12. Kramer, A.D.I., Guillory, J.E., Hancock, J.T.: Experimental evidence of massive-scale emotional contagion through social networks. Proc. Natl. Acad. Sci. United States Am. **111**(29), 8788–8790 (2014). https://doi.org/10.1073/pnas.1320040111
13. Lerner, J.S., Li, Y., Valdesolo, P., Kassam, K.S.: Emotion and decision making. Ann. Rev. Psychol. **66**(1), 799–823 (2015). https://doi.org/10.1146/annurev-psych-010213-115043
14. McCrae, R.R., John, O.P.: An introduction to the five-factor model and its applications. J. Pers. **60**(2), 175–215 (1992)
15. Odić, A., Tkalčič, M., Tasič, J.F., Košir, A.: Predicting and detecting the relevant contextual information in a movie-recommender system. Inter. Comput. **25**(1), 74–90 (2013). https://doi.org/10.1093/iwc/iws003, http://iwc.oxfordjournals.org/content/25/1/74.short, https://academic.oup.com/iwc/article/768060/Predicting
16. Quercia, D., Kosinski, M., Stillwell, D., Crowcroft, J.: Our twitter profiles, our selves: predicting personality with twitter. In: Proceedings - 2011 IEEE International Conference on Privacy, Security, Risk and Trust and IEEE International Conference on Social Computing, PASSAT/SocialCom 2011, pp. 180–185. IEEE (2011). https://doi.org/10.1109/PASSAT/SocialCom.2011.26, http://ieeexplore.ieee.org/lpdocs/epic03/wrapper.htm?arnumber=6113111ieeexplore.ieee.org/xpls/abs_all.jsp?arnumber=6113111
17. Scherer, K.R.: What are emotions? and how can they be measured? Soc. Sci. Inf. **44**(4), 695–729 (2005). https://doi.org/10.1177/0539018405058216
18. Skowron, M., Tkalčič, M., Ferwerda, B., Schedl, M.: Fusing social media cues. In: Proceedings of the 25th International Conference Companion on World Wide Web - WWW 2016 Companion, pp. 107–108. ACM Press, New York (2016). https://doi.org/10.1145/2872518.2889368, http://dl.acm.org/citation.cfm?doid=2872518.2889368 ,
19. Tkalčič, M., Burnik, U., Košir, A.: Using affective parameters in a content-based recommender system for images. User Model. User-Adapt. Inter. **20**(4), 279–311 (2010). https://doi.org/10.1007/s11257-010-9079-z
20. Tkalčič, M., Kunaver, M., Košir, A., Tasič, J.: Addressing the new user problem with a personality based user similarity measure. In: Ricci, F., et al. (eds.) Joint Proceedings of the Workshop on Decision Making and Recommendation Acceptance Issues in Recommender Systems (DEMRA 2011) and the 2nd Workshop on User Models for Motivational Systems: The Affective and the Rational Routes to Persuasion (UMMS 2011) (2011).http://ceur-ws.org/Vol-740/DEMRA_UMMS_2011_proceedings.pdf#page=106

21. Tkalčič, M., Maleki, N., Pesek, M., Elahi, M., Ricci, F., Marolt, M.: A research tool for user preferences elicitation with facial expressions. In: Proceedings of the Eleventh ACM Conference on Recommender Systems - RecSys 2017, no. i, pp. 353–354. ACM Press, New York (2017). https://doi.org/10.1145/3109859.3109978, http://dl.acm.org/citation.cfm?doid=3109859.3109978

22. Tkalčič, M., Maleki, N., Pesek, M., Elahi, M., Ricci, F., Marolt, M.: Prediction of music pairwise preferences from facial expressions. In: Proceedings of the 24th International Conference on Intelligent User Interfaces - IUI 2019, pp. 150–159. ACM Press, New York (2019). https://doi.org/10.1145/3301275.3302266, http://dl.acm.org/citation.cfm?doid=3301275.3302266

23. Tkalcic, M., Odic, A., Kosir, A., Tasic, J.: Affective labeling in a content-based recommender system for images. IEEE Trans. Multimedia **15**(2), 391–400 (2013). https://doi.org/10.1109/TMM.2012.2229970, http://ieeexplore.ieee.org/lpdocs/epic03/wrapper.htm?arnumber=6362231ieeexplore.ieee.org/document/6362231/

24. Vodlan, T., Tkalčič, M., Košir, A.: The impact of hesitation, a social signal, on a user's quality of experience in multimedia content retrieval. Multimedia Tools Appl. **74**(17), 6871–6896 (2015). https://doi.org/10.1007/s11042-014-1933-2

25. Wu, W., Chen, L., He, L.: Using personality to adjust diversity in recommender systems. In: Proceedings of the 24th ACM Conference on Hypertext and Social Media - HT 2013, pp. 225–229 (2013). https://doi.org/10.1145/2481492.2481521, http://dl.acm.org/citation.cfm?doid=2481492.2481521

26. Zheng, Y., Mobasher, B., Burke, R.: Emotions in Context-Aware Recommender Systems, pp. 311–326 (2016). https://doi.org/10.1007/978-3-319-31413-6_15

Nominees for Best Paper Award

The Construction of Action Rules to Raise Artwork Prices

Laurel Powell[1]([✉])(iD), Anna Gelich[2,3], and Zbigniew W. Ras[1,3](iD)

[1] College of Computing and Informatics, University of North Carolina at Charlotte, Charlotte, USA
{lpowel28,ras}@uncc.edu
[2] College of Arts + Architecture, University of North Carolina at Charlotte, Charlotte, USA
agelich@uncc.edu
[3] Polish-Japanese Academy of Information Technology, Warsaw, Poland

Abstract. This work explores the development of action rules for changing the prices of works of contemporary fine art. It focuses on the generation of action rules using LISp-Miner related to artwork profiles and artist descriptions. Additionally, this work explores the use of the dominant color of an artwork as a feature in the generation of action rules for adjusting its prices.

Keywords: Art analytics · Data mining · Action rules

1 Introduction

Artists in the contemporary market for fine art must make many decisions on how to present their work for commercial sale. While these choices are crucial to the artist's long term commercial success, artists are given limited guidance on how to present their work online to its best advantage. Can the price that an artist can ask for regarding a work be altered by changing minor aspects that their potential customers see? What drives the price for a work of fine art?

The challenge of objective valuation is worsened by the complex nature of pricing in the art market. There are a number of theories on how pricing in the art market works. As said in [35],

> In a market that has to reconcile a fierce opposition between commercial and artistic values, and that has to commodify goods whose essence is considered to be non-commodifiable, dealers, artists, and collectors find ways to express and share non-economic values through the economic medium of pricing.

This work proposes a method for developing a set of action rules that can be used to suggest courses of action to artists that will improve the sales potential of their work. It begins by discussing the development of a dataset and feature

D. Helic et al. (Eds.): ISMIS 2020, LNAI 12117, pp. 11–20, 2020.
https://doi.org/10.1007/978-3-030-59491-6_2

construction for the generation of action rules, moves on to contrast the results of different sets of features for the creation of these rules, and it concludes by discussing the future potential of this research avenue. These rules reclassify artworks from one price level to another higher price level. This work discusses possible sets of flexible attributes and proposes a basic set of stable attributes [24]. Stable attributes are generally aspects of the relevant item that are unlikely or difficult to change. In the context of this work, these represent the physical aspects of the artwork and fundamental qualities such as it's medium. Flexible attributes are features that can be altered by an interested party to change its classification. In the context of this work, these represent aspects of how the artist presents themselves to potential customers.

2   Related Work

Applying computing techniques to artworks has received considerable attention from the academic community. Some researchers have addressed the challenge of creating a recommender system that picks out artworks similar to a given one for a user [29], while others are focused on automatically tagging a painting with emotions based on the colors represented [15].

Art price valuation has received limited attention. One interesting project in the realm of price prediction was Hosny's work [13]. As discussed in [7], one notable trend explored in Hosny's work was that certain colors, specifically blacks, whites and grays, are more likely to have a higher sales valuation. Different approaches have been used for price prediction, such as the approach used by [18] to value Rothko paintings. In [11], hedonic regression, a statistical modelling method based on characteristics of the object in question, combined with past sales was used to predict art value.

Action rules were first proposed in [24] as a method of reclassifying objects from one group to another group by changing values of their flexible attributes. They have been used for medical data, such as in [12] and for business purposes, as in [24]. Some have explored expansions on the original methodology, such as considering the cost and feasibility of the implementation of action rules as discussed in [33,34]. The cost of an action rule, which was initially proposed in [34], represents the average cost of changing an attribute from its initial value to another specific value.

This work uses LISp-Miner's Ac4ft-Miner tool, introduced in [25], discussed in [19], and developed by [26]. This is an implementation of the GUHA method for action rule generation [25]. In this method, objects resembling traditional association rules are extracted from the dataset and used to form contingency tables [25]. Two of these objects, with matching stable attributes, are used to construct the action rules. One rule, termed here as the 'before rule', has the stable and flexible attributes of the object before any actions are taken and the other rule, called the 'after rule' here, has the desired set of flexible attributes and matching stable ones. Rules generated in this manner consist of an antecedent and succedent, which are association rules in the form $\varphi \approx \psi$ where the symbol $\approx$

represents the association between φ and ψ [25]. These are then used to create action rules, called G-Action Rules, through the examination of dependencies and similarities across the rules generated in the previous step [25].

3 Dataset

The dataset used in this work is an expanded version of the one used in [21–23]. Our work expands on the ideas discussed in those works by adding action rules to the original dataset and proposed feature sets.

We use a dataset of artworks and artist information collected from the online art sales site Artfinder.com [5]. Artfinder represents artists from all over the world, and has diverse styles and subjects represented. It functions as a platform to allow artists to sell directly to consumers. Using a single source for artworks allows for more consistent definitions of tags used for artworks. The information was scraped using Beautiful Soup [2] to parse the pages, Apache Selenium [4] to work with dynamic webpages and Javascript, and Python to fetch pages and all other major steps. It contains approximately 200,000 artworks from approximately 3,300 artists.

All artworks posted on Artfinder have a dedicated page describing the work. This page consistently has at least one photograph of the work, and has its descriptive information, as well as the tags that the artist selected. These tags are used by potential customers to search for works. These tags are simple and specify the features such as the artistic style, subject or medium of the artwork. Some of the analysis methods discussed below use the primary image of the artwork from the product page. A small number of artwork images could not be retrieved, so they are omitted from the following analysis.

Artists on Artfinder create profiles of themselves discussing their achievements and backgrounds. The profiles have places for artists to provide their biographies, links to their social media and descriptions of achievements in the artist's life. A number of artists have reviews posted from customers, and are rated on a star system with five stars being the best. The majority of artists with one or more reviews have ratings of four out of five stars or higher, and very few have any low reviews. The average of all artists with star ratings score is 4.904. This makes the star ratings received by artists of very low value as predictive features.

4 Methodology

The decision feature considered here, which represents the succedent of the G-Action Rule, is the listed asking price of the artwork. Artists would attempt to make changes to how their artworks are presented to the public in order to increase their sales price, and should be cautious about making changes that could potentially lower their sales price. The prices were discretized into ten groups using LISp-Miner [26]. The cuts were placed to automatically to create partitions containing an approximately equal number of tuples. The price attribute was partitioned into 10 levels, referred to in later sections as levels one through ten. In this work, price is treated as a flexible attribute.

This work defines a set of stable attributes, which are used in the antecedent portion of the G-Action Rule, that can be used to describe a work of art. The set of features used takes inspiration from the sets used in [20]. The chosen medium of the artwork is used as a stable feature. When the artwork was posted the artist tagged it with a medium for search purposes. In order to keep the number of options within reasonable bounds, the mediums were discretized into seven categories and one "NA" category. The artistic style of the work as well as the artist's subject is similarly used as a tag and as a stable attribute. As discussed in Pawlowski [20], the size of a work is very relevant to determining its price. So the length and width of the works were discretized into five bins of roughly equal size using the LISp-Miner [26] discretization tool and used as stable attributes. Additionally, the presence or absence of visible reviews was used as a stable binary feature. The review scores of the artists are consistently very high if they are present at all. This limits the utility of using the score as a quality metric. Instead, the presence or absence of reviews was used as a stable attribute. Lastly, the percentage of edges in the work was utilized as a stable attribute after being discretized. To calculate this, Canny edge detection, which was developed by John Canny in [10], was used from the OpenCV set of tools [32] on the primary artwork image. A small percentage of artworks had errors when an attempt was made to retrieve the photograph, so those tuples were removed from consideration. This algorithm determines if an individual pixel represents an edge or not based on the amount of color variation surrounding it. The number of edge pixels and non-edge pixels are then counted, and the number of edge pixels was used as a stable feature. This can be considered as giving a rough idea of the amount of detail or amount of color variation in the work. The full list of stable attributes is given below.

- Artistic Style
- Artistic Subject
- Medium
- Height
- Width
- Artist Has Visible Reviews
- Percentage of the Artwork Representing Edges

Lastly, the use of color in the artwork was explored as an additional stable feature in one of the rule sets. The pixels in the artwork were clustered using Open-CV [32] across the RGB dimensions using K-Means to determine the 10 most frequently appearing colors. Out of this set of ten, the centroid of the largest cluster was tested against 11 reference colors using the CIEDE2000 color difference measure [30] which was implemented using [31]. The reference colors are based on the idea of universal basic colors which was first proposed in [8]. While this work has been challenged and the colors proposed have been refined such as in [17,28], the concept of basic colors is useful for classification. The 11 reference colors are white (R 255, G 255, B 255), gray (R 128, G 128, B 128), black (R 0, G 0, B 0), red (R 255, G 0, B 0), orange (R 255, G 128, B), yellow (R 255, G 255, B 0), green (R 0, G 255, B 0), blue (R 0, G 0, B 255), purple (R 128, G 0, B 128), pink (R 255, G 192.0, B 203), brown (R 63.8, G 47.9, B 31.9). The RGB values, other than brown, are taken from [3] and the brown comes from [16]. The selected color, referred to as main color, is used as a stable attribute representing the single color that takes up the largest portion of the work.

The set of flexible attributes, part of the antecedent of the rules, explored center around how a work will be perceived by a consumer. How long is the artist's biography? Do they have a presence on social media? What is the tone of their writing? These are easily changeable for an artist hoping to improve their sales. This gives them a very low cost, so stakeholders may be more willing to try recommended rules. The full list of flexible features used is given below.

- Word Count of the Artist Biography (Bio. WC)
- Word Count of the Artwork Description (Desc. WC)
- Social Media (Considered Together as SM)
    - Artist Listed a Facebook Profile
    - Artist Listed a Twitter Profile
    - Artist Listed an Instagram Profile
- Positive Sentiment Level of Biography (Bio. Ps.)
- Negative Sentiment Level of Biography (Bio. Ng.)
- Positive Sentiment Level of Artwork Description (Desc. Ps.)
- Negative Sentiment Level of Artwork Description (Desc. Ng.)

The number of words in the biography of the artist and in the artwork description are used as predictive features. As was discussed in [21–23], the word count does have some utility as a price predicting feature. In other arenas of online sales, the length and wording of a description can have a bearing on the sale ability of an item. In [27], the authors analyzed sales on Ebay, and determined that the length of the description could have an impact on the price.

Similarly, many collectors find artists through social media [1]. It has become an increasingly important tool for helping artists be discovered by collectors. Can adding or removing a link to a profile change the opinion of a collector? Determining the sentiment of the text was done using VADER, the 'Valence Aware Dictionary for sEntiment Reasoning', which is part of the Python Natural Language Toolkit [9,14]. The objective was to determine the polarity of the sentiment of the text. This strategy uses a set of words and phrases to determine the text's sentiment, called an 'opinion lexicon' [6]. VADER uses a combination of an opinion lexicon and a set of rules [14]. The lexicon includes thousands of candidate terms rated by humans on scale, and the rule set considers the impact of capitalization and negation [14]. Each piece of text was given scores as positive, negative or neutral. These scores were discretized and used as flexible attributes. The sentiments found were largely neutral, with only a few pieces of text containing strongly emotional language.

Using the LISp-Miner Ac4ft-Miner tool, sets of rules were generated for combinations of prices and flexible attributes with a single set of stable attributes used throughout for consistency. As the goal is to move prices from lower to higher price points, rules were generated to transition items from lower price levels to higher price levels. Rules generated using this method have support and confidence scores for both the before state association rule and after state association rule that are used together to construct the final action rule.

5   Results

The first set of rules generated had no minimum confidence for a rule and had a requirement that the rule must apply to 50 or more tuples out of the dataset. Rules were generated for each flexible attribute alone, and the social media features alone and in combinations. This work compares the results of different flexible attributes alone to explore which have the greatest impact on the confidence and support of the resulting rules. Each set of rules contains an average of 1700 action rules. At least one rule applies to approximately 97% of the elements at that price level. This is termed the coverage of the rule set and can be used as a measure of its applicability. However, the confidence in these rules is quite low. As all the features discussed here have an extremely low cost, exploring low confidence rules is not a concern. To make the changes these rules suggest, an artist would only need to rewrite their artwork descriptions or change their profile. While some individual rules do have high confidence, the average confidence for the before attribute is approximately 13%.

In response to the low confidence of the initial sets of rules tested, additional rule sets were generated with greater constraints on the generation process. Rules were only generated to move each artwork up to the next higher price level. The social media flexible features (abbreviated in tables as SM) were used to generate rules, and the word counts (abbreviated in tables as WC) of both the artist's biography (abbreviated as Bio.) and description (abbreviated as Desc.) and text polarity features (abbreviated as Ps. for positive sentiment and Ng. for negative sentiment) were all used separately. The minimum support level for a rule was lowered from 50 to 2, but the minimum confidence for a rule to be considered was 60% for the before and after rules. The minimum number of attributes for the stable portion of the rule is one and the maximum number is five. The rule set generated here had a considerably lower average coverage at 9.375%, but a considerably higher average confidence for the before rules at 79.16% and the after rules at 79.2%. Considerably fewer rules were generated, and the average number of rules per group was 1,155. Table 1 displays the coverage of each rule set for the selected price levels and flexible attribute. This represents the percentage of objects at the lower price level that had at least one applicable rule to raise that object to the selected higher price level.

The level of support varies dramatically depending on the price level being addressed. The exact values of the prices are as follows: ($<$12.97–58.28), (58.28–95.90), (95.90–130.04), (130.04–188.13), (188.13–250), (250–350), (350–490), (490–742), (742–1351.24), (1351.24–$>$1,000,000). For example, the coverage of the rules sets that move artworks from the lowest prices, level one, to the next

Table 1. Coverage of rules generated using base stable features

	SM	Bio. WC	Desc. WC	Bio. Ps.	Desc. Ps.	Bio. Ng.	Desc. Ng.
Prices 1 -> 2	19.24	38.89	30.26	37.05	26.56	35.86	19.45
Prices 2 -> 3	4.31	16.35	13.03	17.04	11.97	9.70	5.04
Prices 3 -> 4	3.60	11.91	9.52	11.42	9.40	7.78	3.52
Prices 4 -> 5	2.51	7.87	5.80	8.02	5.67	4.41	2.35
Prices 5 -> 6	2.21	7.31	5.91	7.82	5.54	5.18	1.64
Prices 6 -> 7	1.30	8.20	6.91	7.96	5.77	3.41	3.54
Prices 7 -> 8	1.34	7.28	4.80	6.34	4.41	3.47	1.54
Prices 8 -> 9	1.94	7.79	6.59	8.91	6.03	5.14	2.05
Prices 9 -> 10	4.77	14.25	12.18	13.17	10.87	11.35	5.21

lowest price level, level two, ranges between 19.24% and 38.89% across all the different sets of flexible attributes. Notably, the social media attributes had markedly worse performance than the others. The attributes that changed an artwork's description had less consistently high coverage than the attributes that addressed an artist's biography. At higher values, coverage decreases. This may be due to the size of the shifts necessary to move prices from one tier to another at the higher levels.

To expand on the set of stable attributes, another set of rules was generated that added the dominant color of the work to the list of stable attributes. This set was generated using a randomly selected subset with approximately 100,000 tuples. Lisp-Miner's discretization tools were used to create a new set of partitions similar to those used previously for this set. The same restrictions on rule generation were repeated. Rules must have a minimum confidence of 60% for the before and after rules and rules must have a minimum support of 10. Slightly more rules were generated per set than in the previous variation with an average number of rules per group of 1551. This set did have a slightly higher average coverage of 15.30%, and a similar average before confidence of 78.81% and an after confidence of 79.07%. As with the previous set of rules, the coverage shifts across different price levels. The coverage of each rule group changes depending on the selected flexible attributes and the selected price levels as demonstrated in Table 2.

Table 2. Coverage of rules generated using base stable features and main color

	SM	Bio. WC	Desc. WC	Bio. Ps.	Desc. Ps.	Bio. Ng.	Desc. Ng.
Prices 1 -> 2	25.79	52.14	34.97	52.43	31.83	47.50	24.31
Prices 2 -> 3	9.80	28.21	14.67	28.87	17.96	17.60	9.13
Prices 3 -> 4	7.51	21.34	15.15	20.55	13.29	13.12	7.48
Prices 4 -> 5	5.62	16.48	10.70	16.27	10.49	9.09	5.50
Prices 5 -> 6	3.73	14.87	9.22	14.77	8.31	9.61	3.96
Prices 6 -> 7	4.02	13.92	9.07	12.63	7.67	7.94	4.45
Prices 7 -> 8	3.22	13.47	7.89	13.64	7.26	6.97	3.79
Prices 8 -> 9	4.41	16.72	11.85	18.47	10.01	11.44	5.56
Prices 9 -> 10	12.52	28.39	18.84	27.51	18.43	20.73	10.74

6   Conclusions and Future Work

This work only begins to address the potential for the use of action rules to improve artist sales in the market for contemporary fine art. Many potential avenues for further research exist in the development of features for action rules.

In the initial set of results, the very high rate of coverage is promising. However, the low confidence level, while high for specific rules, is on average quite low. This issue may be attributable to allowing too many low confidence rules. It may also be due to the partitions of the feature sets being overly broad. However, low confidence for the set of G-Action rules is not a barrier to their being useful to an artist. These rules are extremely low cost and easily implemented. Stakeholder's may want to try them on even the small chance that the rules will provide an improvement. The second and third set of rules discussed demonstrates that is possible to significantly raise the average confidence in the rules, but at the expense of lowering the average coverage of the rule sets.

One strong avenue for potential future research is the development of more features for rule generation. The development of more flexible attributes, as well as exploring the attributes discussed here in combination has potential value for stakeholders in the art market. In addition to expanding the list of artwork features, artists have characteristics that may serve as additional stable attributes. A greater exploration of larger scale career changes an artist could make has great potential for research.

Acknowledgement. This research is supported by the National Science Foundation under grant IIP 1749105. Any opinions, findings, and conclusions or recommendations expressed in this material are those of the authors and do not necessarily reflect the views of the National Science Foundation.

References

1. The Hiscox Online Art Trade Report 2018. Technical report, ArtTactic (2018). https://arttactic.com/product/hiscox-online-art-trade-report-2018/
2. Beautiful Soup (2019). https://www.crummy.com/software/BeautifulSoup/
3. Color by name (2019). http://colormine.org/colors-by-name
4. Selenium (2019). https://www.seleniumhq.org/
5. Artfinder.com (2020). https://www.artfinder.com/
6. Aggarwal, C.C.: Machine Learning for Text. Springer, Cham (2018). https://doi.org/10.1007/978-3-319-73531-3
7. Bailey, J.: Machine Learning for Art Valuation. An Interview With Ahmed Hosny, December 2017. https://www.artnome.com/news/2017/12/2/machine-learning-for-art-valuation
8. Berlin, B., Kay, P.: Basic Color Terms: Their Universality and Evolution. University of California Press, Berkeley (1969)
9. Bird, S., Klein, E., Loper, E.: Natural Language Processing with Python, 1st edn. O'Reilly Media, Inc., Sebastopol (2009)
10. Canny, J.: A computational approach to edge detection. IEEE Trans. Pattern Anal. Mach. Intell. **PAMI−8**(6), 679–698 (1986). https://doi.org/10.1109/TPAMI.1986.4767851. http://ieeexplore.ieee.org/lpdocs/epic03/wrapper.htm?arnumber=4767851
11. Galbraith, J., Hodgson, D.: Econometric fine art valuation by combining hedonic and repeat-sales information. Econometrics **6**(3), 32 (2018). https://doi.org/10.3390/econometrics6030032. http://www.mdpi.com/2225-1146/6/3/32
12. Hajja, A.: Object-driven and temporal action rules mining (2013). https://eric.ed.gov/?id=ED564978
13. Hosny, A., Huang, J., Wang, Y.: The Green Canvas (2014). http://ahmedhosny.github.io/theGreenCanvas/
14. Hutto, C., Gilbert, E.: VADER: a parsimonious rule-based model for sentiment analysis of social media text. In: Proceedings of the 8th International Conference on Weblogs and Social Media, ICWSM 2014 (2015)
15. Kang, D., Shim, H., Yoon, K.: A method for extracting emotion using colors comprise the painting image. Multimed. Tools Appl. **77**(4), 4985–5002 (2017). https://doi.org/10.1007/s11042-017-4667-0
16. Labrecque, L.I., Milne, G.R.: Exciting red and competent blue: the importance of color in marketing. J. Acad. Mark. Sci. **40**(5), 711–727 (2012). https://doi.org/10.1007/s11747-010-0245-y
17. Lindsey, D.T., Brown, A.M.: Universality of color names. Proc. Natl. Acad. Sci. **103**(44), 16608–16613 (2006). https://doi.org/10.1073/pnas.0607708103. http://www.pnas.org/cgi/doi/10.1073/pnas.0607708103
18. Liu, D., Woodham, D.: Using AI to Predict Rothko Paintings' Auction Prices (2019). https://www.artsy.net/article/artsy-editorial-ai-predict-mark-rothko-paintings-auction-prices
19. Nekvapil, V.: Using the ac4ft-miner procedure in the medical domain [online] (2009). https://theses.cz/id/0abafc/. Accessed 03 Jan 2020
20. Pawlowski, C., Gelich, A., Raś, Z.W.: Can we build recommender system for artwork evaluation? In: Bembenik, R., Skonieczny, Ł., Protaziuk, G., Kryszkiewicz, M., Rybinski, H. (eds.) Intelligent Methods and Big Data in Industrial Applications. SBD, vol. 40, pp. 41–52. Springer, Cham (2019). https://doi.org/10.1007/978-3-319-77604-0_4

21. Powell, L., Gelich, A., Ras, Z.W.: Developing artwork pricing models for online art sales using text analytics. In: Mihálydeák, T., et al. (eds.) IJCRS 2019. LNCS (LNAI), vol. 11499, pp. 480–494. Springer, Cham (2019). https://doi.org/10.1007/978-3-030-22815-6_37

22. Powell, L., Gelich, A., Ras, Z.W.: Applying analytics to artist provided text to model prices of fine art. In: Appice, A., Ceci, M., Loglisci, C., Manco, G., Masciari, E., Ras, Z.W. (eds.) Complex Pattern Mining. SCI, vol. 880, pp. 189–211. Springer, Cham (2020). https://doi.org/10.1007/978-3-030-36617-9_12

23. Powell, L., Gelich, A., Ras, Z.W.: Art innovation systems for value tagging. In: Encyclopedia of Organizational Knowledge, Administration, and Technologies. IGI Global (2020). https://www.igi-global.com/book/encyclopedia-organizational-knowledge-administration-technology/242894

24. Ras, Z.W., Wieczorkowska, A.: Action-rules: how to increase profit of a company. In: Zighed, D.A., Komorowski, J., Żytkow, J. (eds.) PKDD 2000. LNCS (LNAI), vol. 1910, pp. 587–592. Springer, Heidelberg (2000). https://doi.org/10.1007/3-540-45372-5_70

25. Rauch, J., Šimůnek, M.: Action rules and the GUHA method: preliminary considerations and results. In: Rauch, J., Raś, Z.W., Berka, P., Elomaa, T. (eds.) ISMIS 2009. LNCS (LNAI), vol. 5722, pp. 76–87. Springer, Heidelberg (2009). https://doi.org/10.1007/978-3-642-04125-9_11

26. Rauch, J., et al.: LISp-Miner, October 2019. https://lispminer.vse.cz/index.html

27. Rawlins, C., Johnson, P.: Selling on eBay: persuasive communication advice based on analysis of auction item descriptions. J. Strateg. E-Commerce **5**(1&2), 75–81 (2007)

28. Roberson, D., Hanley, J.: Color vision: color categories vary with language after all. Curr. Biol. **17**(15), R605–R607 (2007). https://doi.org/10.1016/j.cub.2007.05.057. http://www.sciencedirect.com/science/article/pii/S0960982207014819

29. Saleh, B., Elgammal, A.: Large-scale classification of fine-art paintings: learning the right metric on the right feature. arXiv:1505.00855 [cs], May 2015. http://arxiv.org/abs/1505.00855

30. Sharma, G., Wu, W., Dalal, E.N.: The CIEDE2000 color-difference formula: implementation notes, supplementary test data, and mathematical observations. Color Res. Appl. **30**(1), 21–30 (2004). https://doi.org/10.1002/col.20070

31. Taylor, G.: Python-colormath (2014)

32. Team, O.: OpenCV, October 2017

33. Tzacheva, A.A., Bagavathi, A., Ayila, L.: Discovery of action rules at lowest cost in spark. In: 2017 IEEE International Conference on Data Mining Workshops (ICDMW), New Orleans, LA, pp. 87–94. IEEE, November 2017. https://doi.org/10.1109/ICDMW.2017.173. http://ieeexplore.ieee.org/document/8215648/

34. Tzacheva, A.A., Raś, Z.W.: Action rules mining. Int. J. Intell. Syst. **20**(7), 719–736 (2005). https://doi.org/10.1002/int.20092. http://doi.wiley.com/10.1002/int.20092

35. Velthuis, O.: Talking Prices: Symbolic Meanings of Prices on the Market for Contemporary Art. Princeton University Press (2005). http://www.jstor.org/stable/j.ctt4cgd14

Metric-Guided Multi-task Learning

Jinfu Ren, Yang Liu, and Jiming Liu[(⊠)]

Department of Computer Science, Hong Kong Baptist University, Kowloon Tong,
Hong Kong SAR, People's Republic of China
{jinfuren,csygliu,jiming}@comp.hkbu.edu.hk

Abstract. Multi-task learning (MTL) aims to solve multiple related learning tasks simultaneously so that the useful information in one specific task can be utilized by other tasks in order to improve the learning performance of all tasks. Many representative MTL methods have been proposed to characterize the relationship between different learning tasks. However, the existing methods have not explicitly quantified the distance or similarity of different tasks, which is actually of great importance in modeling the task relation for MTL. In this paper, we propose a novel method called Metric-guided MTL (M^2TL), which explicitly measures the task distance using a metric learning strategy. Specifically, we measure the distance between different tasks using their projection parameters and learn a distance metric accordingly, so that the similar tasks are close to each other while the uncorrelated tasks are faraway from each other, in terms of the learned distance metric. With a metric-guided regularizer incorporated in the proposed objective function, we open a new way to explore the related information among tasks. The proposed method can be efficiently solved via an alternative method. Experiments on both synthetic and real-world benchmark datasets demonstrate the superiority of the proposed method over existing MTL methods in terms of prediction accuracy.

Keywords: Multi-task learning · Task relation · Metric learning · Metric-guided multi-task learning

1 Introduction

Multi-task learning (MTL), inspired by human learning behavior and patterns of applying the knowledge and experience learned from some tasks to help learn others, solves multiple learning tasks simultaneously and improves the learning performance of all tasks by exploring the task similarities or commonalities [3,20]. MTL has attracted extensive research interest and has been widely applied to various real-world problems such as human facial recognition and pose estimation [4], traffic flow forecasting [5], climate prediction [6], and disease modeling [14].

One of the key issues in MTL is to characterize the relationship between different learning tasks. Some representative methods have been proposed to

© Springer Nature Switzerland AG 2020
D. Helic et al. (Eds.): ISMIS 2020, LNAI 12117, pp. 21–31, 2020.
https://doi.org/10.1007/978-3-030-59491-6_3

address this challenging issue. They can be categorized into two main categories [20]: feature-based MTL and parameter-based MTL. Feature-based MTL characterizes the relationships among different tasks by introducing constraints on the feature representations of given tasks to model the task similarity. According to the property of weight matrix, feature-based MTL can be further categorized into multi-task feature extraction [1,22] and multi-task feature selection [13]. Parameter-based MTL models the relationships among tasks by manipulating the parameters in each task. It can be further divided into four subcategories: low-rank methods that model the task-relation by constraining the rank of a parameter matrix [18]; task clustering methods that group similar tasks into subsets and share parameters within the cluster [8]; task-relation learning methods that describe the relationships among tasks using a specific criterion such as covariance or correlation [21]; and decomposition methods that characterize task relationships by decomposing the parameter matrix to a set of component matrices and introducing constraints on them [7].

Although the aforementioned methods have explored the task relationship from different aspects, they have not explicitly quantified the distance or similarity of different tasks, which is, of course, very important in modeling the intrinsic correlation of multiple learning tasks. To address this problem, in this paper, we introduce a new MTL method called Metric-guided MTL (M^2TL), aiming at explicitly measuring the task distance via a metric learning strategy. Specifically, we represent the distance between different learning tasks using their projection parameters. Accordingly, we propose to learn a distance metric, under which the similar learning tasks are close to each other while the uncorrelated ones are apart from each other. With the formulated distance metric, we introduce a metric-guided regularizer into the objective function of M^2TL. By jointly optimizing the loss function and the metric-guided regularizer, the learned task relationship is expected to well reflect the explicitly quantified similarity between different tasks.

The rest of the paper is organized as follows. Section 2 briefly reviews the related work on feature-based MTL. Section 3 introduces the proposed M^2TL, including the formulation of task distance, the objective function of M^2TL, the optimization procedure, and the computational complexity analysis. Section 4 validates the effectiveness of M^2TL on both synthetic and real-world benchmark datasets, demonstrating the superiority of the proposed method over existing MTL methods in terms of prediction accuracy. Section 5 draws the conclusion of the paper.

2   Related Work

The proposed M^2TL aims to learn the common feature transformation among different tasks. This section, therefore, briefly reviews some related work in feature-based MTL. Generally, the feature-based MTL approaches, including both feature extraction and feature selection methods, can be formulated under the following regularization framework:

$$\arg \min_{\mathbf{W}} Loss(\mathbf{W}) + \mu \mathcal{R}(\mathbf{W}), \tag{1}$$

where $\mathbf{W} \in \mathbb{R}^{d \times T}$ denotes the weight matrix, which is column stacked by each task's weight vector $\mathbf{w}_i$ $(i = 1, \cdots, T)$. Here d and T denote the dimension of features and the number of tasks, respectively. Moreover, $Loss(\mathbf{W})$ denotes the total loss on T learning tasks that we want to minimize, $\mathcal{R}(\mathbf{W})$ denotes the regularizer that characterize the relationship among different tasks, and $\mu \geq 0$ denotes the balancing parameter that adjusts the importance of $Loss(\mathbf{W})$ and that of $\mathcal{R}(\mathbf{W})$.

Different feature-based MTL approaches adopt their own ways to formulate the loss function $Loss(\mathbf{W})$ and the regularizer $\mathcal{R}(\mathbf{W})$, so that the relationship between different tasks can be modeled and the learning performance of all tasks can be optimized simultaneously. In [1], Argyriou et al. aimed to learn a square transformation matrix for features, which lies in the assumption that transformed feature space is more powerful than the original. They further proposed a convex formulation [2] to solve the optimization problem. In [11,13], the $L_{2,1}$-norm was used as the regularizer to select common features shared across different tasks. A more general form, the $L_{p,q}$-norm, can also be utilized to select the common or shared features as it owns the property of sparsity as well [19]. However, the $L_{p,q}$-norm may perform even worse when the value of shared features are highly uneven [9]. To address this issue, Jalali et al. proposed a method called dirty MTL in which the weight matrix is decomposed into two components, one to ensure block-structured row-sparsity and the other for element-wise sparsity with different regularizers imposed [9].

The regularizers adopted in the above methods have explored the relationship between different tasks from various perspectives. However, none of them has explicitly quantified the distance/similarity between different tasks, which is of vital importance in modeling the task correlation in MTL. This observation motivates us to propose a new MTL method, which can measure the distance between different tasks in an explicit way. Metric learning [16], which aims to learn a distance metric to reflect the intrinsic similarity/correlation between data samples, becomes a natural choice to model the task distance in our case. In recent years, metric learning has been widely used in various applications, such as healthcare [12], person re-identification [17], and instance segmentation [10]. Different from existing metric learning methods that work on data samples, the nature of MTL problem requires us to define the distance metric over tasks, so that the correlation between different tasks can be explicitly measured and integrated into the learning objectives.

3   Proposed Method

In this section, we introduce the proposed M²TL. First, we define the formulation of task distance metric. Based on that, we propose the objective function of M²TL and describe the optimization procedure. Finally, we analyze the computational complexity of the proposed method.

3.1   Task Distance Metric

For each task, the weight vector is a meaningful index to represent the information learned from the corresponding task. Therefore, we define a task distance metric based on weight vectors of different tasks:

$$D(t_i, t_j) = (\mathbf{w}_i - \mathbf{w}_j)^T \mathbf{M} (\mathbf{w}_i - \mathbf{w}_j),\ i, j = 1, \cdots, T, \qquad (2)$$

where t_i and t_j denote the i-th task and the j-th task, respectively; $\mathbf{w}_i$ and $\mathbf{w}_j$ denote the weight vector learned for the i-th task and that for the j-th task, respectively; and $D(t_i, t_j)$ denotes the distance between task i and task j. Similar to the typical distance metric learning, we use the Mahalanobis matrix $\mathbf{M}$ to flexibly adjust the importance of different dimensions. Note that if $\mathbf{M}$ equals to the identity matrix, then the defined distance metric will reduce to the Euclidean distance.

3.2   Objective Function of M²TL

With the above definition on task distance metric, we can formulate the objective function of the proposed M²TL. Given T regression tasks and the corresponding datasets $\{(\mathbf{X}_1, \mathbf{y}_1), (\mathbf{X}_2, \mathbf{y}_2), \cdots, (\mathbf{X}_T, \mathbf{y}_T)\}$, where $\mathbf{X}_i \in \mathbb{R}^{n_i \times d}$ and $\mathbf{y}_i \in \mathbb{R}^{n_i}$ $(i = 1, \cdots, T)$ denote the training samples and the corresponding labels in the i-th task respectively, and n_i denotes the number of samples in the i-th task. In this paper, we use regression tasks as an example to show the formulation of the proposed M²TL. In fact, the idea of the proposed method can be extended to other supervised/unsupervised learning tasks, such as classification tasks and clustering tasks, in a straightforward way. The proposed M²TL specifies the general formulation in Eq. (1) as follows:

$$\arg \min_{\mathbf{W}, \mathbf{M}} Loss(\mathbf{W}) + \frac{\mu}{T^2} \sum_{i=1}^{T} \sum_{j=1}^{T} D(t_i, t_j) + \|\mathbf{M} - \mathbf{Q}\|_F^2. \qquad (3)$$

The first term in Eq. (3) represents the total loss of T tasks as mentioned previously. Without loss of generality, we utilize the least squares formulation as the loss function in our model: $Loss(\mathbf{W}) = \sum_{i=1}^{T} \frac{1}{2n_i} \|\mathbf{X}_i \mathbf{w}_i - \mathbf{y}_i\|_2^2$. Note that $Loss(\mathbf{W})$ in Eq. (3) can be any loss function according to different learning requirements. The second term in Eq. (3) is the task distance formulated in the previous subsection. By minimizing the summation of all task distances, we expect that the commonality/similarity among different tasks, which is generally hidden in the original feature space, can be extracted to the maximum extent. In addition to the loss function and the regularizer, we further introduce the last term in Eq. (3) to avoid the trivial solution on $\mathbf{M}$ by constraining it using a task correlation/covariance matrix $\mathbf{Q}$. Here $\mathbf{Q}$ can be any correlation/covariance matrix that captures the relationship between different tasks. In our paper, we use the Pearson correlation coefficient matrix calculated by the initialization of $\mathbf{W}$ because of its universality.

3.3   Optimization Procedure

To the best of our knowledge, there is no closed-form solution for the optimization problem in Eq. (3). Therefore, we use an alternating method to find the optimal $\mathbf{M}$ and $\mathbf{W}$ iteratively, which guarantees the optimality in each iteration as well as the local optimum of the solution. The detailed optimization procedure is described as follows:

Fix W and update M: With the fixed $\mathbf{W}$, the objective function in Eq. (3) can be rewritten as follows:

$$
\begin{aligned}
&\arg\min_{\mathbf{W},\mathbf{M}} Loss(\mathbf{W}) + \frac{\mu}{T^2}\sum_{i=1}^{T}\sum_{j=1}^{T} D(t_i,t_j) + \|\mathbf{M} - \mathbf{Q}\|_F^2 \\
=\ &\arg\min_{\mathbf{M}} \frac{\mu}{T^2}\sum_{i=1}^{T}\sum_{j=1}^{T}(\mathbf{w}_i - \mathbf{w}_j)^T\mathbf{M}(\mathbf{w}_i - \mathbf{w}_j) + \|\mathbf{M} - \mathbf{Q}\|_F^2 \\
=\ &\arg\min_{\mathbf{M}} \frac{\mu}{T^2}tr(\mathbf{M}\sum_{i,j}^{T}(\mathbf{w}_i - \mathbf{w}_j)(\mathbf{w}_i - \mathbf{w}_j)^T) + \|\mathbf{M} - \mathbf{Q}\|_F^2 \quad\quad (4)\\
=\ &\arg\min_{\mathbf{M}} \frac{2\mu}{T^2}tr(\mathbf{M}(T\sum_{i=1}^{T}\mathbf{w}_i\mathbf{w}_i^T - \sum_{i,j}^{T}\mathbf{w}_i\mathbf{w}_j^T)) + \|\mathbf{M} - \mathbf{Q}\|_F^2 \\
=\ &\arg\min_{\mathbf{M}} \frac{2\mu}{T^2}tr(\mathbf{MWLW}^T) + \|\mathbf{M} - \mathbf{Q}\|_F^2,
\end{aligned}
$$

where $\mathbf{L}$ is a $T \times T$ Lagrange matrix defined as: $\mathbf{L} = T\mathbf{I}_T - \mathbf{1}_T\mathbf{1}_T^T$, with $\mathbf{I}_T$ and $\mathbf{1}_T$ being the T-dimensional identity matrix and all one column vector, respectively. Taking the derivative of the objective function in Eq. (4) with respect to $\mathbf{M}$ and set it to zero, i.e.,

$$
\frac{\partial\left[\frac{2\mu}{T^2}tr(\mathbf{MWLW}^T) + \|\mathbf{M} - \mathbf{Q}\|_F^2\right]}{\partial\mathbf{M}} = 0, \quad\quad (5)
$$

then we can obtain the update of $\mathbf{M}$:

$$
\mathbf{M} = \mathbf{Q} - \frac{\mu}{T^2}\mathbf{WLW}^T. \quad\quad (6)
$$

Algorithm 1: Metric-guided Multi-Task Learning (M^{2}TL)

---

1 **Input**: Training set for T learning tasks: $\{\mathbf{X}_i \in \mathbb{R}^{n_i \times d}, \mathbf{y}_i \in \mathbb{R}^{n_i}\}_{i=1}^T$

2 **Output**: Weight matrix $\mathbf{W} = [\mathbf{w}_1, \cdots, \mathbf{w}_T] \in \mathbb{R}^{d \times T}$

   1: **for** $i \leftarrow 1 : T$ **do**

   2:    Initialize $\mathbf{w}_i$: $\mathbf{w}_i \leftarrow \mathbf{X}_i^T \mathbf{y}_i$;

   3: **end for**

   4: **while** *not convergence* **do**

   5:    Update $\mathbf{M}$ using Eq. (6);

   6:    **for** $i \leftarrow 1 : T$ **do**

   7:      Update $\mathbf{w}_i$ using Eq. (9);

   8:    **end for**

   9: **end while**

Fix M and update W: With the fixed $\mathbf{M}$, the objective function in Eq. (3) can be rewritten as follows:

$$\arg\min_{\mathbf{W},\mathbf{M}} Loss(\mathbf{W}) + \frac{\mu}{T^2} \sum_{i=1}^T \sum_{j=1}^T D(t_i, t_j) + \|\mathbf{M} - \mathbf{Q}\|_F^2$$

$$= \arg\min_{\mathbf{W}} Loss(\mathbf{W}) + \frac{\mu}{T^2} \sum_{i=1}^T \sum_{j=1}^T (\mathbf{w}_i - \mathbf{w}_j)^T \mathbf{M}(\mathbf{w}_i - \mathbf{w}_j)$$

$$= \arg\min_{\mathbf{W}} \sum_{i=1}^T \frac{1}{2n_i} \|\mathbf{X}_i \mathbf{w}_i - \mathbf{y}_i\|_2^2 + \frac{\mu}{T^2} \sum_{i=1}^T \sum_{j=1}^T (\mathbf{w}_i - \mathbf{w}_j)^T \mathbf{M}(\mathbf{w}_i - \mathbf{w}_j)$$

$$= \arg\min_{\mathbf{W}} \sum_{i=1}^T (\frac{1}{2n_i} \|\mathbf{X}_i \mathbf{w}_i - \mathbf{y}_i\|_2^2 + \frac{\mu}{T^2} \sum_{j=1}^T (\mathbf{w}_i - \mathbf{w}_j)^T \mathbf{M}(\mathbf{w}_i - \mathbf{w}_j)).$$

(7)

Note that in the above formulation, each task can be updated individually. Therefore, for the i-th task, we fix the $1, ..., i-1, i+1, ...T$-th tasks, take the derivative with respect to $\mathbf{w}_i$, and set it to zero, then we have:

$$\frac{1}{n_i} \mathbf{X}_i^T (\mathbf{X}_i \mathbf{w}_i - \mathbf{y}_i) + \frac{4\mu}{T^2} \sum_{\substack{j=1 \\ j \neq i}}^T \mathbf{M}(\mathbf{w}_i - \mathbf{w}_j) = 0. \tag{8}$$

From Eq. (8), we can obtain the updating rule of $\mathbf{w}_i$:

$$\mathbf{w}_i = (\frac{1}{n_i} \mathbf{X}_i^T \mathbf{X}_i + \frac{4\mu(T-1)}{T} \mathbf{M})^{-1} (\frac{1}{n_i} \mathbf{X}_i^T \mathbf{y}_i + \frac{4\mu}{T} \mathbf{M} \sum_{\substack{j=1 \\ j \neq i}}^T \mathbf{w}_j). \tag{9}$$

The details of the proposed M^{2}TL are described in Algorithm 1.

3.4 Computational Complexity Analysis

In Algorithm 1, the most time-consuming steps are steps 4–9. The time complexity of step 5, i.e., updating $\mathbf{M}$, is $O(dT^2)$. The time complexity of step 7, i.e., updating $\mathbf{w}_i$, is $O(d^3)$. Assume that t is the number of iterations in the outside *while* loop for convergence, then the total computational complexity of the proposed M^{2}TL is $O(t(dT^2 + d^3T))$.

4 Experimental Results

In this section, we validate the performance of the proposed method on both synthetic and real-world datasets. We use two standard criteria in MTL for performance evaluation: the root mean squared error (RMSE) [23] and the normalized mean squared error (NMSE), which are defined as follows:

$$RMSE = \frac{\sum_{i=1}^m \|\mathbf{X}_i^T \mathbf{w}_i - \mathbf{y}_i\|_2 \times n_i}{\sum_{i=1}^m n_i}, \quad NMSE = \frac{\sum_{i=1}^m MSE_i/var(\mathbf{y}_i) \times n_i}{\sum_{i=1}^m n_i},$$
$$(10)$$

where n_i is the number of samples in i-th task, and $MSE = \frac{1}{n}\sum_{i=1}^n (y_i - \mathbf{x}_i \mathbf{w}_i)^2$ denotes the mean square error. We select the following baselines for performance comparison:

- **STL** [15]: the classical single-task learning method, which learns each task independently without modeling the task relationship. Here we employ Lasso as the STL model.
- **L21** [13]: a typical multi-task feature selection method, which uses $L_{2,1}$-norm to achieve the row sparsity of weight matrix.
- **DirtyMTL** [9]: a representative dirty multi-task learning method, which decomposes the weight matrix into two components and regularizes these two components separately to overcome the shortage of $L_{q,p}$-norm.
- **MTFSSR** [19]: a state-of-the-art multi-task feature selection method with sparse regularization, which extends the $L_{1,2}$-norm regularization to the multi-task setting for capturing common features and extracting task-specific features simultaneously.

4.1 Experiments on Synthetic Dataset

In this subsection, we examine the convergence and the prediction accuracy of the proposed method on a synthetic dataset. We generate the data of T regression tasks. Each task includes N data samples. The data of the i-th task are generated from the normal distribution $\mathcal{N}(i/10, 1)$. The ground truth of the weight matrix, $\mathbf{W}^{(truth)}$, is generated from $\mathcal{N}(1, 1)$. The labels are then generated by: $\mathbf{y}_i = \mathbf{X}_i \mathbf{w}_i^{(truth)} + \epsilon$ accordingly, where ϵ is the Gaussian noise generated from $\mathcal{N}(0, 0.1)$. In the experiments, we assume that $\mathbf{W}^{(truth)}$ is unknown and aim to learn it from the training sets $\{\mathbf{X}_i, \mathbf{y}_i\}$ ($i = 1, \cdots, T$). We set $T = 10$, $d = 10$, and $N = 30$ in our experiments.

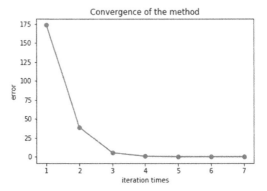

Fig. 1. Convergence speed of the proposed method on the synthetic dataset. The horizontal axis represents the number of iterations while the vertical axis represents the objective value of $\|\mathbf{W}^t - \mathbf{W}^{t-1}\|_F$. The red curve in the figure shows that $\mathrm{M}^2\mathrm{TL}$ can converge to a stable solution quickly. (Color figur online)

We first examine the convergence speed of the proposed method. We determine the stopping point by evaluating the difference between $\mathbf{W}^t$ and $\mathbf{W}^{t-1}$. Specifically, we consider the algorithm as convergent and stop the iteration once the following inequality is satisfied: $\|\mathbf{W}^t - \mathbf{W}^{t-1}\|_F \leq \epsilon$. In this experiment, we set $\epsilon = 0.001$. Figure 1 shows the objective value of $\|\mathbf{W}^t - \mathbf{W}^{t-1}\|_F$ versus the number of iterations. The objective value decreases dramatically and satisfies the stopping criterion within only 7 iterations, demonstrating that the proposed method can converge to a stable solution quickly.

Table 1. The performance (in terms of RMSE) of STL [15], L21 [13], DirtyMTL [9], MTFSSR [19], and the proposed $\mathrm{M}^2\mathrm{TL}$ on the synthetic dataset, with three different training ratios. The best performances are highlighted in bold.

Training No.	Methods				
	STL	L21	DirtyMTL	MTFSSR	$\mathrm{M}^2\mathrm{TL}$
2	60.49 ± 12.45	31.07 ± 9.24	29.56 ± 4.64	38.90 ± 8.95	$\mathbf{25.40 \pm 12.26}$
4	53.36 ± 12.02	25.22 ± 3.23	25.69 ± 5.60	25.55 ± 7.24	$\mathbf{24.93 \pm 13.75}$
6	46.63 ± 19.95	18.70 ± 2.15	20.33 ± 5.53	18.56 ± 4.64	$\mathbf{18.44 \pm 8.39}$

In the second experiment, we compare the performance of the proposed method with that of four aforementioned baselines. We select 2, 4, and 6 samples from the 30 samples for training and use the rest for testing. We repeat the experiment for 10 times on randomly selected training samples and report the average RMSE as well as the standard deviation of each method. Table 1 lists the results of all methods. Obviously, with the exploration on the relationship between different tasks, the MTL methods (including L21, DirtyMTL, MTFSSR, and the proposed $\mathrm{M}^2\mathrm{TL}$) achieve the lower RMSE than the STL method.

By further modeling the task distance in an explicit way, the proposed method outperforms other three MTL methods in all scenarios.

4.2   Experiments on Real-World Multi-task Datasets

In this subsection, we conduct experiments on two real-world multi-task datasets: the School dataset (https://github.com/jiayuzhou/MALSAR/tree/master/data) and the Sarcos dataset (http://www.gaussianprocess.org/gpml/data/). The School dataset is commonly used in many MTL literature. It is provided by the Inner London Education Authority and contains $15,362$ data samples from 139 schools where each data sample has 27 attributes. We treat each school as a task and the learning target is to predict the exam score. The Sarcos dataset is about the inverse dynamic problem. The learning target is to predict the 7 joint torques given the 7 joint positions, 7 joint velocities and 7 joint accelerations. Here we treat prediction of one joint torque as a task so we have 7 tasks in total. For all 7 tasks, the 21 features (7 joint positions, 7 joint velocities and 7 joint accelerations) are used as the input, so the training data for different tasks are the same. We select 200 samples from each task to conduct our experiment. For both School and Sarcos datasets, we 20%, 30%, 40% of data for training and use the rest for testing. Similar to the synthetic experiments, we repeat the experiment for 10 times on randomly selected training samples. We report the average NMSE as well as the standard deviation of each method.

Tables 2 and 3 report the performance of all five methods on the School dataset and the Sarcos dataset, respectively. With the increase of training ratio, the NMSE of all methods decreases (except the L21 method from 20% to 30%), which is consistent with the common observation that providing more training data is generally beneficial to the learning task. Moreover, similar to the observations in the synthetic experiments, the MTL methods perform better than the STL method while our method again achieves the lowest prediction error among all five methods, demonstrating the necessity of exploring the task relationship in MTL and the effectiveness of the proposed formulation.

Table 2. The performance (in terms of NMSE) of STL [15], L21 [13], DirtyMTL [9], MTFSSR [19], and the proposed M^2TL on the School dataset, with three different training ratios. The best performances are highlighted in bold.

Training ratio	Methods				
	STL	L21	DirtyMTL	MTFSSR	M^2TL
20%	2.118 ± 0.104	0.963 ± 0.005	$\mathbf{0.924 \pm 0.001}$	0.929 ± 0.000	0.931 ± 0.001
30%	2.052 ± 0.008	1.061 ± 0.007	0.858 ± 0.001	0.862 ± 0.000	$\mathbf{0.855 \pm 0.007}$
40%	2.013 ± 0.001	1.016 ± 0.001	0.821 ± 0.003	0.820 ± 0.000	$\mathbf{0.809 \pm 0.003}$

Table 3. The performance (in terms of NMSE) of STL [15], L21 [13], DirtyMTL [9], MTFSSR [19], and the proposed M^2TL on the Sarcos dataset, with three different training ratios. The best performances are highlighted in bold.

Training ratio	Methods				
	STL	L21	DirtyMTF	MTFSSR	M^2TL
20%	2.180 ± 0.037	0.309 ± 0.113	0.294 ± 0.001	0.300 ± 0.004	$\mathbf{0.291 \pm 0.004}$
30%	2.052 ± 0.026	0.216 ± 0.001	0.226 ± 0.001	0.216 ± 0.002	$\mathbf{0.211 \pm 0.002}$
40%	2.000 ± 0.091	0.208 ± 0.001	0.202 ± 0.000	0.194 ± 0.000	$\mathbf{0.189 \pm 0.001}$

5   Conclusions

In this paper, we proposed a novel multi-task learning method called Metric-guided Multi-Task Learning (M^2TL), which learns a task distance metric to explicitly measure the distance between different tasks and uses the learned metric as a regularizer to model the multi-task correlation. In the future, we plan to extend our experimental evaluation on large-scale datasets with distributed strategy in order to overcome the issues of quadratic time compelxity with respect to the task number and cubic time complexity with respect to the feature dimension. Moreover, we will investigate more sophisticated ways to model and learn the task distance metric.

Acknowledgement. The authors would like to thank the anonymous reviewers for their valuable comments and suggestions. This work was supported by the Grants from the Research Grant Council of Hong Kong SAR under Projects RGC/HKBU12201318 and RGC/HKBU12201619.

References

1. Argyriou, A., Evgeniou, T., Pontil, M.: Multi-task feature learning. In: Proceedings 19th NIPS, pp. 41–48 (2007)
2. Argyriou, A., Evgeniou, T., Pontil, M.: Convex multi-task feature learning. Mach. Learn. **73**(3), 243–272 (2008)
3. Caruana, R.: Multitask learning. Mach. Learn. **28**(1), 41–75 (1997)
4. Chu, X., Ouyang, W., Yang, W., Wang, X.: Multi-task recurrent neural network for immediacy prediction. In: Proceedings 15th ICCV, pp. 3352–3360 (2015)
5. Deng, D., Shahabi, C., Demiryurek, U., Zhu, L.: Situation aware multi-task learning for traffic prediction. In: Proceedings 17th ICDM, pp. 81–90 (2017)
6. Goncalves, A., Banerjee, A., Zuben, F.V.: Spatial projection of multiple climate variables using hierarchical multitask learning. In: Proceedings 31th AAAI, pp. 4509–4515 (2017)
7. Gong, P., Ye, J., Zhang, C.: Robust multi-task feature learning. In: Proceedings 18th SIGKDD, pp. 895–903 (2012)
8. Jacob, L., Vert, J.P., Bach, F.R.: Clustered multi-task learning: a convex formulation. In: Proceedings 21th NIPS, pp. 745–752 (2009)
9. Jalali, A., Sanghavi, S., Ruan, C., Ravikumar, P.K.: A dirty model for multi-task learning. In: Proceedings 22th NIPS, pp. 964–972 (2010)

10. Lahoud, J., Ghanem, B., Pollefeys, M., Oswald, M.R.: 3d instance segmentation via multi-task metric learning (2019). arXiv preprint arXiv:1906.08650
11. Liu, J., Ji, S., Ye, J.: Multi-task feature learning via efficient l 2, 1-norm minimization. In: Proceedings 25th UAI, pp. 339–348 (2009)
12. Ni, J., Liu, J., Zhang, C., Ye, D., Ma, Z.: Fine-grained patient similarity measuring using deep metric learning. In: Proceedings 26th CIKM, pp. 1189–1198. ACM (2017)
13. Obozinski, G., Taskar, B., Jordan, M.: Multi-task feature selection. University of California, Berkeley, Technical report (2006)
14. Pei, H., B.Yang, Liu, J., Dong, L.: Group sparse Bayesian learning for active surveillance on epidemic dynamics. In: Proceedings 32th AAAI, pp. 800–807 (2018)
15. Tibshirani, R.: Regression shrinkage and selection via the lasso. J. Roy. Stat. Soc. Ser. B (Methodological) **58**(1), 267–288 (1996)
16. Xing, E.P., Jordan, M.I., Russell, S.J., Ng, A.Y.: Distance metric learning with application to clustering with side-information. In: Proceedings 15th NIPS, pp. 521–528 (2003)
17. Yu, H.X., Wu, A., Zheng, W.S.: Cross-view asymmetric metric learning for unsupervised person re-identification. In: Proceedings 16th ICCV, pp. 994–1002 (2017)
18. Zhang, J., Ghahramani, Z., Yang, Y.: Learning multiple related tasks using latent independent component analysis. In: Proceedings 18th NIPS, pp. 1585–1592 (2006)
19. Zhang, J., Miao, J., Zhao, K., Tian, Y.: Multi-task feature selection with sparse regularization to extract common and task-specific features. Neurocomputing **340**, 76–89 (2019)
20. Zhang, Y., Yang, Q.: A survey on multi-task learning (2018). arXiv:1707.08114v2
21. Zhang, Y., Yeung, D.Y.: Multi-task boosting by exploiting task relationships. In: Flach, P.A., De Bie, T., Cristianini, N. (eds.) Proceedings 22th ECML PKDD, pp. 697–710 (2012)
22. Zhang, Z., Luo, P., Loy, C.C., Tang, X.: Facial landmark detection by deep multi-task learning. In: Proceedings 13th ECCV, pp. 94–108 (2014)
23. Zhou, J., Chen, J., Ye, J.: Malsar: multi-task learning via structural regularization. Arizona State University, Technical report (2011)

Sentiment Analysis with Contextual Embeddings and Self-attention

Katarzyna Biesialska[1](✉) (iD), Magdalena Biesialska[1](✉) (iD),
and Henryk Rybinski[2](✉) (iD)

[1] Universitat Politècnica de Catalunya, Barcelona, Spain
{katarzyna.biesialska,magdalena.biesialska}@upc.edu
[2] Warsaw University of Technology, Warsaw, Poland
h.rybinski@ii.pw.edu.pl

Abstract. In natural language the intended meaning of a word or phrase is often implicit and depends on the context. In this work, we propose a simple yet effective method for sentiment analysis using contextual embeddings and a self-attention mechanism. The experimental results for three languages, including morphologically rich Polish and German, show that our model is comparable to or even outperforms state-of-the-art models. In all cases the superiority of models leveraging contextual embeddings is demonstrated. Finally, this work is intended as a step towards introducing a universal, multilingual sentiment classifier.

Keywords: Sentiment classification · Deep learning · Word embeddings

1 Introduction

All areas of human life are affected by people's views. With the sheer amount of reviews and other opinions over the Internet, there is a need for automating the process of extracting relevant information. For machines, however, measuring sentiment is not an easy task, because natural language is highly ambiguous at all levels, and thus difficult to process. For instance, a single word can hardly convey the whole meaning of a statement. Moreover, computers often do not distinguish literal from figurative meaning or incorrectly handle complex linguistic phenomena, such as: sarcasm, humor, negation etc.

In this paper, we take a closer look at two factors that make automatic opinion mining difficult – the problem of representing text information, and sentiment analysis (SA). In particular, we leverage contextual embeddings, which enable to convey a word meaning depending on the context it occurs in. Furthermore, we build a hierarchical multi-layer classifier model, based on an architecture of the Transformer encoder [32], primarily relying on a self-attention mechanism

---

K. Biesialska and M. Biesialska—Contributed equally to this work, which was mostly done at the Warsaw University of Technology.

ⓒ Springer Nature Switzerland AG 2020
D. Helic et al. (Eds.): ISMIS 2020, LNAI 12117, pp. 32–41, 2020.
https://doi.org/10.1007/978-3-030-59491-6_4

and bi-attention. The proposed sentiment classification model is language independent, which is especially useful for low-resource languages (e.g. Polish).

We evaluate our methods on various standard datasets, which allows us to compare our approach against current state-of-the-art models for three languages: English, Polish and German. We show that our approach is comparable to the best performing sentiment classification models; and, importantly, in two cases yields significant improvements over the state of the art.

The paper is organized as follows: Sect. 2 presents the background and related work. Section 3 describes our proposed method. Section 4 discusses datasets, experimental setup, and results. Section 5 concludes this paper and outlines the future work.

2 Related Work

Sentiment classification has been one of the most active research areas in natural language processing (NLP) and has become one of the most popular downstream tasks to evaluate performance of neural network (NN) based models. The task itself encompasses several different opinion related tasks, hence it tackles many challenging NLP problems, see e.g. [16,20].

2.1 Sentiment Analysis Approaches

The first fully-formed techniques for SA emerged around two decades ago, and continued to be prevalent for several years, until deep learning methods entered the stage. The most straight-forward method, developed in [30], is based on the number of positive and negative words in a piece of text. Concretely, the text is assumed to have positive polarity if it contains more positive than negative terms, and vice versa. Of course, the term-counting method is often insufficient; therefore, an improved method was proposed in [10], which combines counting positive and negative terms with a machine learning (ML) approach (i.e. Support Vector Machine).

Various studies (e.g. [31]) have shown that one can determine the polarity of an unknown word by calculating co-occurrence statistics of it. Moreover, classical solutions to the SA problem are often based on lexicons. Traditional lexicon-based SA leverages word-lists, that are pre-annotated with positive and negative sentiment. Therefore, for many years lexicon-based approaches have been utilized when there was insufficient amount of labeled data to train a classifier in a fully supervised way.

In general, ML algorithms are popular methods for determining sentiment polarity. A first ML model applied to SA has been implemented in [21]. Moreover, throughout the years, different variants of NN architectures have been introduced in the field of SA. Especially recursive neural networks [22], such as recurrent neural networks (RNN) [13,28,29], or convolutional neural networks (CNN) [9, 11] have become the most prevalent choices.

2.2   Vector Representations of Words

One of the principal concepts in linguistics states that related words can be used in similar ways [6]. Importantly, words may have different meaning in different contexts. Nevertheless, until recently it has been a dominant approach (e.g. word2vec [19], GloVe [23]) to learn representations such that each and every word has to capture all its possible meanings.

However, lately a new set of methods to learn dynamic representations of words has emerged [5,7,18,24,25]. These approaches allow each word representation to capture what a word means in a particular context. While every word token has its own vector, the vector can depend on a variable-length sequence of nearby words (i.e. context). Consequently, a context vector is obtained by feeding a neural network with these context word vectors and subsequently encoding them into a single fixed-length vector.

ULMFiT [7] was the very first method to induce contextual word representations by harnessing the power of language modeling. The authors proposed to learn contextual embeddings by pre-training a language model (LM), and then performing task-specific fine-tuning. ULMFiT architecture is based on a vanilla 3-layer Long Short-Term Memory (LSTM) NN without any attention mechanism.

The other contextual embedding model introduced recently is called ELMo (Embeddings from Language Models) [24]. Similarly to ULMFiT, this model uses tokens at the word-

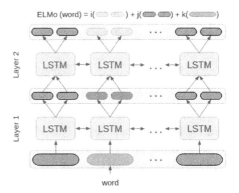

Fig. 1. The architecture of ELMo.

level. ELMo contextual embeddings are "deep" as they are a function of all hidden states. Concretely, context-sensitive features are extracted from a left-to-right and a right-to-left 2-layer bidirectional LSTM language models. Thus, the contextual representation of each word is the concatenation of the left-to-right and right-to-left representations as well as the initial embedding (see Fig. 1).

The most recent model – BERT [5] – is more sophisticated architecturally-wise, as it is a multi-layer masked LM based on the Transformer NN utilizing subword tokens. However, as we are bound to use word-level tokens in our sentiment classifier, we leverage the ELMo model for obtaining contextual embeddings. More specifically, by means of ELMo we are able to feed our classifier model with context-aware embeddings of an input sequence. Hence, in this setting we do not perform any fine-tuning of ELMo on a downstream task.

2.3   Self-attention Deep Neural Networks

The attention mechanism was introduced in [3] in 2014 and since then it has been applied successfully to different computer vision (e.g. visual explanation) and NLP (e.g. machine translation) tasks. The mechanism is often used as an extra source of information added on top of the CNN or LSTM model to enhance the extraction of sentence embedding [15, 26]. However, this scenario is not applicable to sentiment classification, since the model only receives a single sentence on input, hence there is no such extra information [15].

Self-attention (or intra-attention) is an attention mechanism that computes a representation of a sequence by relating different positions of a single sequence. Previous work on sentiment classification has not covered extensively attention-based neural network models for SA (especially using the Transformer architecture [32]), although some papers have appeared recently [2, 14].

3   The Proposed Approach

Our proposed model, called Transformer-based Sentiment Analysis (TSA) (see Fig. 2), is based on the recently introduced Transformer architecture [32], which has provided significant improvements for the neural machine translation task. Unlike RNN or CNN based models, the Transformer is able to learn dependencies between distant positions. Therefore, in this paper we show that attention-based models are suitable for other NLP tasks, such as learning distributed representations and SA, and thus are able to improve the overall accuracy.

The architecture of the TSA model and steps to train it can be summarized as follows:

(a) At the very beginning there is a simple text pre-processing method that performs text clean-up and splits text into tokens.

(b) We use contextual word representations to represent text as real-valued vectors.

(c) After embedding the text into real-valued vectors, the Transformer network maps the input sequence into hidden states using self-attention.

(d) Next a bi-attention mechanism is utilized to estimate the interdependency between representations.

(e) A single layer LSTM together with self-attentive pooling compute the pooled representations.

(f) A joint representation for the inputs is later passed to a fully-connected neural network.

(g) Finally, a softmax layer is used to determine sentiment of the text.

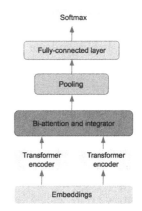

Fig. 2. An overview of the TSA model architecture.

3.1   Embeddings and Encoded Positional Information

Non-recurrent models, such as deep self-attention NN, do not necessarily process the input sequence in a sequential manner. Hence, there is no way they can record the position of each word in a sequence, which is an inherent limitation of every such model. Therefore, in the case of the Transformer, the need has been addressed in the following manner – the Transformer takes into account the order of the words in the input sequence by encoding their position information in extra vectors (so called positional encoding vectors) and adding them to input embeddings. There are many different approaches to embed position information, such as learned or fixed positional encodings (PE), or recently introduced relative position representations (RPR) [27]. The original Transformer used sine and cosine functions of different frequencies.

In this work, we explore the effectiveness of applying a modified attention function to what was proposed in [32]. In the vanilla Transformer a dot-product (multiplicative) attention is used with the scaling factor of $\sqrt{d_z}$. We also propose a different approach to incorporate positional information into the model, namely using RPR instead of PE. Furthermore, we use global average pooling in order to average the output of the last self-attention layer and prepare the model for the final classification layer.

3.2   The Transformer Encoder

The input sequence is combined with word and positional embeddings, which provide time signal, and together are fed into an encoder block. Matrices for a query Q, a key K and a value V are calculated and passed to a self-attention layer. Next, a normalization is applied and residual connections provide additional context. Further, a final dense layer with vocabulary size generates the output of the encoder. A fully-connected feed-forward network within the model is a single hidden layer network with a ReLU activation function in between:

$$FFN(x) = \max\left(0, xW_1 + b_1\right)W_2 + b_2 \tag{1}$$

3.3   Self-attention Layer

The self-attention block in the encoder is called multi-head self-attention. A self-attention layer allows each position in the encoder to access all positions in the previous layer of the encoder immediately, and in the first layer all positions in the input sequence. The multi-head self-attention layer employs h parallel self-attention layers, called heads, with different Q, K, V matrices obtained for each head. In a nutshell, the attention mechanism in the Transformer architecture relies on a scaled dot-product attention, which is a function of Q and a set of K-V pairs. The computation of attention is performed in the following order. First, a multiplication of a query and transposed key is scaled through the scaling factor of $1/\sqrt{d_z}$ (Eq. 2)

$$m_{ij} = \frac{QK^T}{\sqrt{d_z}} \tag{2}$$

Next, the attention is produced using the softmax function over their scaled inner product:

$$\alpha_{ij} = \frac{e^{m_{ij}}}{\sum_{k=1}^{n} e^{m_{ik}}} \qquad (3)$$

Finally, the weighted sum of each attention head and a value is calculated as follows:

$$z_i = \sum_{j=1}^{n} \alpha_{ij} V \qquad (4)$$

3.4 Masking and Pooling

Similar to other sources of data, the datasets used for training and evaluation of our models contain sequences of different length. The most common approach in the literature involves finding a maximal sequence length existing in the dataset/batch and padding sentences that are shorter than the longest one with trailing zeroes. In the proposed TSA model, we deal with the problem of variable-length sequences by using masking and self-attentive pooling. The inspiration for our approach comes from the BCN model proposed in [18]. Thanks to this mechanism, we are able to fit sequences of different length into the final fixed-size vector, which is required for the computation of the sentiment score. The self-attentive pooling layer is applied just after the encoder block.

4 Experiments

4.1 Datasets

In this work, we compare sentiment analysis results considering four benchmark datasets in three languages. All datasets are originally split into training, dev and test sets. Below we describe these datasets in more detail (Table 1).

Table 1. Sentiment analysis datasets with number of classes and train/dev/test split.

Dataset	# Classes	Train	Dev	Test	Domain	Language
SST-2	2	6,920	872	1,821	Movies	English
SST-5	5	8,544	1,101	2,210	Movies	English
PolEmo 2.0-IN	5	5,783	723	722	Medical, hotels	Polish
GermEval	3	19,432	2,369	2,566	Travel, transport	German

Stanford Sentiment Treebank (SST). This collection of movie reviews [28] from the `rottentomatoes.com` is annotated for the binary (SST-2) and fine-grained (SST-5) sentiment classification. SST-2 divides reviews into two groups: *positive* and *negative*, while SST-5 distinguishes 5 different review types: *very* *positive, positive, neutral, negative, very negative*. The dataset consists of 11,855 single sentences and is widely used in the NLP community.

PolEmo 2.0. The dataset [12] comprises online reviews from education, medicine and hotel domains. There are two separate test sets, to allow for in-domain (medicine and hotels) and out-of-domain (products and university) evaluation. The dataset comes with the following sentiment labels: *strong positive, weak positive, neutral, weak negative, strong negative*, and *ambiguous*.

GermEval. This dataset [33] contains customer reviews of the railway operator (Deutsche Bahn) published on social media and various web pages. Customers expressed their feedback regarding the service of the railway company (e.g. travel experience, timetables, etc.) by rating it as *positive, negative*, or *neutral*.

4.2   Experimental Setup

Pre-processing of input datasets is kept to a minimum as we perform only tokenization when required. Furthermore, even though some datasets, such as SST or GermEval, provide additional information (i.e. phrase, word or aspect-level annotations), for each review we only extract text of the review and its corresponding rating.

The model is implemented in the Python programming language, PyTorch[1] and AllenNLP[2]. Moreover, we use pre-trained word-embeddings, such as ELMo [24], GloVe [23]. Specifically, we use the following ELMo models: Original[3], Polish [8] and German [17]. In the ELMO+GloVe+BCN model we use the following 300-dimension GloVe embeddings: English[4], Polish [4] and German[5]. In order to simplify our approach when training the sentiment classifier model, we establish a very similar setting to the vanilla Transformer. We use the same optimizer - Adam with $\beta_1 = 0.9, \beta_2 = 0.98$, and $\epsilon = 10^{-9}$. We incorporate four types of regularization during training: dropout probability $P_{drop} = 0.1$, embedding dropout probability $P_{emb} = 0.5$, residual dropout probability $P_{res} = 0.2$, and attention dropout probability $P_{attn} = 0.1$. We use 2 encoder layers. In addition, we employ label smoothing of value $\epsilon_{ls} = 0.1$. In terms of RPR parameters, we set clipping distance to $k = 10$.

4.3   Results and Discussion

In Table 2, we summarize experimental results achieved by our model and other state-of-the-art systems reported in the literature by their respective authors.

We observe that our models, baseline and ELMo+TSA, outperform state-of-the-art systems for all three languages. More importantly, the presented accuracy

---

[1] https://pytorch.org.
[2] https://allennlp.org.
[3] https://allennlp.org/elmo.
[4] http://nlp.stanford.edu/data/glove.840B.300d.zip.
[5] https://wikipedia2vec.github.io/wikipedia2vec/pretrained.

Table 2. Results of our systems compared to baselines and state-of-the-art systems evaluated on English, Polish and German sentiment classification datasets.

	English		Polish	German
	SST-2	SST-5	PolEmo2.0-IN	GermEval
RNTN [28]	85.4	45.7	–	–
DCNN [9]	86.8	48.5	–	–
CNN [11]	88.1	48.0	–	–
DMN [13]	88.6	52.1	–	–
Constituency Tree-LSTM [29]	88.0	51.0	–	–
CoVe+BCN [18]	90.3	**53.7**	–	–
SSAN+RPR [2]	84.2	48.1	–	–
Polish BERT [1]	–	–	88.1	–
SWN2-RNN [33]	–	–	–	74.9
Our baseline				
ELMo+GloVe+BCN	**91.4**	53.5	88.9	78.2
Our model				
ELMo+TSA	89.3	50.6	**89.8**	**78.9**

scores indicate that the TSA model is competitive and for two languages (Polish and German) achieves the best results. Also noteworthy, in Table 2, there are two models that use some variant of the Transformer: SSAN+RPR [2] uses the Transformer encoder for the classifier, while Polish BERT [1] employs Transformer-based language model introduced in [5]. One of the reasons why we achieve higher score for the SST dataset might be that the authors of SSAN+RPR used word2vec embeddings [19], whereas we employ ELMo contextual embeddings [24]. Moreover, in our TSA model we use not only self-attention (as in SSAN+RPR) but also a bi-attention mechanism, hence this also should provide performance gains over standard architectures.

In conclusion, comparing the results of the models leveraging contextual embeddings (CoVe+BCN, Polish BERT, ELMo+GloVe+BCN and ELMo+TSA) with the rest of the reported models, which use traditional distributional word vectors, we note that the former category of sentiment classification systems demonstrates remarkably better results.

5   Conclusion and Future Work

We have presented a novel architecture, based on the Transformer encoder with relative position representations. Unlike existing models, this work proposes a model relying solely on a self-attention mechanism and bi-attention. We show that our sentiment classifier model achieves very good results, comparable to the state of the art, even though it is language-agnostic. Hence, this work is a step towards building a universal, multi-lingual sentiment classifier.

In the future, we plan to evaluate our model using benchmarks also for other languages. It is particularly interesting to analyze the behavior of our model with respect to low-resource languages. Finally, other promising research avenues worth exploring are related to unsupervised cross-lingual sentiment analysis.

References

1. Allegro: Klej benchmark. https://klejbenchmark.com/. Accessed 20 Jan 2020
2. Ambartsoumian, A., Popowich, F.: Self-attention: a better building block for sentiment analysis neural network classifiers. In: Proceedings of the 9th Workshop on Computational Approaches to Subjectivity, Sentiment and Social Media Analysis, pp. 130–139 (2018)
3. Bahdanau, D., Cho, K., Bengio, Y.: Neural machine translation by jointly learning to align and translate. arXiv (2014)
4. Dadas, S.: A repository of polish NLP resources. Github (2019). https://github.com/sdadas/polish-nlp-resources/. Accessed 20 Jan 2020
5. Devlin, J., Chang, M.W., Lee, K., Toutanova, K.: BERT: pre-training of deep bidirectional transformers for language understanding. In: Proceedings of the 2019 Conference of the North American Chapter of the Association for Computational Linguistics: Human Language Technologies, pp. 4171–4186 (2019)
6. Firth, J.R.: A synopsis of linguistic theory, 1930–1955. Studies in linguistic analysis (1957)
7. Howard, J., Ruder, S.: Universal language model fine-tuning for text classification. In: Proceedings of the 56th Annual Meeting of the Association for Computational Linguistics, pp. 328–339 (2018)
8. Janz, A., Miłkowski, P.: ELMo embeddings for polish (2019). http://hdl.handle.net/11321/690. CLARIN-PL digital repository
9. Kalchbrenner, N., Grefenstette, E., Blunsom, P.: A convolutional neural network for modelling sentences. In: Proceedings of the 52nd Annual Meeting of the Association for Computational Linguistics, pp. 655–665 (2014)
10. Kennedy, A., Inkpen, D.: Sentiment classification of movie reviews using contextual valence shifters. Comput. Intell. **22**, 110–125 (2006)
11. Kim, Y.: Convolutional neural networks for sentence classification. In: Proceedings of the 2014 Conference on Empirical Methods in Natural Language Processing (EMNLP), pp. 1746–1751 (2014)
12. Kocoń, J., Miłkowski, P., Zaśko-Zielińska, M.: Multi-level sentiment analysis of PolEmo 2.0: extended corpus of multi-domain consumer reviews. In: Proceedings of the 23rd Conference on Computational Natural Language Learning (CoNLL), pp. 980–991 (2019)
13. Kumar, A., et al.: Ask me anything: dynamic memory networks for natural language processing. In: Proceedings of the 33rd International Conference on International Conference on Machine Learning, vol. 48, pp. 1378–1387 (2016)
14. Letarte, G., Paradis, F., Giguère, P., Laviolette, F.: Importance of self-attention for sentiment analysis. In: Proceedings of the 2018 EMNLP Workshop BlackboxNLP, pp. 267–275 (2018)
15. Lin, Z., et al.: A structured self-attentive sentence embedding. In: 5th International Conference on Learning Representations (2017)
16. Liu, B.: Sentiment analysis and opinion mining. Synth. Lect. Hum. Lang. Technol. **5**, 1–167 (2012). Morgan & Claypool Publishers

17. May, P.: German ELMo Model (2019). https://github.com/t-systems-on-site-services-gmbh/german-elmo-model. Accessed 20 Jan 2020
18. McCann, B., Bradbury, J., Xiong, C., Socher, R.: Learned in translation: contextualized word vectors. Adv. Neural Inf. Process. Syst. **30**, 6294–6305 (2017)
19. Mikolov, T., Sutskever, I., Chen, K., Corrado, G.S., Dean, J.: Distributed representations of words and phrases and their compositionality. Adv. Neural Inf. Process. Syst. **26**, 3111–3119 (2013)
20. Mohammad, S.M.: Sentiment analysis: detecting valence, emotions, and other affectual states from text. In: Emotion Measurement, pp. 201–237. Elsevier (2016)
21. Pang, B., Lee, L., Vaithyanathan, S.: Thumbs up? sentiment classification using machine learning techniques. In: Proceedings of the 2002 Conference on Empirical Methods in Natural Language Processing (EMNLP), pp. 79–86 (2002)
22. Paulus, R., Socher, R., Manning, C.D.: Global belief recursive neural networks. Adv. Neural Inf. Process. Syst. **27**, 2888–2896 (2014)
23. Pennington, J., Socher, R., Manning, C.: Glove: global vectors for word representation. In: Proceedings of the 2014 Conference on Empirical Methods in Natural Language Processing (EMNLP), pp. 1532–1543 (2014)
24. Peters, M., et al.: Deep contextualized word representations. In: Proceedings of the 2018 Conference of the North American Chapter of the Association for Computational Linguistics: Human Language Technologies, pp. 2227–2237 (2018)
25. Radford, A., Narasimhan, K., Salimans, T., Sutskever, I.: Improving language understanding by generative pre-training (2018)
26. dos Santos, C.N., Tan, M., Xiang, B., Zhou, B.: Attentive pooling networks. arXiv (2016)
27. Shaw, P., Uszkoreit, J., Vaswani, A.: Self-attention with relative position representations. In: Proceedings of the 2018 Conference of the North American Chapter of the Association for Computational Linguistics: Human Language Technologies, pp. 464–468 (2018)
28. Socher, R., et al.: Recursive deep models for semantic compositionality over a sentiment treebank. In: Proceedings of the 2013 Conference on Empirical Methods in Natural Language Processing, pp. 1631–1642 (2013)
29. Tai, K.S., Socher, R., Manning, C.D.: Improved semantic representations from tree-structured long short-term memory networks. In: Proceedings of the 53rd Annual Meeting of the Association for Computational Linguistics and the 7th International Joint Conference on Natural Language Processing, pp. 1556–1566 (2015)
30. Turney, P.: Thumbs up or thumbs down? semantic orientation applied to unsupervised classification of reviews. In: Proceedings of the 40th Annual Meeting of the Association for Computational Linguistics, pp. 417–424 (2002)
31. Turney, P.D., Pantel, P.: From frequency to meaning: vector space models of semantics. J. Artif. Intell. Res. **37**, 141–188 (2010)
32. Vaswani, A., et al.: Attention is all you need. In: Advances in Neural Information Processing Systems 30: Annual Conference on Neural Information Processing Systems, pp. 5998–6008 (2017)
33. Wojatzki, M., Ruppert, E., Holschneider, S., Zesch, T., Biemann, C.: GermEval 2017: shared task on aspect-based sentiment in social media customer feedback. Proc. GermEval **2017**, 1–12 (2017)

Nominees for Best Student Paper Award

Interpretable Segmentation of Medical Free-Text Records Based on Word Embeddings

Adam Gabriel Dobrakowski[1]([✉]), Agnieszka Mykowiecka[2],
Małgorzata Marciniak[2], Wojciech Jaworski[1], and Przemysław Biecek[1]

[1] University of Warsaw, Banacha 2, Warsaw, Poland
ad359226@students.mimuw.edu.pl,
{W.Jaworski,P.Biecek}@mimuw.edu.pl
[2] Institute of Computer Science Polish Academy of Sciences,
Jana Kazimierza 5, Warsaw, Poland
{agn,mm}@ipipan.waw.pl

Abstract. Is it true that patients with similar conditions get similar diagnoses? In this paper we present a natural language processing (NLP) method that can be used to validate this claim. We (1) introduce a method for representation of medical visits based on free-text descriptions recorded by doctors, (2) introduce a new method for segmentation of patients' visits, (3) present an application of the proposed method on a corpus of 100,000 medical visits and (4) show tools for interpretation and exploration of derived knowledge representation. With the proposed method we obtained stable and separated segments of visits which were positively validated against medical diagnoses. We show how the presented algorithm may be used to aid doctors in their practice.

1 Introduction

Processing of free-text clinical records plays an important role in computer-supported medicine [1,13]. A detailed description of symptoms, examination and an interview is often stored in an unstructured way as free-text, hard to process but rich in important information. Although there exist some attempts to process medical notes for English and some other languages, in general, the problem is still challenging [23]. The most straightforward approach to the processing of clinical notes could be their clustering with respect to different features like diagnosis or type of treatment. The process can either concentrate on patients or on their particular visits.

Grouping of visits can fulfil many potential goals. If we are able to group visits into clusters based on interview with a patient and medical examination then we can: follow recommendations that were suggested to patients with similar history to create a list of possible diagnoses; reveal that the current diagnosis is unusual; identify subsets of visits with the same diagnosis but different symptoms. A desired goal in the patients' segmentation is to divide them into groups with

© Springer Nature Switzerland AG 2020
D. Helic et al. (Eds.): ISMIS 2020, LNAI 12117, pp. 45–55, 2020.
https://doi.org/10.1007/978-3-030-59491-6_5

similar properties. In the case of segmentation hospitalized patients one of the most well-known examples are Diagnosis Related Groups [11] which aim to divide patients into groups with similar costs of treatment. Grouping visits of patients in health centers is a different issue. Here most of the information is unstructured and included in the visit's description written by a doctor: the description of the interview with the patient and the description of a medical examination of the patient.

Segmentation (clustering) is a well studied task for structured data such as age, sex, place, history of diseases, ICD-10 code etc. (an example of patients segmentation based only on their history of diseases is introduced in [26]), but it is far from being solved for unstructured free-texts which requires undertaking many decision on how the text and its meaning is to be represented. Medical concepts to be extracted from texts very often are taken from Unified Medical Language System (UMLS, see [4]), which is a commonly accepted base of biomedical terminology. Representations of medical concepts are computed based on various medical texts, like medical journals, books, etc. [5, 8, 10, 21, 22] or based directly on data from Electronic Health Records [6–8]. Other approach for patient segmentation is given in [6]. A subset of medical concepts (e.g. diagnosis, medication, procedures) and embeddings is aggregated for all visits of a patient. This way we get patient embedding that summaries patient medical history.

In this work we present a different approach. Our data include medical records for the medical history, description of the examination and recommendations for the treatment. Complementary sources allow us to create a more comprehensive visit description. The second difference is grouping visits, not patients. In this way a single patient can belong to several clusters. Our segmentation is based on a dictionary of medical concepts created from data, as for Polish does not exist any classification of medical concepts like UMLS or SNOMED. Obtained segments are supplemented with several approaches to visual exploration that facilitate interpretation of segments. Some examples of visual exploration of supervised models for structured medical data are presented in [3, 14, 17]. In this article we deal with a problem of explainable machine learning for unsupervised models.

2   Corpus of Free-Text Clinical Records

The clustering method is developed and validated on a dataset of free-text clinical records of about 100,000 visits. The data set consists of descriptions of patients' visits from different primary health care centers and specialist clinics in Poland. They have a free-text form and are written by doctors representing a wide range of medical professions, e.g. general practitioners, dermatologists, cardiologists or psychiatrists. Each description is divided into three parts: interview, examination, and recommendations.

3   Methodology

In this section we describe our algorithm for visits clustering. The process is performed in the following four steps: (1) Medical concepts are extracted from free-text descriptions of an interview and examination. (2) A new representation of identified concepts is derived with concepts embedding. (3) Concept embeddings are transformed into visit embeddings. (4) Clustering is performed on visit embeddings.

3.1   Extraction of Medical Concepts

As there are no generally available terminological resources for Polish medical texts, the first step of data processing was aimed at automatic identification of the most frequently used words and phrases. The doctors' notes are usually rather short and concise, so we assumed that all frequently appearing phrases are domain related and important for text understanding. The notes are built mostly from noun phrases which consist of a noun optionally modified by a sequence of adjectives or by another noun in the genitive. We only extracted sequences that can be interpreted as phrases in Polish.

To get the most common phrases, we processed 220,000 visits' descriptions. First, we preprocessed texts using Concraft tagger [28] which assigns lemmas, POS and morphological features values. It also guesses descriptions (apart from lemmas) for words which are not present in its vocabulary. Phrase extraction and ordering was performed by TermoPL [19]. The program allows for defining a grammar describing extracted text fragments and order them according to a version of the C-value coefficient [12], but we used the built-in grammar of noun phrases. The first 4800 phrases (all with C-value equal at least 20) from the obtained list were manually annotated with semantic labels. The list of 137 labels covered most general concepts like *anatomy, feature, disease, test*. Many labels were assigned to multi-word expressions (MWEs). In some cases phrases were also labeled separately, e.g. *left hand* is labeled as *anatomy* while *hand* is also labeled as *anatomy* and *left* as *lateralization*. The additional source of information was the list of 9993 names of medicines and dietary supplements.

The list of terms together with their semantic labels was then converted to the format of lexical resources of Categorial Syntactic-Semantic Parser "ENIAM" [15,16]. The parser recognized lexemes and MWEs in texts according to the provided list of terms, then the longest sequence of recognized terms was selected, and semantic representation was created. Semantic representation of a visit has a form of a set of pairs composed of recognized terms and their labels (not recognized tokens were omitted). The average coverage of semantic representation was 82.06% of tokens and 75.38% of symbols in section *Interview* and 87.43% of tokens and 79.28% of symbols in section *Examination*.

Texts of visits are heterogeneous as they consist of: very frequent domain phrases; domain important words which are too infrequent to be at the top of the term list prepared by TermoPL; some general words which do not carry relevant information; numerical information; and words which are misspelled. In

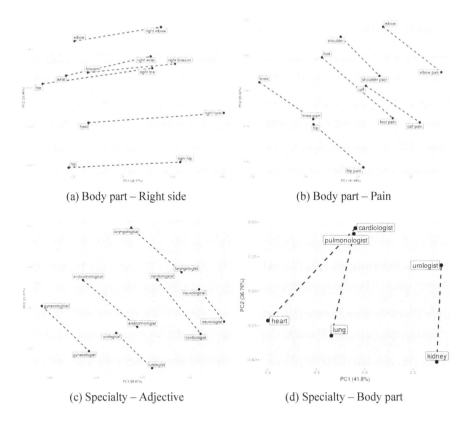

(a) Body part – Right side

(b) Body part – Pain

(c) Specialty – Adjective

(d) Specialty – Body part

Fig. 1. Visualization of analogies between terms. The pictures show term embeddings projected into 2d-plane using PCA. Each panel shows a different type of analogy.

the clustering task we neglect the original text with inflected word forms and the experiments are solely performed on the set of semantic labels attached to each interview and examination.

3.2 Embeddings for Medical Concepts

Operating on relatively large amount of very specific texts, we decided not to use any general model for Polish. In the experiments, we reduce the description of visits to extracted concepts and train on them our own domain embeddings. During creating the term co-occurrence matrix the whole visit's description is treated as the neighbourhood of the concept. Furthermore we choose only unique concepts and abandon their original order in the description (we follow this way due to simplicity).

We compute embeddings of concepts by GloVe [24] for interview descriptions and for examination descriptions separately. Computing two separate embeddings we aim at catching the similarity between terms in their specific context. For example, the nearest words to *cough* in the interview descriptions is *runny nose*, *sore throat*, *fever* but in the examination description it is *rash*, *sunny*, *laryngeal*.

3.3   Visit Embeddings

The simplest way to generate text embeddings based on term embeddings is to use some kind of aggregation of term embeddings such as an average. This approach was tested for example in [2] and [7]. In [9] the authors computed a weighted mean of term embeddings by the construction of a loss function and training weights by the gradient descent method. Thus, in our method we firstly compute embeddings of the descriptions (for interview and examination separately) as a simple average of concepts' embeddings. Then, the final embeddings for visits are obtained by concatenation of two descriptions' embeddings.

3.4   Visits Clustering

Based on Euclidean distance between vector representations of visits we applied and compared two clustering algorithms: k-means and hierarchical clustering with Ward's method for merging clusters [27]. The similarity of these clusterings was measured by the adjusted Rand index [25]. For the final results we chose the hierarchical clustering algorithm due to greater stability.

Table 1. The statistics of clusters for selected domains. The last column shows adjusted Rand index between k-means and hierarchical clustering.

Domain	# clusters	# visits	Clusters' size	K-means - hclust
Cardiology	6	1201	428, 193, 134, 303, 27, 116	0.87
Family medicine	6	11230	3108, 2353, 601, 4518, 255, 395	0.69
Gynecology	4	3456	1311, 1318, 384, 443	0.8
Internal medicine	5	6419	1915, 1173, 1930, 1146, 255	0.76
Psychiatry	5	1012	441, 184, 179, 133, 75	0.81

Table 2. The categories of questions in term analogy task with example pairs.

Type of relationship	# Pairs	Term pair 1		Term pair 2	
Body part – Pain	22	Eye	Eye pain	Foot	Foot pain
Specialty – Adjective	7	Dermatologist	Dermatological	Neurologist	Neurological
Body part – Right side	34	Hand	Right hand	Knee	Right knee
Body part – Left side	32	Thumb	Left thumb	Heel	Left heel
Spec. – Consultation	11	Surgeon	Surgical consult.	Gynecologist	g. consult.
Specialty – Body part	9	Cardiologist	Heart	Oculist	Eye
Man – Woman	9	Patient (male)	Patient (female)	Brother	Sister

Table 3. Mean accuracy of correct answers on term analogy tasks. Rows show different embeddings sizes, columns correspond to size of neighborhoods.

Dim./Context	1	3	5
10	0.1293	0.2189	0.2827
15	0.1701	0.3081	0.4123
20	**0.1702**	0.3749	0.4662
25	0.1667	0.4120	0.5220
30	0.1674	0.4675	0.5755
40	0.1460	**0.5017**	0.6070
50	0.1518	0.4966	**0.6190**
100	0.0435	0.4231	0.5483
200	0.0261	0.3058	0.4410

Table 4. The most common recommendations for each segment derived for gynecology. In brackets we present a percentage of visits in this cluster which contain a specified term. We skipped terms common in many clusters, like: *treatment, ultrasound treatment, control, morphology, hospital, lifestyle, zus (Social Insurance Institution)*.

Cluster	Size	Most frequent recommendations
1	1311	Recommendation (16.6%), general urine test (5.6%), diet (4.1%), vitamin (4%), dental prophylaxis (3.4%)
2	1318	Therapy (4.1%), to treat (4%), cytology (3.8%), breast ultrasound (3%), medicine (2.2%)
3	384	Acidum (31.5%), the nearest hospital (14.3%), proper diet (14.3%), health behavior (14.3%), obstetric control (10.2%)
4	443	To treat (2%), therapy (2%), vitamin (1.8%), diet (1.6%), medicine (1.6%)

For clustering, we selected visits where the description of recommendation and at least one of interview and examination were not empty (some concepts were recognized). It significantly reduced the number of considered visits. Table 1 gives basic statistics of obtained clusters. The last column contains the adjusted Rand index. It can be interpreted as a measure of the stability of the clustering. The higher similarity of the two algorithms, the higher stability of clustering. For determining the optimal number of clusters, for each specialty we consider the number of clusters between 2 and 15. We choose the number of clusters so that adding another cluster does not give a relevant improvement of a sum of differences between elements and clusters' centers (according to so called *Elbow method*).

4   Results

4.1   Analogies in Medical Concepts

To better understand the structure of concept embeddings and to determine the optimal dimension of embedded vectors we use word analogy task introduced in [20] and examined in details in a medical context in [22]. In the former work the authors defined five types of semantic and nine types of syntactic relationship.

We propose our own relationships between concepts, more related to the medical language. We exploit the fact that in the corpus we have a lot of multiword concepts and very often the same words are included in different terms. We would like the embeddings to be able to catch relationships between terms. A question in the term analogy task is computing a vector: $vector(left\ foot) - vector(foot) + vector(hand)$ and checking if the correct $vector(left\ hand)$ is in the neighborhood (in the metric of cosine of the angle between the vectors) of this resulting vector.

We defined seven types of such semantic questions and computed answers' accuracy in a similar way as in [20]: we created manually the list of similar term pairs and then we formed the list of questions by taking all two-element subsets of the pairs list. Table 2 shows the created categories of questions.

We created one additional task, according to the observation that sometimes two different terms are related to the same object. This can be caused for example by the different order of words in the terms, e.g. *left wrist* and *wrist left* (in Polish both options are acceptable). We checked if the embeddings of such words are similar.

We computed term embeddings for terms occurring at least 5 times in the descriptions of the selected visits. The number of chosen terms in interview descriptions was equal to 3816 and in examination descriptions – 3559. Among these there were 2556 common terms for interview and examination. Embeddings of the size 10 to 200 were evaluated. For every embedding of interview terms there was measured accuracy of every of eight tasks. Table 3 shows the mean of eight task results. The second column presents the results of the most restrictive rule: a question is assumed to be correctly answered only if the closest term of the vector computed by operations on related terms is the same as the desired answer. The total number of terms in our data set (about 900,000 for interviews) was many times lower than sets examined in [20]. Furthermore, words in medical descriptions can have a different context that we expect. Taking this into account, the accuracy of about 0.17 is very high and better than we expected. We then checked the closest 3 and 5 words to the computed vector and assumed a correct answer if in this neighbourhood there was the correct vector. In the biggest neighbourhood the majority of embeddings returned accuracy higher than 0.5.

For computing visit embeddings we chose embeddings of dimensionality 20, since this resulted in the best accuracy of the most restrictive analogy task and it allowed us to perform more efficient computations than higher dimensional representations. Figure 1 illustrates PCA projection of term embeddings from four categories of analogies.

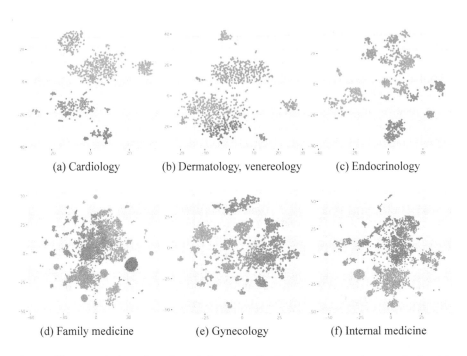

Fig. 2. Clusters of visits for selected domains. Each dot corresponds to a single visit. Colors correspond to segments. Visualization created with t-SNE. (Color figure online)

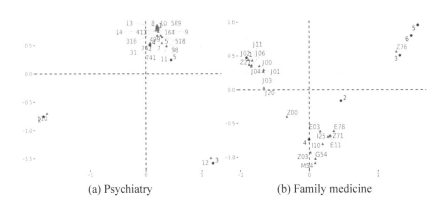

Fig. 3. Correspondence analysis between clusters and doctors' IDs for psychiatry clustering (panel a) and between clusters and ICD-10 codes for family medicine clustering (panel b). Clusters 2 and 3 in panel a are perfectly fitted to a single doctor.

4.2   Visits Clustering

Clustering was performed separately for each specialty of doctors. Figure 2 illustrates two-dimensional t-SNE projections of visit embeddings coloured by clusters [18]. For some domains clusters are very clear and separated (Fig. 2a). This corresponds with the high stability of clustering measured by Rand index.

In order to validate the proposed methodology we evaluate how clear are derived segments when it comes to medical diagnoses (ICD-10). No information about recommendations nor diagnosis is used in the phase of clustering to prevent data leakage.

Figure 3(b) shows correspondence analysis between clusters and ICD-10 codes for family medicine clustering. There appeared two large groups of codes: the first related to diseases of the respiratory system (J) and the second related to other diseases, mainly endocrine, nutritional and metabolic diseases (E) and diseases of the circulatory system (I). The first group corresponds to Cluster 1 and the second to Cluster 4. Clusters 3, 5 and 6 (the smallest clusters in this clustering) covered Z76 ICD-10 code (encounter for issue of repeat prescription). We also examined the distribution of doctors' IDs in the obtained clusters. It turned out that some clusters covered almost exactly descriptions written by one doctor. This happen in the specialties where clusters are separated with large margins (e.g. psychiatry, pediatrics, cardiology). Figure 3(a) shows correspondence analysis between doctors' IDs and clusters for psychiatry clustering.

4.3   Recommendations in Clusters

According to the main goal of our clustering described in Introduction, we would like to obtain similar recommendations inside every cluster. Hence we examined the frequency of occurrence of the recommendation terms in particular clusters.

We examined terms of recommendations related to one of five categories: procedure to carry out by patient, examination, treatment, diet and medicament. Table 4 shows an example of an analysis of the most common recommendations in clusters in gynecology clustering. In order to find only characteristic terms for clusters we filtered the terms which belong to one of 15 the most common terms in at least three clusters.

5   Conclusions and Applications

We proposed a new method for clustering of visits in health centers based on descriptions written by doctors. We validated this new method on a new large corpus of Polish medical records. For this corpus we identified medical concepts and created their embeddings with GloVe algorithm. The quality of the embeddings was measured by the specific analogy task designed specifically for this corpus. It turns out that analogies work well, what ensures that concept embeddings store some useful information.

Clustering was performed on visits embedding created based on word embedding. Visual and numerical examination of derived clusters showed an interesting structure among visits. As we have shown obtained segments are linked with medical diagnosis even if the information about recommendations or diagnosis were not used for the clustering. This additionally convinces that the identified structure is related to some subgroups of medical conditions.

Obtained clustering can be used to assign new visits to already derived clusters. Based on descriptions of an interview or a description of patient examination we can identify similar visits and show corresponding recommendations.

Acknowledgments. This work was financially supported by NCBR Grant POIR.01. 01.01-00-0328/17. PBi was supported by NCN Opus grant 2016/21/B/ST6/02176.

References

1. Apostolova, E., Channin, D.S., Demner-Fushman, D., Furst, J., Lytinen, S., Raicu, D.: Automatic segmentation of clinical texts. In: Proceedings of EMBC, pp. 5905–5908 (2009)
2. Banea, C., Chen, D., Mihalcea, R., Cardie, C., Wiebe, J.: Simcompass: using deep learning word embeddings to assess cross-level similarity. In: Proceedings of SemEval, pp. 560–565 (2014)
3. Biecek, P.: DALEX: explainers for complex predictive models in R. J. Mach. Learn. Res. **19**(84), 1–5 (2018)
4. Bodenreider, O.: The unified medical language system (UMLS): integrating biomedical terminology. Nucleic Acids Res. **32**(suppl-1), D267–D270 (2004)
5. Chiu, B., Crichton, G., Korhonen, A., Pyysalo, S.: How to train good word embeddings for biomedical NLP. In: Proceedings of BioNLP, pp. 166–174 (2016)
6. Choi, E., et al.: Multi-layer representation learning for medical concepts. In: SIGKDD Proceedings, pp. 1495–1504. ACM (2016)
7. Choi, E., Schuetz, A., Stewart, W.F., Sun, J.: Medical concept representation learning from electronic health records and its application on heart failure prediction. arXiv preprint arXiv:1602.03686 (2016)
8. Choi, Y., Chiu, C.Y.I., Sontag, D.: Learning low-dimensional representations of medical concepts. AMIA Summits Transl. Sci. **2016**, 41 (2016)
9. De Boom, C., Van Canneyt, S., Demeester, T., Dhoedt, B.: Representation learning for very short texts using weighted word embedding aggregation. Pattern Recogn. Lett. **80**, 150–156 (2016)
10. De Vine, L., Zuccon, G., Koopman, B., Sitbon, L., Bruza, P.: Medical semantic similarity with a neural language model. In: Proceedings of CIKM, pp. 1819–1822. ACM (2014)
11. Fetter, R.B., Shin, Y., Freeman, J.L., Averill, R.F., Thompson, J.D.: Case mix definition by diagnosis-related groups. Med. Care **18**(2), i-53 (1980)
12. Frantzi, K., Ananiadou, S., Mima, H.: Automatic recognition of multi-word terms: the C-value/NC-value method. Int. J. Digit. Libr. **3**, 115–130 (2000)
13. Ganesan, K., Subotin, M.: A general supervised approach to segmentation of clinical texts. In: IEEE International Conference on Big Data, pp. 33–40 (2014)
14. Gordon, L., Grantcharov, T., Rudzicz, F.: Explainable artificial intelligence for safe intraoperative decision support. JAMA Surg. **154**(11), 1064–1065 (2019)

15. Jaworski, W., Kozakoszczak, J.: ENIAM: categorial syntactic-semantic parser for Polish. In: Proceedings of COLING, pp. 243–247 (2016)
16. Jaworski, W., et al.: Categorial parser. CLARIN-PL digital repository (2018)
17. Kobylińska, K., Mikołajczyk, T., Adamek, M., Orłowski, T., Biecek, P.: Explainable machine learning for modeling of early postoperative mortality in lung cancer. In: Marcos, M., et al. (eds.) KR4HC/TEAAM -2019. LNCS (LNAI), vol. 11979, pp. 161–174. Springer, Cham (2019). https://doi.org/10.1007/978-3-030-37446-4_13
18. Maaten, L.V.D., Hinton, G.: Visualizing data using t-SNE. J. Mach. Learn. Res. 9(Nov), 2579–2605 (2008)
19. Marciniak, M., Mykowiecka, A., Rychlik, P.: TermoPL – a flexible tool for terminology extraction. In: Proceedings of LREC, pp. 2278–2284. ELRA, Portorož, Slovenia (2016)
20. Mikolov, T., Chen, K., Corrado, G., Dean, J.: Efficient estimation of word representations in vector space. arXiv preprint arXiv:1301.3781 (2013)
21. Minarro-Giménez, J.A., Marin-Alonso, O., Samwald, M.: Exploring the application of deep learning techniques on medical text corpora. Stud. Health Technol. Inform. 205, 584–588 (2014)
22. Newman-Griffis, D., Lai, A.M., Fosler-Lussier, E.: Insights into analogy completion from the biomedical domain. arXiv preprint arXiv:1706.02241 (2017)
23. Orosz, G., Novák, A., Prószéky, G.: Hybrid text segmentation for Hungarian clinical records. In: Castro, F., Gelbukh, A., González, M. (eds.) MICAI 2013. LNCS (LNAI), vol. 8265, pp. 306–317. Springer, Heidelberg (2013). https://doi.org/10.1007/978-3-642-45114-0_25
24. Pennington, J., Socher, R., Manning, C.: Glove: global vectors for word representation. In: Proceedings of EMNLP, pp. 1532–1543 (2014)
25. Rand, W.M.: Objective criteria for the evaluation of clustering methods. J. Am. Stat. Assoc. 66(336), 846–850 (1971)
26. Ruffini, M., Gavaldà, R., Limón, E.: Clustering patients with tensor decomposition. arXiv preprint arXiv:1708.08994 (2017)
27. Ward Jr., J.H.: Hierarchical grouping to optimize an objective function. J. Am. Stat. Assoc. 58(301), 236–244 (1963)
28. Waszczuk, J.: Harnessing the CRF complexity with domain-specific constraints. The case of morphosyntactic tagging of a highly inflected language. In: Proceedings of COLING, pp. 2789–2804 (2012)

Decision-Making with Probabilistic Reasoning in Engineering Design

Stefan Plappert$^{(\boxtimes)}$ ⑩, Paul Christoph Gembarski⑩, and Roland Lachmayer⑩

Institute of Product Development, Leibniz University of Hannover, Welfengarten 1a,
30167 Hannover, Germany
{plappert,gembarski,lachmayer}@ipeg.uni-hannover.de

Abstract. The goal of decision making is to select the most suitable option from a number of possible alternatives. Which is easy, if all possible alternatives are known and evaluated. This case is rarely encountered in practice; especially in product development, decisions often have to be made under uncertainty. As uncertainty cannot be avoided or eliminated, actions have to be taken to deal with it. In this paper a tool from the field of artificial intelligence, decision networks, is used. Decision networks utilize probabilistic reasoning to model uncertainties with probabilities. If the influence of uncertainty cannot be avoided, a variation of the product is necessary so that it adjusts optimally to the changed situation. In contrast, robust products are insensitive to the influence of uncertainties. An application example from the engineering design has shown, that a conclusion about the robustness of a product for possible scenarios can be made by the usage of the decision network. It turned out that decision networks can support the designer well in making decisions under uncertainty.

Keywords: Probabilistic reasoning · Decision-making · Engineering design · Decision network · Bayesian network

1 Introduction

For the development of robust products, which are insensitive to uncertainties, it is important to assess possible effects of decisions made in the development process and to consider the relevant influencing factors of tolerances, environments and use cases [1]. Handling uncertainties in this context, e.g. by identifying design rules and learning about sensitivities is a major part in the management of product complexity [2]. Systems that support the designer in the decision-making process have to reduce the uncertainty by providing knowledge or help to estimate possible scenarios, so that the products are robust against possible uncertainties. But almost all computer aids like computer-aided design require discreet parameters [3]. Knowledge-based engineering systems are often used in the late phases of the development process, e.g. to derive variants quickly and check them for validity by tables and rules [4]. Knowledge-based engineering is also beneficial for the conservation of engineering knowledge which is accessible then for later design projects so that uncertainty can be tackled with experience [5].

D. Helic et al. (Eds.): ISMIS 2020, LNAI 12117, pp. 56–65, 2020.
https://doi.org/10.1007/978-3-030-59491-6_6

In this paper, another approach is investigated, targeting at conditional probabilities that arise in the management of requirements that impact the design. In probabilistic reasoning, uncertainties can be represented by probabilities. For this purpose, Bayesian networks and decision networks will be examined in more detail. The structure of this paper is organized as follows: Sect. 2 describes decision making, uncertainty and probabilistic reasoning in engineering design. In Sect. 3 a decision network for the application example of a rotary valve is built. Section 4 provides a summary and describes approaches for further research.

2 Related Work

In the development of products, decisions often have to be made under uncertainty. A tool from the field of artificial intelligence, probabilistic reasoning, offers the possibility to model uncertainties by using probabilities.

2.1 Decision-Making in Product Development

In general, the goal of decision making is to select the most suitable option from a number of possible alternatives [6]. Decisions have to be made by the designer during the entire product development process which may be divided in *task clarification, concept, embodiment design* and *detailed design*. Since it encompasses an initial requirement management, *task clarification* has a major influence on the later stages, especially on *embodiment design* where first geometric considerations are made and the shape is defined [7]. Decisions in the development of a new product must take into account a selection of different criteria [9], which lead to different scopes of change in the product. Due to time pressure, decisions are often made at short notice in practice, which can lead to significant negative consequences, such as delays in deadlines, limitations in functionality, cost overruns and product quality defects [8]. Therefore, tools that support the designer in decision making have to quickly assess possible consequences.

2.2 Uncertainty in Engineering Design

Uncertainty cannot be avoided or eliminated within the product development process, therefore it is necessary to consider and react to uncertainty [10]. Kreye et al. [11] differentiate four types of manifestation of uncertainty: *context uncertainty, data uncertainty, model uncertainty* and *phenomenological uncertainty*. When creating a system, uncertainties arise from the input (data uncertainty), the used model (model uncertainty) and the results of the system (phenomenological uncertainty). Context uncertainty, in contrast, describes the influence of the environment on the system. It can be divided into *endogenous uncertainties*, which arise within the system and can be controlled, and *exogenous uncertainties*, which lie outside the system and typically arise during use of the product [12]. Despite their random occurrence, exogenous variables can be seen as a key to assessing the value of a design, because they reflect the way in which the engineer handles variables that cannot be controlled by himself [6].

According to Chalupnik et al. [13], there are different ways to deal with uncertainty. On the one hand, the uncertainty can be reduced by aiming for an increase in knowledge about the system. On the other hand, the system can be protected from the influence of uncertainty. *Active protection* ensures that the system adapts to uncertain situations. Where, in contrast, *passive protection* ensures that the system withstands the influence of uncertainty and therefore no changes need to be made [13]. Robust products are those that are insensitive to uncontrollable factors [14].

2.3 Modeling Uncertainty with Probabilistic Reasoning

Reasoning with uncertainty given limited resources is part of many technical applications in artificial intelligence [15]. Graph-based models have proven to be an important tool for dealing with uncertainty and complexity, as they build a complex system by combining simpler parts [16].

Bayesian Networks. Probabilities are very suitable for the modelling of reasoning with uncertainty [15]. Bayesian networks use the so-called Bayesian rule (1), because evidence is often perceived as an effect of an unknown cause and the goal is to determine the cause [17].

$$P(\text{cause} \mid \text{effect}) = \frac{P(\text{effect} \mid \text{cause}) \, P(\text{cause})}{P(\text{effect})} \tag{1}$$

The Bayesian rule is explained using a simplified application example for the dosing of powder for the preparation of hot drinks. To support the understanding, the system is reduced to one effect-cause pair. In practice, many effects have very different causes; the multi-causal relationships are discussed in more detail in Sect. 3.2.

In this example the following problem was noticed: the hot drink tastes watery. A possible cause was identified by an insufficient dosage of the powder. For the further solution of the problem it would be helpful to know with which probability the low dosage is the cause for the problem or effect. Based on a statistical analysis, the probability of the effect is $P(\text{watery}) = 0.2$. Since the dosing is carried out automatically, a sensor measures the required powder quantity, which is below its reference value with a probability of $P(\text{low}) = 0.1$. The probability that the hot drink tastes watery if too little powder is dosed is $P(\text{watery} \mid \text{low}) = 0.8$, because the taste is subjective. Based on the information obtained, the Bayesian rule can be used to determine the probability that the low dose of powder is the cause of the watery taste $P(\text{low} \mid \text{watery}) = 0.4$. This leads to the following conclusion that the insufficient dosage of the powder is with a probability of 40% the cause for the watery taste of the hot drink.

According to Russel and Norvig [17], the structure of a Bayesian network can be described by a directed acyclic graph (DAG) in which each node is annotated quantitative probability information. Figure 1 shows a Bayesian network with four nodes, which represents the probabilities for a *watery hot drink* in case of an *incorrect mixing ratio* of a machine for preparing hot drinks due to a *blocked powder supply* or *defective flow sensor* for liquids. The probabilities of the nodes *blocked powder supply* and *defective flow sensor* reflect the probability of their occurrence. The *incorrect mixing ratio* node has the nodes *blocked powder supply* and *defective flow sensor* as parent nodes, so the probability is described in a conditional probability table (CPT) depending on the probability of the parent nodes. The *watery hot drink* node describes the probability of a watery taste of a hot drink depending on the probability of an *incorrect mixing ratio*.

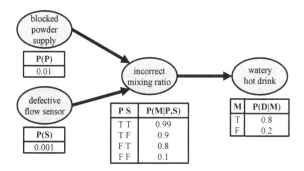

Fig. 1. Simple example for a Bayesian network for a watery hot drink

In a Bayesian network, the direct influences can be displayed by arcs without having to specify each probability manually [17]. The arcs in the DAG specify causal relations between the nodes [18], therefore the Bayesian rule can be applied. The main usage of Bayesian networks is inference, which involves updating the probability distribution of unobserved variables as new evidence or observed variables become available [19].

Decision Networks. A directed acyclic graph (DAG) model that combines chance nodes from a Bayesian network with additional node types for actions and utilities is called a decision network [17]. In general, decision networks can be used for optimal decision making, even if only partial observations of the world are given [20]. According to Zhu [18], a decision network represents the knowledge about an uncertain problem domain, as well as the available actions and desirability of each state. The following three node types, chance nodes, decision nodes and utility nodes form the basic structure of a decision network [17]:

- *Chance Nodes*: random variables as they are used in a Bayesian network, where each node is connected to a conditional distribution indexed by the state of the parent node
- *Decision Nodes*: points where the engineer has a choice of actions to make a decision
- *Utility Nodes*: points with a utility function that describes the preferred outcomes.

Actions are selected based on the evaluation of the decision network for each possible setting of the decision node [17]. Once a decision node is set, the probabilities of the parent nodes of the utility node are calculated by using a standard probabilistic inference algorithm [17]. As a result, the action that has the most added value based on the utility function is selected.

3   Application of Probabilistic Reasoning in Engineering Design

As an application for probabilistic reasoning, an example from engineering design is used to demonstrate a possible handling of uncertainty for the development of robust products.

3.1   Rotary Valve as Application Example

A rotary valve is to be used for the dosing of bulk food for hot drinks. Rotary valves or metering feeders are generally used for metering and conveying free-flowing bulk materials [21]. Figure 2 shows a rotary valve with its components. In this case, the rotary valve is driven via the shaft and receives and transports bulk food through the rotary valve pocket per rotation. To avoid bulk food being drawn in or jammed between the rotary valve and the housing, the gap between the rotary valve and the housing is kept as small as possible.

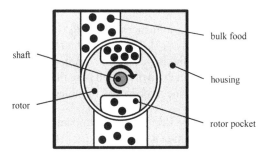

Fig. 2. Rotary valve for the dosing of bulk food

The rotary valve has many advantages in application, as it is easy to handle and provides reproducible results. On the other hand, the dimensioning or design of the rotary valve requires adaptation to the bulk material properties [21]. Especially for discontinuous and quantitative dosing, the rotor pocket size is decisive, which should also be filled as completely as possible during dosing.

The rotary valve in this application example is integrated in a machine for the preparation of various hot drinks. Each hot drink requires a different amount of bulk food to be dosed. In addition, the machine is to be placed at different locations, such as in the home kitchen, the office or a café. These boundary conditions result in uncertainties that cannot be influenced by the engineer. The aim for the engineer is to cover as many possible and probable scenarios with one size of the rotor pocket.

3.2   Modelling of a Decision Network for the Application Example

To support the design engineer in the decision making process for the optimal rotor pocket size, a decision network was established which also represents the given uncertainties due to the boundary conditions.

The decision network (Fig. 3) of the application example consists of a Bayesian network with six chance nodes. The nodes bulk food and place of use form the initial nodes. The nodes weight and density depend on the selected bulk food. Depending on the place of use, a different quantity of liquid is required for the hot drinks, because the number of hot drinks needed is different. The dosing volume depends on the weight and density of the bulk food, and on the required quantity of liquid. The additional chance node filling level represents the uncertainty when filling the rotor pocket size with bulk food. The decision node rotor pocket size represents the different pocket sizes which are available as possible actions. The utility node utility function represents the preferred outcomes, where design conditions are also included.

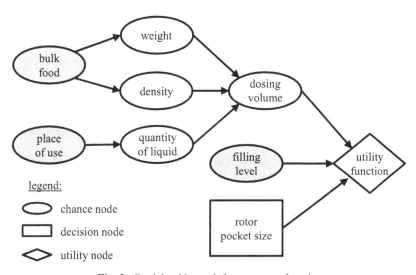

Fig. 3. Decision Network for rotary pocket size

The decision network for the application example was built within Matlab using an open-source package for directed graphical models called Bayes Net Toolbox (BNT). A great strength of BNT is the variety of implemented inference algorithms [20]. In addition, Matlab is very suitable for rapid prototyping, because the Matlab code is high level and easy to read [20].

In the first step, the Bayesian network was represented by six chance nodes (Fig. 4). For the chance nodes *bulk food* and *location* the occurrence probabilities were stored. For the chance nodes *weight*, *density* and *quantity of liquid* the probabilities were stored with conditional probability tables (CPT). CPTs were used, because the application example contains only discrete variables and thus the inference was simplified. The probability values are taken from a similar project and were determined empirically.

For the selection of the possible rotor pocket size, in addition to the possible dosing volumes, the filling level of the rotor pocket size has to be considered. For this purpose, the following assumptions are made that the rotor pocket size has a filling level of 0.9 at 80% and that filling levels 1.0 and 0.8 occur with a probability of 10%. To determine the possible rotor pocket sizes, all divisors with one decimal place of the possible dosing

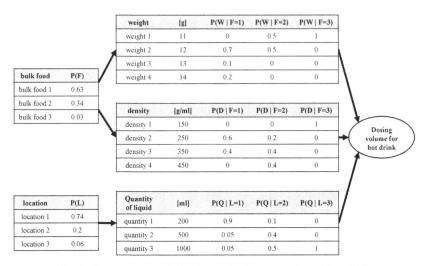

Fig. 4. Bayesian network for dosage in the preparation of hot drinks

volumes were determined. This ensures that all rotor pocket sizes are considered for the required dosing volume.

The general utility function describes the sum of the utilities of all possible outcomes, weighted according to their probability of occurrence [6]. As this general utility function also contains results that lead to an unsuitable layout of the design, the following conditions are also represented in the utility function:

- *High Variability*: One rotor pocket size should be able to cover many different dosing volumes
- *Fast Dosing*: The time for dosing should not take longer than the heating time of the liquid for the hot drink
- *Exact Dosage*: the rotor pocket size should dose exactly the required dosage volume.

The listed conditions lead to conflicts which have to be resolved by the algorithm. For example, high variability leads to the smallest possible rotor pocket size, whereas fast dosing requires the largest possible rotor pocket size. In the following Sect. 3.3 the results of the algorithm for the application example are presented and discussed.

3.3 Results of the Decision Network

The aim of the application example is to support the engineer with a decision network in the selection of the optimum rotor pocket size. Table 1 shows the results of the decision network in Matlab at different information levels. The column known information represents the different levels of evidence or observation for the decision network. The column rotor pocket size shows the optimal rotor pocket size for the given evidence, which is the best choice for dosing the most likely dosing volumes. The column utility probability describes the added probabilities of the dosing volumes, which can be dosed with the rotor pocket size depending on the filling level. For the situation where no further information is available, the optimum rotor pocket size is 10.7 ml with a utility probability of 15.26%. It can also be noticed that a higher information level does not necessarily lead to a higher utility probability, as this depends on the uncertainty or diversity of the probabilities of the chance nodes. This can also be shown by comparing the utility probabilities of no known information with 15.26% and bulk food 2 with 15.16%.

Table 1. Results for the optimal rotary pocket size with utility probabilities

Known information	Rotor pocket size [ml]	Utility probability [%]
No known information	10.7	15.26
Bulk food 1	10.7	23.38
Bulk food 2	1.0	15.16
Bulk food 3	16.3	70.64
Place 1	10.7	19.75
Place 2	1.0	16.84
Place 3	53.3	21.80
Bulk food 1 & Place 1	10.7	30.38
Bulk food 1 & Place 2	1.0	20.84
Bulk food 1 & Place 3	53.3	33.60
Bulk food 2 & Place 1	1.0	17.30
Bulk food 2 & Place 2	0.8	24.00
Bulk food 2 & Place 3	0.8	34.00
Bulk food 3 & Place 1	16.3	76.00
Bulk food 3 & Place 2	16.3	48.00
Bulk food 3 & Place 3	81.5	80.00

In addition, the decision network can support the engineer in making decisions under uncertainty by allowing the utility probability to make a prediction about the robustness of a product:

- *High utility probability*: If a value has a high utility probability, it can be assumed that this value is a suitable solution for as many scenarios as possible and therefore no further changes are necessary. Furthermore, it can be concluded that the described uncertainty within the decision network has little influence on the outcome and therefore it represents a robust product.
- *Low utility probability*: With a low utility probability, only a part of the possible scenarios can be covered with one value. This indicates a high diversity within the decision network, which may be due to a higher influence of uncertainty. For this

reason, the product have to be adaptable to different conditions, i.e. it should have a high variability or modifiability.

The aim of the decision network is to minimize or even avoid the need for design changes at a late stage of product development or in the use phase. If the utility probability is high, the most robust variant is chosen as the solution, because it is insensitive to the assumed uncertainties. If the utility probability is low, a variant cannot cover the whole spectrum of possible results. In this case, a portfolio of variants can be compiled to cover as many scenarios as possible. This portfolio enables a fast reaction to changing conditions.

4   Conclusion and Future Research

Decisions in the product development process often have to be made under uncertainty. Uncertainties can occur during product development or arise from the environment when the product is used. There are two possibilities for dealing with uncertainty. On the one hand, the uncertainty can be reduced by increasing knowledge, on the other hand, the product can be protected from the influence of the uncertainty. To be able to react to the influence of uncertainty, the requirements have to be compared with the behavior of the product during production and use. Especially with a large number of requirements and broad requirement corridors, this requires a great effort from the engineers.

The method presented in this paper supports the engineer in decision-making by representing the uncertainties in a decision network using probabilities. With sensitivities, i.e. the minimum and maximum of a requirement corridor, possible use cases are determined with their probability of occurrence. Based on these, statements about the robustness of the product can be made. Furthermore, they enable a feedback with product management to improve product variety, as possible application scenarios and market segments of the product are already checked.

For future research, it is necessary to investigate how suitable the method with the decision network is for selecting appropriate variants. In addition, the volume of a rotor pocket size was exclusively used as the basis for the application example. For the variation of the rotor, the number of pockets on the circumference or a combination of two pocket sizes could also be interesting. For the illustration of different variants with given uncertainty a coupling between a decision network and a knowledge-based system (KBS) would also be conceivable. Here the requirements with their probabilities of occurrence could be represented in a decision network and the most probable result is transferred to the knowledge-based system for the configuration of the product.

References

1. Suh, N.P.: Complexity: Theory and Applications. Oxford University Press, Oxford (2005). On Demand
2. Gembarski, P.C., Lachmayer, R.: A business typological framework for the management of product complexity. In: Bellemare, J., Carrier, S., Nielsen, K., Piller, F.T. (eds.) Managing Complexity. SPBE, pp. 235–247. Springer, Cham (2017). https://doi.org/10.1007/978-3-319-29058-4_18

3. Antonsson, E.K., Otto, K.N.: Imprecision in engineering design. J. Vibr. Acoust. **117**(B), 25–32 (1995). https://doi.org/10.1115/1.2838671
4. Plappert, S., Gembarski, P.C., Lachmayer, R.: The use of knowledge-based engineering systems and artificial intelligence in product development: a snapshot. In: Świątek, J., Borzemski, L., Wilimowska, Z. (eds.) ISAT 2019. AISC, vol. 1051, pp. 62–73. Springer, Cham (2020). https://doi.org/10.1007/978-3-030-30604-5_6
5. Gembarski, P.C., Bibani, M., Lachmayer, R.: Design catalogues: knowledge repositories for knowledge-based-engineering applications. In: Marjanović, D., Štorga, M., Pavković, N., Bojčetić, N., Škec, S. (eds.) Proceedings of the DESIGN 2016 14th International Design Conference, pp. 2007–2016 (2016)
6. Hazelrigg, G.A.: A framework for decision-based engineering design. J. Mech. Des. **120**(4), 653–658 (1998). https://doi.org/10.1115/1.2829328
7. Pahl, G., Beitz, W.: Konstruktionslehre, 1st edn. Springer, Berlin (1977). https://doi.org/10.1007/978-3-662-02288-7
8. Lenders, M.: Beschleunigung der Produktentwicklung durch Lösungsraum-Management. Apprimus-Verlag (2009)
9. Dostatni, E., Diakun, J., Grajewski, D., Wichniarek, R., Karwasz, A.: Multi-agent system to support decision-making process in design for recycling. Soft. Comput. **20**(11), 4347–4361 (2016). https://doi.org/10.1007/s00500-016-2302-z
10. Wiebel, M., Eifler, T., Mathias, J., Kloberdanz, H., Bohn, A., Birkhofer, H.: Modellierung von Unsicherheit in der Produktentwicklung. In: Jeschke, S., Jakobs, E.M., Dröge, A. (eds.) Exploring Uncertainty, pp. 245–269, Springer Gabler, Wiesbaden (2013). https://doi.org/10.1007/978-3-658-00897-0_10
11. Kreye, M.E., Goh, Y.M., Newnes, L.B.: Manifestation of uncertainty – a classification. In: DS 68-6: Proceedings of the 18th International Conference on Engineering Design (ICED 2011), Impacting Society through Engineering Design, pp. 96–107, Lyngby/Copenhagen, Denmark (2011)
12. De Weck, O., Eckert, C., Clarkson, J.: A classification of uncertainty for early product and system design. In: DS 42: Proceedings of ICED 2007, the 16th International Conference on Engineering Design, pp.159–160, Paris, France (2007)
13. Chalupnik, M.J., Wynn, D.C., Clarkson, P.J.: Approaches to mitigate the impact of uncertainty in development processes. In: DS 58-1: Proceedings of ICED 09, the 17th International Conference on Engineering Design, pp. 459–470, Palo Alto, CA, USA (2009)
14. Ullman, D.G.: Robust decision-making for engineering design. J. Eng. Des. **12**(1), 3–13 (2001). https://doi.org/10.1080/09544820010031580
15. Ertel, W.: Grundkurs Künstliche Intelligenz – Eine praxisorientierte Einführung, 4th edn. Springer, Wiesbaden (2016). https://doi.org/10.1007/978-3-8348-2157-7
16. Jordan, M.I. (eds) Learning in Graphical Models, pp. 1–5. Springer (1999). https://doi.org/10.1007/978-94-011-5014-9
17. Russel, S.J., Norvig, P.: Artificial Intelligence – A Modern Approach, 3rd edn. Pearson, London (2016)
18. Zhu, J.Y., Deshmukh, A.: Application of Bayesian decision networks to life cycle engineering in Green design and manufacturing. Eng. Appl. Artif. Intell. **16**(2), 91–103 (2003). https://doi.org/10.1016/s0952-1976(03)00057-5
19. Moullec, M.L., Bouissou, M., Jankovic, M., Bocquet, J.C.: Product architecture generation and exploration using Bayesian networks. In: DS 70 - Proceedings of DESIGN 2012, the 12th International Design Conference, pp. 1761–1770, Dubrovnik, Croatia (2012)
20. Murphy, K.P.: The Bayes net toolbox for matlab. Comput. Sci. Stat. **33**(2), 1024–1034 (2001)
21. Vetter, G. (eds) Handbuch Dosieren, 2nd edn. Vulkan-Verlag, Essen (2002). https://doi.org/10.1002/cite.330671121

Hyperbolic Embeddings for Hierarchical Multi-label Classification

Tomaž Stepišnik[1,2(✉)] and Dragi Kocev[1,2,3]

[1] Jožef Stefan Institute, Ljubljana, Slovenia
{tomaz.stepisnik,dragi.kocev}@ijs.si
[2] Jožef Stefan International Postgraduate School, Ljubljana, Slovenia
[3] Bias Variance Labs, d.o.o., Ljubljana, Slovenia

Abstract. Hierarchical multi-label classification (HMC) is a practically relevant machine learning task with applications ranging from text categorization, image annotation and up to functional genomics. State of the art results for HMC are obtained with ensembles of predictive models, especially ensembles of predictive clustering trees. Predictive clustering trees (PCTs) generalize decision trees towards HMC and can be combined into ensembles using techniques such as bagging and random forests. There are two major issues that influence the performance of HMC methods: (1) the computational bottleneck imposed by the size of the label hierarchy that can easily reach tens of thousands of labels, and (2) the sparsity of annotations in the label/output space. To address these limitations, we propose an approach that combines graph node embeddings and a specific property of PCTs (descriptive, clustering and target attributes can be specified arbitrarily). We adapt Poincaré hyperbolic node embeddings to obtain low dimensional label set embeddings, which are then used to guide PCT construction instead of the original label space. This greatly reduces the time needed to construct a tree due to the difference in dimensionality. The input and output space remain the same: the tests in the tree use original attributes, and in the leaves the original labels are predicted directly. We empirically evaluate the proposed approach on 9 datasets. The results show that our approach dramatically reduces the computational cost of learning and can lead to improved predictive performance.

Keywords: Hierarchical Multi-label Classification · Hyperbolic embeddings · Ensemble methods · Predictive Clustering Trees

1 Introduction

In the typical supervised learning setting the goal is to predict the value of a single target variable. The tasks differ by the type of the target variable: binary classification deals with predicting a discrete variable with two possible values, multi-class classification deals with predicting a discrete variable with several possible values and regression deals with predicting a continuous variable. In

D. Helic et al. (Eds.): ISMIS 2020, LNAI 12117, pp. 66–76, 2020.
https://doi.org/10.1007/978-3-030-59491-6_7

many real life problems of predictive modelling the target variable is structured. Examples can be labelled with multiple labels simultaneously and some dependencies (e.g., tree-shaped or directed acyclic graph hierarchy) among labels may exist. The former task is called multi-label classification (MLC), while the latter is called hierarchical multi-label classification (HMC). These types of problems occur in domains such as life sciences (finding the most important genes for a given disease, predicting toxicity of molecules, etc.), ecology (analysis of remotely sensed data, habitat modelling), multimedia (annotation and retrieval of images and videos) and the semantic web (categorization and analysis of text and web pages). The most prominent area in a need of efficient HMC models with premium predictive performance is gene function prediction, where the goal is to predict the functions of a given gene. Gene Ontology [3] organizes 45.000 gene functions into a directed acyclic graph. Hence, the task of gene function prediction can be naturally viewed as a task of HMC.

A significant amount of research effort has been dedicated to developing methods for predicting structured outputs. In this sense, the methods for MLC [5] are the most prominent. The methods that consider hierarchical dependencies among the labels during model learning are less abundant. In two overviews of the HMC task [9,13], several methods are analyzed based on the amount of information they exploit from the hierarchy of labels during the learning of the models. The main conclusion is that global models (predicting the complete structure as a whole) generally have better predictive performance than the local models (predicting components of the output and then combining them). The success of the HMC methods is limited by two major factors: computational cost and sparsity of the output space. The number of labels (as well as the number of examples and features) for many domains presents a major *computational bottleneck* for all of the HMC methods and various methods cope differently with this. The *sparsity of the output space* pertains to the fact that the number of labels per example as well as the number of examples per label is very small.

We propose to address the two performance limiting issues by embedding the large hierarchical label space to a smaller space. Learning embeddings of complex data such as text, images, graphs and multi-relational data is currently a highly researched topic in artificial intelligence. Related to HMC are the embeddings of graph nodes (e.g., *Poincaré* embeddings [10], latent space embeddings [7], *NODE2VEC* [4]), as well as the embeddings for multi-relational data for information extraction and completion of knowledge graphs (e.g., *RESCAL, TRANSE, Universal Schema*). We exploit the learned embeddings within the learning of predictive clustering trees (PCTs) – a generalization of decision trees. They support different heuristic functions that guide the tree construction, and different prototype functions that make predictions in the leaves. With different choices of heuristic and prototype functions, they have been used for structured output predictions tasks [8], including HMC [14]. Additionally, PCTs yield state of the art predictive performance for the HMC task [2,6,11] and have been extensively used for gene function prediction [11,12].

The main contributions of this paper are as follows. First, we **learn new embeddings for HMC** by adapting the Poincaré node embeddings to get low dimensional embeddings of label sets assigned to individual examples. Second, we **extend PCTs** so that the heuristic function guiding the tree construction only looks at the embeddings and ignores the original high dimensional label space. This significantly reduces the time needed to construct the trees. The prototype function in the leaves is the same as for standard PCTs when used for HMC and predicts the original labels directly. A mapping from embeddings back to labels or **decoding of the embeddings is not needed**. Third, we **empirically evaluate** our approach on 9 datasets for gene function prediction. The evaluation reveals that it **drastically reduces the computational cost** compared to standard PCTs and, given equal time budget, yields models with **superior predictive performance**.

The remainder of this paper is organized as follows. In Sect. 2, we present our approach and theoretically analyze the reduction in computational cost it offers. Next, we outline the experimental design used to evaluate the performance of the obtained predictive models and discuss the obtained results in Sect. 3. Finally, we conclude and provide directions for further work in Sect. 4.

2 Method Description and Analysis

In this section, we briefly describe the calculation of the HMC embeddings by using the Poincaré hyperbolic node embeddings. We then describe PCTs and their extension so they use the label set embeddings to guide the tree construction.

2.1 Hyperbolic Embedding of Label Sets

A recently proposed approach based on the Poincaré ball model of hyperbolic space was shown to be very successful at embedding hierarchical data into low dimensions [10]. The points in the d-dimensional Poincaré ball correspond to the open d-dimensional unit ball $\mathbb{B}^d = \{x \in \mathbb{R}^d; \|x\| < 1\}$, and the distance between them is given as

$$d(x, y) = \operatorname{arcosh}\left(1 + 2\frac{\|x - y\|^2}{(1 - \|x\|^2)(1 - \|y\|^2)}\right).$$

This means that points close to the center have a relatively small distance to all other points in the ball, whereas the distances between points close to the border (as the denominator approaches 0) is much greater compared to its Euclidean counterpart. This property makes the space well suited for representing hierarchical data, as the norm of the embedding vector can naturally represent the depth of the node in the hierarchy. For example, the root of a tree hierarchy is an ancestor to all other nodes, and as such can be placed near the center, where the distance to all other points is relatively small. On the other hand, leaves of the tree can be placed close to the boundary.

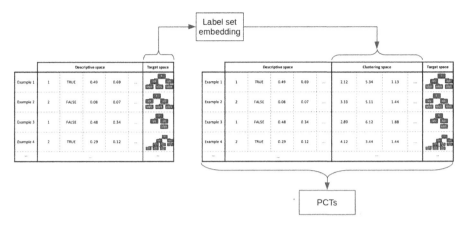

Fig. 1. An overview of our approach. We first calculate the embedding of label sets and add this information to our data. The embeddings are used for the clustering space, which guides the PCT construction, while the input and target space remain the same.

In the HMC task, labels are organized in a hierarchy, which can consist of thousands of nodes. First, we embed the hierarchy into the Poincaré ball, following the proposed method for embedding taxonomies [10]. This way we obtain vectors representing individual labels in the hierarchy. However, each example in a HMC dataset is associated with a set of labels. To get a vector representing the set of labels assigned to an example, we aggregate the vectors representing the individual labels. For the aggregation we have multiple options, for example calculating the component-wise mean vector or the medoid vector.

2.2 Predictive Clustering Trees

Predictive clustering trees (PCTs) are a generalization of decision trees towards predicting structured outputs including hierarchies of labels. In this work, we exploit a unique property of PCTs that allows arbitrary use of the various attributes. Specifically, for the learning of tree models there are three attribute types: descriptive, clustering and target (as illustrated in Fig. 1). The descriptive attributes are used to divide the space of examples; these are the variables encountered in the test nodes. The clustering attributes are used to guide the heuristic search of the best split at a given node. The target attributes are the ones we predict in the leaves.

PCTs are induced with a top-down induction of decision trees algorithm. outlined in Algorithm 1 that takes as input a set of examples and indices of descriptive, clustering and target attributes (can overlap). It goes through all descriptive attributes and searches for a test that maximizes the heuristic score. The heuristic that is used to evaluate the tests is the reduction of impurity caused by splitting the data according to a test. It is calculated on the clustering attributes. If no acceptable test is found (e.g., no test reduces the variance significantly, or the number of examples in a node is below a user-specified

Algorithm 1. Learning a PCT: The inputs are a set of learning examples E, indices of descriptive attributes D, indices of clustering attributes C and indices of target attributes T.

1: **procedure** GROW_TREE(E, D, C, T)
2:     test = best_test(E, D, C)
3:     **if** acceptable(test) **then**
4:         E_1, E_2 = split(E, test)
5:         left_subtree = grow_tree(E1, D, C, T)
6:         right_subtree = grow_tree(E2, D, C, T)
7:         **return** Node(test, left_subtree, right_subtree)
8:     **else**
9:         **return** Leaf(prototype(E, T))
10: **procedure** BEST_TEST(E, D, C)
11:     best = None
12:     **for** $d \in D$ **do**
13:         **for** test $\in$ possible_tests(E, d) **do**
14:             **if** score(test, E, C) > score(best, E, C) **then**
15:                 best = test
16:     **return** best
17: **procedure** SCORE(test, E, C)
18:     E_1, E_2 = split(E, test)
19:     **return** impurity(E, C) - $\frac{1}{2}$ impurity(E_1, C) - $\frac{1}{2}$ impurity(E_2, C)

---

threshold), then the algorithm creates a leaf and computes the prototype of the target attributes of the instances that were sorted to the leaf. The selection of the impurity and prototype functions depends on the types of clustering and target attributes (e.g., variance and mean for regression, entropy and majority class for classification). The support of multiple target and clustering attributes allows PCTs to be used for structured target prediction [8].

In existing uses of PCTs, clustering attributes include target attributes, i.e., the splits minimize the impurity of the target attributes. In this work, we take the attribute differentiation a step further and decouple the clustering and target attributes completely. We propose to use the learned embeddings as clustering attributes to guide the model learning and keep the original label vectors as the target attributes. This reduces the dimensionality and sparsity of the clustering space, which makes split evaluation and therefore tree construction faster. Additionally, we do not need to convert the embeddings to the original label space, since the predictions are already calculated in the label space.

We calculate the prototype in each leaf node as the mean of the label vectors (target variables) of the examples belonging to that leaf [14]. The prototype vectors present label probabilities in the corresponding leaves. The variance function is the same as for learning PCTs for multi-target regression (embeddings are continuous vectors), i.e., the weighted mean of variances of clustering attributes.

Like standard decision trees, the predictive performance of PCTs is typically much improved when used in an ensemble setting. Bagging is an ensemble

Table 1. Properties of the datasets used for the evaluation. The columns show the name of the dataset, the number of examples, the number of attributes describing the examples, and the number of labels in the target hierarchy.

dataset	N	D	L
cellcycle	3751	77	4125
eisen	2418	79	3573
expr	3773	551	4131
gasch1	3758	173	4125
gasch2	3773	52	4131

dataset	N	D	L
hom	3837	47034	4126
seq	3900	478	4133
spo	3697	80	4119
struc	3824	19628	4132

method that constructs base classifiers by making bootstrap replicates of the training set and using each of these replicates to construct a predictive model. Bagging can give substantial gains in predictive performance when applied to an unstable learner, such as tree learners [1]. It reduces the variance component of the generalization error linearly with the number of ensemble members. This means that there is a limit to how much ensembles can improve the performance (the bias component of the error). At a point the ensemble is saturated, and adding additional trees no longer makes a notable difference [8].

The computational cost of learning a PCT is $\mathcal{O}(DN \log^2 N) + \mathcal{O}(CDN \log N)$, where N is the number of examples, D is the number of descriptive attributes and C is the number of clustering attributes [8]. For standard PCTs, C is the number of labels in the hierarchy (L), which in hierarchical classification is typically much greater than $\log N$, making the second term the main contributor.

In our approach, C is instead the dimensionality of the embeddings (E), which can be only a fraction of the number of labels. We must also take into account the time needed to calculate the embeddings. They are optimized with stochastic gradient descent with time complexity $\mathcal{O}(EL)$ [10].

Therefore, we have reduced the time complexity from $\mathcal{O}(LDN \log N)$ using the standard approach, to $\mathcal{O}(EDN \log N) + \mathcal{O}(EL)$ using our approach. For larger datasets with thousands of examples and/or labels in the hierarchy, using our approach with low dimensional embeddings will offer significant speed gains.

3   Evaluation and Discussion

3.1   Experimental Design

For the evaluation of our approach we use 9 benchmark HMC datasets [14], in which examples are *yeast* genes. In different datasets, different features are used to describe the genes. The goal is to predict gene functions as represented by gene ontology terms. Basic properties of the datasets can be found in Table 1.

We optimize the embeddings using a variant of stochastic gradient descent as recommended in [10]. The only difference is that we do not use the burn-in

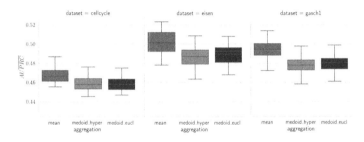

Fig. 2. Performance achieved using different aggregations of label embeddings. For brevity, only three datasets are shown; the results on the other datasets are very similar.

period, as it did not affect our results. We ran the optimization for 100 epochs with batch size 100, which took approximately 10–20 s per dataset using a single Nvidia Titan V graphics card.

In our first set of experiments, we aim to determine the best hyperparameters of our approach: the dimensionality of the embedding vectors and the aggregation function used to combine the embedding vectors of all the labels of an example. Embedding into higher dimensional space makes it easier for the optimization algorithm to find good embeddings, but increases the time required to learn them and learn PCTs on them. We consider dimensionalities from the set $\{2, 5, 10, 25, 50, 100\}$. We also consider three aggregation functions. The simplest one is to calculate the component-wise mean. The other approach is to select the medoid, i.e. the label embedding that is closest to all other label embeddings. Here, we examine both the Euclidean distance and the Poincaré distance as distance between the vectors. Note that the distances discussed here are only used to calculate the medoid of label embeddings. The embeddings themselves are always calculated and optimized in the hyperbolic space. We compare the performance of our approach to two methods: 1) the standard bagging of PCTs and 2) bagging of random PCTs, where at each step a random test to split the data is selected. Bagging of standard PCTs offer state of the art predictive performance for HMC tasks: It is what we would like to achieve or even exceed with our approach, but to learn the trees much faster. For this set of experiments, all ensembles consisted of 50 trees. We used 7 datasets, all but the largest two (hom and struc).

In the second set of experiments, we compare our approach to standard PCTs in a time budgeted manner. We add trees to an ensemble and record the performance and time needed after every couple of trees added. The ensemble is built for one hour or until 250 trees are built (at that point the performance gains are negligible). This evaluation compares our approach to ensembles of standard PCTs given equal time budget. We use the hyperparameters that worked well in the first set of experiments. For all experiments, we use area under the average precision-recall curve ($AU\overline{PRC}$) [14] to measure the predictive performance. Higher values indicate better performance. To estimate $AU\overline{PRC}$ we use 10-fold cross validation and report mean and standard deviation over folds. The entire

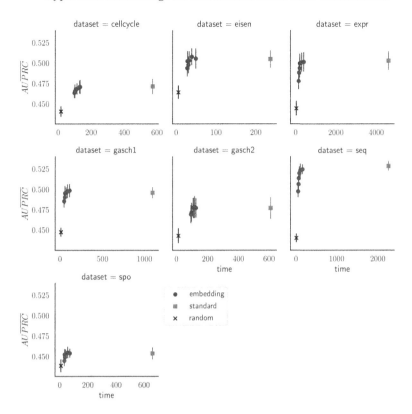

Fig. 3. The comparison of learning time (seconds) and performance of ensembles of standard PCTs, ensembles of random PCTs, and ensembles of PCTs learned on the embeddings. Vertical lines show standard deviation over folds.

experimental pipeline required to reproduce our experiments is available online at http://kt.ijs.si/dragikocev/ISMIS2020/ISMIS2020code.zip.

3.2   Results and Discussion

Figure 2 compares the results using different aggregation functions in the first set of experiments. Both Euclidean and Poincaré distances seem to work equally well for calculating the medoid embedding, in terms of the predictive performance achieved. However, mean aggregation typically offered better performance. Given that mean is also faster to calculate than medoid, especially when examples have many labels, we decided to proceed using mean aggregation.

Figure 3 shows the results obtained with mean aggregation and different embedding dimensionalities. Additionally, it shows the performance of ensembles of standard PCTs and ensembles of random PCTs. First thing to note is that our approach noticeably outperformed ensembles of random PCTs on all datasets.

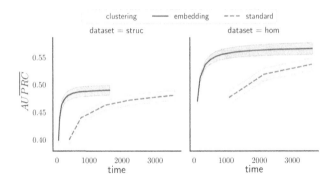

Fig. 4. The relationship between the performance and time (seconds) needed to build the ensemble. With the same time budget, our approach significantly outperforms ensembles of standard PCTs.

This means that the label set embeddings contain useful information about the label hierarchy. We can also see that with enough dimensions, our approach achieves performance that is on par with or exceeds the performance of standard PCTs. Increasing the dimensionality of the embeddings usually improves the performance of the ensemble, but increases the time needed to learn the trees. However, the performance seems to saturate rather quickly, and using more than 25 dimensions rarely results in significant improvement. Most importantly, our approach achieves the performance similar to that of ensembles of standard PCTs in barely a fraction of the time needed to construct them.

Considering the results in Fig. 3, we decided to use the 25-dimensional embeddings for the time-budgeted experiments. The performance improvements with higher dimensions were usually small, whereas the time complexity scales linearly. Figure 4 shows the results of the second set of experiments. On the struc dataset, our approach learned 250 trees well before the hour expired, and on the hom dataset, it learned 225 trees. With the standard approach, we managed to learn only 50 and 20 trees on these datasets, respectively. As discussed in Sect. 2.2, given a high enough time budget, the difference in the number of trees would not make much difference as both ensembles would be saturated. However, the results clearly show that our approach achieves significantly better results much faster. This is especially important when working with large datasets.

4   Conclusions

In this paper, we propose a new approach for solving the HMC task that combines Poincaré hyperbolic node embeddings and PCTs. We aggregate the embeddings of all the labels assigned to an example to obtain low dimensional label set embeddings. We then exploit the property of PCTs that allows us to use the label set embeddings to guide the tree construction, but still predict the original labels directly. Due to the difference in dimensionality between the embeddings and standard vector representations of label sets, our approach constructs PCTs much faster. Because we still predict the original labels directly, we do not need a mapping from the embeddings back to the labels.

We empirically evaluate our approach on 9 benchmark datasets for gene function prediction. First, we compare the aggregation functions used to combine the label-wise embeddings and show that **aggregation with mean works better** than the medoid aggregation. Second, we show that the models learned with our approach are **much more efficient**: the learning time is 5 or more folds faster than learning in the original space. In some cases they even achieve better predictive performance. Third, in time budgeted experiments we show that our approach achieves **premium predictive performance** much sooner than standard ensembles of PCTs. In applications with limited computational resources, it is clear that the models learned on the embeddings should be preferred.

We plan to extend the work along several dimensions. First, we will look into other node embeddings to compare them to the Poincaré variant. Next, we plan to investigate the effect of different optimization criteria on the performance. For example, we could try optimizing embeddings so that the distance between them is similar to the distance between the labels in the graph. Finally, we will investigate the influence of the embeddings on a wider set of domains.

References

1. Breiman, L.: Bagging predictors. Mach. Learn. **24**(2), 123–140 (1996). https://doi.org/10.1023/A:1018054314350
2. Cerri, R., Barros, R.C., de Carvalho, A.C., Jin, Y.: Reduction strategies for hierarchical multi-label classification in protein function prediction. BMC Bioinform. **17**(1), 373 (2016). https://doi.org/10.1186/s12859-016-1232-1
3. Consortium, T.G.O.: The gene ontology resource: 20 years and still GOing strong. Nucleic Acids Res. **47**(D1), D330–D338 (2018)
4. Grover, A., Leskovec, J.: Node2vec: scalable feature learning for networks. In: Proceedings of the 22 ACM SIGKDD Conference (KDD 2016), pp. 855–864. ACM (2016)
5. Herrera, F., Charte, F., Rivera, A.J., del Jesus, M.J.: Multilabel Classification: Problem Analysis, Metrics and Techniques. Springer, Heidelberg (2016). https://doi.org/10.1007/978-3-319-41111-8_2
6. Ho, C., Ye, Y., Jiang, C.R., Lee, W.T., Huang, H.: Hierlpr: decision making in hierarchical multi-label classification with local precision rates (2018)
7. Hoff, P., Raftery, A., Handcock, M.: Latent space approaches to social network analysis. J. Am. Stat. Assoc. **97**(460), 1090–1098 (2002)
8. Kocev, D., Vens, C., Struyf, J., Džeroski, S.: Tree ensembles for predicting structured outputs. Pattern Recogn. **46**(3), 817–833 (2013)
9. Levatić, J., Kocev, D., Džeroski, S.: The importance of the label hierarchy in hierarchical multi-label classification. J. Intell. Inf. Syst. **45**(2), 247–271 (2014). https://doi.org/10.1007/s10844-014-0347-y
10. Nickel, M., Kiela, D.: Poincaré embeddings for learning hierarchical representations. In: Advances in Neural Information Processing Systems 30, pp. 6338–6347. Curran Associates, Inc. (2017)

11. Radivojac, P.: colleagues: a large-scale evaluation of computational protein function prediction. Nat. Methods **10**, 221–227 (2013)
12. Schietgat, L., Vens, C., Struyf, J., Blockeel, H., Kocev, D., Džeroski, S.: Predicting gene function using hierarchical multi-label decision tree ensembles. BMC Bioinform. **11**(2), 1–14 (2010)
13. Silla, C., Freitas, A.: A survey of hierarchical classification across different application domains. Data Min. Knowl. Disc. **22**(1–2), 31–72 (2011). https://doi.org/10.1007/s10618-010-0175-9
14. Vens, C., Struyf, J., Schietgat, L., Džeroski, S., Blockeel, H.: Decision trees for hierarchical multi-label classification. Mach. Learn. **73**(2), 185–214 (2008). https://doi.org/10.1007/s10994-008-5077-3

Natural Language Processing

Joint Multiclass Debiasing of Word Embeddings

Radomir Popović[1(✉)], Florian Lemmerich[1], and Markus Strohmaier[1,2]

[1] RWTH Aachen University, Aachen, Germany
radomir.popovic@rwth-aachen.de
[2] Gesis-Leibniz Institute for Social Sciences, Mannheim, Germany

Abstract. Bias in Word Embeddings has been a subject of recent interest, along with efforts for its reduction. Current approaches show promising progress towards debiasing single bias dimensions such as gender or race. In this paper, we present a joint multiclass debiasing approach that is capable of debiasing *multiple* bias dimensions simultaneously. In that direction, we present two approaches, HardWEAT and SoftWEAT, that aim to reduce biases by minimizing the scores of the Word Embeddings Association Test (WEAT). We demonstrate the viability of our methods by debiasing Word Embeddings on three classes of biases (religion, gender and race) in three different publicly available word embeddings and show that our concepts can both reduce or even completely eliminate bias, while maintaining meaningful relationships between vectors in word embeddings. Our work strengthens the foundation for more unbiased neural representations of textual data.

Keywords: Word embedding · Bias reduction · Joint debiasing · WEAT

1 Introduction

Word Embeddings, i.e., the vector representation of natural language words, are key components of many state-of-the art algorithms for a variety Natural Language Processing tasks, such as Sentiment Analysis or Part of Speech Tagging. Recent research established that popular embeddings are prone to substantial biases, e.g., with respect to gender or race [2,6], which demonstrated in results like *"Man is to Computer Programmer as Woman is to Homemaker"* [2] as results of basic analogy tasks. Since such biases can potentially have an effect on downstream tasks, several relevant approaches for debiasing existing word embedding have been developed. A common deficit of existing techniques is that debiasing is limited to a single bias dimension (such as gender). Thus, in this paper, we propose two new post-processing methods for joint/simultaneous multiclass debiasing, which differ in their trade-off between maintaining word relationships and decreasing bias levels: HardWEAT completely eliminates contained bias as measured by the established Word Embedding Association Test

© Springer Nature Switzerland AG 2020
D. Helic et al. (Eds.): ISMIS 2020, LNAI 12117, pp. 79–89, 2020.
https://doi.org/10.1007/978-3-030-59491-6_8

[3]. SoftWEAT has a stronger and tunable emphasis on maintaining the original relationships between words in addition to bias removal. We demonstrate the effectiveness of our approach on the bias dimensions gender, race and religion on three conventional Word Embedding models: FastText, GloVe and Word2Vec.

2   Background and Related Work

In this section, we introduce key concepts formally and discuss related work on debiasing word embeddings.

We assume a *Word Embedding* with vocabulary size v that maps each word w to a vector representation $\vec{w} \in \mathbf{R}^d$. Protected *classes* c $\in$ C are entities on which bias can exist, e.g., race, religion, gender, or age. *Subclasses* S or *target set of words* refer to directions of a class, e.g., *male, female* for $c = gender$, and are associated with a set of definitional words S_c. The set of *neutral words* N contains all words that should not be associated with any $S_c \, \forall c \in C$. Finally, *attribute sets of words* A (A $\subset$ N) denote word sets that a target set of words can potentially be linked to, e.g., {*pleasant, nice, enjoyable*} or {*science, research, academic*}.

The Word Embedding Association Test. The state-of-the art way of measuring biases in embeddings is the Word Embedding Association Test (WEAT) [3]: It considers two sets of *attribute words* (A, B), e.g., family and career related words, and two target sets (X, Y), e.g., black and white names. The null hypothesis of this test states, that the relative association of target sets' words to both attribute sets' words are equally strong. Thus, rejecting the null hypothesis asserts bias. To examine this, a test statistic s quantifies how strongly X is associated to A in comparison to the association between Y and B. It is computed as $s(X, Y, A, B) = \sum_{\vec{x} \in X} h(\vec{x}, A, B) - \sum_{\vec{y} \in Y} h(\vec{y}, A, B)$, where $h(\vec{w}, A, B) = \text{mean}_{\vec{a} \in A} \cos(\vec{w}, \vec{a}) - \text{mean}_{\vec{b} \in B} \cos(\vec{w}, \vec{b})$ describes the relative association between a single target word $x \in X$ compared to the two attribute sets in a range $[-2, 2]$. Based on s, we assess the statistical significance via an one-sided permutation test through which we acquire p value; Additionally, an effect size d that quantifies the severity of the bias can be calculated as:

$$d(X, Y, A, B) = \frac{\text{mean}_{\vec{x} \in X} h(\vec{x}, A, B) - \text{mean}_{\vec{y} \in Y} h(\vec{y}, A, B)}{\text{std}_{w \in X \cup Y} h(\vec{w}, A, B)}$$

We will use this effect size in our novel debiasing approaches.

Existing Debiasing Techniques. To achieve reduction of bias, we will substantially extend the work of Bolukbasi et al. [2], which describes two ways of debiasing Word Embeddings in a post-processing step: *Hard and Soft Debiasing*. Both rely on identifying gender bias subspace B via Principal Component Analysis computed on gender word pairs differences, such as *he–she, and man–woman*. Hard debiasing employs a *neutralize* operation that removes, e.g., all non-gender related words N from a gender subspace by deducting from vectors their bias subspace projection. Subsequently, an *equalize* operation positions opposing gender pair words (e.g., king, queen) to share the same angle with neutral words. By contrast, Soft Debiasing enables more gradual bias removal by utilizing a tuning parameter λ: An embedding $W \in \mathbb{R}^{d \times |vocab|}$ is being transformed by optimizing a transformation matrix $T \in \mathbb{R}^{d \times d}$:

$$\min_{T} \|(TW)^T(TW) - W^T W\|_F^2 + \lambda\|(TN)^T(TB)\|_F^2$$

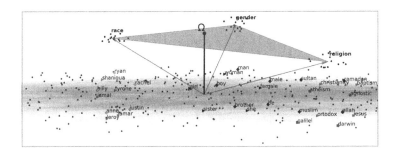

Fig. 1. Visual interpretation of multiclass debiasing via HardWEAT.

Another recent approach by Manzini et al. [10] incorporates these ideas, but expands and evaluates results not only on gender, but separately also on race and religion. It suggests a bias subspace definition for non-binary class environment, which is formulated via PCA of mean shifted k-tuples (k = number of subclasses) of the definitional words. There are also other recently proposed gender bias post-processing [4,5] and pre-processing techniques [16].

In terms of existing joint multiclass debiasing techniques, *Conceptor Debiasing* [7], is based on applying Boolean-like logic operators using soft transformation matrix on a linear subspace where Word Embedding has the highest variance. We will use this technique for comparison in our experiments.

3   Approach

In this chapter, we present our novel debiasing techniques. Our methods substantially extend previous works of Bolukbasi et al. [2] and Caliskan et al. [3].

HardWEAT: To adapt the *neutralize* step from Bolubasi's work [2] jointly in a multiclass debiasing setting, we first define the concept of *class definitional vectors* $\overrightarrow{def_c}$ of classes (e.g., race, gender, ...), which are computed as the top component of a PCA on the vector representations of definitional words $D_i \; \forall i \in \{1, ..., n\}$ for n subclasses (e.g., male, black, ...), cf. [10]. Now, each particular class can vary in terms of bias amount according to WEAT tests with chosen sets of attribute and target words. Thus, we aggregate WEAT effect sizes d for each class c into bias levels δ_c by averaging twice: First, we compute the mean value for all target/subclass pairs within a class. Second, we average those means for all results for class c. Then, we compute a *Centroid* Ω, as the average of the class definitional vectors weighted by their normalized bias levels: $\vec{\Omega} = \sum_{c \in C} \frac{(\delta_c \cdot \vec{def_c})}{\sum_{c \in C} \delta_c}$. We use this centroid to perform *neutralization*, i.e., we re-embed all neutral words such that they perpendicular to it, cf. [2].

To adapt the *equalize* step of Bolubasi's work, we move the definitional words of all subclasses in a way such that there is no relative preference of any attribute set towards a subclass. For that purpose, we generate equally spread out points $e_1, \ldots, e_n$ on some circle with radius r and center $\vec{o}$ as the new center points of the subclasses. In three or more dimensions, this circle can be defined by two vectors $(\vec{v_1}, \vec{v_2})$ perpendicular to $\vec{o}$: $\vec{e_i} = \vec{o} + r\cos(\frac{2\pi i}{n}) \cdot \vec{v_1} + r\sin(\frac{2\pi i}{n}) \cdot \vec{v_2}$.

Given this formula, we use *equidistancing* twice: First, we calculate new temporary central points for each class (e.g., gender) by neutralization (see above), then we determine new temporary central points $\overrightarrow{defs_{S_i}}$ for each subclass (e.g., male) in a circle around that. Then, we compute new embeddings $\vec{w_i}$ for all definitional words D_i in a circle around this subclass vector. Note that r_c, r_{S_c} are randomly generated with the condition $r_{S_c} \gg r_c$, whereas $\vec{v_1}, \vec{v_2}$ are obtained via SVD. Along with an illustration (see Fig. 1), we present the full procedure in Algorithm 1.

For successful equidistancing, i.e., equal dispersion of words, we also require that the length of all subclass/target words within a particular class must be

---

Algorithm 1: HardWEAT algorithm.

Input: Word Embedding matrix $\boldsymbol{W}_{v \times d}$, Words W, set of classes C, set of subclasses S, Bias levels δ_c

1 **for** $c \in C$ **do**   // Generating class definitional vectors

2    $\quad \overrightarrow{def_c} := \text{PCA}\left(\bigcup\limits_{i=1}^{n} \bigcup\limits_{\vec{w} \in D_i} \vec{w} - \mu_i \right)$

3 $\vec{\Omega} = \sum\limits_{c \in C} (\delta_c \cdot \overrightarrow{def_c})$ // Computing centroid

4 $N = W \setminus B$ // Defining neutral words

5 $\boldsymbol{N} := \boldsymbol{N} - \frac{N \cdot \vec{\Omega}}{||\vec{\Omega}||} \vec{\Omega}$ // Neutralizing

6 $\boldsymbol{W} = \frac{W}{||W||}$ // Normalizing vectors

7 **for** $c \in C$ **do**   // Equidistancing

8    $\quad \vec{O_c} := \overrightarrow{def_c} - \frac{\overrightarrow{def_c} \cdot \vec{\Omega}}{||\vec{\Omega}||} \vec{\Omega}$

9    $\quad$ **for** $i \in \{1, ..., |S_c|\}$ **do**

10      $\quad\quad \overrightarrow{defs_{S_i}}\prime = \vec{O_c} + r_c\cos(\frac{2\pi i}{|S_i|}) \cdot \vec{v_1} + r_c\sin(\frac{2\pi i}{|S_i|}) \cdot \vec{v_2}$

11      $\quad\quad$ **for** $w_j \in W_S$ **do**

12        $\quad\quad\quad \vec{w_j} = \overrightarrow{defs}\prime + r_{S_c}\cos(\frac{2\pi j}{|W_S|}) \cdot \vec{v_1} + r_{S_c}\sin(\frac{2\pi j}{|W_S|}) \cdot \vec{v_2}$

13 $\boldsymbol{W} = \frac{W}{||W||}$ // Normalizing vectors

---

equal. Violations of this requirement will result in inadequate WEAT scores. Furthermore, HardWEAT could in theory result in some equidistanced words becoming angle-close to random words: Thus, to counter this, we design this process in an iterative manner, modifying r_{S_c} until there are no neutral words having an angle greater than certain threshold (e.g., we use $45°$ as a default). We discuss factors of success and open issues in more detail in Sect. 4.4. Overall, the HardWEAT method ultimately aims at complete bias elimination, but randomizes the topological structure of subclass words.

SoftWEAT is a more gradual debiasing alternative with a greater focus on quality preservation. It provides the user with a choice on how to debias and to what extent. Let us assume that a particular target set of words S_c is closely related in terms of their angle to some attribute set of words A_i (e.g., A_1 is the set of *pleasant* words, A_2 is the set of *intellectual* words). Hypothetically, bias would be minimized if the subclass words S_c would be perpendicular to the attribute sets A_i. Thus, moving S_c towards such a perpendicular space/vector (null vector), noted as $\vec{n}$, results in bias reduction. Since full convergence towards a vector $\vec{n}$ may result in quality decrease we define a parameter $\lambda \in [0,1]$ as a *level of removal*. Setting $\lambda = 1$ results in maximal angle decrease between S_c and A_i (perpendicular vectors), while $\lambda = 0$ makes no transformation at all. Note, however, that placing vectors from some subclass words S_i to be perpendicular with attribute words A_i may not necessarily result in overall bias level reductions since WEAT tests are relative measures. E.g., they take the relationship between *male* and *intellectual* words and the relationship between *female* and *appearance*-related words into account at the same time. Additionally, higher λ also pose a greater risk of producing new bias. Moving the representation of subclass words away from attribute words A_1 may result in moving them closer to some other A_2 without prior intent. For example, moving *male* away from *science* can get them closer to *aggressive*, resulting in other WEAT test d increase. To address these issues, we choose a nullspace via SVD that minimizes the average WEAT effect size for all tests.

SoftWEAT executes the following steps: Given some S_c and its corresponding related attribute set of words $A_i, \ldots, A_n$, we generate a matrix $\boldsymbol{A}_{n \times d}$, where the n rows consist of the first principal components for each A_i respectively. For $\boldsymbol{A}$, we then find the nullspace vector $\vec{n}$ that decreases WEAT test scores the most. As a goal, we aim to translate our S_c to this space. Since initially, there might be only few target set words S_c, we expand it by adding for all of them the closest, frequently occurring neighbor words (e.g., the 20 closest neighbors with minimum frequency of 200 in the English Wikipedia [1]).) Afterwards, we calculate a transformation vector $\vec{\psi}$ for which it holds that: $\vec{\psi} = \vec{n} - \vec{m}$, where $\vec{m}$ is the mean of S_c words. This transformation vector is then multiplied with a parameter $\lambda \in [0,1]$ followed by conversion into linear translation matrix Ψ. By translating vectors from extended word sets, we preserve relationships of subclass words S_c to these words. Note that we only modify positions of words from the extended neighborhood of subclass words, but not all words from the vocabulary. Finally, as a last optional step, we normalize all vectors.

In SoftWEAT, the user can decide on the λ, but can also select target and attribute sets used for debiasing. By default, we will only take those into account, for which WEAT scores result in an aggregated effect size of $|d| > 0.6$. To compute this, we accumulate attribute sets per each target set that form matrix $\mathbf{A}$ per each S_c. In case of positive d, S_1 gets removed from A_1, and S_2 from A_2. In case of negative d, removal is done other way around. The algebraic operations are formalized as follows: Out of the new matrix $\mathbf{W}'$ with $|S_c|$ rows and $d + 1$ columns, all but the last row are taken as a new vector representation for the given target set of words S_c.

$$\mathbf{A}\vec{n} = 0 \qquad \vec{\psi} = \lambda(\vec{n} - \vec{m})$$

$$\mathbf{W}'_{S_c} = \Psi \begin{bmatrix} (W_{S_c})^T \\ 1 \end{bmatrix} = \begin{bmatrix} 1 & 0 & 0 & \dots & \psi_1 \\ \vdots & \vdots & \ddots & \dots & \vdots \\ 0 & 0 & \dots & 1 & \psi_d \\ 0 & 0 & 0 & \dots & 1 \end{bmatrix} \begin{bmatrix} w_{11} & w_{21} & \dots & w_{n1} \\ \vdots & \vdots & \vdots & \ddots \\ w_{1d} & w_{2d} & \dots & w_{nd} \\ 1 & 1 & \dots & 1 \end{bmatrix}$$

4   Experimental Evaluation

This section presents example results of our methods in practical settings and compares it with Conceptor Debiasing [7] as the current state-of-the-art approach. We measure the bias decrease along with deterioration of embedding quality and show the effects of biased/debiased embeddings in a Sentiment Analysis as a downstream task example. Due to limited space, we also provide an extended set of results in an accompanying online appendix[1].

4.1   Experimental Setup

Datasets: Extending the experimental design from [8], we apply debiasing simultaneously on following target sets/subclasses: *(male, female) - gender*, *(islam, christianity, atheism) - religion* and *(black and white names) - race* with seven distinct attribute set pairs. We collected target, attribute sets, and class definitional sets from literature [3,8,10,11,15,16], see our online appendix for a complete list. As in previous studies [7], evaluation was done on three pretrained Word Embedding models with vector dimension of 300: FastText[2] (English webcrawl and Wikipedia, 2 million words), GloVe[3] (Common Crawl, Wikipedia and Gigaword, 2.2 million words) and Word2Vec[4] (Trained on Google News, 3 million words). For the Sentiment Analysis task, we additionally employed a dataset of movie reviews [9].

---

[1] https://git.io/JvL10.
[2] https://fasttext.cc/docs/en/english-vectors.html.
[3] https://nlp.stanford.edu/projects/glove/.
[4] https://drive.google.com/uc?id=0B7XkCwpI5KDYNlNUTTlSS21pQmM.

Quality Assessment: First, we compared ranked lists of word pairs in terms of their vector similarity to human judgement via Spearman's Rank correlation coefficient [13], by using the collection taken from the Conceptor Debiasing evaluation [7], i.e., *RG65, WS, RW, MEN, MTurk, SimLex, SimVerb*. In addition, we also utilize Mikolov Analogy Test [12].

Methods: Regarding HardWEAT, we specified neutral words through set difference between all words and ones from priorly defined target sets. In terms of SoftWEAT we provide details such as target-attribute sets structure and number of changed words in section F of online appendix. We applied the OR operator in the Conceptor Debiasing, using the same word set of subclasses within the three above defined classes (subspaces in Conceptor Debiasing).

Table 1. Bias levels for gender, race, and religion before debiasing (REG) and after debiasing with Hard/SoftWEAT (HW/SW) or Conceptor (CPT)

		REG	HW	SW $\lambda = 0.2$	SW $\lambda = 0.4$	SW $\lambda = 0.6$	SW $\lambda = 0.8$	SW $\lambda = 1$	CPT
	gender	0.62	0.0**	0.47	0.3	0.28	0.15	0.09*	0.24
GloVe	race	0.71	0.0**	0.6	0.51	0.39	0.3	0.19*	0.43
	religion	0.77	0.0**	0.62	0.46	0.29	0.2	0.16*	0.28
	gender	0.52	0.0**	0.14*	0.17	0.32	0.31	0.32	0.46
FastText	race	0.36	0.0**	0.13*	0.16	0.25	0.3	0.31	0.31
	religion	0.6	0.0**	0.27	0.21*	0.29	0.35	0.42	0.54
	gender	0.63	0.0**	0.48	0.37	0.32	0.2	0.23	0.12*
Word2Vec	race	0.56	0.0**	0.32	0.28	0.2	0.16	0.16*	0.38
	religion	0.47	0.0**	0.31	0.19	0.11*	0.18	0.2	0.28

4.2   Bias Levels and Quality of Word Embeddings

First, we focus on overall bias levels, see Table 1, by measuring WEAT scores before and after debiasing. We observe that HardWEAT removes the measured bias completely as indicated by zero WEAT scores. SoftWEAT also substantially reduces the bias measurements to different degrees for different datasets. In comparison, for example with $\lambda = 1$, SoftWEAT still leads to a stronger reduction in bias compared to the state-of-the-art Conceptor algorithm in all but one measurement.

Regarding quality assessment, Table 2, shows the complete results for seven rank similarity datasets as well as the Mikolov analogy score on the example of the GloVe Word Embedding. As expected, a significant quality drop appears with HardWEAT, most notably on the RG65 dataset. With SoftWEAT, greater λ induce larger modifications of the original embeddings, which corresponds to a greater drop in embedding quality. However, in three out of eight test settings, SoftWEAT with lower λ settings achieved a similar or higher score and also leads to competitive results compared to Conceptor, see Sect. 4.4 for an in-depth discussion. Extended results can be found in the online appendix.

Table 2. Measurements of quality tasks after debiasing for GloVe embeddings.

	REG	HW	SW $\lambda = 0.2$	SW $\lambda = 0.4$	SW $\lambda = 0.6$	SW $\lambda = 0.8$	SW $\lambda = 1$	CPT
RG65	**76.03***	63.42	75.68	75.39	74.34	73.25	71.42	68.73
WS	73.8	69.73	**74.0***	73.96	73.15	71.95	70.34	73.48
RW	46.15	46.09	46.15	46.13	46.08	46.02	45.94	**52.45***
MEN	80.13	77.52	80.34	**80.34***	80.1	79.57	78.74	79.99
MTurk	**71.51***	69.78	71.25	70.78	70.37	69.48	68.64	68.24
SimLex	40.88	40.44	41.46	42.0	42.07	42.21	42.18	**47.36***
SimVerb	28.74	28.74	28.9	29.15	29.27	29.54	29.78	**36.78***
Mikolov	0.65	0.64	**0.65***	0.65	0.65	0.64	0.64	0.63

4.3 Sentiment Analysis

To analyze the effects of debiasing in downstream tasks, we study a sentiment analysis task in the context of movie reviews, i.e., we investigate whether we observe significant differences in predicted sentiments when using biased and debiased Word Embeddings. Modifying setup from Packer et al. [14], we added to the end of each test sentences randomized word from opposing set pairs (e.g., a typical black name or a typical white name), and calculated the difference in the prediction, i.e., the *polarity score* according to a simple neural network that takes the respective pretrained word embeddings as pretrained embedding, which is followed by Flatten-operation and a sigmoid layer. The first two layers were not trainable.

For training, we used a combination of binary-cross entropy loss and Ada-Delta optimizer. We trained our model only on sentences with a maximum length of 50 that did not contain any target sets of words. Shuffling test and training set, we then calculated *polarity score* on 15 different model instances using 6 different kinds of Embeddings: Regular (not debiased), SoftWEAT with $\lambda = 0.1, 0.5, 1$, HardWEAT and Conceptor Debiasing with $\alpha = 2$.

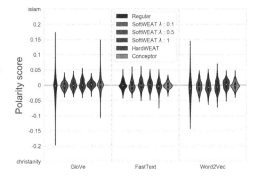

Fig. 2. Classifier Bias on (islam, christianity)

Results for *[christianity, islam]* are shown in Fig. 2. For the non debiased embeddings, we observe that our classifier has no clear trend such that the addition of a bias word in a sentence influences the polarity in a particular direction. However, we can see that bias words have a strong influence on the classifier, leading to a large variance of *polarity scores* in GloVe and Word2Vec-based models. When we apply our debiasing methods, we see that the respective variances are substantially decreased, e.g., around 5.45 times for SoftWEAT already with a small λ of 0.1 in GloVe. This is only the case with Conceptor Debiasing for the Word2Vec-based model, but not for the GloVe-based model.

4.4   Discussion

HardWEAT and SoftWEAT Success Factors: Based on our experiments, see also our online appendix, we could identify a variety of success factors for our methods. Regarding HardWEAT, higher embedding dimensionality and more dispersed vectors are more likely to output more desirable outcome. This is due to the possibility of random words appearing as close neighbors to a target words during the equidistancing procedure, which we counter iteratively as described above. Regarding centroid $(\vec{\Omega})$ neutralization, relevant factors of importance are number of classes, uniformity of bias levels and angles between all def_c: Having more classes with uniform bias levels and distant angles can output non-desirable results, e.g., too large distances between $\vec{\Omega}$ and $\overrightarrow{def_c}$ points. However, the user could also manually define an alternative centroid $\vec{\Omega}$. We acknowledge that *equidistancing* comes with a main drawback, which is the partial loss of relationship between target and non-target words, which is also reflected in a quality drop according to different metrics. Regarding SoftWEAT, decreased angles between target and attribute sets after translation may not necessarily result in lower bias levels, since WEAT is a relative measure. However, we take one nullspace which minimizes these values. Furthermore, we note that in our experiments we used all target and attribute set pairs within each of the WEAT tests, which can be further optimized: We may not want to debias something, which isn't priorly biased. E.g., removing *male* from *science* may be necessary, whereas doing the same for *female* from *art* may not, thus we could exclude this latter pair. Also, we should bear in mind, that attribute sets of words are often correlated (as also shown by our experiments in the online appendix). This implies that by moving subclass words towards a specific set of attribute words, we automatically change the associations also with other attribute sets. Thus, the user plays a crucial role in deciding which sets should be used for debiasing.

Comparison with Conceptor Debiasing: Given the experimental results, we conclude that neither Conceptor Debiasing nor SoftWEAT outperform each other. Yet, SoftWEAT exhibits some distinctive advantages: (i) With Soft-WEAT, relationships within the target set words remain completely the same, whereas in Conceptor Debiasing, overall angle distribution gets more narrow (See online appendix for more details). (ii) We argue that with SoftWEAT, user gets more freedom with choosing on which target/attribute set combination and to which degree debiasing is applied. (iii) Given our method, there is no difference in word representation if one uses small subset of neutral words or complete vocabularies. Nevertheless, we acknowledge that Conceptor Debiasing does succeed in reducing bias equally well by applying it in more global behavior.

5   Conclusion

In this paper, we presented two novel approaches for multi-class debiasing of Word Embeddings. We demonstrated the general viability of these methods for reducing and/or eliminating biases while preserving meaningful relationships of the original vector representations. We also analyzed the effects of debiased representations in Sentiment Analysis as an example downstream task and find that debiasing leads to substantially decreased variance in the predicted polarity. Overall, our work contributes to ongoing efforts towards providing more unbiased neural representations of textual data.

Acknowledgements. We want to thank S. Karve, L. Ungar, L., and J. Sedoc for providing the code for Conceptor Debiasing and their support.

References

1. Arora, S., Liang, Y., Ma, T.: A simple but tough-to-beat baseline for sentence embeddings. In: International Conference on Learning Representations (2017)
2. Bolukbasi, T., Chang, K.W., Zou, J.Y., Saligrama, V., Kalai, A.T.: Man is to computer programmer as woman is to homemaker? Debiasing word embeddings. In: Lee, D.D., Sugiyama, M., Luxburg, U.V., Guyon, I., Garnett, R. (eds.) Advances in Neural Information Processing Systems 29, pp. 4349–4357 (2016)
3. Caliskan, A., Bryson, J.J., Narayanan, A.: Semantics derived automatically from language corpora contain human-like biases. Science **356**(6334), 183–186 (2017)
4. Dev, S., Phillips, J.: Attenuating bias in word vectors. In: International Conference on Artificial Intelligence and Statistics, pp. 879–887 (2019)
5. Font, J.E., Costa-jussà, M.R.: Equalizing gender bias in neural machine translation with word embeddings techniques. In: Proceedings of the First Workshop on Gender Bias in Natural Language Processing, pp. 147–154 (2019)
6. Garg, N., Schiebinger, L., Jurafsky, D., Zou, J.: Word embeddings quantify 100 years of gender and ethnic stereotypes. Proc. Natl. Acad. Sci. **115**(16), E3635–E3644 (2018)
7. Karve, S., Ungar, L., Sedoc, J.: Conceptor debiasing of word representations evaluated on WEAT. In: Workshop on Gender Bias in NLP (2019)
8. Knoche, M., Popović, R., Lemmerich, F., Strohmaier, M.: Identifying biases in politically biased wikis through word embeddings. In: Conference on Hypertext and Social Media, pp. 253–257 (2019)
9. Maas, A.L., Daly, R.E., Pham, P.T., Huang, D., Ng, A.Y., Potts, C.: Learning word vectors for sentiment analysis. In: Annual Meeting of the ACL: Human Language Technologies, pp. 142–150 (2011)
10. Manzini, T., Chong, L.Y., Black, A.W., Tsvetkov, Y.: Black is to criminal as caucasian is to police: detecting and removing multiclass bias in word embeddings. In: Conference of the North American Chapter of the ACL, pp. 615–621 (2019)

11. May, C., Wang, A., Bordia, S., Bowman, S., Rudinger, R.: On measuring social biases in sentence encoders. In: Conference of the North American Chapter of the ACL: Human Language Technologies, pp. 622–628 (2019)
12. Mikolov, T., Chen, K., Corrado, G., Dean, J.: Efficient estimation of word representations in vector space. arXiv preprint arXiv:1301.3781 (2013)
13. Myers, L., Sirois, M.J.: Spearman Correlation Coefficients, Differences Between. American Cancer Society (2006)
14. Packer, B., Mitchell, M., Guajardo-Céspedes, M., Halpern, Y.: Text embeddings contain bias. Here's why that matters. Technical report, Google (2018)
15. Zhao, J., Wang, T., Yatskar, M., Ordonez, V., Chang, K.W.: Gender bias in coreference resolution: evaluation and debiasing methods. In: Conference of the North American Chapter of the ACL, pp. 15–20 (2018)
16. Zhao, J., Zhou, Y., Li, Z., Wang, W., Chang, K.W.: Learning gender-neutral word embeddings. In: Conference on Empirical Methods in Natural Language Processing, pp. 4847–4853 (2018)

Recursive Neural Text Classification Using Discourse Tree Structure for Argumentation Mining and Sentiment Analysis Tasks

Alexander Chernyavskiy$^{(\boxtimes)}$ and Dmitry Ilvovsky

National Research University Higher School of Economics, Moscow, Russia
alschernyavskiy@gmail.com, dilv_ru@yahoo.com

Abstract. This paper considers sentiment classification of movie reviews and two argument mining tasks: verification of political statements and categorization of quotes from an Internet forum corresponding to argumentation (factual or emotional). In the case of the fact-checking problem, justifications can be used additionally in one of its sub-tasks. A strong model for solving these and similar problems still does not exist. It requires the style-based approach to achieve the best results. The proposed model effectively encodes parsed discourse trees due to the recursive neural network. The novel siamese model based on it is suggested to analyze discourse structures for the pairs of texts. In the paper, the comparison with state-of-the-art methods is given. Experiments illustrate that the proposed models are effective and reach the best results in the assigned tasks. The evaluation also demonstrates that discourse analysis improves quality for the classification of longer texts.

Keywords: Argument mining · Fact-checking · Sentiment analysis · Discourse tree structure · Recursive neural network · Siamese model

1 Introduction

Nowadays, social media users are inundated with factual texts about politics, economics, history and so on. Triggered by the fact that some sources can utilize fake statements for their purposes, it would be unwise to trust all of them. For instance, in the United States over a million tweets contained fake information by the end of the presidential election of 2016 (the "Pizzagate" scandal). Thus, the factual text categorization task is significant for public goods.

The analysis of texts from the point of view of psychology and rhetoric is also challenging. There are many papers devoted to analyzing messages from Internet users regarding the existence of factual argumentation, sarcasm and other factors. This area can be used in dialogue systems, sentiment classification and so forth.

© Springer Nature Switzerland AG 2020
D. Helic et al. (Eds.): ISMIS 2020, LNAI 12117, pp. 90–101, 2020.
https://doi.org/10.1007/978-3-030-59491-6_9

The sentiment analysis consists in determining whether the emotional tone (sentiment) of the social-media text is positive or negative. This task has become more popular over the years. It can be applied in e-commerce, marketing and advertising.

In this work, we review the three following tasks:

1. The fact-checking task. The main goal is to categorize given factual text into several pre-defined classes corresponding to its veracity. Two sub-tasks are considered:
   (a) Classification of statements alone
   (b) Classification of pairs of texts (statement and its justification)
2. The detection of emotional and factual argumentation. It is a binary classification task, the main goal of which is to categorize texts from an Internet forum as either factually or emotionally justified.
3. The sentiment classification task. The main objective is to classify movie reviews into positive and negative texts.

This paper aims to develop a universal model capable of effective solving these and similar tasks.

Keyword features seem to be insufficient to this problem but accounting for the writing style of text may be crucial. The idea of this research is to outperform existing methods due to using additional discourse features, which can be constructively represented as trees employing Rhetorical Structure Theory [15].

The developed method encodes discourse trees with the recursive neural network, specifically the binary TreeLSTM model. The analogous model was utilized only for unlabeled syntax trees in the sentiment analysis and semantic relatedness of two sentences task before and achieved top results. We also propose the siamese recursive neural network based on it to categorize pairs of texts which can be effectively used in the sub-task (b) of the fact-checking problem.

Our main contributions are the following:

– The TreeLSTM model was adopted to get the universal and effective model based on the discourse tree structure.
– The siamese recursive model based on discourse trees was proposed to analyze pairs of texts.

We demonstrate the quantitative benefits of these models applied to discourse trees for four popular datasets. Experiments confirmed that discourse analysis suffices to improve quality for relatively long texts. The siamese model was applied in the fact-checking problem where statements with justifications are given. Additionally, we compared two variants of text embeddings in the leaves of the discourse trees.

The remainder of the paper is organized as follows. Initially, related work is observed. This is followed by a description of discourse features and text representation associated with it. Afterwards, the main models are introduced and described in detail. Finally, datasets, implementation details, obtained results and directions for further research are discussed.

2   Related Work

In this paper, we utilize the style-based approach. The main idea is that the writing style of texts in the aforementioned tasks is different. Rashkin and Choi [21] showed that additional lexical features extracted by the LSTM model improve the accuracy of simple models such as Naive Bayes and Logistic Regression. Further, Galitsky et al. [6,7] suggested the usage of discourse trees that are encoded by the Tree kernel SVM model to determine disinformation. Bhatia et al. [2], Ji and Smith [10] considered the same idea and presented structures of the recursive neural networks for these trees, where the latter suggested modification of the former. These models were successfully applied in sentiment analysis and similar tasks. Finally, Tai et al. [23] proposed the N-ry TreeLSTM structure that obtained the best results in different text classification tasks using syntax trees.

Besides recurrent and recursive neural networks, the usage of convolutional neural networks is being researched [8,25]. Deligiannis et al. [4] suggested the graph convolutional neural network which utilizes data about users and publishers for the fake news detection. Huge number of current state-of-the-art approaches consider BERT-based models for text categorization, question answering and other tasks [5,13,14]. For instance, Trautmann et al. [24] showed that BERT achieved top results in the argumentation mining task.

Petz et al. [20] investigated the influence of the quality of social media texts on the performance of popular opinion mining algorithms.

We provide a method based on the discourse structure of texts which combines discourse trees and the recursive neural network. We adopted the method described in [23] for the labeled discourse structure.

3   Methods

3.1   Rhetorical Structure Theory

The main advantage of the proposed method is the usage of additional discourse features. According to the Rhetorical Structure Theory [15] text can be represented as a labeled tree. Its leaves are spans of text named Elementary Discourse Units (denote as EDUs) and connected by corresponding discourse relations. As a result, inner nodes are formed and these nodes contain longer spans. Inner nodes are also connected according to relations and so forth. The most popular discourse relations are Elaboration, Condition, Joint and Attribution.

Spans are said to be Nucleus and Satellite, where Nucleus is more essential for understanding the meaning of the text and Satellite involves some additional information. Figure 1 demonstrates an example of a discourse tree. Arrows are drawn from Nucleus vertices to Satellite vertices.

3.2   EDU Embeddings

We consider two variants of text embedding in EDUs.

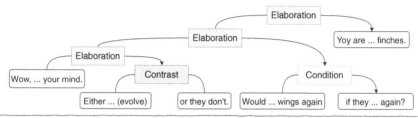

Fig. 1. Discourse tree for text from 4forums.com

1. Mean value of word embeddings
   We utilized 300-dimensional pre-trained GloVe vectors [19] for word embeddings.
2. Sentence embedding
   Universal Sentence Encoder [3] is the open-source model proposed by Google.
   It achieves the best results in different transfer learning tasks. There are two
   modifications of its architecture: Transformer network and Deep Averaging
   Network (DAN), where the former is more qualitative, and the latter is more
   memory- and time-efficient. One benefit of this model to EDU embeddings is
   that it ensures that the semantics of the context will be considered, not just
   the word meaning.

In addition to the aforementioned embeddings, we utilize part-of-speech tags.
We used the sum function to aggregate one-hot vectors based on the corresponding POS-tags. The resulting vector representation of the EDU is a concatenation
of the text and POS-tag embeddings.

3.3   Recursive Neural Model

The recursive neural network encodes the discourse parse tree as a vector. It
calculates embeddings of sub-trees sequentially from the leaves to the root.

Let x_i denote the text embedding corresponding to the node i.

$$
x_i = \begin{cases} \text{EDU embedding} & \text{if } i \text{ is the leaf} \\ \text{Embedding of the empty text} & \text{if } i \text{ is the inner node} \end{cases}
$$

Text Encoder applies a fully-connected layer to this pre-trained vector:

$$\text{Text\_Enc}(i) = \text{FC}(x_i) \tag{1}$$

Let nodes denoted as j and k be children indices for the node i, and r be
the name of the discourse relation that characterizes the link between them. The
vector representation of the input associated with i concatenates four vectors as
follows:

$$t_i = \text{Concat}[\mathbb{I}[j \text{ is Nucleus}], \mathbb{I}[k \text{ is Nucleus}], \text{OneHot}(r), \text{Text\_Enc}(i)] \tag{2}$$

In (2) $\mathbb{I}$ is the indicator function. The OneHot function performs a one-hot encoding of the respective relation's name.

An embedding of the tree which has root in the node i is computed based on embeddings of its left and right sub-trees due to the binary TreeLSTM model [23]. In fact, it is a hidden state of an associated TreeLSTM cell:

$$h_i = \text{TreeLSTM}(t_i, h_j, h_k) \tag{3}$$

In the case of the leaf node, both h_j and h_k are considered to be equal to zero.

We utilize the dropout regularization strategy proposed by Semeniuta et al. [22]. Let c denote a memory cell and α denote a dropout factor. Formally, TreeLSTM with the dropout is expressed with the following equations:

$$\begin{pmatrix} \boldsymbol{i}_i \\ \boldsymbol{f}_{i0} \\ \boldsymbol{f}_{i1} \\ \boldsymbol{o}_i \\ \boldsymbol{u}_i \end{pmatrix} = \begin{pmatrix} \sigma(W_i[t_i, h_j, h_k] + b_i) \\ \sigma(W_{f_0}[t_i, h_j, h_k] + b_{f_0}) \\ \sigma(W_{f_1}[t_i, h_j, h_k] + b_{f_1}) \\ \sigma(W_o[t_i, h_j, h_k] + b_o) \\ D(\tanh(W_u[t_i, h_j, h_k] + b_u), \alpha) \end{pmatrix} \tag{4}$$

$$c_i = c_j * \boldsymbol{f}_{i0} + c_k * \boldsymbol{f}_{i1} + \boldsymbol{i}_i * \boldsymbol{u}_i \tag{5}$$

$$h_i = \boldsymbol{o}_i * c_i \tag{6}$$

Here, σ is the sigmoid function, D is the Dropout function and $*$ is the element-wise multiplication.

The embedding of the discourse tree obtained in its root is utilized in the categorization task. Two fully-connected layers are applied to the output of the recursive neural network. The softmax layer produces the probabilities of the classes followed by the cross-entropy loss function.

We propose the siamese model to classify pairs of texts. Firstly, it applies the recursive model to calculate the embeddings of the discourse trees constructed for the given texts. At this step, the weights of the base model are shared between two branches. Secondly, it concatenates the resulting embeddings and applies two fully-connected layers to get the final predictions. The architecture of this model is shown in Fig. 2.

The main strength of the resulting models is that they are capable of end-to-end learning.

4   Results

4.1   Detection of Factual and Emotional Argumentation

Internet Argumentation Corpus. This dataset contains messages from the Internet forum 4forums.com. These messages were divided into questions and answers and each text was labeled as "factual" and "feeling", depending on whether its argumentation is based on facts or emotions. Oraby et al. [16]

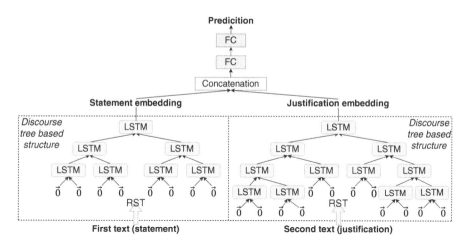

Fig. 2. Siamese model. Here, "LSTM" applies the TreeLSTM cell. Cells with the same color use the same weights. Each cell applied to EDUs receives zero vectors as embeddings of its children.

describe this dataset in detail and consider several simple models to solve the problem. The Internet Argument Corpus (IAC) contains 5848 texts, and 3466 of them are labeled as "factual". The data is split into train, validation and test sets approximately in the ratio 4:2:1.

Implementation Details. In this research, we utilized ALT Document-level Discourse Parser [11]. Output was transformed into the tree's format described in Sect. 3.1. We utilized the python Dynet library to implement the recursive neural network. The size of the hidden layer in LSTM cells was established at 300, the dropout rate α at 0.1, the learning rate at 0.001 and the number of units in the fully-connected layer in the Text Encoder at the dimension of x_i. For the model with GloVe embeddings, the best results are achieved when POS-tag embeddings are not considered. We chose the Adagrad optimizer which is less prone to overfitting for the needed task. The optimal number of epochs is 4–9. The model was trained by mini-batches of 60 texts.

Experiments. The parser identified 18 unique discourse relations. It selects "Elaboration" by default. "Attribution", "Joint" and "Same-Unit" are also popular. "Attribution" usually corresponds to the introductory phrases: "I think that", "I suppose that" and so forth. "Joint" is indicating the frequent use of the unions. "Same-Unit" indicates that the main meaning of the text in both sub-trees is the same.

We investigated the difference between the distributions of discourse relations in the two classes. The most representative relative frequencies are presented in Table 1. It shows that users employ Elaboration and Same-Unit relations more often for the argumentation based on facts and Attribution and Condition relations to express their feelings.

Table 1. The most different relative frequencies of the relations in the IAC classes

Discourse relation	For feeling	For factual
Elaboration	0.449	0.483
Attribution	0.132	0.107
Same-Unit	0.078	0.087
Condition	0.031	0.023

Table 2. Model performance on the IAC test set

Model	Features	Fact. P	Fact. R	Feel. P	Feel. R	Macro avg. F1
Random baseline	–	59.08	59.08	40.59	40.59	49.83
Oraby et al. (2015)	patterns	79.9	40.1	63.0	19.2	41.4
Naïve Bayes	unigrams, binary	73.0	67.0	57.0	65.0	65.0
SVM	unigrams	76.14	74.86	64.31	65.81	70.24
CNN	word2vec	82.58	84.72	76.96	74.06	79.56
	dep. embed.	78.49	77.81	68.18	69.04	73.38
	fact. embed.	76.24	74.93	64.49	66.12	70.43
	all embed.	81.98	81.27	73.14	74.06	77.61
LSTM	word2vec	80.60	77.81	69.32	72.80	75.10
	dep. embed.	78.70	76.66	67.34	69.87	73.12
	fact. embed.	78.77	81.27	71.49	68.20	74.90
	all embed.	77.09	82.42	71.63	64.43	73.75
BERT	–	**84.64**	80.98	74.02	**76.27**	79.51
Recursive model	GloVe	81.61	81.84	73.53	73.22	77.55
	DAN	83.47	**85.88**	**78.60**	75.31	**80.79**

We compared the proposed approach with the CNN-based and RNN-based models that were described in [8] and with other models provided by the authors of the dataset [16]. Despite it, we trained the BERT model. Precision (P), Recall (R) and macro-averaged F1-score were used to compare the results.

Table 2 shows the results of the comparison. According to the F1-score metric, the recursive model with DAN embeddings obtained the best results. The model utilized the GloVe embeddings got lower quality. The fully-connected layer in the Text Encoder is important since it suffers to improve F1-score by 1–2 points. However, the performance is limited by the mistakes in the discourse trees.

Numerous false predictions are observed for the short texts for which discourse trees have only a few nodes. In this case, discourse does not suffice to categorize it. The quality for deeper trees is presented in Fig. 3. It illustrates that for the big trees results will be much better. On average, the BERT model has not such an improvement in quality.

4.2 Fact-Checking Task

LIAR and LIAR-PLUS Datasets. The LIAR dataset [25] contains short statements collected from the website politifact.com. Each of them was rated by experts on a 6-point scale depending on truthfulness. Binary classification is also possible when all labels less than four indicate lie and the rest indicate truth. The dataset contains 12,782 statements which were split into the train, validation and test samples in the ratio of 10:1:1. This dataset is balanced and the accuracy metric can be used to compare results. The LIAR-PLUS dataset [1] additionally contains human-provided justification for each statement evidence.

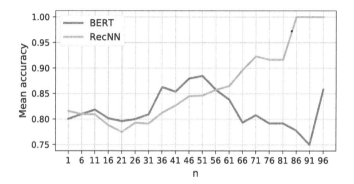

Fig. 3. Performance for the long texts from IAC. Only discourse trees that have at least n nodes are considered

Implementation Details. The implementation details are the same as described in Sect. 4.1 except some hyperparameters. The learning rate was fixed at 0.004 and the size of mini-batches at 150.

Experiments. The discourse parser distinguished 16 unique discourse relations in the LIAR dataset. The most popular relations are "Elaboration", "Same-Unit" and "Attribution". More than a quarter of trees (3,787) contain only one node. We investigated the difference between distributions for the instances labeled as "true" and instances labeled as "pants-fire". The most representative relative frequencies are shown in Table 3. It illustrates that speakers use "Elaboration" and "Contrast" relations more often for the truthful statements whereas "Attribution" and "Enablement" relations are more popular for persuasion in some misleading information.

Table 3. The most different relative frequencies of the relations in classes in the LIAR

Discourse relation	"pants-fire"	"true"
Elaboration	0.377	0.412
Attribution	0.158	0.121
Enablement	0.060	0.043
Contrast	0.015	0.031

The recursive model was compared with the methods discussed in [25] and [17] for the LIAR dataset. Here, we consider models that consume only text data. Table 4 demonstrates the results for the 6-class categorization task. The recursive model with DAN embeddings in EDUs got the highest results. However, the quality is limited by short claims that have trivial discourse structure.

Table 4. Model performance on the LIAR test set

Model	Accuracy
Majority baseline	0.208
SVMs	0.255
Logistic Regression	0.247
LSTM + Attention	0.255
Bi-LSTMs	0.233
CNNs	0.270
MMFD [12]	0.291
BERT	0.288
Recursive model (GloVe)	0.276
Recursive model (DAN)	**0.302**

We also investigate the performance of the proposed model for pairs of texts in the LIAR-PLUS dataset. We applied the siamese recursive model with DAN embeddings in EDUs to predict results in this case. The comparison with methods from [1] with input SJ (statement + justification) is shown in Table 5. The proposed model obtained the best macro-avg. F1-score for both tasks. It has a significant boost in quality when compared to processing the claim alone.

The fully-connected layer in the Text Encoder is crucial since it adds up to 0.02 to accuracy. The usage of the POS-tags embeddings also improves the overall quality approximately by 0.003-0.01.

Table 5. Model performance (macro-avg. F1-score) on the LIAR-PLUS dataset

Model	Binary		Six-way	
	Valid	Test	Valid	Test
LR	0.68	0.67	0.37	0.37
SVM	0.65	0.66	0.34	0.34
BiLSTM	0.70	0.68	0.34	0.31
P-BiLSTM	0.69	0.67	0.36	0.35
Siamese RecNN	**0.71**	**0.69**	**0.40**	**0.40**

4.3 Sentiment Classification

Movies Dataset. The main goal is to categorize positive and negative movie reviews from the corpus constructed by Pang and Lee [18]. The dataset contains 1000 instances for each class. The quality is calculated based on the 10-fold cross-validation.

Implementation Details. The implementation details are almost the same as described in Sect. 4.1. The learning rate was fixed at 0.003. The number of units in the FC layer in the Text Encoder was established at 350.

Experiments. The parser identified 18 unique discourse relations. The most popular relations are "Elaboration", "Same-Unit" and "Joint". It indicates that humans used mainly multi-nuclear relations to give their opinion about movies. Distributions of the relations are not significantly different for the given classes.

The comparison of the suggested model with recent methods works on discourse is shown in Table 6. These methods got state-of-the-art results before. The accuracy metric was used to compare results. The closest work to our research is the paper by Ji and Smith [10]. In this paper, the recursive neural network based on discourse is also proposed but has another structure. Our model outperformed other methods, and DAN embeddings availed to obtain the best quality.

Table 6. Model performance on the Movies dataset

Model	Accuracy
Hogenboom et al. [9]	71.9
Bhatia et al. [2]	82.9
Ji and Smith [10]	83.1
Our model (GloVe emb.)	84.5
Our model (DAN emb.)	**85.2**

5   Conclusion and Future Work

This paper presented an efficient model for text categorization using the style-based approach. The architecture of the proposed recursive neural network is developed to encode parsed discourse trees. Additionally, we suggested the siamese model to classify pairs of texts.

Experiments were performed for the sentiment analysis of movie reviews, the analysis of argumentation from the IAC and the automatic verification of political statements from the LIAR and LIAR-PLUS datasets. The proposed model achieved the best quality in all tasks. It was experimentally confirmed that the quality of the model increases greatly as the tree's size increases for the IAC. The siamese model obtained the highest results for the LIAR-PLUS dataset which contains statements with justifications. The DAN embeddings in EDUs suffered to get better results than GloVe embeddings in all cases.

There are possible directions for future work: modification of the architecture of the siamese model, application of CNNs or BERT to obtain more complex text embeddings in EDUs and investigation of the performance of the proposed models in other various tasks where discourse analysis may be helpful. For instance, the siamese model can be used in question-answering systems.

Acknowledgments. The article was prepared within the framework of the HSE University Basic Research Program and funded by the Russian Academic Excellence Project '5-100'.

References

1. Alhindi, T., Petridis, S., Muresan, S.: Where is your evidence: improving fact-checking by justification modeling. In: Proceedings of the First Workshop on Fact Extraction and VERification (FEVER), pp. 85–90 (2018)
2. Bhatia, P., Ji, Y., Eisenstein, J.: Better document-level sentiment analysis from RST discourse parsing. In: Proceedings of the 2015 Conference on Empirical Methods in Natural Language Processing, pp. 2212–2218 (2015)
3. Cer, D., et al.: Universal sentence encoder. arXiv, abs/1803.11175 (2018)
4. Deligiannis, N., Huu, T.D., Nguyen, D.M., Luo X.: Deep learning for geolocating social media users and detecting fake news. In: NATO Workshop (2018)
5. Devlin, J., Chang, M., Lee, K., Toutanova K.: BERT: pre-training of deep bidirectional transformers for language understanding. In: Proceedings of NAACL-HLT (2018)
6. Galitsky, B., Ilvovsky, D., Kuznetsov, S.: Text classification into abstract classes based on discourse structure. In: RANLP, pp. 200–207 (2015)
7. Galitsky, B., Ilvovsky, D.: Discovering disinformation: discourse level approach. In: 15th National Conference on Artificial Intelligence with International Participation (CAI), pp. 23–32 (2016)
8. Guggilla, C., Miller, T., Gurevych, I.: CNN- and LSTM-based claim classification in online user comments. In: Proceedings of the 26th International Conference on Computational Linguistics: Technical Papers, pp. 2740–2751 (2016)
9. Hogenboom, A., Frasincar, F., de Jong, F., Kaymak, U.: Using rhetorical structure in sentiment analysis. Commun. ACM **58**(7), 69–77 (2015)

10. Ji, Y., Smith, N.: Neural discourse structure for text categorization. In: Proceedings of the 55th Annual Meeting of the Association for Computational Linguistics, pp. 996–1005 (2017)
11. Joty, S., Carenini, G., Ng, R.: A novel discriminative framework for sentence-level discourse analysis. In: Proceedings of the 2012 Joint Conference on Empirical Methods in Natural Language Processing and Computational Natural Language Learning, pp. 904–915 (2012)
12. Karimi, H., Roy, P., Saba-Sadiya, S., Tang, J.: Multi-source multi-class fake news detection. In: Proceedings of the 27th International Conference on Computational Linguistics, pp. 1546–1557 (2018)
13. Lan, Z., et al.: ALBERT: a lite BERT for self-supervised learning of language representations. arXiv, abs/1909.11942 (2019)
14. Liu, Y., et al.: Roberta: a robustly optimized BERT pretraining approach. arXiv, abs/1907.11692 (2019)
15. Mann, W., Thompson, S.: Rhetorical structure theory: a theory of text organization. University of Southern California, Information Sciences Institute (1987). https://www.researchgate.net/publication/319394264_Rhetorical_Structure_Theory_A_Theory_of_Text_Organization
16. Oraby, S., Reed, L., Compton, R., Riloff, E., Walker, M., Whittaker, S.: And that's a fact: distinguishing factual and emotional argumentation in online dialogue. In: Proceedings of the 2nd Workshop on Argumentation Mining (2015)
17. Oshikawa, R., Qian, J., Wang, W.: A survey on natural language processing for fake news detection. arXiv, abs/1811.00770 (2018)
18. Pang, B., Lee, L.: A sentimental education. In: Proceedings of the 42nd Annual Meeting on Association for Computational Linguistics - ACL 2004 (2004)
19. Pennington, J., Socher, R., Manning, C.: Glove: global vectors for word representation. In: Proceedings of the 2014 Conference on Empirical Methods in Natural Language Processing (EMNLP), pp. 1532–1543 (2014)
20. Petz, G., Karpowicz, M., Fuerschuss, H., Auinger, A., Stritesky, V., Holzinger, A.: Computational approaches for mining user's opinions on the Web 2.0. Inf. Process. Manag. 50(6), 899–908 (2014)
21. Rashkin, H., Choi, E., Jang, J., Volkova, S., Choi, Y.: Truth of varying shades: analyzing language in fake news and political fact-checking. In: Proceedings of the 2017 Conference on Empirical Methods in Natural Language Processing, pp. 2931–2937 (2017)
22. Semeniuta, S., Severyn, A., Barth, E.: Recurrent dropout without memory loss. In: Proceedings of the 26th International Conference on Computational Linguistics (COLING), pp. 1757–1766 (2016)
23. Tai, K., Socher, R., Manning, C.: Improved semantic representations from tree-structured long short-term memory networks. In: Proceedings of the 53rd Annual Meeting of the Association for Computational Linguistics and the 7th International Joint Conference on Natural Language Processing (2015)
24. Trautmann, D., Daxenberger, J., Stab, C., Schutze, H., Gurevych, I.: Robust argument unit recognition and classification. arXiv, abs/1904.09688 (2019)
25. Wang, W.: "Liar, Liar Pants on Fire": a new benchmark dataset for fake news detection. In: Proceedings of the 55th Annual Meeting of the Association for Computational Linguistics, vol. 2. arXiv, abs/1705.00648 (2017)

Named Entity Recommendations to Enhance Multilingual Retrieval in Europeana.eu

Sergiu Gordea[1]([✉]), Monica Lestari Paramita[2], and Antoine Isaac[3]

[1] AIT Austrian Institute of Technology Gmbh, Vienna, Austria
sergiu.gordea@ait.ac.at
[2] University of Sheffield, Sheffield, UK
m.paramita@sheffield.ac.uk
[3] Free University Amsterdam, Amsterdam, The Netherlands
aisaac@few.vu.nl

Abstract. In the past years significant research efforts were invested towards the usage of Named Entity recommendation for improving information retrieval in large and heterogeneous data repositories. Such technology is employed nowadays to better understand user's search intention, to improve search precision and to enhance user experience in web portals. Within the current paper we present a case study on recommending Named Entities for enhancing multilingual retrieval in Europe's digital platform for cultural heritage. The challenges of designing an entity auto-suggestion service able to effectively support users searching for information in Europeana.eu are described in the paper together with a preliminary experimental evaluation and the outline indicating the directions for future development.

Keywords: Semantic query suggestion · Named Entities · Knowledge graph

1 Introduction

Europeana.eu currently provides access to more than 50 million (digitized) cultural heritage objects provided by 3700+ partner institutions across Europe. Most often, each individual item is described in one of the European languages, using data structures specific to the individual domains: galleries, libraries, archives and museums. The rich degree of multilinguality within these items create inherent issues for users seeking for particular cultural heritage objects within Europeana.eu web portal. Therefore, it is difficult for visitors to find the most relevant content items when using search terms in their preferred search language (often their native language). While query recommendations are a popular way of supporting users in web portals, the recommendation of Named Entities (e.g., people, places, subject categories) is more appropriate for semantic repositories, in which the information is structured according to a well-defined ontology (i.e.,

D. Helic et al. (Eds.): ISMIS 2020, LNAI 12117, pp. 102–112, 2020.
https://doi.org/10.1007/978-3-030-59491-6_10

Fig. 1. Auto-suggestion service on Europeana.eu

Europeana.eu is a semantic repository using the EDM as the underlying data model[1]).

In this paper, we describe the implementation of the Europeana auto-suggestion service. This feature utilizes a knowledge graph to aid users in formulating their queries by recommending relevant Named Entities (see Fig. 1). These entity recommendations are used to support end users in building search queries and to improve multilingual retrieval. At the same time the service can improve discovery within the web portal, as it may suggest entities that users are unaware of.

The knowledge graph, further referred to as the Europeana Entity Collection (EC), contains a set of entities related to the Cultural Heritage domain that were extracted from a number of Linked Data repositories. We propose a novel ranking algorithm for the auto-suggestion by combining an internal Europeana feature (i.e., Europeana hit count) and an external feature (i.e., Wikidata Page-Rank). By integrating the EC with external knowledge sources, this approach overcomes the cold start problem that is common within recommendation algorithms [10].

The use of Entity Collection (EC) is intended to address the following issues in Europeana.eu:

- the *multilinguality problem* when searching for information within the web portal (i.e. closing the gap between the language used for writing search queries and the language used for describing the cultural heritage objects).

---

[1] See https://pro.europeana.eu/resources/standardization-tools/edm-documentation.

– the need for *semantic enrichment* of cultural heritage objects. Utilizing a knowledge graph in the enrichment will link these objects to EC and, consequently, it will enable semantic search to be employed in the future.

In this study, we aim to answer two research questions:

1. Can the knowledge graph (i.e., extracted from Geonames, DBpedia and Wikidata) be utilized to perform auto-suggestion within the cultural heritage domain?
2. Can the auto-suggestion feature (i.e., utilizing Europeana relevance and Wikidata PageRank) effectively support users in building their search queries?

The challenges of designing an entity auto-suggestion functionality, which is able to effectively support users searching for information in a multilingual data repository, are described in this paper (Sect. 3) together with a preliminary experimental evaluation (Sect. 4), followed by a discussion (Sect. 5) and an outline of future development (Sect. 6).

2   Related Work

An auto-suggestion feature supports users in formulating their queries by suggesting relevant terms that match the prefix entered by users in the search box [8]. In previous work, the suggested terms were computed globally by extracting popular queries from query logs [15]. Others personalized the suggestions by using user-specific information (e.g., recent queries submitted by the user) [1], or demographic-based information [14]. In both approaches, a large amount of data are needed to compute the popular queries, which is often not available. In this work, we investigated the use of a knowledge graph to auto-suggest relevant named entities.

The use of a knowledge graph has been investigated in various areas in Information Retrieval. In previous work, it has been utilized to support users in disambiguating and expanding their query terms [12,16]. It has also been used to improve the indexing process and query representation [6,13], and to improve the ranking and presentation of search results [5,6,13].

The auto-suggestion algorithm proposed in this study utilizes the PageRank algorithm [2], which was originally created for the purpose of improving scalability and precision of web search engines. In fact, the algorithm is an effective measure for computing the popularity of web pages. The development of the Wikipedia platform (i.e., including Wikimedia and Wikidata) as an open multilingual data repository, enabled the usage of PageRank algorithm to compute popularity measures for Named Entities and to use them for retrieval, ranking and disambiguation purposes [5]. Similar to our approach, Prasad et al. [12] employed the PageRank algorithm for named entity recognition and linking based on search click-log data. However, the proposed algorithm is language agnostic and not context aware. The current work proposes a language-aware entity suggestion algorithm, which takes into account also the context of its usage (i.e., Europeana.eu as a semantic repository of cultural heritage objects).

The usage of Europeana access logs for computing recommendations on semantic search shortcuts was investigated by Cecarelli et al. [4]. The authors analyzed the user search sessions recorded from August 2010 to January 2012 and suggested Named Entities from the English version of Wikipedia only. The experimental evaluation was carried out against a subset of manually annotated query logs, using relatedness and diversity as quality metrics. In contrast, the current work uses a different recommendation algorithm based on overall popularity of entities and their relevancy for Europeana.eu. Moreover, the recommendations are language-aware and based on the Europeana Entity Collection, which includes only items relevant for the cultural heritage domain.

3 Entity Suggestions for Enhanced Search Experience

The entity auto-suggestion functionality provides recommendations for writing more precise search queries and therefore improves the user experience in Europeana.eu platform. These suggestions are computed based on the Europeana knowledge graph (i.e., Europeana Entity Collection). The following subsections provide an overview of the Entity Collection and the algorithm used to compute the entity recommendations.

3.1 Europeana Entity Collection

The Europeana Entity Collection was curated for three main purposes: i) to support the semantic enrichment for the description of cultural heritage objects available in Europeana, ii) to improve Europeana search effectiveness (i.e., by using names entities in search queries and by enabling a semantic search in the future), and iii) to enable browsing in Europeana.eu platform by offering alternative entry points, through the entity pages (e.g. like the one of Leonardo Da Vinci for example[2]).

The EC was curated using data extracted from GeoNames, DBpedia, and Wikidata. To avoid redundancy and ambiguous data, only a subset of these resources was selected based on the relevance of entities to Europeana and the cultural heritage domain [9]. Each entity contains the skos:prefLabel property which lists the entity names in different languages. This property is utilized by the auto-suggestion feature, therefore, availability of the property values will enable better suggestion and retrieval across languages. The statistics of the EC are shown in Table 1 indicating the number of available entities by their type and the average multilingual coverage of their labels (i.e., as present in the skos:prefLabel property).

3.2 Europeana Auto-Suggestion Algorithm

Considering the utility of the suggested entities for end users, the entity recommendations can be split in two stages: the search and the ranking of suggestions.

---

[2] http://data.europeana.eu/agent/base/146741.

Table 1. Statistics of the Entity Collection

Entity type	Total	Number of average EU languages available for the skos:prefLabel property
Agent	165,005	2.82 EU language (SD = 3.2, min = 1, max = 23, median = 2)
Concept	1,572	15.30 EU languages (SD = 7.11, min = 1, max = 24, median = 17)
Place*	215,802	0.59 EU languages (SD = 1.8, min = 0, max = 24, median = 0)

* All Place entities contain labels, but most were not associated to any EU languages, hence the low average.

The search stage (i.e., the identification of a candidate set of named entities) is implemented based on string matching over the labels available for Named Entities. Additional filtering by entity type, entity language or use in Europeana, are provided in order to further enhance the precision of recommendations. However, the effectiveness of filtering options was not evaluated in the experiments presented in Sect. 4. Given that the list of recommendations is limited to the top 10 entries, the ranking problem is more challenging. In order to ensure high precision, the ranking function takes into account the overall popularity of the Named Entities and their relevance to the Europeana. repository. The following function is used to rank the list of recommendations:

$$score(i) = ln(relevance_{ER}(i) * popularity(i))|i \in EC$$
$$relevance_{ER}(i) = hits_{ER}(i) * max(hits_{ER}())/max(hits_{ER}(t_i))$$
$$popularity(i) = pageRank(i) * max(pageRank())/max(pageRank(t_i))),$$

where i represents an item from the Entity Collection EC and t_i represents the type of item i (see Table 1). The overall ranking function is computed as the natural logarithm of the product between Europeana relevance $relevance_{ER}$ and the overall popularity $popularity()$, where these two metrics are computed based on the europeana hit count $hits_{er}(i)$ and Wikidata page rank $pageRank(i)$ (see [7]). As the distribution of these metrics over different entity types is quite heterogeneous, both metrics are normalized to facilitate the inclusion of different types of entities within the Top10 recommendations. This normalization is achieved by multiplying the metrics of individual items (i.e. $hits_{ER}(i)$ and $pageRank(i)$) with the ratio between the maximum value of the metric (i.e. $hits_{ER}()$ and $max(pageRank())$ respectively) and the maximum value of the metric for items having the same type (i.e. $max(hits_{ER}(t_i))$ and $max(pageRank(t_i))$ respectively). This normalization ensures that the most relevant item of each type will have the same relevancy score and the most popular item of each type will have the same popularity score.

4 Experiments

Two experiments were conducted to evaluate the performance of the auto-suggestion service to indicate its relevancy and its effectiveness for building enhanced search queries.

4.1 Entity Coverage in Query Logs

Firstly, the auto-suggestion feature is only able to recommend correct entities if the search terms are found in the EC. Therefore, we first evaluated the proportion of Europeana queries that can be mapped to known Named Entities. A reference dataset containing 500 randomly-chosen queries was created by analyzing query logs and extracting the search terms submitted by end users between 1 March and 1 August 2017. Manual annotation was performed on this dataset to indicate if the search queries were involving entities (i.e., Agents, Places or Concepts) or not. For example, the query "Obra compuesta por Lucio Marineo Siculo" (which means "work composed by Lucio Marineo Siculo" in English) was annotated as being related to the person "Lucio Marineo Siculo". The computed coverage indicates that 74.8% of dataset entries (374 queries) were identified as referring to Named Entities. The contents of the EC were shown to be capable of matching 45.7% of these search queries, which represents 34.2% of the entire reference dataset [11].

4.2 Effectiveness of Entity Auto-Suggestion

Whilst the first experiment focused on the feasibility of employing entity suggestions for building search queries, the second experiment measured the effectiveness of the auto-suggestion service. Provided that the relevant named entities do exist in the EC, we evaluated whether or not the service was able to suggest the relevant entities to end users (i.e., by using the top 10 recommendations). To investigate this aspect, a multilingual test dataset was created by evaluating the most frequent queries searched by Europeana users in 26 languages (i.e., 24 official EU languages[3], Norwegian and Russian). The 10 most popular search queries (as reported by Google Analytics for January-February 2018) were collected for each language. Only the top two queries that matched Entity labels for the given language were included in the evaluation dataset. These entries were manually annotated with the IDs of the entity in the EC. The statistics of the resulting dataset are shown in Table 2.

Table 2. Overview of the multilingual test dataset

Number of languages	26 languages
Number of queries	52 queries (2 queries per language)
Type of queries	19 Agents; 29 Places; 4 Concepts
Length of queries (in number of characters)	Mean = 9.10 (SD = 3.92), Min = 4, Max = 18

The performance of the auto-suggestion service (referred to as *Europeana*) is compared to three other methods:

---

[3] See https://europa.eu/european-union/topics/multilingualism_en.

- *Baseline*: standard TF-IDF based ranking which is extensively used in information retrieval (i.e. builtin Solr functionality[4])
- *Relevance*: suggestion ranking based on Europeana relevance function (see 3.2)
- *PageRank*: suggestion ranking based on popularity function (see 3.2)

The auto-suggestion performance was recursively evaluated using the first k characters of each query (entity). Effectiveness was measured using Mean Reciprocal Rank (MRR) and the Success Rate at top-n (SR@10), which indicates the average percentage of queries that retrieve the relevant entity in the top 10 results [3].

Evaluation Results. The experimental results, presented in Fig. 2, indicate that the overall accuracy of the Europeana auto-suggestion algorithm outperforms the TF-IDF baseline. These results are statistically significant ($p < 0.01$). Given the first 3 characters of users' queries, the auto-suggestion was able to recommend the correct entities for more than half of the cases (SR@10 = 0.48), compared to 31% success rate achieved by the TF-IDF ranking (SR@10 = 0.31). Over 95% cases were completed (i.e., the correct entities were recommended by the auto-suggestion in the top 10) when users have typed the first 5 characters (SR@10 = 0.95). On the given test data set, the *Relevance* and *PageRank* methods provide comparable results, which indicate that the evaluated Entities are both popular in broader sense and relevant for Europeana.eu repository. However, there is a large number of items in the EC, which are not very well known by the public users or they are not referenced by a large number of cultural heritage object available in Europeana.eu. These are subject of investigation for future work. The combination of these two scoring functions in the auto-suggestion algorithm proves its effectiveness especially with regard to the MRR metrics, indicating a better ranking of searched entities within the auto-suggestion list.

Overall, the auto-suggestion was able to recommend the correct entities for all 52 queries with an input (on average) of 3.73 characters (SD = 1.33, min = 1, max = 7). The baseline method was able to suggest the correct entities for 34 queries with an average of 3.74 characters (SD = 1.44, min = 1, max = 7). However, it was not able to suggest the correct entities in the Top 10 list for the remaining 18 queries. TF-IDF algorithm appears to be sensitive to the heterogeneous distribution of item labels over different languages and language scripts (see Table 1), term frequency being low for items with low language coverage and document frequency being high in the case of common words, like person first names for example. The drop in the precision encountered at 8th letter in the query string is explained by the low ranking precision for the first 1 to 3 letters in the second word of the query.

The performance of the auto-suggestion features for the different entity types were also evaluated, as shown in Table 3. The results show that the auto-suggestion achieved a high SR@10 for Agents and Places for an input of 5 characters or more (0.72 and 0.92, respectively). In all four cases, the Concepts in

---

[4] https://lucene.apache.org/core/7_6_0/core/org/apache/lucene/search/similarities/TFIDFSimilarity.html.

the dataset were retrieved using the first 3 characters only, but this achievement is also related to the relatively small number of concepts included in the EC. A comparison of the MRR scores also show a better accuracy for Concept queries, followed by Places, and Agents. However, in all cases, high MRR scores were achieved by providing 7 characters or more for all entity types.

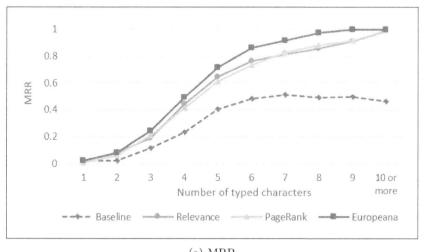

(a) MRR

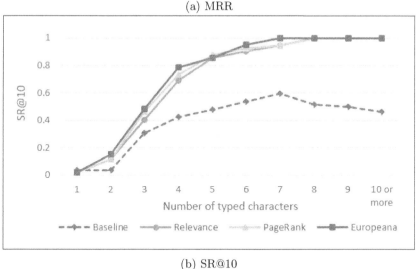

(b) SR@10

Fig. 2. Auto-suggestion accuracy (n = 52)

The overall results show a much higher SR@10 compared to MRR, which suggest that, although the correct entity was found in the top 10, it was not suggested as the first ranked item. After further analysis, we found two reasons

Table 3. Auto-suggestion accuracy by entity type

No of chars	MRR			SR@10		
	Agents	Concepts	Places	Agents	Concepts	Places
3	0.26	0.71	0.17	0.47	1	0.41
5	0.68	1	0.69	0.72	1	0.92
7	0.87	1	0.94	1	1	1
9	1	1	1	1	1	1

that caused the lower MRR scores. The first is the *strict one-to-one matching between the queries and the entities*. In some cases, more than one entities may be relevant for each query. For example, a user searching for "Palermo" may intend to search for the city or the province. We also found cases where duplicate entries were found in the EC (often found in Place entities). For example, "Izola" (a town in Slovenia) was described in two entities. Since the test dataset only allows one entity to be linked for each user query, retrieving multiple entries in the suggestion results may cause the rank of the correct entity to appear to be lower.

Another issue identified is due to the *auto-suggest support for multilingual queries*. To enable retrieval across languages, the auto-suggest function currently retrieves candidate queries by matching user queries against entity labels in all available languages in the EC. In some cases, however, this method also decreased the precision due to finding items that are not relevant to user's queries or preferred language. E.g., users using Dutch language for search, when typing "Maa" with the intention to look for "Maastricht", retrieved the following suggestion "Maatalous" (a Finnish term for "Agriculture") and "Maalikunst" (an Estonian term for "Painting") in higher ranks.

5   Discussion

In this section, we consider the research questions and highlight insights gained through experimentation.

(RQ1) Can the knowledge graph (i.e., extracted from Geonames, DBpedia and Wikidata) be utilized to perform auto-suggestion within the cultural heritage domain? We found that the current coverage of the EC extracted from these sources satisfied less than half of Europeana's entity queries. The remaining out-of-scope entity queries come from genealogy searches, and queries on infrequent entities, e.g., fashion brands, modern artists, history photographs of politicians, which are currently not available in the EC. Our findings also show that the chosen linked data sources have low multilingual availability for the entity labels. As a result, the language filter cannot be used effectively using the current EC, as the recall drops dramatically due to the poor language coverage on entity labels. However, the rapid development on Wikidata may address this data gap in the near future.

(RQ2) Does the auto-suggestion feature (i.e., utilizing Europeana relevance and Wikidata PageRank) effectively support users in building their search queries? Our results indicate that the combination of Europeana relevance and Wikidata PageRank can support the auto-suggestion functionality to retrieve entities that are relevant to users' queries with high accuracy, regardless of the entity type. However, a bigger dataset containing more queries for each entity type and language is needed to further investigate these results. The use of Wikidata rank also overcomes the cold-start problem, due to the lack of user model in the Europeana site to populate the ranking of the items. In some cases, the ranking of the relevant entities can be further improved. A learning-to-rank approach will be investigated in the future work to enable the personalization of auto-suggestion.

6   Conclusion and Future Work

In this study, we have reported on the auto-suggestion service implemented in the Europeana.eu web portal, which uses an entity ranking algorithm combining Wikidata page rank and Europeana relevance scores. Initial results indicate that the EC was able to match 45.7% Entity queries available in a sample of Europeana.eu access logs. This suggests that more work is needed to improve the coverage of the Entity Collection by integrating more knowledge sources, like increasing the size of classification schemes or adding more entity types, like historical events or popular works (e.g. named paintings, sculptures, books, etc.). However, for cases where the entities do exist in the EC, the auto-suggestion functionality was able to suggest entities with high success rate (SR@10) and good ranking (MRR@10) after users typed 5 characters, regardless of the entity types. Overall the experimental results indicate that the Europeana knowledge graph provides effective support for building more precise search queries and improving the multilingual retrieval on Europeana.eu.

Future research will focus on improving the accuracy of suggestions by enabling filtering options (e.g., by Entity Type, or by usage in Europeana) and prioritizing suggestions that match the user's preferred language (e.g., the language selected for the UI). Building a personalized recommendation based on a user model will also be employed by evaluating the effective usage of the auto-suggestion service (e.g., by recording original query, clicked entity and its ranking).

References

1. Bar-Yossef, Z., Kraus, N.: Context-sensitive query auto-completion. In: Proceedings of the 20th International Conference on World Wide Web, WWW 2011, pp. 107–116. ACM, New York (2011). http://doi.acm.org/10.1145/1963405.1963424
2. Brin, S., Page, L.: The anatomy of a large-scale hypertextual web search engine. Comput. Netw. ISDN Syst. **30**(1), 107–117 (1998). http://www.sciencedirect.com/science/article/pii/S016975552980011 0X. Proceedings of the Seventh International World Wide Web Conference

3. Cai, F., De Rijke, M., et al.: A survey of query auto completion in information retrieval. Found. Trends® Inf. Retr. **10**(4), 273–363 (2016)

4. Ceccarelli, D., Gordea, S., Lucchese, C., Nardini, F.M., Perego, R.: When entities meet query recommender systems: semantic search shortcuts. In: Proceedings of the 28th Annual ACM Symposium on Applied Computing, SAC 2013, Coimbra, Portugal, 18–22 March 2013, pp. 933–938 (2013). http://doi.acm.org/10.1145/2480362.2480540

5. Diefenbach, D., Thalhammer, A.: PageRank and generic entity summarization for RDF knowledge bases. In: Gangemi, A., et al. (eds.) ESWC 2018. LNCS, vol. 10843, pp. 145–160. Springer, Cham (2018). https://doi.org/10.1007/978-3-319-93417-4_10. https://2018.eswc-conferences.org/wp-content/uploads/2018/02/ESWC2018_paper_108.pdf

6. Elhadad, M., Gabay, D., Netzer, Y.: Automatic evaluation of search ontologies in the entertainment domain using text classification. In: Applied Semantic Technologies: Using Semantics in Intelligent Information Processing, pp. 351–367 (2011)

7. Eom, Y., Aragón, P., Laniado, D., Kaltenbrunner, A., Vigna, S., Shepelyansky, D.: Interactions of cultures and top people of Wikipedia from ranking of 24 language editions. CoRR abs/1405.7183 (2014). http://arxiv.org/abs/1405.7183

8. Hearst, M.: Search User Interfaces. Cambridge University Press, Cambridge (2009)

9. Isaac, A., Manguinhas, H., Charles, V., Stiller, J.: Task force on evaluation and enrichment: selecting target datasets for semantic enrichment (2015). https://pro.europeana.eu/files/Europeana_Professional/EuropeanaTech/EuropeanaTech_taskforces/Enrichment_Evaluation/EvaluationEnrichment_SelectingDatasets_102015.pdf. Accessed 2 May 2018

10. Lika, B., Kolomvatsos, K., Hadjiefthymiades, S.: Facing the cold start problem in recommender systems. Expert Syst. Appl. **41**(4, Part 2), 2065–2073 (2014). http://www.sciencedirect.com/science/article/pii/S0957417413007240

11. Paramita, M., Clough, P., Hill, T.: D6.3: search improvement report (2017). https://pro.europeana.eu/files/Europeana_Professional/Projects/Project_list/Europeana_DSI-2/Deliverables/d6.3-search-improvement-report.pdf. Accessed 2 May 2018

12. Prasad, S.N.V.A.S.R.K., Gupta, K.G., Manasa, M.: Lightweight multilingual named entity resource extremely extraction and linking using page rank and semantic graphs. IJSRCSEIT **2**(4), 182–188 (2017)

13. Schreiber, G., et al.: Semantic annotation and search of cultural-heritage collections: the MultimediaN E-Culture demonstrator. Web Semant. Sci. Serv. Agents World Wide Web **6**(4), 243–249 (2008)

14. Shokouhi, M.: Learning to personalize query auto-completion. In: Proceedings of the 36th International ACM SIGIR Conference on Research and Development in Information Retrieval, SIGIR 2013, pp. 103–112. ACM, New York (2013). http://doi.acm.org/10.1145/2484028.2484076

15. Whiting, S., Jose, J.M.: Recent and robust query auto-completion. In: Proceedings of the 23rd International Conference on World Wide Web, WWW 2014, pp. 971–982. ACM, New York (2014). http://doi.acm.org/10.1145/2566486.2568009

16. Zhang, L., Färber, M., Rettinger, A.: XKnowSearch!: exploiting knowledge bases for entity-based cross-lingual information retrieval. In: Proceedings of the 25th ACM International on Conference on Information and Knowledge Management, pp. 2425–2428. ACM (2016)

A Deep Learning Approach to Fake News Detection

Elio Masciari[✉], Vincenzo Moscato, Antonio Picariello, and Giancarlo Sperli

University Federico II of Naples, Naples, Italy
{elio.masciari,vincenzo.moscato,antonio.picariello,
giancarlo.sperli}@unina.it

Abstract. The uncontrolled growth of fake news creation and dissemination we observed in recent years causes continuous threats to democracy, justice, and public trust. This problem has significantly driven the effort of both academia and industries for developing more accurate fake news detection strategies. Early detection of fake news is crucial, however the availability of information about news propagation is limited. Moreover, it has been shown that people tend to believe more fake news due to their features [11]. In this paper, we present our framework for fake news detection and we discuss in detail a solution based on deep learning methodologies we implemented by leveraging Google Bert features. Our experiments conducted on two well-known and widely used real-world datasets suggest that our method can outperform the state-of-the-art approaches and allows fake news accurate detection, even in the case of limited content information.

1 Introduction

Social media are nowadays the main medium for large-scale information sharing and communication and they can be considered the main drivers of the Big Data revolution we observed in recent years [1]. Unfortunately, due to malicious user having fraudulent goals fake news on social media are growing quickly both in volume and their potential influence thus leading to very negative social effects. In this respect, identifying and moderating fake news is a quite challenging problem. Indeed, fighting fake news in order to stem their extremely negative effects on individuals and society is crucial in many real life scenarios. Therefore, fake news detection on social media has recently become an hot research topic both for academia and industry.

Fake news detection dates back long time ago [12], for a very long time journalist and scientists fought against misinformation, however, the pervasive use of internet for communication allows for a quicker spread of false information. Indeed, the term *fake news* has grown in popularity in recent years, especially after the 2016 United States elections but there is still no standard definition of fake news [9].

Aside the definition that can be found in literature, one of the most well accepted definition of fake news is the following: *Fake news is a news article that*

D. Helic et al. (Eds.): ISMIS 2020, LNAI 12117, pp. 113–122, 2020.
https://doi.org/10.1007/978-3-030-59491-6_11

is intentionally and verifiable false and could mislead readers [2]. There are two key features of this definition: authenticity and intent. First, fake news includes false information that can be verified as such. Second, fake news is created with dishonest intention to mislead consumers [9].

The content of fake news exhibits heterogeneous topics, styles and media platforms, it aims to mystify truth by diverse linguistic styles while insulting true news. Fake news are generally related to newly emerging, time-critical events, which may not have been properly verified by existing knowledge bases due to the lack of confirmed evidence or claims. Thus, fake news detection on social media poses peculiar challenges due to the inherent nature of social networks that requires both the analysis of their content [6,8] and their social context [3,10].

Indeed, as mentioned above fake news are written on purpose to deceive readers to believe false information. For this reason, it is quite difficult to detect a fake news analysing only the news content. Therefore, we should take into account auxiliary information, such as user social engagement on social media to improve the detection accuracy. Unfortunately, the usage of auxiliary information is a non-trivial task as users social engagements with fake news produce data that are big, noisy, unstructured and incomplete.

Moreover, the diffusion models for fake news changed deeply in recent years. Indeed, as mentioned above, some decades ago, the only medium for information spreading were newspapers and radio/television but recently, the phenomenon of fakes news generation and diffusion take advantage of the internet pervasive diffusion and in particular of social media quick pick approach to news spreading. More in detail, user consumption behaviours have been affected by the inherent nature of these social media platforms: 1) They are more pervasive and less expensive when compared to traditional news media, such as television or newspapers; 2) It is easier to share, comment on, and discuss the news with friends and followers on social media by overcoming geographical and social barrier.

Despite the above mentioned advantages of social media news sharing, there are many drawbacks. First of all, the quality of news on social media is lower than traditional news organizations due to the lower control of information sources. Moreover, since it is cheaper to provide news online and much faster and easier to spread through social media, larger volumes of fake news are produced online for a variety of purposes, such as political gain and unfair competition to cite a few.

Our Approach in a Nutshell. Fake news detection problem can be formalized as a classification task thus requiring features extraction and model construction. The detection phase is a crucial task as it is devoted to guarantee users to receive authentic information. We will focus on finding clues from news contents.

Our goal is to improve the existing approaches defined so far when fake news is intentionally written to mislead users by mimicking true news. More in detail, traditional approaches are based on verification by human editors and expert journalists but do not scale to the volume of news content that is generated in online social networks. As a matter of fact, the huge amount of data to be analyzed calls for the development of new computational techniques. It is worth noticing that, such computational techniques, even if the news is detected as fake, require some sort of expert verification before being blocked. In our framework, we perform an accurate pre-processing of news data and then we apply three different approaches. The first approach is based on classical classification approaches. We also implemented a deep learning approach that leverages neural network features for fake news detection. Finally, for the sake of completeness we implemented some multimedia approaches in order to take into account misleading images. Due to space limitation, we discuss in this paper the deep learning approach.

2   Our Fake News Detection Framework

Our framework is based on news flow processing and data management in a pre-processing block which performs filtering and aggregation operation over the news content. Moreover, filtered data are processed by two independent blocks: the first one performs natural language processing over data while the second one performs a multimedia analysis.

The overall process we execute for fake news detection is depicted in Fig. 1.

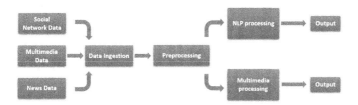

Fig. 1. The overall process at a glance

In the following, we describe each module in more detail.

Data Ingestion Module. This module take care of data collection tasks. Data can be highly heterogeneous: social network data, multimedia data and news data. We collect the news text and eventual related contents and images.

Pre-processing Module. This component is devoted to the acquisition of the incoming data flow. It performs filtering, data aggregation, data cleaning and enrichment operations.

NLP Processing Module. It performs the crucial task of generating a binary classification of the news articles, i.e., whether they are fake or reliable news. It is split in two submodules. The *Machine Learning* module performs classification using an ad-hoc implemented Logistic Regression algorithm after an extensive process of feature extraction and selection TF-IDF based in order to reduce the number of extracted features. The *Deep Learning* module classify data using

Google Bert algorithm after a tuning phase on the vocabulary. It also perform a binary transformation and eventual text padding in order to better analyze the input data.

Multimedia Processing Module. This module is tailored for Fake Image Classification through Deep Learning algorithms, using ELA (Error Level Analysis) and CNN.

Due to space limitation, we discuss in the following only the details of the deep learning module and the obtained results.

2.1   The Software Architecture

The software implementation of the framework described above is shown in Fig. 2.

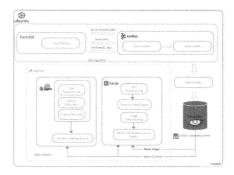

Fig. 2. Our fake news detection framework

Herein: the data ingestion block is implemented by using several tools. As an example for Twitter data we leverage Tweepy, a Python library to access the Twitter API. All tweets are downloaded through this library. Filtering and aggregation is performed using Apache Kafka which is able to build real-time data pipelines and streaming apps. It is scalable, fault-tolerant and fast thus making our prototype well-suited for huge amount of data.

The data crawler uses the Newspaper Python library whose purpose is extracting and curating articles. The analytical data archive stores pre-processed data that are used for issuing queries by traditional analytical tools. We leverage Apache Cassandra as datastore because it provides high scalability, high availability, fast writing, fault- tolerance on commodity hardware or cloud infrastructure. The data analytics block retrieves news contents and news images from Cassandra DB that are pre-processed by the Machine Learning module using Scikit Learn library and by Deep Learning module using Keras library. Image content is processed by the Multimedia Deep Learning module using Keras library.

In the following, we will briefly describe how the overall process is executed. Requests to the Cassandra DB are made through remote access. Each column in Cassandra refers to a specific topic and contains all news belonging to that topic. Among all news, those having a valid *external link* value are selected. In this way, the news content can be easily crawled. As the link for each news is obtained, a check is performed in order to verify the current state of the website. If the website is still running, we perform the article scraping. The algorithm works by downloading and parsing the news article, then, for each article, title, text, authors, top image link, news link data are extracted and saved as a JSON file in Cassandra DB.

Finally, three independent analysis are then performed by three ad-hoc Python modules we implemented. The first two perform text classification, and the last one images classification. Concerning the text analysis, the problem being solved is a binary classification one where class 0 refers to reliable news and class 1 refers to fake ones.

2.2   The Deep Learning Module

The Deep Learning Module computes a binary classification on a text datasets of news that will be labelled as 0 if a news is marked as Real, and as 1 if it is marked as Fake. The Deep Learning Module classifies news content using a new language model called *B.E.R.T.* (Bidirectional Encoder Representations from Transformers) developed and released by Google. Prior to describing the algorithm features in detail, we briefly describe the auxiliary tools being used, while in Sect. 3 we describe the experimental evaluation that lead to our choice on BERT.

Colaboratory. Colab[1] is intended for machine learning education and research, it requires no setup and runs entirely on the cloud. By using Colab it's possible to write and execute code, save and share analytics and it provides access to expensive and powerful computing resources for free by a web interface.

More in detail, Colab's hardware is powered by: Intel(R) Xeon(R) CPU @ 2.00 GHz, nVidia T4 16 GB GDDR6 @ 300 GB/sec, 15 GB RAM and 350 GB storage. This setting is able to speed-up the learning task execution up to 35X and 16X faster in deep learning training compared to a CPU-only server.

Tensor Flow. It is devoted to train and run neural networks for image recognition, word embeddings, recurrent neural networks, and natural language processing. It is a cross-platform tool and runs on CPUs, GPUs, even on mobile and embedded platforms. TensorFlow uses dataflow graphs to represent the computation flow, i.e., these structures describe the data flow through the processing nodes. Each node in the graph represents a mathematical operation, and each connection between nodes is a multidimensional data array called tensor. The TensorFlow Distributed Execution Engine abstracts from the supported devices and provides a high performance-core implemented in C++ for the TensorFlow platform. On top there are Python and C++ frontends. The Layers API provides a simple interface for most of the layers used in deep learning models. Finally,

---

[1] https://research.google.com/colaboratory.

higher-level APIs, including Keras, makes training and evaluating distributed models easier.

Keras. It is a high-level neural network API, implemented in Python and capable of running on top of TensorFlow. It allows for easy and fast prototyping through: 1) User Friendliness as it offers consistent and simple APIs that minimizes the number of user actions required for common use cases; 2) Modularity as neural layers, cost functions, optimizers, initialization schemes, activation functions and regularization schemes are all standalone modules that can be combined to create new models; 3) Extensibility as new modules are simple to add as new classes and functions.

Google BERT. This tool has been developed in order to allow an easier implementation of two crucial tasks for Natural Language Processing (NLP): Transfer Learning through unsupervised pre-training and Transformer architecture. The idea behind Transfer Learning is to train a model in a given domain on a large text corpus, and then leverage the gathered knowledge to improve the model's performance in a different domain. In this respect, BERT has been pre-trained on Wikipedia and BooksCorpus. On the opposite side, the Transformer architecture processes all elements simultaneously by linking individual elements through a process known as attention. This mechanism allows a deep parallelization and guarantee higher accuracy across a wide range of tasks. BERT outperforms previous proposed approaches as it is the first unsupervised, fully bidirectional system for NLP pre-training. Pre-trained representations can be either context-free or context based dependig on user needs. Due to space limitations we do not describe in detail the BERT's architecture and the encoder mechanism.

3   Our Benchmark

In this section we will describe the fake news detection process for the deep learning module and the datasets we used as a benchmark for our algorithms.

3.1   Dataset Description

Liar Dataset. This dataset includes 12.8 K human labelled short statements from fact-checking website Politifact.com. Each statement is evaluated by a Politifact.com editor for its truthfulness. The dataset has six fine-grained labels: pants-fire, false, barely-true, half-true, mostly-true, and true. The distribution of labels is relatively well- balanced [12]. For our purposes the six fine-grained labels of the dataset have been collapsed in a binary classification, i.e., label 1 for fake news and label 0 for reliable ones. This choice has been made due to binary Fake News Dataset feature. The dataset is partitioned into three files: 1) Training Set: 5770 real news and 4497 fake news; 2) Test Set: 1382 real news and 1169 fake news; 3) Validation Set: 1382 real news and 1169 fake news. The three subsets are well balanced so there is no need to perform oversampling or undersampling.

The processed dataset has been uploaded in Google Drive and, then, loaded in Colab's Jupyter as a Pandas Dataframe. It has been added a new column with the number of words for each row article. Using the command *df.describe()* on this column it is possible to print the following statistical information: count 15389.000000, mean 17.962311, std 8.569879, min 1.000000, 25% 12.000000, 50% 17.000000, 75% 22.000000, max 66.000000. These statistics show that there are articles with only one word in the dataset, so it has been decided to remove all rows with less than 10 words as they are considered poorly informative. The resulting dataset contains 1657 less rows than the original one. The updated statistics are reported in what follows: count 13732.000000, mean 19.228663, std 8.192268, min 10.000000, 25% 14.000000, 50% 18.000000, 75% 23.000000, max 66.000000. Finally, the average number of words per article is 19.

FakeNewsNet. This dataset has been built by gathering information from two fact-checking websites to obtain news contents for fake news and real news such as PolitiFact and GossipCop. In PolitiFact, journalists and domain experts review the political news and provide fact-checking evaluation results to claim news articles as fake or real. Instead, in GossipCop, entertainment stories, from various media outlets, are evaluated by a rating score on the scale of 0 to 10 as the degree from fake to real. The dataset contains about 900 political news and 20k gossip news and has only two labels: true and false [14].

This dataset is publicly available by the functions provided by the FakeNews-Net team and the Twitter API. As mentioned above, FakeNewsNet can be split in two subsets: GossipCop and Politifact.com. We decided to analyse only political news as they produce worse consequences in real world than gossip ones. The dataset is well balanced and contains 434 real news and 367 fake news. Most of the news regards the US as it has already been noticed in LIAR. Fake news topics concern Obama, police, Clinton and Trump while real news topics refer to Trump, Republicans and Obama. Such as the LIAR dataset, it has been added a new column and used the command *df.describe()* to print out the following statistical information: count 801, mean 1459.217228, std 3141.157565, min 3, 25% 114, 50% 351, 75% 893, max 17377.

The average number of words per articles in Politifact dataset is 1459, which is far longer than the average sentence length in Liar Dataset that is 19 words per articles. Such a statistics suggested us to compare the model performances on datasets with such different features.

3.2 Pre-elaboration Steps

The above mentioned datasets are available in CSV format and are composed of two columns: text and label. The news text need to be pre-processed for our analysis. In this respect, an ad-hoc Pythin function has been developed for unnecessary IP and URL addresses removal, HTML tags checking and words spell-check. Due to neural features, we decide to maintain some stop words in order to allow a proper context analysis. Thus, to ameliorate the noise problem, we created a custom list of stop words. We leverage Keras Tokenizer for preparing text documents for subsequent deep learning steps. More in detail, we create a vocabulary index based on word frequency, e.g., given the sentence *The cat sat on the mat* we create the following dictionary $word\_index$[the] $= 1$, $word\_index$[cat] $= 2$ so every word gets a unique integer value; the 0 value is reserved for padding. Lower integer means more frequent word. After this encoding step, we obtain for each text a sequence of integers. As BERT needs a more elaborated input than other neural networks wed need to produce a *tsv* file, with four columns, and no header. The columns to be added to dataset are: 1) *guid*, i.e., a row ID; 2)*label*, i.e., the label for the row (it should be an int); 3) *alpha*, a dummy column containing the same letter for all rows, it is not used for classification but it is needed for proper running of the algorithm and 4) *text*, i.e., the news content. The data needs to be converted in InputFeature object to be compatible with Transformer Architecture. The conversion process includes tokenization and converting all sentences to a given sequence length (truncating longer sequences, and padding shorter sequences). Tokenization is performed using WordPiece tokenization, where the vocabulary is initialized with all the individual characters in the language, and then the most frequent/likely combinations of the existing words in the vocabulary are iteratively added. Words that does not occur in the vocabulary are broken down into sub-words in order to search for possible matches in the collection.

4   Evaluation

In order to show that the Google BERT model we implemented outperforms the results of the current performance state of art both on Liar dataset and Polifact dataset, we report in Figs. 3 and 4 the best results obtained for the other approaches commonly used in literature for those datasets.

Neural Network	Accuracy	Precision	Recall	F1	TP	FP	TN	FN	AUC
BERT	0.619	0.583	0.596	0.628	697	472	884	498	0.617
C-HAN	0.557	0.514	0.628	0.565	688	694	735	434	0.574
BI-LSTM	0.586	0.554	0.495	0.523	579	465	917	590	0.607
CNN	0.536	0.489	0.275	0.352	1046	847	322	336	0.539

Fig. 3. Comparison for Google BERT against state of the art approaches on LIAR dataset

We compared the performances on well-established evaluation measure like: Accuracy, Precision, Recall, F1 measure, Area Under Curve (AUC) [5] and the values reported in the obtained confusion matrices for each algorithm, i.e., True Positive (TP), False Positive (FP), True Negative (TN) and False Negative (FN).

We hypothesize that our results are quite better due to a fine hyper parameter tuning we performed, a better pre-processing step and the proper transformation.

For the sake of completeness, we report in Figs. 5 and 6 the detailed confusion matrices obtained for LIAR and Polifact datasets.

Neural Network	Accuracy	Precision	Recall	F1	TP	FP	TN	FN	AUC
BERT	0.588	0.565	0.449	0.628	165	127	306	202	0.578
C-HAN	0.448	0.443	0.795	0.569	67	75	292	367	0.436
BI-LSTM	0.511	0.466	0.455	0.460	167	200	243	191	0.532
CNN	0.540	0.498	0.405	0.447	284	218	149	150	0.520

Fig. 4. Comparison for Google BERT against state of the art approaches on Polifact datataset

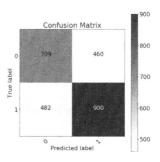

Fig. 5. Confusion Matrix for LIAR dataset

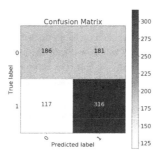

Fig. 6. Confusion matrix for polifact dataset

5   Conclusion and Future Work

In this paper, we investigated the problem of fake news detection by deep learning algorithms. We developed a framework the leverage Google BERT for analyzing real-life datasets and the results we obtained are quite encouraging. As for future work, we would like to extend our analysis by considering also user profiles' features and some kind of dynamic analysis of news diffusion mechanism in our fake news detection model [7].

Acknowledgement. Elio Masciari has been supported by POR Calabria project SPI-DASEC.

References

1. Agrawal, D., et al.: Challenges and opportunities with big data. A community white paper developed by leading researchers across the United States. Technical report, Purdue University, March 2012
2. Allcott, H., Gentzkow, M.: Social media and fake news in the 2016 election. Working Paper 23089, National Bureau of Economic Research, January 2017
3. Cassavia, N., Masciari, E., Pulice, C., Saccà, D.: Discovering user behavioral features to enhance information search on big data. TiiS **7**(2), 7:1–7:33 (2017)
4. Culpepper, J.S., Moffat, A., Bennett, P.N., Lerman, K. (eds.) Proceedings of the Twelfth ACM International Conference on Web Search and Data Mining, WSDM 2019, Melbourne, VIC, Australia, 11–15 February , 2019. ACM (2019)
5. Flach, P.A., Kull, M.: Precision-recall-gain curves: PR analysis done right. In: Cortes, C., Lawrence, N.D., Lee, D.D., Sugiyama, M., Garnett, R. (eds.) Advances in Neural Information Processing Systems 28: Annual Conference on Neural Information Processing Systems 2015, Montreal, Quebec, Canada, 7–12 December 2015, pp. 838–846 (2015)
6. Guo, C., Cao, J., Zhang, X., Shu, K., Yu., M.: Exploiting emotions for fake news detection on social media. CoRR, abs/1903.01728 (2019)
7. Masciari, E.: SMART: stream monitoring enterprise activities by RFID tags. Inf. Sci. **195**, 25–44 (2012)
8. Potthast, M., Kiesel, J., Reinartz, K., Bevendorff, J., Stein, B.: A stylometric inquiry into hyperpartisan and fake news. CoRR, abs/1702.05638 (2017)
9. Shu, K., Sliva, A., Wang, S., Tang, J., Liu, H.: Fake news detection on social media: a data mining perspective. CoRR, abs/1708.01967 (2017)
10. Shu, K., Wang, S., Liu, H.: Beyond news contents: the role of social context for fake news detection. In: Culpepper et al. [4], pp. 312–320
11. Vosoughi, S., Roy, D., Aral, S.: The spread of true and false news online. Science **359**(6380), 1146–1151 (2018)
12. Zhou, X., Zafarani, R., Shu, K., Liu, H:. Fake news: fundamental theories, detection strategies and challenges. In: Culpepper et al. [4], pp. 836–837

Satirical News Detection with Semantic Feature Extraction and Game-Theoretic Rough Sets

Yue Zhou[1], Yan Zhang[1(✉)], and JingTao Yao[2]

[1] School of Computer Science and Engineering, California State University, San Bernardino, San Bernardino, CA, USA
{Yue.Zhou,Yan.Zhang}@csusb.edu
[2] Department of Computer Science, University of Regina, Regina, SK, Canada
jtyao@cs.uregina.ca

Abstract. Satirical news detection is an important yet challenging task to prevent spread of misinformation. Many feature based and end-to-end neural nets based satirical news detection systems have been proposed and delivered promising results. Existing approaches explore comprehensive word features from satirical news articles, but lack semantic metrics using word vectors for tweet form satirical news. Moreover, the vagueness of satire and news parody determines that a news tweet can hardly be classified with a binary decision, that is, satirical or legitimate. To address these issues, we collect satirical and legitimate news tweets, and propose a semantic feature based approach. Features are extracted by exploring inconsistencies in phrases, entities, and between main and relative clauses. We apply game-theoretic rough set model to detect satirical news, in which probabilistic thresholds are derived by game equilibrium and repetition learning mechanism. Experimental results on the collected dataset show the robustness and improvement of the proposed approach compared with Pawlak rough set model and SVM.

Keywords: Satirical news detection · Social media · Feature extraction · Game-theoretic rough sets

1 Introduction

Satirical news, which uses parody characterized in a conventional news style, has now become an entertainment on social media. While news satire is claimed to be pure comedic and of amusement, it makes statements on real events often with the aim of attaining social criticism and influencing change [15]. Satirical news can also be misleading to readers, even though it is not designed for falsifications. Given such sophistication, satirical news detection is a necessary yet challenging natural language processing (NLP) task. Many feature based fake or satirical news detection systems [3,11,14] extract features from word relations given by statistics or lexical database, and other linguistic features. In addition, with

© Springer Nature Switzerland AG 2020
D. Helic et al. (Eds.): ISMIS 2020, LNAI 12117, pp. 123–135, 2020.
https://doi.org/10.1007/978-3-030-59491-6_12

the great success of deep learning in NLP in recent years, many end-to-end neural nets based detection systems [6,12,16] have been proposed and delivered promising results on satirical news article detection.

However, with the evolution of fast-paced social media, satirical news has been condensed into a satirical-news-in-one-sentence form. For example, one single tweet of "If earth continues to warm at current rate moon will be mostly underwater by 2400" by The Onion is largely consumed and spread by social media users than the corresponding full article posted on The Onion website. Existing detection systems trained on full document data might not be applicable to such form of satirical news. Therefore, we collect news tweets from satirical news sources such as The Onion, The New Yorker (Borowitz Report) and legitimate news sources such as Wall Street Journal and CNN Breaking News. We explore the syntactic tree of the sentence and extract inconsistencies between attributes and head noun in noun phrases. We also detect the existence of named entities and relations between named entities and noun phrases as well as contradictions between the main clause and corresponding prepositional phrase. For a satirical news, such inconsistencies often exist since satirical news usually combines irrelevant components so as to attain surprise and humor. The discrepancies are measured by cosine similarity between word components where words are represented by Glove [9]. Sentence structures are derived by Flair, a state-of-the-art NLP framework, which better captures part-of-speech and named entity structures [1].

Due to the obscurity of satire genre and lacks of information given tweet form satirical news, there exists ambiguity in satirical news, which causes great difficulty to make a traditional binary decision. That is, it is difficult to classify one news as satirical or legitimate with available information. Three-way decisions, proposed by YY Yao, added an option - deferral decision in the traditional yes-and-no binary decisions and can be used to classify satirical news [21,22]. That is, one news may be classified as satirical, legitimate, and deferral. We apply rough sets model, particularly the game-theoretic rough sets to classify news into three groups, i.e., satirical, legitimate, and deferral. Game-theoretic rough set (GTRS) model, proposed by JT Yao and Herbert, is a recent promising model for decision making in the rough set context [18]. GTRS determine three decision regions from a tradeoff perspective when multiple criteria are involved to evaluate the classification models [25]. Games are formulated to obtain a tradeoff between involved criteria. The balanced thresholds of three decision regions can be induced from the game equilibria. GTRS have been applied in recommendation systems [2], medical decision making [19], uncertainty analysis [24], and spam filtering [23].

We apply GTRS model on our preprocessed dataset and divide all news into satirical, legitimate, or deferral regions. The probabilistic thresholds that determine three decision regions are obtained by formulating competitive games between accuracy and coverage and then finding Nash equilibrium of games. We perform extensive experiments on the collected dataset, fine-tuning the model by different discretization methods and variation of equivalent classes. The experimental result shows that the performance of the proposed model is superior compared with Pawlak rough sets model and SVM.

2   Related Work

Satirical news detection is an important yet challenging NLP task. Many feature based models have been proposed. Burfoot et al. extracted features of headline, profanity, and slang using word relations given by statistical metrics and lexical database [3]. Rubin et al. proposed a SVM based model with five features (absurdity, humor, grammar, negative affect, and punctuation) for fake news document detection [11]. Yang et al. presented linguistic features such as psycholinguistic feature based on dictionary and writing stylistic feature from part-of-speech tags distribution frequency [17]. Shu et al. gave a survey in which a set of feature extraction methods is introduced for fake news on social media [14]. Conroy et al. also uses social network behavior to detect fake news [4]. For satirical sentence classification, Davidov et al. extract patterns using word frequency and punctuation features for tweet sentences and amazon comments [5]. The detection of a certain type of sarcasm which contracts positive sentiment with a negative situation by analyzing the sentence pattern with a bootstrapped learning was also discussed [10]. Although word level statistical features are widely used, with advanced word representations and state-of-the-art part-of-speech tagging and named entity recognition model, we observe that semantic features are more important than word level statistical features to model performance. Thus, we decompose the syntactic tree and use word vectors to more precisely capture the semantic inconsistencies in different structural parts of a satirical news tweet.

Recently, with the success of deep learning in NLP, many researchers attempted to detect fake news with end-to-end neural nets based approaches. Ruchansky et al. proposed a hybrid deep neural model which processes both text and user information [12], while Wang et al. proposed a neural network model that takes both text and image data [16] for detection. Sarkar et al. presented a neural network with attention to both capture sentence level and document level satire [6]. Some research analyzed sarcasm from non-news text. Ghosh and Veale [7] used both the linguistic context and the psychological context information with a bi-directional LSTM to detect sarcasm in users' tweets. They also published a feedback-based dataset by collecting the responses from the tweets authors for future analysis. While all these works detect fake news given full text or image content, or target on non-news tweets, we attempt bridge the gap and detect satirical news by analyzing news tweets which concisely summarize the content of news.

3   Methodology

In this section, we will describe the composition and preprocessing of our dataset and introduce our model in detail. We create our dataset by collecting legitimate and satirical news tweets from different news source accounts. Our model aims to detect whether the content of a news tweet is satirical or legitimate. We first extract the semantic features based on inconsistencies in different structural parts of the tweet sentences, and then use these features to train game-theoretic rough set decision model.

3.1 Dataset

We collected approximately 9,000 news tweets from satirical news sources such as The Onion and Borowitz Report and about 11,000 news tweets from legitimate new sources such as Wall Street Journal and CNN Breaking News over the past three years. Each tweet is a concise summary of a news article. The duplicated and extreme short tweets are removed. A news tweet is labeled as satirical if it is written by satirical news sources and legitimate if it is from legitimate news sources. Table 1 gives an example of tweet instances that comprise our dataset.

Table 1. Examples of instances comprising the news tweet dataset

Content	Source	Label
The White House confirms that President Donald Trump sent a letter to North Korean leader Kim Jong Un	CNN	0
Illinois Senate plans vote on bills that could become the state's first budget in more than two years	WSJ	0
Naked Andrew Yang emerges from time vortex to warn debate audience about looming threat of automation	TheOnion	1
New study shows majority of late afternoon sleepiness at work caused by undetected carbon monoxide leak	TheOnion	1
Devin Nunes accuses witnesses of misleading American people with facts	BorowitzReport	1

3.2 Semantic Feature Extraction

Satirical news is not based on or does not aim to state the fact. Rather, it uses parody or humor to make statement, criticisms, or just amusements. In order to achieve such effect, contradictions are greatly utilized. Therefore, inconsistencies significantly exist in different parts of a satirical news tweet. In addition, there is a lack of entity or inconsistency between entities in news satire. We extracted these features at semantic level from different sub-structures of the news tweet.

Different structural parts of the sentence are derived by part-of-speech tagging and named entity recognition by Flair. The inconsistencies in different structures are measured by cosine similarity of word phrases where words are represented by Glove word vectors. We explored three different aspects of inconsistency and designed metrics for their measurements. A word level feature using tf-idf [13] is added for robustness.

Inconsistency in Noun Phrase Structures. One way for a news satire to obtain surprise or humor effect is to combine irrelevant or less jointly used attributes and the head noun which they modified. For example, noun phrase such as "rampant accountability", "posthumous apology", "self-imposed mental construct" and other rare combinations are widely used in satirical news, while individual words themselves are common. To measure such inconsistency, we first select all leaf noun phrases (NP) extracted from the trees to avoid repeated calculation. Then for each noun phrase, each adjacent word pair is selected and represented by 100-dim Glove word vector denoted as (v_t, w_t). We define the averaged cosine similarity of noun phrase word pairs as:

$$S_{NP} = \frac{1}{T} \sum_{t=1}^{T} cos(v_t, w_t) \tag{1}$$

where T is a total number of word pairs. We use S_{NP} as a feature to capture the overall inconsistency in noun phrase uses. S_{NP} ranges from -1 to 1, where a smaller value indicates more significant inconsistency.

Inconsistency Between Clauses. Another commonly used rhetoric approach for news satire is to make contradiction between the main clause and its prepositional phrase or relative clause. For instance, in the tweet "Trump boys counter Chinese currency manipulation *by* adding extra zeros to $20 Bills", contradiction or surprise is gained by contrasting irrelevant statements provided by different parts of the sentence. Let q and p denote two clauses separated by main/relative relation or preposition, and $(w_1, w_1, ...w_q)$ and $(v_1, v_1, ...v_p)$ be the vectorized words in q and p. Then we define inconsistency between q and p as:

$$S_{QP} = cos(\sum_{q=1}^{Q} w_q, \sum_{p=1}^{P} v_p)) \tag{2}$$

Similarly, the feature S_{QP} is measured by cosine similarity of linear summations of word vectors, where smaller value indicates more significant inconsistency.

Inconsistency Between Named Entities and Noun Phrases. Even though many satirical news tweets are made based on real persons or events, most of them lack specific entities. Rather, because the news is fabricated, news writers use the words such as "man", "woman", "local man", "area woman", "local family" as subject. However, the inconsistency between named entities and noun phrases often exists in a news satire if a named entity is included. For example, the named entity "Andrew Yang" and the noun phrases "time vortex" show great inconsistency than "President Trump", "Senate Republicans", and "White House" do in the legitimate news "President Trump invites Senate Republicans to the White House to talk about the funding bill." We define such inconsistency as a categorical feature that:

$$C_{NERN} = \begin{cases} 0 & \text{if } S_{NERN} < \bar{S}_{NERN} \\ 1 & \text{if } S_{NERN} \geq \bar{S}_{NERN} \\ -1 & \text{if there's no named entity} \end{cases} \quad (3)$$

S_{NERN} is the cosine similarity of named entities and noun phrases of a certain sentence and $\bar{S}_{NERN}$ is the mean value of S_{NERN} in corpus.

Word Level Feature Using TF-IDF. We calculated the difference of tf-idf scores between legitimate news corpus and satirical news corpus for each single word. Then, the set S_{voc} that includes most representative legitimate news words is created by selecting top 100 words given the tf-idf difference. For a news tweet and any word w in the tweet, we define the binary feature B_{voc} as:

$$B_{voc} = \begin{cases} 1 & \text{if } w \in S_{voc} \\ 0 & \text{otherwise} \end{cases} \quad (4)$$

3.3 GTRS Decision Model

We construct a Game-theoretic Rough Sets model for classification given the extracted features. Suppose $E \subseteq U \times U$ is an equivalence relation on a finite nonempty universe of objects U, where E is reflexive, symmetric, and transitive. The equivalence class containing an object x is given by $[x] = \{y \in U | xEy\}$. The objects in one equivalence class all have the same attribute values. In the satirical news context, given an undefined concept $satire$, probabilistic rough sets divide all news into three pairwise disjoint groups i.e., the satirical group $POS(satire)$, legitimate group $NEG(satire)$, and deferral group $BND(satire)$, by using the conditional probability $Pr(satire|[x]) = \frac{|satire \cap [x]|}{|[x]|}$ as the evaluation function, and (α, β) as the acceptance and rejection thresholds [20–22], that is,

$$POS_{(\alpha,\beta)}(satire) = \{x \in U \mid Pr(satire|[x]) \geq \alpha\},$$
$$NEG_{(\alpha,\beta)}(satire) = \{x \in U \mid Pr(satire|[x]) \leq \beta\},$$
$$BND_{(\alpha,\beta)}(satire) = \{x \in U \mid \beta < Pr(satire|[x]) < \alpha\}. \quad (5)$$

Given an equivalence class $[x]$, if the conditional probability $Pr(satire|[x])$ is greater than or equal to the specified acceptance threshold α, i.e., $Pr(satire|[x]) \geq \alpha$, we accept the news in $[x]$ as satirical. If $Pr(satire|[x])$ is less than or equal to the specified rejection threshold β, i.e., $Pr(satire|[x]) \leq \beta$

we reject the news in $[x]$ as *satirical*, or we accept the news in $[x]$ as *legitimate*. If $Pr(satire|[x])$ is between α and β, i.e., $\beta < Pr(satire|[x]) < \alpha$, we defer to make decisions on the news in $[x]$. Pawlak rough sets can be viewed as a special case of probabilistic rough sets with $(\alpha, \beta) = (1, 0)$.

Given a pair of probabilistic thresholds (α, β), we can obtain a news classifier according to Eq. (5). The three regions are a partition of the universe U,

$$\pi_{(\alpha,\beta)}(Satire) = \{POS_{(\alpha,\beta)}(Satire), BND_{(\alpha,\beta)}(Satire), NEG_{(\alpha,\beta)}(Satire)\} \tag{6}$$

Then, the accuracy and coverage rate to evaluate the performance of the derived classifier are defined as follows [25],

$$Acc_{(\alpha,\beta)}(Satire) = \frac{|Satire \cap POS_{(\alpha,\beta)}(Satire)| + |Satire^c \cap NEG_{(\alpha,\beta)}(Satire)|}{|POS_{(\alpha,\beta)}(Satire)| + |NEG_{(\alpha,\beta)}(Satire)|} \tag{7}$$

$$Cov_{(\alpha,\beta)}(Satire) = \frac{|POS_{(\alpha,\beta)}(Satire)| + |NEG_{(\alpha,\beta)}(Satire)|}{|U|} \tag{8}$$

The criterion coverage indicates the proportions of news that can be confidently classified. Next, we will obtain (α, β) by game formulation and repetition learning.

Game Formulation. We construct a game $G = \{O, S, u\}$ given the set of game players O, the set of strategy profile S, and the payoff functions u, where the accuracy and coverage are two players, respectively, i.e., $O = \{acc, cov\}$.

The set of strategy profiles $S = S_{acc} \times S_{cov}$, where S_{acc} and S_{cov} are sets of possible strategies or actions performed by players acc and cov. The initial thresholds are set as $(1, 0)$. All these strategies are the changes made on the initial thresholds,

$$S_{acc} = \{\beta \text{ no change}, \beta \text{ increases } c_{acc}, \beta \text{ increases } 2 \times c_{acc}\},$$
$$S_{cov} = \{\alpha \text{ no change}, \alpha \text{ decreases } c_{cov}, \alpha \text{ decreases } 2 \times c_{cov}\}. \tag{9}$$

c_{acc} and c_{cov} denote the change steps used by two players, and their values are determined by the concrete experiment date set.

Payoff Functions. The payoffs of players are $u = (u_{acc}, u_{cov})$, and u_{acc} and u_{cov} denote the payoff functions of players acc and cov, respectively. Given a strategy profile $p = (s, t)$ with player acc performing s and player cov performing t, the payoffs of acc and cov are $u_{acc}(s, t)$ and $u_{cov}(s, t)$. We use $u_{acc}(\alpha, \beta)$ and $u_{cov}(\alpha, \beta)$ to show this relationship. The payoff functions $u_{acc}(\alpha, \beta)$ and $u_{cov}(\alpha, \beta)$ are defined as,

$$u_{acc}(s, t) \Rightarrow u_{acc}(\alpha, \beta) = Acc_{(\alpha,\beta)}(Satire),$$
$$u_{cov}(s, t) \Rightarrow u_{cov}(\alpha, \beta) = Cov_{(\alpha,\beta)}(Satire), \tag{10}$$

where $Acc_{(\alpha,\beta)}(Satire)$ and $Cov_{(\alpha,\beta)}(Satire)$ are the accuracy and coverage defined in Eqs. (7) and (8).

Payoff Table. We use payoff tables to represent the formulated game. Table 2 shows a payoff table example in which both players have 3 strategies defined in Eq. (9).

Table 2. An example of a payoff table

		cov		
		α	$\alpha \downarrow c_{cov}$	$\alpha \downarrow 2c_{cov}$
acc	β	$\langle u_{acc}(\alpha,\beta),$ $u_{cov}(\alpha,\beta)\rangle$	$\langle u_{acc}(\alpha-c_{cov},\beta),$ $u_{cov}(\alpha-c_{cov},\beta)\rangle$	$\langle u_{acc}(\alpha-2c_{cov},\beta),$ $u_{cov}(\alpha-2c_{cov},\beta)\rangle$
	$\beta \uparrow c_{acc}$	$\langle u_{acc}(\alpha,\beta+c_{acc}),$ $u_{cov}(\alpha,\beta+c_{acc})\rangle$	$\langle u_{acc}(\alpha-c_{cov},\beta+c_{acc}),$ $u_{cov}(\alpha-c_{cov},\beta+c_{acc})\rangle$	$\langle u_{acc}(\alpha-2c_{cov},\beta+c_{acc}),$ $u_{cov}(\alpha-2c_{cov},\beta+c_{acc})\rangle$
	$\beta \uparrow 2c_{acc}$	$\langle u_{acc}(\alpha,\beta+2c_{acc}),$ $u_{cov}(\alpha,\beta+2c_{acc})\rangle$	$\langle u_{acc}(\alpha-c_{cov},\beta+2c_{acc}),$ $u_{cov}(\alpha-c_{cov},\beta+2c_{acc})\rangle$	$\langle u_{acc}(\alpha-2c_{cov},\beta+2c_{acc}),$ $u_{cov}(\alpha-2c_{cov},\beta+2c_{acc})\rangle$

The arrow $\downarrow$ denotes decreasing a value and $\uparrow$ denotes increasing a value. On each cell, the threshold values are determined by two players.

Repetition Learning Mechanism. We repeat the game with the new thresholds until a balanced solution is reached. We first analyzes the pure strategy equilibrium of the game and then check if the stopping criteria are satisfied.

Game Equilibrium. The game solution of pure strategy Nash equilibrium is used to determine possible game outcomes in GTRS. The strategy profile (s_i, t_j) is a pure strategy Nash equilibrium, if

$$\forall s_i' \in S_{acc}, u_{acc}(s_i, t_j) \geqslant u_{acc}(s_i', t_j), \text{where } s_i \in S_{acc} \wedge s_i' \neq s_i,$$
$$\forall t_j' \in S_{cov}, u_{cov}(s_i, t_j) \geqslant u_{cov}(s_i, t_j'), \text{where } t_j \in S_{cov} \wedge t_j' \neq t_j. \quad (11)$$

This means that none of players would like to change his strategy or they would loss benefit if deriving from this strategy profile, provided this player has the knowledge of other player's strategy.

Repetition of Games. Assuming that we formulate a game, in which the initial thresholds are (α,β), and the equilibrium analysis shows that the thresholds corresponding to the equilibrium are (α^*,β^*). If the thresholds (α^*,β^*) do not satisfy the stopping criterion, we will update the initial thresholds in the subsequent games. The initial thresholds of the new game will be set as (α^*,β^*). If the thresholds (α^*,β^*) satisfy the stopping criterion, we may stop the repetition of games.

Stopping Criterion. We define the stopping criteria so that the iterations of games can stop at a proper time. In this research, we set the stopping criterion as the thresholds are inside the valid range or the increase of one player's payoff is less than the decrease of the other player's payoff.

4   Experiments

There are 8757 news records in our preprocessed data set. We use Jenks natural breaks [8] to discretize continuous variables S_{NP} and S_{QP} both into five categories denoted by nominal values from 0 to 4, where larger values still fall into bins with larger nominal value. Let D_{NP} and D_{QP} denote the discretized variables S_{NP} and S_{QP}, respectively. We derived the information table that only contains discrete features from our original dataset. A fraction of the information table is shown in Table 3.

Table 3. The information table

Id	D_{NP}	D_{QP}	C_{NERN}	B_{voc}	target
1	0	2	0	0	1
2	1	2	0	0	1
3	2	2	0	1	0
4	2	4	1	1	0
5	2	3	0	0	1
6	4	3	−1	1	0
7	2	3	0	0	0
8	3	2	−1	0	1

The news whose condition attributes have the same values are classified in an equivalence class X_i. We derived 149 equivalence classes and calculated the corresponding probability $Pr(X_i)$ and condition probability $Pr(Satire|X_i)$ for each X_i. The probability $Pr(X_i)$ denotes the ratio of the number of news contained in the equivalence class X_i to the total number of news in the dataset, while the conditional probability $Pr(Satire|X_i)$ is the proportion of news in X_i that are satirical. We combine the equivalence classes with the same conditional probability and reduce the number of equivalence classes to 108. Table 4 shows a part of the probabilistic data information about the concept *satire*.

Table 4. Summary of the partial experimental data

	X_1	X_2	X_3	X_4	X_5	X_6	X_7	X_8	X_9	......
$Pr(X_i)$	0.0315	0.0054	0.0026	0.0071	0.0062	0.0018	0.0015	0.0098	0.0009	......
$Pr(Satire|X_i)$	1	0.9787	0.9565	0.9516	0.9444	0.9375	0.9231	0.9186	0.875	......
	......	X_{100}	X_{101}	X_{102}	X_{103}	X_{104}	X_{105}	X_{106}	X_{107}	X_{108}
$Pr(X_i)$	......	0.0121	0.0138	0.0095	0.0065	0.0383	0.0078	0.0107	0.0163	0.048
$Pr(Satire|X_i)$	......	0.0283	0.0248	0.0241	0.0175	0.0149	0.0147	0.0106	0.007	0

4.1 Finding Thresholds with GTRS

We formulated a competitive game between the criteria accuracy and coverage to obtain the balanced probabilistic thresholds with the initial thresholds $(\alpha, \beta) = (1, 0)$ and learning rate 0.03. As shown in the payoff table Table 5, the cell at the right bottom corner is the game equilibrium whose strategy profile is (β increases 0.06, α decreases 0.06). The payoffs of the players are (0.9784,0.3343). We set the stopping criterion as the increase of one player's payoff is less than the decrease of the other player's payoff when the thresholds are within the range. When the thresholds change from (1,0) to (0.94, 0.06), the accuracy is decreased from 1 to 0.9784 but the coverage is increased from 0.0795 to 0.3343. We repeat the game by setting $(0.94, 0.06)$ as the next initial thresholds.

Table 5. The payoff table

		cov		
		α	$\alpha \downarrow 0.03$	$\alpha \downarrow 0.06$
acc	β	$< 1, 0.0795 >$	$< 0.9986, 0.0849 >$	$< 0.9909, 0.1008 >$
	$\beta \uparrow 0.03$	$< 0.9868, 0.2337 >$	$< 0.9866, 0.2391 >$	$< 0.9843, 0.255 >$
	$\beta \uparrow 0.06$	$< 0.9799, 0.3130 >$	$< 0.9799, 0.3184 >$	$\mathbf{< 0.9784, 0.3343 >}$

The competitive games are repeated eight times. The result is shown in Table 6. After the eighth iteration, the repetition of game is stopped because the further changes on thresholds may cause the thresholds lay outside of the range $0 < \beta < \alpha < 1$, and the final result is the equilibrium of the seventh game $(\alpha, \beta) = (0.52, 0.48)$.

Table 6. The repetition of game

	Initial(α, β)	Strategies	Result(α, β)	Payoffs
1	(1, 0)	$(\beta \uparrow 0.03, \alpha \downarrow 0.03)$	(0.94, 0.06)	$< 0.9784, 0.3343 >$
2	(0.94, 0.06)	$(\beta \uparrow 0.03, \alpha \downarrow 0.03)$	(0.88, 0.12)	$< 0.9586, 0.4805 >$
3	(0.88, 0.12)	$(\beta \uparrow 0.03, \alpha \downarrow 0.03)$	(0.82, 0.18)	$< 0.9433, 0.554 >$
4	(0.82, 0.18)	$(\beta \uparrow 0.03, \alpha \downarrow 0.03)$	(0.76, 0.24)	$< 0.9218, 0.6409 >$
5	(0.76, 0.24)	$(\beta \uparrow 0.03, \alpha \downarrow 0.03)$	(0.7, 0.3)	$< 0.8960, 0.7467 >$
6	(0.7, 0.3)	$(\beta \uparrow 0.03, \alpha \downarrow 0.03)$	(0.64, 0.36)	$< 0.8791, 0.8059 >$
7	(0.64, 0.36)	$(\beta \uparrow 0.03, \alpha \downarrow 0.03)$	(0.58, 0.42)	$< 0.8524, 0.8946 >$
8	(0.58, 0.42)	$(\beta \uparrow 0.03, \alpha \downarrow 0.03)$	(0.52, 0.48)	$< 0.8271, 0.9749 >$

4.2   Results

We compare Pawlak rough sets, SVM, and our GTRS approach on the proposed dataset. Table 7 shows the results on the experimental data. The SVM classifier achieved an accuracy of 78% with a 100% coverage. The Pawlak rough set model using $(\alpha, \beta) = (1, 0)$ achieves a 100% accuracy and a coverage ratio of 7.95%, which means it can only classify 7.95% of the data. The classifier constructed by GTRS with $(\alpha, \beta) = (0.52, 0.48)$ reached an accuracy 82.71% and a coverage 97.49%. which indicates that 97.49% of data are able to be classified with accuracy of 82.71%. The remaining 2.51% of data can not be classified without providing more information. To make our method comparable to other baselines such as SVM, we assume random guessing is made on the deferral region and present the modified accuracy. The modified accuracy for our approach is then $0.8271 \times 0.9749 + 0.5 \times 0.0251 = 81.89\%$. Our methods shows significant improvement as compared to Pawlak model and SVM.

Table 7. The comparison results

	(α, β)	Accuracy	Coverage	Modified accuracy
SVM	-	78%	100%	78%
Pawlak	(1, 0)	100%	7.95%	53.98%
GTRS	(0.52, 0.48)	**82.71%**	**97.49%**	81.89%

5   Conclusion

In this paper, we propose a satirical news detection approach based on extracted semantic features and game-theoretic rough sets. In our model, the semantic features extraction captures the inconsistency in the different structural parts of the sentences and the GTRS classifier can process the incomplete information based on repetitive learning and the acceptance and rejection thresholds. The experimental results on our created satirical and legitimate news tweets dataset show that our model significantly outperforms Pawlak rough set model and SVM. In particular, we demonstrate our model's ability to interpret satirical news detection from a semantic and information trade-off perspective. Other interesting extensions of our paper may be to use rough set models to extract the linguistic features at document level.

References

1. Akbik, A., Blythe, D., Vollgraf, R.: Contextual string embeddings for sequence labeling. In: Proceedings of the 27th International Conference on Computational Linguistics, pp. 1638–1649. Springer (2018)

2. Azam, N., Yao, J.T.: Game-theoretic rough sets for recommender systems. Knowl.-Based Syst. **72**, 96–107 (2014)
3. Burfoot, C., Baldwin, T.: Automatic satire detection: are you having a laugh? In: Proceedings of the 2009 International Conference on Natural Language Processing, pp. 161–164. ACL (2009)
4. Conroy, N.J., Rubin, V.L., Chen, Y.: Automatic deception detection: methods for finding fake news. In: Proceedings of the Association for Information Science and Technology, pp. 1–4. Wiley Online Library (2015)
5. Davidov, D., Tsur, O., Rappoport, A.: Semi-supervised recognition of sarcastic sentences in Twitter and Amazon. In: Proceedings of the 14th Conference on Computational Natural Language Learning, pp. 107–116. ACL (2010)
6. De Sarkar, S., Yang, F., Mukherjee, A.: Attending sentences to detect satirical fake news. In: Proceedings of the 27th International Conference on Computational Linguistics, pp. 3371–3380. Springer (2018)
7. Ghosh, A., Veale, T.: Magnets for sarcasm: making sarcasm detection timely, contextual and very personal. In: Proceedings of the 2017 Conference on Empirical Methods in Natural Language Processing, pp. 482–491 (2017)
8. Jenks, G.F.: The data model concept in statistical mapping. Int. Yearb. Cartography **7**, 186–190 (1967)
9. Pennington, J., Socher, R., Manning, C.D.: Glove: global vectors for word representation. In: Empirical Methods in Natural Language Processing, pp. 1532–1543. ACL (2014)
10. Riloff, E., Qadir, A., Surve, P., De Silva, L., Gilbert, N., Huang, R.: Sarcasm as contrast between a positive sentiment and negative situation. In: Proceedings of the 2013 Conference on Empirical Methods in Natural Language Processing, pp. 704–714 (2013)
11. Rubin, V., Conroy, N., Chen, Y., Cornwell, S.: Fake news or truth? using satirical cues to detect potentially misleading news. In: Proceedings of the 2nd Workshop on Computational Approaches to Deception Detection, pp. 7–17. ACM (2016)
12. Ruchansky, N., Seo, S., Liu, Y.: CSI: a hybrid deep model for fake news detection. In: Proceedings of the 2017 Conference on Information and Knowledge Management, pp. 797–806. ACM (2017)
13. Salton, G., McGill, M.J.: Introduction to Modern Information Retrieval. McGraw-Hill, New York (1986)
14. Shu, K., Sliva, A., Wang, S., Tang, J., Liu, H.: Fake news detection on social media: a data mining perspective. ACM SIGKDD Explor. Newsl. **19**(1), 22–36 (2017)
15. Sterling, C.H.: Encyclopedia of Journalism. Sage Publications, New York (2009)
16. Wang, Y., et al.: Eann: event adversarial neural networks for multi-modal fake news detection. In: Proceedings of the 24th SIGKDD International Conference on Knowledge Discovery & Data Mining, pp. 849–857. ACM (2018)
17. Yang, F., Mukherjee, A., Dragut, E.: Satirical news detection and analysis using attention mechanism and linguistic features. arXiv preprint arXiv:1709.01189 (2017)
18. Yao, J.T., Herbert, J.P.: A game-theoretic perspective on rough set analysis. J. Chongqing Univ. Posts Telecommun. **20**(3), 291–298 (2008)
19. Yao, J.T., Azam, N.: Web-based medical decision support systems for three-way medical decision making with game-theoretic rough sets. IEEE Trans. Fuzzy Syst. **23**(1), 3–15 (2015)
20. Yao, Y.Y.: The superiority of three-way decisions in probabilistic rough set models. Inf. Sci. **181**(6), 1080–1096 (2011)

21. Yao, Y.Y.: An outline of a theory of three-way decisions. In: Proceedings of International Conference on Rough Sets and Current Trends in Computing, pp. 1–17. Springer (2012)
22. Yao, Y.Y.: Three-way decisions and cognitive computing. Cognitive Comput. **8**(4), 543–554 (2016)
23. Zhang, Y., Liu, P.F., Yao, J.T.: Three-way email spam filtering with game-theoretic rough sets. In: Proceedings of the 2019 International Conference on Computing, Networking and Communications, pp. 552–556. IEEE (2019)
24. Zhang, Y., Yao, J.T.: Determining three-way decision regions by combining gini objective functions and GTRS. In: Yao, Y., Hu, Q., Yu, H., Grzymala-Busse, J.W. (eds.) RSFDGrC 2015. LNCS (LNAI), vol. 9437, pp. 414–425. Springer, Cham (2015). https://doi.org/10.1007/978-3-319-25783-9_37
25. Zhang, Y., Yao, J.T.: Multi-criteria based three-way classifications with game-theoretic rough sets. In: Kryszkiewicz, M., Appice, A., Ślęzak, D., Rybinski, H., Skowron, A., Raś, Z.W. (eds.) ISMIS 2017. LNCS (LNAI), vol. 10352, pp. 550–559. Springer, Cham (2017). https://doi.org/10.1007/978-3-319-60438-1_54

Deep Learning and Embeddings

Comparing State-of-the-Art Neural Network Ensemble Methods in Soccer Predictions

Tiago Mendes-Neves[1]([✉]) [iD] and João Mendes-Moreira[1,2] [iD]

[1] Faculdade de Engenharia, Universidade do Porto, Porto, Portugal
tiago.m.neves@inesctec.pt
[2] LIAAD-INESC TEC, Porto, Portugal

Abstract. For many reasons, including sports being one of the main forms of entertainment in the world, online gambling is growing. And in growing markets, opportunities to explore it arise. In this paper, neural network ensemble approaches, such as bagging, random subspace sampling, negative correlation learning and the simple averaging of predictions, are compared. For each one of these methods, several combinations of input parameters are evaluated. We used only the expected goals metric as predictors since it is able to have good predictive power while keeping the computational demands low. These models are compared in the soccer (also known as association football) betting context where we have access to metrics, such as rentability, to analyze the results in multiple perspectives. The results show that the optimal solution is goal-dependent, with the ensemble methods being able to increase the accuracy up to +3 % over the best single model. The biggest improvement over the single model was obtained by averaging dropout networks.

Keywords: Sports betting · Neural networks · Ensemble learning

1 Introduction

Gambling was always an interesting concept to human beings. If we ask a person if they want to trade 1€ for 0.95€ they will immediately reject the proposal. Being guaranteed to lose money is something that is not usually accepted without being rewarded. In betting, the reward comes from the existing probability of winning money. Even though in the long term more money is lost than won, the human brain is blinded by the prospect of a big win.

This paper is about soccer betting. Unlike other forms of gambling, in soccer betting, the probabilities are not predefined or easily calculated.

Bookmakers have the advantage of having access to the wisdom of the crowd that when combined with the ability for the market to self regulate, leaves them making a consistent profit regardless of the outcome.

This work is financed by National Funds through the Portuguese funding agency, FCT - Fundação para a Ciência e a Tecnologia, within project UIDB/50014/2020.

© Springer Nature Switzerland AG 2020
D. Helic et al. (Eds.): ISMIS 2020, LNAI 12117, pp. 139–149, 2020.
https://doi.org/10.1007/978-3-030-59491-6_13

The increase of available data on the soccer pitch along time and in terms of information richness allow nowadays the use of more data demanding machine learning algorithms to predict probabilities.

Academia has embraced the problem of predicting soccer matches very heavily. We are going to focus on machine learning approaches. A great amount of work uses Bayesian networks, such as [2] where they attempt to forecast the 2011/2012 season of the English Premier League and evaluate their model using the bookmaker odds. Also in the scope of Bayesian networks, [7] compares Bayesian networks with other state of the art machine learning algorithms, such as decision trees and k-nearest neighbors, concluding that machine learning algorithms had a better performance in the seasons tested (95/96 and 96/97, using Tottenham Hotspurs games on the English Premier League). Another algorithm used was the fuzzy-based models with genetic and neural tuning [11], tested on the Finnish Football League from 1991 to 1993, where the goal was to predict the score difference between both teams. Predictions using neural networks are also present in academia. [9] tested their hypothesis on several sports data, such as rugby and soccer, achieving the significant result of consistently being in the 99th percentile of a tipsters competition, meaning that it was able to beat a great part of human-made predictions. [3] attempts to predict price movements on betting exchanges using Artificial Neural Networks, focusing mainly on horse racing markets and yielding a significant ROI.

In 2017, the 2017 Soccer Prediction Challenge [14] purposed that researchers approached a data set from 52 leagues and seasons from 2000/01 to 2016/17 with more than 200 000 games, known as Open International Soccer Database [13]. Of the best work resulting from this issue, [15] uses Bradley—Terry model and a hierarchical Poisson log-linear model, falling more on the statistical model category. [16] used Bayesian networks that yields a very good results even with constraints of the challenge, since it only allows the usage of the final scores of the matches to build features for the models. It overtakes this obstacle by building dynamic ratings system. The winner, [17], also uses a rating system to generate features used in gradient boosted trees. This challenge has shown that, despite data set restrictions, it was still possible to obtain great results in terms of soccer predictions.

The machine learning algorithm that is of interest is the neural networks. Neural networks are nowadays being used to solve a lot of problems, namely in image data where they dominate over other techniques in terms of published research papers. Neural networks are connection-based models that have a very strong ability to assemble complex models. While neural networks having the ability to generate complex models may look like a plus, it can lead to overfitting, and with that, a bad generalization power, leading to sub-par performance when testing the models in previously unseen instances.

Ensemble models can help to solve this problem. Like the gathering of opinions in the betting markets improves the estimate over a single opinion, ensemble models gather the predictions of several models and by combining them allows the ensemble prediction to be better than any of the individual predictions alone.

Neural network have been used to solve several problems. Neural network ensembles have been extensively explored in the last couple of years. From 2017 to 2019 the number of published papers doubled (Science Direct search engine returns on the query "neural network ensemble"). 2019 has seen 3195 research papers while 2017 had only 1560. Due to its ability to work with image data, the majority comes from the health sector, such as [6] where they attempt to classify skin lesions in order to detect the cancerigenous ones, where they have been able to improve the score of a single network in 8%. Health anomaly detection seems to be the area of most success when using neural networks on image data. Even though image data seems to dominate the research in the area there is some research on quantitative data such the energy sector where [12] attempts to predict the load on the grid to improve systems planning.

What is missing from the academic perspective is a review of the available ensemble methods for neural networks. Neural networks have an extensive problem of overfitting. However, ensemble models have the ability to solve this problem. It is necessary to study what algorithms fit better with this technique, along with the development of ensemble techniques that enhance the advantages of the neural networks while hiding their flaws.

The reason for soccer data to be the ideal environment to test ensemble methods is that there is available a baseline in the bookmakers odds that is also calculated using an ensemble, the wisdom of the crowd.

The paper is organized as following. Section 2 describes the data used in the experiments. Section 3 talks about our experimental setup, focusing on the metrics, feature generation and models used. In Sect. 4 we describe the experiments made and discuss the results. Finally, in Sect. 5, we make our final conclusions and discuss possible future work in the area.

2   Describing the Data

The experiments described in this paper uses data from two sources: fivethirtyeight.com soccer-spi data set [4] and football-data.co.uk [5]. From the first, we retrieve the expected goals metric for every match from the season 2016/2017 to 2018/2019. On the second, we acquire the odds from the matches that we retrieved from the soccer-spi data set. The resulting data set is described in Table 1.

The data set has games from 6 leagues: English Premier League, French Ligue 1, Spanish La Liga, Italian Serie A, German Bundesliga, and Portuguese Primeira Liga. On this last, the 2016/2017 season data is not available.

The training set will be composed of the 2016/2017 and 2017/2018 season data, in a total of 3178 games. For the test set, the full 2018/2019 season is used, amounting to 1656 games.

Table 1. Variables present in the assembled data set.

Variable	Data type	Description
game_id	Symbolic	Internal unique id for integrating the databases
season	Symbolic	Season in which the game occurred
season_day	Numeric	Day of the season (season start set to July 1st)
home_team	Symbolic	Home team identifier
away_team	Symbolic	Away team identifier
ftr	Symbolic	Full time winner (H, D, A)
h_expected_goals	Numeric	Performance of the home team measured in expected goals
a_expected_goals	Numeric	Performance of the away team measured in expected goals
h_odd	Numeric	Average odd from the bookmakers for the home team to win
d_odd	Numeric	Average odd from the bookmakers for the draw
a_odd	Numeric	Average odd from the bookmakers for the away team to win

2.1 Expected Goals

The expected goals metric [10] is a measure of shot quality. It is calculated from the likelihood of a shot ending in a goal, taking into account factors such as distance to the goal, angle of the shot, body part used to make the shot and whether it was a first touch shot or not. The mathematical formulation can be seen in Eq. 1.

$$team\ X\ expected\ goals\ =\ \sum_{for\ all\ team\ X\ shots} P(shot\ leading\ to\ a\ goal) \quad (1)$$

This metric allows us to abstract from the binary goal metric. By incorporating probabilities we can obtain a better representation of how the game was played between both teams, reducing the randomness associated with soccer. This is the reason why this metric is gathering the attention of fans and the media.

3 Experimental Setup

3.1 Performance Metrics

The goal of the experiments is to obtain results that allow us to compare the performance of different ensemble algorithms. We are going to define a set of metrics that will be used, each one focusing on a part of the problem.

The first metric that will be used is the accuracy (Eq. 2). This is a metric that is of standard use in classification problems.

$$accuracy = \frac{correct\ predictions}{number\ of\ predictions} \tag{2}$$

The probabilities generated by the classification algorithms can be evaluated from two points of view: the betting and the regression point of view.

On the betting point of view, the defined metric was the rentability (Eq. 3), that tells us how much money the model made in relation to the stake. For this calculation, the algorithms will be always betting in the predicted favourite and the stake of each bet will be one unit.

$$rentability = \sum_{correct\ predictions} (odd - 1) \quad - \quad number\ of\ incorrect\ predictions \tag{3}$$

From the regression point of view, two metrics were used: bias and estimated standard deviation (referred simply as *estimated stdev* in the rest of the paper). Both bias and estimated stdev are defined in Eqs. 4 and 5, where M is the number of predictions made, γ is the predicted value and y is the real value. The bias is the error caused by the model's simplified assumptions that cause a constant error across different choices of training data. Estimated stdev means that for the same instance, the same model trained in slightly different data will yield different results.

$$bias^2 = \frac{1}{M} \sum_{i}^{M} y_i - \gamma_i \tag{4}$$

$$estimated\ stdev = \frac{1}{M} \sum_{i}^{M} stdev(\gamma_i) \tag{5}$$

The last metric measured is the average training time of the models. No optimization was made in any of the algorithms.

3.2   Feature Set

The feature set will be generated on top of the expected goals metric. For that, we will assemble the expected goals scored and conceded in each of the last 7 games for both teams facing up. This will leave us with 14 features for each team and a total of 28 features for our model.

3.3   Base Learners

There is a wide spectrum of choices when tuning a neural network, and these choices have a high impact in the final results of the neural network. Therefore it is necessary to establish what will be the base learners used in the ensemble models in order to keep them comparable.

We are going to use the Keras API on top of the TensorFlow 1.13 to run these experiments. We have chosen Python 3.6 as the programming language. The parameters where no reference is made in this section revert to the default values that can be found in the Keras documentation.

Through an iteration process we found the base learners described next to yield good results as singular models.

The base learners are sequential models with dense layers. Two architectures were tested. The first uses a single hidden layer with 15 nodes (15,), the second uses two hidden layers of 10 and 5 nodes (10,5). In both cases, the layers use the *softmax* activation function, in order to have predictions in the form of probabilities. These two architectures were the ones that had the better performance when considering also time consumed.

A parameter that will be tested is the early stop. Since it naturally increases the variability of the models it might lead to better performance than the no early stop variant when ensembled.

The other non-default parameters are constant through the models. The learning rate is 0.025 and the batch size is 200. The loss function used is the categorical cross-entropy and the optimizer used is *Adam*. Dropout is used or not depending on the ensemble.

Table 2 summarizes the base learners used in the experiences.

Table 2. Presentation of the base learners.

Model	Architecture	Epochs	Early Stop	Patience
1	(15,)	100	No	
2	(15,)	500	Yes	10
3	(10,5)	100	No	
4	(10,5)	500	Yes	10

3.4 Proposed Algorithms

Bagging. Bagging [1] is perhaps the most common ensemble approach used. The idea is to generate new training data sets from a single data set by sampling, which can be performed with or without replacement.

In our implementation, similarly to what is done in random forest with the random subspace sampling, there is a parameter that allows us to modify the feature subset in which the models are trained. This parameter will be called *feature_ratio*, and it indicates a percentage of the features that will be used at each split. The number of samples from the base data that will be used is indicated by the parameter *sample_ratio*, which is a percentage of the samples that will be used. Samples are drawn without replacement.

Simple Average Dropout Networks (SADN). The SADN is an ensemble in which the goal of each model is to have low estimated stdev. The dropout parameter acts as regularization for the networks, not allowing them to become too complex and overfit. The predicted probabilities from the ensembles' models are then averaged and re-normalized in order to produce the ensemble predictions. On the experiments, the hidden layers will have a 0.3 dropout rate.

Negative Correlation Learning (NCL). In negative correlation learning [8], the approach is to train individual networks in an ensemble and combining them in the same process. All the neural networks in the ensemble are trained simultaneously and interactively through a correlation penalty term in their error function.

The difference from regular neural network training is in the loss function. Subtracted to the regular loss function (a function of the predicted value and the real value) is a percentage (λ parameter) of the loss function calculated between the value predicted by the model and the ensemble predicted value. This can be seen in Eq. 6, where γ is a neural network prediction, ε the ensemble prediction and y is the real value. This incentives models to go in a different direction from the average value of the ensemble when training, creating diversity in the model's opinions and improving the model classification performance.

$$new\ loss\ function\ =\ loss\ function(\gamma, y)\ -\ \lambda * loss\ function(\gamma, \varepsilon) \quad (6)$$

4   Experiments

4.1   Ensemble Hyperparameter Tuning

Before the comparison could be made, the ensemble model parameters need to be tuned. The first parameter defined is that each ensemble will have 50 base learners. While SADN does not need any additional parameter, both bagging and NCL have parameters that need to be tuned. To tune the parameters for the bagging method we do a grid search in order to find the optimal hyperparameters. This can be seen in Table 3.

Table 3. Hyperparameters grid search for bagging in model 1, with results in the format *Accuracy (estimated stdev)*. The results are averages of 10 runs.

		Sample ratio			
		0.25	0.5	0.75	1
Feature ratio	0.25	51.17 (0.0379)	51.07 (0.0455)	51.08 (0.0491)	50.98 (0.0517)
	0.5	51.15 (0.0387)	51.12 (0.0446)	51.04 (0.0484)	50.95 (0.0515)
	0.75	50.97 (0.0536)	50.99 (0.0515)	50.97 (0.0506)	50.92 (0.0513)
	1	50.95 (0.0526)	50.96 (0.0515)	50.95 (0.0509)	50.93 (0.0514)

Table 4. Hyperparameters grid search for NCL. The tests were done on the model 1. The results are averages of 10 runs.

Lambda parameter	0.05	0.10	0.20	0.30	0.40
Accuracy	50.1	50.16	49.6	49.71	47.2
Estimated stdev	0.0252	0.0291	0.0362	0.0494	0.0757

Table 5. Results from the tests with optimal parameters. Note that the expected rentability (calculated by betting in every outcome of every game) is −77.22. The negative number demonstrates the earlier mentioned fact that bookmakers are making consistent profit.

Evaluation metric	Single model		Bagging		SADN		NCL
Early Stop	Yes	No	Yes	No	Yes	No	No
Architecture	(15,)						
Accuracy	48.95	50.18	50.83	51.22	51.70	51.24	50.01
Rentability	−96.23	−77.92	−73.46	−65.85	−43.24	−60.83	−71.39
Bias	0.6284	0.6287	0.6288	0.6287	0.6297	0.6293	0.6293
Estimated stdev	0.2199	0.1049	0.0766	0.0397	0.0104	0.0112	0.0229
Average execution time	4.18	1.65	89.65	24.72	82.19	140.47	102.9
Architecture	(10,5)						
Accuracy	50.21	50.61	51.08	51.35	50.91	51.00	51.31
Rentability	−62.75	−70.44	−52.16	−61.29	−49.66	−52.40	−53.27
Bias	0.6291	0.6286	0.6334	0.6294	0.6326	0.6311	0.6302
Estimated stdev	0.1710	0.0881	0.0528	0.0304	0.0133	0.0095	0.0136
Average execution time	2.56	1.71	43.00	25.51	107.39	170.22	126.46

From Table 3 we can conclude that both feature_ratio and sample_ratio improve the performance when lowered. This can be verified from both accuracy and estimated stdev perspective. The best performing hyperparameters (both parameters equal to 0.25) will be used.

For the NCL we need to set the parameter λ. In the tuning phase, low λ values seemed to perform better, as seen in Table 4. The chosen λ is 0.1.

4.2   Ensemble Methods Comparison

Table 5 shows the results of the experiments. These results are the average of 50 runs of each algorithm.

Accuracy wise, the best performing model was the SADN (15,) with early stop with 51,7%. This is also the only instance where early stopping leads to better accuracy results since neither bagging or single models were able to improve when using early stop. In the (10,5) architecture the SADN with early stopping did not replicate this success. The accuracy performance of the SADN (15,) enabled the rentability to also be the best overall with -43.24, a value nearly 34 units above the expected value.

Table 6. A sample of the odds generated by the best model of each type in comparison with the bookmaker's odd.

Season	Res.	Home	Away	Home					Draw					Away				
				Bookie	NN	BagNN	SADN	NCL	Bookie	NN	BagNN	SADN	NCL	Bookie	NN	BagNN	SADN	NCL
18/19	H	Napoli	Cagliari	1.26	1.83	1.58	1.48	1.72	6	4.45	4.94	5.92	3.99	11.14	4.39	6.12	6.4	5.9
18/19	H	Liverpool	Huddersfield	1.07	1.25	1.24	1.28	1.37	13.61	7.49	7.3	6.22	5.39	36.37	14.73	17.38	17.69	12.02
18/19	A	Nacional	Porto	13.51	10.48	9.18	12	7.02	6.65	6.92	6.86	5.48		1.19	1.28	1.34	1.3	1.48
18/19	H	Atalanta	Genoa	1.51	1.11	1.18	1.15	1.22	4.33	12.03	8.42	9.03	6.9	6.5	50.25	28.52	62.12	30.93
18/19	H	Paris SG	Monaco	1.51	1.48	1.4	1.53	1.38	4.74	4.11	5.78	5.02	5.44	5.69	12.19	8.86	6.81	11.02
18/19	A	Tondela	Santa Clara	1.76	3.78	6.49	4.4	5.5	3.63	2.55	3.62	3.89	4.93	4.57	2.92	1.76	1.94	1.63
18/19	D	Leipzig	Bayern M	4.61	5.02	2.82	2.42	2.18	4.22	2.99	3.19	3.66	3.66	1.69	2.15	3.01	3.19	3.74
18/19	H	Roma	Juventus	2.52	3.58	4.62	3.86	4.8	3.43	4.95	2.85	3.44	4.6	2.79	1.93	2.31	2.22	1.74
18/19	D	Leicester	Chelsea	2.4	4.02	2.58	2.67	2.36	3.57	3.83	3.59	3.43	3.7	2.89	2.04	3	3	3.27

Since these models were focused on estimated stdev reduction, it was expected that the bias did not change considerably. While improvements in bias are scarce, the cost of using ensembles was low, with none of the biases increasing by over 1% over the single model.

On the other side, the estimated stdev was immensely reduced, with some models achieving approximately 95% reduction. The most notable performance here was also obtained by the SADN algorithm, with the NCL being a close contender.

The more complex architecture (10,5) found it harder to improve results. While the performance jumped almost 3% in the best scenario for the (15,) architecture, the (10,5) failed to improve even 1%. However, with the exception of the SADN, all the algorithms managed to perform better on the (10,5) architecture.

Since the (15,) is a less complex model, ensembling with methods that do not induce more variability in predictions (SADN) yielded better results. On the other side, a more complex model, (10,5), needed algorithms that introduced variability in the models (bagging/NCL) to improve the predictions. We know that the most important factor when ensembling is to find the right balance of variability in the ensemble's models. Either too much or too little variance will worsen the results. For this, the combination of simpler models trained with dropout and early stopping seem to be the best solution.

In terms of quickness, bagging is the better option. Without early stopping, bagging is able to only take approximately 15x more time to train than the simple network, while training 50x more models, due to the reduced number of instances and features used.

From the business point of view, all the ensemble methods were able to have a better performance than the single neural network. While the results are still far away from being profitable, we have to take into account the fact that these models are already able to beat the expected rentability of -77.22 units.

In Table 6 we are able to see that the models' predictions are, in some instances, very similar to the bookmakers. This values could potentially be improved since the models do not take into account a lot of information about the games: player availability, how many rest days since last games, among others.

5 Conclusions

Ensembling neural networks improves the results of single models in terms of accuracy up to 3 %. This accuracy improvement can have a massive impact from the business perspective, as it can be seen in the rentability, with the best performing ensemble cutting the losses in half over the single model variant.

In general, ensemble proves itself to be a reliable way to reduce variance in the neural network context. However, in problems that are already very demanding in terms of computation time, this might not be optimal. An increase of 3% in accuracy lead to at least a 600% percent increase in computational time. Therefore, we conclude that the usage of ensemble techniques with neural networks is situational.

5.1 Future Work

While the results look promising, especially for SADN, the tests are only done in one data set. The algorithms need to be tested across multiple data sets to verify if these conclusions are consistent or if they only hold true in the soccer prediction scenario. Even in the same data set, tests with different parameters can be done.

References

1. Aggarwal, C.C.: Neural Networks and Deep Learning: eBook. Springer, Switzerland (2018)
2. Constantinou, A.C., Fenton, N.E., Neil, M.: Profiting from an inefficient association football gambling market: prediction, risk and uncertainty using Bayesian networks. Knowl. Based Syst. **50**, 60–86 (2013)
3. Dzalbs, I., Kalganova, T.: Forecasting price movements in betting exchanges using cartesian genetic programming and ANN. Big Data Res. **14**, 112–120 (2018)
4. FiveThirtyEight. fivethirtyeight.com. Accessed 21st June 2019
5. Football-Data.co.uk. football-data.co.uk. Accessed 21st June 2019
6. Harangi, B.: Skin lesion classification with ensembles of deep convolutional neural networks. J. Biomed. Inform. **86**, 25–32 (2018)
7. Joseph, A., Fenton, N.E., Neil, M.: Predicting football results using Bayesian nets and other machine learning techniques. Knowl.-Based Syst. **19**(7), 544–553 (2006)
8. Liu, Y., Yao, X.: Ensemble learning via negative correlation. Neural Networks **12**(10), 1399–1404 (1999)
9. Mccabe, A., Trevathan, J: Artificial intelligence in sports prediction. In: Fifth International Conference on Information Technology: New Generations, pp. 1194–1197. IEEE Computer Society, Las Vegas (2008)
10. Opta. optasports.com. Accessed 22nd June 2019
11. Rotshtein, A.P., Posner, M., Rakityanskaya, A.B.: Football predictions based on a fuzzy model with genetic and neural tuning. Cybern. Syst. Anal. **41**(4), 619–630 (2005)
12. Ribeiro, G., Mariani, V., Coelho, L.: Enhanced ensemble structures using wavelet neural networks applied to short-term load forecasting. Eng. Appl. Artif. Intell. **82**, 272–281 (2019)

13. Dubitzky, W., Lopes, P., Davis, J., Berrar, D.: The open international soccer database. Mach. Learn. **108**, 9–28 (2019)
14. Berrar, D., Lopes, P., Davis, J., Dubitzky, W.: Guest editorial: special issue on machine learning for soccer. Mach. Learn. **108**(1), 1–7 (2018). https://doi.org/10.1007/s10994-018-5763-8
15. Tsokos, A., Narayanan, S., Kosmidis, I., Baio, G., Cucuringu, M., Whitaker, G., Király, F.: Modeling outcomes of soccer matches. Mach. Learn. **108**(1), 77–95 (2018). https://doi.org/10.1007/s10994-018-5741-1
16. Constantinou, A.C.: Dolores: a model that predicts football match outcomes from all over the world. Mach. Learn. **108**(1), 49–75 (2018). https://doi.org/10.1007/s10994-018-5703-7
17. Hubáček, O., Šourek, G., Železný, F.: Learning to predict soccer results from relational data with gradient boosted trees. Mach. Learn. **108**(1), 29–47 (2018). https://doi.org/10.1007/s10994-018-5704-6

Static Music Emotion Recognition Using Recurrent Neural Networks

Jacek Grekow$^{(\boxtimes)}$ (ID)

Faculty of Computer Science, Bialystok University of Technology,
Wiejska 45A, 15-351 Bialystok, Poland
j.grekow@pb.edu.pl

Abstract. The article presents experiments using recurrent neural networks for emotion detection for musical segments using Russell's circumplex model. A process of feature extraction and creating sequential data for learning networks with long short-term memory (LSTM) units is presented. Models were implemented using the WekaDeeplearning4j package and a number of experiments were carried out with data with different sets of features and varying segmentation. The usefulness of dividing data into sequences as well as the sense of using recurrent networks to recognize emotions in music, whose results have even exceeded the SVM algorithm for regression, were demonstrated. The author analyzed the effect of the network structure and the set of used features on the results of regressors recognizing values on two axes of the emotion model: arousal and valence.

Keywords: Emotion detection · Audio features · Sequential data · Recurrent Neural Networks

1 Introduction

Music is an organization of sounds over time, and one of its more important functions is the transmission of emotions. The music created by a composer is ultimately listened to by a listener. The carriers of emotions are sounds distributed over time, their quantity, pitch, loudness, and their mutual relations. These sounds in music terminology are described by melody, timbre, dynamics, rhythm, and harmony. Before a person notices the emotions in music, he/she must have some time to analyze the listened to fragment [2]; depending on the changes in melody, timbre, dynamics, rhythm, or harmony, we can notice different emotions, such as happy, angry, sad, or relaxed.

The aim of this paper was to imitate the time-related perception of emotions in music by humans through the construction of an automatic emotion detection system using recurrent neural networks (RNN). Just as the human brain is "fed" with subsequent sound information over time, on the basis of which it perceives the emotions in music, similarly, the neural network downloads subsequent information vectors in subsequent time steps to predict the emotion value of the analyzed musical fragment.

© Springer Nature Switzerland AG 2020
D. Helic et al. (Eds.): ISMIS 2020, LNAI 12117, pp. 150–160, 2020.
https://doi.org/10.1007/978-3-030-59491-6_14

Division into categorical and dimensional approach can be found in papers devoted to music emotion recognition (MER). In the categorical approach, a number of emotional categories (adjectives) are used for labeling music excerpts [8,12]. In the dimensional approach, emotion is described using dimensional space, like the 2D model proposed by Russell [13], where the dimensions are represented by arousal and valence [5,6,9,15].

MER task can also be divided into static or dynamic, where static MER detects emotions in a relatively long section of music of 15–20 s [4,6,8], and dynamic MER examines changes in emotions over the course of a composition, for example, every 0.5 or 1 s. Dynamic MER task was conducted by MediaEval Benchmarking Initiative for Multimedia Evaluation, the results of which were presented by Aljanaki et al. [1].

Long-short term memory recurrent neural networks were used in dynamic MER task by Coutinho et al. [5]. Low-level acoustic descriptors extracted using openSMILE and psychoacoustic features extracted with the MIR Toolbox were used as input data. A multi-variate regression using by deep recurrent neural networks was used to model the time-varying emotions (arousal, valence) of a musical piece [15]. In this work, a set of acoustic features extracted from segments of 1 s length were used. Delbouy et al., in [6], used mel-spectrogram from audio and embedded lyrics as input vectors to the convolutional and LSTM networks. Chowdhury et al., in [4], used VGG-style convolutional neural networks to detect 8 emotional characteristics (happy, sad, tender, fearful, angry, valence, energy, tension). For network training perceptual mid-level features (melodiousness, articulation, rhythmic stability, rhythmic complexity, dissonance, tonal stability, modality) were used, and spectograms from audio signals were used as input vector for neural networks.

What distinguishes this work from others is that it uses a different segment length (6 s) than the standard static MER, as well as proposes a method of preparing data for recurrent neural networks, which it tests with various low and mid-level features. Due to the fact that the studied segment is relatively short, a solution of using a sliding window also allows to study changes in emotions throughout the entire composition, i.e. similar to dynamic MER. This paper presents results in relation to previously conducted experiments [9].

The rest of this paper is organized as follows. Section 2 describes the music data set and the emotion model used in the conducted experiments. Section 3 presents the tools used for feature extraction and preparation of data before building the models. Section 4 describes the details of the built recurrent neural networks. Section 5 presents the results obtained while building the models. Finally, Sect. 6 summarizes the main findings.

2   Music Data

A well-prepared database of learning examples affects the results and the correctness of the created models predicting emotions. The advantages of the obtained database are well-distributed examples on the emotion plane as well as congruity between the music experts' annotations.

The data set consisted of 324 six-second fragments of different genres of music: classical, jazz, blues, country, disco, hip-hop, metal, pop, reggae, and rock. The tracks were all 22050 Hz mono 16-bit audio files in .wav format. The training data were taken from the generally accessible data collection project Marsyas[1]. After the selection of samples, the author shortened them to the first 6 seconds, which is the shortest possible length at which experts could detect emotions for a given segment. Bachorik et al. [2] investigated the length of time required for participants to initiate emotional responses to musical samples.

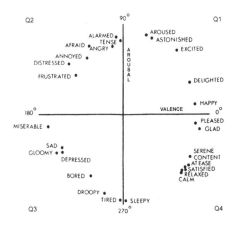

Fig. 1. Russell's circumplex model [13]

Data annotation was done by five music experts with a university musical education. The musical education of the experts, people who deal with the creation and analysis of emotions in music on a daily basis, enables to trust the quality of their annotations. Each annotator annotated all records in the data set - 324 six-second fragments. Each music expert had heard all the examples in the database. As a result during the annotation each annotator was able to see all the shades of emotions in music, which is not always the case in databases with the emotions determined. This had a positive effect on the quality of the received data, which was emphasized by Aljanaki et al. in [1].

During annotation of music samples, we used the two-dimensional arousal-valence Russell's model (Fig. 1) to measure emotions in music, which consists of two independent dimensions of arousal (vertical axis) and valence (horizontal axis). Each music expert making annotations after listening to a music sample had to specify values on the arousal and valence axes in a range from −10 to 10.

Value determination on the arousal-valence axes (A-V) was clear with a designation of a point on the A-V plane corresponding to the musical fragment. The data collected from the five music experts was averaged. After annotation, the amount of examples obtained in the quarters on the A-V emotion plane were: Q1

[1] http://marsyas.info/downloads/datasets.html.

(A:high, V:high): 93; Q2 (A:high, V:low): 70; Q3 (A:low, V:low): 80; Q4 (A:low, V:high): 81.

A well-prepared database, i.e. one suitable for independent regressors predicting valence and arousal, should contain examples where the values of valence and arousal are not correlated. To check if valence and arousal dimensions are correlated in our music data, the Pearson correlation coefficient was used. The obtained value of $r = -0.03$ (i.e. close to zero) indicates that arousal and valence values are not correlated and the music data are a well spread in the quarters on the A-V emotion plane.

All examples in the database were marked by 5 music experts and their annotations had good agreement levels. A good level of mutual consistency was achieved, represented by Cronbach's α calculated for the annotations of arousal ($\alpha = 0.98$) and valence ($\alpha = 0.90$). We can see that the experts' annotations for the arousal value show slightly greater agreement than for the valence value, which is in line with the natural perception of emotions by humans [1]. Details on creating the music data were presented in a previous paper [10].

3   Feature Extraction

3.1   Tools for Feature Extraction

For feature extraction, tools for audio analysis and audio-based music information retrieval, Essentia [3] and Marsyas [14], were used. Marsyas software, written by George Tzanetakis, has the ability to analyze music files and to output the extracted features. The tool enables the extraction of the following features: Zero Crossings, Spectral Centroid, Spectral Flux, Spectral Rolloff, Mel-Frequency Cepstral Coefficients (mfcc), and chroma features - 31 features in total. For each of these basic features, Marsyas calculates four statistic features. The feature vector length obtained from Marsyas was 124.

Essentia is an open-source library, created at the Music Technology Group, Universitat Pompeu Fabra, Barcelona. In the Essentia package, we can find a number of executable extractors computing music descriptors for an audio track: spectral, time-domain, rhythmic, and tonal descriptors. Extracted features by Essentia are divided into three groups: low-level, rhythm, and tonal features. A full list of features is available on the web site[2]. Essentia also calculates many statistic features: the mean, geometric mean, power mean, median of an array, and all its moments up to the 5th-order, its energy, and the root mean square (RMS). The feature vector length obtained from Essentia was 529.

3.2   Preparing Data for RNN

Recurrent neural networks process sequential data and find relationships between the input data sequences and the expected output value. To be able to train the recurrent neural network, it is necessary to enter sequences of the feature vectors.

---

[2] http://essentia.upf.edu/documentation/algorithms_reference.html.

In this paper, to extract correlations with time in the studied music fragments, they were segmented into smaller successive sections. The process of dividing a fragment of music (6 s) into smaller segments of a certain length t (1, 2 or 3 s) and overlap (0 or 50%) is shown in Fig. 2. To split the wav file, the *sfplay.exe* tool from Marsyas toolkit was used. From the created smaller segments of music, feature vectors were extracted, which were used to build a sequence of learning vectors for the neural network. A program was written that allows to select the segmentation option for a music fragment, performs feature extraction, and prepares data to be loaded to a neural network.

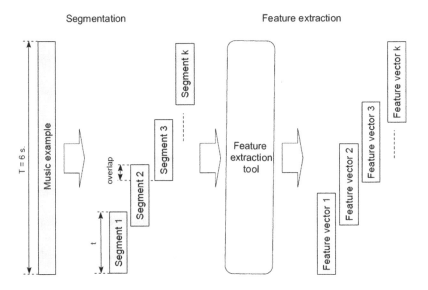

Fig. 2. Creating training data sequences for RNN

4   Recurrent Neural Networks

Long short-term memory (LSTM) units, which were defined in [7], were used to build recurrent networks. LSTM units are special kinds of memory blocks that solve the vanishing gradient problem occurring with simple RNN units. Each LSTM unit consists of a self-connected memory cell and three multiplicative regulators - input, output, and forget gates. Gates provide LSTM cells with write, read, and reset operations, which allows the LSTM unit to store and access information contained in a data sequence that corresponds to data distributed over time. The weights of connections in LSTM units need to be learned during training.

4.1 Implementation of RNN

The WekaDeeplearning4j package [11], which was included with the Weka program [16], was used to conduct the experiments with recurrent neural networks. This package makes deep learning accessible through a graphical user interface. The WekaDeeplearning4j module is based on Deeplearning4j[3], which is a widely used open-source machine learning workbench implemented in Java. Weka with WekaDeeplearning4j package enables users to perform experiments by loading data in the Attribute-Relation File Format (ARFF), configuring a neural network, and running the experiment. To predict emotions in music files, a neural network was proposed with the structure shown in Fig. 3. Input data were given to the network in the form of a sequence set of features, and then processed by a layer consisting of LSTM units (LSTM1–LSTMn). The next layer built of densely connected neurons (1–n) converted the signals received from the LSTM layer and created an output signal.

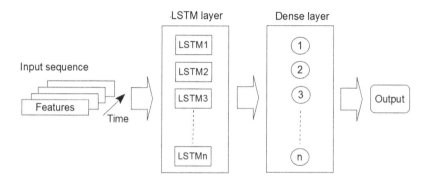

Fig. 3. Recurrent neural network architecture

4.2 Parameters of the RNN

The structure of the neural network was built once with one LSTM layer, once with two layers, and with different amounts of LSTM units (124, 248). A tanh activation function was used for LSTM units. For our regression task (prediction of continuous values of arousal and valence), the identity activation function for a dense layer was used, in conjunction with the mean squared error loss function. For weight initialization, the Xavier method was used. Stochastic gradient descent was used as a learning algorithm with Nesterov updater, which helped to optimize the learning rate. The network was trained with 100 epochs and to avoid overfitting an early stopping strategy was used. The training process was stopped as soon as the loss did not improve anymore for 10 epochs. The loss was evaluated on a validation set (20% of the training data).

---

[3] https://deeplearning4j.org.

5 Experiments and Results

During the conducted experiments, regressors for predicting arousal and valence were built. As baseline for comparing the results of the obtained regressors a simple linear regression model (lr) was chosen. The data were also tested with a fairly good SMOreg algorithm with polynomial kernel, which is an implementation of the support vector machine for regression. The author also tested the usefulness of SMOreg in a previous paper [9]. Both algorithms (SMOreg, lr) were tested on the same music fragments as the neural networks but on non-segmented fragments.

The performance of regression algorithms were evaluated using the 10-fold cross validation technique (CV-10). The coefficient of determination (R^2) and mean absolute error (MAE) were used to assess model efficiency. Before constructing regressors arousal and valence annotations were scaled between $[-0.5, 0.5]$, thus the MAE value also corresponds to the average error percentage. Before providing input data to the neural network, the data was standardized to zero mean and unit variance.

Regressors were built using recurrent neural network RNN (RnnSequence-Classifier [11]) and were tested in 4 variants: RNN1 - 1 layer × 124 LSTM units; RNN2 - 1 layer × 248 LSTM units; RNN3 - 2 layers × 124 LSTM units each; RNN4 - 2 layers × 248 LSTM units each.

5.1 RNN with Marsyas Features

During the testing of RNN efficiency, features obtained from Marsyas tool were divided into 3 sets: (1) all Marsyas features - 124; (2) mfcc features - 13 Mel Frequency Cepstral Coefficients × 4 statistic - 52; (3) chroma features - 54.

Table 1 presents the coefficient of determination (R^2) and mean absolute error (MAE) obtained during building regressors using mfcc and chroma features. The best results for each regressor type (arousal, valence) are marked in bold. From the obtained results, we can see that the usefulness of the chroma features is small compared with the mfcc features. The results for mfcc features far outweigh those for chroma features.

Table 2 presents the results for all Marsyas features. The simple linear regression model and support vector machine for regression (SMOreg) was outperformed by the RNN models, in two cases RNN3, RNN4 for arousal and valence. The best results were obtained with RNN4 (2 layers × 248 LSTM): $R^2 = 0.67$ and $MAE = 0.12$ for arousal, $R^2 = 0.17$ and $MAE = 0.15$ for valence. We see that RNN with two LSTM layers gives better results for both arousal as well as valence. As expected, the results show that the sequential modeling capabilities of the RNN are useful for this task.

The use of all features gives the best results; however, in the case of arousal, the set of mfcc features gives quite comparable results, similar to the whole set of features ($R^2 = 0.66$ and $MAE = 0.12$, Table 1). The best results were obtained at a segment length of 2 s and without overlap, and those are presented here.

Table 1. Results obtained for mfcc and chroma features

	Mfcc features				Chroma features			
	Arousal		Valence		Arousal		Valence	
	R^2	MEA	R^2	MEA	R^2	MEA	R^2	MEA
Linear regression	0.61	0.13	0.07	0.17	0.40	0.17	0.02	0.18
SMOreg	0.60	0.13	0.10	0.17	0.43	0.16	**0.04**	**0.17**
RNN1	0.58	0.14	0.14	0.17	0.43	0.17	0.02	0.18
RNN2	0.61	0.13	0.13	0.16	0.41	0.17	0.02	0.18
RNN3	**0.66**	**0.12**	0.11	0.16	**0.48**	**0.15**	0.03	0.17
RNN4	0.64	0.12	**0.14**	**0.15**	0.46	0.16	0.02	0.18

Table 2. Results obtained for all Marsyas features

	Arousal		Valence	
	R^2	MAE	R^2	MAE
Linear regression	0.56	0.14	0.13	0.17
SMOreg	0.62	0.13	0.15	0.16
RNN1	0.58	0.14	0.12	0.17
RNN2	0.62	0.13	0.12	0.17
RNN3	0.66	0.12	0.16	0.15
RNN4	**0.67**	**0.12**	**0.17**	**0.15**

5.2 RNN with Essentia Features

Experiments with the features obtained from the Essentia package were also conducted. These features include the mfcc and chroma features, which are also in the Marsyas tool, but also contain many higher-level features such as rhythm or harmony. Table 3 shows the results of the experiments. The experiments were expanded by two networks with an increased number of LSTM units, similar to the number of features in the sequence: RNN5 - 1 layer × 529 LSTM units, RNN6 - 2 layers × 529 LSTM units each.

From the obtained results (Table 3) for the Essentia features set, we can see a significant improvement of the results compared with the database algorithms (RNN1-RNN6 for arousal, RNN2-RNN6 for valence). Better features from the Essentia toolkit give better neural network results. The best results were obtained with RNN4: $R^2 = 0.69$ and $MAE = 0.11$ for arousal, $R^2 = 0.40$ and $MAE = 0.13$ for valence. The improvement is also significant for regressors for valence, compared with the results from Marsyas features (Table 2), where the best result was: $R^2 = 0.17$ and $MAE = 0.15$.

In regard to the different numbers of layers and LSTM units, the best results were obtained using the RNN4 network (2 layers × 248 LSTM) for both arousal and valence. Two-layer networks recognized emotions better than one-layer

Table 3. Results obtained for Essentia features

	Arousal		Valence	
	R^2	MAE	R^2	MAE
Linear regression	0.07	0.25	0.06	0.19
SMOreg	0.48	0.18	0.27	0.17
RNN1	0.54	0.14	0.21	0.16
RNN2	0.58	0.14	0.32	0.14
RNN3	0.67	0.12	0.32	0.13
RNN4	**0.69**	**0.11**	**0.40**	**0.13**
RNN5	0.61	0.13	0.29	0.15
RNN6	0.68	0.12	0.36	0.14

networks. What is quite interesting in the case of arousal ($R^2 = 0.69$, $MAE = 0.11$), the results are comparable with the results obtained from the Marsyas package ($R^2 = 0.67$, $MAE = 0.12$, Table 2). Mfcc features are quite good for detecting arousal, and adding new features improved the results only slightly. A significant result of these experiments is that features from the Essentia package, like rhythm and tonal features, significantly improved the detection of valence. In the case of arousal, it is not necessary to use such a rich set of features, which is why the model for arousal is not so complex.

6   Conclusions

This article presents experiments using recurrent neural networks for emotion detection for musical segments. The sequential possibilities of the models turned out to be very useful for this type of task as the obtained results exceeded such algorithms as support vector machine for regression, not to mention the weaker linear regression. In all the built models, the accuracy of arousal prediction exceeded the accuracy of valence prediction. There was more difficulty detecting emotions on the valence axis than arousal. Similar difficulties were noted when music experts were annotating files, which was confirmed during annotation compliance testing.

It is significant that the use of higher-level features (features from Essentia tool) had a very positive effect on the models, especially the accuracy of valence regressors. Interestingly, to predict arousal, even a small set of features (mfcc from Marsyas tool) provided quite good results, similar to those of the large features set from Essentia. Low-level features, like mfcc, are generally sufficient for predicting arousal. In the future, feature selection for the Essentia set could be made and the most useful features chosen, despite that testing them on recurrent neural networks can be very time-consuming.

Acknowledgments. This research was realized as part of study no. KSIiSK in the Bialystok University of Technology and financed with funds from the Ministry of Science and Higher Education.

References

1. Aljanaki, A., Yang, Y.H., Soleymani, M.: Developing a benchmark for emotional analysis of music. PLoS One **12**(3), e0173392 (2017)
2. Bachorik, J., Bangert, M., Loui, P., Larke, K., Berger, J., Rowe, R., Schlaug, G.: Emotion in motion: investigating the time-course of emotional judgments of musical stimuli. Music Percept. **26**, 355–364 (2009)
3. Bogdanov, D., Wack, N., Gómez, E., Gulati, S., Herrera, P., Mayor, O., Roma, G., Salamon, J., Zapata, J., Serra, X.: ESSENTIA: an audio analysis library for music information retrieval. In: Proceedings of the 14th International Society for Music Information Retrieval Conference, pp. 493–498 (2013)
4. Chowdhury, S., Portabella, A.V., Haunschmid, V., Widmer, G.: Towards explainable music emotion recognition: the route via mid-level features. In: Proceedings of the 20th International Society for Music Information Retrieval Conference, ISMIR 2019, Delft, The Netherlands, pp. 237–243 (2019)
5. Coutinho, E., Trigeorgis, G., Zafeiriou, S., Schuller, B.: Automatically estimating emotion in music with deep long-short term memory recurrent neural networks. In: Working Notes Proceedings of the MediaEval 2015 Workshop, Wurzen, Germany (2015)
6. Delbouys, R., Hennequin, R., Piccoli, F., Royo-Letelier, J., Moussallam, M.: Music mood detection based on audio and lyrics with deep neural net. In: Proceedings of the 19th International Society for Music Information Retrieval Conference (ISMIR 2018), Paris, France, pp. 370–375 (2018)
7. Gers, F.A., Schmidhuber, J., Cummins, F.A.: Learning to forget: continual prediction with LSTM. Neural Comput. **12**, 2451–2471 (2000)
8. Grekow, J.: Audio features dedicated to the detection of four basic emotions. In: Saeed, K., Homenda, W. (eds.) CISIM 2015. LNCS, vol. 9339, pp. 583–591. Springer, Cham (2015). https://doi.org/10.1007/978-3-319-24369-6_49
9. Grekow, J.: Music emotion maps in arousal-valence space. In: Saeed, K., Homenda, W. (eds.) CISIM 2016. LNCS, vol. 9842, pp. 697–706. Springer, Cham (2016). https://doi.org/10.1007/978-3-319-45378-1_60
10. Grekow, J.: Human annotation. In: From Content-Based Music Emotion Recognition to Emotion Maps of Musical Pieces. SCI, vol. 747, pp. 13–24. Springer, Cham (2018). https://doi.org/10.1007/978-3-319-70609-2
11. Lang, S., Bravo-Marquez, F., Beckham, C., Hall, M., Frank, E.: WekaDeeplearning4j: a deep learning package for Weka based on Deeplearning4j. Knowl.-Based Syst. **178**, 48–50 (2019)
12. Lu, L., Liu, D., Zhang, H.J.: Automatic mood detection and tracking of music audio signals. Trans. Audio Speech Lang. Proc. **14**(1), 5–18 (2006)
13. Russell, J.A.: A circumplex model of affect. J. Pers. Soc. Psychol. **39**(6), 1161–1178 (1980)
14. Tzanetakis, G., Cook, P.: Marsyas: a framework for audio analysis. Org. Sound **4**(3), 169–175 (2000)

15. Weninger, F., Eyben, F., Schuller, B.: On-line continuous-time music mood regression with deep recurrent neural networks. In: 2014 IEEE International Conference on Acoustics, Speech and Signal Processing (ICASSP), pp. 5412–5416 (2014)
16. Witten, I.H., Frank, E., Hall, M.A., Pal, C.J.: Data Mining: Practical Machine Learning Tools and Techniques, 4th edn. Morgan Kaufmann Publishers Inc., San Francisco (2016)

Saliency Detection in Hyperspectral Images Using Autoencoder-Based Data Reconstruction

Annalisa Appice[1,3]([✉]), Francesco Lomuscio[1], Antonella Falini[1],
Cristiano Tamborrino[2], Francesca Mazzia[1], and Donato Malerba[1,3]

[1] Dipartimento di Informatica, Università degli Studi di Bari Aldo Moro,
via Orabona, 4, 70126 Bari, Italy
{annalisa.appice,antonella.falini,ncesca.mazzia,
donato.malerba}@uniba.it,
f.lomuscio7@studenti.uniba.it
[2] Dipartimento di Matematica, Università degli Studi di Bari Aldo Moro,
via Orabona, 4, 70126 Bari, Italy
cristiano.tamborrino@uniba.it
[3] Consorzio Interuniversitario Nazionale per l'Informatica - CINI, Bari, Italy

Abstract. Saliency detection extracts objects attractive to a human vision system from an image. Although saliency detection methodologies were originally investigated on RGB color images, recent developments in imaging technologies have aroused the interest in saliency detection methodologies for data captured with high spectral resolution using multispectral and hyperspectral imaging (MSI/HSI) sensors. In this paper, we propose a saliency detection methodology that elaborates HSI data reconstructed through an autoencoder architecture. It resorts to (spectral-spatial) distance measures to quantify the salience degree in the data represented through the autoencoder. Finally, it performs a clustering stage in order to separate the salient information from the background. The effectiveness of the proposed methodology is evaluated with benchmark HSI and MSI data.

1 Introduction

Saliency detection has been inspired by the natural visual attention mechanism that identifies an object to be salient when it attracts visual attention more than anything else, i.e., the background. It has been demonstrated that the human vision system is prone to pay more attention to objects having higher contrast with their surroundings [9]. This consideration has paved the way for one of the earliest methodology, the Itti's method [10], that is based on the construction of center-surround contrast models on color images. These models are constructed by computing the Euclidean distance between the considered pixels and their surrounding ones in the color space. While simple, Itti's method has inspired several approaches investigated in the saliency investigation field for RGB images (see [3] for a recent survey). Recent models have also combined Itti's model

© Springer Nature Switzerland AG 2020
D. Helic et al. (Eds.): ISMIS 2020, LNAI 12117, pp. 161–170, 2020.
https://doi.org/10.1007/978-3-030-59491-6_15

with a clustering stage [11] proving that clustering may help in the extraction of salient information. On the other side, in recent years, it has definitely emerged that many applications in various fields (e.g. weather forecasting, military intelligence) may take advantage of the elaboration of MultiSpectral Imaging (MSI) and HyperSpectral Imaging (HSI) data. Using multispectral/hyperspectral sensors, imagery data can be captured with a wide spectrum of light instead of just assigning primary colors (red, green, blue) to each pixel.[1] In general, abundant spectral information, going above and beyond human vision, captures more Earth's features than standard RGB images [2].

Following the growing amount of HSI data collections, increasing attention has been recently devoted to saliency detection in HSI [8,9,13,19]. However, due to the high dimensionality of HSI data, traditional computer vision methods based on gray-scale or color image analysis cannot be directly applied to HSI problems. On the other hand, witnessing the great success of salience estimation models that, similarly to the Itti's method, work in a low dimensional space, a natural research direction consists of combining a saliency detection methodology with a dimensionality reduction technique, in order to handle the abundant spectral information of HSI data [9]. Following this direction, we propose a HSI methodology—AISA (Autoencoding of Hyperspectral Imagery data for Saliency Analysys)—that cascades an autoencoder stage and a clustering stage. The autoencoder stage learns a non-linear encoding and decoding of HSI data. The clustering stage elaborates the autoencoder information, also engineered on surrounding pixels, to separate the salient object from the background, working in a lower dimensional space.

The remainder of this paper is organized as follows. The motivation and contributions are discussed in Sect. 2. The proposed methodology is illustrated in Sect. 3. Section 4 describes the results of the experimental evaluation. Finally, Sect. 5 draws the conclusions and future developments.

2 Motivation and Contributions

The autoencoder is an unsupervised deep neural network that can learn a codification of high dimensional input data so that the decodification resembles the input data as closely as possible [5]. As the codification is commonly used to obtain reduced dimensionality, the encoder representation of HSI data allows us to tackle "the curse of dimensionality" of a high spectrum and to perform data denoising. Recent studies have already proved that autoencoders can learn very interesting HSI data representations compared to other dimensionality reduction techniques [17]. Based on these promising results, autoencoders have been

---

[1] The main difference between multispectral and hyperspectral is the number of bands and how narrow the bands are. MSI technology commonly refers to a small amount of bands, i.e., from 3 to 10, sensed by a radiometer. HSI technology could have hundreds or thousands bands from a spectrometer. In this paper, we generally refer to HSI data defining a methodology that is then evaluated in both HSI and MSI scenario.

recently considered for the dimensionality reduction in HSI, for tasks of both change detection [1] and classification [20]. In this paper, we explore the use of autoencoders in HSI saliency detection tasks.

The rationale of our proposal is that clustering can be used to separate salient pixels, grouped in one cluster, from background pixels, grouped in another cluster. Clustering is performed on the encoder representation of the HSI data. In fact, the encoder should reduce the dimensionality of the spectrum keeping the hidden interactions and dependencies of the input, without loosing relevant information and rejecting noise, as well as unnecessary redundancies [6]. On the other hand, we consider that the classification ability of the cluster algorithm may be improved by making the encoder "aware" on the reconstruction data. To this purpose, we compute the distance between the input HSI data and the data reconstructed in the same input space through the encoding-decoding architecture learned via the autoencoder. The use of the distance is a reasonable choice as the encoding and decoding process should behave similarly on pixels belonging to the same cluster. Consequently, a clustering stage, performed by also accounting for the distance between HSI data and reconstructed data, should contribute to correctly group the salient pixels in the same cluster. We integrate the spatial information into the autoencoder framework by computing a spectral-spatial distance feature that, following the main Itti's idea, aids the identification of salient objects by contrast.

In short, the contributions of our paper are: (1) the use of autoencoder information, derived with both the encoder and the decoder level, in a HSI saliency detection task; (2) the analysis of spectral and spectral-spatial distances, coupled with autoencoder and clustering, for feature engineering; (3) an empirical evaluation of the effectiveness of the proposed methodology using both HSI and MSI data.

3   Methodology

AISA takes an hyperspectral image $\mathbf{Z}$ as input. This image is a scene acquired at a specific time, represented as a tensor of size $m \times n \times K$. In particular, the scene is described by $m \times n$ pixels and K spectral bands. A pixel denotes an area of about a few square meters of the Earth's surface—it is a function of the sensor's spatial resolution—which is unequivocally referenced with spatial coordinates (x, y) with $1 \leq x \leq m$ and $1 \leq y \leq n$, according to the usual matrix representation. Every spectral band is a numeric feature proportional to the surface reflectance. The salience map, given as output by our algorithm, is a binary $m \times n$ matrix, which assigns a binary label ("salient"—1— and "no-salient"—0) to each pixel (x, y) of the scene. Based on these premises, the learning process performed by AISA is, in principle, divided into three stages, (1) training an autoencoder architecture with $\mathbf{Z}$; (2) engineering a new feature space from the autoencoder; (3) performing clustering in the engineered feature space, in order to separate pixels in two clusters (salient cluster vs background cluster).

Stage 1 – Autoencoder. The autoencoder is computed by resorting to an encoder-decoder architecture that is used to map each input $\mathbf{z}$ onto the encoding $\mathbf{h}$ via an encoder network, so that $\mathbf{h} = \sigma(\mathbf{Wz} + \mathbf{b})$, where σ is an element-wise activation function, $\mathbf{W}$ is a weight matrix and $\mathbf{b}$ is a bias vector. Weights and biases are initialized randomly and then updated iteratively during training through backpropagation. The encoding is in turn mapped to the reconstruction $\mathbf{z}'$ by means of a decoder network, so that $\mathbf{z}' = \sigma'(\mathbf{W}'\mathbf{h} + \mathbf{b}')$. In particular, the auto-encoder is trained to minimize the loss through a feedforward neural network that reproduces the input data on the output layer. Both $\mathbf{z}$ and $\mathbf{z}'$ have the same dimension, while the autoencoder has as many layers as needed, placed symmetrically in the encoder and decoder. Every unit located at any of the hidden layers receives several inputs from the preceding layer. The unit computes the weighted sum of these inputs and applies the activation function to produce the output.

In this study, the input of the autoencoder comprises the $m \times n$ pixels spanned on the K spectral bands. We adopt ADAM, an adaptive learning rate optimization algorithm, in order to produce optimal weights and biases by minimizing the mean square error [12]. The encoder consists of six layers with K, $K/2$, $K/3$, $K/4$, $K/5$ and 1 neuron(s), respectively. The decoder starts in the bottleneck layer (the output layer of the encoder) and maps the bottleneck signals back to the input space through five layers with $K/5$, $K/4$, $K/3$, $K/2$ and K neurons, respectively. In addition, we use the sigmoid activation function that is commonly used in binary classification problems. Finally, we set the maximum number of epochs equal to 150 and the learning rate equal to 0.002.

Stage 2 – Feature Engineering. Let us consider the autoencoder trained in stage 1. For each pixel (x, y), let $\mathbf{z}_{xy}$ denote the spectral data acquired on pixel (x, y), $\mathbf{h}_{xy}$ be the representation of $\mathbf{z}_{xy}$ on the bottleneck layer of the autoencoder and $\mathbf{z}'_{xy}$ denote the reconstruction of $\mathbf{z}_{xy}$ computed through the encoder and the decoder. Then we use this information to build a bi-variate representation of the image. In particular, we span every imagery pixel on two features, that is:

1. the encoder feature H that expresses the input data as they are represented at the bottleneck layer of the autoencoder;
2. the decoder feature D that is the distance computed between $\mathbf{z}_{xy}$ and $\mathbf{z}'_{xy}$.

We evaluate the performance of four distance algorithms to build D, that is, Spectral Angle Mapper (SAM), Z-score (ZID), Z-score correction of Spectral Angle Mapper (SAMZID) and Spectral-Spatial Cross Correlation based Distance (SSCCD). SAM, ZID and SAMZID are spectral distance algorithms often used for material identification. They are simple and fast to compute [7]. SAM is independent of the number of spectral bands and insensitive to sunlight [15]. SAMZID minimizes the effect of noise and atmospheric corrections [7]. SSCCD couples the spectral information to the spatial arrangement of the pixels. As discussed in [18], SSCCD is helpful to overcome possible radiometric and dynamic range differences by accounting for the spatial information.

Formally, $SAM(x, y)$ measures the angle between $\mathbf{z}_{xy}$ and $\mathbf{z}'_{xy}$, that is:

$$SAM(x, y) = \arccos \left(\frac{\mathbf{z}_{xy} \cdot \mathbf{z}'_{xy}}{\|\mathbf{z}_{xy}\| \|\mathbf{z}'_{xy}\|} \right), \tag{1}$$

where the "·" denotes the scalar product.
$ZID(x, y)$ is computed as follows:

$$ZID(x, y) = \sum_{k=1}^{K} \left(\frac{(\mathbf{z}_{xy}[k] - \mathbf{z}'_{xy}[k]) - \mu_k}{\sigma_k} \right)^2, \tag{2}$$

where k represents the k-th spectral band so that $\mathbf{z}_{xy}[k] - \mathbf{z}'_{xy}[k]$ denotes the spectral divergence computed by applying the difference operator to the k-th band of both $\mathbf{z}_{xy}$ and $\mathbf{z}'_{xy}$. The symbols μ_k and σ_k represent the mean and standard deviation of the spectral information divergence computed on the k-th spectral band.

The distance $SAMZID$ is computed by combining SAM and ZID through element-wise multiplication and a trigonometric operator. Formally,

$$SAMZID(\mathbf{Z}) = [scale\,(\sin SAM(\mathbf{Z}))] \times [scale\,(\,ZID(\mathbf{Z}))], \tag{3}$$

where $scale()$ is the scale function. In Eq. 3, the combined distance components are scaled to the interval $[0, 1]$.

Finally, the distance algorithm SSCCD is the distance formulation of the spectral-spatial cross correlation measure (SSCC) illustrated in [18]. This correlation metric compares spectral bands of both $\mathbf{z}$ and $\mathbf{z}'$ pairwise by accounting for the spatial arrangement of the spectrum in windows centered at the considered pixel (x, y). Formally, $SSCC(x, y)$ is computed as follows:

$$SSCC(x, y) = \frac{\displaystyle\sum_{(x', y') \in W(x, y)} [\mathbf{z}_{x'y'} - \overline{\mathbf{z}}_{W(x, y)}]^T [\mathbf{z}'_{x', y'} - \overline{\mathbf{z}}'_{W(x, y)}]}{\sqrt{\displaystyle\sum_{(x'y') \in W(x, y)} \|\mathbf{z}_{x', y'} - \overline{\mathbf{z}}_{W(x, y)}\|^2} \sqrt{\displaystyle\sum_{(x'y') \in W(x, y)} \|\mathbf{z}'_{x'y'} - \overline{\mathbf{z}}'_{W(x, y)}\|^2}}, \tag{4}$$

where $\overline{\mathbf{z}}_{W(x, y)}$ and $\overline{\mathbf{z}}'_{W(x, y)}$ denote the mean vectors of the spectral vectors determined band-by-band on all the pixels of $W(x, y)$ in $\mathbf{Z}$ and $\mathbf{Z}'$, respectively. In this study, $W(x, y) = \{(x + I, y + J) | I = \pm 2, J = \pm 2\}$. The distance measure $SSCCD(x, y)$ can be derived from $SSCC(x, y)$ as:

$$SSCCD(x, y) = 1 - scale(SSCC(x, y)), \tag{5}$$

where $scale()$ scales SSCC values in the range $[0, 1]$.

Stage 3 – Clustering. As a clustering algorithm, we consider the popular Gaussian Mixture Model (GMM) algorithm in the version described in [21]. Specifically, we train the estimation network, described in [21], on the bi-variate input constructed at stage 2 (i.e. the input that comprises the imagery data spanned on both the encoder feature H and the distance feature D). This network estimates a probability density function under the GMM framework and predicts to which Gaussian distribution each pixel belongs to. In particular, the estimation network estimates the parameters of the GMM (mixture-component distribution, mixture means and mixture covariance) and evaluates the likelihood/energy of samples by utilizing a multi-layer neural network, in order to predict the mixture membership for each sample. We consider the estimation network as it has been implemented at https://github.com/danieltan07/dagmm. Similarly to [21], we use a threshold to label the samples of high energy as salient pixels (we consider the salient pixels as anomalies with respect to the background). However, differently from [21], where a pre-chosen threshold is considered, we use the Otsu's method [14], in order to automatically determine this threshold. The Otsu's method is a simple algorithm that returns a single intensity threshold to separate pixels into two classes—foreground and background. In particular, this threshold is determined by minimizing the intra-class intensity variance or, equivalently, by maximizing the inter-class variance. In this paper, we use the implementation of threshold_otsu from skimage.filters.

Final considerations concern the fact that the brute application of the described clustering stage may occasionally yield spurious, isolated assignments of pixels to clusters. To avoid this issue, we apply the principle of local autocorrelation congruence of objects [16]—detected clusters are generally expanding over contiguous areas. Based on this principle, we change the assignment of pixels that strongly disagree with the surrounding assignments. This corresponds to perform a spatial-aware correction of the original cluster assignments. Based upon this correction, each pixel may be re-assigned to the cluster (and to its label) that originally groups the most part of its neighboring pixels.

4   Experimental Study

AISA is written in Python 3.7 with Pytorch to build autoencoders. To evaluate AISA, we consider HS-SOD dataset [9]—a collection of 60 hyperspectral images with their respective ground-truth binary saliency images.[2] Several competitors have been already evaluated using these data. We also consider two multispectral images of Madrid collected with Sentinel-2A satellite.

4.1   HS-SOD Data

The images have been collected by accounting for several aspects during the data collection (e.g. variation in object size, number of objects, foreground-background contrast, object position on the image). Data have been acquired with a NH-AIK model hyperspectral camera from fifty scenes at the public parks of Tokyo Waterfront City in Japan in several days between August and September 2017. Spectral bands are collected in the visible spectrum (380–780 nm)

---

[2] The dataset is available at https://github.com/gistairc/HS-SOD.

Table 1. AISA: mean ± standard deviation of AUC-Borji computed on 60 images of HS-SOD by varying the distance among SAM, ZID, SAMZID and SSCCD (column 1); number of images of HS-SOD where AUC-Borji is the highest per distance (column 2).

Distance	AUC-Borji (mean ± stdev)	Count
SAM	0.7152 ±0.1258	9
ZID	0.7057 ±0.1267	12
SAMZID	0.6781 ± 0.1377	6
SSCCD	0.7540 ±0.1244	33
Best distance	0.7727 ±0.1083	–

(a) GT   (b) SAM   (c) ZID   (d) SAMZID   (e) SSCCD

Fig. 1. AISA output on image id = 26 of HS-SOD dataset: the binary ground truth–GT (Fig. 1(a)), the salience maps computed by AISA with SAM (AUC-Borji = 0.8414, Fig. 1(b)), ZID (AUC-Borji = 0.8399, Fig. 1(c)), SAMZID (AUC-Borji = 0.7716, Fig. 1(d)), SSCCD (, AUC-Borji = 0.8509, Fig. 1(e)).

leaving 81 spectral bands. Each image consists of 1024×768 pixels. The saliency ground truth is available for these images. For the quantitative evaluation of the saliency detection performances, we compute the Area Under Curve (AUC) metric. We consider the AUC implementation described in [4] (AUC-Borji) that is commonly used in saliency detection method evaluations [9].

We start this analysis by evaluating the performance of AISA with respect to the distance measure computed in the feature engineering stage. For each distance, the mean and standard deviation of AUC-Borji computed on 60 images is reported in column 1 of Table 1, while the number of times each distance achieves the highest AUC-Borji is reported in column 2 of Table 1. Figure 1 reports an example of saliency maps built by varying the distance. These results highlight that the highest overall performance is achieved by using the SSCCD distance that accounts for the spatial arrangement of the data besides the spectral information. In any case, the accuracy analysis, performed image per image, also shows that although SSCCD achieves the highest AUC-Borji in 33 out of 60 images, SAM, ZID or SAMZID outperform SSCCD in the remaining images. This result suggests that a future development of this research consists in exploring if any dependence exists between the best distance and the image characteristics (e.g. camera settings such as exposure time and gain values, weather condition or salient object material, which may change in the considered images).

We complete the analysis by comparing AISA to competitors reported in [9]. Competitors include: Itti's method computed on spectral data and its variants checking the spectral distances between each spatial region for saliency

Table 2. Competitor analysis. Results of competitors are collected in [9]

Method	AUC-Borji
AISA	0.7760
Itti et al.	0.7694
SED	0.6415
SAD	0.7521
GS	0.7597
SED-OCM-GS	0.7863
SED-OCM-SAD	0.8008
SGC	0.8205

computation by using spectral Euclidean distance (SED) and spectral Angle distances (SAD); the method computing the saliency on the spectral bands divided in groups and calculating the Euclidean distance as color opponency between groups (GS); the methods using the orientation based salient features (OCM) in the combinations SED-OCM-GS and SED-OCM-SAD; the method using the spectral gradient contrast (SGC) in combination with super-pixels extracted by considering both spatial and spectral gradients. Results, reported in Table 2, show that AISA outperforms Itti et al. method, SED, SAD and GS, while it is outperformed by methods including orientation features (SED-OCM-GS and SED-OCM-SAM) and/or super-pixels (SGC). This result suggests that new accuracy can be gained in AISA by augmenting the feature space of the clustering stage with orientation features and/or adding super-pixel in the learning stage. In particular, super-pixels, automatically determined based on spatial-spectral arrangement of data, may aid in revealing higher-level saliency patterns that may significantly help in denoising the data and neglecting details.

4.2 Madrid Data

The Sentinels are a fleet of satellites designed to deliver the wealth of data and imagery that are central to the European Commission's Copernicus programme. Sentinel-2 carries an innovative wide swath high-resolution multispectral images with a swath width of 290 km and 13 spectral bands. We have considered two images 1000 × 1000 in the area of Madrid (Spain). No ground truth is available for these data. RGB rendering of these images has been reported in Figs. 2(a) and 2(c), while the saliency maps computed by AISA with SSCC have been plotted in Figs. 2(b) and 2(d). The algorithm is able to delineate salient areas in the landscape, which roughly correspond to the river, roads and buildings in Madrid 1. The algorithm is also able to track the river path in Madrid 2.

(a) RGB Madrid 1   (b) AISA Madrid 1   (c) RGB Madrid 2   (d) AISA Madrid 2

Fig. 2. RGB rendering and saliency map computed by AISA on Madrid 1 and Madrid2.

5   Conclusion

This paper illustrates AISA—an unsupervised machine learning methodology that addresses saliency detection problems in HSI by resorting to salient information extracted through autoencoders. Spatial data arrangement is also taken into account to improve the accuracy. The experimental study, performed using 60 hyperspectral images and 2 multispectral images, reveal that AISA achieves competitive accuracy compared to recent state-of-the-art techniques disclosing relevant salient regions. Based on the validation results, some directions for further work are still to be explored. Orientation features may be integrated in the elaboration, in order to better delineate the shape of the salient objects detected. On the other hand, super-pixel segmentation may be performed, in order to introduce spectral information aggregated at spatial level and perform salient object detection possibly abstracting on details within the objects.

Acknowledgments. This work fulfills the research objectives of the PON "Ricerca e Innovazione" 2014–2020 project RPASInAir "Integrazione dei Sistemi Aeromobili a Pilotaggio Remoto nello spazio aereo non segregato per servizi" (ARS01_00820), funded by the Italian Ministry for Universities and Research (MIUR). The research of Antonella Falini is founded by PON Project AIM 1852414 CUP H95G18000120006 ATT1. The research of Cristiano Tamborrino is funded by PON Project "Change Detection in Remote Sensing" CUP H94F18000270006. We thank Planetek Italia srl for Madrid data.

References

1. Appice, A., Di Mauro, N., Lomuscio, F., Malerba, D.: Empowering change vector analysis with autoencoding in bi-temporal hyperspectral images. In: CEUR Workshop Proceedings, vol. 2466(2019)
2. Appice, A., Malerba, D.: Segmentation-aided classification of hyperspectral data using spatial dependency of spectral bands. ISPRS J. Photogrammetry Remote Sens. **147**, 215–231 (2019)
3. Borji, A., Cheng, M.-M., Hou, Q., Jiang, H., Li, J.: Salient object detection: a survey. Comput. Vis. Media **5**(2), 117–150 (2019). https://doi.org/10.1007/s41095-019-0149-9

4. Borji, A., Tavakoli, H.R., Sihite, D.N., Itti, L.: Analysis of scores, datasets, and models in visual saliency prediction. In: 2013 IEEE International Conference on Computer Vision, pp. 921–928 (2013)
5. Charte, D., Charte, F., García, S., del Jesus, M.J., Herrera, F.: A practical tutorial on autoencoders for nonlinear feature fusion: taxonomy, models, software and guidelines. Inf. Fusion **44**, 78–96 (2018)
6. Han, W., Wang, G., Tu, K.: Latent variable autoencoder. IEEE Access **7**, 48514–48523 (2019)
7. Hussain, M., Chen, D., Cheng, A., Wei, H., Stanley, D.: Change detection from remotely sensed images: from pixel-based to object-based approaches. ISPRS J. Photogrammetry Remote Sens. **80**, 91–106 (2013)
8. Imamoglu, N., Ding, G., Fang, Y., Kanezaki, A., Kouyama, T., Nakamura, R.: Salient object detection on hyperspectral images using features learned from unsupervised segmentation task. In: IEEE International Conference on Acoustics, Speech and Signal Processing (ICASSP 2019), pp. 2192–2196. IEEE (2019)
9. Imamoglu, N.: Hyperspectral image dataset for benchmarking on salient object detection. In: 2018 Tenth International Conference on Quality of Multimedia Experience (QoMEX), pp. 1–3 (2018)
10. Itti, L., Koch, C., Niebur, E.: A model of saliency-based visual attention for rapid scene analysis. IEEE Trans. Pattern Anal. Mach. Intell. **20**(11), 1254–1259 (1998)
11. Jia, Y., Hao, C., Wang, K.: A new saliency object extraction algorithm based on Itti's model and region growing. In: 2019 IEEE International Conference on Mechatronics and Automation (ICMA), pp. 224–228 (2019)
12. Kingma, D.P., Ba, J.: Adam: a method for stochastic optimization. In: 3rd International Conference on Learning Representations (ICLR 2015), Conference Track Proceedings (2014). arXiv:1412.6980
13. Liang, J., Zhou, J., Bai, X., Qian, Y.: Salient object detection in hyperspectral imagery. In: 2013 IEEE International Conference on Image Processing, pp. 2393–2397 (2013)
14. Otsu, N.: A threshold selection method from gray-level histograms. IEEE Trans. Geosci. Remote Sens. **9**(1), 62–66 (1972)
15. Seydi, S.T., Hasanlou, M.: A new land-cover match-based change detection for hyperspectral imagery. Eur. J. Remote Sens. **50**(1), 517–533 (2017)
16. Wang, J., Liu, S., Zhang, S.: A novel saliency-based object segmentation method for seriously degenerated images. In: 2015 IEEE International Conference on Information and Automation, pp. 1172–1177 (2015)
17. Windrim, L., Ramakrishnan, R., Melkumyan, A., Murphy, R.J., Chlingaryan, A.: Unsupervised feature-learning for hyperspectral data with autoencoders. Remote Sens. **11**(7), 1–19 (2019)
18. Yang, Z., Mueller, R.: Spatial-spectral cross-correlation for change detection : a case study for citrus coverage change detection. In: ASPRS 2007 Annual Conference, vol. 2, pp. 767–777, January 2007
19. Zhang, L., Zhang, Y., Yan, H., Gao, Y., Wei, W.: Salient object detection in hyperspectral imagery using multi-scale spectral-spatial gradient. Neurocomputing **291**, 215–225 (2018)
20. Zhou, P., Han, J., Cheng, G., Zhang, B.: Learning compact and discriminative stacked autoencoder for hyperspectral image classification. IEEE Trans. Geosci. Remote Sens. **57**(7), 4823–4833 (2019)
21. Zong, B., et al.: Deep autoencoding gaussian mixture model for unsupervised anomaly detection. In: International Conference on Learning Representations (2018)

Mesoscale Anisotropically-Connected Learning

Qi Tan, Yang Liu, and Jiming Liu$^{(\boxtimes)}$

Department of Computer Science, Hong Kong Baptist University,
Hong Kong, China
{csqtan,csygliu,jiming}@comp.hkbu.edu.hk

Abstract. Predictive spatio-temporal analytics aims to analyze and model the data with both spatial and temporal attributes for future forecasting. Among various models proposed for predictive spatio-temporal analytics, the recurrent neural network (RNN) has been widely adopted. However, the training of RNN models becomes slow when the number of spatial locations is large. Moreover, the structure of RNN is unable to dynamically adapt to incorporate new covariates or to predict the target variables with varying spatial dimensions. In this paper, we propose a novel method, named Mesoscale Anisotropically-Connected Learning (MACL), to address the aforementioned limitations in RNN. For efficient training, we group the dataset into clusters (which refers to the mesoscale) along the spatial dimension according to the spatial adjacency and develop individual prediction module for each cluster. Then we design an anisotropic information exchange mechanism (i.e., the information exchange is not symmetric), to allow the prediction modules leveraging state information from nearby clusters for enhancing the prediction accuracy. Furthermore, for timely adaptation, we develop a local updating strategy for adapting the learning model to incorporate new covariates and the target variables with varying spatial dimensions. Experimental results on a real-world prediction task demonstrate that our method can be trained faster and more accurate than existing methods. Moreover, our method is flexible to incorporate new covariates and target variables of varying spatial dimensions, without sacrificing the prediction accuracy.

1 Introduction

Predictive spatio-temporal analytics, which aims to model and forecast the data collected across space and time, has attracted increasing research interests recently [14,16]. In predictive spatio-temporal analytics, datasets are generally collected from various spatial locations over a historical period and then fed into the learning model for predicting the target variables in these locations [12]. For example, citywide traffic flow data are collected during the past weeks or months for traffic condition prediction in different district or streets [16]. The nationwide or even global climatic data of the past years or decades are collected for

© Springer Nature Switzerland AG 2020
D. Helic et al. (Eds.): ISMIS 2020, LNAI 12117, pp. 171–180, 2020.
https://doi.org/10.1007/978-3-030-59491-6_16

climate forecast [9]. Moreover, in a dynamic environment, new covariates and new target variables in varying spatial locations arrives sequentially [13]. Given such a large volume of spatio-temporal data in a dynamic environment, on the one hand, it is desired to efficiently train reliable spatio-temporal models. On the other hand, the structure of the designed spatio-temporal models should be flexible to incorporate new covariates and to adapt for making predictions on new dimensions of the target variables (i.e., the new spatial locations), because retraining the entire model for new covariates or the target variables on new spatial locations is computationally expensive or even infeasible.

The recurrent neural network (RNN) based spatio-temporal models achieve the state-of-the-art performance in various predictive spatio-temporal analytics tasks [8,14]. For capturing the temporal correlation among data, the RNN incorporates the covariates in the previous time steps and then generate a final hidden representation for future forecast. For capturing the dependency among different sequences, e.g., the traffic conditions in the nearby districts, conventionally, the RNN concatenates the multiple sequences as multivariate input.

However, there are two limitations in the current RNN-based models. First, when learning with a large-scale dataset, as the network size of a RNN is quadratically related to the hidden size, it is difficult to train an RNN with large hidden size for incorporating all necessary information when the number of spatial locations is large [2,7]. Second, when learning in a dynamic environment, the structure of RNN is unable to dynamically adapt to incorporate new covariates or to predict the target variables with varying spatial dimensions. Therefore, in this paper, we aim to develop a novel RNN-based model for efficient training and flexible adaptation.

1.1 Related Work

Many RNN-based spatio-temporal models are proposed in various applications. Yao et al. [14] employed an RNN for the temporal view, which takes the representations from the spatial view and context features as input. Li et al. [8] developed a novel RNN by integrating the spatial graph in the transition matrix. However, these models cannot be easily adapted to the scenarios with varying covariates and new locations.

Some studies investigated learning with varying feature spaces, such as evolvable features [5] and trapezoidal data [17]. However, they were based on the shallow neural networks, which may limit the learning power of the developed models. In this paper, we investigate the problem of online update under the deep model.

For efficient online updating, Doyen et al. [10] proposed a Hedge Back-Propagation method for refining the parameters of DNN effectively. They modified the DNN structure for fast error back-propagation. On the other hand, factorized LSTMs are studied for reducing the number of parameters to be estimated [1,6], but their input layers are not factorized, which leads to limited computational cost reduction and cannot incorporate new covariate.

1.2 Motivation

As the existing methods cannot address the limitations of RNN in large-scale dynamic predictive spatio-temporal analytics, we attempt to tackle these challenges by exploring the intrinsic property of spatio-temporal data. We observe the fact that the state of one location (i.e., one dimension of the target variable), generally, is closely related to the states of geographically nearby locations, but is less relevant to the states of the faraway locations [11]. Therefore, it may not be necessary to require the temporal sequences at all spatial locations to be used as inputs for predicting the target variables in one location. Similarly, when one new covariate in some locations is collected, it could be the nearby locations that need the new data for the predictive tasks.

In summary, the core question to be answered in this paper is: ***How to design an RNN-based learning model for efficient training and timely adaptation in large-scale predictive spatio-temporal analytics, such that new covariates and target variables of varying spatial dimensions in a dynamic environment can be flexibly incorporated into the model, without sacrificing its prediction accuracy?***

1.3 Our Contributions

In this work, we propose a novel method, named Mesoscale Anisotropically-Connected Learning (MACL), to address the above-mentioned question by utilizing the spatial locality and mesoscale connectivity. We first group the dataset into clusters (which refers to the mesoscale) along the spatial dimension according to the spatial adjacency, so that the target variables of the spatial locations in the same cluster are closely related to each other. We develop individual RNN-based prediction module for each cluster. The features extracted from multiple covariates in one cluster are fused as that cluster's state information. Moreover, the state information in nearby clusters may also benefit the prediction, but the influence of different nearby clusters may not be homogeneous. Therefore, we design an information exchange mechanism, which is anisotropic (i.e., the information exchange is neither symmetric nor homogeneous) to allow the prediction modules leveraging information from nearby clusters to enhance the prediction accuracy. We use different weights to combine information from different nearby clusters. Furthermore, we develop a local updating strategy to adapt the learning model so that new covariates can be incorporated and the target variables with varying spatial dimensions can be predicted. Figure 1 illustrates the procedure of our method. The contributions of this paper are summarized as follows.

1. We investigate an important problem in predictive spatio-temporal analytics, i.e., how to design a deep spatio-temporal recurrent model for efficient training and timely adaptation in a large-scale dynamic environment.
2. We propose a novel learning method MACL to solve the above challenging problem.
3. We perform extensive experiments on a real-world traffic flow dataset, showing the effectiveness and efficiency of the proposed method.

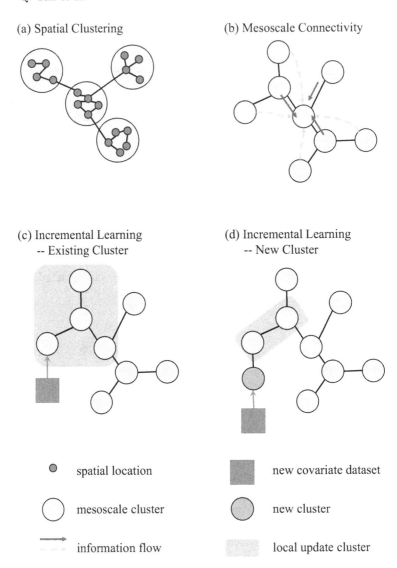

(a) Spatial Clustering

(b) Mesoscale Connectivity

(c) Incremental Learning
-- Existing Cluster

(d) Incremental Learning
-- New Cluster

- spatial location
- mesoscale cluster
- information flow
- new covariate dataset
- new cluster
- local update cluster

Fig. 1. Illustration of the proposed MACL. (a) First, the locations are grouped into several clusters based on the spatial adjacency. Then the mesoscale connectivity among the clusters is constructed by linking the clusters with their K closest clusters. Section 2.2 shows the details of spatial clustering and connecting procedure. (b) During the learning, the features of the data in the same cluster and the state information from the neighborhood clusters are utilized to update the state representation for prediction. The scenario that the state information is forwarded $L = 2$ times is illustrated. Section 2.3 shows the details of the formulations. (c) In a dynamic environment, when incorporating new covariates in the existing cluster, only prediction modules of the clusters within the distance of 2 need to be updated, i.e., training the modules in the pink box with new covariates via error back-propagation, avoiding the retraining of the entire model. (d) In a dynamic environment, when the new covariates are collected in a new cluster, the mesoscale connections are added and the local update is applied. Section 2.4 shows the details of incremental learning procedure.

2 Approach

In this section, we present the proposed MACL. First, we provide the necessary notations and formally define the problem. Then we introduce the structure of the proposed model and the update strategy for learning with new covariates and new target variables in varying spatial dimension.

2.1 Problem Formulation

In this subsection, we formally define the problem of predictive spatio-temporal analytics with large-scale datasets in a dynamic environment, which requires efficient training and timely adaptation. Let $\mathbf{Y} \in \Re^{N \times T}$ be the target data collected in N locations during T time steps, $\mathbf{y}_t \in \Re^N$ be the target data at time step $t, t = 1, \cdots, T$. Let $\mathcal{X}$ be the set of covariate data and $\mathcal{X}_{[t-t^l:t]}$ be the covariate data from time step $t - t^l$ to t. Predictive spatio-temporal analytics task could be regarded as learning the mapping function parameterized by θ, $f_\theta : \mathcal{X}_{[t-t^l:t]} \to \mathbf{y}_t$, where t^l is the time length of the features. For simplicity, we denote $\mathcal{X}_{[t-t^l:t]}$ as $\mathbf{x}_t$ and $\mathbf{x}_t^{(d)}$ as the dth dataset, $d = 1, \dots, D$. The predictive spatio-temporal analytics problem becomes more challenging when predicting with large-scale datasets, i.e. N and $|\mathcal{X}|$ are large. Moreover, in a dynamic environment, new data, $\mathcal{X}^n$ and $\mathbf{Y}^n$, arrive sequentially. We need to timely adapt the model, $f_\theta' : [\mathbf{x}_t, \mathbf{x}_t^n] \to [\mathbf{y}_t, \mathbf{y}_t^n]$, to incorporate the new covariates and to predict the new target variables.

2.2 Spatial Clustering

In order to achieve efficient training, as illustrated in Fig. 1(a), we first group the locations into several clusters and then build a recurrent prediction module for each cluster: we can group the locations based on the spatial adjacency or mutual information correlation. Assume we group N locations into g clusters and construct the adjacency graph, $G = < V, E >, |V| = g$, for these g clusters. There might be multiple covariate datasets collected in one cluster. Thus we use the LSTM [4] to extract the feature for each covariate dataset d:

$$\mathbf{s}_t^{(d)} = \mathrm{LSTM}(\mathbf{x}_t^{(d)}). \qquad (1)$$

where $\mathbf{s}_t^{(d)}$ is the extracted feature from one data set $\mathbf{x}_t^{(d)}$ for predicting $\mathbf{y}_t$. In doing so, we can model the complex interactive among locations within one cluster, as often they are spatially close or semantically similar. We use the hidden state at the last time step as the feature extracted from one dataset. Then we can aggregate the information from the input data for each location:

$$\mathbf{h}_{n,t}^0 = \sum_{d \in \mathcal{S}^n} \mathbf{s}_t^{(d)}, \qquad (2)$$

where $\mathcal{S}^n$ is the set of datasets inputted into the n-th cluster.

2.3   Anisotropically-Connected Learning

In order to improve the prediction accuracy of predictive model for each cluster, we further consider the mesoscale correlation between the clusters, i.e., the data in the nearby clusters is also informative for predicting the target variable. As illustrated in Fig. 1(b), we build up a mechanism for exchanging the information among clusters by passing the message to the neighborhoods:

$$\mathbf{m}_{n,t}^l = \sum_{i \in \mathcal{N}^n} \mathcal{G}^{n,i}(\mathbf{h}_{i,t}^{l-1}), \tag{3}$$

where $\mathcal{N}^n$ is the neighborhoods of cluster n and $\mathcal{G}^{n,i}(\cdot)$ is the specific information mapping from cluster i to cluster n. Then we update the state of one cluster as:

$$\mathbf{h}_{n,t}^l = \mathcal{F}^n([\mathbf{m}_{n,t}^l, \mathbf{h}_{n,t}^{l-1}]), \tag{4}$$

where $\mathcal{F}^n(\cdot)$ is the update function of cluster n. Note that if such message exchange could be operated for L times, one cluster could receive the information from other clusters within the distance of L . Finally, the output for target variable estimation is calculated as:

$$\hat{y}_{n,t} = \mathcal{O}^n(\mathbf{h}_{n,t}^L), \tag{5}$$

where $\mathcal{O}^n(\cdot)$ is the output function for cluster n.

2.4   Incremental Learning

For timely adaptation, we develop an incremental learning strategy based on the mesocale connected structure developed above. The incremental learning strategy consists of the following two steps.

Model Extension. We will introduce the extension strategies in the following two scenarios:

New Covariate From Existing Cluster. In the covariate adaptation, i.e, a new covariate dataset from the existing cluster is collected to facilitate the prediction, we learn a new RNN for the new covariate as shown in Eq. (1). Then we could treat it in the way as other covariates in this cluster and fuse all covariates for prediction, as shown in Eq. (2). Figure1(c) illustrates the procedure of adding new covariate from the existing cluster.

New Covariate and Target Variable From New Cluster. In this case, as illustrated in Fig. 1(d), we first add this cluster into G and link this new cluster to other existing clusters. Then we add and learn new LSTM model, message exchanging function, update function, and output function for the new cluster.

Local Update. In both scenarios, only the nearest L step neighbourhoods need to be updated, i.e., we fix the parameters in other modules while just train the modules of these neighbourhood, which avoids the retraining of the entire model.

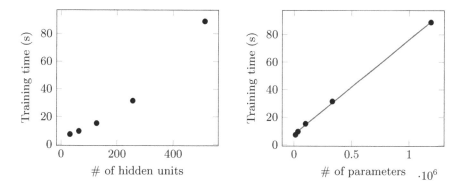

Fig. 2. Running time of LSTM in 1,000 back-propagation iterations with varying numbers of hidden units (left) and parameters (right). The number of parameters in 1-layer LSTM is $4h^2 + 4hd$, where h is the number of hidden units and d is the input size.

Traditional RNN needs to learn $h^2 + hN + hI$ parameters while our model just needs to learn $(2h^2 + hN + hI)/g$ parameters, where g is the number of clusters. In the online update module, we just need to learn $(h^2 + hN + hI)/g^2$ parameters for each influenced module. Thus our approach achieves much faster training speed for new covariate and target variable, compared with traditional RNN.

3   Experiments

In this section, we evaluate the proposed method in efficient training with large-scale datasets and timely adaptation with new covariates and target variables.

We use the traffic crowd flows data[1] in Beijing from Jul. - Oct. 2013 [16] in $1,024$ locations. We use the inflow and outflow in the previous time steps to predict the inflow of each location at the current time step. We use the data at the first 70% time steps for training and the remaining 30% for testing.

We group the locations into 20 clusters and construct the connections between clusters based on spatial coordinates. We link each cluster to $K = 2$ closest clusters. We measure the prediction accuracy using the rooted mean square error (RMSE): $RMSE = \frac{1}{TN} \sum_{t=1}^{T} \sum_{n=1}^{N} ||y_t^n - \hat{y}_t^n||$. We implement the models in Pytorch and conduct the experiments on the cluster with CPU Core i7-4771 3.50 GHz, GPU RTX 2070 and 32 GB Memory.

3.1   Evaluation on Efficient Training

We first show the importance of reducing the number of parameters on the fast learning. Figure 2 shows the training time of LSTM model with varying number of hidden units and parameters. We can see that the training time is

---

[1] Available at https://github.com/lucktroy/DeepST/tree/master/data.

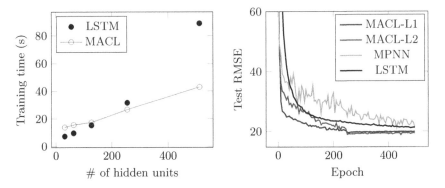

Fig. 3. (Left) Comparison of training time between LSTM and the proposed framework in 1,000 back-propagation iterations with varying number of hidden units. (Right) Comparison of test performance among the proposed framework, MPNN, and LSTM. In our methods, MACL-L1 exchanges the information among the clusters 1 time and MACL-L2 exchanges the information 2 times.

quadratically related to the number of hidden units and linearly correlated with the number of parameters. Thus, reducing the number of parameters and number of hidden units can significantly reduce the computation time of training LSTM.

Then we evaluate the efficiency of the proposed method. Figure 3 shows the training time of our proposed model and LSTM (left) and the testing RMSE of different methods (right). Without loss of generality, all the models are trained with SGD optimizer with the same learning rate. Orthogonal to our work, there are some studies working on using graph convolution neural networks to model spatial correlation [15], among which the message passing network (MPNN) [3] is most similar to our model. In our method, we set the information to be exchanged among clusters for 1 time (MACL-L1) and 2 times (MACL-L2). The results show that our method is trained faster and converges quicker to a better result than the baseline LSTM and the state-of-the-art MPNN. Comparing with LSTM, our method decomposes the large-scale predictive task into several interactive sub-tasks and trains a smaller model for each sub-task, which makes the models can be trained easier. Comparing with MPNN, our method provides more flexibility in modeling the heterogeneous connections by learning an unique weight parameter for each connection.

3.2 Evaluation on Timely Adaptation

In this subsection, we evaluate the adaptation in two scenarios: sequentially arriving new covariates and target variables in new location.

First we test the effectiveness of incorporating new sequentially arriving covariates. We partition the data of each region into 5 groups and input one group of data for each location in certain epoch to simulate the sequentially arriving covariate data. Figure 4 (left) shows the performance of our method

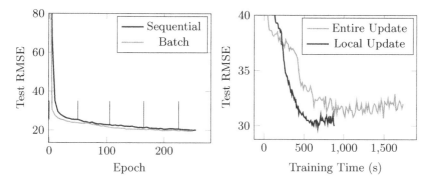

Fig. 4. (Left) The black vertical bars indicate the epochs in which one new covariate dataset is inputted. The performance of the sequential update version, after it converges, is comparable to that of the batch update version. (Right) New locations are added for a trained model. In the entire update version, the entire network is updated after the new location is added; while in the local update version, the partial network is updated. The local update version achieves lower RMSE with shorter training time.

with the sequentially updating way and that with the batch updating. It can be observed from the experimental results that even in the sequentially updating scheme, our model could achieve similar performance as the batch updating version, in which all datasets arrive before the training.

Then we test the efficiency of timely updating when a new target location is explored. We first train the entire network with all clusters except one. Then we adapt the model to incorporate this new cluster and we measure the performance of our method in predicting on this new cluster. Figure 4 (right) shows the performance using the locally updating strategy vs. the entirely updating strategy. The experimental result shows that the locally updating strategy produces better results in a shorter time than the entirely updating strategy.

The above results shows that the spatial mesoscale anisotropically-connected structure is able to achieve fast local adaptation and effective local update, without loss of performance.

4   Conclusion and Future Work

In this paper, we introduced a novel method for efficient training and timely adaptation of RNN model. The results of experimental in the large-scale dynamic environment demonstrated that our model can be trained faster than existing methods while achieving lower prediction error. Moreover, with the local updating and adaptation strategy, our method is flexible to incorporate new covariates and target variables of varying spatial dimensions.

In the future work, we plan to evaluate our framework on larger-scale datasets. Increasing the number of clusters can take the advantage of parallel computation, while this requires more between-cluster connections to improve accuracy, which increases the computational cost. We will further explore how to determine the number of clusters to achieve good efficiency.

Acknowledgement. The authors would like to thank the reviewers for their valuable comments. This work was supported by the grants from the Research Grant Council of Hong Kong SAR under Project RGC/HKBU12201619 and RGC/HKBU12201318.

References

1. Belletti, F., Beutel, A., Jain, S., Chi, E.: Factorized recurrent neural architectures for longer range dependence. In: International Conference on Artificial Intelligence and Statistics, pp. 1522–1530 (2018)
2. Gao, F., Wu, L., Zhao, L., Qin, T., Cheng, X., Liu, T.Y.: Efficient sequence learning with group recurrent networks. In: Proceedings of the 2018 Conference of the North American Association for Computational Linguistics, vol. 1, pp. 799–808 (2018)
3. Gilmer, J., Schoenholz, S.S., Riley, P.F., Vinyals, O., Dahl, G.E.: Neural message passing for quantum chemistry. In: Proceedings of the 34th ICML, pp. 1263–1272 (2017)
4. Hochreiter, S., Schmidhuber, J.: Long short-term memory. Neural Comput. **9**(8), 1735–1780 (1997)
5. Hou, B.J., Zhang, L., Zhou, Z.H.: Learning with feature evolvable streams. In: Advances in Neural Information Processing Systems, pp. 1417–1427 (2017)
6. Kuchaiev, O., Ginsburg, B.: Factorization tricks for LSTM networks. ICLR Workshop (2017)
7. Lei, T., Zhang, Y., Wang, S.I., Dai, H., Artzi, Y.: Simple recurrent units for highly parallelizable recurrence. In: Proceedings of the 2018 Conference on Empirical Methods in Natural Language Processing, pp. 4470–4481 (2018)
8. Li, Y., Yu, R., Shahabi, C., Liu, Y.: Diffusion convolutional recurrent neural network: data-driven traffic forecasting. In: NIPS 2017 Time Series Workshop (2017)
9. Lozano, A.C., et al.: Spatial-temporal causal modeling for climate change attribution. In: Proceedings of the 15th SIGKDD, pp. 587–596. ACM (2009)
10. Sahoo, D., Pham, Q., Lu, J., Hoi, S.C.: Online deep learning: learning deep neural networks on the fly. In: Proceedings of the 27th International Joint Conference on Artificial Intelligence, pp. 2660–2666. AAAI Press (2018)
11. Tobler, W.R.: A computer movie simulating urban growth in the Detroit region. Econ. Geogr. **46**(sup1), 234–240 (1970)
12. Wu, X., Zhu, X., Wu, G.Q., Ding, W.: Data mining with big data. IEEE Trans. Knowl. Data Eng. **26**(1), 97–107 (2014)
13. Yao, H., Liu, Y., Wei, Y., Tang, X., Li, Z.: Learning from multiple cities: a meta-learning approach for spatial-temporal prediction. In: The World Wide Web Conference, pp. 2181–2191. ACM (2019)
14. Yao, H., et al.: Deep multi-view spatial-temporal network for taxi demand prediction. In: Thirty-Second AAAI Conference on Artificial Intelligence (2018)
15. Yu, B., Yin, H., Zhu, Z.: Spatio-temporal graph convolutional networks: A deep learning framework for traffic forecasting. arXiv preprint arXiv:1709.04875 (2017)
16. Zhang, J., Zheng, Y., et al.: Predicting citywide crowd flows using deep spatio-temporal residual networks. Artif. Intell. **259**, 147–166 (2018)
17. Zhang, Q., Zhang, P., Long, G., Ding, W., Zhang, C., Wu, X.: Online learning from trapezoidal data streams. IEEE Trans. Knowl. Data Eng. **28**(10), 2709–2723 (2016)

Empirical Comparison of Graph Embeddings for Trust-Based Collaborative Filtering

Tomislav Duricic[1,2]([⊠]), Hussain Hussain[2], Emanuel Lacic[2], Dominik Kowald[2], Denis Helic[1], and Elisabeth Lex[1]

[1] Graz University of Technology, Graz, Austria
{tduricic,dhelic,elisabeth.lex}@tugraz.at
[2] Know-Center GmbH, Graz, Austria
{hhussain,elacic,dkowald}@know-center.at

Abstract. In this work, we study the utility of graph embeddings to generate latent user representations for trust-based collaborative filtering. In a cold-start setting, on three publicly available datasets, we evaluate approaches from four method families: (i) factorization-based, (ii) random walk-based, (iii) deep learning-based, and (iv) the Large-scale Information Network Embedding (LINE) approach. We find that across the four families, random-walk-based approaches consistently achieve the best accuracy. Besides, they result in highly novel and diverse recommendations. Furthermore, our results show that the use of graph embeddings in trust-based collaborative filtering significantly improves user coverage.

Keywords: Recommender systems · Empirical study · Graph embeddings · Cold-start · Trust networks

1 Introduction

Recommender systems suffer from the well-known cold-start problem [1] that arises when users have rated no or only few items. The cold-start problem is particularly problematic in neighborhood-based recommendation approaches such as collaborative filtering (CF) [2] since the ratings of these users cannot be exploited to find similar users. Trust-based recommender systems (e.g., [3–6]) have been proposed as a potential remedy for the cold-start problem. They alleviate this problem by generating a trust network, i.e., a type of a social network in which nodes usually represent users and edges represent trust connections between users based on their explicitly expressed or implicitly derived trust relationships. Although trust is a complex and ambiguous concept from social sciences, in the context of recommender systems, we use a simple interpretation in which users trust other users in the system if they trust their opinions and ratings on different items [5]. Resulting trust network can be used to find the k-most similar users, whose items are recommended to a target user. Trust networks

© Springer Nature Switzerland AG 2020
D. Helic et al. (Eds.): ISMIS 2020, LNAI 12117, pp. 181–191, 2020.
https://doi.org/10.1007/978-3-030-59491-6_17

are, however, typically sparse [7] since only a fraction of users have trust connections, which makes finding the k-similar users challenging. In the present work, we explore the utility of *graph embeddings* to extract the k-similar users from trust networks. To that end, we conduct experiments on three publicly available benchmark datasets often used in studies on trust-based recommender systems: Epinions [5], Ciao [8], and Filmtrust [9]. We empirically evaluate a range of state-of-the-art graph embedding approaches [10] from four distinct method families, i.e., (i) factorization-based methods, (ii) random-walk-based approaches, (iii) methods based on deep learning, and (iv) the LINE approach [11] that falls in neither of these families, with respect to their ability to deliver accurate, novel, and diverse recommendations [12] for cold-start users.

In our experimental setup, we split each dataset into a validation set (warm-start users) and a test set (cold-start users). For each graph embedding approach, we perform a hyperparameter optimization on each validation set. We then select the hyperparameters which result in highest recommendation accuracy. We generate recommendations for each target user in a CF manner by finding k-similar neighbors using the learned embeddings and ranking their items by similarity scores. Finally, we evaluate the resulting graph embeddings against a corresponding test set with respect to accuracy and beyond accuracy metrics. We compare the graph embedding approaches against five baselines from trust-based recommender systems, commonly used in cold-start settings: (i) Most Popular (MP) recommends the most frequently rated items, (ii) $Trust_{dir}$ extracts trusted users directly from a trust network, (iii) $Trust_{undir}$ ignores edge directions and extracts neighbors from the resulting undirected network, (iv) $Trust_{jac}$ applies the Jaccard coefficient on the explicit trust network, and (v) $Trust_{Katz}$ [13] computes the Katz similarity to infer transitive trust relationships between users. To quantify the algorithmic performance, we evaluate recommendation quality in terms of nDCG, novelty, diversity and user coverage.

We find that as a result of their ability to create a representation of each user in a network, graph embeddings are able to improve user coverage when compared to the baseline approaches. Our experiments also show that random-walk-based approaches, i.e., Node2vec and DeepWalk, consistently outperform other graph embedding methods on all three datasets in terms of recommendation accuracy. Finally, we find a positive correlation between novelty and accuracy in all three datasets suggesting that users in the respective platforms tend to prefer novel content. Summing up, our contributions are three-fold. Firstly, we provide a large-scale empirical study on the efficacy of graph embedding approaches in trust-based recommender systems. Secondly, unlike many previous studies, which evaluated only recommendation accuracy, we compare different approaches with respect to beyond accuracy metrics such as novelty, diversity and user coverage. Lastly, our results provide new insights into user preferences based on correlations between different recommendation quality metrics.

2   Graph Embeddings

In this study, we compare the recommendation performance of graph embedding approaches from four distinct method families [10], i.e., factorization-based methods, random-walk-based approaches, deep-learning-based approaches, and the LINE approach [11] which falls in neither of the first three families.

Factorization-based Approaches. Factorization-based approaches produce node embeddings using matrix factorization. The inner product between the resulting node embedding vectors approximates a deterministic graph proximity measure [14]. In total, we investigate five different factorization approaches:

- *Graph Factorization (GF)* [15] factorizes the adjacency matrix and determines proximity between nodes directly on the adjacency matrix.[1]
- *Laplacian Eigenmaps (LE)* [16] factorizes the normalized Laplacian matrix and preserves the 1st-order proximity.[1]
- *Locally Linear Embedding (LLE)* [17] minimizes the squared difference between the embedding of a node and a linear combination of its neighbors' embeddings, weighted by the edges connecting to them. The solution of this minimization problem reduces to a factorization problem. (see footnote 1)
- *High-Order Proximity preserved Embedding (HOPE)* is able to preserve higher-order proximities and capture the asymmetric transitivity. [18]. (see footnote 1)
- *Graph Representations with Global Structural Information (GraRep)* [19] can handle higher-order similarity as it considers powers of the adjacency matrix.[2]

Random Walk-Based Approaches. RW-based approaches first identify the context of a node with a random walk and then learn the embeddings typically using a skip-gram model [10]. In total, we evaluated three different approaches:

- *DeepWalk* [20] extracts node sequences with truncated random walks and applies a skip-gram model [21] with hierarchical softmax on the node pairs.[3]
- *Node2vec* [22] extends DeepWalk with hyperparameters to configure the depth and breadth of the random walks. In contrast to DeepWalk, Node2vec enables to define flexible random walks, while DeepWalk only allows unbiased random walks over the graph [14].[4]
- *Role2vec* [23] uses attributed random walks to learn embeddings. As Role2vec enables to define functions that map feature vectors to types, it can learn embeddings of *types of nodes*.[5]

---

[1] Implementation used: https://github.com/palash1992/GEM-Benchmark.
[2] Implementation used: https://github.com/benedekrozemberczki/role2vec.
[3] Implementation used: https://github.com/phanein/deepwalk.
[4] Implementation used https://github.com/aditya-grover/node2vec.
[5] Implementation used: https://github.com/benedekrozemberczki/role2vec.

Deep Learning-Based Approaches. Such approaches use deep neural network models to generate node embeddings. In this research paper, we studied three deep learning-based models in total:

- *Deep Neural Networks for Graph Representations (DNGR)* [24] uses random surfing to build a normalized node co-occurrence matrix and employs a stacked denoising autoencoder to learn node embeddings.[6]
- *Structural Deep Network Embedding (SDNE)* [25] finds neighbors by means of 1st and 2nd order proximity and learns node embeddings via autoencoders.[7]
- *Graph sample and aggregate GraphSAGE* [26] is a multi-layered graph convolutional neural network, which represents nodes internally by aggregating their sampled neighborhoods and utilizes a random-walk-based cost function for unsupervised learning. GraphSAGE performs the convolution in the graph space. It uses either mean-based, GCN-based, LSTM-based, mean pooling or max pooling models for aggregation.[8]

Large-Scale Information Network Embedding. *LINE* [11] creates embeddings that preserve 1st-order and 2nd-order proximity which are represented as joint and conditional probability distributions respectively.[9]

3 Preliminaries

3.1 Datasets

We employ three open datasets commonly used when evaluating trust-based recommender systems, i.e., Epinions [5], Ciao [8], and FilmTrust [9]. For all three datasets, we create an unweighted trust network, in which each node represents a user, and each directed edge denotes a trust relationship between two users. The trust network is then an adjacency matrix $\mathbf{A}$ where $\mathbf{A}_{u,v}$ is 1 in case of a trust link between u and v, and 0 otherwise. As a result of preliminary experiments, we found that most of the approaches achieved better accuracy results with an undirected network. One possible explanation is that removing the edge direction increases the average number of edges for each node and reduces the sparsity of the adjacency matrix. Moreover, some approaches are not able to consider link direction by design. Therefore, we convert the trust network to an undirected network in all of our experiments by removing edge direction, thus making $\mathbf{A}$ symmetric. Furthermore, we create a ratings matrix $\mathbf{R}$, where each non-zero entry $\mathbf{R}_{u,i}$ represents a rating given by a user u to an item i. Table 1 shows basic statistics for all three datasets.

Dataset Splits. We split each dataset into two sets: warm-start users, i.e., users with >10 ratings and cold-start users, i.e., users with ≤10 item ratings. While

---

[6] Implementation used: https://github.com/ShelsonCao/DNGR.
[7] Implementation used: https://github.com/suanrong/SDNE.
[8] Implementation used: https://github.com/williamleif/GraphSAGE.
[9] Implementation used: https://github.com/tangjianpku/LINE.

Table 1. Dataset statistics.

Dataset	#Users	#Items	#Edges	#Ratings	Graph density
Epinions	49,288	139,738	487,183	664,824	2×10^{-4}
Ciao	19,533	16,121	40,133	72,665	1.85×10^{-3}
Filmtrust	1,642	2,071	1,853	35,497	2.43×10^{-3}

we use the subset of warm-start users as a validation set for hyperparameter optimization concerning recommendation accuracy, the subset of cold-start users is used as a test set for measuring algorithm performance. Table 2 reports the number of cold-start and warm-start users in our datasets.

Table 2. Number of users per dataset split.

Dataset	Users with ratings		Users with ratings & trust	
	Warm-start	Cold-start	Warm-start (Validation set)	Cold-start (Test set)
Epinions	14,769	25,393	14,769	25,393
Ciao	1,020	16,591	571	2,124
Filmtrust	963	545	499	241

3.2 Experimental Setup

The initial directed trust network is converted to an undirected network by removing edge directions. The resulting undirected symmetric **A** is then used as an input for the graph embedding methods, which, as a result, create a d-dimensional embedding for each node (i.e., user) in the graph.

Recommendation Strategy. After generating the embedding for each node in the graph, a similarity matrix **S** is created based on the pairwise cosine similarity between nodes' embeddings. Recommendations are generated in a kNN manner where we find the k-nearest neighbors N_k (i.e., k most similar users) for the target user u_t using the similarity matrix **S**. We use $k = 40$ across all of our experiments as in [13]. Then, we assign a score for all items the users in N_k have interacted with:

$$score(i, u_t) = \sum_{v \in N_k} S_{u_t, v} \cdot R_v(i), \qquad (1)$$

where $R_v(i)$ corresponds to the rating assigned by the user v to the item i and $S_{u_t, v}$ corresponds to the similarity score in **S** between target user u_t and the neighbor user v from N_k. For each target user u_t with n rated items, we recommend 10 items ranked according to Eq. 1 and compare them with the actual rated items.

Evaluation Metrics. Previous research has shown [27] that accuracy may not always be the only or the best criteria for measuring recommendation quality. Typically, there is a trade-off between accuracy, novelty, and diversity since users also like to explore novel and diverse content depending on the context. Therefore, in our work, we examine both novelty and diversity as well as accuracy. In particular, in our experimental setup, we use the following four accuracy and beyond-accuracy metrics for evaluation.

Normalized Discounted Cumulative Gain (nDCG@n) – a ranking-dependent metric measuring recommendation accuracy based on the Discounted Cumulative Gain (DCG) measure [28].

Novelty@n – corresponds to a recommender's ability to recommend long-tail items that the target user has probably not yet seen. We compute novelty using the Expected Popularity Complement (EPC) metric [29].

Diversity@n – describes how dissimilar items are in the recommendation list. We calculate it as the average dissimilarity of all pairs of items in the recommendation list [30]. More specifically, we use cosine similarity to measure the dissimilarity of two items based on doc2vec embeddings [31] learned using the item vector from **R** where each rating is replaced with the user id.

User Coverage – defined as the number of users for whom at least one item recommendation could have been generated divided by the total number of users in the target set [5].

Baseline Approaches. We evaluate the graph embeddings approaches against five different baselines:

- *Explicit directed trust (Trust$_{dir}$)* is a naive trust-based approach that uses the unweighted, directed trust network's adjacency matrix for finding user's nearest neighbors, i.e. $\mathbf{S} = \mathbf{A}$.

- *Explicit undirected trust (Trust$_{undir}$)* is similar to Trust$_{dir}$ but converts the network to an undirected one by ignoring the edge direction, thus making **A** symmetric, i.e. $\mathbf{S} = \mathbf{A}_{undir}$.

- *Explicit trust with Jaccard (Trust$_{jac}$)* uses the Jaccard index on the undirected trust network $\mathbf{A}_{undir}$. **S** is a result of calculating the pairwise Jaccard index on $\mathbf{A}_{undir}$. The intuition behind this algorithm is that two users are treated as similar if they have adjacent users in common, i.e., trustors and trustees.

- *Explicit trust with Katz similarity (Trust$_{Katz}$)* [13] exploits regular equivalence, a concept from network science by using Katz similarity in order to model transitive trust relationships between users.

- *Most Popular (MP)* is a non-personalized approach in recommender systems, which recommends the most frequently rated items.

4   Results

Table 3 shows our results in terms of nDCG, novelty, diversity and user coverage for $n = 10$ recommendations on cold-start users (test set). The reported results depict those hyperparameter configurations[10], which achieve the highest recommendation accuracy on warm users (validation set).

4.1   Accuracy Results

To ease the interpretation of the evaluation results across all three datasets, we rank the results by nDCG and compute the average of these ranks. Correspondingly, in the $Rank_{nDCG}$ column, we show the resulting average rank for the three datasets for recommendation accuracy.

We can observe that RW-based approaches, especially Node2vec and Deep-Walk, are the best performing approaches on all three datasets. In most cases, approaches based on graph embeddings outperform the baselines, except for $Trust_{jac}$ on Epinions. Contrary to a study conducted in [13], $Trust_{jac}$ achieves higher accuracy in comparison with $Trust_{Katz}$. The reason is that in the present work, we convert the trust network to an undirected network, i.e., do not consider the direction of the trust edge. HOPE and Laplacian Eigenmaps perform best among the factorization-based approaches; LINE shows a good performance on all three datasets concerning all three metrics, and GraphSAGE is the best deep learning approach. SDNE does not perform well in our experiments, which we attribute to not exploring a sufficiently broad range of hyperparameters.

4.2   Beyond-Accuracy Results

Novelty, Diversity, and User Coverage. Being superior in the case of $Rank_{nDCG}$, Node2vec also achieves high novelty and diversity scores. Plus, it performs similarly or better than other RW-based methods across all three datasets. Factorization-based approaches show average performance concerning both novelty and diversity, except for GF, which scores very low on novelty and above average on diversity. DL approaches show average to below-average performance on novelty and average performance on diversity. Trust-based baselines achieve high novelty scores in general and, not surprisingly, MP has a high diversity score and the worst novelty score out of all approaches. Since all graph embedding approaches create a latent representation of each user in a trust network using it to generate a set of item recommendations, there are no differences among them in user coverage. Except for MP, all baselines result in lower user coverage than the graph embedding approaches. Since MP provides the same list of recommendations to all users, it always has a maximum user coverage.

Evaluation Metrics and User Preferences. Table 3 reports only mean values for each of the approaches. However, we store individual nDCG, novelty, and diversity values for each target user and each approach and dataset. By computing the Kendall rank correlation coefficient (Bonferroni corrected, $p < 0.01$) on non-zero metrics values for all approaches, we can get an insight into user

---

[10] Details on the hyperparameter optimization can be found at: https://github.com/tduricic/trust-recommender/blob/master/docs/hyperparameter-optimization.md.

Table 3. Evaluation results on cold-start users for different trust-based CF approaches for $n = 10$ recommendations concerning nDCG, novelty, diversity, and user coverage comparing approaches from five different algorithm families across three different datasets. Values marked with * denote that the corresponding approach was significantly better than every other approach with respect to the appropriate metric according to a Wilcoxon signed-rank test (Bonferroni corrected, $p < 0.01$). Rank$_{nDCG}$ is calculated by summing nDCG-based ranks across the datasets and re-ranking the sums.

Cat.	Approach	Rank $_{nDCG}$	Epinions				Ciao				Filmtrust			
			nDCG	Nov.	Div.	UC	nDCG	Nov.	Div.	UC	nDCG	Nov.	Div.	UC
Baseline	Trust$_{dir}$	15	.0245	.0060	.6006	59.2%	.0140	.0028	.3700	3.9%	.2655	.0313	**.2784***	30.3%
	Trust$_{undir}$	15	.0260	.0063	.5960	97.0%	.0127	**.0045**	.3632	11.4%	.2739	.0284	.2731	42.0%
	Trust$_{jac}$	11	.0373	.0056	.6548	99.9%	.0176	.0027	.3996	12.8%	.3387	**.0369**	.2266	36.1%
	Trust$_{Katz}$	12	.0290	.0046	.6979		.0158	.0026	.3842	13.0%	.3681	.0322	.2185	42.9%
	MP	17	.0134	.0015	**.7621***		.0135	.0012	**.5666**	100%	.3551	.0137	.1672	100%
Factorization	LLE	7	.0309	.0044	.6977		.0239	.0036	.4013		.3649	.0159	.1926	
	LE	3	.0318	.0045	.6961		.0231	.0034	.3962		.3715	.0161	.1853	
	GF	14	.0138	.0023	.7024		.0154	.0022	.3970		.3686	.0154	.1945	
	HOPE	3	.0331	.0047	.6728		.0220	.0033	.3956		.3718	.0158	.1827	
	GraRep	7	.0298	.0042	.6704	100 %	.0209	.0030	.3974	13.1 %	.3694	.0147	.1859	44.2 %
RW	Node2vec	1	.0413	.0064	.6581		.0228	.0036	.4042		**.3904**	.0151	.2235	
	DeepWalk	2	**.0435***	**.0067***	.6707		**.0247***	.0037	.3992		.3654	.0152	.1950	
	Role2vec	6	.0363	.0054	.6910		.0149	.0024	.3933		.3695	.0151	.1919	
DL	DNGR	10	.0353	.0051	.6869		.0197	.0031	.4023		.3583	.0142	.1959	
	SDNE	12	.0184	.0022	.7412		.0175	.0028	.3921		.3687	.0152	.2003	
	GS	7	.0325	.0047	.6810		.0216	.0031	.3963		.3678	.0151	.1883	
LINE	LINE	5	.0407	.0063	.6566		.0222	.0033	.3992		.3667	.0150	.1947	

preferences for each dataset. In this manner, we observe a statistically significant positive mean correlation across all three datasets between nDCG and novelty, ranging from 0.43 on Epinions to 0.36 on Filmtrust. This suggests that users on all three platforms prefer recommendations with higher novelty, especially on Epinions. We also observe a statistically significant mean negative correlation between diversity and novelty on Epinions (-0.15), which suggests that on this platform, more novel content seems to be less diverse.

5   Conclusions and Future Work

In this work, we explored the utility of graph embedding approaches from four method families to generate latent user representations for trust-based recommender systems in a cold-start setting. We found that random-walk-based approaches, (i.e., Node2vec and DeepWalk), consistently achieve the best accuracy. We additionally compared the methods concerning novelty, diversity, and user coverage. Our results showed that again, Node2vec and DeepWalk scored high on novelty and diversity. Thus, they can provide a balanced trade-off between the three evaluation metrics. Moreover, our experiments showed that we can increase the user coverage of recommendations when we utilize graph embeddings in a k-nearest neighbor manner. Finally, a correlation analysis between the nDCG, novelty, diversity scores revealed that in all three datasets, users tend to prefer novel recommendations. Hence, on these datasets, recommender systems should offer a good tradeoff between accuracy and novelty. This could also explain the superior performance of the random-walk based approaches and we plan to investigate this in more detail in follow-up work.

Limitations and Future Work. One limitation of this study is that we treated the trust networks as undirected while, in reality, they are directed. This may have resulted in loss of information, and as such, we aim to further explore how to preserve different properties of trust networks (e.g., asymmetry). Moreover, it is possible that we did not examine an ample enough space of hyperparameters, which might have resulted in a more reduced performance of some of the approaches, e.g., SDNE. Furthermore, we aim to explore node properties of k-nearest neighbors for all methods to find and interpret the critical node properties preserved by the graph embeddings, which impact the recommendation accuracy, thus providing a better understanding of the complex notion of trust. Finally, we aim to incorporate user features obtained from the rating matrix into graph embeddings learned on the trust network.

References

1. Schein, A.I., Popescul, A., Ungar, L.H., Pennock, D.M.: Methods and metrics for cold-start recommendations. In: Proceedings of the 25th Annual International ACM SIGIR Conference on Research and Development in Information Retrieval, pp. 253–260. ACM (2002)
2. Schafer, J.B., Frankowski, D., Herlocker, J., Sen, S.: Collaborative filtering recommender systems. In: Brusilovsky, P., Kobsa, A., Nejdl, W. (eds.) The Adaptive Web. LNCS, vol. 4321, pp. 291–324. Springer, Heidelberg (2007). https://doi.org/10.1007/978-3-540-72079-9_9
3. Fazeli, S., Loni, B., Bellogin, A., Drachsler, H., Sloep, P.: Implicit vs. explicit trust in social matrix factorization. In: Proceedings of the 8th ACM Conference on Recommender Systems, RecSys 2014, pp. 317–320. ACM (2014)
4. Lathia, N., Hailes, S., Capra, L.: Trust-based collaborative filtering. In: Karabulut, Y., Mitchell, J., Herrmann, P., Jensen, C.D. (eds.) IFIPTM 2008. ITIFIP, vol. 263, pp. 119–134. Springer, Boston, MA (2008). https://doi.org/10.1007/978-0-387-09428-1_8
5. Massa, P., Avesani, P.: Trust-aware collaborative filtering for recommender systems. In: Meersman, R., Tari, Z. (eds.) OTM 2004. LNCS, vol. 3290, pp. 492–508. Springer, Heidelberg (2004). https://doi.org/10.1007/978-3-540-30468-5_31
6. O'Donovan, J., Smyth, B.: Trust in recommender systems. In: Proceedings of the 10th International Conference on Intelligent User Interfaces, IUI 2005, pp. 167–174. ACM (2005)
7. Kim, Y.A.: An enhanced trust propagation approach with expertise and homophily-based trust networks. Knowl. Based Syst. **82**, 20–28 (2015)
8. Tang, J., Gao, H., Liu, H.: mTrust: discerning multi-faceted trust in a connected world. In: Proceedings of the Fifth ACM International Conference on Web Search and Data Mining, pp. 93–102. ACM (2012)

9. Guo, G., Zhang, J., Yorke-Smith, N.: A novel Bayesian similarity measure for recommender systems (2013)
10. Goyal, P., Ferrara, E.: Graph embedding techniques, applications, and performance: a survey. Knowl.-Based Syst. **151**, 78–94 (2018)
11. Tang, J., Qu, M., Wang, M., Zhang, M., Yan, J., Mei, Q.: Line: large-scale information network embedding. In: Proceedings of the 24th International Conference on World Wide Web, pp. 1067–1077. International World Wide Web Conferences Steering Committee (2015)
12. Zhang, Y.C., Ó Séaghdha, D., Quercia, D., Jambor, T.: Auralist: introducing serendipity into music recommendation. In: Proceedings of the 5th ACM Conference on Web Search and Data Mining (WSDM-12) (2012)
13. Duricic, T., Lacic, E., Kowald, D., Lex, E.: Trust-based collaborative filtering: tackling the cold start problem using regular equivalence. In: Proceedings of the 12th ACM Conference on Recommender Systems, RecSys 2018, pp. 446–450. ACM (2018)
14. Hamilton, W.L., Ying, R., Leskovec, J.: Representation learning on graphs: methods and applications. IEEE Data Eng. Bull. **40**(3), 52–74 (2017)
15. Ahmed, A., Shervashidze, N., Narayanamurthy, S., Josifovski, V., Smola, A.J.: Distributed large-scale natural graph factorization. In: Proceedings of the 22nd International Conference on World Wide Web, pp. 37–48. ACM (2013)
16. Belkin, M., Niyogi, P.: Laplacian eigenmaps and spectral techniques for embedding and clustering. In: Advances in Neural Information Processing Systems, pp. 585–591 (2002)
17. Roweis, S.T., Saul, L.K.: Nonlinear dimensionality reduction by locally linear embedding. Science **290**(5500), 2323–2326 (2000)
18. Ou, M., Cui, P., Pei, J., Zhang, Z., Zhu, W.: Asymmetric transitivity preserving graph embedding. In: Proceedings of the 22nd ACM SIGKDD International Conference on Knowledge Discovery and Data Mining, pp. 1105–1114. ACM (2016)
19. Cao, S., Lu, W., Xu, Q.: Grarep: learning graph representations with global structural information. In: Proceedings of the 24th ACM International on Conference on Information and Knowledge Management, pp. 891–900. ACM (2015)
20. Perozzi, B., Al-Rfou, R., Skiena, S.: Deepwalk: online learning of social representations. In: Proceedings of the 20th ACM SIGKDD International Conference on Knowledge Discovery and Data Mining, KDD 2014, pp. 701–710. ACM (2014)
21. Mikolov, T., Sutskever, I., Chen, K., Corrado, G.S., Dean, J.: Distributed representations of words and phrases and their compositionality. In: Advances in Neural Information Processing Systems 26, pp. 3111–3119. Curran Associates Inc. (2013)
22. Grover, A., Leskovec, J.: node2vec: scalable feature learning for networks. In: Proceedings of the 22nd ACM SIGKDD International Conference on Knowledge Discovery and Data Mining (2016)
23. Ahmed, N.K., et al.: Learning role-based graph embeddings. arXiv e-prints. arXiv:1802.02896, February 2018
24. Cao, S., Lu, W., Xu, Q.: Deep neural networks for learning graph representations (2016)
25. Wang, D., Cui, P., Zhu, W.: Structural deep network embedding. In: Proceedings of the 22nd ACM SIGKDD International Conference on Knowledge Discovery and Data Mining, pp. 1225–1234. ACM (2016)
26. Hamilton, W., Ying, Z., Leskovec, J.: Inductive representation learning on large graphs. In: Advances in Neural Information Processing Systems, pp. 1024–1034 (2017)

27. Kaminskas, M., Bridge, D.: Diversity, serendipity, novelty, and coverage: a survey and empirical analysis of beyond-accuracy objectives in recommender systems. ACM Trans. Interact. Intell. Syst. **7**, 2:1–2:42 (2016)
28. Järvelin, K., Price, S.L., Delcambre, L.M.L., Nielsen, M.L.: Discounted cumulated gain based evaluation of multiple-query IR sessions. In: Macdonald, C., Ounis, I., Plachouras, V., Ruthven, I., White, R.W. (eds.) ECIR 2008. LNCS, vol. 4956, pp. 4–15. Springer, Heidelberg (2008). https://doi.org/10.1007/978-3-540-78646-7_4
29. Vargas, S., Castells, P.: Rank and relevance in novelty and diversity metrics for recommender systems. In: Proceedings of the Fifth ACM Conference on Recommender Systems, pp. 109–116 (2011)
30. Smyth, B., McClave, P.: Similarity vs. diversity. In: Aha, D.W., Watson, I. (eds.) ICCBR 2001. LNCS (LNAI), vol. 2080, pp. 347–361. Springer, Heidelberg (2001). https://doi.org/10.1007/3-540-44593-5_25
31. Mikolov, T., Sutskever, I., Chen, K., Corrado, G.S., Dean, J.: Distributed representations of words and phrases and their compositionality. In: Advances in Neural Information Processing Systems, pp. 3111–3119 (2013)

Neural Spike Sorting Using Unsupervised Adversarial Learning

Konrad A. Ciecierski[(✉)]

Research and Academic Computer Network, Warsaw, Poland
konrad.ciecierski@gmail.com

Abstract. During the analysis of neurobiological signals acquired during neurosurgical procedures such as Deep Brain Stimulation, one focuses mainly on the signal's background and spiking activity. Spikes, the electrical impulses generated by neuron cells, are events lasting about 1.5 ms with voltage typically well below 200 μV. As the shape of spike derives from the physical construction of the neuron's cell membrane and its distance from the recording point, one can infer that spikes of different shapes were generated by different neuron cells. This information can be beneficial during the analysis of the acquired signals, and the procedure of grouping spikes according to their shapes, is called spike sorting. Typically in this procedure, there are defined various attributes characterizing spike shape, and subsequently, the standard clustering method is used to produce groups of spikes of different shapes. This may cause some difficulties that arise both from the definition of good discriminating attributes as well as from the fact that one does not know beforehand how many clusters of shapes to expect. The approach presented in this paper is different, using the fully unsupervised and explainable AI, it generates groups of spike shapes in a fully automatic way. No attributes have to be defined nor the expected number of shape clusters provided. Spike waveforms are fed into the autoencoder that has defined additional constraints on its latent layers. After training, data obtained from those layers can be used for the purpose of spike shape clustering. Beside shape class, one also receives information on how much given spike shape differs from the mean shape of its class.

Keywords: Spike sorting · Explainable AI · Autoencoder · Unsupervised adversarial learning · Distribution enforcement · Weighted MSE loss

Introduction

In the process of many Deep Brain Stimulation (DBS) surgeries, the brain activity is recorded using thin microelectrodes [1]. Signals acquired by those electrodes help neurosurgeons in precise localization of the target of the surgery. Such signals are mainly analyzed in two ways: first focusing on signal's background and second dealing with neural spikes detected in the signal [2]. As the recording

© Springer Nature Switzerland AG 2020
D. Helic et al. (Eds.): ISMIS 2020, LNAI 12117, pp. 192–202, 2020.
https://doi.org/10.1007/978-3-030-59491-6_18

microelectrode can register electrical activity in a radius of about 50 μ m from its lead [1], the activity of nearest neurons is registered as distinct spikes while a summary of the activity of farther neurons is registered as background noise.

For each neuron cell, the shape of spikes it generates is generally constant as it comes from the concentration and distribution of ion channels in its cell membrane [3]. Spikes registered from neurons being farther from the recording lead have also amplitude lower than those being near [1]. From this comes the assumption that spikes of different shapes do originate in different neuron cells [1]. Information about the number of active cells in the vicinity of the electrode might be beneficial in the identification of the area in which the electrode is located. To provide such information i.e., to cluster spike shapes into classes, the spike sorting is used. There are many papers regarding spike sorting approaches [4,5]. The main difficulty in the classical spike sorting is the definition of attributes that discriminate spike shapes in a good manner without the prior knowledge about the number of spike shape clusters expected in a given recording.

One of the approaches to classification or clustering of data for which there are no known attributes is an autoencoder neural network [6] with an adequately small latent layer. In the purest form, such a network consists of two neural networks. First network – encoder – transforms input data into latent space that has dimension much lower than the dimension of the original data. The second network reconstructs the input data from the latent space. Network in which the output of the encoder is the input for the decoder forms autoencoder. During the training of such networks, the goal is to minimize the difference between input and output. In most cases, this difference, the loss, is calculated as MSE [6]. The latent layer acts as a bottleneck for the information that passes through the network. After training, one might attempt to use data from the latent layer as a set of attributes characterizing data fed into the encoding part of the autoencoder. The difficulty lies in the fact that there are no constraints put on the latent layer. Latent data is self-organized, and only for specific input data, one can find here any naturally occurring clusters.

To improve this situation the variational autoencoders (VAE) were defined [7]. In those autoencoders, the loss is augmented in such way as to reflect how much distribution of the latent layer differs from the normal distribution. During the training process, as the loss is minimized, the distribution of the latent layer shifts towards the normal distribution. This is feasible for data of one kind, where one attempts to represent it by normally distributed latent layer. It shows which data is typical and which one lies on tails of the distribution curve, i.e., which is typical and which one is not. Another limitation of VAE is that it enforces only the normal distribution.

The ability to enforce any chosen distribution upon the latent layer has come with the concept of adversarial autoencoders [8]. Here beside the encoder and decoder, there is another network. This network is trained to discriminate data having specified distribution. The discriminating network takes as input, data of size identical to this given by encoder's output. Discriminator's output produces, as a result, one if its input data has desired distribution, 0 otherwise. For simplicity disregarding the minibatch [6] dimension, the adversarial training works in the following way. a) data with desired distribution is fed into discriminating network which is trained to give one as an output b) input data is fed into encoder; subsequently the encoder's output is fed into discriminator, which is trained to give 0 as an output c) encoder is trained in such way to provide output for which discriminator gives one as output. Summarizing, in step a) discriminator learns to produce 1 for data with desired distribution and in step b) discriminator learns to treat encoder's output as not having desired distribution. In crucial step c), encoder learns to produce output than can fool discriminator, i.e., output that is so close to desired distribution that discriminator outputs 1. Of course, besides described adversarial training, autoencoder consisting of encoder paired with a decoder is still trained to minimize the difference between encoder's input and decoder's output.

Using adversarial autoencoder, one can approach the problem of spike sorting in an unsupervised way. Let us assume that encoder for single input produces output composed of two parts. First part has dimension dim_c while dimension of second part is dim_g. For each of those parts, there is a defined separate discriminator network. The first discriminator enforces the first part of the encoder's output to have categorical distribution [9]. The second discriminator enforces the remaining part of the encoder's output to have the normal distribution. In such a way, it forces the adversarial autoencoder to represent input data in the latent layer in no more than dim_c classes where variability in each class is defined as a normally distributed vector of size dim_g.

1   Construction of the Autoencoder

1.1   Input Data

Waveforms of spikes were obtained from recordings done during DBS surgeries for Parkinson's Disease and Dystonia [1]. Sampling frequency during data acquisition was 24 kHz; each spike waveform was recorded in the form of 48 samples, i.e., as 2 ms of data. The full dataset consists of 330963 spikes. Spikes were divided into training, validation, and test sets in proportions 80:10:10, resulting in training data having 264770 spikes, validation data having 33096 spikes, and finally test data with 33097 spikes. All detected spikes were aligned to have their maximal value at sample 12 so relatively to spike maximal point the registered time span ranges from -0.5 μs to 1.5 μs. Originally the recorded data is in micro-volts and fits in the range $[-500$ μV, 500 μV]. Before being fed into the network, the data is normalized to fit in the range $[-1, 1]$.

1.2   Architecture

In the construction of the autoencoder, it has been assumed that spikes are to be separated into no more than ten classes, and that content of each class is to be described as a normally distributed vector from $\mathbb{R}^3$. Assuming that minibatch size is N, the full autoencoder consists of the following parts.

The encoder is a neural network taking as input tensor of dimensions (N, 48) and producing as an output pair of tensors with dimensions (N, 10) and (N, 3). Let us define first output tensor as categorical head and second as Gaussian head of the encoder.

$$enc(input) = (categorical\_head,\ gaussian\_head)$$

$$where\ input \in \mathbb{R}^{(N,48)},\ categorical\_head \in \mathbb{R}^{(N,10)},\ gaussian\_head \in \mathbb{R}^{(N,3)}$$

The encoder takes input tensor of shape (N, 48) and puts it through two dense layers [6]. Each of the two dense layers, has 100 output neurons and uses ELU [6] activation function. After the second dense layer, there are two dense layers connected to it. Those two layers use the linear activation function. The first layer has ten output neurons and produces tensor of shape (N, 10), which is the categorical head of the encoder. The second one has only three output neurons and produces tensor of shape (N, 3), which is the Gaussian head of the encoder. The architecture of the encoder is shown in Fig. 1.

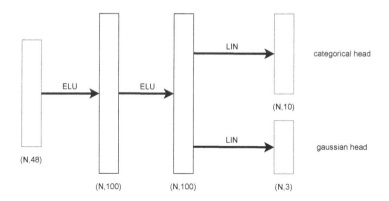

Fig. 1. Architecture of the encoder network

The decoder acts in a way reverse to the encoder. It takes the encoder output in the form of two tensors of shapes (N, 10) and (N, 3) to produce the output of shape identical to encoders input, i.e. (N, 48).

$$dec(categorical\_input,\ gaussian\_input) = output$$

$$where\ categorical\_input \in \mathbb{R}^{(N,10)},\ gaussian\_input \in \mathbb{R}^{(N,3)},\ output \in \mathbb{R}^{(N,48)}$$

The architecture of the decoder (see Fig. 2) is slightly more complicated. There are two dense input networks, one for each of the inputs. The first network takes tensor of shape (N, 10) as an input and outputs tensor of shape (N, 100). The second input network takes tensor of shape (N, 3) as an input and also outputs tensor of shape (N, 100). In the subsequent step, two resulting tensors are added to form a single tensor of shape (N, 100). This tensor is fed into a dense network with 100 output neurons and ELU activation function. Finally, the last layer is also the dense layer but with 48 output neurons and TANH [6] activation function. The choice of the TANH activation function instead of often used Sigmoid [6] function is caused by the fact that original normalized data is in the range [−1, 1] which is identical to the output range of the Tanh function.

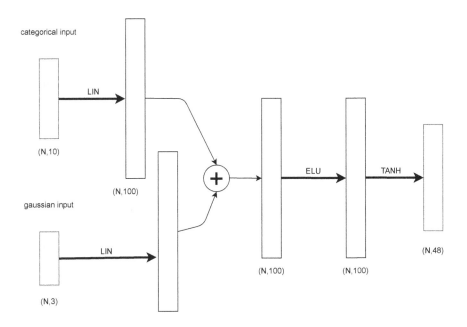

Fig. 2. Layer composition of the decoder

The categorical and Gaussian discriminators are very similar; the only difference between them is the shape of the input tensor. In the case of the categorical discriminator, the shape of the input tensor is identical as the categorical head of the encoder, i.e. (N, 10). Analogously the shape of the input tensor for Gaussian discriminator matches the Gaussian head and is (N, 3). The discriminator consists of three dense layers. The first two layers have 100 output neurons and use ELU activation function. The last layer has two output neurons and is used for one-hot encoding of the discriminator output.

$$dsc_c(categorical\_input) = output$$
$$dsc_g(gaussian\_input) = output$$
$$where\ categorical\_input \in \mathbb{R}^{(N,10)},\ gaussian\_input \in \mathbb{R}^{(N,3)},\ output \in \mathbb{R}^{(N,2)}$$

The architecture of the categorical discriminator is shown in Fig. 3.

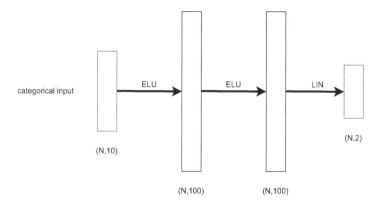

Fig. 3. Layer composition of the categorical discriminator

2   Training

2.1   Training Phases for Single Epoch

For each minibatch, the training consists of three distinct steps.

The Reconstruction Phase. In this phase, autoencoder is trained in a way that is similar to classical autoencoder. For the given input, the loss function is defined as $MSE_w(input, dec(enc(input)))$. The MSE_w being a custom loss function is based upon the standard MSE function. Each spike waveform is 48 samples long, the sampling frequency is 24 kHz, and so the spike waveform span is 2 ms. Not all parts of the spike are equally important for the sorting process [5] let us so define MSE_w for minibatch as follows:

$$in = [x_{i,j}], \; i = 0 \dots N - 1, \; j = 0 \dots 47; \quad out = dec(enc(in))$$

$$SE_p(in, out, b, e) = \sum_{i=0}^{N-1} \sum_{j=b}^{e} (in_{i,j} - out_{i,j})^2$$

$$SE_w(in, out) = SE_p(in, out, 0, 5) + 50 \, SE_p(in, out, 6, 17) + SE_p(in, out, 18, 47)$$

$$MSE_w(in, out) = \frac{SE_w(in, out)}{48 \, N}$$

The coefficient of 50 used in the definition of $SE_w(in, out)$ has been found experimentally. The MSE_w focuses on time span between -250 µs and $+250$ µs as shown in Fig. 4a. The result that is achieved by using such weighted loss function can be seen in Fig. 4b where the main spike shape is recreated very

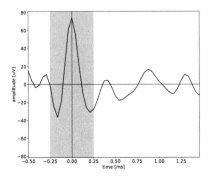

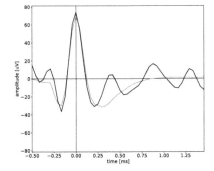

(a) Spike waveform with focus area

(b) Spike original waveform and its recreation by the autoencoder

Fig. 4. Role of spike focus area

closely while the distortions present after time 0.5 ms are replaced by general waveform learned from spikes during the training process.

The Adversarial Phase. In this phase, the discriminators are trained. Firstly for the given *input* of size (N, 48) the *categorical_head* and *gaussian_head* are calculated. In the next four steps, the total discriminatory loss is calculated. The $loss_{HC}$ is calculated as cross-entropy loss between the output of the $dsc_c(categorical\_head)$ and zero filled vector of size N. The $loss_{HG}$ is calculated as cross-entropy loss between the output of the $dsc_g(gaussian\_head)$ and zero filled vector of size N. At this step, two tensors are generated, first tensor tc is of size identical to *categorical_head* and is filled with randomly generated numbers sampled with categorical distribution along horizontal the axis. The second tensor tg is of size identical to *gaussian_head* and is filled with random numbers sampled with the normal distribution. The $loss_{SC}$ is calculated as cross-entropy loss between the output of the $dsc_c(tc)$ and one filled vector of size N. The $loss_{SG}$ is calculated as cross-entropy loss between the output of the $dsc_g(tg)$ and one filled vector of size N. Finally the total discriminatory loss $loss_{HC} + loss_{HG} + loss_{SC} + loss_{SG}$ is used for training of the categorical and Gaussian discriminators. It is expected that discriminators will return class 0 for data produced by the encoder while returning class 1 for the data sampled with desired distributions. cross-entropy loss was calculated using CrossEntropyLoss implementation provided by PyTorch [10].

The Generation Phase. The purpose of this phase is to train encoder to enforce categorical distribution on *categorical_head* and normal distribution on *gaussian_head*. This step is similar to the previous one in this that in each epoch, it uses the same calculated *categorical_head* and *gaussian_head* tensors. The $loss_{HC}$ and $loss_{HG}$ are however calculated differently. The $loss_{HC}$ is calculated as cross-entropy loss between the output of the $dsc_c(categorical\_head)$ and one filled vector of size N. The $loss_{HG}$ is calculated as cross-entropy loss between the output of the $dsc_g(gaussian\_head)$ and one filled vector of size N. Finally, the total generative loss $loss_{HC} + loss_{HG}$ is used for the training of the encoder. The encoder is trained to produce heads that discriminators would recognize as having proper distributions.

2.2   Training of the Autoencoder

For all four networks i.e., encoder, decoder, and both discriminators, for the weight optimization, the Adam [11] algorithm has been chosen. The initial learning rate for encoder and decoder was set to 1e−4, while for discriminators, the 1e−2 value was set. Additionally, for generalization, for the encoder and decoder, the weight decay was set to 0.0005. The stop condition in case of training of adversarial autoencoders is not as clearly defined as in many other applications where the lowest loss for the validation data is the golden standard [6]. Here one of the goals is, of course, the low validation loss on reconstruction phase but the main goal is the achievement of a stable equilibrium between adversarial and generative loss. All those conditions have been achieved during training as can be seen in Fig. 5a and especially in Fig. 5c and d. At epoch 3000, the recreation loss was 0.0020818 validation data (V) and 0.0020843 for test data (T). The respective adversarial losses were 2.3282955 (V) and 2.3300469 (T). Finally, the generative losses were 1.5815901 (V) and 1.5819962 (T). Important information can also be found in Fig. 5b where one can find how many of the classes represented by the categorical head, were actually used by the autoencoder.

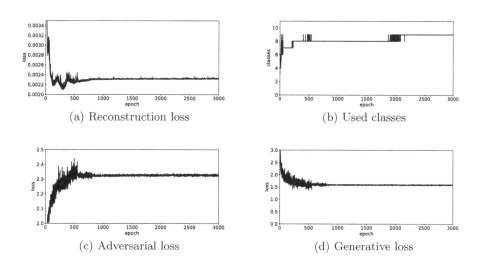

(a) Reconstruction loss   (b) Used classes

(c) Adversarial loss   (d) Generative loss

Fig. 5. Losses and used class count for validation data

3   Evaluation and Summary

During training, only nine out of 10 possible classes were used (see Fig. 5b). Spikes of all those classes were also found in validation and test data. Below, shown are three out of nine found classes. For each class shape, its center is shown with a thick black line; this is the shape for a particular class when all values in the Gaussian input of the decoder are set to 0. Gray shapes were generated by putting random numbers from $\mathcal{N}(0, 0.75)$ into the Gaussian input of the decoder. In fact, the decoder alone has become a GAN network [12] generating various spike shapes.

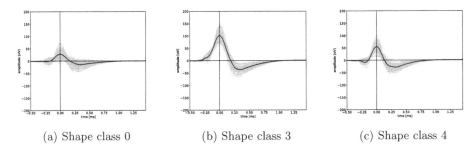

(a) Shape class 0   (b) Shape class 3   (c) Shape class 4

Fig. 6. Selected shape classes found in test data

The question that now has to be answered is if the Gaussian head for the test data has a normal distribution. Table 1 shows the p-values for the D'Agostino and Pearson's test [13].

Table 1. p-values for data in Gaussian head for classes found in test data. Gausian head is represented here as $gh \in \mathbb{R}^3$

Class ID	Count	p-value for gh[0]	p-value for gh[1]	p-value for gh[2]
0	6133	2.042e−07	2.296e−08	3.290e−25
1	4648	9.557e−25	2.173e−09	9.937e−31
2	2965	1.223e−05	5.464e−13	2.833e−22
3	1895	9.965e−07	3.931e−01	2.052e−03
4	5198	1.624e−12	7.758e−11	5.241e−02
5	3581	7.687e−46	5.608e−07	2.595e−05
6	2	–	–	–
7	4221	2.646e−41	3.097e−04	4.114e−32
8	4454	3.731e−55	9.075e−03	8.396e−20

Clearly, for all classes where the calculation of p-vale was feasible, in all but two underlined cases, the p-value is well below 0.005, i.e., the data is normally distributed. Additionally, one case where the p-value is above 0.05 is a narrow miss being 0.052.

It has been shown that adversarial autoencoder can be used for spike shape classification. Additionally, when analyzing spikes detected in a single recording, where their variety is much smaller than in the training set, obtained results might be used as a basis for further research. For example, one might try to find smaller, more specific clusters of spike shapes within classes found by autoencoder.

It must be remarked here that in the architecture of the adversarial autoencoder, setting the sizes of heads sets only the upper limits for the network. As the learning is in this particular case unsupervised, there is no guarantee that the number of classes found will be equal to the size of the categorical head. For example, in this paper, the categorical head has size 10, and only 9 classes were found.

Similarly, setting a certain size of the Gaussian head does not guarantee that all its dimensions will be used. During tests, when the size of the Gaussian head was set to 5, two of those dimensions were found to be very close to zero for all training, validation, and test data. They were left unused by the encoder and what is remarkable, further tests showed that their values had almost no influence on the decoder output.

References

1. Israel, Z., Burchiel, K.J.: Microelectrode Recording in Movement Disorder Surgery. Thieme Medical Publishers (2004)
2. Ciecierski, K., Raś, Z.W., Przybyszewski, A.W.: Foundations of recommender system for STN localization during DBS surgery in Parkinson's patients. In: Chen, L., Felfernig, A., Liu, J., Raś, Z.W. (eds.) ISMIS 2012. LNCS (LNAI), vol. 7661, pp. 234–243. Springer, Heidelberg (2012). https://doi.org/10.1007/978-3-642-34624-8_28
3. Koch, C.: Biophysics of Computation: Information Processing in Single Neurons. Oxford University Press (2004)
4. Lewicki, M.S.: A review of methods for spike sorting: the detection and classification of neural action potentials. Netw. Comput. Neural Syst. **9**(4), R53–R78 (1998)
5. Quiroga, R.Q., Nadasdy, Z., Ben-Shaul, Y.: Unsupervised spike detection and sorting with wavelets and superparamagnetic clustering. Neural Comput. **16**(8), 1661–1687 (2004)
6. Goodfellow, I., Bengio, Y., Courville, A.: Deep Learning. MIT press, New York (2016)
7. Chen, X., et al.: Variational lossy autoencoder. arXiv preprint arXiv:1611.02731 (2016)
8. Makhzani, A., Shlens, J., Jaitly, N., Goodfellow, I., Frey, B.: Adversarial autoencoders. arXiv preprint arXiv:1511.05644 (2015)
9. Agresti, A., Kateri, M.: Categorical Data Analysis. Springer, Heidelberg (2011). https://doi.org/10.1007/978-94-007-0753-5_291
10. Paszke, A., et al.: Automatic differentiation in PyTorch (2017)

11. Kingma, D.P., Ba, J.: Adam: a method for stochastic optimization. arXiv preprint arXiv:1412.6980 (2014)
12. Goodfellow, I., et al.: Generative adversarial nets. In: Advances in Neural Information Processing Systems, pp. 2672–2680 (2014)
13. Pearson, E.S., D"'Agostino, R.B., Bowman, K.O.: Tests for departure from normality: comparison of powers. Biometrika **64**(2), 231–246 (1977)

Digital Signal Processing

Poriferal Vision: Classifying Benthic Sponge Spicules to Assess Historical Impacts of Marine Climate Change

Saketh Saxena[1], Philip Heller[1(✉)], Amanda S. Kahn[2], and Ivano Aiello[2]

[1] San Jose State University, San Jose, CA 95192, USA
philip.heller@sjsu.edu
[2] Moss Landing Marine Laboratories, Moss Landing, CA 95039, USA

Abstract. Sponges and corals are ecologically important members of the marine community. Climate change, while harmful to corals, has historically been favorable to sponges. Sponge population dynamics are studied by analyzing core samples of marine sediment. To date this analysis has been performed by microscopic visual inspection of core cross sections to distinguish spicules (the rigid silica components of sponge skeletons) from the residue of other silica-using organisms. Since this analysis is both slow and error prone, complete analysis of multiple cross sections is impossible.

FlowCam® technology can produce tens of thousands of microphotographs of individual core sample particles in a few minutes. Individual photos must then be classified *in silico*. We have developed a Deep Learning classifier, called Poriferal Vision, that distinguishes sponge spicules from non-spicule particles. Small training sets were enhanced using image augmentation to achieve accuracy of at least 95%. A Support Vector Machine trained on the same data achieved accuracy of at most 86%. Our results demonstrate the efficacy of Deep Learning for analyzing core samples, and show that our classifier will be an effective tool for large-scale analysis.

Keywords: Deep Learning · Sponges · Porifera · Climate

1 Introduction

Sponges (Phylum Porifera) are prolific reef builders [1] and benthic grazers [2, 3]. Many sponges (Phylum Porifera, especially Classes Demospongiae and Hexactinellida) form spicules – the rigid components of their skeletons – composed of silica. Since silica can resist breakdown for millions of years, spicules are part of the geological record, providing insight into the presence and ecological significance of sponges under a variety of climatic conditions. Sponge populations have increased significantly during planetary warm periods, including the Upper Carboniferous period (300 mya) [4], some stages of the Paleozoic (540–250 mya) and Mesozoic (250–65 mya) eras [5], the Triassic/Jurassic transition (200 mya) [6], and the Pliocene warm period (3 mya) [7].

© Springer Nature Switzerland AG 2020
D. Helic et al. (Eds.): ISMIS 2020, LNAI 12117, pp. 205–213, 2020.
https://doi.org/10.1007/978-3-030-59491-6_19

Shallow water reefs are hotspots of marine biodiversity, fringing 1/6 of the world's coastlines [8] and supporting hundreds of thousands of species [9]. Reefs are predominantly coral rather than sponge; the hard carbonate exoskeletons of dead coral polyps accrete to form non-living support structures for living reef components. However, coral reefs are particularly vulnerable to current climate change conditions. Rising temperatures cause bleaching by forcing expulsion of heat-intolerant symbionts [10, 11]. Ocean acidification, driven by increased CO_2 in the atmosphere, slows production of calcium carbonate and accelerates removal of calcium carbonate by dissolution and bioerosion [12–14] in shallow and deep waters; moreover there is mounting evidence that increasing acidity disrupts the life cycle of corals at multiple stages [15]. Goreau et al. have estimated that climate change had destroyed or degraded 25% of all reefs by the year 2000 [16]. The loss of coral reefs can be expected to continue through the century, since atmospheric CO_2 levels are expected to rise to 750 ppm by 2100 [17].

The ongoing degradation of coral reefs, together with the historical success of sponges during warm periods, suggest that sponges may be due for a resurgence in ecological importance, and may replace corals as major reef builders [18]. The ecological and economic importance of reefs motivate research into the mechanics of the historical success of sponges during past warm periods. Although glass sponges do not build significant shallow water reefs, they are similar to reef-building corals in that they form the foundation for unique communities [19], provide three-dimensional habitat [20], are a nursery habitat for many species [21], and enhance local biodiversity [19, 22]. Their historical response to climate change can therefore elucidate present-day coral and sponge population dynamics. Much information can be gained by counting and classifying spicules from smear slides of spicule-bearing core samples (Fig. 1).

Unfortunately, while particle size distribution can be computed mechanically, identification of particles is labor intensive. Core sections must be inspected microscopically, and spicules must be distinguished from the organic and inorganic constituents of the cores, plus refractory detritus of other silica-using organisms such as diatoms and radiolarians (Phylum Retaria). Comprehensive analysis of large numbers of sections is not practical. FlowCam® technology (Fluid Imaging Technologies, Inc.) offers a partial solution to this problem of scale. In a FlowCam device, particles in suspension pass one by one in front of a camera; this approach can generate tens of thousands of photographs from a sample in a matter of minutes. To our knowledge this technology has not yet been systematically applied to core samples, although recent success with cyanobacterial blooms [23] suggests that the approach is promising. However, a FlowCam approach to large-scale spicule analysis requires software classification of large numbers of individual photographs. To facilitate development of a classifier, Fluid Imaging Technologies generously performed a small-scale run on a core sample from the Bering Sea from the Pliocene warm period. 110 photographs of sponge spicules and 113 photographs of non-spicule particles (diatoms and radiolarians) were generated (Fig. 1).

Our experience with computer vision technology suggested that an accurate photograph classifier could be achieved using Artificial Neural Networks (ANNs), which have been shown to be effective in handling large scale image classification problems and are the de-facto standard in image recognition problems [24, 25]. In particular we hypothesized that a deep convolutional neural network (ConvNet) [26] could achieve

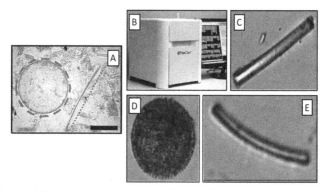

Fig. 1. (A) Smear slide microphotograph showing a diatom (in red dashed circle) and a glass sponge spicule (along dashed red line). (B) FlowCam series 8000 flow imaging microscope. Reprinted with permission from Fluid Imaging Technologies, Inc. (C) Glass sponge spicule. (D) Diatom. (E) Radiolarian. The similarity between (C) and (E) necessitates a sophisticated software classifier. (Color figure online)

acceptable accuracy despite the obvious similarity between glass sponge spicules and radiolarians. However, ConvNets require extensive positive training data, and the available training data consisted of just 110 positive examples and 113 negative examples from the few FlowCam images provided by Fluid Imaging Technologies. Fortunately, training data can be enhanced *in silico*, and we further hypothesized that ConvNet accuracy could benefit from data augmentation [27] via image manipulation of the original photographs. We have trained a deep ConvNet, called "Poriferal Vision", that achieves 95% accuracy with available data. A Support Vector Machine (SVM) classifier [28, 29] was trained using the same data to provide a basis for comparison; the Poriferal Vision model consistently achieved higher accuracy.

2   Methods

To enable rapid *in silico* identification of FlowCam images from marine core samples, a deep convolutional neural network (ConvNet) called "Poriferal Vision" was trained on data from the Bering Sea. Data augmentation was applied to training instances. 20% of the augmented data was withheld from training and used for testing. Withheld data was tested on the Poriferal Vision ConvNet and, for comparison, on a Support Vector Machine.

2.1   Data Collection

Sediments from a section of a Bering Sea core sample (Fig. 3) were provided by the Aiello research group at Moss Landing Marine Laboratories. The Integrated Ocean Drilling Program Expedition 323 to the Bering Sea (IODP Exp 323) discovered that, in the central part of this marginal sea at Bowers Ridge (Site U1340, ~600 m; water depth 1295 m; Fig. 3), sponge spicules comprise a very important component of hemipelagic sedimentation, together with diatom frustules and ice-rafted debris (Fig. 2) [7, 30].

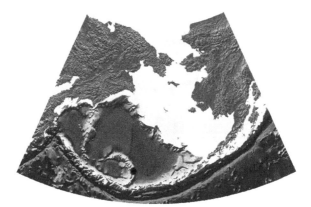

Fig. 2. Bering Sea and adjacent landmasses. The red dot marks sampling site U1340. (Color figure online)

To create positive training and testing sets, 110 sponge spicule photographs were manually selected. Negative example photographs of 80 diatoms and 33 radiolarians were collected. All photographs were randomly split into training and testing sets. To compensate for the small number of training examples, the positive and negative training sets were enhanced by transforming images using the OpenCV software package [31]. Individual images were read into OpenCV in grayscale. In order to preserve the structure of the single objects in the photos and to get more realistic transformations, the polygonal chain approximation method [32] was used to trace highly accurate contours around each object. A rectangular bounding box was drawn around the contours, isolating the object and preserving its local background context. The bounding box was flipped horizontally, vertically, and horizontally + vertically to generate 3 new images from each original. All images were rotated in increments of 15°. Lastly, in order to reduce the probability of misidentification of poorly centered images, 6 perspective transformations were applied to each original image by translating the object by a small distance. To ensure that perspective transformation did not create any blank images, a blurring Gaussian filter [33] was applied to the transformed images; any image with a majority of black pixels or a majority of white pixels was removed from the data set. All image backgrounds were cleaned by removing noise and deleting line artifacts of rotation. Images were saved in .jpg format and, as a quality control check, visually inspected. Accepted images were resized to 138 × 78 pixels (the mean resolution among accepted images).

2.2   ConvNet and Support Vector Machine Design and Training

The Poriferal Vision ConvNet was implemented in TensorFlow [34]; The architecture of the model was conservatively inspired by AlexNet [26], which seems suitable as a pioneering large scale image classification architecture for our specific problem. The motivation behind designing a lean ConvNet was to accommodate for the limited size of the dataset despite data augmentation and learn a generalizable distribution of features from it. It consists of two 2D convolutional layers with a filter size of 64 and kernel size of (3, 3) to convolve over. Considering the relatively small dimension of the transformed images a smaller kernel size would allow learning localized features and differentiate between spicules and radiolarians, due to the high degree of similarity between them. Both layers use relu [35] activation function and max pooling to down sample the input representation.

These are followed by a flatten and dense layer with relu activation to reduce the spatial dimension of the input, finally followed by an output layer with sigmoid activation [36] (Fig. 3). Training proceeded for 14 epochs.

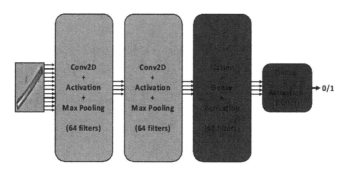

Fig. 3. Poriferal Vision Convolutional Neural Network (CNN) design

To provide a basis for assessing the Poriferal Vision model, a Support Vector Machine classifier was created using the Support Vector Classifier module of SciKit-Learn [37], with a linear kernel. Following the method of R. Sahak et al. [38], training time was reduced by using Principal Component Analysis (PCA) to limit feature dimensionality. The kernel's c and gamma parameters were optimized using SciKit-Learn's GridSearchCV module.

2.3 Evaluation

The Poriferal Vision ConvNet was trained first with the original training data and then with the enhanced data. In each category (sponge spicules, diatoms, and radiolarians), 20% of data was withheld from training and later used for testing (Table 1). All training and experiments performed on Poriferal Vision were also performed on the Support Vector Machine. For all 4 experiments (Poriferal Vision and SVM, original and enhanced data), the classifiers were evaluated by computing test accuracy, f-score, precision, and recall.

Table 1. Training and testing sets.

Data Set	Organism	Original photos	Original testing photos	Original training photos	Flipped images	Rotated images	Perspective transformations	Total selected training images
Positive	Sponges	110	30	80	240	5120	376	5816
Negative	Diatoms	90	22	68	204	4352	408	5032
	Radiolarians	33	8	25	75	1600	150	1850

3 Results

Test accuracy, f-score, precision, and recall for the 4 experiments are given in Table 2. Accuracy and loss over 14 training epochs, for both the original and enhanced data sets, are shown in Fig. 4.

Table 2. Results for SVM and Poriferal Vision, for original and enhanced (i.e. original and transformed) images.

Model	Images	Test accuracy	F-score	Precision	Recall
SVM	Original	0.83	0.83	0.83	0.83
	Enhanced	0.80	0.79	0.86	0.80
Poriferal Vision	Original	0.92	0.92	0.93	0.92
	Enhanced	0.95	0.95	0.95	0.95

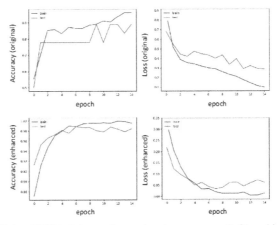

Fig. 4. Poriferal Vision CNN training regimes, showing accuracy (left) and loss (right) with the original (top) and enhanced (bottom) data sets over 14 training epochs. The test set (brown lines) was composed of 20% of each photograph set (spicules, diatoms, radiolarians), selected randomly and withheld from model training. (Color figure online)

4 Discussion

The Poriferal Vision ConvNet, trained with the enhanced data set, achieved 95% accuracy according to all 4 computed statistics (test accuracy, f-score, precision, and recall). Enhancing the positive and negative training data added 2–3% to the statistics. Our hypotheses are therefore confirmed: a ConvNet has been shown to achieve reasonable accuracy, and image enhancement improves accuracy. Significantly, the enhanced-data version of Poriferal Vision is accurate enough to be applied to large-scale sets of FlowCam images when those become available.

The Support Vector Machine classifiers, trained on the same data as the Poriferal Vision models, was less accurate than Poriferal Vision by 5 to 11% points. This may be due to the PCA-based dimension reduction which was applied in order to make model computation time tractable. The time complexity of a linear kernel SVM is $O(n)$, where n is the number of dimensions. Since SVMs are not practical without dimension reduction, they are not appropriate for the kind of classification undertaken here.

The next stage of Poriferal Vision's development will involve taxonomic classification of spicules (and perhaps diatoms and radiolarians). We hope and expect that the insights gained from consequent analysis of the historical record will help the scientific and ocean policy communities to understand the future role of sponges in marine environments.

5 Conclusions

The Poriferal Vision ConvNet, trained with the enhanced data set, achieved at least 95% accuracy according to all 4 computed statistics (test accuracy, f-score, precision, and recall). This result justifies future efforts to expand the training set.

A larger training set, with spicules identified by species or by higher taxonomic level, will enable a more sophisticated classifier that can identify samples by species or by higher taxonomic level. Such a classifier, if deployed on shipboard and receiving FlowCam images in real time, would enable real-time analysis of the local marine environment, thus greatly enhancing the efficiency of sponge analysis at a moment in human history when such analysis is urgent.

References

1. Conway, K.W., Barrie, J.V., Krautter, M.: Geomorphology of unique reefs on the western Canadian shelf: sponge reefs mapped by multibeam bathymetry. Geo-Mar. Lett. **25**(4), 205–213 (2005)
2. Yahel, G., Whitney, F., Reiswig, H.M., Eerkes-Medrano, D.I., Leys, S.P.: In situ feeding and metabolism of glass sponges (Hexactinellida, Porifera) studied in a deep temperate fjord with a remotely operated submersible. Limnol. Oceanogr. **52**(1), 428–440 (2007)
3. Kahn, A.S., Yahel, G., Chu, J.W.F., Tunnicliffe, V., Leys, S.P.: Benthic grazing and carbon sequestration by deep-water glass sponge reefs. Limnol. Oceanogr. **60**(1), 78–88 (2015)
4. West, R.R.: Temporal changes in Carboniferous reef mound communities. PALAIOS **3**(2), 152 (1988)
5. Brunton, F.R., Dixon, O.A.: Siliceous sponge-microbe biotic associations and their recurrence through the phanerozoic as reef mound constructors. PALAIOS **9**(4), 370 (1994)
6. Kiessling, W., Simpson, C.: On the potential for ocean acidification to be a general cause of ancient reef crises: ancient reef crises. Glob. Change Biol. **17**(1), 56–67 (2011)
7. Aiello, I.W., Ravelo, A.C.: Evolution of marine sedimentation in the Bering Sea since the Pliocene. Geosphere **8**(6), 1231–1253 (2012)
8. Birkeland, C. (ed.): Life and Death of Coral Reefs. Chapman and Hall, New York (1997)
9. Reaka-Kudla, M.L., Wilson, D.E., Wilson, E.O.: Biodiversity II. Joseph Henry Press, Washington, D.C. (1997)
10. Glynn, P.W.: Coral reef bleaching: ecological perspectives. Coral Reefs **12**(1), 1–17 (1993). https://doi.org/10.1007/BF00303779

11. Hughes, T.P.: Climate change, human impacts, and the resilience of coral reefs. Science **301**(5635), 929–933 (2003)
12. De'ath, G., Fabricius, K.E., Sweatman, H., Puotinen, M.: The 27-year decline of coral cover on the Great Barrier Reef and its causes. Proc. Natl. Acad. Sci. **109**(44), 17995–17999 (2012)
13. Hoegh-Guldberg, O., et al.: Coral reefs under rapid climate change and ocean acidification. Science **318**(5857), 1737–1742 (2007)
14. Wisshak, M., Schönberg, C.H.L., Form, A., Freiwald, A.: Ocean acidification accelerates reef bioerosion. PLoS ONE **7**(9), e45124 (2012)
15. Albright, R.: Reviewing the effects of ocean acidification on sexual reproduction and early life history stages of reef-building corals. J. Mar. Biol. **2011**, 1–14 (2011)
16. Goreau, T., McClanahan, T., Hayes, R., Strong, A.: Conservation of coral reefs after the 1998 global bleaching event. Conserv. Biol. **14**(1), 5–15 (2000)
17. Nakićenović, N., Intergovernmental Panel on Climate Change (eds.): Special Report on Emissions Scenarios: A Special Report of Working Group III of the Intergovernmental Panel on Climate Change. Cambridge University Press, Cambridge, New York (2000)
18. Bell, J.J., Davy, S.K., Jones, T., Taylor, M.W., Webster, N.S.: Could some coral reefs become sponge reefs as our climate changes? Glob. Change Biol. **19**(9), 2613–2624 (2013)
19. Chu, J., Leys, S.: High resolution mapping of community structure in three glass sponge reefs (Porifera, Hexactinellida). Mar. Ecol. Prog. Ser. **417**, 97–113 (2010)
20. Beaulieu, S.E.: Life on glass houses: sponge stalk communities in the deep sea. Mar. Biol. **138**(4), 803–817 (2001)
21. Marliave, J.B., Conway, K.W., Gibbs, D.M., Lamb, A., Gibbs, C.: Biodiversity and rockfish recruitment in sponge gardens and bioherms of southern British Columbia, Canada. Mar. Biol. **156**(11), 2247–2254 (2009)
22. Guillas, K.C., Kahn, A.S., Grant, N., Archer, S.K., Dunham, A., Leys, S.P.: Settlement of juvenile glass sponges and other invertebrate cryptofauna on the Hecate Strait glass sponge reefs. Invertebr. Biol. **138**(4), e12266 (2019)
23. Graham, M.D., et al.: High-resolution imaging particle analysis of freshwater cyanobacterial blooms: FlowCam analysis of cyanobacteria. Limnol. Oceanogr. Methods **16**(10), 669–679 (2018)
24. Widrow, B., Rumelhart, D.E., Lehr, M.A.: Neural networks: applications in industry, business and science. Commun. ACM **37**, 93–106 (1994)
25. Rawat, W., Wang, Z.: Deep convolutional neural networks for image classification: a comprehensive review. Neural Comput. **29**(9), 2352–2449 (2017)
26. Krizhevsky, A., Sutskever, I., Hinton, G.E.: ImageNet classification with deep convolutional neural networks. Commun. ACM **60**(6), 84–90 (2017)
27. Shorten, C., Khoshgoftaar, T.M.: A survey on image data augmentation for deep learning. J. Big Data **6**(1), 1–48 (2019). https://doi.org/10.1186/s40537-019-0197-0
28. Cortes, C., Vapnik, V.: Support-vector networks. Mach. Learn. **20**(3), 273–297 (1995)
29. Chapelle, O., Haffner, P., Vapnik, V.N.: Support vector machines for histogram-based image classification. IEEE Trans. Neural Netw. **10**(5), 1055–1064 (1999)
30. Schlung, S.A., et al.: Millennial-scale climate change and intermediate water circulation in the Bering Sea from 90 ka: a high-resolution record from IODP Site U1340: Bering Sea climate change since 90 KA. Paleoceanography **28**(1), 54–67 (2013)
31. Bradski, G.R., Kaehler, A.: Learning OpenCV: Computer Vision with the OpenCV Library, 1st edn. O'Reilly, Beijing (2011)
32. Ramer, U.: An iterative procedure for the polygonal approximation of plane curves. Comput. Graph. Image Process. **1**(3), 244–256 (1972)
33. Deng, G., Cahill, L.W.: An adaptive Gaussian filter for noise reduction and edge detection. In: 1993 IEEE Conference Record Nuclear Science Symposium and Medical Imaging Conference (1993)

34. Abadi, M., et al.: TensorFlow: a system for large-scale machine learning, p. 21 (2016)
35. Hahnloser, R.H.R., Sarpeshkar, R., Mahowald, M.A., Douglas, R.J., Seung, H.S.: Digital selection and analogue amplification coexist in a cortex-inspired silicon circuit. Nature **405**, 947–951 (2000)
36. Kilian, J., Siegelmann, H.T.: The dynamic universality of sigmoidal neural networks. Inf. Comput. **128**(1), 48–56 (1996)
37. Pedregosa, F., et al.: Scikit-learn: machine learning in Python. J. Mach. Learn. Res. **12**, 2825–2830 (2011)
38. Sahak, R., Mansor, W., Lee, Y.K., Yassin, A.I.M., Zabidi, A.: Performance of combined support vector machine and principal component analysis in recognizing infant cry with asphyxia. In: 2010 Annual International Conference of the IEEE Engineering in Medicine and Biology (2010)

Experimental Evaluation of GAN-Based One-Class Anomaly Detection on Office Monitoring

Ning Dong[1], Yusuke Hatae[1], Muhammad Fikko Fadjrimiratno[2],
Tetsu Matsukawa[1], and Einoshin Suzuki[1]

[1] ISEE, Kyushu University, Fukuoka 819-0395, Japan
`dongning.ac@gmail.com`, `hatae.yusuke.042@s.kyushu-u.ac.jp`,
`{matsukawa,suzuki}@inf.kyushu-u.ac.jp`
[2] SLS, Kyushu University, Fukuoka 819-0395, Japan
`muhammadfikko@gmail.com`

Abstract. In this paper, we test two anomaly detection methods based on Generative Adversarial Networks (GAN) on office monitoring including humans. GAN-based methods, especially those equipped with encoders and decoders, have shown impressive results in detecting new anomalies from images. We have been working on human monitoring in office environments with autonomous mobile robots and are motivated to incorporate the impressive, recent progress of GAN-based methods. Lawson et al.'s work tackled a similar problem of anomalous detection in an indoor, patrol trajectory environment with their patrolbot with a GAN-based method, though crucial differences such as the absence of humans exist for our purpose. We test a variant of their method, which we call FA-GAN here, as well as the cutting-edge method of GANomaly on our own robotic dataset. Motivated to employ such a method for a turnable Video Camera Recorder (VCR) placed at a fixed point, we also test the two methods for another dataset. Our experimental evaluation and subsequent analyses revealed interesting tendencies of the two methods including the effect of a missing normal image for GANomaly and their dependencies on the anomaly threshold.

Keywords: One-class anomaly detection · Generative Adversarial Networks · Human monitoring

1 Introduction

Monitoring an office environment, especially the humans inside, represents an interesting problem for intelligent systems from both scientific and industrial viewpoints. Detecting anomalies is one of the most fundamental and yet important subproblems, though collecting and even knowing such anomalies beforehand are at the same time laborious and difficult. One-class anomaly detection,

A part of this work is supported by Grant-in-Aid for Scientific Research JP18H03290 from the Japan Society for the Promotion of Science (JSPS).

which takes only normal data in the training stage to detect anomalies in the test stage, solves these shortcomings. Recently Generative Adversarial Networks (GAN) [1], which are deep neural networks capable of learning the probabilistic distribution of the given, originally unlabeled, examples, have shown impressive results in detecting new anomalies from images. For instance, Lawson et al. report that the false positive rate of 4.72% achieved with their previous method [2] dropped to 0.42% with their GAN-based method in an anomaly detection problem by their patrolbot [3].

We have been working on office monitoring including humans inside with autonomous mobile robots, e.g., skeleton clustering [4], facial expression clustering [5], and fatigue detection [6]. Recently our interests are focused on one-class anomaly detection [7,8], though these works adopt non-GAN-based approaches. Motivated to incorporate the impressive, recent progress of GAN-based methods, we test two most relevant methods, which we explain in Sect. 2, on our robotic and VCR datasets.

2   Related Work

Recently GAN-based one-class anomaly detection has attracted considerable attention of the machine learning community. Schlegl et al. [9] proposed AnoGAN to detect anomalies on Optical Coherence Tomography (OCT) data. They assumed that the trained latent space represents the true distribution of the training data. However, their method is time-consuming in finding a latent vector that corresponds to an image that is visually most similar to a given query image [10] in the test stage. To cope with the shortcoming that the parameters need to be updated in the test stage of AnoGAN, Zenati et al. proposed an anomaly detection method [11] which is efficient at test time by leveraging BiGAN [12], which simultaneously learns an encoder with a decoder and a discriminator during training. It can avoid the computationally expensive process during the test stage. Sabokrou et al. proposed a framework for one-class novelty detection which consists of a reconstructor and a discriminator [13]. They added noise to the original normal examples to train the reconstructor network to make it more robust and employed the discriminator as a detector to classify whether the input is abnormal. A deep generative model trained on a single class cannot generate examples belonging to other classes. Perera et al. focused on this problem and proposed OCGAN for novelty detection [14]. They restricted the boundary of the latent space and used a latent discriminator and a visual discriminator to ensure the images generated from any latent vector belong to the same class. Different from the traditional GAN-based encoder-decoder approaches, Akcay et al. [10] employed an encoder-decoder-encoder structure to capture the two latent vectors which show significant differences with an abnormal example. The added encoder aids learning the data distribution for the normal examples. We adopt their GANomaly [10] for evaluation in our experiments as we consider it a relevant cutting-edge method.

We have witnessed a number of developments and applications of autonomous mobile robots in anomaly detection. Chakravarty et al. [15] used modified sparse and dense stereo algorithms to detect anomalies which were never shown during the training stage for a patrolbot. However, light intensity has a great impact on their detection accuracy. Lawson et al. [2] proposed a method with clustering normal features from CNNs in a fixed path with a patrolbot. Abnormal features would produce large distances to these clusters. Later, they extended their work in [3] to find anomalies with an autoencoder-decoder GAN, which achieved much better performance as we stated in the previous section. We extend their method by replacing their autoencoder with a more sophisticated encoder [16], and call the extended method FA-GAN in this paper. We also use FA-GAN in our experiments.

The robotic data and the VCR data we use in our experiments have been introduced in our previous work on detecting anomalous image regions with deep captioning [8]. In the work, our anomaly detector represents each salient region with its image, caption, and position features and uses an incremental clustering method [17] to detect anomalies with these features. The point is to exploit another dataset used in training a combination of Convolutional Neural Network (CNN) [18] and Long Short-Term Memory (LSTM) [19,20] through deep captioning [21]. We will explain the details of our datasets in Sect. 4.1. Since the method [8] uses deep captioning and conducts evaluation on image region level, we leave its comparison with GANomaly and FA-GAN for our future work.

3   Tested Methods

3.1   FA-GAN

Lawson et al. use the DCGAN approach [22], adopting an architecture which is similar to what was proposed in Context Encoders [16]. Unlike DCGAN, they use an autoencoder-style with a bottleneck size of 4096. Considering the more complex nature of our office monitoring problem, we replace their autoencoder with a more sophisticated encoder in [16], as we explained in the previous Section. The resulting FA-GAN has an encoder-decoder architecture, which is shown in Fig. 1(a). The encoder G_E is composed of convolution layers and batch-normalization layers with LeakyReLU activation function. The decoder G_D adopts the structure of DCGAN [22] with deconvolutional layers to generate images from a latent vector. The discriminator D has a similar structure to G_E and uses Sigmoid function to output whether the input is real or generated.

Given a training set X, which consists of N normal images, $X = \{x_1, \ldots, x_N\}$, the generator G first reads an input image x as the input to its encoder G_E to downscale x by compressing it to a latent vector $z = G_E(x)$. Then the decoder G_D tries to reconstruct z to an image $\hat{x}$. To enhance the capability of reconstructing an image, FA-GAN uses the adversarial loss [1] shown in Eq. 1.

$$Loss\_FA = \mathbb{E}_{x \sim p_x} \left[\log D(x) \right] + \mathbb{E}_{x \sim p_x} \left[\log(1 - D(\hat{x})) \right] \tag{1}$$

In the test stage, the generator G produces the latent vectors z and $\hat{z}$ from the original input image and its corresponding generated image through the encoders

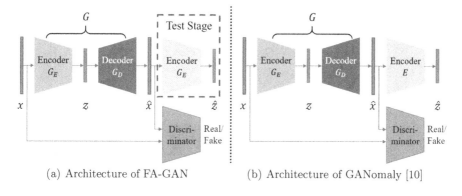

(a) Architecture of FA-GAN   (b) Architecture of GANomaly [10]

Fig. 1. Architectures of FA-GAN (the left one) and GANomaly (the right one)

in the network, respectively. Note that the encoder in the dashed rectangle is not used during the training stage. It is just a copy of the former one after training. Finally, a test example can be determined as normal or abnormal by comparing its abnormal degree A with a user-given threshold ϕ, where $A = \|z - \hat{z}\|_2$ is computed as the distance between the two latent vectors z and $\hat{z}$. If $A > \phi$ then the image is predicted as anomalous, otherwise normal. Since the network is only trained with normal data, the generator cannot reconstruct an anomalous image well, which means there will be a large difference between the two latent vectors.

3.2 GANomaly

GANomaly [10] has a similar architecture to FA-GAN. It also uses an encoder-decoder generator in which the latent vector z of the original input is obtained by $G_E(x) = z$. The difference is that there is one more encoder E to produce $\hat{z}$ after the generated image as shown in Fig. 1 (b). The parameters of the additional encoder E are also optimized during the training stage, in which the distance between z and $\hat{z}$ is considered as a loss. After the second encoder E, the generated image is downscaled to a latent vector $\hat{z} = E(\hat{x})$, which has the same size with z.

The loss function $Loss\_GANomaly$ of GANomaly consists of an adversarial loss L_{adv}, a contextual loss L_{con}, and an encoder loss L_{enc} as in Eq. 2, where w_{adv}, w_{con}, and w_{enc} represent hyper-parameters. In Eq. 3, $f(x)$ represents the value of the activation function on the last CNN layer in the discriminator given image x [10]. To predict whether a test example is abnormal, GANomaly uses the same flow as FA-GAN, which is to compute the distance between z and $\hat{z}$, except $\hat{z}$ is produced by the additional encoder E, and not G_E.

$$Loss\_GANomaly = w_{adv}L_{adv} + w_{con}L_{con} + w_{enc}L_{enc} \qquad (2)$$
$$L_{adv} = \mathbb{E}_{x\sim p_x} \|f(x) - \mathbb{E}_{x\sim p_x} f(G(x))\|_2 \qquad (3)$$
$$L_{con} = \mathbb{E}_{x\sim p_x} \|x - G(x)\|_1 \qquad (4)$$
$$L_{enc} = \mathbb{E}_{x\sim p_x} \|z - \hat{z}\|_2 \qquad (5)$$

Fig. 2. Examples in the robotic dataset (the left two) and the VCR dataset (the right two). From the left, the classes are normal, abnormal, normal, and abnormal.

Table 1. Distributions of the two datasets

	Robotic dataset		VCR dataset	
	Training	Test	Training	Test
Normal	4768	343	16800	684
Abnormal	0	15	0	31

4   Experimental Evaluation

4.1   Experimental Setup

To evaluate the two GAN-based methods on anomaly detection, we conduct experiments with our robotic and VCR datasets, which we introduced in [8]. Figure 2 shows several examples. The robotic dataset is taken in a room by our TurtleBot2 with Kobuki, which is equipped with a Kinect v2. It contains frequent scene changes as the robot moved in an office. The VCR dataset is taken with a VCR placed on a spandrel wall. It has only a few scene changes as the VCR was put on a fixed point and a human occasionally changed its angle. Table 1 shows the distributions of the datasets.

We install GANomaly [10] and implement FA-GAN [3] in PyTorch (v1.3.1 with Python 3.6.9). The networks are optimized by Adam [23] with an initial learning rate of 0.0001, $\beta_1 = 0.5$, and $\beta_2 = 0.999$. We set the batch size to 32 and each network is trained for 70 epochs. The hyper-parameters are set as $w_{adv} = 1$, $w_{con} = 50$, and $w_{enc} = 1$. The size of a latent vector is set to 4096 throughout the experiments. We normalize all the anomaly scores obtained in the test set to the range of [0, 1].

4.2   Results

In this section, we first analyze the dependency of the performance in F1 score on ϕ and then investigate the reasons behind the mistakes by the two methods. The left two plots in Fig. 3 show the results of the dependency. We see that FA-GAN outperforms GANomaly in the robotic dataset but loses to it in the VCR dataset. From Fig. 4 we can see the reasons. In (a), there is a clear boundary to distinguish the normal and abnormal examples with FA-GAN on the robotic dataset. GANomaly succeeds in concentrating the anomaly scores of all normal examples in a small interval on the VCR dataset in (d). The right two plots in Fig. 3 show the results in ROC curve and AUC.

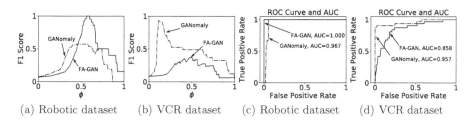

(a) Robotic dataset   (b) VCR dataset   (c) Robotic dataset   (d) VCR dataset

Fig. 3. F1 scores in terms of threshold ϕ (left two plots) and the ROC curve and AUC (right two plots)

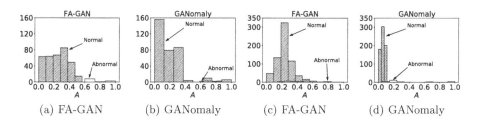

(a) FA-GAN   (b) GANomaly   (c) FA-GAN   (d) GANomaly

Fig. 4. Histogram of the anomaly scores for the test data on the robotic dataset (the left two plots) and the VCR dataset (the right two plots)

We assume that setting ϕ to its best value is possible as long as the office environment does not change drastically. Based on this assumption we conduct our investigation on the best cases in terms of the value of ϕ. We focus on mistakes committed by the two methods, which are summarized in Table 2. In the Table, FN and FP represent the number of false negatives and the number of false positives, respectively. "Same examples" is the number of images wrongly detected by both methods.

On the robotic dataset, we see from Table 2 that FA-GAN made no mistake while GANomaly 18 false positives. Since the 18 examples look all similar, we

Table 2. Statistics of the wrongly detected examples

	Robotic dataset		VCR dataset	
	FN	FP	FN	FP
FA-GAN	0	0	19	21
GANomaly	0	18	4	0
Same examples	0	0	2	0

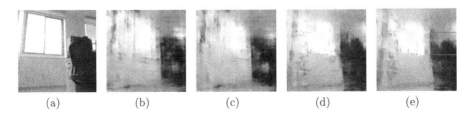

(a)   (b)   (c)   (d)   (e)

Fig. 5. One of the FP examples with GANomaly in the robotic dataset. (a) Original input. (b) Generated image with z by FA-GAN. (c) Generated image with $\hat{z}$ by FA-GAN. (d) Generated image with z by GANomaly. (e) Generated image with $\hat{z}$ by GANomaly.

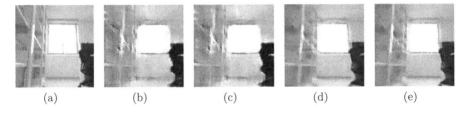

(a)   (b)   (c)   (d)   (e)

Fig. 6. One of the correctly detected examples by the two methods. See the captions of Fig. 5 for (a)–(e).

pick one of them and show it in Fig. 5(a). The anomaly scores A are 0.968 in GANomamly and 0.506 in FA-GAN. We also show the generated images by z and $\hat{z}$ with both methods in Fig. 5(b)–(e) and see that the red rectangle region in Fig. 5(e) accounts for the large A in GANomaly.

We also pick an example which is similar to Fig. 5(a) but correctly classified by both methods and show it in Fig. 6(a). The anomaly scores are 0.266 and 0.106 for FA-GAN and GANomaly, respectively. We see from Fig. 6(b), (c) and (d), (e) that z and $\hat{z}$ are similar in both methods.

Further inspection revealed that no similar image to Fig. 5(a) exists in the training set while all other images in the test set have similar images in the training set. These analyses show the higher generalization capability of FA-GAN to GANomaly for this dataset. To justify our claim, we added random noise to the 18 FP examples to generate 72 additional examples and added them in the training set. We trained GANomaly on this dataset and obtained a perfect result, i.e., no mistake committed and hence AUC = 1.0.

On the VCR dataset, we see from Table 2 that there are only 4 FN examples with GANomaly but 19 FN examples and 21 FP examples with FA-GAN. This dataset was recorded in one corner by the VCR, so the intuitive complexity of the dataset is relatively reduced compared with the first dataset. Moreover, the training set, which consists of 16800 examples, is large, so the second encoder of GANomaly can well learn the distributions of the normal features in this dataset. Figure 7(a) shows an FN example with FA-GAN, which is correctly detected by GANomaly. Hiding under a table[1] is considered as an anomaly because nobody does it in the training set. The subsequent images in (b)–(e) are generated images with z and $\hat{z}$ by the two methods. Unlike in Fig. 5, the anomaly scores A in both methods are small, which is justified by the small differences between (b) and (c) as well as (d) and (e). The different results can be explained by the best values of ϕ. Figure 4 shows the histograms on the anomaly scores A in the both methods. It can be seen from Fig. 4(b) that GANomaly concentrates the anomaly scores of all normal examples between 0 and 0.12. Note that FA-GAN and GANomaly adopted a relatively large and small values for ϕ, which results in a false negative and a true positive, respectively.

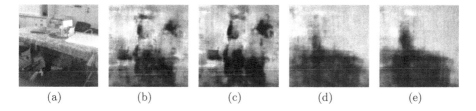

(a)   (b)   (c)   (d)   (e)

Fig. 7. One of the FN examples with FA-GAN. See the captions of Fig. 5 for (a)–(e).

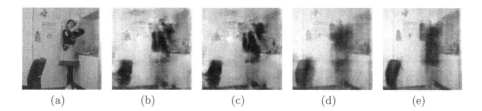

(a)   (b)   (c)   (d)   (e)

Fig. 8. One of the FN examples with the two methods. See the captions of Fig. 5 for (a)–(e).

---

[1] Schools in Japan teach students to take this action under strong shakes during an earthquake.

We also show in Fig. 8 one of the FN examples which were wrongly classified by the two methods. Taking a selfie is considered as an anomaly because nobody does it in the training set. Note that the difference between the images generated by z and $\hat{z}$ is small. Figure 9 shows an FP example with FA-GAN. We see that there is a large difference between (b) and (c), especially on the middle part. However, the last two generated images by GANomaly, which achieved an anomaly score of 0.051, are similar.

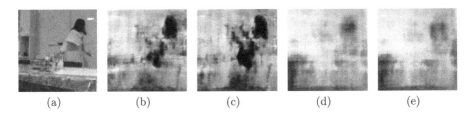

(a)   (b)   (c)   (d)   (e)

Fig. 9. One of the FP examples with FA-GAN. See the captions of Fig. 5 for (a)–(e).

From the experiments above, we see that in overall the two methods show good performance on the two datasets. We conclude that the two GAN-based methods show their ability to solve the problem of anomaly detection on human monitoring. However, the drawback of the methods is also obvious. For some minor anomalies, e.g., the selfie in Fig. 8(a), the latent vector z after the first encoder does not reflect them, which makes z and $\hat{z}$ be quite similar. It results in a small anomaly score for such examples and thus these abnormal examples will be predicted as normal. We can see these results in Fig. 8.

5   Conclusion

We applied two kinds of GAN-based methods, which are FA-GAN and GANomaly, to the datasets collected by our autonomous robot and with a VCR, so that the anomalies can be detected without any supervision. The results show that FA-GAN performs better on the robotic dataset while GANomaly performs better on the VCR dataset, possibly due to the different frequencies of the scene changes. We analyzed the reason behind the dependency of the performance in F1 score on ϕ through the histograms on the anomaly scores. Also, the methods occasionally make wrong detection with minor anomalies in images due to the scarcity of the scene in the training set or the loss of information in the latent vectors. In general, the two GAN-based methods with encoder-decoder architectures should perform well on our autonomous robot and VCR for human monitoring.

Our future work will take image captions into account to improve the performance of anomaly detection. We think that the captions as weak labels can provide additional useful information on anomaly detection, since we already made some progress with a non-GAN-based approach [8].

References

1. Goodfellow, I.J., et al.: Generative adversarial nets. In: Proceedings of the NIPS, pp. 2672–2680 (2014)
2. Lawson, W., Hiatt, L., Sullivan, K.: Detecting anomalous objects on mobile platforms. In: Proceedings of the CVPR Workshop (2016)
3. Lawson, W., Hiatt, L., Sullivan, K.: Finding anomalies with generative adversarial networks for a Patrolbot. In: Proceedings of the CVPR Workshop (2017)
4. Deguchi, Y., Takayama, D., Takano, S., Scuturici, V.M., Petit, J.M., Suzuki, E.: Skeleton clustering by multi-robot monitoring for fall risk discovery. J. Intell. Inf. Syst. **48**(1), 75–115 (2017)
5. Kondo, R., Deguchi, Y., Suzuki, E.: Developing a face monitoring robot for a Deskworker. In: Aarts, E., et al. (eds.) AmI 2014. LNCS, vol. 8850, pp. 226–241. Springer, Cham (2014). https://doi.org/10.1007/978-3-319-14112-1_19
6. Deguchi, Y., Suzuki, E.: Hidden fatigue detection for a desk worker using clustering of successive tasks. In: De Ruyter, B., Kameas, A., Chatzimisios, P., Mavrommati, I. (eds.) AmI 2015. LNCS, vol. 9425, pp. 268–283. Springer, Cham (2015). https://doi.org/10.1007/978-3-319-26005-1_18
7. Fujita, H., Matsukawa, T., Suzuki, E.: Detecting outliers with one-class selective transfer machine. Knowl. Inf. Syst. **62**(5), 1781–1818 (2019). https://doi.org/10.1007/s10115-019-01407-5
8. Hatae, Y., Yang, Q., Fadjrimiratno, M.F., Li, Y., Matsukawa, T., Suzuki, E.: Detecting anomalous regions from an image based on deep captioning. In: Proceedings of the VISIGRAPP, Subvolume for VISAPP, vol. 5, pp. 326–335 (2020)
9. Schlegl, T., Seeböck, P., Waldstein, S.M., Schmidt-Erfurth, U., Langs, G.: Unsupervised anomaly detection with generative adversarial networks to guide marker discovery. In: Niethammer, M., et al. (eds.) IPMI 2017. LNCS, vol. 10265, pp. 146–157. Springer, Cham (2017). https://doi.org/10.1007/978-3-319-59050-9_12
10. Akcay, S., Atapour-Abarghouei, A., Breckon, T.P.: GANomaly: semi-supervised anomaly detection via adversarial training. In: Jawahar, C.V., Li, H., Mori, G., Schindler, K. (eds.) ACCV 2018. LNCS, vol. 11363, pp. 622–637. Springer, Cham (2019). https://doi.org/10.1007/978-3-030-20893-6_39
11. Zenati, H., Foo, C.S., Lecouat, B., Manek, G., Chandrasekhar, V.R.: Efficient GAN-based anomaly detection. arXiv preprint arXiv:1802.06222 (2018)
12. Donahue, J., Krähenbühl, P., Darrell, T.: Adversarial feature learning. arXiv preprint arXiv:1605.09782 (2016)
13. Sabokrou, M., Khalooei, M., Fathy, M., Adeli, E.: Adversarially learned one-class classifier for novelty detection. In: Proceedings of the CVPR, pp. 3379–3388 (2018)
14. Perera, P., Nallapati, R., Xiang, B.: OCGAN: one-class novelty detection using GANs with constrained latent representations. In: Proceedings of the CVPR, pp. 2898–2906 (2019)
15. Chakravarty, P., Zhang, A.M., Jarvis, R., Kleeman, L.: Anomaly detection and tracking for a patrolling robot. In: Proceedings of the Australasian Conference on Robotics and Automation (ACRA) (2007)
16. Pathak, D., Krahenbuhl, P., Donahue, J., Darrell, T., Efros, A.A.: Context encoders: feature learning by inpainting. In: Proceedings of the IEEE Conference on Computer Vision and Pattern Recognition, pp. 2536–2544 (2016)
17. Zhang, T., Ramakrishnan, R., Livny, M.: BIRCH: a new data clustering algorithm and its applications. Data Min. Knowl. Discov. **1**(2), 141–182 (1997)

18. Krizhevsky, A., Sutskever, I., Hinton, G.E.: ImageNet classification with deep convolutional neural networks. In: Proceedings of the NIPS, pp. 1106–1114 (2012)
19. Hochreiter, S., Schmidhuber, J.: Long short-term memory. Neural Comput. **9**(8), 1735–1780 (1997)
20. Zaremba, W., Sutskever, I.: Learning to execute. CoRR abs/1410.4615 (2014)
21. Johnson, J., Karpathy, A., Fei-Fei, L.: DenseCap: fully convolutional localization networks for dense captioning. In: Proceedings of the CVPR, pp. 4565–4574 (2016)
22. Radford, A., Metz, L., Chintala, S.: Unsupervised representation learning with deep convolutional generative adversarial networks. arXiv preprint arXiv:1511.06434 (2015)
23. Kingma, D.P., Ba, J.: Adam: a method for stochastic optimization. arXiv preprint arXiv:1412.6980 (2014)

Ranking Speech Features for Their Usage in Singing Emotion Classification

Szymon Zaporowski[1]($\boxtimes$) ⓘD and Bozena Kostek[2] ⓘD

[1] Faculty of Electronics, Telecommunications and Informatics, Multimedia Systems Department, Gdansk University of Technology, 80-233 Gdansk, Poland
smck@multimed.org
[2] Faculty of Electronics, Telecommunications and Informatics, Audio Acoustics Laboratory, Gdansk University of Technology, 80-233 Gdansk, Poland
bokostek@audioacoustics.org

Abstract. This paper aims to retrieve speech descriptors that may be useful for the classification of emotions in singing. For this purpose, Mel Frequency Cepstral Coefficients (MFCC) and selected Low-Level MPEG 7 descriptors were calculated based on the RAVDESS dataset. The database contains recordings of emotional speech and singing of professional actors presenting six different emotions. Employing the algorithm of Feature Selection based on the Forest of Trees method, descriptors with the best ranking results were determined. Then, the emotions were classified using the Support Vector Machine (SVM). The training was performed several times, and the results were averaged. It was found that descriptors used for emotion detection in speech are not as useful for singing. Also, an approach using Convolutional Neural Network (CNN) employing spectrogram representation of audio signals was tested. Several parameters for singing were determined, which, according to the obtained results, allow for a significant reduction in the dimensionality of feature vectors while increasing the classification efficiency of emotion detection.

Keywords: Mel Frequency Cepstral Coefficients (MFCC) · MPEG 7 low-level audio descriptors · Feature selection · Singing expression classification

1 Introduction

Speech analysis and processing, parametrization as well as automatic classification are the areas being thoroughly researched and developed for the last few decades as their application is of utmost importance in many domains. To name a few [1–4]: telecommunications (VoIP, enhanced IP communication services), automatic speech transcription, automated speech-to-text technologies in video-over IP communications, medical applications such as hearing aids, cochlear implants and speech pathology recognition, language processing for communication services, and more recently human-(intelligent) computer communication based on big data [5, 6]. The last-mentioned application is within the interest of researches as well as commercial usage.

© Springer Nature Switzerland AG 2020
D. Helic et al. (Eds.): ISMIS 2020, LNAI 12117, pp. 225–234, 2020.
https://doi.org/10.1007/978-3-030-59491-6_21

Parametrization is usually the first and often the most crucial block of automatic speech recognition (ASR) in combination with machine learning algorithms. It is only in the last few years that deep learning methodology has forced a different approach to the speech signal processing, in which speech parameters are not retrieved, but the signal in the form of 2D images (i.e., spectrograms, cepstrograms, mel-cepstrograms, chromagrams, *wavenet*-like, etc.) [7–9] is fed at the net input. On the other hand, automatic evaluation of singing quality in the context of its production (e.g., evaluation of the intonation and timbre of the singing voice) is a relatively poorly studied issue comparing to the 'pure' speech area [3, 10]. It should, however, be remembered that singing - like speech - is also a tool for expressing feelings and emotions, thus speech descriptors applied to the singing evaluation should be useful. Also, it is interesting whether deep learning-based methodology may be – in a straightforward way - applied to the singing expression evaluation.

The area of emotion detection in speech is quite well studied, in contrast to the detection of emotion in singing. The article presents issues related to the search for speech signal parameters that may work in the context of automatic evaluation of the quality of expression in singing. For this purpose, a dataset containing recordings of emotional speech and emotionally-singing singing was used, followed by the parametrization of these signals. Some speech descriptors were evaluated for their usage in the feature vector (FV) for singing emotion recognition.

In the next step, the determined parameters were evaluated and reduced using the feature significance algorithm using a Forest of Trees method. Then classification was carried out using the Support Vector Machine (SVM) based on the prepared reduced feature vectors. The final part of the article presents conclusions regarding the development of the proposed methodology to use machine learning methods, including a deep learning approach, to assess the quality of singing expression automatically.

2 Related Work

2.1 Emotion Detection in Speech

The detection of emotions in speech is now very much present in the literature, especially when the possibility of using deep learning for this purpose appeared. Most of the studies describe approaches that use artificial neural networks as classifiers (i.e., convolutional networks, recursive networks, autoencoders), presenting processed spectrograms as input [7, 8, 11]. The use of classical speech signal descriptors (e.g., Mel-Frequency-Cepstral-Coefficients, MFCC) is currently less prevalent in speech research due to the lower accuracy of emotion recognition (approx. 60%) [12, 13]. When 2D image spaces are used as parameters, the classification efficiency can reach over 80% [7, 9]. Such efficiency can also be achieved for some chosen emotions using SVM [8].

2.2 Emotion in Singing

There exist systems that allow automatic assessment of singing and singing quality. The focus of such systems is on assessing the quality of singing individual sounds or a specific singing technique [14, 15]. Classification accuracy can be up to 80% for these types of systems. Another approach researched is to use the fundamental frequency as a parameter to test whether the person singing a given sound or repeating it after the system prompt is able to sing it correctly [16].

3 Parameter Selection

3.1 Dataset

The RAVDESS database of recordings was used to conduct the experiments presented in this paper. This dataset is often used in research studies, thus it may be treated as a benchmark in the area of speech emotion recognition. The database contains recordings of 24 professional actors (12 women, 12 men) speaking and singing two matched English statements with a neutral North American accent. Speech includes expressions of calm, joy, sadness, anger, fear, surprise and disgust, and singing contains the emotions of calm, joy, sadness, anger, and fear. Each expression is sung and pronounced at two levels of emotional intensity (normal, enhanced). Additionally, an emotionally neutral expression was recorded for each phrase. An example of actors' images presenting a set of emotions available in the database is shown in Fig. 1.

Fig. 1. Example of the RAVDESS emotion expression [17]

All emotion recordings are available in three modalities: audio signal (16-bit resolution, 48 kHz sampling frequency, wave format files), audio-video (720 p resolution, H.264 video coding, AAC audio coding, 48 kHz sampling frequency), mp4 file format) and video signal. The database contains 7356 files (24.8 GB), of which 1440 recordings are of speech alone, and 1012 recordings are of singing. The database is available under a Creative Commons license. Only audio files were used in this study.

3.2 Parameter Selection

Two approaches were utilized to parameterize data from the RAVDESS database. The FV in the first scenario consists of 40 consecutive normalized MFCCs. In the second approach, which uses MPEG 7 descriptors and parameters available in the Librosa library [18], FV contains parameters in both time- and frequency-domains. Time-domain parameters include zero crossings (Zero-Crossing, ZC), and signal energy (Root Mean Square Energy, RMS). The following spectral descriptors were used [18]: the spectral center of gravity (Audio Spectrum Centroid, ASC) and the spectral flatness measure (Audio Spectrum Flatness, ASF). Besides, the spectral roll-off parameter built into the Librosa library was employed [18]. This set of parameters is calculated according to the internal settings of the Librosa library. All the above-mentioned descriptors are present in the literature [19, 20]; thus their definitions will not be recalled here.

For the approach based on Convolutional Neural Networks (CNN), spectrograms were calculated for each of the utterances and songs from the RAVDESS corpora. The audio is sampled at 48000 Hz. Each audio frame is windowed using the Kaiser window of the length of 2048. Fast Fourier Transform (FFT) windows of the length of 2048 are then applied on the windowed audio samples with the STFT hop-length as 512 points. As a result of the aforementioned transformations, the bandwidth of the audio signal was reduced to 8 kHz. In total, there were more than 24200 samples for six classes. That means there were more than 4050 examples for each class.

4 Experiments

4.1 Significance Ranking

To reduce the number of parameters used in the classification and, at the same time, increase the accuracy of the classification by leaving only significant descriptors, the Feature Importance algorithm was employed. The authors have successfully utilized this algorithm in earlier publications related to speech classification [21]. The Feature Importance algorithm is based on another algorithm called Extremely Randomized Trees (ERT) [22]. The concept is derived from Random Forest (RT), which provides a combination of tree predictors so that each tree depends on the value of a random vector sampled independently and has the same distribution for all trees in the forest. The error related to generalization for forests is approaching the limit as the number of trees in the forest increases. ERTs generalization error depends on the correlation between trees in the forest and the strength of individual trees in the entire set [21–23].

The conducted experiments used the implementation of the ERT algorithm contained in the scikit-learn library in Python [25]. The ERT algorithm settings were as follows: n_estimators = '40', criterion = 'entropy', min_ samples_split = 2, min_samples_leaf = 1, min_weight_ fraction _leaf = 0.1, max_features = 'auto', min_impurity_ decrease = 0.01, min_impurity_split = None, bootstrap = True, random_state = True, warm_start = True, class_weight = balanced.

4.2 SVM-Based Classifier

The SVM algorithm, using the scikit-learn package in Python, was employed for the classification. The classifier settings were selected experimentally, ultimately the highest accuracy in the classification for all types of emotions studied was obtained using a degree 3 polynomial kernel with the parameter C = 0.1 and 'balanced' mode of adjusting weights of individual classes. For comparing classes with each other, one vs. all approach was used.

4.3 CNN Classifier

The CNN classifier used for this experiment was created using the Tensorflow library. The architecture used for this experiment is shown in Fig. 2. The architecture was created in an empirical approach, adding individual layers, and then examining their impact on classification results. Inspiration for this architecture was research presented in the literature [24, 25]. The created neural network was trained for 200 epochs using a batch size of 32 and data split 60/40 for training and validation set, respectively. Titan RTX graphic processor was employed for training.

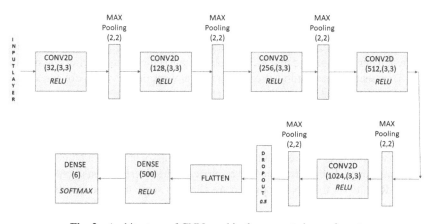

Fig. 2. Architecture of CNN used in the presented experiment

5 Results and Discussions

Below ranking results of the significance of individual parameters and results of emotion classification are shown. Figure 3a) shows the importance of the MFCCs depending on emotions. According to Fig. 3a) coefficient no. 40 is the most versatile. It is the most important feature for several emotions, e.g., joy, calm, sadness, and neutral state. Also, feature no. 1 seems to be most important for anger and fear emotions. Figure 3b) presents the ranking of the parameter importance of speech and singing for all emotions using the MFCC coefficients. As can be seen in Fig. 3b) the most essential parameters for speech and singing are different. The common one is the coefficient no. 1, but the rest differs. For speech, more important are coefficients with the lower numbers of order, in the case of singing, features with the number higher than 30 were indicated. Tables 1, 2 and 3 show several MFCC parameters, the accuracy of classification as well as the mean square error (MSE) for individual emotions using SVM. The MFCCs shown are derived from the Feature Importance algorithm, which indicated the most important features respectively to values presented in Tables 1, 2 and 3. In most cases, the use of four best coefficients provides the best accuracy results. It is worth noticing that reducing FV to only 10 best features in most cases results in a significant increase in the accuracy score. The emotion of anger is an exception here; accuracy values are oscillating all the time around 70%. Table 4 contains the classification results for the second parameterization scenario employing MPEG-7 low-level descriptors. Table 5 presents results for the CNN-based classification approach. The measure of accuracy is understood as the ratio of the number of correct predictions to the total number of input samples [26].

Based on the presented results, it is possible to distinguish a group of MFCC coefficients that are most important in the process of classifying speech and singing within a given emotion. Classifying emotions in speech and singing using these factors is characterized by high accuracy for most emotions (over 88%). In most cases, the feature vector reduced to two descriptors consisted of MFCCs nos. 29 and 40. For anger, these were coefficients nos. 1 and 39. Among the tested coefficients from the second FV variant, the highest result was obtained using spectral centroid (ASC). The low efficiency of the zero-crossing (ZC) parameter and RMS energy is puzzling. Based on the experiments conducted, it can be observed that MFCC coefficients achieve much better classification results. They seem to be a natural direction in further work on the system for assessing the quality of expression in singing. There was also a decrease in the classification accuracy for anger emotions in all the feature vectors used. This is an interesting phenomenon that should be studied based on another database of recordings. Such a difference may result from a significant change in the volume of speech and possible changes in formant frequencies in the case of this emotion. Articulation associated with emotion can also affect the accuracy of classification. The accuracy of the classification of other emotions is similar. Results for the CNN approach are slightly worse than the results for MFCC parameterization. It could be due to the fact that spectrograms bandwidth were limited to only 8 kHz. It is worth noticing that the categorical cross-entropy values indicate that there is room for improvement, however, presented values and architecture were the best from all tested architectures.

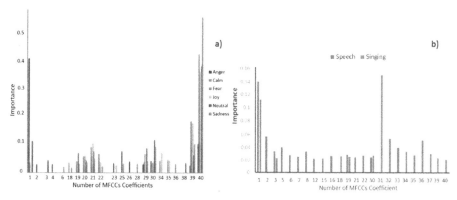

Fig. 3. Classification results for a) The importance of MFCCs in terms of the researched emotion b) for speech and singing for the normalized ranking of MFCCs

Table 1. Classification (speech and singing) results for anger

Quantity of the MFCCs retained	Accuracy [%]	MSE
40	67.7	0.323
20	**66.81**	0.3319
15	68.58	0.3141
10	68.59	0.3142
5	69.47	0.3053
4	**70.35**	**0.2964**
2	68.58	0.3142

Table 2. Classification (speech and singing) results for fear

Quantity of the MFCCs retained	Accuracy [%]	MSE
40	**50.66**	0.493
20	**50.66**	0.4933
15	51.33	0.4867
10	82	0.18
5	90	0.1
4	**90.67**	**0.093**
2	69	0.31

Table 3. Classification (speech and singing) results for neutral emotion

Quantity of the MFCCs retained	Accuracy [%]	MSE
40	**52.7**	0.473
20	53.38	0.4662
15	54.05	0.4595
10	70.95	0.29
5	89.19	0.108
4	91.21	0.0878
2	**97.973**	**0.02**

Table 4. Emotion classification results for speech and singing based on MPEG 7 descriptors

Emotion [%]	ASC	ASF	Roll-off	ZC	RMS
Neutral	**98.52**	48.26	34.93	**34.53**	**68.23**
Joy	97.87	52.13	**38.66**	31.81	67.24
Sadness	95.69	47.42	37.84	30.15	62.51
Anger	**70.47**	27.53	30.92	18.32	47.83
Surprised	93.36	**53.77**	33.36	24.36	53.27
Fear	96.55	43.29	32.67	21.84	56.68
All	79.39	41.23	35.73	27.27	62.26
Average	**90.26**	**44.80**	**34.87**	**26.90**	**59.72**

Table 5. Emotion classification results for speech and singing based on the CNN approach

Emotion	Accuracy [%]	Categorical cross-entropy
Neutral	75.85	0.8693
Joy	**51.33**	7.8441
Sadness	57.83	2.9245
Anger	76.33	0.9483
Surprised	**77.33**	1.9979
Fear	77.00	1.3644
All	**65.96**	1.4873

6 Conclusions

In this paper, an approach to rank speech features based on RAVDESS emotional speech and singing dataset with different approaches to parameterization and classification is presented. Significance of particular MFCC parameters for speech and singing derived from the Feature Importance algorithm is shown. Three different approaches to parameterization using MFCCs, MPEG-7 low-level descriptors, and spectrograms are also demonstrated. The results for each approach are presented and discussed.

In the future, the authors intend to focus on creating parameterization based on all MPEG-7 low-level descriptors and checking their effectiveness in the classification of emotions, both in speech and singing. The next step will also be testing parameterization on sets containing opera singing. The basis of such a system could be the detection of emotions in singing, expanded by a ranking system, using the approach described in the article. However, it seems natural to extend research towards the use of deep learning and 2D representation of signals such as cochleagrams or CQT (Constant-Q) transform.

References

1. Mukesh, K., Shimi, S.: Voice recognition based home automation system for paralyzed people. Int. J. Adv. Res. Electron. Commun. Eng. **4**(10), 2508–2515 (2015)
2. Markoff, J.: From Your Mouth to Your Screen, Transcribing Takes the Next Step (2019). https://www.nytimes.com/2019/10/02/technology/automatic-speech-transcrip tion-ai.html. Accessed 15 Jan 2020
3. Munir, A., Kashif Ehsan, S., Mohsin Raza, S.M., Mudassir, M.: Face and speech recognition based smart home. In: 2019 International Conference on Engineering and Emerging Technologies ICEET 2019, pp. 1–5 (2019)
4. Delić, V., et al.: Speech technology progress based on new machine learning paradigm. Comput. Intell. Neurosci. **2019** (2019). https://doi.org/10.1155/2019/4368036
5. Lei, X., Tu, G.-H., Liu, A.X., Ali, K., Li, C.-Y., Xie, T.: The insecurity of home digital voice assistants – Amazon Alexa as a case study (2017)
6. Kannan, K., Selvakumar, J.: Arduino based voice controlled robot. Int. Res. J. Eng. Technol. **02**(01), 49–55 (2015)
7. Bertero, D., Fung, P.: A first look into a convolutional neural network for speech emotion detection. In: 2017 IEEE International Conference on Acoustics, Speech and Signal Processing (ICASSP), pp. 5115–5119 (2017)
8. Kerkeni, L., Serrestou, Y., Raoof, K., Cléder, C., Mahjoub, M., Mbarki, M.: Automatic speech emotion recognition using machine learning (2019). https://www.intechopen.com/ online-first/automatic
9. Zhao, J., Mao, X., Chen, L.: Speech emotion recognition using deep 1D & 2D CNN LSTM networks. Biomed. Signal Process. Control **47**, 312–323 (2019)
10. Scherer, K.R., Sundberg, J., Tamarit, L., Salomão, G.L.: Comparing the acoustic expression of emotion in the speaking and the singing voice. Comput. Speech Lang. **29**(1), 218–235 (2015)
11. Cibau, N., Albornoz, E., Rufiner, H.: Speech emotion recognition using a deep autoencoder. In: Anales de la XV Reunion de Procesamiento de la Informacion y Control, pp. 934–939 (2013)
12. Sezgin, M.C., Gunsel, B., Kurt, G.K.: Perceptual audio features for emotion detection. EURASIP J. Audio Speech Music Process. **2012**(1), 16 (2012). https://doi.org/10.1186/1687-4722-2012-16

13. Poorna, S.S., Jeevitha, C.Y., Nair, S.J., Santhosh, S., Nair, G.J.: Emotion recognition using multi-parameter speech feature classification. In: 2015 International Conference on Computers, Communications, and Systems (ICCCS), pp. 217–222 (2015)

14. Zwan, P.: Expert system for automatic classification and quality assessment of singing voices. In: Audio Engineering Society - 121st Convention Paper 2006, vol. 1, pp. 446–454, January 2006

15. Amir, N., Michaeli, O., Amir, O.: Acoustic and perceptual assessment of vibrato quality of singing students. Biomed. Signal Process. Control **1**, 144–150 (2006)

16. Półrolniczak, E., Łazoryszczak, M.: Quality assessment of intonation of choir singers using F0 and trend lines for singing sequence. Metod. Inform. Stosow. **4**, 259–268 (2011)

17. Livingstone, S.R., Russo, F.A.: The Ryerson Audio-Visual Database of Emotional Speech and Song (RAVDESS): a dynamic, multimodal set of facial and vocal expressions in North American English. PLoS ONE **13**(5), e0196391 (2018)

18. Ellis, D., et al.: librosa: audio and music signal analysis in Python. In: Proceedings of the 14th Python in Science Conference, no. Scipy, pp. 18–24 (2018)

19. Muhammad, G., Melhem, M.: Pathological voice detection and binary classification using MPEG-7 audio features. Biomed. Signal Process. Control **11**(1), 1–9 (2014)

20. Dave, N.: Feature extraction methods LPC, PLP and MFCC in speech recognition. Int. J. Adv. Res. Eng. Technol. **1**(VI), 1–5 (2013)

21. Zaporowski, S., Czyżewski, A.: Selection of features for multimodal vocalic segments classification. In: Choroś, K., Kopel, M., Kukla, E., Siemiński, A. (eds.) MISSI 2018. AISC, vol. 833, pp. 490–500. Springer, Cham (2019). https://doi.org/10.1007/978-3-319-98678-4_49

22. Geurts, P., Ernst, D., Wehenkel, L.: Extremely randomized trees. Mach. Learn. **63**(1), 3–42 (2006)

23. Louppe, G., Wehenkel, L., Sutera, A., Geurts, P.: Understanding variable importances in forests of randomized trees. In: Advances in Neural Information Processing Systems 26, pp. 431–439 (2013)

24. Svetnik, V., Liaw, A., Tong, C., Christopher Culberson, J., Sheridan, R.P., Feuston, B.P.: Random forest: a classification and regression tool for compound classification and QSAR modeling. J. Chem. Inf. Comput. Sci. **43**(6), 1947–1958 (2003)

25. Pedregosa, F., et al.: Scikit-learn: machine learning in Python. J. Mach. Learn. Res. **12**, 2825–2830 (2012)

26. Goodfellow, I., Bengio, Y., Courville, A.: Deep Learning. The MIT Press, Cambridge (2016)

Leveraging Machine Learning in IoT to Predict the Trustworthiness of Mobile Crowd Sensing Data

Corrado Loglisci[1]([✉]), Marco Zappatore[2], Antonella Longo[2,3],
Mario A. Bochicchio[3], and Donato Malerba[1]

[1] Department of Computer Science, University of Bari, Bari, Italy
{corrado.loglisci,donato.malerba}@uniba.it
[2] Hesplora srl, Lecce, Italy
marco.zappatore@hesplora.it
[3] Department of Engineering Innovation, University of Salento, Lecce, Italy
{antonella.longo,mario.bochicchio}@unisalento.it

Abstract. The advances in Internet-of-things (IoT) have fostered the development of new technologies to sense and monitor the urban scenarios. Specifically, Mobile Crowd Sensing (MCS) represents one of the suitable solutions because it easily enables the integration of smartphones collecting massive ubiquitous data at relatively low cost. However, MCS can be affected by wrong data-collection procedures by non-expert practitioners, which can be make useless (or even counter-productive), if contributed data are not trustworthy. Contextualizing monitored data with those coming from phone-embedded sensors and from time/space proximity can improve data trustworthiness. This work focuses on the development of a machine learning method that exploits context awareness to improve the reliability of MCS collected data. It has been validated on a case study concerning urban noise pollution data and promises to improve the trustworthiness of data generated by end users.

1 Introduction

In the recent years, the Mobile Crowd Sensing paradigm (MCS) [1] has become a widely-exploited solution to observe large-scale events and phenomena thanks to the worldwide diffusion of mobiles. MCS applications constantly grow and diversify but they target mainly environmental monitoring in smart cities where non-professionals can be involved to: 1) improve citizenship's environmental awareness and 2) foster the collaboration with scientists to overcome typical limitations to traditional data collection (expensiveness, equipment scarcity).

MCS initiatives very often refer to environmental monitoring as several benefits from wide citizenship participation can be targeted and, at the same time, numerous advantages can be offered to involved users. Thanks to smartphone sensors (embedded or pluggable), the MCS paradigm allows collecting time-/spatial-referenced quantitative observations about multiple parameters (e.g.,

© Springer Nature Switzerland AG 2020
D. Helic et al. (Eds.): ISMIS 2020, LNAI 12117, pp. 235–244, 2020.
https://doi.org/10.1007/978-3-030-59491-6_22

noise, atmospheric pressure, light intensity, etc.). However, some challenges do exist, especially in terms of data quality. Mobile sensors can exhibit limited sensing capabilities, different accuracy and precision, improper usage patterns by their owner, time and space sparsity. These elements determine heterogeneous levels of adequacy to a specific sensing context and produces biased readings that generate outliers in measurement campaigns. Systematic sensing errors can be coped with effectively in advance, by properly training users and calibrating mobile devices by means of professional metering equipment, but data prediction approaches allow to not rely on user and sensor behaviour solely. More specifically, a machine learning approach can rely on pre-categorized mobile crowd-sensed data in terms of their trustworthiness.

This paper proposes an approach that leverages on the contemporary presence of reliable sensor data (about which everything or quite everything is known, *labelled data*) and of unknown (or partially unknown) sensor readings whose trustworthiness is not known a-priori, (*unlabelled data*). Spatial and temporal auto-correlation of these two data typologies can help to assess the trustworthiness of unlabelled data that are spatially and temporally closed to labelled data. To that purpose, a transductive learning approach has been devised to train a classifier able of inferring categories of trustworthiness for the unlabelled data starting from the labelled ones.

The method has been validated for mobile crowd-sensed noise levels collected via smartphone-embedded microphones in the framework of the APOLLON Project[1]: a research effort granted by Apulia Region (Italy), focused on urban environmental analyses in terms of noise, air pollution and UV-levels, which offers semantic-based and geo-localized near-/real-time monitoring services to citizens and city decision makers.

The paper is so organized. Section 2 summarizes core features as well as data trustworthiness challenges in MCS. Section 3 details the proposed method, whose empirical evidence is described in Sect. 4. Section 5 provides conclusions.

2   Related Works

The MCS paradigm has proven its effectiveness for individuals, communities of interest and city managers in several scenarios, where mobile sensor readings and user-provided data can be profitably gathered to provide context-based innovative services exhibiting both personal (e.g., improved awareness fostering responsible behaviours) and public applicability (increased fine-grained knowledge of city status in local authorities). By exploiting *participatory* (i.e., user-supervised data collection) or *opportunistic* (i.e., once authorized, the device collects data autonomously) usage patterns, MCS can be relied on to address such typical urban issues as traffic monitoring, first-response management and pollution assessment (air, noise, EM fields, etc.) [7,13], provided that a large number of mobiles with suitable applications is used. In this way, traditional

---

[1] http://web.apollon-project.it/.

monitoring campaign can be addressed solely where they are needed, thus reducing costs. Mobile apps for MCS require specific architectural capabilities (e.g., local sensing, storage and pre-processing; cloud-based back-ends to process and aggregate data, etc.) and must offer specific data features (e.g., situated data creation, time continuity, high-spatial resolution, etc.) in order to achieve a realistic representation of the scenario where data collection is performed [8]. When participatory MCS solutions are adopted, user-training elements should be added to mobile apps [14] in order to improve the awareness on the correct data collection procedure as well as to motivate users and improve their involvement rate.

When opportunistic MCS strategies are adopted, instead, intelligent data layers targeting a local enhancement of sensor reading quality are needed, featuring either calibration via regression analysis [3,10] or prediction-based contextualization. Wu et al. [11] investigate a forecasting problem of the trustworthiness by exploiting collaborative filtering. Huang et al. [2] focuses on the device quality and propose a Gompertz function-based classifier to calculate device reputation scores as a reflection of the trustworthiness of data.

However, most part of the methods proposed in the literature can suffer from limited applicability by two main reasons. First, they disregard an intrinsic property of the MCS data, that is, auto-correlation [6], which violates the independence assumption (i.i.d.) on which many machine learning approaches rely on. Second, those methods need user effort in the preparing pre-categorized sensing readings, which requires large collections of labelled data obtained through the manual intervention of authoritative annotators, hence the paradigm *supervised learning*. An approach which inspires the current work, has been studied under the paradigm of *transductive learning* [4] and focuses on the exploitation of unlabelled data in combination with few labelled data, which asks for much less human intervention. This makes unfair any comparison between supervised method and those transductive. In this paper, MCS data refer to a three-level representation [12]. Raw data from physical sensors, virtual sensors (i.e., software applications) or logical sensors (e.g., logs) form the low-level context. The high-level context is inferred through meta-data classified into 1) device context (e.g., connectivity); 2) user context (e.g., user profile); 3) physical context (e.g., temperature); 4) temporal context (i.e., the specific time when the situation occurs). The topmost-level context is the estimation of the user state that can be achieved by combining the two underlying levels.

3   The Method

This section is devoted to the description of the method we design to predict the level (label) of trustworthiness of the MCS data. We first provide basic notions and then explain how the method works.

Let $\mathcal{D}$ be the set of sparsely labelled data which comprises the set $\mathcal{L}$ of labelled data (pre-categorized MCS data) and set of $\mathcal{U}$ of the instances with unknown trustworthiness ($\mathcal{D} = \mathcal{L} \cup \mathcal{U}$). The set $\mathcal{D}$ is spanned on a vector $\mathbf{X}$ of (numeric and discrete) attributes and a discrete attribute Y, which denotes the trustworthiness level. The attributes $\mathbf{X}$ denote the three-level representation introduced in Sect. 2. For the instances of $\mathcal{D}$ included into the set $\mathcal{L}$, the labels are known, while for $\mathcal{U}$, the values of Y are determined by the method.

Following the transductive paradigm, the method inputs both the full information represented by $\mathcal{L}$ and the partially given information represented by $\mathcal{U}$, it learns a classification model and predicts the trustworthiness for the instances of the unlabelled part. This is done through an iterative convergence approach [9] aiming at improving the accuracy of the classification model through a procedure that converges to a configuration of the predictions on $\mathcal{U}$ as accurate as possible. Before the iteration starts, the method performs a feature augmentation step, which generates an extended set of descriptive properties for every instance. It has the aim to make every instance "aware" about the distribution of the values (for each attribute) over the instances which are more correlated to it. These new attributes are updated during the iterative process, in order to "propagate" accurate predictions over correlated instances. To properly do that, we take only the predictions that could truly improve the classification model, so we do consider only the predictions with high confidence, generated for the current iteration, and feed them the learning process of the next iteration.

To identify data correlating to an instance the most, we introduce the notion of "neighborhoods" of instances. So, for every instance m, we build the spatio-temporal neighborhood $N(m, \delta_s, \delta_t)$ composed of the instances whose spatial distance from m does not exceed δ_s and which have been recorded within a time δ_t from m, formally $N(m, \delta_s, \delta_t) = \{p | p \in \mathcal{D}, distance(m.c, p.c) \leq \delta_s, m.t - p.t \leq \delta_t, if\ m.t \geq p.t, p.t - m.t \leq \delta_t, otherwise\}$. The terms $m.c$ and $p.c$ denote the spatial coordinates of the instances m and p respectively, while the terms $m.t$ and $p.t$ denote the recording times. This allows us to account for presence of the spatial autocorrelation and temporal auto-correlation [5] by nicely capturing the typical scenario in which the data recorded by the same device in a short time tend to have the same trustworthiness, as well as, the data, spatially close to each other, which have been recorded in a short time tend to have the same trustworthiness level.

Feature Augmentation. The new attributes are defined to compute variation summarization statistics of two classes. *Type 1:* given the base numeric attribute A, we build two new attributes, $AN(mean)$ and $AN(stDev)$, based on A. Both attributes are computed by aggregating A over the neighborhoods $N(m, \delta_s, \delta_t)$ constructed with maximum spatial distance δ_s and maximum temporal contiguity δ_t. Let m be an instance, $AN(u, mean)$ and $AN(u, stDev)$ are computed as the mean and standard deviation of the values of A falling in the neighborhood $N(m, \delta_s, \delta_t)$. Both the new attributes allow us to summarize average and variance, local to the neighborhoods, of the numeric attributes. *Type 2:* given the base discrete attribute A that takes d distinct values, we build d new attributes. These attributes represent the frequency histogram of A, as it is computed on the neighborhoods $N(m, \delta_s, \delta_t)$. In practice, we build one attribute for every distinct value of A. Let m be an instance, val be a distinct value of A, $AN(m, val)$ is computed as the frequency of val over the neighborhood $N(m, \delta_s, \delta_t)$.

Prediction Reliability. We measure the confidence of the labels predicted at each iteration, in order to select those more confident that are then fed back into the learning process for the next iteration. Intuitively, confident predictions should manifest the property of auto-correlation, so that similar labels can be plausibly propagated to the neighbours. The higher the autocorrelation of the label with neighbour labels, the more confident its prediction. To define the measure, we estimate the presence of the predicted label associated to the instance m over the neighborhood $N(m, \delta_s, \delta_t)$ by quantifying the times in which the prediction on m is identical to the labels of its neighbours included in the set $\mathcal{L}$ (labelled neighbours). The choice of comparing the predicted labels against those original of the set $\mathcal{L}$ is done to provide validity to confidence estimation.

However, the temporal component should have a weight larger, compared to that spatial, because onsets and effects of the urban processes often depend on the timing of human lifestyles and daily periods more than phenomena related to the spatial dislocation. To encode this, we inject the temporal distances between the instance m and its neighbours into the confidence measure and assign weights to the distances dependently on their values. In practice, we build two sorted sets with the neighbours, one set with the instances recorded before the instance m and one set with the instances recorded after m. So, the weights are determined by the number of instances that separate m from the neighbours. The confidence measure for the predicted label done on the instance m is so formulated:

$$\mathcal{R}(m) = \frac{\sum_{p \in \{N(m,\delta_s,\delta_t) \cap \mathcal{L}\}} \left(\sigma(m.t, p.t) \times (equal(\overline{y}, p.y)) \right)}{\sum_{p \in \{N(m,\delta_s,\delta_t) \cap \mathcal{L}\}} \left(\sigma(m.t, p.t) \right)}, \tag{1}$$

where $\sigma(m.t, p.t)$ determines the weight associated to every comparison (that is, temporal distance between m and p), $equal(\overline{y}, p.y)$ is 1 when the prediction $\overline{y}$ equals $p.y$, 0 otherwise. It has values in the range [0, 1], where 1 denotes the highest number of occurrences of the label $\overline{y}$ in the neighborhood and therefore corresponds to the largest confidence, while 0 indicates poor confidence.

Transductive Classification Process. The transductive classification process essentially carries on learning and prediction along two stages, that is, initialization and iteration, as described in the following.

In the initialization stage, it performs three main operations.

(1) For every instance of $\mathcal{D} = \mathcal{L} \cup \mathcal{U}$, it constructs the respective neighborhood with the instances which have spatial distance less than δ_s and temporal distance less than δ_t. This is done by considering the spatial coordinates and recording time of m, as illustrated in the formulation of $N(m, \delta_s, \delta_t)$. The values of δ_s and δ_t are set by the user. Then, for each attribute X of the attribute vector $\mathbf{X}$, it generates new attributes of *Type 1* and *Type 2*, dependently on whether X is numeric or discrete. The computation considers both the labelled instances and unlabelled instances of each neighborhood $N_m \in \mathbf{N_{st}}$ previously determined. Finally, it generates new attributes of *Type 2* for the label-attribute Y with the procedure used for the attributes $\mathbf{X}$. In this case, the computation considers only the labelled instances of each neighborhood $N_m \in \mathbf{N_{st}}$ because the unlabelled instances have no prediction for the attribute Y at the initialization stage. Clearly, all the instances will have the same set of new attributes, while the values are specific per instance m and depend on the data distribution over the respective neighbors N_m.

(2) The algorithm learns a classification model F from the training set $\mathcal{L}$, which is now represented with an augmented feature space $\mathbf{X} \times \mathbf{X}N \times Y \times \mathbf{Y}N$. This allows us inject the auto-correlation into the learning process since the beginning, without making the subsequent computation burden because the new attributes are built once only.

(*3*) The model F is finally used to initialize the unknown labels of the instances $\mathcal{U}$, which are stored as $\hat{\mathcal{U}}$. This way, the predictor F is able to estimate the data trustworthiness by considering additionally the contextual information provided by the nearby MCS data (spatial auto-correlation) and by the readings done by the same devices in the past (temporal auto-correlation), besides of information the devices record in themselves.

In the iteration stage, we aim at improving the predictive accuracy of F and, to this end, we exploit the auto-correlation property from the most confident predictions inferred along the iterations. Basically, the method carries out the following operations.

(*1*) For every instance m previously labelled and stored in the set $\hat{\mathcal{U}}$, it computes the confidence values by comparing the prediction of m against the originally known labels of the instances of $\mathcal{L}$ included in the neighborhood N_m, as illustrated in the formula 1.

(*2*) The confidence values $\mathbf{R_U}$ are sorted and then we pick only the first $size_r$ instances with higher rank, being considered as mostly reputable. The assigned labels (stored in B) will be maintained as such because they will contribute to the subsequent operations, since the instances are now "stabilized". However, these instances will have a role different from the originally labelled instances $\mathcal{L}$, in accordance with the philosophy of the transductive learning and, in fact, they are removed from the target set (unlabelled instances) $\mathcal{U}$ and moved in the set $\hat{\mathcal{U}}$, which is different from $\mathcal{L}$.

(*3*) The new configuration of labels, caused by the reduction of $\mathcal{U}$ and extension of $\hat{\mathcal{U}}$, is propagated over all the instances ($\mathcal{L}$, $\mathcal{U}$, $\hat{\mathcal{U}}$) through the update of the new attributes. It should be noted that only the attributes $\mathbf{Y}N$ are influenced by the update, since those of the set $\mathbf{X}N$ remain unchanged, being derived by the attributes $\mathbf{X}$.

(*4*) In accordance with the transductive learning, the classification model F is (re-)trained on the originally labelled instances $\mathcal{L}$, which are now "aware" about the new labeling scenario. So, the predictor F can leverage the *i)* confidence of the predictions ($\hat{\mathcal{U}}$) and *ii)* reinforced configuration of the descriptive attributes, in order to improve the accuracy of the instances left in $\mathcal{U}$.

This iterative procedure stops when one of the two stopping criteria is satisfied, specifically, either the set $\mathcal{U}$ is empty or the number of iterations completed reaches a user-defined threshold. The depletion of $\mathcal{U}$ is guaranteed as every iteration removes a portion of instances equal to the threshold $size_r$ (user-defined).

4 Empirical Evidence

In order to provide empirical evidence to the proposed solution for trustworthiness assessment, we performed experiments on a MCS-originated dataset of noise level readings recorded by a set of 5 smartphone devices, moving in a $8\,\mathrm{km}^2$ area in Lecce (Apulia, Italy), during the period 2019/05/27–2019/07/01. Each measurement point is time- and geo-referenced and enriched with contextual data provided from additional smartphone-embedded sensors (e.g., accelerometer, proximity sensor, etc.) and device's metadata as well.

Specifically, we have 4335 readings uniformly distributed over the categories (991 instances for *not reliable*, 1782 instances for *poorly reliable* and 1562 for *reliable*), so we have no imbalanced concern for the classification task. For instance, noise readings outside sensor's range are classified as unreliable. Similarly, noise readings with low amplitude (i.e.,]+20;+50] dB(A)), are considered poorly reliable when the proximity sensor indicates low values (i.e., less than 5 cm), as an obstructed microphone affecting noise readings can be inferred.

We arrange experiments aiming at training and testing the classification models. In particular, we performed a quantitative evaluation on the predictive capabilities of the transductive approach in performing accurate inferences on the trustworthiness level of the unknown noise pollution readings. The accuracy was measured in terms of the F-score and averaged over 5 trials executed according to the inverse 5-fold cross validation. More precisely, for each trial, the learner is trained on one fold (which represents $\mathcal{L}$) and tested on the set $\mathcal{U}$ composed of the remaining four folds. We guaranteed that the training set $\mathcal{L}$ was balanced. By following the transductive setting, the set $\mathcal{L}$ contains a smaller part of the whole dataset, and, more precisely, it has a percentage of 10% balanced over the three classes. The accuracy was estimated on three variants of the method, which were built by using three base learners to train the predictor F. Specifically, we integrated the classification algorithms of *Decision Tree* (DT), *Random Forest* (RF) and *Logistic Regression* (RF).

The three variants of the method were tested in two main experimental setups, *i)* size of the set of confident predictions $(size_r)$, and *ii)* size of the neighborhoods in terms of the values of the thresholds δ_s and δ_t. In particular, we considered three different values of $size_r$, that is, 5%, 10%, 15% of $\mathcal{U}$ and three different neighborhood configurations, that is, $\delta_s = 250$ m and $\delta_t = 10$ min (thereafter, n_10_250), $\delta_s = 500$ m and $\delta_t = 5$ min (thereafter, n_5_500), and $\delta_s = 500$ m and $\delta_t = 10$ min (thereafter, n_10_500), which let us build neighborhoods with different sizes (number of instances contained), that is, 4, 5 and 7, on average, respectively. Neighborhoods which, initially had no labelled instance, were extended with the instance of $\mathcal{U}$ which is closest to the samples already present. For a fair comparison, we fix the maximum number of iterations to 10, which allows the experimental trials terminate under different conditions.

The F-score values computed along the iterations with n_10_250 are reported in the Fig. 1, while those computed with n_5_500 are reported in the Fig. 2, finally, the F-score values computed with n_10_500 are reported in Fig. 3.

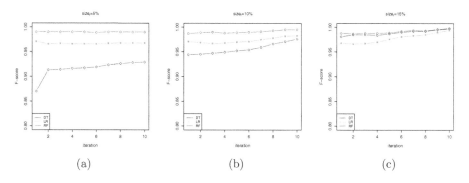

(a)   (b)   (c)

Fig. 1. The values of F-score computed on the unlabelled instances along the iterations when $\delta_s = 250$ m, $\delta_5 = 10$ min.

The first consideration we can do is on the number of the iterations. Regardless of the neighborhood configuration, the predictive accuracy in most cases increases as new iterations are performed. The highest gain accuracy is obtained at the initial iterations,

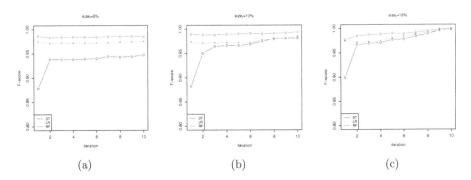

Fig. 2. The values of F-score computed on the unlabelled instances along the iterations when $\delta_s = 500$ m, $\delta_5 = 5$ min.

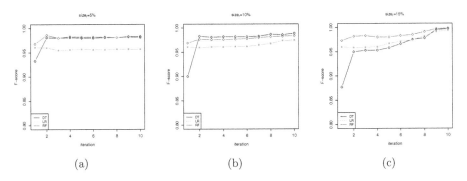

Fig. 3. The values of F-score computed on the unlabelled instances along the iterations when $\delta_s = 500$ m, $\delta_5 = 10$ min.

which indicates the classifier benefits from the best confident predictions since at the early. This confirms the effectiveness of the iterative learning approach. We should also note that acceptable F-score values can be reached even before the execution of 10 iterations. Clearly, this leads benefits from the viewpoint of the running times.

Another consideration deserves the behaviour of the accuracy with respect to the number of confident predictions selected during the iterative process. We see that the lowest value of $size_r$ (5%) guarantees the more stable (less variable) F-score response over the three base learners, meaning that the refinement process of the predictor F allows effectively us to improve the predictions of the instances, which are selected later, instead of removing them from $\mathcal{U}$ at the early. This is confirmed by the higher variance of the F-score when $size_r$ is 15%.

As to the neighborhoods, the indication we can draw is the higher accuracy is obtained with the larger number of neighbours. In fact, we see F-score values greater than 0.95 only for the configuration n_10_500, on the contrary, for n_5_500 and n_10_250, the accuracy is under the threshold of 0.95. This is why the use of greater neighborhoods generally leads to increase the "awareness" of the predictor F about the surrounding instances of a target instance and, consequently, improve the prediction of the trustworthiness levels. This confirms the advantages of the feature augmentation for "contextual" noise readings.

5   Conclusions

Enriching Mobile Crowd Sensing data with contextual details is essential to maximize the effectiveness of contributed data without explicitly requesting additional information to the end-user. This paper proposes to leverage on machine learning to contextualize the gathered observations in a way that accounts for the properties of the crowdsensors, spanning the device characteristics, the end-user's behavior and the environment. We have developed a transductive learning method able to learn on both fully known devices and partially labelled data. A validation session on the recurrent scenarios of urban noise pollution has been conducted to refine the accuracy of the trustworthiness prediction and quantitatively measure the influence of the working conditions on the accuracy.

Acknowledgments. This work was supported in part by the research project "APOLLON - environmentAl POLLution aNalyzer", within the "Bando INNONET-WORK 2017" funded by Regione Puglia (Italy) in the framework of the "FESR - Fondo Europeo di Sviluppo Regionale" - "POR Puglia FESR FSE 2014–2020 - Asse Prioritario 1 - Ricerca, sviluppo tecnologico, innovazione - Azione 1.6".

References

1. Guo, B., et al.: Mobile crowd sensing and computing: the review of an emerging human-powered sensing paradigm. ACM Comput. Surv. **48** (2015). https://doi.org/10.1145/2794400
2. Huang, K.L., Kanhere, S.S., Hu, W.: Are you contributing trustworthy data?: the case for a reputation system in participatory sensing. In: 13th International Symposium on Modeling Analysis and Simulation of Wireless and Mobile Systems, MSWiM 2010, pp. 14–22 (2010). https://doi.org/10.1145/1868521.1868526
3. Karkouch, A., Mousannif, H., Moatassime, H.A., Noël, T.: Data quality in internet of things: a state-of-the-art survey. J. Netw. Comput. Appl. **73**, 57–81 (2016). https://doi.org/10.1016/j.jnca.2016.08.002
4. Loglisci, C., Appice, A., Malerba, D.: Collective regression for handling autocorrelation of network data in a transductive setting. J. Intell. Inf. Syst. **46**(3), 447–472 (2016). https://doi.org/10.1007/s10844-015-0361-8
5. Loglisci, C., Ceci, M., Malerba, D.: Discovering evolution chains in dynamic networks. In: New Frontiers in Mining Complex Patterns - Revised Selected Papers, pp. 185–199 (2012). https://doi.org/10.1007/978-3-642-37382-4_13
6. Loglisci, C., Malerba, D.: Leveraging temporal autocorrelation of historical data for improving accuracy in network regression. Stat. Anal. Data Min. **10**(1), 40–53 (2017). https://doi.org/10.1002/sam.11336
7. Longo, A., Zappatore, M., Bochicchio, M.A.: Collaborative learning from mobile crowd sensing: a case study in electromagnetic monitoring. In: 2015 IEEE Global Engineering Education Conference (EDUCON), pp. 742–750 (2015)
8. Louta, M., Mpanti, K., Karetsos, G., Lagkas, T.: Mobile crowd sensing architectural frameworks: a comprehensive survey. IISA **2016** (2016). https://doi.org/10.1109/IISA.2016.7785385

9. Neville, J., Jensen, D.: Iterative classification in relational data. In: Proceedings of the 17th International Joint Conference on Artificial Intelligence. AAAI Press (2000)
10. Sailhan, F., Issarny, V., Tavares-Nascimiento, O.: Opportunistic multiparty calibration for robust participatory sensing. In: 14th International Conference on Mobile Ad Hoc and Sensor Systems, pp. 435–443, October 2017. https://doi.org/10.1109/MASS.2017.56
11. Wu, C., Luo, T., Wu, F., Chen, G.: EndorTrust: an endorsement-based reputation system for trustworthy and heterogeneous crowdsourcing. In: 2015 IEEE Global Communications Conference, pp. 1–6 (2015). https://doi.org/10.1109/GLOCOM.2014.7417352
12. Yurur, O., Liu, C.H., Sheng, Z., Leung, V.C., Moreno, W., Leung, K.K.: Context-awareness for mobile sensing: a survey and future directions. IEEE Commun. Surv. Tutor. 18(1), 68–93 (2016). https://doi.org/10.1109/COMST.2014.2381246
13. Zappatore, M., Longo, A., Bochicchio, M.A.: Crowd-sensing our smart cities: a platform for noise monitoring and acoustic urban planning. J. Commun. Softw. Syst. 13(2), 53–67 (2017)
14. Zappatore, M., Longo, A., Bochicchio, M., Zappatore, D., Morrone, A., De Mitri, G.: A crowdsensing approach for mobile learning in acoustics and noise monitoring, 04–08 April 2016, pp. 219–224 (2016). https://doi.org/10.1145/2851613.2851699

A Hierarchical-Based Web-Platform for Crowdsourcing Distinguishable Image Patches

Ayman Hajja[✉] and Justin Willis[✉]

Department of Computer Science, College of Charleston, 66 George Street,
Charleston, SC 29424, USA
hajjaa@cofc.edu, willisjk@g.cofc.edu

Abstract. In this work, we present an open-source web-platform for crowdsourcing a unique kind of image labels. This is done by introducing image segments that we refer to by the term "distinguishable patches"; as the name implies, a distinguishable patch is a small segment of an image that identifies a particular object. Although these distinguishable patches will naturally be part of the object; more often than not, these distinguishable patches combined will not cover the entire body of the object, which makes the nature of the data collected distinct and rather different than what would be obtained from a traditional image segmentation system. To minimize human labeling efforts while maximizing the amount of labeled data collected, we introduce a novel top-bottom hierarchical approach to automatically determine the size and the location of the patches our system will present to individuals (referred to by "workers") for labeling, based on previously labeled patches. The three processes of: determining the size and location of the patch, assigning a particular patch to the right worker, and the actual cropping of these patches, all happen in real-time and as the workers are actively using our web-platform. As far as the authors are aware, this unique form of image data has not been collected in the past, and its impact has not been explored, which makes this work highly valuable and important to the research community. One of the many ways that the authors are interested in using these distinguishable patches would be to improve the accuracy of a machine learning image classification system, by providing an enriched dataset of images that not only contain a single label for each image, but rather a spatial distribution of the distinguishability of an object in each image.

Keywords: Crowdsourcing · Labeling · Human computation · Image classification · Machine learning · Web application

1 Introduction and Background

The use of large volumes of labeled data has a direct impact on the accuracy of many machine learning classification algorithms; therefore, researchers and

© Springer Nature Switzerland AG 2020
D. Helic et al. (Eds.): ISMIS 2020, LNAI 12117, pp. 245–254, 2020.
https://doi.org/10.1007/978-3-030-59491-6_23

data science practitioners have been seeking new ways to collect labeled data effectively and accurately. One of the most practical ways of collecting labeled data is through the use of crowdsourcing; one form of crowdsourcing can be defined as the act of engaging many (often low-stake) individuals to collectively collect ground truths for a particular type of data instances (e.g. images, audio segments, video snippets), for the purpose of using these labeled data points to improve the accuracy of a machine learning classifier. It is worth noting here that other forms of crowdsourcing, such as ranking or proposing ideas or solutions, exist; however, crowdsourcing for machine learning purposes is what we are concerned about in this work.

There have been many proposed crowdsourcing systems for labeling images. "Visual Madlibs" [5] is a fill-in-the-blank system that collects labels about people and objects, their appearances, activities, and interactions. "Peekaboom" [4] on the other hand is a game-based image crowdsourcing platform for locating objects in images. Other more specific domains such as medical image analysis have also been interested in creating crowdsourcing platforms to gather ground truth data; one such system is a smartphone app that crowdsources the analysis of bladder cancer TMA core samples [3]. From the domain of audio analysis, Hajja et al. introduced a system to label short audio segments to help improve stutter detection classifiers [2]. Amazon Mechanical Turk (MTurk) [1] is perhaps one of the most known public platforms for crowdsourcing ground truths; it allows "requesters" to create a crowdsourcing study consisting of performing mini-tasks, which need to be undertaken by "workers". With the exception of MTurk, most of these existing platforms were designed to crowdsource labels of a specific dataset and are not publicly available for other researchers to utilize.

Objects in images are generally identified by either assigning a single label to the entire image, or by segmenting the object within the image and assigning that object a label. Crowdsourcing labeled images using the first image-based approach can be extremely effective due to its simple nature; and for that reason, many existing platforms such as MTurk can be utilized to collect such data. The second more specific segment-based approach requires manual segmentation of the object in the image, which tends to be substantially more tedious and time consuming; hence not easily feasible for collecting large sums of labeled data.

In this research paper, we present an open-source web-platform for crowd-sourcing a unique kind of image labels that strikes a delicate balance between the effectiveness benefit of the first approach, and the specificity benefit of the second approach. This is done through the introduction of image segments that we refer to by the term 'distinguishable patches'; as the name implies, a distinguishable patch is a small segment of an image that identifies a particular object. Although these distinguishable patches will naturally be part of the object; more often than not, these distinguishable patches combined will not cover the entire object, which makes the nature of the data collected distinct and rather different than what would be obtained by a traditional image segmentation system. As will be shown below, our web-platform provides a high level of customization for individuals and teams to create their own unique data collection studies or campaigns. Some of the parameters that campaign administrators can set include the sizes of image patches, the desired level of specificity and depth of patch generation hierarchy, and the 'stop threshold' which will be used to determine the optimal patch-worker assignment.

Instead of using the concept of crowdsourcing to collect a single label for each image, our system is designed to collect a distribution of labels for a set of dynamically-generated segments in each image. This is done through an image segmentation process that takes place in real-time using a cloud-based back-end logic implemented as part of our system. The criteria for which the segmentation process takes place can be set and customized by the campaign administrator. We will provide a detailed description of our system and show screenshots of our web-platform in the following section.

It is worth noting here however, that the system we are presenting here serves an entirely different purpose than a system that segments images. Our system identifies patches that can distinguish an object, as opposed to identifying the object itself. We believe that from a machine learning training perspective, that seemingly subtle difference will have a drastic effect on the effectiveness of the classifier training process. We have conducted a simple experiment to demonstrate that distinguishable patches do not always overlap with the entire object in the image. Figure 1 shows a heatmap displaying the object "distinguishability" in two images. Red regions show patches that workers were able to identify the object (i.e. cat) in; these are the patches that our system is interested in crowdsourcing.

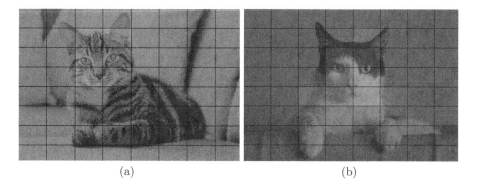

(a)   (b)

Fig. 1. Two sample outputs of our crowdsourcing platform showing a spatial distribution of object "distinguishability" (Color figure online)

2   Our System

We define patch labeling as the process of segmenting an image into different parts, which we refer to as "patches", and labeling each patch individually. The segmentation process, which determines the number (and location) of patches in an image, is established dynamically by our system. The basis behind the dynamic nature of the platform will be guided by the other previously labeled patches for the same image. In this section, we will delve into the logistics of how our system crops and distributes these patches, and we will provide a detailed summary of the parameters and settings that users can specify and tune to create their own unique data collection campaigns.

One of our primary goals for building this platform, is to give researchers the ability to design and administer image crowdsourcing projects according to their needs. Thus, we needed to build a highly-customizable platform that will adapt to the different needs of campaign administrators. In this paper, we will use the term "administrator" or "admin" to refer to an individual (or team) responsible for administering an image labeling project or campaign; although there are no restrictions on who can use the system, we anticipate most of our administrators to be academic researchers and data scientists. The term "worker" on the other hand, will refer to an individual who labels these image segments (referred to by "patches"); again, although anyone can label a public study, we anticipate a good portion of these workers to be research lab members or students.

2.1   The Grid System

Our system starts by dividing the images into an administrator-defined $m \times n$ grid, such that m is the number of rows and n is the number of columns. Each rectangle in the generated grid is referred to by the term "unit". The administrator also needs to define the size of the largest patch set (referred to by "initial", or "L1" patches) that will be labeled by workers. As will be shown below, the concept of patch levels (L1, L2, L3, ...) will be one of the primary key concepts in this research work.

Figure 2(a) shows an example of an 8×8 image grid with a 2×2 initial patch size, resulting in 16 initial patches. Figure 2(b) on the other hand shows a 4×4 grid with a 2×2 initial patch size, which yields 4 different non-overlapping initial patches. These grid lines shown in Fig. 2 are displayed in real-time while the admin is experiment with different values, allowing them to visualize the generated patches and determine the most appropriate grid dimension.

In the next subsection, we will elaborate on the different uses for our grid system and show examples of the multi-level patch system.

2.2   Variable Patch Sizes

As mentioned above, one of the primary reasons for the introduction of our system was to generate an enriched image dataset to enhance the performance of a machine learning image classifier. To accomplish that, we needed to identify as many distinguishable patches as possible in each image, which implies that we are particularly interested in finding the set of smallest distinguishable patches. There are two reasons for that, the first being is that extracting smaller patches will yield a higher number of unique patches, which as a result will provide more training data instances for our classifier to utilize. The second reason for our

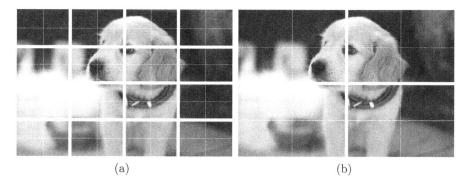

Fig. 2. (a) 8×8 image grid with a 2×2 initial patch size (b) 4×4 image grid with a 2×2 initial patch size.

interest in small patches is due to the information/size ratio; small distinguishable patches occur only when there is a highly distinguishable feature observed in that patch, such as cat whiskers, which can be highly valuable when training an image classification model. Having said that, the size of the smallest distinguishable patches depends on the object in the image, and it cannot be predetermined accurately by campaign administrators. For that reason, we have introduced a new way in which our system can extract these distinguishable patches from workers thorough a top-down hierarchical labeling approach.

Recall that campaign administrators need to specify at least the dimension of the image grid, and the dimension of the initial patch (L1). Each one of the initial patches is displayed to a group of workers (independently), for which they must assign a label. These labels are defined by the campaign administrator for the particular campaign. Figure 3 shows a screenshot of the worker's interface showing one of the four patches from the image shown in Fig. 2(b), along with three labels defined by the campaign administrator.

Campaign administrators also have the option of choosing to divide the set of initial (or L1) patches further. The resulting set of dividing an L1 patch is a set of L2 patches; L2 patches can be further divided into L3, so on and so forth. For example, assuming that the dimension of the image grid is 8×8, the dimension of L1 patches is 4×4, L2 patches is 2×2, and L3 patches is 1×1. Figure 4 shows the set all patches that could potentially get generated from our image. Note here that although Fig. 4(a) shows a 2×2 grid, the size of each cell in that grid is 4×4 (L1 dimension) provided that the image grid dimension is defined as 8×8.

As you may have guessed, there are multiple sets of dimensions that will result in the same patches; for example, an L1 division of 9×9 and an L2 division of 3×3 is identical to an L1 division of 3×3 and an L2 division of 1×1 (assuming L2 is the highest level generated patches). Our system allows the administrator to specify any valid set of dimensions; in addition to that, our system will display the number of potentially generated patches for each level.

Fig. 3. A screenshot from the worker's interface showing one of the four patches from the image shown in Fig. 2(b), along with three labels defined by the campaign administrator

Figure 5 shows a screenshot of the page on which the administrators specify the grid size of each level. Note here that these patches are only potentially generated since this will be determined based on the labels provided by previous workers, as will be explained in the next subsection.

2.3   Hierarchical Patch Generation

To minimize human labeling efforts, we have decided to adopt a top-down approach for generating the patches. Our system will start by asking workers to label all the L1 patches first, and based on the obtained labels for each patch, our system will either mark these patches with "unknown" or "distinguishable" label. Next, our system will divide all distinguishable L1 patches into their L2 sub-patches, and present them to workers, so on and so forth. This process will continue until we go through all levels (defined by campaign admin), or until all patches have been labeled with "unknown", and hence no further division will be required. To provide few examples to demonstrate the idea of hierarchical patch generation, we would have to define three key terms:

1. **Stop threshold**: The stop threshold can be defined as the minimum percentage of agreeability that campaign administrators require for any patch to be marked "distinguishable". As soon as a particular patch satisfies this agreeability threshold, and as long as the number of responses for that patch is within the minimum and maximum range (defined below), our system will not present this patch for additional labeling.
2. **Minimum number of responses for a patch**: As the name implies, this value is used to define a lower bound for the number of responses obtained from workers for a particular patch. For example, if this value is set to 3,

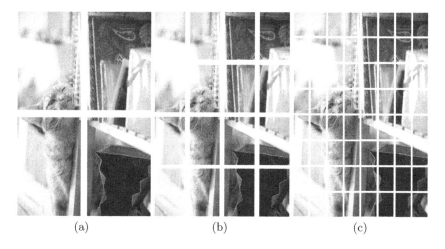

(a)   (b)   (c)

Fig. 4. (a) L1 patches (b) L2 patches (c) L3 patches

Fig. 5. Screenshot of what the admins see when setting the dimensions of the patch levels.

that simply means that every patch needs to be presented to at least three different workers.

3. **Maximum number of responses for a patch**: This value is used to define an upper bound for the number of responses obtained from workers for a particular patch. As we will demonstrate in the next few examples, this value will often not be reached, since the stop threshold and minimum number constraints may be satisfied prior to reaching this value.

Next, we will show an example for demonstrating the hierarchical patch generation in action. In this example, we will use the levels defined in Fig. 4. Let us assume the following:

1. the stop threshold is 70%,
2. the minimum responses is 3 and the maximum responses is 7, and
3. the labels provided to workers are "Cat", "Dog", and "I don't know".

Let us also agree on a patch numbering system that we will use in this example. If the patch is an L1 patch, then its ID will be L_m such that m is the number of the patch, counting from top left to bottom right; for example, L_1 would be the top left patch in Fig. 3(a), L_2 would be the top right patch in Fig. 3(a), L_3 would refer to the bottom left patch, and L_4 would be the bottom right patch in Fig. 3(a). If the patch is an L2 patch, then we will use two subscripts, the first one referring to the L1 patch that it was generated from, and the second subscript will be used to indicate the number of the patch counting from top left to bottom right; for example, the top right patch in Fig. 3(b) is referred to by $L_{2,2}$ since it is part of patch L_2 and it is the second patch; the bottom right patch in Fig. 3(b) would be identified by $L_{4,4}$ since it is part of patch L_4, and it is the fourth and last patch (within that sub-patch).

Table 1 below shows a valid patch-worker assignment, with potential labels provided by workers; we will use this table to demonstrate this example. In Table 2, we provide a description of how our system will behave based on the labels provided in Table 1.

Table 1. A hypothetical example of possible labels provided by workers

Time	Patch ID	Worker ID	Label provided
0	L_1	15	Cat
1	L_1	9	Cat
2	L_2	9	I don't know
3	L_1	12	Cat
4	L_2	15	I don't know
5	L_2	5	I don't know
6	L_3	5	Cat
7	L_4	12	I don't know
8	L_4	5	I don't know
9	L_3	9	I don't know
10	L_4	9	I don't know
11	L_3	12	Cat
12	$L_{1,2}$	26	I don't know
13	L_3	29	Dog
14	...	...	...

Table 2. A detailed description of how our system behaves based on the labels provided by workers shown in Table 1

Time	System behavior
3	L_1 satisfies the three criteria (3 is more than or equal to minimum responses, 3 is less than or equal to maximum responses, and 3 workers provided the label "Cat" out of the 3 labels provided, which means that the agreeability is 100% which is higher than or equal to the stop threshold 70%). Since L_1 satisfied the three conditions, we will crop it into its L_2 level patches and repeat
5	L_2 now has been labeled "I don't know" by three workers. Here, we would stop asking workers to label L_2 and we do not divide it into its Level 2 ($L2$) patches. The reason for this decision is because even if the rest of workers (4, since the maximum responses is 7) agree on a label (say "Cat"), the agreeability will be 4/7 which is less than the stop threshold
10	Similar to L_2, L_4 has been labeled with "I don't know" by three workers, so we stop asking users to label L_4, and we do not divide that patch any further
11	Here, L_3 has been labeled "Cat" by two workers and "I don't know" by one worker. The three criteria have not been satisfied yet since the stop threshold (70%) has not been met. That being said, there is still a possibility that if we ask more workers, we will satisfy all three conditions
12	We can observe two things here: 1. Our system is now asking workers to label patches from the second level (since for patch L_1, it has been agreed that it exhibits the label "Cat"), and 2. We notice that the worker ID is now different, and that is because the four workers with ID's 5, 9, 12, and 15 have seen L_1 patch (which contains $L_{1,2}$) and therefore may bias their response
13	After the label provided at Time 13, we have 2 "Cat" labels for L_3, 1 "I don't know" label, and 1 "Dog" label. Our system will automatically stop asking workers to label this patch since there is no way that we will reach agreeability higher than or equal to the stop threshold

3   Conclusion and Future Work

In this paper, we described an open-source web-platform designed to provide researchers and data scientists with the ability to efficiently and effectively crowdsource labeled patches from images. The nature of the crowdsourced distinguishable patches presented in this paper is unique, and as far as the authors are aware has not been explored in the past.

The system that we have designed adopts a top-bottom hierarchical-based approach to automatically determine the size and the location of the patches presented to workers; therefore minimizing human labeling efforts while maximizing the amount of labeled data collected, without adding human bias.

One of the areas in which future work can be conducted involves exploring ways to improve the training of image classification models by utilizing these collected distinguishable patches; either by treating each distinguishable patch as a unique labeled instance, or by augmenting the spatial distribution of workers' responses to existing single-label image datasets.

Another area of future work involves extending our platform to crowdsource labels from sets of many isolated patches within an image. This idea of multi-patch labeling is particularly interesting since it may reveal interesting findings about how combining independent (and indistinguishable) patches could result in meaningful groups of image patches.

References

1. Buhrmester, M., Kwang, T., Gosling, S.D.: Amazon's Mechanical Turk: a new source of inexpensive, yet high-quality data? (2016)
2. Hajja, A., Hiers, G.P., Arbajian, P., Raś, Z.W., Wieczorkowska, A.A.: Multipurpose web-platform for labeling audio segments efficiently and effectively. In: Ceci, M., Japkowicz, N., Liu, J., Papadopoulos, G.A., Raś, Z.W. (eds.) ISMIS 2018. LNCS (LNAI), vol. 11177, pp. 179–188. Springer, Cham (2018). https://doi.org/10.1007/978-3-030-01851-1_18
3. Smittenaar, P., et al.: Harnessing citizen science through mobile phone technology to screen for immunohistochemical biomarkers in bladder cancer. Br. J. Cancer **119**(2), 220–229 (2018)
4. Von Ahn, L., Liu, R., Blum, M.: Peekaboom: a game for locating objects in images. In: Proceedings of the SIGCHI Conference on Human Factors in Computing Systems, pp. 55–64 (2006)
5. Yu, L., Park, E., Berg, A.C., Berg, T.L.: Visual Madlibs: fill in the blank description generation and question answering. In: Proceedings of the IEEE International Conference on Computer Vision, pp. 2461–2469 (2015)

Performing Arithmetic Using a Neural Network Trained on Digit Permutation Pairs

Marcus D. Bloice[1]([✉]), Peter M. Roth[2], and Andreas Holzinger[1,3]

[1] Institute for Medical Informatics, Statistics, and Documentation,
Medical University of Graz, Graz, Austria
{marcus.bloice,andreas.holzinger}@medunigraz.at
[2] Institute of Computer Graphics and Vision, Graz University of Technology,
Graz, Austria
p.m.roth@ieee.org
[3] xAI Lab Alberta Machine Intelligence Institute, T6G 2H1 Edmonton, Canada

Abstract. In this paper a neural network is trained to perform simple arithmetic using images of concatenated handwritten digit pairs. A convolutional neural network was trained with images consisting of two side-by-side handwritten digits, where the image's label is the summation of the two digits contained in the combined image. Crucially, the network was tested on permutation pairs that were not present during training in an effort to see if the network could learn the task of addition, as opposed to simply mapping images to labels. A dataset was generated for all possible permutation pairs of length 2 for the digits 0–9 using MNIST as a basis for the images, with one thousand samples generated for each permutation pair. For testing the network, samples generated from previously unseen permutation pairs were fed into the trained network, and its predictions measured. Results were encouraging, with the network achieving an accuracy of over 90% on some permutation train/test splits. This suggests that the network learned at first digit recognition, and subsequently the further task of addition based on the two recognised digits. As far as the authors are aware, no previous work has concentrated on learning a mathematical operation in this way.

1 Introduction

The aim of this study is to attempt to find experimental evidence that would suggest that a network can be trained to perform the task of addition, when supplied with image data containing two digits that should be summed. To ensure that the network has indeed learned this, and is not simply mapping images to labels, a constraint was applied whereby the network is tested with a held back test set of previously unseen permutation pairs. This forces the network to learn more than simply a mapping between individual images and labels as it is tested using digit combination pairs that it has not seen, meaning a direct mapping from an image or shape to a label would not function.

© Springer Nature Switzerland AG 2020
D. Helic et al. (Eds.): ISMIS 2020, LNAI 12117, pp. 255–264, 2020.
https://doi.org/10.1007/978-3-030-59491-6_24

Table 1. Example input images and their corresponding labels. Each image consists of two side-by-side MNIST digits merged into one single image where each image's label is the summation of the two digits. The label contains no information as to the individual digits contained within the image, nor is there any indication given to the network prior to training that each image consists of two digits or otherwise.

Image	Interpretation	Label
5 0	5 + 0	5
8 5	8 + 5	13
9 9	9 + 9	18

To test this hypothesis, a network was trained with data generated using 90 of the possible 100 combinations of the digits 0–9 up to length 2. Once trained, the network was tested by inputting images based on the remaining 10 permutation pairs, and was required to predict the summations for them. Some examples of the generated samples can be seen in Table 1. 1,000 samples were generated for each permutation pair. This was repeated 10 times for 10 different permutation pair train/test splits.

2   Background Work

There has been previous work relating to this experiment. As opposed to most work, however, the goal of this study was not to recognise digits, extract their numerical values from the images, and then perform (after the network's character recognition procedure) some mathematical function on the numerical values. The task of this experiment was to learn if the network could learn the logical task of the mathematical operation itself using an end-to-end approach. For example, [3] experimented with computer generated image data, however as output the network was trained to produce images containing the summations. Their work concentrated on the visual learning of arithmetic operations from images of numbers. In contrast, the work presented here outputs its predictions as a real number. Their approach used numbers of longer lengths and were therefore also able to generate many thousands of training samples, despite not using hand written digits. The input consisted of two images, each showing a 7-digit number and the output, also an image, displayed a number showing the result of an arithmetic operation (e.g., addition or subtraction) on the two input numbers. The concepts of a number, or of an operator, are not explicitly introduced. Their work, however, was more akin to the learning of a transformation function, rather than the task of learning a mathematical operation. Other operations, such as multiplication, were not learnable using this architecture. Some tasks were not learnable in an end-to-end manner, for example the addition of Roman numerals, but were learnable once broken into separate sub-tasks: first perceptual character recognition and then the cognitive arithmetic sub-task.

Similarly, a convolutional neural network was used by [6] to recognise arithmetic operators, and to segment images into digits and operators before performing the calculations on the recognised digits. This again is different to the approach described here, as it is not an attempt to learn the operation itself, but to learn to recognise the operator symbols and equations (and perform the mathematics on the recognised symbols).

In [8], the authors addressed the task of object counting in images where they applied a learning approach in which a density map was estimated directly from the input image. They employed convolutional neural networks with layered boosting and selective sampling. It would be possible to create an experiment based on their work, that would perform arithmetic by counting the values of domino tiles, for example.

Until now, as far as the authors are aware, no work has concentrated on learning the actual mathematical operation itself. Previous work tends to concentrate on first recognising digits and operators within images, and then to perform the mathematical operations separately, after this extraction has been carried out. In the case of this work, an end-to-end algorithm has been developed that performs the digit recognition, representation learning of the values of the digits, and performs the arithmetic operation.

3   Experiment

A convolutional neural network was trained to perform arithmetic on images consisting of two side-by-side hand-written digits. Each image's corresponding label is the sum of the two digits, and the network was trained as a regression problem. For the digits 0–9, up to length 2, there are 100 possible permutation pairs. For each permutation pair, 1,000 unique images were generated using the MNIST hand written digit database [4]. MNIST was chosen due to its familiarity in machine learning, and because LeNet5 (used as the basis for the network trained in this work) clearly functions well on this dataset. Also, MNIST continues to be a focus of research and is used as a benchmarking dataset to this day [2,7].

Training was performed on images generated from a random 90-long subset of the possible permutations, and testing was performed on images based on the remaining 10 permutations. A number of example input images and their labels are shown in Table 1 where it can be seen that a single input image consists of two MNIST digits side-by-side, and the image's label is the summation of the two digits. For each permutation, 1,000 combined images are generated resulting in 100,000 samples that are separated into a training and test set based on the permutation pairs.

The task of the experiment was to train a neural network with data generated from a subset of the possible 100 permutations, that when presented with a new samples, generated from the unseen permutation pairs, the network would be required to predict the correct summation. This was done in order to ascertain whether a network could learn a simple arithmetic operation such as addition,

given only samples of images and their summations and no indication as to
the value of each individual digit contained within the image, while only being
trained on a subset of all possible permutation pairs and tested on the remaining
pairs.

In summary, the experimental setting is as follows:

- By permutations, it is meant all possible combinations of the digits 0–9 of
  length 2. Formally, if the set of digits $D = \{0, 1, 2, 3, 4, 5, 6, 7, 8, 9\}$, all possible
  permutations is the Cartesian product of $D \times D$, which we define as P, so
  that $P = \{(0,0), (0,1), (0,2), \ldots, (9,8), (9,9)\}$.
- Of the 100 possible permutations pairs P, a random 90 are used as a basis to
  train the network and the remaining 10 pairs are used as a basis to test the
  network. These are the training permutations, P_t, and the test permutations,
  P_v. This permutation train/test split was repeated 10 times as a 10-fold cross
  validation.
- For each permutation, 1,000 samples are generated. So, for each of the permutations in P, 1,000 concatenated images are generated using random MNIST
  digits M (appropriate for that permutation).
- By appropriate this means that, for example, generating an image for the
  permutation pair $(3, 1)$ a random MNIST digit labelled 3 is chosen and a
  random MNIST digit labelled 1 is chosen and these images are concatenated
  to create a single sample for this permutation pair. This means each sample
  is likely unique (likely, as each image is chosen at random with replacement,
  see Table 2). The generated images are contained in a matrix X, where X_t
  are the training samples and X_v are the test samples.
- For the generation of the training set images, X_t, only images from the
  MNIST training set, M_t, are used.
- For the test permutation images, X_v, only images from the MNIST test set,
  M_v, are used.
- The network is not given any label information regarding each individual digit
  within the concatenated images, only the summation is given as label data.
- The permutations pairs in the test set are not seen during training. This
  means the training set and test set are distinct in two ways: they contain
  different permutations pairs that do not overlap, and the individual MNIST
  images used to generate each permutation sample do not overlap between the
  training set and test set.

The decision to train the network as a regression problem is done for the
following reasons. First, the number of output neurons would change depending on the train/test split. For example, the sum of $(9, 9)$ is 18 and cannot be
made by any other permutation, meaning a possible discrepancy between the
number of possible output neurons of the training set and test set. Second, when
training on permutations of longer length, an ever increasing number of output
neurons would be required. Last, for measuring how well the network performs,
classification poses problems which are mitigated by using regression and mean
squared error loss.

Table 2. For each permutation pair, random MNIST digits are used for generating each sample. For example, for the permutation pair (0, 2), each sample that is generated uses a random digit 0 combined with a random digit 2 obtained from MNIST.

Sample 1	Sample 2	...	Sample 1000
		...	

The following sections describe the experiment itself, beginning with a description of the dataset and how it was created. Then, the neural network's architecture is described as well as the training strategy. Last, the results of the experiment are discussed, and the paper is concluded with a discussion.

4 Dataset

The dataset used for the creation of the concatenated image data was MNIST, a 70,000 strong collection of labelled hand written digits. As per the original dataset, 60,000 digits belong to the training set and 10,000 belong to the test set. Images in the MNIST dataset are 8-bit greyscale, 28×28 pixels in size. The generated images are therefore 28×56 pixels in size as they are the concatenation of two MNIST digits placed side-by-side and stored as a single image (see Table 3, for example). Each generated image's label is the summation of the two individual digits' labels (see Table 1 for several examples). For each permutation, one thousand samples are generated, and the MNIST digits are chosen at random in order to create distinct samples.

4.1 Train/Test Split

As mentioned previously, for the digits 0–9, with a maximum of length of $l = 2$, there are n^l or $10^2 = 100$ possible permutation pairs. To generate the training and testing data, the permutations pairs are split into a permutation training set and a permutation test set at random, so that: $P = P_t \cup P_v$ and $P_t \cap P_v = \emptyset$. For training, 90% of the permutations were used to generate the training samples, X_t and the remaining 10% were used for generating the test set samples X_v, meaning $|P_t| = 90$ and $|P_v| = 10$ while $|X_t| = 90,000$ and $|X_v| = 10,000$ for any particular run. The experiment was performed using a 10-fold cross validation, based on the permutations pairs, and the loss was averaged across the 10 runs.

In terms of the generated samples, the generated training set images and generated test set images honoured the MNIST training set and test set split. This means that the generated training set samples are distinct from the generated test set samples both in terms of the permutation pairs and the images used to create each sample. It should be noted that the images were chosen from MNIST randomly *with replacement*, meaning images could appear twice in different permutation pairs within the training set or test set, but not between both.

Therefore, the generated data set matrix X contains 100,000 samples, 10,000 for each permutation pair. This also means that X_t contains the corresponding samples for the permutations P_t, and X_v contains the samples for P_v. A label vector y contains the labels, which are the summations of the two digits in the sample. Similarly, $y = y_t \cup y_v$.

4.2   Network Architecture and Training Strategy

The neural network used was a multilayer convolutional neural network similar to the original LeNet5, which was first reported by [5]. However, rather than treating the problem as a classification problem, the network is trained as a regression problem. The network was evaluated using mean squared error loss and optimised with ADADELTA [10]. All experimentation was performed with Keras using TensorFlow [1] as its back-end, running under Ubuntu Linux 14.04.

A LeNet5-type network was chosen due to its association with MNIST, having been optimised and developed for this dataset, and has repeatedly been shown to work well at the general task of character recognition. As the inputs to this network are similar to the original MNIST images, having twice the width in pixels but having the same height in pixels, the only other modification that was made was with the output of the network. Instead of a 10 neuron, fully connected output layer with *Softmax*, the final layer was replaced with a single output neuron and trained as a regression problem, optimising mean squared error loss.

4.3   Data Generation

The data generation procedure algorithm is shown in Algorithm 1. During data generation, the permutations, P, are iterated over and $m = 1000$ samples are generated for each permutation. A sample consists of two random MNIST images, corresponding to the labels in the current permutation, concatenated together as one image (this is represented by the function ConcatenateImages). As well as this, the label vector y is generated, which contains the sum of the two digit labels that make up each individual MNIST image used to create the sample. Note that in Algorithm 1 the symbol $\Leftarrow$ represents append, as is the case for the matrix X and its corresponding label vector y.

The procedure shown in Algorithm 1 is repeated for the train set permutation pairs, P_t, and the test set permutation pairs, P_v. It is important to note that when generating the data for the training set permutation images, the MNIST training set is used, and conversely when generating the test set permutation images, the MNIST test set is used. This ensures no overlap between the training set or test set in terms of the permutation pairs or the data used to generate the samples.

Algorithm 1. Data Generation

Input: permutation pairs P, MNIST images M, labels y', size $m \leftarrow 1000$.
Initialise $X \leftarrow [\,]$.
Initialise $y \leftarrow [\,]$.
for all (p_1, p_2) **in** P **do**
    for 1 **to** m **do**
        $r^1 \leftarrow$ random index from M with label p_1
        $r^2 \leftarrow$ random index from M with label p_2
        $X \Leftarrow \text{ConcatenateImages}(M_{r^1}, M_{r^2})$
        $y \Leftarrow y'_{r^1} + y'_{r^2}$
    end for
end for

5   Results

Averaged across the different training/test splits of a 10-fold cross validation of
the permutation pairs, mean squared error was generally under 1.0 and averaged
0.85332, as shown in Table 7. Tables 3, 4, and 5 show a number of examples of
a trained network's predictions on permutations from a test set. Table 3 shows
a number of sample inputs from the test set and their predictions, as well as
their true labels. It is interesting to note that the network learned to deal with
permutations with images in reverse order, as is the case for (6, 4) and (4, 6) or
(1, 3) and (3, 1) in Table 4. In some cases, three different permutation pairs exist
in the test set which sum to the same number, and these were also predicted
correctly, as seen in Table 5.

Table 3. Example results for permutation samples from the test set passed through
the trained network. As can be seen, all samples use distinct MNIST digits.

Input Image	Prediction	Actual
	11.1652	11
	10.3215	10
	5.01775	5
	8.99357	9
	5.99666	6
	8.6814	9

Although the network was trained as a regression problem, accuracy can
also be measured by rounding the predicted real number output to the nearest
integer and comparing it to the integer label. When rounding to the nearest digit,
accuracy was as high as 92%, depending on the train/test split and averaged

Table 4. Example results of correct predictions for test set permutations that have the same label but consist of different pairs of digits such as (6, 6) and (4, 8) or (9, 2) and (3, 8). Note also that the model was also able to deal with digits in swapped order, as is the case for (1, 3) and (3, 1). Table 5 shows a further example of this.

Input Image	Prediction	Actual
6 6	11.8673	12
4 8	11.8703	12
9 2	10.8862	11
3 8	10.8827	11
1 3	3.7308	4
3 1	4.23593	4

Table 5. Example of correct predictions for three permutations from the same test set that sum to the same value.

Input Image	Prediction	Actual
8 2	9.84883	10
4 6	9.9731	10
6 4	10.0746	10

70.9%. Accuracy increases, if the predicted value is used with a floor or ceiling function, and both values compared to the true value, achieving an accuracy of approximately 88% across a 10 fold cross validation. When allowing for an error of ±1, the accuracy, of course, increases further. Table 6 shows the accuracy of the trained network over a 10 fold cross validation. The accuracies are measured across all samples of all test set permutations—a total of 10,000 images. The accuracies are provided here merely as a guide, for regression problems it is of course more useful to observe the average loss of the predictions versus the labels. Errors presented here are likely the result of misclassifications of the images themselves rather than the logic of the operator learned, as the network was trained to optimise the loss and not the accuracy. Also, even for folds with low accuracy there are always correct predictions for all permutation pairs, again a further reason why it is not entirely useful to report accuracies. The mean squared error loss is provided as a truer measure of the network's performance, and in order to provide as accurate a loss as possible a 10-fold cross validation was performed. The results of a 10-fold cross validation of the permutation pairs can be seen in Table 7. The average mean squared error loss over the 10-fold cross validation was 0.853322.

Table 6. Accuracy of each run of a 10-fold cross validation. For each permutation there are always correct predictions, even for poorly performing folds, such as folds 2 and 4. By using floor and ceiling functions on the predicted values for each of the images, the accuracy increases significantly. Note that for the results here 2,000 samples per permutation were generated.

Fold	Rounding	Floor/ceiling	±1
1	81.19%	85.51%	94.93%
2	43.00%	90.89%	96.23%
3	80.95%	86.95%	95.12%
4	55.08%	71.21%	86.56%
5	82.35%	93.56%	95.28%
6	92.23%	95.49%	96.76%
7	54.32%	86.94%	95.38%
8	74.86%	94.71%	96.67%
9	87.59%	94.33%	95.91%
10	57.48%	85.53%	96.23%
Avg.	70.91%	88.51%	94.90%

Table 7. Results of each run of a 10-fold cross validation. The average mean squared error (MSE) across 10 runs was ≈ 0.85 on the test set.

Fold	Test set MSE	Train set MSE
1	1.1072	0.0632
2	0.6936	0.0623
3	0.7734	0.0661
4	0.7845	0.0607
5	0.9561	0.0694
6	0.7732	0.0553
7	1.2150	0.0803
8	0.7278	0.0674
9	0.9464	0.0602
10	0.5556	0.0709
Avg.	0.8533	0.0656

6   Conclusion

In this work, we have presented a neural network that achieves good results at the task of addition when trained with images of side-by-side digits labelled with their summations, and tested with digit combination pairs it has never seen. The network was able to predict the summation with an average mean squared error of 0.85 for permutation pairs it was not trained with. By testing the network on a distinct set of digit combinations that were unseen during training, it suggests the network learned the task of addition, rather than a mapping of individual images to labels. A number of further experiments would be feasible using a similar experimental setup. Most obviously, the use of three digits per image could be performed using permutations up to length 3, or higher. Furthermore, other arithmetic operations could also be tested, such as subtraction or multiplication. More generally, the applicability of this method to other datasets in other domains needs to be investigated more thoroughly, for example whether there is an analogous experiment which could be performed on a dataset that does not involve arithmetic but involves the combination and interpretation of unseen combinations of objects in order to make a classification, such as through the use of the Fashion-MNIST [9] or ImageNet datasets.

References

1. Abadi, M., et al.: TensorFlow: large-scale machine learning on heterogeneous distributed systems. arXiv:1603.04467 (2016)
2. Delahunt, C.B., Kutz, J.N.: Putting a bug in ML: the moth olfactory network learns to read MNIST. Neural Netw. (2019)
3. Hoshen, Y., Peleg, S.: Visual learning of arithmetic operations. In: Thirtieth AAAI Conference on Artificial Intelligence, pp. 3733–3739 (2016)
4. LeCun, Y., Bottou, L., Bengio, Y., Haffner, P.: Gradient-based learning applied to document recognition. Proc. IEEE **86**(11), 2278–2324 (1998)
5. LeCun, Y., et al.: Handwritten digit recognition: applications of neural network chips and automatic learning. IEEE Commun. Mag. **27**(11), 41–46 (1989)
6. Liang, Z., Li, Q., Liao, S.: Character-level convolutional networks for arithmetic operator character recognition. In: International Conference on Educational Innovation through Technology, pp. 208–212 (2016)
7. Phong, L.T., Aono, Y., Hayashi, T., Wang, L., Moriai, S.: Privacy-preserving deep learning via additively homomorphic encryption. IEEE Trans. Inf. Forensics Secur. **13**(5), 1333–1345 (2018)
8. Walach, E., Wolf, L.: Learning to count with CNN boosting. In: Leibe, B., Matas, J., Sebe, N., Welling, M. (eds.) ECCV 2016. LNCS, vol. 9906, pp. 660–676. Springer, Cham (2016). https://doi.org/10.1007/978-3-319-46475-6_41
9. Xiao, H., Rasul, K., Vollgraf, R.: Fashion-MNIST: a novel image dataset for benchmarking machine learning algorithms. arXiv:1708.07747 (2017)
10. Zeiler, M.D.: ADADELTA: an adaptive learning rate method. arXiv:1212.5701 (2012)

Modelling and Reasoning

CatIO - A Framework for Model-Based Diagnosis of Cyber-Physical Systems

Edi Muškardin$^{(\boxtimes)}$, Ingo Pill, and Franz Wotawa

Christian Doppler Laboratory for Quality Assurance Methodologies
for Cyber-Physical Systems Institute for Software Technology, Graz University
of Technology, Inffeldgasse 16b/II, 8010 Graz, Austria
{edi.muskardin,ipill,wotawa}@ist.tugraz.at

Abstract. Diagnosing cyber-physical systems is often a challenge due to the complex interactions between its individual cyber and physical components. With CatIO (From 'Causarum Cognitio', Latin for *"(seek)* knowledge of causes"), we propose a framework that supports a designer in developing corresponding diagnostic solutions that utilize either abductive or consistency-based diagnosis for detecting and localizing faults at runtime. Employing an interface to tools of the modeling language Modelica, a designer is able to simulate a cyber-physical system's detailed behavior, and based on the observed data she can then assesses the diagnostic solution(s) under development and explore the trade-offs of individual solutions. For the abductive reasoning variant, CatIO supports also in coming up with the required abductive diagnosis model via an automated concept based on fault injection and the simulation of corresonding Modelica models.

Keywords: Model-based diagnosis · Modelica · Cyber-physical system · Co-simulation.

1 Introduction

For an intelligent cyber-physical system (CPS), it is crucial to be aware of its current status and also that of its environment. This allows the CPS to respond intelligently to environmental inputs and changes, and also to make the right choices when having to deal with issues like faults. Model-based diagnosis [4,10, 18] (MBD) is a powerful means to assess an observed situation, and furthermore supports an engineer in isolating the root cause(s) of some encountered issue(s). Deploying MBD in practice is, however, usually not a trivial task, and this is especially true for CPSs [20]. That is, providing an appropriate model for diagnostic reasoning concerning a CPS—where a computation core monitors and controls an aggregation of physical and cyber components—is a quite complex and cumbersome task.

Authors are listed in alphabetical order.

D. Helic et al. (Eds.): ISMIS 2020, LNAI 12117, pp. 267–276, 2020.
https://doi.org/10.1007/978-3-030-59491-6_25

For consistency-based MBD [10, 13], we need a specific type of system model that describes a CPS's expected behavior, and when considering fault modes [5] we need to cover also faulty behavior. Choosing the right abstraction level and description format for this model in order to arrive at an attractive trade-off between scalability and diagnostic preciseness is not an easy task.

For abductive diagnostic reasoning [4, 7], we need to come up with a knowledge-base (KB) describing dependencies between faults and their effects considering the system's observable behavior. Such a KB then allows us to abductively reason backwards from experienced symptoms to possible root causes. While such a KB is usually more abstract than a detailed model of a system's behavior, we still have to consider details like multiple faults occurring simultaneously [16]. For more information on modeling for consistency-based and abductive diagnosis we refer the interested reader to [22].

While a variety of MBD algorithms, e.g., [9, 10, 13, 18, 21], corresponding libraries [17], and also solutions for automatically generating abductive diagnosis models [15, 16] have been proposed, we are in need of frameworks that assist a designer in exploring the options available for some application scenario. In addition, several tools and languages for supporting designers in coming up with diagnostic systems have been proposed, including DiKe [6] and Rodelica [3].

With CatIO, we are proposing a framework that connects Modelica [8] with diagnostic reasoning in order to allow an engineer to simulate a system's behavior and directly assess the diagnostic solutions she is developing in the context of such a simulation. Modelica is a flexible and intuitive programming language for modeling a system, and there is commercial as well as free tool support[1] available. One can draw furthermore on a variety of libraries, e.g., for digital circuits, fluids, or mechanics, so that Modelica is a flexible tool for modeling a CPS. Implementing the concept proposed in [15], via this Modelica interface, CatIO can support an engineer also in the development of an abductive diagnosis model based on fault injection and simulation (see Sect. 2.1).

As we will show in Sect. 2, CatIO offers a special interface for connecting an external component to a simulation. All data computed in the framework is available via this interface, and the external component is allowed to dynamically change the input values of the simulation. The motivation for this *controller* interface was to enable connecting a system's control logic to a simulation, so that a designer can assess and verify a system's reactions in fault scenarios.

Summing up CatIO's features, an engineer can (i) simulate a CPS in Modelica to create data for testing a diagnostic solution under development, (ii) compare and assess MBD solutions in the context of different abstraction levels and drawing on either MBD concept, (iii) automatically generate an abductive diagnosis model, (iv) connect an external controller that uses the diagnostic information.

The remainder of our manuscript is organized as follows: After proposing the details of our framework in Sect. 2, we will show how CatIO supports designers in the context of a simple example in Sect. 3. In Sect. 4 we will draw a brief summary and will depict future work.

---

[1] https://www.modelica.org/modelicalanguage.

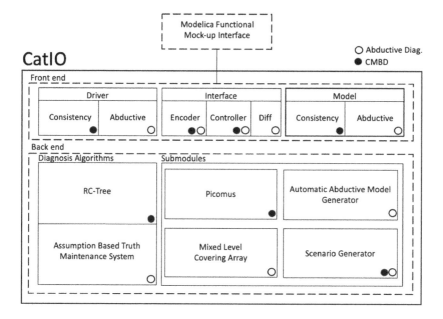

Fig. 1. CatIO's high level architecture.

2   The CatIO Framework and Its Architecture

In Fig. 1, we illustrate CatIO's architecture. The front end aggregates all the parts a user interacts with. This includes (i) the *driver* module for configuring the diagnostic features, (ii) the *interface* module that connects to Modelica and where we specify the *encoder* for matching the signals and translating between the simulation and the diagnostic models, the *controller* which is an external component we can connect to the simulation as described in the introduction, *diff* which we use to detect deviations between the observed and expected behavior (for the abductive variant, as will be explained later), and (iii) the *model* module which encapsulates the diagnostic model in either variant. As back end, we use a set of diagnosis algorithms complemented by a set of sub-modules.

2.1   The Back-End: Diagnosis Algorithms and Related Sub-modules

For consistency-based MBD (CMBD), we use RC-Tree [13]. RC-Tree is a variant of HS-DAG [9] and Reiter's original algorithm [18] that avoids all the redundancy in its search for diagnoses. The computation in this MBD variant is based on isolating and resolving the conflicts between observed and expected behavior via computing the minimal hitting sets of those conflicts [18]. RC-Tree allows for an on-the-fly computation of the conflicts which is of advantage if we are just interested in diagnoses up to a certain size. While the architecture allows the integration of any solver, for our ongoing Java implementation we rely on SAT models as system description and use picomus [1] to isolate the conflicts. We could connect also, e.g., constraint solvers though—or any other solver that can generate conflicts (ideally minimal conflicts). There is also another algorithmic concept for implementing CMBD where we compute the diagnoses directly in the solver as outlined in [11]. Our performance comparison of [12] did not show a substantial advantage in one or the other direction, so that we currently focus on RC-Tree. Depending on the diagnosis engine configuration, RC-Tree can compute both persistent faults and intermittent faults, and we can combine results from multiple simulation scenarios as was proposed in [14].

Abductive diagnosis in CatIO relies on an Assumption Based Truth Maintenance System (ATMS) [19], were we use a specific knowledge-base encoding our knowledge about a system's fault-and-effect dependencies and then reason backwards from observed symptoms to probable causes as explanations for the symptoms [15]. Coming up with such a knowledge-base is a complex task, where CatIO helps a designer in its creation via an automation option proposed in [16]. The underlying principle there is that we test the system and faulty system versions (that we can create via fault injection) for collecting possible symptoms caused by faults and also the corresponding fault-and-effect dependency rules for aggregating the knowledge base (in CatIO, these rules use a Prolog-like syntax). In this approach, combinatorial testing (CT) allows us to explore the input space in a controlled way such that we adequately cover all possible fault/input scenario combinations. As CT engine, we use the ACTS [23] library for generating mixed level covering arrays (MCAs) from a simulation's input variables and their domains. These MCAs then define the set of simulations conducted for generating the abductive diagnosis model [15,16].

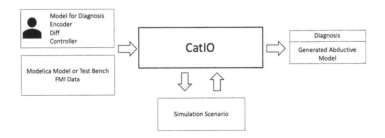

Fig. 2. Inputs and outputs of CatIO.

2.2   Front-End: Modelica Models and Simulations

In Fig. 2, we can see a user's view of the in- and outputs of our framework. A central part is the Modelica model that allows the simulation of a system's behavior. As usual, we can consider a simulation as a function that computes a system's outputs based on the inputs as defined in a simulation scenario. There are three categories of such inputs, namely *parameters*, *inputs* and *mode assignment variables*. System parameters describe values that are usually constant during a simulation, but which depend on a concrete scenario—like resistor values for an electrical circuit or the distance between a robot's differential drive's wheels in our example in Sect. 3. Inputs, on the other hand, are the dynamic inputs that connect a system to its environment—like the signals a robot's differential drive needs for controlling its left and right wheel. Mode assignment variables are special in that they concern the behavior of a system component, so that we can, e.g., describe whether a resistor is *OK*, *broken*, or *shorted*.

Modelica models can be converted to Functional Mock-up Interfaces (FMIs) [2], which an external program—like our CatIO framework—can then use to conduct and control a corresponding simulation. Unless we configure variables as input variables in Modelica, we can't change them during the simulation though. Thus, at least the inputs and mode-assignment variables have to be configured as inputs for an FMI. This allows us to support the dynamic control of a simulation (like for the robot control logic example mentioned in the introduction), and to simulate a CPS's behavior for test cases/simulations defined in a test bench. As we will see in Sect. 3, CatIO offers a graphical user interface (see Fig. 4) for (i) an easier extraction of the necessary data from FMIs and (ii) supporting a designer in the manual creation of simulation scenarios.

When we would like to use observations from a Modelica simulation in our diagnostic reasoning, we have to employ abstraction. In particular, since we have that the simulated signal values in a simulation are expressed in Modelica data types (like Boolean, integer or float), we have to map them to Boolean variables or propositional variables as used in the CMBD diagnosis model or the abductive KB respectively. For this purpose, CatIO uses the *encoder* function that has to be provided by an engineer. For each time step, this function reads the desired signal values of a simulation and performs the checks and comparisons defined in the function in order to derive observations in the correct format.

A central part in the concept for automating the generation of an abductive diagnosis model via Modelica simulations [15, 16] (see also Sect. 2.1) is the *diff* function. This function (also to be provided by the engineer) takes observations encoded with respect to the model, and identifies as well as encodes discrepancies between correct and faulty behavior. It so to say acts as deviation detector and encoder. Alternatively, *diff* can interpret values also directly from a simulation, i.e., without *encoder* mapping them to the desired format beforehand.

2.3 Front End: Further Inputs and Interfaces

CatIO's *controller* interface allows an engineer to connect an external function to a simulation by specifying a class encapsulating the external component to be interfaced. Such a component might describe a system's control logic that takes diagnostic information into account for coming up with repair and compensation actions, or might implement more general concepts for affecting/controlling a simulation. An example scenario for the case of a robot would be to connect its control logic, so that a designer can assess and verify the correct exploitation of diagnostic information.

For either diagnosis variant the user has to come up with the required diagnosis model/KB and the artifacts described before. Then he or she has to specify the length of a simulation and the time-step size, as well as the desired diagnosis parameters (persistent vs. intermittent faults, single or multiple scenarios, ...).

3   Example

Let us consider a robot's differential drive as depicted in Fig. 3. For this example, we have the distance d between the wheels as parameter, the voltages u_R, u_L for the left and right wheel as system inputs, and we have mode assignment variables m_L, m_R for the wheels whose domain is {*ok, faster, slower, stuck*}. Let us assume that, as suggested in Sect. 2.2, all changing variables in the Modelica model are considered as inputs for the FMI (Fig. 4).

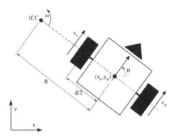

Fig. 3. Mobile robot with differential drive.

The first step now is to select those Modelica signals/variables which we're going to consider in the diagnostic process (i.e. those relevant for the diff and encoder functions). For our example, these are the inputs and outputs for the two wheels. For our current Java implementation of CatIO, we can see the corresponding graphical user interface for this step in Fig. 4. At the same time, the user can also define an input scenario for a simulation (in the bottom half of the figure) either manually, or via defining possible variable domains (defined in the column labeled "values" in the top table in the Fig. 4) to be considered for an automated generation of scenarios via computing mixed-level-covering arrays— a concept that is used also when generating an abductive knowledge base as described in Sect. 2.1.

In the following listing, we show a part of the abductive KB that we might generate for our example and which would contain also background knowledge like the mutual exclusiveness of some propositions. In particular, it is physically impossible for a robot to move straight forward and to move along a curve to the left at the same time. Note that *nominal, slower* and *faster* are propositions relating to a wheel's spinning speed with respect to the other wheel, while the robot's motion direction is denoted by *straight, left_curve, right_curve*.

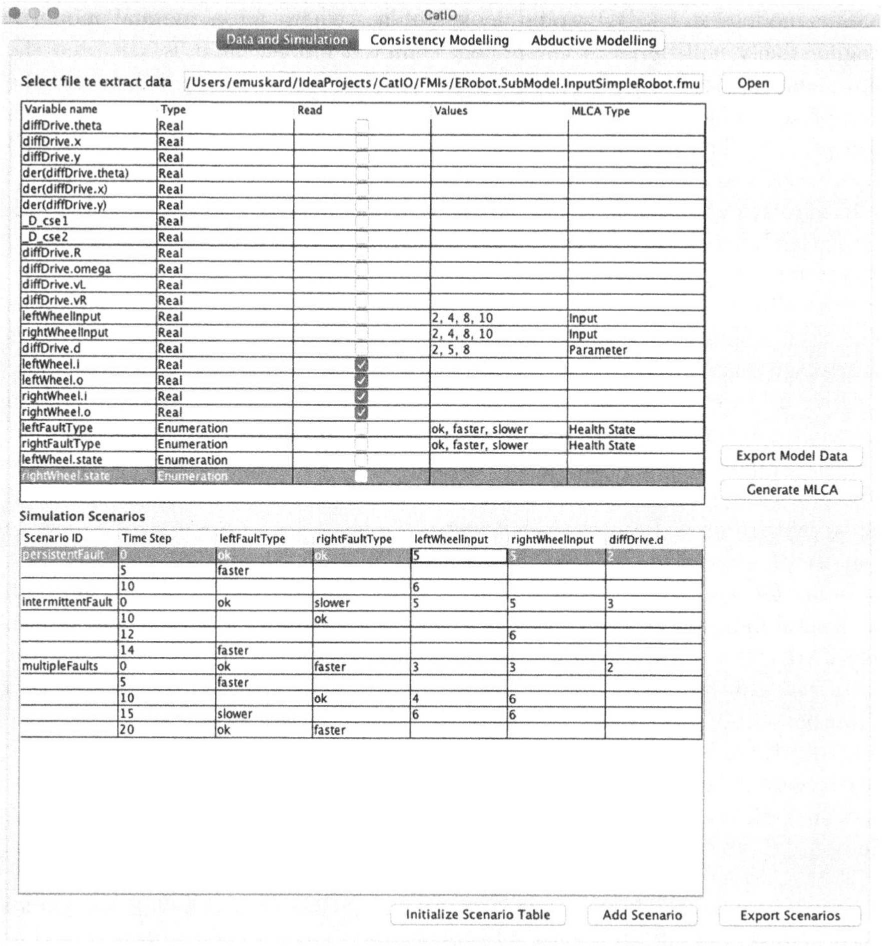

Fig. 4. Graphical user interface depicting FMI data extraction and scenario creation.

```
. . .
wheel(left).
wheel(right).

wheel(left), Expected(left) -> nominal(left).
wheel(left), Reduced(left) -> slower(left).
wheel(left), Increased(left) -> faster(left).
nominal(left), nominal(right) -> straight.
nominal(left), faster(right) -> left_curve.
faster(left), nominal(right) -> right_curve.
. . .
```

The model for CMBD would look similar, where for a manual model a designer likely will derive a component-centered model. That is, describe the components individually via arbitrary Boolean formulae first and then add the appropriate connections between the components. Please note that this Boolean description is then automatically converted to conjunctive normal form (CNF) which is the usual input format for SAT solvers.

In Fig. 4, we can see that we selected the in- and output voltages for both wheels (*leftWheel.i, rightWheel.i, leftWheel.o, leftWheel.o*) in the column labeled "Read" for being interfaced to the diagnostic process. For each time step, the corresponding evalues are passed to the *encoder* in order to map the observations to desired propositions, and in the *diff* function, we check whether the wheels are behaving as expected (such that the input signal equals the output signal).

Obviously we could now also connect the robot's control logic via our special *controller* interface as described in Sect. 2.3. This could be a Java class that alters the input voltages for the wheels if the robot drives a curve to the left instead of going straight ahead, or where we decide to turn 270 degrees to the left instead of turning 90 degrees to the right if the latter is not possible due to some diagnosed fault. Let us assume though, that we just aim to run our diagnostic reasoning for a simulation scenario, so that we run the diagnostic process with the desired parameters—like diagnosing persistent faults for a single simulation with CMBD.

As we discussed in Sect. 2.1, CatIO offers an implementation of an approach for automatically generating abductive KBs that was proposed in [15,16]. The basic idea there is to inject faults, simulate and compare faulty behavior against the correct one, and identify deviations in the system behavior in order to encode them in fault-and-effect dependency rules that we then aggregate in the KB. Via the diff function mentioned in Sect. 2.2, CatIO detects and encodes such deviations, which are then added to simple rules of the form $fault \to deviation$. The model automatically generated for our example contains the following rules concerning these fault-and-effect dependencies:

```
. . .
faster(leftWheel) -> right_curve.
faster(rightWheel) -> left_curve.
slower(leftWheel) -> left_curve.
slower(rightWheel) -> right_curve.
faster(rightWheel), slower(leftWheel) -> left_curve.
. . .
```

4   Summary and Future Work

In this manuscript, we propose an architecture for a framework supporting designers in the implementation of model-based diagnosis for cyber-physical system applications. CatIO's architecture was designed with a direct interface to Modelica simulation models that allow an engineer to cover a multitude of system domains via Modelica's versatile modeling options. The framework allows her to explore both abductive and consistency-based MBD for offering enhanced flexibility, she can conduct corresponding experiments for assessing and comparing developed solutions, and she is supported in the creation of an abductive knowledge base via an automated generation approach. An intrinsic feature is the option to connect an external controller (e.g., a robot's control logic) that can consider all available data and can react to them via dynamically defining the future inputs to the simulation.

Currently we are in the process of implementing and testing our CatIO architecture as described in this manuscript. In future work we will add more of the mentioned MBD algorithm variants, e.g., by interfacing the PyMBD library [17] and we will add more system description formats for CMBD like constraint representations and the corresponding solvers. Also interfaces allowing CatIO to run multiple simulations simultaneously (possibly on multiple machines) for exploiting today's PC's multi-core designs would be an interesting feature for a practical application. While the architecture has already been designed with flexibility on our minds, tuning the (graphical) interfaces for the specific tasks will also be subject to future work such as to foster an intuitive deployment and efficient use in practice.

Acknowledgment. The financial support by the Austrian Federal Ministry for Digital and Economic Affairs and the National Foundation for Research, Technology and Development is gratefully acknowledged.

References

1. Biere, A.: Picosat essentials. J. Satisfiability Boolean Modeling Comput. JSAT (2008)
2. Blochwitz, T., et al.: Functional mockup interface 2.0: the standard for tool independent exchange of simulation models, 09 2012. https://doi.org/10.3384/ecp12076173
3. Bunus, P., Lunde, K.: Supporting model-based diagnostics with equation-based object oriented languages. In: EOOLT. Linköping Electronic Conference Proceedings, vol. 29, pp. 121–130. Linköping University Electronic Press (2008)
4. Console, L., Torasso, P.: Integrating models of correct behavior into abductive diagnosis. In: Proceedings of the European Conference on Artificial Intelligence (ECAI), pp. 160–166. Pitman Publishing, Stockholm, August 1990
5. De Kleer, J., Williams, B.C.: Diagnosis with behavioral modes. In: 11th International Joint Conference on Artificial Intelligence - Volume 2, pp. 1324–1330 (1989)
6. Fleischanderl, G., Schreiner, H., Havelka, T., Stumptner, M., Wotawa, F.: DiKe - a model-based diagnosis kernel and its application. In: Proceedings of the Joint German/Austrian Conference on Artificial Intelligence (KI), Vienna, Austria (2001)
7. Friedrich, G., Gottlob, G., Nejdl, W.: Hypothesis classification, abductive diagnosis and therapy. In: Gottlob, G., Nejdl, W. (eds.) Expert Systems in Engineering Principles and Applications. LNCS, vol. 462, pp. 69–78. Springer, Heidelberg (1990). https://doi.org/10.1007/3-540-53104-1_32
8. Fritzson, P.: Principles of Object-oriented Modeling and Simulation with Modelica 3.3: A Cyber-physical Approach. Wiley (2014)

9. Greiner, R., Smith, B.A., Wilkerson, R.W.: A correction to the algorithm in Reiter's theory of diagnosis. Artif. Intell. **41**(1), 79–88 (1989)
10. de Kleer, J., Williams, B.C.: Diagnosing multiple faults. Artif. Intell. **32**(1), 97–130 (1987)
11. Metodi, A., Stern, R., Kalech, M., Codish, M.: Compiling model-based diagnosis to Boolean satisfaction. In: AAAI Conference on Artificial Intelligence, pp. 793–799 (2012)
12. Nica, I., Pill, I., Quaritsch, T., Wotawa, F.: The route to success - a performance comparison of diagnosis algorithms. In: International Joint Conference on Artificial Intelligence (IJCAI), pp. 1039–1045 (2013)
13. Pill, I., Quaritsch, T.: RC-Tree: a variant avoiding all the redundancy in Reiter's minimal hitting set algorithm. In: IEEE International Symposium Software Reliability Engineering Workshops (ISSREW), pp. 78–84 (2015). https://doi.org/10.1109/ISSREW.2015.7392050
14. Pill, I., Wotawa, F.: Exploiting observations from combinatorial testing for diagnostic reasoning. In: International Workshop on Principles of Diagnosis (DX) (2019, in press). (accepted)
15. Pill, I., Wotawa, F.: Fault detection and localization using Modelica and abductive reasoning, pp. 45–72 (2018)
16. Pill, I., Wotawa, F.: On using an I/O model for creating an abductive diagnosis model via combinatorial exploration, fault injection, and simulation. In: 29th International Workshop on Principles of Diagnosis (DX 2018) (2018). http://ceur-ws.org/Vol-2289/paper9.pdf
17. Quaritsch, T., Pill, I.: PyMBD: a library of MBD algorithms and a light-weight evaluation platform. In: International Workshop on Principles of Diagnosis (DX), pp. 1–5 (2014)
18. Reiter, R.: A theory of diagnosis from first principles. Artif. Intell. **32**(1), 57–95 (1987)
19. Reiter, R., Kleer, J.: Foundations of assumption-based truth maintenance systems: preliminary report, pp. 183–189 (1987)
20. Sayed-Mouchaweh, M. (ed.): Diagnosability, Security and Safety of Hybrid Dynamic and Cyber-Physical Systems. Springer, Cham (2018). https://doi.org/10.1007/978-3-319-74962-4
21. Wotawa, F.: A variant of Reiter's hitting-set algorithm. Inf. Process. Lett. **79**, 45–51 (2001)
22. Lughofer, E., Sayed-Mouchaweh, M. (eds.): Predictive Maintenance in Dynamic Systems. Springer, Cham (2019). https://doi.org/10.1007/978-3-030-05645-2
23. Yu, L., Lei, Y., Kacker, R.N., Kuhn, D.R.: Acts: a combinatorial test generation tool. In: 2013 IEEE Sixth International Conference on Software Testing, Verification and Validation, pp. 370–375, March 2013. https://doi.org/10.1109/ICST.2013.52

Data Publishing: Availability of Data Under Security Policies

Juba Agoun[✉] and Mohand-Saïd Hacid

Université de Lyon, Université Lyon 1, CNRS UMR 5205, Villeurbanne, France
{juba.agoun,mohand-said.hacid}@univ-lyon1.fr

Abstract. In this paper, we study the problem of data publishing under policy queries. We consider the privacy-aware in the context of data publishing where given a database instance, a published view, and a policy query that specifies what information is sensitive and should be protected. Our approach exploits necessary and sufficient conditions for a view to comply with a policy query. We formulate a preliminary work that consists on a data-independent method to revise the non-privacy-preserving views. With this revising process, we strike a balance between data privacy and data availability.

Keywords: Privacy-preserving · Data publishing · Data availability

1 Introduction

With the rapid growth of stored information, governments and companies are often motivated to derive valuable information. Enterprises and organizations have begun to employ advanced techniques to analyze the data and extract *"useful"* information. Data publishing has fueled significant interest in the database community [3]. It is a widespread solution to make the data available publicly and enable data sharing. It consists on exporting or publishing information of an underlying database through views to be used by peers. At the same time, the publisher tries to ensure that sensitive information will not leak.

There are great opportunities associated with data publishing, but also associated risks [16]. Indeed, it is complex to find a trade-off between preventing the inappropriate disclosure of sensitive information and guarantee the availability of non-sensitive data: Removing of all the data exported by a violating view achieves perfect privacy, but it is a total uselessness, while publishing the entire data, is at the other extreme.

In this paper, we focus on a variant problem of data availability referred also as utility of the data [13], and we specifically consider data privacy in database publishing. The owner of a given database D wishes to publish a set of views $\mathcal{V}$

This research is performed within the scope of the DataCert (Coq deep specification of privacy aware data integration) project that is funded by ANR (Agence Nationale de la Recherche) grant ANR-15-CE39-0009 - http://datacert.lri.fr/.

under a set of restrictions $\mathcal{S}$. We assume that the restrictions on the views are expressed as a set of queries, called policy queries [11,12]. Inspired by prior work on privacy-preservation [18], we define a view to be *safe* w.r.t. a policy query if they don't return tuples in common, in other words, the view and the policy query are disjoint. We provide a revision algorithm for the privacy-preservation protocol to ensure data availability. Indeed, instead of neutralizing the unsafe view, we propose a rewriting technique that exploits the interaction between a view and a policy query to enrich the view with relevant information for ensuring the availability of data. Our approach exploits the concept of residue introduced in [2].

We illustrate our data publishing setting by the following running example. Let D be a database schema of a company consisting of two relations:

$$Employee(id\_emp, name, dept)$$
$$PayrollService(id\_emp\_ps, accountNum, salary)$$

The relation *Employee* stores, for each employee her/his identifier, her/his name and her/his department. The relation *PayrollService* stores information related to employees, in particular, their identifier, their account number, and their salary.

Consider the following set $\mathcal{S}$ of policy queries. The policy queries define the information that is sensitive to make secret to public.

$$S_1(i, n) : -Employee(i, n, d)$$
$$S_2(n, s) : -Employee(i, n, d), \ PayrollService(i, a, s), \ s > 3000$$

Query S_1 projects the *identifier* and the *name* of an employee. It states that both attributes are sensitive when they are returned simultaneously. S_2 states that the *name* and the *salary* of employees with a salary over \$3 000 are sensitive if returned together.

Now, consider the following set of published views $\mathcal{V}$. The view V_1 projects the *name* and the *salary* of the employees in department "*info501*" whereas V_2 projects the *name* and the *salary* of the employees with a salary under \$10 000.

$$V_1(n, s) : -Employee(i, n, d), \ PayrollService(i, a, s), \ d =' info501'$$
$$V_2(n, s) : -Employee(i, n, d), \ PayrollService(i, a, s), \ s < 10000$$

The policy query S_1 is not violated by any of the views in $\mathcal{V}$ since none of them returns the *identifier* and the *name* of the employees. The view V_1 discloses some sensitive information since $S_2(I)$ and $V_1(I)$ overlap for some database instances, such as $I = \{Employee\{\langle'p552','Jhon','info501'\rangle\}$, PayrollService $\{\langle'p552','FRB1015', 4500\rangle\}\}$. Regarding the view V_2 and the query S_2, one can notice that there could be some overlapping tuples since the selection conditions in the two queries are both satisfiable for some values of s.

The questions addressed in our paper are the following: **Are the published views $\mathcal{V}$ safe w.r.t. the policy queries $\mathcal{S}$? If the views are not safe w.r.t.**

the policy queries, how could we revise them to make available the subset of tuples which are not in the set of tuples identified by S?

The paper is organized as follows. Section 2 discusses related work. Section 3 discusses the basic concepts and notions relevant to our approach. Section 4 presents the privacy preservation protocol. Section 5 describes a mechanism for revising the non-privacy-preserving views, while Sect. 6 concludes our paper.

2 Related Works

Access control [1] is considered as traditional database security. It prevents unauthorized access to the database, therefore it usually offers a control at the physical level. However, access controls are the most common solution in several systems. View-based access control is a mechanism for implementing database security [15]. It is mostly based on defining the rights of authorized subjects (e.g., users) to read or modify data. Some rewritten techniques (e.g., [7]) are used to preform access control [14] by answering queries using only information contained in authorized views. Some research works enforce the security in data publishing over XML documents through an access control using cryptography [10]. However, our approach does not deal with authorization methods, it differs since the secret data is specified at the logical level rather than at the physical level.

Perfect-Security is the main method to determine leakage of private data that could occur in a publishing database. The authors of [11] presented a query-view security model based on a probabilistic formal analysis [4]. In this approach, the authors assume that they know the probability distribution of an instance and they model it as the knowledge of an adversary. Then, it compares a prior knowledge of an adversary with the knowledge of the adversary after data has been published. Although involving probability distribution, the main results of [11] shows that it can be converted to a logical test. Our approach differs from the setting considered in [11]. Indeed, the perfect-security is based on a probabilistic distribution on the instances whereas our approach is data-independent.

Determining when two given queries are disjoint was firstly introduced in [5]. The topic of disjoint queries has found application mostly in the *irrelevant updates* domain [6].

The issue of detecting privacy violations in data publishing has been investigated in [19]. The goal of this work is to prevent inferring sensitive information in XML documents. The authors developed algorithms for finding partial documents of a published document that do not leak sensitive information. In addition to the problem of detecting privacy violation. There are other techniques to avoid privacy violations such k-anonymity [17], t-closeness [8], and l-diversity [9]. They rely on the generalization of data which aims at transforming the published data. Thus, specific attribute values are transformed into less specific attribute. Depending on the nature of the attribute, different generalization methods could be considered. For example, category value could be generalized using a hierarchy, and this could replace the category value by its ancestor according to the hierarchy. Our approach does not prevent violation by transforming the published data. Nevertheless, the views are revised to return only non-sensitive information, thus, ensure availability of data.

3   Preliminaries

In this section, we introduce the relevant concepts to our framework. We consider a relational setting. A *database schema* D consists of a finite set of relation schemas $R_1, ..., R_n$, where each relational schema (or just *relation*) consists of a unique name and a finite set of attributes. The set of attributes in a relational schema R_i is denoted by $att(R_i)$. A *tuple* for a relational schema R_i, where $att(R_i) = \{X_1, ..., X_k\}$ is an element consisting of a set of k constants. A relation r defined over a relational schema R_i, denoted by $r(R_i)$ (*or simply by* r_i *if* R_i *is understood*) consists of a finite set of tuples defined over R_i. A database instance defined over a schema $D = R_1, ..., R_n$, denoted by $I(D)$ (*or simply by* I *if* D *is understood*) is a finite set of relations $\{r_1(R_1), ..., r_n(R_n)\}$. We consider nonrecursive DATALOG queries with inequalities defined as follows. For the simplicity of the presentation, we assume that Boolean queries are not allowed[1]. Hence, the head of the query contains only variables and at least one variable. Despite the fact that constants do not appear in the head of the query, the essentials of our results could be extended for this case.

Definition 1. *We assume a set of variable names* $\mathcal{N}$, *and the function* typ: $\mathcal{N} \rightarrow att(R_1) \cup ... \cup att(R_n)$, *where* $D = \{R_1, ..., R_n\}$. *A conjunctive query* Q *is defined by:*

$$Q(\bar{X}) : -t_1, ..., t_n, C_{n+1}, ..., C_{n+m}$$

where $t_1, ..., t_n$ *are terms,* $C_{n+1}, ..., C_{n+m}$ *are inequalities, and* $\bar{X} \in \mathcal{N}$. *A term is of the form* $R(X_1, ..., X_n)$, *where* R *is a relational schema in* D *and* $\{X_1, ..., X_n\} \subseteq \mathcal{N}$. *An inequality has one of the following form:*

1. $X \odot c$, *where* $X \in \mathcal{N}$, c *is a constant (i.e., numeric or string),* $\odot \in \{=, \leq, \geq, <, >, \neq\}$.
2. $X \odot Y$, *where* $\{X, Y\} \subseteq \mathcal{N}$, $\odot \in \{=, \leq, \geq, <, >, \neq\}$.

$Q(\bar{X})$ *is called the head of the query* Q, *and* $t_1, ..., t_n, C_{n+1}, ..., C_{n+m}$ *is called the body of* Q. $\bar{X}$ *is called the schema of* Q *and is also denoted by* $att(Q)$. *The queries have the following restrictions:*

a. *We assume that a variable can appear at most once in the head of a query.*
b. *If a variable* X_i *appears as the* i^{th} *variable in a term* $R(, ..., Xi, ...)$ *then* $typ(X_i) = A_i$, *where* A_i *is the* i^{th} *attribute of* R;
c. $\bar{X} \neq \emptyset$.

---

[1] Policy queries specify tuples to keep private. Thus, a policy query with a set of variables empty in the head is useless.

We recall that the views are considered to be conjunctive queries in our framework. Condition (a) above does not prevent join and self-join queries from being defined in our framework. Condition (c) mentions clearly that Boolean queries are not allowed. We also assume that the queries are range restricted (safety of a DATALOG query), *i.e.*, every variable X in $\bar{X}$ also appears in some term in the body of Q, or there exists a variable Y such that $C_{n+1}, ..., C_{n+m}$ imply that $X = Y$ and Y appears in some term in the body of Q.

Next we will introduce some notations needed to compare the schema of queries based on the types of the variables.

Definition 2. *Given queries $S(X_1, ..., X_n)$ and $V(X'_1, ..., X'_m)$, $att(S)$ is contained in $att(V)$, denoted by $att(S) \preceq att(V)$, if $n \leq m$ and $typ(X_i) = typ(X'_i)$ for all $i \in \{1, ..., n\}$. The schemas $att(S)$ and $att(V)$ are equivalent, denoted by $att(S) \equiv att(V)$, if $att(S) \preceq att(V)$ and $att(V) \preceq att(S)$. The difference between $att(V)$ and $att(S)$, denoted by $\underset{typ}{\setminus}$ is defined as:*

$$att(V)\underset{typ}{\setminus}att(S) = \{X'_i | \nexists X_j(typ(X'_i) = typ(X_j))\}.$$

4 Privacy Preservation

Our notion of privacy preservation is build upon the protocol introduced in [18]. Below, we formalize the notion of privacy preservation as defined in [18] and in the next section we extend it for the query revision. First, we summarize the notion of disjoint queries.

Definition 3. *If I is an instance of a database schema D and S and V are queries that have the same schema, then S and V are defined to be disjoint if for every I in D, $Q(I) \cap V(I) = \emptyset$.*

The adopted approach for defining privacy consists in specifying the information that is private by a query S, and then the notion of a user query being legal translates to the requirement of our approach that the user query must be disjoint from S for all possible database states. The privacy violation occurs only when the same tuple is returned by both the policy query and the user query, for some database instance. The basic semantic unit of information is a tuple, not an attribute value therefore the intersection is at the tuple level and not the attribute level.

The privacy preservation we want to achieve in this paper can be determined using only the structure of S and V, and so can be checked at compile time. We define privacy preservation as follows.

Definition 4. *Query V is privacy-preserving with respect to a policy query S if either:*

a. $att(S)\underset{typ}{\setminus}att(V) \neq \emptyset$;or

b. $att(S) \preceq att(V)$ and S and V' are disjoint, where V' is the query defined by: $V'(att(S)) : -V$

Essentially, the case (a) is the situation where there are some attributes of S that are not in V, for which we do not consider to be a violation. It could be illustrated by V_1 and S_1 from our running example. As it is previously discussed, we do not consider V_1 to be a privacy violation for S_1. In the situation of case (b) every variable in $att(S)$ has a matching variable in $att(V)$, in which case we require that the projection of V on $att(S)$ be disjoint from S.

The main result established in [18] characterizes when a published query V preserves the privacy of a secret query S. So, given S and V two DATALOG queries over a database scheme D such that $\bar{C}_S$ and $\bar{C}_V$ are separately satisfiable, then V preserves the privacy of S if either:

1. $att(S) \setminus_{typ} att(V) \neq \emptyset$; or
2. the set of inequalities $\pi_{att(S)}(\bar{C}_S) \cup \pi_{att(S)}(\bar{C}_V)$ is unsatisfiable.

Next, we illustrate by an example the detection of privacy violations of a published view $w.r.t.$ a policy query.

Example 1. We demonstrate over a simplified version of the running example, where the set of policy queries $\mathcal{S} = \{S_2\}$ and the set of published views $\mathcal{V} = \{V_3\}$.

$$V_3(n, a, s) : -Employee(i, n, d),\ PayrollService(i, a, s),\ s \leq 5000,\ d =' info501'$$

We note that $\bar{C}_{V_3} = \{s \leq 5000,\ d =' info501'\}$ and $\bar{C}_{S_2} = \{s > 3000\}$. Then $\pi_{att(S_2)}(\bar{C}_{S_2}) = s > 3000$ and $\pi_{att(S_2)}(\bar{C}_{V_3}) = \{s \leq 5000\}$. We say that V_3 is not privacy-preserving w.r.t. policy query S_2 since $att(S_2) \setminus_{typ} att(V_3) = \emptyset$ and $\pi_{att(S_2)}(\bar{C}_{S_2}) \cup \pi_{att(S)}(\bar{C}_V) = \{s > 3000, s \leq 5000\}$ which is satisfiable.

5   Revising Privacy Violating Views

In the previous section we introduced the privacy preservation and described how to detect the violating views. In this Section, we present a technique for revising the views w.r.t. policy queries, over all database instances.

Our approach to revise the views is based on the explorations of residues resulting after the unification process. This technique is used for semantic query optimization [2]. The unification process allows to capture residues from policy queries and will be associated with published views.

Revising the views can be described informally as the process of transforming the non-privacy-preserving view to comply with the requirements of the policy query. When the schema of the view and the query is the same (*see* Definition 4), this means that bodies are unsatisfiable. In this case, we propose using the partial subsumption technique, to extract the residue of a policy query and associate it with the view. A policy query S partially subsumes a view V if a subclause[2] of S subsumes the body of V.

In the following, the process is going to be illustrated by an example. Consider a policy query S and a view to be published V over a database with two relations $R_1(a, b, c)$ and $R_2(a, c)$. The policy query S states that the attribute association $\{a, b\}$ is sensitive for tuples in R_1 that could be join with R_2 on the attribute a and where c is greater than 5. The view V project the attribute a' and b' for tuples in R_1 join R_2 on a' and where a' is under 4.

$$S(a, b) : -R_1(a, b, c), R_2(a, c), c > 5$$
$$V(a', b') : -R_1(a', b', c'), R_2(a', c'), a' < 4$$

Since V is not privacy-preserving to S we proceed to the revision the view. It consists on generating the residue from the policy query S using the V using the partial subsumption technique. First, the body of V is negated and, thereby, the set $\{\neg R_1(a', b', c'), \neg R_2(a', c'), a' \geq 4\}$ is obtained.

Then, we construct a linear refutation tree (*see* Fig. 1) by considering the body of S as the root. At each step, an element of the negated body of V is unified using an element in the root: The first step of the refutation tree, unify $\neg R_1(a', b', c')$ and $R_1(a, b, c)$ with the following substitutions $\{a'/a, b'/b, c'/c\}$. In the next step, $\neg R_2(a', c')$ of the negated body V is unified with $R_2(a', c')$ in S according to the substitutions of the previous step.

Finally, we obtain at the bottom of the tree the residue: $c' > 5$, which could interpreted as "c' *cannot be over* 5". The residue is then associated to the view V:

$$V^r(a', b') : -R_1(a', b', c'), R_2(a', c'), a' < 4, \{c' > 5\}$$

According to the interpretation before, the meaning of the previous view is as follows:

$$V^r(a', b') : -R_1(a', b', c'), R_2(a', c'), a' < 4, c' \leq 5$$

Revising views depends on the type of resulting residue. For instance, if an empty set is obtained at the bottom of the refutation tree called *null residue*, we would associate to the view an empty residue $\{\ \}$ which is then replaced by a *false* term. A false term in a query leads to directly exclude the view from the set of published views $\mathcal{V}$. We recall that we omit the case of unsatisfied residue since we consider only the non-privacy-preserving views for revision.

Above we assumed V and S having the same schema. Now, we consider the case where some elements of the schema $att(S)$ appears in $att(V)$(*i.e.*, $att(S)$ $\setminus att(V) = \emptyset$ and $att(S) \preceq att(V)$). Same as previously, the residue would be
$_{typ}$
computed using the body of the view V and S and added to the given view. In case the result of the refutation tree is empty, instead of incorporating the

---

[2] C is a subclause of D if every literal in C is also in D. A literal could be either a term or an inequality.

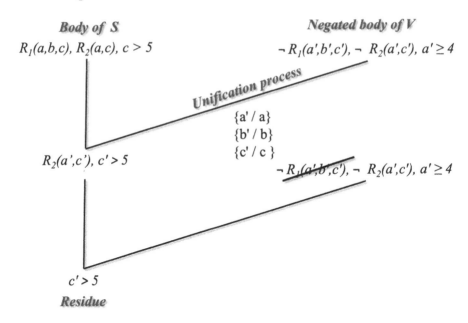

Fig. 1. Refutation tree

term *false*, which implies removing the entire view as seen before, we propose to keep the view and put the sensitive attributes as secret. Indeed, the attributes of $att(S)$ in the view V are replaced by $NULL$. Putting the sensitive attributes to $NULL$ prevents the disclosure and helps to redesign the views for a better utility.

Example 2. Consider the view V_1 from our running example and the following policy query S_3:

$$S_3(n) : -Employee(i, n, d), PayrollService(i, a, s)$$

The result of the refutation tree would be a *null residue* and we note that $att(S_3)$ \ $att(V_1) = \emptyset$ and $att(S_3) \preceq att(V_1)$. In this case, we propose to rewrite V_1 into V^r having the following form:

$$V^r{}_1(NULL, s) : -Employee(i, n, d), PayrollService(i, a, s), d =' info501'.$$

6   Conclusion

We addressed the problem of privacy-preserving and data availability in data publishing under policy queries. We proposed a preliminary method to revise views. The method is based on the concept of residue usually adopted in semantic query optimization. Our work aims to strike the balance between data privacy and data availability. Indeed, instead of only neutralizing the non-privacy-preserving views, we proposed a data-independent process for revising the views *w.r.t.* policy queries that sanitize the views from sensitive information. Our revising model returns changed views that do not guarantee the correctness since some information could be hidden. However, it provides safe access to the published data and considerable availability.

Future work will address more complex data-publishing scenarios. Firstly, we have presented the revising of one view *w.r.t.* to a policy query. We are extending our approach to accommodate data dependencies and complex interactions between views. When there are multiple published views, a violation could potentially occur from a combination of views which, individually, are privacy-preserving *w.r.t.* a set of policy queries. We could pursue our revising process by considering the interplay between multiple views. Finally, we could envision an application in other contexts (*e.g.,* data integration ...).

References

1. Bertino, E., Jajodia, S., Samarati, P.: Database security: research and practice. Inf. Syst. **20**(7), 537–556 (1995)
2. Chakravarthy, U.S., Grant, J., Minker, J.: Logic-based approach to semantic query optimization. ACM Trans. Database Syst. (TODS) **15**(2), 162–207 (1990)
3. Clifton, C., et al.: Privacy-preserving data integration and sharing. In: Proceedings of the 9th ACM SIGMOD Workshop on Research Issues in Data Mining and Knowledge Discovery, pp. 19–26. ACM (2004)
4. Dalvi, N., Suciu, D.: Management of probabilistic data: foundations and challenges. In: Proceedings of the Twenty-Sixth ACM SIGMOD-SIGACT-SIGART Symposium on Principles of Database Systems, pp. 1–12 (2007)
5. Elkan, C.: A decision procedure for conjunctive query disjointness. In: Proceedings of the Eighth ACM SIGACT-SIGMOD-SIGART Symposium on Principles of Database Systems, pp. 134–139 (1989)
6. Gupta, A., Mumick, I.S., et al.: Maintenance of materialized views: problems, techniques, and applications. IEEE Data Eng. Bull. **18**(2), 3–18 (1995)
7. Levy, A.Y., Mendelzon, A.O., Sagiv, Y.: Answering queries using views. In: Proceedings of the Fourteenth ACM SIGACT-SIGMOD-SIGART Symposium on Principles of Database Systems, pp. 95–104 (1995)
8. Li, N., Li, T., Venkatasubramanian, S.: t-Closeness: privacy beyond k-anonymity and l-diversity. In: 2007 IEEE 23rd International Conference on Data Engineering, pp. 106–115. IEEE (2007)
9. Machanavajjhala, A., Kifer, D., Gehrke, J., Venkitasubramaniam, M.: l-diversity: privacy beyond k-anonymity. ACM Trans. Knowl. Discov. Data (TKDD) **1**(1), 3-es (2007)
10. Miklau, G., Suciu, D.: Controlling access to published data using cryptography. In: Proceedings 2003 VLDB Conference, pp. 898–909. Elsevier (2003)
11. Miklau, G., Suciu, D.: A formal analysis of information disclosure in data exchange. J. Comput. Syst. Sci. **73**(3), 507–534 (2007)
12. Nash, A., Deutsch, A.: Privacy in GLAV information integration. In: Schwentick, T., Suciu, D. (eds.) ICDT 2007. LNCS, vol. 4353, pp. 89–103. Springer, Heidelberg (2006). https://doi.org/10.1007/11965893_7
13. Rastogi, V., Suciu, D., Hong, S.: The boundary between privacy and utility in data publishing. In: Proceedings of the 33rd International Conference on Very Large Data Bases, pp. 531–542. Citeseer (2007)

14. Rizvi, S., Mendelzon, A., Sudarshan, S., Roy, P.: Extending query rewriting techniques for fine-grained access control. In: Proceedings of the 2004 ACM SIGMOD International Conference on Management of Data, pp. 551–562 (2004)
15. Rosenthal, A., Sciore, E.: Content-based and view-based access control (2011)
16. Shay, R., Blumenthal, U., Gadepally, V., Hamlin, A., Mitchell, J.D., Cunningham, R.K.: Don't even ask: database access control through query control. ACM SIGMOD Rec. **47**(3), 17–22 (2019)
17. Sweeney, L.: k-anonymity: a model for protecting privacy. Int. J. Uncertainty Fuzziness Knowl.-Based Syst. **10**(05), 557–570 (2002)
18. Vincent, M.W., Mohania, M., Iwaihara, M.: Detecting privacy violations in database publishing using disjoint queries. In: Proceedings of the 12th International Conference on Extending Database Technology: Advances in Database Technology, pp. 252–262. ACM (2009)
19. Yang, X., Li, C.: Secure XML publishing without information leakage in the presence of data inference. In: Proceedings of the Thirtieth International Conference on Very Large Data Bases-Volume 30, pp. 96–107 (2004)

Matrix Factorization Based Heuristics Learning for Solving Constraint Satisfaction Problems

Seda Polat Erdeniz[✉], Ralph Samer, and Muesluem Atas

Graz University of Technology, Inffeldgasse 16B/2, 8010 Graz, Austria
{spolater,rsamer,muatas}@ist.tugraz.at
http://ase.ist.tugraz.at

Abstract. In configuration systems, and especially in Constraint Satisfaction Problems (CSP), *heuristics* are widely used and commonly referred to as variable and value ordering heuristics. The main challenges of those systems are: producing high quality configuration results and performing real-time recommendations. This paper addresses both challenges in the context of CSP based configuration tasks. We propose a novel learning approach to determine transaction-specific variable and value ordering heuristics to solve configuration tasks with high quality configuration results in real-time. Our approach employs matrix factorization techniques and historical transactions (past purchases) to learn accurate variable and value ordering heuristics. Using all historical transactions, we build a sparse matrix and then apply matrix factorization to find transaction-specific variable and value ordering heuristics. Thereafter, these heuristics are used to solve the configuration task with a high prediction quality in a short runtime. A series of experiments on real-world datasets has shown that our approach outperforms existing heuristics in terms of runtime efficiency and prediction quality.

1 Introduction

Configuration systems [5] can find solutions for problems which have many variables and constraints. For example, a configuration problem can be *the customization of cars* where many hardware and software components exist and all should work together without any conflicts. Therefore, a conflict-free solution should be found. Many scheduling and configuration problems can be defined as constraint satisfaction problems (*CSPs*) [12], in which there are a set of variables, domains (a set of possible values of each variable), and a set of constraints. A solution of a CSP is a complete set of variable assignments where all constraints are satisfied [4].

Various search techniques can be applied in order to improve the performance of CSP solvers. Variable and value ordering heuristics are common intelligent search techniques which are used for solving many kinds of problems such as *configuration, job shop scheduling,* and *integrated circuit design* [10].

D. Helic et al. (Eds.): ISMIS 2020, LNAI 12117, pp. 287–297, 2020.
https://doi.org/10.1007/978-3-030-59491-6_27

Selecting the most appropriate (the best performing) heuristics within constraint solvers is challenging due to the following reasons:

- *Runtime.* Solving a configuration task on a finite domain is an NP-complete problem with respect to the domain size and the complexity of the given constraints. The runtime performance of the configurator becomes crucial for real-time configuration tasks.
- *Accuracy.* When there are too many solutions for a given CSP, it would be very hard for a user to select a preferred solution out of a very long solutions list. A successful configuration system should be able to provide a configuration result with high probability to be *accepted/preferred by the user* as the first solution.

In this paper, we propose a novel learning approach to find variable and value ordering search heuristics for configuration systems to overcome the aforementioned two challenges. Our approach uses historical transactions which are composed of past user interactions and stored on the configuration system. Using all historical transactions, our approach builds a sparse matrix and then applies matrix factorization to learn transaction-specific variable and value ordering heuristics. Thereafter, these heuristics are used to solve the configuration task with a high prediction accuracy in a short runtime.

2   The Example Domain: An Online Bike Shop

To demonstrate our approach on a small example, we use an online personalized bike shop example on the basis of a real world knowledge base[1]. This online bike shop offers a variety of bikes which are personalized according to the user requirements and product constraints of the available bike modules.

In configuration systems, a user may specify own preferences. In addition to that, the products themselves have constraints. Given these user requirements and product constraints, the configuration system is then used to determine a configuration which satisfies both the user requirements and the product constraints.

We classify user transactions into three as follows.

Complete Transaction (CT): CT represents a complete and historical transaction (i.e., a complete past purchase). A transaction of this type can later serve as valuable input for configurations in the future.

Incomplete Transaction (IT): Another possible scenario refers to the situation where a user may leave the online shop without having purchased a configuration result. Such transactions are incomplete and denoted as IT. These transactions represent stored incomplete transactions which only contain user requirements but not a complete specification of a product. Another scenario refers to situations where new product features/services are introduced to existing products for which some complete and incomplete transactions already exist.

---

[1] https://www.itu.dk/research/cla/externals/clib/bike2.cp.

With the introduction of new product features/services, all existing CTs are converted into ITs.

Active Transaction (AT): Whenever a new user starts a new configuration session, he or she will receive instant configurations while he or she is defining requirements in the online bike shop. This scenario is referred to as an active transaction. In sharp contrast to IT, the transaction is active (i.e., ongoing) and the user can receive instant configurations based on the requirements provided by him or her.

In our online bike shop example, there are five users: *Alice, Bob, Tom, Ray,* and *Joe* as shown in the historical transactions table (see Table 1). These users interacted with the online personalized bike shop in the past.

Table 1. Historical transactions of the online bike shop

User name:	Alice	Bob	Tom	Ray	Joe
frame_biketype	3	4	0		2
frame_internal	1	0		1	
extra_propstand	1	1	1		
gear_internal	1	0	1	1	1
Type:	CT	CT	IT	IT	IT

In the same bike shop example, Lisa is an active user who has not left the online shop yet and has defined own preferences (REQ_{Lisa}) as shown in Table 2.

Table 2. A configuration satisfaction problem: CSP_{Lisa}

V_{bike}	$frame\_biketype = \{0, .., 4\},$
	$frame\_internal = \{0, .., 1\},$
	$extra\_propstand = \{0, .., 2\},$
	$gear\_internal = \{0, .., 1\}$
C_{bike}	$c_1 : (frame\_biketype = (1 \lor 2)) \implies (frame\_internal = 1)$
	$c_2 : (frame\_biketype = 4) \implies (gear\_internal = 0)$
	$c_3 : (frame\_internal = gear\_internal)$
REQ_{Lisa}	$c_4 : (frame\_biketype = 2)$
	$c_5 : (gear\_internal = 1)$

3   The Proposed Method

State-of-the-art variable and value ordering heuristics lack of possibilities to consider both *runtime performance* and *prediction accuracy* for configuration systems. The motivation of our approach is to decrease the runtime in accordance with the increase of prediction quality for configuration systems.

Our approach calculates transaction-specific variable and value ordering heuristics to solve two kind of configuration tasks which are based on: *incomplete historical transactions (IT)* and *active transactions (AT)*.

In order to find configurations for IT based configuration tasks, we generate a sparse matrix using the historical transactions and decompose it into user and value factors. Then, we obtain a dense matrix which holds similar values of the sparse matrix. In addition, it has the estimated values for the missing values in the sparse matrix. Therefore, in this dense matrix, we have predictions for the missing values of ITs. These predictions represent the probabilities of the corresponding values of ITs. Therefore, we start the configuration search with the highest probability values to find a configuration for an IT.

However, ATs are not included in these matrices since the online computation of such matrix factorization can be time taking. Thus, to find configurations for AT based configuration tasks, we use the calculated heuristics of the most similar historical transaction of AT and use them to solve the configuration task of an AT.

In our working example, M_TRX_{bike} is a sparse bitmap matrix (see Table 3-(a)) which is composed of transactions in Table 1. The bitmap matrix contains only 0s and 1s (i.e., bits). In a row (transaction), 1s mean that the variables hold as values the *corresponding* values of these columns. In return, all of the other domain values of these variables are set to 0. The blank values represent the undefined variables in the user requirements. For example, *Alice* has specified *frame_biketype*=3 which is the fourth value in its domain, so in the *Alice*'s row in the bitmap matrix (M_TRX_{bike}), the fourth bit ($b4$) is set to 1, and all the rest domain values of *frame_biketype* (bits: $b1, b2, b3, b5$) are set to 0.

3.1   Configurations for ITs

When a user revisits the online shop, the corresponding pre-calculated configuration result is offered to him/her according to his/her incomplete historical transaction (IT). This configuration is calculated using the estimated historical transaction in the dense matrix as variable and value ordering heuristics to guide the search of configuration.

Matrix Factorization. Matrix factorization based collaborative filtering algorithms [1,6–8] introduce a rating matrix R (a.k.a., user-item matrix) which describes preferences of users for the individual items the users have rated.

We use a bitmap matrix [3,9] to hold the transactions as shown in Table 3-(a). Each row of the matrix represents a *transaction*. Each column of the matrix represents a domain value of a variable. In terms of *matrix factorization*, the sparse matrix M_TRX_{bike} is decomposed into a $m \times k$ *user-feature matrix* and a $k \times n$ *value-feature matrix* which both contain the relevant information of the sparse matrix. Thereby, k is a variable parameter which needs to be adapted accordingly depending on the internal structure of the given data.

As shown in Table 3, the dense matrix, and user-feature and value-feature matrices are calculated by applying matrix factorization based on singular-value decomposition (SVD) [2][2].

Table 3. The factorization result of the sparse matrix M_TRX_{bike}. Columns represent frame_biketype, frame_internal, extra_propstand, and gear_internal.

(a) Sparse matrix M_TRX_{bike}

	0	1	2	3	4	0	1	0	1	2	0	1
	b1	b2	b3	b4	b5	b6	b7	b8	b9	b10	b11	b12
Alice	0	0	0	1	0	0	1	0	1	0	0	1
Bob	0	0	0	0	1	1	0	0	1	0	1	0
Tom	1	0	0	0	0			0	1	0	0	1
Ray						0	1				0	1
Joe	0	0	1	0	0						0	1

(b) Dense matrix $M\_TRX'_{bike}$

	0	1	2	3	4	0	1	0	1	2	0	1
	b1	b2	b3	b4	b5	b6	b7	b8	b9	b10	b11	b12
Alice	0.5	-0.3	0.3	0.4	0.1	0.3	1.3	-0.2	2	-0.2	-0.3	2
Bob	0.4	-0.3	0.3	0	0.6	0.9	0.6	-0.2	1.9	-0.2	-0.3	2
Tom	0.5	-0.3	0.2	0.3	0.3	0.5	1.1	-0.3	2.1	-0.3	-0.4	2.1
Ray	0.6	-0.5	0.5	0.5	0.1	0.5	1.7	-0.4	2.7	-0.4	-0.5	2.9
Joe	0.3	-0.4	0.6	0.2	0.2	0.6	1.1	-0.3	2.1	-0.3	- -0.4	2.3

Calculating the Heuristics. Using the matrix factorization outputs (user factors and value factors), we find variable and value ordering for searching configurations for an IT. To find heuristics for CT and ITs, we use the dense matrix (see Table 3-(b)) whereas to find heuristics for ATs, we use the value factors matrix.

For example, for *Ray* we sort the probabilities of the dense matrix $M\_TRX'_{Ray}$. We take the first bit in the sorted dense matrix which is $b12$. It represents the assignment *gear_internal = 1*. Therefore, we set *gear_internal* as the first variable to assign in the variable ordering and the value ordering of gear_internal starts with 1 as shown in Table 4-(b). Then, we take the next bit in the dense matrix which is $b9$. It represents the assignment *extra_propstand = 1*. Thus, we set *extra_propstand* as the second variable in the variable ordering and its value ordering starts with the value 1. Then, we complete the variable and value ordering heuristics for all CTs and ITs as shown in Table 5.

Table 4. Learning heuristics for CSP_{Ray} from the dense matrix. Columns represent frame_biketype, frame_internal, extra_propstand, and gear_internal.

(a) Dense matrix of M_TRX'$_{Ray}$

0	1	2	3	4	0	1	0	1	2	0	1
b1	b2	b3	b4	b5	b6	b7	b8	b9	b10	b11	b12
0.6	-0.5	0.5	0.5	0.1	0.5	1.7	-0.4	2.7	-0.4	-0.5	2.9

(b) Variable and Value Ordering for CSP_{Ray}

1	0	1	0	2	1	0	0	2	3	4	1
b12	b11	b9	b8	b10	b7	b6	b1	b3	b4	b5	b2

---

[2] We set the parameters in SVD as $latent factors = 6$ and $iterations = 1000$.

Searching for a Configuration. After calculating the variable and value ordering heuristics by sorting the values in the dense matrix, we solve the configuration tasks of ITs using these heuristics. We provide the variable and value ordering heuristics to the CSP solver as an input. Then the solver searches for a consistent configuration based on the orders in the given heuristics.

We store the calculated heuristics and configuration results for historical transactions as shown in Table 5. We use these configuration results directly. For example, when *Ray* revisits the online bike shop, the pre-calculated personalized bike configuration (see the column *CONF* in Table 5) is directly provided without any online calculations. The heuristics are used to guide the search for configurations for similar ATs. That's why, we also calculate the heuristics for CT even though they do not need heuristics since they already consist a complete personalized bike settings.

Table 5. Calculated heuristics and configurations

Type	User	Variable Ordering	Value Ordering					*CONF*
CT	Alice	1. extra_propstand	1	0				
		2. gear_internal	1	0	2			
		3. frame_internal	1	0				
		4. frame_biketype	0	3	2	4	1	
CT	Bob	1. gear_internal	1	0				
		2. extra_propstand	1	0	2			
		3. frame_internal	0	1				
		4. frame_biketype	4	0	2	3	1	
IT	Tom	1. extra_propstand	1	0	2			1
		2. gear_internal	1	0				1
		3. frame_internal	1	0				1
		4. frame_biketype	0	3	4	2	1	0
IT	Ray	1. gear_internal	1	0				1
		2. extra_propstand	1	0	2			1
		3. frame_internal	1	0				1
		4. frame_biketype	0	2	3	4	1	0
IT	Joe	1. gear_internal	1	0				1
		2. extra_propstand	1	0	2			1
		3. frame_internal	1	0				1
		4. frame_biketype	2	0	3	4	1	2

3.2 Configurations for ATs

In order to calculate a variable and value ordering heuristics for ATs, we cannot use the dense matrix since ATs do not yet exist in the transaction matrix. If we would include an AT in the dense matrix and again decompose it, the runtime would not be feasible for a real-time configuration task. Therefore, for an AT, we use the information of k most similar ITs and CTs from the dense matrix of transactions to calculate heuristics for ATs.

Calculating the Heuristics. We use Euclidean Distance [13] based similarity to find the most similar historical transactions to an AT as shown in Formula 1. $M\_TRX'_{HT}$ is a historical transaction (CT or IT) from the dense matrix and M_TRX_{AT} is the bit-mapped requirements of the active user. b_i represents the ith bit of the bitmap matrix. The bits of M_TRX_{AT} and $M\_TRX'_{HT}$ are compared on the basis of the instantiated bits in the M_TRX_{AT}. Therefore, i stands for the index of the instantiated bits in M_TRX_{AT}.

$$\sqrt{\sum_{b_i \in AT.REQ} \|M\_TRX_{AT}.b_i - M\_TRX'_{HT}.b_i\|^2} \tag{1}$$

For the working example, to find the most similar transactions, first we need to convert REQ_{Lisa} (see Table 2) into a bitmap form (M_TRX_{AT}) as shown in Fig. 1.

$$M\_TRX_{\text{Lisa}} = \begin{bmatrix} 0\ 0\ 1\ 0\ 0 & \_\ \_\ \_\ \_\ \_ & 0\ 1 \end{bmatrix}$$

Fig. 1. The active transaction of Lisa (M_TRX_{Lisa}).

In order to find the most similar ITs and CTs to M_TRX_{Lisa}, we find euclidean distance between M_TRX_{Lisa} and each row of the dense matrix $M\_TRX'_{bike}$ based on the instantiated bits (the first five and the last two) in M_TRX_{Lisa}. According to the calculated distances, k most similar transactions to M_TRX_{Lisa} are selected as M_TRX_{Alice} and M_TRX_{Bob}, since k is set to 2 for this working example and they have the shortest distances to M_TRX_{Lisa}.

Thus, during the search for a consistent configuration for $Lisa$, we use the variable and value ordering heuristics of $M\_TRX'_{Lisa}$ which is calculated by multiplying the aggregated user features matrix of the most similar transaction with the item features matrix. The predicted user-feature matrix of Lisa $M\_UF'_{Lisa}$ is multiplied with the value-feature matrix M_VF_{bike} in order to find the estimated transaction matrix for Lisa as shown in Table 6. Using the sorted values of the estimated transaction matrix, we obtain the variable and value ordering heuristics to solve CSP_{Lisa}.

Table 6. The estimated transaction for CSP_{Lisa}

b1	b2	b3	b4	b5	b6	b7	b8	b9	b10	b11	b12
0.4	−0.2	0.7	0.3	0	0.8	0.7	−0.1	1.7	−0.1	−0.2	1.9

4   Experiments

In this section, we first describe the experimental settings, the knowledge bases, evaluation criteria and comparative approaches. Then, we demonstrate the run-time and accuracy performance of our approach compared to the combinations of existing variable and value ordering heuristics.

Our experiments are executed on a computer with an Intel Core i5-5200U, 2.20 GHz processor, 8 GB RAM and 64 bit Windows 7 Operating System and Java Run-time Environment 1.8.0. For constraint satisfaction, we used a Java based CSP solver *choco-solver*[3]. For decomposing the bitmap matrix, we have used the *SVDRecommender* of Apache Mahout library [11] with a latent factor $k=100$, number of iterations $=1000$. We present all other parameters and test cases in Table 7. We have used real-world constraint-based knowledge bases from a *Configuration/Diagnosis Benchmark in Choco (CDBC)*[4]. The test parameters based on knowledge bases and corresponding transactions are in Table 7.

Table 7. Parameters of experimental tests

Experimental knowledge bases and datasets			
	Bike KB	PC KB	Camera KB
# of CT	80*	50*	150**
# of IT	20*	50*	50**
# of AT	50*	50*	64**
# of variables	31	45	10
# of constraints	32	42	20
# of variable ordering heuristics	10	10	10
# of value ordering heuristics	7	7	7
# of test cases for each configuration task	70	70	70
# of configuration tasks based on ATs	50	50	50
# of configuration tasks based on ITs	20	50	50

*synthetic transactions dataset,
**real world transactions dataset

We evaluate the performance of our approach based on the following two performance criteria: time (τ) and accuracy (π) which are calculated as explained in the following paragraphs.

Runtime (τ). Runtime, as one of our performance indicators, represents the time spent in calculating the configuration result (n represents the number of configuration tasks, see Formula 2).

$$\tau = \frac{1}{n} \times \sum_{i=1}^{n} runtime(CSP_i.solve()) \qquad (2)$$

---

[3] http://www.choco-solver.org/.
[4] https://github.com/CSPHeuristix/CDBC.

Table 8. Performance results of variable ordering (on the left column) and value ordering (on the top row) combinations. Our proposed method is the combination of MF_var (MF based learned variable ordering heuristics) and MF_val (MF based learned value ordering heuristics).

(a) Runtime to solve AT based configuration tasks (given in ms)

	Int-Domain-Max	Int-Domain-Min	Int-Domain-Median	Int-Domain-Middle	Int-Domain-Random	Int-Domain-Random-Bound	MF_val
Smallest	14.96	5.46	5.49	7.05	5.48	4.78	6.37
Largest	5.57	4.91	7.47	4.19	5.23	4.37	5.20
FirstFail	6.89	6.71	8.25	7.67	6.60	7.47	3.99
Anti- FirstFail	6.81	7.06	6.94	5.88	6.78	7.93	4.67
Occurance	4.37	4.62	3.28	3.01	3.72	3.38	3.76
Input Order	4.88	4.32	4.77	6.51	7.42	6.16	3.91
DomOver Dweg	25.50	3.83	3.99	3.91	4.17	5.16	4.01
Generalized MinDomain	4.23	3.15	4.21	3.88	4.02	3.97	4.05
Random	3.90	4.14	3.77	4.27	4.15	4.50	4.34
MF_var	3.45	3.77	3.55	3.24	3.26	3.35	**2.36**

(b) Prediction Accuracies for AT based configuration tasks

	Int-Domain-Max	Int-Domain-Min	Int-Domain-Median	Int-Domain-Middle	Int-Domain-Random	Int-Domain-Random-Bound	MF_val
Smallest	0.69	0.67	0.69	0.67	0.67	0.64	0.67
Largest	0.69	0.64	0.72	0.72	0.72	0.64	0.72
FirstFail	0.72	0.64	0.69	0.61	0.67	0.64	0.67
Anti- FirstFail	0.69	0.67	0.67	0.69	0.67	0.67	0.69
Occurance	0.72	0.64	0.67	0.69	0.69	0.75	0.69
Input Order	0.72	0.64	0.67	0.69	0.67	0.67	0.69
DomOver- Dweg	0.69	0.64	0.67	0.64	0.64	0.61	0.58
Generalized- MinDomain	0.72	0.64	0.69	0.61	0.75	0.69	0.67
Random	0.67	0.61	0.67	0.72	0.72	0.75	0.69
MF_var	0.61	0.64	0.64	0.61	0.58	0.61	

(c) Prediction Accuracies for IT based configuration tasks

	Int-Domain-Max	Int-Domain-Min	Int-Domain-Median	Int-Domain-Middle	Int-Domain-Random	Int-Domain-Random-Bound	MF_val
Smallest	0.69	0.67	0.69	0.67	0.67	0.64	0.69
Largest	0.69	0.64	0.72	0.72	0.72	0.64	0.74
FirstFail	0.72	0.64	0.69	0.61	0.67	0.64	0.69
Anti- FirstFail	0.69	0.67	0.67	0.69	0.67	0.67	0.72
Occurance	0.72	0.64	0.67	0.69	0.69	0.75	0.75
Input Order	0.72	0.64	0.67	0.69	0.67	0.67	0.72
DomOver- Dweg	0.69	0.64	0.67	0.64	0.64	0.61	0.61
Generalized- MinDomain	0.72	0.64	0.69	0.61	0.75	0.69	0.69
Random	0.67	0.61	0.67	0.72	0.72	0.75	0.72
MF_var	0.64	0.64	0.67	0.67	0.67	0.66	**0.88**

Accuracy (π). The configuration result ($CONF$) can be purchased by the user or not. If the purchased product is the same as $CONF$, then $CONF$ is considered as an accurate configuration $Accurate\_CONF$, otherwise an inaccurate configuration. The prediction quality of a configuration approach is calculated by dividing the number of $Accurate\_CONF$s by the total number of solved CSPs (see Formula 3).

$$\pi = \frac{\#(\text{Accurate\_CONF})}{\#(\text{CSP})} \tag{3}$$

Runtime performance is not critical for IT based configuration tasks since their calculations are executed when the corresponding user is offline. Therefore, we have not included its runtime performance into evaluations. For the real-time (AT-based) configuration tasks, we have not used matrix factorization to decrease the runtime. Instead of that we have selected the most similar transactions from the dense matrix of historical transactions. By this way, our approach outperforms the compared heuristics combinations in terms of runtime performance as shown in Table 8-(a). On the other hand, our approach outperforms the compared variable and value ordering heuristics for both IT and AT based configuration tasks in terms of prediction accuracy as shown in Table 8-(b) and Table 8-(c). IT based configuration tasks are solved with better prediction accuracy results rather than AT based configuration tasks. This is because, we have used matrix factorization based heuristics estimation for IT-based configuration tasks since the trade-off factor runtime is not critical in these offline calculations.

5   Conclusion

The quick generation of configurations within a reasonable time frame (i.e., before users leave the web page) can be very challenging especially for configuration results in a high prediction accuracy.

In this paper, we have proposed a novel learning approach for variable and value-ordering to guide the configuration search to address the runtime and prediction accuracy challenges. According to our experimental results on the basis of real-world configuration benchmarks, our approach outperforms the compared combinations of variable and value ordering heuristics in terms of *runtime efficiency* and *prediction quality* for AT based (real-time) configuration tasks. Moreover, for also IT based (offline) configuration tasks, our approach outperforms the compared approaches in terms of *prediction quality*.

References

1. Baltrunas, L., Ludwig, B., Ricci, F.: Matrix factorization techniques for context aware recommendation. In: Proceedings of the Fifth ACM Conference on Recommender Systems, pp. 301–304. ACM (2011)
2. Boutsidis, C., Gallopoulos, E.: Svd based initialization: a head start for nonnegative matrix factorization. Pattern Recogn. **41**(4), 1350–1362 (2008)

3. Chan, C.Y., Ioannidis, Y.E.: Bitmap index design and evaluation. In: ACM SIG-MOD Record, vol. 27, pp. 355–366. ACM (1998)
4. Eiben, A., Ruttkay, Z.: Constraint satisfaction problems (1997)
5. Felfernig, A., Hotz, L., Bagley, C., Tiihonen, J.: Knowledge-Based Configuration: From Research to Business Cases, 1st edn. Morgan Kaufmann Publishers Inc., San Francisco (2014)
6. Gemulla, R., Nijkamp, E., Haas, P.J., Sismanis, Y.: Large-scale matrix factorization with distributed stochastic gradient descent. In: Proceedings of the 17th ACM SIGKDD International Conference on Knowledge Discovery and Data Mining, pp. 69–77. ACM (2011)
7. Koren, Y., Bell, R., Volinsky, C.: Matrix factorization techniques for recommender systems. Computer **42**(8), 30–37 (2009)
8. Mnih, A., Salakhutdinov, R.R.: Probabilistic matrix factorization. In: Advances in Neural Information Processing Systems, pp. 1257–1264 (2008)
9. Papagelis, M., Plexousakis, D.: Qualitative analysis of user-based and item-based prediction algorithms for recommendation agents. Eng. Appl. Artif. Intell. **18**(7), 781–789 (2005)
10. Sadeh, N., Fox, M.S.: Variable and value ordering heuristics for the job shop scheduling constraint satisfaction problem. AI J. **86**(1), 1–41 (1996). https://doi.org/10.1016/0004-3702(95)00098-4
11. Schelter, S., Owen, S.: Collaborative filtering with apache mahout. In: Proceedings of ACM RecSys Challenge (2012)
12. Tsang, E.: Foundations of Constraint Satisfaction. Academic Press, Cambridge (1993)
13. Weinberger, K.Q., Blitzer, J., Saul, L.K.: Distance metric learning for large margin nearest neighbor classification. In: Advances in Neural Information Processing Systems, pp. 1473–1480 (2006)

Explaining Object Motion Using Answer Set Programming

Franz Wotawa$^{(\boxtimes)}$ ⓘ and Lorenz Klampfl

Christian Doppler Laboratory for Quality Assurance Methodologies
for Cyber-Physical Systems, Institute for Software Technology,
Graz University of Technology, Inffeldgasse 16b/2, 8010 Graz, Austria
{wotawa,lklampfl}@ist.tugraz.at

Abstract. Identifying objects and their motion from sequences of digital images is of growing interest due to the increasing application of autonomous mobile systems like autonomous cars or mobile robots. Although the reliability of object recognition has been increased significantly, there are often cases arising where objects are not classified correctly. There might be objects detected in almost all images of a sequence but not all. Hence, there is a need for finding such situations and taking appropriate measures to improve the overall detection performance. In this paper, we contribute to this research direction and discuss the use of logic for identifying motion in sequences of images that can be used for this purpose. In particular, we introduce the application of diagnosis and show an implementation using answer set programming.

Keywords: Spatial reasoning · Qualitative reasoning · Diagnosis · Application of answer set programming

1 Introduction

With scientific advancements already achieved in artificial intelligence there is also a growing trend of using the resulting methods and techniques in applications. In the automotive industry more and more automated and autonomous functions have been made available for customers including automated emergency braking or lane assist. All these functions depend on certain sensors like cameras, laser scanners, or radar and have to interpret the sensor data to finally identify scenarios where a function has to be executed, i.e., reducing speed because of a braking car in front. When gaining more and more autonomy such cars must take actions under tight time constraints so that under no circumstances it harms people but also be reliable assuring a smooth and non-obstructive driving behavior. Hence, what we want are dependable and trustworthy systems.

In this paper, we contribute to this vision of trustworthy and dependable autonomous systems. In particular, we discuss steps towards utilizing logics for

---

Authors appear in reverse alphabetical order.

© Springer Nature Switzerland AG 2020
D. Helic et al. (Eds.): ISMIS 2020, LNAI 12117, pp. 298–307, 2020.
https://doi.org/10.1007/978-3-030-59491-6_28

identifying object motion an autonomous system may have to deal with. Using logics is appealing in this context because it allows to explain decisions drawn from the underlying knowledge base. Its declarative nature allows us to state what we want and not requires us to come up with an implementation. Tracing object movements is important because it allows us to classify objects in being potential dangerous or not. The former may require additional actions to prevent the autonomous system from crashes or other not wanted interactions. Let us have a look at an example. In Fig. 1 we depict a typical view a driver has through the front window of a car (and an autonomous vehicle has as well).

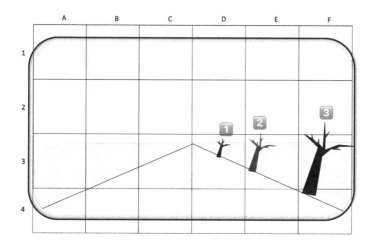

Fig. 1. View from the front window of a driving car recognizing a passing tree on the right side considering three different time steps.

In this figure, we see a tree on the right at three points in time, i.e., **1**, **2**, and **3**. The tree is coming closer when we drive and also its visual appearance increases. However, it seems to move to the right and thus we would not expect any crash or other dangerous situation (assuming a certain uniformity of nature, where we do not consider that the tree might fall on the street when we are passing with our car). Hence, identifying movements of objects over time in our visual view is important and there are many application scenarios, e.g., (i) distinguishing driving scenarios as being critical or not, (ii) identifying spurious objects caused by wrong classifications of the used sensor, or (iii) keeping track of objects that might not be always correctly seen by the sensor. The later application may deal with a situation where the tree from Fig. 1 is seen by the sensor at time **1** and **2** but not **3**. In this case, we are able to predict that the tree has to be at position **F3** at time **3** because an object like a tree cannot vanish.

In the following, we focus on identifying and explaining motion of objects, i.e., going from left to right, using answer set programming (ASP) [6]. We follow ideas outlined in Suchan et al. [12] but give attention to finding a certain movement using concepts obtained from model-based diagnosis [5,8,10]. In particular, we show how model-based diagnosis can be easily implemented using ASP and how we obtain a model for identifying objects' motion.

2 Basic Foundations

Diagnosis is an activity aiming at providing an explanation for certain symptoms, i.e., behaviors deviating from expectation. In artificial intelligence diagnosis has been considered early, finally leading to the field of model-based diagnosis (MBD) where the idea is to utilize models of systems directly for obtaining the diagnoses. Davis [5] was one of the first introducing the basic concepts of MBD followed by work from Reiter [10] and de Kleer and Williams [8] mainly dealing with formalizing MBD, its algorithms, and probing. The original formalization of MBD is tailored towards diagnosis of systems comprising connected components. Because we are interested in finding explanations, we are going to adapt the formalization accordingly. Instead of models we make use of a logical theory, and instead of considering components we are considering assumptions. We start the formalization stating the parts necessary to formulate a diagnosis problem comprising a theory, assumptions, and given observations, and where we interested in obtaining those assumptions necessary and keep them in order to explain the given observations.

Definition 1 (Diagnosis Problem). *A tuple (Th, A, O) is called a diagnosis problem where Th is a model describing the system's behavior or structure, or a knowledge base, A is a set of assumptions, and O is a set of observations.*

Let us use this definition, to formulate the Nixon diamond as a diagnosis problem. We know that quakers are usually pacifists. We further assume that republicans are usually not pacifists, and we also know as facts that the former president of the U.S. Nixon was both a quaker and a republican. For assumptions like usually quakers or republicans behave as expected, we introduce the formal assumptions nab_quaker and $nab\_republican$ respectively. Hence, the corresponding diagnosis problem can be formalized as follows:

$$Th = \left\{ \begin{array}{l} nab\_quaker \wedge quaker \rightarrow pacifist, \\ nab\_republican \wedge republican \rightarrow \neg pacifist \end{array} \right\}$$
$$A = \{nab\_quaker, nab\_republican\}$$
$$O = \{quaker, republican\}$$

When we put together $Th \cup A \cup O$, i.e., assume that all assumptions are true, we obviously obtain an inconsistent sentence, and we further need to find out the reason behind. The following definition of diagnosis allows to characterize explanations for such unsatisfiable logical sentences formally.

Definition 2 (Diagnosis). *Given a diagnosis problem (Th, A, O). A set $\Delta \subseteq A$ is a diagnosis if and only if the logical sentence $Th \cup \Delta \cup \{\neg a | a \in A \backslash \Delta\} \cup O$ is consistent. A diagnosis is minimal if no proper subset is a diagnosis.*

For the Nixon diamond, we obtain two minimal diagnosis: $\Delta_1 = \{nab\_quaker\}$ and $\Delta_2 = \{nab\_republican\}$. Hence, either Nixon was an ordinary quaker or an ordinary republican but not both. It is worth noting that if we know that Nixon was not a pacifist, i.e., adding $\neg pacifist$ to the set of observations O, we receive only the one minimal diagnosis Δ_1 and know that he was not behaving like an ordinary quaker.

For more details about MBD including algorithms we refer the interested reader to [14]. In the following we briefly outline how to use ASP [6] to implement MBD. Informally an answer set can be characterized as follows: First we assume a logical propositional theory comprising facts, rules, and integrity constraints. The latter is a logical rule that allows to derive a contradiction, which is used to state that the given propositions on the left side of a rule cannot be true at the same time. A logical model is a set of propositions used in the rules and integrity constraints that satisfies all of them. A logical model is an answer set if every proposition has an acyclic logical derivation from the given logical theory.

For example, given the theory of the Nixon diamond together with its observations, i.e., $Th \cup O$, we obtain the one answer set $\{quaker, republican\}$ because nothing else can be directly derived. If we add the fact nab_quaker, the answer set is $\{quaker, nab\_quaker, pacifist, republican\}$. In order to allow an answer set solver to implement MBD, we need a way such that the solver can decide a truth value for nab_quaker and $nab\_republican$ by itself. This can be achieved via introducing the following rules:

$$\neg\, nab\_quaker \rightarrow ab\_quaker$$
$$\neg\, ab\_quaker \rightarrow nab\_quaker$$

for nab_quaker (and similar ones for $nab\_republican$). These rules allow the answer set solver to select either nab_quaker or ab_quaker to be true, but not both.

Hence, in general we only need to come up with an answer set theory that comprises $Th \cup O$ and rules

$$\neg\, a \rightarrow n\_a$$
$$\neg\, n\_a \rightarrow a$$

for all assumptions $a \in A$ to have an answer set solver representation of a diagnosis problem. Using this representation, we can compute diagnoses using the answer set solver directly. For our implementation we rely on the Potassco answer set solver[1], which makes use of Prolog syntax and where $\neg$ is represented as **not**.

3   Modeling Object Motion

The Artificial Intelligence (AI) community today, has widely recognized the possibilities and significance of **qualitative physics**. Unfortunately codifying qualitative knowledge about the physical world has turned out to be surprisingly

---

[1] See https://potassco.org/.

difficult [13]. As stated in [4], a qualitative representation of the world with the goal to later on qualitatively reason about this world is not new. Many subareas, like qualitative spatial reasoning, qualitative temporal reasoning and qualitative simulation, where established over time. The idea in this paper focuses on the field of qualitative spatial reasoning (QSR) which provides a calculus that allows us to model object motion without the need of precise quantitative entities. The representation of objects in the model is done by stating spatial qualitative knowledge on which later on reasoning is applied with decision-making methods and techniques.

In AI systems like robotics, geographical information systems and medical analysis systems, spatial information plays a crucial role where spatial reasoning can be a potential application [2]. Cohn et al. [3] state that in most cases not a complete a priori quantitative knowledge but qualitative abstractions drive spatial reasoning. Furthermore they indicate that the challenge of QSR lies in providing a calculi which allows a machine to represent and reason with spatial entities but without using the traditional quantitative techniques which are for instance common in the computer graphics or computer vision communities. Since the concept of space in commonsense knowledge and spatial reasoning is quite complex, due to its multi-dimensionality, most work in this area has focused on single aspects of space such as topology, orientation and distance [11].

In the example given in the introduction, we approach the underlying problem by explaining object motion with the single aspect of topology. In topological approaches qualitative spatial reasoning in most cases describe relationships between spatial regions rather than points that are usually used in the case of approaches dealing with orientation [11]. The best known approach in the topological domain is the Region Connection Calculus (RCC) by Randell, Cui and Cohn [9] and includes the eight basic relations DisConnected (DC), Externally Connected (EC), Partially Overlapping (PO), EQual (EQ), Tangential Proper Part (TPP), Non-Tangential Proper Part (NTPP) and the inverse of the latter two TPP^{-1} and NTPP^{-1} which are shown in Fig. 2. In addition to this basic relations, Bennett [1] presented a method for reasoning about spatial relationships on the basis of entailments in propositional logic and showed that reasoning on spatial relations represented in a logical representation is decidable.

In the following, we outline the use of **MBD for spatial reasoning using ASP** using the example discussed in the introduction for illustration purposes. In the `clingo` source code given after this paragraph we depict a simplified ASP model together with necessary integrity constraints. This ASP model considers only movements along the x-axis. With this simplification we can reduce our possible states to one dimension which results in three possible explanations of an object movement, namely `stable`, `toLeft` and `toRight`. `stable` states that the object in the following time step is still in the same grid region, which may indicate that the object is moving towards our direction which could result in a crash or dangerous situation. `toLeft` and `toRight` indicate that the object is moving to the left or right field in our projected grid during the transition from

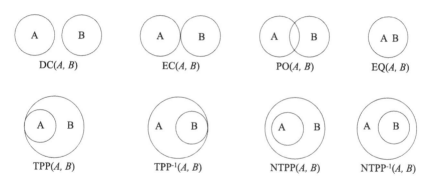

Fig. 2. A representation of the basic relations in the RCC theory by Randell et al. [9].

time step T to T+1, which is usually an indicator that the object is consistent in the environment and we would not expect any crash or other dangerous situation.

In source code line 3–10 the above mentioned explanations for a movement on the x-axis are defined. In addition the predicate `stable` is satisfied if the predicate `n_stable` (not stable) is not satisfied and vice versa. Similar this is done for `toRight` and `toLeft`. Further on the predicate `no_expl(N)` in line 14 counts the number of explanations for given observations and with its following integrity constraint in line 16 it is ensured that we only search for a single explanation. Line 20–28 define the basic definitions which are used if an object of interest is in the `next` or `prev` field of our projected grid. The knowledge for stating that an actual position change from time step T to time step T+1 based on the defined action was performed is implemented in line 33–35.

```
1   % defining explanations
2
3   stable :- not n_stable.
4   n_stable :- not stable.
5
6   toRight :- not n_toRight.
7   n_toRight :- not toRight.
8
9   toLeft :- not n_toLeft.
10  n_toLeft :- not toLeft.
11
12  % counting the number of explanations using predicate
         no_expl
13
14  no_expl(N) :- N = #count {1:stable; 2:toRight; 3:toLeft}.
15
16  :- not no_expl(1). % Search for single explanations only
17
18  % basic definition of next and previous operator
19
```

```
20  next(outofsight,a).
21  next(a,b).
22  next(b,c).
23  next(c,d).
24  next(d,e).
25  next(e,f).
26  next(f,outofsight).
27
28  prev(X,Y) :- next(Y,X).
29
30  % knowledge base stating that an object position change
31  % from time step 1 to 2 based on the action performed.
32
33  object(Xn,2) :- object(X,1), toRight, next(X,Xn).
34  object(Xp,2) :- object(X,1), toLeft, prev(X,Xp).
35  object(X,2)  :- object(X,1), stable.
```

KB for explaining motion

In addition, we require an integrity constraint to assure that an object cannot be at two different positions at the same time step T.

```
36  :- object(X,T), object(Y,T), not X = Y.
```

Used integrity constraint

The described ASP model allows for capturing the case of identifying an object moving along the x-axis. For example, we might call clingo using additional observations that an object (like the tree in Fig. 1) is at position $d3$ at time step 1 and at position $e3$ at time step 2. In order to allow clingo computing diagnoses, i.e., explanations for this movement, we have to add the following facts (representing the observations) to the ASP source code (where we ignore the y-axis):

```
37  object(d,1).
38  object(e,2).
```

Observations for our running example

When calling clingo using the described file, we obtain the following result:

```
Solving...
Answer: 1
next(outofsight,a) next(a,b) next(b,c) next(c,d) next(d,e
    ) next(e,f) next(f,outofsight) prev(a,outofsight)
    prev(b,a) prev(c,b) prev(d,c) prev(e,d) prev(f,e)
    prev(outofsight,f) object(d,1) object(e,2) toRight
    n_toLeft n_stable no_expl(1)
SATISFIABLE

Models      : 1
Calls       : 1
```

```
Time              : 0.002s (Solving: 0.00s 1st Model: 0.00s
     Unsat:  0.00s)
CPU Time          : 0.002s
```

Result obtained using <code>clingo</code>

From the result we see that the object is moving to the right (toRight is true) because there is only one valid model.

After discussing the basic spatial model, we further want to discuss its potential application in practice. In Fig. 3 we illustrate a possible **application scenario of qualitative spatial reasoning** on real-world data. With the same idea as in the example outlined in the introduction, we are now looking at three frames of real-world data extracted from the well-known KITTI raw dataset [7]. On the left side the possible output of an object detector performing car detection within each frame is depicted. For simplicity we only consider one object of interest, the parked car on the right side of the street which is correctly detected in the first two frames. Let us assume that in the third frame the object detector is not able to detect the car due to an occurring sensor error. In this case the autonomous car may decides to drive too far on the right side which could result in an accident or dangerous situation. In such situations, spatial reasoning and the methods described can serve as a supplementary system capable of performing sanity checks in addition to the object detector. With spatial reasoning and the previous observations of the object detector, we are able to predict that the parked car has to be at position E3 in frame three due to the fact that the car cannot vanish within one time step. With this prediction we are now able to revise the initial output of the detector to a correct detection of the car which furthermore resolves the possible dangerous situation, and which can also result in an improvement of the overall detection performance.

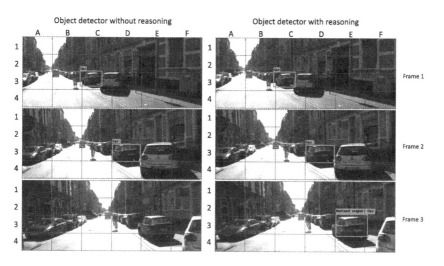

Fig. 3. *Left:* an object detector recognizing the object of interest in the first two frames but missing the object in the third frame. *Right:* revised output at frame three after the application of spatial reasoning in addition to the object detector.

For simplicity reasons only movements along the x-axis were considered. Nevertheless it should be noted that with similar rules as used for movements along the x-axis, the approach can be extended to recognize movements also along the y-axis. Therefore it is also possible to extend the approach to larger problems.

Besides improving detection performance when coupling logical reasoning with object recognition, there are other application scenarios possible. We might come up with rules evaluating traffic situations as being dangerous or not as a first application. For example, whenever there is an object that remains in the center of our view but becomes bigger as well, there is danger of a crash and appropriate actions are required. We also might handle the case of total occlusion of a detected object due to other traffic participants via introducing additional spatial rules following the approach of Suchan et al. [12].

4   Conclusion

In this paper, we introduced the use of model-based diagnosis implemented using answer set programming for spatial reasoning. In particular, we discussed logical rules allowing to identify the movement of objects considering two consecutive images in time comprising the same object but at different positions. The obtained direction of movement can be used for improving the detection performance of objects and other application scenarios. To illustrate improving object detection, we made use of an example scenario obtained from a dataset used in the domain of autonomous driving.

In order to bring the approach into practice, we intend to further work on directly coupling object recognition with logical reasoning requiring to come up with the right time steps that allow for mapping the quantitative world into an abstract qualitative representation. We want to further work on adding rules for identifying critical scenarios where we need to consider object movements that may lead to a crash situation. Combining quantitative approaches like distances with qualitative ones seems to be a promising choice for future applications. In addition, we want to carry out an experimental evaluation showing that the presented approach really improves detection considering adversarial attacks (like adding rain or other disturbances) to some of the images in the sequence.

Acknowledgement. The financial support by the Austrian Federal Ministry for Digital and Economic Affairs and the National Foundation for Research, Technology and Development is gratefully acknowledged.

References

1. Bennett, B.: Spatial reasoning with propositional logics. In: Doyle, J., Sandewall, E., Torasso, P. (eds.) Principles of Knowledge Representation and Reasoning. The Morgan Kaufmann Series in Representation and Reasoning, pp. 51–62. Morgan Kaufmann (1994). https://doi.org/10.1016/B978-1-4832-1452-8.50102-0, http://www.sciencedirect.com/science/article/pii/B9781483214528501020
2. Bennett, B.: Modal logics for qualitative spatial reasoning. Logic J. IGPL **4**(1), 23–45 (1996). https://doi.org/10.1093/jigpal/4.1.23
3. Cohn, A., Hazarika, S.: Qualitative spatial reasoning. Fundamenta Informaticae **46**(1–2), (2001)
4. Dague, P.: Qualitative reasoning: a survey of techniques and applications. AI Commun. **8**(3/4), 119–192 (1995)
5. Davis, R.: Diagnostic reasoning based on structure and behavior. Artif. Intell. **24**, 347–410 (1984)
6. Eiter, T., Ianni, G., Krennwallner, T.: Answer set programming: a primer. In: Tessaris, S., et al. (eds.) Reasoning Web 2009. LNCS, vol. 5689, pp. 40–110. Springer, Heidelberg (2009). https://doi.org/10.1007/978-3-642-03754-2_2
7. Geiger, A., Lenz, P., Stiller, C., Urtasun, R.: Vision meets robotics: the kitti dataset. Int. J. Rob. Res. (IJRR) **32**, 1231–1237 (2013)
8. de Kleer, J., Williams, B.C.: Diagnosing multiple faults. Artif. Intell. **32**(1), 97–130 (1987)
9. Randell, D., Cui, Z., Cohn, A.: A spatial logic based on regions and connection. KR **92**, 165–176 (1992)
10. Reiter, R.: A theory of diagnosis from first principles. Artif. Intell. **32**(1), 57–95 (1987)
11. Renz, J., Nebel, B.: Qualitative spatial reasoning using constraint calculi. In: Aiello, M., Pratt-Hartmann, I., Van Benthem, J. (eds.) Handbook of Spatial Logics, pp. 161–215. Springer, Dordrecht (2007). https://doi.org/10.1007/978-1-4020-5587-4_4
12. Suchan, J., Bhatt, M., Varadarajan, S.: Out of sight but not out of mind: an answer set programming based online abduction framework for visual sensemaking in autonomous driving. In: Proceedings of the Twenty-Eighth International Joint Conference on Artificial Intelligence (2019). https://doi.org/10.24963/ijcai.2019/260
13. Weld, D., de Kleer, J. (eds.): Readings in Qualitative Reasoning about Physical Systems. Morgan Kaufmann, Burlington (1989)
14. Wotawa, F.: Reasoning from first principles for self-adaptive and autonomous systems. In: Lughofer, E., Sayed-Mouchaweh, M. (eds.) Predictive Maintenance in Dynamic Systems, pp. 427–460. Springer, Cham (2019). https://doi.org/10.1007/978-3-030-05645-2_15

The GraphBRAIN System for Knowledge Graph Management and Advanced Fruition

Stefano Ferilli[1([⊠])] and Domenico Redavid[2]

[1] University of Bari, Bari, Italy
stefano.ferilli@uniba.it
[2] Artificial Brain S.r.l, Bari, Italy
redavid@abrain.it

Abstract. The possibility of inter-relating different information items is crucial in the perspective of enhanced storage, handling and fruition of knowledge. GraphBRAIN is a general-purpose tool that allows to design and collaboratively populate knowledge graphs, and provides advanced solutions for their fruition, consultation and analysis. Its functionalities are also provided as Web services to other applications. A peculiarity of GraphBRAIN is its fusion of methods and tools coming from different research areas: ontologies to describe such variated knowledge, collaborative tools to collect the knowledge scattered across many people, graph databases to store the knowledge base, data mining and social network analysis tools for personalized fruition of the collected knowledge. It is currently used as the knowledge management platform in a tourism-related project.

1 Introduction

To better support the information needs of scholars, practitioners, and even non-expert end users, considering also the context of the information items they handle is of utmost importance to put them in perspective and grasp a deeper understanding thereof. This requires a step up from simple 'information' handling to 'knowledge' handling. Collecting, storing and using knowledge is not trivial. It might be scattered and spread across many people with different expertise, culture, background and perspectives. Its effective exploitation might involve complex information patterns and aggregates, that might be domain- or user-dependent. From a technological perspective, a switch from databases to knowledge bases is required, so as to enable high-level analysis and reasoning tasks. In turn, knowledge bases require suitable schemes to represent and organize the knowledge, often in the form of ontologies, but a proper formalization might be unavailable in mainstream research for some domains. Last but not least, different domains might be inter-related, and cross-fertilization among them should be enforced.

Tackling all these issues together requires a suitable infrastructure, including advanced data representation and storage facilities, and advanced tools and

D. Helic et al. (Eds.): ISMIS 2020, LNAI 12117, pp. 308–317, 2020.
https://doi.org/10.1007/978-3-030-59491-6_29

algorithms for knowledge handling. The solution we propose is **GraphBRAIN**, a general-purpose system aimed at supporting all stages and tasks in the lifecycle of a knowledge base—from knowledge base design, to knowledge acquisition, to knowledge organization and management, to (personalized) knowledge fruition and delivery—, also providing an on-line tool for interactive exploitation of these functionalities[1]. Its main peculiarities and innovative features include: the mix of an ontology-based approach with a graph database, so as to take advantage of both the efficiency of the latter and the flexibility and power of the former for storing and handling knowledge; the cooperation among different ontologies/domains, obtained through shared entities and relationships, and through a general ontology that interconnects all the domain-specific ones, yielding a single knowledge base in which the various domains cross-fertilize each other; the integration of data mining and social network analysis tools to provide high-level functionalities for knowledge handling and fruition aimed at finding relevant, personalized and non-trivial information; an interactive interface for collaborative knowledge enrichment and personalized consultation.

This paper is organized as follows. The next section describes the main features and interface of GraphBRAIN. Then, Sect. 3 discusses its current use and relevance, also by relating it to other works in the literature. Finally, Sect. 4 concludes the paper and outlines future work issues.

2 GraphBRAIN

This section will highlight the most relevant architectural and functional features of GraphBRAIN, and how they concur in providing its advanced support to knowledge management.

The knowledge base is implemented as a graph database, using the Neo4j [13] DBMS. Neo4j implements the Labeled Property Graphs (LPG) model, in which both nodes and arcs in the graph may have associated attribute-value maps; one or many labels (usually representing classes) may be associated to nodes (representing class instances), while each arc (representing a relationship) may be labeled with one type only. Neo4j is schema-free: the user may apply any label and/or attribute to each single node or arc. While ensuring great flexibility, this does not allow to associate a clear semantics to the graph items. To overcome this, and enable high-level reasoning on the available knowledge, GraphBRAIN imposes the use of pre-specified data schemes, expressed in the form of ontologies, so that only data that are compliant to the ontologies may be added to the graph. In this way, it brings to cooperation a database management system for efficiently handling, mining and browsing the individuals, with an ontology level that allows it to carry out formal reasoning and consistency or correctness checks on the individuals.

GraphBRAIN's administrators may build and maintain several ontologies by specifying for each the hierarchies of classes and relationships to be considered, each with its attributes and associated datatypes. The universal class is

---

[1] A demo of the system is available at http://193.204.187.73:8088/GraphBRAIN/.

implicit, and common to all ontologies. Such ontologies act as DB schemes that drive and support all functionalities: knowledge base creation and enrichment; advanced tools for searching and browsing the knowledge base; mining, analysis and knowledge extraction tools that may be used interactively by end users or provided as services to other systems for obtaining selective and personalized access to the stored knowledge. Some classes and relationships may appear in different ontologies, possibly with different attributes, in order to reflect different perspectives on them. This is another innovative feature, that allows cross-fertilization among, and knowledge reuse across, different domains: individuals of shared classes act as bridges, allowing the users of a domain to reach information coming from other domains. Instead of taking external ontologies and relying on ontology alignment techniques, which may return wrong associations, GraphBRAIN requires administrators to define their ontologies from inside the GraphBRAIN environment. When defining their ontologies, they may see the classes and relationships of existing ontologies, and may reuse them, possibly extending or refining them, or defining super- or sub-classes thereof.

In particular, GraphBRAIN provides a top-level ontology, defining very general and highly reusable concepts (e.g., **Person**, **Place**) and relationships (e.g., **Person.wasIn.Place**). It acts as a hub and plays a crucial role to interconnect the domain-specific ontologies, ensuring that a single, overall connected, knowledge graph underlies all the available domains. Particularly interesting is class **Category**, aimed at storing items from different taxonomies as individuals in the knowledge graph. This allows to handle them within the graph. Currently, it includes the WordNet lexical ontology [12] (also using class **Word** for its lexical part) and the standard part of the Dewey Decimal Classification (DDC) system [2]. So, Category and Word nodes may be linked by arcs to individuals of other classes (e.g., documents, persons, places) and used as tags to express information about them. Specifically, words may be used for lexically tagging other items (e.g., this paper, as a Document node, might be linked to 'graph', 'ontology', etc.), while categories may be used to semantically tag them (e.g., this paper might be linked to 'Computer Science', 'Artificial Intelligence', etc.) without formalizing thousands of classes in the ontologies. Category and Word nodes are also related to each other, thus enabling a form of reasoning as graph traversal and allowing to find non-trivial paths between instances of other classes.

The ontologies are saved in a proprietary XML format, purposely designed to be used as a schema for the graph database. However, the tool may also export them into standard Semantic Web formats, to make them publicly available for reuse. Currently, serialization to Ontology Web Language (OWL)[2] format is provided, so that they can be published and exploited for ensuring semantic access to the knowledge base and make it interoperable with other resources.

Information is fed into the knowledge base by interaction with users or by automatic knowledge extraction from documents and other kinds of resources (e.g., the Internet). The interactive interface, shown in Fig. 1, includes two form-based tabs, one for entities (Fig. 1, left) and one for relationships (Fig. 1, right),

---

[2] http://www.w3c.org/owl.

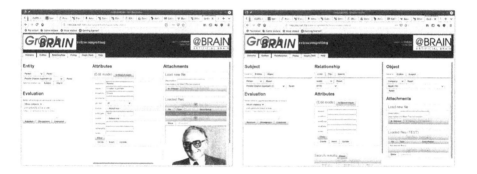

Fig. 1. GraphBRAIN interface for managing and consulting the knowledge base.

allowing the user to insert/update/remove instances and/or their attribute values. The same form-based interfaces can be used to query the knowledge base for instances of entities and relationships. The retrieved instances may be graphically displayed in another tab, as nodes and arcs in the graph. This allows the user to continue his search in a less structured way, by directly browsing the graph (e.g., expanding or compressing node neighbors). This is useful to explore the available knowledge without a pre-defined goal in mind, but letting the data themselves drive the search. Thus, serendipity in information retrieval is supported, and the users may find unexpected information that is relevant to their information needs.

The forms are automatically generated by the system from the XML format specification of the ontologies: top-level classes and relationships are shown in drop-down menus; sub-classes are shown as a tree after the top-level class has been selected; class and relationship attributes are reported in a form where names are labels and values are in textboxes, dropdown menus or other widgets depending on their type. Available types are Integer, Real, Boolean, Date, Enumeration, Instance (another instance in the knowledge base).

Albeit GraphBRAIN may handle several ontologies, each specifying a different domain, the form-based interface for data management and querying requires the user to select one of the available domains in order to load the corresponding scheme/ontology to be used. While the knowledge base content may be published as linked open data (LOD) [8], it is not available in its entirety as LOD. It is accessible only through the querying and graph browsing facilities in the on-line interface, or through pre-defined tools exposed as services, that, based on their input parameters, return relevant portions of the graph serialized as RDF.

Users may also manage (add, show, delete) attachments for each instance. In this way GraphBRAIN also acts as an archive, whose content is indirectly organized according to formal ontologies, and thus may foster interoperability with other systems. Finally, users may add comments, or approve/disapprove, each entity or relationship instance, and even each single attribute value thereof. Using the comments, the users may also provide suggestions to improve and extend the ontologies. Through the approval/disapproval mechanism, the system may establish a trust mechanism for the users that supports 'distributed' quality assurance on the content of the knowledge base. Users are encouraged to provide high-quality knowledge, because using a combination of their number of contributions and trust they are assigned 'credits' that they may spend in using advanced features provided by GraphBRAIN. Interactions of users are tracked in order to build models of their preferences to be used for personalization purposes.

Finally, GraphBRAIN provides several analysis, mining and information extraction functionalities, including tools to:

- perform consistency checks and various kinds of reasoning on the knowledge base (currently: ontological reasoning, using OWL reasoners, by providing in OWL format and giving them access to the individuals contained in the graph in RDF; or multi-strategy (deduction, abduction, abstraction, induction, argumentation, probabilistic inference, analogy) reasoning based on an inference engine implemented in Prolog);
- assess relevance of nodes and arcs in the graph, and extract the most relevant ones (currently: Closeness centrality, Betweenness centrality, PageRank, Harmonic centrality, Katz centrality);
- extract a portion of the graph that is relevant to some specified starting nodes and/or arcs (currently: Spreading Activation and a proprietary variation thereof);
- extract frequent patterns and associated sub-graphs (currently using the Gspan algorithm, and a variation thereof that is bound to include specific nodes in the graph);
- predict possible links between nodes (currently: Resource Allocation, Common Neighbors, Adamic Adar, and a proprietary approach);
- retrieve complex patterns of nodes and relationships (currently based on the application of Prolog deduction, using user-defined or automatically learned rules, on selected portions of the knowledge graph, suitably translated into Prolog facts);
- translate selected portions of the graph into natural language (using a Prolog-based engine to organize the selected information into coherent and fluent sentences).

Some of the underlying algorithms are reused from the literature; others have been purposely extended to improve their ability to return personalized outcomes that may better satisfy the user's information needs, which is another novelty introduced by GraphBRAIN. For instance, since the graph is too large to be entirely displayed, when opening the graph tab, a connected neighborhood of the most relevant nodes is shown. If a user model is available, based on statistics collected about his previous interaction with the system, the starting nodes may be those more related to his interests, preferences, aims, background, etc. Of course, the displayed portion of the graph may also be the result of a specific user query. For instance, Fig. 2 shows a portion of the graph automatically selected starting from 4 nodes selected by the user (indicated by arrows). Different colors of nodes are associated to different classes, and a clear predominance of some classes (those reported as more relevant to the user by his user model) is visible. Also, several clusters of instances (some of which indicated by circles) clearly emerge, that may be analyzed by the user to find potentially interesting but possibly unexpected information.

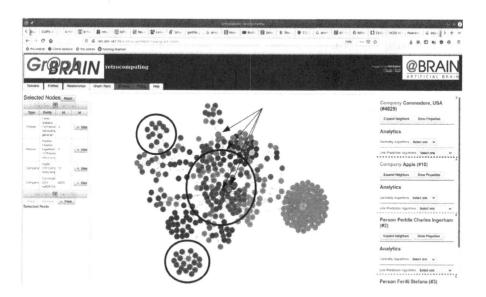

Fig. 2. Portion of GraphBRAIN's knowledge base.

3   Discussion and Related Work

We believe that GraphBRAIN is the first example of a methodology by which many different, currently separate, Artificial Intelligence tasks, techniques and approaches can be brought to cooperation for improving knowledge management and (personalized) fruition by users: database technology, ontologies, data mining, machine learning, automated reasoning, natural language processing, personalization and recommendation, social interaction. The relevance of this proposal is in these approaches being really integrated, and not simply juxtaposed, so that each of them takes directly or indirectly advantage from all the others. E.g., a social approach is used to build and integrate ontologies; user models are used to guide data mining; ontologies are used to guide database interaction and interface generation; data mining is used to filter a manageable and relevant portion of a huge graph on which carrying out automated reasoning, etc.

A prototype of GraphBRAIN is currently in use as part of a larger ongoing project [5], aimed at providing advanced support to end-users, entrepreneurs and institutions involved in touristic activities. It currently includes the following domain-specific ontologies: **tourism** (concerning history, cultural heritage items, points of interest, logistics and services, etc.); **food** (concerning typical dishes and beverages from specific regions); **computing** (concerning computing devices and their history); **lam** (concerning libraries, archives and museums). While this paper provided a detailed account on the general structure and functionality of GraphBRAIN, [4,6] focused on the development and population of the **computing** and **lam** ontologies, respectively. Table 1 reports statistics on the current ontologies and knowledge base content of GraphBRAIN. Obviously, the vast majority of knowledge items is in the **general** subgraph, including items automatically loaded from WordNet and DDC. Next comes the **lam** subgraph, also mostly automatically loaded from the records of a private collection, including 4295 titles. Then, with much less items, come the other ontologies, whose knowledge items were manually entered using the on-line collaborative interface. There are less class instances than relationship instances, indicating a quite connected graph, which is important for interlinking the knowledge and enabling effective graph browsing by the users. The **general** subgraph is the most connected. As expected, the average number of attributes per instance is larger for class instances than for relationship instances. Indeed, relationships are by themselves information carriers. The 'information density' is different between classes and relationships for the various domains. Specifically, much relevant information in the **lam** domain is in the relationships rather than in the attributes, which makes sense considering the strict interplay among documents, authors, publishers, places, categories, series.

Table 1. Statistics on ontologies in GraphBRAIN

Ontology	Classes		Attributes		Relationships		Attributes	
General	17 + 27	333 020	79	1 744 116	88	488 639	23	39 186
Tourism	9 + 60	250	64	1173	49	318	13	54
Food	9 + 24	181	40	405	23	65	3	0
Computing	15 + 97	551	111	2 096	117	739	21	343
Lam	12 + 63	9 902	51	31 615	51	13 649	24	10 614
Total	62 + 271	343 904	345	1 779 405	328	503 410	84	50 197

Concerning the ontology development functionality, several tools have been proposed in the literature, each pursuing specific objectives as regards the construction, editing, annotation and merging of ontologies [1]. The most popular and mature one is protégé[3], based on the OWL-API, which is fully compliant with the OWL specifications by W3C[4]. GraphBRAIN adopted the same OWL-API for its ontology export functionality, so that the generated ontologies are

---

[3] https://protege.stanford.edu.
[4] http://owlcs.github.io/owlapi.

fully compliant with the standard and may be edited using protégé. We developed a specific ontology definition and handling tool for several reasons. First, it had to be embedded into GraphBRAIN's interface, so that the administrators could seamlessly and collaboratively build and refine the ontologies. Second, while existing tools are mainly aimed at defining formal ontologies starting from an RDF knowledge base model, our motivation was in the need to define a schema for the graph DB, and the translation in standard ontology format was a consequential objective in order to enable OWL reasoning capabilities. Third, also due to the use of Neo4j as the DBMS infrastructure, our tool allows one to develop ontologies in which also relationships may have attributes, which is not allowed by current ontological standards.

Of course, the availability in GraphBRAIN of an ontology editor does not prevent the reuse of the ontologies and/or knowledge graphs already available in the literature or in the practice. Indeed, standard ontologies exist for describing many domains (e.g., the DCMI [9] for the library/archive domain, or Cultural-ON [11] and its evolution ArCo[5] for cultural heritage). However, to the best of our knowledge, nothing exists aimed at expanding the area of interest to a broader, 'contextual' perspective that may leverage different related domains and be attractive also for non-specialized users.

On the methodological side, a few works analyze the possibilities of cooperation between ontologies and graph DBs. Some, taken from research literature, are more theoretical. [3] outlines the potential of applying graph DBMSs to an ontological context in order to create essentially an ontological tensor, and assesses its complexity. [10] discusses technical issues that might limit the impact of symbolic Knowledge Representation on the Knowledge Graph area, and summarizes some developments towards addressing them in various logics. Some other, more practical, were specifically developed for Neo4j (https://neo4j.com/blog). One approach consists in just expressing ontologies as nodes and arcs in Neo4j, and mapping some kinds of ontological reasoning onto graph queries. We also provide this opportunity in the **general** ontology, but additionally allow to link the ontological items to the instance nodes. Another approach is driven by an ontology in determining the graph content, but this is done 'manually' without a system module that ensures that nothing not compliant with the ontology is entered, as we do.

More in general, some NoSQL DBMSs already support the integration with ontologies. The semantic graph database GraphDB is based on the RDF graph[6] model. So, it does not allow to represent node and relationship attributes, which causes a significant increase in nodes and relationships to represent, and thus a decrease in efficiency compared to the LPG model. Also triplestores can be considered as database management systems aimed at managing knowledge graphs. Compared to triplestores, our aim is keeping separate instance representation (owl:Individual) from representation of owl:Class, owl:ObjectProperty and owl:DatatypeProperty in the ontology, so as to be able to leverage the

---

[5] http://wit.istc.cnr.it/arco/index.php?lang=en.
[6] https://db-engines.com/en/system/GraphDB%3BNeo4j.

advantages of LPG (scalability, small size, etc.) on Individuals and the advantages of OWL Reasoners (main reasoning tasks over ontologies: consistency of the ontology, concept and role consistency, concept and role subsumption, instance checking, instance retrieval, query answering) on the ontological part. In particular, we may use a number of available Semantic Web reasoners[7] that provide implementations for all or part of the inferences. When performing ontological reasoning that involves instances, we may still make available the LPG part as RDF thanks to the existence of a formal mapping between RDF and LPG [7], that can be leveraged (e.g., in the form of a Neo4j plugin) to keep the representation of Node and Relation properties in LPG and render instances in RDF. More in detail, the issue with RDF is that it cannot express datatype properties on relationships, and that datatype properties on classes must be also specified as triples, this way scattering the a single entity into many nodes of the graph. By representing the ontology within the graph, one needs to explicitly specify node and relationship attributes as specific relationships. In order to exploit the ontology to keep control over the data that are inserted into the graph, one must necessarily make explicit which should be considered node and attribute relationships. This amounts to representing all RDF triples as if a classical RDF store were used (some estimate that an increase in triples with respect to nodes of up to one order of magnitude might be required[8]): as a consequence, functionality that might be used on LPG graphs, such as centrality and link prediction, would at least become inefficient.

4   Conclusions and Future Work

The possibility of inter-relating different information items is crucial in the perspective of enhanced storage, handling and fruition of knowledge, that goes beyond 'simple' information retrieval based on lexical content or metadata. GraphBRAIN is a general-purpose tool that allows to design and collaboratively populate knowledge graphs, and provides advanced solution for their fruition, consultation and analysis. Its functionality are also provided as Web services to other applications. A peculiarity of GraphBRAIN is its strategy for cooperation of several components coming from different research areas: ontologies to describe such variated knowledge, collaborative tools to collect the knowledge scattered across many people spread all over the world, and to store it in a knowledge base, data mining and social network analysis tools for personalized fruition of the collected knowledge by interested stakeholders and end users. A prototype of GraphBRAIN is currently used as the knowledge management layer of a platform for providing advanced support to stakeholders in turistic applications.

Based on the feedback of users of the on-line prototype, we are currently extending and improving the functionality and content of GraphBRAIN. We are also developing specific analysis and mining algorithm that can leverage the

---

[7] http://owl.cs.manchester.ac.uk/tools/list-of-reasoners/.

[8] https://neo4j.com/blog/rdf-triple-store-vs-labeled-property-graph-difference/.

features and peculiarities of the system, aimed at improved fruition by end-users and client systems. Another direction for future extension is the inclusion of the possibility to find solutions via Question Answering.

References

1. Abburu, S., Babu, G.S.: Survey on ontology construction tools. Int. J. Sci. Eng. Res. **4**, 1748–1752 (2013)
2. Dewey, M.: A classification and subject index for cataloguing and arranging the books and pamphlets of a library, Amherst, Mass (1876)
3. Drakopoulos, G., Kanavos, A., Mylonas, P., Sioutas, S., Tsolis, D.: Towards a framework for tensor ontologies over neo4j: representations and operations. In: 8th International Conference on Information, Intelligence, Systems & Applications, IISA 2017, Larnaca, Cyprus, 27–30 August 2017, pp. 1–6. IEEE (2017)
4. Ferilli, S., Redavid, D.: An ontology and knowledge graph infrastructure for digital library knowledge representation. In: Ceci, M., Ferilli, S., Poggi, A. (eds.) IRCDL 2020. CCIS, vol. 1177, pp. 47–61. Springer, Cham (2020). https://doi.org/10.1007/978-3-030-39905-4_6
5. Ferilli, S., De Carolis, B., Buono, P., Di Mauro, N., Angelastro, S., Redavid, D.: Una piattaforma intelligente per la gestione integrata del settore turistico. In: Primo Convegno Nazionale CINI sull'Intelligenza Artificiale - Workshop on AI for Cultural Heritage, p. 2 (2019). http://www.ital-ia.it/submission/163/paper. (in Italian)
6. Ferilli, S., Redavid, D.: An ontology and a collaborative knowledge base for history of computing. In: Proceedings of the 1st International Workshop on Open Data and Ontologies for Cultural Heritage (ODOCH-2019), at the 31st International Conference on Advanced Information Systems Engineering (CAiSE 2016). Central Europe (CEUR) Workshop Proceedings, vol. 2375, pp. 49–60 (2019)
7. Hartig, O.: Foundations to query labeled property graphs using SPARQL. In: Kaffee, L., et al. (eds.) Joint Proceedings of the 1st International Workshop On Semantics For Transport and the 1st International Workshop on Approaches for Making Data Interoperable Co-located with 15th Semantics Conference (SEMANTiCS 2019), Karlsruhe, Germany, 9 September 2019. CEUR Workshop Proceedings, vol. 2447 (2019). CEUR-WS.org
8. Heath, T., Bizer, C.: Linked Data: Evolving the Web into a Global Data Space. Morgan & Claypool Publishers, Synthesis Lectures on the Semantic Web (2011)
9. ISO/TC 46/SC 4 Technical Committee: Information and documentation - the dublin core metadata element set - part 1: Core elements. Technical report, ISO 15836–1:2017 (2017)
10. Krötzsch, M.: Ontologies for knowledge graphs? In: Artale, A., Glimm, B., Kontchakov, R. (eds.) Proceedings of the 30th International Workshop on Description Logics, Montpellier, France, 18–21 July 2017, CEUR Workshop Proceedings, vol. 1879. CEUR-WS.org (2017). http://ceur-ws.org/Vol-1879/invited2.pdf
11. Lodi, G., et al.: Semantic web for cultural heritage valorisation. In: Hai-Jew, S. (ed.) Data Analytics in Digital Humanities. MSA, pp. 3–37. Springer, Cham (2017). https://doi.org/10.1007/978-3-319-54499-1_1
12. Miller, G.A.: Wordnet: a lexical database for English. Commun. ACM **38**, 39–41 (1995)
13. Robinson, I., Webber, J., Eifrem, E.: Graph Databases. O'Reilly Media, 2nd edn. (2015)

Mining Exceptional Mediation Models

Florian Lemmerich$^{(\boxtimes)}$, Christoph Kiefer, Benedikt Langenberg,
Jeffry Cacho Aboukhalil, and Axel Mayer

RWTH Aachen University, Aachen, Germany
florian.lemmerich@gmail.com

Abstract. In statistics, mediation models aim to identify and explain the direct and indirect effects of an independent variable on a dependent variable. In heterogeneous data, the observed effects might vary for parts of the data. In this paper, we develop an approach for identifying interpretable data subgroups that induce exceptionally different effects in a mediation model. For that purpose, we introduce mediation models as a novel model class for the exceptional model mining framework, introduce suitable interestingness measures for several subtasks, and demonstrate the benefits of our approach on synthetic and empirical datasets.

1 Introduction

Mediation analysis is a classical statistical technique that aims for explaining the relationship between an independent variable and an (dependent) outcome variable by integrating a third, so-called mediator variable. The model suggests that the independent variable not only influences the outcome directly, but it also has an effect on the mediator variable, which in turn influences also the outcome. By doing so, it allows for distinguishing between a direct effect of the independent variable on the outcome, and an indirect effect via the mediator. Mediation analysis is a key technique predominantly used in (but not limited to) social and behavioral research, and psychology. However, in its original form, the mediation model assumes constant direct and indirect effects, i.e., it does not take heterogeneity in the data into account.

In this paper, we propose a novel approach that enables exploratory analysis of subgroups with exceptional mediation effects by integrating mediation models as a novel model class into the *exceptional model mining* framework [12]. Extending the well known subgroup discovery task [8,11], exceptional model mining aims for identifying describable subgroups in the data with significantly different model parameters compared to the other instances in a pre-specified model of a specific class. We introduce mediation models as a novel model class for exceptional model mining, discuss multiple options, how exceptional model mining with mediation models can lead to valuable insights, and present respective interestingness measures that can identify the most interesting subgroups in the data. By doing so, our research allows for the detection of interpretable subgroups, in which the direct, indirect, or total effect in the mediation analysis is specifically strong or weak. In our experimental evaluation, we show that

D. Helic et al. (Eds.): ISMIS 2020, LNAI 12117, pp. 318–328, 2020.
https://doi.org/10.1007/978-3-030-59491-6_30

our approach can successfully recover subgroups with exceptional effects in synthetic data, and demonstrate the benefits of the approach in an example on social survey data.

2 Background and Related Work

Next, we introduce the necessary background on exceptional model mining and mediation analysis, and discuss existing research most closely related to ours.

2.1 Exceptional Model Mining

In this paper, we assume a dataset D to be a multiset of data instances $i \in I$ described by a set of attributes A. In this set, we distinguish between describing attributes $A_D \subset A$ and model attributes $A_M \subset A$. In a given dataset, a *subgroup* is defined by a *subgroup description* $p : D \rightarrow \{true, false\}$ that selects instances from the dataset in a way that can be understood by humans. As it is common practice, we use in this paper conjunctions of selection conditions on single describing attributes, i.e., attribute-value pairs in the case of a nominal attribute, or intervals in the case of numeric attributes (such as $Age < 18 \wedge Gender = Male$). However, this is not a necessary condition. We call the set of all instances $c(p) = \{i \in I | p(i) = true\}$ that are described by a subgroup description p the *subgroup cover* and the set of all instances not covered by the subgroup description as its *subgroup complement*.

In exceptional model mining [12], we are now interested in finding interesting subgroups, i.e., subgroups for which the parameters of the model, which is picked a priori by the analyst, differ substantially depending on the model being fitted on the subgroup cover or subgroup complement instances. For example, consider we could choose in a population survey the simple correlation model class and the two model attributes religiosity and happiness as a model. A (fictitious) finding of exceptional model mining could then be: *"While overall there is a positive correlation between religiosity and happiness ($\rho = 0.1$), the subgroup of women younger than 18 years shows a negative correlation ($\rho = -0.15$)"*. To extract interesting subgroups from the (exponentially) large set of describable candidates, an interestingness measure (quality) q is employed that can assign a score (as a real number) to each candidate based on the statistical properties of the subgroup and the model fitted on its instances. Then, a search algorithm can identify the candidate subgroups with the highest scores. Unfortunately, solving tasks based on new model classes into exceptional model mining requires the development of tailored interestingness measures. As a key contribution, this paper accomplishes this for several use cases related to the mediation model.

2.2 Mediation Model

Mediation analysis is of interest whenever researchers assume that the effect of a putative cause X on an outcome Y of interest might be transmitted through

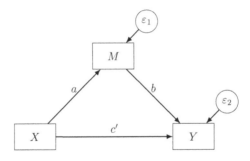

Fig. 1. Basic mediation model

an intermediate variable M [15]. A basic linear mediation analysis, cf. Baron and Kenny [1], involves three steps: First, regressing the outcome variable Y on the cause variable X to compute the total effect c of X. Second, regressing the mediator variable M to the cause variable X to obtain the first part of the indirect effect, a, i.e., the effect of X on M. Third, regressing the outcome variable Y on the cause variable X (i.e., c', the direct effect) and the mediator variable M (i.e., b, the second part of the indirect effect). Note that in such linear models, the total effect from the first step is given by $c = c' + a \cdot b$, that is, it is decomposed into the direct effect c' and the indirect effect $a \cdot b$ (from the second and third step), see also Sect. 3. Using regression notation for mean centered variables, the basic mediation model is given by the following two equations:

$$M = a \cdot X + \varepsilon_1 \tag{1}$$
$$Y = b \cdot M + c' \cdot X + \varepsilon_2 \tag{2}$$

The two regression equations of a mediation model can be estimated simultaneously in a structural equation modeling (SEM) framework. Based on the parameters of the model, different interpretable effects can be computed as described above. These computations hold for linear models. For extensions to non-linear models and causal assumptions needed in these models, see existing work by Imai et al. [9], Pearl [16], and Mayer et al. [15].

   Figure 1 shows a path diagram, which is a graphical representation of the basic mediation model. Structural equation models [2] provide a lot of flexibility to examine structural relations among a variety of variables. For example, covariates and additional mediators can be added. Or measurement models can be added for latent variables. Or the model can be estimated as a multi-group model, where the effects of a categorical variable (e.g., gender, ethnicity) can be investigated by using a distinct model for both males and females allowing for different magnitudes and directions of effects. The latter option will be important for our proposed approach to find interesting subgroups. It is also possible to set equality constraints on specific parameters among groups if needed.

2.3   Related Approaches

Exceptional model mining has been applied in a wide range of applications and for diverse model classes including classification [12], regression [4], and sequential data [14] models. In this context, Duivestijn et al. investigate Bayesian networks for exceptional model mining [5]. While these could in principle also be used to model the dependency structure of mediation, the authors focus on the difference of dependency structures derived from subgroup instances. By contrast, in this paper we study a setting, in which a specific dependency structure (a mediation) is the explicit interest of the research task.

In research on structural equation modeling, van Kesteren and Oberski [10] propose a method for detecting potential mediators from a larger search space in an exploratory way. In the presented research, we consider the mediator as fixed, and identify circumstances with extraordinary effects associated with it. Brandmaier et al. introduce structural equation model trees [3], combining the decision tree paradigm with structural equation models (which also include mediation models). These partition the dataset with respect to different parameterizations of a SEM. While individual paths in those could also be interpreted as rule-like patterns, it differs from the research presented here by the typical, well-explored differences between rule- based and decision tree- based data mining, including a different focus of local patterns vs global models, potential overlap and redundancy in the results, and different modularity of results.

While the basic mediation model assumes constant effects, there exists some extensions for investigating differential effects. In so-called *moderated mediation* analysis [17], the total, direct, and indirect effects may differ depending on categorical variables (e.g., in multigroup models), or depending on values of continuous variables, that is, the effects-of-interest can be modeled as decreasing/increasing depending on the moderator's value.

3   Mining Exceptional Mediation Models

Next, we describe how we can integrate mediation models into the exceptional model mining framework in order to find subgroups with exceptional effects.

3.1   GeneralaApproach

We assume that a mediation model with independent, dependent, and mediation variables has been specified and selected by domain experts for a more fine-grained analysis. We then identify subgroups with exceptional effects with exceptional model mining. Overall, we employ a branch-and-bound search over the space of all potential candidate subgroups. For that purpose, we start by forming selection expressions over the describing attributes in our data set and then considering all possible conjunctions over these expressions as candidates. Then, any exhaustive (e.g., search depth-first-search or apriori) or heuristic (e.g., beam search) search algorithm for subgroup discovery can be employed to enumerate and individually evaluate the candidates. The evaluation of a candidate assigns a score from a user chosen interestingness measure. These interestingness measures are task dependent and will be discussed below. The top-K subgroups with the highest scores are finally returned as the results set.

3.2   Effects of Interest

For many other model classes that are used with exceptional model mining, parameters are symmetrical to each other. For example, in a Markov chain transition model, all parameters specify a probability of a transition from one specific state to another specific state. As a result, interestingness measures in these scenarios are treated all equally, and the exceptionality and thus interestingness of subgroups can be quantified by computing a distance measure (e.g., a Manhattan distance) with equal importance of all parameters. This is not the case for mediation models: Here, all parameters have distinct meanings and are used to determine different types of effects that are interpretable by domain specialists. We can employ exceptional model mining to unveil subgroups, in which a specific effect type (eff) is specifically weak or strong. These include:

- The *direct effect* quantifies, by how much the dependent variable changes per unit increase of the independent variable assuming a fixed value of the mediator. It is directly given by the value of the parameter $eff_{dir} = c'$.
- In linear models, the *indirect effect* is computed as the product of the parameters $eff_{indir} = a \cdot b$. It measures how much the dependent variable changes when the mediator increases as much as it would change for a unit increase of the independent variable.
- The *total effect* describes the unconditional dependency of the dependent on the independent variable. In linear models, it is calculated as the sum of the direct and the indirect effect: $eff_{total} = a \cdot b + c'$.

3.3   Interestingness Measures

To find subgroups that are exceptional with respect to a specific effect type, we propose two groups of interestingness measures. In both cases, we determine parameters that maximize the joint likelihood of the model in the subgroup instance and in its complement. Based on that, we compute the respective effect of interest (direct, indirect, or total effect) in the subgroup and in the complement, denoted by eff^{sg} and eff^{compl}.

For the first group of interestingness measures, which allows for a statistically guided selection of subgroups, we then conduct a Wald Test with the null hypothesis that the effect is the same in the subgroup and its complement. We then use the test statistic of this test as the score for subgroup selection, i.e., we aim to find those subgroups, in which the effect is most significantly different between the subgroup and its complement. Note that the resulting p-value of the test could in principle also be used as an alternative, but can quickly lead to arithmetic underflow in large datasets. The Wald Test directly allows to assert the statistically significance, but the resulting p-values have to be adjusted for multiple comparison, e.g., using Bonferroni corrections.

The second group of interestingness measures is designed for a more exploratory setting. In analogy to popular measures for subgroup discovery and exceptional model mining, cf. [11,12], we suggest to compute the score as $n^a \cdot (eff^{sg} - eff^{compl})$, where n is the number of instances covered by a subgroup and a an user chosen parameter. This parameter enables the data analyst to trade-off preferences for larger subgroups (in terms of number of instances) and stronger deviations in the effects. Due to this option, this class of interestingness measures is specifically suited for iterative and interactive mining. In its proposed form, the measures identify subgroup with a specifically strong positive effect, extension for weak or negative effects are straightforward.

3.4 Redundancy Reduction

A well-known issue in exceptional model mining is redundancy in the result set, i.e., a high similarity in the top-k found subgroups. For example, if the subgroup A induces an exceptionally strong direct effect and thus achieves a high score, then also the subgroup $A \wedge B$ with any random noise selector B can be expected to also imply an unusual model and receive a high score. To limit these effects, we suggest to use the discussed quality measures with a generalization-aware modification [7]. That is, instead of using the interestingness measures discussed above directly, we compute the score difference between a subgroup and the maximum score in any of its direct generalizations and use it as a final score. For example, assuming the subgroup $A \wedge B$ has a base score of 100, and the subgroups A and B have score 40 and 70 respectively, then the final (adapted) score of $A \wedge B$ is $100 - max(40, 70) = 30$. Efficient search for this type of interestingness measure can (preferably) be achieved by using a level-wise search strategy such as apriori, or by extensive caching of subgroup scores.

4   Experimental Evaluation

We demonstrate the benefits of the proposed approach in a study with synthetically generated data, and a case study on real world data.

4.1   Setup and Implementation

For our evaluation, we implemented the mediation model class as an extension of the exceptional model mining Python library *pysubgroup* [13]. For fitting the mediation models and calculating Wald Test score we employed the well-known *lavaan* package with the *nlminb* [18] solver in R, which we connected to pysubgroup with the *rpy2* R-Python bridge. As a result of connecting these two specialized highly efficient implementations, all discussed experiments could be finished within seconds on a standard desktop machine.[1]

---

[1] Code is available at http://florian.lemmerich.net/mediation-emm.

Table 1. Results for the synthetic data with an interestingness measure based on the Wald Test on the direct effect. The p-values p_{WT} are Bonferroni corrected.

Rank	Score	p_{WT}	Description	# Instances	eff_{dir}	eff_{ind}
1	218.32	$< 10^{-14}$	A=A1 AND B=B1	1000	0.049	0.754
2	72.49	$< 10^{-14}$	A=A1	2000	0.558	0.287
3	72.49	$< 10^{-14}$	A=A2	2000	0.859	0.117
4	66.68	$3.30 \cdot 10^{-13}$	B=B1	2000	0.555	0.331
5	66.68	$3.30 \cdot 10^{-13}$	B=B2	2000	0.844	0.091
6	7.58	> 0.5	N17=1	227	0.536	0.248
7	7.58	> 0.5	N17=0	3773	0.733	0.186
8	7.25	> 0.5	N11=1 AND N15=0	20	0.171	0.147

4.2 Experiments on Synthetic Data

To evaluate our interestingness measures, we test if they are able to recover relevant subgroups among noise. For that purpose, we created synthetic data as follows: First, we defined two mediation models with manually chosen parameters. While we tried different settings with equivalent results, we report here on results for one model with a strong mediator effect $((a, b, c') = (0.9, 0.8, 0.1))$, and one with a strong direct effect $((a, b, c') = (0.3, 0.4, 0.8))$. The residual variances were fixed at one. Then, we create the model attributes of 1000 instances by the first model, and of 3000 instances by the second model. Next, we generate two binary descriptive variables A and B such that the instances from the first model are assigned $(A = A1, B = B1)$, and the instances from the second model in equal parts $(A = A1, B = B2)$, $(A = A2, B = B1)$, and $(A = A2, B = B2)$. Additionally, we generate 20 binary variables with random noise with a random probability $p \in [0, 1]$. The exceptional model mining task then should recover the subgroup $A = A1 \wedge B = B1$ as the one with significantly different effects, while subgroups with noise-based descriptions should not appear in the top subgroups.

Table 1 shows the results, i.e., the top-8 subgroups, of our approach for a representative task with a maximum search depth of two and using the Wald Test for the direct effect as interestingness measure. We can observe that we successfully recovered the correct subgroup $A = A1 \wedge B = B1$ as the one with the most exceptional effect by a wide margin w.r.t to its score. The following subgroups are the generalizations $B = A1$, $B = B2$, $A = A1$, and $A = A2$ of this main pattern, which –by construction– also exhibit an unusual direct effect compared to the subgroup complement. Only starting from position 6, subgroups described by noise attributes follow. Note that the top 5 subgroups can be shown to be highly significant by the Wald test, but all noise-based subgroups are not statistically significant according to this test. Results for other Wald Test-based interestingness measure equally recover the correct subgroup as the most interesting one. We tested these results for several generated datasets with different effect parameterizations and could identify the right subgroups

Table 2. Results for the European social survey data with a wald test-based interestingness measures.

Rank	Score	p_{WT}	Subgroup description	# Instances	eff_{dir}	eff_{ind}
1	23.538	< .001	hincfel = Living comfortably on present income	12427	0.070	0.043
2	21.903	< .001	edulvla = Tertiary education completed (ISCED 5–6)	12233	0.080	0.041
3	20.986	< .001	edulvla = Less than lower secondary education (ISCED 0–1)	6127	−0.121	0.112
4	12.083	< .001	maritala = Married	23933	0.072	0.047
5	11.899	< .001	$28 \leq$ agea < 41	9831	0.072	0.040

consistently. Thus, in summary, we show that our approach is in principle capable to recover a subgroup with a different model compared to the remaining data reliably.

4.3   Example: European Social Survey Data

Next, we illustrate the benefits of our approach in an empirical example inspired by a psychological case study [19]. In that work, the authors investigated – roughly speaking– differences between religious and non-religious individuals with respect to social recognition and potential effects on their subjective well-being. For their analysis, the authors analyzed a large-scale dataset from the European Social Survey (third round) [6] and utilized a mediation model between the variables *religiosity* (independent variable), *happiness* (dependent variable), and *social recognition* (mediator variable). Here, we show how the contributions of our paper can be used to study this model with the same data on a more fine-grained level. This demonstration, however, should not be regarded as a full scale empirical case study since some simplifying assumptions have been made, e.g., additional covariates, mediators, and weights on the survey answers have been ignored.

For data preprocessing, we performed the following steps: From the full survey data ($N = 47,099$), we first selected relevant attributes. Then, we performed a two-dimensional confirmatory factor analysis with the latent variables *religiosity* and *social recognition*. Religiosity was measured by two indicators self-identification (*rlgdgr*) and attendance of religious services (*rlgatnd*), and social recognition was measured by three indicators perceived respect, perceived unfair treatment and perceived recognition. If necessary, indicators were reverse-coded for consistency. Based on the confirmatory factor analysis, factor scores were computed and used in the further analysis.

Table 3. Results for the European Social Survey Data with an exploratory interestingness measure $q = $ sg size$^{0.1} \cdot |eff_{ind_{sg}} - eff_{ind_{\overline{sg}}}|$ and $sg_{size} > 20$, i.e., $a = 0.1$, minimum 20 instances.

Rank	score	p_{WT}	Subgroup description	# Inst.	eff_{dir}	eff_{ind}
1	1.201	0.030	rlgdnm = Islamic ∧ age ≥ 65.0	44	−0.389	0.955
2	1.099	0.037	maritala = Never married ∧ edulvla = Other	20	0.092	0.962
3	0.647	0.141	rlgdnm = Islamic ∧ maritala = Widowed	32	0.316	0.593
4	0.636	0.011	trstlgl = 1 ∧ uempli = Marked	60	−0.148	0.625
5	0.560	0.060	agea < 28 ∧ rlgdnm = Eastern rel.	26	−0.547	−0.539

Since the original study had a focus on the indirect effect, we now set out to find circumstances (subgroups), under which the indirect effect in the mediation model was significantly different compared to the complement. As the search space for candidate subgroup description, we used the variables age, marital status(*maritala*), unemployment status (*uempli, uempla*), feeling about household's income (*hincfel*), highest education level (*edulvla*), trust in the legal system (*trstlgl*), religion or denomination at present (*rlgdnm*), ever belonging to a religion or denomination (*rlgblge*) and years living in current country (*livecntr*).

The indirect effect for the model corresponding to the overall data was 0.065. We employed exceptional model mining analysis with maximum search depth of two, the Wald test for the indirect effect as interestingness measure, and the redundancy reduction described above. Results are shown in Table 2. Top subgroups include those living comfortably with their income (rank 1), highly educated people with tertiary education completed (rank 2), lower educated people with less than secondary education completed (rank 3), married people (rank 4), and people in the age range between 28 and 41 years (rank 5). For all these subgroups except lower educated people the indirect effect of religiosity on happiness via social reputation is smaller compared to the complementary group. In contrast, for those with low education (rank 3) this indirect effect is higher. However, while for all these subgroups the change in indirect effect is highly significant due to large subgroup sizes, the changes in effect sizes are only moderate.

To find smaller subgroups with stronger effects, we used in a second exceptional model mining run an exploratory interestingness measure with setting $a = 0.1$ and a subgroup size (number of instances) constraint of 20. As expected, this indeed leads to results, cf. Table 3, with much smaller subgroup sizes, but stronger changes in effects. For example, the top subgroup now describes elderly muslims, for which the size of the indirect effect is substantially increased. Note that in this case the results can not immediately be confirmed as significant, but can rather be seen as candidates for more detailed investigations in future studies.

In summary, with exceptional model mining on mediation models, we could find several interesting subgroups w.r.t. to the mediation model between religiosity, social reputation, and happiness in the analyzed survey data. We also found indications that the interestingness measures proposed in this paper lead to results with their desired properties.

5 Conclusions

In this paper, we integrated mediation analysis as a novel model class into the exceptional model mining framework. This enabled us to mine for subgroups in the data that induce exceptionally different model effects compared to the rest of the data set. We introduced statistically principled and exploratory interestingness measures for this task and showed its benefits in synthetic and empirical application examples.

In the future, we will investigate the applicability of exceptional model mining towards more complex structural equation models such as latent growth models or confirmatory factor analysis. Furthermore, we look for ways to improve the runtime of the suggested approach, e.g., by introducing pruning into the search.

Acknowledgement. Funded by the Excellence Initiative of the German federal and state governments.

References

1. Baron, R.M., Kenny, D.A.: The moderator-mediator variable distinction in social psychological research: conceptual, strategic and statistical considerations. J. Pers. Soc. Psychol. **51**(6), 1173–1182 (1986)
2. Bollen, K.A.: Structural Equation Modeling with Latent Variables. Wiley, New York (1989)
3. Brandmaier, A.M., von Oertzen, T., McArdle, J.J., Lindenberger, U.: Structural equation model trees. Psychol. Methods **18**(1), 71 (2013)
4. Duivesteijn, W., Feelders, A., Knobbe, A.: Different slopes for different folks: mining for exceptional regression models with cook's distance. In: International Conference on Knowledge Discovery and Data Mining (KDD), pp. 868–876 (2012)
5. Duivesteijn, W., Knobbe, A., Feelders, A., van Leeuwen, M.: Subgroup discovery meets Bayesian networks-an exceptional model mining approach. In: International Conference on Data Mining (ICDM), pp. 158–167 (2010)
6. ESS3: ESS Round 3: European Social Survey Round 3 Data (2006)
7. Grosskreutz, H., Boley, M., Krause-Traudes, M.: Subgroup discovery for election analysis: a case study in descriptive data mining. In: Pfahringer, B., Holmes, G., Hoffmann, A. (eds.) DS 2010. LNCS (LNAI), vol. 6332, pp. 57–71. Springer, Heidelberg (2010). https://doi.org/10.1007/978-3-642-16184-1_5
8. Herrera, F., Carmona, C.J., González, P., Del Jesus, M.J.: An overview on subgroup discovery: foundations and applications. Knowl. Inf. Syst. **29**(3), 495–525 (2011)

9. Imai, K., Keele, L., Tingley, D.: A general approach to causal mediation analysis. Psychol. Methods **15**, 309–334 (2010)
10. van Kesteren, E.J., Oberski, D.L.: Exploratory mediation analysis with many potential mediators. Structural Equation Modeling, 1–14 (2019)
11. Klösgen, W.: Explora: a multipattern and multistrategy discovery assistant. In: Advances in Knowledge Discovery and Data Mining, pp. 249–271. American Association for Artificial Intelligence (1996)
12. Leman, D., Feelders, A., Knobbe, A.: Exceptional model mining. In: Daelemans, W., Goethals, B., Morik, K. (eds.) ECML PKDD 2008. LNCS (LNAI), vol. 5212, pp. 1–16. Springer, Heidelberg (2008). https://doi.org/10.1007/978-3-540-87481-2_1
13. Lemmerich, F., Becker, M.: pysubgroup: easy-to-use subgroup discovery in python. In: Brefeld, U., et al. (eds.) ECML PKDD 2018. LNCS (LNAI), vol. 11053, pp. 658–662. Springer, Cham (2019). https://doi.org/10.1007/978-3-030-10997-4_46
14. Lemmerich, F., Becker, M., Singer, P., Helic, D., Hotho, A., Strohmaier, M.: Mining subgroups with exceptional transition behavior. In: International Conference on Knowledge Discovery and Data Mining (KDD), pp. 965–974 (2016)
15. Mayer, A., Thoemmes, F., Rose, N., Steyer, R., West, S.G.: Theory and analysis of total, direct and indirect causal effects. Multivar. Behav. Res. **49**(5), 425–442 (2014)
16. Pearl, J.: Direct and indirect effects. In: Conference on Uncertainty in Artificial Intelligence, pp. 411–420. Morgan Kaufmann, San Francisco (2001)
17. Preacher, K.J., Rucker, D.D., Hayes, A.F.: Addressing moderated mediation hypotheses: theory, methods, and prescriptions. Multivar. Behav. Res. **42**, 185–227 (2007)
18. Rosseel, Y.: lavaan: an R package for structural equation modeling. J. Stat. Softw. **48**(2), 1–36 (2012)
19. Stavrova, O.: Fitting in and Getting Happy: How Conformity to Societal Norms Affects Subjective Well-Being, vol. 4. Campus Verlag (2014)

Machine Learning Applications

Multivariate Predictive Clustering Trees for Classification

Tomaž Stepišnik[1,2]([✉]) [iD] and Dragi Kocev[1,2,3] [iD]

[1] Jožef Stefan Institute, Ljubljana, Slovenia
{tomaz.stepisnik,dragi.kocev}@ijs.si
[2] Jožef Stefan International Postgraduate School, Ljubljana, Slovenia
[3] Bias Variance Labs, d.o.o., Ljubljana, Slovenia

Abstract. Decision trees are well established machine learning models that combined in ensembles produce state-of-the-art predictive performance. Predictive clustering trees are a generalization of standard classification and regression trees towards structured output prediction and semi-supervised learning. Most of the research attention is on univariate decision trees, whereas multivariate decision trees, in which multiple attributes can appear in a test, are less widely used. In this paper, we present a multivariate variant of predictive clustering trees, and experimentally evaluate it on 12 classification tasks. Our method shows good predictive performance and computational efficiency, and we illustrate its potential for performing feature ranking.

Keywords: Predictive clustering trees · Multivariate decision trees · Classification · Multi-label classification

1 Introduction

In predictive modeling, the task is to use a set of learning examples to construct a model that can be used to make accurate predictions for unseen examples. The examples are described with features and associated with a target variable. The model uses the features to predict the value of the target variable. Depending on the type of the target, we differentiate between different predictive modelling tasks. In this paper, we focus on classification, where the target (or targets) has a finite set of possible values. When there are only two possible target values, the task is called binary classification. If there are more than two possible values, it is called multi-class classification. Furthermore, in multi-label classification, a single example can have multiple labels, and in hierarchical multi-label classification, the labels are also organized in a hierarchy (e.g., biological taxonomy).

Decision trees [3] are a well known and widely used machine learning method. However, most of the research focus is on univariate trees, i.e., trees in which the tests for splitting the examples use a single feature. Multivariate decision trees, where the tests in the internal nodes can use more than one attribute, have received much less attention. More flexible splits can lead to smaller trees that

© Springer Nature Switzerland AG 2020
D. Helic et al. (Eds.): ISMIS 2020, LNAI 12117, pp. 331–341, 2020.
https://doi.org/10.1007/978-3-030-59491-6_31

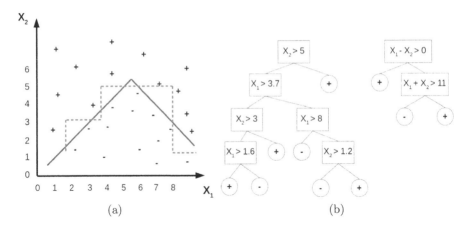

Fig. 1. A toy dataset (a) with drawn decision boundaries learned by a univariate (red, dashed) and multivariate (blue, solid) decision tree (b). (Color figure online)

provide better generalization, but broadening the space of possible splits also complicates the search for the optimal split. Figure 1 demonstrates the advantage of multivariate trees compared to univariate trees on a toy dataset. Some existing multivariate decision trees have been proposed [3, 7–9], however they are computationally inefficient [3, 8] or focused on a very specific task [9].

Predictive clustering trees (PCTs) [6] are a generalization of standard decision trees. When used in a standard classification or regression setting, they work the same as classification or regression trees [3]. However, they support different splitting heuristics and prototype functions in the leaves, and can be used for a wider variety of predictive modeling tasks. When used in ensembles, they offer state-of-the-art performance [6], and they have been successfully applied to a variety of real world problems [4, 10, 11].

In this paper, we present a multivariate variant of predictive clustering trees, that retains the flexibility of the original predictive clustering trees. As in univariate PCTs, splits in multivariate PCTs are optimized to minimize the impurity of the clustering (target) variables on each side of the hyperplane. We engineer a differential criterion function that allows for efficient optimization with gradient descent methods. We experimentally evaluate our method and show that it often outperforms standard PCTs, and is more computationally efficient on datasets with many features or labels.

We first describe our method, then theoretically analyze its time complexity and compare it to the top-down induction of univariate PCTs. We present an experimental evaluation of our method both in single tree and ensemble setting on different classification tasks (binary, multi-class and multi-label). We also perform parameter sensitivity analysis and show how the learned multivariate trees can be interpreted via feature importance scores.

2   Method Description

2.1   Tree Induction Overview

Tree based models, including predictive clustering trees, are most often built using a greedy top-down induction algorithm [3], outlined in Algorithm 1. Examples are recursively partitioned until no acceptable test is found. At that point a leaf is created, where the prototype (majority class or mean value/vector) of the remaining examples is stored for prediction. When searching for a test, one goes through all the possible splits for each individual attribute and finds the one that produces the partitions with the lowest impurity. The impurity can depend on the task, but PCTs by default use the sum of target variances to measure the impurity of a set of examples [6]. The sum can be weighted, if we want to prioritize reducing the impurity of certain targets (e.g., in hierarchical multi-label classification, we can prioritize labels higher in the hierarchy).

The prototype function also depends on the task. Typically, the mean values of the target variables are stored in the leaves. For classification, they are interpreted as the probability that an example in that leaf has the corresponding labels. To make a prediction, an example is passed through the tree according to the tests in internal nodes until it reaches a leaf. In binary and multi-class classification, the label with the highest probability is predicted (majority class in the leaf), whereas in (hierarchical) multi-label classification, all labels with probabilities greater than some user-defined threshold (usually 0.5) are predicted.

Individual trees can be useful, because they are easily interpretable models. However, to achieve state of the art predictive performance, they are most often used in an ensemble setting [1,2,6] (bagging ensembles and random forests).

2.2   Learning Multi-variate Splits

Our proposed method follows the greedy top-down induction method, but changes the split search procedure. Instead of only considering univariate splits, we allow any linear combination of features to be a split. This means that split boundaries are no longer only parallel to the coordinate axes, but can be any hyperplane in the feature space (as shown in Fig. 1).

Algorithm 1. Top-down decision tree induction: The inputs are matrices of features $X \in \mathcal{R}^{N \times D}$ and targets $Y \in \mathcal{R}^{N \times T}$.

```
1: procedure GROW_TREE(X, Y)
2:     test = find_test(X, Y)
3:     if acceptable(test) then
4:         X_1, Y_1, X_2, Y_2 = split(X, Y, test)
5:         left_subtree = grow_tree(X_1, Y_1)
6:         right_subtree = grow_tree(X_2, Y_2)
7:         return Node(test, left_subtree, right_subtree)
8:     else
9:         return Leaf(prototype(Y))
```

Let $X \in \mathcal{R}^{N \times D}$ be the matrix containing the D features of the N examples in the learning set. Let $Y \in \{0,1\}^{N \times L}$ be the binary matrix denoting which of the L labels are present for each of the N examples in the learning set. For example, in binary classification $L = 1$, whereas for multi-class and (hierarchical) multi-label classification L is the number of possible labels. In multi-class classification, each example only has one label, i.e., the sum of each row in Y equals 1, whereas in multi-label classification there can be multiple 1s in each row. Below, $M_{i.}$ will refer to the i-th row of the matrix M and $M_{.i}$ to the i-th column, where M will be either X or Y. We wish to learn a vector of weights $w \in \mathcal{R}^D$ and a bias term $b \in \mathcal{R}$ that define a hyperplane, which splits the learning examples into two subsets. We will refer to the subset where $X_{i.}w + b \geq 0$ as the positive subset, and the subset where $X_{i.}w + b < 0$ as the negative subset. We want the examples in the same subset to have similar labels.

First, we calculate the vector $s = \sigma(Xw + b) \in [0,1]^N$, where the sigmoid function σ is applied component-wise. The vector s contains values from the $[0,1]$ interval, and we treat it as a fuzzy membership indicator. Specifically, the value s_i tells us how much the i-th example belongs to the positive subset, whereas the value $1 - s_i$ tells us how much it belongs to the negative subset.

To measure the impurity of the positive subset, we calculate the weighted variance of each column (label) in Y, and we weigh each row (example) with its corresponding weight in s. To measure the impurity of the negative subset, we calculate the weighted variances with weights from $1 - s$. Weighted variance of a vector $v \in \mathcal{R}^n$ with weights $a \in \mathcal{R}^n$ is defined as

$$\mathtt{var}(v, a) = \frac{\sum_i^n a_i (v_i - \mathtt{mean}(v, a))^2}{A} = \mathtt{mean}(v^2, a) - \mathtt{mean}(v, a)^2,$$

where $A = \sum_i^n a_i$ is the sum of weights and $\mathtt{mean}(v, a) = \frac{1}{A} \sum_i^N a_i v_i$ is the weighted mean of v.

The final impurity is the weighted sum of weighted variances over all the labels. The impurity of the positive subset is $\mathtt{imp}(Y, p, s) = \sum_j^L p_j \mathtt{var}(Y_{.j}, s)$, and similarly $\mathtt{imp}(Y, p, 1-s)$ is the impurity of the negative subset. As mentioned above, weights $p \in \mathcal{R}^L$ enable us to give different priorities to different labels. The final objective function we want to minimize is

$$f(w, b) = S * \mathtt{imp}(Y, p, s) + (n - S) * \mathtt{imp}(Y, p, 1 - s),$$

where $s = \sigma(Xw + b)$ and $S = \sum_i^N s_i$. The terms S and $n - S$ represent the sizes of positive and negative subsets, and are added to incentivize balanced splits.

For examples that are not close to the hyperplane, s_i is close to 0 or 1. Weighted variances are therefore approximations of the variances of the subsets, that the hyperplane produces, and the optimization function is almost equivalent to the one used by the original PCTs. The only difference is that we use the weighted variances instead of directly calculating the subset variances. This makes the objective function differentiable and enables us to use the efficient Adam [5] gradient descent optimization method. For each split the w is initialized randomly, and then b is set so that initially the examples are split in half.

Like standard PCTs, multivariate PCTs can also be used in ensemble setting, by using different bootstrapped replicates of the learning sets for different ensemble members (bagging) and/or using random subsets of features for individual splits (random forests).

2.3   Time Complexity Analysis

The time complexity of splitting a node in standard PCTs is $\mathcal{O}(DN \log N) + \mathcal{O}(LDN)$, where N is the number of examples in the node, D the number of features and L the number of labels [6]. The first term is the result of sorting the examples by each feature, when searching for the optimal split by that feature. The second term comes from the evaluation of the splits.

In multivariate trees, we perform iterative optimization of the splits. The most costly parts of each iteration (epoch) are the matrix-vector products of X with w ($\mathcal{O}(ND)$) and Y with s ($\mathcal{O}(NL)$), required to calculate the objective function and its derivative. The time complexity is therefore $\mathcal{O}(EN(D + L))$, where E is the number of optimization epochs. This shows that our approach has an efficiency edge when D and L are large.

3   Evaluation and Discussion

3.1   Experimental Design

To evaluate our method, we compare it to standard PCTs in terms of predictive performance, model size and learning time. For benchmarking, we use 3 datasets for each of binary classification, multi-class classification, multi-label classification and hierarchical multi-label classification. All datasets are publicly available online[1,2] and their properties are presented in Table 1.

Table 1. Properties of the datasets used for the evaluation: N is the number of examples, D is the number of features and L is the number of labels.

dataset	N	D	L	task	dataset	N	D	L	task
banknote	1372	4	2	binary	birds	645	260	19	multi-label
OVA_breast	1545	10936	2	binary	scene	2407	294	6	multi-label
diabetes	768	8	2	binary	bibtex	7395	1836	159	multi-label
balance	625	4	3	multi-class	diatoms	1098	200	80	hierarchical
imagesegment	2310	19	7	multi-class	clef07d	11006	80	46	hierarchical
gasdrift	13910	128	6	multi-class	enron	1648	1001	56	hierarchical

For standard PCTs, we use their java implementation in the CLUS framework[3]. We implemented the multivariate PCTs in a python package *spyct* – it is available online[4].

---

[1] www.openml.org.
[2] http://kt.ijs.si/DragiKocev/PhD/resources/doku.php?id=phd_thesis_datasets.
[3] http://source.ijs.si/ktclus/clus-public/.
[4] https://gitlab.com/TStepi/spyct.

To measure the performance, we use F1 score for binary and multi-class classification, with macro averaging over the labels for the latter. For (hierarchical) multi-label classification, we use ranking loss [12], which computes the average number of label pairs that are incorrectly ordered (by their predicted probability). In hierarchical multi-label classification we use the default label weighting strategy from CLUS: label has a weight of 0.75^d, where d is the depth of that label in the hierarchy ($d = 0$ for root nodes). We estimate the performance with 10-fold cross validation. We evaluate both single trees and bagging ensembles of 50 trees. The splits in multivariate PCTs are optimized for a maximum of 100 epochs with learning rate 0.1. The optimization is stopped early, if the objective function does not decreases after an epoch. Before learning multivariate PCTs we also standardize each feature to 0 mean and standard deviation of 1. We include the standardization in the times reported below. All experiments were performed on the same computer with Intel i7-6700K processor. Ensembles used 4 cores to learn trees in parallel.

3.2 Results

In this section, we present and discuss the results from the evaluation. Model sizes and learning times are reported only for the ensembles, because they are more reliable than single tree results. Figure 2 presents the results for binary and multi-class datasets. The most notable difference in F1 score is on the balance dataset, where multivariate trees significantly outperform univariate trees. The same holds for bagging ensembles. On the other datasets, the differences in single tree performances are small, and for ensembles they are barely noticeable. In terms of model sizes, multivariate trees are mostly much smaller. This was expected, because multivariate splits are much more expressive than univariate. Multivariate trees are also faster to learn on datasets with many features (OVA_breast, gasdrift) datasets, and slower on datasets with few features. These results support the theoretical time complexity analysis from Sect. 2.3.

Figure 3 presents the results for multi-label and hierarchical datasets. For single trees, our method outperforms standard PCTs on most datasets, and the same is true for ensembles. Interesting cases are bibtex and enron datasets, our method performs poorly in single tree setting, but very well in ensemble setting. This indicates that overfitting, was the reason for poor performance of single trees. Tree sizes are consistently smaller over all datasets, and learning times are also strongly in favor of our method. The reason is that these datasets mostly have many possible labels, and learning time for univariate trees scales with $D \times L$, whereas for our method it only scales with $D + L$.

3.3 Parameter Sensitivity Analysis

We also performed parameter sensitivity analysis. We evaluated our method with different combinations of learning rate and maximum numbers of optimization epochs. The results of single trees on two datasets are presented in Fig. 4.

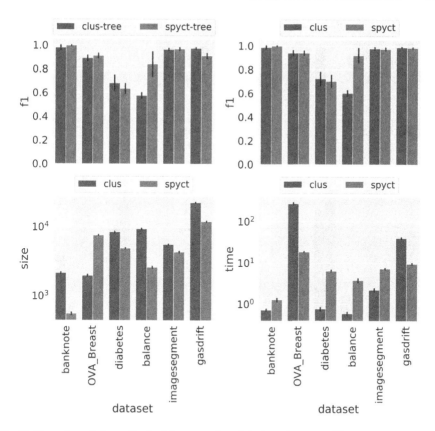

Fig. 2. Experimental results for binary and multi-class datasets. Top row presents the predictive performance (f1, higher is better), bottom row shows efficiency in terms of model size (number of nodes) and learning time (in seconds). Note that times and sizes are on logarithmic scale.

As might be expected, increasing the maximum number of epochs generally leads to better predictive performance, but also comes with greater learning times. On the other hand, there is no general rule for learning rates. While increasing the learning rate might make finding a good split faster (e.g., image-segment dataset), too large learning rate can also prevent the algorithm from finding a good solution (e.g., gasdrift dataset). With higher learning rates, early stopping comes into effect quickly (objective function stops improving), so increasing the maximum number of epochs has little effect on learning time.

Our selection of maximum 100 epochs at 0.1 learning rate is a sensible default, that proved to work well overall. However, it is unlikely to be optimal for any particular dataset. To achieve optimum predictive performance on a given dataset, the parameters (especially learning rate) should be fine-tuned for it specifically.

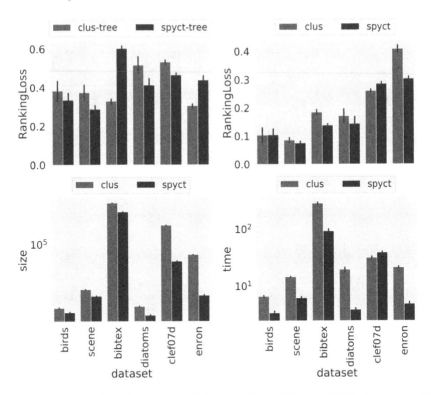

Fig. 3. Experimental results for multi-label and hierarchical multi-label datasets. Top row presents the predictive performance (ranking loss, lower is better), bottom row shows efficiency in terms of model size (number of nodes) and learning time (in seconds). Note that times and sizes are on logarithmic scale.

3.4   Interpretability

One of the advantages of decision trees is their interpretability. Multivariate decision trees are harder to interpret than univariate, especially for datasets with many attributes. However, large decision trees and ensembles thereof are also difficult to interpret. In those cases, feature importance scores are often calculated from the model and used to gain insights into its inner working. We illustrate the ability of our method to produce meaningful feature importance scores.

To obtain the feature importance vector from a tree, we calculate the weighted sum of vectors $|w|$ learned in the split nodes. Each vector w is normalized and weighted by the share of the dataset that was split in that node. This way, splits closer to the top of the tree, that influence more examples, have a bigger impact on feature importance. To get feature importance vector from an ensemble of trees, we simply calculate the average of feature importances of individual trees.

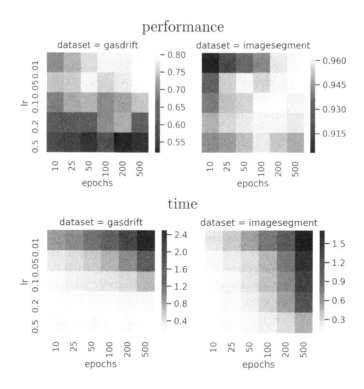

Fig. 4. Heatmaps showing the F1 scores (top row) and learning time (bottom row) depending on the parameter selection on the gasdrift and imagesegment datasets. Lighter color indicates better values.

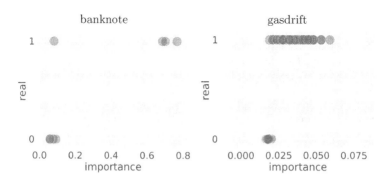

Fig. 5. Feature importances calculated on two datasets. The x-axis shows the importance score, and y-axis indicates whether the feature was real (1) or random (0).

To evaluate our feature importance scoring, we add random features (for each real feature we add one random feature) to a dataset, and compare their feature importance to the importance of real features. Results for two datasets are presented in Fig. 5. We can clearly see that the added random features have very low scores, meaning that the model does not rely on them. There are some real features that also have low scores, but that is to be expected. Datasets often include features that are not useful for predicting the target. This experiment confirms that feature importances obtained with our method are meaningful.

4   Conclusion

We present a multivariate variant of predictive clustering trees, that retains the flexibility of the original framework. We evaluate our proposed method on binary, multi-class, multi-label and hierarchical multi-label classification datasets from various domain. The results show that our method outperforms univariate PCTs and generally produces smaller models. Both theoretical and experimental analyses show that learning time is smaller on datasets with many features or labels. We also perform parameter sensitivity analysis, and demonstrate the ability of our method to produce meaningful feature ranking.

For future work, we plan to evaluate our method on other tasks supported by standard PCTs, such as (multi-target) regression and semi-supervised learning. We plan to evaluate our method in random forest setting and explore its potential advantages in exploiting sparse data using sparse matrix multiplication.

References

1. Breiman, L.: Bagging predictors. Mach. Learn. **24**(2), 123–140 (1996)
2. Breiman, L.: Random forests. Mach. Learn. **45**(1), 5–32 (2001)
3. Breiman, L., Friedman, J., Olshen, R., Stone, C.J.: Classification and Regression Trees. Chapman & Hall/CRC, San Francisco (1984)
4. Debeljak, M., Squire, G.R., Kocev, D., Hawes, C., Young, M.W., Džeroski, S.: Analysis of time series data on agroecosystem vegetation using predictive clustering trees. Ecol. Model. **222**(14), 2524–2529 (2011)
5. Kingma, D.P., Ba, J.: Adam: a method for stochastic optimization (2014)
6. Kocev, D., Vens, C., Struyf, J., Džeroski, S.: Tree ensembles for predicting structured outputs. Pattern Recogn. **46**(3), 817–833 (2013)
7. Menze, B.H., Kelm, B.M., Splitthoff, D.N., Koethe, U., Hamprecht, F.A.: On oblique random forests. In: Gunopulos, D., Hofmann, T., Malerba, D., Vazirgiannis, M. (eds.) ECML PKDD 2011. LNCS (LNAI), vol. 6912, pp. 453–469. Springer, Heidelberg (2011). https://doi.org/10.1007/978-3-642-23783-6_29
8. Murthy, S.K., Kasif, S., Salzberg, S.: A system for induction of oblique decision trees. J. Artif. Intell. Res. **2**, 1–32 (1994)
9. Prabhu, Y., Varma, M.: FastXML: a fast, accurate and stable tree-classifier for extreme multi-label learning. In: ACM-Association for Computing Machinery (2014)

10. Slavkov, I., Gjorgjioski, V., Struyf, J., Džeroski, S.: Finding explained groups of time-course gene expression profiles with predictive clustering trees. Mol. BioSyst. **6**(4), 729–740 (2010)
11. Struyf, J., Džeroski, S., Blockeel, H., Clare, A.: Hierarchical multi-classification with predictive clustering trees in functional genomics. In: Bento, C., Cardoso, A., Dias, G. (eds.) EPIA 2005. LNCS (LNAI), vol. 3808, pp. 272–283. Springer, Heidelberg (2005). https://doi.org/10.1007/11595014_27
12. Tsoumakas, G., Katakis, I., Vlahavas, I.: Mining multi-label data. In: Maimon, O., Rokach, L. (eds.) Data Mining and Knowledge Discovery Handbook, pp. 667–685. Springer, Boston (2010). https://doi.org/10.1007/978-0-387-09823-4_34

Comparison of Machine Learning Methods to Detect Anomalies in the Activity of Dairy Cows

Nicolas Wagner[1,2(✉)], Violaine Antoine[1], Jonas Koko[1],
Marie-Madeleine Mialon[2], Romain Lardy[2], and Isabelle Veissier[2]

[1] UCA, LIMOS, UMR 6158, CNRS, Clermont-Ferrand, France
nicolas.wagner@uca.fr
[2] UCA, INRAE, UMR Herbivores, 63122 Saint-Genès-Champanelle, France

Abstract. Farmers need to detect any anomaly in animals as soon as
possible for production efficiency (e.g. detection of estrus) and animal
welfare (e.g. detection of diseases). The number of animals per farm
is however increasing, making it difficult to detect anomalies. To help
solving this problem, we undertook a study on dairy cows, in which their
activity was captured by an indoor tracking system and considered as
time series. The state of cows (diseases, estrus, no problem) was manually
labelled by animal caretakers or by a sensor for ruminal pH (acidosis).
In the present study, we propose a new Fourier based method (FBAT)
to detect anomalies in time series. We compare FBAT with the best
machine learning methods for time series classification in the current
literature (BOSS, Hive-Cote, DTW, FCN and ResNet). It follows that
BOSS, FBAT and deep learning methods yield the best performance but
with different characteristics.

Keywords: Machine learning · Deep learning · Time series
classification · Detection of anomalies · Precision livestock farming

1 Introduction

Precision livestock farming is based on the use of smart technologies (mainly
sensors) to monitor closely the animals or their environment. The aim is to optimize the production and reduce farmers work load. The increase in computers
storage capacity and in the precision of sensors makes possible to record a high
quantity of data which requires automatic processing to be used by farmers.
Machine learning tools is beginning to be employed to extract relevant information from these massive data. For example, machine learning has been used to
determine grass growth from satellite and weather data [10] or to predict the
quantity of manure to be spread on pastures or crops as fertilizer [11].

Farmers need to detect any anomaly in animals as soon as possible both for
milk production efficiency and animal welfare. Such a detection seems possible

D. Helic et al. (Eds.): ISMIS 2020, LNAI 12117, pp. 342–351, 2020.
https://doi.org/10.1007/978-3-030-59491-6_32

through the analysis of the animals' activities. For instance, [15] found that dairy cows' activity varies according to a circadian cycle which significantly changes if the cow is about to be sick or in estrus.

Furthermore, time series classification (TSC) or anomaly detection are among the most challenging problem in machine learning [6,12,18] and are present in many fields of science like sensor-based human activity recognition [16], credit card fraud detection [1], electroencephalogram and electrocardiogram analysis [3], geo-distributed networks [5], etc. TSC differs from classical machine learning problems since it deals with data listed in time order. Some algorithms were developed for TSC like Dynamic Time Warping (DTW) [4], Bag of SFA Symbols (BOSS) [14] or Hive-Cote [8]. Most recently, deep learning neural networks as FCN and ResNet were also used and found to outperform the other algorithms [6, 17].

In this paper, we employed algorithms of time series classification (BOSS, DTW, Hive-Cote FCN and ResNet) considered as the best ones. In addition, because the activity of cows follows a circadian cycle, we proposed and tested a new method based on Fourier transformations (Fourier Based Approximation with Thresholding or FBAT).

The first section describes the most popular TSC classifier. The Sect. 2 details our new FBAT method. Section 3 first describes the data set, i.e. time series of activities of dairy cows, then explains the experimental protocol and finally presents the results. Perspectives of the work are given in a conclusion.

2 Time Series Classifier

This section presents the current best algorithms for TSC [2,6,17] used as baseline to compare the FBAT method.

2.1 Dynamic Time Warping

DTW [4] is a method that measures the similarity between two time series. It is often used as a distance with the one Nearest Neighbor algorithm (1-NN). Although combining 1-NN and DTW gives good results in practice, however, DTW is not a distance function. Indeed, it does not respect all mathematical properties of a distance especially the triangle inequality [13].

The difference between DTW and standard distance measures is the following: standard distances assume that the ith of a series is aligned with the ith point of an other series while DTW is designed to minimize the effects of shifting and distortion in time series. The DTW method have many advantages. It is easy to employ, it can be used with many algorithms like k-NN and when combined with 1-NN it is one of the best algorithms for time series classification [2]. This makes it an interesting baseline. Its quadratic time complexity is a disadvantage but it remains faster than other algorithms like Hive-Cote.

2.2 Hive-Cote

Hive-Cote [8] is an improved version of the algorithm Cote (or Flat-Cote). Flat-Cote consists in using 35 classifiers of time series classification. Each classifier produces a result and the final decision is based on a vote of all classifiers. The vote is weighted by the training accuracy of each classifier. One problem with Flat-Cote is the flat architecture. It means that all classifiers vote independently. However, some algorithms pertain to the same category and probably give similar results. To solve this problem, Hive-Cote gathers the algorithms into groups called modules. Each module computes the probability for each class to be the solution. This probability is computed with the weighted results of the algorithms for each module. Then, the final solution is the class that has the highest probability across all module's outputs.

Hive-cote is composed of five modules: elastic ensemble, shapelet transform ensemble, BOSS, time series forest and random interval features.

2.3 Fully Convolutional Networks

Fully Convolutional Networks (FCNs) [9] are similar to Convolutional Neural Networks (CNNs) excepted that they contain local pooling layer so as to keep the same dimensionality input through the convolutional layers. In addition, a standard CNN generally ends by a Fully Connected (FC) layer that is replaced by a Global Average Pooling (GAP) in the FCNs. The architecture used in this paper was proposed by [17]. It consists of three parts that are composed of a convolution layer, a Batch Normalization (BN) layer and a ReLu activation layer. These three parts are followed by a GAP layer and a classical softmax layer. The three convolution blocks contain 128, 256 and 128 filters with a filter length of respectively 8, 5 and 3. The stride is set to one with a zero padding that enables to preserve the same length of time series across the network. This architecture has the advantage to remain stable according to the length of the time series (excepted for the last softmax layer). This allows us to use exactly the same network used in [6,17]. FCN is the best deep learning algorithm on the 44 data sets analyzed in [17].

2.4 ResNet

A Residual Network [7] is close to CNN. The difference lies in shortcuts added from the input of convolution blocks to their output. These shortcuts inject the information that may be lost by the convolutional block. The ResNet proposed by [6,17] is composed of three convolutional blocks. All blocks are composed of three convolutional layers with respectively a filter's length set to 8, 5 and 3. Each convolutional layer is followed by a BN and Relu layer. The convolutional layers of the first block are composed of 64 filters and the convolutional layers of the second and last block are composed of 128 filters. The three blocks are followed by a global pooling and a softmax layer. This architecture has the same advantage as FCN: it does not vary with the length of the time series. ResNet is the best deep learning algorithm on the 85 data sets tested in [6].

2.5 Bag of SFA Symbols

BOSS [14] is a method that combines the advantage of the Fourier transform and the bag of words model. It allows to reduce noise and to handle variable lengths.

First, a sliding window of size w with a step of 1 is applied over each time series. The obtained time windows are converted into sequences (or words) of symbols of length l with an alphabet size of c using the Symbolic Fourier Approximation (SFA) algorithm. A time series is then represented by the sequences of each window. Finally, a histogram is built using sequences as modal class. The last step consists in using 1-NN algorithm with the BOSS distance function. Given two histograms B_1 and B_2, the formula of the BOSS distance function is:

$$dist(B_1, B_2) = \sum_{a \in B_1 ; B_1(a) > 0} [B_1(a) - B_2(a)]^2, \qquad (1)$$

where a is a word and $B_i(a)$ the number of occurrences of a in the ith histogram.

3 Fourier Based Approximation with Thresholding

We propose a Fourier Based Approximation with Thresholding (FBAT) method to classify time series by measuring the variations of the cyclic components. It is made to classify time series as normal or abnormal by assuming that an abnormal series includes a break on the cycle. Thus, if the variations of the cyclic components are high, the algorithm classifies the time series S_i as abnormal. The algorithm starts by extracting two sub-series A and B of size p and delayed of q from the input series with $p < |S_i|$ and $p + q < |S_i|$. A Fourier transform is applied on both sub-series to extract their harmonic decomposition. With these harmonics, a new model $m(t)$ is computed for each sub-series following the formula:

$$m(t) = \sum_{f=-z}^{z} |h_f| cos(2\pi f \frac{t}{p} + arg(h_f)), \ z = 0...\lceil \frac{p-1}{2} \rceil, \qquad (2)$$

where h_f is the harmonic corresponding to the frequency f and z is the number of harmonics to keep in the model. Note that h_f is a complex number with $|h_f|$ its modulus and $arg(h_f)$ its argument. Moreover, the two sub-series A and B are delayed by q. As a consequence, it is necessary to synchronize the models of A and B by applying a temporal shift to the model of B. This shift is performed by adding a delay $-\frac{q}{p}2\pi$ in the formula of the model of B.

A L2-norm distance d_{L2} is then computed between the two models. This distance reflects the variation of the cyclic component of the input time series. A high distance means a high variation and vice versa.

To classify the input time series as normal or abnormal, the algorithm needs to compute a threshold τ for the distance. If $d_{L2} > \tau$, the time series is classified as abnormal. To compute this threshold, all distances d_{L2} are computed for each time series that belongs to the training set. Then, s samples are computed between the minimum and the maximum obtained distances. The accuracy of the training set is computed for each sample and the sample that yields the best accuracy is chosen to be the threshold.

4   Experiments and Results

4.1   Data Set

The data were collected on 28 Holstein cows during a two month experimentation in which a subacute ruminal acidosis, a metabolic disease common in ruminants, was induced.

Data Construction. The raw data consist of the record of the location of each cow every second with an indoor tracking system (CowView, GEA Farm Technologies, Bönen, Germany). Three activities were identified: *eating* if the cow was located next to the trough, *resting* if it was in a cubicle (resting place) and *in alleys* if the cow was in an alley. These activities were aggregated in a new variable called level of activity. The procedure is described in [15]. We thus obtained for our study time series consisting in the evolution over time of the level of activity of each cow estimated per hour. All anomalies, such as acidosis, oestrus, etc. were noted. The acidosis was detected by a sensor that measured the pH in the rumen.

A set of 28 time series, corresponding to the 28 cows for two months was available. To build the data set, a sliding window of 36 h was applied on each cow to extract sub-series (see justification in next section). The obtained data set was divided into two parts: one for the training and one for test. Half of series labelled as *abnormal* were used for the training set and the other half for the test set, except for the series related to acidosis that were all placed in the test set. Indeed, this specific anomaly was induced by experimenters. The idea is to avoid the perturbation of a classifier during the learning phase with an unnatural anomaly. Finally, to balance the training data set, a reduction of the *normal* series was performed by randomly selecting few ones. The same number of normal series were randomly chosen for the test data set. The training data set is composed of 1088 *normal* series and 972 *abnormal* series. The test set is composed of 1408 *normal* series and 4212 *abnormal* series (including 3180 series with acidosis).

Data Properties. The first property of this data set is that each cow has its own natural daily rhythm based on a circadian cycle with a low activity during the night and a higher activity during the day [15]. This change of activity between nights and days can be modified if the cow is sick or under stress. We decided to work with series of 36 h in order to observe a cycle of more than one day and to be able to detect anomalies with precision (e.g. a normal cycle of 24 h followed by 12 h abnormal).

Figure 1 illustrates the activity of two cows during two normal days. Both cows are more active during the day than at night, in line with the fact that the rhythm is circadian. This figure also illustrates that the rhythm of normal activity can differ between cows and days. Figure 2 illustrates the level of activity of a cow under lame during three days. This shows how the activity of a cow can be modified by an anomaly. Given that there exists normal variations between days (as illustrated by the Cow b, Fig. 1), the difficulty lies in discriminating between changes due to an anomaly and those due to spontaneous variations.

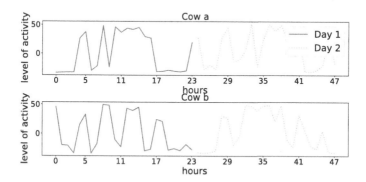

Fig. 1. 48 h of normal activity for two different cows.

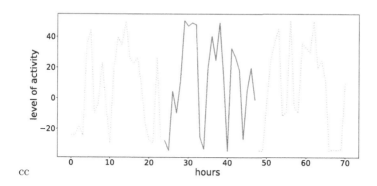

Fig. 2. Level of activity of a cow during three consecutive days: the solid line represents the day detected as lame.

4.2   Experiments

We compare the algorithms described in Sect. 2 with the FBAT method presented Sect. 3. The code of the five methods comes from the github repository of the original authors [2,6]. The code of the FBAT method is available on the github repository https://github.com/nicolas-wagner/FBAT.

For BOSS we use the same parameters as in [2]: the word length $l = [8, 10, 12, 14, 16]$, the alphabet size $c = 4$ and the window size, $w = [10, 12, ..., 36]$.

For FBAT we set the time window p to 24, the delay q to 12 and the number of samples s to 10000. We test all possibilities of the number of harmonics z, i.e. from 0 to 12 harmonics.

As in [6], the deep learning methods were run 10 times to train them with 10 different initializations of parameters. The results presented in this paper are the average over these 10 runs.

We use the same train and test data set for all methods tested. The train data set is composed of 1088 time series labelled as *normal* (negative) and 972 time series labeled as *abnormal* (positive). The test set is composed of 1408 *normal* time series and 4212 *abnormal*.

We define normal label as the negative class and abnormal label as the positive class. For each classifier it is possible to count the number of True Positive (TP), True Negative (TN), False Positive (FP) and False Negative (FN). We then calculated the overall accuracy as well as the precision and the recall for positive and the negative classes as follows:

$$accuracy = \frac{TP + TN}{TP + TN + FP + FN}, \tag{3}$$

$$precision_- = \frac{TN}{TN + FN} \quad and \quad recall_- = \frac{TN}{TN + FP}, \tag{4}$$

$$precision_+ = \frac{TP}{TP + FP} \quad and \quad recall_+ = \frac{TP}{TP + FN}. \tag{5}$$

The accuracy is used as a single value to measure and compare the performance of the methods. The precision and the recall measured for each class help to understand the behavior of each classifier in detail. A high $recall_-$ (resp. $precision_-$) means that the majority of time series labelled as normal (resp. classified as normal) are classified as normal (resp. labelled as normal) and inversely for a high $recall_+$ (resp. $precision_+$).

The CPU time (in hour) is also retrieved from the experiments (training + test time). For the deep learning methods, a GPU mode is available so, FCN and ResNet were run on CPU and GPU. The first machine were composed of CPUs Intel Xeon 2.4GHz with 80 cores and 1 TB of RAM. The second were composed of CPUs Intel Xeon 2.4GHz with 10 cores and 62.5 GB of RAM with a GPU NVIDIA Quadro P5000 (16 GB of GDDR5 memory and 2560 cores). All algorithms were run in a sequential mode using only one core.

4.3   Results

The performances of all methods are summarized Table 1.

All of these methods are intended to be used by livestock farmers with a personal computer. Consequently, the CPU time is an important characteristic

Table 1. Results of all classifiers

		DTW	Hive-Cote	BOSS	FBAT	FCN	ResNet
Accuracy		0.54	0.63	**0.72**	0.60	0.66	0.67
Precision$_-$		0.31	0.38	**0.43**	0.38	0.40	0.40
Recall$_-$		0.68	0.73	0.36	**0.90**	0.73	0.71
Precision$_+$		0.82	0.87	0.80	**0.94**	0.88	0.87
Recall$_+$		0.49	0.60	**0.84**	0.50	0.64	0.65
Time (h)	CPU	1 h10	28 h	0 h38	**0 h06**	19 h28	16 h36
	GPU	–	–	–	–	1 h16	2 h13

to take into account and we notice that Hive-Cote has a too large CPU time (28 h) to be used in real conditions.

DTW obtains the worst performance in terms of accuracy (0.54) and its CPU time is rather high. DTW is faster than Hive-Cote but slower than BOSS and FBAT and similar to the GPU time of the deep learning models. Therefore, we estimate that DTW does not obtain enough satisfying results to be kept as a solution for this problem.

The performances of the neural networks are similar, excepted in terms of GPU time where FCN is almost two times faster than ResNet. They are a compromise between BOSS and FBAT in terms of $recall_+$ and $recall_-$. However, FCN and ResNet are considered as expensive solutions for livestock farmers since they need a GPU to be used in a real application.

BOSS produces the best accuracy results and obtains a low recall of the negative class. This means that among all time series labelled as *normal*, most of them are incorrectly classified as *abnormal*. If a farmer decides to use BOSS as a tool for detecting anomalies in dairy cows, he/she will then receive a high number of false alerts. These wrong detections may overshadow the correct ones and the method can become worthless. On the opposite, FBAT has the higher recall for the negative class. This would lead to a low number of false alerts for the farmer. The recall of abnormal days $(recall_+)$ metric however decreases from 0.84 for BOSS to 0.50 for FBAT. As a matter of fact, a low $recall_+$ score may be due to the data set construction. Indeed, thanks to previous observations [15], we chose to label all time series included between two days before and one day after an anomaly as abnormal because the behavior of an animal can be disturbed shortly before and after clinical symptoms are detected. But all anomalies may not last for four consecutive days and this can lower the $recall_+$. We checked if for each anomaly, at least one of the four days is detected as abnormal by FBAT. The FBAT method detected at least one day among the four consecutive ones labelled as abnormal in 83% of the lameness cases, 61% of the acidosis and 100% of the estrus. These results seem adequate for an on-farm use and a test by farmers is necessary to decide if the $recall_-$ is satisfactory.

Another advantage of the FBAT method in a farm application is the threshold. Indeed, we proposed a solution to automatically set the threshold between normal vs. abnormal time series. Nevertheless, the threshold can be adjusted. If a farmer thinks that the method does not detect enough anomalies, the threshold can be decreased. On the opposite, if the method detects too much false positive series, the threshold can be increased. Moreover, a threshold can be defined for each cow. If a cow is particularly insensitive to anomalies, its threshold can be decreased and inversely for a cow with a higher sensitivity.

The last advantage of FBAT is its CPU time, which is the best one of our experiments with 6 min. We also tested FBAT on personal computers (instead of big and expensive servers) and the CPU time didn't exceed 30 min.

5 Conclusion

The early detection of anomalies is very important for a farmer. Thanks to tools developed for precision livestock farming, it is possible to collect data in real time that can be analyzed by machine learning methods. In this study, we proposed a method based on Fourier transforms (FBAT) that we tested with the best algorithms and deep learning models available in the current literature for time series classification (BOSS, Hive-Cote, DTW, FCN and ResNet). The results showed that FBAT and BOSS are the two best solutions to solve the problem of anomaly detection in dairy cow activity. BOSS gives the best recall in the negative class whereas FBAT gives the best recall in positive class. They both obtain the best CPU time and they are both easy to implement. FBAT has the advantage to employ a threshold that can be adjusted to each cow. Testing these methods in real conditions, that is by farmers themselves, should help to choose the best method for the purpose. As other perspective, we propose to consider the labels as fuzzy. Indeed, the anomalies are detected when clinical signs are well visible, but it is reasonable to assume that anomalies gradually appear and disappear. We expect to increase the performances of the classifiers by better defining the labels. Finally, we plan to study the robustness of the algorithms to noisy labels. Indeed, label noise often occurs when humans are involved. In our application, caretakers are detecting and labeling anomalies but they easily have imperfect evidence.

Acknowledgment. This collaborative work was made possible thanks to the French Government IDEX-ISITE initiative 16-IDEX-0001 (CAP 20–25). The PhD grant for N. Wagner was provided by INRA and Université Clermont Auvergne. We thank the HERBIPOLE staff, B. Meunier, Y. Gaudron and M. Silberberg for data.

References

1. Adewumi, A.O., Akinyelu, A.A.: A survey of machine-learning and nature-inspired based credit card fraud detection techniques. Int. J. Syst. Assurance Eng. Manag. 8(2), 937–953 (2017)
2. Bagnall, Anthony., Lines, Jason., Bostrom, Aaron., Large, James, Keogh, Eamonn: The great time series classification bake off: a review and experimental evaluation of recent algorithmic advances. Data Min. Knowl. Disc. 31(3), 606–660 (2016). https://doi.org/10.1007/s10618-016-0483-9

3. Berkaya, S.K., et al.: A survey on ECG analysis. Biomed. Signal Process. Control **43**, 216–235 (2018)

4. Berndt, D.J., Clifford, J.: Using dynamic time warping to find patterns in time series. In: KDD Workshop, Seattle, WA, vol. 10, pp. 359–370 (1994)

5. Corizzo, R., Ceci, M., Japkowicz, N.: Anomaly detection and repair for accurate predictions in geo-distributed big data. Big Data Res. **16**, 18–35 (2019)

6. Fawaz, H.I., et al.: Deep learning for time series classification: a review. Data Mining Knowl. Discov. **33**(4), 917–963 (2019)

7. He et al.: Deep residual learning for image recognition. In: Proceedings of the IEEE Conference on Computer Vision and Pattern Recognition, pp. 770–778 (2016)

8. Lines, J., Taylor, S., Bagnall, A.: Hive-cote: the hierarchical vote collective of transformation-based ensembles for time series classification. In: 2016 IEEE 16th International Conference on data Mining (ICDM), pp. 1041–1046. IEEE (2016)

9. Long, J., Shelhamer, E., Darrell, T.: Fully convolutional networks for semantic segmentation. In: Proceedings of the IEEE Conference on Computer Vision and Pattern Recognition, pp. 3431–3440 (2015)

10. Marwah, R., Cawkwell, F., Hennessy, D., Green, S.: Improved estimation of grassland biomass using machine learning and satellite data. In: 9th ECPLF 2019, pp. 174–179. Teagasc (2019)

11. Mollenhors, H., de Haan, M., Oenema, J., Hoving-Bolink, A., Veerkamp, R., Kamphuis, C.: Machine learning to realize phosphate equilibrium at field level in dairy farming. In: 9th ECPLF 2019, pp. 41–44. Teagasc (2019)

12. Munir, M., Siddiqui, S.A., Dengel, A., Ahmed, S.: Deepant: a deep learning approach for unsupervised anomaly detection in time series. IEEE Access **7**, 1991–2005 (2018)

13. Ruiz, E.V., Nolla, F.C., Segovia, H.R.: Is the DTW "distance" really a metric? An algorithm reducing the number of DTW comparisons in isolated word recognition. Speech Commun. **4**(4), 333–344 (1985)

14. Schäfer, P.: The BOSS is concerned with time series classification in the presence of noise. Data Min. Knowl. Disc. **29**(6), 1505–1530 (2015)

15. Veissier, I., Mialon, M.M., Sloth, K.H.: Early modification of the circadian organization of cow activity in relation to disease or estrus. J. Dairy Sci. **100**(5), 3969–3974 (2017)

16. Wang, J., Chen, Y., Hao, S., Peng, X., Hu, L.: Deep learning for sensor-based activity recognition: a survey. Pattern Recogn. Lett. **119**, 3–11 (2019)

17. Wang, Z., Yan, W., Oates, T.: Time series classification from scratch with deep neural networks: a strong baseline. In: 2017 International Joint Conference on Neural Networks (IJCNN), pp. 1578–1585. IEEE (2017)

18. Yang, Q., Wu, X.: 10 challenging problems in data mining research. Int. J. Inf. Technol. Decision Making **5**(04), 597–604 (2006)

Clustering Algorithm Consistency
in Fixed Dimensional Spaces

Mieczysław Alojzy Kłopotek[1]($\boxtimes$)(iD) and Robert Albert Kłopotek[2](iD)

[1] Institute of Computer Science, Polish Academy of Sciences, Warsaw, Poland
mieczyslaw.klopotek@ipipan.waw.pl
[2] Faculty of Mathematics and Natural Sciences, School of Exact Sciences,
Cardinal Stefan Wyszyński University in Warsaw, Warsaw, Poland
r.klopotek@uksw.edu.pl

Abstract. Kleinberg introduced an axiomatic system for clustering
functions. Out of three axioms, he proposed two (scale invariance and
consistency) are concerned with data transformations that should produce the same clustering under the same clustering function. The so-
called consistency axiom provides the broadest range of transformations
of the data set. Kleinberg claims that one of the most popular clustering
algorithms, k-means does not have the property of consistency. We challenge this claim by pointing at invalid assumptions of his proof (infinite
dimensionality) and show that in one dimension in Euclidean space the
k-means algorithm has the consistency property. We also prove that in
higher dimensional space, k-means is in fact inconsistent. This result is
of practical importance when choosing testbeds for implementation of
clustering algorithms while it tells under which circumstances clustering
after consistency transformation shall return the same clusters.

Keywords: Cluster analysis · Consistency axiom · Consistency
transformation · Fixed dimensional euclidean space consistency ·
k-Means algorithm

1 Introduction

In his heavily cited paper [4], Kleinberg introduced an axiomatic system for
clustering functions. Out of three axioms, he proposed two are concerned with
data transformations that should produce the same clustering under the same
clustering function. We can speak here about "clustering preserving transformations" induced by these axioms. The so-called *consistency axiom*, mentioned
below, shall be of interest to us here as it provides the broadest range of transformations.

Property 1. *Let Γ be a partition of S, and d and d' two distance functions on
S. We say that d' is a Γ-transformation of d if (a) for all $i, j \in S$ belonging to
the same cluster of Γ, we have $d'(i, j) \leq d(i, j)$ and (b) for all $i, j \in S$ belonging
to different clusters of Γ, we have $d'(i, j) \geq d(i, j)$. The clustering function f has*

© Springer Nature Switzerland AG 2020
D. Helic et al. (Eds.): ISMIS 2020, LNAI 12117, pp. 352–361, 2020.
https://doi.org/10.1007/978-3-030-59491-6_33

the consistency *property if for each distance function d and its Γ-transformation d' the following holds: if f(d) = Γ, then f(d') = Γ*

The validity or no-validity of any clustering preserving axiom for a given clustering function is of vital practical importance, as it may serve as a foundation for a testbed of the correctness of the function. Any modern software developing firm creates tests for its software in order to ensure its proper quality. Generators providing versatile test data are therefore of significance because they may detect errors unforeseen by the developers. Thus the consistency axiom may be used to generate new test data from existent one knowing a priori what the true result of clustering should be.

Note that Kleinberg [4, Section 2] defines clustering function as:

Definition 1. *A clustering function is a function f that takes a distance function d on [set] S [of size n ≥ 2] and returns a partition Γ of S. The sets in Γ will be called its clusters.*

where the distance is understood by him as

Definition 2. *With the set $S = \{1, 2, \ldots, n\}$ [...] we define a distance function to be any function $d : S \times S \rightarrow \mathbb{R}$ such that for distinct $i, j \in S$ we have $d(i, j) \geq 0, d(i, j) = 0$ if and only if $i = j$, and $d(i, j) = d(j, i)$.*

Note that his distance definition is not a Euclidean one and not even metric, as he stresses. This is of vital importance because based on this he formulates and proves a theorem (his Theorem 4.1)

Theorem 1. *Theorem 4.1 from [4]. For every $k \geq 2$ and every function g [...] and for [data set size] n sufficiently large relative to k, the (k; g)-centroid clustering function [this term encompassing k-means]* [1] *does not satisfy the Consistency property.*

which we claim is wrong with respect to k-means for a number of reasons as we will show below. The reasons are:

- The objective function underlying k-means clustering is *not* obtained by setting $g(d) = d^2$ contrary to Kleinberg's assumption (k-medoid is obtained).
- k-means always works in fixed-dimensional space while his proof relies on unlimited dimensional space.
- Unlimited dimensionality implies a serious software testing problem because the correctness of the algorithm cannot be established by testing as the number of tests is too vast.

---

[1] Kleinberg defined the centroid function as follows: for any natural number $k \geq 2$, and any continuous, non-decreasing, and unbounded function $g : \mathbb{R}^+ \rightarrow \mathbb{R}^+$, the $(k; g)$-centroid clustering consists of: (1) choosing the set of k centroid points $T \subseteq S$ for which the objective function $\Delta_d^g(T) = \sum_{i \in S} g(d(i, T))$ is minimized, where $d(i, T) = \min_{j \in T} d(i, j)$. (2) a partition of S into k clusters is obtained by assigning each point to the element of T closest to it. He claims that the objective function underlying k-means clustering is obtained by setting $g(d) = d^2$.

– The consistency property holds for k-means in one-dimensional space.

The last result opens the problem of whether or not the consistency also holds for higher dimensions.

2 k-Means Algorithm

The popular clustering algorithm, k-means [5] strives to minimize the partition quality function (called also *partition cost function*)

$$J(U, M) = \sum_{i=1}^{m} \sum_{j=1}^{k} u_{ij} \|\mathbf{x}_i - \boldsymbol{\mu}_j\|^2 \tag{1}$$

where $\mathbf{x}_i$, $i = 1, \ldots, m$ are the data points, M is the matrix of cluster centers $\boldsymbol{\mu}_j$, $j = 1, \ldots, k$, and U is the cluster membership indicator matrix, consisting of entries u_{ij}, where u_{ij} is equal to 1 if among all of cluster (gravity) centers $\boldsymbol{\mu}_j$ is the closest to $\mathbf{x}_i$, and is 0 otherwise.

It can be rewritten in various ways while the following are of interest to us here. Let the partition $\Gamma = \{C_1, \ldots, C_k\}$ b a partition of the data set onto k clusters $C_1, \ldots, C_k$. Then

$$J(\Gamma) = \sum_{j=1}^{k} \sum_{\mathbf{x}_i \in C_j} \|\mathbf{x}_i - \boldsymbol{\mu}(C_j)\|^2 \tag{2}$$

where $\boldsymbol{\mu}(C) = \frac{1}{|C|} \sum_{\mathbf{x}_i \in C} \mathbf{x}_i$ is the gravity center of the cluster C. The above can be presented also as

$$J(\Gamma) = \frac{1}{2} \sum_{j=1}^{k} \frac{1}{|C_j|} \sum_{\mathbf{x}_i \in C_j} \sum_{\mathbf{x}_l \in C_j} \|\mathbf{x}_i - \mathbf{x}_l\|^2 \tag{3}$$

The problem of seeking the pair (U, M) minimizing J from Eq. (1) is called *k-means-problem*. This problem is known as NP-hard. We will call *k-means-ideal* such an algorithm that finds a pair (U, M) minimizing J from equation (1). Practical implementations of k-means usually find some local minima of $J()$. There exist various variants of this algorithm. For an overview of many of them, see e.g., [8]. An algorithm is said to be from the k-means family if it has the structure described by Algorithm 1. We will use a version with random initialization (randomly chosen initial seeds) as well as an artificial one initialized close to the true cluster center, which mimics k-means-ideal.

3   Kleinberg's Proof of Theorem 1 and Its Unlimited Dimensionality Deficiency

Kleinberg's proof, delimited to the case of $k = 2$ only, runs as follows: Consider a set of points $S = X \cup Y$ where X, Y are disjoint and $|X| = m$, $|Y| = \gamma m$, where

---

Data: the data points $\mathbf{x}_i$, $i = 1, \ldots, m$, the required number of clusters k
Result: $\boldsymbol{\mu}_1, \ldots, \boldsymbol{\mu}_k$
[1] Initialize k cluster gravity centers $\boldsymbol{\mu}_1, \ldots, \boldsymbol{\mu}_k$;
while *a stop criterion (no change of cluster membership, or no sufficient improvement of the objective function, or exceeding some maximum number of iterations, or some other criterion) not reached* **do**

  [2] Assign each data element $\mathbf{x}_i$ to the cluster C_j identified by the closest $\boldsymbol{\mu}_j$
  ;

  [3] Update $\boldsymbol{\mu}_j$ of each cluster C_j as the gravity center of the data elements in C_j;
end

Algorithm 1: Structure of a practical algorithm from k-means-family

$\gamma > 0$ is "small". $\forall_{i,j \in X} d(i,j) = r$, $\forall_{i,j \in Y} d(i,j) = \epsilon < r$, $\forall_{i \in X, j \in Y} d(i,j) = r + \delta$ where $\delta > 0$ and δ is "small". By choosing $\gamma, \epsilon, r, \delta$ appropriately, the optimal choice of k = 2 centroids will consist of one point from X and one from Y. The resulting partition is $\Gamma = \{X, Y\}$. Let divide X into $X = X_0 \cup X_1$ with X_0, X_1 of equal cardinality. Reduce the distances so that $\forall_{c=1,2} \forall_{i,j \in X_c} d'(i,j) = r' < r$ and $d' = d$ otherwise. If r' is "sufficiently small", then the optimal choice of two centroids will now consist of one point from each X_c, yielding a different partition of S. But d' is a Γ-transform of d so that a violation of consistency occurs. So far the proof of Kleinberg of the Theorem 1.

The proof cited above is a bit excentric because the clusters are heavily unbalanced (k-means tends to produce rather balanced clusters). Furthermore, the distance function is awkward because Kleinberg's counter-example would require an embedding in a very high dimensional space, non-typical for k-means applications.

We claim in brief:

Theorem 2. *Kleinberg's proof of [4, Theorem 4.1] that k-means (k = 2) is not consistent, is not valid in $\mathbb{R}^p$ for data sets of cardinality $n > 2(p + 1)$.*

Proof. In terms of the concepts used in the Kleinberg's proof, either the set X or the set Y is of cardinality $p + 2$ or higher. Kleinberg requires that distances between $p + 2$ points are all identical which is impossible in $\mathbb{R}^p$ (only up to $p + 1$ points may be equidistant).

Furthermore Kleinberg's minimized target function $\Delta_d^g(T) = \sum_{i \in S} g(d(i, T))$, where $d(i, T) = \min_{j \in T} d(i, j)$, differs significantly from the formula (3). For the original set X, the formula (3) would return $\frac{1}{2}(m - 1)r^2$, while Kleinberg's would produce $(m - 1)r^2$. For a combination of a elements from X and b elements from Y in one cluster we get $\frac{a(a-1)r^2/2 + b(b-1)\epsilon^2/2 + ab(r+\delta)^2}{a+b}$ from (2) or the minimum of $(a-1)r^2 + b(r+\delta)^2$ and $(b-1)\epsilon^2 + a(r+\delta)^2$ for Kleinberg's $\Delta_d^g(T)$. The discrepancy between these formulas is shown in Fig. 1.

We see immediately that

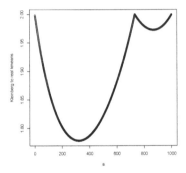

Fig. 1. Quotient of Kleinberg's k-means target and the real k-means target.

Theorem 3. *Kleinberg's target function does not match the real k-means target function.*

4 The Impact of Dimensionality of Consistency Property

Theorem 4. *k-means is consistent in one dimensional Euclidean space.*

Proof. Consider two alternative partitions in one dimensional space:

- the partition $\Gamma_1 = \{C_{1.}, \ldots, C_{k.}\}$ which will be base for the Γ-transform
- and the competing partition $\Gamma_2 = \{C_{.1}, \ldots, C_{.k'}\}$.

Due to the nature of k-means let each cluster of each partition after Γ-transform be represented as an interval not intersecting with any other cluster of the same partition. For Γ_1, it holds before the transform; therefore, it holds afterward. Γ_2 shall be the competing optimal transform; therefore, it holds for sure afterward. We intend to demonstrate that under the Γ transformation, assuming that the actual partition is Γ_1, the target function of k-means for Γ_1 will decrease not less than that for Γ_2. For simplicity, assume that the indices of clusters grow with the growing value of the cluster center.

For this purpose assume that $C_{ij} = C_{i.} \cap C_{.j}$ are non-empty intersections of clusters $C_{i.} \in \Gamma_1$, $C_{.j} \in \Gamma_2$, of both partitions. Define $minind(C_{i.})$, resp. $maxind(C_{i.})$ as the minimal/maximal index j such that C_{ij} is not empty. The $Q(\Gamma_1)$ will be the sum of centered sums of squares over all C_{ij} plus the squared distances of centers of all C_{ij} to the center of $C_{i.}$ times cardinality of C_{ij}.

$$Q(\Gamma_1) = \sum_{C_{i.} \in \Gamma_1} \sum_{j; C_{ij} \neq \emptyset} \left(|C_{ij}|(\mu(C_{ij}) - \mu(C_{i.}))^2 + \sum_{x \in C_{ij}} (x - \mu(C_{ij}))^2 \right)$$

$$= \left(\sum_{i,j; C_{ij} \neq \emptyset} \sum_{x \in C_{ij}} (x - \mu(C_{ij}))^2 \right) + \left(\sum_{C_{i.} \in \Gamma_1} \sum_{j; C_{ij} \neq \emptyset} |C_{ij}|(\mu(C_{ij}) - \mu(C_{i.}))^2 \right)$$

Please note that

$$\sum_{j;C_{ij}\neq\emptyset}|C_{ij}|(\mu(C_{ij})-\mu(C_{i.}))^2 = \sum_{j;C_{ij}\neq\emptyset}|C_{ij}|(\mu(C_{ij})-\sum_{j';C_{ij'}\neq\emptyset}\frac{|C_{ij'}|}{|C_{i.}|}\mu(C_{ij'}))^2$$

$$= \sum_{j;C_{ij}\neq\emptyset}|C_{ij}|(\sum_{j';C_{ij'}\neq\emptyset}(\frac{|C_{ij}|}{|C_{i.}|}\mu(C_{ij})-\frac{|C_{ij'}|}{|C_{i.}|}\mu(C_{ij'})))^2$$

$$= \sum_{j;C_{ij}\neq\emptyset}|C_{ij}|(\sum_{j';C_{ij'}\neq\emptyset}(\frac{|C_{ij'}|}{|C_{i.}|}\mu(C_{ij})-\mu(C_{ij'})))^2$$

$$= 0.5\sum_{j;C_{ij}\neq\emptyset}\sum_{j';C_{ij'}\neq\emptyset}\frac{|C_{ij}|\cdot|C_{ij'}|}{|C_{i.}|}(\mu(C_{ij})-\mu(C_{ij'})))^2$$

The $Q(\Gamma_2)$ can be computed analogously, but let us follow a bit distinct path.

$$Q(\Gamma_2) = \sum_{C_{.j}\in\Gamma_2}\frac{1}{|C_{.j}|}\sum_{x\in C_{.j}}\sum_{y\in C_{.j}}(x-y)^2$$

$$= \sum_{C_{.j}\in\Gamma_2}\frac{1}{|C_{.j}|}\left(\left(\sum_{i;C_{ij}\neq\emptyset}\sum_{x\in C_{ij}}\sum_{y\in C_{ij}}(x-y)^2\right)\right.$$

$$\left.+\left(\sum_{i',i'';C_{i'j}\neq\emptyset,C_{i''j}\neq\emptyset}\sum_{x\in C_{i'j},y\in C_{i''j}}(x-y)^2\right)\right)$$

$$= \sum_{C_{.j}\in\Gamma_2}\frac{1}{|C_{.j}|}\left(\left(\sum_{i;C_{ij}\neq\emptyset}|C_{ij}|\sum_{x\in C_{ij}}(x-\mu(C_{ij}))^2\right)\right.$$

$$\left.+\left(\sum_{i',i'';C_{i'j}\neq\emptyset,C_{i''j}\neq\emptyset}\sum_{x\in C_{i'j},y\in C_{i''j}}(x-y)^2\right)\right)$$

$$= \left(\sum_{i,j;C_{ij}\neq\emptyset}\frac{|C_{ij}|}{|C_{.j}|}\sum_{x\in C_{ij}}(x-\mu(C_{ij}))^2\right)$$

$$+ \sum_{C_{.j}\in\Gamma_2}\frac{1}{|C_{.j}|}\left(\sum_{i',i'';i'\neq i'',C_{i'j}\neq\emptyset,C_{i''j}\neq\emptyset}\sum_{x\in C_{i'j},y\in C_{i''j}}(x-y)^2\right)$$

Both summands of $Q(\Gamma_1)$, that is $\left(\sum_{i,j;C_{ij}\neq\emptyset}\sum_{x\in C_{ij}}(x-\mu(C_{ij}))^2)\right)$ and $\left(\sum_{C_{i.}\in\Gamma_1}\sum_{j;C_{ij}\neq\emptyset}(|C_{ij}|(\mu(C_{ij})-\mu(C_{i.}))^2\right)$ will decrease upon Γ_1 based consistency transformation. $(x-\mu(C_{ij}))^2$ decreases because the distance to each of elements of C_{ij} decreases as they are all in the same cluster $C_{i.}$. Each $(\mu(C_{ij})-\mu(C_{ij'}))^2$ decreases because all the elements constituting C_{ij} and $C_{ij'}$

belong to the same cluster C_i. Hereby there is always an extreme data point $P_{ij} \in C_{ij}$ separating it from $C_{ij'}$. As the points of both C_{ij} and $C_{ij'}$ get closer to P_{ij} consistency transformation, so the centers of both C_{ij} and $C_{ij'}$ will get closer to P_{ij}, so that they will move closer to each other. As summands of $Q(\Gamma_2)$ are concerned, the first, will also decrease upon consistency transformation. $\left(\sum_{i,j;C_{ij}\neq\emptyset} \frac{|C_{ij}|}{|C_{\cdot j}|} \sum_{x\in C_{ij}} (x - \mu(C_{ij}))^2 \right)$ but not by the same absolute value as $\left(\sum_{i,j;C_{ij}\neq\emptyset} \sum_{x\in C_{ij}} (x - \mu(C_{ij}))^2 \right)$, because always $|C_{ij}| \leq |C_{\cdot j}|$. But $\sum_{C_{\cdot j}\in\Gamma_2} \frac{1}{|C_{\cdot j}|} \left(\sum_{i',i'';i'\neq i'',C_{i'j}\neq\emptyset,C_{i''j}\neq\emptyset} \sum_{x\in C_{i'j},y\in C_{i''j}} (x - y)^2 \right)$ will increase because x, y stem from different clusters of Γ_1. Therefore, if Γ_1 was the optimal clustering for k-means cost function prior to consistency transformation, it will remain so afterward.

But what about higher dimensions? Let us illustrate by a realistic example (balanced, in Euclidean space) that inconsistency of k-means in $\mathbb{R}^m$ is a real problem - see Theorems 5 and 6.

Theorem 5. *k-means in 3D is not consistent.*

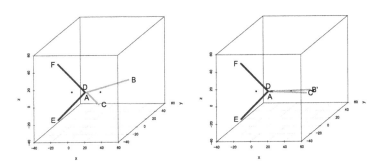

Fig. 2. Inconsistency of k-means in 3D Euclidean space. Left picture - data partition before consistency transform. Right picture - data partition after consistency transform.

Proof. Let A, B, C, D, E, F be points in three-dimensional space (see Fig. 2) with coordinates: $A(1,0,0)$, $B(33,32,0)$, $C(33,-32,0)$, $D(-1,0,0)$, $E(-33,0,-32)$, $F(-33,0,32)$. Let $S_{AB}, S_{AC}, S_{DE}, S_{DF}$ be sets of say 1000 points randomly uniformly distributed over line segments (except for endpoints) AB, AC, DE, EF resp. Let $X = S_{AB}\cup S_{AC}\cup S_{DE}\cup S_{EF}$. k-means with $k = 2$ applied to X yields a partition $\{S_{AB}\cup S_{AC}, S_{DE}\cup S_{DF}\}$. Let us perform a Γ transformation consisting in rotating line segments AB, BC around the point A in the plane spread by the first two coordinates towards the first coordinate axis so that the angle between this axis and AB' and AC' is say one degree. Now the k-means with $k = 2$ yields a different partition, splitting line segments AB' and AC'. [2]

With this example not only *consistency violation* is shown, but also *refinement-consistency violation*. Not only in 3D, but also in higher dimensions. So what about the case of two dimensions - 2D?

Theorem 6. *k-means in 2D is not consistent.*

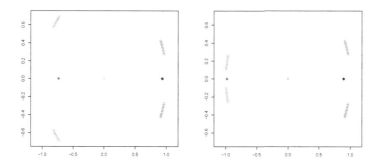

Fig. 3. Inconsistency of k-means in 2D Euclidean space. Left picture - data partition before consistency transform. Right picture - data partition after consistency transform. Cluster elements are marked with blue and green. Red points indicate cluster centers.

Proof. The proof of Theorem 6 uses a less realistic example than in Theorem 5, hence Theorem 5 was worthy considering in spite of the fact that it is implied by Theorem 6. Imagine a unit circle with data points arranged as follows (Fig. 3 left): one data point in the center, and the remaining points arranged on the circle with the following angular positions with respect to the circle center. Set $A = \{16°, 17°, \dots, 25°, -16°, -17°, \dots, -25°\}$. Set $B = \{136°, 137°, \dots, 145°, -136°, -137°, \dots, -145°\}$. k-means with $k = 2$ will merge points of the set B and the circle middle point as one cluster, and the set A as the other cluster. After a consistency-transformation (Fig. 3 right) let A turn to A' identical with A and let B change to B' $B' = \{165°, 166°, \dots, 174°, -165°, -166°, \dots, -174°\}$, while the point in the center of the circle remains in its position. Now k-means with $k = 2$ yields one cluster consisting of points of the set B' and the second cluster consisting of the circle middle point and the set A'. The center point of the circle switches the clusters upon consistency transformation.

---

[2] In a test run with 100 restarts, in the first case we got clusters of equal sizes, with cluster centers at $(17, 0, 0)$ and $(-17, 0, 0)$, (between_SS/total_SS = 40%) whereas after rotation we got clusters of sizes 1800, 2200 with centers at $(26, 0, 0)$, $(-15, 0, 0)$ (between_SS/total_SS = 59%).

5   Experiments

Experiments have been performed to check whether or not the Theorem 4 that denies Kleinberg's findings for one-dimensional space really holds. Samples were generated from uniform distribution (sample size 100, 200, 400, 1000, 2000, 4000, 10000) for each $k = 2, \ldots, floor(\sqrt{samplesize})$. Then the respective sample was clustered into k clusters ($k = 2, \ldots, floor(\sqrt{samplesize})$) and k-means clustering (R package) was performed with $100k$ restarts. Subsequently, Γ transformation was performed where the distances within a cluster were decreased by a randomly chosen factor (a separate factor for each pair of neighboring data points), and at the same time the clusters were moved away so that the distance between cluster elements of distinct clusters is not decreased. Then k-means clustering was performed with $100k$ restarts in two variants. The first variant was with random initialization. The second variant was with the initialization of the midpoint of the original (rescaled) cluster interval. Additionally, for control purposes, the original samples were reclustered. The number of partitions was counted for which errors in restoring the original clustering was observed. Experiments were repeated ten times. Table 1 presents the average results obtained.

Table 1. Validation of the Theorem 4.

Sample size	100	200	400	1000	2000	4000	10000
Max. k	10	14	20	31	44	63	100
Errors variant 1	0	0	0	2.4	10.2	21.5	62.4
Errors variant 2	0	0	0	0	0	0.5	2.0
Errors reclustering	0	0	2.3	14.1	30.2	49.5	86.2

In this table, looking at the errors for variant 1, we see that with the increasing sample size (and hence increasing maximum of k), more errors are committed. This contrasts with the variant 2 where the number of errors is negligible. The second variant differs from the first in that seeds are distributed so that there is one in each intrinsic cluster.

Clearly the Theorem 4 holds (as visible from the variant 2). At the same time, however, the table shows that k-means with random initialization is unable to initialize properly for a larger number k of clusters in spite of a large number of restarts (variant 1). This is confirmed by the experiments with reclustering original data.

This study also shows how a test data generator may work when comparing variants of k-means algorithm (for one-dimensional data).

6   Conclusions

In this paper, we have provided a definite answer to the problem of whether or not k-means algorithm possesses the consistency property. The answer is negative except for one-dimensional space. Settling this problem was necessary because the proof of Kleinberg of this property was inappropriate for real application areas of k-means that it is a fixed-dimensional Euclidean space. The result precludes usage of consistency axiom as a generator of test examples for k-means clustering function (except for one-dimensional data) and implies the need to seek alternatives.

Kleinberg's consistency was subject of strong criticism and new variants were proposed like Monotonic consistency [6] or MST-consistency [9]. See also criticism in [2,3]. The mentioned new definitions of consistency are apparently restrictions of Γ-consistency, and therefore the Theorem 4 would be valid. The Monotonic consistency seems not to impose restrictions on Kleinberg's proof on k-means violating consistency. Therefore in those cases, the consistency of k-means under higher dimensionality needs to be investigated. Note that we have also challenged the result [7], who claims that Kleinberg's consistency may be achieved by k-means with random initialization (see our Theorem 5). The shift of axioms from clustering function to quality measure [1] was suggested to the problems with consistency, but this approach fails to tell what the outcome of clustering should be, which is not useful for the mentioned test generator application.

References

1. Ben-David, S., Ackerman, M.: Measures of clustering quality: a working set of axioms for clustering. In: Proceedings Advances in Neural Information Processing Systems, vol. 21, pp. 121–128 (2008)
2. Carlsson, G., Mémoli, F.: Characterization, stability and convergence of hierarchical clustering methods. J. Mach. Learn. Res. **11**, 1425–1470 (2010)
3. Correa-Morrisa, J.: An indication of unification for different clustering approaches. Pattern Recogn. **46**, 2548–2561 (2013)
4. Kleinberg, J.: An impossibility theorem for clustering. In: Proceedings NIPS, vol. 2002, pp. 446–453 (2002). http://books.nips.cc/papers/files/nips15/LT17.pdf
5. MacQueen, J.: Some methods for classification and analysis of multivariate observations. In: Proceedings Fifth Berkeley Symposium on Mathematical Statistics and Probability, vol. 1, pp. 281–297. University of California Press (1967)
6. Strazzeri, F., Sánchez-García, R.J.: Morse theory and an impossibility theorem for graph clustering (2018). https://arxiv.org/abs/1806.06142
7. Wei, J.H.: Two examples to show how k-means reaches richness and consistency. DEStech Trans. Comput. Sci. Eng. (2017). https://doi.org/10.12783/dtcse/aita2017/16001
8. Wierzchoń, S., Kłopotek, M.: Modern clustering algorithms. Stud. Big Data **34** (2018)
9. Zadeh, R.: Towards a principled theory of clustering (2010). http://stanford.edu/rezab/papers/principled.pdf

Estimating the Importance of Relational Features by Using Gradient Boosting

Matej Petković[1,2(✉)] [iD], Michelangelo Ceci[1,3] [iD], Kristian Kersting[4] [iD], and Sašo Džeroski[1,2] [iD]

[1] Jozef Stefan Institute, Jamova 39, Ljubljana, Slovenia
[2] Jozef Stefan Postgraduate School, Jamova 39, Ljubljana, Slovenia
{matej.petkovic,saso.dzeroski}@ijs.si
[3] Università degli Studi di Bari Aldo Moro, via E. Orabona 4, Bari, Italy
michelangelo.ceci@uniba.it
[4] CS Department, TU Darmstadt, Hochschulstrasse 1, Darmstadt, Germany
kersting@cs.tu-darmstadt.de

Abstract. With data becoming more and more complex, the standard tabular data format often does not suffice to represent datasets. Richer representations, such as relational ones, are needed. However, a relational representation opens a much larger space of possible descriptors (features) of the examples that are to be classified. Consequently, it is important to assess which features are relevant (and to what extent) for predicting the target. In this work, we propose a novel relational feature ranking method that is based on our novel version of gradient-boosted relational trees and extends the Genie3 score towards relational data. By running the algorithm on six well-known benchmark problems, we show that it yields meaningful feature rankings, provided that the underlying classifier can learn the target concept successfully.

Keywords: Feature ranking · Relational trees · Gradient boosting

1 Introduction

One of the most frequently addressed tasks of machine learning is predictive modeling, where the goal is to build a model by using a set of known examples, which are given as pairs of target values and vectors of feature values. The model should generalize well to previously unseen combinations of feature values and accurately predict the corresponding target value. In this paper, we limit ourselves to classification, i.e., to the case when the target can take one of finitely many nominal values. For example, when modeling the genre of a movie, the possible values might be `thriller`, `drama`, `comedy` and `action`.

In the simplest and most common case, the data comes as a single table where rows correspond to examples and columns to features, including the target.

This is financially supported by the Slovenian Research Agency (grants P2-0103, N2-0128, and a young researcher grant to MP).

D. Helic et al. (Eds.): ISMIS 2020, LNAI 12117, pp. 362–371, 2020.
https://doi.org/10.1007/978-3-030-59491-6_34

However, when predicting the genre of a movie, it might be beneficial to not only know the properties of the movie (i.e., the feature values), but also to have access to the properties of ratings of the movie (e.g., number of stars and date) on some website. As a consequence, the data is now represented by two tables: one describing the movies and the other one describing the ratings. A link between them might then be the relation *belongsTo* (rating, movie) that tells which movie a given rating rates. The actual `movies` dataset used in our experiments (shown in Fig. 1) contains five additional tables and four relations among them.

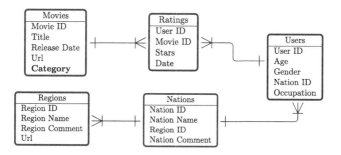

Fig. 1. The `movies` dataset consists of 5 tables and 4 relations among them. Single and multiple ends on the connections among the tables represent one-to- (or -to-one) and many-to- (or -to-many) relations, respectively. The task at hand is to predict the property Category (genre) of a movie (shown in bold).

In the era of abundance of data, such complex datasets are more and more frequent in various domains, e.g., sports statistics, businesses, epidemiology [14], among others. In biology, for instance, protein networks [18] encode complex data about proteins and their functions, together with different protein-protein inter-actions. Concrete examples are given in Sect. 5. In all these cases, relational data are very much of interest because they offer a powerful representation language where a broad spectrum of predictive modeling tasks can be applied. A notable example is link prediction where the goal is to predict whether two examples are in a given relation or not [4], say, two researchers are co-authors of a paper [16]. It is also possible to predict links of multiple types [3]. Another prominent task is classification. For example, movies can be categorized into different genres. Again, predicting multiple targets is possible [12,14].

The main focus of the present paper, however, is not predictive modeling, but rather feature ranking. In the simplest case when the data is given as a single table, the task of feature ranking is to estimate the importance of each descriptive feature when predicting the target value. The features are then ranked with respect to their importance. A possible motivation for performing feature ranking is dimensionality reduction. Discarding the less relevant features from the dataset (before proceeding to predictive modeling) results in lower time and space complexity of learning the models, which may also be more accurate since a lower number of features reduces overfitting. Moreover, feature ranking can explain black box models such as ensembles of decision trees [1,5] or neural networks, which may be of crucial importance in domains such as medicine.

Feature ranking on relational data, however, is not a trivial task. In fact, with two tables or more (and relations between them) at hand, the notion of a feature is a more general concept. It can refer to a relation between tables or to a descriptive attribute, which is part of the description of the objects in a given table. Furthermore, the existing relations and descriptive attributes can be combined into new ones. For example, the schema in Fig. 1 implicitly defines a feature whose value for a given movie is the name of the nation which contributes the highest number of ratings for the movie. Consequently, the feature space can be extremely large, and discovering only the most relevant features for a given predictive modeling task might have a high value. A feature ranking algorithm can estimate the importance of the features explicitly given (such as release date in the movies dataset), or it can use some heuristic to construct new features from the ones existing in the descriptive relations.

In this paper, we propose a feature ranking algorithm that focuses on the second approach, i.e., heuristically introduces new features. The novel heuristic for building relational trees starts the search for relevant features at those which are explicitly given, and gradually proceeds towards more and more complex ones if needed. To improve the quality (and stability) of feature rankings, we use gradient boosting ensembles [5] instead of a single tree. Once the ensemble is built, feature ranking is seamlessly computed out of it by using the adaptation of the Genie3 score [9].

The remainder of the paper is organized as follows. Section. 2 explains the necessary background and notation, Sect. 3 reviews related work, Sect. 4 introduces our extension of gradient boosting to relational data and our novel feature ranking algorithm, Sect. 5 gives the experimental setup, while Sect. 6 the results are discussed. Finally, Sect. 7 concludes the work.

2   Background on Relational Predictive Modeling

Let $\mathcal{X}$ be a generic domain that identifies an object type and x be an example (instance) of such type. Every object $x \in \mathcal{X}$ is represented in terms of its ID (object identifier) and a list of attribute values. A relation r of arity $t \geq 1$ is defined as a subset of a Cartesian product of the domains $\mathcal{X}_i$ of relation's arguments X_i, $1 \leq i \leq t$. The fact that $x_1 \in \mathcal{X}_1$, ..., $x_t \in \mathcal{X}_t$ are in a relation r, is denoted either as $(x_1, \ldots, x_t) \in r$ or $r(x_1, \ldots, x_t)$.

If $t = 1$ the relation describes a property of a single object, e.g., $isMale$(user). If $t = 2$, the relation is said to be binary. A binary relation is one-to-many if each $x_1 \in \mathcal{X}_1$ is in a relation with *at least* one $x_2 \in \mathcal{X}_2$, and each x_2 is in a relation with *exactly* one x_1. The other three cases (one-to-one, many-to-one and many-to-many) are defined analogously.

The goal of the relational classification task is thus to learn a model that uses descriptive relations to predict the target relation $r(X_0, Y)$ where the first component corresponds to the example to be classified and the last component corresponds to the target values. For the link prediction task, the target relation is of form $r(X_0^1, X_0^2, Y)$ where both X_0^1 and X_0^2 correspond to examples and the last component corresponds to the target values. Thus, a relational representation of the data elegantly unifies different classification tasks into the same framework. However, in this work, we focus on the $r(X_0, Y)$ target relations only.

A dataset is typically stored as a group of tables and relations between them. For example, the `movies` dataset in Fig. 1 contains five tables that describe five different types of objects: movies, ratings, users, nations, and regions. Every movie is given as a 5-tuple, consisting of movie ID, title, release date, url, and category. Similarly, ratings are given as 4-tuples of user ID (the author of the rating), movie ID (the movie that rating refers to), stars and date.

Since both the fist component of movies and the second component of ratings are of the same type (movie ID), the dataset also implicitly defines the one-to-many relation via which we can find, for example, all ratings of a given movie. Similar links exist between users and ratings (connected via user ID columns), between users and nations, and between nations and regions. Should there be a relation among users, e.g., *isARelative*, we would need an additional table, e.g., relatives, that would contain two columns (both of the type user ID) and would explicitly list all the pairs of relatives.

Prior to applying our algorithm (and also the majority of those discussed in the related work), the data at hand should be converted into a pure relational representation, coherent with the formal description above, where all facts are given as relation elements $r(x_1, \ldots, x_t)$, e.g., $gender(\text{Ana}, \text{female})$.

3 Related Work

Since our feature ranking is embedded into a classifier [7], it is closely related to relational ensemble techniques. Gradient boosting of relational trees is proposed in [11]. It takes relations (given as sets of tuples) as the input and builds gradient-boosted regression trees. As usual, multi-class problems demand 1-hot encoding of the target variable, which converts the original dataset into a series of 1-versus-all classification problems, and an ensemble is built for each of them separately. In turn, a tree induction in an ensemble bases on the TILDE learner [17] and its predecessor FOIL [13], which results in two possible limitations of the method.

First, it allows only for the nominal descriptive features, thus the numeric ones (e.g., age of a user) should first be discretized into bins which results in the loss of ordering of values. The second possible limitation are candidate tests in the internal nodes of the trees. Without loss of generality, we assume that the variable X_0 that corresponds to the example ID whose target values is to be predicted, always appears as the first component of a relation r. In this case, the candidate splits are of two types. First, a split can be a conjunction of predicates

$$r_1(X_0, x_1^2, \ldots) \wedge \cdots \wedge r_j(X_0, x_j^2, \ldots) \wedge r_{j+1}(x_{j+1}^1, x_{j+1}^2, \ldots) \wedge \cdots \wedge r_\ell(x_\ell^1, x_\ell^2, \ldots)$$

where $\ell \geq 1$, and x_i^k is the value of the variable at position k for relation r_i. Second, some of the variables X_i^k may not be grounded yet (i.e., their value is not determined). In that case, a split is of the form

$$\exists\, X_{i_1}^{k_1} \ldots \exists\, X_{i_n}^{k_n} \,:\, r_1(X_0, x_1^2, \ldots) \wedge \cdots \wedge r_\ell(x_\ell^1, x_\ell^2, \ldots) \tag{1}$$

where n is the number of non-grounded variables. When the actual example ID x_0 reaches a split, X_0 takes its value, the split is evaluated, and the example follows the YES or NO branch accordingly. That means that having splits like *Is the average age of users that rated a movie, larger than 60?* is not possible. That was overcome by introducing aggregates into TILDE [17]. There, also constrained aggregation is possible, e.g., *Is the average age of users that have contributed at least 5 ratings in total, and have rated the given movie, larger than 60?*. For the exact formulation of the possible split tests, we refer the reader to [17] where the extension of the method to relational random forests is described.

Regarding relational feature ranking methods, there is only the FARS method [8], which belongs to the group of filters [7], i.e., no predictive model is needed for computing the ranking. As such, it cannot be used for explaining the decisions of classifiers. It is suitable for classification datasets and is based on propagation of the class values from the table that contains the target attribute to the other tables. The method supports neither estimation of numeric attributes nor the estimation of implicitly defined features.

4   Our Method

Here, we describe our two contributions. We first present the proposed test split generation for relational trees (and boosting ensembles). Afterward, the proposed feature ranking approach is introduced.

4.1   Relational Gradient Boosting with Aggregates

Relational trees are built with the standard top-down induction procedure [2] whose main part is greedily finding the optimal test (according to a heuristic h that is based on an impurity measure, e.g., Gini index [2]) that splits the data into two subgroups. The point at which relational tree induction differs from the standard one is how the candidate splits are created. Indeed, one option is using TILDE with aggregates. However, also from the feature ranking perspective, we find the following feature-value back-propagation splits more appropriate, since by doing so (in contrast to TILDE), we can directly link the values of a given feature to the given example ID, no matter which table is the feature present in.

Consider Fig. 2, which depicts our candidate test generation. Each test is generated in two stages. First (as shown in Fig. 2a), we start from an example ID x_0, and follow any relation r_1 where x_0 can appear in. The group g_1 of all tuples $\boldsymbol{x}_1^i \in r_1$, which x_0 is part of, is thus found. Then, for each of the tuples $\boldsymbol{x}_1^i$, we recursively repeat the search from $\boldsymbol{x}_1^i$, thus finding the group of examples g_2^i by following any relation r_2 that shares at least one input domain $\mathcal{X}$ with relation r_1. The search is finished after at most ℓ steps which is a user-defined parameter. In Fig. 2a, we have $\ell = 2$. For example, following the schema of the movie data set in Fig. 1, we might start from $x_0 = $ titanic, and find the group g_1 of all pairs $x_1^i = ($titanic, ratingID$^i)$. For each such pair, a (singleton) group g_2^i of all pairs $($ratingID^i, stars$)$ is found.

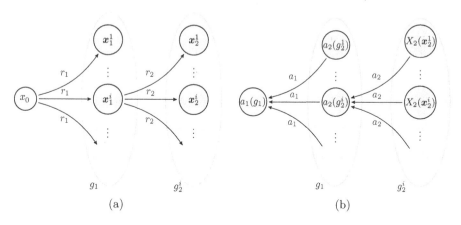

(a)   (b)

Fig. 2. Candidate test generation: finding related tuples by following descriptive relations (a), and back-propagation of the feature values by aggregation (b).

After the search is finished, the back-propagation of feature values by aggregation starts, by choosing one of the variables that were introduced in the last step of the search. Let this be X_ℓ, i.e., X_2. Its type defines possible starting aggregates. We can always use *count* or *countUnique*. Additionally, we can use *max*, *min*, *mean*, and *sum* if X_2 is numeric, and *mode* if X_2 is nominal. The data type returned by the first aggregate in turn defines the possible options for the next one in the chain. In general, we proceed form right to left (Fig. 2b).

By using aggregate a_2, we aggregate every group g_2^i over X_2. Following the example above, the variable at hand would be the number of stars. Then, the tuple $x_1^i \in g_1$ is effectively replaced by the aggregated value $a_2(g_2^i)$. After this is done for all tuples x_1^i, the group g_1 is similarly aggregated into the final value $a_1(g_1)$. If we set a_2 to *mean* and a_1 to *max*, in the above example, we compute the maximal rating of a movie (the mean of a single value is the value itself).

Finally, the procedure for generating candidate splits proceeds to finding the optimal threshold ϑ the aggregated values are compared against, and chooses the best test among the candidates. This covers also the existentially quantified split tests in Eq. (1) if $\vartheta = 0$ and the aggregates are set to *count*. This motivates the idea to allow the algorithm to continue the search for a good split in the YES-child at any of the steps from its parent node.

Therefore, the evaluation of the tests becomes more time-consuming deeper in the tree, so gradient boosting [5] where the trees are shallower is more efficient than, e.g., bagging [1] where bias-variance decomposition of the error reveals that trees should be grown to a full depth.

Using aggregates is necessary since the preliminary experiments (not part of this work) show that this increases the expressiveness of the splits which reflects in substantially improved predictive power of the models.

4.2 Feature Ranking

Now, we are ready to introduce our novel relational feature ranking. Let (R, A, ϑ) be a triplet denoting a test in a node $\mathcal{N}$ in a tree $\mathcal{T}$, where R and A are lists of relations and aggregates used in the test. The size s of a test is defined as $s = |R| (= |A|)$. Let $E(\mathcal{N})$ be the set of examples that reach the node $\mathcal{N}$, and $h(\mathcal{N})$ the heuristic value of the split in $\mathcal{N}$ [2]. We write $r \in \mathcal{N}$ if $r \in R$. Then, a natural extension of the Genie3 score [9] for the already existing relations r is

$$
importance(r) = \sum_{\mathcal{T}} \sum_{\mathcal{N} \in \mathcal{T}} \frac{\mathbf{1}[r \in \mathcal{N}]}{s(\mathcal{N})} h(\mathcal{N}) |E(\mathcal{N})|, \tag{2}
$$

where $\mathbf{1}$ is the indicator function. The definitions says that all relations that appear in a given node are rewarded equally and proportionally to the heuristic value. The term $|E(\mathcal{N})|$ assures that relations that appear higher in a tree (and influence more examples) receive bigger award.

Please note that Eq. (2) is naturally extended to (parts of) lists R of relations by summing up the importances of their atomic parts. Note also that the relations and combinations thereof that do not appear in the ensemble, have the importance score with the value 0.

5 Experimental Setting

In order to investigate the performance of our relational feature ranking methods, we computed feature rankings for six well-known datasets. In addition to the movie dataset (shown in Fig. 1), these were: basket [10] (basketball players, coaches, teams, etc.), IMDB [6] (movies, actors, directors, etc.), Stack [15] (user posts, users and comments), and Yelp [19] (different businesses, their reviews, users etc.). Additional statistics for the data are given in Table 1.

In our experiments, the quality of the underlying predictive model will be used as a proxy for the quality of the ranking, thus 10-fold cross-validation (CV) is performed. For each training set, internal 3-fold CV was performed to tune the boosting parameters via grid-search. The parameters (and their possible values) that were optimized are shrinkage ($\{0.05, 0.2, 0.4, 0.6\}$), proportion of chosen examples ($\{0.6, 0.8, 1.0\}$), proportion of the evaluated tests in a node ($\{0.2, 0.4, \ldots, 1.0, \mathrm{sqrt}\}$), and maximal depth of trees ($\{2, 4, 6, 8\}$). Ensemble size was set to 50 and the size of splits was $\ell \leq 2$.

6 Results and Discussion

The experimental results are summarized in Fig. 3. It shows the feature importances, averaged over the 10-folds of CV, and the three most relevant features, for each dataset. We observe two qualitatively different results: for the datasets basket, Stack and UWCSE (and to some extend Yelp), it is evident that feature importance score have mostly converged to their final values, meaning that the Gini heuristic in the splits goes down to 0.0 and the trees with higher indices do not influence the ranking or predictions of the model. This is confirmed by the accuracy of the corresponding models: 0.98 (basket), 0.95 (Stack), 0.83 (basket) and 0.88 (Yelp). Since accuracy on the training sets are even higher, this means that only a few training examples are being miss-classified and the target values at 50-th iteration are mostly close to 0.0.

The two most prominent members of the second class of results are IMDB and movie where the feature importance scores are still noticeably growing. On the other hand, the order of the features is mostly fixed, so trees are similar to each other, but unable to fully solve the predictive problem. Indeed, the accuracy of the corresponding two models is 0.63 (IMDB) and 0.57 (movie) which is closer to the default accuracy than in the previous cases (see Table 1).

The next observation is that for all six datasets, a group of 1–3 most important features is established. The difference between this group and the other features is most notable for the UWCSE dataset. Here, the goal is to predict, which discipline a given course belongs to, and the most important relation is $taughtBy$(course, person, session). This is not surprising as it is also the link to the $advisedByDiscipline$(person, person, discipline), ranked second. Since professors typically work only in one discipline, the discipline a professor advises someone in, is likely equal to a discipline that the professor teaches.

A similar difference between the top feature and the others is visible also in the cases of the Yelp dataset (as well as for basketball). In the case of Yelp, the goal is to predict the category of a business (e.g., restaurant, Health&Medical etc.), and counting in how many tuples of the *attributes* relation a business appears, is a good indication of the target value. For example, restaurants tend to have a lot of attributes such as classy, hipster, romantic, dessert etc. whereas Health&Medical places are sometimes described only by ByAppointmentOnly.

Table 1. Data characteristics: number of tables, number of relations in the final representation, sum of relation sizes (number of descriptive facts), number of target facts, number of classes, and the proportion of the majority class.

Dataset	Tables	Relations	Descriptive facts	Target facts	Classes	Majority class
basket	9	118	630038	95	2	0.70
IMDB	21	57	614662	8816	4	0.58
movie	5	16	183469	1422	4	0.51
Stack	7	52	383040	5855	5	0.36
UWCSE	12	15	1961	115	5	0.25
Yelp	9	51	3348181	24959	4	0.57

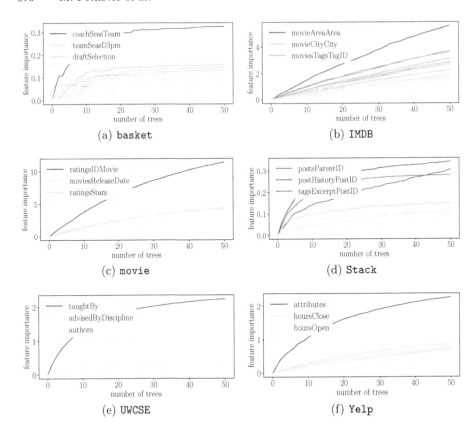

Fig. 3. Development of feature importance values with the number of trees, for different datasets. Every line corresponds to a feature that is present in an ensemble. It's color corresponds to the final feature importance at 50 trees. Additionally, the three most relevant features are listed.

In the cases where the models are not that accurate, feature ranking is a way of seeing whether the model overfits and if some relations should be excluded from the descriptive space. This happens in the case of the IMDB dataset where the goal is to predict the genre of a movie but the most relevant feature is the area where the movie was taken.

7   Conclusions and Future Work

We have proposed an adaptation of the Genie3 feature ranking to relational data, using our adaptation of gradient boosting as the underlying ensemble, and evaluated its appropriateness empirically. The main motivation for choosing boosting was that the trees learned in boosting can be quite shallow as compared to those learned in bagging. However, parameter-tuning for boosted relational trees takes a considerable amount of time, so we plan to extend the proposed approach to other ensemble methods, such as bagging and random forests (and parallelize them). Also, the feature-ranking-motivated definition of the possible split tests will be evaluated in a predictive modeling scenario. Finally, we plan to compare the different relational ensembles against each other and against other learners.

References

1. Breiman, L.: Bagging predictors. Mach. Learn. **24**(2), 123–140 (1996)
2. Breiman, L., Friedman, J., Olshen, R., Stone, C.J.: Classification and Regression Trees. Chapman & Hall/CRC, Boca Raton (1984)
3. Davis, D., Lichtenwalter, R., Chawla, N.V.: Multi-relational link prediction in heterogeneous information networks. In: 2011 International Conference on Advances in Social Networks Analysis and Mining, pp. 281–288 (2011)
4. Dong, Y., et al.: Link prediction and recommendation across heterogeneous social networks. In: 2012 IEEE 12th International Conference on Data Mining, pp. 181–190 (2012)
5. Friedman, J.H.: Greedy function approximation: a gradient boosting machine. Ann. Stat. **29**(5), 1189–1232 (2001)
6. GroupLens Research: Imdb dataset. https://grouplens.org/datasets/hetrec-2011/
7. Guyon, I., Elisseeff, A.: An introduction to variable and feature selection. J. Mach. Learn. Res. **3**, 1157–1182 (2003)
8. He, J., Liu, H., Hu, B., Du, X., Wang, P.: Selecting effective features and relations for efficient multi-relational classification. Comput. Intell. **26**, 258–281 (2010)
9. Huynh-Thu, V.A., Irrthum, A., Wehenkel, L., Geurts, P.: Inferring regulatory networks from expression data using tree-based methods. PLOS ONE **5**(9), 1–10 (2010). https://doi.org/10.1371/journal.pone.0012776
10. Moore, A.W.: Basket dataset. http://www.cs.cmu.edu/~awm/10701/project/data.html
11. Natarajan, S., Kersting, K., Khot, T., Shavlik, J.: Boosted Statistical Relational Learners. SCS. Springer, Cham (2014). https://doi.org/10.1007/978-3-319-13644-8
12. Pio, G., Serafino, F., Malerba, D., Ceci, M.: Multi-type clustering and classification from heterogeneous networks. Inf. Sci. **425**, 107–126 (2018)
13. Quinlan, J.R.: Boosting first-order learning. In: Arikawa, S., Sharma, A.K. (eds.) ALT 1996. LNCS, vol. 1160, pp. 143–155. Springer, Heidelberg (1996). https://doi.org/10.1007/3-540-61863-5_42
14. Serafino, F., Pio, G., Ceci, M.: Ensemble learning for multi-type classification in heterogeneous networks. IEEE Trans. Knowl. Data Eng. **30**(12), 2326–2339 (2018)
15. Stack Exchange: Stack dataset. https://archive.org/details/stackexchange
16. Sun, Y., Barber, R., Gupta, M., Aggarwal, C.C., Han, J.: Co-author relationship prediction in heterogeneous bibliographic networks. In: 2011 International Conference on Advances in Social Networks Analysis and Mining, pp. 121–128 (2011)
17. Vens, C.: Complex aggregates in relational learning. Ph.D. thesis, Faculteit Ingenieurswetenschappen, Katholieke Univeristeit Leuven (2007)
18. Škrlj, B., Kralj, J., Lavrač, N.: Targeted end-to-end knowledge graph decomposition. In: Riguzzi, F., Bellodi, E., Zese, R. (eds.) ILP 2018. LNCS (LNAI), vol. 11105, pp. 157–171. Springer, Cham (2018). https://doi.org/10.1007/978-3-319-99960-9_10
19. Yelp: Yelp dataset. www.yelp.com/dataset_challenge

Multi-objective Discrete Moth-Flame Optimization for Complex Network Clustering

Xingjian Liu[1], Fan Zhang[1], Xianghua Li[1], Chao Gao[1], and Jiming Liu[2(✉)]

[1] College of Computer and Information Science, Southwest University, Chongqing, China
[2] Department of Computer Science, Hong Kong Baptist University, Kowloon, Hong Kong
jiming@comp.hkbu.edu.hk

Abstract. Complex network clustering has been extensively studied in recent years, mostly through optimization approaches. In such approaches, the multi-objective optimization methods have been shown to be capable of overcoming the limitations (e.g., instability) of the single-objective methods. Nevertheless, such methods suffer from the shortcoming of incapability of maintaining a good tradeoff between exploration and exploitation, that is, to find better solutions based on the good ones obtained so far. In this paper, we present a new nature-inspired heuristic optimization method, called multi-objective discrete moth-flame optimization (DMFO) method, which achieves such a tradeoff. We describe the detailed algorithm of DMFO that utilizes the Tchebycheff decomposition approach with an l_2-norm constraint on the direction vector (2-Tch). Furthermore, we show the experimental results on synthetic and several real-world networks that verify that the proposed DMFO and the algorithm are both effective and promising for tackling the task of complex network clustering.

Keywords: Complex network clustering · Multi-objective optimization · Discrete moth-flame optimization · Decomposition

1 Introduction

Complex networks are ubiquitous in the real world, some examples of which include biological networks [1], social networks [2], and transportation networks [3]. Computationally speaking, a complex network can be viewed as a graph in which nodes and links correspond, respectively, to the objects and the relationships between them in a certain complex system. Of great interest in analyzing a complex network is to detect its intrinsic cluster structure, such as those found in social networks [2]. Here, a cluster refers to a subset of closely related or linked nodes, while their connections with the nodes in other communities are sparse [4]. The network clustering aims to discover such community structures

© Springer Nature Switzerland AG 2020
D. Helic et al. (Eds.): ISMIS 2020, LNAI 12117, pp. 372–382, 2020.
https://doi.org/10.1007/978-3-030-59491-6_35

in a network, which can also be viewed as a task of dividing the network into various subsets of nodes according to their characteristics.

Network clustering, in essence, can be modeled as an optimization problem [5]. Several optimization algorithms for solving this problem have been proposed in recent years. Traditional single-objective methods first select an optimal objective (such as the modularity Q [6] or the community score [7]) and apply different strategies to optimize such an objective. For example, GA-Net optimizes the community score using a genetic algorithm for network clustering in social networks [7]. DCRO optimizes the modularity by simulating a chemical reaction process in order to achieve a better clustering result [8]. Generally, one of the key short-comings in the above-mentioned methods lies in that they consider only a single structural attribute in the formulation of their objective functions, even though network clustering is required to address multiple underlying structural attributes. To overcome this limitation, multi-objective methods that mine multiple attributes of a network simultaneously would be desired so as to improve the accuracy and efficiency of network clustering [9–12]. Along this direction, several multi-objective optimization methods for network clustering have been developed that simultaneously make tradeoffs among multiple contradictory objectives. For instance, a multi-objective network clustering method, called MOGA-Net, applies the non-dominated sorting genetic algorithm-II (NSGA-II) to optimize both the community score and community fitness simultaneously [13]. Two contradictory optimization functions, kernel K-means (KKM) and ratio cut (RC), are used in developing the discrete multi-objective particle swarm optimization (MODPSO) method for network clustering [9].

While they represent an improvement over the single-objective methods, the existing multi-objective optimization methods also have certain limitations, the most notable of which is known as the problem of instability when performing network clustering, that is, the processes of exploration and exploitation cannot be well balanced. Such an imbalance may cause inefficient search of the solution space, and as a result, some solutions may fall into local optima [14].

The moth-flame optimization (MFO) algorithm is a new stochastic optimization algorithm [15]. Combined with the simple flame generation (SFG) and the spiral flight search (SFS) strategy, the MFO is capable of achieving a good balance between exploration and exploitation. By incorporating these characteristics, MFO has been successfully utilized to deal with many real-word scientific and engineering optimization problems. However, the existing MFO variants cannot be applied to network clustering problems well. In this paper, we aim to further develop a novel multi-objective discrete moth-flame optimization algorithm (DMFO) for network clustering. On the basis of the label-based representation of discrete individuals, a discrete MFO variant is first extended by redesigning the SFG and SFS strategies. Moreover, a multi-objective optimization framework based on the two contradictory optimization functions (i.e., KKM and RC) is presented for network clustering. Experimental results show that our DMFO method obtains the outstanding performance.

The rest of this paper is organized as follows. The problem formulation is given in Sect. 2. Section 3 presents the proposed method for network clustering. Extensive experiments are carried out to show the performance of DMFO in Sect. 4. Finally, basic concluding remarks are discussed in Sect. 5.

2   Problem Formulation

A complex network can be modeled by an undirected graph $G = (V, E)$, where V and E represent the sets of nodes and links, respectively. A community of complex network is defined as a group of nodes, which have more intra-links than inter-links [4]. The task of network clustering in a complex network is to obtain densely connected subgraphs $G_i(i = 1, \cdots, k)$ in G, where G obeys $\cup_{1 \le i \le k} G_i = G$ and $\cap_{1 \le i \le k} G_i = \varnothing$.

The process of network clustering can be formulated as a multi-objective optimization problem (MOP) [9] in which several criteria are proposed to measure the densities of intra-links and inter-links. In this paper, we select the kernel K-means (KKM) and the ratio cut (RC) as two conflicting objective functions to maximize the internal connections and to minimize the external connections, respectively [9]. Given a network G consisting of n nodes and m links, the adjacent matrix of G is $A = (A_{ij})_{n \times n}$. If G has a cluster partition $C = \{C_1, \cdots, C_k\}$ in which C_i is the node set of subgraph $G_i(i = 1, \cdots, k)$, the multi-objective optimization problem can be formulated as Eq. (1).

$$\min \begin{cases} f_1 = KKM = 2(n - k) - \sum_{i=1}^{k} \frac{L(C_i, C_i)}{|C_i|} \\ f_2 = RC = \sum_{i=1}^{k} \frac{L(C_i, \overline{C_i})}{|C_i|} \end{cases} \tag{1}$$

where $L(C_i, C_i) = \sum_{i \in C_i, j \in C_i} A_{ij}$ and $L(C_i, \overline{C_i}) = \sum_{i \in C_i, j \in \overline{C_i}} A_{ij}$ mean the densities of internal and external links for nodes in $C_i(i = 1, \cdots, k)$, respectively.

To evaluate the quality of network partitions, we use the modularity Q [6] and the normalized mutual information (NMI) [16] as evaluation metrics, which are defined in Eq. (2) and Eq. (3), respectively. If the real community partition of a network is known, we evaluate the performance of the algorithm using both NMI and Q. Otherwise, only Q is used to estimate the clustering results.

$$Q = \sum_{s=1}^{N_c} \left(\frac{l_s}{2m} - \left(\frac{d_s}{2m} \right)^2 \right) \tag{2}$$

where N_c is the number of detected clusters in a network. l_s and d_s denote the total number of links and degrees in the cluster s, respectively. Specifically, the larger the Q value is, the higher the clustering quality of the algorithm is.

$$NMI = \frac{-2 \sum_{i=1}^{c_A} \sum_{j=1}^{c_B} C_{ij} \log (C_{ij} n / C_{i.} C_{.j})}{\sum_{i=1}^{c_A} C_{i.} \log (C_{i.}/n) + \sum_{j=1}^{c_B} C_{.j} \log (C_{.j}/n)} \tag{3}$$

where $c_A(c_B)$ denotes the number of clusters in partition $A(B)$. C represents the confusion matrix. $C_{i.}(C_{.j})$ means the sum of the elements of C in the ith row

(j^{th} column). The larger the NMI value is, the closer the obtained division is to the intrinsic partition. The value of NMI is between 0 and 1.

3 The Proposed Method

In this section, we first introduce the representation and initialization of discrete individuals. Then, we redesign two strategies of DMFO (i.e., SFG and SFS) to achieve a good balance of exploration and exploitation in the clustering process. Finally, an improved Tchebycheff decomposition method is applied to optimize the multi-objective network clustering problem.

3.1 Representation and Initialization of Discrete Individuals

The representation problem of discrete position is essential for MFO-based optimization. It is a key step for an algorithm to encode the scheme of a solution. This paper adopts the label-based representation strategy whose label denotes a cluster value of a node [9]. Although there are two populations of moths and flames in MFO, their positions, in essence, represent a division of a network. Therefore, we adopt the same representation strategy. More specifically, if there are D nodes in a network, the positions of the ith moth and the ith flame are redefined based on Eq. (4).

$$M_i = \{m_{i1}, \cdots, m_{ij}, \cdots, m_{iD}\}, F_i = \{f_{i1}, \cdots, f_{ij}, \cdots, f_{iD}\} \qquad (4)$$

where $m_{ij}(f_{ij})$ denotes the group value of the jth node in the ith moth (flame). j is a random integer in the range of [1, D]. If $m_{ij}(f_{ij})$ and $m_{ik}(f_{ik})$ are equal, then the jth and k^{th} node of the ith moth (flame) belong to the same cluster.

Figure 1 presents an illustration of the mechanism for moth (of flame) encoding and decoding, based on its discrete position representation. To speed up the convergence of the proposed DMFO, we apply a label propagation strategy, as defined in [9], to initialize the moth population randomly.

3.2 SFG Strategy of DMFO

In the proposed DMFO method, the aim of SFG strategy is redesigned to avoid falling into local optima during the clustering process. More specifically, we first select the best individuals in the current two generations as flames to guide the moths to find the optimal solution. Since the flames play a vital role in the discrete variant, the neighborhood-based two-way crossover and neighbor-based mutation (NBM) [9] are adopted in the SFG strategy in order to obtain the well-performed flames while avoiding the flames falling into local optima. Then, the obtained flames are merged with the moths obtained in this iteration, and the well-performed individuals are further selected as the next-generation flames based on the modularity Q.

The core of neighborhood-based two-way crossover is as follows. For the ith individual $F_i(i = 1, \cdots, N)$, an individual $F_j(j = 1, \cdots, H)$ is first selected

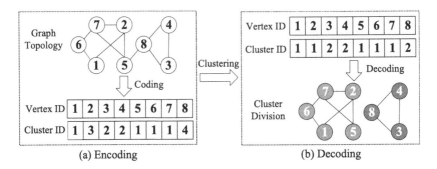

Fig. 1. An illustration of discrete position representation and the mechanism for a discrete individual encoding and decoding. (a) shows the process of encoding a graph into a discrete individual based on a label-based strategy. (b) shows how to decode a discrete individual into cluster label after clustering a discrete individual using the network clustering method. The Cluster ID denotes the cluster label of each node.

randomly in its neighborhood H. The neighborhood where each moth selects the crossover object is chosen by the distances among their aggregation multi-objective weight vectors. Then, F_i and F_j will serve as both the target individual and source individual of the crossover. Finally, two new individuals will be generated by the two-way crossover simultaneously. F_i' represents the better of the two new individuals. If F_i' is better than F_i, replace F_i, otherwise, keep it unchanged.

3.3 SFS Strategy of DMFO

The SFS strategy is redesigned in order to efficiently and extensively search for better results in a solution space. Specifically, the distance U moves when the position of a moth is updated based on Eq. (5).

$$U_i = R_i \cdot Sig\left(\left|e^{bt} \cdot \cos 2\pi t\right|\right) \tag{5}$$

where b is a constant and t represents a random number in the range of $[-1, 1]$. $R_i = M_i \oplus F_j$ represents the distance between the moth M_i and the flame F_j. Specifically, $\oplus$ is a XOR operator. If the corresponding node cluster value between M_i and F_j is the same, R_i is 0, otherwise 1. Suppose $Y = (y_1, \cdots, y_D)$, $X = (x_1, \cdots, x_D)$. In Eq. (5), the function $Y = Sig(X)$ is defined in Eq. (6).

$$y_i = \begin{cases} 1 & \text{for } \mathrm{rand}(0,1) < sigmoid\,(x_i) \\ 0 & \text{for } \mathrm{rand}(0,1) \geq sigmoid\,(x_i) \end{cases} (i = 1, \cdots, D) \tag{6}$$

where $sigmoid(x) = 1/(1 + e^{-x})$ [9].

The redefined U function in Eq. (5) is used to determine if the moth needs to be updated. Here, we assume that the kth and the $(k+1)^{th}$ generation moth are $M_i(k) = (m_{i1}^k, m_{i2}^k, \cdots, m_{iD}^k)$ and $M_i(k+1) = (m_{i1}^{k+1}, m_{i2}^{k+1}, \cdots, m_{iD}^{k+1})$, respectively, and $U_i = (u_1, u_2, \cdots, u_D)$. The update rule of moth position is redefined as $M_i(k + 1) = U_i \odot M_i(k)$. k is the number of iterations and the operator $\odot$ is the specific update strategy of the moth based on Eq. (7).

$$m_{id}^{k+1} = \begin{cases} m_{id}^k & \text{if } u_d = 0 \\ Nbest_d^k & \text{else} \end{cases} \tag{7}$$

where $Nbest_d^k$ represents the cluster value owned by the most of neighbors of the dth node of a moth.

In addition, in the redefined SFS strategy, the number of flames automatically decrease with the iterative process. That is, $F_{no} = $ round $(N - (k(N - 1))/K)$, where the round($\cdot$) operation is used to obtain the nearest integer of $\cdot$. N is the maximum number of flames. K is the iterative maximum number. F_{no} denotes the adaptive number of flame. Figure 2 plots the position update process of a moth according to the redefined SFS strategy.

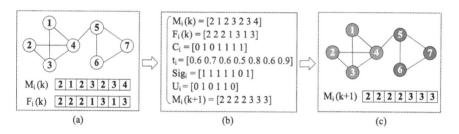

(a)   (b)   (c)

Fig. 2. An illustration of a moth using redefined SFS strategy for position updates. (a) The discrete representation of the ith moth and flame (i.e., M_i and F_i) at the kth generation. (b) The regeneration process of the ith moth. (c) A new i^{th} moth and its graph structure.

3.4 The l_2-norm Constraint

In our DMFO method, an improved Tchebycheff decomposition method with l_2-norm constraint on direction vectors (2-Tch) [17] is used to decompose the multi-objective optimization problem into a set of scalar optimization sub-problems. The 2-Tch method is defined in Eq. (8).

$$\min_{\mathbf{x} \in \Omega} g^{ptch}\left(\mathbf{F}(\mathbf{x}) | \mathbf{w}, \mathbf{z}^*\right) = \max_{1 \le i \le q} \left\{ \frac{f_i(\mathbf{x}) - z_i^*}{w_i} \right\} \tag{8}$$

where $w = (w_1, \cdots, w_q)$ with $\|w\|_2 = 1$ and $w_1, \cdots, w_q \ge 0$ is a weight vector of a subproblem. q is the number of objective functions and $z_i^* = (z_1^*, \cdots, z_q^*)$ is an ideal objective vector with $z_i^* = min\{f_i(x) | x \in \Omega\}(i = 1, \cdots, q)$. Algorithm 1 presents the DMFO method framework for network clustering.

Algorithm 1: Framework of the proposed DMFO

Input: The adjacency matrix A of a network, max generation: Max_iter, moths or flames size: $popsize$, mutation probability: pm, crossover probability: pc, neighborhood size: $niche$.

Output: Pareto front solutions: PF, each solution represents a cluster division.

1 **Initialization:**
2 Initialize $M(0)$;
3 Generate a uniformly distributed weight vector $w = (w_1, \cdots, w_{popsize})$;
4 Initialize the neighborhood of each moth(flame) based on the Euclidean distance of weight vectors;
5 Initialize reference point z^*.
6 **Main loop:**
7 **For** $k = 1 : Max\_iter$ **do**
8     $F(k) \leftarrow$ execute SFG strategy;
9     $M(k+1) \leftarrow$ execute SFS strategy;
10   **If** $rand(0, 1) < pm$ **then**
11   Execute the mutation operation for each $M_i(k+1)$;
12   Evaluate all moths $M(k+1)$;
13   **If** $M_i(k+1)$ is better than $M_i(k)$ **then**
14       $M_i(k+1) \leftarrow M_i(k+1)$;
15   **Else**
16       $M_i(k+1) \leftarrow M_i(k)$;
17   Update reference point z^*;

---

4 Experiments

4.1 Datasets and Parameters

Synthetic Datasets. The extended Girvan-Newman (GN) benchmark networks [6] are selected to verify the performance of proposed DMFO method. These networks contain 128 nodes, which are equally divided into four clusters with 32 nodes, and the average degree of each node is 16. μ denotes a mixing parameter in the range of $[0, 1]$. More specifically, a fraction of $(1 - \mu)$ links of each node is shared with the nodes in the same cluster and the rest of μ links are shared with the other nodes in other clusters. In general, as μ increases, the community structure gradually becomes less evident. When the mixing parameter $\mu < 0.5$, the number of neighbors of each node belonging to its cluster is more than the number of other neighbors in other clusters. Thus, the networks have a clear community structure. When $\mu > 0.5$, the community structure will be very blurred. Therefore, the mixing parameter μ ranges from 0.0 to 0.5, with a spacing of 0.05, resulting in a total of 11 synthetic networks in our experiments. This paper uses Q as the clustering evaluation metrics for the synthetic networks.

Real-World Datasets. Four real-world networks are used to test the performance of DMFO, which include the Zachary's Karate Club network (G_1) [18], the Bottlenose Dolphin network (G_2) [13], the American College Football Network (G_3) [6], and the Books about US Politics network (G_4) [9]. Specifically, G_1 is a social network of friendships among 34 members of a karate club. This network is divided into two groups. G_2 is an animal social network consisting of 62 bottlenose dolphins. G_3 is a football network, which consists of 115 nodes and 616 links representing the football teams and the regular-season matches between different teams, respectively. The whole network is divided into 12 clusters. G_4 is a book network, consisting of 105 nodes and 441 links. We utilize Q and NMI simultaneously to evaluate the clustering performance of the algorithm on four real-world datasets.

Parameter Setting. In DMFO, the number of maximum of iteration for all datasets is set to 80, and the population size is customized according to the size of datasets. Specifically, the size of population for G_1 and G_2 is 150, and the rest is set to 100. In addition, moths are mainly updated according to the flames in DMFO, and it is easy to fall into local optima. Therefore, we increase the mutation rate and crossover rate, i.e., $pm = 0.3$, $pc = 0.9$, for avoiding trapping in the local optimization.

To evaluate the performance of different methods, we select six state-of-the-art network clustering methods, namely, GA-Net [7], MOGA-Net [13], MODPSO [9], QDM-PSO [10], MOPSO-Net [11], MOCD-ACO [12], to compare with the proposed DMFO. The parameter settings for other algorithms are based on their corresponding references. Besides, the averaged results are obtained by running 10 times independently for each experiment in order to obtain more reliable and accurate results.

4.2   Experimental Results

Comparison Results on Synthetic Networks. Figure 3 provides the statistical results of the NMI for each method on synthetic networks with varying mixing parameters from 0 to 0.5. More specifically, DMFO can also detect the true clustering structures even when the mixing parameter μ is 0.45 in which the community structure is blurred. GA and MOGA-Net methods only detect the true cluster division of networks when μ is below 0.15. When $\mu > 0.35$, NMI of GA-Net is 0 which means that such method cannot cluster a network effectively. For the MODPSO, QDMPSO, and MOPSO-Net, the performance of three methods is similar. All of them can discover the real division of a network when $\mu < 0.45$. In summary, our DMFO method achieves outstanding performance for clustering the extended GN benchmark datasets.

Comparison Results on Real-World Networks. Figure 4 illustrates the histogram of the comparison results in terms of Q and NMI on four real-world datasets. As we can see from Fig. 4(a), the optimal average value of Q can be obtained by our DMFO method on four real-world networks. It can be seen from Fig. 4(b) that the proposed DMFO can obtain $NMI = 1$ on both G_1 and G_2

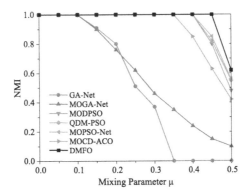

Fig. 3. Comparison results in terms of NMI on the GN extended benchmark datasets. The figure shows that our DMFO outperforms other methods on synthetic datasets.

networks, which means that DMFO can obtain true cluster partition results. In addition, on G_3 and G_4, our DMFO method can also get the best average value of NMI. Therefore, we can conclude that the proposed DMFO can effectively find the optimal cluster partition results compared with other algorithms.

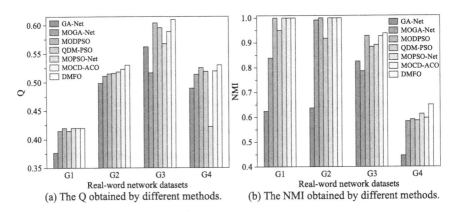

(a) The Q obtained by different methods.   (b) The NMI obtained by different methods.

Fig. 4. Comparison results in terms of Q and NMI on the four real-word datasets. From the figure we can obtain that our DMFO method can achieve the optimal network clustering results on four real-world datasets.

5   Conclusion

In this article, a decomposition-based multi-objective discrete moth-flame optimization (DMFO) method is developed for network clustering. More specifically, the novelty of our proposed method is to achieve a good balance between exploration and exploitation in the clustering process by redefining the two update strategies of MFO algorithm, which allow for better clustering to be obtained. Moreover, the improved Tchebycheff method (2-Tch) is used in the proposed method to decompose the network clustering problem into a family of sub-problems. Experimental results in synthetic and real-world datasets show that our proposed method can obtain the outstanding solutions than other comparison algorithms.

Acknowledgment. This work was supported by the National Natural Science Foundation of China (Nos. 61976181, 11931015) and Natural Science Foundation of Chongqing (Nos. cstc2018jcyjAX0274, cstc2019jcyj-zdxmX0025).

References

1. Chiti, F., Dobson, C.M.: Protein misfolding, amyloid formation, and human disease: a summary of progress over the last decade. Annu. Rev. Biochem. **86**, 27–68 (2017)
2. He, K., Li, Y., Soundarajan, S., Hopcroft, J.E.: Hidden community detection in social networks. Inf. Sci. **425**, 92–106 (2018)
3. Strano, E., Viana, M.P., Sorichetta, A.: Mapping road network communities for guiding disease surveillance and control strategies. Sci. Rep. **8**(1), 1–9 (2018). Article no. 4744
4. Radicchi, F., Castellano, C., Cecconi, F.: Defining and identifying communities in networks. Proc. Natl. Acad. Sci. **101**(9), 2658–2663 (2004)
5. Newman, M.E.J.: Fast algorithm for detecting community structure in networks. Phys. Rev. E **69**(6), 066133 (2004)
6. Girvan, M., Newman, M.E.J.: Community structure in social and biological networks. Proc. Natl. Acad. Sci. **99**(12), 7821–7826 (2002)
7. Pizzuti, C.: GA-Net: a genetic algorithm for community detection in social networks. In: Rudolph, G., Jansen, T., Beume, N., Lucas, S., Poloni, C. (eds.) PPSN 2008. LNCS, vol. 5199, pp. 1081–1090. Springer, Heidelberg (2008). https://doi.org/10.1007/978-3-540-87700-4_107
8. Chang, H., Feng, Z., Ren, Z.: Community detection using dual-representation chemical reaction optimization. IEEE Trans. Cybern. **47**(12), 4328–4341 (2017)
9. Gong, M., Cai, Q., Chen, X., Ma, L.: Complex network clustering by multiobjective discrete particle swarm optimization based on decomposition. IEEE Trans. Evol. Comput. **18**(1), 82–97 (2014)
10. Li, L., Jiao, L., Zhao, J., Shang, R., Gong, M.: Quantum-behaved discrete multiobjective particle swarm optimization for complex network clustering. Pattern Recogn. **63**, 1–14 (2017)
11. Rahimi, S., Abdollahpouri, A., Moradi, P.: A multi-objective particle swarm optimization algorithm for community detection in complex networks. Swarm Evol. Comput. **39**, 297–309 (2018)
12. Ji, P., Zhang, S., Zhou, Z.P.: A decomposition-based ant colony optimization algorithm for the multi-objective community detection. J. Ambient Intell. Humaniz. Comput. **11**(1), 173–188 (2019). https://doi.org/10.1007/s12652-019-01241-1
13. Pizzuti, C.: A multiobjective genetic algorithm to find communities in complex networks. IEEE Trans. Evol. Comput. **16**(3), 418–430 (2012)
14. Chen, X., et al.: A new evolutionary multiobjective model for traveling salesman problem. IEEE Access **7**, 66964–66979 (2019)

15. Mirjalili, S.: Moth-flame optimization algorithm: a novel nature-inspired heuristic paradigm. Knowl.-Based Syst. **89**, 228–249 (2015)
16. Danon, L., Díaz-Guilera, A., Duch, J., Arenas, A.: Comparing community structure identification. J. Stat. Mech. Theory Exp. **2005**(09), P09008 (2005)
17. Ma, X., Zhang, Q., Tian, G., Yang, J., Zhu, Z.: On Tchebycheff decomposition approaches for multiobjective evolutionary optimization. IEEE Trans. Evol. Comput. **22**(2), 226–244 (2018)
18. Duch, J., Arenas, A.: Community detection in complex networks using extremal optimization. Phys. Rev. E **72**(2), 027–104 (2005)

Predicting Associations Between Proteins and Multiple Diseases

Martin Breskvar[1,2]([✉]) [iD] and Sašo Džeroski[1,2] [iD]

[1] Department of Knowledge Technologies, Jožef Stefan Institute, Ljubljana, Slovenia
[2] Jožef Stefan International Postgraduate School, Ljubljana, Slovenia
{Martin.Breskvar,Saso.Dzeroski}@ijs.si

Abstract. We formulate the task of predicting protein-disease associations as a multi-label classification task. We apply both problem transformation (binary relevance), i.e., local approaches, and algorithm adaptation methods (predictive clustering trees), i.e., global approaches. In both cases, methods for learning individual trees and tree ensembles (random forests) are used. We compare the predictive performance of the local and global approaches on one hand and different feature sets used to represent the proteins on the other.

Keywords: Protein-disease associations · Multi-label classification · Predictive clustering trees · Random forests · Network embeddings

1 Introduction

Mutations and other alterations in many genes and gene products (proteins) in the cell can manifest themselves as distinct disease prototypes [10]. Since proteins interact with each other through protein-protein interaction networks (PPIs), the impact of altering one protein can spread along the links of the PPI and affect neighboring proteins. The sets of proteins associated with a given disease and their interactions are known as disease pathways.

Studying a single disease protein in isolation cannot fully explain most human diseases. On one hand, each disease can be associated with many proteins. Computational methods have thus been developed to predict which proteins are associated with a given disease and to bring them together into pathways. Typically, a binary classification problem is formulated for the disease at hand, with each protein corresponding to one instance/example. On the other hand, a protein can be associated with more than one disease. In this context, each disease gives rise to a binary classification problem. Considering the association between proteins and multiple diseases simultaneously corresponds to a multi-label classification (MLC) problem.

Diseases can be divided and subdivided into categories and subcategories, following the Human Disease Ontology [16]. This means that the space of labels (diseases) is organized into a hierarchy. Given that each gene can be associated to

© Springer Nature Switzerland AG 2020
D. Helic et al. (Eds.): ISMIS 2020, LNAI 12117, pp. 383–392, 2020.
https://doi.org/10.1007/978-3-030-59491-6_36

multiple diseases (labels) and that the diseases are organized into a hierarchy, the task at hand becomes a task of hierarchical MLC, i.e., predicting the association between proteins and multiple diseases simultaneously, while taking into account the Disease Ontology.

Proteins (and the genes that encode them) can be described in many different ways. Here we choose to use three different sources of information about the proteins/genes. First, we use the protein-protein interaction network. In particular, we use features from the embedding of the nodes in the PPI network, as proposed by Agrawal et al. [1]. We also use functional annotations of the genes/proteins from the Gene Ontology (GO) [2,7,9]. Finally, we use information on the known pathways in which genes/proteins are involved from the Kyoto Encyclopedia of Genes and Genomes (KEGG) [12].

In this paper, we apply machine learning methods to the task of learning to predict associations between proteins and multiple diseases. We consider this task as a MLC task and apply both problem transformation (binary relevance), i.e., local approaches, and algorithm adaptation methods (predictive clustering trees), i.e., global approaches. In both cases, methods for learning individual trees and tree ensembles (random forests) are used. On one hand, we compare the predictive performance of the local and global approaches. On the other hand, we compare the different feature sets used to represent proteins.

The remainder of the paper is organized as follows. Section 2 describes the different datasets for predicting associations between proteins and multiple diseases, including the used data sources and the feature engineering process. Section 3 briefly summarizes the applied machine learning methods. Section 4 describes the experimental setup. In Sect. 5, we present the results of applying the machine learning methods to the prepared MLC datasets. Section 6 concludes with a discussion and an outline of directions for further work.

2   Data

We use five different sources of information to construct the MLC datasets we consider, three of which have been used by Agrawal et al. [1]. Examples/instances in our datasets correspond to proteins and labels/targets to diseases. To construct the label part of the data, we use DisGeNET [15] and the Human Disease Ontology. For the feature part of the data, we use a human PPI network [8,14], the Gene Ontology and the Kyoto Encyclopedia of Genes and Genomes.

2.1   Data Sources

To derive the label part of our datasets, we use a database of protein-disease associations, which contains pairs $a(p, d)$, indicating that alteration of protein p is linked to disease d. Protein-disease associations were taken from Agrawal et al. [1], who in turn took them from DisGeNET [15]. Agrawal et al. [1] consider 519 diseases that each have at least 10 disease proteins.

Of the 519 diseases, we consider the 302 which can be found in the Disease Ontology. Namely, for each of the 519 diseases, represented with UMLS (Unified Medical Language System) [5] codes, only 302 diseases map to the codes in the Disease Ontology.

We use the human PPI network compiled by Menche et al. [14] and Chatr-Aryamontri et al. [8], as provided by Agrawal et al. [1]. The network contains physical interactions between proteins experimentally confirmed in humans. This includes metabolic enzyme-coupled interactions and signaling interactions. The network is unweighted and undirected with $n = 21,557$ proteins and $m = 342,353$ experimentally confirmed interactions. Proteins are mapped to genes and the largest connected component of the PPI network is used.

The Gene Ontology (GO) knowledge base is the world's largest source of information on the functions of genes. Terms in GO describe the molecular functions, cellular locations, and processes gene products may carry out. Genes (and proteins) are annotated with terms from GO and we use these annotations to generate features describing proteins.

The Kyoto Encyclopedia of Genes and Genomes (KEGG) is a resource for understanding high-level functions and utilities of biological systems, such as the cell and the organism, from genomic and molecular-level information. KEGG is an integrated database resource consisting of eighteen databases. In this paper, we use one of them, namely the KEGG PATHWAY database.

2.2 Feature Engineering

The first set of features that we used to describe proteins was derived from the PPI network mentioned above. In particular, an embedding of the nodes in the network was generated by using the *node2vec* [11] approach: This is a neural embedding approach that learns a vector representation for each protein using a single-layer neural network and random walks. The default parameter values of *node2vec* were used, generating an embedding of 128 dimensions (features).

An alternative representation of the proteins is obtained by using functional annotations from the Gene Ontology (GO). In principle, each term from the GO gives rise to one binary feature, which is true for a given gene/protein if the latter is annotated with the term at hand. We filter the resulting features by excluding those terms that are not assigned to any of the considered proteins (e.g., are false throughout). The total number of generated GO features is 11,695.

In a similar manner, we use KEGG to derive new features describing the proteins. Each term from KEGG corresponds to one binary feature. The feature is true for a given protein if the protein is annotated with the term, i.e., is known to belong to the pathway corresponding to the KEGG term. The total number of generated KEGG features is 222.

2.3 Disease Groups

It is known that a single protein can influence several diseases and that a single disease can be affected by several proteins. Therefore it may be beneficial to

Table 1. Properties of the generated datasets. The datasets share a common set of features (embeddings and GO/KEGG terms) and have different numbers of labels (diseases) and examples. The features from network embeddings are numeric, whereas the GO/KEGG features are binary. The column l_c denotes label cardinality, i.e., average number of labels per example.

	Examples	Features	Labels	l_c
Embeddings				
All diseases	5,255	128	302	2.398
Cardiovascular system diseases	951		33	1.221
Cancer	2,214		68	1.486
Immune system diseases	1,055		21	1.144
Nervous system diseases	1,063		44	1.277
GO/KEGG				
All diseases	5,199	11,917	302	2.410
Cardiovascular system diseases	946		33	1.222
Cancer	2,203		68	1.488
Immune system diseases	1,049		21	1.145
Nervous system diseases	1,051		44	1.279

narrow the set of considered diseases by focusing the analysis and model learning on specific subsets of it. We study four groups of diseases, defined by the Disease Ontology. We generate subsets of the original datasets by examining the second-level of the ontology. This level consists of several categories. Here, we focus only on the following four: cardiovascular system diseases (33), various types of cancers (68), immune system diseases (21) and nervous system diseases (44).

For each of the considered disease groups, new variants of the original datasets are constructed, containing features generated from neural embeddings as well as GO/KEGG features. This results in 8 new variants of the original two datasets. Considering only a subset of the diseases can result in unlabeled examples. This problem is less prominent when examples in the dataset are densely labeled. However, in our case, the label space is particularly sparse. In both original datasets, on average, each example is labeled only with less than 3 out of the 302 possible labels. The resulting 8 generated variants consequently contain considerably fewer examples than the original two datasets. Table 1 summarizes all generated datasets, which are available online[1]. The entry "all diseases" refers to the two original/complete datasets.

3   Methods

Local vs. Global Approaches to MLC. The tasks we address in this paper are tasks of MLC. Such tasks can generally be approached in two ways: locally or

---

[1] http://kt.ijs.si/martin_breskvar/data/ismis2020.

globally. The local approaches learn a model for each label separately, whereas the global approaches learn one model for all labels simultaneously. In this paper, we use both.

Local Approach. The local approach, known under the name of binary relevance, addresses the MLC task by splitting it up into many binary classification tasks. This approach is often used and performs well. Its advantage is the ability to use any readily available binary classification method. This approach yields as many models as there are labels. Its disadvantage is its inability to capture potential relations in the target space.

Global Approach. The global approaches require specially crafted methods in order to capture the potential relations in the target space. Such approaches result in one model, which is able to make predictions for all considered labels. Such approaches have been shown to reduce overfitting and computational effort. The disadvantage of using global approaches is the need to use specially designed methods that can handle more complex target spaces.

Single Trees and Tree Ensembles for MLC. In this paper we build tree models and ensembles thereof. In particular, we use predictive clustering trees (PCTs) [4] as individual models and Random forests of PCTs [13] as model ensembles to address the MLC task.

Single Trees. A PCT is a generalized decision tree, obtained with the standard top-down induction (TDI) algorithm for decision trees. The induction algorithm for PCTs is general in the sense that it allows for different impurity and proto-type functions. The impurity function is used to calculate the homogeneity of a given set of data instances, whereas the prototype function is applied to calculate representative target values for the set that can be used as predictions. With this, PCTs can address many different tasks, including regression and classification, (hierarchical) multi-target regression and (hierarchical) multi-label classification, and time series prediction. The CLUS software package[2] contains implementations of all mentioned approaches.

To address the MLC task, the PCT induction algorithm uses the *impurity function* $IM(E) = \frac{1}{|T|} \sum_{a \in T} \text{Gini}_a(E)$, where T is the set of all considered labels and *Gini* is the Gini index. The *prototype function* is defined as a vector $\hat{y}$, where the components $(\hat{y}_a)$ represents the predictions for target attributes (labels) a. The presence of a label in a leaf node is predicted if the majority of training data instances, that arrive to that node, are labeled with it.

Tree ensembles. An ensemble is a set of models, called base predictive models, where each model contributes to the overall prediction. It is a common practice to use ensembles to lift the predictive performance of the base predictive models and thus obtain better predictive performance. Ensembles make predictions by combining the predictions of their base models (we use probability distribution voting [3] that has been extended towards MLC). We build our ensembles

---

[2] http://source.ijs.si/ktclus/clus-public/.

with the Random forest algorithm [6] that has been extended to multi-target prediction [13], including MLC, and uses PCTs as base predictive models.

4   Experiments

Experimental Setup. We consider two datasets, with the same examples (proteins) and targets (disease associations), composed as previously described. They differ in the set of features (attributes/independent variables). The first dataset comprises the features obtained via the neural embedding (NE) of the nodes of the PPI. The second dataset comprises the features obtained by using the GO and the KEGG databases. In Table 1, these two datasets are marked with "all diseases". In addition to these two datasets, we also consider their variants, generated according to the disease groups described in Sect. 2.3.

We apply PCTs and random forests thereof in four different ways. We learn global single trees, global ensembles, local single trees (one tree per label) and local ensembles (one ensemble per label). When building single trees (global or local), we apply F-test pruning, with the optimal value of the F parameter determined by internal 3-fold cross-validation.

Method Parameters. For single trees, we search for the optimal F-test value considering the following values: $0.125, 0.1, 0.05, 0.01, 0.005, 0.001$. Random forests use 100 base predictive learners (PCTs), which are left unpruned and evaluate the candidate splits for 25% of all input attributes.

Evaluation Measures. We evaluate our models by calculating two measures of predictive performance: the area under the average receiver operating characteristic (AU$\overline{\text{ROC}}$) and the area under the average precision-recall curve (AU$\overline{\text{PRC}}$). Standard AUROC and AUPRC values are calculated for each target (disease) separately and then averaged across all targets (302 diseases, in case of "all diseases" datasets). We selected to monitor these two measures because they are threshold independent (unlike the accuracy, precision and recall measures). Although AUROC is widely used, it can give overly optimistic evaluation scores in cases where a large skew in the class distribution is present. This happens because correctly predicting the absence of a label is rewarded. To address this issue, we also report AUPRC, which handles this deficiency appropriately.

We estimated the predictive performance of the obtained models by using 10-fold cross-validation. The results which we report were obtained on the testing folds. In addition to the estimated predictive performance, we also report the *overfitting score*: a measure of how much a model overfits the training data. We calculate the overfitting score (OS) for AU$\overline{\text{ROC}}$ as follows:

$$\text{OS} = (\text{AU}\overline{\text{ROC}}_{train} - \text{AU}\overline{\text{ROC}}_{test})/\text{AU}\overline{\text{ROC}}_{train}.$$

The OS for AU$\overline{\text{PRC}}$ is calculated analogously. Smaller values are better, because they indicate less overfitting.

5   Results

Table 2 gives the results, i.e., estimates of predictive performance on unseen data, in terms of AUROC and AUPRC scores, respectively. Table 3 gives the overfitting scores, calculated for the AUROC and AUPRC metrics. We investigate the predictive performance along the following dimensions:

1. Which approach performs better (local or global)?
2. Which method performs better (single tree or ensembles)?
3. Which input data is more informative?
4. Which models overfit less?

Table 2. AUROC and AUPRC results for local and global models as well as single tress and ensembles for the considered datasets.

	AUROC				AUPRC			
	Embeddings		GO/KEGG		Embeddings		GO/KEGG	
	Local	Global	Local	Global	Local	Global	Local	Global
All diseases								
Single tree	0.405	0.335	0.469	0.332	0.013	0.006	0.053	0.006
Random forest	0.633	0.622	0.730	0.712	0.042	0.042	0.121	0.092
Cancer								
Single tree	0.461	0.348	0.487	0.346	0.032	0.017	0.081	0.017
Random forest	0.623	0.607	0.705	0.696	0.077	0.072	0.150	0.119
Cardiovascular system diseases								
Single tree	0.473	0.315	0.483	0.307	0.050	0.029	0.165	0.029
Random forest	0.637	0.619	0.715	0.732	0.107	0.118	0.267	0.212
Nervous system diseases								
Single tree	0.506	0.320	0.532	0.313	0.061	0.021	0.146	0.021
Random forest	0.640	0.640	0.741	0.721	0.132	0.129	0.240	0.189
Immune system diseases								
Single tree	0.491	0.358	0.504	0.347	0.059	0.044	0.114	0.044
Random forest	0.607	0.578	0.714	0.711	0.097	0.089	0.199	0.177

The results suggest that the single tree models, obtained with the local approach, outperform the global single tree models for both the dataset that contains neural embeddings as input features and the dataset constructed from GO terms and KEGG pathways. The differences in performance are substantial, although the overall performances of both local and global single tree models do not seem encouraging. For ensemble models, the differences in performance are much smaller and performances can be considered comparable.

Table 3. Overfitting scores calculated from AUROC and AUPRC for local and global models as well as single trees and ensembles for the considered datasets.

	AUROC				AUPRC			
	Embeddings		GO/KEGG		Embeddings		GO/KEGG	
	Local	Global	Local	Global	Local	Global	Local	Global
All diseases								
Single tree	0.363	0.330	0.304	0.336	0.948	0.229	0.837	0.232
Random forest	0.367	0.378	0.270	0.288	0.958	0.958	0.879	0.908
Cancer								
Single tree	0.437	0.304	0.322	0.308	0.947	0.212	0.814	0.229
Random forest	0.377	0.393	0.295	0.303	0.923	0.928	0.850	0.881
Cardiovascular system diseases								
Single tree	0.439	0.370	0.363	0.385	0.926	0.220	0.696	0.223
Random forest	0.363	0.381	0.285	0.268	0.893	0.882	0.733	0.788
Nervous system diseases								
Single tree	0.419	0.360	0.302	0.374	0.915	0.261	0.728	0.277
Random forest	0.360	0.360	0.259	0.279	0.868	0.871	0.759	0.810
Immune system diseases								
Single tree	0.466	0.284	0.347	0.305	0.926	0.185	0.792	0.200
Random forest	0.393	0.422	0.286	0.289	0.903	0.911	0.801	0.823

Although predictive superiority of ensembles was expected, the poor performance of single trees was somewhat surprising. It turned out that the low performance of single trees was due to extensive F-test pruning. The results also suggest that the GO/KEGG input features carry more information as compared to the input features obtained with neural embeddings from the PPI network. The performance scores show substantial improvements when using GO/KEGG instead of neural embeddings.

In terms of overfitting, the results are as follows. When using the neural embeddings as input features, single tree global models tend to overfit less (in case of AUPRC, the difference is substantial) w.r.t. local single trees. However, in the ensemble setting, global ensembles overfit *slightly more* w.r.t. local ensembles in terms of AUROC and no difference is visible in terms of AUPRC. Similar observations can be made for the GO/KEGG dataset. Global single trees overfit more than the local single trees in terms of AUROC and considerably less in terms of AUPRC.

Models built for the described disease groups exhibit improved predictive performance in terms of AUPRC as compared to the models built on datasets that contain all diseases. Improvement of predictive performance on the same datasets is comparable in terms of AUROC. The predictive performance advantage of using GO/KEGG features over neural embeddings is retained on datasets that contain only labels of a specific disease group. Global models do not improve over local ones and perform comparably as on the datasets with all diseases.

In terms of $\overline{\text{AUPRC}}$, the disease group models overfit slightly less than the models for all diseases. The overfitting scores, in terms of $\overline{\text{AUROC}}$, seem comparable to the ones obtained with models built on datasets with all diseases.

6   Conclusions and Further Work

In this paper, we have presented results of learning predictive models for associating proteins with 302 diseases. We consider two groups of datasets, where each group consists of five datasets: a dataset with all diseases and 4 variants of it, one for each considered disease group. The first group contains features obtained by using neural embeddings on a PPI network. The second group contains features constructed from Gene Ontology and KEGG pathways. We have presented our results in terms of two evaluation measures, namely $\overline{\text{AUROC}}$ and $\overline{\text{AUPRC}}$. We have also reported overfitting scores which were calculated based on the aforementioned evaluation measures. We were learning local and global models of single predictive clustering trees and ensembles thereof. We show that the GO/KEGG datasets contain more informative features for predicting the considered diseases.

Further work is possible along several directions. We will consider another group of datasets, with both feature sets (neural embeddings and GO/KEGG features). Along this direction, we will also generate embeddings of higher dimensions (i.e., vectors of 256, 512, 1024 features). We also believe that using a hierarchy in the output space can be beneficial. In particular, we will reformulate the learning task into a hierarchical MLC task and take advantage of the Disease Ontology in order to obtain better predictive models. Given the low label cardinality of all considered datasets, we believe it would be beneficial to apply the hierarchical single-label classification (HSC) approach [17]. Feature ranking will also be performed. It is expected that many of the GO/KEGG features will be deemed uninformative: This will speed up the learning process and potentially improve predictive performance. We will also consider evaluation measures only for the top N predicted labels, ordered descending by the predicted probabilities. Examples of such measures are Recall-at-100 and Precision-at-100.

Acknowledgements. We acknowledge the support of the Slovenian Research Agency (grants P2-0103 and N2-0128), the European Commission (grant HBP, The Human Brain Project SGA2), and the ERDF (Interreg Slovenia-Italy project TRAIN). The computational experiments were executed on the computing infrastructure of the Slovenian Grid (SLING) initiative.

References

1. Agrawal, M., Žitnik, M., Leskovec, J.: Large-scale analysis of disease pathways in the human interactome. Pac. Symp. Biocomput. **23**, 111–122 (2018)

2. Ashburner, M., et al.: Gene ontology: tool for the unification of biology. Nat. Genet. **25**(1), 25 (2000)
3. Bauer, E., Kohavi, R.: An empirical comparison of voting classification algorithms: bagging, boosting, and variants. Mach. Learn. **36**(1), 105–139 (1999)
4. Blockeel, H., Raedt, L.D., Ramon, J.: Top-down induction of clustering trees. In: Proceedings of the 15th International Conference on Machine Learning, pp. 55–63. Morgan Kaufmann (1998)
5. Bodenreider, O.: The unified medical language system (UMLS): integrating biomedical terminology. Nucleic Acids Res. **32**(suppl-1), D267–D270 (2004)
6. Breiman, L.: Random forests. Mach. Learn. **45**(1), 5–32 (2001). https://doi.org/10.1023/a:1010933404324
7. Carbon, S., et al.: Amigo: online access to ontology and annotation data. Bioinformatics **25**(2), 288–289 (2008)
8. Chatr-Aryamontri, A., et al.: The biogrid interaction database: 2015 update. Nucleic Acids Res. **43**(D1), D470–D478 (2014)
9. Consortium, G.O.: The gene ontology resource: 20 years and still going strong. Nucleic Acids Res. **47**(D1), D330–D338 (2018)
10. Creixell, P., et al.: Pathway and network analysis of cancer genomes. Nat. Methods **12**(7), 615 (2015)
11. Grover, A., Leskovec, J.: node2vec: Scalable feature learning for networks. In: Proceedings of the 22nd ACM SIGKDD International Conference on Knowledge Discovery and Data Mining, pp. 855–864. ACM (2016)
12. Kanehisa, M., Furumichi, M., Tanabe, M., Sato, Y., Morishima, K.: Kegg: new perspectives on genomes, pathways, diseases and drugs. Nucleic Acids Res. **45**(D1), D353–D361 (2016)
13. Kocev, D., Vens, C., Struyf, J., Džeroski, S.: Tree ensembles for predicting structured outputs. Pattern Recogn. **46**(3), 817–833 (2013). https://doi.org/10.1016/j.patcog.2012.09.023
14. Menche, J., et al.: Uncovering disease-disease relationships through the incomplete interactome. Science **347**(6224), 1257601 (2015)
15. Piñero, J., et al.: Disgenet: a discovery platform for the dynamical exploration of human diseases and their genes. Database **2015** (2015)
16. Schriml, L.M., et al.: Human disease ontology 2018 update: classification, content and workflow expansion. Nucleic Acids Res. **47**(D1), D955–D962 (2018). https://doi.org/10.1093/nar/gky1032
17. Vens, C., Struyf, J., Schietgat, L., Džeroski, S., Blockeel, H.: Decision trees for hierarchical multi-label classification. Mach. Learn. **73**(2), 185 (2008)

Short Papers

Exploiting Answer Set Programming for Building explainable Recommendations

Erich Teppan[1](✉) and Markus Zanker[2](✉)

[1] Universität Klagenfurt, Klagenfurt, Austria
erich@ifit.uni-klu.ac.at
[2] Free University of Bolzano, Bolzano, Italy
mzanker@unibz.it

Abstract. The capability of a recommendation system to justify its proposals becomes an ever more important aspect in light of recent legislation and skeptic users. Answer Set Programming (ASP) is a logic programming paradigm aiming at expressing complex problems in a succinct and declarative manner. Due to its rich set of high level language constructs it turns out that ASP is also perfectly suitable for realizing knowledge and/or utility-based recommendation applications, since every aspect of such a utility-based recommendation capable of producing explanations can be specified within ASP. In this paper we give an introduction to the concepts of ASP and how they can be applied in the domain of recommender systems. Based on a small excerpt of a real life recommender database we exemplify how utility based recommendation engines can be implemented with just some few lines of code and show how meaningful explanations can be derived out of the box.

Keywords: Recommender systems · Explanations · Answer set programming

1 Introduction

The ability of a recommendation system to explain to users why a specific item is recommended has reached considerable research momentum, supported also by recent legislation like GDPR[1] that even codifies a right for explanations of the outcomes of algorithmic decision makers. Although knowledge-based recommendation strategies [9] are a niche topic compared to predominant machine learning based approaches, they are nevertheless in actual use for high-involvement product domains like financial services [3] or consumer products where many variables and aspects are typically considered during decision making processes. Utility-based recommender systems can be considered to be a specific variant of knowledge-based systems, where the matching between user preferences or needs and item properties or features is realized via utility functions. The recent

---

[1] See https://eugdpr.org/ for reference.

© Springer Nature Switzerland AG 2020
D. Helic et al. (Eds.): ISMIS 2020, LNAI 12117, pp. 395–404, 2020.
https://doi.org/10.1007/978-3-030-59491-6_37

revived attention for knowledge-based approaches [1] is actually based on three pillars: the wide availability of structured and unstructured content that can be exploited for knowledge extraction, the promise of achieving beyond accuracy goals [11] like transparency, validity and explainability as well as the development of ever more efficient computational mechanisms for processing declarative knowledge. The answer set programming (ASP) paradigm under the stable model semantics [7] must be seen in the context of the latter pillar. It is a successful logic programming (LP) paradigm that has evolved from an academic discipline rooted in deductive databases to an approach that is practically applicable in many different domains [6]. It turns out that ASP, due to its rich set of different language constructs, seems to be perfectly appropriate for also expressing recommendation problems, particularly those ones that can be naturally modeled based on a knowledge or utility-based recommendation paradigm. One big advantage of encoding a recommendation engine in ASP is the compact high level representation that eases maintenance. Another advantage is that explanations for recommendations can be derived quite naturally as a side product of recommendation calculation based on logic rules.

In this paper, we introduce for the first time how the ASP mechanism can be used to build the complete logic of a utility-based recommender system. Hereby, we focus particularly on how to automatically derive meaningful explanations for recommendations. As a first proof of concept we use an excerpt from a real world knowledge base from the financial service domain [4] and show how a corresponding recommendation engine, in particular calculations of item utilities and explanations, can be expressed in ASP in a succinct way.

2 Related Work

To the best of our knowledge there exist only two papers employing ASP in the context of recommender systems. In [10] the authors propose dynamic logic programming, an extension of ASP, in order to provide a user the means to specify changes of their user profile in order to support the recommendation engine in producing better recommendations. Clearly, the focus of this approach is totally different as it is not concerned with building any aspect of the recommendation engine itself.

Not so for the second paper identified. The system described in [8] consists of two ASP based components. The first component is used for automatic extraction of relevant information from touristic offers contained in leaflets produced by tour operators. This information is in turn added to a tourism ontology, i.e. the second component. The second component is responsible for answering the question which touristic offers match a certain customer profile that is also added to the ontology. Thus, the system described in [8] can be seen as a raw content-based recommender. In contrast to that, the approach discussed herein is different in several ways. First, we do not build upon ontologies, but directly harness the ASP knowledge representation. Second, beyond making solely the binary choice, whether an item is matching the user preferences or not, we calculate fine grained utilities to rank items. Third, we demonstrate meaningful explanations following the taxonomy of [5], where explanations actually reason on all three information categories: user preferences, item properties as well as comparisons with alternatives.

3 Working Example

In order to illustrate the applicability of ASP for knowledge-based recommendation approaches we develop a motivating example. It builds on a utility-based recommendation scenario, where the domain knowledge is encoded by relating user needs and item properties via utility values on different dimensions based on multi-attribute utility theory (MAUT) [2]. Due to the clarity of the dependencies we build on a published example [4] from the domain of financial services that actually constitutes a small excerpt from a real world recommender system in use by a European financial services provider.

The core of MAUT specifications are utility dimensions. These dimensions are abstract concepts which contribute to the overall utility of an item as perceived by a user. In our working example there are two dimensions: *profit*, i.e. the financial return of an investment, and *availability*, i.e. how quickly an investment can be converted back to cash. Note, that these utility dimensions are abstract concepts capable to model and map also indirect relationships between user and/or product attributes. In our example, the simplified user profiles contain just two attributes (or explicit preferences), namely the *duration* of an investment and the personal *goal* of a user with respect to some investment. For the investment *duration* let us assume three distinct values *long*, *medium* and *short* term. Analogously, for the *goal* description we again assume three values: *savings* for a rainy day, profitable *growth* and *venture*. Similarly, properties also describe product items. Let us assume that financial products can be described based on the portions of *shares* a product contains (0%, $\leq 30\%$, $\leq 60\%$, $\leq 80\%$, $> 80\%$) as well as expected price *fluctuation* (*low*, *medium* and *high*).

The connection between utility dimensions and attribute values can be given by so called scoring values (or just scores for short) that specify how much user preferences and product properties contribute to the utility dimensions. The higher the score for a dimension based on a given attribute value the more this attribute contributes to fulfilling this dimension. In this example we use scores between 0 and 10 where a score of 0 signifies that an attribute value does not positively contribute to a dimension. Figure 1 depicts the scores for the user attribute values and resulting utility dimension weights for an example user profile. Conforming to MAUT, the weight that is given to a dimension is estimated as the sum of user attribute value scores on that dimension. Figure 2 depicts the scores for the product attribute values, attribute values for a small example set of four financial products and the resulting product contributions. For a given product, its contribution to a dimension is calculated as the sum of its attribute value scores on that dimension. Finally, the overall utility of each product for a specific user is defined by the sum of weighted dimension contributions. The computations for our toy example are given in Fig. 3.

Hence, for this example scenario, the top ranked product for user *Simon* would be *bondBX* followed by *mutual fund AXP*. These overall utilities can be utilized already for a very basic justification for each recommendation, such as 'Item X promises the highest overall utility u_x'. A closer look to the dimensional contributions and weights provides us means for further argumentation, for instance, 'The top ranked *bond BX* scores also strongest in terms of *availability*, which is also considered most relevant given your preferences p_u'. On the other hand, *mutual fund AXP* scores second best on both dimensions, i.e. it can be justified as being a good compromise. Although *mutual fund CXP* is the worst overall it shows the best contributions to *profit*. For *bond RX* there is, for instance, a clear *negative* explanation why it must not be recommended since it is dominated by *mutual fund AXP* that offers the same utility score in terms of *availability*, but even more in terms of *profit*. Thus, there is no rational reason for opting towards *bond RX* [12–14].

duration	profit	availability		goal	profit	availability
long	10	2		savings	2	8
medium	7	6		growth	6	4
short	3	10		venture	10	2

profile	duration	goal		profile	profit	availability
Simon	medium	savings		Simon	7+2=9	6+8=14

Fig. 1. User scores, example user profile and resulting dimension weights

shares	profit	availability		fluctuation	profit	availability
0%	2	7		high	7	4
1 − 30%	4	6		medium	5	6
31 − 60%	5	5		low	1	8
61 − 80%	8	2				
81 − 100%	10	1				

product	shares	fluctuation		product	profit	availability
mutual fund AXP	31-60%	medium		mutual fund AXP	5+5=10	5+6=11
bond BX	0%	medium		bond BX	2+5=7	7+6=13
bond RX	0%	high		bond RX	2+7=9	7+4=11
mutual fund CXP	61-80%	high		mutual fund CXP	8+7=15	2+4=6

Fig. 2. Product scores, attribute values and resulting utility contributions

4   ASP for Recommenders

In this section we show how utility based recommendation and explanation can be expressed based on the answer set programming (ASP) paradigm in form of

user	product	utility
Simon	mutual fund AXP	9*10+14*11=244
	bond BX	9*7+14*13=245
	bond RX	9*9+14*11=235
	mutual fund CXP	9*15+14*6=219

Fig. 3. Product utilities for the example user profile

first order logic facts and rules. Since it is not possible at this point to give a complete introduction to ASP, we explain the ASP code snippets on an intuitive level[2].

Listing 1.1 shows how the user profile data from Fig. 1, i.e. the actual user requirements, is encoded by ASP logic. Analogously, Listing 1.2 depicts how the scoring and product data in Fig. 2 can be expressed by logic facts.

```
profile("Simon",duration,medium).
profile("Simon",goal,savings).
```

Listing 1.1. User profile given as logic facts

```
user_score(duration,long,profit,10).
user_score(duration,medium,profit,7).
user_score(duration,short,profit,3).
user_score(duration,long,availability,2).
user_score(duration,medium,availability,6).
user_score(duration,short,availability,10).
user_score(goal,savings,profit,2).
user_score(goal,growth,profit,6).
user_score(goal,venture,profit,10).
user_score(goal,savings,availability,8).
user_score(goal,growth,availability,4).
user_score(goal,venture,availability,2).
product_score(shares,"0",profit,2).
product_score(shares,"1-30",profit,4).
product_score(shares,"31-60",profit,5).
product_score(shares,"61-80",profit,8).
product_score(shares,"81-100",profit,10).
product_score(shares,"0",availability,7).
product_score(shares,"1-30",availability,6).
product_score(shares,"31-60",availability,5).
product_score(shares,"61-80",availability,2).
product_score(shares,"81-100",availability,1).
product_score(fluctuation,high,profit,7).
product_score(fluctuation,medium,profit,5).
product_score(fluctuation,low,profit,1).
product_score(fluctuation,high,availability,4).
product_score(fluctuation,medium,availability,6).
product_score(fluctuation,low,availability,8).
product("mutual fund AXP",shares,"31-60").
product("mutual fund AXP",fluctuation,medium).
product("bond BX",shares,"0").
product("bond BX",fluctuation,medium).
product("bond RX",shares,"0").
product("bond RX",fluctuation,high).
product("mutual fund CXP",shares,"61-80").
product("mutual fund CXP",fluctuation,high).
```

Listing 1.2. Scoring and product data given as logic facts

---

[2] For an in-depth introduction to ASP please refer to [6].

For calculating the contributions of all products for all utility dimensions only the single ASP rule given in Listing 1.3 is needed. Like in other logic programming languages (e.g. Prolog) the ':-' operator stands for left implication. To put it very simple, the left hand side of the rule (i.e. left from ':-') specifies the atoms that are to be added to the solution, which in ASP is called answer set, based on the calculations done on the right hand side. The core of the calculation is performed by a #*sum* aggregate that sums up all scoring values for a product *Productname* on a dimension *Dimension*. Terms beginning with capital letters like *Productname* or *Dimension* stand for logic variables. Consequently, this rule 'fires' once for each combination of products and dimensions.

```
contribution(Productname,Dimension,Total):-
  product(Productname,_,_), product_score(_,_,Dimension,_),
  Total=#sum{Score,Attribute:product(Productname,Attribute,Value),
                    product_score(Attribute,Value,Dimension,Score)}.
```

Listing 1.3. Calculation of product contributions in ASP

The atoms produced by the rule in Listing 1.3 and included in the answer set are the following:

```
contribution("mutual fund AXP",profit,10)
contribution("bond BX",profit,7)
contribution("bond RX",profit,9)
contribution("mutual fund CXP",profit,15)
contribution("mutual fund AXP",availability,11)
contribution("bond BX",availability,13)
contribution("bond RX",availability,11)
contribution("mutual fund CXP",availability,6)
```

```
weight(Username,Dimension,Total):-
  profile(Username,_,_), user_score(_,_,Dimension,_),
  Total=#sum{Score,Attribute:profile(Username,Attribute,Value),
                    user_score(Attribute,Value,Dimension,Score)}.
```

Listing 1.4. Calculation of user weights in ASP

Similarly, all user weights based on the profile and scoring facts are calculated by the ASP rule defined in Listing 1.4. This rule produces the following two solution atoms:

```
weight("Simon",profit,9)
weight("Simon",availability,14)
```

```
utility(Username,Productname,Total):-
  profile(Username,_,_), product(Productname,_,_),
  Total=#sum{W*C,Dimension:weight(Username,Dimension,W),
                    contribution(Productname,Dimension,C)}.
```

Listing 1.5. Utility calculation expressed in ASP

Finally, the rule in Listing 1.5 calculates the resulting utilies based on the product contributions and user weights. The solution atoms produced by the rule in Listing 1.5 are:

```
utility("Simon","mutual fund AXP",244)
utility("Simon","bond BX",245)
utility("Simon","bond RX",235)
utility("Simon","mutual fund CXP",219)
```

At this point we want to emphasize that the three rules depicted in Listing 1.3–1.5, which can be seen as a logic representation of the core of a utility-based recommendation engine, are totally generic and do not have to be changed in case of extending the set of products, product or user attributes, attribute values or scoring values.

Building on such an ASP implementation of a recommendation engine, additional rules can be easily added in order to produce solution atoms for explanations. For instance, if we want to support the top ranked item by a corresponding explanation, we can add the rule in Listing 1.6. This rule basically expresses that a product *Productname* with a utility *U* is top ranked if there is no other product *Productname1* with a higher utility *U1*.

```
top_ranked(Username,Productname,U):- utility(Username,Productname,U),
   #count{Productname1: utility(Username,Productname1,U1),U1>U}=0.
```

Listing 1.6. Producing an argument for the top ranked product

The rule in Listing 1.6 produces the following solution atom:

```
top_ranked("Simon","bond BX",245)
```

```
top_in_dimension(Dimension,P,C):- contribution(P,Dimension,C),
   #count{P1: contribution(P1,Dimension,C1),C1>C}=0.
```

Listing 1.7. Calculating the top item in a dimension

Similarly, we can add a rule for identifying whether a product is best in some dimension. Listing 1.7 shows such a rule, which produces the following atoms:

```
top_in_dimension(profit,"mutual fund CXP",15)
top_in_dimension(availability,"bond BX",13)
```

```
domination(P,"dominated by",P1):- product(P,_,_),product(P1,_,_),
   #count{Dimension: contribution(P,Dimension,C),
      contribution(P1,Dimension,C1),C>C1}=0,
   #count{Dimension: contribution(P,Dimension,C),
      contribution(P1,Dimension,C1),C1>C}>=1.
```

Listing 1.8. Calculating totally dominated alternatives

We can also produce counter arguments, for example, on totally dominated product items, i.e. those where there exists an item that is at least equally good in all dimensions and better in at least one dimension. The rule given in Listing 1.8 achieves these *negative* explanations and adds the following atom to the answer set:

```
domination("bond RX","dominated by","mutual fund AXP")
```

5 Evaluation

We evaluated the applicability of our proposed ASP approach with respect real life recommender system scenarios. A typical online recommendation scenario comprises two steps: (1) Based on a given user profile, the recommendation engine first must calculate utilities for all product items. (2) An evolved set of items with the highest utilities is then presented on some form of result page allowing the comparison and explanation of the different proposals. The number of presented items depends on the web page and is commonly between 5 and 20 items. Databases comprising some ten-thousands of product items subject to recommendation can be considered as (very) large yet possible cases. For instance, the streaming platform Netflix offers more than 10000 movies, series and documentations and Amazon Prime more than 20000, and increasing[3].

Consequently, we produced two benchmark data sets, one for step 1 and one for step 2, reflecting the size of real recommender system scenarios. The first benchmark dataset consists of 48 generated problem instances comprising product item databases, a random user profile and random item and user scoring rules. The problem instances differ in (a) the number of products (1000, 10000 or 100000), (b) the number of utility dimensions (2, 4, 8 or 16), and (c) the number of user/item attributes (2, 4, 8 or 16). We assumed numeric product item attributes taking values in the range of $\{0, \ldots, 10\}$. The benchmark dataset for recommendation step 2, i.e. the explanation of an evolved product item set, consists of 32 instances and is similar to the first dataset. However, the numbers of products are significantly smaller, that is 10 or 20, as there are hardly more items in an evolved set that is presented on the result/comparison page.

We solved each problem instance for benchmark 1 with an encoding consisting of Listing 1.3–1.5. We refer to this encoding as the *utility encoding*. Each problem instance of benchmark 2 (i.e. targeting at explanations for evolved sets) we solved with an encoding that consists of Listings 1.3–1.8. We refer to this latter encoding as *explanation encoding*. For each of the test cases we measured the total time and the peak main memory consumption that was needed by the ASP process for coming up with the solution (i.e. answer set). We ran the experiments on an AMD EPYC 7551 with 32 cores @ 2 GHz (turbo core max 3 GHz) and 256 GByte of main memory. As an ASP solver, we used Clingo 4.5.4 in single-core mode.

Summarizing the test cases related to benchmark 2 comprising only up to 20 evolved product items, we can say that calculation times are always below 0.1 s such that also many more types of explanations would never be a problem also in online scenarios. Also concerning benchmark 1 targeting at the calculation of all utilities for all products for a given profile the measured calculation times allow online scenarios. For the most extreme case, that is 100000 products, 16 dimensions, 16 item attributes and 16 user attributes we measured a calculation time of 78 s and a peak memory consumption of roughly 3 GByte. Note that 100000 product items is about five times more than Amazon Prime

---

[3] https://www.techradar.com/news/netflix-vs-amazon-prime-video-which-streaming-service-is-best-for-you.

offers movie and series items in total. Also the maximal number of 16 utility dimensions is very high, and it is hard to imagine a product category where this would be exhausted. For more common scenarios calculation times and memory consumption are much lower, e.g. for the 10000 product test case involving 4 utility dimensions, 8 item attributes and 8 user attributes we measured a calculation time of 1.3 s and a peak memory consumption of 71 MByte. Thus, ASP represents an attractive alternative to efficiently represent the complete recommendation logic in a declarative and highly compact way.

References

1. Anelli, V.W., et al.: Knowledge-aware and conversational recommender systems. In: Proceedings of the 12th ACM Conference on Recommender Systems, pp. 521–522. ACM (2018)
2. Dyer, J.S.: Multiattribute utility theory (MAUT). In: Greco, S., Ehrgott, M., Figueira, J.R. (eds.) Multiple Criteria Decision Analysis. ISORMS, vol. 233, pp. 285–314. Springer, New York (2016). https://doi.org/10.1007/978-1-4939-3094-4_8
3. Felfernig, A., Friedrich, G., Jannach, D., Zanker, M.: An integrated environment for the development of knowledge-based recommender applications. Int. J. Electron. Commer. **11**(2), 11–34 (2006)
4. Felfernig, A., Teppan, E., Friedrich, G., Isak, K.: Intelligent debugging and repair of utility constraint sets in knowledge-based recommender applications. In: Proceedings of the International Conference on Intelligent User interfaces (IUI), pp. 217–226. Springer, Berlin Heidelberg (2008). https://doi.org/10.1145/1378773.1378802
5. Friedrich, G., Zanker, M.: A taxonomy for generating explanations in recommender systems. AI Mag. **32**(3), 90–98 (2011)
6. Gebser, M., Kaminski, R., Kaufmann, B., Schaub, T.: Answer set solving in practice. Synth. Lect. Artif. Intell. Mach. Learn. **6**(3), 1–238 (2012). Morgan and Claypool Publishers
7. Gelfond, M., Lifschitz, V.: The stable model semantics for logic programming. In: Kowalski, R., Bowen, K. (eds.) Proceedings of the Fifth International Conference and Symposium of Logic Programming (ICLP 1988), pp. 1070–1080. MIT Press, Cambridge (1988)
8. Ielpa, S.M., Iiritano, S., Leone, N., Ricca, F.: An ASP-based system for e-tourism. In: Erdem, E., Lin, F., Schaub, T. (eds.) LPNMR 2009. LNCS (LNAI), vol. 5753, pp. 368–381. Springer, Heidelberg (2009). https://doi.org/10.1007/978-3-642-04238-6_31
9. Jannach, D., Zanker, M., Felfernig, A., Friedrich, G.: Recommender Systems: An Introduction. Cambridge University Press, Cambridge (2010)
10. Leite, J., Ilić, M.: Answer-set programming based dynamic user modeling for recommender systems. In: Neves, J., Santos, M.F., Machado, J.M. (eds.) EPIA 2007. LNCS (LNAI), vol. 4874, pp. 29–42. Springer, Heidelberg (2007). https://doi.org/10.1007/978-3-540-77002-2_3
11. McNee, S.M., Riedl, J., Konstan, J.A.: Being accurate is not enough: how accuracy metrics have hurt recommender systems. In: CHI 2006 Extended Abstracts on Human Factors in Computing Systems, pp. 1097–1101. ACM (2006)
12. Teppan, E., Felfernig, A.: Impacts of decoy elements on result set evaluations in knowledge-based recommendation. Int. J. Adv. Intell. Paradigms **1**, 358–373 (2009)

13. Teppan, E.C., Felfernig, A.: Calculating decoy items in utility-based recommendation. In: Chien, B.C., Hong, T.P., Chen, S.M., Ali, M. (eds.) Next-Generation Applied Intelligence. IEA/AIE 2009. Lecture Notes in Computer Science, vol. 5579, pp. 183–192. Springer, Berlin, Heidelberg (2009). https://doi.org/10.1007/978-3-642-02568-6_19
14. Teppan, E.C., Zanker, M.: Decision biases in recommender systems. J. Internet Commer. **14**(2), 255–275 (2015)

Tailoring Random Forest
for Requirements Classification

Andreas Falkner[(✉)] ⓘ, Gottfried Schenner ⓘ, and Alexander Schörghuber ⓘ

Siemens AG Österreich, Vienna, Austria
{andreas.a.falkner,gottfried.schenner,alexander.schoerghuber}@siemens.com

Abstract. Automated and semi-automated classifications of requirements (type and topics) are important for making requirements management more efficient. We report how we tailored a random forest approach
in the EU funded project OpenReq, aiming for sufficient quality for practical use in bid projects. Evaluation with thirty thousand requirements
in English from nine tender documents for rail automation systems in
various countries show that user expectations are hard to meet.

Keywords: Random forest · Evaluation · Industrial application ·
Requirement classification · Text categorization

1 Introduction

Requirements management for large projects is a time-consuming and error-
prone task which can be supported by artificial intelligence [10]. In the Horizon
2020 project OpenReq[1], we developed several solution approaches and evaluated
them with data from bid projects in the domain of railway safety systems.

Requests for proposal (RFP) or tenders for large infrastructure systems are
typically issued by national authorities and comprise natural language documents of several hundred pages with requirements of various kind (domain
specific, physical, non-functional, references to standards and regulations, etc.).
Preparing a proposal (bid) to answer a tender requires (1) to identify the requirements in the tender text and (2) to assign experts to assess the company's compliance to those requirements. The difficult part is the classification of the (real)
requirements w.r.t. predefined topics (which are covered by the experts).

For both tasks, it is important to achieve a very high true positive rate
(recall), because requirements which are not detected or are assigned to the
wrong experts will not be assessed correctly and may lead to high non-compliance
cost. On the other hand, the true negative rate shall also be high, so that unnecessary work is reduced.

The contribution of this work is twofold: Firstly, a new way to tailor the
well-known random forest approach [5] by optimizing the model's configuration

---

[1] http://openreq.eu/.

© Springer Nature Switzerland AG 2020
D. Helic et al. (Eds.): ISMIS 2020, LNAI 12117, pp. 405–412, 2020.
https://doi.org/10.1007/978-3-030-59491-6_38

to requirements classification in general. Secondly, its evaluation in the domain of rail automation (30,000 real-world requirements, 50 topics).

The remainder of this paper is structured as follows: In Sect. 2 we list previous approaches to solve this and similar problems. After presenting our solution in Sect. 3, we report on the evaluation results in Sect. 4. Section 5 summarizes the main outcome and its impact on the users.

2   Related Work

Many approaches to automatic text classification are not specific to requirements management. In the past they tended to be rule-based, but lately (supervised) machine learning has become increasingly popular [13].

Approaches specific to requirements classification vary in the preprocessing NLP pipeline and in their choice of used classifiers. For example, a micro-service for requirements classification developed in the OpenReq project uses Naïve Bayes classifiers [9]. [18] describes a NLP pipeline for extracting requirements from prescriptive documents and uses a SVM classifier to classify the requirements into disciplines.

[14] is an early paper on automatic topic categorization of requirements written in natural language using a bootstrapping approach with machine learning (Naïve Bayes). [17] uses automatic requirement categorisation in an industrial setting to support the review of large natural language specifications in the automotive domain.

Semantic approaches for text classification incorporate not only syntactic but also semantic information, e.g.., provided by systems for automatic information and relation extraction [11]. For a survey of such approaches see [3]. [16] discusses the use of ontologies and semantic technologies in requirements management.

In contrast to new approaches which use pre-trained models, e.g. BERT [8], this work relies solely on traditional machine learning approaches and a model which has been in industrial use for two years.

3   Solution

Text categorization labels paragraphs of natural language documents with pre-defined categories (or classes). It is a typical application of supervised learning which relies on an initial set of labelled instances used for training [1]. We use binary classification for *type classification* (whether an instance is a requirement or not) and multi-label classification for *topics* (an instance can be assigned to either zero or one or several topics). For example, an input instance to classification is the paragraph "The power supply shall consist of the two sources: one main and one for backup." and the corresponding output could be $requirement = yes$ for binary type classification and $topics = \{Power, Diesel\}$ for multi-label topic classification. Internally, we implemented multi-label classification as multiple isolated binary classification problems [21].

Our solution comprises: (1) a *Random Forest* approach which is a proven classifier for text categorization [1], (2) text preprocessing such as tokenization, n-grams, stop word removal and reduction of word inflections, (3) a feature engineering stage which includes calculating feature weights and selecting relevant features [2], and (4) various sampling strategies to overcome problems with imbalanced data [12].

For tailoring this solution, we evaluated various combinations for these steps and identified the most promising model configuration for application to bid projects (for more details, see [19]):

- As a sampling strategy, we analyzed random under-sampling (RUS) [12], SMOTE [6] and no rebalancing. RUS showed superior performance. For training, we apply RUS ten times with 10 different random seeds, resulting in ten training sets. One model is trained per training set and all ten models are finally aggregated using majority vote.
- To remove word inflection, lemmatization (StanfordNLP [15]) [20] and stemming (Porter Stemmer) [20] were compared. Although evaluation revealed that using the lemmatizer increases performance, we decided on the stemmer because of its less restrictive software licence.
- We evaluated usage of tokens based on n-grams, $n \in \{1\}$, $n \in \{1,2\}$, ..., $n \in \{1,2,3,4,5\}$. This parameter has no significant influence on performance – therefore uni-grams are used to keep feature space small.
- Different feature weights were compared: set of words [21], term frequency (TF) [20], TF-IDF [20] and (R)TF-IGM [7]. As this parameter did not show significant influence on performance, we selected TF because of its algorithmic simplicity.
- Using a stop-word list [20] from the Natural Language Toolkit[2] increased performance.
- The following common feature selection methods were evaluated: information gain (IG), χ^2 and term frequency (TF) [21]. TF showed good results and a fast runtime. The algorithm is configured to keep 1,300 features.

Additionally to the described configurations above, a user can set a threshold for positive classification before starting the predictor. Only if the built model's probability for a requirement is greater than or equal to the given threshold, the requirement is classified as positive instance. By that, the priority of true positives versus true negatives can be decided [23].

4   Evaluation

After having tailored the random forest approach including text preprocessing, we evaluated it using previous bid projects provided by the bid group. In addition to a quantitative evaluation, we did a small field study with three experts (unstructured interviews, application to a new, yet unlabelled bid project).

---

[2] http://www.nltk.org/, accessed 09.01.2020.

4.1   Data Set

The data set used for evaluation comprises the text paragraphs of nine tender documents. All of them were written in English (most of them translated from a native language). Each entry was labelled by experts as a requirement (or non-requirement) and assigned to relevant topics (mostly between one and three, out of 52 potential topics). Thereafter, we randomly chose six documents as training data, resulting in a 47% test data split. Table 1 lists the numbers of requirements, non-requirements, and assigned topics for training data as a whole and for test data separately for each project[3] and in total.

Concerning type classification, 14,714 out of 17,556 potential requirements are labeled as requirement, leading to a prevalence of 84%.

From 52 potential topics, 50 occur in the training data, and 34 occur in the test data. Depending on prevalence in the training data, we selected three groups of 5 topics each: A (5/6) comprises all topics which occur more than 1,000 times in the training data (and at least once in the test data). For B (5/17), we chose – from all topics which occur more than 200 times in the training data – those which occur at least 500 times in the test data or in all three test projects. For

Table 1. Test data – numbers of types and topics

		Training data	Test data	Project 1	Project 2	Project 3
Total		17,556	8,256	978	6,465	813
Req		14,714	6,923	867	5,336	720
Non-Req		2,842	1,333	111	1,129	93
52 topics		20,288	10,479	867	8,892	720
PM	A	4,710	1,663	683	432	548
IXL		2,073	338	0	338	0
MMI		1,590	1,287	0	1,287	0
Power		1,448	405	47	299	59
BidManager		1,141	56	0	56	0
LED	B	507	361	116	132	113
SCADA		476	1,317	0	1,317	0
Diagnosis		422	820	0	820	0
SystemMgmt		400	548	0	548	0
Engineering		203	1,328	0	1,328	0
GSMR	C	172	181	13	168	0
Commissioning		39	70	0	70	0
PIS		27	223	0	223	0
Diesel		8	135	0	135	0
EHS		8	195	0	195	0

---

[3] Project names are confidential – therefore we use numbers 1 to 3.

C (5/27), we selected – from all topics which occur less than 200 times in the training data – those which occur most often in the test data.

4.2 Metrics

We use (standard) metrics which help to directly judge user benefit: recall (sensitivity, true positive rate, TPR), specificity (true negative rate, TNR) [22], receiver operating characteristics (ROC) curve analysis [4] and custom metrics for estimating time savings. For bid projects, a high recall is very important in order to reduce risk of high non-compliance cost due to ignorance of information. Specificity, on the other hand, is important to avoid unnecessary work due to wrongly assigned topics. All metrics are micro-averaged, i.e., summing all quantities and then calculating the metrics on the sum. This leads to a combined metric for all test data and a combined metric for multi-labels per topic.

For estimating the time savings, we compare our solution to the decisions by a requirements manager, using the metrics defined by Eqs. 1 and 2, which are based on the time to comprehend a requirement ($t_{analyze}$), the time to change a label (t_{change}) and standard evaluation quantities: true positives (TP), true negatives (TN), false positives (FP), and false negatives (FN). Assuming that an expert does not make any mistakes, he or she needs to analyze each requirement and set only the positive labels. In the automated approach, true positives need not be analyzed nor set, but additional work is necessary: For type classification (td_{type}), true and false negatives are still analyzed by the requirements manager and false negatives are set to positive in order to get TPR high – this has no effect in Eq. 1. False positives must be changed to negative by topic experts. For topic classification (td_{topic}), false positives and negatives are corrected by the topic experts during assessment. If the values for $t_{analyze}$ and t_{change} are known (e.g., as seconds per requirement on average) then the difference in hours can be calculated.

$$td_{type} = (FP - TP) \times t_{change} - (TP + FP) \times t_{analyze} \qquad (1)$$

$$td_{topic} = FP \times t_{analyze} + (FP - TP) \times t_{change} \qquad (2)$$

4.3 Type Classification

We used various thresholds (20%, 30%, ..., 80%) to get a feeling of the balance of TPR and TNR – see the ROC curve in Fig. 1a. Projects 1 and 3 perform very well, probably because they have similar properties as projects in the training set (same author, different stations on the same railway line).

In our interviews, the requirements managers turned out to be very risk-averse. Therefore, they prefer a very high TPR (e.g., 99%, as achieved with threshold 20%) and accept the relatively low TNR of 72% compared to threshold 50% (with fairly balanced TPR of 91% and TNR of 87%) – see Table 2. Assuming $t_{analyse} = 30\,$s and $t_{change} = 5\,$s, the time savings are more than 60 working hours for the three test projects. This equates to savings of approximately one working day for each thousand requirements which was confirmed in a field study with a new bid project.

Table 2. Evaluation results – TPR, TNR and estimated time savings

	Threshold 50%				Threshold 20%			
	Micro-avg	P1	P2	P3	Micro-avg	P1	P2	P3
TPR (%)	91	99	89	98	99	100	99	100
TNR (%)	87	98	85	98	72	97	68	94
td_{type} (h)	−62	−8	−47	−7	−69	−8	−54	−7

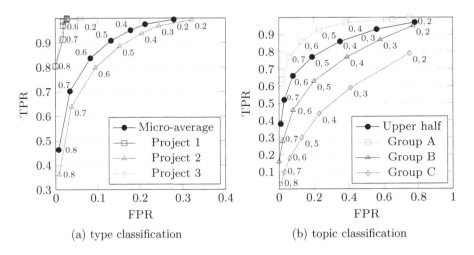

(a) type classification   (b) topic classification

Fig. 1. ROC curves

4.4 Topic Classification

Prevalence of topics is very low – the average in the training data was lower than 2% and even for the 5 most common topics only around 6%. The ROC curve in Fig. 1b shows that prediction quality on average (for the upper half of topics) is not as good as for type classification. The groups B and C from Table 1 perform much worse than group A (with comparably higher prevalence).

In our interviews, the requirements managers preferred a threshold of 50% - see Table 3. However, they judged the achieved TPR of 73% (micro-average of all topics occurring in at least one test project) as too low for practical use. Even the TPR of 77% (and TNR 81%) for the upper half of topics was not sufficient, as nearly 2,100 topic assignments are missing and more than 16,300 assignments are wrong. Again, projects 1 and 3 perform much better than the more typical project 2. The new project from the field study performed similar to the latter.

Although the requirements managers do not need to spend any time for topic assignment, the metric td_{topic} estimates an additional effort of 127 h for the necessary manual adjustments by topic experts. Only for project 3 or with a high threshold can time savings be achieved.

Table 3. Evaluation results – TPR and TNR at threshold 50%

	TPR				TNR			
	Micro-avg	P1	P2	P3	Micro-avg	P1	P2	P3
34 topics in test	73	98	69	100	81	92	81	98
Upper half	77	99	73	100	81	97	80	98
5 topics A	93	100	89	100	77	96	75	97
5 topics B	63	99	61	100	80	99	78	99
5 topics C	30	38	30		87	76	87	

5 Conclusion

We chose the random forest approach for requirements classification because it is easier to maintain and deploy than advanced deep learning solutions. Training is less expensive and can be done on local servers.

The results were fairly good for type classification and topics with a prevalence $> 5\%$ (better than, e.g.., an alternative approach based on Naïve Bayes). Application in a field study showed a high potential for reducing efforts for requirements managers (e.g.., 80% of the time for type classification). However, improvements – especially for topics with a low prevalence $< 3\%$ – are necessary to fulfil the users' demand for high TPR and TNR, i.e., both $>> 95\%$.

Acknowledgments. The work presented here has been conducted in the scope of the Horizon 2020 project OpenReq, supported by the European Union under Grant Nr. 732463.

References

1. Aggarwal, C.C.: Machine Learning for Text. Springer, Cham (2018). https://doi.org/10.1007/978-3-319-73531-3
2. Allahyari, M., et al.: A brief survey of text mining: classification, clustering and extraction techniques. CoRR abs/1707.02919 (2017)
3. Altınel, B., Ganiz, M.C.: Semantic text classification: a survey of past and recent advances. Inf. Process. Manag. **54**(6), 1129–1153 (2018)
4. Bekkar, M., Djemaa, H.K., Alitouche, T.A.: Evaluation measures for models assessment over imbalanced datasets. J. Inf. Eng. Appl. **3**(10), 27–38 (2013)
5. Breiman, L.: Random forests. Mach. Learn. **45**(1), 5–32 (2001)
6. Chawla, N.V., Bowyer, K.W., Hall, L.O., Kegelmeyer, W.P.: Smote: synthetic minority over-sampling technique. J. Artif. Intell. Res. **16**, 321–357 (2002)
7. Chen, K., Zhang, Z., Long, J., Zhang, H.: Turning from TF-IDF to TF-IGM for term weighting in text classification. Expert Syst. Appl. **66**, 245–260 (2016)
8. Devlin, J., Chang, M.W., Lee, K., Toutanova, K.: BERT: pre-training of deep bidirectional transformers for language understanding. arXiv preprint arXiv:1810.04805 (2018)

9. Falkner, A., Palomares, C., Franch, X., Schenner, G., Aznar, P., Schoerghuber, A.: Identifying requirements in requests for proposal: a research preview. In: Knauss, E., Goedicke, M. (eds.) REFSQ 2019. LNCS, vol. 11412, pp. 176–182. Springer, Cham (2019). https://doi.org/10.1007/978-3-030-15538-4_13

10. Fucci, D., et al.: Needs and challenges for a platform to support large-scale requirements engineering. A multiple case study. CoRR abs/1808.02284 (2018)

11. Gupta, P., Schütze, H., Andrassy, B.: Table filling multi-task recurrent neural network for joint entity and relation extraction. In: Proceedings of COLING 2016, the 26th International Conference on Computational Linguistics: Technical Papers, pp. 2537–2547 (2016)

12. Haixiang, G., Yijing, L., Shang, J., Mingyun, G., Yuanyue, H., Bing, G.: Learning from class-imbalanced data: review of methods and applications. Expert Syst. Appl. **73**, 220–239 (2017)

13. Kadhim, A.I.: Survey on supervised machine learning techniques for automatic text classification. Artif. Intell. Rev. **52**(1), 273–292 (2019)

14. Ko, Y., Park, S., Seo, J., Choi, S.: Using classification techniques for informal requirements in the requirements analysis-supporting system. Inf. Softw. Technol. **49**(11–12), 1128–1140 (2007)

15. Manning, C.D., Surdeanu, M., Bauer, J., Finkel, J., Bethard, S.J., McClosky, D.: The Stanford CoreNLP natural language processing toolkit. In: Association for Computational Linguistics (ACL) System Demonstrations, pp. 55–60 (2014)

16. Moser, T., Winkler, D., Heindl, M., Biffl, S.: Requirements management with semantic technology: an empirical study on automated requirements categorization and conflict analysis. In: Mouratidis, H., Rolland, C. (eds.) CAiSE 2011. LNCS, vol. 6741, pp. 3–17. Springer, Heidelberg (2011). https://doi.org/10.1007/978-3-642-21640-4_3

17. Ott, D.: Automatic requirement categorization of large natural language specifications at Mercedes-Benz for review improvements. In: Doerr, J., Opdahl, A.L. (eds.) REFSQ 2013. LNCS, vol. 7830, pp. 50–64. Springer, Heidelberg (2013). https://doi.org/10.1007/978-3-642-37422-7_4

18. Pinquié, R., Véron, P., Segonds, F., Croué, N.: Requirement mining for model-based product design. Int. J. Product Lifecycle Manag. **9**(4), 305–332 (2016)

19. Schörghuber, A.: Classification of requirements in the tender process. Master's thesis, University of Technology Vienna, Vienna (2019)

20. Schütze, H., Manning, C.D., Raghavan, P.: Introduction to Information Retrieval, vol. 39. Cambridge University Press, Cambridge (2008)

21. Sebastiani, F.: Machine learning in automated text categorization. ACM Comput. Surv. (CSUR) **34**(1), 1–47 (2002)

22. Sokolova, M., Lapalme, G.: A systematic analysis of performance measures for classification tasks. Inf. Process. Manag. **45**(4), 427–437 (2009)

23. Tosun, A., Bener, A.: Reducing false alarms in software defect prediction by decision threshold optimization. In: Proceedings of the 2009 3rd International Symposium on Empirical Software Engineering and Measurement, pp. 477–480 (2009)

On the Design of a Natural Logic System for Knowledge Bases

Troels Andreasen[1]([✉]), Henrik Bulskov[1], and Jørgen Fischer Nilsson[2]

[1] Computer Science, Roskilde University, Roskilde, Denmark
{troels,bulskov}@ruc.dk
[2] Mathematics and Computer Science, Technical University of Denmark,
Lyngby, Denmark
jfni@dtu.dk

Abstract. Natural logics are logics that take form of stylized natural language sentences within a selected fragment of natural language. Natural Logics are at the same time formal logics with a well-defined syntax and semantics. Therefore, natural logics may be advanced as knowledge base logics enhancing explainability of query answers. This paper is concerned with a natural logic, NATURALOG, having been proposed as a deductive knowledge base language. The paper briefly reviews and brings together in compact form the main points in our previously but separately published design proposals, systems functionalities and implementation principles.

Keywords: Natural logic · Knowledge base systems · Deductive querying · Life science applications

1 Introduction

In a historical perspective there are two development lines in logic, that is logic-of-language and mathematical logic. The logic-of-language tradition dates back to Aristotle and went through developments during the medieval times until the end of the 19th century. This development line was halted and largely abandoned by the advent of quantified predicate logic due mainly to G. Frege and B. Russell. Predicate logic and related logics, then, have become foundational concepts as well as important tools in in computer science, in particular in computational logic, e.g. with logic programming.

However, the logic-of-language tradition recently has attracted renewed interest in connection with attempts to ease communication with computers. This paper discusses natural logics [9,12] that are rooted in the logic-of-language tradition and describes undertakings aimed at adopting and adapting natural logics for logical knowledge base systems. The following sections briefly surveys and discusses the various aspects of design principles and implementation methods described in the here chronologically listed range of papers [1–5,10,11] followed up by [6,7] forthcoming in 2020.

© Springer Nature Switzerland AG 2020
D. Helic et al. (Eds.): ISMIS 2020, LNAI 12117, pp. 413–421, 2020.
https://doi.org/10.1007/978-3-030-59491-6_39

Our design concern throughout is to obtain a useful trade-off between expressivity and computational tractability, having in mind also the requirements for potentially useful application domains in the design. Although the proposed natural logic is meant as a general purpose specification language for real-world domains, we have had in mind in particular applications within the life sciences as it appears in the mentioned publications.

2   Designing a Natural Logic

Our natural logic proposal, termed NATURALOG, takes as point of departure syllogistic logic from the Aristotelian tradition, cf. [8,11]. This means that the basic so-called categorical sentence forms are every C is D and some C is D, where C and D are class- or concept terms. In the simplest cases C and D are common nouns representing classes aka concepts. Accordingly, these forms known as copula forms express, respectively, a subclass relationship and a class-class overlap relationship. In the following, for the copula "is" we follow the conventions in computer science and write "isa". With the form every C isa D (or in convenient short form simply C isa D), one can specify hierarchically- as well as non-hierarchically structured formal ontologies as partial orders, with the isa relationship being transitive.

In addition to these affirmative sentence forms the old syllogistic logic also comprises negative forms as mentioned in [11], also known from the square-of-opposition. However, at present we refrain from admitting negative statements in a knowledge base itself, resorting instead to negation-as-non-provability as known from databases and logic programming.

2.1   Beyond Copula Forms: General Relationships

In addition to the copula isa in NATURALOG one can introduce transitive verbs (i.e. verbs taking a linguistic object) as one pleases. To this end in [2,3,11] we introduced the more general sentence forms with a verb R

$$(\text{every} \mid \text{some}) \ C \ R \ (\text{every} \mid \text{some}) \ D$$

giving four determiner constellations. The verb R represents a binary relationship between the two concepts represented by the subject and object. As convenient default form for the most common sentences in knowledge bases we propose

$$C \ R \ D \quad \text{for the full form} \quad \text{every } C \ R \text{ some } D$$

Example: persons like pets is shorthand for every person likes some pet and some persons drink beer is shorthand for some persons drinks some beer. Actually, in NATURALOG we ignore linguistic inflection rules as seen in the following examples. For the some form we have

$$\text{some } C \ R \ D \quad \text{for} \quad \text{some } C \ R \text{ some } D$$

In predicate logic every C R some D is construed as

$$\forall x (Cx \to \exists y (Rxy \wedge Dy))$$

and some C R some D is

$\exists x (Cx \wedge \exists y (Rxy \wedge Dy))$, which is equivalent to $\exists x \exists y (Cx \wedge Dy \wedge Rxy)$

However, we stress that NATURALOG sentences are not translated into predicate logic in our systems proposal as explained in Sect. 3 and 4.

2.2 Compound Terms

In addition to the concepts given by common nouns in NATURALOG one can form expressions for creation of new concepts by attachment of restrictive modifiers to common nouns as in the sample sentence

betacell isa cell that produce insulin

The restrictive modifiers may take form of relative clauses as in the compound concept term cell that produce insulin or prepositional phrases such as in cell in gland. Both forms semantically consist of a relationship given as a verb or as a preposition followed by a concept. Provision is made for nesting of such constructs reflecting the usual recursive syntax for modifiers in natural language phrases.

There are other forms of restrictive nominal modifiers in natural language, in particular adjectives (including participles such as "increased") and noun-noun-compounds such as "heart disease". The incorporation of these modifiers into natural logic is more problematic and is postponed since they, unlike the above ones, do not directly provide a modifying relationship. As a temporary solution noun-noun compounds may be rewritten using a relative clause, so that for instance "bacteria infection" would become infection that is-caused-by bacteria.

Verbs may also be modified restrictively using an adverbial prepositional phrase as in the verb form produce in gland as a restriction of the verb produce. Incorporation of this useful feature, which falls outside the simple predicate logical explanation in Sect. 2.1, is thoroughly discussed in [6].

The syntax for the current version of NATURALOG is specified in the form of a BNF grammar in [6]. In order to ensure that there be no structural ambiguities, parentheses are enforced in one production rule for stipulating the intended recursive phrase structure.

In [3] we discuss various language extensions intended for promoting the usability of NATURALOG by approaching some common forms in natural language. Examples are appositions and conjunctions. We distinguish semantically conservative extensions and non-conservative ones. The former ones do not extend the semantic coverage. The order of the natural logic sentences in a knowledge base is logically irrelevant; the sentences are syntactically independent of each other.

2.3   Active-Passive Voice and Existential Import

Now, consider the active voice sentence betacell produce insulin or in full form every betacell produce some insulin. The corresponding passive voice sentence is some insulin is-produced-by some betacell, where is-produced-by represents the inverse relation of produce. Although this latter sentence follows intuitively, the sentence does not follow logically, because in predicate logic the denotation of a monadic predicate may well be empty. This problem is overcome by appealing to the existential import principle known from the Aristotelian logic tradition, cf. [6,11]. This principle declares that all mentioned classes be non-empty without being specific about any member entity. As a special consequence, the presence of a copula sentence of the form [every] C isa D implies availability also of the weakened converse sentence form some D isa C.

2.4   Remarks on Natural Logic and Description Logic

Todays most common logics for ontologies and knowledge bases are presumably the various description logic dialects. Both NATURALOG and description logics are examples of variable free logics covering small but useful fragments of predicate logic. A key difference between these two logics is that description logics offer sentences in copula form only (at the so-called T-box level of concepts), which seems awkward from the point of conventional use in natural language. This is in disagreement with common formulations in natural language. Another difference is that description logics have to resort to awkward reformulations for sentences beginning with the determiner "some", cf. the discussion of active-passive forms in the previous section. A further comparison of the two logics is given in [6].

The endorsing in the natural logic of non-copula sentences (with verbs fetched from the target application) agrees well with an entity-relationship model view of a knowledge base: A NATURALOG knowledge base typically takes the form of an ontology formed by the stated copula sentences extended with non-copula sentences connecting concepts across the ontology with relations expressed by transitive verbs. This view further invites the introduction of a distinction between definitional and observational (i.e. empirical) statements mentioned in the next section.

3   The Metalogic Framework for NATURALOG

So far, NATURALOG may be conceived simply as a "sugared" fragment of predicate logic for describing the application domain of discourse. As a next step appropriate proof rules admitting computational derivation of logical consequences as NATURALOG sentences are to be introduced. Importantly, these rules are to be applied directly to natural logic sentences and terms, rather than to their would-be predicate logical translations.

Such rules enable deductive querying of the knowledge base giving answer results in the form of classes and more generally compound terms. In principle deductive querying is achievable by introduction of variables ranging over the terms in the natural logic as the metalogical variable X the sample query form

X isa cell that produce hormone

supposed to give the answer betacell. These term variables should not be confused with the quantified entity- or individual variables of predicate logic that range over entities in the application domain of discourse. We account formally for these variables by introduction of a metalogic in which NATURALOG becomes embedded.

As metalogic we can choose a "domesticated" form of predicate logic: In [10, 11] we suggested DATALOG to this end, and [6] gives elaborate description of the metalogic inference engine, succeeding a more compact presentation in [5]. Recall that DATALOG consists of definite clauses without compound terms and enjoys decidability.

Let us exemplify the encoding of our natural logic into DATALOG: The sentence betacell produce insulin becomes the atomic metalogic clause

proposition(every, betacell, produce, insulin)

where the natural logic terms formally appear as, and are treated as, constants.

The sentence betacell isa cell that produce insulin becomes in the metalogic representation

proposition(every, betacell, isa, cell-that-produce-insulin)

where cell-that-produce-insulin is a new simple concept term that becomes defined by the following pair of defining metalogic clauses

definition(cell-that-produce-insulin, isa, cell)
definition(cell-that-produce-insulin, produce, insulin)

The decomposition applies recursively to nested concept terms.

The distinction between definitional and non-definitional contributions in the decomposition is internal to the system. However, [11] hints at further introducing an external epistemic distinction between definitional and observational sentences in the knowledge base.

3.1 The Encoded Knowledge Base as Graph

The metalogic knowledge base representation may be conceived as a labeled graph whose nodes are concept terms, and whose directed arcs represent relationships with accompanying determiners (quantifiers). The graph picture further supports the conception of the knowledge base as an extended ontology. The concept terms present in the KB sentences are uniquely represented as nodes in the graph across sentences. Moreover the graph view helps visualizing pathway querying, cf. Sect. 5. In the decomposition of sentences into clauses there is no loss of information.

4   Design of Inference Engine

Natural logics use high level inference rules reflecting "intuitive" rules applied by humans when reasoning with descriptions in natural language. This adds to the explainability of the deductive reasoning and hence the query processing. These rules are now to be formalized in the metalogic, exploiting the decomposed and encoded NATURALOG sentences.

As a key principle, answers to queries stated to the knowledge are computed by use of the inference rules. We refer to [5,6] for the rules we apply for NATURA-LOG. These papers also contain references to the background literature dealing with deductive reasoning in natural logics.

We present here using DATALOG only a few rules. As key rules there are the so-called monotonicity rules:

proposition(every,$C, R, Dsuper$) $\leftarrow$
    proposition(every,C, R, D) $\wedge$ proposition(every,D,isa,$Dsuper$)
proposition(every,$Csub, R, D$) $\leftarrow$
    proposition(every,C, R, D) $\wedge$ proposition(every,$Csub$,isa,C)

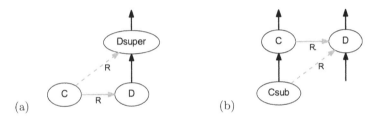

Fig. 1. Monotonicity rules: (a) inheritance and (b) generalization. Dashed relations are inferred.

The graphs for these are shown in Fig. 1. One observes that the latter rule provides "property inheritance". Further, one may observe that the transitivity rule for isa obtains with the special case of R being instantiated to isa. As another distinctive feature we introduce a rule for obtaining the corresponding passive voice sentence from a given active voice sentence. This implies in particular that the sentence form C isa D gives rise to some D isa C as mentioned in Sect. 2.3 besides giving the weakened some C isa D by means of appropriate rules. In Sect. 5.1 we mention the potential for extension with "non-logical" application-specific inference rules.

4.1   Materialization of Deductive Closure

In [7] we propose that the part of the deductive closure of a knowledge base being relevant for query answers of a knowledge base is computed and stored in advance in a compilation process jointly involving the sentences. This means that sentences and terms potentially appearing in a query answer is made present in advance in the compiled knowledge base.

Furthermore, in [7] (forthcoming 2020) we elaborate a version of the inference engine where DATALOG is replaced by relational database query operations. This enables use of a database system for efficient retrieval of sentences in the knowledge base, and in addition inference computations are made algorithmically more efficient by "bulk processing" applying database join operations.

The recursive NATURALOG syntax generally admits infinitely many terms. However, only a finite subset of these are known to have a non-empty denotation in the form of subconcepts in the knowledge base – namely either by their being explicitly present or by being a superclass of such a mentioned term.

5   Systems Functionalities

A range of systems functionalities can be obtained on basis of the relevant deductive closure computed by the inference rules. First of all there are answers in the form of sets of terms from instantiation of metalogical variables in query forms exemplified by

$$\text{proposition}(\text{every}, X, \text{produce}, \text{hormone})$$

supposed to give as answer cells that produce hormone, such as betacells. Notice here that computation of such answers in general draws on inference rules, say, for combining the sentence betacell produce insulin with the sentence insulin isa hormone in a monotonicity inference rule. Query answer terms may well be compound terms stemming from a given sentence or having been computed in the compilation process.

So far, we accept only affirmative sentences in a knowledge base. Negative sentences may be accepted as query sentences in the form no C R D being logically contrary to every C R D and contradictory to some C R D, with a supporting inference rule appealing to negation by non-provability.

The graph conceptualization of a NATURALOG knowledge base as (usually) one coherent graph invites path-finding operations for retrieving shortest paths between two given terms as discussed and examplified in [1,3,4]. Path-finding is of particular interest for tracing pathways in life-science knowledge bases.

5.1   Application-Specific Query Inference Rules

In our two level logic setup the natural logic level describes the application domain of discourse and the metalogic level prescribes the computing with natural logic sentences and terms. This opens for introducing application specific rules in the metalogic. For instance, one easily introduces a rule providing a verb, say, "causes" with the property of transitivity of the underlying relation.

As another example, [6] describes an additional metalogic rule for computing the commonalities of two given terms. When asking for instance about common properties of the two concepts alphacell and betacell, the deduced answer may comprise informative compound terms such as cell that produce hormone. More sophisticated general rules may afford the computing of analogies, asking for instance which concept is related to alphacell as insulin is related to betacell.

Inference rules may be introduced to verify ad hoc consistency requirements formulated as rules expected to yield empty query answers in case of consistency fulfilment as known from logic programming.

6   Conclusion and Open Problems

The described natural logic with the accompanying realization principles attempts to strike a balance between on one hand language expressivity and interesting computational functionalities and on the other hand an acceptable computational tractability. NATURALOG covers basic essential application domain demands within the considered life-science domains, but additional useful features are to be included in coming versions. Among the possible semantical extensions let us just mention exception handling for non-monotonic blocking of unrestricted inheritance of properties, and introduction of generalized quantifiers such as "most" and "few".

It remains to be verified that computational tractability can be obtained with the suggested relational database implementation, when scaling up to interesting large size knowledge bases.

An interesting but highly challenging problem is to conduct a computer-assisted, if not completely automatic, translation of essential parts of given natural language descriptive texts into NATURALOG. This complex and difficult problem of computationally extracting natural logic sentences from descriptions in free natural language is touched in [1,3,4]. To this end a syntactic-semantic analysis using NATURALOG as target language might be ameliorated by inductive machine learning methods. Eventually, more expressive versions of natural logic may come into use directly as logical specification languages in natural science domains.

References

1. Andreasen, T., Bulskov, H., Nilsson, J.F., Jensen, P.A.: Computing pathways in bio-models derived from bio-science text sources. In: Proceedings of the IWB-BIO International Work-Conference on Bioinformatics and Biomedical Engineering, Granada, April, pp. 217–226 (2014)
2. Andreasen, T., Bulskov, H., Nilsson, J.F., Jensen, P.A.: A system for conceptual pathway finding and deductive querying. Flexible Query Answering Systems 2015. AISC, vol. 400, pp. 461–472. Springer, Cham (2016). https://doi.org/10.1007/978-3-319-26154-6_35
3. Andreasen, T., Bulskov, H., Nilsson, J.F., Jensen, P.A.: Partiality, Underspecification, and Natural Language Processing, chapter A Natural Logic for Natural-Language Knowledge Bases. Cambridge Scholars, Newcastle upon Tyne (2017)

4. Andreasen, T., Bulskov, H., Jensen, P.A., Nilsson, J.F.: Pathway computation in models derived from bio-science text sources. In: Kryszkiewicz, M., Appice, A., Ślęzak, D., Rybinski, H., Skowron, A., Raś, Z.W. (eds.) ISMIS 2017. LNCS (LNAI), vol. 10352, pp. 424–434. Springer, Cham (2017). https://doi.org/10.1007/978-3-319-60438-1_42

5. Andreasen, T., Bulskov, H., Jensen, P.A., Nilsson, J.F.: Deductive querying of natural logic bases. In: Cuzzocrea, A., Greco, S., Larsen, H.L., Saccà, D., Andreasen, T., Christiansen, H. (eds.) FQAS 2019. LNCS (LNAI), vol. 11529, pp. 231–241. Springer, Cham (2019). https://doi.org/10.1007/978-3-030-27629-4_22

6. Andreasen, T., Bulskov, H., Nilsson, J.F., Jensen, P.A.: Natural logic knowledge bases and their graph form, p. 34. (2020). Submitted to journal (under review)

7. Andreasen, T., Bulskov, H., Nilsson, J.F.: A natural logic system for large knowledge bases. In: The 30th International Conference on Information Modelling and Knowledge Bases, Ejc 2020, Hamburg, Germany, 8–12 June 2020 (2020)

8. Klima, G.: Natural logic, medieval logic and formal semantics. Magyar Filózfiai Szemle **54**(4), 58–75 (2010)

9. Moss, L.S.: Syllogistic logics with verbs. J. Log. Comput. **20**(4), 947–967 (2010)

10. Fischer Nilsson, J.: Querying class-relationship logic in a metalogic framework. In: Christiansen, H., De Tré, G., Yazici, A., Zadrozny, S., Andreasen, T., Larsen, H.L. (eds.) FQAS 2011. LNCS (LNAI), vol. 7022, pp. 96–107. Springer, Heidelberg (2011). https://doi.org/10.1007/978-3-642-24764-4_9

11. Nilsson, J.F.: In pursuit of natural logics for ontology-structured knowledge bases. In: The Seventh International Conference on Advanced Cognitive Technologies and Applications (2015)

12. van Benthem, J.: Essays in logical semantics. In: Studies in Linguistics and Philosophy, vol. 29. D. Reidel, Dordrecht (1986)

Evaluation of Post-hoc XAI Approaches Through Synthetic Tabular Data

Julian Tritscher[1($\boxtimes$)], Markus Ring[2], Daniel Schlör[1($\boxtimes$)], Lena Hettinger[1($\boxtimes$)], and Andreas Hotho[1($\boxtimes$)]

[1] University of Würzburg, Würzburg, Germany
{tritscher,schloer,hettinger,hotho}@informatik.uni-wuerzburg.de
[2] University of Coburg, Coburg, Germany
markus.ring@hs-coburg.de

Abstract. Evaluating the explanations given by post-hoc XAI approaches on tabular data is a challenging prospect, since the subjective judgement of explanations of tabular relations is non trivial in contrast to e.g. the judgement of image heatmap explanations. In order to quantify XAI performance on categorical tabular data, where feature relationships can often be described by Boolean functions, we propose an evaluation setting through generation of synthetic datasets. To create gold standard explanations, we present a definition of feature relevance in Boolean functions. In the proposed setting we evaluate eight state-of-the-art XAI approaches and gain novel insights into XAI performance on categorical tabular data. We find that the investigated approaches often fail to faithfully explain even basic relationships within categorical data.

Keywords: Explainable AI · Evaluation · Synthetic data

1 Introduction

Black box classifiers such as deep neural networks (DNNs) have been established as state-of-the-art in many machine learning areas. Even though they give strong predictions, their models contain lots of non-linear dependencies, causing their decisions to become untraceable. As a result, a branch of research in explainable artificial intelligence (XAI) has developed, aiming to give local explanations for single predictions of trained black box models in a post-hoc fashion [6].

Problem. While there are many approaches to acquire such explanations, no unified evaluation method has been proposed so far. While easily comprehensible domains like image and text classification use simple presentation of explanations [3] and user studies [6], XAI behavior on tabular data is largely unexplored.

Objective. Post-hoc XAI explanations are inherently approximations, giving simplified but not necessarily faithful insights into highly complex models [11].

© Springer Nature Switzerland AG 2020
D. Helic et al. (Eds.): ISMIS 2020, LNAI 12117, pp. 422–430, 2020.
https://doi.org/10.1007/978-3-030-59491-6_40

Therefore, assessing the limits of these approaches is a central point of research interest. We take a first step to investigate post-hoc XAI performance on DNNs trained on tabular data, by developing a test setting for categorical tabular data, where feature relationships can often be expressed via Boolean functions.

Approach and Contribution. In this paper, we design synthetic datasets reflecting typical relationships of real-world data such as logical AND, OR and XOR connections between categorical attributes. We propose a definition of feature importance in Boolean functions in the context of XAI to generate gold-standard explanations for our datasets. Using this data, we present an evaluation setting for XAI approaches, that allows for evaluation of data with underlying complex feature relationships. This setting is used to analyze and compare eight state-of-the-art XAI approaches. Evaluation on an expert-annotated real dataset suggests that our results translate well to real data. We publish our datsets to facilitate comparison of XAI approaches in a standardized evaluation setting.

2 Related Work

Aside from subjective image explanations [3] and resource intensive user studies [6], several approaches have been used to evaluate XAI performance.

In [5,9] model faithfulness of their approaches is evaluated by explaining predictions of inherently explainable linear models and comparing obtained explanations directly to the model. Performance of XAI approaches when explaining non-linear classifiers, however, can not be assessed in this evaluation setting.

Additionally, [9] also evaluate the faithfulness of their local approximation model by measuring if it corresponds to changed inputs in the same way as the classifier it approximates. This evaluation can, however, only be performed on XAI approaches that train a simplified classifier as a local approximation.

Perturbation-based evaluations, e.g. as used in [12], iteratively remove features with the highest relevance from data and re-classify. Explanations are rated higher, the faster the classification error increases. While perturbation-based evaluation can be applied to classification tasks where the removal of single features is expected to gradually impair the performance of the classifier, this assumption does not hold for categorical tabular data in general.

3 Investigated XAI Approaches

Perturbation-based explanation approaches mask or remove input features from data samples to observe the change in classifier output. While they pose no architectural constraints on the classifier, they are computationally intensive. LIME (Local Interpretable Model-agnostic Explanations) [9] uses perturbations to explore the classifier outputs locally around a given input. It trains a local linear classifier and uses the weights as scores of input feature relevance. Shapley Value sampling [4] is based on the Shapley value from cooperative game theory, a unique solution for distributing an achieved score onto cooperating players, under a list of desirable criteria. Shapley value sampling approximates this NP-complete problem with a sampling approach, evaluating the output of possible feature value combinations through perturbation. (Kernel) SHAP (Kernel SHapley Additive exPlanation) [6] uses a custom kernel for the LIME XAI approach, in order to adapt LIME to approximate Shapley values.

Gradient-based XAI approaches use the gradient of a gradient descent based classifier to approximate explanations through few backpropagations, considerably saving runtime in comparison to perturbation-based approaches. For our evaluation, we use the implementations of [1]. Saliency maps [14] highlight the most influential pixels using a first order approximation of the absolute gradient of the predicted output with respect to the input for a specific data sample. Gradient×Input [13] builds on the Saliency approach, multiplying the signed result of Saliency with the corresponding input feature. Integrated Gradients [15] computes the average output gradients with respect to different inputs. Gradients are computed for values on a linear path between the data sample and an uninformative baseline input. ϵ-LRP (ϵ-Layerwise Relevance Propagation) [3] defines the relevance of a neuron as all influence it has on the neurons of the next layer, multiplied by these neurons' activations for a specific data sample. In this work we use the reformulated implementation by [1]. DeepLIFT [12], like LRP, computes the relevance of a neuron by measuring the influence on neurons of the next layer, additionally subtracting the influence of an uninformative baseline.

4   Data Generation Approach

Since categorical attributes can be binarized (e.g. one-hot encoding) to Boolean features, we will focus on feature relationships modeled as Boolean functions.

In [7] a feature is considered influential in a sample if changing its value would also change the function output. While this is intuitive on some inputs, on others it assigns no influence to any feature. For example, consider the Boolean function $y = 0 \wedge 0 = 0$, where no single feature can be changed to change y. In the context of XAI, this would not allow to differentiate between the 0-inputs involved in the function and irrelevant features. To address this, we adapt the influence definition on basic Boolean operations (AND, OR, XOR) to assign influence to both features, if no single feature can be changed to change the function output. For more complex functions, we proceed as follows:

Definition 1. *Let the Boolean function y be represented by a Boolean binary expression tree [8]. For each data sample, we consider a child node c relevant to the explanation of its parent operation-node o, iff the value of the subtree formed by c, evaluated with respect to the sample, is relevant to the operation-node o.*

We thereby distribute the relevance of a complex function by decomposing it into its basic operations. We calculate their intermediate results for each data sample, propagating the relevance of the entire function through all basic operations down to its input features. The resulting explanations contain the input features that most determine the output of complex Boolean functions.

When assessing XAI performance, we have to take into account that non-matching explanations might be correct explanations of a weak classifier, instead of a poorly performing XAI approach. For this, we train our classifiers in a 5-fold stratified cross-validation setting, using only classifiers that reliably achieve 100% accuracy on training- and test-sets. Further, we generate synthetic datasets including every permutation of categorical attributes exactly once. This guarantees that the test-sets contain permutations not seen during training, ensuring that the classifier learned to perfectly generalize to the unseen test-data without sensitivity to irrelevant inputs.

Following these restrictions, we expect XAI approaches to give the highest scores to the relevant features. Thus, we consider a data sample to be correctly explained, if the top scoring features given by an XAI approach match the relevant features of the ground truth explanation.

5   Experiments

The following setup is used throughout all of our experiments.

Datasets are generated following the criteria of Sect. 4. We set a fixed dataset size of n binary features, for which we include every permutation once in the dataset, giving 2^n data samples. We then generate the label for each data sample with a Boolean function and generate the explanation of every data sample according to Sect. 4. Features that were not used in the generation of the label hereby act as noise that XAI approaches may falsely consider relevant. All following experiments use 12 binary features and $2^{12} = 4096$ data samples.

Classifier & XAI setup also follow Sect. 4. We encode our Boolean input data with the values 1 for True and -1 for False, and train a feed forward neural network with 5 layers, 20 neurons per layer, and ReLU activations in a 5-fold stratified cross-validation setting. The classifiers reliably achieve 100% accuracy on training- and test-sets for all evaluated datasets. For each cross-validation fold, we compute the explanations of the test-data. We repeat the evaluation 10 times per dataset, reporting the average over the results.

Baseline Some XAI approaches replace classifier input features with uninformative values to observe classifier behavior with missing information. We let LIME and SHAP extract their own baseline from the cross-validation training data, using k-means clustering with $k = 20$ for SHAP. For the gradient-based approaches, we use 0 as baseline value, as discussed in [2].

5.1   Evaluation of Basic Boolean Operations

We initially evaluate the behavior of XAI approaches on datasets where two features are linked by common Boolean operations AND ($\wedge$), OR ($\vee$) or XOR ($\otimes$) in Table 1. We find that most XAI approaches fail to fully explain even the linear Boolean AND and OR operations, with only LIME and SHAP finding the most relevant features for each sample. Results on the XOR dataset show that LIME, due to training a local linear model around the sample, fails to give good explanations when the underlying local function is non-linear. SHAP appears to improve on LIMEs behavior, correctly matching the gold standard explanations with its kernel-based Shapley value adaptation of LIME.

Detailed Analysis. We take a closer look at the input permutations causing problems to the XAI approaches. We find that falsely explained samples for all gradient-based approaches and Shapley sampling on the AND and OR datasets are caused by issues with $y = 0 \wedge 0$ and $y = 1 \vee 1$. In this case, the mentioned approaches consider one of the two features as irrelevant, even though both features are equally important to the label. Additionally, the Saliency approach shows issues on unequal inputs, where one feature speaks against the prediction outcome. Since the gradient of the output with respect to this feature is negative, the Saliency's absolute gradient causes this negatively influencial feature to overshadow the relevant feature. On the non-linearly separable XOR dataset, all approaches show a similar amount of errors for each input permutation.

Table 1. XAI performance in percent correctly explained samples after Definition 1.

Approach	$\wedge$	$\vee$	$\otimes$	$(\otimes) \wedge (\otimes)$	Synthetic	Real
LIME	**100.00**	**100.00**	43.76	10.12	**100.00**	84.78
Shapley sampling	99.83	99.92	69.10	56.34	79.79	77.15
SHAP	**100.00**	**100.00**	**100.00**	**95.01**	98.12	**93.34**
Saliency	95.81	95.69	85.99	55.55	59.86	59.12
Gradient×input	97.66	97.06	75.81	57.31	69.47	69.49
Integrated gradients	99.64	99.56	73.02	57.73	76.46	76.27
ϵ-LRP	97.66	97.06	75.84	57.34	69.48	69.71
Deeplift	99.67	99.59	73.28	58.83	75.48	76.77

5.2 Evaluation of Boolean Functions with Multiple Variables

Using our relevance definition (Definition 1), we investigate XAI performance on complex Boolean functions. Results of the function $y = (f_1 \otimes f_2) \wedge (f_3 \otimes f_4)$, shown as $(\otimes) \wedge (\otimes)$ in Table 1, indicate that XAI performance deteriorates with an increased number of relevant features involved. To test this, we create similar datasets with increasing numbers of variables that may impact the output label.

Linearly Separable Boolean Functions. Since XAI performance on basic operations suggests different XAI behavior on linear and non-linear Boolean functions, we first investigate XAI performance with increasing function complexity on linear functions. For this, we generate eight datasets using the function $y = ((((f_1 \wedge f_2) \vee f_3) \wedge f_4) \vee ...)$, appending 3 to 10 relevant features as label.

To validate that the used datasets are linearly separable, we ensure that a linear Support Vector Machine can perfectly separate each dataset.

The average results on each dataset are shown in Fig. 1a. We find LIME to be able to fully explain all samples of datasets with up to 6 relevant features. Both LIME and SHAP are capable of explaining a large amount of samples in all tested datasets. Shapley sampling and all gradient-based methods show difficulties with explaining functions with more then 2 variables involved, with performance declining further with more than 3 variables. The small inclines in explanation score with increased function complexity may be explained by all datasets consisting of a total of 12 variables. This means that when 10 variables are involved in the function, randomly assigning the 2 non relevant variables the lowest scores may occur more often then with 6 relevant and irrelevant variables.

Non-linearly Separable Boolean Functions. We also evaluate performance on eight non-linearly separable datasets generated using the function $y = ((((f_1 \otimes f_2) \wedge f_3) \otimes f_4) \wedge ...)$ with 3 to 10 relevant features used for label generation. As seen in Fig. 1b, all approaches show lower performance compared to the non-linear XOR dataset (see Table 1). While SHAP still maintains stronger performance than other approaches on non-linear datasets throughout the experiment, its performance deteriorates when more than 3 features influence the label. All other approaches show poor performance on all datasets. The fluctuation between scores with increasing complexity may be caused by the alternating label-generation: If the last operator in the outermost brackets of the function is an AND, then for all samples that evaluate to $x \wedge 0$ with the entire previous term $x = 1$, the only relevant variable for this sample is the last 0. Therefore XAI approaches only have to find the last variable 0 as explanation, simplifying the problem down to a basic AND operation for several input permutations.

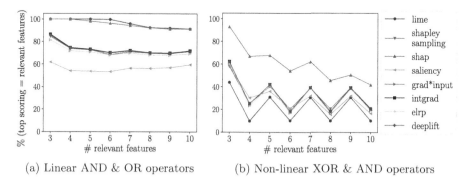

(a) Linear AND & OR operators   (b) Non-linear XOR & AND operators

Fig. 1. XAI performance on datasets with multiple relevant features. Performance in percent of correctly explained samples according to Definition 1.

5.3 Application Scenario

Next, we choose a setting from the intrusion detection domain, to show that our findings can be applied to realistic settings. A method to synthetically create flow-based network traffic is proposed in [10], where a flow describes a network connection between two hosts and contains attributes like transport protocol and TCP-flags. In this setting, only flows which represent TCP traffic are allowed to set any TCP-flags, creating the task of validating whether samples resemble valid network traffic. While transport protocol and TCP-flags can be represented as binary attributes with a set of predefined rules, the complexity of the problem is low enough to create expert-annotated labels. In this experiment, we use N-WGAN-GP from [10] to create 2048 correct and 2048 incorrect flows. Each flow is represented by six categorical (*TCP flags*) and two categorical, one-hot-encoded features (*weekday, protocol*), six numerical features (*bytes, packets, duration, time, source- and destination-port*) and two values encoded with multiple numeric features (*IP-addresses*). The six *TCP flags* and the *protocol* are considered as relevant. To create a comparable synthetic setting, we generate a dataset using the function $y = (f_1 \vee -(f_2 \vee f_3 \vee f_4 \vee f_5 \vee f_6 \vee f_7))$. We then evaluate the XAI approaches on both datasets.

The results, marked as "synthetic" and "real" in Table 1, indicate a similar ranking of the XAI approaches with respect to their performance on both datasets. We observe that, while all perturbation-based approaches perform worse on real data, LIME achieves considerably better results on synthetic data in comparison to the real setting. This may be due to its local linear approximation that benefits more from the equal distribution of different sample permutations in the synthetic data. This experiment suggests that XAI performance on our synthetic datasets closely resembles real world application scenarios.

6 Discussion

Evaluation of eight post-hoc XAI approaches shows that many approaches fail to give satisfactory explanations even on basic categorical tabular data. The gradient-based approaches used in this work all show weaker performance than perturbation-based methods. While the approaches LIME and SHAP are both capable of well explaining basic linearly separable Boolean functions, only SHAP is capable of explaining non-linearly separable functions with up to 3 variables. Overall, we find that investigated approaches struggle to explain more complex, as well as non-linear Boolean functions. The datasets generated for these experiments may be used to gain first insights into XAI performance on categorical tabular data and will therefore be made available as benchmark datasets[1].

7 Conclusion

In this paper, we investigated XAI performance on categorical tabular data, proposing a setting in which XAI approaches can be evaluated independently

---

[1] http://www.dmir.uni-wuerzburg.de/projects/deepscan/xai-eval-data/.

of classifier performance using synthetic datasets with gold standard explanations. We generated benchmark datasets containing typical relationships between binary attributes such as AND, OR and XOR, as well as explanations according to a novel definition of relevance of features in Boolean functions.

Using these datasets, we empirically evaluated eight state-of-the-art XAI approaches. We found that many approaches fail to capture simple feature relationships such as non-linear XOR connections, with performance decreasing with increasing relationship complexity. Overall, we found the tested gradient-based approaches to yield worse results than the perturbation-based methods. By evaluating an expert-annotated dataset from the intrusion detection domain and comparing the results to explanations from synthetic data, we showed that the findings on our synthetic datasets can be applied to realistic data.

Acknowledgement. The authors acknowledge the financial support by the Federal Ministry of Education and Research of Germany as part of the DeepScan project (01IS18045A).

References

1. Ancona, M., Ceolini, E., Öztireli, C., Gross, M.: A unified view of gradient-based attribution methods for deep neural networks. In: NIPS 2017 - Workshop on Interpreting, Explaining and Visualizing Deep Learning. ETH Zurich (2017)
2. Ancona, M., Ceolini, E., Öztireli, C., Gross, M.: Gradient-based attribution methods. In: Samek, W., Montavon, G., Vedaldi, A., Hansen, L.K., Müller, K.-R. (eds.) Explainable AI: Interpreting, Explaining and Visualizing Deep Learning. LNCS (LNAI), vol. 11700, pp. 169–191. Springer, Cham (2019). https://doi.org/10.1007/978-3-030-28954-6_9
3. Bach, S., Binder, A., Montavon, G., Klauschen, F., Müller, K.R., Samek, W.: On pixel-wise explanations for non-linear classifier decisions by layer-wise relevance propagation. PloS one **10**(7), e0130140 (2015)
4. Castro, J., Gómez, D., Tejada, J.: Polynomial calculation of the shapley value based on sampling. Comput. Oper. Res. **36**(5), 1726–1730 (2009)
5. Kindermans, P.J., et al.: Learning how to explain neural networks: Patternnet and patternattribution. In: International Conference on Learning Representations (2018)
6. Lundberg, S.M., Lee, S.I.: A unified approach to interpreting model predictions. In: Advances in Neural Information Processing Systems, pp. 4765–4774 (2017)
7. O'Donnell, R.: Analysis of Boolean Functions. Cambridge University Press, Cambridge (2014)
8. Preiss, B.: Data Structures and Algorithms with Object-Oriented Design Patterns in Java (1999)
9. Ribeiro, M.T., Singh, S., Guestrin, C.: Why should i trust you?: explaining the predictions of any classifier. In: 22nd ACM SIGKDD International Conference on Knowledge Discovery and Data mining, pp. 1135–1144. ACM (2016)
10. Ring, M., Schlr, D., Landes, D., Hotho, A.: Flow-based network traffic generation using generative adversarial networks. Comput. Secur. **82**, 156–172 (2019)
11. Rudin, C.: Stop explaining black box machine learning models for high stakes decisions and use interpretable models instead. Nat. Mach. Intell. **1**(5), 206 (2019)

12. Shrikumar, A., Greenside, P., Kundaje, A.: Learning important features through propagating activation differences. In: 34th International Conference on Machine Learning, vol. 70, pp. 3145–3153. JMLR. org (2017)
13. Shrikumar, A., Greenside, P., Shcherbina, A., Kundaje, A.: Not just a black box: Learning important features through propagating activation differences. arXiv preprint arXiv:1605.01713 (2016)
14. Simonyan, K., Vedaldi, A., Zisserman, A.: Deep inside convolutional networks: visualising image classification models and saliency maps. In: Bengio, Y., LeCun, Y. (eds.) ICLR (Workshop Poster) (2014)
15. Sundararajan, M., Taly, A., Yan, Q.: Axiomatic attribution for deep networks. In: 34th International Conference on Machine Learning, vol. 70, pp. 3319–3328. JMLR. org (2017)

SimLoss: Class Similarities in Cross Entropy

Konstantin Kobs[(✉)], Michael Steininger[(✉)], Albin Zehe[(✉)],
Florian Lautenschlager[(✉)], and Andreas Hotho[(✉)]

Julius-Maximilians University Würzburg, Würzburg, Germany
{kobs,steininger,zehe,lautenschlager,hotho}@informatik.uni-wuerzburg.de

Abstract. One common loss function in neural network classification tasks is Categorical Cross Entropy (CCE), which punishes all misclassifications equally. However, classes often have an inherent structure. For instance, classifying an image of a rose as "violet" is better than as "truck". We introduce SimLoss, a drop-in replacement for CCE that incorporates class similarities along with two techniques to construct such matrices from task-specific knowledge. We test SimLoss on Age Estimation and Image Classification and find that it brings significant improvements over CCE on several metrics. SimLoss therefore allows for explicit modeling of background knowledge by simply exchanging the loss function, while keeping the neural network architecture the same. Code and additional resources are available at https://github.com/konstantinkobs/SimLoss.

Keywords: Cross entropy · Class similarity · Loss function.

> *Roses are red, violets are blue,*
> *both are somehow similar, but the classifier*
> *has no clue.*
>
> (Common proverb)

1 Introduction

One common loss function in neural network classifiers is Categorical Cross Entropy (CCE). CCE tries to maximize the assigned target class probability and punishes every misclassification in the same way, independent of other information about the predicted class. Often, however, classes have a special order or are similar to each other, such as different flowers in image classification. Including class similarities using the inherent class structure (e.g., class order), class properties (e.g., class names) or external information about the classes (e.g., knowledge graphs) in the training procedure would allow the classifier to make less severe mistakes as it learns to predict similar classes.

In this work, we modify Categorical Cross Entropy and propose Similarity Based Loss (SimLoss) as a way to explicitly introduce background knowledge

© Springer Nature Switzerland AG 2020
D. Helic et al. (Eds.): ISMIS 2020, LNAI 12117, pp. 431–439, 2020.
https://doi.org/10.1007/978-3-030-59491-6_41

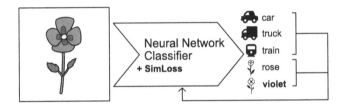

Fig. 1. SimLoss includes knowledge about class relations in the loss function.

into the training process, as visualized in Fig. 1. For this, we augment CCE with a matrix containing class similarities and propose two techniques in order to prepare such matrices that exploit certain class relations: class order and general class similarities. We show on two tasks, Age Estimation (exploiting class order) and Image Classification (exploiting semantic similarities using word embeddings), that SimLoss can significantly outperform CCE. We also show that tuning the hyper-parameters of both generation techniques influences the model's performance on metrics measuring either more or less specific predictions.

Our contribution is twofold: First, we introduce a drop-in replacement for CCE that incorporates class similarities to support the training of neural network classifiers. Second, we describe two techniques to convert task-specific knowledge into matrices that can be used in the proposed loss function.

2   Related Work

Previous work on including task-specific knowledge in classification is mostly designed for specific use cases, requires modifications to the model architecture or training procedure, or implicitly learns the information while training. Sukhbaatar et al. *implicitly* learn a probability instead of a similarity matrix (we provide an analysis of the relationship in the online material) that indicates the chance of a falsely assigned class label in order to compensate for noise [13]. This, however, requires changes in the network architecture and a special training procedure. An analysis of the relation between probability and similarity based matrices is further analyzed in the online resources. Related to tasks with similar classes are tasks where classes have a taxonomic structure, which is called hierarchical classification. Specifically designed loss functions and/or model architectures use the fact that classes that belong to the same category are more similar than others [1,14]. Izbicki et al. exploit the geospatial relation between areas on earth to automatically geotag input photos [5]. Their model learns to predict a mixture of densities that spread across multiple areas instead of specific classes/areas. A number of task-specific methods try to use the inherent class order of so-called ordinal classification tasks [3,4]. For example, Niu et al. use multiple binary classifications each indicating whether the value is greater than the class value [11]. Model architectures incorporating semantic similarities using word embeddings were shown to usually predict more similar classes if they fail compared to models without similarity information [2,12]. In contrast to the related work, the use of SimLoss does not require special model architectures and works on any common neural network classifier. This makes it easy to explicitly support the training procedure with background knowledge.

3 Similarity Based Loss

Our proposed Similarity Based Loss (SimLoss) is based on the Categorical Cross Entropy (CCE). CCE assumes that only one class is correct and is defined as $L_{\mathrm{CCE}} = -\frac{1}{N} \sum_{i=1}^{N} \log(\boldsymbol{p}_i[y_i])$, where N is the size of the dataset and $\boldsymbol{p}_i[y_i]$ is the probability vector output of the network at the target index y_i for the ith example. To model additional knowledge, SimLoss adds a matrix $\boldsymbol{S}$, which gives

$$L_{\mathrm{SimLoss}} = -\frac{1}{N} \sum_{i=1}^{N} \log \left(\sum_{c=1}^{C} \boldsymbol{S}_{y_i,c} \cdot \boldsymbol{p}_i[c] \right), \qquad (1)$$

where $\boldsymbol{S} \in [0,1]^{C \times C}$ encodes class relations. $\boldsymbol{S}_{i,j}$ is the similarity between classes i and j. $\boldsymbol{S}_{i,j} = 1$ if and only if classes i and j are identical or interchangeable.

SimLoss is equal to CCE if $\boldsymbol{S} = I_c$ (identity matrix). Non zero values lead to smaller losses when the network gives a high score to classes similar to the correct one. For misclassifications, this leads the network to predict similar classes.

Matrix Generation. We now propose two techniques to generate the matrix $\boldsymbol{S}$, which explicitly captures background knowledge about class relations. Our techniques allow the modeling of class order and general class similarities.

Class Order: If classes have an inherent order, we can calculate class similarities based on the distance between the class indices. As classes lying next to each other are more similar, we construct the similarity matrix $\boldsymbol{S}$ as follows: Assuming the same distance between neighboring classes, we define the reduction factor $r \in [0, 1)$ to be the rate at which the similarity will get smaller given the distance to the correct class. The similarity matrix is then

$$\boldsymbol{S}_{i,j} = r^{|i-j|} \quad \forall i,j \in \{1,\dots,C\}. \qquad (2)$$

The smaller the reduction factor, the faster the entries converge to 0 with increasing distance to the target class. If the reduction factor is set to 0, the matrix becomes the identity, resulting in the CCE loss. The reduction factor is a hyperparameter of this technique, which can be tuned using a validation dataset to optimize the model for different metrics, as we show in Sect. 4. As SimLoss is equivalent to CCE when $r = 0$ (assuming $0^0 = 1$), an optimized r will always perform at least as good as CCE unless we overfit.

General Class Similarity: For some classification tasks, a similarity between classes, such as class names, is available or can be defined. Then, we can use an appropriate similarity measure $sim : C \times C \to [0, 1]$ that returns the similarity for two classes $i, j \in \{1, \dots, C\}$ and calculate all entries of the similarity matrix S. Such similarity measures can be manual, semi- or fully-automatic. Additionally, we define a lower bound $l \in [0, 1)$ as a hyper-parameter that controls the minimal class similarity that should have an impact on the network punishment. We cut all similarities below l and then scale them such that l becomes 0:

$$S_{i,j} = \frac{max(0, sim(i,j) - l)}{1 - l} \; \forall i, j \in \{1, ..., C\}. \tag{3}$$

Assuming only the diagonal of S are ones, converging $l \to 1$ leads to the CCE loss, as only the ones in the diagonal are preserved by the lower bound cut-off.

4   Experiments

In the following, we compare SimLoss to CCE by applying them to the same neural network model with the same hyper-parameters for Age Estimation and Image Classification. *Age Estimation* is an ordinal classification task with the goal of predicting the age of a person given an image of their face. The classes have an inherent order: two classes are more similar if they represent similar ages. A misclassification is thus less harmful for nearer classes. In *Image Classification*, the goal is to recognize an object shown in an image. Here, we use class name word embeddings to model semantic class similarities. For example, classifying an image of a rose as "violet" is less harmful than classifying it as "truck".

Datasets and Resources. For *Age Estimation*, we train neural networks on the UTKFace [15] and AFAD [11] datasets, both containing human face images annotated with their age. For UTKFace, we use all images for ages 1 to 90, while AFAD has 61 age classes. We randomly sample training/validation/test sets using 60/20/20 splits. For *Image Classification*, we use the CIFAR-100 dataset [7]. We also use word embeddings from a word2vec model pretrained on Google News [9] to calculate the semantic similarity between class names. Four class names do not yield a word embedding and are therefore eliminated. Each remaining class has 450 training, 50 validation, and 100 test examples.

Evaluation Metrics. To evaluate our method, we employ task-dependent evaluation metrics that focus both on correct predictions and the similarity of predicted and target class. For *Age Estimation:* Accuracy (Acc), Mean Absolute Error (MAE), and Mean Squared Error (MSE). Accuracy captures exact predictions, while MAE and MSE capture the distance to the target class, thus considering class order.: *Image Classification:* Accuracy, Superclass Accuracy (SA), and Failed Superclass Accuracy (FSA). Every example in the CIFAR-100 dataset has a main class and a superclass (e.g., classes "rose" and "orchid" have the superclass "flower"). Superclass Accuracy is the fraction of examples that are correctly put into the corresponding superclass. This value is always at least as high as Accuracy, as a correctly assigned class implies the correct superclass. Failed Superclass Accuracy only observes misclassified examples, thus measuring the similarity of misclassifications compared to the target class. A high FSA means that if the model predicts the wrong class, the predicted class is at least similar to the correct class. Accuracy only counts exact predictions, while SA and FSA focus on the semantic similarity of the prediction to the target class.

Generating the Similarity Matrix. Since *Age Estimation* has equidistant classes, the similarity matrix can be built using Eq. (2) without any modifications. In *Image Classification*, we define the similarity matrix as the cosine similarity $sim_{cos} : w \rightarrow [-1, 1]$ between class name embeddings, where $sim_{cos}(w, w) = 1$. To ensure compatibility with the definition in Sect. 3, we set $sim(i, j) = \max(0, sim_{cos}(w_i, w_j))$ in Eq. (3).

Experimental Setup. Since SimLoss is a drop-in replacement for CCE, we investigate the effects of changing the loss function on our example tasks. Recall that we do not focus on task specific models, but rather on the evaluation of SimLoss as a general loss function which can be used on various tasks. Both classification tasks are typical examples for using CCE. For *Age Estimation*, we take the Convolutional Neural Network (CNN) from [11] and change the output size to be the dataset's number of classes. The input images are resized to 60 px by 60 px and the values of all color channels are standardized. We use the softmax function and apply the SimLoss loss function using the similarity matrix introduced above. We study the effect of the reduction factor r by performing grid search for $r \in \{0.0, 0.1, \ldots, 0.9\}$ on the validation set. Optimizing the network using Adam [6] with a learning rate of 0.001 and a batch size of 1024, we employ early stopping [10] with a patience of 10 epochs on the validation MAE. We smooth random differences (e.g., by weight initialization) by averaging over 10 runs. For *Image Classification*, the LeNet CNN [8] is used. Global standardization is applied to the color channels of the input images. We stop early if the Accuracy on the validation set plateaus for 20 epochs of the Adam optimizer with a learning rate of 0.001, and a batch size of 1024. We optimize the matrix generation technique's lower bound $l \in \{0.0, 0.1, \ldots, 0.8, 0.9, 0.99\}$ with grid search and average 10 runs per configuration. $l = 0.99$ makes the loss equivalent to CCE, cutting all similarities except the diagonal.

5   Results

Table 1 shows the resulting mean metrics for the validation and test sets given a reduction factor r for both *Age Estimation* datasets. The best performing reduction factors on the validation and test set are always higher than 0.0, meaning that SimLoss outperforms CCE. Choosing the reduction factor then depends on the metric to optimize for. For UTKFace, a reduction factor of 0.3 leads to the best validation Accuracy, while 0.8 or 0.9 optimize MAE and MSE, respectively. For AFAD, $r = 0.5$ yields the best validation result on Accuracy, while $r = 0.7$ results in the best MAE and MSE. Overall, choosing a smaller reduction factor $r \approx 0.4$ optimizes the Accuracy, while larger $r \approx 0.8$ optimizes MAE and MSE. This is because large r lead to higher matrix values and thus smaller punishments for estimating a class near the correct age. A model optimized for that is favored by metrics that accept approximate matches, such as MAE or MSE.

Table 1. Validation and test results averaged over 10 runs on UTKFace and AFAD. Accuracy (Acc) is given in percent. Best validation values are written in bold. Statistically significantly different test values are marked by $+$ or $-$, if they are on average better or worse than CCE (i.e. $r = 0.0$).

r	UTKFace						AFAD					
	Validation			Test			Validation			Test		
	Acc	MAE	MSE	Acc	MAE	MSE	Acc	MAE	MSE	Acc	MAE	MSE
0.0	15.23	7.09	122.12	14.47	7.39	131.65	11.17	4.05	32.61	11.22	4.10	33.64
0.1	15.43	7.06	119.87	14.48	7.29	127.18	11.21	4.06	32.75	11.30	4.10	33.73
0.2	15.94	7.06	121.28	14.57	7.27	127.13	11.40	4.09	33.52	11.37	4.15^-	34.60^-
0.3	**16.25**	6.95	117.67	15.17^+	7.19^+	125.70	11.34	4.10	33.53	11.38^+	4.16^-	34.53^-
0.4	16.13	6.95	117.52	15.46^+	7.18^+	125.74	11.33	4.10	33.44	11.45^+	4.16^-	34.56^-
0.5	16.10	6.89	115.59	15.09	7.18^+	123.94	**11.44**	4.06	33.02	11.49^+	4.13	34.21
0.6	15.62	6.83	112.85	14.34	7.09^+	120.34^+	11.26	4.01	31.99	11.31	4.05^+	32.84^+
0.7	14.39	6.79	110.12	13.07	7.08^+	121.19^+	11.22	**3.95**	**31.17**	11.11	4.02^+	32.36^+
0.8	13.50	**6.74**	108.80	12.57^-	7.01^+	117.99^+	8.58	4.58	38.69	8.55^-	4.64	39.78
0.9	9.69	6.90	**106.23**	9.16^-	7.18^+	117.62^+	6.55	5.09	44.87	6.47^-	5.15^-	45.82^-

A Wilcoxon-Signed-Rank-Test with a confidence interval of 5 % shows that optimizing the reduction factor always leads to significant improvements over CCE. Sometimes, however, choosing the reduction factor based on a specific metric also results in a trade-off between the chosen and other metrics.

For the *Image Classification* task, Table 2 shows the results for the validation and test set of the CIFAR-100 dataset given a lower bound l. On average, the best performing model always has a lower bound of less than 0.99, again showing that SimLoss outperforms CCE. Also, a statistical test reveals that $l = 0.9$ gives significantly better results on the test set in terms of Accuracy and Superclass Accuracy. Smaller lower bounds tend to reduce the Accuracy as the loss function hardly punishes any misclassification. For $l \approx 1$, the loss is equivalent to CCE, forcing the network to predict the correct class, thus increasing Accuracy. In between, the network is guided to predict the correct class but is also not punished severely for misclassifications of similar classes. This improves Superclass Accuracy, which pays attention to more similar classes.

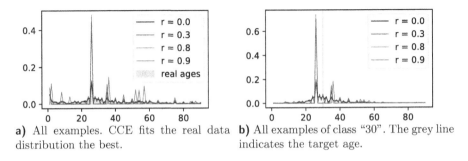

a) All examples. CCE fits the real data distribution the best.   b) All examples of class "30". The grey line indicates the target age.

Fig. 2. Mean probability distribution output for different r. High reduction factors lead the network to choose only few representative classes.

Table 2. Validation and test results over 10 runs with early stopping on the modified CIFAR-100 dataset. Best validation values are written in bold. Statistically significantly different test values are marked by $+$ or $-$, if they are on average better or worse than CCE (i.e. $l = 0.99$).

l	Validation			Test		
	Accuracy	SA	FSA	Accuracy	SA	FSA
0.99	46.89 %	55.78 %	16.73 %	39.51 %	49.22 %	16.05 %
0.90	**47.42 %**	56.32 %	16.95 %	40.15 %$^+$	49.93 %$^+$	16.36 %
0.80	46.37 %	55.38 %	16.80 %	39.49 %	49.32 %	16.22 %
0.70	46.95 %	55.92 %	16.90 %	39.86 %	49.63 %	16.25 %
0.60	47.28 %	**56.44 %**	17.39 %	40.00 %	50.00 %	16.67 %$^+$
0.50	46.36 %	56.18 %	18.28 %	39.26 %	49.40 %	16.70 %$^+$
0.40	38.03 %	50.58 %	20.28 %	32.18 %$^-$	44.58 %$^-$	18.30 %$^+$
0.30	28.65 %	43.76 %	**21.18 %**	24.43 %$^-$	38.90 %$^-$	19.13 %$^+$
0.20	21.66 %	37.97 %	20.80 %	18.54 %$^-$	33.68 %$^-$	18.58 %$^+$
0.10	16.40 %	31.68 %	18.31 %	14.25 %$^-$	28.70 %$^-$	16.85 %
0.00	2.80 %	8.37 %	5.77 %	2.53 %$^-$	8.06 %$^-$	5.71 %$^-$

Analysis. To understand the effect of SimLoss, we focus on Age Estimation whose one dimensional classes are easy to visualize. We compare the best models for UTKFace trained using SimLoss and CCE, i.e. $r \in \{0.0, 0.3, 0.8, 0.9\}$. For each r, we plot the mean output distribution for all examples in the dataset as well as the real age distribution, which is shown in Fig. 2a. CCE ($r = 0.0$) resembles the real age distribution the best, while higher reduction factors tend to aggregate groups of multiple age classes. With a higher reduction factor, the number of spikes decreases and the distances between them increase: The model chooses representative classes to which it mainly distributes the output probability mass. This becomes apparent in Fig. 2b, where we plot the mean output distribution for all examples of age 30. The network with $r = 0.9$ focuses its probability output to the two nearest representative classes, in this case "26" and "35". The Accuracy of the network decreases, as the output probability mass is not on the correct class, but the distance of the prediction to the correct class is smaller than for CCE. Representative classes are apparently chosen such that frequent items receive more probability mass from the model. A higher reduction factor therefore leads to a coarser class selection. This can be explained by the optimization objective of the loss function. The loss should be smaller for misclassifications of similar classes than for dissimilar classes. Representing multiple similar classes as one class and predicting it more often for similar classes does not lead to the smallest possible loss value. However, the loss gets smaller compared to predicting dissimilar classes, as the punishment should be smaller for classifying a similar class. In the case of Age Estimation, predicting an age that lies close to the correct age will decrease the Accuracy, but perform better than CCE on MAE and MSE. In Image Classification, selecting one or multiple representative classes leads to smaller Accuracy but to higher Superclass Accuracy and Failed Superclass Accuracy than CCE. Higher similarities in the matrix thus guide the network to make coarser predictions, improving metrics that accept predictions of similar classes. The results from Sect. 4 also show that keeping the loss near CCE by choosing the similarity matrix conservatively can improve on specific prediction metrics such as Accuracy as well.

6 Conclusion

In this work, we have presented SimLoss, a modified Categorical Cross Entropy loss function that incorporates background knowledge about class relations in form of class similarities. We have introduced two techniques to prepare similarity matrices to exploit class order and general class similarity that can be used to significantly improve the performance of neural network classifiers on different metrics. Also, SimLoss helped with predicting more similar classes if the model misclassified an example. In our analysis, we found that SimLoss forced the model to focus on choosing representative classes. The number of representative classes can be implicitly tuned by a hyper-parameter. While finding the best hyper-parameter and similarity metric can be computationally expensive and non-trivial, SimLoss can incorporate arbitrary similarity metrics into a classifier.

References

1. Cesa-Bianchi, N., Gentile, C., Zaniboni, L.: Incremental algorithms for hierarchical classification. J. Mach. Learn. Res. **7**, 31–54 (2006)
2. Frome, A., Corrado, G.S., Shlens, J., Bengio, S., Dean, J., Mikolov, T., et al.: Devise: a deep visual-semantic embedding model. In: NIPS (2013)
3. Fu, Y., Huang, T.S.: Human age estimation with regression on discriminative aging manifold. IEEE Trans. Multimed. **10**(4), 578–584 (2008)
4. Guo, G., Mu, G., Fu, Y., Huang, T.S.: Human age estimation using bio-inspired features. In: CVPR. IEEE (2009)
5. Izbicki, M., Papalexakis, E.E., Tsotras, V.J.: Exploiting the earth's spherical geometry to geolocate images. In: Brefeld, U., Fromont, E., Hotho, A., Knobbe, A., Maathuis, M., Robardet, C. (eds.) ECML PKDD 2019, vol. 11907, pp. 3–19. Springer, Cham (2019). https://doi.org/10.1007/978-3-030-46147-8_1
6. Kingma, D.P., Ba, J.: Adam: a method for stochastic optimization. arXiv preprint arXiv:1412.6980 (2014)
7. Krizhevsky, A., Hinton, G.: Learning multiple layers of features from tiny images. Technical report, Citeseer (2009)
8. LeCun, Y., Bottou, L., Bengio, Y., Haffner, P., et al.: Gradient-based learning applied to document recognition. Proc. IEEE **86**(11), 2278–2324 (1998)
9. Mikolov, T., Chen, K., Corrado, G., Dean, J.: Efficient estimation of word representations in vector space. arXiv preprint arXiv:1301.3781 (2013)
10. Morgan, N., Bourlard, H.: Generalization and parameter estimation in feedforward nets: some experiments. In: NIPS (1990)

11. Niu, Z., Zhou, M., Wang, L., Gao, X., Hua, G.: Ordinal regression with multiple output CNN for age estimation. In: CVPR (2016)
12. Norouzi, M., et al.: Zero-shot learning by convex combination of semantic embeddings. arXiv preprint arXiv:1312.5650 (2013)
13. Sukhbaatar, S., Bruna, J., Paluri, M., Bourdev, L., Fergus, R.: Training convolutional networks with noisy labels. arXiv preprint arXiv:1406.2080 (2014)
14. Wu, C., Tygert, M., LeCun, Y.: Hierarchical loss for classification. arXiv preprint arXiv:1709.01062 (2017)
15. Zhang, Z., Song, Y., Qi, H.: Age progression/regression by conditional adversarial autoencoder. In: CVPR (2017)

Efficient and Precise Classification of CT Scannings of Renal Tumors Using Convolutional Neural Networks

Mikkel Pedersen[1], Henning Christiansen[1(✉)], and Nessn H. Azawi[2,3,4]

[1] Roskilde University, Roskilde, Denmark
{mikkped,henning}@ruc.dk
[2] Zealand University Hospital, Roskilde, Denmark
nesa@regionsjaelland.dk
[3] University of Copenhagen, Copenhagen, Denmark
[4] Odense University Hospital, Odense, Denmark

Abstract. We propose a new schema for training and use of deep convolutional neural networks for classification of renal tumors as benign or malign from CT scanning images. A CT scanning of a part of the human body produces a stack of 2D images, each representing a slice at a certain depth, and thus comprising a 3D mapping. An additional temporal dimension may be added by injection of contrast fluid with CT scannings performed at certain time intervals. We reduce dimensionality – and thus computational complexity – by ignoring depth and temporal information, while maintaining an ultimate accuracy. Classification of a given scan is done by majority voting over the classifications of all its 2D images. Images are divided into training and validation sets on a patient basis in order to reduce overtraining. Current experiments with scans for 369 patients, yielding almost 20,000 2D images, demonstrate an accuracy of 93.3% for single images and 100% for patients.

1 Introduction

A renal tumor may be benign or malign, and in the latter case, immediate surgery is likely needed. Biopsy procedures are currently the only way for precise classification of renal tumors, but non-invasive methods are preferable. The decision of whether or not to operate a given patient is typically done by manual inspection of CT scanning images, perhaps complemented by other medical tests. However, this practice has a high error rate, and especially false positives are problematic as they result in unnecessary surgery, leading to lost life quality for the patients and waste of resources in the health sector. Recent studies indicate rates of 11–17% of surgery for patients with renal tumors, turning out to be benign [7,10,12]. A precise automatic classification from CT scan images may, thus, become an important decision support tool.

In the recent years, deep learning methods for convolutional neural networks (CNN) have lead to numerous reports of impressive results in medical image

© Springer Nature Switzerland AG 2020
D. Helic et al. (Eds.): ISMIS 2020, LNAI 12117, pp. 440–447, 2020.
https://doi.org/10.1007/978-3-030-59491-6_42

analysis; CNN and deep learning are able to identify subtle but indicative features that are difficult to recognize for a human eye as, e.g., indicated above.

A CT scanning produces a 3D mapping of the tumor, consisting of a stack of 2D images, each focused at a specific depth. An additional temporal dimension may be added by injection of contrast fluid with scannings performed at certain time intervals, leading to so-called multiphase images. We present an experiment of training and testing a CNN on a substantial number of CT images collected from 369 Danish patients, yielding more than 20,000 2D images. Compared with other approaches, we reduce dimensionality – and thus computation time – by ignoring depth and temporal information while maintaining a very high accuracy.

For classification of single images, our accuracy amounts 93.3%, and by a majority voting over all 2D images for a given patient, we obtain 100% accuracy. The selection procedure for separation the total set of images is important for these results. We can show that the separation of images into training and validation sets should be done on the level of patients, rather that randomly over all 2D images as a whole, which results in a significant overtraining.

Section 2 gives an account on related work, focusing specifically on CNN in relation to CT scanning. In Sect. 3, we explain more about the datasets produced by CT scannings, as well as the dataset available for our experiments. Our specific CNN model is explained in Sect. 4 and test results presented in Sect. 5. Section 6 gives a summary and our directions for future work.

2 Related Work

Convolution Neural Networks (CNNs), introduced by LeCun et al in 1989 [8] (for a recent overview, see [4, chap. 9]), have developed into a standard for medical image analysis. The migration into standard equipment for medical usage is still in its infancy, but a lot of research effort is invested into the field all over the world. Even when restricted to CT scan images, the DBLP bibliography reports more that 300 scientific papers since 2015 on the topic.[1] CNNs are supported by several software platforms, including the TensorFlow [1] program library and its high level API Keras [3] that we have used for our experiments.

There are some recent works on classification of renal tumors. Pan et al [9] apply a complex network structure based on CNN, involving explicit phases of segmentation ("where is the tumor?") and classification, where we rely on CNN's ability automatically to identify the image features that are important for classification. They test out different CNN structures for scans of a total of 131 patients for training and test, each given by 300 2D images; accuracy up to 100% is reported for the best setup. The referenced paper emphasizes the importance of using multi-phasic CT images, but it is not clear whether or how this information is employed in their model and in training and test.

Han et al [5] analyze multiphase CT images with three phases, using manual segmentation and manual selection of one "best" 2D image from each phase;

---

[1] DBLP, The Computer Science Bibliography, https://dblp.uni-trier.de; searches for "convolution tomography" and "convolution CT"; March 2020.

each triplet of matching phase images is then combined into one in an ad-hoc manner. A total of 169 patients were used for training and test, and an accuracy of 85% is reported from the experiment.

In comparison, our model is trained and tested on a larger set of 369 patients, but compared with [9], a smaller number of 2D images (20–100, vs. their fixed 300) for each patient. Moreover, our approach has removed all time and phasic information related to the CT images, reducing the complexity of our network structure, shown in Fig. 1, and thus also training and classification time and hardware cost. The fact that we can do with perhaps as few as 20 images for classification may further speed up computation times. Unfortunately, none of the mentioned experiments (incl. our own) have reported figures for computation times, so we cannot make a precise comparison.

The mentioned approaches, including our own, involve known CNN structures adapted with new classification layers for the purpose. Pre-trained weight may be used unchanged (and only the classification layers are trained with CT scan images), as initial with weights for a retraining, or the network may be trained from scratch. We have used ResNet50V2 [6] as it performed best among those we tried out.

3   Data Set

Our datasets consist of CT scans for 369 patients (20% benign, 80% malign) in total performed at different Danish hospitals collected from 2015 until beginning of 2020; the possible influence of regional differences and changes in equipment and clinical procedures over the years are not considered in the present study.

True labelings as benign/malign were determined by subsequent surgery or other medical tests. No records are maintained of whether a scan involves different phases (nor how many). The number of 2D images per patient ranges from 20 to 100, yielding more that 20,000 2D-images. Original images with three colour channels of 512×512 pixel images are downscaled to 224×224 (for easy fit with the ResNet network structure).

We split data into 70% for training, 10% for validation (to check convergence during training) and 20% for independent test (of the trained model). To form a balanced training set (50% benign, 50% malign) we applied oversampling[2] to the original 20% benign ones.

Considering the set of 2D images from a single CT scan, and thus related to one patient, we note that 2D images at adjacent or close depths may be nearly identical, and the downscaling emphasizes this phenomenon. This increases the risk of over-training that might not be detected by a standard comparison of accuracy (or, alternatively, loss) of a trained model on its training data vs. the disjoint sets of validation and test data: if the splitting is done completely at random over all 2D images, each validation/test image will likely have several

---

[2] See [2] for an overview; overtraining is a potential disadvantage of oversampling, but shown by our tests, this is not the case in our experiments.

almost identical companions in the training set. To eliminate this phenomenon, we perform the random splitting at the patient level, rather that on individual images. Thus each such group of "similar" images belongs entirely to either the training set or to the test set. After that, each set is viewed as a collection of arbitrary images, each carrying a true classification as benign/malign but no information about which patient or possible phase. These decisions obviously throw away some information, dimensionality is reduced and so is the time complexity, and our test results, shown in Sect. 5, below, shows that we still obtain very good results.

4   Experiment and Model Structure

We use a modified version of the ResNet50V2 [6] CNN implemented with the TensorFlow and Keras software running on standard, affordable hardware (AMD Ryzen 2700x CPU with 16GB RAM, GeForce TTX 980ti 4GB and 1050ti 2GB GPUs).

Figure 1 shows the original ResNet50V2 structure together with our version. While ResNet50V2 classifies images into 1000 different categories according to the object depicted, we are interested in a binary classification into benign/malign, and thus the final, fully connected network layers can be much simpler; the convolutional layers are reused. We use transfer learning from ResNet50V2, applying its trained weights as initial one (rather than random weights), and the new fully connected layers for classification need to be trained from scratch.

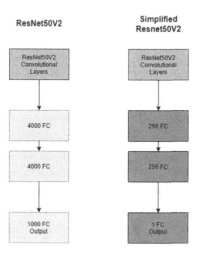

Fig. 1. ResNet50V2 and our simplied version. The convolutional layers are identical to ResNet50V2 [6]; the fully connected layers (FC) are reduced from 4000 nodes to 256 nodes and the final output layer from 1000 nodes to 1 node.

5   Results

Our adaptation of ResNet50V2 has been trained on the data explained above, with no overlap between images from single patients across the training, validation and test set. Here we report the results for (1a) the randomly per patient chosen 20%, and (1b) a small set of 12 patients collected in 2019 independently from the other data. To check the hypothesis that splitting per patient rather than randomly over all 2D images, we retrained the model with this selection methods and tested with (2a) 20% images selected in a similar manner, and (2b) the same 12 patients as in (1b). In addition, we performed in both cases, what we refer to as complete test, where the training and validation data are combined (1c); comparing these with (1a) and (2a) may also give indication of possible over-training.

Tests (1a) and (1b) (and (1c) as well) showed a 100% accuracy when each patient were classified as benign/malign by a majority voting of individual image classifications. In the following, we concentrate on the finer detail of the image classification, showing ROC curves and AUC values for the different tests. Figure 2 shows the results for (1a) and (1c), and Fig. 3 for (1b). Test (1a) indicates an impressive AUC of 0.973, and (1b) an AUC of 0.992, indicating a small and insignificant overtraining. Test (1b) indicates, as one would expect, a slightly lower, but still satisfactory, results on single images, but not enough to degrade the majority voting's 100% accuracy. However, this 12 patient set is too small for drawing any firm conclusions, and it motivates our planned future studies of influences of regional and historical variations.

To check our hypothesis presented in Sect. 3 that overtraining is reduced by our splitting principle, of separating the 2D image sets into training and validation parts randomly on a patient basis rather that on all images, we re-did training and tests using the latter principle for data set splitting, i.e., with a random

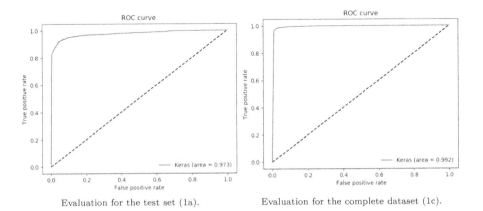

Evaluation for the test set (1a).   Evaluation for the complete dataset (1c).

Fig. 2. ROC curve and AUC value for the (1a) test set and the complete dataset (1c). The accuracy amounts to 93.3% for (1a) and 97.7% for (1c).

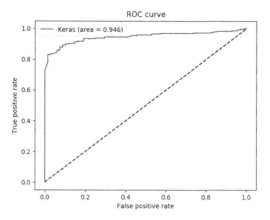

Fig. 3. ROC curve and AUC value for the (1b) test with an accuracy of 90.4%.

70/30 splitting at image level. ROC curves and AUC values for (2a–c) are shown in Fig. 4. As we expected, the performance on a test set – in which each image statistically will have a number of similar companions in the training set – is "too good", and for unseen 12 patient data set, the performance degrades heavily. It appeared that the patient classification based on majority voting (not shown in the graphs) now classified 2 out of 12 patients incorrectly.

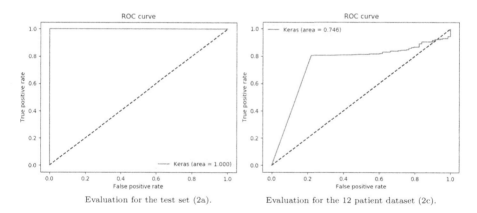

Evaluation for the test set (2a).   Evaluation for the 12 patient dataset (2c).

Fig. 4. Test (2a–2b) for the traditional splitting method that we do not recommend for these specific data. There is a severe disparity in indicated performance versus tests (1a–b).

6   Conclusion and Future Work

We have shown our first experiment of training and using a Convolutional Neural Network based on ResNet50V2 for automatic classification of renal tumors from multi-phasic CT scan images. We used a substantially larger dataset than those related works we have compared with, cf. Sect. 2, and we tested on independent tests sets not included in the training and validation sets. In contrast to the compared works, we discard information about phases as well as depth information of each 2D image; as it appears in our tests, this does not affect the accuracy, so no essential information is lost. On the other hand, it reduces complexity of the neural network and thus time complexity for training and classification. We used oversampling to obtain balanced training sets as there were about four times as many malign patients as benign ones.

We presented a hypothesis that the overtraining can be reduced by separating the 2D image sets into training and validation parts randomly on a patient basis rather that on all images; this was clearly verified by our test results.

Classification using the trained model yielded an accuracy per image ranging from 90.4% to 97.7% in the different tests, leading to a consistent 100% classification accuracy per patient using majority voting. All in all, we consider the method we have developed as a worthy candidate to be matured into an effective and efficient decision support tool for medical experts, giving an instant and reliable proposal for a diagnosis.

Our plans for future work include collecting a much larger dataset based on all available CT scan images from Danish hospitals for a critical assessment of the observed 100% classification, and we need also study the details of possible influences regional difference and historical developments in CT scanning equipments. We plan a detailed assessment and interpretation of the feature maps extracted from the image data through network training, aiming at a visualization, e.g., along the lines of Selvaraju et al [11]. From a clinical standpoint, interpretation an visualization of the feature maps may provide some clarity to areas of interest on CT imaging that are not currently understood when diagnosing renal cancer, and it may diminish the magic blackbox flavour of CNN-based classification and promote its acceptance as a practical decision support tool.

References

1. Abadi, M., et al.: Tensorflow: a system for large-scale machine learning. In: 12th USENIX Symposium on Operating Systems Design and Implementation (OSDI 2016), pp. 265–283 (2016), https://www.usenix.org/system/files/conference/osdi16/osdi16-abadi.pdf

2. Buda, M., Maki, A., Mazurowski, M.A.: A systematic study of the class imbalance problem in convolutional neural networks. Neural Networks **106**, 249–259 (2018). https://doi.org/10.1016/j.neunet.2018.07.011

3. Chollet, F., et al.: Keras (2015). https://keras.io

4. Goodfellow, I., Bengio, Y., Courville, A.: Deep Learning. MIT Press (2016). http://www.deeplearningbook.org

5. Han, S., Hwang, S.I., Lee, H.J.: The classification of renal cancer in 3-Phase CT images using a deep learning method. J. Digit. Imaging **32**(4), 638–643 (2019). https://doi.org/10.1007/s10278-019-00230-2

6. He, K., Zhang, X., Ren, S., Sun, J.: Identity mappings in deep residual networks (2016)

7. Johnson, D.C., et al.: Preoperatively misclassified, surgically removed benign renal masses: a systematic review of surgical series and United States population level burden estimate. J. Urol. **193**(1), 30–35 (2015). https://doi.org/10.1016/j.juro.2014.07.102

8. LeCun, Y., et al.: Backpropagation applied to handwritten zip code recognition. Neural Comput. **1**(4), 541–551 (1989). https://doi.org/10.1162/neco.1989.1.4.541

9. Pan, T., et al.: A multi-task convolutional neural network for renal tumor segmentation and classification using multi-phasic CT images. In: 2019 IEEE International Conference on Image Processing, ICIP 2019, Taipei, Taiwan, 22–25 September, 2019, pp. 809–813. IEEE (2019). https://doi.org/10.1109/ICIP.2019.8802924

10. Pedersen, C.L., Winck-Flyvholm, L., Dahl, C., Azawi, N.H.: High rate of benign-histology in radiologically suspect renal lesions. Danish Med. J. **61**(10) (2014)

11. Selvaraju, R.R., Cogswell, M., Das, A., Vedantam, R., Parikh, D., Batra, D.: Grad-CAM: visual explanations from deep networks via gradient-based localization. Int. J. Comput. Vis. **128**(2), 336–359 (2019). https://doi.org/10.1007/s11263-019-01228-7

12. Srougi, V., Kato, R.B., Salvatore, F.A., Ayres, P.P.M., Dall'Oglio, M.F., Srougi, M.: Incidence of benign lesions according to tumor size in solid renal masses. Int. Braz. J. Urol. **35**(4), 427–431 (2009). https://doi.org/10.1590/S1677-55382009000400005

Deep Autoencoder Ensembles
for Anomaly Detection on Blockchain

Francesco Scicchitano, Angelica Liguori, Massimo Guarascio[⊠],
Ettore Ritacco, and Giuseppe Manco

ICAR-CNR, Via P. Bucci, 8/9c, Rende, Italy
{francesco.scicchitano,angelica.liguori,massimo.guarascio,
ettore.ritacco,giuseppe.manco}@icar.cnr.it

Abstract. Distributed Ledger technologies are becoming a standard
for the management of online transactions, mainly due to their capability to ensure data privacy, trustworthiness and security. Still, they
are not immune to security issues, as witnessed by recent successful
cyber-attacks. Under a statistical perspective, attacks can be characterized as anomalous observations concerning the underlying activity. In
this work, we propose an Ensemble Deep Learning approach to detect
deviant behaviors on Blockchain where the base learner, an encoder-
decoder model, is strengthened by iteratively learning and aggregating
multiple instances, to compute an outlier score for each observation. Our
experiments on historical logs of the Ethereum Classic network and synthetic data prove the capability of our model to effectively detect cyber-
attacks.

Keywords: Blockchain · Anomaly detection · Sequence to sequence
models · Encoder-decoder models · Ensemble learning

1 Introduction

There is a growing interest in adopting *Distributed Ledger Technology* (DLT) due
to its capability to ensure privacy, trustworthiness and security in online transactions and payments. In particular, the Blockchain represents the most known
and widespread DLT [5] and allows to save data under the form of permanent
and verifiable transactions between two parties.

However, as discussed in [11], Blockchain is not immune to security issues,
therefore the early detection of in-progress attacks represents a challenging and
important problem. In particular, all processes within the blockchain are logged
and, since the ledger is open, it is natural to ask whether these logs can be
exploited to the early detection challenge. In this paper we focus on the information collected by the activities of *Ethereum Classic*[1] (ETC), a public blockchain.
ETC blockchain has experienced two significant attacks: [1] reports an attack on

---

[1] http://ethereumclassic.org/.

© Springer Nature Switzerland AG 2020
D. Helic et al. (Eds.): ISMIS 2020, LNAI 12117, pp. 448–456, 2020.
https://doi.org/10.1007/978-3-030-59491-6_43

18 June 2016 (referred as DAO from now on) and in January 2019 researchers confirmed a successful 51% attack[2]. Our purpose is then to identify the core features from ETC logs that allow to early detect attacks.

The current literature focused on the machine learning techniques as a powerful tool to identify cyber-attacks and detect anomalous behaviors real-time or for post-incident analysis. Notably, both supervised and unsupervised machine learning algorithms have been successfully employed to support intrusion detection and prevention systems, as well as to detect system misuses and security breaches. However, these techniques were seldom applied to blockchain, with few exceptions where preliminary machine learning based approaches have been proposed to improve the security on blockchain [4]. A visual analytical approach of attack discovery is proposed in [2], where a set of statistics, collected from the Ethereum blockchain, is used as input to an unsupervised anomaly detection system: the work shows an anomalous peak close to the corresponding DAO attack date. This approach has been extended in [10], where authors propose an encoder-decoder deep learning model to detect ETC attacks. Notably, several anomaly detection techniques has been proposed in literature for different types of scenarios [3,9], but we mainly focused on the current approaches adopted to early detect anomalies on blockchain.

Specifically, in this paper we investigate the adoption of autoencoder ensembles to the analysis of ETC activities. Ensembles can strengthen the capabilities of the basic autoencoders which tend to overfit, especially in noisy contexts (like the ETC logs). Our strategy, based on sequence-to-sequence ensemble, consisting in by progressively training a weak learner with a Snapshot Procedure [8], i.e., a cyclic alteration of the learning rate. As shown in [8], this approach allows to guarantee a better exploration of the search space by a combination of multiple local minima, without affecting the training performance.

Our main contributions can be summarized as follow: *(i)* an unsupervised neural network architecture is devised for the early detection of anomalies. Basically, multiple sequence-to-sequence models are discovered by exploiting a Snapshot procedure and combined according to a suitable strategy, *(ii)* an ETC real dataset is used to assess the outlier detection capability of the proposed model. In addition, a robustness analysis is performed on synthesized data.

The paper is organized as follows: Sect. 2 is devoted to describe the ensemble architecture and the snapshot procedure; in Sect. 3, first, we illustrate the experimentation performed on a real (from ETC blockchain) and a synthetic dataset. Finally, in Sect. 4, we discuss relevant open issues and future works.

2   Methodology

In this section we describe the machine learning approach proposed to detect attacks in progress on blockchain-based systems and, more in general, in cyber-security scenarios. Due to the lack of labeled data, an unsupervised method is

---

[2] See http://tiny.cc/fri3iz.

adopted to identify outliers: the underlying assumption is that successful attacks represents extremely rare events and, typically, do not share common patterns.

As a consequence, frequently supervised techniques fail to detect new incoming/in-progress attacks, resulting in poor performances.

Our proposal adapts the snapshot ensemble method defined in [8] to a sequential unsupervised scenario exploiting an encoder-decoder model as a base model. In detail, it aims at replicating an input sequence by producing a reconstructed copy from the compressed representation of the input, therefore the reproduction error can be used as anomaly indicator. Although the idea to use the reconstruction error as anomaly score to identify deviant behaviors is not new itself[3], it has been not fully explored to monitor malicious behaviors on blockchain and new research lines can be analyzed. In [10], the authors proposed a preliminary approach able to recognize attacks on blockchain. However this method could be affected by the concept drift problem, while our approach, by considering different models, is able to handle smooth changes.

In our framework we instantiate several base models, which outlierness score is finally averaged. These models are generated according to a simple algorithm, which takes advantage of the iterative gradient-descent optimization procedure of the neural network learning phase. The latter assumes that, at each step, the weights of the underlying network are updated in the direction opposite to the gradient, in order to progressively approach a local minimum. The learning rate determines the speed of convergence and it is progressively adapted to guarantee convergence. The local optimum obtained by the Stochastic Gradient Descent procedure and its variants depends both on the initialization and the choice of the learning rate. As noticed in [8], a cyclic reset of the latter has the same effect of re-initializing the network and restarting the optimization from another spot in the search space, allowing to identify different variants of the model.

Formally, we modeled data as temporally-sorted multi-dimensional events (i.e. each event is composed of several features). Let $X = \{x_1, \ldots, x_N\}$ be the sequence of events observed in a window with length N, and x_t the feature vector of the t-th occurrence in the sequence X. An anomaly x_t over X is an abnormal event significantly different from close events. The encoder-decoder is composed by two sub-networks. The encoder Θ compresses the input x into a latent space $\Theta(x) = z \in \mathbb{R}^K$, generating an embedding of the original input into a latent vector of size K. By converse, the decoder Φ, given a K-dimensional vector z, aims at generating an output $\Phi(z) = y$ as close as possible to the original input. Θ and Φ are modeled as Recurrent Neural Networks (RNNs) [6], which represent a natural choice to handle sequential data: by iterating over the sequence a recurrent network is able to store (partial) memory of each event. In our implementation we use Long Short-Term Memories (LSTM) [7]. Thus, given an input sequence $\mathcal{I} = \{x_1, \ldots, x_n\}$, a single encoder-decoder learner computes

---

an output sequence $\mathcal{O} = \{\boldsymbol{y}_1, \ldots, \boldsymbol{y}_n\}$ as follows:

$$
\begin{aligned}
\boldsymbol{h}_t^{(e)} &= \mathtt{RNN}_\theta(\boldsymbol{x}_t, \boldsymbol{h}_{t-1}^{(e)}) \\
\boldsymbol{z} &= \mathtt{mlp}_\vartheta(\boldsymbol{h}_t^{(e)}) \\
\boldsymbol{h}_t^{(d)} &= \mathtt{RNN}_\phi(\boldsymbol{z}, \boldsymbol{h}_{t-1}^{(d)}) \\
\boldsymbol{y}_t &= \mathtt{mlp}_\varphi(\boldsymbol{h}_t^{(d)})
\end{aligned}
\tag{1}
$$

where $\mathtt{RNN}_\theta$ and $\mathtt{mlp}_\vartheta$ represent the encoder, with internal state $\boldsymbol{h}_t^{(e)}$ given the t-th event; symmetrically, $\mathtt{RNN}_\phi$ and $\mathtt{mlp}_\varphi$ represent the decoder, with inner state $\boldsymbol{h}_t^{(d)}$. Further, $\mathtt{mlp}_\vartheta$ and $\mathtt{mlp}_\varphi$ represent multilayer networks parameterized by θ and ϕ, respectively. Since the main purpose of the autoencoder is to reconstruct the input from a compact representation, the model can be trained by considering a reconstruction loss:

$$
\ell(\mathcal{I}, \mathcal{O}) = \frac{1}{n}\sum_{t=1}^{n}\|\boldsymbol{x}_t - \boldsymbol{y}_t\|_2
\tag{2}
$$

Input subsequences are obtained from X through a sliding window mechanism. Each timestep within X is associated with a subsequence $W_t = \{\boldsymbol{x}_{t-m+1}, \ldots, \ldots, \boldsymbol{x}_t\}$, where m is the window size. As shown in [10], the autoencoder can be trained on a set $\{W_m, \ldots, W_N\}$ of subsequences that can be obtained from X, by learning to reconstruct them in a way that minimizes the specified loss. The distance between the input and the output is used to measure the outlierness of the analyzed sequence. The final score is computed as the average of the outlierness scores of all the involved windows.

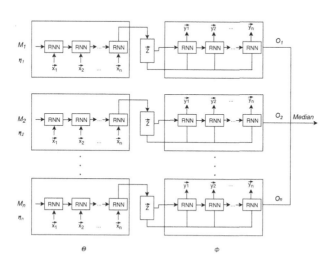

Fig. 1. Snapshot ensemble encoder-decoder model.

The Snapshot Ensemble Encoder-Decoder (SEED) is trained as follows. An autoencoder is randomly initialized and devised as M_1. Then, the procedure iteratively learns model M_i by re-training M_{i-1} with the initial learning rate η for a fixed number of epochs. At each epoch η is progressively lowered. M_i is then collected in the ensemble and the learning rate is reinitialized. The final architecture is shown in Fig. 1, where the different encoder-decoder weak learners $\{M_1, M_2, \ldots, M_n\}$ are shown. In the prediction stage, each sequence is passed as input to all instances and the reconstructed sequences are obtained. The final outlierness score is the median of the n reconstruction errors $\{O_1, \ldots, O_n\}$ produced by the instances.

3   Experimental Evaluation

In this section, first, we specify the values of the main parameters characterizing the DNN architecture, then the experimentation on the real and the synthetic data are shown. In detail, two LSTM layers (each one composed of 32 cells) are employed for encoding and decoding the input data and a *Hyperbolic Tangent* is adopted as activation function in each cell of these layers. The reconstruction error is measured in terms of *Mean Squared Error* (MSE) which is the loss function minimized by the optimizer (in our case we used the Adaptive Moment Estimation (Adam) algorithm). Finally, we set 200 epochs to learn the base models composing the ensemble.

3.1   Analysis of the Ethereum Classic Network

In this section, we apply SEED to the ETC network to identify attacks. As discussed in Sect. 1, ETC has experienced two known attacks: DAO (18 June 2016) and 51% attack. Documentation concerning the latter, is scarce but the reports state that it occurred in a period within the interval 5–8 January 2019. Our goal is thus to adopt SEED to highlight possible anomalies in the given time intervals. Our experiments[4] have been performed on a four year sample of ETC blockchain by using preprocessing steps and data split (Fig. 2) like in [10]. At the end of the preprocessing phase, we obtained the following subset of relevant features, computed on a daily basis: (i) `block_size_average`, the average size (in bytes) of a block; (ii) `provided_gas_average`, referring to the average *provided gas* needed to perform the transaction; (iii) `block_difficulty_average`, the average effort necessary to validate a block; (iv) `transaction_average_per_block`, the average number of transactions contained in a block; (v) `gas_used_sum`, the total amount of employed gas; (vi) `transactions_number`, the total number of transactions in all blocks.

---

[4] https://github.com/francescoscicchitano/Anomaly_Detection_On_Blockchain.

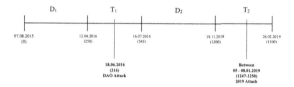

Fig. 2. Data split according to the attacks, as indicated in [10].

We consider two different training sets, namely D_1 (relating to the period prior DAO) and D_2 (covering the period post DAO and prior the 51% attack). T_1 and T_2 (the periods within a range of about two months from the attacks, exact dates are listed in Fig. 2) are used as test sets, for evaluating the outlierness scores. We performed three different tests by training two instances of SEED model. The first two use D_1 as training set and scored all the events within T_1 and T_2, respectively. In the third experiment SEED is trained on $D_1 \cup D_2$ and scores events in T_2. We used 10 weak learners, obtaining scores in the Fig. 3.

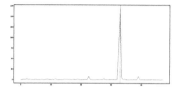

(a) Prediction on T_1, red line represents the confirmed DAO attack.

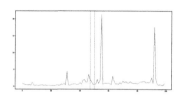

(b) Prediction on T_2, red lines represent the interval of the reported 2019 attack.

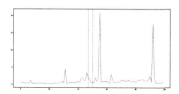

(c) Prediction on T_2, SEED was trained on $D_1 \cup D_2$, red lines represent the interval of the reported 2019 attack.

Fig. 3. Outlierness score on ETC. (Color figure online)

Figures 3a and 3b plot the outlierness scores computed by our model on both T_1 and T_2. We can see that the DAO is perfectly detected, as shown in Fig. 3a. In fact, outlierness score exhibits a peak on day 316, corresponding to DAO. The result is consistent with the findings of [2] but in our case the model is capable of perfectly detect the exact day of the attack.

In Figs. 3b and 3c we can see that the peak is translated of a few days respect to vertical red lines (the presumed starting and ending days of the 2019 attack): in fact, outlierness score changes its pattern as the attack period approaches; later, many companies frozen all activities on ETC network and blockchain didn't register a core amount of transactions, which were, instead, restarted in the forthcoming days[5,6,7], triggering the registered peak in day 55. The 51% attack highlighted by SEED was also confirmed in [10].

However, the selected features seem not fully sufficient to detect some types of attacks occurring on blockchain but we figure out that integrating data from other sources (as show in [2]) could improve the detection capability of SEED.

3.2 Experiments on Synthetic Data

In this section we perform a sensitivity analysis of SEED in a controlled scenario where were both the dimensionality of the data and the number of outliers is tuned. Data were generated according to the following procedure: *(1)* first a sequence D (as a matrix $1440 \times nFeat$) is generated; then *(2)* n points are randomly selected and a candidate feature is extracted for each point; finally

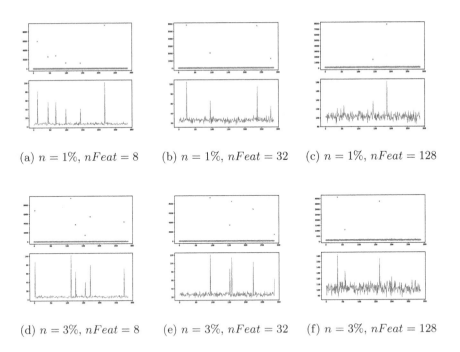

(a) $n = 1\%$, $nFeat = 8$ (b) $n = 1\%$, $nFeat = 32$ (c) $n = 1\%$, $nFeat = 128$

(d) $n = 3\%$, $nFeat = 8$ (e) $n = 3\%$, $nFeat = 32$ (f) $n = 3\%$, $nFeat = 128$

Fig. 4. Outlierness score on synthetic data. (Color figure online)

---
[5] http://shorturl.at/giz14.
[6] https://bit.ly/30Qs9P8.
[7] http://cryptonomist.ch/2019/01/07/ethereum-classic-attacco-del-51/.

(3) an uniform noise is injected in the candidate feature. In our experiments, n ranges into $\{1\%, 3\%\}$ of the entire dataset and $nFeat \in \{8, 32, 128\}$, while we set the number of weak learners to 10, $|D| = 1440$ as in ETC and $\gamma = 200$. The evaluation is performed by adopting an hold-out validation protocol (80%–20%). The results are highlighted in Fig. 4. For each experiment we show two graphs: respectively, the topmost one represents, for each time-sorted event in the test set, the anomaly score assigned by the model; blue dots are regular events, while red crossed dots are the detected anomalies; while, the bottom graph, shows the overlap of all features, thus highlighting the peaks where we injected noise. These results highlight the capability of SEED to detect outliers in different settings.

4   Concluding Remarks

In this paper we proposed an unsupervised ensemble deep architecture for anomaly detection. The ensemble schema was obtained by exploiting the Snapshot Procedure, defined in [8] as a strategy to improve classifiers based on neural networks. In particular, we defined a snapshot ensemble encoder-decoder model to identify anomalies in sequential data, without providing to it any prior knowledge about outliers. An extensive evaluation on real and synthetic data proves the capability of the approach in detecting incoming attacks.

As future work, we plan to study new effective methods to select the weak learners composing the ensemble, so to improve the overall performances of the approach.

Acknowledgment. This work has been partially supported by MIUR - PON Research and Innovation 2014–2020 under project Secure Open Nets.

References

1. Atzei, N., Bartoletti, M., Cimoli, T.: A survey of attacks on Ethereum smart contracts (SoK). In: Maffei, M., Ryan, M. (eds.) POST 2017. LNCS, vol. 10204, pp. 164–186. Springer, Heidelberg (2017). https://doi.org/10.1007/978-3-662-54455-6_8
2. Bogner, A.: Seeing is understanding: anomaly detection in blockchains with visualized features. In: UbiComp 2017, pp. 5–8. ACM (2017)
3. Corizzo, R., Ceci, M., Japkowicz, N.: Anomaly detection and repair for accurate predictions in geo-distributed big data. Big Data Res. **16**, 18–35 (2019)
4. Dey, S.: Securing majority-attack in blockchain using machine learning and algorithmic game theory: a proof of work. CoRR, abs/1806.05477 (2018)
5. El Ioini, N., Pahl, C.: A review of distributed ledger technologies. In: Panetto, H., Debruyne, C., Proper, H.A., Ardagna, C.A., Roman, D., Meersman, R. (eds.) OTM 2018. LNCS, vol. 11230, pp. 277–288. Springer, Cham (2018). https://doi.org/10.1007/978-3-030-02671-4_16
6. Graves, A.: Supervised Sequence Labelling with Recurrent Neural Networks. SCI, vol. 385. Springer, Heidelberg (2012). https://doi.org/10.1007/978-3-642-24797-2
7. Hochreiter, S., Schmidhuber, J.: Long short-term memory. Neural Comput. **9**(8), 1735–1780 (1997)

8. Huang, G., Li, Y., Pleiss, G., Liu, Z., Hopcroft, J.E., Weinberger, K.Q.: Snapshot ensembles: train 1, get M for free. In: 5th International Conference on Learning Representations, ICLR 2017, Toulon, France, 24–26 April 2017, Conference Track Proceedings (2017)

9. Munir, M., Siddiqui, M.S.A., Dengel, A., Ahmed, S.: DeepAnT: a deep learning approach for unsupervised anomaly detection in time series. IEEE Access **7**, 1991–2005 (2018)

10. Scicchitano, F., Liguori, A., Guarascio, M., Ritacco, E., Manco, G.: A deep learning approach for detecting security attacks on blockchain. In: Italian Conference on Cybersecurity, ITASEC (2020). CEUR-WS, http://ceur-ws.org/Vol-2597/paper-19.pdf

11. Ye, C., Li, G., Cai, H., Gu, Y., Fukuda, A.: Analysis of security in blockchain: case study in 51%-attack detecting. In: 2018 5th International Conference on Dependable Systems and Their Applications (DSA), pp. 15–24 (2018)

A Parallelized Variant of Junker's QUICKXPLAIN Algorithm

Cristian Vidal Silva[1]([✉]), Alexander Felfernig[3], Jose Galindo[2], Müslüm Atas[3], and David Benavides[2]

[1] Catholic University of the North, Antofagasta, Chile
cristian.vidal@ucn.cl
[2] University of Sevilla, Seville, Spain
{jagalindo,benavides}@us.es
[3] Graz University of Technology, Graz, Austria
{afelfernig,muatas}@ist.tugraz.at

Abstract. Conflict detection is used in many scenarios ranging from interactive decision making to the diagnosis of potentially faulty hardware components or models. In these scenarios, the efficient identification of conflicts is crucial. Junker's QUICKXPLAIN is a divide-and-conquer based algorithm for the determination of preferred minimal conflicts. Motivated by the increasing size and complexity of knowledge bases, we propose a parallelization of the original algorithm that helps to significantly improve runtime performance especially in complex knowledge bases. In this paper, we introduce a parallelized version of QUICKXPLAIN that is based on the idea of predicting and executing parallel consistency checks needed by QUICKXPLAIN.

1 Introduction

Conflict detection is used in many applications of constraint-based representations (and beyond). Examples thereof are *knowledge-based configuration* [13] where users define requirements and conflict detection is in charge of figuring out minimal sets of potential changes to the given requirements in order to restore consistency (if the configurator is not able to identify a solution), *recommender systems* [3,10], and many other applications of model-based diagnosis [9]. Especially in interactive settings, there is often a need of identifying preferred conflicts [7,11], for example, users of a car configurator or a camera recommender who have strict preferences regarding the upper price limit, are more interested in relaxations related to technical features (e.g., related to the availability of a skibag in a car or a wide-aperture lens in a pocket camera).

Conflict detection helps to find combinations of constraints in the knowledge base that are responsible for an inconsistency. QUICKXPLAIN is such a conflict detection algorithm which is frequently used and works for constraint-based representations, description logics, and SAT solvers [7]. The algorithm is based on a divide-and-conquer approach where consistency analysis operations are based

© Springer Nature Switzerland AG 2020
D. Helic et al. (Eds.): ISMIS 2020, LNAI 12117, pp. 457–468, 2020.
https://doi.org/10.1007/978-3-030-59491-6_44

on the division of a constraint set $C = \{c_1..c_m\}$ into two subsets $C_a = \{c_1..c_k\}$ and $C_b = \{c_{k+1}..c_m\}$ assuming, for example, $k = \lfloor \frac{m}{2} \rfloor$. If C_b is *inconsistent*, the consideration set C can be reduced by half since C_a must not be analyzed anymore (at least one conflict exists in C_b). Depending on the QUICKXPLAIN variant, either C_a or C_b is checked for consistency.

Conflict detection is typically applied in combination with conflict resolution which helps to resolve all existing conflicts. In this context, the minimality (irreducibility) of conflict sets is important since this allows to resolve each conflict by simply deleting one of the elements in the conflict set. The elements to be deleted to restore global consistency are denoted as *hitting set* and can, for example, be determined on the basis of a hitting set directed acyclic graph [9]. Due to the increasing size and complexity of knowledge bases, there is an increasing need to further improve the performance of solution search and conflict detection/hitting set calculation [2,4,5,8]. Parallelizations of algorithms in these scenarios have been implemented in different contexts. Approaches to parallelization have, for example, been proposed on the *reasoning level* [2] where the determination of a solution is based on the idea of identifying subproblems which can be solved to some degree independently by the available cores. Due to today's multicore CPU architectures, such parallelization techniques become increasingly popular in order to be able to better exploit the offered computing resources.

A similar motivation led to the development of parallelization techniques in model-based diagnosis [9]. J. Marques-Silva et al. [6] propose a parallelization approach for hitting set determination where Reiter's approach to model-based diagnosis is parallelized by a level-wise expansion of a breadth-first search tree with the goal of computing minimal (cardinality) diagnoses. On each level, (minimal) conflict sets are determined in parallel, however, the determination of individual conflict sets is still a sequential process (based on QUICKXPLAIN [7]). In diagnosis search, the efficient determination of minimal conflicts is a core requirement [6]. Especially in constraint-based reasoning scenarios, the identification of minimal conflict sets is frequently based on QUICKXPLAIN [7]. Compared to iterative approaches of removing elements from inconsistent constraint sets [1], QUICKXPLAIN follows a divide-and-conquer strategy that helps to reduce the number of needed consistency checks. Although the algorithm is often used in interactive settings with challenging runtime requirements, up-to-now no parallelized version has been proposed. In this paper, we propose an algorithm that enables efficient parallelized minimal conflict detection and thus helps a.o. to significantly improve the runtime performance of conflict detection in interactive applications.

The contributions of this paper are the following. *First*, we show how to parallelize conflict detection with look-ahead strategies that scale with the number of available computing cores. *Second*, we show how to integrate our approach with the QUICKXPLAIN algorithm that is often used in constraint-based applications. *Third*, we show the applicability and improvements of our approach on the basis of performance evaluations. Finally, we point out in which way the proposed approach can help to improve the performance of existing diagnosis approaches. The remainder of the paper is organized as follows. In Sect. 2, we introduce the basic idea of Junker's QuickXPlain using a working example. Thereafter, in Sect. 3 we introduce a parallelized variant of the algorithm. In Sect. 4, we analyze the proposed approach and report the results of a performance evaluation which shows significant improvements compared to standard QuickXPlain. The paper is concluded with Sect. 5.

2   Calculating Minimal Conflicts

In the remainder of this paper, we introduce our approach to parallelized conflict detection on the basis of constraint-based knowledge representations [14]. A *conflict set* can be defined as a minimal set of constraints that is responsible for an inconsistency, i.e., a situation in which no solution can be found for a given constraint satisfaction problem (CSP) (see Definitions 1–2).

Definition 1. *A* Constraint Satisfaction Problem *(CSP) is a triple (V,D,C) with a set of variables $V = \{v_1..v_n\}$, a set of domain definitions $D = \{dom(v_1)..dom(v_n)\}$, and a set of constraints $C = \{c_1..c_m\}$.*

Definition 2. *Assuming the inconsistency of C, a* conflict set *can be defined as a subset $CS \subseteq C : CS$ is inconsistent. CS minimal if $\neg \exists CS' : CS' \subset CS$.*

Examples of a CSP and a conflict set are the following (see Examples 1–2).

Example 1. An example of a CSP for car configuration is the following: $V = \{cartype, fuel, pdc, color, skibag\}$, $D = \{dom(cartype) = [s,c], dom(fuel) = [p,d], dom(pdc) = [y,n], dom(color) = [b,r,g], dom(skibag) = [y,n]\}$, and $C = \{c_1 : fuel = d, c_2 : skibag = y, c_3 : color = b, c_4 : cartype = c, c_5 : pdc = y, c_6 : skibag = y \rightarrow cartype = s, c_7 : cartype = c \rightarrow color = b\}$.

Note that PCD refers to the Park Distance Control implemented in recent vehicles It is convenient to distinguish between a *consistent background knowledge B* of constraints that cannot be relaxed (in our case, $B = \{c_6, c_7\}$) and a *consideration set C* of relaxable constraints (in our case, requirements $C = \{c_1..c_5\}$).

Example 2. In Example 1, there is one minimal conflict which is $CS = \{c_2, c_4\}$, since $\{c_2, c_4\} \subseteq C$ and inconsistent (CS). Furthermore, CS is minimal since there does not exist a CS' s.t. $CS' \subset CS$. The execution trace of QuickXPlain for this working example is depicted in Fig. 1.

QuickXPlain [7] (a variant is shown in Algorithms 1 and 2) supports the determination of minimal (irreducible) conflicts in a given set of constraints (C). QuickXPlain is activated if the background knowledge B (often assumed to be empty or a consistent set of constraints) is *inconsistent* with the set of constraints C (we assume this consistency check to be performed by a direct solver call). The core algorithm is implemented in the function QX (Algorithm 2) that determines a minimal conflict which is a minimal subset of the constraints in C with the conflict set property.

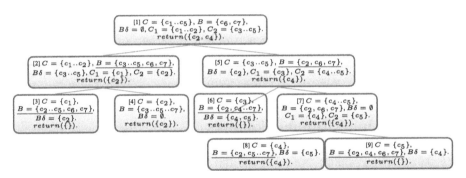

Fig. 1. QX execution trace for $C = \{c_1..c_5\}$ and $B = \{c_6, c_7\}$ assuming a minimal conflict set $CS = \{c_2, c_4\}$. Underlined Bs denote QX consistency checks. For example, in the incarnation [2] of the QX function, the consistency check activated is $\{c_3..c_5, c_6, c_7\}$.

The function QX (Algorithm 2) focuses on isolating those constraints that are part of a minimal conflict. If C includes only one element ($C = \{c_\alpha\}$), this element can be considered as element of the conflict - this is due to the invariant property $inconsistent(C \cup B)$. If the B is consistent and C has more than one element, C is divided into two separate sets, where (in our QX variant) the second part (C_b) is added to B in order to analyse further elements of the conflict. The function QX activates a consistency check (INCONSISTENT) to figure out whether the considered background knowledge is inconsistent, i.e., no solution exists. $B\delta$ indicates constraints added to B.

Algorithm 1. QUICKXPLAIN$(C, B) : CS$

1: **if** CONSISTENT$(B \cup C)$ **then**
2:     $return$('no conflict')
3: **else if** $C = \emptyset$ **then**
4:     $return(\emptyset)$
5: **else**
6:     $return(\text{QX}(C, B, \emptyset))$
7: **end if**

Algorithm 2. QX$(C = \{c_1..c_m\}, B, B\delta) : CS$

1: **if** $B\delta \neq \emptyset$ and INCONSISTENT(B) **then**
2:     return$(\emptyset)$
3: **end if**
4: **if** $C = \{c_\alpha\}$ **then**
5:     return$(\{c_\alpha\})$
6: **end if**
7: $k = \lfloor \frac{m}{2} \rfloor$
8: $C_a \leftarrow c_1...c_k; C_b \leftarrow c_{k+1}...c_m;$
9: $\Delta_2 \leftarrow \text{QX}(C_a, B \cup C_b, C_b); \Delta_1 \leftarrow \text{QX}(C_b, B \cup \Delta_2, \Delta_2);$
10: return$(\Delta_1 \cup \Delta_2)$

In many of the mentioned application scenarios, there exists an exponential number of conflicts and ways to resolve a conflict [7]. Especially in interactive scenarios, it is extremely important to identify *preferred conflicts*, i.e., conflicts with a high probability of being the basis of a relaxation acceptable for the user. For example, if a user is strongly interested in low-priced digital cameras, a conflict set that includes a price limit might be of low relevance for the user (since the user is not willing to change the price limit). QUICKXPLAIN [7] supports the determination of *preferred conflicts*. Although our discussions focus on constraint-based representations, the approach can be applied to any kind of satisfiability problem such as propositional satisfiability (SAT) and description logics (DL). We assume *monotonic satisfiability* (see Proposition 1).

Proposition 1. *If a solution for a CSP satisfies all constraints $c_i \in C$ then it also satisfies every proper subset $C' \subset C$.*

QUICKXPLAIN determines one *minimal preferred conflict* at a time which includes constraints one might be willing to relax. In the line with [7], an explanation X is preferred over an explanation Y under the following condition (see Definition 3).

Definition 3. *We define a total order $<$ on the constraints in $C = \{c_1..c_m\}$ which is represented as $[c_1 < c_2 < .. < c_m]$. If $c_i < c_j$, i.e., the importance of c_i is higher than c_j, then $i < j$. If X and Y are two lexicographical constraint orderings of $c_1..c_n$, Y is preferred over X ($Y >_{lex} X$) iff $\exists k : c_k \in X - Y$ and $X \cap \{c_{k-1}..c_1\} = Y \cap \{c_{k-1}..c_1\}$.*

Example 3. Given the constraint ordering $[c_3 < c_4 < c_5 < c_6]$ and two binary conflict sets $X = \{c_3, c_4\}$ and $Y = \{c_4, c_5\}$, Y is preferred over X since $c_3 \in X - Y$ with $X \cap \{c_2, c_1\} = Y \cap \{c_2, c_1\}$.

3   Parallelizing QUICKXPLAIN

Our approach to parallelize the consistency checks in QX substitutes the *direct* solver call INCONSISTENT(B) in QX with the activation of a lookahead function (QXGEN) in which consistency checks are not only triggered to directly provide feedback to QX requests, but also to be able to provide fast answers for consistency checks potentially relevant in upcoming states of a QX instance. In the parallelized variant of QUICKXPLAIN, consistency checking is activated by QX with INCONSISTENT($C, B, B\delta$) (see Algorithm 3). This also activates the QXGEN function (see Algorithm 4) that starts to generate and trigger (in a parallelized fashion) further consistency checks that might be of relevance in upcoming QX phases. For the description of QXGEN, we employ a two-level *ordered set* notation which requires, for example, to embed the QX B into $\{B\}$, etc. In QXGEN, C, $B\delta$, and B are interpreted as *ordered sets*.

QXGEN-generated consistency checking tasks are stored in a LOOKUP table (see, e.g., Table 1). Thus, in the parallelized variant, QX has to activate the consistency check with INCONSISTENT($C, B, B\delta$). In contrast to the original QUICKXPLAIN approach, C and $B\delta$ are needed as additional parameters to conduct

---

Algorithm 3. INCONSISTENT$(C, B, B\delta)$:*Boolean*

1: **if** $\neg$EXISTSCONSISTENCYCHECK(B) **then**
2:    QXGEN$(\{C\}, \{B\delta\}, \{B - B\delta\}, \{B\delta\}, 0)$
3: **end if**
4: $return(\neg$LOOKUP$(B))$

---

inferences about needed future consistency checks. While in the standard QX version $B\delta \subseteq B$, we assume $B\delta$ and B to be separate units in QXGEN.

Table 1. LOOKUP table indicating the consistency of individual constraint sets. The consistency checking tasks have been generated by ADDCC in the QXGEN function (see Fig. 2) and are executed in parallel. The '-' entry for the constraint set $\{c_4, c_5, c_6, c_7\}$ indicates that the corresponding consistency check is still ongoing or has not been started up to now. Algorithm 3 uses the LOOKUP function to test the consistency of a constraint set.

Node-id	Constraint set	Inconsistent
1	$\{c_3, c_4, c_5, c_6, c_7\}$	*true*
1.1	$\{c_2, c_3, c_4, c_5, c_6, c_7\}$	*false*
1.1.1	$\{c_1, c_6, c_7\}$	*true*
1.2.1	$\{c_4, c_5, c_6, c_7\}$	-

The QXGEN function (see Algorithm 4) predicts future potentially relevant consistency checks needed by QX and activates individual consistency checking tasks in an asynchronous fashion using the ADDCC (*add consistency check*) function. The ADDCC function triggers an asynchronous service that is in charge of adding consistency checks (parameter of ADDCC) to a LOOKUP table and issuing the corresponding solver calls. The global parameter $lmax$ is used define the maximum search depth of one activation of QXGEN.

In QXGEN, $|f(X)|$ denotes the number of constraints c_i in X (it is introduced due to the subset structure in C). Furthermore, SPLIT(C, C_a, C_b) splits C at position $\lfloor \frac{|C|}{2} \rfloor$ if $|C| > 1$ or C_1 (the first element of C) at position $\lfloor \frac{|C_1|}{2} \rfloor$ if $|C| = 1$ and $|C_1| > 1$ into C_a and C_b. Otherwise, no split is needed (C_1 is a singleton).

The first inner condition of QXGEN ($|f(\delta)| > 0$) generates a consistency check if this is needed. A consistency check is needed, if $B\delta$ gets extended from C or a singleton C has been identified which then extends B. The function ADDCC is used to add consistency check tasks which can then be executed asynchronously in a parallelized fashion. Thus, a consistency check in the LOOKUP table can be easily identified by an ordered constraint set that has also been used as parameter of ADDCC, for example, ADDCC$(\{\{c_4, c_5\}, \{c_6, c_7\}\})$ results in the LOOKUP table entry $\{c_4, c_5, c_6, c_7\}$ which is internally represented with 4567.

Algorithm 4. QXGEN($C, B\delta, B, \delta, l$)
$C = \{C_1..C_r\}$ ... *consideration set* (*subsets* $\mathbf{C}_\alpha$)
$B\delta = \{B\delta_1..B\delta_n\}$ ... *added knowledge* (*subsets* $\mathbf{B}\delta_\beta$)
$\mathbf{B} = \{\mathbf{B}_1..\mathbf{B}_o\}$ ... *background* (*subsets* $\mathbf{B}_\gamma$)
$\delta = \{D_1..D_p\}$ ... *to be checked* (*subsets* D_π)
l ... *current lookahead depth*

1: **if** $l < lmax$ **then**
2:   **if** $|f(\delta)| > 0$ **then**
3:     ADDCC($B\delta \cup B$)
4:   **end if**
5:   $\{B\delta \cup B$ assumed consistent$\}$
6:   **if** $|f(C)| = 1 \wedge |f(B\delta)| > 0$ **then**
7:     QXGEN($B\delta, \emptyset, B \cup \{C_1\}, \{C_1\}, l+1$)
8:   **else if** $|f(C)| > 1$ **then**
9:     SPLIT(C, C_a, C_b)
10:    QXGEN($C_a, C_b \cup B\delta, B, C_b, l+1$)
11: **end if**
12: $\{B\delta \cup B$ assumed inconsistent$\}$
13: **if** $|f(B\delta)| > 0 \wedge |f(\delta)| > 0$ **then**
14:    QXGEN($\{B\delta_1\}, B\delta - \{B\delta_1\}, B, \emptyset, l+1$)
15: **end if**
16: **end if**

If $B\delta \cup B$ is assumed to be *consistent*, additional elements from C have to be included such that an inconsistent state can be generated (which is needed for identifying a minimal conflict). This extension of $B\delta$ can be achieved by dividing C (if $|f(C)| > 1$, i.e., more than one constraint is contained in C) into two separate sets C_a and C_b and to add C_b to $B\delta$. If $|f(C)| = 1$, this singleton can be added to B which is responsible of collecting constraints that have been identified as being part of the minimal conflict. This is the case due to the already mentioned invariant property, i.e., $C \cup B\delta \cup B$ is inconsistent. If C contains only one constraint, i.e., C_1 is a singleton, it is part of the conflict set. If $B\delta \cup B$ is assumed to be *inconsistent*, it can be reduced and at least one conflict element will be identified in the previously added $B\delta_1$. If δ does not contain an element, no further recursive calls are needed since $B\delta - B\delta_1$ has already been checked.

The QXGEN function (Algorithm 4) is based on the idea of issuing recursive calls and adapting the parameters of the calls depending on the two possible situations 1) *consistent*($B\delta \cup B$) and 2) *inconsistent*($B\delta \cup B$). In Table 2, the different parameter settings are shown in terms of FOLLOW sets representing the settings of $C, B\delta, B$, and δ in the next activation of QXGEN.

Optimizations. To further improve the performance of QX, we have included mechanisms that help to identify irrelevant executions of consistency checks (see Table 3). If a consistency check has been completed, we are able to immediately decide whether some of the still ongoing or even not started ones can be canceled since they are not relevant anymore. For example, if the result of a consistency check is $\{c_3..c_7\}$ is *true* (see Fig. 2), the check $\{c_4..c_7\}$ does not have to be

Table 2. FOLLOW sets of the QXGEN function. Depending on the assumption about the consistency of $B\delta \cup B$, the follow-up activations of QXGEN have to be parameterized differently.

COND.	FOLLOW SETS			
	C	$B\delta$	B	δ
$\|f(C)\| = 1$	$B\delta$	$\emptyset$	$B \cup \{C_1\}$	$\{C_1\}$
$\|f(C)\| > 1$	C_a	$C_b \cup B\delta$	B	C_b
$\|f(B\delta)\| > 0$	$\{B\delta_1\}$	$B\delta - \{B\delta_1\}$	B	$\emptyset$

executed anymore (see also Proposition 1) or has to be canceled since QUICKX-PLAIN will not need this check (see Fig. 2). The deletion criteria for ongoing or even not started consistency checks can be pre-generated for $lmax$. An example for $lmax = 3$ is provided in Table 3.

Table 3. Optimizing the execution of consistency checks by detecting irrelevant ones. If the result of a specific check is known, LOOKUP (Algorithm 3) triggers the corresponding delete (*del*) operation. Nodes without associated consistency check are ignored.

NODE-ID	RESULT OF CONSISTENCY CHECK	
	true	*false*
1	$del(1.2.x)$	$del(1.1.x)$
1.1	$del(1.1.2.x)$	$del(1.1.1.x)$
1.2	$del(1.2.2.x)$	$del(1.2.1.x)$

In Table 3, the used *node-ids* are related to QXGEN instances, for example, in Fig. 2 the consistency check $\{c_3..c_7 = true\}$ has the *node-id* 1.

4   Analysis

QX *Complexity.* Assuming a splitting $k = \lfloor \frac{m}{2} \rfloor$ of $C = \{c_1..c_m\}$, the worst case time complexity of QUICKXPLAIN in terms of the number of consistency checks needed for calculating one minimal conflict is $2k \times log_2(\frac{m}{k}) + 2k$ where k is the minimal conflict set size and m represents the underlying number of constraints [7]. Since consistency checks are the most time-consuming part of conflict detection, the runtime performance of the underlying algorithms must be optimized as much as possible.

QXGEN *Complexity.* The number (*nc*) of different possible minimal conflict sets that could be identified with QXGEN for an inconsistent constraint set C consisting of m constraints is represented by Formula 1.

$$nc(C) = \binom{m}{1} + \binom{m}{2} + ... + \binom{m}{m} \tag{1}$$

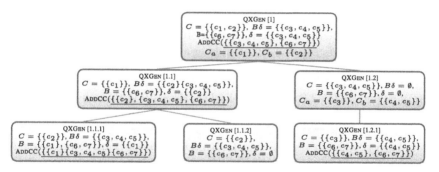

Fig. 2. QXGEN execution trace for $C = \{\{c_1..c_5\}\}$, $B\delta = \emptyset$, $B = \{\{c_6, c_7\}\}$, $\delta = \emptyset$, and $lmax = 3$. The consistency checks $\{c_3, c_4, c_5, c_6, c_7\}$ and $\{c_2, c_3, c_4, c_5 c_6, c_7\}$ (flattened list generated by ADDCC) can be used by the QuickXPlain instance of Fig. 1. The QXGEN nodes $\{[1], [1.1], [1.1.2]\}$ represent the first part of the QuickXPlain search path in Fig. 1.

The upper bound of the space complexity in terms of recursive QXGEN calls for $lmax = n$ is in the worst case $2^{n-1} - 1$. Due to the combinatorial explosion, only those solutions make sense that scale $lmax$ depending on the available computing cores (see the reported evaluation results).

If the non-parallelized version of QuickXPlain is applied, only sequential consistency checks can be performed. The approach presented in this paper is more flexible since $\#processors$ consistency checks can be performed in parallel. Assuming a maximum QXGEN search depth of $lmax = 4$, the maximum number of generated consistency checks is $\frac{2^{lmax-1}}{2}$ due to the binary structure of the search tree, i.e., 4 in our example. Out of these 4 checks, a maximum of 3 will be relevant to QX, the remaining ones are irrelevant for identifying the conflict in the current QX session. Thus, the upper bound of relevant consistency checks generated by QXGEN is $lmax$, i.e., one per QXGEN search level (see the outer left search path in Fig. 2), the minimum number of relevant consistency checks is $\lceil \frac{lmax}{2} \rceil$, i.e., 2 in our example. Assuming $\#processors = 16$, our approach can theoretically achieve a performance boost of factor 3 since max. 3 relevant consistency checks can be performed in parallel. It is important to mention, that within the scope of these upper and lower bounds, a performance improvement due to the integration of QXGEN can be guaranteed independent of the underlying knowledge base.

Termination of QXGEN. If the parameter $lmax = n$, recursive calls of $QXGen$ stop at level $n - 1$. In every recursive step, based on the inconsistency invariant between $C \cup B\delta \cup B$, *either* 1) C is reduced to C_a and $B\delta$ gets extended with C_b (if $B\delta \cup B$ is consistent) or to $B \cup C$ if C is a singleton, *or* 2) $B\delta$ is further reduced (if $B\delta \cup B$ is inconsistent). Obviously, in the second case, the constraints in C are not relevant anymore, since a conflict can already be found in $B\delta$ (which is inconsistent with B).

QX-*conformance of* QXGEN. QXGEN correctly predicts QX consistency checks. It follows exactly the criteria of QX. If $|f(C)| = 1$, i.e., C includes only one constraint c_α, $B\delta \cup B$ is consistent and - as a consequence of the inconsistency invariant - c_α is a conflict element and therefore has to be added to B. The issued consistency check is $B \cup C$ - if inconsistent, a conflict has already been identified. If $|f(C)| > 1$, $C_b \cup B\delta \cup B$ has to be checked, since $B\delta \cup B$ is consistent and by adding C_b we follow the goal of restoring the inconsistency of $B\delta \cup B$. Finally, if $B\delta \cup B$ is inconsistent, no check has to be issued since $B\delta - \{B\delta_1\} \cup B$ has already been checked in the previous $QXGen$ call and obviously was considered consistent. Thus, QXGEN takes into account all QX states that can occur in the next step and exactly one of the generated consistency checks (if needed) will be relevant for QX. Finally, the generated consistency checks are irredundant since each check is only generated if new constraints from C are added to $B\delta$ or a new constraint (singleton C) is added to B.

Runtime Analysis. The following evaluations have been conducted on the basis of a *Java 8* based implementation of the parallelized QUICKXPLAIN (QX) version presented in this paper. For the implementation, we applied the *ForkJoin* framework for running parallel tasks in *Java*. This, while being less automatic than newer java implementations allowed us to fully control when threads are created and destroyed. For representing our test knowledge bases and conducting the corresponding consistency checks, we have used the SAT solving environment *SAT4j* (see www.sat4j.org). All experiments reported in the following have been conducted using an *Intel Xeon multi-core (16 cores) E5-2650 of 2.60* GHz computer *and 64* GB *of RAM* that supports hyperthreading simulation of 64 cores which would allow us to run up to lmax = 6. We have selected configuration knowledge bases (feature models) from the publicly available BETTY toolsuite [12], which allows for a systematic testing of different consistency checking and conflict detection approaches for knowledge bases. The knowledge base instances (represented as background knowledge B in QUICKXPLAIN) that have been selected for the purpose of our evaluation, range from 500 to 5.000 binary variables and also vary in terms of the number of included constraints (B 25-1.500 binary constraints). On the basis of these knowledge bases, we randomly generated requirements ($c_i \in C$) that covered 30%−50% of the variables included in the knowledge base. These requirements have been generated in such a way that conflict sets of different cardinalities could be analyzed (see Table 4). In order to avoid measuring biases due to side effects in thread execution, we repeated each evaluation setting 3×.

The results of our QXGEN performance analysis are summarized in Table 4. On an average, the runtime needed by standard QUICKXPLAIN (*lmax* = 1) to identify a preferred minimal conflict of cardinality 1 is 3.2× higher compared to a parallelized solution based on $QXGen$ (*lmax* = 5). In Table 4, each entry represents the average runtime in *msec* for all knowledge bases with a preferred conflict set of cardinality n, where the same set of knowledge bases has been evaluated for *lmax* sizes 1–6 (*lmax* = 1 corresponds to the usage of standard QX without QXGEN integration). In this context, we have introduced an additional baseline version $6(rd)$, where only a randomly selected set of QXGEN consistency checks has been evaluated. It can be observed that with an increasing $lmax$ the performance of QX increases. Starting with $lmax = 6$, a performance deterioration can be observed which can be explained by the fact that the number of pre-generated consistency checks starts to exceed the number of physically available processors. In the line of our algorithm analysis, the number of relevant consistency checks that can be performed with $lmax = 5$ is between 5 and 3. Taking into account the overheads for managing the parallelized consistency checks, the shown results support our theoretical analysis of QXGEN.

Table 4. Avg. runtime (in $msec$) of parallelized QX when determining minimal conflicts.

$lmax$	Conflict cardinality				
	1	2	4	8	16
1	**103993.0**	**286135.4**	**11006.4**	**30995.3**	**177354.7**
2	69887.3	193354.2	9528.0	26258.7	154093.7
3	49823.5	130657.5	11094.6	28639.6	154155.0
4	50042.8	136150.8	7577.0	19753.1	120992.0
5	**32481.4**	**88963.7**	**7750.0**	**18975.6**	**98242.7**
6	34946.2	94783.0	6367.6	17743.7	95816.3
6(rd)	105678.3	195987.5	112546.8	28676.1	179876.2

5 Conclusions

We have introduced a parallelized variant of the QUICKXPLAIN algorithm that is used for the determination of minimal conflict sets. Example applications are the model-based diagnosis of hardware designs and the diagnosis of inconsistent user requirements in configuration and recommender applications. Current approaches to the detection of minimal conflicts do not take into account the capabilities of multi-core architectures. Our parallelized variant of QUICKXPLAIN provides efficient conflict detection especially when dealing with large and complex knowledge bases.

Acknowledgements. This work has been partially funded by the EU FEDER program, the MINECO project OPHELIA (RTI2018-101204-B-C22); the TASOVA network (MCIU-AEI TIN2017-90644-REDT); and the Junta de Andalucia METAMORFOSIS project.

References

1. Bakker, R., Dikker, F.: Diagnosing and solving over-determined constraint satisfaction problems. In: 13th International Joint Conference on Artificial Intelligence (IJCAI 1993), Chambéry, France, pp. 276–281 (1993)

2. Bordeaux, L., Hamadi, Y., Samulowitz, H.: Experiments with massively parallel constraint solving. In: 21st International Joint Conference on Artificial Intelligence, pp. 443–448. Morgan Kaufmann Publishers, San Francisco (2009)

3. Felfernig, A., Burke, R.: Constraint-based recommender systems: technologies and research issues. In: ACM International Conference on Electronic Commerce (ICEC 2008), Innsbruck, Austria, pp. 17–26 (2008)

4. Gent, I., et al.: A review of literature on parallel constraint solving. Theory Pract. Log. Program. **18**(5–6), 725–758 (2018)

5. Hamadi, Y., Sais, L. (eds.): Handbook of Parallel Constraint Reasoning. Springer, Cham (2018). https://doi.org/10.1007/978-3-319-63516-3

6. Jannach, D., Schmitz, T., Shchekotykhin, K.: Parallelized hitting set computation for model-based diagnosis. In: 29th AAAI Conference on Artificial Intelligence, pp. 1503–1510. AAAI Press, Austin (2015)

7. Junker, U.: QuickXPlain: preferred explanations and relaxations for over-constrained problems. In: 19th National Conference on Artificial Intelligence, pp. 167–172. AAAI Press, San Jose (2004)

8. Marques-Silva, J., Heras, F., Janota, M., Previti, A., Belov, A.: On computing minimal correction subsets. In: 23rd International Joint Conference on Artificial Intelligence, Beijing, China, pp. 615–622 (2013)

9. Reiter, R.: A theory of diagnosis from first principles. Artif. Intell. **23**(1), 57–95 (1987)

10. Ricci, F., Rokach, L., Shapira, B., Kantor, P.B. (eds.): Recommender Systems Handbook. Springer, Boston (2011). https://doi.org/10.1007/978-0-387-85820-3

11. Rossi, F., Venable, K., Walsh, T.: A Short Introduction to Preferences: Between Artificial Intelligence and Social Choice. Morgan & Claypool Publishers, San Rafael (2011)

12. Segura, S., Galindo, J., Benavides, D., Parejo, J., Ruiz-Cortés, A.: BeTTy: benchmarking and testing on the automated analysis of feature models. In: 6th International Workshop on Variability Modeling of Software-Intensive Systems, VaMoS 2012, pp. 63–71. ACM, New York (2012)

13. Stumptner, M.: An overview of knowledge-based configuration. AI Commun. **10**(2), 111–125 (1997)

14. Tsang, E.: Foundations of Constraint Satisfaction. Academic Press, London (1993)

Author Index

Printed in the United States
By Bookmasters

MANUAL OF BIOLOGICAL MARKERS OF DISEASE

Edited by

W. J. VAN VENROOIJ

University of Nijmegen,
The Netherlands

and

R. N. MAINI

The Mathilda and Terence Kennedy Institute of Rheumatology,
London, UK

KLUWER ACADEMIC PUBLISHERS
DORDRECHT / BOSTON / LONDON

A C.I.P. Catalogue record for this book is available from the Library of Congress.

hardback
ISBN 0-7923-4242-9

paperback
ISBN 0-7923-4243-7

Neither Kluwer Academic Publishers nor any person acting on its behalf is responsible for the use which might be made of the information contained herein.

Published by Kluwer Academic Publishers,
P.O. Box 17, 3300 AA Dordrecht, The Netherlands.

Kluwer Academic Publishers incorporates
the publishing programmes of
D. Reidel, Martinus Nijhoff, Dr W. Junk and MTP Press.

Sold and distributed in the U.S.A. and Canada
by Kluwer Academic Publishers,
101 Philip Drive, Norwell, MA 02061, U.S.A.

In all other countries, sold and distributed
by Kluwer Academic Publishers Group,
P.O. Box 322, 3300 AH Dordrecht, The Netherlands.

Printed on acid-free paper

Contents

Contents

Introduction

The origins of this volume, which will be part of a series, are intertwined with the work of the European Consensus Workshop on Autoantibodies in Rheumatic Diseases. The impetus for the setting up of a real 'Workshop' on detection of autoantibodies involving the analysis of sera from patients by several research laboratories in Europe arose in context of a meeting of the European Workshop for Rheumatological Research held under the Chairmanship of Harry Moutsopoulos in Corfu in 1988. Here, the increasing range of autoantibody specificities being detected in the sera of patients became apparent, as did the disparity in the prevalence of autoantibodies in rheumatological disorders reported by different groups. This raised questions about the factors which may have contributed to the inconsistent results. We concluded that reproducibility of the laboratory tests and heterogeneity of patient populations seemed to be the most obvious variables.

With the support and encouragement of Josef Smolen who was chairing the European Research Meeting in Vienna the following year, we decided to set up the first consensus workshop on autoantibodies. The Workshop's aim was to examine the level of agreement in autoantibody specificities reported by laboratories who wished to participate in the study and the first exercise involving 19 laboratories and ten sera took place. The conclusions and lessons learnt from this and subsequent workshops has been reported by the organizers (P. Charles, R.N. Maini, and W.J. van Venrooij) in the literature (J. Immunol. Methods 1991, **140**, 181–189 and Clin. Exp. Rheumatol., 1992, **10**, 507–511).

As organisers of these Workshops, it became apparent to us that the Workshop (which has now grown to involve 35–40 laboratories) was both popular and strongly desired for a variety of reasons. These included the need to validate and compare newer techniques such as ELISA and Western blotting with previously established methods such as immunofluorescence, immunodiffusion and counter-immunoelectrophoresis. It is of interest to note, for example, that the number of laboratories using ELISA and Western blotting has grown from 5 and 8 to 19 and 21 respectively in four years. An equally compelling reason for the popularity of the Workshop was the opportunity it created for laboratories to gain access to (and experience with) detecting rare specificities in sera, such as anti-rRNP and anti-PCNA or newly described antibodies such as anti-RA33, which few laboratories would have encountered previously. The Workshop also introduced participating laboratories to relatively unknown methods for important autoantibodies in clinical work, an example being the techniques used by Allan Wiik and Cees Kallenberg for detecting ANCA. This was consolidated by the inclusion of sera with ANCA reactivity in ensuing Workshops. Since the detection of

ANCA has become valuable in the diagnosis and classification of vasculitis, the educational value of the exercise to groups previously interested in rheumatic disease was widely acknowledged as having been most valuable.

A significant milestone in the life of the Workshop was reached when 'experts' in our consensus groups agreed to submit their 'cookbook' methods towards a collection of papers which were circulated to all participating laboratories. We urged participants to use these methods whenever possible and were gratified to receive encouraging feedback on the value of the methods detailed therein. Such enthusiasm seemed to be matched by improved results in the subsequent Workshops. Armed with this knowledge we decided to put together for publication this current volume of methods for autoantibody detection. We have chosen a loose-leaf format so that it will be possible for us to update information on a regular basis, probably biennially. We would encourage our readers to contact us with comments, criticisms, or suggestions about any additions which should be included so that suitable amendments and additions can be made in the next edition.

The scope of our venture has grown since we embarked on producing this volume. It has become apparent to us that there is a need for a subsequent volume which deals with the biochemical and molecular structure of autoantigens and the clinical significance of corresponding antibodies. Preparations are well advanced for this 'sister-book' to appear within the next twelve months. It is our hope that this publication and subsequent volumes will be available widely to immunologists, molecular biologists and physicians interested in the research and applied aspects of autoantibody detection.

We gratefully acknowledge the strong support we have received from the European League against Rheumatism and the WHO/IUIS/ILAR Committee on Autoantibodies, chaired by Dr. Eng Tan, to publish this manual.

The Editors   W.J. VAN VENROOIJ
                                                   R.N. MAINI

Manual of Biological Markers of Disease **A1**: 1–5, 1993.
© 1993 *Kluwer Academic Publishers. Printed in the Netherlands.*

International cooperative activities in standardization of antinuclear antibodies

E. M. TAN

W.M. Keck Autoimmune Disease Center, The Scripps Research Institute, 10666 North Torrey Pines Road, La Jolla, California 92037, U.S.A.

Since 1980, clinical investigators in the United States have been engaged in efforts to standardize reference sera for antinuclear antibody determinations. This activity has continued and more recently, it has expanded to become an international cooperative effort involving laboratories in the United States, Europe, Japan and Australia.

In 1980, the Arthritis Foundation in the United States in collaboration with the Centers for Disease Control established a Committee on Antinuclear Antibody Serology. The charge to this committee was to establish a repository of ANA reference sera which should be made available to research and diagnostic laboratories in the United States and in foreign countries. These ANA reference reagents were to serve as standards defining ANA specificities and used by individual laboratories to establish their own secondary reference reagents. Sera from patients undergoing therapeutic plasmapheresis were screened for the desired antinuclear reactivities and certain sera selected as possible reference reagents. These sera were processed at the Centers for Disease Control in Atlanta, Georgia and aliquots of 0.50 ml were dispensed into borosilicate ampoules, freeze-dried and sealed under reduced pressure. The reference reagents were determined by the United States Pharmacopiea-approved methods of sterility testing and shown to be free of hepatitis-B surface antigen and negative for rheumatoid factor. The standard requirements for establishing reference reagents – such as studies to determine mean dry weights, water content and stability properties – were performed and data were provided with the ampoules. The reference reagents before and after processing were tested in the laboratories of members of the Committee on Antinuclear Antibody Serology. In 1982, five reference reagents were made available [1]. These five reference sera are described in Table 1. The five reference sera provided reference reagents for antinuclear antibodies defined in the following terms:

A. Three patterns of fluorescent antinuclear antibodies: homogeneous/rim pattern, and two different speckled nucleoplasmic patterns.

B. Four different antinuclear antibodies identified by immunodiffusion analysis: antibodies to native DNA, SS-B/La, nuclear RNP, and Sm.

Table 1. AF/CDC ANA Reference Sera, 1982.

			Specificities of reference sera				
			CDC1	CDC2	CDC3	CDC4	CDC5
Assay	Method	Assay readout	FANA and anti-nDNA	FANA and anti-SS-B/La	FANA only	Anti-nRNP only	Anti-Sm only
ANA	Indirect immunofluorescence	Pattern	H/R	Sp	Sp		
		Titer	1:512 (1:160–1:640)	1:160 (1:80–1:320)	1:320 (1:180–1:640)		
Anti-nDNA	Millipore filter	% DNA bound	74% (64–82)				
	Farr	DNA bound	85% (75–95)				
	Crithidia	Titer	1:90 (1:180–1:320)				
Anti-SS-B/La	Immunodiffusion	Titer		1:40(1:16–1:80)			
Anti-nRNP	Immunodiffusion	Titer				1:64 (1:32–1:80)	
Anti-Sm	Immunodiffusion	Titer					1:64 (1:32–1:64)

* FANA = fluorescence ANA; H/R = homogeneous/rim; Sp = speckled pattern. Results show the median value with modified range (see text) indicated in parenthesis.

Following this initial activity of the AF/CDC Committee on Antinuclear Antibody Serology, several international organizations expressed support for expansion of these activities. These organizations were the International League Against Rheumatism (ILAR), the International Union of Immunological Societies (IUIS) and the World Health Organization (WHO). The bank of reference reagents at the CDC were expanded to include five more reference ANAs [2]. The total number of reference reagents available in the CDC depository now comprised ten different sera all of which had been analyzed and processed in the same manner as the first five reference reagents described above. These reference reagents were made available in 1988 and are shown in Table 2. This depository of ten reference reagents provides ANA specificities covering the following spectrum:

A. Five fluorescent ANA patterns: homogeneous pattern, two speckled patterns, nucleolar pattern and centromere pattern.
B. Seven ANA specificities: anti-native DNA, anti-SS-B/La, anti-U1-RNP, anti-Sm, anti-SS-A/Ro, anti-Scl-70, and anti-Jo-1.

A recent analysis of the usage of this bank of reference reagents at the CDC shows that there have been, on the average, 250 requests annually for reference reagents. These requests come form research investigators and diagnostic laboratories. Some of the requests are for individual reference

Table 2. AF/CDC ANA Reference Sera, 1988.

Reagent	Immunologic features
AF/CDC1	Homogeneous-pattern ANA (doubles as anti-native DNA)
AF/CDC2	Anti-La (SS-B). Doubles as speckled pattern ANA
AF/CDC3	Speckled pattern ANA
AF/CDC4	Anti-U1 RNP (U1 small nuclear RNP)
AF/CDC5	Anti-Sm (U1, U2, U5, U4/6 small nuclear RNP)
AF/CDC6	Nucleolar-pattern ANA
AF/CDC7	Anti-Ro (SS-A)
AF/CDC8	Centromere pattern ANA
AF/CDC9	Anti-Scl-70 (DNA topoisomerase I)
AF/CDC10	Anti-Jo-1 (histidyl-transfer RNA synthetase)

reagents while others are for multiple reagents. These reagents are supplied by the CDC free of charge. At the present time, some of the reagents are running low and policies have been instituted to reserve a certain number of ampoules for future reference purposes, if new sera of similar specificities should replace the original reagents.

For several years, laboratories in Europe have established active collaborations with the purpose of standardizing ANA methodology. This group has been spearheaded by R.N. Maini (London), W.J. van Venrooij (Nijmegen), J.S. Smolen (Vienna), and S. Bombardieri (Pisa). Four annual workshops have taken place since 1989 and, at the last workshop, 35 European laboratories participated. It has been noted that over this period of time, there has been a continuous improvement in laboratory performance so that incorrect assignments have fallen from 21% in the first workshop to 6% in the most recent workshop. The experience of the "European Consensus Study Group" has been reported recently [3].

In October 1992, the Committee on Antinuclear Antibody Serology in the United States was reorganized and is now called the IUIS-ANA Subcommittee. This subcommittee is a standing committee of the IUIS Standardization Committee. Participating organizations in the IUIS-ANA Subcommittee include the Centers for Disease Control (Atlanta), the Arthritis Foundation (Atlanta), the World Health Organization, the International League Against Rheumatism, and the European League Against Rheumatism. The committee includes members from the old committee on antinuclear antibody serology and members of the European Consensus Study Group, as well as one representative each from Japan and Australia. The current initiatives of the IUS-ANA subcommittee are the following:

A. Define the fine specificities of ANA reference reagents in the CDC reference bank as well as those in WHO ANA reference repository. This will include determination of RNA and protein antigenic determinants recognized by reference sera and semi-quantitation of fluorescent ANA titers.

B. Support and participate in the production of the ANA laboratory manuals currently being organized by the European Consensus Study Group.
C. Expand reference reagents to include ANAs of other specificities.
D. Quallity oversight and spot testing of commercial ANA test kits.

These activities have been successful largely because of the enthusiasm and voluntary contribution of time and resources donated by each member who has served on committees in the past and at the present. The importance of these activities cannot be over-emphasized, because ANA testing plays an important role in the diagnosis and care of patients with many immunologic diseases.

References

1. Tan, E.M., Fritzler, M.J., McDougal, J.S. *et al.* (1982) Reference sera for antinuclear antibodies. I. Antinuclear to native DNA, Sm, nuclear RNP and SS-B/A. Arthr Rheum 25:1003-1005
2. Tan, E.M., Feltkamp, T.E.W., Alarcon-Segovia, D. *et al.* (1988) Reference reagents for antinuclear antibodies. Arthr Rheum 31:1331
3. Reports of the European Consensus Study Groups 1992 Clin Exp Rheumatol 10:505-554

Manual of Biological Markers of Disease **A2**: 1–16, 1993.

Detection of antinuclear antibodies by immunofluorescence

R.L. HUMBEL

Centre Hospitalier de Luxembourg, 4, rue Barblé, L-1210 Luxembourg

Abbreviations

ANA: antinuclear antibodies; CIE: counter-immunoelectrophoresis; hnRNP: heterogeneous nuclear RNP; IFA: indirect immunofluorescence assay; JCA: juvenile chronic arthritis; RA: rheumatoid arthritis; RNP: ribonucleoprotein; SLE: systemic lupus erythematosus; snRNP: small nuclear RNP; SS: Sjögren's syndrome

Key words: Immunofluorescence, autoantibodies, autoantigens, antinuclear antibodies

1. Introduction

The indirect immunofluorescence assay (IFA) is the most commonly used technique for the detection of antinuclear antibodies (ANA)[1]. It is a sensitive, reproducible and simple assay that can be performed in almost every laboratory. Because IFA enables the detection of a wide range of autoantibodies, it is generally used as a primary screening test for ANA [2].

A variety of substrates can be used, including frozen tissue sections of murine liver or kidney and more recently cell cultures [3]. Compared to tissue sections, cell lines have several advantages. The cells are grown directly on the glass slide and, thus lying in a flat plane, it allows a good recognition of both nuclear and cytoplasmic structures [4]. In addition, cultured cells form a homogeneous population and are not embedded in too much extracellular matrix. When maintained in an asynchronous culture, antigens that are only expressed during certain stages of the cell cycle are represented as well. This allows detection of antibodies that are directed to cell cycle specific antigens which are not or are poorly expressed in tissue sections.

At present, cultures of HEp-2 epithelial cells (human larynx carcinoma, American Type Culture Collection CCL-23) are the most widely used. Both nuclei and nucleoli in these cells are large, allowing a good recognition of the many different nuclear and nucleolar staining patterns. As HEp-2 cells are rapidly dividing, many mitotic figures are present in the preparation, providing additional information on the nature of the antibodies and allowing detection of antibodies against structures unique to mitotic cells.

2. The immunofluorescence assay

2.1 The commercial substrate

HEp-2 cells (ATCC-CCL 23) can be grown in the laboratory (see 2.2) but can also be purchased from commercial suppliers. Many of the available HEp-2 kits produce reliable results in ANA screening. There is, however, a wide range in the sensitivity for some ANA, mainly due to variations in the fixation procedure. All the kits provide in their instruction leaflets some data pertaining to their sensitivity. Proper standardization and quality assurance of any IFA-tests depend on calibrated reference standards to compare sensitivities and specificities of the various antigen substrates. Reference sera are available from the World Health Organization (WHO) (see [5] and the chapter of T.E.W. Feltkamp in this manual) and the joint committee of CDC-Arthritis Foundation [6].

2.2 The culturing of HEp-2 cells

Steps in the procedure
1. Seed HEp-2 cells on a slide (approx. 1×10^5 cells per slide) in 5 ml DMEM containing 10% newborn calf serum per slide.
2. Grow the cells at 37 °C in 5% CO_2 for 40–48 h.
3. Fix the cells when cell culture on slide is 80–90% confluent.

2.3. Fixation of the cells

Generally, commercial cells have been fixed by methods which are not easily disclosed. For the fixation of your own cultured cells the following two procedures are recommended.

2.3.1 Methanol-acetone fixation

Steps in the procedure
1. Wash the cells 3 times with cold PBS.
2. Immerse the slide on which the cells were grown in −20 °C methanol for 5 min.
3. Dip the fixed cells in acetone (room temperature).

4. Repeat step 3 with a second batch of acetone.
5. Air-dry the cells for about 15 min. They can be stored for several months at -20 °C in a closed box.
6. Immediately prior to use the slides should be thawed quickly at room temperature. If necessary use a cold fan.

2.3.2 Paraformaldehyde-Triton fixation

Steps in the procedure
1. Wash the cells 3 times with cold PBS.
2. Fix the cells in paraformaldehyde (3% in PBS) for 15 min at room temperature.
3. Wash the cells 3 times with PBS.
4. Immerse the cells in Triton X-100 (0.2% in PBS) for 5 min at room temperature.
5. Wash the cells 3 times with PBS.

Notes to the procedures
1. Paraformaldehyde fixed cells should not be stored, but used immediately.
2. Different fixation procedures may produce different ANA patterns. For example, with anti-SS-A/Ro antibodies the paraformaldehyde fixation mostly produces a weak cytoplasmic staining in addition to the bright nuclear fluorescence. In methanol-acetone fixed cells this cytoplasmic staining is mostly absent.
3. Antigens may have different sensitivities to different fixation procedures.

2.4 Preparation of the serum sample

Steps in the procedure
1. Collect the blood sample without any additives to prevent lysis of blood cells.
2. Centrifuge the sample at 2,500 g for 10 min.
3. Collect the upper layer of serum, preventing contamination with the pelleted blood cells.
4. If not used immediately, store serum at -70 °C. Addition of sodium azide (100–300 μg/ml) is recommended.
5. Dilute serum in PBS supplemented with 0.5 g/l Tween 20 and 0.1 g/l sodium azide (PBS-T). ANA-screening on HEp-2 cells is usually performed at serum dilutions of 1:40 or 1:80.

Notes to the procedure
ad. 4. When a serum is analyzed within one week, storage at 4 °C is sufficient. Storage for longer periods (months/years) at -20 °C may cause some lyophilization resulting in changes in ANA titer. Frequent thawing and freezing the sample may lead to a decrease or loss of antibody activity.
ad. 5. Bovine serum albumin (20 g/l) can be added to the dilution buffer to reduce nonspecific binding of serum globulins to the substrate. The use of Tween 20 is not essential.

2.5 The conjugate antibodies

Polyvalent goat or rabbit anti-human IgG conjugated with fluorescein-isothiocyanate (F/P ratio about 3) is the best reagent for screening ANA and is available from many suppliers. The use of an anti-γ-globulin conjugate will also allow the detection of IgM and IgA ANA. Such antibodies, however, are very common and therefore may produce unwanted nonspecific background fluorescence and many positive normal sera!

Dilute the conjugate with PBS-T and determine the appropriate working dilutions by chessboard titrations [7]. The undiluted conjugate is stable for at least 6 months when stored at 4 °C in light-tight containers.

In commercial kits the reagent is supplied prediluted and matched for optimal reactivity and sensitivity with the appropriate substrate.

2.6 The incubation procedure

Steps in the procedure
1. Place 10–20 μl of the diluted serum on the substrate well.
2. Incubate the slides in a humid chamber for 30 min at laboratory temperature.
3. Rinse the slides briefly with PBS-T buffer (room temperature) and place them immediately in PBS-T. Wash for 10 min with agitation in two charges of PBS-T.
4. Remove the adhering buffer from the slides with absorbing paper and immediately add 1 drop (10–20 μl) of FITC-conjugate onto each well. Incubate again for 30 min at room temperature in the humid chamber.
5. Rinse as indicated under step 3.
6. If counterstaining is desirable, add EVANS BLUE to the last washing buffer.
7. Place 2–3 drops of mounting medium (70% glycerol in PBS pH 8.3) on each slide and cover carefully with a coverslip.
8. Observe the wells through an epi-fluorescence microscope mounted with a combination of filters optimal for FITC-detection. A 400 × magnification gives patterns which allow easy identification of various standard patterns [1].

Notes to the procedure
ad. 2. This operation must be carried out in such a way as to ensure that the slide wells do not dry out.
ad. 6. The concentration of EVANS BLUE that can be used is about 250 μg per ml.
ad. 7. Various chemicals can be added to the mounting medium to improve retention of fluorescence upon illumination. We recommend DABCO (1,4 diazobicyclo [2,2,2]-octane, Aldrich Chemical Co, Inc.) 2.5 g and sodium azide 1.0 g per 100 ml of mounting medium. Carefully adjust the pH to 8.3 with concentrated HCl and seal coverslips with nail varnish to avoid oxygenation of DABCO [8].

3. Interpretation of the results

The intensity and the patterns of fluorescence are assessed and recorded. A serum is considered to be positive when the observed fluorescence is significant. The intensity can be expressed according to a scale of values compatible with the guidelines established by the reference centres (e.g. CDC, Atlanta) as + to + + + + or as negative. A semi-quantitative evaluation can be obtained by performing serial dilutions of the test serum to endpoint fluorescence (titer = reciprocal of dilution).

In general, a titer higher than 1:80 is suggestive of a connective tissue disease [9].

A multitude of antibodies directed to nuclear as well as cytoplasmic antigens can be detected using HEp-2 cells. Fluorescence is seen in different structural sizes and shapes and may be localized in different cell compartments.

Notes to the interpretation of immunofluorescence patterns
1. ANA patterns are only indicative and precise determination of the antibody specificities always needs confirmation by other techniques, such as CIE, immunoblotting, RNA or protein precipitation or ELISA. This is especially true because sera mostly contain a mixture of antibody activities in different concentrations.
2. Mixed patterns occur when the patients' serum contains more than one antibody specificity, which is mostly the case. For example when anti-snRNP antibodies are associated with anti-DNA antibodies in a serum of an SLE patient, a homogeneous nuclear pattern will be observed and the chromosomal region of mitotic cells will be strongly positive. However, after serial dilution the typical, speckled pattern for anti-snRNP antibodies may appear. Different serum dilutions thus may produce different fluorescence patterns!
3. We have tried to indicate how frequent the various immunofluorescence patterns tend to occur in patients with a connective tissue disease (common = more than 10% of the patients; fairly common = between 5 and 10% ; rare = between 1 and 5% ; very rare = less than 1%).

3.1 Nuclear fluorescence patterns

3.1.1 Nuclear membrane patterns (rare)
It is possible to distinguish two types of membranous patterns, namely a homogeneous ring-like pattern and a punctate membrane pattern.

The former pattern normally shows a homogeneous ring-like fluorescence of the fine inner nuclear membrane and a homogeneous staining of the entire nucleus and the external structure of the membrane. In telophase the newly formed nuclei are surrounded by a fine membranous fluorescence. In mitosis the fluorescence is diffusely localized in the cytoplasm while the chromatin is negative (Fig. 1). This pattern is found when antibodies to components associated with the nuclear membrane such as the lamins A, B and C are present [10, 11].

The other pattern is characterized by a discontinuous punctate fluorescence along the nuclear envelope. On focusing through the nucleus, the punctate staining can be seen on the surface of the entire nuclear membrane, giving the appearance of a finely granular nuclear pattern. In mitosis the fluorescence is diffusely localized throughout the cytoplasm (Fig. 2). This pattern is observed with antibodies to the nuclear pore complexes (for example antibody directed to the gp 210 protein) or to the lamin B receptor of the inner nuclear membrane [11, 12].

3.1.2 The homogeneous pattern (common in SLE, RA, JCA, drug induced LE)

A uniform diffuse fluorescence covering the entire nucleus can be associated with different patterns [4, 13]. In some patterns the nucleoli are not stained, in some others they are. These patterns are obtained when antibodies to histones (Fig. 3) or other abundant chromatin-associated antigens such as DNA or DNA topoisomerase I are present. In some other cases a more intense staining of the large inner edge of the nucleus (nuclear rim) can be seen. This may indicate that antibodies to dsDNA are present (Fig. 4) although anti-lamin antibodies may produce a somewhat similar pattern in interphase cells.

3.1.3 The speckled pattern (common)

This pattern is characterized by speckles or dots of fluorescence scattered throughout the nucleoplasm. The size of these speckles may vary from large to very fine and the shape may be uniform or highly irregular. Speckles may number from a few to numerous, the latter leading to a dense grainy, almost homogeneous staining pattern. No fluorescence in the chromosomal region of mitotic cells is observed.

Several patterns can be distinguished :

3.1.3.1 Large speckled pattern (very rare). Variable large speckles in a sponge-like network (figure 5). This pattern is observed with antibodies directed to components of the nuclear matrix, e.g. heterogeneous nuclear ribonucleoproteins (hnRNP) [14].

3.1.3.2 Coarse speckled pattern (common in SLE and SLE-overlap disease). Densely distributed, intermediate sized speckles, generally associated with large speckles (Fig. 6). This pattern is typical for anti-snRNP or anti-Sm antibodies [13]. The antigens are contained in the small nuclear ribonucleoproteins (snRNPs) [15, 16].

3.1.3.3 Fine speckled (granular) pattern (common in primary SS, subacute cutaneous lupus and hyper-γ globulinaemic purpura). Fine grainy staining in a uniform distribution, sometimes very dense so that an almost homogeneous pattern is obtained (Fig. 7). This pattern is common to many different nuclear proteins, the best studied being SS-B/La and SS-A/Ro

[13, 16]). The precise identity of these antibody activities should be tested by other techniques.

Notes
1. Some anti-Ro/SS-A sera do not produce a fluorescent pattern (ANA-negative). This can be caused for example by the fact that the epitopes recognized by the antibodies are hidden or masked as a consequence of fixation.
2. Anti-topoisomerase I antibodies (Fig. 8) produce a fine speckled pattern that often has the appearance of a homogeneous pattern.

3.1.3.4 Heterogeneous, pleomorphic speckled pattern (PCNA-like pattern; very rare, mostly in SLE). Fine, as well as coarse irregular speckles in the nuclei of the cells in S-phase (10 to 50% of the cells, depending on the cell preparation). In some cells the nucleoli are stained as well (Fig. 9). This antibody activity recognizes the proliferating cell nuclear antigen (PCNA or cyclin [13, 16]) which has been found to be identical to an auxiliary protein of DNA polymerase δ, a multisubunit particle which is required for DNA replication [17].

3.1.3.5 Discrete speckled (similar to anti-centromere pattern; common in limited systemic sclerosis, fairly common in primary Raynaud's phenomenon). Uniform discrete speckles located throughout the entire nucleus, usually in multiples of 46 (40–60 speckles per nuclei). Mitotic cells show these speckles in the condensed chromosomal material (Fig. 10). This pattern is consistent with antibodies to the chromosome centromeres (kinetochores) and is rather specifically found in the CREST variant of scleroderma (limited systemic sclerosis) [13, 16].

3.1.3.6 Nuclear dots (rare). A few discrete nuclear speckles of variable size are brightly stained. The antigens recognized by the antibodies which produce such patterns can be several but are thought to be contained in somewhat larger nuclear substructures. In contrast to anti-centromere antibodies, no staining of chromosomes in mitotic cells is seen. One can distinguish at least two types of patterns :

Type 1. 2–6 dots / nucleus.

This is a pattern typically produced by anti-Coiled Body antibodies (anti-p80 coilin [18]) (Fig. 11). The dots are located randomly in the nucleoplasm, although they are often seen in close proximity to the nucleolus.

Type 2. 5–10 dots / nucleus (similar to multiple nuclear dots).

This pattern has been found with antibodies that recognize a soluble acidic nuclear protein of 100 KDa refered to as Sp100 [19], although it is not known if it is restricted only to this antibody specificity (Fig. 12).

3.2 Nucleolar patterns (common in progressive systemic sclerosis)

The neoplastic HEp-2 cell usually has 2 or more nucleoli. Different patterns of nucleolar fluorescence can be observed [20, 21]:

3.2.1 Homogeneous pattern
This pattern shows fluorescence of the entire nucleolus. It can be observed with antibodies directed to nucleolin [22] or to the PM/Scl antigen [23] (Fig. 13).

3.2.2 Clumpy pattern
Brightly clustered larger granules corresponding to decoration of the fibrillar centres of the nuclei. Antibodies are directed to fibrillarin, one of the proteins associated with the U3 small nucleolar ribonucleoprotein U3 (Fig. 14) [24].

3.2.3 Speckled pattern
Small discrete speckles in the centre of the nucleolus. This punctate pattern might be caused by antibodies directed to RNA polymerase I [25] (Fig. 15).

3.2.4 Punctate pattern with multiple dots
In interphase a few small single dots of irregular size can be seen in the nucleolus. This typical pattern can be seen with antibodies to the nucleolar organizing region (NOR, [26, 27]) (Fig. 16).

3.3 Spindle apparatus fluorescence patterns (very rare in connective tissue disease)

3.3.1 The mitotic spindle pattern
The mitotic spindle is clearly deliniated by the fluorescent stain and both chromosomal (kinetochore to pole) and interpolar (pole to pole) fibers are apparent. A weak cytoplasmic filamentous fluorescence is apparent in interphase cells. This pattern characterizes antibodies to tubulin (not shown).

3.3.2 The centriole pattern
The metaphase spindle displays discrete fluorescent spots at each of the spindle poles lying perpendicular to one another (Fig. 17). One or two bright spots closely apposed to the nucleus are seen in the cytoplasm of interphase cells. This pattern is seen with antibodies to the centriole, very rare in connective tissue disease but fairly common in mycoplasma-pneumonia infections [28].

3.3.3 The NuMA pattern
The IF staining is solely concentrated at the spindle poles without staining of

the interpolar fibers. At metaphase the spindle poles are fully formed and the chromosomes positioned at the cell equator are unstained. This pattern is obtained with anti-NuMA (Nuclear Mitotic Apparatus protein, [29]) antibodies (Fig. 18). Such antibodies are very rarely found in connective tissue disease patients.

When spindle fibers are decorated as well, we are most probably dealing with anti-tubulin antibodies (see 3.3.1).

3.3.4 MSA-2 and MSA-3

In case of MSA-2 (Mitotic Spindle Antigen) the interphase cells are not stained. In S and G2 phases discrete speckles, changing into patchy speckles, appear. In prophase and metaphase the fluorescence is localized in the chromosomal region, whereas in telophase the staining is restricted to the cleavage furrow and the narrow connecting midbody between cells completing cytokinesis. This antibody activity is referred to as anti-Midbody [30] (Fig. 19).

In case of MSA-3 a very fine dense nuclear speckled pattern is seen in some interphase cells. A dense punctate decoration of the chromosomes appears in prophase cells, whereas in metaphasic cells the pattern is constituted by 2 sets of dense large granules surrounding the chromosomal metaphase plate as a grip. No staining of telophase cells is observed [30, 31] (Fig. 20).

3.4 Cytoplasmic fluorescence patterns

3.4.1 The speckled patterns

3.4.1.1 Fine speckled pattern (common in polymyositis). Fine granules condensed around the nucleus diminishing towards the periphery of the cytoplasm (Fig. 21) is evocative of antibodies to aminoacyl-tRNA synthetases (such as Jo1, PL7 or PL12) [32].

3.4.1.2 The ribosomal pattern (fairly common in SLE). A very fine dense granular to homogeneous staining or cloudy pattern covering all the cytoplasm is obtained with serum of patients (mostly SLE) containing antibodies to ribosomal RNP (Fig. 22). In most cases also the nucleolus, center of ribosome synthesis, is stained [33].

3.4.1.3 Large cytoplasmic organelles (very rare in connective tissue disease; common in primary biliary cirrhosis). Larger irregular granules extending from the nucleus throughout the cytoplasm in a filamentous pattern are stained with anti-mitochondrial antibodies (Fig. 23). One of the major antigens that can be recognized is M2, a cluster of four major mitochondrial inner membrane proteins [34, 35].

However, antibodies to the signal recognition particle (SRP) appear to produce a similar staining pattern [36]. A fluorescent pattern with larger granules condensed in the cytoplasm more pronounced at one site of the nucleus can be obtained with antibodies to the endoplasmic reticulum.

Irregular large organelles distributed throughout the cytoplasm can be stained with antibodies supposed to be directed to organelles like lysosomes or peroxysomes.

A polar paranuclear arrow surrounding one part of the nucleus and composed of irregular large granules is observed with antibodies to the Golgi apparatus [37] (Fig. 24).

3.4.2 The fibrous patterns

3.4.2.1 The actin pattern (rare in patients with connective tissue disease; common in chronic active hepatitis). Stress fibers that span the whole length of the cell and lying in a simple focal plane just underneath the plasma membrane (Fig. 25). This pattern corresponds to antibodies against microfilaments and mainly to actin cables [38]. Shorter stress fibers with membrane ruffling and dendritic-like extensions are decorated with antibodies to tropomyosin.

3.4.2.2 The vimentin pattern (rare in patients with connective tissue disease; often caused by crossreaction with anti-α helix antibodies).
Abundant fine fibers extending through the cytoplasm in a more or less radial arrangement. In some cells the fibers are aggregated in bundles around the nucleus in a coiled form. This pattern is typical for antibodies directed to vimentin (Fig. 26). These antibodies are detected in all kinds of (chronic) inflammatory diseases. Their clinical significance is unknown.

4. Solutions and suppliers

PBS: 13 mM Na_2HPO_4, 3 mM KH_2PO_4, 140 mM NaCl, pH 7.4
PBS-T: PBS containing Tween 20 (0.5 mg per ml) and sodium azide (0.1 mg per ml).

4.1 Solutions for culturing the cells

The composition of most solutions and media can be found in the catalogue of the supplier.

Penicillin-Streptomycin solution
The solution contains 5000 U/ml Pen and 5000μg/ml Strep and is sold as a lyphilized product (Gibco BRL 061–05075).

Dulbecco's MEM : Mix the following sterile solutions:
400 ml distilled water
50 ml Dulbecco's MEM 10x (Gibco BRL 042-2501)

24 ml Na-bicarbonate 7.5% (Gibco BRL 043-05080)
 5 ml L-glutamin 200 mM (Gibco BRL 043-05030)
 5 ml Na-pyruvate 100 mM (Gibco BRL 043-01360)
0.2 ml Penicillin/Streptomycin solution (Gibco BRL 061-05075)
50 ml Newborn Calf Serum

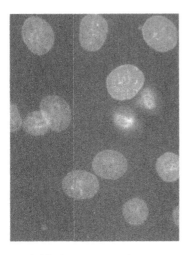

Fig. 1. Membranous pattern 1.
Homogeneous fluorescence of the nuclear
membrane (anti-lamin).

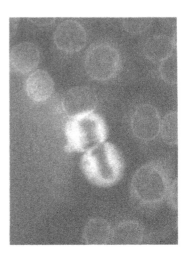

Fig. 2. Membranous pattern 2.
Punctate fluorescence of the nuclear
membrane (anti-nuclear pore complexes).

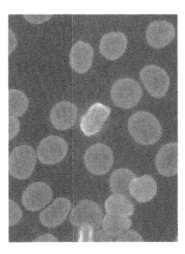

Fig. 3. Homogeneous pattern.
Uniform fluorescence of the nucleus.
Strong fluorescence of chromatin in
mitotic cells (anti-histone).

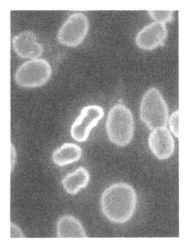

Fig. 4. Peripheral pattern.
Fluorescence is more intense at the inner
edge of the nucleus (nuclear rim).

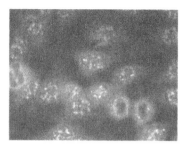

Fig. 5. Speckled pattern 1.
Large speckles (anti-hnRNP, anti-nuclear matrix).

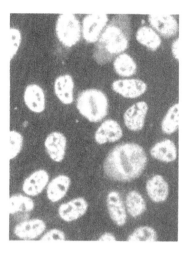

Fig. 6. Speckled pattern 2.
Coarse speckles (anti-snRNP, anti-Sm).

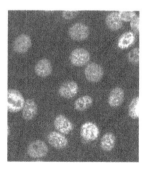

Fig. 7. Speckled pattern 3.
Fine speckles (granular) (anti-SSA/Ro).

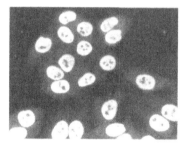

Fig. 8. Speckled pattern 4.
Fine speckled, almost homogeneous pattern (anti-Scl70 = anti-DNA Topoisomerase I).

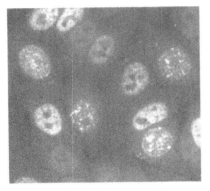

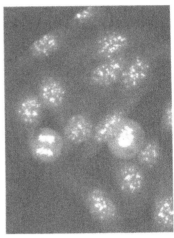

Fig. 9. Pleomorphic speckled, anti-PCNA pattern.
Only nuclei of cells in the S phase are stained in different patterns (anti-PCNA/Cyclin).

Fig. 10. Anti-centromere pattern.
In the mitotic cells the antigen is lined up.

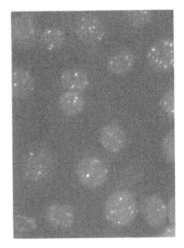

Fig. 11. Nuclear dots; anti-coilin pattern.
A few nuclear dots, mitotic cells are negative.

Fig. 12. Multiple nuclear dots.
Anti-Sp100 staining pattern.

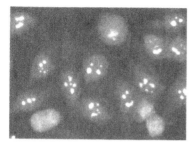

Fig. 13. Nucleolar pattern, homogeneous. Homogeneous staining of nucleoli (anti-nucleolin, anti-PM/Scl).

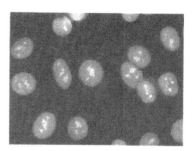

Fig. 14. Nucleolar pattern, clumpy. Anti-fibrillarin.

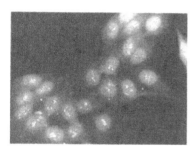

Fig. 15. Nucleolar pattern, speckled. Anti-RNA polymerase I.

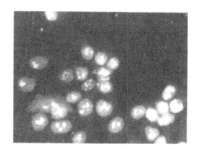

Fig. 16. Nucleolar pattern, nucleolar dots. Punctate pattern, also seen in metaphase cells (anti-NOR).

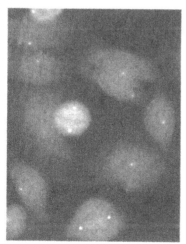

Fig. 17. Antibodies to the spindle apparatus 1.
Antibodies to the centriole.

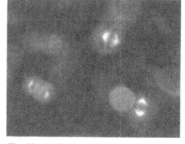

Fig. 18. Antibodies to the spindle apparatus 2.
Antibodies to NuMA.

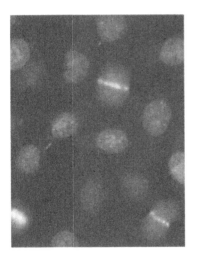

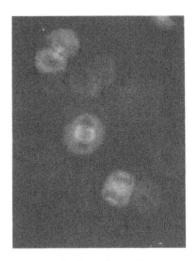

Fig. 19. Antibodies to the spindle
apparatus 3.
Anti-midbody.

Fig. 20. Antibodies to the spindle
apparatus 4.
Large granule surrounding the
chromosomal metaphase plate (type
MSA3).

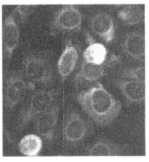

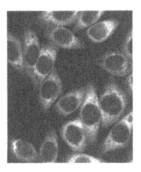

Fig. 21. Cytoplasmic pattern 1.
Fine speckled, condensed around the
nucleus (anti-Jo1, anti-tRNA synthetase
such as anti-PL7 and anti-PL12).

Fig. 22. Cytoplasmic pattern 2.
Fine dense staining, cloudy pattern (anti-
ribosomes). Some nucleoli may be stained.

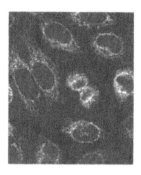

Fig. 23. Cytoplasmic pattern 3.
Large speckles in the cytoplasm (anti-mitochondria).

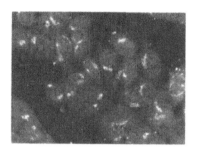

Fig. 24. Cytoplasmic pattern 4.
Polar paranuclear arrow and large
granular pattern surrounding the nucleus.
Anti-Golgi apparatus pattern.

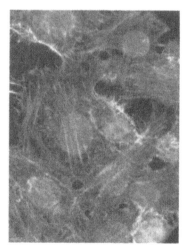

Fig. 25. Cytoplasmic pattern 5.
Stress fibres (anti-actine).

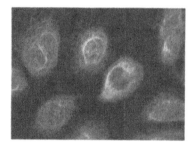

Fig. 26. Cytoplasmic pattern 6.
Fine fibres aggregated around the nucleus
(anti-vimentine).

References

1. McCarty GA, Valencia DW & Fritzler MJ (1984) Antinuclear antibodies: contemporary techniques and clinical application to connective tissue diseases, pp. 1–96. Oxford Univ. Press., N.Y.
2. Fritzler MJ & Tan EM (1985) Antinuclear antibodies and the connective tissue diseases. In: Laboratory diagnostic procedures in rheumatic diseases, (pp. 207–247). Grune & Stratton Edt.
3. Fritzler MJ (1986) Autoantibody Testing: Procedures and significance in systemic rheumatic diseases. Meth Archiv Exp Pathol 12:224–260. Karger, Basel
4. Humbel RL & Kadusch P (1985) Use of HEp-2 cells in the IF test for detection of antinuclear antibodies. Protides of the Biological Fluids, Proc. 35th Colloquium, H. Peeters Edit. Pergamon Press Ldt. 33:385–387
5. Feltkamp TEW, Klein F & Janssens MBJA (1988) Standardisation of the quantitative determination of antinuclear antibodies (ANAs) with a homogeneous pattern. Ann Rheum Dis 47:906–909
6. Tan EM, Feltkamp TEW, Alarcon-Segovia D et al.(1988) Reference reagents for antinuclear antibodies. Arthr Rheum 31:1331
7. Nakamura RM & Tan EM (1986) Recent advances in laboratory tests and the autoantibodies to nuclear antigens in systemic rheumatic diseases. Rec Adv Immunopath 6:41–43
8. Krenik KD, Kephart GM, Offord KP, Dunnette SL & Gleich GJ (1989) Comparison of antifading agents used in immunofluorescence. J Immunol Meth 117:91–97
9. Fritzler MJ, Pauls JD, Douglas-Kinsella T, Bowen TJ (1985) Anti-nuclear, anti-cytoplasmic and anti-Sjögren's syndrome antigen A (SSA/Ro) antibodies in female blood donors. Clin Immunol Immunopathol 36:120–128
10. Lassoued K, Guilly MN, Danon F, Chantal A, Dhumeaux D, Clauvel J-P, Brouet J-C, Seligmann M & Courvalin J-C (1988) Antinuclear autoantibodies specific for lamins. Ann Int Med 108:829–833
11. Senecal JL & Raymond Y (1991) Autoantibodies to DNA, lamins and pore complex proteins produce distinct peripheral fluorescent antinuclear antibody patterns on the HEp-2 substrate. Arthr Rheum 34:249–251
12. Worman HJ & Courvalin JC (1991) Autoantibodies against nuclear envelope proteins in liver disease. Hepatology 14:1270–1279
13. Tan EM (1982) Autoantibodies to nuclear antigens (ANA): Their immunobiology and medicine. Adv Immunol 33:167–240
14. Fritzler MJ, Ali R & Tan EM (1984) Antibodies from patients with mixed connective tissue disease react with heterogeneous nuclear ribonucleoproteins or ribonucleic acid (hnRNP/RNA) of the nuclear matrix. J Immunol 132:1216–1222
15. Van Venrooij WJ & Sillekens PTG (1989) Small nuclear RNA associated proteins: autoantigens in connective tissue diseases. Clin Exp Rheum 7:635–645
16. Tan EM (1989) Antinuclear antibodies: Diagnostic markers for autoimmune diseases and probes for cell biology. Adv Immun 44:93–152
17. Bravo R, Frank R, Blundell PA, MacDonald-Bravo H (1987) Cyclin/PCNA is the auxiliary protein of DNA polymerase δ. Nature 326:515–517
18. Andrade LEC, Chan EKL, Raska I, Peebles CL, Roos G & Tan EM (1991) Human autoantibody to a novel protein of the nuclear Coiled Body: Immunological characterization and cDNA cloning of p80-coilin. J Exp Med 173:1407–1419
19. Szostecki C, Krippner H, Penner E & Bautz FA (1987) Autoimmune sera recognize a 100 kD nuclear protein antigen (Sp-100). Clin Exp Immunol 68:108–116
20. Bernstein RM, Steigerwald JC & Tan EM (1982) Association of antinuclear and antinucleolar antibodies in progressive systemic sclerosis. Clin Exp Immunol 48:43–51
21. Imai H, Ochs RL, Kiyosawa K, Furuta S, Nakamura RM & Tan EM (1992) Nucleolar

antigens and autoantibodies in hepatocellular carcinoma and other malignancies. Am J Pathol 140:859-870

22. Minota S, Jarjour WN, Suzuki N, Nojima Y, Roubey RAS, Mimura T, Yamada A, Hosoya T, Takaku F & Winfield JB (1991) Autoantibodies to nucleolin in systemic lupus erythematosus and other diseases. J Immunol 146:2249-2252

23. Blüthner M & Bautz FA (1992) Cloning and characterization of the cDNA coding for a polymyositis-scleroderma overlap syndrome-related nucleolar 100 kDa protein. J Exp Med 176

24. Okano Y, Steen VD & Medsger TA jr (1992) Autoantibody to U3 nucleolar ribonucleoprotein (fibrillarin) in patients with systemic sclerosis. Arthritis Rheum 35:95-100

25. Reimer G, Rose KM, Scheer U & Tan EM (1987) Autoantibody to RNA polymerase I in scleroderma sera. J Clin Invest 79:65-72

26. Courvalin JC, Hernandez-Verdun D, McCarty MC, Maunoury R & Bornens M (1985) A 80 Kd protein is associated with the nucleolus organizer (NOR) of the human cell lines. Protides of the Biological Fluids, Proc. 35th. Colloquium, H. Peeters Edit. Pergamon Press Ltd, pp. 221-224

27. Rodriguez JL, Gelpi C, Juarez C & Hardin JA (1987) Anti-NOR 90: A new autoantibody in scleroderma that recognizes a 90-kDa component of the nucleolus-organizing region of chromatin. J Immunol 139:2579-2584

28. Rattner JB, Martin L, Waisman DM, Johnstone SA & Fritzler MJ (1991) Autoantibodies to the centrosome (centriole) react with determinants present in the glycolytic enzyme enolase. J Immunol 146:2341-2344

29. Price CM, McCarty GA & Pettijohn DE (1984) NuMA protein is a human autoantigen. Arthr Rheum 27:774-779

30. Fritzler MJ, Ayer LM, Gohill J, O'Connor C, Laxer RM & Humbel R-L (1987) An antigen in metaphase chromatin and the midbody of mammalian cells binds to scleroderma sera. J Rheumatol 14:291-294

31. McCarty GA, Valencia D & Fritzler MJ (1984) Antibody to the mitotic spindle apparatus: Immunologic characterics and cytologic studies. J Rheumatol 11:213-218

32. Saito E, Yoshimoto Y, Oshima H, Yoshida H & Kinoshita M (1989) Fluorescent antibodies in polymyositis using cultured human skin fibroblasts: Granular perinuclear cytoplasmic staining pattern by sera from patients with polymyositis and pulmonary fibrosis. J Rheumatol 16:47-54

33. Bonfa E & Elkon KB (1986) Clinical and serologic associations of the antiribosomal P protein antibody. Arthr Rheum 29:981-985

34. Mackay IR & Gershwin ME (1989) Molecular basis of mitochondrial autoreactivity in primary biliary cirrhosis. Immunol Today 10:315-318

35. Coppel RL, McNeilage LJ, Surh CD, Van De Water J, Spithill TW, Whittingham S & Gershwin ME (1988) Primary structure of the human M2 mitochondrial autoantigen of primary biliary cirrhosis: Dihydrolipoamide acetyltransferase. Proc Natl Acad Sci USA 85:7317-7321

36. Reeves WH, Nigam SK & Blobel G (1986) Human autoantibodies reactive with the signal recognition particle. Proc Natl Acad Sci USA 83:9507-9511

37. Rodriguez JL, Gelpi C, Thompson TM, Real FJ & Fernandez J (1982) Anti-Golgi complex autoantibodies in a patient with Sjögren's syndrome and lymphoma. Clin Exp Immunol 49:579-586

38. Kurki P & Vitanen I (1984) The detection of human antibodies against cytoskeletal components. J Imm Methods 67:209-223

Manual of Biological Markers of Disease A3: 1–12, 1993.

Counterimmunoelectrophoresis and immunodiffusion for the detection of antibodies to soluble cellular antigens

C. BUNN[1] and T. KVEDER[2]

[1] Department of Immunology, Royal Free Hospital, Pond Street, London NW3 2QG, U.K.;
[2] Department of Rheumatology, University Medical Centre, SLO-61107, Ljubljana, NW3 2QG Slovenia

Abbreviations

CIE: counterimmunoelectrophoresis; EEO: electroendosmosis; ENA: extractable nuclear antigen; ID: immunodiffusion; PBS: phosphate buffered saline; HIV: human immunodeficiency virus

Key words: Autoantibodies, autoimmunity, immunodiffusion, counterimmunoelectrophoresis

Introduction

The detection of autoantibodies by indirect immunofluorescent screening techniques has been supplemented by laboratory tests that characterise individual reactions, thereby providing clinicians with important diagnostic and prognositic information [1]. Counterimmunoelectrophoresis (CIE) and double diffusion (Ouchterlony) are two of the methods available for identifying specific antibodies in complex antigen mixtures [2, 3] and have found particular application with respect to the detection of antibodies to soluble cellular antigens, sometimes referred to as extractable nuclear antigens (ENA).

Both methods are able to resolve multiple autoantibody specificities by the formation of precipitin lines in agarose gel using a cell extract as a source of antigen. The saline soluble cell extracts are made from fresh tissue or from commercially available acetone powders. The latter are a rich source of the majority of common nuclear antigens but are notably deficient in the Ro antigen. For this reason a second extract usually made from human spleen obtained at necropsy is employed as the preferred source of this antigen although cytoplasmic preparations from bovine tissue are also adequate.

The diffusion technique has the advantages of requiring no rigorous attention to pH or ionic strength and needs no specialist electrophoresis equipment. CIE, on the other hand, is more rapid and can detect as little as 20 μg/ml of specific antibody, approximately a ten fold improvement in sensitivity over immunodiffusion (ID). The constraints of CIE require that the antigen is negatively charged and that there is an electroendosmotic flow. Under these conditions antibody and antigen are forced together to form a precipitin line at equivalent proportions [4, 5]. Originally termed immunoelectrosmophoresis, CIE has been used as a microscreening method for the detection of the α2-migrating hepatitis B surface antigen [6], and in an adapted two-stage form to detect anti-DNA antibodies. In this modification the relatively slow migration of γ-globulin is allowed to proceed before the introduction of the rapidly moving DNA antigen [7]. The technique was first used to detect ENA antibodies by Kurata and Tan (1976) [8] and is now an established method in the clinical immunology laboratory [9–12].

The descriptions of antigen preparations and conditions for diffusion and electrophoresis presented in the literature contain hundreds of variations. In this manual we have attempted to describe workable methods and to draw attention to some of the alternatives.

1. Antigen extract preparations

It should be noted that all the procedures detailed in this section should be carried out as close to 0 °C as possible in order to minimise proteolysis.

Thymus acetone powders from calf and rabbit have been successfully used as the source of most extractable antigens with the exception of Ro. The most frequently used and widely referenced source of these products is Pelfreez Biologicals Inc. (Rogers, Arkansas). The products of suppliers such as C-six (Mequon, Wisconsin), Immunovision (Springdale, Arkansas) and Sigma (Poole, UK) remain relatively untried but may be considered for use. It is also possible to buy commercial antigen mixtures or, at less cost, to make soluble extracts from fresh slaughterhouse material. The Ro antigen is usually prepared from human spleen obtained at post-mortem, however there is a growing reluctance to work with human tissue that may be unknowingly infected with HIV. The alternatives appear to be to make Ro from primate, bovine or even canine spleen.

Most laboratories employ two extracts to cover the whole range of ENAs. A rabbit thymus extract, supplemented by a human spleen extract for Ro. If calf thymus preparations are used a weak precipitin attributable to anti-Ro is sometimes seen. We believe that this can be confusing particularly in sera that give multiple lines and we recommend the use of a separate extract when testing for anti-Ro.

1.1 Extraction from acetone powders [12]

Steps in the procedure
1. 0.5 g of rabbit thymus acetone powder is added to 10 ml of ice cold phosphate buffered saline (PBS) and the antigens extracted by stirring the mixture for 4 h in the cold.
2. The mixture is then centrifuged to obtain a clear supernatant at 10,000 g for 30 min.
3. Dialyse the supernatant against 20% polyethylene glycol 6000 in PBS until the concentration is 30-40 mg/ml. Alternatively concentrate by filtration through an Amicon PM10 filter under nitrogen pressure.
4. Test the antigen against all desired prototype antisera prior to use.
5. Aliquot the extract and freeze at $-70\ ^{\circ}$C.

Notes to the procedure
ad 1. Overnight extraction does not have any deleterious effects.
ad 5. The majority of antigens are stable at $-20\ ^{\circ}$C, however there is a noticeable deterioration in amounts of Scl-70 and XR present after 3–4 months. Antibodies to the latter, undefined antigen, are found in patients with liver disease [13].

1.2 Whole cell extract [4, 14]

Steps in the procedure
1. Fresh tissue from the abattoir is placed in a plastic bag and kept on ice until received in the laboratory.
2. Remove the outer membranes and blood vessels.
3. Add an equal volume (w/v) of ice cold PBS and homogenise in a blender.
4. Centrifuge at 10,000 g for 60 min.
5. Centrifuge the supernatant at 76,000 g for 2 h.
6. Dialyse against PBS.
7. Concentrate to 30–40 mg/ml.
8. Test and store the extract as described above.

Notes to the procedure
ad 1. Calf thymus is a suitable source of nuclear proteins, bovine speen for Ro and sheep liver for cytoplasmic antigens such as aminoacyl-tRNA synthetases. Tissues can be stored frozen at $-20\ ^{\circ}$C or $-70\ ^{\circ}$C prior to use.
ad 3. A Waring blender or kitchen food processor is adequate.
ad 5. This step may be omitted.

1.3 Spleen extract for Ro [15]

Steps in the procedure
1. Human (or bovine) spleen less than 24 h post mortem trimmed of membranes and blood vessels is minced in a tissue press.
2. Mix with an equal volume of PBS.
3. Homogenise in a motor driven Potter homogeniser with teflon pestle at 2,500 rpm for 10 min.
4. Centrifuge at 27,000 g for 30 min.
5. Pass supernatant through two layers of gauze to remove lipid.
6. Centrifuge at 105,000 g for 90 min.
7. Test, aliquot and store as described above.

Notes to the procedure
ad 4. Post mitochondrial pellet.
ad 5. Post microsomal pellet, this step may be omitted.

1.4 Liver extract for aminoacyl-tRNA synthetases [16]

Steps in the procedure
1. Sheep liver obtained from the slaughterhouse is mixed with 2 volumes of 0.35 M Sucrose in 100 mM Tris, 60 mM KCl, 10 mM $MgCl_2$
2. Blend in a Waring blender or food processor for 1 min.
3. Homogenise in a motor driven Potter homogeniser with teflon pestle at 2,500 rpm for 10 min.
4. Centrifuge at 27,000 g for 30 min.
5. Pass supernatant through two layers of gauze to remove lipid.
6. Centrifuge at 105,000 g for 90 min.
7. Dialyse against PBS.
8. Test, aliquot and store as described above.

1.5 Nuclear extract [17, 18]

Steps in the procedure
1. Trim and remove membranes and blood vessels.
2. Homogenise gently in 0.88 M sucrose with 1.5 mM calcium chloride using a Potter homogeniser with teflon pestle.
3. Raise the sucrose concentration to 1.6 M and layer on to 2.2 M sucrose containing 1.5 mM calcium chloride.
4. Centrifuge at 76,000 g for 90 min.
5. Wash the nuclear pellet in 0.25 M sucrose with 0.5 mM calcium chloride.

6. Resuspend the pellet in 9 volumes of 150 mM sodium chloride.
7. Homogenise with a 45 second burst in a Virtis macerator and stand at 4 °C overnight.
8. Centrifuge at 106,000 g for 2 h.
9. Test and store as previously described.

Notes to the procedure
ad 2. Calcium ions prevent nuclei clumping.
ad 5. 0.25 M Sucrose is isotonic.
ad 8. This step can be omitted (microsome removal is not essential).

1.6 Ion-Exchange Chromatography [4, 19]

Spleen extracts for Ro can be heavily contaminated with haemoglobin and some workers prefer to remove it by ion exchange chromatography in order to improve the clarity of the plates. Fractions can be collected that are devoid of most other antigens, and this is considered advantageous if the extract is only for the detection of anti-Ro. If a range of antibodies is to be detected, fractions should be tested and pooled as necessary. Ku and XH [13] are abundant antigens that can be eliminated by this procedure.

Steps in the procedure
1. Dialyse the extract against PBS.
2. Make a suspension of DE52 (Whatman) in PBS and bring the pH to 7.5 using a 0.5 M solution of sodium dihydrogen phosphate. Use several changes of cold PBS to equilibrate the mixture.
3. Add 2 ml of extract for every 1 ml of slurry and leave to stir in the cold for 2 h.
4. Pour slurry into a column and allow to settle.
5. Elute antigens from the column by adding PBS with 0.4 M sodium chloride added. Collect fractions and assay them against reference antisera for the presence of the desired antigens.
6. Pool the required fractions, dialyse them against PBS and store at −70 °C.

Notes to the procedure
ad 4. Column technology is not essential, although it is more efficient. Alternatively, centrifuge the slurry and discard the supernatant, wash the DE52 with cold PBS and again discard the supernatant following centrifugation. Elute the antigen from the resin by adding 5 volumes of PBS with 0.4 M sodium chloride. Stir for 3 h in the cold. Centrifuge and retain the supernatant, dialyse against cold PBS and concentrate to 30–40 mg/ml.
ad 5. Fractions are collected after the haemoglobin has been excluded.

2. Procedures

2.1 Immunodiffusion [17]

Steps in the procedure
1. Dissolve 0.6 g of agarose in 100 ml of PBS pH 7.5 by heating the mixture on a hot plate stirrer.
2. Pour 10 ml of molten agarose on to an alcohol cleaned 80 × 80 mm glass plate on a level surface and allow to harden prior to cutting.
3. With the aid of a paper template or a predrilled plastic template use a cylindrical metal cutter to punch a central 7 mm well separated by 4 mm from six peripheral 7 mm wells arranged in a circle around the central well (Fig. 1).
4. Remove the agarose plugs and fill the central well with antigen and the peripheral wells with serum.
5. Plates are placed in a moist chamber at room temperature and read at 24 and 48 h. Staining with Coomassie blue as described below can be used to produce a permanent record.
6. Positive reactions can be characterised by placing prototype sera in adjacent wells and examining the plates for lines of immunological identity.

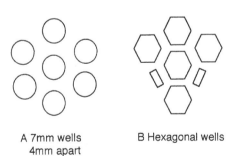

A 7mm wells
4mm apart

B Hexagonal wells

Fig. 1. Well arrangement for Immunodiffusion. A) 7 mm wells 4 mm apart, central well for antigen, peripheral wells for antisera. B) Hexagonal central well for antigen, upper and lower wells for reference serum. Remaining hexagonal wells for testing two unidentified antisera. The rectangular wells are filled with the same unidentified sera giving an effective dilution.

Notes to the procedure
ad 1. 0.6% agarose is the commonest medium, other workers have used lower (0.4%) and higher (0.8%) gel concentrations [20–22]. Although agar is also a suitable gel medium, agarose is almost exclusively used. 150 mM NaCl can also be used in place of PBS. It is usual to add 0.1% sodium azide as a preservative.

Once dissolved the agarose can be stored aliquoted at 4 °C and liquified by heating in a boiling water bath.

ad 2. This gives approximately a 1.5 mm depth of agarose on the plate; alternatively pour 25 ml of agarose into a 100 mm diameter petri dish to give a 3 mm depth.

ad 3. Many variations of this pattern have been used such as 8 mm wells 3 mm apart, 4 mm wells 3 mm apart, a 7 mm central well separated from 4 mm peripheral wells by 2 mm or 3 mm. An interesting variation, also shown in Fig. 1, uses a hexagonal central well with similarly shaped peripheral wells interspersed with smaller rectangular wells that act to build in a 1/4 dilution factor thus reducing any prozoning effects that could possibly occur (Inova Diagnostics Inc., San Diego). Reference sera are used in the wells at the top and the bottom.

ad 4. Agarose plugs are easiest removed if the plated are chilled. They can be removed by suction on a vacuum line or using a needle.

2.2 Counterimmunoelectrophoresis (CIE) [12]

Steps in the procedure

1. Dissolve 1 g of Agarose in 100 ml of 0.065 M barbital buffer pH 8.3 by heating on a hot plate stirrer.
2. Pour 10 ml of molten agarose on to an alcohol cleaned 80 × 80 mm glass plate on a level surface and allow to harden prior to cutting.
3. A perspex template is highly desirable in order to cut the maximum number of wells in each plate. Cut three parallel rows of 3 mm wells over the length of the plate, each well in the row is separated from its neighbour by the minimum practical distance (usually approx. 1 mm). To the cathodal side of each row cut a 2 mm trough separated from the row of wells by 3 mm (see Fig. 2). The wells are for the antisera and the trough for the antigen. The agarose plugs should be removed from the wells at this stage but not from the troughs.
4. Load 10 μl of heat inactivated serum into each well, transfer the plate to an electrophoresis tank containing 0.065 M barbital buffer pH 8.3 and electrophorese the samples at a constant current of 12 mA per plate (40 V/cm) for 15 min.
5. Turn off the current, remove the gel from the antigen trough and fill it with freshly thawed cold antigen (150–200 μl) and continue electrophoresis for a further 40 min.
6. Wash plates overnight in cold PBS.
7. Staining with Coomassie blue, as described below, can be used to obtain a permanent record.

Notes to the procedure

ad 1. Electroendosmosis (EEO) is an important consideration when choosing which agarose to use. In publications where a preference has been stated Agarose type 1 (EEO 0.1–0.15) (Sigma, Poole, UK) Indubiose A37 (EEO 0.17–0.19) (IBF, Gennevilliers, France) and Agarose B (EEO 0.25) (Pharmacia, Milton Keynes, UK) have all been used with satisfactory results. However, it should be noted that if EEO is high it can reverse the direction of electrophoretic migration, moving antigens

A. Well Arrangement   B. Plate Arrangement

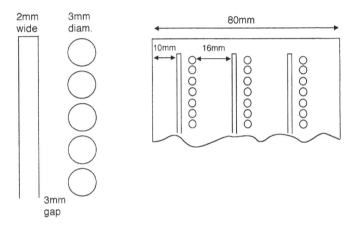

Fig. 2. Counterimmunoelectrophoresis plate. Wells are filled with antisera, troughs filled with antigen after the initial first stage electrophoresis.

with little negative charge to the cathode rather than towards the anode. The pH of the buffer is also important, at less than pH 8.0 some antigens will lose their negative charge. Scl-70 is at risk here as it is a basic protein at physiological pH, and carries little charge at pH 8.0. At pH 8.8 and above, antigens such as Jo-1 may be denatured or not detected for other reasons. Therefore pH 8.3 is a compromise within the safe window. The addition of sodium azide as a preservative may require some alteration of conditions.

ad 2. This gives approximately a 1.5 mm depth of agarose on the plate.

ad 3. The arrangement described allows three rows of 16–20 wells to be cut per plate and therefore as many as 60 samples can be tested. Other methods use larger wells separated by greater distances i.e. 4 mm wells separated by 5–6 mm [9]. Fewer samples requiring more serum can be tested at any one time. The use of an antigen trough as opposed to placing the antigen in a well facilitates the identification of antibody by allowing the formation of lines of identity between sera in adjacent wells.

ad 4. Sera are heat inactivated at 56 °C for 30 min to reduce non-specific reactions. The technique described is a two-stage electrophoresis, a development designed to position the precipitin line away from the antibody well when using fast moving antigens such as DNA. (Kurata and Tan originally described a single stage electrophoresis [8]). The exact amperage and timing depends on the thickness of agarose, sizes of plate, types of wick, electrophoresis tank and power supply and the published values show considerable variation (1.5 mA/cm to 4 mA/cm). The important factor, at the end of the first stage, is that the serum can be seen to have migrated 3–4 mm.

ad 5. During this time the precipitin lines will have formed and further electrophoresis will neither benefit or harm the process. The serum albumin will have travelled approximately 15 mm, the distance to the next trough.

ad 6. The majority of precipitin lines are visible immediately but weaker lines are more readily observed after 30 min in PBS and finally after overnight soaking prior to staining with coomassie blue. Soaking the plates in 5% sodium citrate (for between 30 min and several days) has been recommended by some workers as a method of improving definition.

2.3 Staining with Coomassie Blue [2]

Steps in the procedure

1. Place the plates on the bench and flood them with PBS.
2. Overlay with wet filter paper (Whatman No.1) excluding air bubbles that would cause the gel to split.
3. Place several layers of absorbent towel on top of the filter paper followed by a glass plate and heavy weight (2–3 kg).
4. Squash the gel under the weight for 10–15 min then remove filter paper etc. and dry the gel using a hair dryer.
5. Stain the gel for 10–15 min in mature 0.1% Coomassie blue and destain as required until the background is clear.

Notes to the procedure

ad 1. The gel can be stained on the original glass plate or transferred to gel-bond plastic film.

ad 5. Coomassie blue ripens with age and should be replaced when destaining becomes difficult. The destaining process can be speeded up using warm destain.

2.4 Test planning and interpretation

After negative sera have been screened out, positive reactions can be characterised by re-running the test using sera of defined specificity in adjacent wells and demonstrating lines of immunological identity. The most efficient and economical use of sometimes rare reference sera should take into account the diagnosis and the immunofluorescent staining pattern produced by the antiserum on HEp-2 cells. Using this strategy the most appropriate antibodies are tried first. Some examples are given below.

The most common specificities in patients with Scleroderma or Polymyositis are Scl-70, PM-Scl and Jo-1. They give nuclear, nucleolar and cytoplasmic staining respectively by indirect immunofluorescence on HEp-2 cells and should be tested appropriately.

In SLE and Sjogrens syndrome the four most common antibodies are anti-nRNP, anti-Sm, anti-Ro and anti-La. They are found in various combinations, but frequently anti-nRNP is present with anti-Sm and anti-La with anti-Ro. The anti-RNP/ Sm pair correlate with the variable large speckled staining pattern whereas the anti-Ro/La pair give a fine speckled or diffuse pattern by indirect immunofluorescence on HEp-2 cells. Positive reactions that cannot be attributed to any of the antibodies already mentioned should be tested against other less common reference sera, anti-Ku anti-SL (Ki), anti-rRNP and anti-PCNA.

Solutions and suppliers

Solutions
- 0.15 M PBS pH 7.2
    - 8 g Sodium chloride
    - 2 g Potassium chloride
    - 1.15 g Disodium hydrogen phosphate (anhydrous)
    - 0.2 g Potassium dihydrogen phosphate
    - Make to 1 litre with distilled water

- 0.065 M Barbitone Buffer pH 8.3
    - Dissolve 3.1 g barbitone in 200 ml distilled water with heating
    - Add 9.8 g sodium barbitone and make to 1 litre with distilled water

- Coomassie blue stain
    - 1 g Coomassie brilliant blue R
    - 100 ml glacial acetic acid
    - 450 ml ethanol or methanol
    - 450 ml distilled water

- Destain solution
    - 100 ml glacial acetic acid
    - 250 ml ethanol or methanol
    - 650 ml distilled water

Suppliers of acetone powders and antigen extracts
Pel-Freez Biologicals Inc., PO Box 68, Rogers, AR 72756, USA.
C-Six Diagnostics Inc., 9653, North Granville Road, Mequon, Wisconsin 53092, USA. Distributed by Saxon Europe Ltd., P.O. Box 28, Newmarket, CB8 0BA, UK.
Immunovision Inc., 1506 Ford Ave., Springdale, AR72764, USA. Distributed by Imperial Laboratories (Europe) Ltd., West Portway, Andover, SP10 3LF, UK.
Inova Diagnostics Inc., 10451, Roselle St., San Diego, CA 92121, USA. Distributed by The Binding Site Ltd., Vincent Drive, Birmingham. B15 2SQ, UK.
Biodiagnostics Ltd., Rectory Road, Upton-upon-Severn, WR8 0XL, UK.

References

1. Tan EM (1989) Antinuclear antibodies: diagnostic markers for autoimmune disease and probes for cell biology. Adv Immunol 44:93–151
2. Ouchterlony O & Nilsson L-A (1978) Immunodiffusion and Immunoelectrophoresis. In: Weir DM (ed) Handbook of Experimental Immunology (3rd Ed) Vol. 1. Blackwell Scientific Publications, Oxford
3. Williams CA (1971) Immunoelectrophoretic Analysis. In: Williams CA & Chase MW (eds) Methods in Immunology and Immunochemistry Vol. 3, pp. 234–279. Academic Press, London
4. Isenberg DA & Maddison PJ (1987) Detection of antibodies to double stranded DNA and extractable nuclear antigen. J Clin Pathol 40:1374–1381
5. McCarty GA, Valencia DW & Fritzler MJ (1984) Antinuclear antibodies: contemporary techniques and clinical application to connective tissue diseases, pp. 70–80. Oxford University Press, Oxford
6. Prince AM & Burke K (1970) Serum Hepatitis antigen (SH), rapid detection by high voltage immunoelectrosmophoresis. Science 169:593–597
7. Johnson GD, Edmonds JP & Holborow EJ (1973) Precipitating antibody to DNA detected by two-stage electroimmunodiffusion. Lancet (ii) 883–885
8. Kurata N & Tan EM (1976) Identification of antibodies to nuclear acidic antigens by counterimmunoelectrophoresis. Arthr Rheum 19:574–579
9. Keiser HD & Weinstein J (1980) Detection and identification of antibodies to saline extractable nuclear antigens by counterimmunoelectrophoresis. Arthr Rheum 23:1026–1035
10. Wasicek CA & Reichlin M (1982) Clinical and serologic difference between systemic lupus erythematosus patients with antibodies to Ro versus patients with antibodies to Ro and La. J Clin Invest 69:835–843
11. Bunn CC, Gharavi AE & Hughes GRV (1982) Antibodies to extractable nuclear antigens in 173 patients with DNA-binding positive SLE: an association between antibodies to RNP and Sm antigens observed by counterimmunoelectrophoresis. J Clin Lab Immunol 8:13–17
12. Bernstein RM, Bunn CC & Hughes GRV (1982) Identification of antibodies to acidic antigens by counterimmunoelectrophoresis. Ann Rheum Dis 41:554–555
13. Bernstein RM, Neuberger JM, Bunn CC, Callender ME, Hughes GRV & Williams R (1984) Diversity of autoantibodies in primary biliary cirrhosis and chronic active hepatitis. Clin Exp Immunol 55:553–560
14. Mimori T, Akizuki M, Yamagata H, Inada S, Yoshida S & Homma M (1981) Characterisation of a high molecular weight acidic nuclear protein recognised by autoantibodies in sera from patients with polymyositis-scleroderma overlap. J Clin Invest 68:611–620
15. Clark G, Reichlin M & Tomasi TB (1969) Characterisation of a soluble cytoplasmic antigen reactive with sera from patients with systemic lupus erythematosus. J Immunol 102:117–122
16. Deuscher MP (1967) Rat liver glutamyl ribonucleic acid synthetase. J Biol Chem 242:1123–1131
17. Akizuki M, Powers R & Holman HR (1977) A soluble acidic protein of the cell nucleus which reacts with serum from patients with systemic lupus erythematosus and Sjogrens syndrome. J Clin Invest 59:264–272
18. Maggio R, Siekevitz P Palade GE (1963) Studies on isolated nuclei. J Cell Biol 18:267–302
19. Venables PJW, Smith PR & Maini RN (1983) Purification and characterisation of the Sjogrens syndrome A and B antigens. Clin Exp Immunol 54:731–738
20. Takano M, Agris PF & Sharp GC (1980) Purification and biochemical characterisation of nuclear ribonucleoprotein antigen using purified antibody from a patient with mixed connective tissue disease. J Clin Invest 65:1449–1456

21. Northway JD & Tan EM (1972) Differentiation of antinuclear antibodies giving speckled staining patterns in immunofluorescence. Clin Immunol Immunopathol 1:140–154
22. Nishikai M & Reichlin M (1980) Heterogeneity of precipitating antibodies in polymyositis and dermatomyositis. Arthr Rheum 23:881–888

Manual of Biological Markers of Disease **A4**: 1–25, 1993.

Protein blotting

R. VERHEIJEN[1], M. SALDEN[2] and W.J. VAN VENROOIJ[1]
[1] *Department of Biochemistry, University of Nijmegen, The Netherlands;*
[2] *Euro-Diagnostica, Apeldoorn, The Netherlands*

Key words: Autoantibody, autoantigen, autoimmune disease, protein blotting

1. Introduction

In rheumatic diseases several clinical entities are distinguished, many being syndromes with overlapping clinical features. For this reason, patients can evolve in the course of their disease from one diagnosis to another. In sera of many patients with rheumatic diseases highly specific autoantibodies to cellular macromolecules are detectable. Some of these autoantibodies are marker antibodies for certain diseases, e.g. antibodies reactive with double stranded DNA or the Sm antigen in systemic lupus erythematosus (SLE) [1–3]. Other examples are anti-DNA topoisomerase I (anti-Scl-70) antibodies present in sera from patients with diffuse systemic sclerosis [4], anti-centromere antibodies present in sera from patients with limited systemic sclerosis [4], and anti-Jo-1 antibodies present in sera from patients suffering from polymyositis/dermatomyositis [2, 5]. Some other autoantibodies are found in more than one syndrome, e.g. anti-U1 RNP in mixed connective tissue disease (MCTD) [1, 6] and SLE [7, 8], anti-Ro60/SS-A and anti-La/SS-B in Sjögren's syndrome and SLE [9, 10], and anti-histone in both SLE and drug-induced SLE [2]. Therefore, the determination of serological autoantibodies reactive with cellular antigens located in the nucleus or the cytoplasm can be helpful in the classification of patients with rheumatic diseases [11].

Over the years, polyacrylamide gel electrophoresis (PAGE) has proved to be a very powerful technique for analyzing complicated protein mixtures. The analytical power of this method has been greatly expanded by the protein blotting technique, in which proteins are first resolved by gel electrophoresis, followed by transfer to synthetic membrane supports. Thereafter, detection of proteins of interest on the blot is performed with specific antibodies or ligand reagents. One of the great advantages of protein blotting over gel techniques is that proteins on a blot are readily accessible to antibodies or other probes. Another important advantage is that the resolution of protein bands obtained during electrophoresis is retained on the membrane and not lost by diffusion during subsequent incubation steps. Furthermore, immunoblot assays require small reagent and volumes, relatively short processing times, and are generally easy to perform. Many applications and methodologies of protein blotting have been reviewed previously [12–20].

A typical blotting experiment consists of five steps:

1. Immobilization of proteins on a membrane, either by electrophoretic transfer from a gel or by direct application of protein solutions.
2. Blocking of unoccupied sites on the membrane to prevent non-specific binding of antibodies, being proteins themselves, to the membrane.
3. Probing of the blot for proteins of interest with specific, primary antibodies.
4. Labeling of the antibody-antigen complexes with secondary antibodies or ligand reagents, specific for the primary antibody type and conjugated to markers, i.e. enzymes or gold particles.
5. Visualisation of the enzyme-labeled protein bands by incubation with appropriate substrates to form insoluble, coloured products at the proteins' locations.
   In case of incubation with gold-conjugated reagents, the protein bands are directly visible without this last step.

Currently, the most frequently used methods for screening anti-nuclear antibodies (ANAs) are indirect immunofluorescence, double immuno-diffusion and counter-immunoelectrophoresis. They all have important technical drawbacks as poor methodological standardisation and lack of sensitivity, substrate variability, instability, insolubility and operator dependancy. The immunoblot is a useful alternative for simultaneously detecting all autoantibodies in a patients' serum in a highly sensitive and specific test that is easy to perform.

It is beyond the scope of this manual to exhaustively describe all possible methods for antigen detection on immunoblots. We will merely describe our own experience in making good and reliable immunoblots for the detection and identification of autoantibodies in human serum directed to nuclear or cytoplasmic antigens. Because of this limitation, some additional reading on new and improved detection methods remains recommendable. A comprehensive updated source of information for blotting experiments is offered by Bio-Rad Laboratories in a manual called "Protein blotting: a guide to transfer and detection".

2. The antigens

Crucial in the search for autoantibodies in human sera is the quality of the antigens on the immunoblot. A frequently used antigen source is a protein extract of HeLa S3 (human cervix carcinoma) cells. HeLa S3 cells can be grown in suspension at 37 °C at densities ranging between $0.2-1.0 \times 10^6$ cells/ml on Suspension Minimal Essential Medium (Flow Laboratories, U.K.; cat. no. 11–170–22) supplemented with 10%(v/v) newborn calf serum (Gibco, U.K.; cat. no. 021–06010 M), glutamine, vitamins and antibiotics.

Because of their low abundance in nuclear extracts and/or because of their instability, most of the nucleolar autoantigens are hardly detectable on conventional nuclear immunoblots. Significantly better results in screening patient sera for anti-nucleolar antibody specificities, however, can be obtained by immunoprecipitation from a sonicated HeLa cell extract which has been labeled with [^{35}S] methionine. Such labeling can be performed by incubating the cells for 16 h at 37 °C with $5-10\mu$Ci/ml [^{35}S] methionine (Amersham U.K.; cat. no. SJ 1015; $\pm$ 37 TBq/mmol = $\pm$ 1000 Ci/mmol) at densities of $1-2 \times 10^6$ cells/ml. For the first 2–3 h the cells are incubated in Minimal Essential Medium without methionine and glutamine (Flow Laboratories, U.K.; cat. no. 16–222–49) containing 5% (v/v) dialyzed fetal bovine serum (Hyclone Laboratories, Logan, U.K.; cat. no. A-1115-L) and [^{35}S] methionine. Subsequently, 0.1 volume of complete medium (medium plus 10% (v/v) newborn calf serum) is added.

Notes

– HeLa cells, nuclei and cytoplasm in large quantities are also commercially available from: Prof. Dr. A. Miller, Laboratoire de Biochimie Moléculaire et Unité de Biotechnologie Appliqée, Faculté de Médecine, Université de l'Etat, Avenue de Champ de Mars 24, 7000 Mons, Belgium. Tel. (65) 373549/50/51.
– Generally, several types of blots are made in order to simplify the antibody profiles of the sera.
   Nuclear blots: contain an extract of HeLa cell nuclei and can be used for the detection of antibodies against RNP, Sm, La (SS-B), DNA topoisomerase I (Scl-70), centromere proteins, NOR-90, PM-Scl, Ku, histones, NuMA and other antigens.
   Cytoblots: contain a cytoplasmic extract of HeLa cells and can be used for the detection of antibodies against Ro60 (SS-A), Ro52, La (SS-B), Jo-1, PCNA, PDH, ribosomal RNP and other antigens.

2.1 Preparation of nuclear extracts

Use only chemicals of analytical grade.

Cell fractionations should be performed in the presence of 0.5 mM phenylmethylsulfonyl fluoride (PMSF) to reduce proteolytic degradation. This agent can be stored as a 125 mM stock solution in 2-propanol (see Appendix).

The solution of ribonuclease A (RNase A; Sigma Chemical Co, München) should be pre-incubated for 15 min at 100 °C to reduce possible protease activity.

Unless indicated otherwise, centrifugation steps are carried out for 5 min at 800 g and +4 °C.

Steps in the procedure
1. Cells (1 × 10⁹) are harvested by centrifugation and washed twice with isotonic NKM solution (130 mM NaCl, 5 mM KCl, 1.5 mM MgCl$_2$) to remove all culture medium.
2. The cells are resuspended in 22.5 ml buffer A (10 mM NaCl, 10 mM Tris-HCl pH7.4, 1.5 mM MgCl$_2$), and 2.5 ml of a freshly prepared mixture of 5% (w/v) sodium deoxycholate (DOC)/10% (v/v) Tween 40 in buffer A is added (i.e. add 1.25 ml 10% (w/v) DOC to 1.25 ml 20% (v/v) Tween 40, mix and add this to the resuspended cells).
3. Subsequently, the cells are dounced 7 times using an S-pestle (1 time is up and down). Check under a microscope if all cells are damaged. If there are still many intact cells, continue douncing.
4. The nuclei are pelleted (10 min at 3000 g and +4 °C), washed twice with 50 ml buffer A and incubated with 500 μl DNase I (Sigma, 2 mg/ml in buffer A), 40 μl RNase A (5 mg/ml in buffer A), 625 μl buffer B (110 mM NaCl, 10 mM Tris-HCl pH7.5, 1.5 mM MgCl$_2$) and 10 μl PMSC for 30 min at 20 °C.
5. After incubation, the DNA-depleted nuclei are dissolved by adding 2 ml of 2× concentrated sample buffer (4% (w/v) sodium dodecyl sulphate (SDS), 10% (v/v) 2-mercaptoethanol, 20% (v/v) glycerol, 0.01% (w/v) Bromophenol Blue, 125 mM Tris-HCl pH6.8), followed by heating the suspension for 3 min at 100 °C.
6. After centrifugation for 10 min at 3000 g and 20 °C, the supernatant is used as nuclear extract. From such an extract, 250 μl (± 60 × 10⁶ nuclei) should be used on one 14 × 13 × 0.15 cm SDS-polyacrylamide gel.

Note
– In order to obtain a higher yield of nucleolar antigens, the isolated nuclei can be sonicated instead of treated with DNase I and RNase A. In that case the dounced and washed nuclei are resuspended in 3 ml buffer A and sonicated 5 times (30 s) on ice with a Branson sonifier (Branson Instr. Inc., Connecticut, U.S.A.) with a microtip at level 4. Subsequently, the sonicated nuclei are dissolved in 4× concentrated sample buffer.

2.2 Preparation of cytoplasmic extracts

Steps in the procedure
1. Cells (1 × 10⁹) are harvested by centrifugation and washed twice with isotonic NKM buffer (130 mM NaCl, 5 mM KCl, 1.5 mM MgCl$_2$) and once with hypotonic HEPES buffer (10 mM HEPES pH7.9, 10 mM KCl, 1.5 mM MgCl$_2$, 0.5 mM dithiotreitol (DTT)).

2. The cells are swollen for 30 min at + 4 °C in two volumes of HEPES buffer and subsequently homogenized by douncing 7 times using an S-pestle. Check under a microscope if all cells are damaged.
3. Subsequently, 0.1 volume of isotonic buffer is added (300mM HEPES pH7.9, 1.4 mM KCl, 30 mM MgCl$_2$). The nuclei are pelleted by centrifugation for 5 min at 3000 g and + 4 °C.
4. To the supernatant containing the cytoplasmic antigens 1/3 volume of a 4x concentrated sample buffer (8% (w/v) SDS, 20% (v/v) 2-mercaptoethanol, 40% (v/v) glycerol, 0.02% (w/v) Bromophenol Blue, 250 mM Tris-HCl pH6.8) is added, followed by heating for 3 min at 100 °C.
5. After centrifugation for 10 min at 3000 g and 20 °C, the supernatant is used as cytoplasmic extract. From such an extract, 250 μl (cytoplasmic proteins of $\pm$ 60 $\times$ 10^6 cells) should be used on one 14 $\times$ 13 $\times$ 0.15 cm SDS-polyacrylamide gel.

3. SDS polyacrylamide gel electrophoresis

Nuclear proteins can be best separated on 13% or 15% polyacrylamide gels, whereas the separation of most cytoplasmic antigens is best on 10% polyacrylamide gels. The volumes of reagents used to cast two polyacrylamide separation gels are indicated in Table 1.

Table 1. Volumes of reagents used to cast two polyacrylamide gels ($14 \times 13 \times 0.15$cm)

Reagents	polyacrylamide concentration		
	10%	13%	15%
40% (w/v) acrylamide : 1.07% (w/v) bisacrylamide	17.3ml	22.3ml	26.0ml
1.5M Tris-HCl pH8.8	19.8ml	19.8ml	19.8ml
H$_2$O	31.2ml	26.2ml	22.5ml
100% glycerol	1.0ml	1.0ml	1.0ml
10% (w/v) SDS	800μl	800μl	800μl
10% (v/v) TEMED	250μl	250μl	250μl
10% (w/v) ammonium persulfate	160μl	160μl	160μl

Acrylamide (crystallized 2 $\times$: Serva, Heidelberg, Germany; cat. no. 10675) and bisacrylamide (N,N'-methylene-bisacrylamide: Bio-Rad; cat. no. 161–0200/161–0201) are dissolved as a mixture. The solution is filtrated and stored in the dark at + 4 °C.

After casting, some saturated 1-butanol is layered on top of the gel. Just before use, the butanol is removed and a stacking gel of about 0.5 cm high is put on top of the polymerized separation gel. The volumes of reagents used to cast two 4% polyacrylamide stacking gels are indicated in Table 2.

Table 2. Volumes of reagents used to cast two stacking gels (14 × 0.5 × 0.15 cm)

Reagents	4%
40% (w/v) acrylamide : 1.07% (w/v) bisacrylamide	1.0ml
0.5M Tris-HCl pH6.8	2.4ml
H$_2$O	5.6ml
100% glycerol	700μl
10% (w/v) SDS	100μl
10% (v/v) TEMED	100μl
10% (w/v) ammonium persulfate	50μl

3.1 Buffer selection

The most common buffer system for SDS-polyacrylamide gel electrophoresis is the Laemmli buffer system [21]. The composition of these buffers is:

Lower buffer compartment (anode): 192 mM glycine, 25 mM Tris-base
Upper buffer compartment (cathode): 192 mM glycine, 25 mM Tris-base, 0.1% (w/v) SDS

3.2 Power conditions

Under normal running conditions the stacking gel is passed using a constant current of 15 mA per gel. When the Bromophenol Blue front enters the separation gel, the current can be increased to 30–40 mA per gel. When using such high currents the gel system needs to be cooled with running tap water. The total running time will vary between 4–6 h.

When the run is performed overnight during at least 16 h, a constant current of about 7–10 mA per gel is sufficient. In this case cooling of the gel sytem with running tap water is not necessary.

Separation of the proteins is completed when the Bromophenol Blue front has reached the bottom of the gel.

4. Protein blotting

Once a protein mixture is separated on a polyacrylamide gel, the components can be transferred to a membrane. One way to do this is by using an electric field. Electrophoretic transfer is fast and efficient, maintaining the high resolution obtained with gel electrophoresis. There are two main types of electroblotting apparatus:

1. 'Semi-dry' transfers with horizontally placed electrodes.
2. Tanks of buffer with vertically placed wire or plate electrodes.
   For characteristics and differences between semi-dry blotting and tank blotting systems we refer to the protein blotting guide of Bio-Rad.

4.1 Membrane selection

A variety of membranes is available for immunoblotting applications. The physical properties and performance characteristics of a membrane should be evaluated in selecting the appropriate transfer conditions.

A suitable general-purpose membrane is nitrocellulose. This type of membrane with a binding capacity of 80–100 μg protein/cm^2 is available in two pore sizes. We routinely use the nitrocellulose membrane of Schleicher & Schuell: BA 85, 0.45 μm (Dassel, Germany; cat. no. 401196).

Notes
- Supported nitrocellulose (Bio-Rad; cat. no. 162-0090/162-0097) consists of pure nitrocellulose cast on an inert synthetic support. The great advantage of this membrane is the increased mechanical strength as compared to ordinary nitrocellulose.
- Protein binding on nitrocellulose is achieved mainly on the basis of hydrophylic interactions. Binding of proteins on PVDF (polyvinylidene difluoride) membrane (Bio-Rad; cat. no. 162-0180/162-0185) is achieved mainly on the basis of hydrophobic interactions and is under certain circumstances preferable [22]. PVDF membrane has a high mechanical strength and chemical stability, and a binding capacity of 170–200 μg/cm^2.

4.2 Blotting filter paper

Blotting filter paper, made of 100% cotton fiber, provides a uniform current through the gel. This paper contains no additives that might interfere with the transfer process. Very suitable blotting paper is available from Bio-Rad (thin, thick, extra thick) and from Schleicher & Schuell (GB 002).

4.3 Buffer selection

The composition of the most commonly used transfer buffer, originally described by Towbin [23], is:

192 mM glycine, 25 mM Tris-base, 20% (v/v) methanol, pH8.3. Addition of 0.015% (w/v)–0.1% (w/v) SDS improves the elution of high molecular weight proteins, but may interfere with the binding of the proteins to the membrane. When SDS is used, the buffer needs to be extensively stirred during the transfer.

Alternative buffers of Bjerrum and Schafer-Nielsen [24] (39 mM glycine, 48 mM Tris-base, 20% (v/v) methanol, pH9.2), and Dunn [25] (10 mM NaHCO$_3$, 3 mM NaCO$_3$, 20% (v/v) methanol, pH9.9), at a higher alkaline pH and lower conductivity than the Towbin buffer, are recommended for semi-dry transfers. Our experience with the Towbin buffer system in semi-dry transfer, however, is also very good.

4.4 Semi-dry blotting

In our laboratory, semi-dry blotting is performed using the Towbin buffer system (192 mM glycine, 25 mM Tris-base, 20% (v/v) methanol, 0.1% (w/v) SDS, pH8.3) and the Novablot electrophoresis transfer kit of Pharmacia (cat. no. 18 1016 86). In this apparatus, and other comparable semi-dry blotting devices, the lower graphite plate is the anode and the upper the cathode (see Fig. 1).

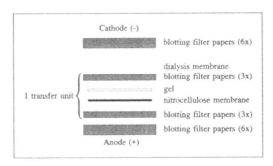

Fig. 1. Schematic representation of the composition of a transfer unit in semi-dry blotting.

Blotting procedure
1. Rinse the graphite plates with deionized water.
2. Place 6 layers of blotting filter paper (15.5 × 15 cm) which have been soaked in blotting buffer on the anodic graphite plate.
3. Assembly of a transfer unit: Place 3 layers of soaked blotting filter papers (15.5 × 15 cm) on top of the 6 layers, followed by a wet nitrocellulose membrane (15.5 × 15 cm). The gel (14 × 13 cm) is then placed carefully to the nitrocellulose membrane. Avoid air-bubbles. When air-bubbles occur, wet the gel with some blotting buffer, and gently push out the bubbles by rolling a glass tube (∅ 2–4 cm) over the gel. Place three soaked filter papers on top of the gel to complete the transfer unit.
4. If more than one gel is to be transferred during the run, rinse a dialysis membrane (20 × 20 cm) (Pharmacia; cat. no. 80 1129 38) in water and place it on top of the filter papers. The dialysis membrane prevents any possible cross-contamination between the individual transfer units. Continue to assemble another transfer unit, starting with three soaked filter papers. When all the transfer units have been assembled, place six soaked filter papers on top of the stack of transfer units.
5. Finally, cover the apparatus with its lid to which the cathode is attached.
6. The run is performed with a constant current of 0.8 mA/cm² of gel (= 150 mA/6V) for 1 h at room temperature.

Notes
- When assembling the transfer units, always wear gloves.
- To obtain optimal transfer of proteins from the gel, it is extremely important to avoid trapping of air-bubbles between any of the components of the transfer units.
- The applied current is irrespective of the number of transfer units.
- It is our experience that a simultaneous transfer of more than three gels in the semi-dry blotting technique may lead to a considerable loss of quality of the blots.

4.5 Tank blotting

In tank blotting the transfer unit is held vertically in a porous, non-conducting cassette. Composition of one transfer unit (see Fig. 2):
- a plastic mould
- a fiber pad
- a sheet of blotting filter paper (17 × 16 cm)
- the gel (14 × 13 cm)
- a sheet of nitrocellulose membrane (15.5 × 15 cm)
- a sheet of blotting filter paper (17 × 16 cm)
- a fiber pad
- a plastic mould

Prior to assembling the transfer unit, the fiber pads, blotting filter paper and nitrocellulose membrane should be soaked in blotting buffer.

The transfer units are submerged in buffer between platinum electrodes mounted on opposite walls of the tank with the nitrocellulose membrane

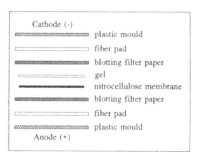

Fig. 2. Schematic representation of the composition of a transfer unit in tank blotting. During blotting the transfer units are placed vertically.

facing the anode and the gel facing the cathode. The amount of transfer buffer varies between 3–6 l.

In our laboratory, tank blotting is performed in 6 l Towbin buffer with platinum wire electrodes placed 10 cm apart. Using this system, three gels can be blotted simultaneously. With this large amount of transfer buffer, cooling is not necessary.

Transfer in the absence of SDS is carried out for approximately 16 h at a constant current of 250 mA. In the presence of 0.015% (w/v) SDS, the transfer time can be reduced to 1–3 h.

The same buffer can be used for several transfers.

Notes
See notes in 4.4.

4.6 Total protein staining

Total protein staining of a membrane provides a visual image of the complete protein pattern. This will give valuable information on the quality of the blot concerning the degree of separation of the proteins, the effect of "smiling" of the gel and the presence of bad sections where air-bubbles have obstructed an accurate transfer of proteins.

From the many reagents used for total protein staining we mention:
– Anionic dyes, i.e. amido black, Coomassie brilliant blue R-250 and fast green FCF.
– Colloidal gold. With this method all proteins are stained by binding of colloidal gold particles to the proteins, revealing them as reddish bands.
– Biotin. In this case, all proteins are biotinylated with NHS-biotin (N-hydroxy-succinimide ester of biotin), followed by the use of the avidin horseradish peroxidase conjugate system to detect the biotinylated proteins.

For comparison of the staining techniques using these reagents we once again refer to the blotting guide of Bio-Rad.

A very useful, yet not well-known dye for staining all proteins on a blot is Ponceau S. Although the staining by Ponceau S is not as sensitive as staining by colloidal gold or biotin, the procedure is fast and very suitable for checking the quality of the blot.

Ponceau S staining solution is prepared by mixing 20 ml 2% (w/v) Ponceau S concentrate in 30% (w/v) trichloroacetic acid and 30% (w/v) sulfosalicylic acid (Sigma; cat. no. P 7767) with 180 ml deionized water. This solution is stable for at least 1 month when stored at room temperature.

Staining of the blot is performed by shaking it in the Ponceau S solution for 1 to 2 min. Thereafter, the blot is destained by washing it in tap water until the protein bands have the proper reddish appearance. When the blot is washed in 5% (v/v) acetic acid, Ponceau S will be more stably bound to the proteins and cannot be easily removed from the blot again.

4.7 Blocking reagents

Following transfer and total protein staining, unoccupied binding sites on the blot must be blocked to prevent non-specific binding of the primary antibodies. Failure to adequately saturate the membrane leads to high backgrounds. Suitable blocking solutions are:
– 3% (w/v) BSA (fraction V), 350 mM NaCl, 10 mM Tris-HCl pH7.5
– 5% (w/v) non-fat dry milk, 500 mM NaCl, 20 mM Tris-HCl pH7.5
– 3% (w/v) gelatin, 500 mM NaCl, 20 mM Tris-HCl pH7.5

Notes
– A blocking solution consisting of 2% (w/v) non-fat dry milk in 50 mM Tris, 150 mM NaCl, 0.5% (v/v) Tween 20, pH10.3 is recommended when using the alkaline phosphatase (AP) system for detection of the antibody-antigen complexes.
– Our experience with 0.05% (v/v) Tween 20 in phoshate buffered saline (PBS) pH7.4 is that this solution is only suitable as blocking reagent when using (mouse) monoclonal antibodies.

4.8 Serum samples

The blot should be probed with serum. Lipemic, hemolyzed or icteric serum should not be used. Also, bacterially contaminated specimens should not be used. Serum can be decomplemented by heating for 30 min at 56 °C. Serum samples can be stored at + 2–8 °C for up to 3 days. For prolonged storage, freeze at − 70 °C. Avoid repeated freezing and thawing.

Note
– Freezing for prolonged times at − 20 °C may introduce lyophilization effects.

4.9 Immunoblot assay

For analyzing numerous serum samples on a single blot, a blot-incubator is a very useful, but not essential tool. Blot-incubators for blots of 15.5 × 15 cm are available from Euro-Diagnostica (Apeldoorn, The Netherlands) or from Immunetics (Cambridge, U.S.A.). When such an incubator is unavailable at your lab, the blot can also be cut into strips of approximately 0.5 cm in width, whereafter each strip can be incubated separately with the serum of interest.

Optimal performance of any specific detection of antibody-antigen complexes requires the appropriate reagents at each step of the process. We shall focus on three detection systems:

- ^{125}I conjugated reagents
- Horseradish peroxidase (HRP) system
- Alkaline phosphatase (AP) system

Assay procedure

1. Pre-incubate the blot in blocking buffer for 0.5–1 h at room temperature with gentle agitation.
2. Mount the wet blot in the blot-incubator with the antigen-bearing side facing the channels of the incubator. Make sure that the blot covers the full-length of each channel to be used. If all channels of the incubator will not be used, the blot may be cut to any width.
3. In a routine screening, when the antibody titers are unknown, a serum dilution of 1:30 in radioimmunoassay (RIA) buffer (0.3% (w/v) BSA, 150 mM NaCl, 10 mM Tris-HCl pH7.5, 1% (v/v) Triton X-100, 0.5% (w/v) DOC, 0.1% (w/v) SDS) is suitable to start with. When using reference sera with known high antibody titers, however, higher dilutions will often give better results.
4. Pipet the diluted serum samples in the channels of the blot-incubator and incubate for 1 h at room temperature with gentle agitation on a rocking platform.
5. The sera can be discarded from the lanes by vacuum aspiration. By using the special detachable washing devices, all channels can be washed simultaneously. The washing solutions for the various methods are:
   - RIA buffer in the ^{125}I system.
   - TP buffer (0.5% (v/v) Triton X-100 in PBS) in the HRP system.
   - TNT buffer (50mM Tris-HCl pH10.3, 150 mM NaCl, 0.5% (v/v) Tween 20) in the AP system.
6. Remove the blot from the incubator using tweezers, place in a plastic tray and wash twice for 5 min with gentle agitation.
7. Remove the washing buffer and incubate the blot with the appropriate secondary antibody for 1 h at room temperature with gentle agitation. *In the ^{125}I system:*
   - ^{125}I-labeled sheep anti-human Ig (whole antibody) (Amersham U.K.; cat. no. IM133; 28-111TBq/mmol = 750–3000 Ci/mmol) in a 1:1000 dilution in RIA buffer or:
   - ^{125}I-labeled sheep anti-human Ig (F(ab')$_2$ fragment) (Amersham U.K.; cat. no. IM1330; 19-74Tbq/mmol = 500–2000 Ci/mmol) in a 1:1000 dilution in RIA buffer.

In the HRP system:
- Goat anti-human peroxidase-conjugated Ig (IgG, IgA, IgM, Fc and Fab) (Nordic Immunological Laboratories, Tilburg, The Netherlands) in a 1:1000 dilution in TP buffer with 0.5% (w/v) BSA or:
- Isotype specific anti-human peroxidase-conjugated antibodies (Nordic) in 1:1000 dilutions in TP buffer with 0.5% (w/v) BSA.

In the AP system:
- Rabbit anti-human alkaline phoshatase-conjugated antibodies (IgA, IgG, IgM) (DAKOpatts, Glostrup, Denmark; cat. no. D342) in a 1:500 dilution in TNT buffer with 0.33% (w/v) non-fat dry milk or:
- Isotype specific anti-human alkaline phosphatase-conjugated antibodies (DAKOpatts) in 1:500 dilutions in TNT buffer with 0.33% (w/v) non-fat dry milk.

8. Remove the secondary antibody solution and wash the blot three times for 10 min with the same washing buffer as indicated in step 5. When TP buffer is used, a fourth wash of 5 min in PBS is recommended.

9. When using the ^{125}I system the blot can be dried between filter paper, whereafter the antigen-antibody complexes are visualized by autoradiography.

When using the HRP or AP system, the staining reaction is performed as follows:

Staining with 4-chloro-1-naphtol in the HRP system:
- Dissolve 50 mg 4–chloro-1-naphtol (Merck, Darmstadt, Germany; cat. no. 11952.0010) in 1ml 100% ethanol and mix with 99 ml deionized water and 40 μl 30% (w/w) H_2O_2.
- Clarify the solution by filtration and use immediately for staining.
- The reaction is terminated by washing the blot in deionized water.
- The blot is air-dried between filter paper and must be stored in the dark as the blue-purple stain is light-sensitive.

Staining with diaminobenzidine in the HRP system:
- Dissolve 40 mg 3,3' diaminobenzidine (Sigma; cat. no. D-8001) in 100 ml 100 mM sodium citrate buffer pH5.0.
- Add 30 μl 30% (w/w) H_2O_2 and use the solution immediately for staining.
- The reaction is terminated by washing the blot in deionized water.
- The blot is air-dried between filter paper.

Staining with BCIP/NBT in the AP system:
- Prepare the following substrate stock solutions:
  25 mg/ml BCIP (5-bromo-4-chloro-3-indolyl phosphate, p-toluidine salt: Research Organics Inc. Cleveland, U.S.A.; cat. no. 1181 B) in 100% dimethylformamide (DMF)
  50 mg/ml NBT (p-nitro blue tetrazolium chloride: Research Organics Inc.; cat. no. 0414 N) in 70% (v/v) DMF.
  Store both solutions in small aliquots in the dark at -20 °C.

- Mix 330 μl of the NBT stock solution with 50 ml TNM buffer (100 mM Tris-HCl pH9.5, 100 mM NaCl, 5 mM MgCl$_2$) and then add 330 μl of the BCIP stock solution. Use immediately.
- The reaction is terminated by washing the blot in deionized water.
- The blot is air-dried between filter paper.

Notes
- There are several excellent alternatives for the secondary antibodies indicated in step 7 from, for example, DAKOpatts, Kirkegaard & Perry Laboratories (Gaithersburg, Maryland, U.S.A.) and Bio-Rad Laboratories for both the HRP and AP system. When using these antibodies, please follow the instruction manual for appropriate dilutions.
- To reduce background staining, the colour development should be performed in a clean container which is not contaminated with secondary antibodies.
- The staining reaction in the AP system is rather slow, but the blot can be incubated for a long period (overnight) without risk of overstaining. Using the HRP system, the staining reaction responds more rapidly with a possibility of overstaining.
- Azide is a potent inhibitor of horseradish peroxidase. For this reason, all buffers and solutions should be azide-free.

5. Antigen appearance on the immunoblot

5.1 Nuclear antigens

Sm and RNP proteins. Anti-Sm antibodies are directed to the B (28kD), B' (29kD) and D (16kD) proteins (Fig. 3A and 3B, lanes 3). In some anti-Sm sera, also, antibodies to the E (12kD), F (11kD) and G (9kD) proteins can be found [26]. Occasionally, a 27kD degradation product of B/B' is visible as well. In conventional polyacrylamide gels the D protein is seen as a single band, sometimes accompanied by the so-called D' protein which migrates slightly faster than the D protein. Recently, Lehmeier et al [27] have shown that when using an increased TEMED concentration of 0.4% (v/v) instead of the usual 0.04% (v/v), the D' band is clearly separated into two bands. All three proteins, designated D$_1$ (16kD), D$_2$ (16.5kD) and D$_3$ (18kD) appear to be antigenic. On nuclear immunoblots, anti-Sm sera often decorate an Sm core protein of approximately 70kD which is not identical to the U1–70K protein [28]. It probably is identical to the newly detected p69 Sm antigen (R. Lührmann, personal communication). In cytoplasmic extracts anti-Sm antibodies detect an additional 59kD protein, in this paper referred to as the p59 Sm protein.

The Sm autoantibody specificity is often found in association with low titers of anti-RNP antibodies [2, 8].

Anti-RNP antibodies are directed to one or more of the U1 RNP specific proteins U1–70K (70kD), U1-A (34kD) and U1-C (22kD) (Fig. 3A and 3B, lanes 1) [6,8]. On immunoblots, the U1–70K protein is often seen as a triplet [28], whereas the U1-C band may have a very broad appearance, sometimes separating into at least two bands.

Besides anti-Sm and anti-RNP antibody systems, another autoantibody specificity reacting with snRNP proteins has been described. These so-called anti-U1/U2 RNP sera contain antibodies against the U2 RNP specific U2-B" protein (28.5kD) and, in some cases, also antibodies against the U2 RNP specific U2-A' protein (31kD) (Fig. 3A and 3B, lanes 2). The anti-U1 reactivity in such sera is due to cross-reactivity of anti-(U2-B") antibodies with the U1-A protein. This antibody specificity is found in approximately 10% of all anti-RNP sera [8].

Centromere proteins. Anti-centromere antibodies (ACA) are directed to at least three immunologically related centromere antigens, CENP-A (17–19kD), CENP-B (80kD), and CENP-C (140kD) [29–31]. A fourth antigen, CENP-D (50kD) has been described as well [32].

CENP-A has been shown to be a centromere-specific histone H3 variant [33]. CENP-B is a highly acidic DNA-binding protein that recognizes a specific 17bp sequence in α-satellite DNA [34, 35]. The function of CENP-C, a component of the inner kinetochore plate, is still unknown [36]. CENP-D is probably identical to the mitosis specific autoantigen described by Hadlaczky *et al.* [37].

The most prominent band on nuclear immunoblots is that of CENP-A (Fig. 3A, lane 6). Because of the low abundancy of the other CENP antigens in nuclear extracts, these proteins are hardly detectable on nuclear blots when using either the HRP- or the AP-detection system. Considerably better results in the detection of anti-CENP-B antibodies are obtained when ^{125}I-labeled secondary antibodies are used [38], but even then CENP-C is almost or entirely undetectable.

DNA-topoisomerase I. DNA topoisomerase I (topo I) is an enzyme that catalyses interconversions between different topological forms of DNA, by creating a transient single-stranded nick in the DNA backbone, passing the unbroken strand of the DNA through the nick and resealing the original scission [39–41].

Autoantibodies to topo I have been described by several investigators [42–47]. The reported apparent molecular weight of the antigen varies from 86kD [45] to 95–100kD [42, 48]. Because of the discrete immunoreactive degradation product of 70kD, which is also often found on nuclear immunoblots, the antigen was formerly designated as the Scl-70 antigen (Fig. 3A, lane 7) [49]. Although anti-topo I antibodies of the IgG type are most common, IgA autoantibodies are frequently found as well [50, 51].

It is the general experience that anti-topo I antibodies rarely coexist with anti-centromere antibodies [52].

Ku (p70/p80) antigen. The Ku autoantigen is a DNA-binding nuclear protein complex consisting of one 80kD (p80) and one 70kD (p70) subunit [53–55]. Ku is identical to nuclear factor IV (NF IV), providing evidence for a possible role in DNA replication, repair or recombination [56].

The abundance of Ku in nuclear extracts and the fact that both proteins of the heterodimer are antigenic make the appearance of a 70kD/80kD

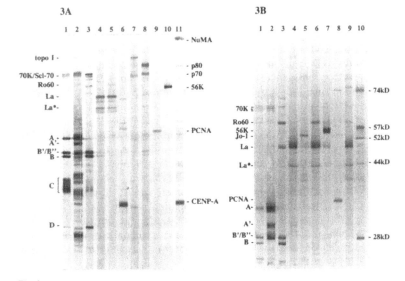

Fig. 3.

A. Analysis of human autoimmune sera on a nuclear blot.

HeLa nuclear proteins were separated on a 13% polyacrylamide gel and transferred to nitrocellulose. Subsequently, the blot was probed with various patient sera (dilution 1:30). Detection of the antibody-antigen complexes was performed using the AP detection system.

lane 1: anti-U1 RNP + anti-Sm serum H129, recognizing the U1 RNP specific proteins 70K (70kD), A (34kD) and C (22kD), as well as the Sm antigens B (28kD), B' (29kD) and D (16kD). Notice the extremely broad appearance of the C protein.

lane 2: anti-U1/U2 RNP serum V26, recognizing the U1 RNP specific proteins 70K and A, as well as the U2 RNP specific proteins A' (31kD) and B'' (28.5kD). Several undefined proteins are stained as well.

lane 3: anti-Sm serum C45, recognizing the Sm antigens B, B' and D. A weak staining reaction with the C protein is also seen. The detected antigen of ~70kD is not the U1–70K protein, but most probably the p69 Sm antigen [R. Lührmann, personal communication]

lane 4: anti-La serum W10, recognizing the La protein (50kD) and its 43kD degradation product (La*). The 28kD degradation product of La is not seen.

lane 5: anti-La + anti-Ro serum S67, recognizing La, La* and Ro60 (60kD).

lane 6: anti-centromere serum H33, recognizing CENP-A (17–19kD). Both CENP-B (80kD) and CENP-C (140kD) are not seen on this blot.

lane 7: anti-DNA topoisomerase I serum Z3, recognizing the intact topo I (110kD) and the 70kD degradation product (Scl-70).

lane 8: anti-Ku serum M4, recognizing both p70 (70kD) and p80 (80kD) of the Ku complex.

lane 9: anti-PCNA serum, recognizing PCNA (36kD).

lane 10: anti-56K serum P217, recognizing the 56K protein.

lane 11: anti-NuMA + anti-centromere serum F1, recognizing both the NuMA protein (250kD) and CENP-A (17–19kD).

B. Analysis of human autoimmune sera on a cytoblot.

HeLa cytoplasmic proteins were separated on a 10% polyacrylamide gel and transferred to nitrocellulose. Subsequently, the blot was probed with various patient sera (dilution doublet on nuclear immunoblots very characteristic for anti-Ku autoantibodies (Fig. 3A, lane 8). However, sera with autoantibodies to each of the individual protein components of the Ku complex also occur.

Proliferating Cell Nuclear Antigen (PCNA). PCNA or cyclin is a 36kD protein which can be detected on both nuclear and cytoplasmic immunoblots (Fig. 3A, lane 9; Fig. 3B, lane 8). Expression of the antigen increases 2 to 3-fold in late G and S phase. The antigen has been shown to be identical to an auxiliary protein for DNA polymerase delta [57, 58].

Histones. Histones are small, highly basic proteins. On SDS polyacrylamide gels they migrate slower than would be expected on the basis of their known molecular weights. This is due to a reduced overall-negative charge, a result of the high proportion of basic amino acids in histones. The practical consequence of this phenomenon is that the molecular weights of histones are over-estimated if calculated solely by comparison of histone mobility with that of standard, non-basic proteins in SDS-PAGE.

When using 18% SDS polyacrylamide gels, five major histones can be resolved, i.e. H1 (30kD), H3 (17kD), H2B (16.5kD), H2A (16kD) and H4 (14kD). The H1 antigen is mostly seen as a doublet, which on 13% SDS polyacrylamide gels migrates even slower than the 34kD U1-A protein.

The frequency of autoantibodies to the individual histones varies according to the type of autoimmune disease but, in general, anti-H1 antibodies are the most frequent, followed by anti-H2B, anti-H2A, anti-H3 and anti-H4 antibodies, respectively [59–61, reviewed in 62].

1:30). Detection of the antibody-antigen complexes was performed using the AP detection system.

lane 1: anti-U1 RNP + anti-Sm serum H129 (see also Fig. 3A, lane 1) Notice that on this blot the U1–70K protein is seen as a triplet.

lane 2: anti-U1/U2 RNP serum V26 (see also Fig. 3A, lane 2).

lane 3: anti-Sm serum C45 (see also Fig. 3A, lane 3).
The antigen of ~ 60kD is not Ro60 but an Sm protein, known as the p59 Sm antigen.
Notice that the protein of ~ 50kD migrates slightly faster than La, whereas the characteristic 43kD degradation product of La is not present. This indicates that this 50kD antigen is an as yet undefined protein.

lane 4: anti-La serum W10 (see also Fig. 3A, lane 4).

lane 5: anti-Jo-1 serum K160, recognizing the 54kD Jo-1 antigen.

lane 6: anti-La + anti-Ro serum S67 (see also Fig. 3A, lane 5).

lane 7: anti-56K serum P217 (see also Fig. 3A, lane 10). A weak anti-La activity is also detected.

lane 8: anti-PCNA serum, recognizing PCNA (36kD) and an undefined 74kD protein (see also Fig. 3A, lane 9).

lane 9: anti-La + anti-Ro serum B125, recognizing La, La*, Ro60 and some undefined high molecular weight proteins.

lane 10: anti-mitochondrial protein serum, recognizing four proteins of 74kD, 57kD, 52kD and 44kD, respectively. An undefined 28kD protein is stained as well.

Nuclear Mitotic Apparatus (NuMA) protein. The NuMA protein is a non-histone chromosomal phosphoprotein of approximately 250kD that, during mitosis, becomes associated with the mitotic spindle poles [63, 64]. The protein may have a role in post-mitotic reassembly [64].

Anti-NuMA antibodies have been described in sera of patients having different rheumatic diseases [63] and can be best detected on nuclear immunoblots (Fig 3A, lane 11).

5.2 Nucleolar antigens

U3 RNP. The U3 RNP particle is involved in ribosomal RNA processing [65, 66]. The complex is composed of U3 RNA (217nt) and at least 6 proteins (74kD, 59kD, 34kD, 30kD, 13kD, 12,5kD) with the 34kD, termed fibrillarin, being the antigenic target [65, 67]. Fibrillarin can be best detected on nuclear immunoblots.

Anti-U3 RNP antibodies rarely coexist with either anti-centromere antibodies or anti-topo I antibodies.

Th RNP. The human mitochondrial RNA processing (MRP) RNP complex (also termed Th RNP, To RNP or 7–2 RNP) is composed of an RNA component (267nt) and at least 6 proteins (120kD, 40kD, 30kD, 29kD, 23kD, 18kD) with the 40kD polypeptide being the antigenic target [4, 68–71]. MRP RNP appears to be localized in the mitochondria ($\sim$ 1%) as well as in the granular component of nucleoli ($\sim$ 99%). In mitochondria the complex is responsible for an endoribonucleic cleavage in mitochondrial primer RNA, which is thought to be involved in replication of mitochondrial DNA [72, 73]. In the nucleolus, Th RNP is believed to be involved in processing of preribosomal RNA.

Interestingly, sera with anti-Th activity also immunoprecipitate RNase P, an RNP endonuclease that processes the 5' termini of pre-tRNAs [74]. This indicates that Th RNP and RNase P share at least one common or antigenically related polypeptide. The RNA components of both RNases show sequence similarity to each other in a few short blocks [69].

The Th antigen is not detectable on nuclear immunoblots.

RNA polymerase I. RNA polymerase I is an enzyme complex selectively transcribing the nucleolar genes that code for precursor RNA of 28S, 18S and 5.8S ribosomal RNA. The complex is composed of at least 13 polypeptides with molecular weights ranging from 12.5kD to 210kD with the 210kD polypeptide being the antigenic target [75]. However, sera with anti-RNA polymerase I activity that recognized other proteins of the RNA polymerase I complex have been described as well [76]. The antigens are hardly detectable on nuclear immunoblots.

PM-Scl antigen. Certain patients with polymyositis/scleroderma (PM-Scl) overlap syndromes produce autoantibodies to a nucleolar particle termed PM-Scl [4]. The structure and function of this particle is unknown, but its localization within the granular component of the nucleolus suggests a role in ribosome biogenesis and/or transport.

The PM-Scl particle is composed of 11–16 proteins with molecular weights ranging from 110kD to 20kD of which an 80kD and 110kD component have been identified as autoantigens [68, 77, 78]. Reactivity with the 110kD antigen was reported with all anti-PM-Scl positive sera examined [77], while only some of the sera recognized the 80kD antigen [77]. The antigens can be detected on nuclear immunoblots.

NOR-90 antigen. Recently, autoantibodies to a 90kD protein that is localized in the nucleolus-organizing regions (NORs) in dividing cells and in nucleoli in interphase have been reported [79]. The antigen, originally designated as the NOR-90 antigen, has been shown to be identical to the ribosomal RNA transcription factor hUBF (human upstream binding factor) [80]. The antigen can be detected on nuclear immunoblots.

5.3 Cytoplasmic antigens

Ro (SS-A) and La (SS-B). Anti-Ro antibodies are directed to the Ro60 (60kD) (Fig. 3A, lane 5; Fig. 3B, lane 6) and/or Ro52 (52kD) proteins, both components of the Y1-Y5 RNP particles [reviewed in 10]. Moreover, in approximately 60% of anti-Ro positive sera, antibodies to the La protein (50kD) can be detected as well (Fig. 3A, lanes 4 and 5; Fig. 3B, lanes 4 and 6). In conventional SDS polyacrylamide gel systems, Ro52 and La usually co-migrate; this is the reason why it is difficult to distinguish between anti-Ro52 or anti-La antibody specificities. However, by using an alternative acrylamide-bisacrylamide ratio, e.g. 40% (w/v) : 0.53% (w/v), the antigens can be separated from each other [81–83]. Anti-Ro antibodies can be best detected on cytoblots.

Discrete antigenic degradation products of the La protein have apparent molecular weights of 28kD and 43kD, respectively. The 43kD fragment is often seen as a doublet. These degradation products are helpful in identifying anti-La activity. The La protein can be detected on both nuclear- and cytoblots.

tRNA synthetases. In polymyositis (PM)/dermatomyositis (DM) certain species of tRNA synthetases act as autoantigen [reviewed in 2]. These include Histidyl-tRNA synthetase (Jo-1; 54kD) (Fig. 3B, lane 5) [5], Threonyl-tRNA synthetase (PL-7; 80kD) [84] and Alanyl-tRNA synthetase (PL-12; 110kD) [85]. Recently, the two synthetases for isoleucine (139kD) and glycine (77kD) have been identified as autoantigens [86, 87]. Anti-tRNA synthetase antibodies can be best detected on cytoblots.

56K antigen. Although the 56K (56kD) autoantigen has not been well characterized yet, it has been shown that Ro60, Ro52, La, Jo-1, calreticulin and 56K all represent distinct polypeptides [83]. The intracellular localization of the 56K antigen is not clear. Biochemical fractionation of cells reveals that most of the antigen is present in the cytoplasmic fraction [88, compare also Fig. 3A, lane 10 with Fig. 3B, lane 7]. However, by using the more elegant method of cell enucleation to obtain nuclear and cytoplasmic cell fractions, approximately equal amounts of the antigen were shown to be present in each fraction [83]. The autoantigen is not stably associated with either RNA or other proteins [83, 88].

Mitochondrial proteins. On cytoblots, anti-mitochondrial antibodies recognize groups of polypeptides, designated M2, with apparent molecular weights of 70–74kD, 56kD, 52kD and 46kD (Fig. 3B, lane 10) [89–91]. The 70kD M2 autoantigen has been identified as the E2 component (lipoate acetyltransferase) of the pyruvate dehydrogenase complex [89] and is recognized by virtually every patient with primary biliary cirrhosis (PBC) [89–91]. Autoantibodies to the other mitochondrial proteins are only occasionally found in addition to the 70kD reactivity [91].

Sera from patients with anti-mitochondrial antibodies may also contain antibodies to the centromere antigens.

Ribosomal proteins. P-proteins (phosphorylated ribosomal proteins) are generally present in multiple copies on the ribosome and have isoelectric points in the range of 3 to 5, in contrast to most ribosomal proteins which are single copy and basic. A typical profile of an anti-ribosomal P-protein antibody specificity is composed of 3 antigens of 37kD (protein P0), 18kD (protein P1) and 15kD (protein P2) [92]. An additional band of approximately 20kD is often present as well. P0, P1 and P2 represent the eukaryotic analogs of the *E. coli* proteins L10, L7 and L12, respectively [93]. The 20kD protein is the analog of *E. coli* protein S10. Although the P0 protein is an easy detectable antigen on cytoblots, anti-ribosomal RNP antibodies are best detected on a blot containing an extract of purified ribosomes.

Notes
- Antibody identification is based on reactivity with antigens designated according to apparent molecular weight. Single band reactivities should be interpreted with care and confirmed by additional reference methodologies [94].
- The immunoblot only serves as an aid to diagnosis and should not be interpreted as diagnostic in itself.
- Band colour intensities do not necessarily correlate with antibody titers as determined with other reference methodologies.
- Occasionally, autoantibody activities as described above can be detected in human control sera. For this reason it is advised to use pooled normal human serum as a negative control.

Appendix: PMSF and protease inhibition.

During cell fractionation and protein purification it is necessary to inhibit protease activity. Phenylmethylsulfonyl fluoride (PMSF) is widely used as a protease inhibitor. Stock solutions of > 100 mM PMSF in 100% 2-propanol have a shelf-life at room temperature of >1 year. Under certain conditions, however, PMSF may have a half-life as short as 35 min (see Table 3).

High salt conditions (3M), may reverse the inhibition by PMSF. Thus, protease inhibition by PMSF alone may be insufficient to protect proteins from degradation during isolation and purification. This is especially true when proteins are isolated from tissues that contain a high protease activity, i.e. liver and spleen.

If protease activity is suspected or detected, additional PMSF may be added. If the protease activity still remains, try other inhibitors. Boehringer Mannheim Biochemicals advises the use of a cocktail of Leupeptin, EDTA, Pepstatin and PMSF as a starting point.

In order to verify the effectiveness of protease inhibition after each purification step, one can also use the Endoproteinase Test Kit of Boehringer (cat. no. 582 433) for the determination of endoproteinases in the ng and μg range (fibrin method).

Table 3. Half-life of PMSF in aqueous solution [95]

Temperature	pH7.0	pH7.5	pH8.0
+ 4°C	± 25 hr	± 16 hr	± 3 hr
+25°C	110 min	55 min	35 min

References

1. Tan EM (1982) Adv Immunol 33: 167–240
2. Tan EM (1989) Adv Immunol 44: 93–151
3. Smeenk R, Brinkman K, van den Brink H, Termaat R-M, Berden J, Nossent H & Swaak T (1990) Clin Rheumatol 9 (Suppl 1): 100–110
4. Reimer G (1990) In: LeRoy EC (Ed) Rheumatic disease clinics of North America Vol 16(1) (pp 169–183) WB Saunders Company, Philadelphia
5. Yoshida S, Akizuki M, Mimori T, Yamagata H, Inada S & Homma M (1983) Arthr Rheum 26: 604–611
6. Sharp GC & Alspaugh MA (1985) In: Gupta S and Talal N (Eds) Immunology of Rheumatic Diseases (pp 197–219) Plenum Publ. Corporation
7. Reichlin M & van Venrooij WJ (1991) Clin Exp Immunol 83: 286–290
8. van Venrooij WJ & Sillekens PTG (1989) Clin Exp Rheumatol 7: 635–645
9. Chan EKL & Tan EM (1989) Curr Opin Rheumatol 1: 376–381
10. Slobbe RL, Pruijn GJM & van Venrooij WJ (1991) Ann. Med. Interne 142: 592–600
11. van Venrooij WJ & van de Putte LB (1991) Seminars Clin Immunol 3: 27–32
12. Gershoni JM (1985) Trends Biochem Sci 10: 103–106
13. Bers G & Garfin D (1985) BioTechniques 3: 276–288
14. Beisiegel U (1986) Electrophoresis 7: 1–18
15. Gershoni JM (1987) In: Grambach A, Dunn MJ and Radola BJ (Eds) Adv. Electrophoresis Vol 1 (pp 141–175) VCH, Weinheim
16. Tovey ER & Baldo BA (1987) Electrophoresis 8: 452–463
17. Gershoni JM (1988) In: Glick D (Ed) Methods Bioch Anal Vol 33 (pp 1–58) Wiley, New York
18. Garfin D & Bers G (1989) In: Baldo BA and Tovey ER (Eds) Protein blotting: methodology, research and diagnostic applications (pp 5–41) Karger AG, Basel
19. Hoch SO (1989) In: Baldo BA and Tovey ER (Eds) Protein blotting: methodology, research and diagnostic applications (pp 140–164) Karger AG, Basel
20. Stott DI (1989) J Immunol Methods 119: 153–187
21. Laemmli UK (1970) Nature (Lond.) 227: 680–685
22. Mozdzanowski J, Hembach P & Speicher DW (1992) Electrophoresis 13: 59–64
23. Towbin J, Staehelin T & Gordon J (1979) Proc Natl Acad Sci USA 76: 4350–4354
24. Bjerrum OJ & Schafer-Nielsen C (1986) In: Dunn MJ (Ed) Electrophoresis '86 (pp 315–327) VCH, Weinheim
25. Dunn SD (1986) Anal Biochem 157: 144–153
26. Reuter R, Rothe S, Habets W, van Venrooij WJ & Lührmann R (1990) Eur J Immunol 20: 437–440
27. Lehmeier T, Foulaki K & Lührmann R (1990) Nucleic Acids Res 18: 6475–6484
28. van Venrooij WJ (1987) J Rheumatol 14 (Suppl 13): 78–82
29. Earnshaw WC & Rothfield NF (1985) Chromosoma (Berlin) 91: 313–321
30. Earnshaw W, Bordwell B, Marino C & Rothfield N (1986) J Clin Invest 77: 426–430
31. Guldner HH, Lakomek H-J & Bautz FA (1985) Clin Exp Immunol 58: 13–20
32. Kingwell B & Rattner JB (1987) Chromosoma (Berlin) 95: 403–407
33. Palmer DK, O'Day K, Trong HL, Charbonneau H & Margolis RL (1991) Proc Natl Acad Sci USA 88: 3734–3738
34. Masumoto H, Masukata H, Muro Y, Nozaki N & Okazaki T (1989) J Cell Biol 109: 1963-1973
35. Pluta AF, Saitoh N, Goldberg I & Earnshaw WC (1992) J Cell Biol 116: 1081–1093
36. Saitoh H, Tomkiel J, Cooke CA, Ratrie H, Maurer M, Rothfield NF & Earnshaw WC (1992) Cell 70: 115–125
37. Hadlaczky G, Praznovszky T, Rasko I & Kereso J (1989) Chromosoma (Berlin) 97: 282–288

38. Verheijen R, de Jong BAW, Oberyé EHH & van Venrooij WJ (1992) Mol Biol Rep 16: 49–59
39. Wang JC (1985) Annu Rev Biochem 54: 665–697
40. Wang JC (1987) Biochim Biophys Acta 909: 1–9
41. Osheroff N (1989) Pharmacol Ther 41: 223–241
42. Guldner HH, Szosteki C, Vosberg H-P, Lakomek HJ, Penner E & Bautz FA (1986) Chromosoma 94: 132–138
43. Maul GG, French BT, van Venrooij WJ & Jiminez SA (1986) Proc Natl Acad Sci USA 83: 5145–5149
44. Shero JH, Bordwell B, Rothfield NF & Earnshaw WC (1986) Science 231: 737–740
45. van Venrooij WJ, Stapel SO, Houben H, Habets WJ, Kallenberg CGM, Penner E & van de Putte LB (1985) J Clin Invest 75: 1053–1060
46. Shero JH, Bordwell B, Rothfield NF & Earnshaw WC (1987) J Rheumatol 14 (Suppl 13): 138–140
47. Kumar V, Kowalewski C, Koelle M, Qutaishat S, Chorzelski TP, Beutner EH, Jarzabek-Chorzelska M, Kolacinska Z & Jablonska S (1988) J Rheumatol 15: 1499–1505
48. Alderuccio F, Barnett AJ, Campbell JH, Pedersen JS & Toh BH (1986) Clin Exp Immunol 64: 94–100
49. Douvas AS, Achten M & Tan EM (1979) J Biol Chem 254: 10514–10522
50. Hildebrandt S, Weiner E, Senécal J-L, Noell S, Daniels L, Earnshaw WC & Rothfield NF (1990) Arthr Rheum 33: 724–727
51. Verheijen R, de Jong BAW & van Venrooij WJ (1992) Clin Exp Immunol 89: 456–460
52. Steen VD, Powell DL & Medsger TA Jr (1988) Arthr Rheum 31: 196–203
53. Mimori T & Hardin JA (1986) J Biol Chem 261: 10375–10379
54. Reeves WH (1987) J Rheumatol 14 (Suppl 13): 97–105
55. de Vries E, van Driel W, Bergsma WG, Arnberg AC & van der Vliet PC (1989) J Mol Biol 208: 65–78
56. Stuiver MH, Ceonjaerts FEJ & van der Vliet PC (1990) J Exp Med 172: 1049–1054
57. Bravo R, Frank R, Blundell PA & MacDonald-Bravo H (1987) Nature (Lond.) 326: 515–517
58. Prelich G, Tan KK, Kostura M, Mathews MB, So AG, Downey KM & Stillman B (1987) Nature (Lond.) 326: 517–520
59. Gohill J, Cary PD, Couppez M & Fritzler MJ (1985) J Immunol 135: 3116–3121
60. Costa O & Monier JC (1986) J Rheumatol 13: 722–725
61. Cohen MG, Pollard KM & Webb J (1992) Ann Rheum Diseases 51: 61–66
62. Chou C-H, Satoh M, Wang J & Reeves WH (1992) Mol Biol Rep 16: 191–198
63. Price CM, McCarty GA & Pettijohn DE (1984) Arthr Rheum 27: 774–779
64. Price CM & Pettijohn DE (1986) Exp Cell Res 166: 295–311
65. Parker KA & Steitz JA (1987) Mol Cell Biol 7: 2899–2913
66. Kass S, Tyc K, Steitz JA & Sollner-Webb B (1990) Cell 60: 897–908
67. Okano Y, Steen D & Medsger TA Jr (1992) Arthr Rheum 35: 95–100
68. Kipnis RJ, Craft J & Hardin JA (1990) Arthr Rheum 33: 1431–1437
69. Gold HA, Topper JN, Clayton DA & Craft J (1989) Science 245: 1377–1380
70. Reddy R, Tan EM, Henning D, Nohga K & Busch H (1983) J Biol Chem 258: 1383–1386
71. Reimer G, Raska I, Scheer U & Tan EM (1988) Exp Cell Res 176: 117–128
72. Chang DD & Clayton DA (1987) EMBO J 6: 409–417
73. Chang DD & Clayton DA (1987) Science 235: 1178–1184
74. Altman S, Baer MF, Guerrier-Takada C & Vioque (1986) Trends Biochem Sci 11: 515–518
75. Reimer G, Rose KM, Scheer U & Tan EM (1987) J Clin Invest 79: 65–72
76. Stetler DA, Rose KM, Wenger ME, Berlin CM & Jacob ST (1982) Proc Natl Acad Sci USA 79: 7499–7503
77. Reimer G, Scheer U, Peters J-M & Tan EM (1986) J Immunol 137: 3802–3808
78. Gelpi C, Alguero A, Angeles Martinez M, Vidal S, Juarez C & Rodriquez-Sanchez (1990) Clin Exp Immunol 81: 59–64

79. Rodriquez-Sanchez JL, Gelpi C, Juarez C & Hardin JA (1987) J Immunol 139: 2579–2584
80. Chan EKL, Imai H, Hamel JC & Tan EM (1991) J Exp Med 174: 1239–1244
81. Byon JP, Slade SG, Chan EKL, Tan EM & Winchester R (1990) J Immunol Methods 129: 207–210
82. Slobbe RL, Pruijn GJM, Damen WGM, van der Kemp JWCM & van Venrooij WJ (1991) Clin Exp Immunol 86: 99–105
83. Pruijn JM, Božič B, Schoute F, Rokeach LA & van Venrooij WJ (1992) Mol Biol Rep 16: 267–276
84. Mathews MB, Reichlin M, Hughes GRV & Bernstein RM (1984) J Exp Med 160: 420–434
85. Targoff IN & Arnett FC (1990) Am J Med 88: 241–251
86. Targoff IN (1990) J Immunol 144: 1737–1743
87. Targoff IN, Trieu EP, Plotz PH & Miller FW (1992) Arthr Rheum 7: 821–830
88. van Venrooij WJ, Wodzig KW, Habets WJ, de Rooij DJ & van de Putte LB (1989) Clin Exp Rheumatol 7: 277–282
89. Yeaman SJ, Fussey SPM, Danner DJ, James OFW, Mutimer DJ & Bassendine MF (1988) Lancet i: 1067–1070
90. Mackay IR & Gershwin ME (1989) Immunol Today 10: 315–318
91. Zurgil N, Bakimer R, Kaplan M, Youinou P & Shoenfeld Y (1991) J Clin Immunol 11: 239–245
92. Takehara K, Nojiman Y, Kikuchi K, Igarashi A, Soma Y, Tsuchida T & Ishibashi Y (1990) Arch Dermatol 126: 1184–1186
93. Rich BE & Steitz JA (1987) Mol Cell Biol 7: 4065–4074
94. van Venrooij WJ, Charles P & Maini RN (1991) J Immunol Methods 140: 181–189
95. James GT (1978) Anal Biochem 86: 574–579

Manual of Biological Markers of Disease **A5**: 1–23, 1993.

Enzyme-linked immunosorbant assay in the rheumatological laboratory

P.J. CHARLES and R.N. MAINI

Immunology and Immunogenetics Laboratory, Department of Rheumatology, Charing Cross Hospital, and Kennedy Institute of Rheumatology, London, W6 8RF, U.K.

Abbreviations

OD: optical density; ELISA: enzyme-linked immunosorbent assay; DIB: dot immunobinding assay; CIE: counterimmuno electrophoresis; OPD: ortho-phenylene diamine; TMB: 3,3',5,5'-tetramethyl-benzidine; ABTS: 2,2'-azino-dia(3'-ethylbenzthiazoline-6-sulfonic acid); ONPG: ortho-nitro-phenyl-B-D-galactosidase

Key words: Autoantibodies, autoantigens, enzyme immunoassay, enzyme linked immunosorbant assay

1. Introduction

The increasing popularity of enzyme-linked immunosorbant assay (ELISA) for the detection of autoantibodies of value in the diagnosis of rheumatological diseases has led to the development of assay systems which are proving to be of great utility. Unfortunately, in practice, the results do not always match the high standards which are appropriate for use in routine diagnosis or application in research. In this paper we offer guidelines on practical aspects of employing ELISA to detect autoantibodies and describe methods and an approach which should assure a good quality of result.

The principle of ELISA requires an initial step of immobilization of the antigen on a solid phase, either directly or via binding to an antibody. In a subsequent step, bound antibody or antigen is then detected by means of an immunochemical reaction involving the use of a chromogenic substrate.

The first point that has to be addressed is the question 'why use ELISA?'. In general ELISA is a safe, rapid, sensitive and specific technique for the detection of antibodies to well-characterised antigens. Each assay detects only antibodies to one specific antigen and many assays may be required for the detection of antibodies of multiple specificity in a serum sample, which in the case of autoimmune diseases is the rule. Attempts have been made to establish screening assays to multiple constituent antigens using crude tissue extracts or combinations of purified proteins as the antigen. To date, these have not been very successful. However, given the experience now established, it should be possible to develop a good screening system, which can be used before proceeding on to antibody identification by antigen specific assays.

2. General notes

2.1 Solid phase

The solid phase onto which the antigen is immobilized consists of either polystyrene or polyvinyl chloride. This can take the form of tubes or beads. These are best when small numbers of samples are to be assayed, as washing can be time-consuming. In recent years the most commonly used format is microtitre plates, which allow large numbers of sera to be screened and can be used in conjunction with automated instruments such as diluters and washers.

The nature of the solid phase can be of importance to the assay performance. As a general rule polystyrene takes up less protein, and the background is consequently lower. Polyvinyl chloride has a high uptake, but this in turn may lead to a higher absorption of any contaminating proteins in the coating solution. Polyvinyl chloride would obviously be the choice when small quantities of antigen are available and maximum absorption is required. Another advantage of polyvinyl chloride is that antigen coating can be carried out in a short period, sometimes as short as one hour at room temperature.

One disadvantage of microtitre plates was that they only were manufactured in a 96-well format. However, recently polystyrene and polyvinyl chloride plates have become available in a variety of strip formats consisting of 8 or 16 wells, allowing for a more purposeful and efficient use of plates.

More recently, new solid phase materials have been developed for use in immunoassays. An example of this is the Covalink plate designed for use with synthetic peptides that can be co-valently linked directly to the plate [1].

2.2 Conjugates

The conjugate in an ELISA assay is usually an immunoglobulin linked by a chemical reaction to an enzyme capable of reacting with a chromogenic substrate. A number of different enzymes are available as the active agent for conjugates (see Table 1), and the choice is often dictated by the nature of the assay system and the reaction conditions of its substrate. The most commonly used system is that employing alkaline phosphatase. One of the major advantages of this enzyme is that no colour developing stage is required, and thus, the reaction can be allowed to continue whilst the optical density can be monitored until it develops to the required level. Urease conjugates are used when the results of the assay are read by direct vision rather than using an instrument. Other conjugates used include peroxidase, β- galactosidase, and glucose oxidase which may be used when the antigen has phosphatase activity.

Table 1. Detection systems used in enzyme immunoassays

Enzyme	Substrate (see par. 8.4)	Stop solution (see par. 8.5)	Absorbance wavelength
1. Alkaline phosphatase	p-Nitrophenyl phosphate	not required	405 nm
2. Peroxidase	OPD	A	492 nm
	TMB	B	450 nm
	ABTS	not required	405 nm
3. β-galactosidase	ONPG	C	410 nm
4. Glucose oxidase	Glucose-OPD	D	490 nm
5. Urease	Bromocresol purple	not required	read macroscopically

As an alternative to using enzyme conjugated to immunoglobulin, the enzyme can be conjugated to protein A. This is only useful if the antibody to be detected is of a species and immunoglobulin isotypes which react with protein A. More recently, a novel IgG binding protein, Protein H, has been suggested for use as this does not react with bovine serum immunoglobulin and has no binding to other human immunoglobulin classes. This means that it is likely to be more specific than Protein A, especially in systems involving products of cell culture systems [2].

2.3 Substrates

Chromogenic substrates are compounds which are colourless in the native form and produce a coloured product when degraded by enzymatic activity. Most chromogenic substrates require a defined reaction time and some require the addition of stop solution to visualise the coloured product. This can present difficulties when the reaction time may be variable, or during the setting up of an assay, when the optimum conditions are unknown. To overcome these problems most assays use the alkaline phosphatase – p-Nitrophenyl phosphate detection system as this does not require a developing agent to visualise the end product.

Detection systems are outlined in Table 1.

2.4 Blocking agents

Once the antigen is bonded to the solid phase, free binding sites on the solid phase are blocked to prevent the non-specific binding of patient immunoglobulin to these sites, which would then be detected by the assays detection system. These blocking agents are protein solutions against which immune reactions are not usually found. Numerous different protein agents have been used including:
a) 2% Casein in coating buffer
b) 3% Bovine serum albumin in coating buffer
c) 1% Tween 20 in coating buffer
d) 2% Foetal calf serum in coating buffer
e) 2% Glycerol in coating buffer
f) 1% Gelatin in coating buffer

Blocking agents must contain no substances to which antibodies can be directed, e.g. immunoglobulins. The use of such agents may cause false positive results.

2.5 Antigens and antigen presentation

When considering our protocol for ELISA, the first questions concern the target antigen and how it will best be presented on the solid phase. To assess the best presentation method it is necessary to know the answer to four questions:
1. Is the antigen available in a biochemically pure form?
   If so, this relieves the investigator from the problems involved in the purification of the antigenic material. These may include the selection of an antibody through which to purify it, and the determination of a suitable elution protocol that will ensure there is no denaturation of the antigen. The investigator has only to ensure that the material contains no other substances which are likely to interfere with or give rise to false positive reactions in the assay system.
2. Is the antigen available commercially?
   When antigens are available commercially the investigator has only to be assured of the purity of the antigen provided. Although the standard of commercially produced antigens is usually very good, this cannot be assumed and should be checked before the material is used in an assay system.
3. If the answer to question 2 is no, can the antigen be purified using affinity chromatography or other biochemical techniques?
   The investigator who does not have pure antigen, but has either a polyclonal or monoclonal antibody directed against the antigen, can produce an antigenic preparation using affinity chromatography.
4. Are antibodies specifically directed against the antigen available?

If purification of the antigen is not practical (if, for example, the antigen becomes degraded during the elution from an affinity column), then the antibody may be used in an antigen capture ELISA.

If the answer to any of the first three questions is yes, then it is likely that a solid phase ELISA (see paragraph 3.1) is the technique of choice, whereas if the answer to question 1 to 3 is no, but the answer to question 4 is yes, then a sandwich or antigen capture ELISA technique (see paragraph 3.3) will be required.

An example of the evolution of an antigen and its use in assay systems is the La antigen, which was originally purified from crude tissue extracts using polyclonal antisera derived from patients with autoimmune diseases. Subsequently the development of monoclonal antibodies led to their use in affinity chromatographic columns to purify the antigen. Further studies have enabled the production of recombinant proteins and synthetic peptides, which are currently being evaluated as substrates for ELISA systems.

Most antigens will either be:
1. Whole protein, either biochemically or immunologically prepared.

   It is most important to realize that not all antigen-containing preparations will be suitable or optimal for assay use. If using commercially prepared antigens it is advisable to try a number of different suppliers before choosing the source for your assay, and to be aware that there may be batch-to-batch variations even from the same supplier. Thus some form of internal quality control for the antigen preparation is necessary.
2. Synthetic peptides.
3. Recombinant proteins.

Most assays described to date have used native protein molecules. The only synthetic peptide that has been regularly used is the 22 amino acid peptide from the ribosomal ribonucleoprotein (rRNP) or P protein described by Elkon [3]. This short peptide provides an excellent substrate for the detection of anti P-antibodies. More recently, recombinant proteins have been used. These have a major advantage in the ease of mass production. However, when fusion proteins are used they may lack the original conformational structure of the antigen, and this conformational aspect may be of importance in some autoreactive epitopes.

3. Assay methods

3.1 Solid phase ELISA

Establishment of optimal assay conditions
Reagents required:
Antigen preparation

Immune sera containing antibodies to target antigen
Normal non-immune sera
Blocking agents
Detection system

Steps in the procedure
1. Plate out the antigen diluted at various concentrations, in coating buffer, at concentrations in the range of 1–10 μg/ml. The volume added to each well is 50 –100 μl. The plate should be covered to prevent evaporation.
2. The antigen-coated plates are incubated at 4 °C for 16 hours to allow antigen to bind to the plate.
3. After coating, excess antigen/buffer is removed by inverting the plate, and 200μl of washing buffer is then added to each well and discarded by inverting the plate. This is repeated 3–6 times. After the final wash the plate is inverted to discard buffer and tapped on absorbent paper to remove residual fluid.
4. The remaining free binding sites on the solid phase are blocked with an excess of an immunologically inert protein. Examples of these are listed in paragraph 2.4. The volume of blocking agent is usually twice that of the antigen solution, i.e. if 50μl of antigen was added, the volume of blocking agent used should be 100μl. The plate is incubated at 22 °C for two hours.
5. Excess protein is washed off, as described in step 3.
6. Serially diluted samples of positive and negative sera are added to the appropriate wells. The volume of sera added should be equivalent to the volume of the antigen-coating solution. The plate is incubated at 22 °C for two hours.
7. Unbound proteins are washed off as in step 3.
8. Anti-immunoglobulin conjugated to a chromogenic enzyme is added to each well. The specificity of this conjugate should be directed against the immunoglobulin class or classes of the positive control sera. The volume should be equivalent to that of the sera. The plate is incubated at 22 °C for 2 hours.
9. Unbound conjugate is washed off, as described in step 3.
10. The appropriate substrate is then added to the wells, developed at 37 °C and read at suitable time intervals until there is maximal optical density readings with the lowest dilution of the positive sera. The volume of the substrate solution should be equivalent to that of the conjugate solution.
11. If necessary the reaction is stopped using the appropriate stop solution (see paragraph 10.5).

Notes to the procedure
ad 2. Antigen binding can also be carried out at 37 °C for one hour. However this does not work with all antigen systems and 4 °C coating is safer for use in initial experiments.
ad 4. Temperatures quoted are approximate. In most laboratories a temperature of 22 °C is equivalent to incubation on the bench top. If the temperature within a laboratory varies from this by more than 5 °C, then either incubations should be carried out in a controlled-temperature cabinet, or incubation times should be adjusted to take these variations into account.

3.2 The chequerboard titration

An alternative to the method outlined in paragraph 3.1.1 is the use of the chequerboard titration. This is especially useful if a wide range of sera and antigen dilutions need to be tested. In this method the antigen is diluted across the plate, column 1 containing the highest concentration and column 6 the lowest. These dilutions are then repeated across columns 7–12. The plate is divided into two halves, that with columns 1–6 is tested against dilutions of the positive sera, and columns 7–12 against the negative sera. The plate layout is shown in Table 2. The assay is then performed as described in paragraph 3.1.1.

Table 2. The plate layout for the performance of a chequerboard titration.

	\multicolumn{12}{c}{Antigen concentration (ug/ml)}										
COL	1	2	3	4	5	6	7	8	9	10	11	12
	10	5	2.5	1.2	0.6	0.3	10	5	2.5	1.2	0.6	0.3
ROW												
1	POSITIVE SERUM 1/100						NEGATIVE SERUM 1/100					
2	POSITIVE SERUM 1/200						NEGATIVE SERUM 1/200					
3	POSITIVE SERUM 1/400						NEGATIVE SERUM 1/400					
4	POSITIVE SERUM 1/800						NEGATIVE SERUM 1/800					
5	POSITIVE SERUM 1/1600						NEGATIVE SERUM 1/1600					
6	POSITIVE SERUM 1/3200						NEGATIVE SERUM 1/3200					
7	POSITIVE SERUM 1/6400						NEGATIVE SERUM 1/6400					
8	POSITIVE SERUM 1/12800						NEGATIVE SERUM 1/12800					

3.3 Antigen capture (sandwich) ELISA

Reagents required:
Antibody directed against target antigen
Material containing antigen
Immune sera containing antibodies to target antigen

AMAN-A5/7

Normal non-immune sera
Blocking agent
Detection system

Steps in the procedure
1. The antigen-capturing antibody is diluted in coating buffer. The optimum concentration for this antibody is usually in the region of 5–15μg/ml. The plate is covered with a sealer or plate lid and incubated at 4 °C for 16 hours.
2. The plate is washed by adding 200μl of wash buffer to each well and then discarded by inverting the plate. This is repeated 3–6 times. After the final wash the plate is then tapped against some layers of absorbent paper to remove the residual liquid.
3. Free binding sites are blocked with 200μl per well of a suitable blocking agent. The plate is covered and incubated at 22 °C for 2 hours.
4. Excess material is removed by inverting the plate, and then washed as in step 2.
5. The antigen-containing material is then added to the wells. The aim must be to saturate the antigen-binding sites of the capture antibody. The plate is covered and left for 16 hours at 4 °C to ensure maximal binding.
6. Excess material is removed by inverting the plate, and then washed as in step 2.
7. Serial dilutions of positive or negative sera are added to the wells, and the plates are covered and incubated at 22 °C for two hours.
8. Excess sera is removed by inverting the plate, and then washing as in step 2.
9. Anti immunoglobulin conjugated to a chromogenic enzyme is added to each well. The specificity of this conjugate should be directed against the immunoglobulin class or classes of the positive control sera. The volume should be equivalent to that of the sera. The plate is incubated at 22 °C for two hours.
10. Unbound conjugate is washed off, as described in step 2.
11. The appropriate substrate is then added to the wells and developed at 37 °C and read at suitable time intervals until there is maximal optical density readings with the lowest dilution of the positive sera. The volume of the substrate solution should be equivalent to that of the conjugate solution.
12. If necessary the reaction is stopped using the appropriate stop solution.

ad 1. If human antibodies are to be detected using a sandwich technique, then the capture antibody needs to be of a species other than human, e.g. rat or mouse.
ad 3. Temperatures quoted are approximate. In most laboratories a temperature of 22 °C is equivalent to incubation on the bench top. If the temperature within a laboratory varies from this by more than 5 °C, then either incubations should be carried out in a controlled-temperature cabinet, or incubation times should be adjusted to take these variations into account.
ad 9. The conjugate used in the detection system should be tested to show that it does not bind to the capture antibody. This can be simply done by running a blank well in the assay, which contains no human sera. Any positivity in this well indicates direct binding of the conjugate to the capture antibody. An alternative conjugate should than be sought.

3.4 Dot immunobinding (dot blot) assay

A recent development has been the use of dot immunobinding assays (DIB) [4]. The capture antibody or antigen is bound to a micropourous membrane such as nitrocellulose, and the specimen containing the analyte is allowed to flow over the capture antibody or antigen. After washing, the conjugate is added and subsequently, after further washing the colour is developed. The results of these assays are only semi-quantitative as the coloured product cannot be read in a spectrophotometer. It has been claimed that this technique increases binding of the analyte to the capture phase and decreases non-specific binding.

4. Evaluation and improvement of the assay system

4.1 Evaluation

The data for each antigen concentration should be plotted as shown in Fig. 1. The optimum antigen concentration is that which gives the greatest differentiation between the optical density of the positive and negative sera. This can be assessed by measuring the area between the two graphs (Fig. 2). Once the antigen concentration is set, this data also supplies the optimum serum dilution for testing, this being the dilution at which there is maximal difference in optical density. In the example shown in Fig. 2, this would be at a serum dilution of 1:40.

4.2 Improvement of the assay system

If there is insufficient sensitivity, then the following modifications can be made in order to improve the performance.
1. Change solid phase medium.

CURVES OF POSITIVE AND NEGATIVE SERA

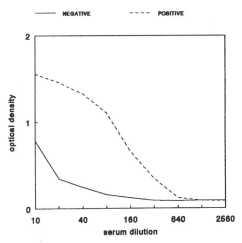

Fig. 1. Comparison of dilution curves for positive and negative sera. The optimum serum dilution is the point at which there is greatest difference between the two curves (in this example at dilution 1:40).

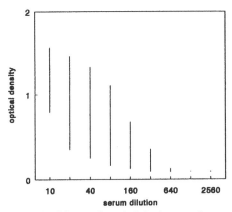

Fig. 2. Graph showing the difference in optical density at a given serum dilution for the positive and negative sera in an ELISA assay for the dilution curves shown in Fig. 1. The optimum differential occurs in this example at 1:40.

Consider use of alternative material or format.
2. Adjustment of blocking agent.
   Some blocking agents have the ability to inhibit binding of antibodies to certain antigens.
3. Alter incubation temperature from 22 °C to 37 °C or 4 °C.
   This is usually accompanied by a change in incubation time.
4. The alteration of incubation times, especially of serum incubation. In some systems this may require an incubation up to 18 h.
5. Change specificity of conjugate.
   If the antibodies present in the sera are only of IgG class, it is possible that a better differentiation between positive and negative will be obtained if an anti-IgG conjugate is used rather than an anti-polyvalent immunoglobulin conjugate.
6. Use other detection systems.
   If the signal received from the assay is weak, it is possible to increase the sensitivity of the reaction by the use of an enhanced detection system. The most commonly used is the biotin-avidin system. Instead of the detecting immunoglobulin being bound directly to the enzyme, it is conjugated to biotin and the enzyme is conjugated to avidin. Thus, following the incubation of the detecting immunoglobulin-biotin, the plate is washed and avidin-enzyme is added to the wells. After incubation and washing the chromogenic substrate is added as normal.
   A further modification of this is the use of biotin labelled immunoglobulin and enzyme using avidin as a bridge (Antigen-Antibody-immunoglobulin/biotin-avidin-biotin/enzyme-substrate).
   These systems may be hampered by non-specific binding of antibodies to avidin. This can be overcome by pre-incubation of the sample with immobilized avidin.
7. Adjust antigen dilution.
   Whilst the dilution of most protein antigens lies in the range 1–10 μg/ml, it is possible that antigens with low binding capacities will need to be coated at a higher concentration in order to ensure that sufficient antigenic material absorbs onto the solid phase.
8. Pre-incubation of the sample with murine immunoglobulin.
   With the recent development of monoclonal antibody therapy, the presence of antibodies to murine immunoglobulin may interfere in an assay system. In order to remove these antibodies, the serum can be pre-incubated with murine immunoglobulin.
9. Shaking the plates during incubation.
   This ensures optimal mixing, and may increase antibody-antigen binding and decrease the quantities of reagents required.
10. Adjustment of conjugate dilution.
    The dilution of the conjugate can have an effect on the sensitivity of the assay and the incubation time required. An increase in the

concentration of the conjugate used may enhance the performance of the assay.

5. Validation of the assay system

Once the optimum conditions have been established, the assay needs to be standardised. The best way to do this is to establish a standard curve drawn from positive sera diluted to give a range of values to which a numerical value can be assigned. If this curve is then run on every plate, it provides a comparative reference point to which results drawn from different assays can be compared. To establish a standard curve, a strongly positive serum is titrated until the optical density is equivalent to that given by a normal sera at the same dilution. This point is assigned the value of 1 unit of antibody activity, and the other dilutions are assigned values accordingly, i.e. if the dilution at which the OD of the positive matches that of the negative is 1:40000, then the positive sera has a value of 2 units at 1:20000, 4 units at 1:10000, 32 units at 1:1200, and 128 units at 1:300. Once this curve has been established, the OD of the test sample at the selected screening dilution is read off of the curve drawn from the OD values of the standard dilutions.

In some cases it may be possible to use an international standard to convert the arbitrary unit calculated from the assay into international units to allow comparison of levels obtained from different laboratories. Although primary standards are available for most antinuclear antibody specificities from the Center for Disease Control, Atlanta, USA, these are only qualitative standards and, as yet, the only quantitative standards available are for dsDNA (available from WHO,CLB laboratory, Amsterdam, The Netherlands). Once the standard curve has been established the assay can be validated with reference to specificity and sensitivity.

Disease specificity is determined by the number of patients outside the target disease group who are positive for the antibody. Disease sensitivity is the ratio of patients within the target disease group who are positive for the antibody, relative to patients in the disease group who are negative for the antibody. Thus, if an antibody is present in 23 out of 25 patients in the target disease group and only 4 out of 400 sera from patients outside the disease group it has a high sensitivity ($23/25 = 92\%$) and a high specificity ($396/400 = 99\%$). Sera for testing should be chosen from disease groups expected to be positive for the antibody and from disease groups which form part of a differential diagnosis, and sera from groups likely to give false positive results. This latter group most commonly includes sera from rheumatoid arthritis (antiglobulins or rheumatoid factors), and from patients with bacterial or viral infections (polyclonal activation). These sera are tested according to the conditions established in previous experiments.

Ideally, the perfect assay has a high sensitivity and specificity. However, in a diagnostic system an assay with high specificity and a sensitivity of 30–50% will still be of use.

6. Validation of a commercial assay system

This will follow the same protocol as validation of an in-house assay, although financial constraints may necessitate that a smaller sera panel will be used. If such a validation is not possible, it is still vital that the 'end user' obtains from the manufacturer published comparison studies or names of users willing to discuss the advantages and disadvantages of the system. It cannot be stressed strongly enough that it is the responsibility of the end user to properly validate any commercial assay before using it in clinical practice.

7. Summary

The investigator wishing to use ELISA assays in his laboratory has first to make the decision whether he is able or wishes to make use of commercially available diagnostic kits, or whether to proceed with the establishment of an in-house assay. There are advantages to both. Once the commercial kit has been validated with the laboratory of the user, the majority of the time-consuming work of plate preparation and batch-to-batch quality control becomes the province of the manufacturer, the end user has only to run sufficient quality control to ensure the continued good performance of the assay. In addition, commercially produced assays are often more rapid than in-house methods, with a consequent saving in staff time and decrease in turnaround. The arguments in favour of in-house systems fall into three categories. First, that they can be more economical. Lower antigen concentrations can be used, although this may lead to an increased assay time. Second, the specificity and sensitivity of an in-house assay system can be set to suit the needs of the user. The user may find that the sensitivity of commercially available assays is not sufficient for his needs and he will be able to tune an in-house system to his own requirement. Third, there may not yet be a commercially available assay system for the analyte to be assayed. This is usually the case with all new specificities until their clinical and commercial viability has been proven and the commercial producers feel confident enough to move into production. Having considered all these points the user will then have to decide whether to proceed with the validation of a commercial assay or to formulate his plans for an in-house assay. When choosing assays for our laboratories, a number of factors need to be taken into consideration. These include the specificity and sensitivity of the assay, the suitability of the technology and its ease of use in the laboratory. ELISA, when properly set up, provides a system of analysis which is sensitive, specific and rapid. It is easy to handle, and need not require expensive equipment to read the results. The reagents are inexpensive and stable for long storage.

There has been a rapid and explosive rise in the use of enzyme immunoassays in medical and other laboratories over the last decade, and these developments look certain to continue at an even greater pace over the forthcoming decade. New analytes, new technologies and modifications to existing technology will make their appearance, and may require us to change our approach to the management and diagnosis of patients with rheumatic disease.

8. Solutions and materials

8.1 Materials

ELISA plates
a) Dynatech Laboratories Ltd, Daux Rd, Billinghurst, RH14 9SJ,UK.
b) ICN Flow, Eagle House, Business Park, Gomm RD, High Wycombe, Buckinghamshire, HP13 7DL, UK.
c) Greiner Labortechnik Ltd, GesmbH, Greinerstrasse 70, PO Box 6, A-4550, Kremsmunster, Austria.

Immunochemicals
a) Sigma chemical co Ltd, Fancy Rd, Poole, Dorset, BH17 7BR, UK.
b) Sera-lab Ltd, Crawley Down, Sussex, RH10 7FF, UK.
c) ICN Flow, Eagle House, Business Park, Gomm RD, High Wycombe, Buckinghamshire, HP13 7DL, UK.
d) DAKO Ltd, 16 Manor Courtyard, Hughendon Ave, High Wycombe, Bucks, HP13 5RE, UK.

Chemicals
a) Sigma chemical co Ltd, Fancy Rd, Poole, Dorset, BH17 7BR, UK.
b) ICN Flow, Eagle House, Business Park, Gomm RD, High Wycombe, Buckinghamshire, HP13 7DL, UK.
c) BDH Ltd, PO Box 15, Freshwater Rd, dagenham, Essex, RM8 1RF, UK.
d) Gibco BRL, Life Technologies Inc, 8400 Helgewrman Ct, Gaithersburg MD 20877 USA.

8.2 Coating buffers

a) 0.01 M Phosphate buffer 0.145 M NaCl pH 7.2
Na_2HPO_4, $12H_2O$ 2.680 g
NaH_2PO_4, H_2O 0.345 g

NaCl 8.474 g
H₂O Add 1 litre
Check pH
Store at 4 °C.
b) 0.05 M Carbonate buffer pH 9.6
Na₂CO₃ 1.59 g
NaHCO₃ 2.93 g
NaN₃ 0.20 g
H₂O Make up to 1 litre
Store at 4 °C
c) Glycine pH 10.4
Glycine 7.50 g
MgCl₂ 0.203 g
Zn(SO₄)₂ 0.288 g
H₂0 Make up to 1 litre pH to 10.4

8.3 Dilution/Washing buffer

0.01 M Phosphate buffer 0.145 M NaCl pH 7.2
Made as coating buffer. To this can be added 1% Tween-20 to reduce background binding. Tween may however have an interference effect in certain systems.

8.4 Chromogenic substrate solution

a) p-Nitrophenyl phosphate
   5 mg p-Nitrophenyl phosphate dissolved in 5 ml glycine buffer.
b) ortho-phenylenediamine (OPD)
   8 mg of OPD is dissolved in 12 ml of 0.1 M citric acid-phosphate buffer pH 5.0. Allow to dissolve, then add 5 ul 30% H_2O_2. This is stable for 1 h if protected from light. Avoid contact with skin or eyes.
c) 3,3',5,5'-tetramethyl-benzidine (TMB)
   Dissolve 10 mg of TMB in 1 ml of dimethylsulfoxide. Mix slowly into 100 ml of 0.1M sodium acetate-citric acid buffer pH 6.0. Add 20 μl 30% H_2O_2. Use within 30 min.
d) 2,2'-azino-di-(3-ethylbenzthiazoline-6-sulfonic acid) (ABTS)
   Dissolve 5 mg of ABTS in 50 ml 0.1M citric acid- phosphate buffer pH 5.0.Add 20 μl of 30% H_2O_2. Use within 30 min.
e) ortho-Nitrophenyl-β-D-galactosidase (ONPG).
   Dissolve 90 mg ONPG in 10ml phosphate buffered saline pH7.2 (Heat to 50 °C to dissolve). This stock solution is stable for several days at room temperature in the dark. Before use dilute stock solution 1:10 with PBS containing 10 mM MgCl₂ and 0.1M 2-Mercaptoethanol.

8.5 Stop solutions

A) 1 M Sulphuric acid
   55 ml 95% H_2SO_4
   900 ml Distilled water
   NB Add acid very slowly to the larger volume of distilled water. Cool reaction. Adjust volume to 1 litre with distilled water.
B) 2 M Sulphuric acid
C) 1 M Na_2CO_3
D) 2.5 M Hydrochloric acid

9. Laboratory notes on ELISA systems for the detection of anti-nuclear, anti-cytoplasmic and anti-phospholipid antibodies

9.1 Antinuclear antibodies

Attempts have been made to produce a screening ELISA for antibodies to nuclear antibodies, using cell nuclei or combinations of antigens. Some of these systems appear to lack sensitivity [5]. Attempts to produce screening assays using a mixture of purified antigens have been more successful [6], although it remains to be seen whether they will replace the fluorescent antinuclear antibody test as the initial screening test for ANA.

9.2 Anti dsDNA

Native dsDNA does not readily bind to polystyrene plates unless contaminated by protein or regions of single stranded (ssDNA) or denatured DNA. This means that dsDNA ELISAs require the use of a ligand such as poly-l-lysine or methylated bovine serum albumin to be used to couple the DNA to the plate (see the article of R. Smeenk in this issue). Any regions of denatured DNA can be removed using digestion with S1 nuclease to remove regions of single stranded DNA. Some assays provide unreliable results [7] and this may be due to the detection of low avidity antibodies [8], but other assays give good correlations between the ELISA and crithidia luciliae immunofluorescent test [9], and between the ELISA and the Farr assay [10]. There is disagreement about the specificity of anti-dsDNA ELISA and SLE [9], but it is likely that these disagreements are due to the individual assays, and it may well be that the assays which do not have good specificity are detecting low avidity antibodies, other non-specific binding proteins, or anti-ssDNA [11, 12].

9.3 Anti-histone

There has been little published work on anti-histone ELISAs. However it is important to check that all 5 histone polypeptides (H1,H2A,H2B,H3, and H4) are present in the antigen preparation, and that the antigen is not contaminated with DNA. ELISA for anti-histone is more sensitive than immunofluorescent assays and is not complicated by the presence of other autoantibody specificities. Anti-histone ELISA assays were compared to immunoblotting and have been found to be slightly less sensitive [13]. Quantitative measurements of anti-histone by ELISA have been reported as correlating with disease activity in SLE [14].

9.5 Anti-Ro/SS-A

ELISA assays for anti-Ro are generally considered to be more sensitive than counter-immunoelectrophoresis (CIE) [15–19]. ELISA has been used to detect IgA anti Ro in the saliva of patients with Sjögrens syndrome, and has been found in patients who do not have circulating antibodies [20]. It is important to know which of the Ro peptides are present in the assay system, as this may have an effect on the performance of the system.

Recently, recombinant antigens for Ro60 protein have been used as antigens in an ELISA system and have been found to give comparable results to systems using native protein [21].

9.6 Anti-La

ELISA assays for anti-La are more sensitive than CIE [17, 21]. Some sera have been reported which are negative by ELISA, but positive by CIE [17], and this may represent a reaction with an epitope which is concealed or denatured on binding to the solid phase. IgA antibodies to La have been detected in saliva in the absence of circulating antibodies [20].

Recombinant La antigen has been used in ELISA for the detection of antibodies and gives comparable results to those obtained using the native protein [21–23].

9.7 Anti-nRNP

ELISA assays using native protein have been reported as giving comparable results to CIE [24] and to immunoblotting [25], and in some cases have a greater sensitivity than CIE [17]. Assays using recombinant proteins have been reported as having a lower specificity than those using the native protein [26]. Some studies have measured antibodies to the individual peptides of the nRNP molecule [25, 27, 28].

9.8 Anti-Sm

ELISA assays for the detection of antibodies to Sm have produced contradictory results with some groups reporting an increased sensitivity compared to CIE [24, 29], whilst others have reported it to be comparable with CIE [15], and immunoblotting [25]. Recombinant Sm-B peptide has been used in ELISA and has been reported as giving comparable results to CIE [30].

9.9 Anti-ribosomal RNP (P protein)

Since this antibody specificity is difficult to detect using gel precipitation techniques [31], it has only been possible to investigate its role in SLE since the description of an ELISA using the 22 amino-acid synthetic P protein which contains the reactive epitope [3].

9.10 Anti-Scl70 (Topoisomerase 1)

A good correlation has been reported between ELISA for anti-Scl70 and CIE [32], and between ELISA and immunoblotting [33]. Some authors have reported a higher frequency of the antibody using ELISA compared with gel diffusion techniques [34]. More recently an ELISA system has been described using a purified recombinant non-fusion protein [35].

9.11 Anti-Ki

An ELISA has been described for the detection of anti-Ki, which indicates that these antibodies are more common than previously reported by gel precipitation [36].

9.12 Anti-Jo-1

An ELISA has been described for the detection of antibodies to Jo-1, which is more sensitive than the gel precipitation techniques such as CIE [37], and gives comparable results to those obtained with immunoblotting.

9.13 Anti-Cardiolipin

ELISA assays are the most commonly used detection system for the detection of anti-cardiolipin antibodies. There have been a number of reports on factors affecting the performance of this system [38–43], which seems more sensitive to changes in assay conditions than most other ELISA systems. Indeed, as some inter-laboratory programmes have shown, even when different laboratories use the same commercial assay, vastly different results can be obtained.

9.14 Anti-Centromere

An ELISA system has been produced using the CENP – B recombinant protein. Initial results show that this is a promising substrate for future studies [44].

Acknowledgements

We thank Ms. Pat Mumford for helpful discussion during the preparation of this manuscript.

References

1. Sondergaard-Andersen J, Lauritzen E, Lind K, Holm A (1990) Covalently linked peptides for enzyme-linked immunosorbent assay. J Immunol Methods, 131:99–104.
2. Akesson P, Cooney J, Kishimoto F, Bjorek L (1990) Protein-H. Anovel IgG binding bacterial protein. Mol Immunol 27:523–531.
3. Elkon KB, Bonfa E, Skelly S, Weissbach H, Brot N (1987) Ribosomal protein autoantibodies in systemic lupus erythaematosus. Bioassays 7:258–61
4. Smith LH, Yin A, Beiber M, Teng NNN (1987) Detection of antibodies to HIV viral protein. J Immunol Meth 105:263–7
5. Jitsukawa T,Nakajima S, Usui J,Watanabe H (1991) Detection of anti-nuclear antibodies from patients with systemic rheumatic diseases by ELISA using Hep-2 nuclei. J Clin Lab Anal 5:49–53.
6. Monce NM, Bogusky RT, Cappel NN (1991) An enzyme immunoassay screening test for the detection of total antinuclear antibodies. J Clin Lab Analysis 5:439–442.
7. Brinkman K, Termat R, Van den Brink H, Berden J, Smeenk R (1991) The specificity of the anti dsDNA ELISA. A closer look. J Immunol Methods 139:91–100.
8. Nossent JC, Huysen V, Smeenk RJT, Swaak AJG (1989) Low avidity antibodies to double stranded DNA in systemic lupus erythaematosus: a longitudinal study of their clinical significance. Ann Rheum Dis 48:677–82.
9. Mohan TC, Jalil HA, Nadarajah M, Sng EH (1989) Evaluation of an ELISA method for the measurement of antibodies to dsDNA. Singapore Med J 30(3):242–5
10. Tipping PG, Buchanan RC, Riglar AS, Dimech WJ, Littlejohn GO, Holdsworth SR (1991) Measurement of anti-DNA antibodies by ELISA: a comparative study with Crithidia and a farr assay. Pathology 23(1):21–4
11. Kroubouzas G, Tosca A, Konstadoulakis MM, Varelzidis A (1992) Poly-l-lysine causes false positive results in ELISA methods detecting anti-dsDNA antibodies. J Immunol Methods 148(1–2):261–3
12. Kadlubowski M, Jackson M, Yap PL, Niell G (1991) Lack of specificity for antibodies to double stranded DNA found in four commercial kits. J Clin Path 44(3) 246–50
13. Brinet A, Fournel C, Faure JR, Venet C, Monier JC (1988) Anti Histone antibodies (ELISA and immunoblot) in canine lupus erythaematosus. Clin Exp Immunol 74(1):105–9
14. Muller S, Barakat S, Watts R, Joubaud P, Isenberg D (1990) Longitudinal analysis of antibodies to histones, Sm-D peptides, and ubiquitin in the serum of patients with systemic lupus erythaematosus, rheumatoid arthritis and tuberculosis. Clin Exp Rheumatol 8(5):445–53
15. Merryman P, Louis P (1991) Comparison of assay systems for the detecting antibodies to nuclear ribonucleoproteins. J Clin Pathol 44(8):685–9
16. Provost TT,Levin LS, Watson RM, Mayo M, Ratrie H (1991) detection of anti-Ro (SS-A) antibodies by gel double diffusion and a sandwich ELISA in systemic and sub acute cutaneous lupus erythaematosus and Sjogrens syndrome. J Autoimmun 4(1) 87–96
17. Charles PJ, Tyler SZ, Dawda P, Maini RN (1990) Antibodies to nRNP, Sm, Ro and La: comparison of gel diffusion and ELISA assays. Clin Rheum 9:546
18. Yagamata H, Harley JB, Reichlin M (1984) Molecular properties of the Ro/SS-A antigen and enzyme linked immunosorbent assay for the quantitation of antibody. J Clin Invest 74:625–33
19. Sontheimer RD, McCauliffe DP, Zappi E, Targoff I (1992) Antinuclear antibodies: clinical correlations and biologic significance. Adv Dermatol 7:3–52
20. Charles PJ, Foster H, Shattles W, Griffiths I, Maini RN (1991) Salivary antibodies to Ro and La in primary Sjogrens syndrome. Hung Rheum 32(supp):275
21. Veldhoven CH, Meilof JF, Huisman JG, Smeenk RJ (1992) The development of a quantitative assay for the detection of anti- Ro/SS-A and anti-La/SS-B autoantibodies using purified recombinant proteins. J Immunol Methods 151(1–2):177–89

22. St Clair EW, Pisetsky DS, Reich CF, Chambers JC, Keene JD (1988) Quantitative immunoassay of anti-La antibodies using purified recombinant La antigen. Arthr Rheum 31:506-14.

23. Whittingham S, Naselli G, McNeilage LJ, Coppel RL, Sturgess AD (1987) Serological diagnosis of primary Sjogrens syndrome by means of human recombinant La(SS-B) as a nuclear antigen. Lancet 2(8549):1-3

24. Blazek M, Naser W, Bayer-Beck J, Lakomek HJ, Werle E, Fiehn W (1992) Enzyme immunoassay for the detection of autoantibodies to nRNP and Sm: a rapid and sensitive alternative to current procedures. Z Rheumatol 51:87-93

25. Takeda Y, Wang GS, Wang RJ, Anderson SK, Petterson I, Amaki S, Sharp GC (1989) Enzyme-linked immunosorbent assay using isolated (U) small nuclear ribonucleoprotein polypeptides as antigens to investigate the clinical significance of autoantibodies to these polypeptides. Clin Immunol Immunopathol 50(2):213-30

26. Barka NE, Agopian MS, Peter JB (1992) Comparison of two commercial ELISA kits for the detection of anti snRNP and anti-Sm autoantibodies using immunoblot as reference method. Lupus 1 (supp 1) 65

27. Ter-Borg EJ, Horst G, Limburg PC, van Venrooij WJ, Kallenberg CG (1991) Changes in levels of antibodies against the 70kDa polypeptide of the U1RNP complex in relation to exacerbations of systemic lupus erythaematosus. J Rheumatol 18(3):363-7

28. Williams DG, Charles PJ, Maini RN (1988) Preparative isolation of p67,A,B,B', and D from nRNP/Sm antigens by reverse-phase chromatography. Use in a polypeptide-specific ELISA for independent quantitation of anti-nRNP and anti-Sm antibodies. J Immunol Methods 113(1):25-35

29. Field M, Williams DG, Charles PJ, Maini RN (1988) Anti Sm antibodies by ELISA are specific for systemic lupus erythaematosus. Increased sensitivity of detection using purified peptide antigens. Ann Rheum Dis 47:820-825

30. Hines JJ, Danho W, Elkon KB (1991) Detection and quantification of anti-Sm antibodies using synthetic peptide and recombinant Sm B antigens. Arthritis Rheum 34(5):572-9

31. Kveder T, Bozic B, Lestan B, Rozman B (1991) Is counterimmunoelectrophoresis an appropriate method for the detection of antibodies against ribosomal RNP. Clin Rheum 9:547

32. Ghirardello A, Ruffatti A, Calligaro A, Del Ross T,Gambari PF. (1992) ELISA detection anti ENA antibodies:comparison with counterimmunoelectrophoresis Clin Rheum 11:126

33. Merety K, Cebecauer L, Bohm U, Kozakova D, Brozik M (1989) Study of anti Scl-70 (topoisomerase-1) autoantibodies in rheumatic patients. Clin Rheum 8:20

34. Hildebrandt S, Weiner ES, Senecal J-L, Noell GS, Earnshaw WC, Rothfield NF (1990) Autoantibodies to topoisomerase 1 (Scl-70): analysis by gel diffusion, immunoblot, and enzyme-linked immunosorbent assay. Clin Immunol Immunopathol 57:399-410.

35. Verheijen R, de Jong BAW, van Venrooij (1992) A recombinant topoisomerase 1 ELISA:screening for IgG, IgM and IgA anti-topo 1 autoantibodies in human sera. Clin Exp Immunol 89:456-460

36. Sakamoto M, Takasaki Y, Yamanaka K (1989) Purification and characterization of KI antigen and detection of anti-Ki antibody by enzyme linked immunosorbent assay in patients with systemic lupus erythematosus. Arthr Rheum 32:1554-62.

37. Walker EJ, Tymms KE, Webb J, Jeffrey PD (1987) Improved detection of anti Jo-1 antibody, a marker for myositis, using purified histidyl-tRNA synthetase. J Immunol Methods 96:149-56.

38. Cowchuck S, Fort J, Munoz S, Norberg R, Maddery W (1988) False positive ELISA tests for anti-cardiolipin antibodies in sera from patients with repeated abortion, rheumatologic disorders and primary biliary cirrhosis: correlation with elevated polyclonal IgM and implications for patients with repeated abortion. Clin Exp Immunol 73:289-94.

39. Peter JB (1990) Cardiolipin antibody assays. Lancet 335:1405

40. Hughes TP, Jones P, Rozenberg MC. (1989) Hypergammaglobulinaemia and the

anticardiolipin antibody test. Arthr Rheum 32:813

41. Matsura E, Igarashi Y, Fujimoto M, Ichikawa KT, Koike T (1990) Anticardiolipin cofactors and differential diagnosis of autoimmune diseases. Lancet 336:177–8
42. Rupin A, Gruel Y, Leroy J, Bardos P (1992) Methodological approach for the detection of IgM anticardiolipin antibodies. Immunol Invest 21(2):159–67
43. Kumar V, Bartholomew W,Shieh MT, Pierce C, Kaul N (1991) Standardization of ELISA for the detection of anti-cardiolipin antibodies- effect of non-specific IgG binding. Immunol Invest 20(7):583–93
44. Verheijen R, de Jong BAW, Oberye EHH, van Venrooij WJ (1992) Molecular cloning of a major CENP-B epitope and its use for the detection of anticentromere autoantibodies. Mol Biol Rep 16:49–59

Manual of Biological Markers of Disease **A6**: 1–13, 1993.

Immunoprecipitation of labelled proteins*

E. M. TAN and C. L. PEEBLES

W.M. Keck Autoimmune Disease Center, The Scripps Research Institute, La Jolla, California 92037, U.S.A.

Abbreviations

SLE: systemic lupus erythematosus; SDS: sodium dodecyl sulphate; rRNP: ribosomal ribonucleoprotein

Key words: Autoantigen, autoantibodies, immunoprecipitation

1. Introduction

The majority of autoantibodies to nuclear and other intracellular antigens have been shown to react with specific antigens which in the native state are components of subcellular particles (for review see [1]). These subcellular particles frequently comprise aggregates of several different proteins, with or without different species of RNA. Specific immunoprecipitation using a defined autoantibody, therefore results in the precipitation of not only the target antigen reacting with antibody but also other molecular components complexed with the target antigen. In this manner, the entire molecular composition of certain intracellular particles can be elucidated, although the autoantibody is reactive only with one component of the particle.

A widely used approach currently employs the biosynthetic labeling of cellular components with radioisotope tagged essential amino acid [^{35}S]-methionine. After disruption of the cells, desired antigens are recovered by specific immune precipitation with autoantibody. Protein antigens which do not have good turnover rates or are not synthesized during the period of incubation with [^{35}S]-methionine do not lend themselves to this method of detection. The isolation technique described here is based on the principle that protein A-bearing strains of Staphylococcus aureus bind specifically and strongly to the Fc regions of many mammalian IgG subclasses. Several advantages of using Staphylococcus protein A over second antibody as the

* Supported by NIH grant AR32063.

immunoprecipitating agent are the rapidity of the technique (which minimizes degradation of antigens), the preferential binding of protein A with IgG in the form of immune complexes (rather than with free IgG) and lower levels of background radioactivity (due to other proteins which may be non-specifically precipitated with second antibody) [2]. A known disadvantage of the protein A procedure is its poor reactivity with IgG3 and with IgM and IgA immunoglobulins.

In 1979 Lerner and Steitz [3] adapted the protein A immunoprecipitation procedure of Kessler [2] for the analysis of sera from patients with SLE. Patients with antibodies to Sm and U1-RNP precipitated several proteins in snRNP particles. Different RNP particles contain a certain number of core protein components as well as proteins which are unique to the particular RNP particle. The major components have been defined as A, B/B', C, D, E, F and G. The distinction between Sm and RNP specificity could not be determined using this method alone. When these sera were tested using an immunoblotting technique, it was demonstrated that antibodies to U1-RNP react primarily with a 68 kD protein which could not be detected by standard [^{35}S]-methionine immunoprecipitation and proteins A and C. Antibodies to Sm, on the other hand, react with the proteins B/B', D and occasionally E. Using this same technique, others have demonstrated the precipitation of other protein/RNA complexes, including those of SS-A/Ro , SS-B/La, rRNP, To (or Th), Jo-1 and other tRNA synthetases. With few exceptions, the antibody binding site resides on the protein moiety of the complex.

The procedure involves the mixing of patient serum with radiolabeled antigens and insoluble particles (Sepharose beads coated with Staphylococcus protein A). The particle suspension is then mixed to allow for a specific antigen/antibody reaction to occur and the subsequent attachment of the antigen-antibody complexes to the protein A-coated beads. The resulting labelled antigens bound by the antibodies are then separated by electrophoresis and the antigens identified on exposure to radiographic film.

The advantages of immunoprecipitation include a higher sensitivity and specificity than can be obtained with the standard immunodiffusion assay. The method detects antigens in their native conformation and also allows identification of other molecules associated with the specific antigen. The disadvantages include the use of radioisotopes and the inability in some cases to determine the specific antigenic peptide due to some of the reasons described above.

Interpretation of the results requires an understanding of the patterns of proteins precipitated by defined antibody specificities. When a complex is precipitated, the combination of bands becomes characteristic for the antibody specificity. Care must be taken when evaluating results if only one protein is precipitated, as multiple proteins may occur at the same location in a one-dimensional gel.

The following procedure is for determination of proteins using a soluble extract of [^{35}S]-methionine labeled HeLa cells for immunoprecipitation.

2. Protein immunoprecipitation procedures

2.1 Preparation of [^{35}S]-methionine radiolabeled extract

2.1.1 Stock HeLa cell culture
HeLa cells are grown in a T-175 flask to subconfluence in the standard culture medium.

Steps in the procedure
1. Just prior to transfer add 10 ml trypsin solution (0.05% in Ca^{2+}/Mg^{2+}-free Tyrode buffer containing 0.02% EDTA), shake gently to spread over the entire surface of the cells and incubate at 37 °C for 15 min or until cells are detached.
2. Under sterile conditions, transfer cells to a 50 ml conical tube and centrifuge for 5 min at 1500 rpm in a tabletop centrifuge at room temperature.
3. Resuspend cells in 50 ml fresh sterile medium.
4. Add 10 ml of cell suspension to each of five tissue culture dishes.
5. Add 10 ml additional medium to each dish. Swirl gently to give an even cell distribution.
6. Incubate at 37 °C until cell growth is 50% confluent.

2.1.2. Isotope labeling of HeLa cells and preparation of the soluble extract

Steps in the procedure
1. Replace the culture medium from the tissue culture dishes with 10 ml of the medium minus methionine.
2. Add 1 mCi [^{35}S]-methionine to each of the tissue culture dishes. Maintain culture dishes in a horizontal position for even cell growth.
3. Incubate overnight in an incubator with 10% CO$_2$.
4. After overnight incubation, appropriately dispose of radioisotope containing media.
5. Wash cells with 10 ml of cold PBS and dispose of as radioisotope wash.
6. Add 1 ml of *cold* buffer A to each plate.
7. *Carefully* scrape off cells with a disposable cell scraper.
8. Transfer supernatant to Eppendorf tubes. Centrifuge at 10,000 g for 15 min at 4 °C. Transfer supernatant to new microcentrifuge tubes.
9. Preclear extract by adding 100 μl protein A-Sepharose stock/ml of extract. Mix 10 min at 4 °C.

11. Centrifuge for 5 min.
12. Transfer supernatant to new microcentrifuge tubes. Store at -70 °C. Discard microcentrifuge tubes as radioactive waste.

2.1.3 Sample preparation

Two methods are given for immunoprecipitation. The first method is a simple one-step method where all reactants are added at the same time. The second involves precoating of the protein A-Sepharose beads with antibody and the subsequent addition of antigen in a two-step method. In most instances the simple one-step method will give adequate results. For those instances when the sera may contain substances with adverse effects on the antigens such as RNases, the two-step method is advised.

2.1.3.1 Method 1

Label a 1.5 ml microcentrifuge for each sample to be tested. Controls should include a normal human serum and one or more known positive sera. All reagents and procedures are on ice or at 4 °C.

Steps in the procedure

1. Add 500 μl NET-2, 100 μl protein A-Sepharose bead suspension, 10 μl patient serum sample and 25–40 μl of labeled HeLa antigen extract to the appropriate tubes.
2. Rotate at 4 °C for 1 hour.
3. Centrifuge at 10,000 g in a microcentrifuge for 5 min.
4. Remove the supernatant into a suction flask, add 1 ml of NET-2, vortex briefly to resuspend and repeat centrifugation.
5. Repeat step 4 at least 5 times.
6. After the last wash, remove the wash to leave only a small amount of buffer over the beads.
7. Add 20 μl of 2× gel sample buffer. The samples may be stored frozen at -20 °C until ready to use.

2.1.3.2 Method 2

Label a 1.5 ml microcentrifuge for each sample to be tested. Controls and precautions are the same as in Method 1 (see 2.1.3.1).

Steps in the procedure

1. Add 500 μl NET-2, 100 μl protein A-Sepharose bead suspension and 10 μl patient serum sample to the appropriate tubes.
2. Rotate at 4 °C for 1 hour.
3. Centrifuge at 10,000 g in a microcentrifuge for 5 min.
4. Remove the supernatant into a suction flask, add 1 ml of NET-2, vortex briefly to resuspend and repeat centrifugation.
5. Repeat step 4 at least 3 times.

6. After last wash, add 500 μl NET-2 and 25–40 μl of the radiolabeled extract.
7. Rotate the tubes at 4 °C for 1 hour.
8. Centrifuge and wash as described in step 4 for a total of 5 washes.
9. At the last wash, remove the supernatant down to the surface of the pellet. Add 20 μl 2 × gel sample buffer and mix well. Samples may be stored at −20 °C for future evaluation or processed immediately.

Notes to the procedures
ad 6. The level of the SS-A/Ro antigen is usually low. If this is the primary antigen being studied, a greater amount of antigen is suggested. The lower amounts of antigen are satisfactory for the detection of most of the other common extractable antigens.

2.2 Preparation of the gels

2.2.1 Preparation of separating gels
The following gel protocol is modified from the Laemmli's method [4, 5] with the major difference being the lower ratio of bisacrylamide to acrylamide [6]. This modification, using a 15% gel system, is more effective in the separation of antigens with molecular weights between 40,000 and 60,000 due to the increased gel pore size. The chart below gives the recipe for several different percentages of separating gels which are 90 × 130 × 1.5 mm (height × length × width). To use gels which are 1 mm in width use 0.67 × the amounts. To prepare gels that are 210 mm in length use 5 × the amounts.

Reagent		10%	12.5%	15%
30% Acrylamide	ml	10.00	12.50	15.00
2% bis-acrylamide	ml	2.00	1.65	1.30
1M Tris pH 8.7	ml	11.25	11.25	11.25
H₂0	ml	6.65	4.65	2.35
20% SDS	μl	150	150	150
TEMED	μl	30	30	30
10% Ammonium persulfate	μl	105	105	105

2.2.2. Preparation of the 5% stacking gel
Mix together 1.7 ml 30% acrylamide, 0.7 ml 2% bis-acrylamide, 1.25 ml 1 M Tris pH 6.9, 6.35 ml distilled water, 50 μl 20% SDS, 25 μl TEMED. Just prior to adding to gel (see par. 2.2.3, step 6), add 50 μl 10% ammonium persulfate.

2.2.3. Preparation of gel for use

Steps in the procedure
1. Prepare the separating gel and pour to the 90 mm height.
2. Remove the trapped bubbles by gentle tapping of the glass plates. Overlay with n-butanol stored saturated over water.
3. Allow gel to polymerize for at least 45 min before adding the stacking gel.
4. When ready to use, remove the n-butanol and rinse well with water to remove any residual. Blot out remaining water.
5. Place a toothed comb between the gel plates and adjust to have 10–20 mm distance between the bottom of the teeth and the top of the separating gel.
6. Add the ammonium persulfate to the stacking gel and add the mixture immediately as the stack gel polymerizes very rapidly.
7. Allow stacking gel to polymerize for 15 min.

2.2.4 Running the gels

Steps in the procedure
1. Heat samples (beads with bound labeled proteins in sample buffer) 5 min at 100 °C.
2. Centrifuge to pack beads in the bottom of the tube.
3. Add clarified sample solution to each lane. Use at least one lane for molecular weight markers (see 'notes to the procedure').
4. The gels are run at 100 V for 4–5 h at room temperature until the bromphenol-blue dye reaches the bottom of the gel.
5. The gel is fixed overnight.
6. After fixation the gel is rinsed briefly in water and placed in DMSO for 30 min on a rotator.
7. Two more changes of DMSO are used to insure the removal of all of the water from the gel.
8. The gel is then placed in DMSO/PPO for at least 3 h.
9. At the end of this time the DMSO/PPO is removed and water is added. This results in precipitation of the PPO fluor causing the gel to become opaque. The gel is placed on a rotator for 20–30 min.
10. The gel is dried at 60 °C under vacuum.
11. When dry, it is placed directly onto an X-ray film for exposure. The following example (Figure 1) demonstrates the results of known prototype antibodies.

Notes to the procedure
ad 3. The sample volume added to each lane will vary according to its width. Using a gel comb with tooth width of 7mm allows about 15–20 μl of a supernatant sample to be added to each lane.

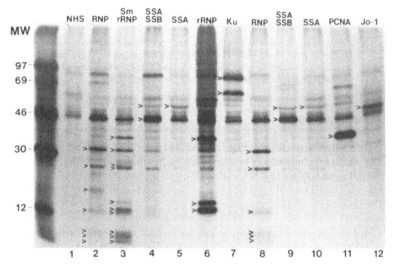

Fig. 1. Protein patterns obtained after immunoprecipitation of [³⁵S]-methionine labelled HeLa cell extracts by human autoantibodies.

Immunoprecipitation was carried out as described in paragraph 2. The precipitated complexes were analyzed by SDS polyacrylamide gel electrophoresis (par.2.4).

The sera used were:

lane 1 : Normal human serum (NHS). No specific bands were precipitated.

lane 2 : Anti-nRNP serum. The bands marked with an arrowhead are U1-A (33kD), B/B' (28kD), U1-C (22 kD), D (13 kD) and the E-F-G triplet (9–11 kD). The nature of the 46 kD band has not been determined but it could be actin.

lane 3 : Anti-Sm serum containing anti-rRNP activity. The anti-Sm activity precipitates the same proteins as the anti-nRNP activity shown in lane 2. The anti-rRNP activity precipitates three additional bands migrating at 38, 16 and 15 kD.

lane 4 : Serum containing anti-La/SS-B and anti-Ro/SS-A activity. Bands at 60kD (the Ro60 protein) and 46 kD (the La antigen) are marked.

lane 5 : Serum containing anti-Ro/SS-A activity. Precipitated are the Ro60 protein (60 kD) and the La antigen (46 kD) which is tightly associated with the Ro antigen.

lane 6 : Serum containing anti-rRNP activity. The typical rRNP bands of 38, 16 and 15 kD are marked. Some of the other bands are presumably ribosomal proteins.

lane 7 : Serum containing anti-Ku activity. The typical Ku bands (p70 and p80) are marked.

lanes 8–10: See lanes 2, 4 and 5, respectively.

lane 11 : Serum containing anti-PCNA activity. The precipitated protein marked is the PCNA antigen (34 kD).

lane 12 : Serum containing anti-Jo-1 activity. The precipitated antigen is the His-tRNA synthetase (54 kD).

The molecular weight markers used should include [¹⁴C]-labeled myosin (200 kD), phosphorylase B (97.4 kD), bovine serum albumin (69 kD), ovalbumin (45 kD), carbonic anhydrase (30 kD), and cytochrome-C (12.3 kD) (Dupont/NEN, Boston, MA).

ad 4. The addition of Coomassie blue dye to the fixative will allow visualization of the immunoglobulin heavy and light chains as well as any other proteins present in large quantity.

ad. 8. Care should be taken in handling the DMSO/PPO as the PPO is carcinogenic. Numerous other fluoro-enhancement methods are commercially available.

ad 11. In running unknown samples, the controls used should include several lanes using individual or pooled samples of known antibody specificities.

3. Interpretation

Figure 1 demonstrates patterns obtained between [^{35}S]-methionine labelled HeLa cells and human sera with autoantibodies to Sm, U1-RNP, SS-A/Ro, SS-B/La, rRNP, Ku, PCNA and Jo-1.

Notes

1. As described previously, the antibodies react with intact complexes and not just the individual components as is seen in immunoblotting techniques. This is best seen in the reactions of the Sm and U1-RNP.
2. The normal human serum control lane contains precipitin bands common to all of the sera. These bands are disregarded in the evaluation.
3. Multiple bands are commonly observed in patients with SLE due to the presence of multiple antibody specificities in their sera. These complex patterns may be evaluated by comparison to monospecific antibody patterns.
4. Arrowheads in each lane of Figure 1 point to the bands which are specifically associated with the antibody. They are described in order from the uppermost arrowhead.

3.1 Sm and U1RNP

The complex of proteins immunoprecipitated by patients with antibodies to Sm and U1-RNP (Fig. 1, lanes 2 and 8) are designated as A, B, C, D, E, F, and G. The molecular weights assigned to these proteins are 33, 28, 22, 16, 13, 11, and 9 kD respectively. Lane 3 containing the Sm proteins also contains proteins of the rRNP complex (compare with Fig. 1, lane 6).

3.2 SS-B/La and SS-A/Ro

Antibodies to SS-B/La immunoprecipitate a 46 kD protein (Fig. 1, lanes 4 and 9, second arrowhead). The antigen is very labile so characteristic breakdown products may be noted on the gels. These include bands at 43 kD and 28 kD. These antibodies rarely occur alone and are usually accompanied by antibodies to SS-A/Ro. SS-A/Ro is seen as a weakly labeled band of 60 kD (Fig. 1, lanes 4, 5, 9, 10, first arrowhead). The 52 kD protein component of Ro observed by immunoblotting [9] is not visualized by immunoprecipitation.

3.3 PCNA, Jo-1, rRNP and Ku

Antibodies to PCNA and Jo-1 produce single bands by immuno-precipitation (Fig. 1, lanes 11 and 12) while antibodies to rRNP and Ku are examples of antibodies that immunoprecipitate small complexes of proteins (Fig. 1, lanes 6 and 7). The rRNP complex contains 3 proteins of 38, 16, 15 kD, respectively [9]. The Ku complex using this gel system contains 2 proteins of 66 and 86 kD [10]. Using a different gel system the same components have been designated p70/p80 [11].

Note
1. There are numerous other proteins which may be demonstrated by this technique. As sera may contain multiple specificities (lane 3 with Sm and rRNP), the interpretation should be made carefully by comparison with known antibody patterns.

4. Materials

4.1 Reagents and equipment

4.1.1. Chemicals
All chemicals are obtained from Sigma Chemical Co. (St. Louis, MO) unless otherwise stated. The acrylamide and bisacrylamide are from Biorad (Richmond, CA). The chemicals used for the preparation of the polyacrylamide gel and electrophoresis buffers are all electrophoresis-grade. DMSO is purchased from Fisher (Tustin, CA), DMSO/PPO is obtained from Dupont/NEN (Boston, MA). The protein A-Sepharose CL-4B beads are obtained from Pharmacia (Piscataway, NJ). The [^{35}S] radiolabeled methionine is obtained as [^{35}S]Trans-label from ICN (Irvine, CA). While the major component is [^{35}S]-methionine, there is also a minor labeled cysteine component.

4.1.2 Tissue culture materials
Tissue culture flasks and dishes for culturing and radiolabeling of HeLa or suspension stock cells are obtained from Falcon.

4.1.3 Electrophoretic apparatus
The electrophoresis system used in this procedure was made in house. A comparable vertical gel system with power supply is available from numerous manufacturers.

4.1.4 Miscellaneous
Microcentrifuge tubes, microcentrifuge, vacuum flask/suction apparatus for the removal of supernatants during washing steps, vortex mixer,

glass dishes for processing gels after electrophoresis, Whatman 3MM filter paper sheets or comparable, autoradiography X-Ray film XAR-5 (Eastman Kodak Co., Rochester, NY), X-Ray cassettes, cell scrapers.

5. Solutions

5.1 Tissue culture media

5.1.1 Media for stock HeLa cells
Take 500 ml of Dulbecco's minimum essential medium with high glucose (e.g. Whittaker Bioproducts, Inc. (Walkersville, MD), and add:
– 5 ml of L-glutamine (200 mM stock), (Sigma, St. Louis, MO)
– 125 μl of a 10 mg/ml solution of gentamicin sulfate (Sigma)
– 50 ml of calf serum (e.g. Gemini Products, Calabasas, CA)
  The medium can be stored at 4 °C for days. It is warmed to 37 °C prior to use.

5.1.2 Medium for labelling cells
To prepare media for 5 tissue culture plates, take:
– 50 ml of methionine free minimum essential medium (Flow/ICN, Irvine, CA) and add
– 1 ml calf serum
– 0.5 ml L-glutamine (200mM stock)
– 12.5 μl of a 10 mg/ml solution of gentamicin sulfate.
  The medium is usually prepared just prior to use but may be stored at −20 °C. Medium is warmed to 37 °C prior to use.

5.2 Acrylamide solutions and electrophoresis buffers

Acrylamide solutions: Stock solutions of 30% acrylamide and 2% N,N'-methylenebisacrylamide (bisacrylamide) are made in water and stored in dark bottles at 4 °C to maintain stability.

Note: Acrylamide and bisacrylamide are strong neurotoxins in both the powder and liquid form but are nontoxic when polymerized.

SDS: Prepare a 20% SDS stock solution (wt/vol) in deionized water. Store at room temperature.

Ammonium persulfate: Prepare a fresh 10% solution (wt/vol) in water. The stock may be stored at 4 °C for two weeks.

Gel sample buffer (2x):
To prepare 100 ml, mix:
- 6 g SDS
- 20 ml glycerol
- 10 ml β-mercaptoethanol
- 12.5 ml of 1 M Tris.Cl pH 6.8
- 200 mg bromophenol blue
- add water to bring volume to 100 ml.
Store frozen at -20 °C in 5 ml aliquots.

Gel electrophoresis running buffer (10x):
To prepare 8 L of buffer, mix:
- 242.4 g Tris base
- 1,154 g glycine
- 80 g SDS
- add water to 8 liters.
Store at room temperature.

Fixative:
To prepare 8 L of fixative, mix:
- 3.6 L methanol
- 3.6 L distilled water
- 720 ml glacial acetic acid.

5.3 Other buffers and solutions

Trypsin solution in Tyrode buffer:
To prepare 1 L of Tyrode buffer, mix:
- 8 g NaCl
- 0.2 g KCl
- 0.05 g $NaH_2PO_4.H_2O$
- 1 g $NaHCO_3$
- 2 g glucose
Adjust pH to 7.4
Add EDTA (0.2 g) and trypsin (0.5 g) per liter Tyrode buffer.

PMSF: A 1 M stock solution is prepared by dissolving 0.87 g of phenylmethylsulfonyl fluoride (PMSF) (Calbiochem, La Jolla, CA) in dimethyl sulfoxide (Eastman Kodak) to a final volume of 5 ml. Store frozen at -20 °C in 0.5 ml aliquots.

Note: PMSF is highly toxic. It also has a short half-life in aqueous solutions.

Cell lysis buffer A:
To prepare 100 ml of buffer, dissolve:
- 870 mg NaCl
- 30 mg $MgCl_2.6H_2O$
- 127 mg Tris-HCl pH 6.0
- 24 mg of Tris base
- 100 μg pepstatin
- 100 μg leupeptin in about 90 ml water.
Then add:
- 0.5 ml NP-40
- 0.5 ml of aprotonin
and adjust with water to 100 ml.

The solution should be pH 7.4. Store at $-20\ °C$ in 5 ml aliquots. Immediately before use, add PMSF to a final concentration of 1 mM using the 1 M stock solution.

Tris.HCl solutions: Prepare 1 M stock solutions of Tris base (pH 9) and Tris.HCl (pH 6) separately and mix to obtain the desired 1 M stock buffers of pH 6.9 and 8.7. Store at room temperature.

NET-2 buffer:
To prepare 1 liter of working solution, mix:
- 5 ml NP-40
- 30 ml 5M NaCl
- 10 ml 500mM EDTA (pH 8.0)
- 50 ml 1M Tris (pH 7.4)
- 5 ml 20% SDS
- 200 mg Sodium Azide
- 5g deoxycholic acid
- add distilled water to 1 liter.
   Filter into a sterile flask. Store at 4 °C or for long term storage at $-20\ °C$.

Protein A-Sepharose Cl-4B bead suspension:
To prepare a 10% stock, add 1.5 g beads (one vial) of protein A-Sepharose CL-4B, 100 mg bovine serum albumin and NET-2 to a final volume of 50 ml. Mix well and store at 4 °C.

References

1. Tan EM (1989) Antinuclear antibodies: diagnostic markers for autoimmune disease and probes for cell biology. Adv Immunol 44: 93–151
2. Kessler SW (1975) Rapid isolation of antigens from cells with a Staphylococcal protein A-antibody adsorbent: Parameters of the interaction of antibody-antigen complexes with protein A J Immunol 115: 1617–1624
3. Lerner MR & Steitz JA (1979) Antibodies to small nuclear RNAs complexed with proteins are produced by patients with systemic lupus erythematosus. Proc Natl Acad Sci USA 76: 5495–5497
4. Takacs B (1979) Electrophoresis of proteins in polyacrylamide slab gels. In: Lefkovits I & Benvenuto P (Ed) Immunological methods (pp. 81–105). Academic Press, New York
5. Laemmli UK (1970) Cleavage of structural proteins during the assembly of the head of bacteriophage T4. Nature (London) 227: 680–685
6. Buyon JP, Slade SG, Chan EKL, Tan EM & Winchester R (1990) Effective separation of the 52 kDa SSA/Ro polypeptide from the 48 kDa SSB/La polypeptide by altering conditions of polyacrylamide gel electrophoresis. J Immunol Methods 129: 207–210
7. Francoeur AM, Chan EKL, Garrels JI & Mathews MB (1985) Characterization and purification of the lupus antigen La. Mol Cell Biol 5: 586–590
8. Ben-Chetrit E, Fox RI & Tan EM (1990) Dissociation of immune responses to the SS-A (Ro) 52 kDa and 60 kDa polypeptides in systemic lupus erythematosus and Sjögren's syndrome. Arthr Rheum 33: 349–355
9. Francoeur AM, Peebles CL, Heckman KJ, Lee JC & Tan EM (1985) Identification of ribosomal protein autoantigens. J Immunol 135: 2378–2384
10. Francoeur AM, Peebles CL, Gomper PT & Tan EM (1986) Identification of Ki (Ku, p70/p80) autoantigens and analysis of anti-Ki reactivity. J Immunol 136: 1648–1653
11. Reeves WH (1985) Use of monoclonal antibody for the characterization of novel DNA-binding proteins recognized by human autoimmune sera. J Exp Med 161: 18–39

Manual of Biological Markers of Disease A7: 1–13, 1993.

Analysis of autoimmune sera by immunoprecipitation of cellular RNPs

G. ZIEVE[1], M. FURY[1] and E.J.R. JANSEN[2]

[1] Department of Pathology, SUNY Stony Brook, Stony Brook, NY 11794-8691, U.S.A.;
[2] Department of Biochemistry, University of Nijmegen, NL-6500 HB Nijmegen, The Netherlands

Abbreviations

RNP: ribonucleoprotein; ribRNP: ribosomal RNP; snRNP: small nuclear RNP; SLE: systemic lupus erythematosus; MCTD: mixed connective tissue disease

Key words: Autoantigen, autoantibody, immunoprecipitation, ribonucleo-protein, RNA precipitation

1. Introduction

Rheumatic diseases are often accompanied by autoimmune antibodies directed towards cellular proteins. Many of these autoantigens are protein components of cytoplasmic and nuclear ribonucleoprotein (RNP) particles (reviewed in [1]). Patients with systemic lupus erythematosus (SLE), mixed connective tissue disease (MCTD) and Sjögren's syndrome often develop antibodies against the snRNP, Ro and La ribonucleoprotein particles, respectively (Table 1). Other RNP particles are recognized with lower frequency [2]. The close correlation of these antibody activities with a disease makes them valuable diagnostic markers. The RNA components of the particles are of unique low molecular weight and often abundant, making them easy to identify (Table 2). Therefore the immunoprecipitation of cellular RNPs with autoimmune sera and analysis of their RNA components is a sensitive assay for autoimmune activities which tests the ability of antibodies to recognize native antigens with high affinity [3–5].

   The Sm antigens, recognized by autoantibodies in up to 30% of the patients with SLE, are the protein components of the nucleoplasmic snRNP particles. The snRNP particles are a family of six abundant particles, U1 – U6, and a growing number of less abundant particles in both the nucleoplasm and nucleolus. The four major nucleoplasmic particles in

Table 1. Frequency of anti-RNP antibodies in rheumatic diseases

	SLE	MCTD	Sjögren's Syndrome
Sm	30%		
U1(RNP)	40%	> 95%	
Ro	30%		50
La	10%		30
U2 snRNP	10%		

Table 2. RNAs precipitated by anti-RNP antibodies

		Nucleotides
Sm	U1	165
	U2	188/189
	U4	143–145
	U5	116/117
	U6	106/107
	other less abundant snRNAs	
U1(RNP)	U1	165
Ro	hY1	112
	hY3	101
	hY4	93
	hY5	82
La polymerase III products	hY RNAs	83–112
	U6	106/107
	pre-tRNA	approx. 80–90
	5S rRNA	approx. 123
	7SL	300/301
	7SK	331

human cells (U1, U2, U4 and U5) share a common core of at least eight proteins (B/B', D1, D2, D3, E, F and G) in addition to a unique RNA component and several snRNA specific proteins (for reviews, see [6,7]). The anti-Sm autoantibody usually recognizes determinants on the B and D family of proteins of the common core [8] and will immunoprecipitate the entire family of nucleoplasmic snRNP particles. The U6 snRNP particle, which, like the nucleolar U3 snRNP, lacks an Sm core, is precipitated because it is associated with U4 in a single U4/U6 particle. Because of their higher abundance, the U1, U2, U4, U5 and U6 RNAs are usually the only RNAs detected in Sm immunoprecipitates.

The U1(RNP) autoantibody response is a diagnostic marker for SLE-overlap syndromes like MCTD [1]. These antibodies usually recognize the U1–70kD and the U1-A protein and, less often, the U1-C protein. These sera specifically immunoprecipitate the U1 snRNP particle. Anti-U1RNP antibodies are often found in conjunction with anti-Sm activity in SLE sera which makes it difficult to distinguish this activity in the immuno-precipitation assay.

The anti-La (or SS-B) antibodies recognize a 48 kD protein found associated with RNAs transcribed by RNA polymerase III (for a review, see [9]). The protein has been described to function in the termination of polymerase III transcription and binds to a poly-uridine stretch at the 3' end of the transcript. La binding is usually lost after the uridines are removed during RNA maturation [10]. The La associated RNAs are found in the nucleus and in the cytoplasm. Several of these RNAs, including pre-tRNA, are nuclear but leak from nuclei prepared by aqueous extraction and, thus, appear in the cytoplasmic fractions [11]. The La antibodies immunoprecipitate a disperse set of low molecular weight RNAs including pre-tRNA, pre-5S rRNA, pre-7S RNA and the Y RNAs. However, the fraction of these RNAs associated with the La antigen is often low, except for the Y RNAs [9, 12]. U1 snRNA which is not transcribed by RNA polymerase III is also reported to be transiently associated with the La antigen and immunoprecipitable by anti-La antibody during its maturation in the cytoplasm [13].

The anti-Ro (or SS-A) antibodies recognize a 52 kD (Ro52) and a 60 kD (Ro60) protein associated with the Y RNAs in the cytoplasm [14]. The Y family of RNAs in humans, hY1, hY3, hY4 and hY5, range in size from 83–112 nucleotides (for a review, see [9]). These RNAs, of unknown function, are of low abundance and are difficult to detect in whole cell fractions. The RNAs are transcribed by RNA polymerase III and, therefore, are also bound to the La antigen. If anti-Ro and anti-La antibodies are both present, the immunoprecipitation assay is not a sensitive approach to distinguish these activities.

The immunoprecipitation of cellular extracts with autoimmune sera is a sensitive assay for the detection of antibodies to the snRNP, La and Ro particles. Several studies have shown that most, although not all, autoimmune sera with the anti-Sm, anti-(U1)RNP, anti-Ro and anti-La antibodies will immunoprecipitate their target RNP particles (4,5). RNAs can be radioactively labeled to high specific activity which provides a sensitive assay. The manipulations required for immunoprecipitation, however, require in most cases extensive handling of radioactive materials, including cell labeling, cell fractionation and gel electrophoresis which makes it difficult to process a large number of samples. However, there are non radioactive alternatives such as the sensitive silver staining procedure. The high sensitivity of the immunoprecipitation assay makes it especially valuable for diagnostic tests when other procedures are inconclusive.

2. Preparation of cellular extracts

The principle of the method described in this chapter is that RNP complexes, precipitated by the autoantibodies, are analysed via their RNA content. For this purpose one can start with an extract in which the RNAs are radiolabeled *in vivo* (par. 2.1). An alternative method is to use an unlabeled extract and demonstrate the presence of precipitated RNAs by a sensitive staining method such as silver staining (par. 4.3) or by probing with specific, radiolabeled probes (not discussed in this chapter). Another possibility is to endlabel the precipitated RNAs with 5'^{32}P-pCp [15].

2.1 Radioactive labeling of RNAs

Cellular RNAs are radiolabeled by incubating HeLa cells or other cultured mammalian cells with either ^{3}H-uridine or ^{32}P-orthophosphate.

2.1.1 ^{32}P labeling of suspension cells

Steps in the procedure
1. Grow cells to a density of 0.5×10^6 cells/ml in normal growth medium (Standard Minimum Essential Medium plus 10% Newborn Bovine Serum).
2. Spin cells down and wash them twice with phosphate free medium (Minimum Essential Medium, Eagle without phosphate, Flow Laboratories, Irvine, Scotland).
3. Resuspend cells in phosphate free medium to a density of 0.5×10^6 cells/ml.
4. Add ^{32}P orthophosphate to a concentration of 10–50 μCi/ml.
5. Incubate for 2 h at 37 °C.
6. Add 1/10th volume of normal complete medium.
7. Incubate overnight.

2.1.2. ^{3}H-uridine labeling of cells

Steps in the procedure
1. To label stable RNA species, 5 μCi/ml of ^{3}H-uridine is added directly to the cultured cells under normal growth conditions.
2. Labeled cultures are then maintained for 16–24 h under normal growth conditions before fractionation.

Notes to the procedure
1. To label newly synthesized RNA species, suspension cells are concentrated 10 times or monolayers are incubated in the minimum medium that will cover the cells. 50 μCi/ml ^{3}H-uridine is then added to the cells in normal medium. Cells are maintained

under normal growth conditions for 30 min to 2 h before fractionation.

2. The labeling conditions may inhibit cell growth and if dividing cells are required, the concentration of label must be reduced to 1 μCi/ml ^{3}H-uridine or 0.4 μCi/ml ^{32}P-phosphate.

3. ^{32}P provides a constant rate of incorporation during the entire labeling period; however, ^{3}H-uridine is rapidly taken up by cells. In long labels ^{3}H-uridine provides a labeling period of several hours followed by an effective cold chase. Short lived RNA species are not labeled as well after a 12-hour label with ^{3}H-uridine.

4. Cytoplasmic and nuclear fractions or a total extract from cells labeled for 16 h can be used to assay for Ro, La, Jo-1, Sm or U1(RNP) antisera, respectively (see Fig. 1) Pulse labeled nuclear and cytoplasmic fractions are appropriate to assay for anti-La antibodies.

5. ^{32}P has a 14.3 day half-life and can be stored until it decays (greater than ten half-lives) and then disposed of as normal waste. ^{3}H with a 12.3 year half-life must be disposed of as radioactive waste.

6. It is also possible to prepare unlabeled cell extracts and use this for immunoprecipitations. The precipitated RNAs can then be detected by 3′ end-labeling with 5′^{32}P-pCp [15], by silver staining (par. 4.3) or by probing with radiolabeled oligonucleotides (not discussed in this chapter).

7. If a CO_2 incubator is not available, add 10 mM Hepes pH 7.1 to the medium to maintain the proper pH during labeling.

2.2. Cell fractionation

Cells can be fractionated into cytoplasmic and nuclear fractions or a total cell extract can be prepared.

2.2.1 Preparation of cytoplasmic and nuclear fractions

Steps in the procedure

1. Following isotopic labeling, cells are collected by centrifugation and rinsed with PBS (145 mM NaCl, 2.6 mM NaH$_2$PO$_4$, 7.4 mM Na$_2$HPO$_4$ pH 7.2). All procedures are carried out at 4 °C unless stated otherwise.

2. Cells are resuspended in CSK buffer (100 mM NaCl, 1.5 mM MgCl$_2$, 10 mM Pipes pH 6.9) containing 0.5% Triton X-100 and a battery of protease and RNAase inhibitors (see notes) at a ratio of 1 ml of buffer per 5 × 10^6 cells.

3. Cells are vortexed vigorously for at least 15 s to assist the extraction of the cytoplasm.

4. Extracted cells are collected by centrifugation at 200 × g for 2 min. The supernatant, representing the cytoplasmic fraction is collected.

5. The nuclear pellet is extracted in a high salt buffer (0.3 M NaCl, 2 mM MgCl$_2$, 20 mM Tris.Cl pH 8.0) based on the procedure of Dignam et al. [16] to solubilize the nucleoplasmic snRNP particles. The nuclear pellet is resuspended in the high salt buffer at a ratio of 0.5 ml buffer per 5 × 10^6 cells. The suspension is agitated by either rocking or rotation for 15 min at room temperature.

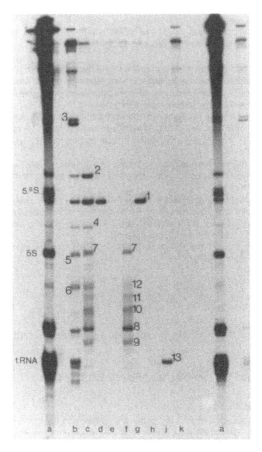

Fig. 1. *Analysis of RNPs precipitated by various autoantibodies.* HeLa cells were labeled with
^{32}P orthophosphate as described in paragraph 2.1.1. Total cell extracts were prepared as
described in paragraph 2.2.2. Immunoprecipitated RNAs were isolated as described in
paragraph 3.1 and analysed on a 10% polyacrylamide-8.3 M urea gel (see paragraph 4.1).

The sera used contained the following activities:

lane a: total cellular RNA. Note the 5.8S RNA which appears in this gel system as a double
band just above U1RNA.

lane b: serum of an SLE patient containing anti-Sm, anti-U1 RNP and anti-ribRNP
activity. The precipitated RNAs which can be identified are bands 1 (U1RNA), 2 (U2RNA),
4 (U4RNA), 5 (U5RNA) and 6 (U6RNA). The heavy band at the top is probably the small
ribosomal RNA (see also lane k). Band 3 is the 7SL/7SK doublet which is precipitated by an
unidentified antibody activity. This complex serum also contains antibody to tRNA species.

lane c: serum of an SLE patient containing anti-Sm, anti-U1 RNP and anti-La/Ro
activities. Precipitated RNAs that can be identified are the U RNAs (bands 1, 2, 4 and 6 are
clearly seen), the pre-5S RNA (band 7), the pre-tRNA (band 8) and the Y RNAs (bands 12,
11, 10 and 9). Note that an RNA precipitation analysis cannot easily detect anti-Ro activity
when anti-La is present. After longer exposures it can be seen that anti-La antibodies also

6. The nuclear suspension is centrifuged at 1000 × g for 4 min and the supernatant, containing the solubilized snRNP particles, is collected.

2.2.2 Preparation of total cell extract

Steps in the procedure
1. Collect cells by centrifugation (550 × g, 5 min, 4 °C) and wash twice with PBS to remove unincorporated label.
2. Resuspend cells in NP40 buffer (10 mM Tris pH 7.5, 150 mM NaCl, 0.5% NP40) at 10 × 10⁶ cells/ml.
3. Lyse cells by freeze-thawing 5 times. Extract the cells by extensive vortexing during thawing.

Notes to the procedures
1. The CSK buffer containing Triton X-100 solubilizes the membrane lipids and releases many cellular components, including the newly synthesized snRNPs, into the soluble phase. The buffer conditions minimize the leakage of the mature snRNP particles out of the nucleus.
2. To minimize degradation, cell fractionation should be done quickly and at 4 °C. Protease inhibitors (Aprotinin (0.03 TIU/ml) or PMSF (1 mM final concentration)) and RNAase inhibitors (placental ribonuclease inhibitor (75 units/ml)) or vanadate-ribonucleoside complex (5 mM) [17]) can be added to the extraction buffer to further reduce degradation.
3. An aliquot of each cell fraction should be saved to run on the final gels as markers.
4. The Dignam extraction procedure is very reproducible and elutes over 75% of the snRNP particles from the nucleus. The 0.3M NaCl used in this procedure reduces nonspecific interactions between proteins in the extract, resulting in lower backgrounds in the immunoprecipitations.

---

precipitate a small amount of (pre-) 7SL/7SK RNPs. This serum probably contains a weak anti-ribRNP activity (compare lanes b and k) as well.

lane d: serum of an SLE-overlap patient containing anti-U1RNP activity. The only RNA precipitated is U1 RNA (band 1).

lane e: serum from a scleroderma patient containing anti-topoisomerase I antibodies. No RNA is precipitated.

lane f: serum of a Sjögren's syndrome patient containing anti-Ro and anti-La activities. RNAs precipitated are the Y RNAs (hY1, hY3, hY4, hY5 can be seen as bands 12, 11, 10 and 9, respectively) and pre-5S and pre-tRNA (bands 7 and 8, respectively). After longer exposures the presence of the 7SK/7SL doublet (band 3) is visible.

lane g: Another anti-U1RNP serum. The only RNA precipitated is U1 RNA (band 1).

lane h: control normal human serum. No RNA is precipitated.

lane j: serum of a polymyositis patient with anti-Jo1 antibodies. The only RNA precipitated is the His-tRNA (band 13) which was associated with the antigenic protein (i.e. his-tRNA synthetase).

lane k: serum of an SLE patient with anti-ribRNP antibodies. The upper band probably is the 18S ribosomal RNA.

5. An alternative procedure for solubilizing the nuclear snRNP particles is to resuspend the detergent resistant nuclear pellet in CSK buffer and to disrupt it by sonication. Five to 10s bursts of a probe tip sonicater (setting 5) with the material maintained *on ice* are effective at disrupting the nuclei and solubilizing the snRNP particles. Care must be taken not to overheat the material during the sonication. Insoluble material is collected by centrifugation at 10,000 × g for 10 min and the supernatant, containing the snRNP particles, is collected.

3. Immunoprecipitation

3.1. Binding of antibodies to agarose beads and immunoprecipitation of RNP complexes

For the immunoprecipitation of the RNPs, antibodies of the human serum are bound to protein A or protein G agarose beads prior to their addition to the cellular fractions. All reagents are maintained at 4°C.

Steps in the procedure
1. Twenty μl of patient serum diluted with 500 μl of IPP150 (10 mM Tris.Cl pH 7.5, 150 mM NaCl, 0.1% NP40) is incubated with 20 μl of a 50% suspension of protein A agarose.
2. Rotate the suspension 1 hour at room temperature or overnight at 4 °C.
3. Wash the protein A agarose-antibody complex 3 times with IPP500 (10 mM Tris.Cl pH 7.5, 500 mM NaCl, 0.05% NP40).
4. Resuspend the antibody-protein A agarose beads in 450 μl IPP150.
5. Remove insoluble material from the cell fractions by centrifugation at 12,000 × g for 10 min at 4 °C.
6. Mix 50 μl of total cell extract or nuclear fraction or 100 μl of the cytoplasmic fraction (equivalent of 0.5 × 10^6 cells) with antibody conjugated protein A agarose and rotate for 1–5 h at 4 °C.
7. Pellet the agarose beads by brief centrifugation, remove the supernatant and wash the beads three to five times with 1 ml of IPP150.
8. Add 100 μl IPP150/0.5% SDS/ 2 μg total yeast RNA to the agarose pellet and incubate for 3 min at 65 °C.
9. Extract the total sample with 1 volume of phenol/chloroform (1:1).
10. Collect the water phase and add 0.1 volume Na-Acetate pH: 5.2. Mix well.
11. Add 2.5 volumes ethanol and mix well again.
12. Incubate for at least 1 hour at −20 °C to precipitate the RNA.
13. Collect the RNA by centrifugation (20 min, 12,000 × g, 4 °C).
14. Wash pellet with 500 μl of 80% ethanol and dry it under vacuum.

Notes to the procedure
1. Protein G has a wider binding specificity than does protein A. However, protein A has the advantage that it does not bind fetal calf antibodies and this may be convenient if contamination with calf serum components from the medium can be expected. Protein G has the advantage that the efficiencies of antibody binding are almost 100%, while protein A typically binds only 50% of the input antibodies. Antibody binding to protein A can be enhanced by adjusting the incubation mixture to pH: 8.
   When using protein A, the unbound supernatant can be rebound a second time to recover further antibodies. Ratios of agarose beads to antibody can be optimized for each antibody depending on the titer of the serum. Generally, 20 μl of a 50% suspension of protein A agarose per 10 μl of serum is most effective. If the resulting pellets are small, normal agarose beads can be added to increase the size of the pellets so they can be monitored visually during the rinsing procedures.
2. The antibody-protein A agarose bead conjugates remain stable with full activity for extended periods of time when stored at 4 °C.
3. Pre-clearing the extract by incubating with normal agarose beads removes cellular material that will adhere non-specifically to the beads.
4. Immunoprecipitations are conveniently performed in microcentrifuge tubes.
5. Caution should be used not to aspirate the agarose beads while removing the supernatants.
6. Phenol extraction of the final pellets is not always necessary. The rinsed immunoprecipitates can be resuspended directly in sample buffer. However, whole cell fractions require phenol extraction before analysis.

4. RNA Gel Electrophoresis

The snRNAs and other small RNAs can be resolved on a variety of gel systems. Two established systems are a 10% acrylamide gel system containing urea to denature the RNAs (see Fig. 1) and a 6–15% gradient gel system [6, 11].

4.1 10% acrylamide/8.3 M urea gel

Steps in the procedure
1. Resuspend the RNA pellet in loading buffer (80% formamide, 90 mM Tris-borate pH 8.3, 2 mM EDTA, 0.025% bromophenol-blue, 0.025% Xylene blue).
2. Separate the sample on an acrylamide/urea gel (10% acrylamide/ bisacrylamide (19:1), 90 mM Tris-borate pH 8.3, 2 mM EDTA, 8.3 M urea).
3. Gels containing ^{32}P are dried directly after electrophoresis, covered with Saran Wrap (Dow Corning) and exposed to X-ray film with an intensifying screen for periods of one to five days. Gels containing ^{3}H are processed with Enhance fluorographic enhancer (NEN), dried and exposed to X-ray film at -70 °C.

4.2 6–15% gradient acrylamide gels

Steps in the procedure
1. Resuspend either the RNA pellet or the final rinsed immuno-precipitate in 100 μl of sample buffer (10 mM Tris.Cl pH 6.8, 10% glycerol, 1% SDS, 0.05% bromophenol-blue, 0.05% Xylene cyanol).
2. Pour a 6–15% polyacrylamide gradient gel (6% and 15% acrylamide:bisacrylamide (30:0.8), 370 mM Tris.Cl pH 8.8, 0.1% SDS).
3. Load the samples and run gels until bromophenol-blue runs off the bottom of the gel (Running buffer is 192 mM glycine, 25 mM Tris, 0.1% SDS, pH will equal 8.8).
4. Gels containing ^{32}P are dried directly after electrophoresis, covered with Saran Wrap (Dow Corning) and exposed to X-ray film with an intensifying screen for periods of one to five days. Gels containing ^{3}H are processed with Enhance fluorographic enhancer (NEN) dried and exposed to X-ray film at -70 °C. Exposures are from one day to two weeks.

Note to the procedures
1. The advantage of a non-denaturing gel system is that the 5.8S rRNA hydrogen bonded to the 28S rRNA is not denatured and released under these conditions. The 5.8S rRNA migrates with similar mobility to U1 (see Fig. 1) and can interfere with the identification of this species under denaturing conditions.

4.3 Silver staining procedure

Steps in the procedure
1. Fix the gel for 30 min in fixation solution (40% v/v methanol, 10% v/v acetic acid).
2. Rinse the gel for 15 min in rinsing solution (10% v/v ethanol, 5% v/v acetic acid).
3. Repeat step 2.
4. Wash the gel with distilled water for 1 min.
5. Stain the gel for 30 min at room temperature in 12 mM AgNO$_3$ solution.
6. Wash the gel with distilled water for *maximal* 1 min.
7. Develop the colour in two changes of 50 ml developer solution (0.28 M Na$_2$CO$_3$, 0.018% v/v formaldehyde solution).

5. Reagents and suppliers

Protein A Sepharose, Sigma – P-3391
GammaBind Plus Sepharose, Pharmacia LKB #17–0886–02 or Kem-en-Tec, Copenhagen, Denmark (cat.nr. 1060 D).
Placental ribonuclease inhibitor, Promega #N2112
Enhance, Dupont/NEN – NEF-981

.

References

1. Tan EM (1989) Antinuclear antibodies: Diagnostic markers for autoimmune diseases and probes for cell biology. Adv Immunology 44: 93–151
2. Craft J, Mimori T, Olsen TL & Hardin JA (1988) The U2 small nuclear ribonucleoprotein particle as an auto antigen. Analysis with sera from patients with overlap syndromes. J Clin Invest 81: 1716–1724
3. Lerner MR & Steitz JA (1979) Antibodies to small nuclear RNAs complexed with proteins are produced by patients with systemic lupus erythematosus. Proc Natl Acad Sci USA 76: 5495 – 5499
4. Pettersson I, Hinterberger M, Mimori T, Gottlieb E & Steitz JA (1984) The structure of mammalian small nuclear ribonucleoproteins: Identification of multiple protein components reactive with anti-(U1)RNP and anti-Sm autoantibodies. J Biol Chem 259: 5907–5914
5. Meilof JF, Bantjes I, De Jong J, Van Dam VP & Smeenk RJT (1990) The detection of anti-Ro/SS-A and anti- La/SS-B antibodies. J Immunol Meth 133: 215–226
6. Zieve GW & Sauterer RA (1990) Cell biology of the snRNP particles. CRC Crit Rev Biochem Mol Biol 25: 1–46
7. Lührmann R, Kastner B & Bach M (1990) Structure of spliceosomal snRNPs and their role in pre-mRNA splicing. Biochim Biophys Acta 1087: 265–292
8. Lehmeier T, Foulaki K & Lührmann R (1990) Evidence for three distinct D proteins, which react differentially with anti-Sm autoantibodies, in the cores of the major snRNPs U1, U2, U4/U6 and U5. Nucl Acids Res 18: 6475–6484
9. Pruijn GJM, Slobbe RL & Van Venrooij WJ (1991) Structure and function of Ro and La RNPs. Mol Biol Rep 14: 43–48
10. Gottlieb E & Steitz JA (1989) Function of the mammalian La protein: evidence for its action in transcription termination by RNA polymerase III. EMBO J 8: 841–850
11. Zieve GW, Sauterer RA & Feeney RJ (1988) Newly synthesized snRNAs appear transiently in the cytoplasm. J Mol Biol 199: 259–267
12. Rinke J & Steitz JA (1982) Precursor molecules of both human 5S ribosomal RNA and transfer RNAs are bound by a cellular protein reactive with the anti-La lupus antibodies. Cell 29: 149–159
13. Madore SJ, Wieben ED & Pederson T (1984) Eukaryotic small ribonucleoproteins: Anti-La human autoantibodies react with U1 RNA-protein complexes. J Biol Chem 259: 1929–1933
14. Slobbe RL, Pluk W, Van Venrooij WJ & Pruijn GJM (1992) Ro ribonucleoprotein assembly in vitro. Identification of RNA-protein and protein-protein interactions. J Mol Biol 227: 361–366
15. England TE & Uhlenbeck OC (1978) 3'-terminal labeling of RNA with T4 ligase. Nature 275: 560–561
16. Dignam JD, Lebovitz RM & Roeder RG (1983) Accurate transcription initiation by RNA polymerase II in a soluble extract from isolated mammalian nuclei. Nucl Acids Res 11: 1475–1489
17. Berger SL & Birkenmeirer CS (1979) Inhibition of intractable nucleases with ribonucleoside – vanadyl complexes. Biochemistry 18: 5143–5149

Manual of Biological Markers of Disease **A8**: 1–12, 1993.

Measurement of antibodies to DNA

R. SMEENK
Department of Autoimmune Diseases, Section Immunoreumatology, Central Laboratory of the Netherlands Red Cross Bloodtransfusion Service (C.L.B.), Plesmanlaan 125, 1066 CX Amsterdam, The Netherlands

Abbreviations

SLE: systemic lupus erythematosus; ELISA: enzyme linked immunosorbent assay; PEG: polyethylene glycol; IFT: immunofluorescence technique; PBS: phosphate buffered saline; BSA: bovine serum albumin; HRP: horse radish peroxidase; RIA: radioimmuno assay; FITC: fluorescein isothiocyanate

Key words: Autoantibodies, anti-dsDNA, Elisa, FARR assay

1. Introduction

Antibodies to DNA can be found in the circulation of the majority of patients with Systemic Lupus Erythematosus (SLE). They are quite specific for this disease, which makes their detection an important diagnostic aid to the clinician. Fluctuations in the level of anti-dsDNA in an individual patient generally parallel the clinical status of that patient. Furthermore, the presence of anti-dsDNA may precede the diagnosis of SLE by more than a year.

For the measurement of antibodies to DNA many different assays have been developed. Each of these detects different parts of the spectrum of anti-DNA antibodies present in an individual patient. Four methods of relevance to the detection of anti-dsDNA antibodies will be discussed in this chapter: the ELISA, the indirect immunofluorescence test on *Crithidia luciliae*, the PEG assay and the Farr assay.

Comparing these methods using sera of patients with defined SLE, we obtained good indices of correlation (Table 1). Discrepancies between the assays first came to light when the assays were used for screening purposes, i.e. when assaying sera referred to our institute for diagnostic reasons (Table 2). Actually, the latter screening is a more realistic approach, since this is the way anti-DNA assays are bound to be used mostly, i.e. to confirm or refute a possible diagnosis of SLE. Adopting this latter approach, we

Table 1. Correlations between 5 anti-dsDNA assays. Anti-dsDNA levels of sera of defined SLE patients were determined by (in-house) Farr assay, Amersham Kit, PEG assay, Crithidia test and ELISA. All levels were expressed in IU/ml. Correlations were calculated using linear regression analysis and Kendall's rank correlation test. Data have been compiled from [5] and [23].

anti-DNA assay	Amersham Kit	PEG assay	Crithidia test	ELISA
Farr assay	$r = 0.93$[a] $t = 0.73$[b] ($p < 0.0001$)	$r = 0.67$ $t = 0.63$ ($p < 0.0001$)	$r = 0.85$ $t = 0.69$ ($p < 0.0001$)	$r = 0.67$ $t = 0.53$ ($p < 0.0001$)
Amersham Kit	--	$r = 0.93$ $t = 0.57$ ($p < 0.001$)	$r = 0.52$ $t = 0.59$ ($p < 0.001$)	n.d.
PEG assay	--	--	$r = 0.83$ $t = 0.69$ ($p < 0.0001$)	$r = 0.62$ $t = 0.63$ ($p < 0.0001$)
Crithidia test	--	--	--	$r = 0.70$ $t = 0.66$ ($p < 0.0001$)

[a] Coefficient of correlation by linear regression analysis.
[b] Kendall's rank correlation coefficient.

Table 2. Comparison of results obtained on 289 sera referred to our institute for routine anti-DNA determination (taken from ref. [24]).

PEG assay	ELISA	Crithidia test	Farr assay	sera (n)
+[a]	+	+	+	22
+	+	+	−	22
−	+	+	+	7
+	+	−	−	15
+	−	+	−	2
−	+	+	−	3
−	+	−	+	1
+	−	−	−	33
−	+	−	−	23
−	−	+	−	3
−	−	−	−	158
94	93	59	30	289

[a] A + indicates a positive result, a − a negative result with the indicated assay.

initially noticed a large discrepancy between the Farr assay and the IFT on *Crithidia luciliae* [1]. Most of the deviating sera turned out to contain anti-DNA of low avidity only, detected with the Crithidia test and the PEG assay but not with the Farr assay [1–3]. Yet, discrepancies between the Crithidia test and the PEG assay were also found. In the first place, not all Crithidia-positive sera were found PEG-positive; furthermore, a significant number

AMAN-A8/2

of sera were found to be PEG-positive but Crithidia-negative [1,3]. Introduction of the anti-DNA ELISA to these comparisons introduced even more discrepancies [4,5].

Of the methods described in this chapter, the ELISA is the most sensitive method, whereas the Farr assay is the most specific for SLE. Using the latter method, only antibodies of a relative high avidity for DNA are detected. Mild forms of SLE, where patients only have anti-dsDNA of a low avidity in their circulation, may easily be missed by this technique. On the other hand, if screening for the presence of anti-DNA takes place using an assay that is not selective for anti-DNA of high avidity, then a positive assay result does not always indicate that the patient has SLE: anti-DNA of lower avidity seems to occur in other diseases than SLE as well (Table 3). An evaluation of the diagnostic value of low avidity anti-dsDNA has shown that of patients whose anti-dsDNA was PEG-positive but Farr-negative, 52% had SLE [6]. When high avidity anti-dsDNA was present as well, 86% of the patients were found to have SLE. Therefore, screening using a "broad spectrum" method should be followed by an assay selective for high avidity anti-dsDNA.

Table 3. Comparison of the disease specificity of 4 anti-dsDNA assays.

| Disease | percentage of sera positive in assay: | | | |
	Farr assay	PEG assay	Crithidia test	ELISA
active SLE	98	96	96	100
Rheumatoid arthritis	1	5	0	3
Sjögren's syndrome	0	7	0	20
Scleroderma	0	0	0	30
Autoimmune hepatitis	0	20	0	15
Myasthenia gravis	0	32	0	20
Autoimmune thyroiditis	0	7	0	13
Autoimmune gastritis	0	0	0	18
Morbus Addison	0	0	0	0
Autoimmune haemolytic anemia	0	10	0	8
Normal donors	0	0	0	0

Because the Farr assay and PEG assay may be readily compared as they only differ with respect to the way DNA/anti-dsDNA complexes are precipitated, a relative avidity index can be acquired by calculating the ratio between results (in IU/ml) of both assays. An optimal discrimination between low and high avidity DNA was obtained using a cut-off Farr/PEG-ratio of 5 [7]. Using this approach we found that the anti-dsDNA avidity of patients with nephritis was significantly higher than that of patients with CNS involvement [8].

2. Procedures

2.1 ELISA

Steps in the procedure
1. Coat ELISA plates for 2 hours at room temperature with 200 μl per well of a solution of protamine sulphate (0.5 mg/ml in distilled water).
2. Wash the plates thoroughly with PBS (5 times) and distilled water (3 times).
3. Coat the plates overnight at room temperature with a solution of 2 μg/ml DNA in PBS (200 μl per well).
4. Wash the plates with PBS (5 times) and distilled water (3 times).
5. Samples to be tested (sera or (monoclonal) antibody preparations) are serially diluted in PBS containing 2% BSA and 0.02% Tween 20 (PTB). Start with a dilution of 1:50 or 1:100.
6. Add 100 μl of the relevant dilution to a coated well and incubate the plates for 2 hours at room temperature.
7. Wash the plates with PBS (5 times) and distilled water (3 times).
8. Add 100 μl of a solution of horseradish peroxidase (HRP) conjugated anti-immunoglobulin antiserum (made in PTB) to each well. Incubate the plates for 2 hours at room temperature.
9. Wash the plates with PBS (5 times) and distilled water (3 times).
10. Add 100 μl per well of the substrate solution (TMB) and allow the colour development to take place for 15 min.
11. Stop the reaction by addition of an equal volume of 2 N H_2SO_4.
12. Read the plates in a Titertek Multiskan at 450 nm.

Notes to the procedure

ad 1–3. Doublestranded DNA binds very weakly to plastic, yet dsDNA is to be preferred above singlestranded DNA in this ELISA. Therefore, a "trick" is used: the plates are precoated with a positively charged protein such as protamine (both sulphate and chloride will do) or poly-L-lysine in order to facilitate the binding of DNA to the plates. An important drawback of such a precoat is that complexes with affinity for protamine, for example, will score positive in the ELISA. In practice, this means that DNA/anti-DNA complexes, nucleosome/anti-nucleosome complexes, or nucleosome/anti-histone complexes will also score positive in this ELISA. It is estimated that between 10 and 20% of the sera found positive with the anti-DNA ELISA are false-positive. Therefore, a positive ELISA result should *always* be confirmed using another anti-DNA assay.

An alternative procedure is to use (photo)biotinylated DNA and precoat the plates with streptavidin. Problems inherent to this approach are, for instance, the putative presence of anti-avidin antibodies in otherwise healthy individuals.

ad 4. Addition of Tween-20 to the washing buffer at this stage leads to elution of bound DNA. For similar reasons Tween is also omitted from the washing buffer of step 2.

ad 8. The conjugate to be used may either be directed against all immunoglobulin classes or against only one particular (sub)class. IgG, IgM and other isotypes of anti-DNA can thus be measured separately. The optimal dilution of the conjugated antibody used is to be established experimentally for each batch of antiserum.

ad 10. The bound enzyme (HRP) catalyses the oxidation of TMB to a (blue) coloured reaction product.

ad 11. The colour now changes into yellow.

ad 12. Units are defined as the reciprocal titer of the serum dilution that gives an absorption of 1.0. To express anti-dsDNA levels in International Units (IU)/ml, sera are compared with the W.H.O reference serum (Wo/80; [20]).

2.2 *IFT on* Crithidia luciliae

This method is an indirect immunofluorescence technique (IFT), originally described by Aarden *et al.* [9] and later subjected to some modifications by Aarden and Smeenk [10]. *Crithidia luciliae* microorganisms, hemo-flagelates containing a giant mitochondrion which contains large amounts of dsDNA, are used as a substrate. Immunofluorescence of this so-called kinetoplast therefore indicates the unequivocal presence of antibody to dsDNA. An example is given in Figure 1.

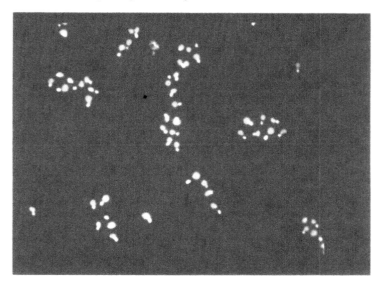

Fig. 1. Positive immunofluorescent staining of Crithidia luciliae by anti-DNA. The yellow-green immunofluorescence of kinetoplasts indicate the presence of antibodies to dsDNA. Nuclei are counterstained with ethidium bromide.

Steps in the procedure

1. Slides, containing spots of *Crithidia luciliae* organisms, are usually kept in the freezer at −20 °C. They are briefly thawed just prior to the assay.

2. Samples to be tested are (serially) diluted in PBS, starting at a 1:10 dilution.
3. About 50 μl of each dilution are incubated with a spot of *Crithidia luciliae* for 30 min at room temperature.
4. Slides are briefly rinsed with tap water and then washed for 30 min with PBS (3 changes).
5. Bound antibodies are detected by means of a fluorescein isothiocyanate (FITC) conjugated anti-immunoglobulin antiserum, diluted in PBS. Incubation is allowed to take place for 30 min at room temperature.
6. Slides are briefly rinsed with tap water and then washed for 30 min with PBS (3 changes).
7. The slides are mounted using a solution of 65% sucrose in PBS *of pH 8.2*, in which 0.5 μg/ml propidium iodide is dissolved.
8. Slides are read using a standard fluorescence microscope (at 500x), equipped with incident illumination. A KP560 barrier filter (Leitz) can be used to separate propidium iodide fluorescence (red) from FITC fluorescence (green). The last dilution giving a positive fluorescence is given as the titer. Using the Wo/80 reference serum, titers can be expressed in IU/ml.

Notes to the procedure

ad 1. Slides containing spots of *Crithidia luciliae* can either be obtained commercially, or prepared yourself [9]. *Crithidia luciliae* organisms are available from the American Type Culture Collection, 12301 Parklawn drive, Rockville, Maryland, USA.
Slides can be kept at −20 °C for at least one year. Just prior to their use in the anti-DNA test they are thawed using a fan (to prevent condensation of moisture).

ad 4. Extreme care should be taken to make sure that the slides do not dry up after the start of the procedure. Drying may lead to a local increase of the salt concentration and, consequently, to elution of antibody. The interaction of anti-DNA with DNA is electrostatical and therefore sensitive to dissociation by increased salt concentrations.

ad 5. The conjugate to be used may either be directed against all immunoglobulin classes or against one particular (sub)class, as described for the anti-DNA ELISA. The optimal dilution of the used conjugated antiserum is to be established experimentally for each batch.

ad 7. The propidium iodide or ethidium bromide in the mounting solution acts as a counterstain to facilitate the localization of the kinetoplasts during microscopy. After mounting, the slides must be kept in the dark for at least 30 min before reading them. Prepared this way, slides can be kept in the dark for several days at room temperature or several weeks at −20 °C.

ad 8. Nuclei, of *Crithidia luciliae* or other cells, always contain a lot of antigens other than DNA and therefore, cannot be used to detect anti-dsDNA. Therefore, it is of extreme importance to score only kinetoplast fluorescence in this assay.

2.3 PEG assay

The PEG assay is a radioimmunoassay (RIA) in which radiolabelled DNA is used as antigen. For this purpose we use circular DNA from the bacteriophage PM2. DNA/anti-DNA complexes are separated from free radio-labelled DNA by means of a 3.5% polyethylene glycol (PEG) precipitation. Details of this assay can also be found elsewhere [1–3,11].

Steps in the procedure

1. Mix 50 μl of (a dilution of) the sample to be assayed, 50 μl of a solution of 1.6 mg/ml human gammaglobulin (HGG), 50 μl of a solution of 0.2 mg/ml dextran sulphate, 50 μl of a solution of 2 μg/ml of [^{3}H]-PM2-DNA and 200 μl of a solution of 7% polyethylene glycol (PEG 6000) in small glass tubes. Always add a control (in duplicate) in which PBS instead of serum is assayed.
2. Incubate the tubes first for 1 hour at 37°C, then for 2 hours on ice (0°C).
3. Centrifuge the tubes for 15 minutes at 3000 g in a refrigerated centrifuge (4°C).
4. Dissolve 200 μl of the supernatant in 8 ml of NE-260$_{sp}$ scintillation fluid (see par. 3.3) and count the radioactivity (in counts per minute, cpm) in a liquid scintillation counter.
5. Calculate the percentage of binding via the formula: [1-(cpm sample/cpm control)]x100%.
   The outcome can also be expressed in units per ml; preferably, anti-dsDNA levels should be expressed in International Units (IU)/ml.

Notes to the procedures

ad 1. PM2 DNA is a circular doublestranded bacteriophage DNA molecule of 5.8 x 10^6 D. It is prepared by culturing PM2 bacteriophages, in the presence of ^{3}H-labelled thymidine, on their host the bacterium Pseudomonas strain BAL31. It is isolated from the lysate by standard DNA purification methods (for details see ref [10]). Instead of PM2 DNA any doublestranded DNA of homogeneous length and not larger than about 10 x 10^{6}D can be used. Commercial kits (such as from Amersham International or DPC) generally employ iodine labelled plasmid DNA, which works very well. If you iodinate DNA yourself make sure that you don't introduce nicks in the DNA that would lead to exposure of singlestranded regions. This assay is very sensitive to nonspecific binding of DNA by serum proteins. Because there is no inherent check on the immunoglobulin nature of the DNA-binding material, any molecule with sufficient affinity for DNA may in principle cause a false-positive reaction. This nonspecific binding can, however, easily be eliminated by addition of an excess of the polyanion dextran sulphate to the incubation mixture.

Normally, sera are screened undiluted (i.e. 50 μl per test), in duplicate. If necessary, positive sera can be titrated (we test 20μl, 10μl, 5 μl, 2μl, 1μl etc; as far down as necessary to get into the 30% binding range).

ad 2. In the originally published procedure [11] the incubation takes place without PEG for the first hour at 37 °C; then an equal volume of PEG is added and incubation is continued for 2 hours on ice. However, the procedure described above works equally well.

DNA is not precipitated by 3.5% PEG, neither is IgG nor IgM. Complexes must have a certain minimum size before they are precipitated by 3.5% PEG.

ad 3. This precipitates the DNA/anti-DNA complexes. The pellets are, however, not very solid, so care should be taken not to disrupt them when taking the tubes out of the centrifuge.

ad 5. Actually, the amount of radioactivity *not* precipitated is counted. By dividing this by the amount of radioacivity not precipitated in the PBS control, one obtains the percentage of DNA *not* bound by the sample. Subtracting this from 100 gives the percentage of bound DNA.

In the past, we defined 1 unit of anti-DNA as the amount that binds 30% of the input DNA under the given conditions. It is preferable to express anti-DNA levels in International Units, by comparing sera with the WHO (Wo/80) reference serum [20]. The observation that some sera are PEG-negative but react positive in the other three assays may be explained by the finding that complexes need to be large to be detected by the PEG assay [2].

2.4 Farr assay

This technique was introduced by Wold *et al.* in 1968 [12], and employs ammonium sulphate precipitation to separate DNA/anti-DNA complexes from free (radiolabeled) DNA. This radioimmunoassay has been studied quite extensively [10, 13–18] and was shown to be very specific for SLE, provided that purely doublestranded DNA is used as antigen [14, 19]. In our assay we use circular DNA from the bacteriophage PM2.

Steps in the procedure

1. Mix 50μl of (a dilution of) the sample to be assayed, 50μl of a solution of 16 mg/ml HGG, 50 μl of PBS with 50μl of a solution of 2μg/ml [^{3}H]-labelled PM2 DNA, in large glass tubes (16 x 100 mm).
2. Incubate the tubes for 1 hour at 37 °C.
3. Add 5 ml of a *cold* 50% saturated solution of ammonium sulphate (in PBS) to each tube.
4. Allow the precipitate to form for 30 min at 4 °C.
5. Centrifuge the tubes for 15 min at 3000 g in a refrigerated centrifuge (at 4 °C).
6. Wash the precipitate twice with 50% saturated ammonium sulphate solution at 4 °C (5 ml per wash; centrifuge again for 15 min at 3000 g).
7. Dissolve the pellet in 1 ml Soluene-100 (Packard) by thorough mixing.
8. Add 10 ml scintillation fluid (e.g. Instafluor II from Packard) and count the tubes for radioactivity in a liquid scintillation counter. Always include at least 2 tubes in which 50 μl of the used PM2 DNA solution is dissolved in Soluene-100 and scintillation fluid (the "input" DNA). This allows calculation of the amount of DNA bound by the samples.

9. The percentage of DNA binding is calculated via the formula: [(cpm sample)/(cpm input)]x100%. Anti-dsDNA is preferably expressed in IU/ml, by comparison with the first international standard for anti-dsDNA (Wo/80) [20].

Notes to the procedure

ad 1. In the Farr assay the HGG solution used has a 10 times higher concentration than the one used in the PEG assay. HGG is necessary to facilitate the precipitation of the serum IgG: especially when highly diluted sera are tested, precipitation is improved by addition of an excess of (precipitable) protein.
The choice of assay conditions employed may be of great influence to the outcome of the Farr assay [19, 21].
The use of iodine-labelled calf-thymus DNA makes the assay more prone to nonspecific DNA-binding [14, 22], but iodinated plasmid DNA such as that used in commercial anti-dsDNA kits (Amersham, DPC) serves very well [23].

ad 3. This precipitates most proteins, including immunoglobulin. If DNA is bound to immunoglobulin it will co-precipitate. Because the employed ammonium sulphate solution is about 3 M, DNA/anti-DNA interactions of low avidity will be dissociated by this precipitation step. This explains the frequently occurring discrepancies between results obtained with Farr and PEG assay.

ad 6. The washing procedure is necessary to remove non-specifically precipitated DNA.

ad 7. Be sure that not too much water remains with the precipitate, because water and Soluene do not mix very well.

3. Reagents and solutions

3.1 ELISA

1. ELISA plates: NUNC maxisorb.
   Protamine sulphate can be obtained from Organon, Oss, The Netherlands.
2. PBS: phosphate buffered saline: 0.14 M NaCl, 0.01 M NaH_2PO_4/K_2HPO_4, pH 7.4.
3. DNA may be calf thymus DNA (e.g. from Worthington Biochemicals, Harrow, U.K.), or, preferably, purified plasmid DNA (e.g. pUC 9).
5. PTB: PBS containing 2% BSA and 0.02% Tween 20.
   BSA: bovine serum albumin (e.g. from Povite, Oss, the Netherlands).
   Tween 20: e.g. from J.T. Baker Chemicals, Deventer, the Netherlands.
10. TMB:3,5,3',5'-tetramethylbenzidine (TMB, Merck, Darmstadt, F.R.G.)
    TMB solution: 100 μg/ml TMB in 0.11 M sodium acetate buffer, pH 5.5, supplemented with 0.003% H_2O_2.

3.2 IFT on Crithidia luciliae

7. Propidium iodide and ethidium bromide can be obtained from Calbiochem, San Diego, CA. Concentration most often used is 0.5 ug per ml.

3.3 PEG assay

1. [³H]-PM2 DNA: is prepared as described in [10]. Stock solutions can be kept in the refrigerator ($+4\,^{\circ}$C.) for over one year. Add a few drops of chloroform to prevent bacterial growth.
HGG: human gammaglobulin, Cohn fraction II (can be obtained commercially, but can also be prepared yourself). A solution of 1.6 mg/ml is prepared in PBS. Stock solutions can be kept frozen for prolonged periods of time.
DXS: dextran sulphate, obtained from Pharmacia Fine Chemicals, Uppsala, Sweden. A stock solution of 2 mg/ml in PBS is prepared and stored at $+4\,^{\circ}$C; a 1:10 dilution (in PBS) is made freshly from the stock.
PEG: polyethylene glycol 6000, obtained from Koch Light Laboratories, Colnbrook, U.K. A solution of 7% is made in PBS and can be kept at $+4\,^{\circ}$C.
4. NE-260$_{sp}$ is prepared by mixing 1 part NE-260 (New England Nuclear Corp., Boston, MA) with 9 parts Instafluor II scintillation fluid (Packard Instrument Co., Downers Grove, IL).

3.4 Farr assay

1. [³H]-PM2 DNA: see par.3.3 PEG assay, point 1.
3. A 50% saturated ammonium sulphate solution is prepared in PBS of pH 7.4; please check and, if necessary, adjust the pH of the final 50% saturated solution.
7. Soluene-100 can be obtained from Packard (Packard Instrument Co., Downers Grove, IL).
8. For scintillation fluid we use Instafluor II from Packard.

References

1. Smeenk R, van der Lelij G & Aarden LA (1982) Measurements of low avidity anti-dsDNA by the Crithidia luciliae test and the PEG-assay. Clin Exp Immunol 50: 603–610
2. Smeenk R & Aarden LA (1980) The use of polyethylene glycol precipitation to detect low avidity anti-DNA antibodies in Systemic Lupus Erythematosus. J Immunol Methods 39: 165–180
3. Smeenk R, van der Lelij G & Aarden LA (1982) Avidity of antibodies to dsDNA. Comparison of IFT on Crithidia luciliae, Farr-assay and PEG-assay. J Immunol 128: 73–78
4. Smeenk R (1986) Detection of antibodies to DNA by enzyme-linked immunosorbent assay (ELISA). In: Pal SB (Ed)Immunoassay Technology, volume 2 (pp. 121–144). Walter de Gruyter, Berlin-New York
5. Smeenk R (1986) A comparison of four different anti-DNA assays. In: Pal SB (Ed)Immunoassay Technology, volume 2 (pp. 145–166). Walter de Gruyter, Berlin-New York
6. Nossent JC, Huysen V, Smeenk RJT & Swaak AJG (1989) Low avidity antibodies to dsDNA as a diagnostic tool. Ann Rheum Dis 48: 748–752
7. Swaak AJG & Smeenk R (1985) Clinical differences between SLE patients in relation to the avidity of their anti-dsDNA. In: Peeters H (Ed) Protides of the biological fluids (pp. 317–320). Pergamon Press, London
8. Smeenk R, Rooyen A van & Swaak AJG (1988) Dissociation studies of DNA/anti-DNA complexes in relation to anti-DNA avidity. J Immunol Methods 109: 27–35
9. Aarden LA, De Groot ER & Feltkamp TEW (1975) Immunology of DNA III. Crithidia Luciliae, a simple substrate for the determination of anti-dsDNA with the immunofluorescence technique. NY Acad Sci USA 254: 505–515
10. Aarden LA & Smeenk R (1981) Measurements of antibodies specific for DNA. In: Lefkovits I & Pernis B (Ed) Immunological methods, volume II (pp. 75–82). Academic Press, New York
11. Riley RL, McGrath HJr & Taylor RP (1979) Detection of low avidity anti-DNA antibodies in systemic lupus erythematosus. Arthr Rheum 22: 219–225
12. Wold RT, Young FE, Tan EM & Farr RS (1968) Deoxyribonucleic acid antibody: a method to detect its primary interaction with deoxyribonucleic acid. Science 161: 806–807
13. Feltkamp TEW (1975) The significance of the determination of anti-DNA and DNA/anti-DNA complexes. Scand J Rheumatol Supplement 11
14. Maini RN & Holborow EJ (1977) Detection and measurement of anti-DNA antibodies. Ann Rheum Dis 36S
15. Aarden LA, Lakmaker F & Feltkamp TEW (1976) Immunology of DNA. I. The influence of reaction conditions on the Farr-assay as used for the detection of anti-DNA. J Immunol Methods 10: 27–37
16. Monier JC, Perraud M, Ricard F & Gioud M (1978) Comparison of three techniques for the detection of antibodies to dsDNA: immunofluorescence on Trypanosoma gambiense, immunofluorescence on Crithidia luciliae and radioimmunoassay using the Farr-assay. Ann Immunol (Inst. Pasteur) 129C: 550–556
17. Ballou SP & Kushner I (1979) Immunochemical characteristics of antibodies to DNA in patients with active SLE. Clin Exp Immunol 37: 58–64
18. Rubin RL, Lafferty J & Carr RI (1978) Re-evaluation of the ammonium-sulphate assay for DNA antibody. Arthr Rheum 21: 950–954
19. Aarden LA, Lakmaker F & Feltkamp TEW (1976) Immunology of DNA. II. The effect of size and structure of the antigen on the Farr-assay. J Immunol Methods 10: 39–48
20. Feltkamp TEW, Kirkwood TBL, Maini RN & Aarden LA (1988) The first international standard for antibodies to dsDNA. Ann Rheum Dis 47: 740–746
21. Smeenk R, Duin T & Aarden LA (1982) Influence of pH on the detection of low- and high-

avidity anti-dsDNA. J Immunol Methods 55: 361–373

22. Aarden LA & Smeenk R (1982) Immunochemical properties of antibodies to DNA and their influence on detection methods. In: Kalden JR & Feltkamp TEW (Ed) Antibodies to nuclear antigens (pp. 23–28). Excerpta Medica, Amsterdam

23. Smeenk RJT, Van den Brink HG, Brinkman K, Termaat R–M, Berden JHM & Swaak AJG (1991) Anti-dsDNA: choice of assay in relation to clinical value. Rheumatol Int 11: 101–107

24. Smeenk R, Brinkman K, Van den Brink H & Swaak T (1990) A comparison of assays used for the detection of antibodies to DNA. Clin Rheumatol 9 Suppl.1: 63–72

Manual of Biological Markers of Disease **A9**: 1–14, 1993.

Methods to detect autoantibodies to neutrophilic granulocytes

A. WIIK[1], N. RASMUSSEN[2] and J. WIESLANDER[1]

[1] *Department of Autoimmunology, Statens Seruminstitut, DK-2300 Copenhagen S., Denmark;*
[2] *Department of Oto-rhino-laryngology, Rigshospitalet, DK-2100 Copenhagen Ø., Denmark*

Abbreviations

ANCA: anti-neutrophil cytoplasmic antibodies; cANCA: cytoplasmic ANCA; IIF: indirect immunofluorescence; EIA: enzyme immuno assay; MPO: myeloperoxidase; PR-3: proteinase-3; GS-ANA: granulocyte-specific antinuclear antibodies

Key words: Anti-neutrophil cytoplasmic antibodies (ANCA), granulocyte-specific anti-nuclear antibodies (GS-ANA), proteinase-3, myeloperoxidase, immunofluorescence, enzyme immuno-assay

1. Introduction

Autoantibodies to neutrophilic granulocytes have been known under the name 'granulocyte-specific antinuclear antibodies' (GS-ANA) since the mid 1960s [1, 2]. Although recognized as early as 1973 [3], autoantibodies to cytoplasmic antigens of neutrophils were not described in literature until 1982 [4]. The morphology and content of a granulocyte is given in Fig. 1. Autoantibodies to surface components of neutrophils are considered to be extremely rare, and neutropenias found in relation to autoimmune diseases are most likely induced by circulating immune complexes interacting with and thereby activating neutrophils [5, 6, 7].

   Autoantibodies staining the nuclei or the perinuclear zone of neutrophils by indirect immunofluorescence (IIF) are commonly directed to cationic cytoplasmic components of human neutrophils and monocytes (such as lysozyme, myeloperoxidase, elastase, and lactoferrin), which redistribute to anionic components of the nucleus upon removal of lipids from granule membranes by ethanol fixation (see A 2.2). Such fixation is widely used for cell smears to make intracellular compartments accessible to antibodies. Since antibodies to such artifactually distributed cytoplasmic antigens cannot be distinguished from antibodies to granulocyte nuclei (true GS-ANA), the terminology perinuclear anti-neutrophil cytoplasmic antibody/granulocyte specific ANA (pANCA/GS-ANA) will be used in

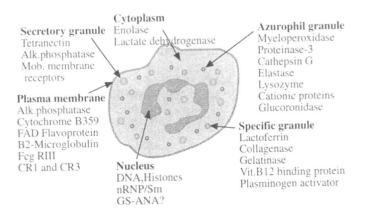

Fig. 1. Content of a neutrophilic granulocyte. In the cytoplasm of neutrophilic granulocytes a number of granules is seen. They are called azurophilic, specific and secretory granules. The content of each of these is indicated in the figure along with other important molecules.

this article. Antibodies giving a clear-cut cytoplasmic fluorescence on neutrophils and monocytes, but not on lymphocytes or HEp-2 cells, will be designated cytoplasmic ANCA (cANCA) [8].

The IIF technique has been the preferred screening technique for demonstrating pANCA/GS-ANA and cANCA [9,10]. This technique was adopted as the international reference method in 1988 [11] at the first International Workshop on ANCA. Since this method is only qualitative or semiquantitative and extensive experience is required for interpreting the results, quantitative techniques have been developed such as enzyme immunoassays (EIA) and solid phase radio-immuno-assays (SPRIA) using crude [12, 13] or purified antigens [14, 15] as targets. Although some concordance between different EIA for cANCA demonstration already existed in 1989 [16], such assays need to be thoroughly standardized, whereas the standard IIF seems to be a reliable screening method for all ANCAs [17]. This is also the conclusion reached in a European multicenter study of ANCA assay standardization [18].

2. Demonstration of neutrophil-reactive autoantibodies by IIF

2.1 Steps in the procedure

1. Ten ml of blood from a healthy donor is drawn into a 10 ml syringe with no anticoagulant added.
2. The blood is immediately transferred into a conical flask and is defibrinated with 10 glass beads by circular movements for about 5–7 min.

3. The defibrinated blood is transferred into a polystyrene tube containing 250 IU of heparin and the blood is thoroughly mixed.
4. 2 ml aliquots are layered onto 5 ml of a dextran-metrizoate gradient (see below) in 10 ml polystyrene tubes and left to stand at room temperature for 45 min.
5. The leucocyte containing plasma layers are pipetted from each tube into one 10 ml polystyrene tube.
6. This tube is centrifuged at 200 × g for 10 min at room temperature to sediment the leucocytes.
7. The cell pellet is resuspended in 10 ml of PBS, pH 7.4, containing 1% human serum albumin.
8. Steps 6, 7 and 6 are repeated to remove serum constituents from the cells.
9. The leucocyte pellet is freed from washing solution by pipetting and only 150–200 μl of fluid is left over the pellet.
10. The leucocytes are resuspended in this fluid by tapping the tube several times with a finger to obtain an evenly turbid suspension without visible cell aggregates.
11. A fine-tipped Pasteur pipette is drawn out over a gas flame and broken to obtain a very fine pipette tip.
12. The suspension is soaked into the pipette and immediately cells are deposited as 1 mm diameter droplets on carefully defatted objective slides.
13. No more than 10 s later the suspension droplets are smeared using ground edged glass slides.
14. The smears are left to air dry.
15. Just after drying, the slides are fixed in 99% cold ethanol (4 °C) for 5 min.
16. After drying, each cell smear is encircled using a glass marker.
17. These preparations should be used as antigen source within 30 min for IIF (or be wrapped, airtight, to be stored at − 20 °C for use within a week).
18. Patient serum diluted in PBS, pH 7.4 (usually 1:20) is added as a drop covering the encircled cell smear, negative control serum at the same dilution to another smear, positive cANCA serum to a third smear and positive pANCA/GS-ANA serum to a fourth smear.
19. The smears are left at room temperature for 30 min in a humid incubation chamber.
20. Each smear is individually washed with excess of PBS and all smears are submersed in the chamber in PBS for 10 min to remove non-specifically attached serum.
21. After removal of all PBS from the slides, the submersion of slides in PBS is repeated for another 10 min.
22. Excess PBS is removed around the circle of each slide, but a minute amount of PBS is left in the circle to avoid drying of the sample.

23. Fluorescein isothiocyanate-labelled IgG fraction of rabbit anti-serum to human IgG (Dako, Glostrup, Denmark), diluted appropriately in PBS (usually 1:25–1:100), is added as a drop onto the encircled cell smear of each slide.
24. The slides are again incubated for 30 min at room temperature in a humid chamber.
25. Steps 20, 21 and 22 are repeated.
26. A drop of a 2:1 glycerol/PBS mixture is then applied to each smear and a cover cover glass is applied.
27. The smears should either be read directly in a fluorescence microscope or left at 4 °C in the dark until reading.
28. Reading is best done in an epi-illumination microscope with 400 × magnification using an objective with a high numerical aperture.
29. Interpretation of results (see Interpretation § 2.3).

Notes to the procedures

ad 1 + 2. The volume of donor blood can be adapted to the requirement of slides (5 ml blood plus 5 glass beads, 20 ml blood plus 20 glass beads).

ad 3. Heparinization is essential to avoid clotting due to incomplete defibrination.

ad 10. If cell aggregates are found in the leucocyte suspension, activation and partial degranulation of leucocytes has taken place, and the cells are unsuitable for use as cell substrate.

ad 12 + 13. Extra hands may be required to carry out the smearing of cells just after deposition of each suspension droplet on the slide to avoid any drying of the minute amount of fluid before smearing.

ad 18. It is important to reach step 18 as quickly as possible to avoid artifacts by leakage of antigens out of the cells or from one cell to another (see below).

ad 23. Use FITC-conjugates with a low mean molar fluorescein/protein ratio (< 2.5) (Dako, Glostrup, Denmark) since higher ratios will cause non-specific staining of cationic proteins of eosinophils and neutrophils.

ad 27. If slides are read immediately or after a few days, it may be an advantage to add an anti-fading reagent, such as paraphenylene diamine 1 mg/ml (Sigma, St. Louis, MS), to the glycerin/PBS mixture to decrease fluorescence fading during reading or photomicrography.

2.2 Solutions

- Dextran/metrizoate solution:
  - 24.7– ml dist. H_2O
  - 89.4 ml 6% dextran T-500 (Pharmacia, Uppsala, Sweden)
  - 20.0 ml metrizoate sodium 75% w/v (Nycomed, Oslo, Norway)
- Phosphate-buffered saline, pH 7.4 (PBS)
  - 8.0 g NaC1
  - 200 mg KC1
  - 1.44 g Na_2HPO_4, 2 H_2O
  - 200 mg KH_2PO_4
  - dist. H_2O to 1000 ml

- PBS/albumin 1% w/v
  - PBS as above, desired volume
  - add human serum albumin to 1%
- Ethanol 99%
- Glycerol/PBS mixture
  - PBS as above, desired volume
  - add glycerol to 66% mixture

2.3 Interpretation, artifacts, control cells

The smears are rich in neutrophils and red cells. Lymphocytes and eosinophils should therefore be clearly distinguishable.

2.3.1 cANCA
c-ANCA is typically seen as a granular cytoplasmic fluorescence which is more intense close to the nucleus of neutrophils than in the periphery (Fig. 2a). Monocytes may be positive showing a less granular pattern. Eosinophils and lymphocytes should show a negative cytoplasm, and, in case of doubt, other cells of human origin, such as isolated human lymphocytes or HEp-2 cells should be studied separately to show clearly whether the cytoplasm is negative or positive. Eosinophils should be negative except for a brownish autofluorescence. If they show green fluorescence, the reason most often is a non-specific binding of too highly labelled conjugate to cationic proteins of eosinophilic granules. If cytoplasmic staining of isolated lymphocytes or HEp-2 cells is found, cANCA often cannot be demonstrated and solid phase assays using isolated proteinase-3 as antigen are then necessary. A clearly higher titer on neutrophil cytoplasm than on lymphocyte cytoplasm may indicate presence of both cANCA and a non-organ-specific cytoplasmic antibody. Low intensity non-granular cytoplasmic/perinuclear fluorescence of neutrophils may be seen in certain sera where circulating immune complexes are present, most likely due to the interaction of complexes with Fcγ-receptors on the cell surface. Also, in such cases, a specific test for anti-proteinase-3 antibodies is preferable.

2.3.2 pANCA/GS-ANA
pANCA/GS-ANA is seen as a perinuclear/nuclear fluorescence in neutrophils (Fig. 2b) and in some monocytes, whereas the majority of lymphocytes and eosinophils should be negative. However, this pattern of reactivity is often the result of a fixation artifact as explained in Fig. 3. The ethanol makes cellular membranes of the smeared cells penetrable to Ig molecules and fixes the cells to the glass surface. However, it removes much of the membrane lipids so that molecules can move freely around in the cell or even make contact with neighbouring cells. Cationic molecules such as

lysozyme, myeloperoxidase, leucocyte elastase, cathepsin G and lactoferrin attach to oppositely charged nuclear components of the cell itself, but may also bind to the nuclei of neighbouring lymphocytes. In that case, autoantibodies to one of these constituents may be misinterpreted as non-organ-specific ANA. Therefore, another cell substrate must be used as a control. Isolated human lymphocytes from the same donor are especially well suited since they are comparable in cell size to neutrophils. If lymphocytes for routine use are not available, HEp-2 cells may be used instead. However, these cells are much larger, and titer differences between pANCA/GS-ANA reactivity and HEp-2 cell reactivity are less meaningful due to different fluorescence intensities at comparable serum dilutions. This means that pANCA/GS-ANA can be said to be present if reactivity is seen solely with the neutrophils and monocytes, or if such reactivity is present in a significantly higher titer ($\geq$ 2 titer steps) than that of a human substrate for ANA demonstration [9], such as isolated lymphocytes. Such titer comparisons are not justified when HEp-2 cells are used as ANA substrate except when ANA is distinctly different from the pANCA/GS-ANA pattern, i.e. anti-nucleolar or anti-centromere antibodies.

It has been proposed to use formaldehyde for cross-linking proteins in situ followed by acetone penetration of membranes to avoid the artifactual

A   **B**

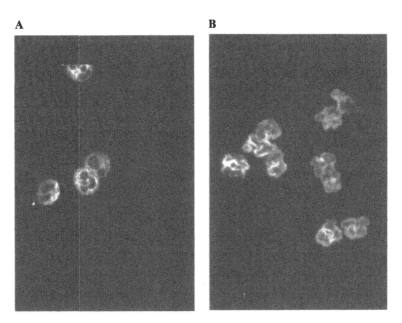

Fig. 2. A. Cytoplasmic ANCA pattern as seen by IIF. B. Perinuclear ANCA pattern as seen by IIF.

IF patterns

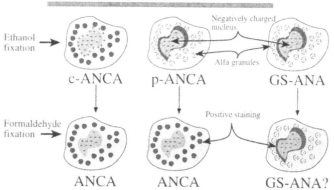

Fig. 3. Ethanol fixation gives rise to the two patterns seen in Fig. 2. It is not possible to distinguish between true GS-ANA and pANCA morphologically. Granule membranes are made penetrable and positively charged protein can migrate to the nucleus to interact with negatively charged components there, giving rise to nuclear/perinuclear staining as if antibody reacted with true nuclear antigens. Formaldehyde cross-links proteins in situ resulting in cANCA pattern in all situations where antibodies are directed to granule constituents. Cross-linking of nuclear proteins should not alter the nuclear staining pattern of true GS-ANA (see reservations about utility in the text). GS-ANA is indicated with ? in the figure since it is not really known if a true GS-ANA exists or if some of the staining called GS-ANA could be pANCA.

redistribution of cationic antigens illustrated in Fig. 3. However, there are some problems involved with this technique: cross-linking of the proteins leads to augmented autofluorescence and, at the same time, antigens may lose some reactivity with autoantibody – both factors increasing the problems with interpretation of positivity. Non-organ-specific ANA found in prototype sera usually do not react with nuclear antigens of cells fixed in this way. This means that true GS-ANA may not be detectable on formalin fixed neutrophils and a negative result cannot be interpreted. Thus formalin/acetone fixation is not suitable for routine ANCA testing.

2.3.3 Alternative substrates
HL-60 cells (a promyelocytic leukemic cell line) have been used as cell substrate for ANCA testing instead of normal donor leucocytes in several laboratories [19,20]. However, these cells tend to lose granules as they go through many cultivation passages; they only harbour α-granules, and their content of nuclear and cytoplasmic constituents is unpredictable. All these facts impose restrictions on the use of these cells in screening procedures for ANCA. They may, however, be a suitable substrate for PR-3 ANCA or MPO-ANCA testing, if their contents of these constituents are regularly checked with known monoclonal or human monospecific antibodies.

3. EIA procedures

3.1 Preparation of α-granules as described by Borregaard et al. [21]

3.1.1 Steps in procedure
1. Order buffy coats from the local hospital.
2. Measure the volume of the buffy coats and add an equal volume of 2% Dextran T-500 (Pharmacia, Uppsala, Sweden). Let it stand and sediment for 1 h at room temperature.
3. Collect the leucocyte rich upper layer in 50 ml centrifuge tubes and centruge 200 × g for 10 min at 4^0 C.
4. Remove the supernatant. The pellet is resuspended by adding 25 ml 0.9% NaCl to each tube at room temperature.
5. Layer 15 ml Lymfoprep (Nycomed, Oslo, Norway) under the resuspended cells and centrifuge 400 × g for 30 min.
6. Resuspend the pellet in 10 ml ice-cold distilled water kept on ice. Shake for 30 s and then add 10 ml 1.8% NaCl. Mix again for 30 s.
7. Combine the tubes two and two and centrifuge 200 × g for 8 min. Collect the pellets and resuspend them in 10 ml ice-cold 0.2% NaCl.
8. Add 10 ml ice-cold 1.8% NaCl and centrifuge 200 × g for 8 min.
9. Collect the pellet and resuspend it in relaxation buffer (1 × conc.). ATP should be added to the relaxation buffer just prior to use.
10. Put the suspended pellet into a 50 ml centrifuge tube in the nitrogen bomb (Parr Instrument Company, Moline, Il, USA).
11. Connect the bomb to the nitrogen flask. Carefully open the valve on the bomb and increase pressure to 375 psi.
12. Leave pressure on for 20 min. Place a 50 ml plastic centrifuge tube containing 600 ml 100 mM EGTA solution (if 40 ml cells have been used) at the outlet to collect the disrupted cells.
13. Prepare the Percoll (Pharmacia, Uppsala, Sweden) gradient. Prepare 90 ml density 1.120 and 90 ml density 1.050 solutions according to the recipe of the supplier.
14. Prepare the gradient in tubes for the ultracentrifuge. Pour 14 ml, density 1.050, in the tube, bring carefully 14 ml of density 1.12 solution under the 1.050 solution. Apply 8–10 ml cavitated cells on the top of this discontinuous gradient.
15. Centrifuge at 48 000 × g for 30 min at +6 °C.
16. After centrifugation there should be 3 bands visible; from the bottom α, β and γ with the supernatant on top (see Fig. 4).
17. Collect the α band with a Pasteur pipette and remove Percoll by centrifuging at 100000 × g for 180 min at +6 °C.
18. Discard the supernatant and mix the pellet with about 5 ml relaxation buffer (1 ×) without ATP.
19. Add 10 μl Triton X-100/ml of the fraction. Incubate for 1 h and centrifuge 2000 × g on the tabletop centrifuge.

20. The supernatant is now ready for coating to microtiterplates at a dilution around 1/1000. The ELISA is then performed as described [13].

Percoll gradient of granules

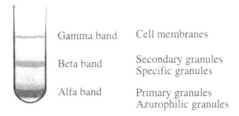

Fig. 4. Percoll gradient of granules from neutrophils. The centrifugation of disrupted neutrophils on the Percoll gradient separates them into α, β and γ fractions. The α fraction is used for ELISA.

3.1.2 Solutions
- Relaxation buffer (1 ×)
  - KCl 100 mM
  - NaCl 3 mM
  - Pipes 10 mM
  - $MgCl_2.6\ H_2O$ 3.5 mM
  - ATP $(Na)_2$ 1.0 mM

3.2 EIA for quantification of ANCA using α-granules

3.2.1 Steps in the procedure
1. Microtiterplates are coated with 100 μl per well of the Triton X-100 extract of α-granules, optimally diluted (1:500–1:1000) in carbonate buffer, pH 9.5–9.7.
2. Incubate overnight at room temperature.
3. Wash three times by filling and emptying the wells, the last time by knocking the plates upside-down on an absorbant tissue.
4. Leave plates for air drying at room temperature.
5. Cover plates with protective film and store at + 4 °C if plates are not used immediately.
6. Dilute test sera 1:50 in PBS/BSA/Tween incubation buffer. Dilute positive and negative control sera and an internal standard serum in the same buffer.
7. Add 100 μl of these sera to each well, and remember to leave some blank wells with incubation buffer alone.
8. Incubate for 60 min at room temperature on a shaker.

9. Wash three times in 0.9% NaCl/Tween washing solution using the same technique as in step 3.
10. Dilute alkaline phosphatase (AP)-labelled anti-human IgG conjugate (Dako, Glostrup, Denmark) to about 1:1000–1:2000 in incubation buffer.
11. Add 100 μl of AP-conjugate to each well.
12. Incubate for 60 min on a shaker at room temperature.
13. Prepare substrate solution by dissolving five p-nitrophenyl phosphate (PNP) tablets (Sigma, St. Louis, MS) with 25 ml of substrate buffer.
14. Add 100 μl of substrate to each well.
15. Incubate for 60 min at room temperature.
16. Read the plates in a microtiterplate reader at 405 nm.

Notes to the procedures

ad 1. Use NUNC Maxisorp plates or their equivalents from other firms.
ad 10. Do not use horseradish peroxidase conjugates due to the myeloperoxidase present in the α-granules.
ad 16. After reading, the results should be compared with those on an internal standard serum. Values can then be given as Strong, Intermediate or Weak relative to this standard. It is important to define the cut-off level based on patients sera, for example from immunoinflammatory conditions, because these sera give a slightly higher background than donor sera.

3.2.2 Solutions
– Coating buffer
  – 1.125 g Na_2CO_3
  – 3.30 g $NaHCO_3$
  – 200 mg NaN_3
  – dist. H_2O to 1000 ml
  – adjust pH to 9.5–9.7
– Washing buffer
  – 45 g NaCl
  – 2.5 ml Tween 20
  – dist. H_2O to 5000 ml
– Incubation buffer
  – 8.0 g NaCl
  – 200 mg KCl
  – 1.136 g $Na2HPO_4$, 2 H_2O
  – 200 mg KH_2PO_4
  – 200 mg NaN_3
  – dist. H_2O to 1000 ml
  – adjust pH to 7.4
  – add 0.5 ml Tween 20
  – add 2.0 g BSA

- Substrate buffer
  - 97 ml diethanolamine (should be colourless!)
  - 800 ml dist. H_2O
  - mix and adjust pH to 9.79.9 with HCl
  - 0.101 g $MgCl_2$, 6 H_2O
  - 200 mg NaN_3
  - dist. H_2O to 1000 ml
  - check pH again

3.3 EIA for quantification of Proteinase-3 (PR-3) ANCA

3.3.1 Steps in the procedure
1. Microtiterplates are coated with PR-3 dissolved in carbonate buffer pH 9.5–9.7, 100 μl per well at a protein concentration of about 0.5–1 μg/ml. Then follow steps 2–16 of § 3.2.1.

Notes to the procedures
ad 1. Use NUNC Maxisorp plates or their equivalents from other firms. Use PR-3 isolated by methods described [22, 23, 24]. The optimal PR-3 preparation for EIA determination of cANCA cannot yet be selected, but comparative studies are currently being performed [25].

3.3.2 Solutions
- Coating buffer
  - 1.125 g Na_2CO_3
  - 3.30 g $NaHCO_3$
  - 200 mg NaN_3
  - dist. H_2O to 1000 ml
  - adjust pH to 9.5–9.7
- Washing buffer
  - 45 g NaCl
  - 2.5 ml Tween 20
  - dist. H_2O to 5000 ml
- Incubation buffer
  - 8.0 g NaCl
  - 200 mg KCl
  - 1.136 g $Na2HPO_4$, 2 H_2O
  - 200 mg KH_2PO_4
  - 200 mg NaN_3
  - dist. H_2O to 1000 ml
  - adjust pH to 7.4
  - add 0.5 ml Tween 20
  - add 2.0 g BSA
- Substrate buffer

- 97 ml diethanolamine (should be colourless!)
- 800 ml dist. H_2O
- mix and adjust pH to 9.79.9 with HCl
- add 0.101 g $MgCl_2.6\ H_2O$
- add 200 mg NaN_3
- dist. H_2O to 1000 ml
- check pH again

3.4 EIA for quantification of myeloperoxidase (MPO) ANCA

3.4.1 Procedures

Use myeloperoxidase just as described above for PR-3, at the same protein concentration but with known positive and negative control sera for MPO ANCA as well as an internal standard serum. All the steps are identical to those for PR-3 ANCA (see 3.3). Calculate values in arbitrary units and set limit for clearly abnormal values after running both normal donor sera and disease control sera. Some SLE sera may contain MPO ANCA according to the literature.

Notes to the procedure

MPO can be obtained commercially (Calbiochem, La Jolla, CA) or produced by standard methods [26].

References

1. Faber V, Elling P, Norup G, Mansa B, Nissen NI (1964) An antinuclear factor specific for leucocytes. Lancet ii: 344–345
2. Elling P (1967) On the incidence of antinuclear factors in rheumatoid arthritis. A comparative study with special reference to the reaction with polymorphonuclear granulocytes. Acta Rheum Scand 13: 101–112
3. Rasmussen N, Wiik A (1985) Autoimmunity in Wegener's granulomatosis. In Veldman JE, McCabe BF, Haizing EH, Mygind N (Eds) Immunobiology, Autoimmunity, Transplantation in Otorhinolaryngology (pp. 231–236). Amsterdam: Kugler Publications
4. Davies DJ, Moran JE, Niall JF, Ryan GB (1982) Segmental necrotizing glomerulonephritis with anti-neutrophil antibodies: possible arbovirus aetiology? Br Med J 285: 606
5. Wiik A (1975) Circulating immune complexes involving granulocyte-specific antinuclear factors in Felty's syndrome and rheumatoid arthritis. Acta Pathol Microbiol Scand, sect C 83: 354–364
6. Hurd ER, Andreis M, Ziff M (1977) Phagocytosis of immune complexes by polymorphonuclear leucocytes in patients with Felty's syndrome. Clin Exp Immunol 28: 413–425
7. Petersen J, Wiik A (1983) Lack of evidence for granulocyte-specific membrane-directed autoantibodies in serum from patients with Felty's syndrome and rheumatoid factor-positive autoimmune neutropenia. Acta Path Microbiol Immun Scand, sect C 91: 15–19
8. Wiik A, van der Woude FJ (1990) The new ACPA/ANCA nomenclature. Neth J Med 36: 107–108
9. Wiik A, Jensen E, Friis J (1974) Granulocyte-specific antinuclear factors in synovial fluids and sera from patients with rheumatoid arthritis. Ann Rheum Dis 33: 515–522
10. Rasmussen N, Wiik A (1989) Indirect immunofluorescence for IgG ANCA in sera submitted for the 1st International workshop on ANCA, 1988. APMIS 97: suppl. 6, 16–20
11. Wiik A (1989) Delineation of a standard procedure for indirect immunofluorescence detection of ANCA. APMIS 97: suppl. 6, 12–13
12. Savage COS, Winearls CG, Jones S, Marshall PD, Lockwood CM (1987) Prospective study of radioimmunoassay for antibodies against neutrophil cytoplasm in diagnosis of systemic vasculitis. Lancet i: 1389–1393
13. Rasmussen N, Sjölin C, Isaksson B, Bygren P, Wieslander J (1990) An ELISA for the detection of anti-neutrophil cytoplasm antibodies (ANCA). J Imm Meth 127: 139–145
14. Falk RF, Jennette JC (1988) Anti-neutrophil cytoplasmic antibodies with specificity for myeloperoxidase in patients with systemic vasculitis and idiopathic necrotizing and crescentic glomerulonephritis. N Engl J Med 318: 1651–1657
15. Lüdemann J, Utecht B, Gross WL (1988) Detection and quantification of anti-neutrophil cytoplasmic antibodies in Wegener's granulomatosis by ELISA using affinity purified antigen. J Imm Meth 114: 167–174
16. Rasmussen N, Lüdemann J, Utecht B (1989) ELISA examination for IgG ANCA in sera submitted for the 1st International workshop on ANCA. APMIS 97: suppl. 6, 21–22
17. Daha MR, Rasmussen N (1990) Presentation of solid phase assays from several laboratories. Neth J Med 36: 137–142
18. Hagen EC, Andrassy K, Chernok E, Daha MR, Gaskin G, Gross W, Lesavre P, Lüdemann J, Pusey CD, Rasmussen N, Savage COS, Sinico A, Wiik A, van der Woude FJ. The value of indirect immuno-fluorescence and solid phase techniques for ANCA detection. A report on the first phase of an International cooperative study on the standardization of ANCA assays. J Imm Meth (in press)
19. Charles LA, Falk RF, Jennette JC (1989) Reactivity of anti-neutrophil cytoplasmic autoantibodies with HL-60 cells. Clin Immunol Immunopathol 53: 243–253
20. Wheeler FB, Saluta G, Wise CM, Semble EL, Pisko EJ (1991) Detection of anti-neutrophil cytoplasmic autoantibodies using the promyelocytic HL-60 cell-line. Clin Exp Rheumatol 9: 569–580

21. Borregaard N, Heiple JM, Simons ER, Clark RA (1983) Subcellular localization of the b-cytochrome component of the human neutrophil microbicidal oxidase: Translocation during activation. J Cell Biol 97: 52–61
22. Kao RC, Wehner NG, Skubitz KM, Gray BH, Hoidal JR (1988) Proteinase 3: A distinct human polymorphonuclear leucocyte proteinase that produces emphysema in hamsters. J Clin Invest 82: 1963–1973
23. Lüdemann J, Utecht B, Gross WL (1990) Antineutrophil cytoplasm antibodies in Wegener's granulomatosis recognize an elastinolytic enzyme. J Exp Med 171: 357–362
24. Wieslander J, Rasmussen N, Bygren P (1989) An ELISA for ANCA and preliminary studies of the antigens involved. APMIS 97: suppl. 6, 42
25. Hagen EC, Andrassy K, Csernok E, Daha MR, Gaskin G, Gross W, Lesavre P, Lüdemann J, Pusey CD, Rasmussen N, Savage COS, Sinico A, Wiik A, van der Woude F (1993) The value of indirect immunofluorescence and solid phase techniques for ANCA detection. A report on the first phase of an International cooperative study on the standardization of ANCA assays. J Immunol Methods 159: 1–16
26. Olsson I, Olofsson T, Odeberg H (1972) Myeloperoxidase-mediated iodination in granulocytes. Scand J Haematol 9: 483–491

Manual of Biological Markers of Disease **A10**: 1–8, 1993.

Detection of antiperinuclear factor and antikeratin antibodies

R.M.A. HOET

Department of Biochemistry, University of Nijmegen, P.O. Box 9101, NL 6500 HB Nijmegen, The Netherlands

Abbreviations

APF: antiperinuclear factor; AKA: antikeratin antibody; RA: rheumatoid arthritis; RF: rheumatoid factor; WHO: World Health Organization

Key words: APF, AKA, autoantibodies, autoantigens, RA

1. Introduction

The diagnosis of rheumatoid arthritis (RA) is initially based on clinical manifestations. Serological support for such a diagnosis is not very well established and depends mainly on the presence of rheumatoid factors (RF). A positive RF test has a predictive value [1] and is related to disease with a more severe outcome [2]. However, RF is also present in other (autoimmune) diseases and in control sera from healthy people [3]. Therefore, testing for a second RA-specific antibody is very useful, especially for the management of seronegative RA patients. Both the antiperinuclear factor (APF) (reviewed in [4]) and antikeratin antibodies (AKA) (reviewed in [4] and [5]) are helpful in this respect.

The APF originally described by Nienhuis and Mandema [6], can be found in 49–91% of sera from RA-patients [4, 6–19]. The antibodies, mostly of the IgG type, are directed against a protein component present in the 0.5–4 μm spherical keratohyalin granules [6–8] in the cytoplasm of human buccal mucosa cells. A typical immunofluorescence staining pattern of a buccal mucosa cell from an APF-positive RA serum is shown in Fig. 1.

The so-called antikeratin antibodies were first described by Young *et al.* [20] and can be found in 36–59% of sera from RA patients. This antibody specificity also has been found to be specific for RA [5, 8, 13, 16, 20–30]. A typical immunofluorescence staining pattern of AKA on a rat oesophagus section is shown in Fig. 2. It should be mentioned here that the AKA are most probably not directed to cytokeratin polypeptides, but to an as yet unknown protein component in this epithelial layer of the oesophagus. For a detailed discussion see [4].

In the following paragraphs we will critically deal with the methods employed for the detection of APF and AKA.

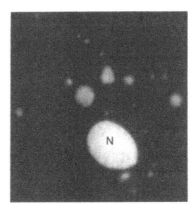

Fig. 1. Typical immunofluorescence staining pattern of a buccal mucosa cell with an APF-positive RA serum. The immunofluorescence staining was performed as described in the text. The nucleus (N) is counterstained with ethidiumbromide.

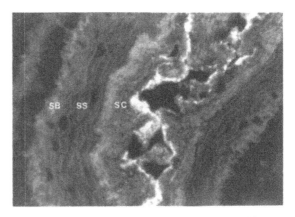

Fig. 2. Typical immunofluorescence staining pattern of an AKA-positive serum on a rat oesophagus cryostat section. The immunofluorescence staining was performed as described in the text. The laminar staining of the stratum corneum (SC) of rat oesophagus is specific for RA. SB, stratum basale; SS, stratum spinosum; SC, stratum corneum.

2. Detection of the APF

2.1 Antigen substrate and their donors

As substrate for the APF test, buccal mucosa cells (cheek cells) from so-called 'positive donors' are used. The availability of a good donor is essential to the performance of the method.

Buccal mucosa cells from at least 10 normal, healthy potential donors (e.g. laboratory workers) should be tested in an immunofluorescence assay as described in the protocol below using a positive reference serum (for example, WHO standard rheumatoid arthritis serum [31], available at the WHO branch of the CLB, P.O. Box 9190, 1006 AD Amsterdam, the Netherlands).

Important considerations for finding a good donor are:
- At least 20% of the cells should be positive for APF using the WHO reference serum.
- Disturbing contaminations with bacteria or other micro-organisms should be as low as possible [8, 12–15, 18]. It is recommended to have at least two good donors at your disposal who are readily available.

Note
Although the proportion of antigen-positive donors in a normal population is high (at least 70–80%) [4], most authors report that only a small percentage (about 10%) of the donors have sufficient antigen present in their buccal mucosa cells to allow easy detection of the APF.

2.2 Preparation of buccal mucosa cells for immunofluorescence

Steps in the procedure
1. The donor should brush his/her teeth and rinse his/her mouth with water several times.
2. Remove buccal mucosa cells from the inner cheek of the donor with a small piece of sterile foam plastic sponge (not too rigid and not too soft).
3. Bring the cells into suspension by rinsing the foam sponge with about 10 ml phosphate buffered saline (PBS) pH 7.4.
4. Wash the cells twice with 5 ml PBS (5 min centrifugation at 800 g).
5. Wash the cells with 5 ml PT-buffer (0.5% Triton-X100 in PBS). Centrifuge for 5 min at 800 g.
6. Wash the cells with 5 ml PBS. Centrifuge for 5 min at 800 g.
7. Suspend the cells in 1–5 ml PBS at a concentration of about 1.10^5 cells per ml.
8. Spot 10 μl cell suspension (about 1000 cells) per hole on a microscopic multi-hole slide and air-dry the cells under a cold föhn (e.g. 10 or 12 holes multi-hole slides).

9. Cells are now ready for use or can alternatively be stored for up to 2 weeks at −70 °C.

Notes to the procedure
ad 5. Treatment of the buccal mucosa cells with 0.5% Triton-X100 increases the sensitivity of the APF test, probably because this treatment renders the cells more permeable to the antibodies and because it helps in reducing background staining by removing most soluble cellular proteins and residual bacterial contamination [8].
ad.8. The cells should be used without fixation because both acetone and methanol have a traumatic effect on the antigenicity of the perinuclear factor [6, 8].

2.3 Serum dilution and immunofluorescence procedure

Steps in the procedure
1. Dilute the serum samples 1:5 in PBS and centrifuge 5 min at 13,000 g.
2. Incubate the slides containing the buccal mucosa cells with the diluted serum samples for 90 min at room temperature in a 100% humid atmosphere.
3. Wash the cells 3 times for 10 min with PBS.
4. Incubate for 30 min with purified fluorescein isothiocyanate labelled rabbit antibody against human IgM, IgG and IgA (Dakopatts F200, dilution 1:100 in PBS).
5. Wash the cells 3 times for 10 min with PBS.
6. Mount the preparations in a glycerol/PBS (1:1) solution with ethidium-bromide (0.5 μg/ml) as a nuclear counterstain.
7. The slides can now be read under a fluorescence microscrope.

Notes to the procedure
ad 1. Serum dilution seems to be a critical variable in the test protocol. Although in earlier studies undiluted sera were used [4], at present 1:5 diluted sera are used most often because of the higher sensitivity obtained [12, 13, 18]. Westgeest and co-workers [32] found a positive APF result in 6% of 123 control blood donors and in 41% of 123 RA patients when undiluted sera were used. When the same sera were tested in a 1:10 dilution, 12% of the controls and 70% of RA patients showed a positive result. There was no difference in sensitivity between 1:10 or 1:5 diluted sera (A.M.Th. Boerbooms, personal communication). The sensitivity of the test is thus increased and the specificity decreased when the serum is diluted.
ad 4. Kataaha *et al.* [16] found IgG-APF activity in all 16 sera tested, although in four sera additional IgM could be detected and in three sera, additional IgA activity. Therefore, the use of a total anti-Ig conjugate preparation is recommended.

2.4 Criteria for positivity and reproducibility of the test

1. A serum is considered to be positive when the typical perinuclear staining as shown in Fig. 1 is seen in at least a few cells.
2. Each test should be accompanied by a reference serum. The WHO

rheumatoid arthritis reference serum contains (per definition) 100 IU/ml. By comparing the titre of the WHO rheumatoid arthritis reference serum with the titre of an unknown serum it is possible to calculate the APF-activity in the latter serum sample (in IU/ml).

Notes to the procedure
ad 2. In an interlaboratory assay, 5 sera from RA patients were independently tested using the WHO reference RA serum (100 IU/ml) and a reference anti-Ig conjugate as standards. The sera contained APF activity ranging between 0 (control serum) and 400 IU/ml. The results showed that although all five participating laboratories used their own donors and their own criteria for positivity, the APF concentrations obtained were very comparable [33, 34].

2.5 Reagents and solutions

– WHO standard rheumatoid arthritis serum [31] is available at the WHO branch of the CLB, P.O. Box 9190, 1006 AD Amsterdam, the Netherlands), and contains per definition 100 IU/ml APF-activity.
– Purified fluorescein isothiocyanate labelled rabbit antibody against human IgM, IgG and IgA can be obtained from Dakopatts (Dakopatts F200, Glostrup, Denmark).
– Phosphate buffered saline (PBS).

3. Detection of antikeratin antibodies (AKA)

3.1 Antigen substrate

Steps in the procedure
1. Isolate the middle third of a rat oesophagus (e.g. of a Wistar rat), freeze it in liquid nitrogen, and store at $-70\,°C$.
2. Make 4–5 μm thick cryostat sections of rat oesophagus (can be stored at $-70\,°C$ for up to one month).
3. Treat these sections for 5 min at room temperature with PT-buffer (PBS containing 0.5% Triton-X100).
4. Wash for 5 min with PBS.
5. Sections can now be used as substrate for the AKA test.

Notes to the procedure
ad 1. For detection of AKA by immunofluorescence (first reported by Young and co-workers [20]) unfixed rat oesophagus cryostat sections (4–5 μm) are used. Johnson et al. [13] found that the specificity and sensitivity of the test depends on the part of the oesophagus that is used to prepare the cryosections. Low oesophagus provided the best discrimination between RA and controls, and cardia of the stomach gave the highest incidence of staining in all patient groups (RA, other autoimmune

diseases, and healthy controls). In most studies the middle third of the oesophagus is used as substrate for the AKA test.

3.2 Serum dilution and immunofluorescence procedure

Steps in the procedure
1. Dilute your serum sample 1:10 in PBS and centrifuge 5 min at 13,000 g.
2. Incubate oesophagus substrate with diluted serum for 30 min at room temperature.
3. Wash 3 times for 5 min with PBS.
4. Incubate 30 min at room temperature with FITC-conjugated rabbit immunoglobulins to human IgG (F_c-fragment) (Dakopatts F123, dilution 1:100 in PBS, Dakopatts, Glostrup, Denmark).
5. Wash 3 times for 5 min with PBS.
6. Mount preparations in glycerol/PBS (1:1).
7. The slides can now be read under a fluorescence microscope.

Notes to the procedure
ad 1. In nearly all studies, a 1:10 dilution of the serum in PBS is used.
ad 4. Most AKA are of the IgG-type and, according to Vincent and co-workers [5], only these IgG-AKA are specific for RA. Therefore, the use of an IgG-specific second antibody is recommended.

3.3 Criteria for positivity of the AKA test

A serum sample is considered AKA positive if an intense linear laminated labeling pattern restricted to the stratum corneum is observed (see Fig. 2).

Note
Most authors [5, 13, 16, 21, 23, 28–30] regard only the distinct laminar staining of the keratin layer as a positive reaction (see Fig. 2), while others [22, 25] also consider a speckled fluorescence labeling pattern positive. Vincent *et al.* [5] studied different labeling patterns on the rat oesophagus epithelium and found only the intense, linear laminated labeling restricted to the stratum corneum to be highly specific for RA, while the weak, diffuse labeling of the three epithelial compartments (stratum basale, stratum spinosum, and stratum corneum) was not specific.

3.4 Reagents and solutions

FITC-conjugated rabbit immunoglobulins to human IgG (F_c-fragment) can be obtained from Dakopatts, Dakopatts F123, Glostrup, Denmark.

References

1. Valkenburg HA, Ball J, Burch TA *et al.* (1966) Rheumatoid factors in a rural population. Ann Rheum Dis 25: 497–508
2. Kellgren JH, O'Brien WH (1962) On the natural history of rheumatoid arthritis in relation to sheep cell agglutination test (SCAT). Arthr Rheum 5: 115
3. Waller MV, Toone EC, Vaughan E (1964) Study of rheumatoid factor in a normal population. Arthr Rheum 7: 513–520
4. Hoet RMA, van Venrooij WJ (1992) The antiperinuclear factor (APF) and antikeratin antibodies (AKA) in rheumatoid arthritis. In Smolen JS, Kalden JR, Maini RN (Eds) Rheumatoid Arthritis, Recent Research Advances (pp. 299–327). Springer Verlag Berlin Heidelberg New York
5. Vincent C, Serre G, Lapeyre F, Fournié B, Ayrolles C, Fournié A, Soleilhavoup J-P (1989) High diagnostic value in rheumatoid arthritis of antibodies to the stratum corneum of rat oesophagus epithelium, so-called "antikeratin antibodies". Ann Rheum Dis 48: 712–722
6. Nienhuis RLF, Mandema E (1964) A new serum factor in patients with rheumatoid arthritis. The antiperinuclear factor. Ann Rheum Dis 23: 302–305
7. Smit JW, Sondag-Tschroots IRJM, Aaij C, Feltkamp TEW, Feltkamp-Vroom TM (1980) The antiperinuclear factor. II. A light-microscopical and immunofluorescence study on the antigenic substrate. Ann Rheum Dis 39: 381–386
8. Hoet RMA, Arends M, Boerbooms AMTh, Ruiter DJ, van Venrooij WJ (1991) Antiperinuclear factor, a marker autoantibody for rheumatoid arthritis. Colocalisation of the perinuclear factor and profilaggrin. Ann Rheum Dis 50: 611–619
9. Visconti A, Cava L, Fontana G (1964) Ricerca del fattore anti-perinucleare (APF) nell' artrite reumatoide ed in altre malattie. Osp Magg 59: 1357
10. Marmont AM, Damasio EE, Bertorello C, Rossi F (1967) Studies on the antiperinuclear factor. Arthr Rheum 10: 117–128
11. Roques MT (1969) Les anticorps dits périnucléaires. Thesis, Paris
12. Sondag-Tschroots IRJM, Aaij C, Smit JW, Feltkamp TEW (1979) The antiperinuclear factor. 1. The diagnostic significance of the antiperinuclear factor for rheumatoid arthritis. Ann Rheum Dis 38: 248–251
13. Johnson GD, Carvalho A, Holborow EJ, Goddard DH, Russell G (1981) Antiperinuclear factor and antikeratin antibodies in rheumatoid arthritis. Ann Rheum Dis 40: 263–266
14. Cassani F, Ferri S, Bianchi FB, Zauli D, Pisi E (1983) Antiperinuclear factor in an Italian series of patients with rheumatoid arthritis. Ric Clin Lab 13: 347–352
15. Youinou P, Le Goff P, Miossec P (1983) Untersuchungen zur beziehung zwischen antiperinucleären factoren, anti-keratin-antikörpern und dem agglutinierenden und nicht agglutinierenden rheumafactor bei der chronischen polyarthritis. Z Rheumatol 42: 36–39
16. Kataaha PK, Mortazavi-Milani SM, Russel G, Holborow EJ (1985) Anti-intermediate filament antibodies, antikeratin antibody, and antiperinuclear factor in rheumatoid arthritis and infectious mononucleosis. Ann Rheum Dis 44: 446–449
17. Westgeest AAA, Boerbooms AMTh, Jongmans M, Vandenbroucke JP, Vierwinden G, van de Putte LBA (1987) Antiperinuclear factor: Indicator of more severe disease in seronegative rheumatoid arthritis. J Rheumatol 14: 893–897
18. Janssens X, Veys EM, Verbruggen G, Declercq L (1988) The diagnostic significance of the antiperinuclear factor for rheumatoid arthritis. J Rheumatol 15: 1346–1350
19. Vivino FB, Maul GG (1990) Histologic and electron microscopic characterization of the perinuclear factor antigen. Arthr Rheum 33: 960–969
20. Young BJJ, Mallya RK, Leslie RDG, Clark CJM, Hamblin TJ (1979) Anti-keratin antibodies in rheumatoid arthritis. Br Med J ii: 97–99
21. Scott DL, Delamere JP, Jones LJ, Walton KW (1981) Significance of laminar antikeratin antibodies to rat oesophagus in rheumatoid arthritis. Ann Rheum Dis 40: 267–271
22. Ordeig J, Guardia J (1984) Diagnostic value of antikeratin antibodies in rheumatoid

arthritis. J Rheumatol 11: 602–604

23. Hajiroussou VJ, Skingle J, Gillett AP, Webley M (1985) Significance of antikeratin antibodies in rheumatoid arthritis. J Rheumatol 12: 57–59

24. Mallya RK, Young BJJ, Pepys MB, Hamblin TJ, Mace BEW, Hamilton EBD (1983) Antikeratin antibodies in rheumatoid arthritis: frequency and correlation with other features of the disease. Clin Exp Immunol 51: 17–20

25. Kirnstein H, Mathiesen FK (1987) Antikeratin antibodies in rheumatoid arthritis. Scand J Rheumatology 16: 331–337

26. Youinou P, Le Goff P, Colaco CB, Thivolet J, Tater D, Viac J, Shipley M (1985) Antikeratin antibodies in serum and synovial fluid show specificity for rheumatoid arthritis in a study of connective tissue diseases. Ann Rheum Dis 44: 450–454

27. Meyer O, Fabregas D, Cyna L, Ryckewaert A (1986) Les anticorps anti-kératine. Un marqueur des polyarthrites rhumatoïdes évolutives. Rev Rhum Mal Osteoartic 53: 601–605

28. Quismorio PF, Kaufman RL, Beardmore T, Mongan ES (1983) Reactivity of serum antibodies to the keratin layer of rat esophagus in patients with rheumatoid arthritis. Arthritis Rheum 26: 494–499

29. Paimela L, Gripenberg M, Kurki P, Leirisalo-Repo M (1992) Antikeratin antibodies: diagnostic and prognostic markers for early rheumatoid arthritis. Ann Rheum Dis 51: 743–746

30. Miossec P, Youinou P, Le Goff P, Moineau MP (1982) Clinical relevance of antikeratin antibodies in rheumatoid arthritis. Clin Rheumatol 1: 185–189

31. Anderson SAG, Bentzon MW, Houba V, Krag P (1970) International reference preparation of rheumatoid arthritis serum. Bull. Wld Hlth Org 42: 311–318

32. Westgeest AAA, Boerbooms AMTh, van de Putte LBA (1990) The influence of serumdilution on finding of antiperinuclear factor prevalence in rheumatoid arthritis. Arthr Rheum 33: 759–760

33. Feltkamp TEW, Boerbooms AMTh, De Keyser F, Dumais M, Hoet RM, van Venrooij WJ, Verbruggen G, Veys EM, Youinou P (1990) Antiperinuclear factor – standardization program. Clin Rheumatol 9: 112–113

34. Feltkamp TEW, Boerbooms AMTH, Geertzen HGM, Hoet RMA, De Keyser F, van Venrooij WJ, Verbruggen G, Veys EM, Youinou P (1993) Interlaboratory variability of the antiperinuclear factor (APF) test for rheumatoid arthritis. Clin Rheumatol 11: 57–59

Manual of Biological Markers of Disease **A11**: 1–12, 1993.

Standards and reference preparations

T.E.W. FELTKAMP

Central Laboratory of the Netherlands Red Cross Blood Transfusion Service, PO Box 9190, 1006 AD Amsterdam; the Netherlands and the Netherlands Ophthalmic Research Institute, PO Box 12141, 1100 AZ Amsterdam, the Netherlands

Abbreviations

ANA: antinuclear antibody, APF: antiperinuclear factor CDC: Centers for Disease Control, IFT: immunofluorescence technique; IU: international unit; IUIS: International Union of Immunological Societies; RF: rheumatoid factor; WHO: World Health Organization

Key words: Centers for Disease Control, International Union of Immunological Societies, International units, Reference preparation, Standardization, World Health Organization

1. Introduction

All laboratories want the results of their tests to be reproducible and comparable to the results of the same tests performed in other laboratories. Since the methods to demonstrate the antibodies mentioned in this manual are, at most, about the same but never exactly equal to each other in distinct laboratories, the need for standardization is often urgently felt as soon as a new method becomes generally used [1–9]. Often the first idea is then to standardize the method. This, however, has never appeared to be practical, since most laboratory workers are individualists [10]. But even if they would like to submit their work to rules, they would soon be confronted with the fact that such rules hamper the development of small improvements of the tests which nobody would like to miss.

2. Standards

A considerable help in this respect is the production of so-called standard sera or, if the sera are submitted to certain treatments and are freeze-dried, standard preparations [8,9]. Such standard sera are given by definition a certain amount of (international) units. They serve as a yardstick to compare the results obtained with the serum to study.

It should be underlined that standard preparations try to serve the performance of certain tests. They are not at all developed to acquire pure antibodies to be measured by weight. On the contrary, some standards represent mixtures of several sera to represent as much as possible the situation that will occur in the daily practice of a laboratory for autoimmune serology. Many standards contain several antibodies and of course many other proteins. In short, they represent the best that could be done at the moment of preparation.

Since the amount of each standard serum is limited, "secondary standards" are often prepared in comparison to the worldwide accepted, so called "primary standard" [10]. Such secondary standards serve the needs of a certain country or area. Finally, each laboratory has to prepare his own, usually "tertiary standard" in comparison to the secondary standard.

The primary standards for the tests mentioned in this manual are prepared under the initiative and guidance of the Standardization Committee of the International Union of Immunological Societies (IUIS). This Committee nominates a Subcommittee for the development of a certain standard or group of standards. The final product is then offered to the World Health Organization (WHO) to be adapted as a WHO standard and to serve as a primary standard on a worldwide scale.

3. Reference preparations

If a certain serum or preparation cannot (yet) be used completely as a standard for quantitative methods, but nevertheless fulfills a great need as a guide in recognizing a certain type of antibody, the preparation can be accepted by the IUIS and WHO as a reference preparation. The development of secondary and tertiary reference preparations is the same as that of the standards.

About ten years ago the American Arthritis Foundation, together with the Centers for Disease Control (CDC), made available reference sera for the distinct types of nuclear fluorescence seen at ANA determination. A committee of active American investigators, under the chairmanship of Dr. Eng Tan, was appointed. They produced the reference preparations listed in Table 1, in addition to others (see article by Tan in this issue). These were all offered for further validation by expert investigators under the aegis of WHO, IUIS and the International League against Rheumatism (ILAR). For unknown reasons these sera never reached the status of WHO reference preparations. Nevertheless they are considered to be of great value to the workers in the field [11, 12].

4. How to use a standard

The use of the standards is extremely simple: Both the standard and the serum to study are subjected to the same test on the same day. The results are expressed in the usual way. For example, if you expressed your results as titres, then you will obtain two sets of data, one for the standard and one for the serum under study. If the standard, in the situation in which you started your test, was defined to contain, for instance, 200 IU/ml, and if the serum under study showed an endpoint two steps in the twofold dilution series greater than the standard, the serum under study contains 800 IU/ml.

It seems obvious (but from my experience this is not always the case) that for the determination of a titre the titration series should end with at least one completely negative result. For example an ending with titre > or = to 1:256 could never be used for the calculation of the amount of IU/ml.

If the results of your local test are expressed in local units, as is often the case in performing ELISA or Farr assays, the principle of converting local units to IUs is the same. If the result of the serum under study is p local units, and the result obtained with the standard is s local units, and if the standard is defined as b IU/ml, then the serum under study contains p/s x b IU/ml.

If we use this formula for the example of the experiment above, the serum under study contains: p/s x 200 IU/ml. Now p represents the reciprocal titre of the serum to study and s the reciprocal titre of the standard. Since in the example p = 2 x 2 x s, the serum contains 4s/s x 200 IU/ml or 800 IU/ml.

From the above we learn that at a national level p is the proposed national standard and s the WHO standard. At the level of a local laboratory, p is the proposed laboratory standard and s the national standard. However, in the daily practice of such a laboratory, p will be the serum of the patient under study and s the laboratory standard.

5. Why are standards often not used?

Although most of the leading laboratories readily agree that the use of standards is a great help in increasing reproducibility between laboratories and in decreasing the interlaboratory variability, one of the greatest problems concerning standardization is its implementation. The production of standards takes time and by the time they are ready, the laboratories and clinicians are used to their way of expression in titres or local units instead of IU/ml [13].

In the Netherlands, the members of the Working Group of the Foundation Reference Laboratory for Rheuma-Serology (RELARES) had no real problems defining and introducing national standards for rheumatoid factor (RF), antiperinuclear factor (APF), antinuclear antibodies (ANA) and anti-DNA antibodies [8]. It was surprising to see that even the introduction of the RF standard, and thus the switch from titres to IUs, was readily accepted [14].

Sometimes there are good reasons to stick to the old habit of expressing the result in titres. This is so for the ANA determination with the indirect immunofluorescence technique (IFT). Since only a standard for the homogeneous type is in existence, it would be confusing for a clinician if one result of an ANA determination is expressed in IU/ml and another in a titre. Another reason is that the performance of the indirect IFT of dilution series is time-consuming. Many laboratories prefer, therefore, to obtain semi-quantitative results by testing only one or two serum dilutions – for example 1:10 and 1:100 – further quantifying the antibody content of the serum by indicating the intensity of the fluorescence, e.g. weak, moderate or strong.

In this respect it is interesting to follow the experience obtained by Bonifacio et al. [5] and Hollingworth et al. [6], who observed that if a dilution series is made of a local standard serum, which is referred to the WHO standard for ANA of the homogeneous type, the intensity of the fluorescence obtained with the serum to study can be compared with the intensity of a certain dilution of the standard. The result can then be expressed in IUs without performing a dilution series. They suggest that this is also true for non-homogeneous types of fluorescence.

6. Examples of the use of standards

To augment the reproducibility within a laboratory, the daily use of laboratory standards is of course already a great help, even if such a standard is not compared to a national standard. To decrease the interlaboratory variability the use of national standards is a good tool [2, 7–9, 15]. This is best illustrated by the following figures: Fig. 1 shows the results of the demonstration of the RF with the latex fixation test on a serum tested by several laboratories expressed both in titres and in IU/ml. Fig. 2 shows the same for the RF tested with the Waaler-Rose test, Fig. 3 for the APF tested with the indirect IFT on buccal mucosa cells, Fig. 4 for ANA of the homogeneous type tested with the indirect IFT on Hep-2 cells, and Fig. 5 for anti-DNA tested with the indirect IFT on Crithidia luciliae. These figures were based on the results given in previous articles: Fig. 1 [7, 14], Fig. 2 [14], Fig. 3 [9], Fig. 4 [8, 12] and Fig. 5 [12].

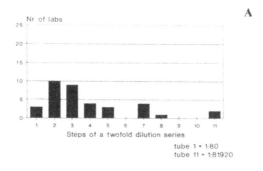

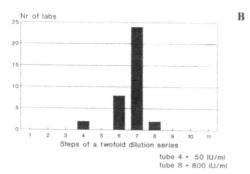

Fig. 1. A. Titres found by 18 laboratories, testing one serum twice with the Latex test for RF.
B. The same results expressed in IU/ml.

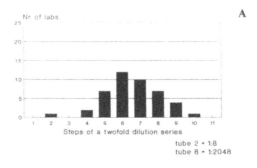

tube 2 = 1:8
tube 8 = 1:2048

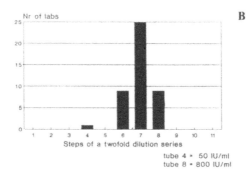

tube 4 = 50 IU/ml
tube 8 = 800 IU/ml

Fig. 2. **A.** Titres found by 22 laboratories, testing one serum twice with Waaler-Rose test for RF.
B. The same results expressed in IU/ml.

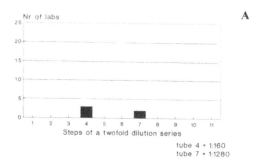

tube 4 = 1:160
tube 7 = 1:1280

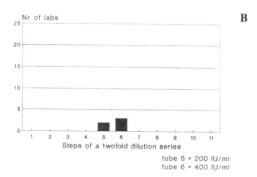

tube 5 = 200 IU/ml
tube 6 = 400 IU/ml

Fig. 3. A. Titres found by 5 laboratories, testing one serum with the IFT for APF.
B. The same results expressed in IU/ml.

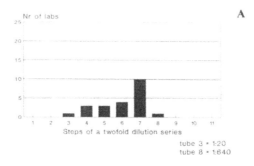

tube 3 = 1:20
tube 8 = 1:640

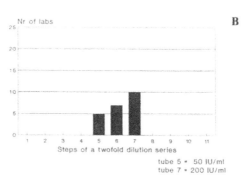

tube 5 = 50 IU/ml
tube 7 = 200 IU/ml

Fig. 4. A. Titres found by 11 laboratories, testing one serum twice for ANA (homogeneous type).
B. The same results expressed in IU/ml.

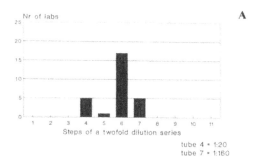

tube 4 = 1:20
tube 7 = 1:160

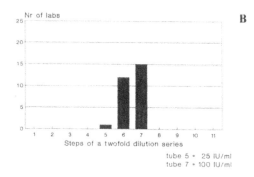

tube 5 = 25 IU/ml
tube 7 = 100 IU/ml

Fig. 5. A. Titres found by 8 laboratories, testing one serum with IFT for anti-dsDNA.
B. The same results expressed in IU/ml.

7. Characteristics of standards and reference preparations

Table 1 lists the most important characteristics of the available standards and reference preparations. Details about these standards can be found in the literature mentioned in this table.

Table 1. International standards and reference preparations

Antibody	Code	IU/ml	Lit.Ref.	Available from
RF	WHO-64/1	100/ampoule	[2, 15]	CLB
APF	WHO-64/1	100/ampoule	[9, 15]	CLB
ANA homogeneous	WHO-66/233	100/ampoule	[8, 18]	CLB
ANA speckled	AF/CDC 3	ref. prep.	[11, 17]	CDC
ANA nucleolar	AF/CDC 6	ref. prep.	[11, 17]	CDC
ANA centromere	AF/CDC 8	ref. prep.	[11, 17]	CDC
FITC anti Hu Ig	WHO-480010	100/ampoule	[8, 10, 12]	CLB
anti-nRNP	WHO-αNRNP	ref. prep.	[12, 19]	CLB
anti-Sm	AF/CDC 5	ref. prep.	[11, 17]	CDC
anti-Ro/SSA	AF/CDC 7	ref. prep.	[11, 17]	CDC
anti-La/SSB	AF/CDC 2	ref. prep.	[11, 17]	CDC
anti Scl-70	AF/CDC 9	ref. prep.	[11, 17]	CDC
anti-Jo-1	AF/CDC 10	ref. prep.	[11, 17]	CDC
anti-dsDNA	WHO-Wo/80	200/vial	[20]	CLB

The only standard which requires special attention is the standard FITC labelled sheep anti-human Ig conjugate, known as 480 010 [8, 10, 12]. This standard conjugate should be used to standardize national conjugates. Such national conjugates can then be used to adjust laboratory conjugates. This has to be done by performing block or chessboard titrations each time a new substrate or conjugate is introduced in the laboratory [10, 16].

8. How to obtain the standards?

The international standards and reference preparations in the field covered by this manual are listed in Table 1. For organizations who would like to improve the serology at a national level, all these standards are available from: Dept. of Reagents, CLB, P.O. Box 9190, 1006 AD Amsterdam, The Netherlands.

The CDC reference preparations are also listed in Table 1 [11, 12, 17]. They can be ordered by national organizations from: AF/CDC ANA Ref. Lab., Immunology 1-1202 A25, CDC, Atlanta, GA 30333, USA.

References

1. Bozsoky S (1963) The problem of standardization in rheumatoid arthritis serology. Arthr Rheum 6:641–649
2. Fulford KM, Taylor RN, Przybyszewski VA (1978) Reference preparations to standardise results of serological tests for rheumatoid factor. J Clin Microbiol 7:434–441
3. Molden DP, Nakamura RM, Tan EM (1984) Standardization of the immunofluorescence test for autoantibody to nuclear antigens (ANA): Use of reference sera of defined antibody specificity. Amer J Clin Pathol 82:57–66
4. Nakamura RM, Rippey JH (1985) Quality assurance and proficiency testing for autoantibodies to nuclear antigen Arch Pathol Lab Med 109:109–114
5. Bonifacio E, Hollingworth PN, Dawkins RL (1986) Antinuclear antibody: Precise and accurate quantitation without serial dilution. J Immununol Meth 91:249–255
6. Hollingworth PN, Bonifacio E, Dawkins RL (1987) Use of a standard curve improves precision and concordance of antinuclear antibody measurement. J Clin Lab Immunol 22:197–200
7. Klein F, Janssens MBJA (1987) Standardization of serological tests for rheumatoid factor measurement. 46:647–680
8. Feltkamp TEW, Klein F, Janssens MBJA (1988) Standardization of the quantitative determination of antinuclear antibodies (ANAs) with a homogeneous pattern. Ann Rheum Dis 47:906–909
9. Feltkamp TEW, Berthelot JM, Boerbooms AMT *et al.* (1993) Interlaboratory variability of the antiperinuclear factor (APF) test for rheumatoid arthritis. Clin Exp Rheumatol 11:57–59
10. Johnson GD, Chantler S, Batty I, Holborow EJ (1978) Use and abuse of international reference preparations in immunofluorescence. In: Dumonde C, Steward MW (eds): Laboratory and clinical standardization in rheumatoid arthritis, Part I (pp. 93–100). Lancaster: MTP Press
11. Tan EM, Feltkamp TEW, Alarcon-Segovia D *et al.* (1988) Reference reagents for antinuclear antibodies. Arthr Rheum 31:1331
12. Feltkamp TEW (1990) Standards for ANA and anti-DNA. Clin Rheumatol 9: 74–81
13. Hay FC, Nineham LJ (1979) Standardization of assays for rheumatoid factors and antiglobulins. In: Dumonde DC, Steward MW (eds) Laboratory tests in rheumatic diseases: Standardization in laboratory and clinical practice (pp. 101–105). Lancaster: MTP Press
14. Feltkamp TEW (1988) Expression of rheumatoid factors in titres or units? Scand J Rheumatology. Suppl 75:54–57
15. Anderson SG, Bentzon MW, Houba V, Krag P (1970) International reference preparation of rheumatoid arthritis serum. Bull Wld Hlth Org 42: 311–318.
16. Beutner EH, Holborow EJ, Johnson GD (1967) Quantitative studies of immunofluorescent staining I. Analysis of mixed immunofluorescence. Immunology 12: 327.
17. Tan EM, Fritzler MJ, McDougal JS *et al.* (1982) Reference sera for antinuclear antibodies. Arthr Rheum 25:1003–1005
18. Anderson SG, Addison IE, Dixon HG (1971) Antinuclear-factor serum (homogeneous): An international collaborative study of the proposed research standard 66/233. Ann NY Acad Sci 177: 337–345
19. WHO Expert Committee on Biological standardization (1984) In: WHO Technical Report Series. 700:22
20. Feltkamp TEW, Kirkwood TBL, Maini RN, Aarden LA (1988) The first international standard for antibodies to double stranded DNA. Ann Rheum Dis 47:740–746

Introduction to Section B

Our introduction to Section A gave a historical perspective to the genesis of this manual and the anticipated publication of Section B. Our concept was to proceed from detection methods described in the first section to a detailed molecular description of antigens recognised by autoantibodies present in the sera of patients. Section B is the outcome of the progression of our plan.

Two main criteria guided us in deciding which 'autoantigens' to include in the manual. Firstly, we thought, the autoantigens should be well-characterised molecular entities; and secondly, presence of cognate autoantibodies in sera should be of clinical relevance. As a result of the need for good molecular information we soon realised that the information would be greatly enhanced if its presentation conformed to a common template. Thus, we developed the following sequential theme: definition of the antigen followed by biochemical characteristics; methods for detection of antibody binding; its tissue (or cellular) distribution; function, genetic structure and gene (DNA) sequence (if known); relationship to other molecules; and, finally, information on epitopes recognised by B and T cells.

In planning the layout of chapters we decided against a system based on classification of antigens according to molecular, functional or tissue localisation, although each of these had points to commend it. Instead, we chose to use a sub-division based on clinical diagnosis since the approach emphasised the clinical importance of the study of autoimmunity. Thus, in this initial print, we chose autoantigens of interest in rheumatic diseases, e.g., rheumatoid arthritis, SLE, overlap syndromes, Sjögren's syndrome, scleroderma, myositis and vasculitis. The tentative, but positive, decision to include mitochondrial antigens and p80 coilin attests to our unwillingness to be constrained solely by a clinical speciality or by diagnostic utility. We decided that the study of autoimmunity as a biological phenomenon is of itself a sufficient justification, provided that the antigens were well-characterised.

Our evolving philosophy has several consequences. For one thing we now see the need to follow-on from a description of molecular aspects of autoimmunity in non-organ specific autoimmune disease to another section on 'tissue specific' autoimmunity. This is in the planning stage. In addition, we are well advanced in the preparation of the 'clinical part' related to Section B. The clinical part (Section C) will deal with an assessment of the indication for measuring autoantibodies and interpreting the data in a diagnostic, monitoring and therapeutic context.

A final word of thanks to our expert contributors whose excellence, patience and personal involvement in the subject led to the completion of the difficult task of providing an in-depth review and the information enshrined in this book. We hope that future interest of readers in the product (this manual) will in time reward them and make all the team feel that we have prompted a further increase in our knowledge of understanding of disease. Therein must lie the key to conquering its consequences of disability and premature death.

The Editors   W.J. VAN VENROOIJ, NIJMEGEN
                                                       R.N. MAINI, LONDON
                                                       1994

Manual of Biological Markers of Disease **B1.1**: 1–27, 1994.

The RF antigen

RIZGAR A. MAGEED

Kennedy Institute of Rheumatology, 6 Bute Gardens, Hammersmith, London W6 7DW, U.K.

Abbreviations

Ag: antigen; Cγ: IgG heavy chain constant region; C$_H$: Ig heavy chain constant region domain; C$_L$: light chain constant region domain; Cμ: IgM constant region; DAG: diacylglycerol; DEAE: diethyl amino ethyl; ELISA: enzyme-linked immunosorbent assay; ER: endoplasmic reticulum; Fab: fragment antigen binding; Fc: fragment crystalline; FcγR: IgG Fc receptor; Gm: IgG allotypic marker; HC: heavy chain; Ig: immunoglobulin; IP3: inositol 1,4,5-triphosphate; kD: kilo dalton; LC: light chain; MHC: major histocompatibility complex; mIg: membrane-bound immunoglobulin; mRNA: messenger RNA; PIP2: phosphatidyl inositol 4,5-biphosphate; PKC: phosphokinase C; PLC: phospholipase C; RA: rheumatoid arthritis; RF: rheumatoid factor; RER: rough endoplasmic reticulum; RIA: radio-immunoassay; S: Svedberg units (a unit for measuring sedimentation coefficient of molecules); SPA: Staphylococcal protein A; SRBC: sheep red blood cell; V$_H$: immunoglobulin heavy chain variable region domain; V$_L$: immunoglobulin light chain variable region domain

Key words: RF antigen, rheumatoid arthritis, autoantigen, immuno-globulin, rheumatoid factor

1. The antigen

It is now more than 70 years since rheumatoid factors (RFs) were discovered and nearly 50 years since they were first associated with rheumatoid arthritis (RA) [1, 2]. During that time, a vast amount of work has been done on the incidence, nature and specificity of RF in many laboratories. These studies established that RFs react with determinants expressed on the Fc of IgG from human and other primates [3]. Thus the term RF is generally used to describe autoantibodies to IgG Fc. The term antiglobulins (which has been used as a synonym for RF in some studies) refers to any antibody reactive with determinants other than those on the Fc of IgG, or on immunoglobulin isotypes, such as IgE and IgA [4, 5]. Some RFs are also known to cross-react with other non-IgG related antigenic determinants [6].

AMAN-B1.1/1

2. Synonyms

Globulin, γ-globulin when RF = antiglobulin.

3. The antibody

The early studies of RF distribution and properties were focused on IgM RFs. This was primarily because of the efficiency of detection of IgM RF in agglutination assays (the only assays available then). However, with the design of new solid-phase assays and isotype specific reagents, IgG RFs, IgA RFs and IgE RFs were found in the serum and synovial fluid of RA patients [7–9]. Perhaps the most important of these isotypes (from the point of view of immunopathology) is IgG RF, which via self association and binding their own Fc to IgM RF, can form large immune complexes resulting in complement activation and phagocytosis. IgA RFs have been shown to be linked with the presence of high levels of IgM RF and to exist in both dimeric and monomeric forms mainly as IgA1 subclass [10]. IgE RF have also been detected in the serum of patients with extra articular features of RA and vasculitis in particular [11].

4. Biochemical characteristics

It is necessary to appreciate the structural and functional properties of IgG in order to describe the antigenic determinants in the Fc of IgG heavy chains. The structure of IgG (presented in Fig. 1) consists of three functional units. Two of the units are identical and are involved in binding to antigen, "fragment antigen binding" (Fab) while the third unit, "fragment crystalline" (Fc) is involved in binding to effector molecules such as complement components and Fcγ receptors (FcγR). The Fab units contain regions of sequence of great variability (variable regions) between different antibody molecules. These regions confer on IgG its unique antigen binding specificity [12]. The three regions are contributed to by a four-chain structure consisting of two identical heavy chains of approximate molecular weight of 50 kD spanning the Fab and Fc and two identical light chains of approximate molecular weight of 25 kD associated with the Fab. The Fab arm is linked to the Fc via the hinge region which is highly sensitive to proteolytic attacks [13].

The heavy chains can be grouped into 4 different subclasses having heavy chains $\gamma1$, $\gamma2$, $\gamma3$ and $\gamma4$ and giving rise to IgG1, IgG2, IgG3 and IgG4 subclasses, respectively (Fig. 2). These subclasses have very similar primary heavy chain sequences, with the greatest differences being in the hinge region [13]. IgG3, which has an additional 41 amino acids in the lower hinge, has a molecular weight of 170 kD, while IgG1, 2 and 4 each have a molecular

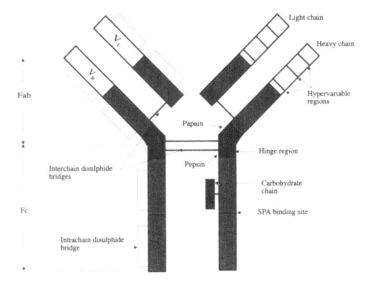

Fig. 1. A schematic representation of the basic IgG architecture showing the two heavy and light chains and the three functional units (2 Fab and one Fc unit) that make up the IgG molecule. The variable (V) and constant (C) region domains within both heavy and light chains are shown as clear and hatched areas, respectively. The light chains exist in two forms, kappa ($\varkappa$) and lambda (λ) light chain isotypes. The heavy chains can be grouped into 4 different subclasses, IgG1, IgG2, IgG3 and IgG4. In a single molecule, the two heavy chains are identical as are the two light chains. IgG is organised into 12 homology regions, or domains, each possessing an internal (or intrachain) disulphide bond. All of these domains with the exception of the $C\gamma2$ domain are in close lateral association with another domain (domain pairing or trans-interaction). The interacting forces of the domains are predominantly hydrophobic. The Fab arrangement is further stabilised by a disulphide bond between the $C\gamma1$ and CL domains. The V_H-$C\gamma1$ and V_L-C_L cis-interactions are very limited allowing flexibility about the V-C switch region (elbow bending). The $C\gamma2$ domains have two N-linked branched carbohydrate chains interposed between them. The location of one N-linked carbohydrate chain is shown as a black rectangle. Regular arrangements of inter- and intrachain disulphide bridges are indicated. Length of the hinge region and number of the interchain disulphide bridges presented is typical of IgG1. Also depicted is the site of Staphylococcal protein A (SPA) binding and pepsin and papain cleavage sites. The hinge region (depicted in black) is highly flexible to allow divalent recognition of variably spaced antigenic determinants. The hinge region is structurally divided into three subregions. The upper subregion allows flexibility of the Fab arms relative to one another (Fab-Fab flexibility). The middle subregion of the hinge contains the interheavy cysteine disulphide bridges and a high content of proline. This subregion adopts a relatively rigid double-stranded structure. The lower hinge subregion is probably responsible for the flexibility of Fc relative to Fab (Fc-Fab flexibility) [16]. There is an association between hinge flexibility and effector functions of IgG.

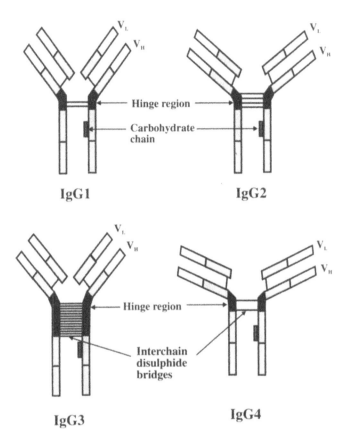

Fig. 2. A schematic diagram of the four IgG subclasses.
The subclasses differ in the length of the hinge region and the number and arrangement of the interchain disuphide bridges. The hinge region is extremely elongated with 14 disulphide bridges in IgG3. The length of the hinge and arrangement of the disulphide bridges have major influence on the orientation the Fab arms. This can be observed for the IgG2 and IgG4 subclass in the Figure. The orientation of the Fab arms and length of the hinge region are particularly relevant to the efficiency of IgG3 in binding to Fcγ receptors and in complement activation. The orientation of the Fab arms renders accessible epitopes involved in binding to C1q and FcγR [14].

weight of 146 kD. IgG1, 2 and 3 have slow electrophoretic mobility in the γ region, while IgG4 has a fast electrophoretic mobility in the same region. Amino acid sequence comparisons show that IgG can be organised into 12 homology regions or domains, each possessing an internal (or intrachain) disulphide bond [14]. Crystallographic studies show that each domain has a common pattern of polypeptide chain folding (Fig. 3) [15]. Each light chain

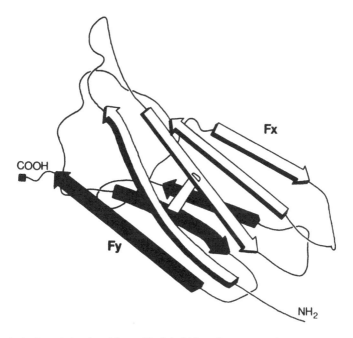

Fig. 3. A schematic drawing of the peptide chain folding of a constant region domain (Cγ1, Cγ2 or Cγ3).

The Fx (white) and Fy (black) segments form two roughly parallel faces of the anti-parallel β-pleated sheets linked by an intra-chain disulphide bond. The pattern of chain orientation consists of two twisted stacked β-pleated sheets enclosing an internal volume of tightly packed hydrophobic amino acid residues. The arrangement is stabilised by an internal disulphide bond linking the two sheets in a central position. The β-pleated sheets are linked by helices or bends. The amino acids that compose the β-pleated sheets are highly conserved while changes are seen in the bends or helices.

consists of two domains while each heavy chain consists of four domains. The Cγ2 domains have two N-linked branched carbohydrate chains covalently attached. All four IgG subclasses have the same amount of carbohydrates attached (2–3% of total IgG mass). The carbohydrate chains of the IgG molecule (attached to asparagine at position 297) are composed of a complex set of different combinations of oligosaccharide structures (about 20 different combinations). IgG molecules can vary in the presence or absence of galactose or sialic acid on the outer parts of the carbohydrate chain. The carbohydrate chain is important in maintaining the Cγ2-Cγ2 orientation [14].

Analysis of the solvent water accessibility of amino acid residues in the Fc region has enabled the identification of hydrophobic patches likely to be involved in IgG interaction with other proteins (Fig. 4) [13].

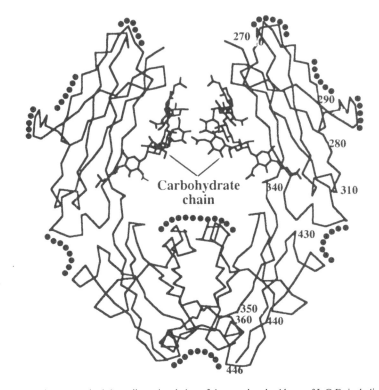

Fig. 4. A computerised three-dimensional view of the α-carbon backbone of IgG Fc including the Cγ1 and Cγ2 domains (RF binding sites).

The location of the N-linked carbohydrate chains are indicated. The approximate locations of the hydrophobic patches are highlighted by black dots. The largest hydrophobic patch on the inside of the Cγ3 domains although accessible to solvent is unlikely to be accessible to large macromolecules because of the proximity of the Cγ2 domains and the carbohydrate chains. The large patch on the outside of the Cγ3 domains is accessible to macromolecules. The third patch forms, with residues in the Cγ2 domain, the general RF and Staphylococcal protein A (SPA) binding sites. The Cγ2 residues of this side show a good deal more sequence conservation between species and subclasses than their counterparts in the Cγ3 domain. (Modified from Burton 1985 [13] with permission).

5. RF detection

The incidence of RF in a population depends on the assay used and the titre chosen to separate positive and negative individuals. The titre of RF usually behaves as a continuous variable that differs among various ethnic groups. With increasing age, both the percentage of individuals with a particular titre and the mean titre in a population increases. In most populations, the distribution of titres among men and women is similar. Some studies have shown that the prevalence of RF in the general population tends to decline beyond the age of 70 [17, 18].

5.1 Immunofluorescence

The basic technique (i.e. binding of antibodies to cells or tissue sections and revealing with fluorescence labelled antiserum) has not been used to detect RF. However, fluorescence-labelled aggregated IgG have been used to detect RF producing cells in tissue sections [19]. Also fluorescence-labelled anti-Ig heavy chains have been used to detect RF binding in agglutination, immunoprecipitation or solid-phase based assays [20, 21]. These methods are difficult to use for quantification of RFs and the interference of IgM RF in the detection of other isotypes cannot be ruled out.

5.2 Counter-immunoelectrophoresis/immunodiffusion

Counter-immunoelectrophoresis has not been used to detect RF. Immuno-diffusion has been used to detect RF in a few studies [22]. However, the method is usually isotype non-specific. In combination with a prior separation procedure (such as ultracentrifugation or affinity purification) to separate the different isotypes the method can be made isotype-specific. In some studies, alteration in the ability of IgG from RA sera, compared with normal IgG, to diffuse in agarose gels has been used to detect self-associated IgG RF [23]. Enzyme- or fluorescence-labelled antisera have been used to detect RF isotype precipitated by immunodiffusion. The method suffers from many technical drawbacks such as binding of the indicator reagents to IgM RF.

5.3 Immunoblotting

For general techniques see chapter A4 of this manual.

Immunoblotting protocols devised to detect RF involve the binding of IgG or denatured IgG Fc to nitrocellulose membranes and application of sera or culture supernatants to the membrane (Fig. 5) [24]. The detection systems have varied between simple enzyme-labelled isotype or light chain-specific antisera to using enhancing systems such as the avidin-biotin system. The assay can detect different RF isotypes and RF of low binding affinity. Immunoblotting has proved useful in detecting IgG RF in culture supernatants [25] but its application to patients' sera may require prior separation of the different RF isotypes [26].

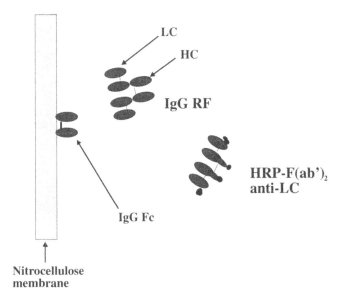

LC

HC

IgG RF

HRP-F(ab')₂
anti-LC

IgG Fc

Nitrocellulose
membrane

Fig. 5. The general technique of immunoblotting involves the prior separation of the antigen in an analytical gel, such as sodium dodecyl sulphate polyacrylamide gel (SDS-PAGE) or isoelectric focusing gel, before transferring electrophoretically to a nitrocellulose membrane. For the detection of RF, a nitrocellulose membrane can be directly sensitised with intact IgG or IgG Fc (for IgG RF detection). The membrane is then blotted with isolated RF or culture supernatant, and washed. Bound RFs are revealed with radio- or enzyme-labelled (e.g. horseradish peroxidase, HRP) F(ab')₂ anti-IgM or IgA or anti light chain (LC). IgG RFs in fractionated serum or culture supernatants can be detected using radio- or enzyme labelled anti-IgG Fd antibodies.

5.4 Protein immunoprecipitation

For general technique see chapter A6 of this manual.

Immunoprecipitation methods to detect RF are usually isotype non-specific. However, the method was used together with analytical centrifugation [27], polyethylene glycol precipitation [28] or nephelometry [29]. The latter method, which is used in routine tests in some hospitals, relies on complex formation between IgG and RFs leading to optical density changes resulting from light scattering by the complexes [29, 30]. These methods usually detect IgM RF. Precipitation may be enhanced by reducing temperature and ionic strength to allow for the interaction of low affinity RF.

5.5 Agglutination assays

RF were originally defined by their ability to agglutinate sheep red blood cells (SRBCs) sensitised with rabbit anti-SRBC anti-serum (the Rose-Waaler test) (Fig. 6). Since then many modifications of the original method have been described. These include the use of carrier red cells from other species [31], tanning of the cells [32] or trypsin treatment and passive chromic chloride coupling of IgG from various species to SRBCs [33]. These modifications improved the sensitivity and specificity of the test for RA patients, but the simpler Rose-Waaler test remains in general use for diagnostic purposes. Other carriers have also been used for agglutination tests including bentonite [34] and latex particles [35]. Cross-linking of the human IgG-coated latex or bentonite particles by IgM RF in serum produces a visible flocculus. The latex flocculation test, which is similar in principle to the Rose-Waaler test except for the use of human IgG attached to latex particles as the indicator, was also adapted to a rapid slide method which remained the most widely used in small hospitals for some time [36].

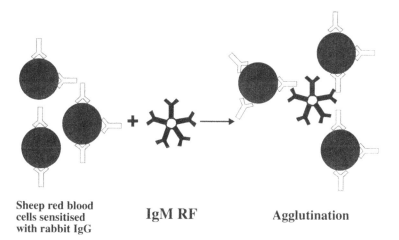

Sheep red blood cells sensitised with rabbit IgG   **IgM RF**   **Agglutination**

Fig. 6. The Rose-Waaler haemagglutination assay involves the use of sheep red blood cells (SRBCs) sensitised with rabbit antibodies to SRBCs. Multivalent IgM RF in serum, added in serial dilutions (usually in doubling dilutions) to tubes or microtitre plate wells, cross-link the rabbit IgG-sensitised SRBCs and sink as a mat to the bottom of the tube or well. In the latex flocculation test, latex particles coated passively with human IgG are used as the detecting agent. IgM RFs cross-link the IgG-coated latex particles to produce a visible flocculus.

Agglutination techniques usually detect IgM RFs, as IgM antibodies are more efficient in agglutination reactions, but the exact contribution of IgM RFs, IgG RFs and IgA RFs and the effect of affinity in these tests is not known. The quantity of IgM RF are usually expressed as the highest dilution of serum yielding detectable agglutination. Reagents for these assays are available commercially from a number of suppliers with varying degrees of sensitivity and accuracy [37].

In general terms, a positive Rose-Waaler test may be more specific for RA than the bentonite or latex flocculation assays, since the latter are coated with human IgG and can detect anti-allotypic antibodies. However, because only a minor proportion of RF cross-react with rabbit IgG, titres in the Rose-Waaler test are usually lower than those detected in systems that employ human IgG as antigen.

5.6 Ultracentrifugation

IgM RFs can combine with monomeric autologous IgG *in vivo* to form soluble 22S immune complexes. These immune complexes have been visualised in many rheumatoid sera by analytical ultracentrifugation [38]. The rationale for using this approach is based on the increase in the molecular size of IgM or IgG after complex formation. IgG RF have also been detected by their characteristic sedimentation profile as intermediate size complexes in the analytical ultracentrifuge [39]. The method is, however, laborious and requires relatively large volumes of serum and expensive equipment.

5.7 Affinity adsorption-based techniques

These methods were devised to measure RF of all immunoglobulin isotypes. However, the methods may be better suited for qualitative assessment of RF isotypes. The techniques depend upon adsorption to, and subsequent elution from, insolubilised immunosorbents (Fig. 7). This step is followed by specific detection of RF isotypes by immunodiffusion in agarose gels using specific anti-immunoglobulin reagents. The techniques were originally devised for the detection of IgG RF [40]. However, non-specific adsorption of normal IgG to the immunosorbent material or IgG complexes with IgM RF may interfere with the detection of IgG RF. The method is laborious and uses a large volume of serum for each test.

5.8 Solid-phase based assays

For general techniques see chapter A5 of this manual.

ELISA and RIA protocols were used as alternative but essentially similar methods to affinity adsorption. Hay *et al.* [8] developed a method where rabbit IgG was bound to plastic tubes (and later, microtitre plate wells) and serum FR were allowed to react with the antigen. Bound RF isotypes were

Serum or plasma

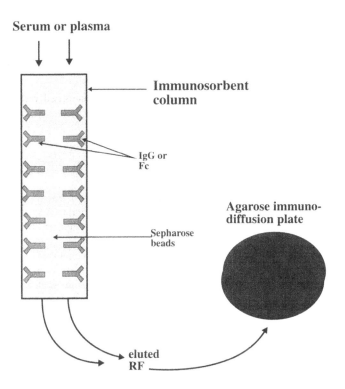

Immunosorbent column

IgG or Fc

Agarose immuno-diffusion plate

Sepharose beads

eluted RF

Fig. 7. Affinity adsorption-based techniques depend on adsorption to, and subsequent elution from, insolubilised IgG or IgG Fc of RF from plasma or serum followed by specific detection of RF isotypes by immunodiffusion in agarose gels. Plasma or serum containing RFs is applied to the immunosorbent column under physiological conditions. RFs are bound to the IgG while unbound proteins are washed through. Bound RFs are then eluted using elution buffer (e.g. acetate pH 3–4; 3M guanidine or thiocyanate) and dialysed back to normal physiological buffers before detecting RF isotypes in immunodiffusion.

detected using radiolabelled heavy chain isotype-specific antisera. Modified methods using human IgG Fc as antigen and anti-human IgG Fd (Fd is composed of the heavy chain variable region and first constant region domains) specific antiserum to detect IgG RF have been described [9]. Faith *et al.* [41] used enzyme-labelled, rather than radiolabelled antisera to detect different isotypes. One significant advantage of RIA or ELISA methods is that they can detect RF in rheumatoid sera diluted 1,000- to 100,000-fold.

Solid-phase methods are now in general use to measure the different isotypes of RF. However, it is still uncertain whether these methods accurately measure amounts of RF, particularly in the case of IgG RF. IgM RF may bind both the solid-phase antigen and IgG in the revealing antiserum, whatever its specificity, leading to false positive results for IgG and IgA RF. This problem may be overcome by using $F(ab')_2$ fragments of IgG as the labelled indicator [42]. IgG RF may be overestimated due to IgG-IgM RF complexes, or underestimated as a result of self-association. Prozones (areas of weak or negative reaction when the antibody is tested at high concentration) may also be seen when measuring low affinity RF [42]. A major problem of these assays is the lack of a recognised method of standardisation to enable comparison of results between different laboratories.

5.9 General notes

1. Pre-treatment of sera before testing has been used to increase sensitivity and specificity of the various methods. For example, "hidden" IgM RF or "undetectable" self-associated IgG RF can be revealed by acid or 1 M sodium chloride solution pretreatment [43, 44].

2. The ability to detect RF is greatly influenced by the manner of antigen (IgG) presentation and RF affinity. Therefore, the method of testing and pre-treatment may be critical in the interpretation of the results of RF measurement. Generally, studies using different methods may not be comparable.

3. The interference by complexes of normal IgG with IgM RF in IgG RF assays can be avoided by reduction and alkylation [45], pepsin digestion [46] or preparation of different isotypes of immunoglobulin on DEAE cellulose [40].

4. The specificity of RF for RA is increased if positivity on two or more consecutive occasions, high titre, reactivity with both human and rabbit IgG and distribution among the IgM, IgG and IgA isotypes is observed.

6. Cellular localisation

Following the successful rearrangement and transcription of the IgG genes, mature mRNA is translated, separately, into functional Ig heavy chains (which express the RF antigen) and light chains on the polyribosomes of the rough endoplasmic reticulum (RER) (Fig. 8) [47]. In B-lymphocytes most of the newly synthesised Igs are incorporated into surface membrane but some are shed rapidly (half-life of 2–4 hours). In mature plasma cells, soluble Igs are held intracellularly until secreted at the late stage of plasma cell maturation.

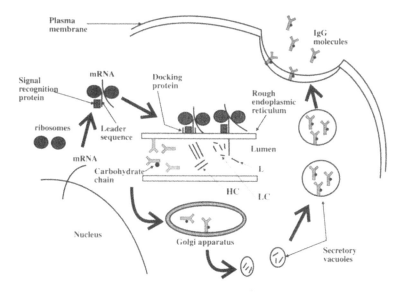

Fig. 8. Messenger RNA (mRNA) for Ig heavy and light chains are transcribed from the DNA in the nucleus and leave to the cytoplasm where it binds to ribosomes.

The heavy chains (which express the determinants for RF) and light chains are synthesised on separate polyribosomes. The leader sequence (L) is the first to be translated and binds to signal recognition protein which blocks further mRNA translation. The signal recognition protein complex with the ribosomes migrate to the rough endoplasmic reticulum (RER) where the complex binds to the docking protein. Translation now proceeds and the completed nascent chains are released into the lumen, the leader sequence is cleaved, the free chains fold and heavy and light chains assemble. The synthesised heavy chains form dimers and subsequently combine with light chains to form Igs. The heavy chains are then glycosylated (black dots) by the transfer of the core oligosaccharides to the acceptor site on the heavy chain. The heavy and light chains are then S-S bonded, the carbohydrate portion first trimmed, then terminal sugars added as the Ig travels from the RER to the Golgi apparatus. Two forms of Ig are synthesised, membrane-bound (mIg) and secretory Ig. Membrane-bound and secreted Igs are synthesised from mRNA that have been spliced differently. Membrane Igs, but not secreted Igs, have an additional tail made of hydrophobic amino acids at the C-terminal end of the Fc region that helps to anchor the Ig molecule in the plasma membrane. Completed Ig molecules are either released from the ER membrane (secretory Ig) or remain embedded in the membrane (membrane Ig) before being transported in vacuoles (or vesicles), which bud off the Golgi, to the plasma membrane. Secretory Ig are transferred by simple physical entrapment within the secretory vacuoles, whereas membrane Ig are transferred attached to the membrane. Fusion of the secretory vacuoles with the plasma membrane results in the membrane Ig being inserted in the plasma membrane while the soluble contents of the vesicles are disgorged to the exterior of the cell. In mature plasma cells, soluble Igs are held intracellularly until secreted at the late stage of plasma cell maturation.

7. Cellular function

Membrane-bound IgG (mIgG) plays a crucial role in the ultimate fate of B-lymphocytes. The precise role of the RF antigen (IgG Fc) in mIgG cellular functions has not been resolved. However, IgG Fc provides the link between binding to antigen through the Fab regions and cellular signalling via the Fc associated proteins (IgG-α and β proteins) (Fig. 9) [48]. Binding of antigen to mIgG results in stimulation and expansion or tolerance of B-lymphocytes, depending on their state of maturation and level and nature of antigen [49]. Endocytosis of membrane-IgG, can also occur after binding to Ag. This allows the internalisation of antigen and its subsequent processing and presentation to T-lymphocytes (Fig. 10). The role of the Fc in this process may be secondary to the binding of IgG to antigen.

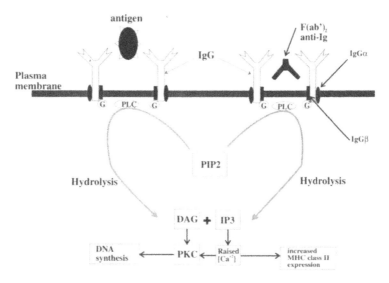

Fig. 9. A cartoon depicting the possible steps involved in cellular signalling of B-lymphocytes involving mIgG.

Membrane IgGs (mIgG) have short cytoplasmic tails and exist on the cell surface non-covalently associated with several proteins. These proteins are involved in anchoring IgG through its Fc in the membrane and signalling that result from cross-linking of mIgG. Antigen or $F(ab')_2$ of anti-IgG, which can mimic antigen in cross-linking membrane IgG may trigger B-lymphocyte into DNA synthesis. This involves phospholipase C (PLC) mediated hydrolysis of phosphatidyl inositol 4,5-biphosphate (PIP2) to produce inositol 1,4,5-triphosphate (IP3) and diacylglycerol (DAG) which mediate the increase in calcium ions $[Ca^{+2}]$ and phosphokinase C (PKC) induction required for cell activation. The coupling of PLC to mIg involves a regulatory guanosine triphosphate (GTP) binding protein (G-protein). The hydrolysis of PLC, production of IP3 and subsequent rise of $[Ca^{+2}]$ induce an increased expression of major histocompatibility complex class II antigens (MHC class II). Activation of PKC results in cell activation and DNA synthesis [51].

Secreted IgG can induce signals by binding through its Fc region to specific receptors (FcγR). This results in signal transduction via the Fcγ receptor with different effector functions depending on the cells expressing such receptors [50].

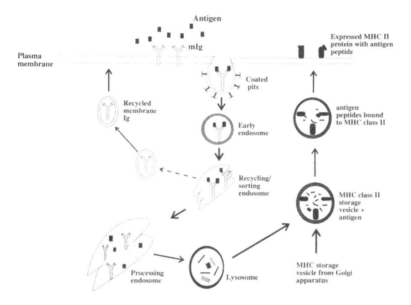

Fig. 10. A diagram of the possible endocytosis and processing of antigen bound to membrane IgG.

Following antigen binding to mIgG, the complex is endocytosed in plasma membrane coated pits. Antigen complexed with mIgG is then delivered to a site where recycling of mIgG may occur. The process of mIgG recycling is dependent upon antibody affinity for antigen. Antibodies with high affinity are not recycled resulting in the processing of antigen as a complex with mIgG. Antigen-IgG complexes are transported to a proteolytically active site (processing endosome) where the complex is processed shortly after endocytosis (~30–40 min). Processing of mIgG may also take place but an intermediate stage when processed antigen is bound to an intact mIg has been identified in lysosomes [52]. MHC class II molecules, synthesised in the rough endoplasmic reticulum and passed through the Golgi apparatus, associate with antigen peptides in acidic compartments before being expressed on the membrane.

8. Presence of IgG in other species

Recent recombinant DNA studies have shown that vertebrates evolutionarily as far apart as reptiles, amphibians, birds and mammals all have IgG (or a related isotype) [53]. However, the precise evolutionary conservation of the RF antigens is not clear. The general RF antigen (Ga determinant), or a similar determinant, is expressed on IgG from many mammals [54]. The determinant recognised on baboon IgG is the most homologous determinant to the human Ga determinant [55]. Rabbit IgG, on the other hand, appears to have a similar but not identical determinant [56]. Some studies have suggested that goat, sheep and chicken IgG do not express any Ga- related determinants [55]. Interpretation of these data however, is difficult because of the lack of adequate sequences. IgG from gorilla, chimpanzee, orangutan, baboon, macacus an other Old World monkeys also express some of the Gm allotypic determinants of IgG recognised by a fraction of RFs [57].

9. cDNA

Comparison of cDNA from vertebrates reveals that the IgG heavy chains did not evolve as single entities [58]. Rather the individual domains appear to

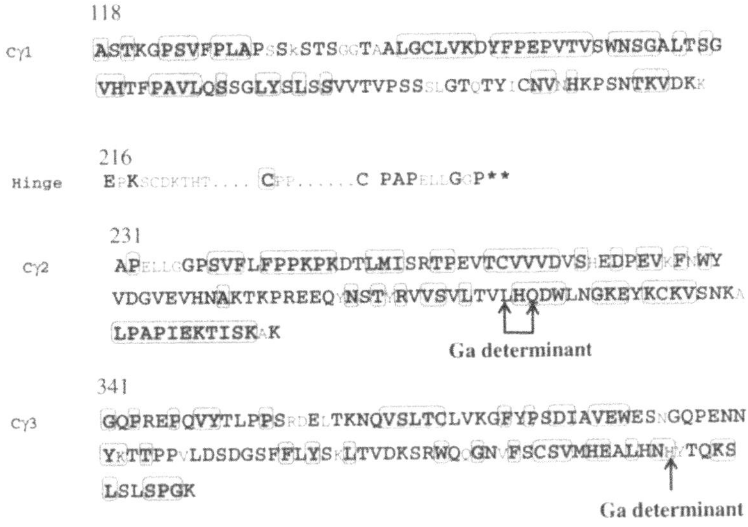

118

Cγ1   ASTKGPSVFPLAP SkSTSGGTAALGCLVKDYFPEPVTVSWNSGALTSG
      VHTFPAVLQSSGLYSLSSVVTVPSSLGTQTYICNVNHKPSNTKVDKK

216

Hinge   EPKSCDKTHT.... CPP......C PAPELLGGP**

231

Cγ2   APELLGGPSVFLFPPKPKDTLMISRTPEVTCVVVDVSHEDPEVKFNWY
      VDGVEVHNAKTKPREEQYNSTFRVVSVLTVLHQDWLNGKEYKCKVSNKA
      LPAPIEKTISKAK
                                    Ga determinant

341

Cγ3   GQPREPQVYTLPPSRDELTKNQVSLTCLVKGFYPSDIAVEWESNGQPENN
      YKTTPPVLDSDGSFFLYSKLTVDKSRWQQGNVFSCSVMHEALHNHYTQKS
      LSLSPGK
                                    Ga determinant

Fig. 11. Amino acid sequence of the three constant region domains of human IgG1 and the hinge region.
Amino acid numbering is according to the IgG1 Eu protein [54]. Residues that are shared between human IgG1, IgG2, IgG3 and IgG4 are given in bold large capital letters. Differences are given as normal small capital letters. Amino acids shared between IgG1 subclass proteins in human, rabbit, guinea pig, rat and mouse are outlined by shaded boxes. Amino acids are given using the single letter coding: A = alanine; C = cysteine; D = aspartic acid; E = glutamic acid; F = phenylalanine; G = glycine; H = histidine; I = isoleucine; K = lysine; L = leucine; M = methionine; N = asparagine; P = proline; Q = glutamine; R = arginine; S = serine; T = threonine; V = valine; W = tryptophan; and Y = tyrosine. Amino acids involved in the general Ga determinants are indicated with arrows. **IgG3 has an extra 41 amino acids in the hinge region.

have emerged from common ancestral genes. Examination of cDNA and amino acid sequences of human IgG shows that there are a number of positions that are conserved between all five isotypes of Ig (IgM, IgG, IgA, IgD and IgE) [54]. The homology between any two of the immunoglobulin isotypes is ~30%. This is in contrast to the subclasses of IgG, where the homology is approximately 95% (Fig. 11). Inspection of the Fc sequences for the five isotypes of Ig shows that the distribution of homologous residues is not random. Residues that are highly conserved tend to be clustered within segments of β-pleated sheets, especially around the two cystines that form the intrachain disulphide bonds, and tryptophan that is nearby in the three-dimensional structure (Fig. 3). There are also similarities of IgG with IgG from other species, especially mammalians. Amino acids involved in the human Ga determinant are partly conserved between human, rabbit, guinea pigs, mice and rats (Fig. 11). The number of the subclasses varies widely from species to species suggesting that the number of IgG genes fluctuates rapidly during evolution.

10. The gene

The IgG molecule results from the recombination of a number of genes at different gene loci through a process of gene recombination (juxtaposition of genes to be recombined to encode a functional protein, and the deletion of intervening DNA) [59, 60]. Heavy and light chain variable region domains are encoded by the recombination of variable region (V), diversity (D, for the heavy chain) and joining region (J) genes that are located on separate mini loci on the same chromosome [61]. Each constant region domain, as well as the hinge region of the heavy chain, is encoded by a separate exon in the DNA (Fig. 12) [62]. Heavy and light chain genes are located on three separate chromosomes, with the human heavy chain genes being on chromosome 14, $\varkappa$ chain genes on 2 and λ chain genes on 22 [63, 64]. Within the DNA region encoding IgG constant region domains, there are four complex functional genes each encoding one IgG subclass. The coding sequences (or exons) for these four genes are interrupted by intervening non-coding sequences (or introns) and there are conserved residues at the various exon and intron boundaries that are involved in the rearrangement processes [61].

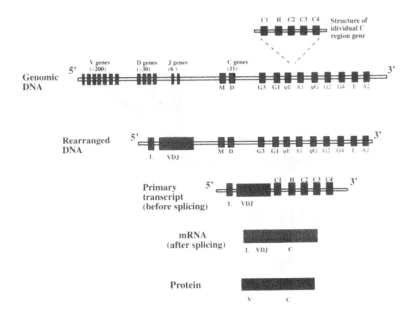

Fig. 12. A cartoon showing the organisation and rearrangement of heavy chain genes on human chromosome 14.

There are about 200 V region genes within the V_H minilocus, 30 D minigens and 6 J region genes. The C region complex is composed of 11 genes, of which two are pseudogenes (ΨG and ΨE). Each C region gene is composed of either 4 (for IgG, IgA, and IgD) or 5 exons (for IgM and IgE) separated by introns. Three or four, respectively, of the exons encode for constant region domains (C) while the H exon encodes for the hinge region. The structure of the four IgG subclass genes is generally similar. Nevertheless, major differences are found around the hinge exon, and in the case of IgG3, the hinge region is encoded by four separate exons [14]. The exons are interrupted by intervening non-coding sequences (or introns) and there are conserved residues at the various exon and intron boundaries that are involved in the rearrangement processes.

DNA for membrane and secretory Igs have two different transcription stop signals and two polyadenylation sites. This gives rise to two different mRNAs, resulting from the differential splicing of the primary mRNA, for membrane and secretory Igs. A key feature of the generation of a functional gene for both heavy and light chains is the recombination of genes. Any one of a number of genes from the mini gene loci (3 for light chain and 4 for the heavy chain) can recombine to form a functional Ig gene. In the original stages, one J gene is moved into a position next to D (for the heavy chain) or to a V gene (for the light chain). The intermediate DJ DNA is then combined with one gene from the V region. The VJ or VDJ segments and one gene from the C gene complex are then transcribed into pre-mRNA. The pre-mRNA thus contains the contiguous VJ or VDJ coding sequences, the interconnecting non-coding sequence and the fragmented coding sequences of one of the C-gene complexes. In the next stage, all the non-coding sequences are excised from the pre-mRNA and a mRNA molecule with contiguous VDJC or VJC is formed. This molecule is then translated into the polypeptide chain.

11. Relation with nucleic acids

There is no known relationship between IgG and nucleic acids. However, IgG is known to have affinity for nucleic acids (e.g., anti-DNA and anti-RNA autoantibodies).

12. Relation with other proteins

Membrane-bound IgG (about 1% of peripheral blood B-lymphocytes) and soluble IgG bind to a number of proteins with important biological functions [65]. Membrane Ig have short cytoplasmic tails and exist on the plasma membrane non-covalently associated with several proteins. Two of these proteins are well characterised, the Ig-α and β [48]. In the absence of the Igα, membrane-IgGs are retained in the endoplasmic reticulum and degraded intracellularly [66]. The relative quantity required and manner of association of Ig-α and β proteins for the different isotypes may be different [48].

Soluble IgG (in complex or aggregated form) can bind to the first component of the complement system [14]. This happens through the interaction of the subcomponent of C1, C1q, with the Cγ2 domain of IgG. The human IgG subclasses bind to C1q with the order of binding constants IgG3>IgG1>IgG2>IgG4 [67]. Proximity of the Fab arm to the C1q binding region may modulate the interaction (Fig. 2). The other two subcomponents of C1, C1s and C1r are also suggested to bind to, as yet unidentified, sites on IgG [68]. Activated forms of C3 and C4 (C3b and C4b) also bind to IgG via covalent interactions with residues in the heavy chain Fab [69].

IgG also binds to Fc receptors on a number of cell types [50]. These interactions are associated with a number of functions including phagocytosis (monocytes, macrophages, neutrophils), antibody-dependent cellular cytotoxicity (monocytes, macrophages, lymphocytes), maternofoetal transport (trophoblast) and immunomodulation (lymphocytes). There are at least three different types of Fc receptors involved in IgG binding (FcγRI, II and III). IgG1 and IgG3 bind with equal order, while IgG4 binds weakly and IgG2 does not.

In addition to these proteins, IgG can also bind to Staphylococcal protein A (SPA), which is a major cell component of most strains of Staphylococcus aureus. SPA binds to the Fc region of Igs of a variety of species and subclasses with varying affinity. As shown by crystallographic studies, SPA binds to a determinant located between the Cγ2 and Cγ3 domains.[15].

13. B-lymphocyte determinants

Within the context of reactivity with IgG, the term RF is generic for antibodies with specificity for determinants in the Fc region of IgG. Many different specificities have been described to different RFs.

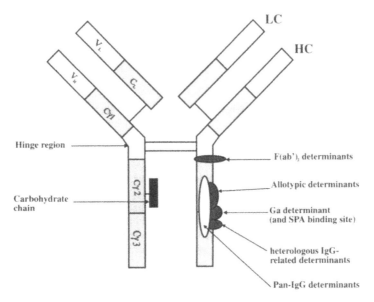

Fig. 13. A schematic diagram showing the location and physical relationship of the antigenic determinants on IgG Fc as recognised by RFs.

RFs bind predominantly to the Ga determinant, located in the $C\gamma2$-$C\gamma3$ region. This determinant is in the COOH-terminal end of the $C\gamma2$ domain. The Staphylococcal protein A (SPA) binding site is also located to the same region. Histidine at position 435 is the contact residue for SPA binding and is also central for the main Ga determinant. The determinants recognised on heterologous IgG (mainly rabbit IgG) by RFs appear to overlap partly with the Ga determinants. The pan-IgG determinants are expressed on all four human IgG subclasses irrespective of allotypic markers. There are recent experimental evidence that this determinant is also partly overlapping with the Ga determinant. The Gm (allotypic) markers appear to be immunologically distinct determinants but physically close to the Ga determinant. These determinants have been associated with amino acid sequences in the $C\gamma2$ and $C\gamma3$ domains.

13.1 IgM RF

IgG determinants recognised by human IgM RFs can be categorised into: 1) isotypic antigens, found on all or some of the four subclasses of human IgG; 2) genetically defined alloantigens on IgG heavy chain (Gm allotypes); 3) cross-reactive antigens shared by human and other species' (heterologous) IgG; 4) species-specific antigens found in human but not heterologous IgG; and 5) neo-antigens expressed on aggregated, denatured, or enzymatically digested IgG (Fig. 13).

Differences in IgM RF reactivity has been correlated, in some instances, with known amino acid interchanges and the topographical distribution of epitopes determined. There are basically three predominant isotypic epitopes with which monoclonal and polyclonal IgM RF have been shown to react. These are: epitopes expressed on all IgG proteins (or pan-IgG determinants); epitopes expressed on IgG1, IgG2 and IgG4 but not IgG3 proteins (or the Ga determinant); and epitopes expressed on one or more but not all IgG subclasses.

Extensive studies have shown that reactivity with the Ga determinant is the main (or the general) specificity known for RF (Table 1) [7, 43, 55, 70–75]. The Ga determinant corresponds to the binding of IgG to Staphylococcal protein A (SPA) (Fig. 13). The overall impression from studies of IgM RF specificity has been that the ability to bind all IgG subclasses (including IgG3 proteins) by IgM RF was not common [7, 42, 71, 74, 76].

Table 1. Spectrum of IgG determinants recognised by polyclonal RF and RF paraproteins in patients' sera

RF isotype	Source of RF	Main IgG determinant recognised	Other IgG determinants recognised	Reference
IgM	RA	Ga	ND	43
IgM	RA	Ga	Gm	72
IgM	EMC	Ga	Pan IgG isotype	97
IgM	RA	Ga	Rabbit IgG	70
IgM	RA	Ga	Pan IgG isotype	7
IgM	EMC	Ga	Pan IgG isotype	7
IgG	RA	Ga	Rabbit IgG	71
IgM	RA	ND	Gm	98
IgM	RA	Ga	Pan IgG isotype	73, 99
IgM	EMC	Ga	Pan IgG isotype	100
IgM	EMC	New Ga	Pan IgG isotype	74

The Table summarises results of some of the main studies aimed at identifying the site of RF reactivity using human IgG paraproteins of different subclasses and allotypes and rabbit IgG. Numerous other studies have either established similar patterns of IgM RF reactivity or revealed minor differences. For the sake of brevity and simplicity the Table is focused on results obtained from a small number of these studies. RA refers to rheumatoid arthritis patients; EMC, refers to patients with essential mixed cryoglobulinaemia. RF from the latter group of patients are monoclonal in nature while those in RA are polyclonal. Ga determinant is expressed on human IgG1, 2 and 4 subclasses but not IgG3 proteins. New Ga determinant is expressed on IgG1, 2 and 4 subclasses and IgG3 of the G3m(s,t) allotypes which have histidine at position 435, but not IgG3 of the G3m(b) and G3m(g) allotypes which have arginine at position 435. Gm determinant refers to allotypic determinants expressed on IgG heavy chains. Pan IgG determinant is expressed on IgG of all 4 subclasses irrespective of allotypic markers.
ND: not determined.

IgM RF with anti-Gm (IgG allotype) specificity in RA sera are usually with specificities for markers on IgG1 and IgG3 [G1m(a) and (x), and G3m(b), (g) and (s)] [77]. By far the most frequent allotype reactivity of RF from RA patients is with the G1m(a) and G3m(g) allotypic markers [78, 79].

Table 2. Spectrum of IgG determinants recognised by monoclonal RF produced by B-lymphocyte clones established from peripheral blood mononuclear cells (PBMC) or synovial membrane (SM) of patients with RA

Isotype of monoclonal RF	Original source of lympho-cyte	Main IgG determinant recognised (no. of clones tested)	Other IgG determinants recognised (no. of clones tested)	Reference
IgM	PBMC	New Ga (1)	Rabbit and baboon IgG	101
IgM	SM	Pan IgG (2)	Rabbit IgG (1)	85
IgM	SM	Ga (9)	Rabbit IgG (3)	85
IgM	SM	New Ga (2)	Rabbit IgG (1)	85
IgM	SM	Gm (1)	None	85
IgM	SM	New Ga (1)	None	102
IgM	SM	Pan IgG (3)	Rabbit IgG (1)	103
IgM	SM	IgG3 (1)[a]	Rabbit IgG	103
IgM	SM	Ga (1)	Rabbit IgG	103
IgM	SM	IgG1 and 2 (1)	Monkey IgG[b]	104
IgA	PBMC	New Ga (1)	Rabbit IgG	90
IgG	SM	Pan IgG (5)	Rabbit IgG (5)	105

The Table summarises recent studies of monoclonal RF produced by B-lymphocyte clones established by hybridoma technology or Epstein-Barr virus (EBV) transformation of B-lymphocytes from patients with RA.
[a] The authors of this study investigated the reactivity of RF with IgG3 of the G3m(b) and (g) only but not IgG3 of the G3m(s,t) allotypes.
[b] The authors of this study did not specify the precise monkey subspecies.

An intriguing feature of RF with anti-Gm specificity has been their specificity for allotypic markers not present on the individual's own IgG (heteroclitic specificity) [80, 81].

Studies of IgM RF reactivity with heterologous IgG have shown that the epitopes can be divided broadly into three categories: 1) cross-reactive epitopes expressed on human and heterologous IgG; 2) epitopes expressed on human but not heterologous IgG; and 3) epitopes expressed on heterologous but not human IgG [70, 82–85].

In a number of studies IgM RF have been shown to react with neo-determinants expressed on pepsin digested IgG, aggregated, complexed or denatured but not native IgG. The question has always been whether this is a reflection of the appearance of neo-antigenic determinants or due to the more favourable energy of binding of RF that result from the local concentration of antigenic determinants on aggregated IgG. It is believed by some investigators (using circular dichroism and small angle X-ray scattering) that the appearance of neo-determinants on IgG might be the result of conformational changes [86, 87].

13.2 IgG RF

The specificity of IgG RFs in RA have not been studied in so much detail as IgM RF. This has been primarily due to the lack, until recently, of monoclonal IgG RFs. So far, the antigenic specificity of polyclonal IgG RF from only two RA patients has been extensively studied [7, 71]. These studies suggested that the specificity of IgG RF in patients' sera is perhaps similar to the general IgM RF specificity to the Ga determinant. More recently monoclonal RF producing B-lymphocyte hybridomas have been successfully produced in a number of laboratories. Detailed study of the fine specificity of these IgG RF producing hybridomas suggest that monoclonal IgG RFs may have reactivity patterns different from the general IgM RF specificity (Ga determinant) [25].

13.3 IgA RF

The molecular specificity of IgA RF has not been studied in great detail. However, the presence and levels of IgA RF have been shown to be associated with that of IgM RF [88, 89]. Furthermore, there is evidence showing that in most patients with RA, IgA RFs appear to be predominantly dimeric [10]. This would suggest that in the original characterisation of IgM RF, specificity studies – which have used agglutination studies – may have inadvertently examined IgA RF specificity, too. This may be evidence that, in the main, IgA RF may be reactive with the general Ga determinant on IgG. Recently, the specificity of a monoclonal IgA RF-producing B-lymphocyte clone, established from blood of an RA patient, was shown to be for a new Ga-like specificity (Table 2) [90].

13.4 IgE RF

The specificity of IgE RF has not been studied, primarily because of the low level of this isotypes of RF in patients sera.

Antiglobulins reactive with determinants in the antibody binding region (Fab region) of IgG are also known to occur [91–94]. Whether or not such antiglobulins can be described within the context of RF activity is debatable. Antiglobulins reactive with determinants in, or close to, the binding site of IgG antibodies in normals have been described as "natural" antibodies [95, 96]. These antiglobulin can be polyreactive or monoreactive and exist as IgM, IgG or IgA isotypes.

13.5 Note: Cross-reactions

Studies of polyclonal RF isolated from serum have shown that some IgM RF can cross-react with a variety of substances other than IgG. These include nitrophenyl groups, DNA, nucleosomes, DNA-nucleoproteins, histones, non-histone nuclear proteins, bromelain-treated erythrocytes and β_2-microglobulin [6, 106–113]. The nuclear antigen involved in RF cross-reactivity was identified as a mononucleosome containing both DNA and histone. Several recent reports have shown that certain RF cross-react with a nuclear antigen found in normal cells obtained from a wide variety of species, and with histone and other basic polycations when bound to an immobilised tyrosine-glutamic acid ployanionic co-polymer [114]. Using histone-derived synthetic peptides, Tuaillon *et al.* [114] demonstrated that the binding of RF to histones involved residues 1–16 and 204–218 of H1, 1–20 and 65–85 of H2A and 1–21 of H3.

14. T-lymphocyte determinants

There is some evidence in favour of the existence of specific mouse T- and B-lymphocyte interactions involving Ig or Fc fragments [115, 116]. Some of these interactions are thought to be based on T-lymphocyte recognition of idiotypic determinants on membrane Ig.

In humans, there is no convincing evidence for the existence of T-lymphocyte epitopes on IgG Fc or fragments thereof. Nevertheless, the finding that most human RF and anti-IgG monoclonal antibodies produced in mice react with the Ga determinant may argue for the existence of some T-lymphocyte epitopes on IgG Fc.

Acknowledgements

I would like to express my gratitude to Professor Roy Jefferis (University of Birmingham) for providing Figure 3, Mr Robert Maziak for providing a computer printout of the α carbon backbone of IgG Fc and Mr Richard Wyatt for his advice in drawing the figures presented in this chapter. The financial support of the Arthritis and Rheumatism Council of Great Britain is greatly acknowledged.

References

1. Meyer K (1922) Immunitatsforsch Exp Ther 34:229–234
2. Waaler E (1940) Acta Pathol Microbiol Scand 17:172–188
3. Osterland CK, Harboe M & Kunkel HG (1963) Vox Sang 8:133–152
4. Shakib F, Boulstridge L & Smith SJ (1993) Allergologia et Immunopathologia 21:20–24
5. Natvig JB, Harboe M, Fausa O & Tveit A (1971) Clin Exp Immunol 8:229–236
6. Rauch J, Tannenbaum H, Straaton K, Massicotte H & Wild J (1986) J Clin Invest 77:106–112
7. Sasso EH, Barber CV, Nardella FA, Yount WJ & Mannik M (1988) J Immunol 140:3098–3107
8. Hay F, Nineham LJ & Roitt IM (1975) Br Med J 3:203–204
9. Carson DA, Lawrance S, Catalano MA, Vaughan JH & Abraham G (1977) J Immunol 119:295–300
10. Schrohenloher RE, Koopman WJ, Alarcon GS (1986) Arthitis Rheum 29:1194–1202
11. Merety K, Falus A, Erhardt CC & Maini RN (1982) Ann Rheum Dis 41:405–408
12. Frangione B, Milstein C & Pink JRL (1969) Nature 221:145–148
13. Burton DR (1985) Mol Immunol 22:161–206
14. Burton DR (1987) In: Calabi F & Neuberger MS, eds, Molecular Genetics of Immunoglobulin, pp 1–50. Elsevier Science Publishers, Amsterdam, The Netherlands
15. Deisenhofer J (1981) Biochem 20:2361–2370
16. Feinstein A, Richardson NE & Taussig MJ (1986) Immunol Today 7:169–174
17. Hooper B, Whittingham S, Mathews JD, Mackay IR, & Curnow DH (1972) Clin Exp Immunol 12:79–87
18. Carson DA (1993) In: Kelly WN, Harris ED, Ruddy S & Sledge CB, eds, Textbook of Rheumatology, 4th ed, vol 1, pp 155–163. WB Sunders Co., Philadelphia, PA, USA
19. Natvig JB & Munthe E (1975) Ann NY Acad Sci 256:88–95
20. Johnson GD & Holborow EJ (1963) Lancet 2:1142–1143
21. Estes D, Atra E & Peltier A (1973) Arthritis Rheum 16:59–65
22. Vaughan JH (1963) Arthritis Rheum 6:446–466
23. Nardella FA, Gilliland BC & Mannik M (1979) Arthritis Rheum 22:141–144
24. Newkirk MM (1992) J Immunol Methods 148:93–99
25. Newkirk MM, Rauch J & Commerford K (1992) J Rheumatol Suppl 32:54–5
26. Aucouturier P, Maillochon JP, Joseph-Theodore G, Duarte F & Preud'homme JL (1991) J Clin Lab Analysis 5:378–381
27. Harboe M (1961) Ann Rheum Dis 20:363–368
28. Santoro F, Duquesnoy B, Wattre P, Delcambre B & Capron A (1977) Biomedicine 27:230–233
29. Pritchard MH & Jobbins K (1981) J Clin Path 34:396–399
30. Knight RK & Pritchard MH (1982) Ann Rheum Dis 41:426–429
31. Cohen E, Neter E, Mink I & Norcross RM (1958) Am J Clin Path 30:32–34
32. Ansell BM, Holborow EJ, Zutshi D, Reading A & Epstein WV (1969) Ann NY Acad Sci 168:21–29
33. Koritz TN, Gurner BW, Chalmers DG & Coombs RRA (1980) J Immunol Methods 32:1–9
34. Bozicevich J, Bunim JJ, Freund J & Ward SB (1958) Proc Soc Exp Biol & Med 97:180–188
35. Singer JM & Plotz CM (1956) Am J Med 21:888–892
36. Waller M (1969) Ann NY Acad Sci 168:5–16
37. Jaspers JPM, Van Oers RJM & Leekes B (1988) J Clin Chem Clin Biochem 26:863–871
38. Franklin EC, Holman HR, Muller-Eberhard HJ & Kunkel HG (1957) J Exp Med 105:425–438
39. Pope RM, Teller DC & Mannik M (1975) Ann NY Acad Sci 256:82–87
40. Torrigiani G & Roitt IM (1967) Ann Rheum Dis 26:334–341

41. Faith A, Pontesilli O, Unger A, Panayi GS & Johns P (1982) J Immunol Methods 55:169–177
42. Taylor RB, Clarke LJ & Elson CJ (1979) J Immunol Methods 26:25–37
43. Allen JC & Kunkel HG (1966) Arthritis Rheum 9: 758–768
44. Winblad S, Hansen A & Jansen R (1969) Acta Rheumatol Scand 15:200–205
45. Pope RM & McDuffy SJ (1979) Arthritis Rheum 22:988–998
46. Powell RJ, Leyland AM, Pound JD & Bossingham DH (1985) J Rheumatol 12:427–431
47. Tonegawa S (1983) Nature 302:575–581
48. Venkitaraman AR, Williams GT, Dariavach P & Neuberger MS (1991) Nature 352:777–781
49. Nossal GJV (1983) Annu Rev Immunol 1:33–62
50. Burton DR & Woof JM (1992) Adv Immunol 51:1–84
51. Klaus GGB, Bijsterbosch M, O'Garra A, Harnett M & Rigley KP (1987) Immunol Rev 99:19–38
52. Watts C, Reid PA, West MA & Davidson HW (1990) Seminars Immunol 2:247–253
53. Pasquier LD (1993) Curr Opinion Immunol 5:185–193
54. Kabat EA, Wu TT, Perry HM, Gottesman KS & Foeller C (1991) Sequences of proteins of immunological interest. Department of Health and Human Services, Public Health Services, National Institute of Health.
55. Jones MG, Shipley ME, Hearn JP & Hay FC (1990) Ann Rheum Dis 49:757–762
56. Stewart GA, Hunneyball IM & Stanworth DR (1975) Immunochem 12:657–662
57. Schanfield MS & van Loghem E (1986) In: Weir DM, ed, Handbook of Experimental Immunology, vol 3, Genetics and Molecular Immunology, 4th ed, pp 94.1–94.18, Blackwell Scientific Publications, Oxford, UK
58. Marchalonis JJ & Schluter SF (1989) The FASEB J 3:2469–2479
59. Honjo T (1983) Annu Rev Immunol 1:499–528
60. Croce CM, Shander M, Martinis J, Cicurel L, D'Ancona GG, Dolby TW & Koprowski H (1979) Proc Natl Acad Sci USA 76:3416–3419
61. Brüggemann M (1987) In: Calabi F & Neuberger MS, eds, Molecular Genetics of Immunoglobulin, pp 51–80. Elsevier Science Publishers, Amsterdam, The Netherlands
62. McBride OW, Hieter PA, Hollis GF, Swan D, Otey MC, & Leder P (1982) J Exp Med 155:1480–1490
63. Lötscher E, Grzeschik K-H, Bauer HG, Pohlenz H-D, Straubinger B & Zachau HG (1986) Nature 320:456–458
64. Taub RA, Hollis GF, Hieter PA, Korsmeyer S, Waldmann TA & Leder P (1983) Nature 304:172–174
65. Banchereau J & Rousset F (1992) Adv Immunol 52:125–262
66. Sitia R, Neuberger MS & Milstein C (1987) EMBO J 6:3969–3977
67. Burton DR, Gregory L & Jefferis R (1986) Monogr Allergy 19:7–35
68. Hughes-Jones NC & Gorick BD (1982) Mol Immunol 19:1105–1112
69. Reid KBM (1983) Biochem Soc Trans 11:1–12
70. Gaarder PI & Michaelsen TE (1974) Acta Pathol Microbiol Scand – Section B, Microbiol Immunol 82B:733–741
71. Nardella FA, Teller DC, Barber CV & Mannik M (1985) J Exp Med 162:1811–1824
72. Gaarder PI & Natvig JB (1970) J Immunol 105:928–937
73. Henney CS (1969) Ann NY Acad Sci 168:52–62
74. Mageed RA, Carson DA & Jefferis R (1988) Scand J Immunol 28:233–240
75. Johnson PM & Faulk WP (1976) Clin Immunol Immunopathol 6:414–430
76. Nardella FA, Oppliger IR, Stone GC, Sasso EH, Mannik M, Sjöquist J, Schröeder AK, Christensen P, Johansson PJ & Björck L (1988) Scand J Rheumatol – Supplement 75:190–198
77. Gran JT, Gaarder PI, Husby G & Thorsby E (1985) Scand J Rheumatol 14:144–148
78. Grubb R (1956) Acta Pathol Microbiol Scand 39:195–197
79. Grubb R, Matsumoto H & Sattar MA (1988) Arthritis Rheum 31:60–62

80. Grubb R & Laurell AB (1956) Acta Pathol Microbiol Scand 39:390-398
81. Williams RC Jr, Malone CC & Casali P (1992) J Immunol 149:1817-1824
82. Hunneyball IM & Stanworth DR (1976) Immunol 30:881-894
83. Michaelsen TE, Aase A & Gaarder PI (1988) Scand J Rheumatol - Suppl 75:164-171
84. Artandi SE, Calame KL, Morrison SL & Bonagura VR (1992) Proc Natl Acad Sci USA 89:94-98
85. Randen I, Thompson KM, Natvig JB, Førre Ø & Waalen K (1989) Clin Exp Immunol 78:13-18
86. Johnson PM, Watkins J, Scopes PM & Tracey BM (1974) Ann Rheum Dis 33:366-370
87. Hansson U-B, Ohlin M & Lindstrom FD (1985) Scand J Immunol 22:27-32
88. Carpenter AB & Smailer S (1989) Immunological Investigations 18:765-773
89. Gioud-Paquet M, Auvinet M, Raffin T, Girard P, Bouvier M & Lejeune E (1987) Ann Rheum Dis 46:65-71
90. Mierau R, Gause A, Kuppers R, Michels M, Mageed RA, Jefferis R & Genth E (1992) Rheumatol Int 12:23-31
91. Osterland CK, Harboe M & Kunkel HG (1963) Vox Sang 8:133-152
92. Kunkel HG & Tan EM (1964) Adv Immunol 4:351-395
93. Birdsal HH, Lidsky MD & Rossen RD (1983) Arthritis Rheum 26:1481-1492
94. Heimer R, Wolfe LD & Abruzzo JL (1982) Arthritis Rheum 25:1298-1306
95. Nasu H, Chia DS, Knutson DW & Barnett EV (1980) Clin Exp Immunol 42:378-386
96. Birdsal HH & Rossen RD (1983) Clin Exp Immunol 53:497-504
97. Johnston SL & Abraham GN (1979) Immunol 36:671-683
98. Jones VE, Puttick AH, Williamson EA & Mageed RA (1988) Clin Exp Immunol 71:451-458
99. Normansell DE & Young CW (1975) Immunochem 12:187-188
100. Kunkel HG, Agnello V, Joslin FG, Winchester RJ & Capra JD (1973) J Exp Med 137:331-342
101. Jefferis R, Nik Jaafar MI & Steinitz M (1984) Immunol Lett 7:191-194
102. Brown CMS, Plater-Zyberk C, Mageed RA, Jefferis R & Maini RN (1990) Clin Exp Immunol 80:366-372
103. Robbins DL, Kenny TP, Snyder LL, Ermel RW & Larrick JW (1993) Arthritis Rheum 36:389-393
104. Robbins DL, Kenny TP, Coloma MJ, Gavilondo-Cowley JV, Soto-Gil RW, Chen PP & Larrick JW (1990) Arthritis Rheum 33:1188-1195
105. Randen I, Thompson KM, Thorpe SJ, Førre Ø & Natvig JB (1993) Scand J Immunol 37:668-672
106. Hannestad K (1969) Ann NY Acad Sci 168:63-75
107. Hannestad K & Stollar BD (1978) Nature 275:671-673
108. Hannestad K, Pekvig OP & Husebekk A (1981) Springer Semin Immunopathol 4:133-160
109. Agnello V, Arbetter A, Ibanez de Kasep G, Powell R, Tan EM & Joslin F (1980) J Exp Med 151:1514-1527
110. Hobbs RN, Lea DJ, Phua KK & Johnson PM (1983) Ann Rheum Dis 42:435-438
111. Tan EM, Robinson J & Robitaille P (1976) Scand J Immunol 5:811-818
112. Mason JC, Venables PJ, Smith PR & Maini RN (1985) Ann Rheum Dis 44:287-293
113. Cunliffe DA & Cox KO (1980) Nature 286:720-722
114. Tuaillon N, Martin T, Knapp AM, Pasquali JL & Muller S (1992) J Autoimm 5:1-14
115. Eichmann K, Falk I & Rajewsky K (1978) Eur J Immunol 8:853-857
116. Sy MS, Lowy A, Hayclass K, Janeway CA Jr, Gurish M, Greene MI & Benacerraf B (1984) Proc Natl Acad Sci USA 81:3846-3850

Manual of Biological Markers of Disease **B1.2**: 1–9, 1994.

The antiperinuclear factor

JEAN-MARIE BERTHELOT[1,2], CHRISTIAN VINCENT[3],
GUY SERRE[3] and PIERRE YOUINOU[1]

[1] *Laboratory of Immunology, Brest University Medical School Hospital, BP 824, F 29609, Brest cedex, France;* [2] *Unit of Rheumatology, Nantes University Medical School Hospital, BP 1005, F 44035, Nantes cedex, France;* [3] *Department of Cell Biology and Histology, Toulouse-Purpan School of Medicine, University of Toulouse, F 31059, Toulouse cedex, France*

Abbreviations

APF: antiperinuclear factor; PNA: perinuclear antigen; RA: rheumatoid arthritis; KHG: keratohyaline granules; AKA: "Antikeratin" antibodies

Key words: Autoantigen, autoantibody, rheumatoid arthritis, APF, antiperinuclear factor, keratohyaline granule

1. The antigen

During an investigation of the occurrence of antinuclear antibodies by means of Coon's immunofluorescent technique, using epithelial cells from human buccal mucosa as substrate [1], Nienhuis and Mandema noticed in 1964 that sera from patients with rheumatoid arthritis (RA) gave a cytoplasmic fluorescence. Since the round cytoplasmic components recognized were mainly located in a perinuclear fashion, the corresponding autoantibodies were designated as an antiperinuclear factor (APF).

2. Synonyms

There is no synonym widely accepted.

3. Biochemical characteristics

The true nature of the antigen (perinuclear antigen: PNA) remains unknown. However, the staining characteristics of the components recognized by APF (Table 1) are reminiscent of those of buccal keratohyaline granules (KHG). Moreover, an exact colocalisation of PNA and filaggrin has been

Table 1. Staining characteristics of perinuclear keratohyaline granules of buccal mucosa cells

Stain	Substrate	Result
May-Grunwald Giemsa	Acidic and basic structures	Basophilic
Hematoxylin and eosin	Acidic and basic structures	Basophilic
Papanicolaou's	Acidic and basic structures	Basophilic
Toluidine blue	Nonspecific	Positive
Methylene blue	Nonspecific	Positive
Methyl green pyronin	DNA and RNA	Negative
Feulgen's	DNA	Negative
Hoechst bisbenzimidine	DNA	Negative
Periodic acid-Schiff	Glycogen	Positive
Sudan black	Lipids	Negative
Congo red	Amyloid	Negative
Acid phosphatase	Lysosomes	Negative
Oil red	Lipids/lipofuscins	Negative
Alcian blue	Acid mucosubstances	Negative
Azur A	Acid mucosubstances	Negative

References [2] and [8] and unpublished results

Table 2. Effects of various treatments on the antigenicity of the perinuclear granules

Treatment	Antigenicity after treatment
Enzymes	
● Trypsin 0.35%	−
● Pepsin 1,000 U/ml	−
● Achromopeptidase 100 U/ml	−
● Hyaluronidase 0.1 mg/ml	+
● Chondroitinase ABC, 1 U/ml	+
● RNAse 100 μg/ml	+
● DNAse 100 μg/ml	+
Detergents	
● Nonidet P-40 (1%)	+
● Triton X-100 (%)	+
● Tween 20 (2%)	+
● Sodium desoxycholate 0.5%	+
Urea	
● Urea 2M	+
● Urea 4M	−
● Urea 8M	−
Miscellaneous	
● β Mercapto-ethanol	−
● HCl 0.1 M	−
● NaOH 0.1 M	−
● Freezing/thawing (3 times)	±
● Acetone	±
● Methanol	±
● Dithiothreitol 50 mM	+

Reference [2] and unpublished results

AMAN-B1.2/2

demonstrated [2] whilst (pro)filaggrin is the major component of human KHG. PNA is thus associated with human buccal KHG. The effects of the various treatments (Table 2) suggest that it is a protein which is rather insoluble.

4. Antibody detection

APF is primarily detected by the indirect immunofluorescence technique (Fig. 1) (for technical details, see this Manual, chapter A.10).

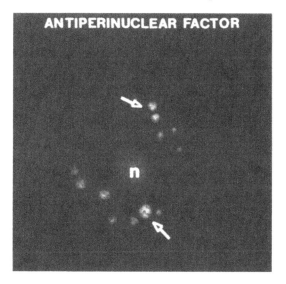

Fig. 1. Immunofluorescence pattern (arrows) produced by the binding of the antiperinuclear factor to the keratohyalin granules located within buccal mucosa cells obtained from normal donors.
Buccal cells are scraped with a wooden tongue depressor from the inner side of both cheeks of healthy volunteers. Serum was tested diluted 1/80 and applied to the cells for 90 minutes in a moisture chamber. After three washes, the preparation was incubated for 30 minutes with fluorescein isothiocyanate-conjugated F(ab')$_2$ goat anti-human IgG.

4.1 Immunofluorescence

As discussed previously [reviewed in 3], several factors greatly influence the results, among which is the dilution of sera. Indeed, while Westgeest *et al.* showed that specificity of APF for RA was increased without simultaneous loss of specificity at 1/10 dilution instead of 1/5 or undiluted sera [4], other investigators demonstrated that 1/80 dilution was even better for this purpose, especially when the criteria for positivity was the presence of at least 10% of cells containing fluorescent granules [5].

4.2 Counter-immunoelectrophoresis

It cannot be used for detection of APF.

4.3 Immunoblotting

So far, no specific band has been disclosed using human buccal cells as a substrate.

4.4 Elisa

Not available (PNA has not been identified yet).

5. Cellular localization

PNA is found within the KHG of human buccal mucosa cells. KHG only appear in the "granulous layer" (5 to 8 sheets of cells) of keratinizing epithelia, among which lie the endojugal mucosa of most of the donors.

KHG is a general term for round cytoplasmic aggregates of various sizes, shapes, chemical compositions and ultrastructural patterns, according to the species and/or epithelium studied. This heterogeneity accounts for some confusion in the literature. In mammals, three main types of KHG have been described: L-KHG, in which loricrin accumulates, F-KHG with (pro)filaggrin as major constituent, and CG (composite granules) that appear to consist of L-granules embedded in large F-granules [6]. PNA colocalizes with (pro)filaggrin in F-granules which, in jugal cells, appear as electron dense inhomogeneous material resembling aggregated rough endoplasmic reticulum [7]. Table 3 shows the reactivity of the F- granules from human endojugal mucosa with a variety of monoclonal and polyclonal antibodies.

6. Cellular function

The cellular function of PNA is unknown. In mammals, KHG appears to be an organel containing proteins which are necessary for the final stage of maturation of keratinizing epithelia [6].

Table 3. Reactivity of keratohyaline granules with various monoclonal and polyclonal antibodies

Antibody specificity	Staining of keratohyaline granules in buccal cells
Ck 5+8	−
Epidermal Ck	−
Ck 18	−
Ck 10	−
Ck 4	−
Ck 7	−
Ck 13+16	−
Vimentin	−
Lamins A, B and C	−
Loricrin	−
Involucrin	−
U1A/U2B'' (anti-sn RNP)	−
U2B''	−
Sm	−
Topoisomerase 1	−
Human heat-shock protein (HSP) 65	−
HSP 72	−
HSP 90	−
Ubiquitin	−
Ck 1,2,10,11+filaggrin	+++
Human (pro)filaggrin	+++

Ck: cytokeratin, HSP: heat-shock protein
Reference [2] and unpublished results

7. Presence in other species

KHG has been noticed in keratinized epithelia of all mamalian species studied so far, for example, in rabbit [8]. PNA, however, has been demonstrated only in rabbit oral and oesophagal mucosa, although these substrates are less sensitive and specific for the detection of APF than human buccal cells.

8. Cdnas

Sequences of the PNA are unknown.

9. Genes

The genes of PNA have not been analysed yet.

10. Relation with other proteins

An exact colocalisation of PNA and profilaggrin in human buccal cells has been noted [9]. However, this does not apply to all keratinizing epithelia. For instance, APF positive sera do not recognize KHG from the human skin, while antifilaggrin antibodies do [2]. This could suggest either that PNA and (pro)filaggrin are different molecules tightly associated within KHG from buccal cells, or that APF recognizes only some determinants of profilaggrin not inexpressed or inaccessible in human skin. The relationship between profilaggrin and filaggrin is schematically depicted in Figure 2.

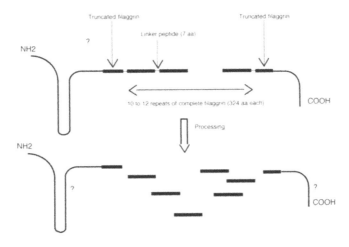

Fig. 2. Relationships between profilaggrin and filaggrin.
Profilaggrin, the insoluble precursor of filaggrin [10], is an histidine-rich, insoluble protein consisting of 10 to 12 repeats of filaggrin arranged in tandem and separated by a short heptapeptide linker sequence. This precursor accumulates in a non-functional and highly phosphorylated form in the granular layer of keratinizing epithelia, before being dephosphorylated and cleaved by excision of the linker sequence to release the functional and highly basic polypeptide filaggrin. Dephosphorylation may be the key event in the processing of profilaggrin. *In vitro* this is very rapidly achieved, resolving into peptides of lower molecular weights [11]. Profilaggrin aggregation inside the granules and further processing could depend on the calcium concentration, since its amino-terminus shows great homology with the S-100 family of calcium-binding proteins [11].

11. Relation with the AKA system

As underlined by several authors, APF and AKA not only recognize similar epithelia but, also, are strongly correlated to each other, i.e. most sera positive for APF are also positive for AKA (Table 4).

Table 4. Correlations of the antiperinuclear factor to the antikeratin activity in sera from patients with rheumatoid arthritis

Authors (year)	Antiperinuclear factor	"Antikeratin" antibody		Correlation
		Positive	Negative	
Johnson et al. (1981) [12]	positive	40	35	$p = 0.068$
	negative	1	7	
Miossec et al. (1982) [13]	positive	28	35	$p = 0.178$
	negative	10	23	
Youinou et al. (1983) [14]	positive	29	47	$p = 0.424$
	negative	7	17	
Hoet et al. (1991) [2]	positive	34	3	$p = 0.004$
	dubious	2	3	
	negative	1	3	
Aho et al. (1993) [15]	positive	6	4	$p = 0.003$
	negative	3	29	

12. B-cell epitopes

Have not been analysed so far.

13. T-cell epitopes

Have not been analysed so far.

References

1. Nienhuis RLF & Mandema E (1964) Ann Rheum Dis 23:302–305
2. Hoet RMA, Boerbooms AMT, Arends M, Ruiter DJ & van Venrooij WJ (1991) Ann Rheum Dis 50:611–618
3. Hoet RMA & van Venrooij WJ (1992) In: Smolen JS, Kalden JR, Maini RN (Eds) Rheumatoid arthritis, pp 299–318. Springer-Verlag, Berlin, Heidelberg
4. Westgeest AAA, Boerbooms AMT & van de Putte LBA (1990) Arthritis Rheum 33:759–760
5. Youinou P, Le Goff P, Dumay A, Lelong A, Fauquert P & Jouquan J (1990) Clin Exp Rheumatol 8:259–264
6. Steven AC, Bisher ME, Roop DR & Steinert PM (1990) J Struct Biol 104:150–162
7. Vivino FB & Maul GG (1990) Arthritis Rheum 33:960–969
8. Smit JW, Sondag-Tschroots IRJM, Aaij C, Feltkamp TEW & Feltkamp-Vroom TM (1980) Ann Rheum Dis 39:381–386
9. Hoet RM, Voorsmit RACA & van Venrooij WJ (1991) Clin Exp Rheumatol 84:59–65
10. Gan SQ, Mc Bride OW, Idler WW, Markova N & Steinert PM (1990) Biochemistry 29:9432–9440
11. Presland RB, Haydock PV, Fleckman P, Nirunsuksiri W & Dale BA (1992) J Biol Chem 267:23772–23781
12. Johnson GD, Carvalho A, Holborow EJ, Goddard DH & Russell G (1981) Ann Rheum Dis 40:263–266
13. Miossec P, Youinou P, Le Goff P & Moineau MP (1982) Clin Rheum 1:185–189
14. Youinou P, Le Goff P, Miossec P, Moineau MP & Ferec C (1983) Z Rheumatol 42:36–39
15. Aho K, von Essen R, Kurki P, Paluoso T & Heliovaara M (1993) J Rheumatol 20:1278–1281

Manual of Biological Markers of Disease **B1.3**: 1–9, 1994.

◆

The A2 protein of the heterogeneous nuclear ribonucleoprotein (hnRNP-A2)/RA33

GÜNTER STEINER

Ludwig Boltzmann-Institute for Rheumatology and Balneology, Second Department of Medicine, Lainz Hospital, Wolkersbergenstrasse 1, A-1130-Vienna, Austria

Abbreviations

RA: Rheumatoid arthritis; hnRNP: heterogeneous nuclear ribonucleoprotein; hnRNA: heterogeneous nuclear RNA; RBD: RNA binding domain; RRM: RNA recognition motif

Key words: Autoantigen, hnRNP-A2, RA-33, rheumatoid arthritis, autoantibody

1. The antigen

In 1989 Hassfeld *et al.* described autoantibodies in sera from rheumatoid arthritis (RA) patients, which were directed to a protein of approximately 33 kD contained in HeLa cell derived nuclear extracts [1]. Since initially such autoantibodies were almost exclusively detected in sera from RA patients, the antigen was given the name RA33. Partial sequencing of highly purified RA33 revealed that it was identical to the A2 protein of the heterogeneous nuclear ribonucleoprotein (hnRNP) complex [2] whose cDNA sequence had been published by Burd *et al.* in 1989 [3]. Studies employing purified antigen showed that autoantibodies to this protein were not confined to RA patients but could also be detected in sera from MCTD and SLE patients. In addition, it was demonstrated that most anti-A2/RA33 sera were also reactive with the proteins B1 and B2 of the hnRNP complex (hnRNP-B1 and B2).

2. Synonyms

Since the biochemical name of the antigen is hnRNP-A2, the name RA33 can be considered a synonym. However, it must be noted that most anti-A2/RA33 sera are also reactive with the two hnRNP-B proteins, and therefore the term anti-RA33 may be used to designate reactivities to the three hnRNP proteins A2, B1, B2. This is analogous to other autoantibody systems: thus the term anti-Sm designates reactivities directed primarily to the snRNP core proteins B, B′, D, and the term anti-U1RNP defines reactivities directed against the U1-snRNP specific proteins 70K, A, C.

3. Biochemical characteristics

The calculated molecular weight of hnRNP-A2 as deduced from the amino acid sequence is 36 kD [3], which is slightly higher than the molecular weight estimated (for RA33) by SDS polyacrylamide gel electrophoresis [1]. A2 is a basic protein with an isoelectric point of 8.4 [4] and at least a portion of it is phosphorylated *in vivo* [5]. It belongs to a family of nucleic acid binding proteins which are characterized by the presence of one or several copies of a relatively conserved RNA binding domain (RBD, also called RNA recognition motif (RRM) or RNP-80) [6]. A2 contains two RBDs which are followed by a glycine rich section at the carboxy terminus (see also section 8). The protein can be purified easily from nuclear extracts by heparin Sepharose chromatography and ion exchange or by affinity chromatography on single-stranded DNA, since it binds with intermediate to high affinity to single-stranded nucleic acids and heparin [2, 4]. However, the purified protein is insoluble in aqueous buffers, and strong detergents such as 6M urea or 1% SDS are needed to keep it in solution. Similar to other ribonucleoprotein antigens the antigenicity of A2 resides in the protein moiety and is not dependent on the presence of RNA or DNA [1]. The protein is relatively resistant to proteases and can be isolated from nuclear extracts even in the absence of protease inhibitors. Limited digestion with trypsin yields a 24 kD fragment which contains the amino terminal part of the protein lacking the glycine rich auxiliary domain (7).

4. Antibody detection

4.1 Immunofluorescence

No characteristic staining pattern has been observed with anti-A2/RA33 positive sera [1]. However, some sera may produce a faint homogeneous staining.

4.2 Immunodiffusion/counterimmunoelectrophoresis

Anti-RA33 cannot be detected by this method since the sera do not form precipitins.

4.3 Immunoblotting

Antibodies to A2/RA33 were initially detected on immunoblots employing crude nuclear extracts as antigen source. However, interpretation of such "nuclear blots" requires much experience and may sometimes be difficult. Therefore the use of semi-purified antigen is recommended. Partially purified antigen can be prepared by heparin-Sepharose chromatography of crude nuclear extracts essentially as described [2]. When using the semi-purified material a characteristic triplet can be observed with most sera (Fig. 1). Anti-

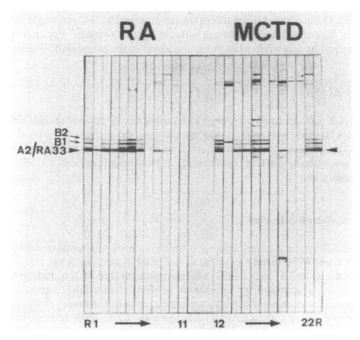

Fig. 1. Anti-RA33 detected by immunoblotting in sera from RA and MCTD patients.
A partially purified preparation of A2/RA33 obtained by heparin Sepharose chromatography was separated on 12% SDS gels and blotted onto nitrocellulose membranes. Sera were diluted 1:25 in phosphate buffered saline containing 3% non-fat dried milk and incubated for 40 min with the blotted antigen. Detergents such as Triton-X-100 or Tween 20 should be avoided in the incubation buffer since they may cause false positive results. For immunodetection the use of alkaline phosphatase conjugated anti-human antibodies is recommended since some peroxidase-labelled antibodies may produce high background staining, particularly of the A2/RA33 band. Note the double band above the RA33 band which is stained by most anti-RA33 positive sera; this double band corresponds to the hnRNP proteins B1 and B2 [2, 3]. The A2/B1/B2 triplet greatly facilitates identification of anti-RA33 positive sera.
(R) Reference serum, [1–11] sera from rheumatoid arthritis patients, [12–22] sera from MCTD patients.

A2/RA33 antibodies are detectable by immunoblotting in approximately 35–40% of RA and MCTD sera and in 20% of SLE sera, but usually not in other rheumatic diseases [2].

4.4 Immunoprecipitation

Immunoprecipitation using even the strongest anti-A2/RA33 sera from RA patients has always given very unsatisfactory results. In contrast, monoclonal antibodies to other hnRNP proteins such as A1 or the C proteins easily co-precipitate A2 and the other components of the hnRNP complex [4, 8, 9]. The reason for this discrepancy remains unclear, but one might assume that human autoantibodies are directed against "hidden" epitopes which may not be accessible when A2 is contained in the native hnRNP particle.

4.5 ELISA

Anti-RA33 can be detected by ELISA using highly purified antigen derived from HeLa nuclear extracts. For this purpose, the antigen can be purified by heparin-Sepharose chromatography followed by chromatofocusing or by cation exchange on CM sepharose [2, 7]. This procedure yields a > 95% pure preparation.

5. Cellular localization

The A2 protein is a nuclear protein which is almost as abundant as histones. A2 seems to be mainly localized in the hnRNP particle: newly transcribed RNA, which is also called heterogeneous nuclear RNA (hnRNA), is complexed with a large number of proteins. Upon mild digestion with ribonucleases discrete particles are generated which sediment at 35–40 S in sucrose density gradients. These particles have been termed hnRNP complexes. They are composed of hnRNA and approximately 30 different proteins which can be separated by two dimensional gel electrophoresis [4]. Six of these proteins (A1, A2, B1, B2, C1, C2) are considered the "core" proteins of this particle, and A2 is presumably one of its major components [10]. The core proteins as well as most of the hnRNP proteins whose cDNAs have been sequenced so far contain one or several RBDs. The detailed molecular structure of the hnRNP particle is not known nor is its function fully understood [11, 12]. In the nucleus at least some of the hnRNP proteins are found at sites of active transcription, and there is some indication for a specific transcript dependent binding of certain hnRNP proteins to pre-mRNA [13, 14].

6. Cellular function

Only very little is known about the function of A2. Like many other hnRNP proteins, A2 is involved in hnRNA packaging [5, 10–12] and thus could play an important role in pre-mRNA processing, i.e. the generation of intron-free and polyadenylated mature mRNA. Moreover, there is considerable evidence that both A1 and A2 are involved in the transport of mature mRNA to the cytoplasm [15]. In addition, A2 could be involved in the structural organization of genomic DNA since it was recently shown to bind to telomere repeat sequences [16]. Thus, taken together, hnRNP-A2 may fulfil multiple functions, particularly during mRNA processing and transport but only few experimental data supporting this assumption have been published so far.

7. Presence in other species

Since hnRNP proteins are highly conserved in evolution, proteins homologous to human A2 are presumably present in all vertebrates. This assumption is supported by the recent identification of the genes for both hnRNP-A1 and A2 in Xenopus [17, 18]. Moreover, proteins homologous to the vertebrate hnRNP-A and B proteins have been identified in Drosophila [19–21], and a protein with considerable sequence homology to human A2 was described in maize [22]. Interestingly, in Drosophila most hnRNP proteins characterized so far resemble the mammalian A/B proteins.

8. The CDNA and derived sequences

The cDNA sequence of hnRNP-A2 was published by Burd *et al.* in 1989 [3]. The deduced amino acid sequence characterizes a protein of 341 amino acids (Fig. 2). A variant of this protein, presumably generated by alternate splicing of the A2 pre-mRNA, contains a 12 amino acid insertion near the amino terminus and is identical to the hnRNP-B1 protein.

9. The gene

To date, the gene structure for human hnRNP-A2 has not been published. However, preliminary data seem to indicate that it is structurally similar to the gene of hnRNP-A1 [23, 24].

A

```
                                          10
      Met Glu Arg Glu Lys Glu Gln Phe Arg Lys Leu Phe Ile Gly Gly Leu Ser Phe
      ATG GAG AGA GAA AAG GAA CAG TTC CGT AAG CTC TTT ATT GGT GGC TTA AGC TTT
   20                                      30
Glu Thr Thr Glu Glu Ser Leu Arg Asn Tyr Tyr Glu Gln Trp Gly Lys Leu Thr Asp Cys
GAA ACC ACA GAA GAA AGT TTG AGG AAC TAC TAC GAA CAA TGG GGA AAG CTT ACA GAC TGT
   40                                      50
Val Val Met Arg Asp Pro Ala Ser Lys Arg Ser Arg Gly Phe Gly Phe Val Thr Phe Ser
GTG GTA ATG AGG GAT CCT GCA AGC AAA AGA TCA AGA GGA TTT GGT TTT GTA ACT TTT TCA
   60                                      70
Ser Met Ala Glu Val Asp Ala Ala Met Ala Ala Arg Pro His Ser Ile Asp Gly Arg Val
TCC ATG GCT GAG GTT GAT GCT GCC ATG GCT GCA AGA CCT CAT TCA ATT GAT GGG AGA GTA
   80                                      90
Val Glu Pro Lys Arg Ala Val Ala Arg Glu Glu Ser Gly Lys Pro Gly Ala His Val Thr
GTT GAG CCA AAA CGT GCT GTA GCA AGA GAG GAA TCT GGA AAA CCA GGG GCT CAT GTA ACT
   100                                     110
Val Lys Lys Leu Phe Val Gly Gly Ile Lys Glu Asp Thr Glu Glu His His Leu Arg Asp
GTG AAG AAG CTG TTT GTT GGC GGA ATT AAA GAA GAT ACT GAG GAA CAT CAC CTT AGA GAT
   120                                     130
Tyr Phe Glu Glu Tyr Gly Lys Ile Asp Thr Ile Glu Ile Ile Thr Asp Arg Gln Ser Gly
TAC TTT GAG GAA TAT GGA AAA ATT GAT ACC ATT GAG ATA ATT ACT GAT AGG CAG TCT GGA
   140                                     150
Lys Lys Arg Gly Phe Gly Phe Val Thr Phe Asp Asp His Asp Pro Val Asp Lys Ile Val
AAG AAA AGA GGC TTT GGC TTT GTT ACT TTT GAT GAC CAT GAT CCT GTG GAT AAA ATC GTA
   160                                     170
Leu Gln Lys Tyr His Thr Ile Asn Gly His Asn Ala Glu Val Arg Lys Ala Leu Ser Arg
TTG CAG AAA TAC CAT ACC ATC AAT GGT CAT AAT GCA GAA GTA AGA AAG GCT TTG TCT AGA
   180                                     190
Gln Glu Met Gln Glu Val Gln Ser Ser Arg Ser Gly Arg Gly Gly Asn Phe Gly Phe Gly
CAA GAA ATG CAG GAA GTT CAG AGT TCT AGG AGT GGA AGA GGA GGC AAC TTT GGC TTT GGG
   200                                     210
Asp Ser Arg Gly Gly Gly Gly Asn Phe Gly Pro Gly Pro Gly Ser Asn Phe Arg Gly Gly
GAT TCA CGT GGT GGC GGT GGA AAT TTC GGA CCA GGA CCA GGA AGT AAC TTT AGA GGA GGA
   220                                     230
Ser Asp Gly Tyr Gly Ser Gly Arg Gly Phe Gly Asp Gly Tyr Asn Gly Tyr Gly Gly Gly
TCT GAT GGA TAT GGC AGT GGA CGT GGA TTT GGG GAT GGC TAT AAT GGG TAT GGA GGA GGA
   240                                     250
Pro Gly Gly Gly Asn Phe Gly Gly Ser Pro Gly Tyr Gly Gly Gly Arg Gly Gly Tyr Gly
CCT GGA GGT GGC AAT TTT GGA GGT AGC CCC GGT TAT GGA GGA GGA AGA GGA GGA TAT GGT
   260                                     270
Gly Gly Gly Pro Gly Tyr Gly Asn Gln Gly Gly Gly Tyr Gly Gly Gly Tyr Asp Asn Tyr
GGT GGA GGA CCT GGA TAT GGC AAC CAG GGT GGG GGC TAC GGA GGT GGT TAT GAC AAC TAT
   280                                     290
Gly Gly Gly Asn Tyr Gly Ser Gly Asn Tyr Asn Asp Phe Gly Asn Tyr Asn Gln Gln Pro
GGA GGA GGA AAT TAT GGA AGT GGA AAT TAC AAT GAT TTT GGA AAT TAT AAC CAG CAA CCT
   300                                     310
Ser Asn Tyr Gly Pro Met Lys Ser Gly Asn Phe Gly Gly Ser Arg Asn Met Gly Gly Pro
TCT AAC TAC GGT CCA ATG AAG AGT GGA AAC TTT GGT GGT AGC AGG AAC ATG GGG GGA CCA
   320                                     330
Tyr Gly Gly Gly Asn Tyr Gly Pro Gly Gly Ser Gly Gly Ser Gly Gly Tyr Gly Gly Arg
TAT GGT GGA GGA AAC TAT GGT CCA GGA GGC AGT GGA GGA AGT GGG GGT TAT GGT GGG AGG
   340
Ser Arg Tyr
AGC CGA TAC
```

Fig. 2A. cDNA and amino acid sequence of hnRNP-A2 [3].
The amino terminal part contains two conserved RNA binding domains between amino acids 1–86 (RBD-I) and 92–177 (RBD-II). Amino acids within these domains are in bold and the two most conserved sequence elements within each domain are underlined. This part of the molecule is highly homologous (> 80% identity) to the amino terminal part of hnRNP-A1 [30]. The carboxy terminal half (starting around position 190) – the so-called auxiliary domain – contains about 50% glycin residues and shows some sequence similarities to other glycine rich structures, such as collagen, keratin or EBNA-1. It also shares about 40% sequence homology with the glycin rich auxiliary domain of hnRNP-A1. The hnRNP-B1 protein is identical with A2 except for a stretch of 12 amino acid inserted at the amino terminus between amino acids two and three.

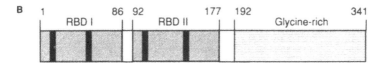

Fig. 2B. B. Structural features of hnRNP-A2.
The three structural domains of A2 are shown. RBD: RNA binding domain; black bars symbolize the two most highly conserved sequences within each domain.

10. Relation with nucleic acids

Little is known about the interaction of A2 with nucleic acids. The protein binds with intermediate to high affinity to single-stranded DNA and RNA and there is strong evidence for a direct binding to hnRNA [25]. Recently it was demonstrated that A2 binds with high affinity to the telomeric DNA sequence TTAGGG and even stronger to the analogous RNA sequence UUAGGG [15]. It is assumed that the amino terminal portion which contains two RBDs is involved in interaction(s) with nucleic acids but studies with recombinant A1 have demonstrated that the glycine rich auxiliary domain may also contribute to the interaction with nucleic acids [26–28].

11. Relation with other proteins

There are no data available on interactions of A2 with other proteins. However, it is assumed that particularly the glycine-rich carboxy terminal part is involved in interactions with other proteins. Candidates are the other hnRNP core proteins, especially the B proteins, but so far there is no direct evidence for such an interaction. It has been proposed that within the hnRNP particle A2 may be organized in a tetrameric structure composed of three molecules A2 and one molecule B1 [29].

12. B-cell epitopes

So far, only preliminary data on B-cell epitopes have been published. These indicate that a major epitope resides in the amino terminal part of the antigen [30].

13. T-cell epitopes

No published data.

References

1. Hassfeld W, Steiner G, Hartmuth K, Kolarz G, Scherak O, Graninger W, Thumb N & Smolen JS (1989) Arthritis Rheum 32:1515-1520
2. Steiner G, Hartmuth K, Skriner K, Maurer-Fogy I, Sinski A, Thalmann E, Hassfeld W, Barta A & Smolen JS (1992) J Clin Invest 90:1061-1066
3. Burd CG, Swanson MS, Görlach M & Dreyfuss G (1989) Proc Natl Acad Sci USA 86:9788-9792
4. Pinol-Roma S, Choi YD, Matunis MJ & Dreyfuss G (1988) Genes Dev 2:215-227
5. Wilk HE, Werr H, Friedrich D, Kiltz HH & Schaefer KP (1985) Eur J Biochem 146:71-81
6. Bandziulis RJ, Swanson MS & Dreyfuss G (1989) Genes Dev 3:431-437
7. Kumar A, Williams KR & Szer W (1986) J Biol Chem 261:11266-11273
8. Choi YD & Dreyfuss G (1984) Proc Natl Acad Sci USA 81:7471-7475
9. Pinol-Roma S, Choi YP & and Dreyfuss G (1990) Methods Enzymol 181:317-326
10. Beyer AL, Christensen ME, Walker BW & LeStourgeon WM (1977) Cell 11:127-138
11. Dreyfuss G, Swanson MS & Pinol-Roma S (1988) Trends Biochem Sci 13:86-91
12. Dreyfuss G, Matunis MJ, Pinol-Roma S & Burd CG (1993) Annu Rev Biochem 62:289-321
13. Bennett M, Pinol-Roma S, Staknis D, Dreyfuss G & Reed R (1992) Mol Cell Biol 12:3165-3175
14. Matunis EL, Matunis MJ & Dreyfuss G (1993) J Cell Biol 121:219-228
15. Pinol-Roma S & Dreyfuss G (1992) Nature 355:730-732
16. McKay SJ & Cooke H (1992) Nucleic Acids Res 20:6461-6464
17. Kay BK, Sawhney RK & Wilson SH (1990) Proc Natl Acad Sci USA 87:1367-1371
18. Good PJ, Rebbert ML & Dawid IB (1993) Nucleic Acids Res 21:999-1006
19. Haynes SR, Raychaudhuri G & Beyer AL (1990) Mol Cell Biol 4:1104-1114
20. Raychaudhuri G, Haynes SR & Beyer AL (1992) Mol Cell Biol 12:847-855
21. Matunis EL, Matunis MJ & Dreyfuss G (1992) J Cell Biol 116:257-269
22. Gomez J, Sánchez-Martinez D, Stiefel V, Rigau J, Puigdomènech P & Pagès M (1988) Nature 334:262-264
23. Biamonti G, Bassi MT, Cartegni L, Mechta F, Buvoli M, Cobianchi F & Riva S (1993) J Mol Biol 230: 77-89
24. Riva S, Biamonti G, Buvoli M, Cartegni L & Cobianchi F (1993) Third EMBO Workshop on Eukaryotic RNPs (abstract)
25. Economidis IV & Pederson T (1983) Proc Natl Acad Sci USA 80:1599-1602
26. Pontius BW & Berg P (1990) Proc Natl Acad Sci USA 87:8403-8407
27. Kumar A & Wilson SH (1990) Biochemistry 29:10717-10722
28. Munroe SH & Dong X (1992) Proc Natl Acad Sci USA 89:895-899
29. Barnett SF, Northington SJ & LeSturgeon WM (1990) Methods Enzymol 181:293-307
30. Steiner G, Skriner K, Sommergruber WH & Smolen JS (1993) Arthritis Rheum 36 (supplement):S242

Manual of Biological Markers of Disease **B1.4**: 1–18, 1994.
© 1994 *Kluwer Academic Publishers. Printed in the Netherlands.*

Collagen II

DAVID G WILLIAMS

Kennedy Institute of Rheumatology, 6 Bate Gardens, Hammersmith London, U.K.

Abbreviations

CII: collagen type II; CB1: Cyanogen Bromide cleavage fragment number 1 derived from collagen type II; CNBr: Cyanogen Bromide; RA: Rheumatoid arthritis

Key words: cartilage, autoantigen, rheumatoid arthritis, collagen, autoantibody

1. The antigen

Collagens are the most abundant animal protein family and the major structural component of connective tissue. They are glycoproteins with a characteristic trimeric structure of 3 intertwined left-handed helices (Fig. 1), each wound in a right-handed superhelix with two other helices. Immune responses to collagens result in antibodies to the native helical form or to linear epitopes in heat-denatured isolated polypeptide constituents (α-chains). This review concentrates on collagen type II which is the major form in articular cartilage, and to which autoimmune responses have been detected in rheumatoid arthritis.

2. Synonyms

None.

3. Biochemical characteristics

The triple helical structure (Fig. 1) is completely dependent on the imino acid sequence of each polypeptide chain [Gly (G) X Y]$_n$ because only the glycine side chain hydrogen atom is small enough to fit into the interior space of the triple helix. On average, 30% of the X and Y positions are occupied by proline or hydroxyproline. The relative rigidity of the peptide backbone contributed by the fixed bond angle about the C-N bond of these imino acids contributes to the stabilisation of the triple helical structure.

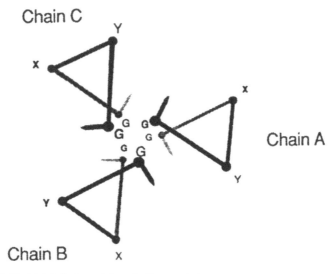

Fig. 1. The triple helical association of collagen α chains.
A transverse view of the collagen triple helix, looking towards the amino terminus of the 3 intertwined α chains (A, B & C), showing the arrangement of the G, X and Y amino acids. (This figure was kindly taken from [61] with permission.)

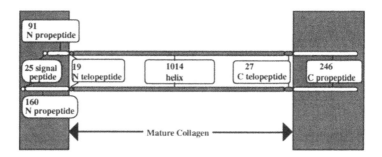

Fig. 2. The domains of collagen type II.
Sizes (in number of amino acids) of each domain in procollagen. The terminal domains within the shaded boxes are removed during post-translational processing. The two forms, with different N-propeptides, represent products of alternative mRNA splicing.

There are at least 14 types of collagen each distinguished by a Roman numeral e.g. collagen type II. At least 5 types are expressed in cartilage (Table 1). Each collagen triplex contains either identical or different α chains, 27 of which have been described so far. Each α chain is distinguished by the chain number (Arabic numeral and the collagen type (Roman) e.g. α1(XI) is polypeptide chain number 1 from collagen type eleven.

Table 1. Collagen types in cartilage

Type	Formula	M_r ($\times 10^{-3}$)	Length of helix (nm)	% of total collagen	Distribution in cartilage
II	$\alpha 1(II)_3$	290	300	95%	Throughout
XI	$\alpha 1(XI),\alpha 2(XI),\alpha 3(XI)$	300	320	3%	Throughout
VI	$\alpha 1(VI),\alpha 2(VI),\alpha 3(VI)$	550	105	0–1%	Pericellular
IX	$\alpha 1(IX),\alpha 2(IX),\alpha 3(IX)$	250	170	1%	Throughout
X	$\alpha 1(X)_3$	170	140	1%	Growth plate and basal calcified zone

List of collagen types (Roman numerals), their abundance and distribution detected in articular cartilage, together with the length of the triplex and names of their component α-chains (from [55])

3.1 The nascent structure of collagen

The nascent product (protocollagen) translated from collagen mRNA is devoid of hydroxyamino acids and glycosylation and bears a signal peptide, required for secretion, and terminal propeptides, which are both absent in the mature fibril (Fig. 2).

3.2 Post-translational modifications

Hydroxylation and glycosylation occur prior to helix assembly. Analysis of the amino acid composition shows that human and chick collagen II contains about 218 prolines, 45% of which are hydroxylated, whereas of the 36 lysine residues, 66% are hydroxylated [1]. Hydroxylated residues occur at specific positions, and may be further glycosylated. A disaccharide is formed on specific hydroxylysine residues to form 2-O-α-D-glucopyranoside-O-β-D-galactopyranosyl-hydroxylysine [2]. In addition, interchain links are formed in the extracellular molecule to stabilise the fibril structure (see Section 10).

3.3 The terminal propeptides

The nascent, immature form of collagen, procollagen, bears terminal propeptides which are cleaved during maturation by specific proteases. The C-propeptides in fibrillar collagens possess 3 characteristic features: the C-protease cleavage position; Cysteinyl residues, important for disulphide formation; and the N-linked glycosylation site. The C-propeptides are highly conserved structures which initiate collagen trimer assembly by inter- and intra-chain disulphide bonding. After cleavage, the C-propeptide stays

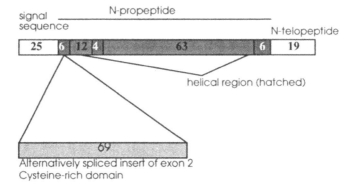

Fig. 3. Structure of the N-terminal of procollagen II.
Size, in number of amino acids, of the signal peptide, the alternatively spliced Cys-rich globular region, the N-propeptide, with its 2 helical regions, and the non-helical N-telopeptide.

associated with the collagen fibril [3] and was identified as chondrocalcin in the calcifying zone of growth plate.

The N-propeptide exists in 2 forms, the product of alternative mRNA splicing (Fig. 2). N-propeptides also possess a Cysteine-rich region, together with a triple helical domain and an N-terminal 25 amino acid signal sequence, cleaved during the secretion process (Fig. 3). Unusually amongst the fibrillar collagens (collagens I, II, III, V & XI), type II N-propeptide is expressed in 2 alternatively-spliced variants. These 2 forms differ in the Cys-rich globular domain encoded by exon 2 [4, 5], which is differentially expressed in various tissues [6] and during embryonic development [7]. The dominant form of collagen in adult articular cartilage lacks this Cys-rich domain [8]. The function of the N-propeptide has been suggested to include regulation of fibrillogenesis and feedback regulation of collagen biosynthesis [9].

4. Antibody detection

4.1 Indirect immunofluorescence

Was used on cartilage in early studies on the specificity of serum autoantibodies in RA patients [10]. Most, but not all, of the type II collagen molecule, as represented by the major CNBr cleavage fragments is available to bind antibody in native cartilage (Table 2). Indirect immunofluorescence on cartilage is however, much less sensitive than ELISA and radio-immunoassay in detecting anti-collagen antibodies and is not generally used.

Table 2. Availability of individual collagen determinants in normal cartilage

Monoclonal antibodies binding to collagen fragment	Number which bind to mouse cartilage
CB8	5/5
CB9	4/4
CB10	5/5
CB11	12/16
CB12	2/2

Monoclonal antibodies derived from immunised DBA/1 mice were characterised by binding to collagen CNBr fragments and tested for their ability to bind to normal mouse cartilage (data from [56])

4.2 Counterimmunoelectrophoresis and immunodiffusion

Have not been used for anti-collagen antibody detection. The native collagen molecule is a large, glycosylated trimer, which will diffuse slowly, limiting the utility of gel diffusion assays.

4.3 ELISA

Collagen-coated polystyrene microwells have been used to detect and quantitate anti-collagen antibody from patients with rheumatic diseases [11–18]. Bovine or chick collagen has been used, but comparison with data using human collagen has suggested that antibodies to xenogeneic collagen are present in some healthy individuals as well as patients with RA [14]. The purity of collagen is of crucial importance in this assay. Collagen is routinely isolated by guanidine-washing of powdered cartilage to remove proteoglycan and other soluble matrix molecules. Pepsin digestion liberates the collagen triple helix from the cross-linked fibril. Salt fractionation of the solubilised digest precipitates collagen [19]. The removal of contaminants such as pepsin and residual proteoglycan by ion exchange chromatography results in a more specific ELISA, indicating that such molecules bind antibodies from healthy individuals [20].

4.4 Haemagglutination

The agglutination of collagen-coated erythrocytes by anti-collagen antibody [21] cannot be used as an antibody detection system because fibronectin, present in serum, by directly binding to collagen, will produce false positive agglutination [22].

4.5 Radioimmunoassay

This technique has been used with higher sensitivity than haemagglutination or indirect immunofluorescence as an alternative to the ELISA [23–29]. Some assays described rely on the radioiodination of lysine residues in the collagen using the Bolton Hunter reagent, which thereby potentially may alter immunoreactive sequences in the antigen. Other assays using radioiodinated anti-immunoglobulin or protein-A do not suffer from this drawback.

4.6 Immunoblotting

This technique has not been used to detect anti-collagen antibodies, but has been used with CNBr peptides to map the specificities of polyclonal sera [30–32] from patients with systemic lupus erythematosus and rheumatoid arthritis and collagen-immunised mice.

5. Cellular location

Collagens are components of the extracellular matrix (see Table 1).

6. Cellular function

Collagen functions in cartilage matrix as a high tensile strength fibre matrix in which is enmeshed the highly charged and solvated aggrecan and hyaluronan.

7. Presence in other species

Collagens are found in most multicellular animals including insects [33, 34] and nematode worms [35, 36] They are not described in organisms devoid of extracellular matrix.

8. cDNA's and derived amino acid sequences

The relative sizes of telopeptides and helical regions are similar in collagens I, II and III (Table 3).

Sequence comparisons between murine, rat, bovine, chick and human collagen II are of particular interest because of the use of xenogeneic collagens in rodents to induce autoimmunity and collagen-induced arthritis.

Table 3. Comparison of collagen domain sizes

	Number of amino acid residues		
	$\alpha 1(I)$	$\alpha 1(II)$	$\alpha 1(III)$
N-telopeptide	17	19	14
Helical region	1014	1014	1029
C-telopeptide	26	27	25

Size, in amino acid residues, of the helical and globular domains of $\alpha 1$ chains from the 3 major fibrillar collagens (taken from [8])

Table 4. Predicted sequence of human collagen type II and its CNBr fragments

SIGNAL PEPTIDE + N-PROPEPTIDE:
(Amino acids: 1–62, numbering from start of N-propeptide)

MIRLGAPQSLVLLTLLVAAVLRCQGQDVRQPGPKGQKGEPGDIKDIVGPK
GPPGPQGPAGEQGPRGDRGDKGEKGAPGPRGRDGEPGTLGNPGPPGPPGP
PGPPGLGGNFAA

MATURE COLLAGEN SEQUENCE:

Starting at number: 1 (# -18)
*QMAGGFDEKAGGAQLGVM*QGPMGPMGPRGPPGPAGAPGPQGFQGNPGEPG
EPGVSGPM

CB 12   Starting at number: 59 (# 40)
GPRGPPGPPGKPGDDGEAGKPGKAGERGPPGPQGARGFPGTPGLPGVKGH
RGYPGLDGAKGEAGAPGVKGESGSPGENGSPGPM

CB 11   Starting at number: 143 (# 124)
GPRGLPGERGRTGPAGAAGARGNDGQPGPAGPPGPVGPAGGPGFPGAPGA
KGEAGPTGARGPEGAQGPRGEPGTPGSPGPAGASGNPGTDGIPGAKGSAG
APGIAGAPGFPGPRGPPDPQGATGPLGPKGQTGKPGIAGFKGEQGPKGEP
GPAGPQGAPGPAGEEGKRGARGEPGGVGPIGPPGERGAPGNRGFPGQDGL
AGPKGAPGERGPSGLAGPKGANGDPGRPGEPGLPGARGLTGRPGDAGPQG
KVGPSGAPGEDGRPGPPGPQGARGQPGVM

CB 8   Starting at number: 422 (# 403)
GFPGPKGANGEPGKAGEKGLPGAPGLRGLPGKDGETGAEGPPGPAGPAGE
RGEQGAPGPSGFQGLPGPPGPPGEGGKPGDQGVPGEAGAPGLVGPRGERG
FPGERGSPGAQGLQGPRGLPGTPGTDGPKGASGPAGPPGAQGPPGLOGM

CB 10   Starting at number: 571 (# 552)
PGERGAAGIAGPKGDRGDVGEKGPEGAPGKDGGRGLTGPIGPPGPAGANG
EKGEVGPPGPAGSAGARGAPGERGETGPPGTSGIAGPPGADGQPGAKGEQ
GEAGQKGDAGAPGPQGPSGAPGPQGPTGVTGPKGARGAQGPPGATGFPGA
AGRVGPPGSNGNPGPPGPPGPSGKDGPKGARGDSGPPGRAGEPGLQGPAG
PPGEKGEPGDDGPSGAEGPPGPOGLAGQRGIVGLPGQRGERGFPGLPGPS
GEPGQQGAPGASGDRGPPGPVGPPGLTGPAGEPGREGSPGADGPPGRDGA
AGVKGDRGETGAVGAPGAPGPPGSPGPAGPTGKQGDRGEAGAQGPM

AMAN-B1.4/7

Table 4. Continued

CB 9, 7   Starting at number: 917 (#898)
GPSGPAGARGIQGPQGPRGDKGEAGEPGERGLKGHRGFTGLQGLPGPPGP
SGDQGASGPAGPSGPRGPPGPVGPSGKDGANGIPGPIGPPGPRGRSGETG
PAGPPGNPGPPGPPGPP*GPGIDM*

C-PROPEPTIDE:
Amino acids: 28–273, numbering from start of C-terminal telopeptide

DQAAGGLRQHDAEVDATLKSLNNQIESIRSPEGSRKNPARTCRDLKLCHP
EWKSGDYWIDPNQGCTLDAMKVFCNMETGETCVYPNPANVPKKNWWSSKS
KEKKHIWFGETINGGFHFSYGDDNLAPNTANVQMTFLRLLSTEGSQNITY
HCKNSIALDEAAGNLKKALLIQGSNDVEIRAEG

Translated from EMBL accession number: X16468. This sequence lacks the N-terminal propeptide Cys-rich globular domain, produced by alternative mRNA splicing. Mature collagen was numbered either: from the start of mature collagen; or (#) from the first glycine in the central helical region. Non-helical regions of mature collagen are shown in italics. CNBr fragments are indicated CB11, CB12, etc.

However, only the complete sequences of human (Table 4) and murine collagen II are published as yet. The overall nucleotide sequence identity human vs mouse CII is 89.1% and the amino acid similarity is 95.2% (only 37 changes within the mature CII sequence) [5]. Sequence conservation is highest in the C-telopeptide/C-propeptide domain (90.2% nucleotide similarity). The bovine collagen II amino acid sequence is identical to human at amino acids 4–36 and differs at only 3 residues at positions 37–302 and at 4 residues at positions 422–681 (8). Comparison of the sequence of the human CB11 fragment with the murine and bovine equivalent (Table 5) shows 28 replacement mutations, only one mutation being at the same position in both the bovine and mouse sequences.

Table 5. Sequence comparison of human, bovine and murine CB11

GPRGLPGERGRTGPAGAAGARGNDGQPGPAGPPGPVGPAGGPGFPGA
PGAKGEAGAPGARGPEGAQGSRGEPGNPGaPGPAGAaGN
PGaDGIPGAKGSAGAPGIAGAPGFPGaRGPPGPtGAsGPLGPK
GQAGePGIAGFKGDQGPKGETGPAGvQGAPGPAGEEGKRGARG
EPGGa/VGPaGPPGERGAPGsRGFPGQDGiAGPKGpPGERGspGavG
PKGspGeaGRPGEaGLPGAkGLTGRPGDAGPQGKVGPSGAPGEDGRP
GPPGPQGARGQPGVM

The predicted sequence of human CB11 with bovine and murine sequence variations: bovine differences are in lower case [57], murine differences are underlined [5].

Table 6. The genomic structure of collagen type II

Exon	Exon size	Codons (1 = first G of central helical region)	Intron size
1	90	− 131.. − 104	4400
2	206	− 104 −	
2B	17	− 103.. − 98	195
3	33	− 97.. − 87	104
4	33	− 86.. − 76	105
5	54	− 75.. − 58	163
5B	102	− 57.. − 24	979
6	78	− 23..3	630
7	45	4.. 18	111
8	54	19.. 36	400
9	54	37.. 54	774
10	54	55.. 72	374
11	54	73.. 90	131
12	54	91.. 108	306
13	45	109.. 123	535
14	54	124.. 141	3062
15	45	142.. 156	479
16	54	157.. 174	1513
17	99	175.. 207	296
18	45	208.. 222	92
19	99	223.. 255	189
20	54	256.. 273	524
21	108	274.. 309	365
22	54	310.. 327	84
23	99	328.. 360	138
24	54	361.. 378	440
25	99	379.. 411	393
26	54	412.. 429	406
27	54	430.. 447	348
28	54	448.. 465	243
29	54	466.. 483	242
30	45	484.. 498	146
31	99	499.. 531	234
32	108	532.. 567	342
33	54	568.. 585	278
34	54	586.. 603	376
35	54	604.. 621	372
36	54	622.. 639	254
37	108	640.. 675	491
38	54	676.. 693	444
39	54	694.. 711	746
40	162	712.. 765	197
41	108	766.. 801	170
42	108	802.. 837	354
43	54	838.. 855	172
44	108	856.. 891	165
45	54	892.. 909	181
46	108	910.. 945	244

Table 6. Continurf

Exon	Exon size	Codons (1 = first G of central helical region)	Intron size
47	54	946.. 963	443
48	108	964.. 999	357
49	289	1000..1095	454
50	188	1096..1158	343
51	243	1159..1239	535
52	144	1240..1287	

Size in base pairs of exons and introns of the human collagen type II gene (taken from [58])

9. The gene

The locus for collagen $\alpha 1$(II), resides on chromosome 12. The gene is 30 kb long. The helical region, typically for fibrillar collagens, contains about 44 exons of 54 bp, 99 bp, 108 bp or 162 bp which start with a codon for glycine and end with a codon for the "Y" amino acid. Only 3 of the 54 exons in mouse collagen type II begin with a split codon: exons 2, 3 and 50 [5]. The commonest exon size of 54 bp encodes 18 amino acids, corresponding to 1.5 turns of the helix. Introns vary in size between 3799 and 84 bp.

The gene organisation is shown in Figures 4 and 5, and in Table 6.

10. Relation with nucleic acids

A non-sequence-specific affinity of collagens I, II, IV and V for both single-stranded and double-stranded DNA has been described. This interaction may be of pathogenic significance for instance in the tissue deposition of

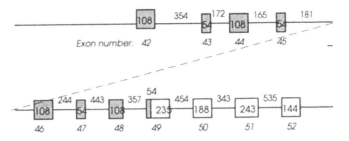

Fig. 4. Exon structure of the collagen type II gene encoding the carboxyl terminal region. Exon numbers are italicised. Other numbers refer to sizes of introns or exons (boxed) in numbers of base pairs. The end of the central helical region is shaded. (From [62].)

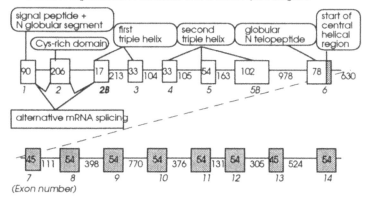

Fig. 5. Exon structure of the collagen type II gene encoding the amino terminal region and its relationship to the protein domains.

Exon numbers are italicised. Other numbers refer to sizes of introns or exons (boxed) in numbers of base pairs. The start of the central helical region is shaded. (Taken from [58, 63, 64].)

DNA-anti-DNA immune complexes in patients with systemic lupus erythematosus [37, 38].

11. Relation with other proteins

Although aggrecan is the major non-collagenous component of cartilage essential for maintaining the resilience of cartilage to compressive forces, no specific molecular interaction can be demonstrated between aggrecan and collagen fibres. The tensile strength of cartilage is little affected by the removal of proteoglycan [39]. On the other hand, collagen α-chains, as a structural component of articular cartilage collagen fibrils make interactions with other polypeptides within the collagen fibril. These interactions impart tensile strength to the fibril to withstand the osmotic swelling pressure in cartilage, induced by the highly charged polysulphated cartilage proteoglycans.

Type II collagen interacts with the minor articular collagens types IX and XI, being present at ratios of 8:1:1 in chick cartilage [40]. The amino and carboxyl telopeptides are the sites of covalent cross-linking between α-chains. Pyridinoline cross-links the CB12 sequence at residue 87; and the CB9,7 sequence at residue 930, to telopeptides in two neighbouring α-(1)II chains (17C-telopeptide and 9N-telopeptide respectively) [41]. Collagen II telopeptides are also cross-linked to collagen IX which decorates the fibril [41]. Interactions with collagen XI in the interior of the fibril appear to be non-covalent [42].

12. B-cell epitopes

12.1 Polyclonal antibodies

Many studies have aimed at mapping arthritogenic or protective/suppressive determinants relative to the large CNBr-derived fragments of collagen type II. Most such studies use either the Wistar rat model immunised with bovine CII or the DBA/1 mouse model immunised with chick CII. In general, all fragments of collagen are antigenic in both models, and are exposed in normal cartilage, allowing antibody binding *in vivo* and *in vitro*. However, the arthritogenic determinants are restricted to CB11, whereas protective determinants are found on CB10 and CB11 (Table 7). Rats immunised with chick, bovine or human CII develop antibodies primarily to CB11, CB9-7, CB12, with less reactivity to CB8, CB10 and CB6 [43, 44] (Table 7). The antibodies to CB10 do not cross-react with rat CII unlike the antibodies to other CB fragments. In contrast, antibodies to CB11, CB9-7 and CB12 from rats immunised with bovine collagen bind directly to rat cartilage, indicating the expression of these autoepitopes in normal tissue [45].

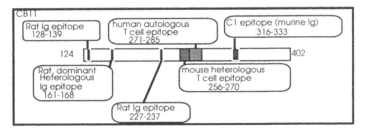

Fig. 6. Antibody and T-cell determinants on collagen type II fragment CB 11.
Antibody epitopes were mapped with polyclonal rat sera [65] and with a murine monoclonal antibody [49], and T-cell determinants were mapped using murine T-cell clones [49, 54, 66].

Mice immunised with chicken collagen show a pattern of antibody reactivities different to rats. In order of decreasing binding affinity, antibodies to CB11, CB10, and CB8 with only weak binding to CB 9, 7, 12, 6 were observed [46]. It was also found that immunisation with CB11 but not CB8 or CB10 was able to induce arthritis in rats and mice (Table 7). On the other hand, both CB11 and CB12 from chick collagen can prevent collagen-induced arthritis (Table 7).

The epitopes on bovine CB11 recognised by polyclonal serum antibodies from collagen-arthritic Wistar rats have been mapped using synthetic octapeptides attached to plastic supports [47] (Fig. 6). Of the 21 epitopes recognised,

all sera bound to	**PAGGPGFP**	CB11: (161–168)
5/6 sera also bound	**IAGAPGFPGAR**	CB11: (227–237)
5/6 sera also bound	**LPGERGRTGPAG**	CB11: (128–139)

Table 7. Mapping important immune determinants using CNBr fragments of type II collagen

CNBr fragment	12	11	8	10	9,7
Binding activity of antibodies from chick-CIA DBA/1 mice to chick fragment [46]	+/−	+++	+	++	+/−
% of hybridomas from CIA-DBA/1 mice binding to chick fragment [56]	15	36	17	14	17
% of bovine-CIA Wistar rat polyclonal anti-CII binding to bovine fragment [59]	24	6	24	15	27
Induction of passive arthritis in DBA/1 mice by IgG antibody purified on bovine fragment [59]	−	+	−	−	−
Induction of arthritis in Wistar rats by immunisation with bovine fragment [46]	ND	+	−	−	ND
Suppression of chick-CIA in DBA/1 mice, by chick fragment or antibodies to chick fragment [60]	+	+	−	−	ND
Reaction of serum IgG, from bovine-CIA Wistar rats, with renatured rat fragment [45]	+++	++	+	−	++++
Reaction of femur cartilage-bound antibody, from bovine-CIA Wistar rat, to rat fragment [45]	++	++++	+	+	++

This table summarises many papers comparing the CNBr-fragments or antibodies to the fragments in immunogenicity, or antigenicity or effectiveness at causing or preventing collagen-induced arthritis in rodents

12.2 Monoclonal antibodies

A group of 20 monoclonal antibodies from DBA/1 mice immunised with rat collagen type II were grouped into 7 sets on the basis of binding to interacting or overlapping epitopes on the native helical domain. Representative antibodies from each group bound with similar affinity to chick, mouse, rat, bovine and human collagen II in 5 cases, but with 2–10 times lower affinity to mouse collagen in 2 cases [48]. The epitopes were not defined by reaction with particular peptide sequences in most cases. Antibody C1 and D3 were shown to bind to conformational determinants on CB11 and 10 respectively [49] (Table 8).

Table 8. Sequence specificity of murine monoclonal antibody C1

GFAGQAGPAGATGAPGRP	= CHICK
GFPGQDGLAGPKGAPGER	= HUMAN
GFPGQDGLAGPKGAPGER	= BOVINE
GFDGQDGLAGPPGPVGPA	= RAT
GF. GQ. G. AG. . G. . G. .	= CONSENSUS

Antibody C1 which also binds to a conformational determinant on collagen, was shown to bind to the CB11 sequence: 316–333 [31]. Because the glycine residues are not exposed, it is likely that the critical C1-antibody binding residues are F-( ) ( )-Q (from: [31])

From a different group of monoclonal antibodies, 3 epitopes (1–5; 2–15; 2–25) were shown to bind antibodies from both collagen-arthritic mice and patients with RA or relapsing polychondritis [50]. Of these 3 antibodies, rabbit antiidiotype to 1–5 was found to induce high levels of anti-collagen antibody and induce arthritis in sensitised mice [51]. The sequence specificity of these antibodies are unknown.

13. T-cell determinants on collagen II

13.1 Mouse T cells

A peptide from chicken collagen CB11: residues 245–270, when given to mice intravenously as a soluble free peptide, was found to induce tolerance against collagen immunisation and protect against collagen-induced arthritis [52]. The T-cell determinant was localised to residues 260–270 (Fig. 6): IAGFKGEQGPK, by measuring the effect of amino acid substitutions in this sequence on γ-interferon release from mouse lymphocytes *in vitro* [53].

Table 9. Sequence specificity of a human T-cell clone TC9 which reacts to human CB11 SEQUENCE: 271–285

G E P G P A G P Q G A P G P A	= HUMAN: stimulation
. . T	= MOUSE : no stimulation
. . . . V	= BOVINE : stimulation
. . . K . . . . E R	= BOVINE 324–333: INHIBITS

Comparison of the T-cell stimulation caused by human, murine and bovine CB11: 271–285 showed the importance of threonine 273, the insensitivity of position 279 and the ability of a related sequence which overlaps 276–285, to inhibit stimulation.

13.2 Human T cells

A CD4+ T-cell clone, TC9, was isolated from a healthy human donor [54] which recognised human and bovine but not mouse collagen peptides containing CB11: 271–285 (Table 9, Figure 6) in the context of MHC DR7. Monoclonal anti-collagen antibody C1, but not D3, inhibited collagen-induced activation of TC9. It was suggested that C1 may protect a collagen peptide, normally destroyed during antigen processing, allowing the peptide to compete with the 271–285 determinant for binding to the MHC or T-cell receptor [49, 54]. In support of this hypothesis, it was shown that the peptide 324–333 overlaps the C1 epitope; that it shares 7 of 10 residues with the TC9 determinant; and that the free 324–333 peptide inhibited TC9 activation (Table 9).

References

1. Miller EJ & Lunde LG (1973) Biochem 12:3153–3159
2. Kuhn K (1987) In: Mayne R and Burgeson D (Eds) Structure and Function of Collagen Types, pp 1–42. Academic Press, Orlando, Florida.
3. Ruggiero F, Pfaffle M, von der Mark K & Garrone R (1988) Cell Tissue Res 252:619–624
4. Ryan MC & Sandell LJ (1990) J Biol Chem 265:10334–10339
5. Metsaranta M, Toman D, de Crombrugghe B & Vuorio E (1991) J Biol Chem 266:16862–16869
6. Sandell LJ, Morris N, Robbins JR & Goldring MB (1991) J Cell Biol 114:1307–1319
7. Nah HD & Upholt WB (1991) J Biol Chem 266:23446–23452
8. Baldwin CT, Reginato AM, Smith C, Jimenez SA & Prockop DJ (1989) Biochem J 262:521–528
9. Bornstein P & Sage H (1988) Proc Natl Acad Sci USA 85:67–106
10. Beard HK, Ryvar R, Skingle J & Greenbury CL (1980) J Clin Pathol 33:1077–1081
11. Gioud M, Meghlaoui A, Costa O & Monier JC (1982) Collagen & Related Res 2:557–564
12. Pesoa SA, Vullo CM, Onetti CM & Riera CM (1989) Autoimmunity 4:171–179
13. Fujii K, Tsuji M, Murota K, Terato K, Shimozuru Y & Nagai Y (1989) J Immunol Meth 124:63–70
14. Terato K, Shimozuru Y, Katayama K, Takemitsu Y, Yamashita I, Miyatsu M, Fujii K, Sagara M, Kobayashi S, Goto M et al. (1990) Arthritis Rheum 33:1493–1500
15. Ferris J, Cooper S, Roessner K & Hochberg M (1990) J Rheumatol 17:880–884
16. Morgan K, Clague RB, Collins I, Ayad S, Phinn SD & Holt PJ (1987) Ann Rheum Diseases 46:902–907
17. Choi EK, Gatenby PA, McGill NW, Bateman JF, Cole WG & York JR (1988) Ann Rheum Diseases 47:313–322
18. Collins I, Morgan K, Clague RB, Brenchley PE & Holt PJ (1988) J Rheumatol 15:770–774
19. Furuto DK & Miller EJ (1987) Meth Enzymol 144: 41–61
20. Williams RO, Williams DG & Maini RN (1992) J Immunol Meth 147:93–100
21. Andriopoulos NA, Mestecky J, Miller EJ & Bradley EL (1976) Arthritis Rheum 19:613–617
22. Beard HK, Lea DJ & Ryvar R (1979) J Immunol Meth 31:119–128
23. Clague RB, Firth SA, Holt PJ, Skingle J, Greenbury CL & Webley M (1983) Ann Rheum Diseases 42:537–544
24. Clague RB & Moore LJ (1984) Arthritis Rheum 27:1370–1377
25. Clague RB, Shaw MJ & Holt PJ (1981) Ann Rheum Diseases 40:6–10
26. Charri'ere G, Hartmann DJ, Vignon E, Ronzi'ere MC, Herbage D & Ville G (1988) Arthritis Rheum 31:325–332
27. Clague, RB Shaw, MJ & Holt PJ (1980) Ann Rheum Diseases 39:201–206
28. Stuart JM, Huffstutter EH, Townes AS & Kang AH (1983) Arthritis Rheum 26:832–840
29. Rowley M, Tait B, Mackay IR, Cunningham T & Phillips B (1986) Arthritis Rheum 29:174–184
30. Choi EK, Gatenby PA, Bateman JF & Cole WG (1990) Immunol Cell Biol 68:27–31
31. Burkhardt H, Holmdahl R, Deutzmann R, Wiedemann H, von der Mark H, Goodman S & von der Mark K (1991) Euro J Immunol 21:49–54
32. Rowley MJ, Mackay IR, Brand CA, Bateman JF & Chan D (1992) Rheum Intl 12:65–69
33. Ashhurst DE & Bailey AJ (1980) Euro J Biochem ·103:75–83
34. Francois J, Herbage D & Junqua S (1980) Euro J Biochem 112:389–396
35. Leushner JR, Semple NL & Pasternak J (1979) Biochim Et Biophys Acta 580:166–174
36. Ouazana R & Herbage D (1981) Biochim Et Biophys Acta 669:236–243
37. Rosenberg AM & Prokopchuk PA (1986) J Rheum 13:512–516
38. Gay S, Losman MJ, Koopman WJ & Miller EJ (1985) J Immunol 135:1097–1100
39. Broom ND & Silyn-Roberts H (1990) Arthritis Rheum 33:1512–1517

40. Vaughan L, Mendler M, Huber S, Bruckner P, Winterhalter KH, Irwin MI & Mayne R (1988) J Cell Biol 106:991–997
41. Eyre DR, Apone S, Wu JJ, Ericsson LH & Walsh KA (1987) Febs Lett 220:337–341
42. Mendler M, Eich-Bender SG, Vaughan L, Winterhalter KH & Bruckner P (1989) J Cell Biol 108: 191–197
43. Wooley PH, Luthra HS, Stuart JM & David CS (1981) J Experim Med 154:688–700
44. Watson WC, Thompson JP, Terato K, Cremer MA & Kang AH (1990) J Experim Med 172:1331–1339
45. Cremer MA, Terato K, Watson WC, Griffiths MM, Townes AS & Kang AH (1992) J Immunol 149:1045–1053
46. Terato K, Hasty KA, Cremer MA, Stuart JM, Townes AS & Kang AH (1985) J Experim Med 162:637–646
47. Worthington J, Brass A & Morgan K (1991) Autoimmunity 10:201–207
48. Holmdahl R, Rubin K, Klareskog L, Larsson E & Wigzell H (1986) Arthritis Rheum 29: 400–410
49. Burkhardt H, Yan T, Broker B, Beck-Sickinger A, Holmdahl R, von der Mark K & Emmrich F (1992) Euro J Immunol 22:1063–1067
50. Iribe H, Kabashima H, Ishii Y & Koga T (1988) Clin Experim Immunol 73:443–448
51. Iribe H, Kabashima H & Koga T (1989) J Immunol 142:1487–1494
52. Myers LK, Stuart JM, Seyer JM & Kang AH (1989) J Experim Med 170:1999–2010
53. Myers LK, Terato K, Seyer JM, Stuart JM & Kang AH (1992) J Immunol 149:1439–1443
54. Yan T, Burkhardt H, Ritter T, Broker B, Mann KH, Bertling WM, von der Mark K & Emmrich F (1992) Euro J Immunol 22:51–56
55. Eyre DR (1991) Seminars Arthritis Rheum 21:2–11
56. Terato K, Hasty KA, Reife RA, Cremer MA, Kang AH & Stuart JM (1992) J Immunol 148:2103–2108
57. Seyer JM, Hasty KA & Kang AH (1989) Euro J Biochem 181:159–173
58. Ala-Kokko L & Prockop DJ (1990) Genomics 8:454–460
59. Englert ME, Ferguson KM, Suarez CR, Sapp TM, Oronsky AL & Kerwar SS (1987) Cell Immunol 105:447–453
60. Kobayashi S, Terato K, Harada Y, Moriya H & Taniguchi M (1991) Arthritis Rheum 34:48–54
61. Petit B, Freyria A, van der Rest M & Herbage D (1992) In: Adolphe M (Ed) Biological Regulation of the Chondrocyte (pp 33–84). CRC Press, Boca Raton.
62. Cheah KS, Stoker NG, Griffin JR, Grosveld FG & Solomon E (1985) Proc Natl Acad Sci USA 2555–2559
63. Huang MC, Seyer JM, Thompson JP, Spinella DG, Cheah KS & Kang AH (1991) Euro J Biochem 195:593–600
64. Su MW, Benson-Chanda V, Vissing H & Ramirez F (1989) Genomics 4:438–441
65. Morgan K, Turner SL, Reynolds I, Hajeer AH, Brass A & Worthington J (1992) Immunology 77:609–616
66. Michaelsson E, Andersson M, Engstrom A & Holmdahl R (1992) Euro J Immunol 22:1819–1825

Manual of Biological Markers of Disease **B2.1**: 1–15, 1994.

DNA as antigen in SLE

RUUD J.T. SMEENK

Department of Autoimmune Diseases, Central Laboratory of the Netherlands Red Cross Blood Transfusion Services, Amsterdam, The Netherlands

Abbreviations

(m)Ab: (monoclonal) antibodies; BSA: bovine serum albumin; (ds/ss)DNA: (double-stranded/single-stranded) desoxyribonucleic acid; ELISA: enzyme-linked immunosorbent assay; GVHD: graft-versus-host disease; IFT: immunofluorescence test; IgG/M: immunoglobulin G/M; MW: molecular weight; RIA: radio immunoassay; RNA: ribonucleic acid; RNP: ribonucleo-protein; SLE: systemic lupus erythematosus; UV: ultraviolet

Key words: DNA, anti-DNA, SLE

1. The antigen

Desoxyribonucleic acid (DNA) is an important antigen in the field of autoimmune diseases [1–3]. DNA as antigen may either be double-stranded (dsDNA) or single-stranded (ssDNA); the influence of DNA strands will be further discussed in paragraph 14. Furthermore, DNA as antigen will *in vivo* almost always occur in the form of nucleosomes, i.e. closely associated with histones.

Apart from DNA, ribonucleic acid (RNA) also constitutes an antigen in SLE. Antibody reactivity towards RNA may either be based on epitopes shared with DNA or form a separate entity [4, 5].

Because the epitopes situated on DNA are – at least in part – based on the repetitive negative charge of the molecule, synthetic polynucleotides are often also recognized by anti-DNA antibodies.

2. Synonyms

Desoxyribonucleic acid; DNA; polynucleotide(s).

3. Antibody detection

A multitude of assays has been developed to detect the presence of antibodies to DNA. Currently, the most commonly used assays are immunofluorescent techniques (e.g. IFT on *Crithidia luciliae*), radio immunoassays (RIAs, e.g. Farr assay and PEG assay) and enzyme-linked immunosorbent assays (ELISAs) [6–10]. The technical details of these methods can be found in part A of this Manual [11]. All methods can be performed either by using commercial kits or as in-house assays. Different assay systems are not always comparable, for the following reasons:
1. The source of antigen differs; DNA may be from eukaryotic or prokaryotic origin, may be double-stranded or single-stranded, poly-disperse in size or homogenous, etc.
2. The way DNA is presented to the antibody may differ; in RIAs it is generally in solution, in ELISAs it is coated on plastic. In fluorescence methods the cellular DNA is mostly presented in an intact form.
3. Reaction conditions of the various assays may be different; e.g., due to the employed ammonium sulphate precipitation step employed in the Farr assay, anti-dsDNA of low avidity is missed using this method. In second antibody techniques such as IFT and ELISA, the choice of conjugated antibody is of importance; often, only IgG anti-DNA is measured with these techniques.

Regarding the source of antigen, it is difficult to decide which antigen preparation should be preferred. However, some general remarks can be made.
1. DNA such as the oft-used calf thymus DNA may be contaminated by proteins, especially histones. It is crucial to avoid such contaminations, since the measurement of anti-DNA is easily disturbed by the detection of other specificities of antibodies or even immune complexes [12, 13].
2. dsDNA is preferred above ssDNA, since several publications have shown that anti-dsDNA is more specific to SLE than anti-ssDNA [1, 14, 15].
3. The DNA should be large enough to present antigenic epitopes in a proper way. This also implies that synthetic polynucleotides may not always present (all) the relevant epitopes.

3.1 Immunofluorescence on Crithidia luciliae

The IFT on *Crithidia luciliae* is a method that couples strong sensitivity with a high disease specificity and, therefore, is one of the preferred methods [16–18]. A potential drawback could be the presence of histones in the kinetoplast. However, histones have not yet been detected in this structure. The method specifically detects antibodies to dsDNA. An example of the binding of anti-dsDNA to *Crithidia luciliae* is given in Figure 1.

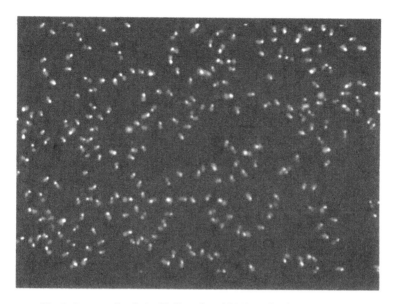

Fig. 1. An example of the binding of anti-DNA antibodies to *Crithidia luciliae*; fluorescence of kinetoplasts indicates the presence of antibodies to DNA. For technical details, see this Manual, chapter A8 [11].

3.2 Radio immunoassays

In RIAs the choice of antigen is again of great importance. The DNA employed should be larger than 10^5 but smaller than 10^7 kD [15]. Furthermore, the DNA must be double-stranded and, to allow quantitation of antibody reactivity, monodisperse in size [19]. This indicates that circular double-stranded bacteriophage DNA (such as from PM2) or plasmids (such as pUC9) are preferred. For technical details, see this manual, chapter A8 [11].

3.3 ELISAs

Regarding the coating of DNA on plastic, generally, it has been found to be difficult to directly coat dsDNA on plastic [20]. In contrast, ssDNA sticks to plastic very easily [20, 21]. To obtain a good coating of dsDNA on plastic, several approaches have been used. The most commonly used approach, probably, is to effect the coating of DNA via intermediate molecules such as poly-L-lysine, protamine or methylated BSA [22–24]. Although very effective, such precoats introduce problems related to binding to the plates (via the intermediate molecule) of immune complexes and/or immuno-globulins not directed against DNA [25, 26]. An alternative is to coat biotinylated DNA to the plates via Streptavidin [27]. Another approach that can be followed is to coat DNA to UV-irradiated plates [28].

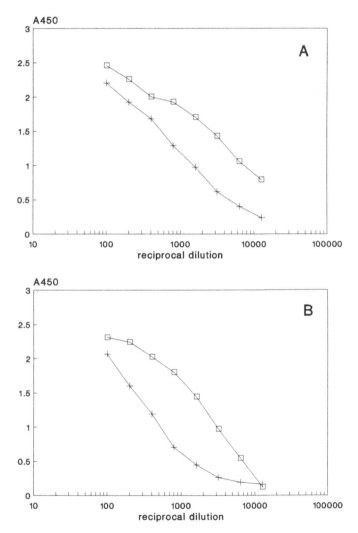

Fig. 2. Binding curves obtained for two SLE sera (+ = serum 1; □ = serum 2) tested in ELISAs on calf thymus DNA coated to the plate via protamine (A) and on pUC9 DNA, photobiotinylated and coated to the plate via Streptavidin (B).

An example of the binding curves obtained for two SLE sera, tested on calf thymus DNA, coated via protamine (Fig. 2A), and on pUC9 DNA, photobiotinylated and coated via Streptavidin (Fig. 2B), illustrates the above.

4. Cellular localization

DNA can, of course, be found in all prokaryotic and nucleated eukaryotic cells. In eukaryotic cell nuclei there are also many proteins present. Immunofluorescence of cell nuclei (as in the so-called ANF test) therefore cannot be used to detect anti-DNA antibodies. *Crithidia luciliae* contain a giant mitochondrion, the kinetoplast, composed of purely dsDNA, not "contaminated" with proteins. Therefore, fluorescence of kinetoplasts is indicative of the presence of anti-dsDNA (Fig. 1).

5. Cellular function

The cellular functions of DNA are numerous and will not be covered in this chapter.

6. Presence in other species

DNA, of course, is found in all species. Whether DNA from various species behaves differently from the viewpoint of antigenicity has not been studied in great detail. However, differences clearly exist. Studying species specificity of anti-DNA antibodies, Stollar *et al.* found that (human) serum antibodies bound DNA of all species tested, although not to the same extent [29]. Comparable results have been reported for monoclonal antibodies to DNA (see paragraphs 12 and 13) [30].

7. cDNA(s) and derived amino acid sequences

Not relevant to DNA as antigen.

8. The gene(s)

Not relevant to this antigen.

9. Relation to nucleic acids

Not relevant to this antigen.

10. Relation to proteins

DNA of eukaryotic cells is organised in the form of chromatin, a compact structure composed of a string of subunits, the nucleosomes (Fig. 3). A nucleosome contains about 200 basepairs of DNA wound around a histone octamer. This octamer consists of two copies each of H2A (MW = 14 kD), H2B (MW =13.8 kD), H3 (MW = 15.3 kD) and H4 (MW = 11.3 kD). Nucleosomal DNA can be divided into two regions: the core DNA with an invariant length of 146 basepairs, which is relatively resistant to digestion by nucleases, and the linker DNA, the length of which may vary between 8 to 114 basepairs per nucleosome associated with this linker DNA. Recently,

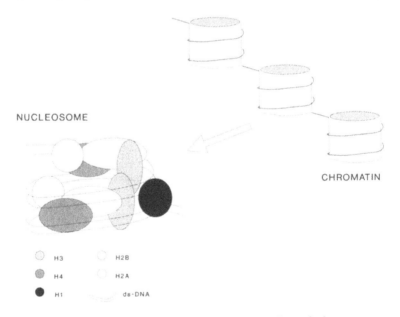

Fig. 3. Schematic drawing of a (string of) nucleosome(s). Chromatin has a compact organization, composed of a string of subunits, which have the same type of design in all eukaryotes, i.e. the nucleosomes. A nucleosome contains about 200 basepairs of DNA coupled to a histone octamer. The octamer consists of two copies each of H2A (MW = 14 kD), H2B (13.8 kD), H3 (15.3 kD) and H4 (11.3 kD). These are known as core histones. Associated with each nucleosome is a single molecule of H1 (23 kD). The nucleosomal DNA can be divided into two regions: the core region with an invariant length of 146 basepairs and the linker DNA, of which the length varies between 8 and 114 basepairs per nucleosome.

evidence has been presented that DNA circulating in the blood is always in the form of nucleosomes [31, 32]. Furthermore, autoantibodies specific for nucleosomes have been described [33].

11. Biochemical characteristics

Not further detailed in this chapter.

12. Murine models of SLE

Various strains of mice spontaneously develop an autoimmune disorder, which closely resembles SLE [34]. Of these, the NZB/W F_1 and MRL/lpr have been most profoundly studied. Autoantibodies produced by these mice include antibodies to DNA as well as to other antigens (anti-histone, anti-Sm, anti-nRNP, anti-rRNP) [35–37]. Another animal model claimed to be comparable with human SLE is obtained by inducing chronic graft-versus-host disease (GVHD) in mice, through the injection of F_1 mice with parental spleen and/or lymph node cells [38, 39]. The validity of chronic GVHD as a model for SLE is based upon clinical as well as serological considerations. A spectrum of autoantibodies, amongst which are anti-DNA, can be found in such mice [40]. When we analyzed the reaction patterns of monoclonal antibodies to DNA, derived from MRL/lpr, NZB/W F_1 and GVHD mice, we found that anti-DNA of all three models behaved comparably in 4 different anti-DNA assays [41, 42]. Furthermore, no differences could be found with regard to cross-reactivity of the antibodies or their isotypes.

13. Monoclonal antibodies

Monoclonal antibodies (mAb) to DNA have been prepared from both SLE mice and SLE patients [4, 30, 43–47]. Generally, such mAb reflect the antibody profile that can be found in the circulation of diseased animals or humans, i.e. the mAb also differ in characteristics such as isotype, avidity, cross-reactivity, specificity, ss/ds DNA reactivity, etc. Human monoclonal antibodies to DNA might be obtained from Stollar and Schwartz [48], Winkler and Kalden [46, 49] and Isenberg [47]; murine monoclonal anti-DNA from Eilat [4, 50], Staines [51], Smeenk [44] and Voss [45].

Sequencing studies of mAb [49, 52–54] have led to the initial conclusions that
1. the anti-DNA response is an immune response like any other (no specific Vh-gene usage, no particularities in somatic mutations); and
2. the anti-DNA response is antigen-driven.

14. B-cell epitopes

The specificity of anti-DNA antibodies is diverse, and conformational and sequential, as well as backbone, determinants can be the targets of anti-DNA antibody recognition.

14.1 Backbone determinants

The potential epitopes on either single-stranded or double-stranded DNA are short regions of DNA helix or short nucleotide sequences. The interaction between B-cell paratope and dsDNA epitope appears to be based on electrostatic interactions, since this binding is extremely sensitive to salt concentration and pH [55–58]. Especially in the case of high avidity, anti-DNA antibodies, secondary hydrogen bonding is also likely to play a role [59]. Most likely, such dsDNA epitopes are constituted by the sugar-phosphate backbone of the DNA. Specificity of autoantibodies for such epitopes might also explain the extent of anti-DNA cross-reactivity previously described [60–63]. Backbone epitopes may occur on dsDNA from any species. Indeed, SLE serum antibodies react with DNA of animal, bacterial, viral and plant origin [64].

14.2 Sequence determinants

Not all DNA molecules are recognized to the same extent [30]. So there also seems to be selective recognition of DNA sites, variably expressed on different DNAs. Such binding seems more pronounced in the case of single-stranded DNA and differs from simple recognition of a sugar-phosphate backbone. Thus, anti-DNA antibodies may also bind selectively to synthetic or natural polynucleotides, presumably on the basis of their nucleotide content [30, 65–67].

14.3 Conformational determinants

There are no firm data available to sustain an important role of conformational epitopes in the reactivity of antibodies with DNA.

14.4 ssDNA epitopes

Although anti-DNA antibodies specific for ssDNA may indeed exist as a separate entity, most of what is generally called anti-ssDNA reactivity is actually anti-dsDNA of low avidity. When dsDNA is denatured, the strands of DNA become more flexible so that, upon cooling, internal duplex formation over short stretches of DNA can occur. Stollar and co-workers have shown that reactivity of anti-dsDNA antibodies with ssDNA is mainly due to this kind of internal duplex formation [68]. One could therefore hypothesize that binding of antibodies to ssDNA is not due to a reaction with the exposed bases in the DNA, but to an interaction with internal duplex structures. In ssDNA these epitopes are exposed in a way completely different from dsDNA. The difference lies in the flexibility of the DNA backbone, which is of extreme importance in terms of allowing multi-point attachment (and thus high avidity binding) of antibodies to DNA. Therefore, the greater flexibility of ssDNA will lead to higher avidity binding. Indeed, Pisetsky and Reich recently reported that small fragments of DNA are not antigenic when coated to an ELISA plate, but are effective in inhibition assays [67]. This can be explained in terms of the restricted flexibility of coated DNA fragments.

14.5 Multivalency

Although the actual combining site of an anti-DNA autoantibody with a DNA sequence encompasses only about 6 nucleotides [69], most anti-DNA antibodies require DNA fragments from 40 to several hundred basepairs in length for stable interaction. The size dependency, however, differs very much among antibodies [70, 71]. These findings suggest that both Fabs of an anti-DNA antibody need to be bound for a stable interaction via (monogamous) bivalent interactions with antigenic sites distributed along the DNA molecule.

15. Cross-reactivity

DNA as antigen has been as reported cross-reactive with phospholipids, cell wall structures and glycosaminoglycans [60, 72–74]. Not only were antibodies to DNA found to cross-react with cardiolipin, antibodies prepared against cardiolipin also reacted with DNA [75]. A comparable relation was found with respect to glycosaminoglycans: not only was anti-DNA reported to cross-react with heparan sulphate, antibodies elicited against heparan sulphate also cross-react with DNA [76].

Such cross-reactivities, in general, have been attributed to the reactivity of antibodies with backbone epitopes, in particular, against repetitive negatively charged moieties (Fig. 4). From our own studies we would conclude the following with regard to anti-DNA cross-reactivity:
1. cross-reactivity of anti-DNA is inversely related to its avidity [44, 72];
2. selection by solid phase systems (i.e. ELISA) preferentially leads to obtainment of low avidity antibodies to DNA; such antibodies poorly reflect the antibody population found in the circulation of patients [77];

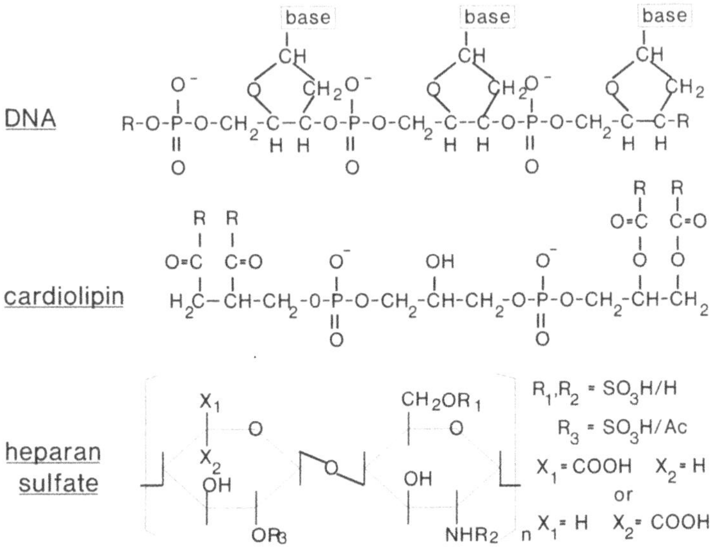

Fig. 4. Structure of DNA in comparison with cardiolipin and heparan sulphate. Based on comparable repetitive negative charges, antibodies to DNA have been supposed to cross-react with cardiolipin and heparan sulphate. More recently, it has been shown that cross-reactivity with heparan sulphate does not exist as such: binding to heparan sulphate is based on the interaction of nucleosomes, complexed to anti-DNA, with heparan sulphate.

3. cross-reactivity with heparan sulphate does not exist; this reactivity is based on the presence of immune complexes composed of nucleosomes and antibodies in sera; such complexes may react via their histone-part with heparan sulphate or via their DNA-part to the employed precoat in the anti-heparan sulphate ELISA (protamine) [13, 78].

16. T-cell epitopes

T-cells with specificity for DNA have not yet been identified. Recently Rajagopalan *et al.* were able to clone autoimmune T-helper (Th) cells from patients as well as from mice with lupus nephritis [79]. Less than 15% of these "autoreactive" T-cell clones had the functional ability to selectively induce the production of pathogenic anti-DNA autoantibodies [79, 80]. Histone-derived peptides from nucleosomes probably form the T-cell stimulus [81].

Acknowledgements

The author wishes to thank Machteld Hylkema and Tom Swaak for stimulating discussions and critical readings of the manuscript. The financial support of "Het Nationaal Reumafonds" of the Netherlands for the research of RJTS is greatfully acknowledged.

References

1. Tan EM (1982) Adv Immunol 33:167–240
2. Tan EM, Chan EKL, Sullivan KF & Rubin RL (1988) Clin Immunol Immunopathol 47:121–141
3. Stollar BD (1973) In: Sela M (Ed) The Antigens, vol. I (pp 1–85). Academic Press, New York
4. Eilat D, Hochberg M, Pumphrey J & Rudikoff S (1984) J Immunol 133:489–494
5. Field AK, Davies ME & Tyrell AA (1980) Proc Soc Exp Biol Med 164:524–529
6. Isenberg DA, Dudeney C, Williams W, Addison I, Charles S, Clarke J & Todd-Pokropek A (1987) Ann Rheum Dis 46:448–456
7. Tipping PG, Buchanan RRC, Riglar AG, Dimech WJ, Littlejohn GO & Holdsworth SR (1988) Br J Rheumatol 27:206–210
8. Sanford DG & Stollar BD (1992) Methods in Enzymology 212:355–371
9. Ward MM, Pisetsky DS & Christenson VD (1989) J Rheumatol 16:609–613
10. Smeenk R, Brinkman K, Van den Brink H & Swaak T (1990) Clin Rheumatol 9 (suppl. 1):63–72
11. Smeenk RJT (1993) In: van Venrooij WJ & Maini RN (Eds) Manual of Biological Markers of Disease (pp A8/1–A8/12). Kluwer Academic Publishers, Dordrecht/Boston/London
12. Subiza JL, Caturla A, Pascual-Salcedo D, Chamorro MJ, Gazapo E, Figueredo MA & de la Concha EG (1989) Arthritis Rheum 32:406–412
13. Brinkman K, Termaat R-M, de Jong J, Van den Brink HG, Berden JHM & Smeenk RJT (1989) Res Immunol 140:595–612
14. Aarden LA & Smeenk R (1982) In: Kalden JR & Feltkamp TEW (Eds) Antibodies to Nuclear Antigens (pp 23–28). Excerpta Medica, Amsterdam
15. Aarden LA, Lakmaker F & Feltkamp TEW (1976) J Immunol Methods 10:39–48
16. Aarden LA, De Groot ER & Feltkamp TEW (1975) NY Acad Sci USA 254:505–515
17. Smeenk R, van der Lelij G & Aarden LA (1982) J Immunol 128:73–78
18. Smeenk R, van der Lelij G & Aarden LA (1982) Clin Exp Immunol 50:603–610
19. Aarden LA, Lakmaker F & De Groot ER (1976) J Immunol Methods 11:153–163
20. Smeenk R (1986) In: Pal SB (Ed) Immunoassay Technology, vol. 2 (pp 121–144). Walter de Gruyter, Berlin-New York
21. Lacy MJ & Voss EW Jr (1989) J Immunol Methods 116:87–98
22. Smeenk R (1986) In: Pal SB (Ed) Immunoassay Technology, vol. 2 (pp 145–166). Walter de Gruyter, Berlin-New York
23. Preudhomme JL (1987) Immunol Today 8:74
24. Tzioufas AG, Manoussakis MN, Drosos AA, Silis G, Gharavi AE & Moutsopoulos HM (1987) Clin Exp Rheumatol 5:247–253
25. Brinkman K, Termaat R-M, van den Brink HG, Berden JHM & Smeenk RJT (1991) J Immunol Methods 139:91–100
26. Kroubouzos G, Tosca A, Konstadoulakis MM & Varelzidis A (1992) J Immunol Methods 148:261–263
27. Emlen W, Jarusiripipat P & Burdick G (1990) J Immunol Methods 132:91–101
28. Zouali M & Stollar DB (1986) J Immunol Methods 90:105–110
29. Stollar BD, Levine L, Lehrer HI & Van Vunakis H (1962) Proc Natl Acad Sci USA 48:874–880
30. Wu DP, Gilkeson GS, Armitage J, Reich CF & Pisetsky DS (1990) Clin Exp Immunol 82:33–37
31. Rumore PM & Steinman CR (1990) J Clin Invest 86:69–74
32. Rumore P, Muralidhar B, Lin M, Lai C & Steinman CR (1992) Clin Exp Immunol 90:56–62
33. Losman MJ, Fasy TM, Novick KE & Monestier M (1992) J Immunol 148:1561–1569
34. Theofilopoulos AN & Dixon FJ (1985) Adv Immunol 37:269–390

35. Kotzin BL, Lafferty JA, Portanova JP, Rubin RL & Tan EM (1984) J Immunol 133:2554–2559
36. Eisenberg RA, Tan EM & Dixon FJ (1978) J Exp Med 147:582–587
37. Bonfa E, Marshak-Rothstein A, Weissbach H, Brot N & Elkon K (1988) J Immunol 140:3434–3437
38. Gleichmann E, van Elven EH & Van Der Veen JPW (1982) Eur J Immunol 12:152–158
39. Gleichmann E, Pals ST, Rolink AG, Radaszkiewicz T & Gleichmann H (1984) Immunol Today 5:324–332
40. van Elven EH, Rolink AG, van der Veen FM, Issa Ph, Duin ThM & Gleichmann E (1981) J Immunol 127:2435–2438
41. Brinkman K, van Dam A, Van den Brink H, Termaat R-M, Berden J & Smeenk R (1990) Clin Exp Immunol 80:274–280
42. van Dam AP, Meilof JF, Van den Brink HG & Smeenk RJT (1990) Clin Exp Immunol 81:31–38
43. Karounos DG & Pisetsky DS (1987) Immunology 60:497–501
44. Smeenk RJT, Brinkman K, Van den Brink HG & Westgeest AAA (1988) J Immunol 140:3786–3792
45. Smith RG & Voss EW, Jr (1990) Mol Immunol 27:463–470
46. Winkler TH, Jahn S & Kalden JR (1991) Clin Exp Immunol 85:379–385
47. Ehrenstein M, Longhurst C & Isenberg DA (1993) Clin Exp Immunol 92:39–45
48. Fleming JO & Pen LB (1988) J Immunol Methods 110:11–18
49. Winkler TH, Fehr H & Kalden JR (1992) Eur J Immunol 22:1719–1728
50. Eilat D (1982) Mol Immunol 19:943–955
51. Morgan A, Buchanan RRC, Lew AM, Olsen I & Staines NA (1985) Immunology 55:75–83
52. Shlomchik MJ, Aucoin AH, Pisetsky DS & Weigert MG (1987) Proc Natl Acad Sci USA 84:9150–9154
53. Eilat D, Webster DM & Rees AR (1988) J Immunol 141:1745–1753
54. Van Es JH, Aanstoot H, Gmelig-Meyling FHJ, Derksen RHWM & Logtenberg T (1992) J Immunol 149:2234–2240
55. De Groot ER, Lamers MC, Aarden LA, Smeenk RJT & van Oss CJ (1980) Immunol Comm 9:515–528
56. Smeenk RJT, Aarden LA & van Oss CJ (1983) Immunol Comm 12:177–188
57. Smeenk R, Duin T & Aarden LA (1982) J Immunol Methods 55:361–373
58. Smeenk R, van Rooyen A & Swaak AJG (1988) J Immunol Methods 109:27–35
59. van Oss CJ, Smeenk RJT & Aarden LA (1985) Immunol Invest 14:245–253
60. Lafer EM, Rauch J, Andrzejewski C Jr, Mudd D, Furie B, Schwartz RS & Stollar DB (1981) J Exp Med 153:897–909
61. Eilat D, Zlotnick AY & Fischel R (1986) Clin Exp Immunol 65:269–278
62. Shoenfeld Y, Rauch J, Massicotte M, Stollar BD & Schwartz RS (1983) N Engl J Med 308:414–420
63. Raz E, Ben-Bassat H, Davidi T, Shlomai Z & Eilat D (1993) Eur J Immunol 23:383–390
64. Stollar BD (1991) Mol Immunol 28:1399–1412
65. Casperson GF & Voss EW Jr (1983) Mol Immunol 20:581–589
66. Lee JS, Lewis JR, Morgan AR, Mossman TR & Singh B (1981) Nucl Acids Res 9:1707–1725
67. Pisetsky DS & Reich CF (1994) J Immunol Methods. In Press
68. Stollar BD & Papalian M (1980) J Clin Invest 66:210–218
69. Stollar DB, Zon G & Pastor RW (1986) Proc Natl Acad Sci USA 83:4469–4473
70. Stollar BD & Papalian M (1980) J Clin Invest 66:210–218
71. Ali R, Dersimonian H & Stollar BD (1985) Mol Immunol 22:1415–1422
72. Smeenk RJT, Lucassen WAM & Swaak AJG (1987) Arthritis Rheum 30:607–617
73. Jacob L, Lety MA, Louvard D & Bach J-F (1985) J Clin Invest 75:315–317
74. Andreason GL & Evans GA (1988) Biotechniques 6:650–660
75. Rauch J, Tannenbaum H, Stollar BD & Schwartz RS (1984) Eur J Immunol 14:529–534

76. Shibata S, Harpel P, Bona C & Fillit H (1993) Clin Immunol Immunopathol 67:264–272
77. Eilat D (1986) Immunol Today 6–4:123–127
78. Termaat R-M, Brinkman K, van Gompel F, van de Heuvel LPWJ, Veerkamp JH, Smeenk RJT & Berden JHM (1990) J Autoimmunity 3:531–538
79. Rajagopalan S, Zordan T, Tsokos GC & Datta SK (1990) Proc Natl Acad Sci USA 87:7020–7024
80. Adams S, Zordan T, Sainis K & Datta SK (1990) Eur J Immunol 20:1435–1443
81. Mohan C, Adams S, Stanik V & Datta SK (1993) J Exp Med 177:1367–1381

Manual of Biological Markers of Disease **B2.2**: 1–28, 1994.
© 1994 *Kluwer Academic Publishers. Printed in the Netherlands.*

Histones

RUFUS W. BURLINGAME and ROBERT L. RUBIN

W.M. Keck Autoimmune Disease Center, Department of Molecular and Experimental Medicine, The Scripps Research Institute, 10666 North Torrey Pines Road, La Jolla, CA, U.S.A. 92037

Abbreviations

SLE: systemic lupus erythematosus; PIL: procainamide-induced lupus; ELISA: enzyme-linked immunosorbent assay; HIL: hydralazine-induced lupus; PAGE: polyacrylamide gel electrophoresis.

Key words: Chromatin structure; nucleosome; anti-histone antibody; anti-nucleosome antibody

1. The antigens

The histones are a set of basic proteins that organize the chromosomal DNA in all eukaryotes. The 4 core histones, H2A, H2B, H3 and H4 are complexed as an $(H2A-H2B-H3-H4)_2$ octamer *in vivo*. Approximately 146 basepairs of DNA are wrapped 1 3/4 turns around the octamer to form the nucleosome core particle. Between 20 and 60 basepairs of DNA comprise the linker region between adjacent core particles in chromatin, and histone H1 binds outside the core particle to this region of DNA to form the nucleosome, the repeating unit of chromatin [1]. Thus, the nucleosome consists of the histone octamer, 200 basepairs of DNA and H1.

2. Synonyms

The original nomenclature for histones adopted in the late 1950's was based on their order of fractionation in cation exchange chromatography. Attempts to improve these awkward names finally resulted in the mid 1970's with the currently accepted classification (Table 1). However, some publications still use the older terminology.

Table 1. Histone nomenclature and synonyms

		—————From 1958 to 1974—————		After 1975
lysine rich	f1	I	KAP	H1
slightly lysine rich	f2a2	IIb1	ALK	H2A
slightly lysine rich	f2b	IIb2	KSA	H2B
arginine rich	f3	III	ARK	H3
arginine rich	f2a1	IV	GRK	H4

The nomenclature of the histones was originally based on their fractionation behavior. Differential salt extraction yielded three fractions: lysine rich, slightly lysine rich and arginine rich. Separation on CM-cellulose yielded 3 different fractions: f1 (fraction 1), etc. Fraction 2 was eventually found to contain 3 polypeptides: f2a2, f2b, f2a1. Elution from Amberlite CG-50 yielded 4 peaks, I–IV; peak II was asymmetric and contained 2 proteins. When it was recognized that there were 5 main classes of histones, they were named with the 1 letter abbreviations for their 3 most common amino acids, i.e. KAP (lysine, alanine, proline). Finally, in a meeting in 1974 [58] it was agreed to replace the diverse and confusing histone nomenclature with the one accepted today: histones H1, H2A, H2B, H3 and H4.

3. Biochemical characteristics

3.1 Chromatin

Eukaryotic chromosomes are made up of 40% DNA, 40% histones, and 20% other proteins, RNA and carbohydrates. Chromatin, the biochemically isolated form of chromosomes, is insoluble at ionic strength equivalent to 0.15 M NaCl. Chromatin is soluble in distilled water, but it is difficult to manipulate in a test tube because of its high viscosity. Additionally, it has a complex macromolecular composition. Thus, a number of biochemical treatments have been developed to solubilize chromatin and to reduce its complexity. Mild treatment of chromatin with micrococcal nuclease yields a mixture of soluble poly- and mononucleosomes. When chromatin is washed in 0.5 M NaCl, histone H1, most non-histone proteins and RNA are removed and it becomes soluble. This is called H1-stripped chromatin. Brief digestion of H1-stripped chromatin with micrococcal nuclease cuts the linker region DNA and yields nucleosome core particles consisting of the histone octamer and DNA as described above [2]. When chromatin is extensively digested with micrococcal nuclease and treated with 3M urea, subnucleosome particles can be isolated. They consist of H1 with 60–70 basepairs of DNA, H2A-H2B with 50–60 basepairs of DNA, and (H3-H4)$_2$ with 70–80 basepairs of DNA [3].

3.2 Denatured histones

Histones can be extracted from chromatin with dilute mineral acid (0.4 N H_2SO_4) and then separated from each other by column chromatography [4] (Figs. 1a and b). The individual histones isolated by this technique are considered denatured [5] and tend to aggregate in high salt and neutral pH.

3.3 Native histone complexes

The $(H2A-H2B-H3-H4)_2$ histone octamer and H1 are extracted from chromatin at neutral pH by 2 M NaCl, and can be separated from the DNA by ultracentrifugation [6]. The octamer can then be dissociated into 2 H2A-H2B dimers and an $(H3-H4)_2$ tetramer by adjusting the pH or ionic strength, and these subunits can be separated by ion exchange or gel filtration chromatography (Fig. 1c). Under conditions of high salt and neutral pH, the octamer, dimer and tetramer are soluble and considered to be native by a number of criteria [7]. The H2A-H2B dimer and the $(H3-H4)_2$ tetramer can be reconstituted with DNA to form native histone-DNA subnucleosome complexes [8].

3.4 Nucleosome structure

Supporting the significance of the biochemical studies, the structure of the histone octamer is tripartite (Fig. 2), with two H2A-H2B dimers flanking the central $(H3-H4)_2$ tetramer [9]. Thus, the histone-DNA complex in chromatin can be separated into a limited number of biochemically relevant structures.

4. Antibody detection

Autoantibodies to histones have been found in a number of diseases such as systemic lupus erythematosus (SLE), drug-induced lupus, adult and juvenile rheumatoid arthritis, undifferentiated rheumatic diseases and murine models of lupus [10, 11].

4.1 Immunofluorescence

Some antihistone antibodies can be detected by their characteristic immunofluorescent properties on a set of 3 specially prepared cells. The first substrate is untreated, the second is extracted with dilute acid to remove histones, and the third is extracted with dilute acid and reconstituted with purified total histones [13]. Sera with certain anti-histone antibodies are

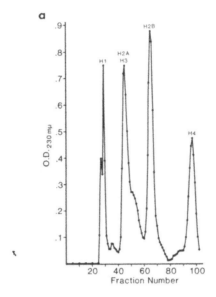

Fig. 1a. Preparation and composition of *bovine* total histones, individual histones and histone-histone complexes.

a. Chromatogram of acid extracted histones separated by exclusion chromatography on Biogel P-60 [4]. H2A and H3 can be further separated by chromatography on Sephadex G-100.

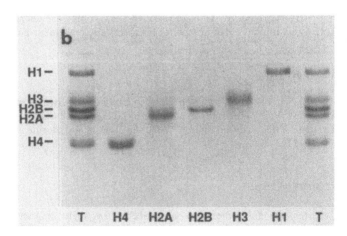

Fig. 1b. Acid urea polyacrylamide gel electrophoresis (PAGE) [60] of chromatographically purified individual histones. The first and last lanes contain total histones (T). Note that proteins are separated by a combination of net charge and molecular size in this type of PAGE.

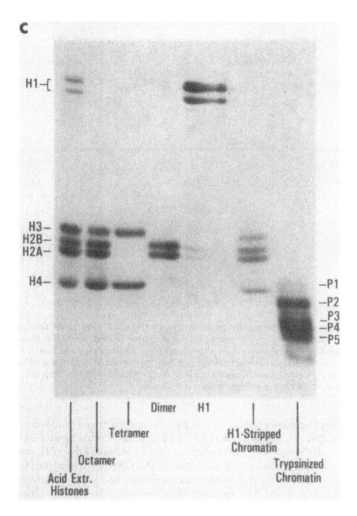

Fig. 1c. Histone composition of various chromatin preparations used in ELISA and prepared as described [20]. The samples applied to the SDS polyacrylamide gel (18% acrylamide, 0.24% bis, as in [61]) were: lane 1, total acid extracted histones (H1a = 22,180 Da, H1b = 21,730 Da, H3 = 15,320 Da, H2B = 13,760 Da, H2A = 13,940 Da, H4 = 11,280 Da; lane 2, the core histone octamer; lane 3, the (H3-H4)$_2$ tetramer; lane 4, the H2A-H2B dimer; lane 5, H1; lane 6 H1-stripped chromatin; lane 7 trypsinized H1-stripped chromatin (see Table 2). Note that histone H1 runs substantially slower than expected in SDS PAGE, and H2A runs faster than H2B even though H2B is 4 amino acids shorter. These data are for bovine main histones.

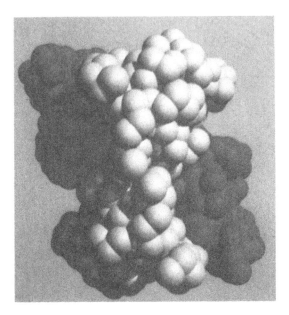

Fig. 2. View of the structure of the histone octamer looking down the molecular two-fold axis as determined by x-ray crystallography (reproduced from [9] with permission from E.N. Moudrianakis). The C-alpha positions contain space filling models of the amino acid backbone. The tripartite structure is obvious from this view, with the two dark H2A-H2B dimers flanking the central (H3-H4)$_2$ tetramer. Compare this with Figures 7 and 13 to see how DNA is wrapped around the octamer to form the nucleosome core particle.

positive with a homogeneous nuclear pattern on the first and third substrates, but negative on the second (Fig. 3). These anti-histone antibodies are common in patients with SLE and procainamide-induced lupus (PIL) and, as described later, recognize regions of the (H2A-H2B)-DNA complex that are exposed in chromatin [12]. If the serum contains antibodies to extractable nuclear antigens, the first substrate shows immunofluorescence, while the next 2 are negative [13]. Sera with antibodies to native DNA are positive on all 3 substrates. The advantage of this technique is its specificity for certain anti-histone antibodies, while its disadvantage is that it is labor intensive.

Other anti-histone antibodies yield a changing pattern of immuno-fluorescence on cells extracted with various amounts of NaCl. Sera from patients with undifferentiated rheumatic diseases [14] are negative on the unextracted HEp-2 cell substrate, display a variable large speckled nuclear pattern on cells washed with 0.5 M NaCl, but are negative again on the 1.5 M NaCl washed cells (Figs. 4D–F) [15]. Interestingly, these sera show a variable large speckled pattern on unextracted mouse kidney cells (Fig. 4G). The pattern is retained on cells washed with 0.5 M NaCl, but is negative on

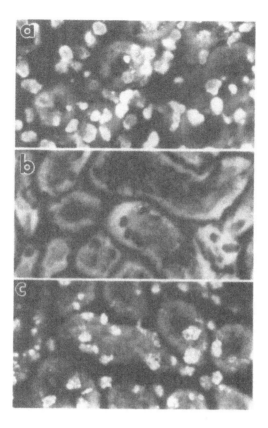

Fig. 3. Detection of antihistone antibodies by the histone reconstitution of immunofluorescence assay. Thin sections of mouse kidney fixed for 5 minutes in 3:1 acetone:methanol at $-20\,^{\circ}C$ were incubated with diluted serum from a patient with PIL and detection of bound IgG was performed with fluorescein conjugated goat anti-human IgG. a. untreated substrate; b. substrate washed with 0.4 N H_2SO_4 for 10 minutes, rinsed with PBS; c. substrate washed with 0.4 N H_2SO_4 for 10 minutes, rinsed with PBS, and incubated with total histones in PBS [13].

mouse kidney cells washed with 1.5 M NaCl (Figs. 4H–I). The variable large speckled pattern is caused by antibody binding to centromeric heterochromatin [15], and these antibodies bind to DNA-free H3 and H4 [14, 15]. In contrast, antibody from a patient with PIL reacting with the (H2A-H2B)-DNA complex displays a strong homogeneous nuclear staining pattern on untreated or 0.5 M NaCl washed HEp-2 cells, but becomes nearly negative on cells washed with 1.5 M NaCl (Figs. 4A–C). Loss of antibody reactivity is due to extraction of histones from the nucleus at 1.5 M NaCl.

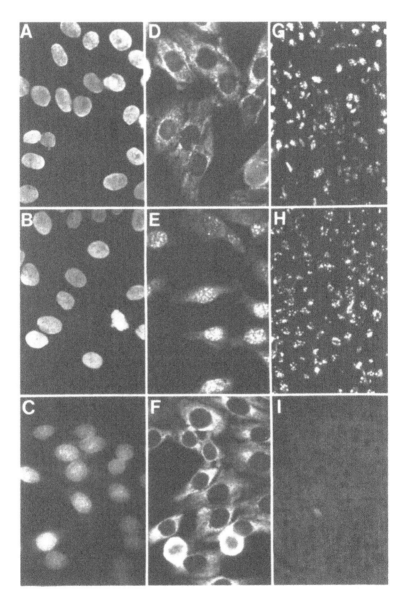

Fig. 4. Detection of antihistone antibodies producing variable large speckled immuno-fluorescence pattern.

Panels A–F. Commercial HEp-2 cells (Bion, Park Ridge, IL).

Panels G–I. Commercial mouse kidney cells (Kallestad, Chaska, MN).

The slides in A, D and G were washed with 0.15 M NaCl; those in B, E and H were washed

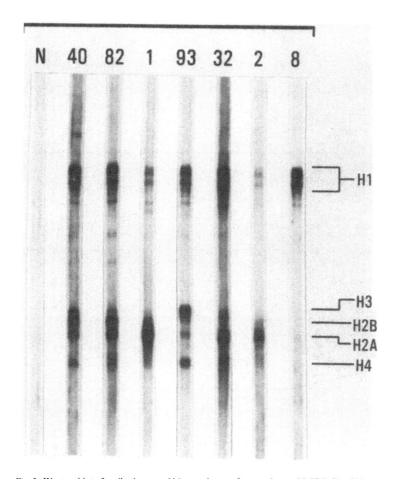

Fig. 5. Western blot of antibody to total histones by sera from patients with SLE. Total histones were separated by SDS PAGE (see legend to Fig. 1 for details) and transferred to nitrocellulose [62]. Antibody binding by SLE sera was revealed with ^{125}I anti-human IgG. Patients are identified with numbers. "N" is a normal serum. On the far right side the positions of the individual histones are marked.

with 0.5 M NaCl; those in C, F and I were washed in 1.5 M NaCl, all at neutral pH. A–C were incubated with a serum from a patient with PIL and detected with fluorescein conjugated goat anti-human IgG. D–I were incubated with a serum from a patient with an undifferentiated rheumatic disease yielding the variable large speckled pattern [14, 15], and bound IgM was detected with fluorescein conjugated goat anti-human IgM.

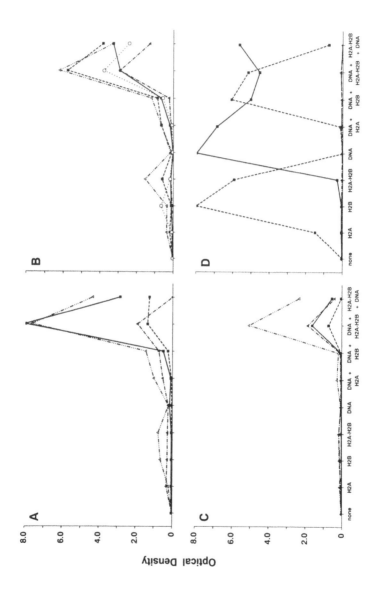

4.2 Counterimmunoelectrophoresis

Counterimmunoelectrophoresis and immunodiffusion have not been used to study anti-histone antibodies. These techniques need both the antigen and antibodies to be in relatively high concentrations. Acid extracted individual histones are not soluble at high concentration in physiologic ionic strength. Native histone complexes are soluble, but at elevated concentrations they bind nonspecifically to human immunoglobulins (unpublished observation), presumably because of the high positive charge of the histones. Whole chromatin is insoluble at physiologic ionic strength. However, it is possible that special preparations such as nucleosomes and H1-stripped chromatin could be used as antigens in immunodiffusion.

4.3 Immunoblotting

Anti-histone antibodies can also be detected by Western blot [16–19]. Sera from some patients with SLE bind strongly to all histones (Fig. 5, patients 40, 82, 93 and 32), while other sera bind to only one (Fig. 5, patient 8) or a few histones (Fig. 5, patients 1 and 2). Some sera from SLE patients are negative on these substrates (not shown). The advantages of the Western blot technique are that total acid extracted histones are commercially available and that antibodies to each of the 5 histones can be measured simultaneously. The disadvantage is that only reactivities to denatured individual histones, not native histone-histone or histone-DNA complexes, can be measured (but see [17] for a possible exception).

4.4 Immunoprecipitation

If proper care were taken to keep the antigen soluble, it should be possible to study anti-histone antibodies by Staphlococcus protein A-mediated im-

---

Fig. 6. ELISA of IgG autoantibodies to various forms of H2A, H2B and DNA. The substrates for the ELISA are listed under the lower panels. H2A and H2B were prepared by acid extraction of chromatin [4]. H2A-H2B is the native dimer. DNA + H2A, DNA + H2B, and DNA + (H2A-H2B) were complexes formed directly on the ELISA plate with DNA bound first and the indicated protein added afterward. (H2A-H2B) + DNA was prepared by binding the native dimer to the ELISA plate first. Each continuous line connects the points corresponding to the antibody activity in one serum.
Panel A. IgG binding from patients with lupus induced by isoniazid ($\triangle$–····–$\triangle$), penicillamine (■————■), methyldopa (▲–·–▲), acebutolol (■–––■).
Panel B. IgG binding from five patients with PIL.
Panel C. IgG binding from four patients with quinidine-induced lupus.
Panel D. Various control sera from patients with anti-native DNA (■————■), anti-H2B ■–––■ and a representative normal (▲–·–▲). All the sera from patients with drug-induced lupus show predominant reactivity with the (H2A-H2B)-DNA complex (from [50]).

munoprecipitation. However, to our knowledge this technique has not been used to study anti-histone antibodies.

4.5 ELISA

Individual histones, native histone-histone complexes, histone-DNA subnucleosome complexes, and various forms of chromatin have been employed as antigens for enzyme-linked immunosorbent assays (ELISA) [20]. These are quantitative assays that allow a detailed mapping of both linear epitopes and those requiring quaternary structure (Fig. 6). The disadvantage to this approach is that the substrates are not presently commercially available.

5. Cellular localization

The histones organize eukaryotic chromosomal DNA, and as such are restricted to the nucleus (Figs. 7a and b). Mitochondrial, chloroplast and kinetoplast DNA are not complexed with histones. An exception to nuclear localization occurs when histones are translated from mRNA in the cytoplasm. Additionally, there is a large reservoir of histone mRNA and protein in the oocytes of animals with very rapid early embryogenesis such as *Drosophila melanogaster* and *Xenopus laevis* [21]. However, this storehouse of histones is not found in mammalian eggs. As expected, the histones are an integral part of metaphase chromosomes in mitotic cells (Fig. 7c).

6. Cellular function

Histones are responsible for the first two levels of compaction of chromosomal DNA and are also involved in the regulation of transcription. By wrapping around a histone octamer, the effective length of DNA is reduced about 8-fold, from 496 Å (146 basepairs × 3.4 Å per basepair) to 60 Å, the width of the nucleosome core particle [22]. A series of core particles is the first level of compaction of chromosomal DNA (Fig. 8, right). These "beads on a string" are coiled into a 300 Å diameter solenoid by interaction with H1 [23, 24], yielding the second order of compaction (Fig. 8, middle). Except for regions controlling the initiation of transcription and certain sites of very active transcription such as the genes encoding ribosomal RNA, it is thought that all the DNA in eukaryotic chromosomes is organized as a series of nucleosomes [24]. Thus a chromosome containing 20,000,000 basepairs of DNA is organized into 100,000 nucleosomes in the form of a solenoid. The chromosomal DNA undergoes many further levels of compaction in order to become a metaphase chromosome (Figs. 7c and 8, left).

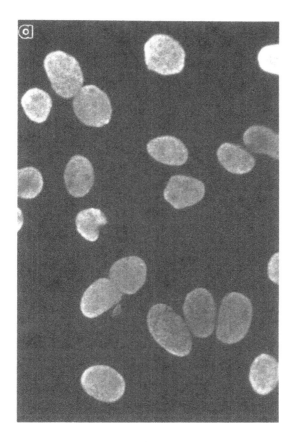

Fig. 7. Immunofluorescence of cells and mitotic chromosomes incubated with anti-(H2A-H2B)-DNA autoantibodies.

a. HEp-2 cells (Bion) incubated with diluted serum from a patient with PIL and bound antibodies detected with goat anti-human IgG. Note that the nuclei but not the mitochondria are immunofluorescent.

At one time it was thought that histones were non-specific inhibitors of transcription. It is now believed that their effects are more subtle. *In vitro*, RNA polymerase II can transcribe DNA bound with histone octamers, but not octamers plus H1 [25]. However, histone octamers bound to the promoter region strongly inhibit transcription [26]. *In vivo*, the nuclease hypersensitive sites around active genes are probably nucleosome-free regions of DNA, bound by sequence-specific DNA binding proteins. RNA polymerase can bind to chromosomal DNA at these sites in preparation for the start of transcription [24]. Thus the competition between histone

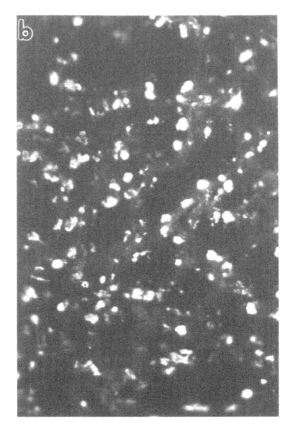

Fig. 7b. Mouse kidney cells (Kallestad) processed in the same way as above.

Table 2. Trypsin resistant domains within the histones[a]

Designation	Residue #'s of parent molecule (total #)
P1	27–129 of H3 (135)
P2′	12–118 of H2A (129)
P2	21–125 of H2B (125)
P3	24–125 of H2B
P4	18–102 of H4 (102)
P5	20–102 of H4
P0 (G–H1)	36–121 of H1 (212–222)

[a] Histones in the form of chromatin are digested with trypsin. Amino acid residue numbers refer to calf thymus histones. The trypsin-resistant domains are arranged in order of increasing electrophoretic mobility. Data derived from Bohm and Crane-Robinson [59].

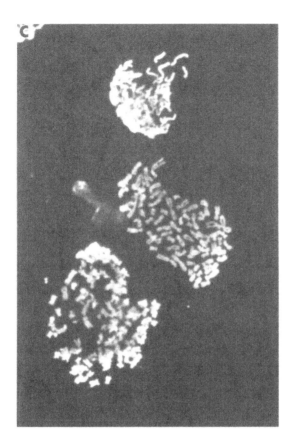

Fig. 7c. Hela cells in mitosis were obtained by selective detachment from the tissue culture container and metaphase chromosomes were prepared [63]. The mitotic spread was incubated as above. The chromosomes are evenly stained, demonstrating that histones are located throughout the mitotic chromosome.

octamers and activator proteins for binding to these DNA sequences influences the transcriptional activity of specific genes.

In addition, the N-terminal regions of histones that are post-translationally modified [1] are involved in chromatin transcription and compaction. Trypsinized H1-stripped chromatin (in which exposed mobile N-terminal and C-terminal peptides are removed, Table 2) cannot be re-condensed by reconstitution with H1 [27]. In an elegant series of *in vivo* studies, it was demonstrated that the 4 lysines subject to reversible acetylation in the N-terminus of yeast H4 were essential for growth [28], and that the N-terminus of H4 in yeast is essential for promoter activation [29].

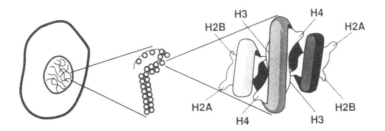

Fig. 8. Expanded view of the cellular localization and higher order structure of histones. On the right is a diagrammatic representation of the nucleosome core particle in which 1 3/4 turns of DNA wrap around an octamer of histones consisting of two H2A-H2B dimers flanking an (H3-H4)$_2$ tetramer [9, 22]. Separation of dimers from the tetramer is exaggerated for clarity, since the native core particle is a compact structure stabilized by noncovalent bonds [6] (see Figs. 2 and 13). Association with H1 completes the nucleosome, and an array of polynucleosomes coils into a solenoid-like structure (middle diagram). Supercoiling into higher order structures takes place so the DNA stays untangled in the nucleus (left side). The tightest compaction occurs during mitosis when chromatin fibers eventually are visible in the light microscope as mitotic chromosomes.

7. Presence in other species

Histones are found in all eukaryotes but not in any prokaryotes. However, the protista such as yeast do not have H1. The amino acid sequences of the 4 core histones are highly conserved throughout nature. Histone H4 is one of the most conserved of all proteins with only one difference known among mammals and only 2 conservative substitutions between pea and cow [30, 31]. It is more difficult to compare the other histone sequences in detail because most species have a number of minor histone variants in addition to the major histones synthesized in S-phase [30, 31]. The major variants of H3 from calf and pea differ by only 4/135 amino acids [30]. Bovine H3.1 is identical to that from human and mouse, and cow, mouse and chicken contain a variant in which the only change is serine in place of cysteine at position 96 [31]. Both human and chicken contain a variant with the same 5 changes from the H3.1 sequence. There are only 5 amino acid differences between human, bovine, rat and chicken H2A.1. Each species also has a variant called H2A.Z that displays greater than 50 amino acid changes from H2A.1, yet is nearly identical to the H2A.Z from the other species [31]. Although there is very high identity in the C-terminal half of H2B, considerable variation in the N-terminal half of H2B has been observed in the animal kingdom [30, 31]. However, human, mouse and bovine H2B.1 are identical except for 2 amino acid substitutions. The extreme conservation of core histone sequences both in the major and minor variants suggests the importance of all portions of these proteins in the biology of the cell. In summary, calf thymus core histones can generally be considered as valid antigens for the study of human and mouse autoantibodies.

In contrast to the core histones, the H1 class is heterogeneous. A globular internal segment of about 70 residues displays extensive homology among the animal H1 histones sequenced, but the N-terminal 40 residues and the C-terminal half show substantial divergence. There are numerous variants and alternative forms of H1 within an animal and between species [30, 31]. This complexity causes difficulty in separating and identifying H1 variants, and complicates the interpretation of studies involving H1-reactive antibodies.

8. Human cDNA and derived amino acid sequences

A substantial increase in both the rate of transcription and the half-life of the major histone mRNAs during S-phase leads to at least a 60-fold increase in the synthetic rate of the major histone polypeptides in synchrony with the replication of DNA [32, 33]. The amino acid sequences for the human S-phase core histones and one H1 variant are shown in Figure 9. A cDNA sequence for each protein is also displayed, but as noted below, there are many different DNA sequences for each histone gene in mammals.

The minor histone variants make up less than 10% of the histones synthesized during S-phase [33]. However, at other times of the cell cycle the minor histone variants are the main histones synthesized and they become the dominant histones in the chromosome as a cell ages without replication [30]. The sequences of many variant histones are known [30, 31].

9. The genes

9.1 Humans

In mammals, histone genes are moderately repetitive, with between 10 and 40 copies of each histone [21]. The genes are grouped randomly; sometimes genes for all 5 histones are found close together, sometimes just 2 are together [34]. Human histone genes map to at least 3 different chromosomes – 1, 6 and 12 [35]. For mammals, the DNA regions flanking the protein coding regions of the different gene clusters are highly divergent [31, 36]. The DNA sequences of the coding regions are much less variable, presumably because they are constrained to code for the same amino acid sequences [21, 31]. The mRNAs for the S-phase histones are different from all other translated RNAs in that they are not polyadenylated, there are no intervening sequences to be spliced out, and they have a unique 3' end structure [21]. In contrast, the genes for the constitutively expressed histone variants are similar to normal genes. They have intervening sequences, they are probably present at one copy per haploid genome, and the transcribed mRNAs are polyadenylated [37].

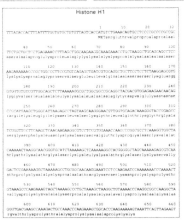

Fig. 9. The *human* cDNA nucleotide sequences (capital letters, above) and amino acid sequences (lower case, below) of the 4 major S-phase histones [38] and an H1 variant [64]. The bases are numbered according to the start of the clone [38], or the start of the coding region [64]. Only a portion of the non-coding regions are shown. For histone H4, the glutamine in capital letters below bases 204–206 replaces a histidine in the original paper. No histidine is known in this position from other H4 sequences, so it is either a rare variant or a sequencing error. ALA is placed below residues 237–239 (which could not be sequenced) to agree with protein sequencing data. For H2A the serine deduced from the nucleotide sequence at position 378–380 is an alanine based on protein sequence data. Glycine-asparagine at positions 369–374 were also derived from

9.2 Other species

There are only 2 copies of each core histone gene in yeast. In sea urchins and fruit flies, there are 100–300 of each histone gene per haploid genome, and the genes are arranged in a strikingly regular tandemly repeated motif in which both the flanking and coding DNA sequences are virtually identical for each repeat. However, the motifs and sequences are different in the two species [21]. Different sets of histone genes are expressed early and late in sea urchin embryogenesis, but this is not the case in mammals. However, see reference [38] for a possible exception. The histone gene family is more diverse in structure among different species than any other multi-gene family thus far discovered [21].

10. Relation with nucleic acids and other proteins

The structure of the histone octamer is known to atomic resolution by x-ray crystallographic studies [9]. It is a wedge-shaped disk approximately 70 Å in diameter and 60 Å at its widest point (Fig. 2). The x-ray structure of the nucleosome core particle is known to low resolution [22]. The DNA (20 Å in diameter) wraps along the outside of the histone octamer, forming a disk 110 Å in diameter and 60 Å long (see Figs. 8 and 13). Thus only 7–10 Å separates each turn of DNA in the nucleosome core particle. The N-terminal tails of the histones are highly positively charged yet they are mobile and do not bind to DNA within a nucleosome [39], but rather to other nucleosomes in the solenoid [27]. The structure of H1 is not known to atomic resolution, but it has flexible N- and C-termini, with a central globular core [40]. Histone H1 is thought to bind to the linker DNA at the points where the DNA enters and exits the nucleosome.

11. B-cell epitopes

11.1 Individual histones and epitopes on peptides

Anti-histone antibodies from patients with a number of diseases have been characterized extensively on individual histones [10, 11]. Studies often differ concerning the most reactive histone or most reactive set of histones in any particular disease. This is due in part to the substantial patient-to-patient variability, especially in SLE (Figs. 5 and 10). However, there is general agreement that in SLE and PIL [16, 18, 19, 41] the major linear epitopes on

---

the protein sequence. After post-translational processing, none of the mature histone polypeptides begins with methionine; the first residue after the methionine is the amino terminus.

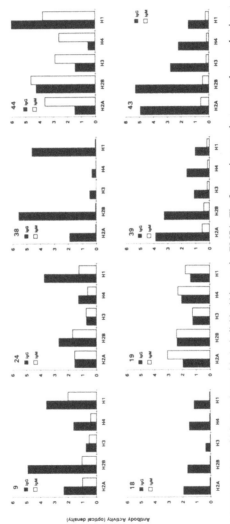

Fig. 10. IgG and IgM reactivity of 8 SLE patients with the individual histones by ELISA. The four patients on the top row show predominant reactivity with H2B and H1, but the four patients on the bottom row show no consistent pattern. In most patients IgG reactivity (solid bars) is greater than IgM (open bars) [65].

the core histones are found in their trypsin sensitive regions (Fig. 11). This is also true in murine lupus [42], directly demonstrated with the N-terminus of H2B [43]. The mobile N-terminal amino acids of all histones as well as 6 to 10 C-terminal residues of H3 and H2A are removed by trypsin digestion of chromatin (Table 2), facilitating the immunologic analysis of these peptides (Fig. 11). In hydralazine-induced lupus (HIL), there are prominent epitopes on the trypsin resistant histone peptides, particularly of H3 and H4 [17, 19]. Since the trypsin sensitive regions of histones are exposed in chromatin, these results are consistent with chromatin being the autoimmunogen inducing anti-histone antibodies in SLE and PIL, but not HIL.

11.2 Epitopes exposed in chromatin

The homogeneous nuclear staining by indirect immunofluorescence of SLE and PIL sera also indicate that at least some epitopes recognized by these autoantibodies are exposed in the nucleus (Figs. 3a and 7a). In addition, immunofluorescence can be removed by absorption with chromatin (Fig. 12), indicating that these epitopes are also exposed in native chromatin isolated from the cell [44].

11.3 Native structures and epitopes requiring quaternary interactions

Antibodies to linear epitopes on histones represent only a portion of the histone-reactive antibodies. Recently, quaternary histone-histone and histone-DNA complexes have been used as antigens. Antibodies from patients with SLE [48, 49], PIL [12], and lupus induced by 4 other drugs [50], as well as antibodies from murine lupus [51, 52], all bound prominently to a structural epitope on the (H2A-H2B)-DNA complex (Fig. 13). Some patients with SLE also bound native DNA and the $(H3-H4)_2$-DNA subnucleosome. In murine lupus, antibodies to (H2A-H2B)-DNA were found early in disease, before antibodies to native DNA and $(H3-H4)_2$-DNA arose [51]. Adsorption with chromatin removed most of the antibody reactivity to both the (H2A-H2B)-DNA and $(H3-H4)_2$-DNA subnucleosome structures, indicating that regions of subnucleosome particles buried in chromatin were not antigenic in SLE, PIL or murine lupus. Antibody reactivity to native DNA was also removed by adsorption with chromatin so that anti-DNA antibodies can also be viewed as a subset of the wide variety of anti-chromatin autoantibodies found in human and murine SLE [49, 51]. These autoantibody specificities could be explained by autoimmunization with chromatin accompanied by sequential loss of tolerance first to the (H2A-H2B)-DNA region and then to $(H3-H4)_2$-DNA and native DNA.

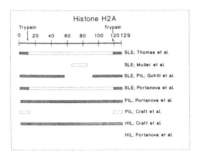

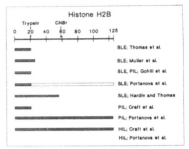

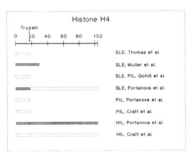

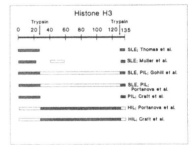

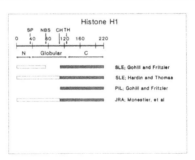

Fig. 11. Linear B cell epitopes on individual histones for human autoantibodies. The number of amino acids in the protein is depicted above the top solid black bar, and arrows mark the sites of the limit trypsin digest of each of the 4 core histones in chromatin (also see Table 2), plus a cyanogen bromide (CNBr) cleavage site of H2B. For histone H1, arrows mark the site of cutting by submaxillary protease (SP), N-bromosuccinimide (NBS), chymotrypsin (CH) and thrombin (TH) as described [45]. There is an additional line for H1 depicting its N-terminal region, globular domain, and C terminal tail [40]. In each row underneath, an open rectangle indicates the region of the protein containing an uncommon epitope, a crosshatched rectangle indicates the region of a common epitope, and the absence of a rectangle indicates no or very rare reactivity to that region of the protein. On the right of each line, the disease that was analyzed is listed first, followed by the reference. The abbreviations for the diseases are: SLE, systemic lupus erythematosus; PIL, procainamide-induced lupus; HIL, hydralazine-induced lupus; JRA, juvenile rheumatoid arthritis. The references are: Thomas *et al.* [16]; Muller *et al.* [41]; Gohill *et al.* [18]; Portanova *et al.* [17]; Craft *et al.* [19]; Gohill and Fritzler [45]; Hardin and Thomas [46]; Monestier *et. al.* [47].

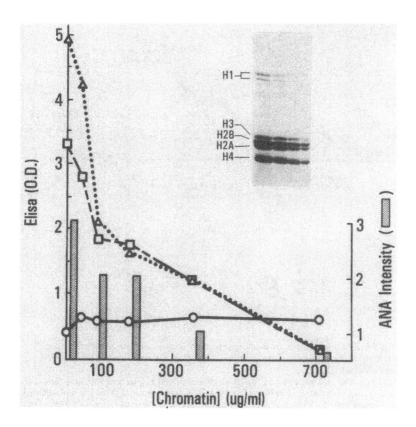

Fig. 12. Adsorption of immunofluorescent activity of antihistone antibodies in procainamide-induced lupus by chromatin. Incubation with increasing amounts of chromatin stoichiometrically removed the antinuclear staining properties of these antihistone antibodies, in parallel with the loss of the antihistone (■ --- ■) and anti-chromatin (△···△) but not anti-tetanus (○———○) activities as measured by ELISA. The insert shows SDS PAGE of serial dilutions of the chromatin immunoadsorbent.

11.4 *Epitopes not exposed in chromatin*

In contrast to PIL and SLE, IgM antibodies from patients taking hydralazine or chlorpromazine [12] had the reciprocal reactivity (Fig. 13), as did IgG from patients with juvenile rheumatoid arthritis [53]. These antibodies bound strongest to DNA-free histones, less to subnucleosome structures, and very little or not at all with chromatin. Adsorption with chromatin could not remove reactivity to subnucleosome structures, consistent with the direct binding studies, and indicating that regions exposed in chromatin were not

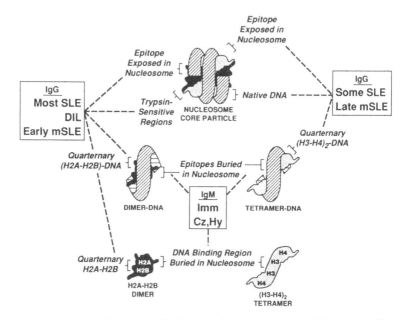

Fig. 13. Relationship between B-cell epitopes and nucleosome core particle structure. The drawing of the nucleosome core particle is based on the structures determined by x-ray crystallography [9, 22].

Left side. The epitopes common to almost all human and murine sera with anti-chromatin antibodies are bracketed: i.e. regions of the H2A-H2B dimer exposed in chromatin and requiring quaternary protein-DNA interactions; the histone tails sensitive to digestion with trypsin; and regions of the dimer requiring quaternary protein-protein interactions [12, 44, 49–52].

Right side. The epitopes recognized by the subset of human and murine sera that also bind to native DNA are bracketed: i.e. native DNA itself and the part of the (H3-H4)$_2$-DNA complex exposed in chromatin [49, 51]. Note that the DNA-free (H3-H4)$_2$ tetramer is not bracketed, since it is not normally antigenic in these sera.

Center. In the middle of the picture the regions of the nucleosome that are not antigenic in lupus sera are bracketed: i.e. the regions of (H2A-H2B)-DNA and (H3-H4)$_2$-DNA that are buried in chromatin by protein-protein interactions; and the regions of H2A-H2B and (H3-H4)$_2$ covered by DNA in chromatin. As explained in the text these regions are recognized by IgM antibodies from patients taking hydralazine or chlorpromazine [12], IgG in JRA sera [53], and antibodies induced by immunization of normal mice [54].

Abbreviations: DIL: drug-induced lupus, specifically antibodies induced by procainamide, quinidine, acebutalol, isoniazid or penicillamine [12, 50]; mSLE: murine lupus; Imm: immunization; Cz: chlorpromazine; Hy: hydralazine.

antigenic in patients taking these two drugs. This is the same pattern of antibody binding found when normal mice were immunized with histones [54].

From these studies it is clear that a full understanding of histone antigenicity and the origin of anti-histone autoantibodies encompasses not only the biochemistry of histone primary structure but also of the histone-histone and histone-DNA interactions that determine the natural organization of histones in the cell.

12. T-cell epitopes

T-cell epitopes for histones have not as yet been defined. There is evidence that T-cell lines and T-cell hybridomas can help anti-DNA production *in vitro* when antigen presenting cells are exposed to chromatin, but not DNA [55]. However, activated B cells from autoimmune mice required macrophage but not T cells to secrete anti-chromatin autoantibodies *in vitro*, in contrast to the secretion of anti-Sm which did require T cells [56]. Peripheral B cells from humans with SLE also did not require the co-presence of T cells to secrete anti-histone antibodies *in vitro* [57].

References

1. McGhee JD & Felsenfeld G (1980) Ann Rev Biochem 49:1115–1156
2. Lutter LC (1978) J Mol Biol 124:391–420
3. Nelson DA, Mencke AJ, Chambers SA, Oosterhof DK & Rill RL (1982) Biochem 21:4350–4362
4. Bohm EL, Strickland WN, Strickland M, Theraits BH, Von der Westhuizen DR & Von Holt C (1973) FEBS Lett 34:217–221
5. Beaudette NV, Fulmer AW, Okabayashi H & Fasman GD (1981) Biochem 20:6526–6535
6. Eickbush TH & Moudrianakis EN (1978) Biochem 17:4955–4964
7. Burlingame RW, Love WE, Eickbush TH & Moudrianakis EN (1986) In: Sarma RH & Sarma MH (Eds) Biomolecular Stereodynamics III. Proceedings of the Fourth Conversation in the Discipline Biomolecular Stereodynamics (pp 11–33). Adenine Press, Guilderland
8. Oohara I & Wada A (1987) J Mol Biol 196:389–397
9. Arents G, Burlingame RW, Wang B-C, Love WE & Moudrianakis EN (1991) Proc Natl Acad Sci USA 88:10148–10152
10. Rubin RL (1992) In: Lahita RG (Ed) Systemic Lupus Erythematosus, 2nd ed (pp 247–271). John Wiley, New York
11. Monestier M & Kotzin BL (1992) Rheum Dis Clin N Amer 18:415–436
12. Burlingame RW & Rubin RL (1991) J Clin Invest 88:680–690
13. Fritzler MJ & Tan EM (1978) J Clin Invest 62:560–567
14. Molden DP, Klipple GL, Peebles CL, Rubin RL, Nakamura RM & Tan EM (1986) Arthritis Rheum 29:39–43
15. Burlingame RW & Rubin RL (1992) Hum Antibod Hybridomas 3:40–47
16. Thomas JO, Wilson CM & Hardin JA (1984) FEBS Lett 169:90–96
17. Portanova JP, Arndt RE, Tan EM & Kotzin BL (1987) J Immunol 138:446–451
18. Gohill J, Cary PD, Couppez M & Fritzler MJ (1985) J Immunol 135:3116–3121
19. Craft JE, Radding JA, Harding MW, Bernstein RM & Hardin JA (1987) Arthritis Rheum 30:689–694
20. Burlingame RW & Rubin RL (1990) J Immunol Methods 134:187–199
21. Maxson R, Cohn R, Kedes L & Mohun T (1983) Ann Rev Genet 17:239–277
22. Richmond TJ, Finch JT, Rushton B, Rhodes D & Klug A (1984) Nature 311:532–537
23. Langmore JR & Paulson JR (1983) J Cell Biol 96:1120–1131
24. Pederson DS, Thoma F & Simpson RT (1986) Annu Rev Cell Biol 2:117–147
25. Laybourn PJ & Kadonaga JT (1991) Science 254:238–245
26. Lorch Y, LaPointe JW & Kornberg RD (1987) Cell 49:203–210
27. Allan J, Harborne N, Rau D & Gould H (1982) J Cell Biol 93:285–297
28. Megee PC, Morgan BA, Mittman BA & Smith MM (1990) Science 247:841–845
29. Durrin LK, Mann RK, Kayne PS & Grunstein M (1991) Cell 65:1023–1031
30. Wu RS, Panusz HT, Hatch CL & Bonner WM (1986) CRC Crit Rev Biochem 20:201–261
31. Wells D & McBride C (1989) Nuc Acids Res 17S:r311–r350
32. Stein GS, Stein JL, Lian JB, Van Wijnen AJ, Wright KL & Pauli U (1989) Cell Biophys 15:201–223
33. Wu RS & Bonner WM (1981) Cell 27:321–330
34. Sierra F, Lichtler A, Marashi F *et al.* (1982) Proc Natl Acad Sci USA 79:1795–1799
35. Tripputi P, Emanuel BS, Croce CM, Green LG, Stein GS & Stein JL (1986) Proc Natl Acad Sci USA 83:3185–3188
36. Liu T-J, Liu L & Marzluff WF (1987) Nuc Acids Res 15:3023–3039
37. Wells D & Kedes L (1985) Proc Natl Acad Sci USA 82:2834–2838
38. Zhong R, Roeder RG & Heintz N (1983) Nuc Acids Res 21:7409–7425
39. Smith RM & Rill RL (1989) J Biol Chem 264:10574–10581
40. Allan J, Hartman PG, Crane-Robinson C & Aviles FX (1980) Nature 288:675–679

41. Muller S, Bonnier D, Thiry M & Van Regenmortel MHV (1989) Int Arch Allergy Appl Immunol 89:288–296
42. Portanova JP, Arndt RE & Kotzin BL (1988) J Immunol 140:755–760
43. Portanova JP, Cheronis JC, Blodgett JK & Kotzin BL (1990) J Immunol 144:4633–4640
44. Rubin RL, Reimer G, McNally EM, Nusinow SR, Searles RP & Tan EM (1986) Clin Exp Immunol 63:58–67
45. Gohill J & Fritzler MJ (1987) Mol Immunol 24:275–285
46. Hardin JA & Thomas JO (1983) Proc Natl Acad Sci USA 80:7410–7414
47. Monestier M, Losman JA, Fasy TM *et al.* (1990) Arthritis Rheu 33:1836–1841
48. Burlingame RW, Tan EM & Rubin RL (1990) Arthritis Rheum 33 (Supplement):S100
49. Burlingame RW, Boey ML, Starkebaum, G & Rubin RL (1994) J Clin Invest 94, in press
50. Rubin RL, Bell SA & Burlingame RW (1992) J Clin Invest 90:165–173
51. Burlingame RW, Rubin RL, Balderas RS & Theofilopoulos AN (1993) J Clin Invest 91:1687–1696
52. Losman MJ, Fasy TM, Novick KE & Monestier M (1992) J Immunol 148:1561–1569
53. Burlingame RW, Rubin RL & Rosenberg AM (1993) Arthritis Rheum 36:836–841
54. Rubin RL, Tang F-L, Tsay G & Pollard KM (1990) Clin Immunol Immunopathol 54:320–332
55. Mohan C, Adams S, Stanik V & Datta SK (1993) J Exp Med 177:1367–1381
56. Fisher CL, Shores EW, Eisenberg RA & Cohen PL (1989) Clin Immunol Immunopathol 50:231–240
57. O'Dell JR, Bizar-Schneebaum A & Kotzin BL (1988) Clin Immunol Immunopathol 47: 343–353
58. Bradbury EM (1975) In: The Structure and Function of Chromatin, Ciba Foundation Symposium 28 (pp 1–4). Associated Scientific Publ, Amsterdam, The Netherlands
59. Bohm L & Crane-Robinson C (1984) Biosci Rep 4:365–386
60. Panyim S & Chalkley R (1969) Arch Biochem Biophys 130: 337–346
61. Laemmli UK (1970) Nature (London) 227: 680–685
62. Towbin H, Staehelin T & Gordon J (1979) Proc Natl Acad Sci USA 76:4350–4354
63. Stenman S, Rosenqvist M & Ringertz NR (1975) Exptl Cell Res 90:87–94
64. Eick S, Nicolai N, Mumberg D & Doenecke D (1989) Eur J Cell Biol 49:110–115
65. Suzuki T, Burlingame RW, Casiano CA, Boey ML & Rubin RL (1994) J Rheumatol 21:1081–1091

Manual of Biological Markers of Disease **B2.3**: 1–11, 1994.
© 1994 *Kluwer Academic Publishers. Printed in the Netherlands.*

Ubiquitin

SYLVIANE MULLER

Institut de Biologie Moléculaire et Cellulaire, UPR 9021 – CNRS, 15 rue Descartes, 67084 Strasbourg Cedex, France

Abbreviations

CEP: carboxyl extention protein; ELISA: enzyme-linked immunosorbent assay; U-H2A: ubiquitinated histone H2A; Ub: ubiquitin

Key words: autoantigen, autoantibody, ubiquitin, histone

1. The antigen

Ubiquitin is a small stress-protein. It consists of a single 8.5 kD polypeptide chain of 76 amino acid residues [1–3]. In the nucleus, ubiquitin can be found covalently joined via an isopeptide linkage to approximately 10% of the nucleosomal histone H2A [4]. Ubiquitinated form of H2A (U-H2A) represents the antigenic target recognized by a distinct subset of autoantibodies [5].

2. Synonyms

Ubiquitin was discovered in 1975 by Goldstein *et al.* [6]. Early nomenclature was, for example, UBIP, APFI, HMG-20 and non-histone protein S [4]. The term "ubiquitin" has been definitely adopted.

3. Ubiquitin antibody detection

3.1 Immunofluorescence

(Manual, part I, Section A2): no application known with autoantibodies.

3.2 Counterimmunoelectrophoresis

(Manual, part I, Section A3): no application known with autoantibodies.

3.3 Immunoblotting

(Manual, part I, Section A4): Autoantibodies can be detected by Western blot with commercially available ubiquitin [5, 7]. Because of the low molecular weight of ubiquitin, the use of SDS / 12–20% polyacrylamide gradient gel and Immobilon affinity hydrophobic membranes (Millipore) is recommended. No application known in Western blot for autoantibodies to U-H2A except their fortuitous detection found by using a commercial preparation of purified ubiquitin contaminated by complexes of ubiquitinated H2A [5].

3.4 Protein immunoprecipitation

(Manual, part I, Section A6): no application known with autoantibodies.

3.5 ELISA

(Manual, part I, Section A5): Autoantibodies to ubiquitin can be detected by ELISA using commercial ubiquitin and the synthetic peptide 22-45 of ubiquitin [7]. Autoantibodies reacting specifically with U-H2A can be studied using a branched octapeptide of U-H2A containing residues of both ubiquitin and the histone [5].

4. Cellular localisation

Ubiquitin is an abundant protein (2 to 100 $\mu g/g$ tissue) and is found in apparently all eukaryotic cells. It has been isolated from a wide variety of mammalian tissues including reticulocytes, erythrocytes, thymus, liver, testis, pituitary, parathyroid and hypothalamus. It is found throughout the cell either as free protein or as a covalent adduct to a wide variety of proteins in the nucleus, cytosol, cytoskeleton and plasmalemma. By using affinity-purified anti-ubiquitin antibodies and colloidal gold immunoelectron microscopy, unconjugated ubiquitin has been found within the cytoplasm, nucleus, the microvilli, autophagic vacuoles and lysosomes of hepatoma cells [8]. The distribution of ubiquitin conjugates has also been studied; for example, ubiquitinated conjugates were localized in Z-bands of normal and atrophic skeletal muscles as well as in Z-bands and intercalated discs of cultured rat cardiomyocytes [9]. Importantly, ubiquitin has been found in the paired helical filaments which are the principal constituents of the neurofibrillary tangles found in patients with Alzheimer's disease [10, 11].

5. Cellular functions

Ubiquitin modification plays an important role in many cellular processes [12]. Among these are intracellular proteolysis (Fig. 1), regulation of gene expression, involvement in the cellular stress response, modification of cell surface receptor and biogenesis of mitochondria and ribosomes [13–15]. One

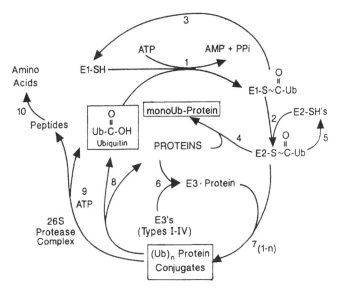

Fig. 1. Functions of ubiquitin.
One important role of ubiquitination is to participate in the selective, rapid degradation of abnormal proteins and certain short-lived normal proteins. The proposed sequence of events in conjugation and degradation of proteins by the ubiquitin system is as follows [13]:
1. Activation of ubiquitin (Ub) by E1 (*ubiquitin-activating enzyme*)
2. Transfer of high-energy ubiquitin intermediate to E2 (*ubiquitin carrier protein*)
3. Recycling of E1
4. Conjugation of protein to ubiquitin by E2
5. Recycling of E2
6. Formation of E3 (*ubiquitin-protein ligase*)-protein complex
7. Conjugation of multiple molecules of ubiquitin to the protein substrate
8. Recycling of an intact sustrate and ubiquitin; mediated by isopeptidase(s)
9. ATP-dependent degradation of conjugates into peptides by the 26S protease complex containing a 20S degradative particle called the proteosome
10. Release of amino acids from peptides

of the genes responsible for programmed-cell death has been shown to be a polyubiquitin gene [16]. Thus, ubiquitin as a polypeptide-chain binding protein has been said to act as a "localized detergent", "foldase" or "chaperonine" [17].

6. Presence in other species

The presence of ubiquitin seems to be restricted to eukaryotes, since no related polypeptides have been detected in either eubacteria or archaebacteria. Ubiquitin is present in all eukaryotic cells, including various mammalian tissues, trout testis, Mediterranean fruit fly eggs and, also, in plants, yeast and a number of parasites [18].

7. The cDNA and amino acid sequence

Ubiquitin is the most highly conserved protein yet identified with only four amino acid changes (in residues 19, 24, 28 and 57) between the yeast, oat and human ubiquitin primary structures. The sequence of ubiquitin from various sources is known either by sequencing the protein or from the cDNA nucleotide sequence [19, 20] (Fig. 2).

→

Fig. 2. Primary sequence of ubiquitin.
Nucleotide sequence of the coding region of the human ubiquitin gene (A) [20] and amino acid sequences of ubiquitin from various species compared with the human ubiquitin sequence (B). Only those residues which differ from the human protein are shown. The branchpoint lysine for multi-ubiquitination is indicated (Lys48). Most of the amino acid substitutions are conservative. A few non-conservative mutations are however observed at positions 14, 19, 22, 28 and 57. The amino acid sequence of ubiquitin from a parasite *Entamoeba histolytica* which causes amebiasis in man has been deduced from a cDNA clone [21]. This sequence deviates at six positions from the consensus of all other known ubiquitin ranging from *Trypanosoma cruzi* to *homo sapiens*. In the structure of ubiquitin known by X-ray diffraction at 1.8 Å resolution, most of the amino acid changes are clustered in two small patches on one surface of the molecule [21]. Out of the C-terminal Gly-Gly sequence involved in the branching with targeted proteins and the Cys48 which can be involved in the conjugation with the C-terminus Gly76 of the neighboring ubiquitin, there is no particular motif or feature in ubiquitin which lacks Cys and Trp residues and contains one residue of each Met (N-terminal), His and Tyr.
[a] Human ubiquitin is identical to ubiquitin of cow, swine, chicken, *X. laevis, D. melanogaster, M. sexta, S. frugiperda* and *C. congregatus.*
[b] *Arabidopsis* ubiquitin is identical to ubiquitin of *H. annus, H. vulgare, G. max, L. polyphyllus* and *C. reinhardii* [14, 21].

A. Nucleotide sequence of the coding region of the human ubiquitin gene

```
          1                                              15
          Met Gln Ile Phe Val Thr Lys Thr Leu Thr Gly Lys Thr Ile Thr Leu Glu Val Glu Pro Ser Asp Thr Ile Glu Asn Val Lys Ala Lys Ile Gln Asp Lys Glu Gly Ile Pro Pro
          1                                                                                              30                                                          114
          ATG CAG ATC TTC GTG ACT AAG ACT CTG ACC GGT AAG ACC ATC ACC CTC GAG GTG GAG CCC AGT GAC ACC ATC GAG AAT GTC AAG GCA AAG ATC CAA GAT AAG GAA GGC AAT CCT CCT

                          45                                             60                                                          75
          Asp Gln Arg Leu Ile Phe Ala Gly Lys Gln Leu Glu Asp Gly Arg Thr Leu Ser Asp Tyr Asn Ile Gln Lys Glu Ser Thr Leu His Leu Val Leu Arg Leu Arg Gly Gly
          115                                                                                                                        228
          GAT CAG AGG TTG ATC TTT GCC GGA AAA CAG CTG GAA GAT GGT CGT ACC CTG TCT GAC TAC AAC ATC CAG AAA GAG TCC ACC TTG CAC CTG GTA CTC CGT CTC AGA GGT GGG
```

B. Ubiquitin Proteins

```
                    1        10        20        30        40      48 50       60        70      76
                    |        |         |         |         |        |-|         |         |       |
H. sapiens[a]       MQIFVKTLTGKTITLEVEPSDTIENVKAKIQDKEGIPPDQQRLIFAGKQLEDGRTLSDYNIQKESTLHLVLRLRGG
C. elegans          ———————————————————————————————————————————————————————————————————————————
T. cruzi            ——————————————A————————————————————————————————A——————————————————————————
A. thaliana[b]      ——————————————A————————————————————————————————A——————————————————————————
A. sativa           ——————————————————S——D———————————————————————A——————————————————————————
S. cerevisiae       ——————————————————S——D——S—————————————————————————————————————————————————
N. crassa           ——————————————————————————Q————————————————————————————————————————————————
D. discoideum       —————————————————————G—N————————————————————————————————————————————————————
T. pyriformis       ——————————————D—A——————————————————————————————————————————————————————————
E. histolytica      ——————————————N—S—DAI————E——————————————————————————————————E—K————————————
```

8. Ubiquitin genes

Ubiquitin is encoded in the eukaryotic genome by a multigene family that is organised in two modes (Fig. 3). The first organization comprises polygenes of lengths varying from 2 to more than 50 ubiquitin genes arranged in head-to-tail spacerless repeats. The final repeat at the 3′ end of the polyubiquitin gene is usually followed by a codon encoding a non-ubiquitin C-terminal amino acid which is not conserved among species. The length of the polyubiquitin genes, the number of loci and the amino acid employed to block the C-terminal glycine vary as a function of species [3, 14]. The human genome contains at least two polyubiquitin genes, a nine repeat and a three repeat.

The second family of ubiquitin genes contains the reading frame of a ribosomal protein fused directly to the 3′ end of a single ubiquitin coding region [22–24]. Two types of C-terminal extension genes have been identified; the first type encodes ubiquitin fused to a carboxyl extension protein (CEP)

Polyubiquitin   genes

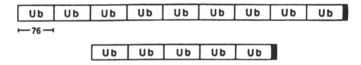

Ubiquitin   Carboxyl   Extension   Protein   genes

Fig. 3. Organization of the ubiquitin gene family in eukaryotic cells.
Two classes of genes exist: polyubiquitin genes and ubiquitin carboxyl extension genes. Polyubiquitin genes encode tandem head-to-tail repeats of ubiquitin whereas extension-encoding genes code for a single ubiquitin fused to CEP sequences encoding a protein of 52 amino acids or 76–80 amino acids. The number of genes encoding polyubiquitin and the number of ubiquitin-encoding units in each gene is highly variable among different organisms. In Drosophila, polyubiquitin genes reside on chromosome 3 at the early-ecdysone and heat shock puff region 63F. In yeast, genes UB11 and UB12 encoding both for the UbCEP52, gene UB13 encoding UbCEP76 and UB14 encoding the pentamer polyubiquitin reside on chromosome 9 (UB11), 11 (UB12) and 12 (UB13, UB14), respectively (from [3]).

of 52 amino acids, a ribosomal protein of the large subunit, whereas the second encodes an extension, depending upon the species, of 76–80 amino acids. The latter protein is a constituent of the small subunit of the ribosome (ribosomal protein S37 of yeast, S27a of rat). To produce functional ubiquitin, the initial translation products of both types of ubiquitin fusion genes are processed to release the ubiquitin C-terminal carboxyl group. Several enzymes responsible for the cleavage of ubiquitin linked to other proteins via α peptide bonds have been identified in yeast, mammals and higher plants.

9. Relation with nucleic acids

Despite having 15% basic residues, ubiquitin binds only very weakly to DNA [25].

10. Relation with other proteins

See Section 5. Figure 4 (and Fig. 1) show the branched regions of U-H2A and U-H2B [26], two stable ubiquitin-protein conjugates. Other known stable conjugates have been characterized with actin, PGDF receptor, the growth hormone-receptor and the lymphocyte homing receptor gp90[MEL-14]. For the latter conjugates, the precise residue of the protein linked to the ubiquitin moiety is not known.

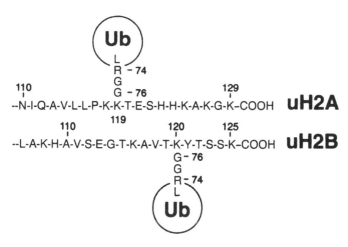

Fig. 4. Branched regions of U-H2A and U-H2B.
Data are taken from [26, 31].

11. Biochemical characteristics

Ubiquitin is extremely stable; it can be boiled without loss of activity. It is also reversibly denatured by urea, guanidin hydrochloride and alcohol. The protein is also quite stable to a variety of proteases but is rapidly inactivated by trypsin and tryptic-like activity which cleave the carboxyl-terminal Gly-Gly motif involved in the binding with protein substrates. Its isoelectric point is neutral (6.7) and the molecule remains folded over a pH range of 1–13 [25]. Ubiquitin is not glycosylated.

12. B-cell epitopes

As expected from its remarkable conservation along evolution, ubiquitin alone in adjuvant is not immunogenic. However, when ubiquitin is conjugated to carrier proteins or by using ubiquitin synthetic peptides, antibodies to ubiquitin can be readily obtained [7, 27]. By using these antibodies induced in rabbits or in mice, a number of continuous B-epitopes of ubiquitin have been identified. They cover residues 6–19 [28], 22–45 [7], 34–48 [29], 50–65 and 64–76 [30] (Fig. 5). Autoantibodies reacting with ubiquitin and synthetic peptide 22–45 of ubiquitin have been found in human and murine lupus [7, 31, 32]. A number of lupus sera react also with a branched octapeptide of U-H2A [5].

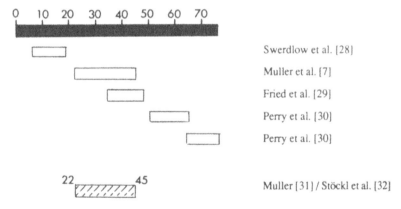

Fig. 5. B-cell epitopes of ubiquitin.
The open blocks delineate epitopes located using anti-peptide antibodies and monoclonal antibodies reacting with ubiquitin synthetic peptides. The hatched block delineates an autoepitope identified with antibodies from both autoimmune patients and lupus mice.

13. Monoclonal antibodies (mab)

Several Mabs have been described:

- #1-2 H11 binds to purified Ub in solid-phase microtiter assay and on immunoblots but does not recognize soluble, free Ub. It recognizes specific Ub-protein conjugates on immunoblots and by immunoprecipitation. The epitope recognized by this MAb is contained in the amino acid sequence of Ub, residues 34–54 [29, 33, 34].
- #5-2 E6, 2-3 D7, 3-3 G6, 4-3 H8, 5-2 F3, 7-2 E7 and 4-2 D8 recognize epitopes located between residues 34 and 53 of Ub [29, 33, 34].
- #5-25 and 3-39 reconize epitopes in the regions 5–25 and 50–65 of Ub respectively [35]. These MAbs have been obtained from mice immunised against paired helical filaments (PHF) from the brain of patients with Alzheimer's disease. They react more strongly with PHF than with free Ub.
- #DF2 (IgM) from a mouse immunised with PHF recognises an epitope which has been proposed by deduction to reside in residues 40–44 of Ub [10].

14. T-cell epitopes

Not known.

References

1. Rechsteiner M (1988) In: Rechsteiner M (ed) Ubiquitin 346 pp. Plenum Press, New York
2. Schlesinger M & Hershko A (1988) The Ubiquitin System. Current Communications in Molecular Biology, Cold Spring Harbor Laboratory Press, New York
3. Monia BP, Ecker DJ & Crooke ST (1990) Biotech 8:209–215
4. Busch H (1984) Meth Enzymol 106:239–262
5. Plaué S, Muller S & Van Regenmortel MHV (1989) J Exp Med 169:1607–1616
6. Goldstein G, Scheid M, Hammerling U, Boyse EA, Schlesinger DH & Niall HD (1975) Proc Natl Acad Sci USA 72:11–15
7. Muller S, Briand JP & Van Regenmortel MHV (1988) Proc Natl Acad Sci USA 85:8176–8180
8. Schwartz AL, Ciechanover A, Brandt RA & Geuze HJ (1988) EMBO J 7:2961–2966
9. Hilenski LL, Terracio L, Haas AL & Borg TK (1992) J Histochem Cytochem 40:1037–1042
10. Mori H, Kondo J & Ihara Y (1987) Science 235:1641–1644
11. Lowe J, Mayer RJ & Landon M (1993) Brain Pathol 3:55–65
12. Jentsch S (1992) Annu Rev Genet 26:179–207
13. Ciechanover A (1993) Brain Pathol 3:67–75
14. Jentsch S, Seufert W, Sommer T & Reins H-A (1990) Trends Biochim Sci 15:195–198
15. Goldberg AL & Rock KL (1992) Nature 357:375–379
16. Schwartz LM (1991) BioEssays 13:389–395
17. Rothman JE (1989) Cell 59:591–601
18. Wilkinson KD (1988) In: Rechsteiner M (ed) Ubiquitin (pp 5–38). Plenum Press, New York
19. Jentsch S, Seufert W & Hauser H-P (1991) Biochim Biophys Acta 1089:127–139
20. Wiborg O, Pedersen MS, Wind A, Berglund LE, Marcker KA & Vuust J (1985) EMBO J 4:755–759
21. Wöstmann C, Tannich E & Bakker-Grundwal T (1992) FEBS Lett 308:54–58
22. Lund PK, Moats-Staats BM, Simmons JG, Hoyt E, D'ercole AJ, Martin F & Van Wyk JJ (1985) J Biol Chem 260:7609–7613
23. Redman KL & Rechsteiner M (1989) Nature 338:438–440
24. Finley B, Barte B & Varshavsky A (1989) Nature 338:394–401
25. Cary PD, King DS, Crane-Robinson C, Bradbury EM, Rabbani A, Goodwin GH & Johns EW (1980) Eur J Biochem 112:577–586
26. Thorne AW, Sautière P, Briand G & Crane-Robinson C (1997) EMBO J 6:1005–1010
27. Haas AL & Bright PM (1985) J Biol Chem 260:12464–12473
28. Swerdlow PS, Finley D & Varshavsky A (1986) Anal Biochem 156:147–153
29. Fried VA, Smith HT, Hildebrandt E & Weiner K (1987) Proc Natl Acad Sci USA 84:3685–3689
30. Perry G, Mulvihill P, Fried VA, Smith HT, Grundke-Iqbal I & Iqbal K (1989) J Neurochem 52:1523–1528
31. Muller S (1992) Medecine/Science 8:223–232
32. Stöckl F, Muller S, Batsford S, Schmiedecke T, Waldherr R, Andrassy K, Sugisaki Y, Nakabayashi K, Nagasawa T, Rodriguez-Iturbe B, Donini U & Vogt A (1994) Clin Nephrol 41:10–17
33. Murti KG, Smith HT & Fried VA (1988) Proc Natl Acad Sci USA 85:3019–3023
34. Manetto V, Perry G, Tabaton M, Mulvihill P, Fried VA, Smith HT, Gambetti P & Autilio-Gambetti L (1988) Proc Natl Acad Sci USA 85:4501–4505
35. Wang GP, Khaton S, Iqbal K & Grundke-Iqbal I (1991) Brain Res 566:146–151

Manual of Biological Markers of Disease **B2.4**: 1–29, 1994.

The Sm antigens

SALLIE O. HOCH
The Agouron Institute, 505 Coast Blvd. South, La Jolla, CA 92037, U.S.A.

Abbreviations

ELISA: enzyme-linked immunosorbent assay; kDa: kilodalton; mAb: monoclonal antibody; PBS: phosphate buffered saline; snRNP: small nuclear ribonucleoprotein

Key words: Sm antigen, autoantibodies, systemic lupus erythematosus, small nuclear ribonucleoprotein

1. The antigen

The Sm family of antigens is described in Table 1. The B and D proteins are the major autoantibody targets.

2. Synonyms

The Sm antigen is now synonymous with at least nine different polypeptides. These proteins are also referred to as the *common* or *core* proteins of the Sm snRNP (small nuclear ribonucleoprotein) particles [1].

3. Relation between Sm and RNP antigens

Sm antisera react with all of the so-called Sm snRNP particles, each of which contains one or two discretely sized low molecular weight snRNAs and a number of polypeptides. The shared element in each snRNP is the Sm core complex of proteins which are the actual targets of the anti-Sm antibodies. (U1)RNP antisera recognize a subset of these snRNP particles designated the (U1)snRNP because each contains the U1 snRNA. The specific U1-associated proteins ((U1)RNP-70K, A and C) are the targets of the anti-RNP antibodies (see this Manual, chapter B 3.1). Thus, anti-Sm and anti-RNP antibodies distinguish non-overlapping polypeptide targets, although these proteins may be contained within the same snRNP particle.

AMAN-B2.4/1

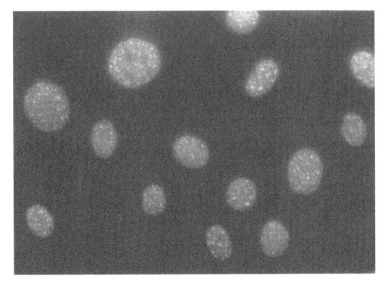

Fig. 1. Indirect immunofluorescent staining with an anti-Sm monoclonal antibody. Nil 8 hamster fibroblasts were stained with Y12 anti-Sm mAb [40] followed by a fluorescein-conjugated second antibody.

A. Nuclear staining. Cells grown on coverslips were permeabilized with 0.5% Triton X-100 in PBS prior to fixation in 3.7% formaldehyde. Anti-RNP antibodies exhibit a similar pattern. This is not an unexpected finding in light of the association of these proteins within the snRNP particle (see this Manual, par. A2/6).

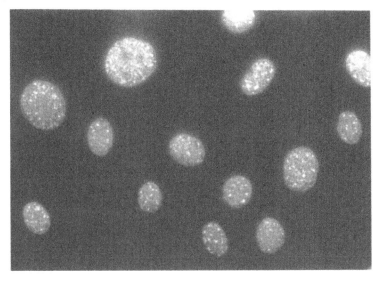

Table 1. Characteristics of the Sm antigens[a]

Sm Polypeptide	SDS-PAGE Estimated m.w.[b]	cDNA Predicted m.w.	Comments
B3(N)	27.5–29.5	24.6	Shares 93% amino acid homology with B2. Restricted pattern of expression relative to cell source and stage of development.
B2(B′)	27–29	24.6	B2/B1 are alternative processing products of the same gene. Variable expression of B2 according to species.
B1(B)	26–28	23.7	Ubiquitous expression across species and tissue sources. Immunoreactive C-terminus of B3/B2/B1 includes PPPGMRPP octamer motif.
D3	18	–	D3/D2 are fractionated from D1 by high TEMED SDS-PAGE.
D2	16.5	–	Least reactive of Sm-D antigens on protein blot.
D1	13–16	13.3	Conformational or discontinuous epitopes predominate. Immunoreactive C-terminus includes 9-fold GR repeat.
E	12	10.8	While B and D proteins are considered the major Sm antigens, E, F and G are each recognized by some rheumatoid autoantibodies.
F	11	–	Only acidic Sm polypeptide.
G	9	–	Sm protein that interacts with Sm binding site in the U1 snRNP. May be a doublet.

[a] See text for details and references.
[b] Reported estimates vary slightly depending on gel concentration, molecular weight standards, etc. routinely used in a particular laboratory.

4. Antibody detection

4.1 Immunofluorescence

Anti-Sm antibodies exhibit a characteristic punctate staining pattern throughout the nucleus; only the nucleolar regions are generally unstained (Fig. 1A). The cytoplasm of the interphase cell has historically been described as unstained by anti-Sm antibodies, although cell fractionation experiments

←
Fig. 1B. Nuclear and cytoplasmic staining. Cells were fixed in 3.7% formaldehyde prior to permeabilization by sequential incubations in 50% acetone, 100% acetone, 50% acetone and PBS. Cytoplasmic staining is dependent on the method of cell fixation, i.e. permeabilization after fixation. (Micrographs and protocols courtesy of G.W. Zieve, State University of New York Stony Brook.)

have clearly demonstrated the presence of the Sm proteins in this cell compartment [2]. Under defined conditions, cytoplasmic staining can be demonstrated (Fig. 1B).

4.2 Counterimmunoelectrophoresis (CIE)

CIE is a more rapid and sensitive alternative to the immunodiffusion assay; both rely on the visualization of a precipitin line versus a known prototype sera (see this Manual, chapter A3). Typically, anti-RNP reactivity is distinguished from anti-Sm reactivity by heat treatment of the antigen source or by incubation with RNase, both of which treatments eliminate the anti-RNP precipitin line (Fig. 2).

4.3 Immunoprecipitation

Assays such as this and the immunoblot have a more molecular orientation in discriminating anti-Sm antibodies. Anti-Sm sera precipitate a number of

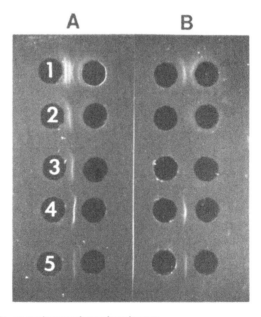

Fig. 2. Anti-Sm counterimmunoelectrophoresis assay.
The samples were as follows: A. untreated extract of calf thymus acetone powder (CTE) (Immunovision); B. CTE heat treated at 56 °C for 30 min. The sera used were: 1, anti-Sm and anti-RNP; 2 and 3, anti-RNP; 4 and 5, anti-Sm.

snRNP complexes, the most abundant of which are designated U1, U2, U4/U6 and U5 snRNP on the basis of their associated snRNA (Fig. 3). The precipitated snRNAs can be readily visualized by gel electrophoresis. The observed *snRNA* pattern (*not* the protein pattern) distinguishes Sm from RNP reactivity. The isolated Sm snRNP contain the U1, U2, U4-U6 snRNA, and the anti-RNP antibodies will precipitate only the (U1)snRNP containing the U1 snRNA. When, however, a serum contains both anti-Sm and anti-RNP antibodies, the immunoprecipitation assay will not discriminate the anti-RNP snRNA pattern against the Sm background.

4.4 Immunoblotting

The protein blot is the assay of choice to distinguish anti-Sm and anti-RNP sera on the basis of discrete *protein* targets. The snRNP proteins are fractionated by SDS-PAGE, electrotransferred to a nitrocellulose or PVDF (polyvinylidene difluoride) membrane, and reacted with antibody (for procedures see this Manual, chapter A4). Anti-Sm sera will target the B, D, E, F and G polypeptides in a multiplicity of combinations with the most frequently recognized being the B and D proteins [3] (Fig. 4). Anti-RNP sera will target the polypeptides unique to the individual snRNP (see also this Manual Chapter B).

4.5 Elisa

The use of the ELISA assay to quantitate anti-Sm reactivity is yet another alternative, with a number of options as to the antigen source. Immunoaffinity-purified Sm snRNP from rabbit thymus or HeLa cells are the starting point for the isolation of the B'/B and D polypeptides by HPLC reverse-phase chromatography [4] or by gel electroelution after SDS-PAGE [5]. Such protocols are necessary because the Sm snRNP polypeptides can not be fractionated out under physiological buffer conditions. Nonetheless, these isolates retain their immunoreactivity and provide a bona fide source of Sm antigens. Affinity-purified Sm snRNP are the prototype RNP/Sm antigen complex. The separation of "free" Sm from the Sm/RNP complex has been reported in a number of ways, including the use of RNase and trypsin. However, if the goal is the maintenance of the intact polypeptide antigens, one of the more efficient approaches is the use of differential immunoaffinity chromatography in which the (U1)snRNP or Sm/RNP complex are isolated on an anti-RNP column, and the unbound fraction is passed over an anti-Sm column to isolate "free" Sm [6]. The latter fraction will still contain the less abundant U2, U4-U6-associated RNP proteins, but as described, anti-RNP reactivity is generally associated with the (U1)RNP-70K, A and C proteins. Thus, these two fractions, Sm/RNP and "free" Sm facilitate a sensitive

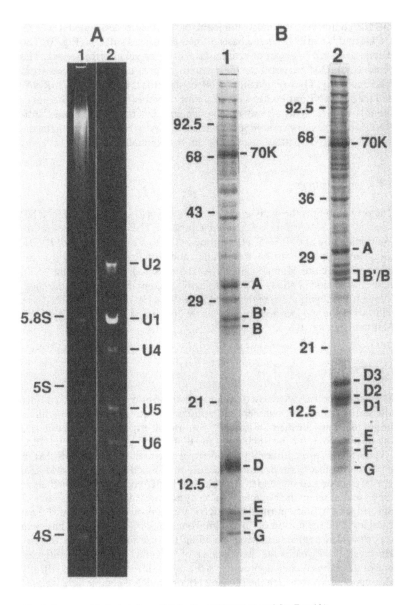

Fig. 3. Immunoaffinity isolation of HeLa Sm snRNP (7.13 anti-Sm-D mAb).
A. snRNA pattern. The snRNA in the eluate were extracted with phenol-chloroform and analyzed on a Tris-borate gel containing 7 M urea (lane 2). The gel was stained with ethidium bromide. Size markers are indicated in lane 1. (From Billings PB *et al.* (1985) J Immunol 135:428.

ELISA assay for the respective RNP and Sm antibodies [7]. Because of the difficulties in separating out the Sm core proteins, the availability of recombinant Sm proteins and/or synthetic peptides representing Sm determinants would appear to be a promising alternative. However, until the level of reactivity and specificity of recognition of the rSm-D1 antigen is identical to the bona fide polypeptide, any ELISA using recombinant Sm-D1 should be supplemented with native Sm-D1. In summary, to extrapolate from the conclusions of the European consensus workshops on ANA detection, no *one* assay is ideal yet in terms of sensitivity, quantitation and spectrum in measuring anti-Sm reactivity [8].

5. Cellular localization

During mitosis, from prophase to late anaphase, the Sm proteins appear in a diffuse distribution pattern throughout the cell plasm (except for the nucleolus) [9], although immunoelectron microscopy has suggested a perichromosomal location for a subset of Sm snRNP [10]. As the nuclear envelope reforms, the characteristic Sm speckled pattern reemerges in telophase. Throughout these mitotic translocations, there is strong evidence that the preformed Sm snRNP particles remain intact [9]. Assembly and transport of the Sm snRNP can be viewed along a continuum (Fig. 5).

6. Cellular function

A functional role for the Sm snRNP was first proposed in 1980 when two groups of investigators suggested independently that the snRNAs, in particular the U1 species, might play a pivotal role in the critical cellular event of RNA splicing, i.e. the processing of the primary RNA transcript (pre-mRNA) [11, 12] (Fig. 6). It is now known that the abundant snRNP

Copyright 1985, Journal of Immunology.) Alternatively, the snRNA can be silver stained [52] or can be radiolabelled with ^{32}P orthophosphate and visualized by autoradiography (see this Manual, chapter A7).
B. Coomassie blue stain of snRNP proteins.
Lane 1, standard 12.5% Laemmli SDS gel; lane 2, 12.5% SDS gel with 8-fold excess TEMED. Molecular weight markers (in kDa) are indicated to the left. The polypeptide originally designated Sm-D by standard SDS-PAGE, can be discriminated into multiple bands by manipulation of the electrophoresis conditions. Thus, Sm-D has been characterized as a doublet designated D and D' on SDS-urea gels [53], and as shown here, as three polypeptides designated D1, D2 and D3 on SDS gels containing an excess of the cross-linking agent TEMED [39]. Alternatively, the snRNP proteins can be radiolabelled with ^{35}S-methionine and visualized by autoradiography (see this Manual, chapter A6). Anti-Sm and anti-RNP sera are *not* distinguished on the basis of the protein patterns displayed after immunoprecipitation or immunoaffinity chromatography (see Section 3). Both sera immunoprecipitate intact snRNP; in both cases one will see the Sm core proteins and the major U1-snRNA-associated proteins.

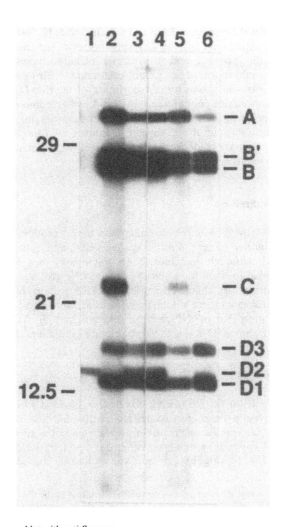

Fig. 4. Immunoblot with anti-Sm sera.
HeLa Sm snRNP were immunoaffinity-purified using anti-Sm SLE antibodies. The Sm snRNP were fractionated on a 10% SDS gel with 8-fold excess TEMED and blotted. Lane 1, anti-Sm-D2 (rabbit anti-electroeluted Sm-D); lanes 2–6, human anti-Sm sera. Molecular weight markers (in kDa) are indicated to the left. Reactivity with the B and D polypeptides is the most representative pattern of anti-Sm antibodies [3, 54]. All three of the Sm-D polypeptides resolved on high TEMED gels are reactive with anti-Sm sera as shown. The one confounding factor in these observations is that some anti-Sm sera also recognize the (U1)RNP-A and-C proteins. These two proteins and the Sm-B polypeptides display a similar proline-rich motif that is both antigenic and accessible to antibody on immunoblots of the individual proteins [35, 55, 56].

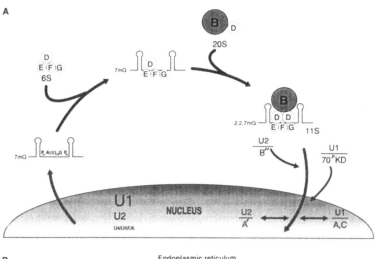

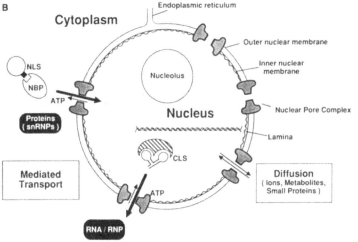

Fig. 5. Assembly and transport of the snRNP particles.
A. Cytoplasmic snRNP assembly.
There is a cytoplasmic pool of RNA-free Sm proteins that will be assembled into the common core of the Sm snRNP as determined in a mouse model [2, 53]. These proteins can be found in three different size intermediate particles: (i) 2–5S complexes containing B protein homo-oligomers; (ii) 6S complexes consisting of D, E, F, G proteins; (iii) 20S complexes containing D and B oligomers. Newly transcribed snRNAs move from the nucleus to the cytoplasm where they assemble with the D through G Sm proteins of the 6S particle, followed by the incorporation of additional D and B proteins. The assembled Sm core particle enters the interphase nucleus by a process of active transport and the snRNA-specific RNP proteins are added to produce the individual, mature snRNP complexes. (Cartoon courtesy of G.W. Zieve, State University of New York Stony Brook.)

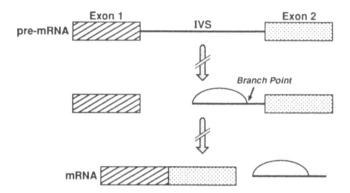

Fig. 6. Pre-mRNA splicing.
In brief, the nascent hnRNA transcript in higher organisms consists of exons (or coding regions) and introns (or non-coding regions). The introns (or intervening sequences (IVS)) must be excised and the exons assembled in the correct order and joined one to the other to produce an mRNA ready for translation into protein. Each of the U snRNP alone or in concert (as in the U4/U6 snRNP) have been identified with specific steps in the multi-step process that is RNA splicing. Yet a specific activity for each of the snRNP proteins, including the Sm core domain, remains obscure. Current models of splicing suggest that the splicing enzymatic activities (simplistically cutting and ligating) may be solely attributable to the RNA moiety akin to a *self-splicing* mechanism [59]. The snRNP proteins would ensure an adequate scaffolding for catalysis and interaction between the components of the splicing machinery – making the proteins obligatory, but leaving their mode of action as yet undefined.

(complexes of proteins and the U1, U2, U4, U5 and U6 snRNAs) are an integral part of the spliceosome, the large macromolecular complex containing an array of protein and RNA components, that facilitates the production of the mature messenger RNA. Of the less abundant Sm snRNP, the best characterized to date of these minor snRNP are associated with the U7, U11 and U12 snRNA (estimated at 10^3–10^4 copies/cell). The U7 snRNP are identified with the 3'-end processing of the histone mRNA [13]. The functions of the U11 and U12 snRNP are unknown [14].

B. snRNP nuclear import.
The transport of protein across the nuclear pore requires suitable nuclear location sequences (NLSs) and the Sm core particle is no exception. There appear to be a diversity of signals which include the Sm binding site on the snRNA which is integral to the assembly of the core protein-RNA particle [29], and the trimethylguanosine (m_3G) cap structure on the snRNA which is created by hypermethylation of the 5' end of the RNA after assembly into the core particle [57]. Once assembled, there is recent evidence to suggest that there is a nuclear location signal in the Sm core domain that is independent of the cap structure, but whose precise physicochemical nature has yet to be defined [58]. NLS, nuclear localization signal; NBP, NLS-binding protein; CLS, cytoplasmic localization signal. (With permission from Nigg EA *et al.* (1991) Cell 6:15.)

7. Presence in other species

Early reports on the number and size of the various Sm polypeptides, as reviewed by MacGillivray et al. [15], differed widely depending upon the tissue or cell source, the form of the tissue (fresh, frozen or acetone-extracted), the method of extraction and purification (if any). Table 2 summarizes a spectrum of non-human sources from which the Sm antigens have currently been identified. The characterization of the Sm polypeptides from a number of the sources is not yet complete, as reflected in the Sm descriptions ranging from partial lists of Sm proteins to mere reactivity with Sm antibodies. Yet repetitively one sees a pattern consisting of the three smallest polypeptides (E, F, G), 2–3 Sm-D proteins and Sm-B(B1), with the variable unit being the Sm-B'(B2) and –N(B3) proteins (see Fig. 7). Where it has not been possible to directly visualize the individual Sm proteins, there are alternative approaches, such as the recent identification of the Sm-D1 homolog in yeast through cDNA cloning [16]. The most recent additions to this list are the two prokaryotes, *S. leopoliensis* and *B. subtilis*, which raise intriguing questions as to evolutionary implications in this conservation of the Sm splicing proteins [17].

Table 2. Non-Human Sources of Sm Proteins

Source	Sm Description	Reference
Calf	B' B D E/F	[42]
Goat	B D	[43]
Pig	anti-Sm reactive (IP)[a]	[15]
Dog	anti-Sm reactive (CIE)	[44]
Rabbit	B' B D3 D2 D1 E F G	[45]
Mouse (FM3A cells)	B D3 D2 D1 E F G	[39]
Frog (*Xenopus laevis*)	B D E F G	[46]
Nematode (*Caenorhabditis elegans*)	anti-Sm reactive (IP)	[47]
Insect (*Drosophila melanogaster*)	B^U B^L D^U D^L E F G	[48]
Protozoa (*Plasmodium falciparum*)	anti-Sm reactive (IP)	[49]
Yeast/fungi (multiple species)	anti-Sm reactive (IP)	[50]
Cyanobacteria (*Synechococcus leopoliensis*)	anti-Sm reactive (IP)	[17]
Eubacteria (*Bacillus subtilis*)	anti-Sm reactive (IP)	[17]
Plant (broad bean)	B' B D doublet E F G	[51]

[a] IP, immunoprecipitation; CIE, counterimmunoelectrophoresis; IFA, indirect immunofluorescence assay

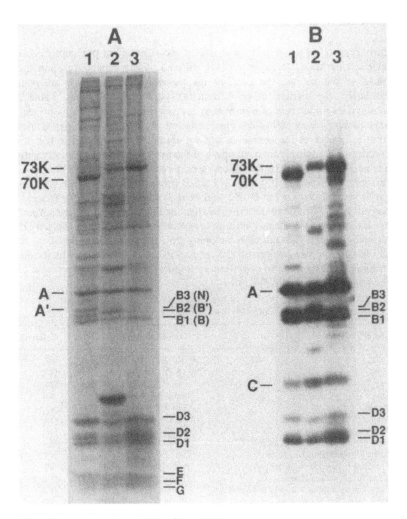

Fig. 7. Tissue and species variability of Sm snRNP.
Sm snRNP were affinity-purified on an anti-m₃G column. The cell sources were as follows: lane 1, human HeLa cells; lane 2, rabbit brain acetone powder; lane 3, rabbit thymus acetone powder. The samples were blotted with a pool of anti-Sm/RNP sera. The distribution of the Sm-B(B1) protein can be described as ubiquitous. The presence of the other B proteins (B'(B2) and N(B3)) is species- and tissue-specific. For example, Sm-B'(B2) is absent in Sm snRNP isolated from murine cells, and is found in higher animals with the proportion of B'(B2) to B(B1) varying depending upon the source (see rabbit thymus and HeLa cells in this figure). Sm-N(B3) on the other hand is associated with certain cell types, as one example, having been demonstrated in mouse, rabbit and human brain.

8. The cDNAs and derived amino acid sequences

8.1 B/B'/N or B1/B2/B3

The Sm-B family of proteins expanded when cDNA cloning revealed there were at least three distinct polypeptide members. To simplify the various alphabetical acronyms, these proteins have been given numerical designations following the format established for the Sm-D polypeptide. These are in order of apparent decreasing molecular weight (by SDS-PAGE): Sm-B3 (= B'/B or N), Sm-B2 (B'), and Sm-B1 (B) [18]. Figure 8 illustrates the

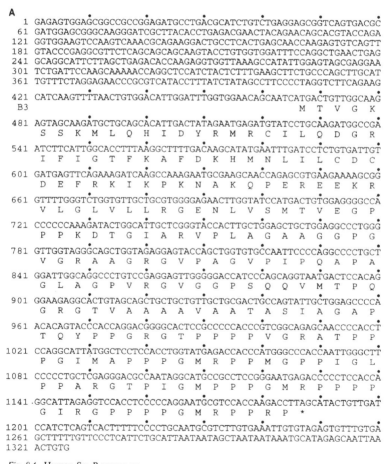

```
A
    1 GAGAGTGGAGCGGCCGCCGGAGATGCCTGACGCATCTGTCTGAGGAGCGGTCAGTGACGC
   61 GATGGAGCGGGCAAGGGATCGCTTACACCTGAGACGAACTACAGAACAGCACGTACCAGA
  121 GGTGGAAGTCCAAGTCAAACGCAGAAGGACTGCCTCACTGAGCAACCAAGAGTGTCAGTT
  181 GTACCCGAGGCGTTCTCAGCAGCAGCAAGTACCTGTGGTGGATTTCCAGGCTGAACTGAG
  241 GCAGGCATTCTTAGCTGAGACACCAAGAGGTGGTTAAAGCCATATTGGAGTAGCGAGGAA
  301 TCTGATTCCAAGCAAAAACCAGGCTCCATCTACTCTTTGAAGCTTCTGCCCAGCTTGCAT
  361 TGTTTCTAGGAGAACCCGCGTCATACCTTTATCTATAGCCTTCCCCTAGGTCTTCAGAAG

  421 CATCAAGTTTTAACTGTGGACATTGGATTTGGTGGAACAGCAATCATGACTGTTGGCAAG
  B3                                             M   T   V   G   K

  481 AGTAGCAAGATGCTGCAGCACATTGACTATAGAATGAGATGTATCCTGCAAGATGGCCGA
       S   S   K   M   L   Q   H   I   D   Y   R   M   R   C   I   L   Q   D   G   R

  541 ATCTTCATTGGCACCTTTAAGGCTTTTGACAAGCATATGAATTTGATCCTCTGTGATTGT
       I   F   I   G   T   F   K   A   F   D   K   H   M   N   L   I   L   C   D   C

  601 GATGAGTTCAGAAAGATCAAGCCCAAAGAATGCGAAGCAACCAGAGCGTGAAGAAAAGCGG
       D   E   F   R   K   I   K   P   K   N   A   K   Q   P   E   R   E   E   K   R

  661 GTTTTGGGTCTGGTGTTGCTGCGTGGGGAGAACTTGGTATCCATGACTGTGGAGGGGCCA
       V   L   G   L   V   L   L   R   G   E   N   L   V   S   M   T   V   E   G   P

  721 CCCCCCAAAGATACTGGCATTGCTCGGGTACCACTTGCTGGAGCTGCTGGAGGCCCTGGG
       P   P   K   D   T   G   I   A   R   V   P   L   A   G   A   A   G   G   P   G

  781 GTTGGTAGGGCAGCTGGTAGAGGAGTACCAGCTGGTGTGCCAATTCCCCAGGCCCCTGCT
       V   G   R   A   A   G   R   G   V   P   A   G   V   P   I   P   Q   A   P   A

  841 GGATTGGCAGGCCCTGTCCGAGGAGTTGGGGGACCATCCCAGCAGGTAATGACTCCACAG
       G   L   A   G   P   V   R   G   V   G   G   P   S   Q   Q   V   M   T   P   Q

  901 GGAAGAGGCACTGTAGCAGCTGCTGCTGTTGCTGCGACTGCCAGTATTGCTGGAGCCCCA
       G   R   G   T   V   A   A   A   V   A   A   T   A   S   I   A   G   A   P

  961 ACACAGTACCCACCAGGACGGGGCACTCCGCCCCCACCCGTCGGCAGAGCAACCCCACCT
       T   Q   Y   P   P   G   R   G   T   P   P   P   P   V   G   R   A   T   P   P

 1021 CCAGGCATTATGGCTCCTCCACCTGGTATGAGACCACCCATGGGCCCACCAATTGGGCTT
       P   G   I   M   A   P   P   P   G   M   R   P   P   M   G   P   P   I   G   L

 1081 CCCCCCTGCTCGAGGGGACGCCAATAGGCATGCCGCCTCCGGGAATGAGACCCCCTCCACCA
       P   P   A   R   G   T   P   I   G   M   P   P   P   G   M   R   P   P   P   P

 1141 .GGCATTAGAGGTCCACCTCCCCCAGGAATGCGTCCACCAAGACCTTAGCATACTGTTGAT
       G   I   R   G   P   P   P   P   G   M   R   P   P   R   P   *

 1201 CCATCTCAGTCACTTTTTCCCCTGCAATGCGTCTTGTGAAATTGTGTAGAGTGTTTGTGA
 1261 GCTTTTTGTTCCCTCATTCTGCATTAATAATAGCTAATAATAAATGCATAGAGCAATTAA
 1321 ACTGTG
```

Fig. 8A. Human Sm-B sequences.
Nucleotide sequence of Sm-B3(N) [34] (GenBank accession No. J04615) and deduced amino acid sequence.

cDNA clones and deduced amino acid sequences. Sm-B3 and –B2 are both 240 residues in length with calculated molecular weights of 24.6 kDa. Sm-B1 is 231 residues in length with a calculated molecular weight of 23.7 kDa.

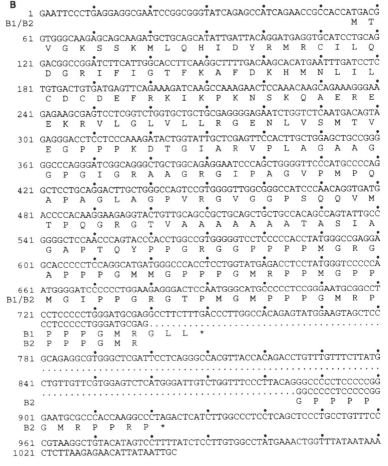

Fig. 8B. Nucleotide sequences of Sm-B2(B′) and –B1(B) [20] (GenBank accession No. X17567) and deduced amino acid sequences.
The cDNA encoding Sm-B1 contains an additional insert of 146 nucleotides as indicated here from position 740 through 885.

```
B3 (N)   MTVGKSSKMLQHIDYRMRCILQDGRIFIGTFKAFDKHMNLILCDCDEFRKIKPKNAKQPE  60
B2 (B')  ...........................................................S..A.
B1 (B)   ...........................................................S..A.

B3 (N)   REEKRVLGLVLLRGENLVSMTVEGPPPKDTGIARVPLAGAAGGPGVGRAAGRGVPAGVPI  120
B2 (B')  ....................................I............I......M
B1 (B)   ....................................I............I......M

B3 (N)   PQAPAGLAGPVRGVGGPSQQVMTPQGRGTVAAAAVAATASIAGAPTQYPPGRGTPPPPVG  180
B2 (B')  .............................A...................G.....M.
B1 (B)   .............................A...................G.....M.

B3 (N)   RATPPPGIMAPPPGMRPPMGPPIGLPPARGTPIGMPPPGMRPPPPGIRGPPPPGMRPPRP  240
B2 (B')  .GA....M.G...........M.I..G....M..............M
B1 (B)   .GA....M.G...........M.I..G....M..............M..LL
```

Fig. 8C. Comparison of amino acid sequences.
The amino acid sequences were derived from cDNA clones from the following human cell sources: **B3**, Raji/B lymphoblastoid (B'/B), Rokeach *et al.* [34]; cerebellar (N), Schmauss *et al.* [60]; pituitary gland (B/B'), Renz *et al.* [61]; HeLa/epitheloid carcinoma (B/B'), Sharpe *et al.* [62]. **B2**, HL60/peripheral blood leukocyte (B'), van Dam *et al.* [20]; HeLa (B'), Elkon *et al.* [63]. **B1**, HL60 (B), van Dam *et al.* [20]; thyroid carcinoma (B), Ohosone *et al.* [64, 65]; HeLa (B), Elkon *et al.* [63]. The letters in parentheses are the original designations. Sm-B3 and –B2 share 93% identity at the amino acid level. There are 17 amino acid differences between the two proteins, all of which can be considered conservative changes. B3 and B2 are rich in proline (20%) and glycine (15–17% respectively). Full-length cDNA clones encoding rodent homologs of human B3 [60, 66] and B1 [67] have been described and exhibit near identity (one residue change in the murine B1). Reflective of the Sm antigenic conservation as one moves down the evolutionary scale is the truncated B1 clone of *D. melanogaster* (65% amino acid sequence identity, 80% similarity) [68].

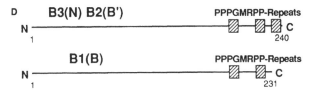

Fig. 8D. Structural features of Sm-B polypeptides.
Numbers refer to the amino acids of the respective proteins. N is the amino-terminus; C, the carboxy-terminus. The most striking structural feature is a 3-fold octamer repeat (PPPGMRPP) found in the carboxy-terminus. Similar sequences are found in the snRNP proteins (U1)RNP-A and –C, as well as the hnRNP C protein, which can be formulated in the consensus motif PP/a PGMR/iPP [34].

8.2 *DI/D2/D3*

Figure 9 illustrates the cDNA sequence and the deduced amino acid sequence for Sm-D1. Sm-D1 is 119 residues in length with a calculated molecular weight of 13.3 kDa. cDNA clones encoding Sm-D2 and Sm-D3 have yet to be reported.

A
```
  1 GAATTCCCCCCCCCCCCCCCAGTGCTCCGCGCGCTCTTGACGTCCGGAGCCCCTGGAGTAG
 61 GCGCTTCCGGCCATTCATACTGCAGTCGGTCAGTGTTCGGTTGAAGGATTCTGTGTGCTG
121 TCGGACCCAGAGGGTGACGGCGCCGCTAGGATGAAGCTCGTGAGATTTTTGATGAAATTG
                                      M  K  L  V  R  F  L  M  K  L
181 AGTCATGAAACTGTAACCATTGAATTGAAGAACGGAACACAGGTCCATGGAACAATCACA
     S  H  E  T  V  T  I  E  L  K  N  G  T  Q  V  H  G  T  I  T
241 GGTGTGGATGTCAGCATGAATACACATCTTAAAGCTGTGAAAATGACCCTGAAGAACAGA
     G  V  D  V  S  M  N  T  H  L  K  A  V  K  M  T  L  K  N  R
301 GAACCTGTACAGCTGGAAACGCTGAGTATTCGAGGAAATAACATTCGGTATTTTATTCTA
     E  P  V  Q  L  E  T  L  S  I  R  G  N  N  I  R  Y  F  I  L
361 CCAGACAGTTTACCTCTGGATACACTACTTGTGGATGTTGAACCTAAGGTGAAATCTAAG
     P  D  S  L  P  L  D  T  L  L  V  D  V  E  P  K  V  K  S  K
421 AAAAGGGAAGCTGTTGCAGGAAGAGGCAGAGGAAGAGGAAGAGGAAGAGGACGTGGCCGT
     K  R  E  A  V  A  G  R  G  R  G  R  G  R  G  R  G  R  G  R
481 GGCAGAGGAAGAGGGGGTCCTAGGCGATAATGTCTCTCAAGATTTCAAAGTCATATGAGA
     G  R  G  R  G  G  P  R  R  *
541 TTTGGGATATTTTTTGTACAGGTTGTGTTTGTTTATGTCAGTTTTTAATAAACATAAATG
601 TGGGACAGAGCTGTCTATTTAGTATATCAAAGTTTTAGTAGTTTCCTCCACATTCACGAA
661 ATTACCACAGTGAGAGCTAAGCATTTCTACTGGGCAGTTTCATTTTTAGTTGATCAGGTT
721 TTAAGTTTTTGAACTAAAATTTTTCTTTTTCTTTTTATGATGAATAAGGTTAAAATAAAA
781 GCCTTAGACAAATTAAATTTGGCAGAGTTTAATTGAGCAAAGGACAATTCACAAATCAGG
841 TAGCCCCTGAACCATAATAGGCTCAGAGGCTTCAGCCCAGCTGCATAGTTGAAGATTTAT
901 GGACAGAAGGAAAGTGATGTATGGAAATGGAAGTGAGATACAGCAACAGCCGGATTAGT
961 TACAGTTCAGCGTTTGCCTTATTTGAATATGGTTTGAACAGTTCGCTGTCTTTGGTTGGC
1021 TGAAACTTAGTGATTGCCACAAGAGTAGGGTACCGTCTGTTTACACGTCCAGTTAGGCTA
1081 CAGTTCTATGTACTGAGAAACCTTTAAGCTGAACTTGAGATATGTAAAGAGACTTTAGGC
1141 TAAACTTAACAATATATATAGGAATATATCCCTTCTACTTCACATGCACTGAATATGCAT
1201 TTTATTGCTTTACTCTTCATTCTGTGGCACCTACCCACAGGGGAAGTAAGAAGTTTGTTT
1261 TGGTATTTCGGAAACTAAAGTCCTTATGGGATGGGGTCTAGAATTGATTCTCCTTTCCTG
1321 AGTTTTACTCCACGGAGTCTTAGGTACCTGGTAAAAAGTTGTCTTCTAAATTAAGGGTCA
1381 TTGCTTTGTTGTCTAGCTGCTAATGTCTTACTTTTGTTTCTTTTGCTTTTTAATCAGTTC
1441 TTAATAGGATATAGTTTTATGTTTTCCAAGTTATAACTTGGAGTTAATGGTCACTAGATT
1501 ATCAGTTATGAGCAGTGTTAAAATCCTCCTATTAATGTGTAATGTACCTGTCAGTGCCTCC
1561 TTTATTAAGGGGTTCTTTGAGAATAAAAGAGAAAAGACCTACTTTATTTGACAGCAAAAA
1621 AAAAAAGGAATTC
```

Fig. 9A. Human Sm-D1 sequences.
cDNA sequence of Sm-D1 [69] (GenBank accession No. J03798) and deduced amino acid sequence.
The amino acid sequence was derived from cDNA clones from the following human cell sources: B-lymphoblastoid (Raji) [69]; pituitary gland [61] libraries; and from a murine B cell lymphoma library [70]. The most salient features of the deduced amino acid sequence are: i) a 9-fold repeated glycine-arginine (Gly-Arg)₉ motif between residues 97–114 (see Fig. 13); ii) a lysine-rich hydrophilic region located between residues 82–93; iii) a region 114-RGGPRR-119 with strong similarity to protamines. The yeast homolog of Sm-D1 has now been identified [16] and shares 40% identity overall with the human protein, but there are two blocks of 28 and 22 residues that average 76% identity. These conserved regions may give insights into functionally-relevant domains. The deletion of the Sm-D1 gene is lethal to the yeast cell, but substitution of the human gene (using a construct of the hSm-D gene with a yeast promoter) complemented the deletion, suggesting the hSm-D gene is encoding functional protein in yeast. Moreover, complementation analyses of yeast deletion mutants with the hSm-D argued strongly for the conservation of the basic eukaryotic splicing apparatus [71].

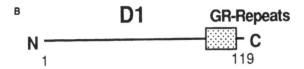

Fig. 9B. Structural features of Sm-D1.

8.3 E/F/G

The remaining Sm cDNA clone reported to date is that of Sm-E, and was in fact the first Sm cDNA to be described. The full-length cDNA encodes a protein 92 residues in length with a calculated molecular weight of 10.8 kDa [19] (Fig. 10).

9. The genes

9.1 B/B'/N or B1/B2/B3

Sm-B1 and B2 are produced from the same gene by alternative RNA splicing [20–22]. The cDNA encoding the shorter of the two proteins, Sm-B1, contains a 146 bp insert at the 3' end of the open reading frame (ORF) encoding the first 228 residues of B1 and B2. The fourth in-frame codon in this insert is a termination codon such that the 231-residue B1 protein is

```
  1 GCTCTCAGAGGCAGCGTGCGGGTGTGCTCTTTGTGAAATTCCACCATGGCGTACCGTGGC
                                                  M   A   Y   R   G
 61 CAGGGTCAGAAAGTGCAGAAGGTTATGGTGCAGCCCATCAACCTCATCTTCAGATACTTA
     Q   G   Q   K   V   Q   K   V   M   V   Q   P   I   N   L   I   F   R   Y   L
121 CAAAATAGATCGCGGATTCAGGTGTGGCTCTATGAGCAAGTGAATATGCGGATAGAAGGC
     Q   N   R   S   R   I   Q   V   W   L   Y   E   Q   V   N   M   R   I   E   G
181 TGTATCATTGGTTTTGATGAGTATATGAACCTTGTATTAGATGATGCAGAAGAGATTCAT
     C   I   I   G   F   D   E   Y   M   N   L   V   L   D   D   A   E   E   I   H
241 TCTAAAACAAAGTCAAGAAAACAACTGGGTCGGATCATGCTAAAAGGAGATAATATTACT
     S   K   T   K   S   R   K   Q   L   G   R   I   M   L   K   G   D   N   I   T
301 CTGCTACACAAAGTGTCTCCAACTAGAAATGATCAATGAAGTGAGAAATTGTTGAGAAGGAT
     L   L   Q   S   V   S   N   *
361 ACAGTTTGTTTTTAGATGTCCTTTGTCCAATGTGAACATTTATTCATATTGTTTTGATTA
421 CCCTCGTGTTACTACAAGATGGCAATAAATACTATGGGATTGTTTGTATTAAAAAAATTTA
481 CATTGCTTCTTAAAAAAAAAA
```

Fig. 10. Human Sm-E sequences.
cDNA sequence of Sm-E [19] (GenBank accession Nos. X12466 and X13772) and deduced amino acid sequence. Again, there is evidence of strong evolutionary conservation when the human predicted amino acid sequence was compared to that of mouse and chicken (100% identity) [72] and *Xenopus* (94% identity) [73].

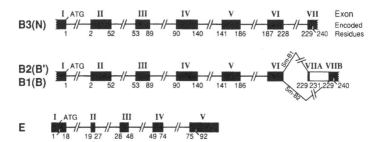

Fig. 11. Schematic of Sm gene organization.
The exon coordinates are derived from the following reports: B3, Schmauss *et al.* [23]; B2/B1, van Dam *et al.* [22], Chu and Elkon [21]; E, Stanford *et al.* [24].

produced. Analysis of the untranslated sequence at the 3' end of this insert reveals a consensus sequence for an intron-exon junction and the splice site for the B2 coding sequence that encompasses residues 229–240 (see the alternative exons VIIA and VIIB in Fig. 11). Isolation of the Sm-B1/B2 gene confirmed this interpretation of the cDNA sequence. Sequence analysis of the intron-exon junctions in this gene demonstrated a total of seven exons, with the sixth intron containing the two alternative 3' acceptor splice sites.

Isolation of the gene encoding Sm-B3(N) confirmed the existence of two distinct genes that encode Sm-B1/B2 and Sm-B3 respectively [23]. The organization of the Sm-B3 gene is similar, but not identical, to that of Sm-B2 comprising seven coding exons and six introns. The 5' untranslated region of the Sm-B1/B2 mRNA is contained in its entirety within the coding exon I of B1/B2, but the B3 untranslated region is split between two exons [22, 23]. Using human/hamster somatic cell hybrids, the Sm-B1/B2 gene was mapped to human chromosome 20, and that of Sm-B3 to chromosome 4. Overall, the similar structural organization of the two genes for B1/B2 and B3 and the near homology of the encoded proteins would suggest that they share a common ancestral gene.

9.2 D1/D2/D3

There is no published information as yet on the organization of the human Sm-D genes.

9.3 E/F/G

The human gene encoding the Sm-E protein (> 10 kilobases) has been sequenced almost in its entirety (exception is intron 4) and found to comprise five exons and four introns (Fig. 11) [24]. Wieben and co-workers have identified at least five additional processed pseudogenes (without introns) in this multigene family [25]. The expressed E protein gene has been mapped to human chromosome 1 (specifically 1q25–1q43) using somatic cell and *in situ* hybridization, and linkage analysis [26]. Comparison of the genomic 5′ untranslated region and the immediate 5′ upstream region in both human and murine clones reveals a high degree of conservation in the basal promoter structure at the nucleotide level.

10. Sm protein interactions

10.1 Relation with nucleic acids

The so-called *Sm binding site* was first identified as a structural motif shared by the U1, U2, U4 and U5 snRNA [27]. It was described as a single-stranded oligonucleotide comprising the sequence PuA(U)nGPu where n > 3. Its relationship to the Sm snRNP was clarified when it was found that this motif was within the snRNA fragments protected from nuclease digestion in the presence of the core snRNP proteins (DEFG) or intact snRNP particles [28]. Deletion of the Sm binding site resulted in the loss of anti-Sm immunoprecipitability, suggesting this sequence was critical to snRNP assembly and transport into the nucleus [29]. Likewise, the same deletion significantly reduced the efficiency of the trimethylation reaction, i.e. the addition of the 5′ m_3G cap in the maturation of the snRNA [30]. Considering the size of the Sm binding site, it seemed likely that only one of the core proteins was directly interacting with this site. Using reconstituted and UV-cross-linked U1 snRNP particles, the Sm polypeptide linked to the U1 snRNA was identified as Sm-G [31] based on: (i) one- and two-dimensional gel electrophoresis; (ii) immunoprecipitation with a monospecific anti-Sm-G serum. Fingerprint analysis demonstrated the site of interaction as the 5′ AAU sequence within the U1 Sm binding site $(AA(U)_3GUGG)$.

10.2 Relation with other proteins

Information is available on the general organization of the Sm proteins into higher order aggregates as described in Section 5 (Cellular localization).

11. Biochemical characteristics

With the exception of Sm-F (est. pI ≤ 3.3), the Sm antigens are a group of basic proteins (est. pI for B2, 10.7; B1, 10.6; D, 10.5; E, 10.3; G, 8.8) [32]. Despite both the overall charge of these proteins and the presence of regions of localized charge (from the predicted amino acid sequence), the molecular weights estimated by SDS-PAGE are in reasonable agreement with the calculated molecular weight determined from the primary sequence (see Table 1). There is no indication as yet of significant post-translational modification of the Sm proteins, but the variables examined have been limited. None of the Sm proteins of known sequence exhibit any distinctive functional motifs, including the so-called RNA recognition motif (RRM) [33].

12. B-cell epitopes

A recent review of the Sm B-cell epitopes is available for more detail [18].

12.1 B/B'/N or B1/B2/B3

The importance of the proline-rich carboxy terminus of the B antigens was predicted from the primary sequences [20, 34, 35]. Epitope mapping, using both recombinant protein and synthetic peptides, confirmed the immunoreactivity of this region. A clear consensus has not emerged as to other domains in the Sm-B antigens (see Fig. 12). Certainly potential contributing factors arise from the diversity of (i) the antigen sources being utilized – recombinant, including *in vivo* expression – and *in vitro* translation-products, and synthetic peptides; (ii) the assay systems including protein blot, immunoprecipitation and ELISA; and (iii) ethnic and racial differences in the patient populations from which sera are collected. Thus, the epitope mapping continues as a work in progress.

12.2 D1/D2/D3

B-cell epitope data is available for Sm-D1 (Fig. 13). Rokeach *et al.* [36] describe two general patterns of anti-Sm reactivity: antibodies that recognize only the full-length antigen; and antibodies that recognize the carboxy terminus which embodies an extended/charged structure – pointing to a preponderance of *conformational* or *discontinuous* epitopes in describing the Sm-D1 profile. A consensus has not emerged as to the relative importance of amino – and carboxyl – terminal epitopes in Sm-D1 [36–38].

Antibodies, affinity-purified against the three Sm-D polypeptides, were

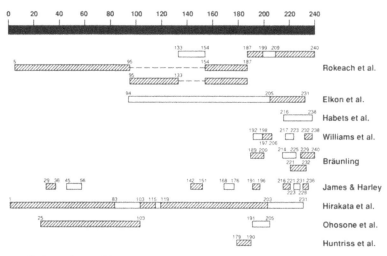

Fig. 12. Autoepitopes of Sm-B.
Location of deduced Sm-B autoepitopes are indicated in amino acid coordinates as defined in the following reports: B3, Rokeach et al. [55], Bräunling [74]; B2/B1, Elkon et al. [63], Habets et al. [35], Williams et al. [56], James and Harley [75], Ohosone et al. [76], Hirakata et al. [38], Huntriss et al. [77]. The open and hatched blocks delineate individual epitopes. (Updated from Rokeach LA & Hoch SO (1992) Mol Biol Rep 16:165; with permission.) Rokeach et al. [55] identified six B3 epitopes, five of which overlapped the C-terminus. There was not a simple recognition pattern, as exemplified by several sera that reacted with the fragment (187–199) containing one copy of the PPPGMRPP octamer, but not with a longer fragment (209–240) that contained two copies. Elkon et al. [63] reported similar observations for the B2 and B1 polypeptides. These observations underscored the influence of spacial location and/or neighboring residues in addition to the primary sequence of a reactive region. An alternative approach was to use overlapping peptides to scan the B protein domains and these studies reiterated the PPPGMRPP motif as an immunodominant epitope [56, 74, 75]. This octamer was manipulated by substitution and deletion. James and Harley [75] concluded that the prolines at positions 7 and 8 are not critical, but that substitution of the arg residue in the GMR sequence with a non-basic residue resulted in a substantial loss of antibody binding. Bräunling [74] concluded that the order of the GMR sequence is essential, and that the proline content *per se* is not as essential as the position of these residues, particularly relative to the N-terminus of the GMR core sequence. Considerable attention has also been paid to the proline-rich C-terminus of the Sm-B proteins as the basis of the cross-reactivity observed with the (U1)RNP-A and-C polypeptides [35, 55, 56, 75, 76]. The amino-terminus has been reported to be immunoreactive [38, 65, 75] and non-immunoreactive [63, 74] using both recombinant antigen and synthetic peptides. Similar discrepant reports relate to the central portion of the polypeptides. There are hints of polypeptide-specific epitopes, such as one attributable to the B1 C-terminal residues relative to B2/B3 [63], or a B1/B2-specific sequence (residues 179–190) relative to B3 [77]. There also remain the questions of conformational epitopes at the polypeptide level [55] and of epitope exposure within a particular snRNP particle (e.g. U1 snRNP vs. all Sm snRNPs) [76].

IMMUNOREACTIVE DOMAINS OF Sm-D

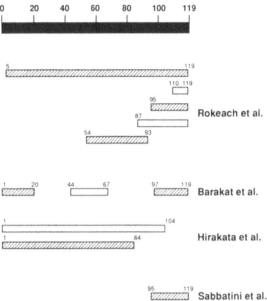

Fig. 13. Immunoreactive domains of Sm-D.
Sm-D1 regions reactive with patient anti-Sm autoantibodies are indicated in amino acid coordinates as defined in the following reports: Rokeach *et al.* [36], Barakat *et al.* [37], Hirakata *et al.* [38], Sabbatini *et al.* [78]. The open and hatched blocks delineate distinct domains. (Updated from Rokeach LA & Hoch SO (1992) Mol Biol Rep 16:165; with permission.) Recognition of the C-terminus centered on a region originally described from the primary sequence in terms of four domains: a basic hydrophilic region (residues 82–93); a flexible (gly-arg)$_9$ repeat (residues 97–114); a protamine-like sequence (residues 114–119); a combined domain (residues 96–119) with striking similarity to a region in EBNA1 (Epstein Barr nuclear antigen) [36]. A recent study with synthetic branched peptides encompassing D1 residues 96–119 and the EBNA homolog (residues 35–58) reinforced the concept of a shared epitope recognized by a specific population of anti-Sm antibodies [78]. Contrasting results were found by Barakat *et al.* [37] using overlapping synthetic peptides in an ELISA assay. Three peptides, spanning residues 1–20, 44–67 and 97–119 were recognized by SLE sera (Sm$^+$ on immunoblot) at frequencies of 75%, 89% and 37% respectively. In a second longitudinal study of 12 SLE patients, 70% of the bleeds had antibodies reactive with peptide 1–20 although there was no correlation with serum anti-Sm antibodies [79]. Using immunoprecipitation assays with peptides produced by *in vitro* translation, Rokeach *et al.* [36] and Hirakata *et al.* [38] did not observe such strong amino-terminal reactivity.

used to explore their relationship. In a survey of 38 antisera, the reactivity patterns were divided into three groups: I (at 55% incidence), sera reacted with D1, D3 and B1/B2; II (37%), reacted with D1, D2, D3 and B1/B2; and III (8%), reacted strongly with D2 and very weakly with D1 and D3 [39].

12.3 E/F/G

No data is presently available on B-cell epitopes of these proteins.

12.4 Mapping of shared Sm epitopes

Sm-D1 is known to share at least one epitope with the Sm-B proteins based on monoclonal antibodies with multiple specificities [40, 41], e.g. Y12 mAb reacts with the B proteins and D1 and D3 [39]. Yet there is no obvious amino acid sequence homology between these polypeptides.

13. T-cell epitopes

There are no published data about Sm specific T-cells.

References

1. Lerner MR & Steitz JA (1979) Antibodies to small nuclear RNAs complexed with proteins are produced by patients with systemic lupus erythematosus. Proc Natl Acad Sci USA 76:5495–5499
2. Zieve GW & Sauterer RA (1990) Cell biology of the snRNP particles. CRC Rev Biochemistry Mol Biol 25:1–46
3. Craft J (1992) Antibodies to snRNPs in systemic lupus erythematosus. Rheum Dis Clin North Am 18:311–335
4. Williams DG, Charles PJ & Maini RN (1988) Preparative isolation of p67, A, B, B′ and D from nRNP/Sm and Sm antigens by reverse-phase chromatography. Use in a polypeptide-specific ELISA for independent quantitation of anti-nRNP and anti-Sm antibodies. J Immunol Methods 113:23–35
5. Takeda Y, Wang GS, Wang RJ, Anderson SK, Pettersson I, Amaki S & Sharp GC (1989) Enzyme-linked immunosorbent assay using isolated (U) small nuclear ribonucleoprotein polypeptides as antigens to investigate the clinical significance of autoantibodies to these polypeptides. Clin Immunol Immunopathol 50:213–230
6. Billings PB & Hoch SO (1984) Characterization of U small nuclear RNA-associated proteins. J Biol Chem 259:12850–12856
7. Field M, Williams DG, Charles P & Maini RN (1988) Specificity of anti-Sm antibodies by ELISA for systemic lupus erythematosus: increased sensitivity of detection using purified peptide antigens. Ann Rheum Dis 47: 820–825
8. van Venrooij WJ, Charles P & Maini RN (1991) The consensus workshops for the detection of autoantibodies to intracellular antigens in rheumatic diseases. J Immunol Methods 140:181–189
9. Verheijen R, Kuijpers H, Vooijs P, van Venrooij W & Ramaekers F (1986) Distribution of the 70K U1 RNA-associated protein during interphase and mitosis. Correlation with other U RNP particles and proteins of the nuclear matrix. J Cell Sci 86:173–190
10. Spector DL & Smith HC (1986) Redistribution of U-snRNPs during mitosis. Exp Cell Res 163:87–94
11. Lerner MR, Boyle JA, Mount SM, Wolin SL & Steitz JA (1980) Are snRNPs involved in splicing? Nature 283:220–224
12. Rogers J & Wall R (1980) A mechanism for RNA splicing. Proc Natl Acad Sci USA 77:1877–1879
13. Birnstiel ML & Schaufele FJ (1988) Structure and function of minor snRNPs. In: Birnstiel ML (Ed) Structure and Function of Major and Minor Small Nuclear Ribonucleoprotein Particles (pp 155–182). Springer-Verlag, New York
14. Gilliam AC & Steitz JA (1993) Rare scleroderma autoantibodies to the U11 small nuclear ribonucleoprotein and to the trimethylguanosine cap of U small nuclear RNAs. Proc Natl Acad Sci USA 90:6781–6785
15. MacGillivray AJ, Carroll AR, Dahi S, Naxakis G, Sadaie MR, Wallis CM & Jing T (1982) The composition of the nuclear antigens Sm and RNP of human rheumatic and connective tissue diseases and the relevance of their autoantibodies as probes for RNA processing mechanisms. FEBS Lett 141:139–147
16. Rymond BC (1993) Convergent transcripts of the yeast *PRP38-SMD1* locus encode two essential splicing factors, including the D1 core polypeptide of small nuclear ribonucleoprotein particles. Proc Natl Acad Sci USA 90:848–852
17. Kovacs SA, O'Neil J, Watcharapijarn J, Moe-Kirvan C, Vijay S & Silva V (1993) Eubacterial components similar to small nuclear ribonucleoproteins: identification of immunoprecipitable proteins and capped RNAs in a cyanobacterium and a gram-positive eubacterium. J Bacteriol 175:1871–1878
18. Rokeach LA & Hoch SO (1992) B-cell epitopes of Sm autoantigens. Mol Biol Rep 16:165–174

19. Stanford DR, Kehl M, Perry CA, Holicky EL, Harvey SE, Rohleder AM, Rehder K Jr, Lührmann R & Wieben ED (1988) The complete primary structure of the human snRNP E protein. Nucleic Acids Res 16:10593–10605

20. van Dam A, Winkel I, Zijlstra-Baalbergen J, Smeenk R & Cuypers HT (1989) Cloned human snRNP proteins B and B' differ only in their carboxy-terminal part. EMBO J 8:3853–3860

21. Chu J-L & Elkon KB (1991) The small nuclear ribonucleoproteins, SmB and B', are products of a single gene. Gene 97:311–312

22. van Dam AP, Winkel I, Smeenk R & Cuypers HT (1990) The snRNP proteins B and B' are alternative splicing variants encoded by the same gene. In: van Dam AP (Ed) Autoantibodies as Diagnostic Markers in SLE (pp 125–133). PhD Thesis, University of Amsterdam, The Netherlands

23. Schmauss C, Brines ML & Lerner MR (1992) The gene encoding the small nuclear ribonucleoprotein-associated protein N is expressed at high levels in neurons. J Biol Chem 267:8521–8529

24. Stanford DR, Perry CA, Holicky EL, Rohleder AM & Wieben ED (1988) The small nuclear ribonucleoprotein E protein gene contains four introns and has upstream similarities to genes for ribosomal proteins. J Biol Chem 263:17772–17779

25. Stanford DR, Holicky EL, Perry CA, Rehder JK, Harvey SE, Rohleder AM & Wieben ED (1991) The snRNP E protein multigene family contains five pseudogenes with common mutations. DNA Sequence 1:357–363

26. Neiswanger K, Stanford DR, Sparkes RS, Nishimura D, Mohandas T, Klisak I, Heinzmann C & Wieben ED (1990) Assignment of the gene for the small nuclear ribonucleoprotein E (SNRPE) to human chromosome 1q25-q43. Genomics 7:503–508

27. Branlant C, Krol A, Ebel JP, Lazar E, Haendler B & Jacob M (1982) U2 RNA shares a structural domain with U1, U4 and U5 RNAs. EMBO J 1:1259–1265

28. Liautard JP, Sri-Widada J, Brunel C & Jeanteur P (1982) Structural organization of ribonucleoproteins containing small nuclear RNAs from HeLa cells. Proteins interact closely with a similar structural domain of U1, U2, U4 and U5 small nuclear RNAs. J Mol Biol 162:623–643

29. Hamm J, Darzynkiewicz E, Tahara SM & Mattaj IW (1990) The trimethylguanosine cap structure of U1 snRNA is a component of a bipartite nuclear targeting signal. Cell 62:569–577

30. Mattaj IW (1986) Cap trimethylation of U snRNA is cytoplasmic and dependent on U snRNP protein binding. Cell 46:905–911

31. Heinrichs V, Hackl W & Lührmann R (1992) Direct binding of small nuclear ribonucleoprotein G to the Sm site of small nuclear RNA. Ultraviolet light cross-linking of protein G to the AAU stretch within the Sm site (AAUUUGUGG) of U1 small nuclear ribonucleoprotein reconstituted in Vitro. J Mol Biol 227:15–28

32. Woppmann A, Patschinsky T, Bringmann P, Godt F & Lührmann R (1990) Characterisation of human and murine snRNP proteins by two-dimensional gel electrophoresis and phosphopeptide analysis of U1-specific 70K protein variants. Nucleic Acids Res 18:4427–4438

33. Kenan DJ, Query CC & Keene JD (1991) RNA recognition: towards identifying determinants of specificity. TIBS 16:214–220

34. Rokeach LA, Jannatipour M, Haselby JA & Hoch SO (1989) Primary structure of a human Sm small nuclear ribonucleoprotein polypeptide as deduced by cDNA analysis. J Biol Chem 264:5024–5030

35. Habets WJ, Sillekens PTG, Hoet MH, McAllister G, Lerner MR & van Venrooij WJ (1989) Small nuclear RNA-associated proteins are immunologically related as revealed by mapping of autoimmune reactive B-cell epitopes. Proc Natl Acad Sci USA 86:4674–4678

36. Rokeach LA, Jannatipour M, Haselby J & Hoch SO (1992) Mapping of the immunoreactive domains of a small nuclear ribonucleoprotein-associated Sm-D autoantigen. Clin Immunol Immunopathol 65:315–324

37. Barakat S, Briand J-P, Weber J-C, Van Regenmortel MHV & Muller S (1990) Recognition of synthetic peptides of Sm-D autoantigen by lupus sera. Clin exp Immunol 81:256–262
38. Hirakata M, Craft J & Hardin JA (1993) Analysis of domains recognized by the Y12 monoclonal anti-Sm antibody and by patient sera. J Immunol 150:3592–3601
39. Lehmeier T, Foulaki K & Lührmann R (1990) Evidence for three distinct D proteins, which react differentially with anti-Sm autoantibodies, in the cores of the major snRNPs U1, U2, U4/U6 and U5. Nucleic Acids Res 18:6475–6484
40. Lerner EA, Lerner MR, Janeway CA & Steitz JA (1981) Monoclonal antibodies to nucleic acid-containing cellular constituents: probes for molecular biology and autoimmune disease. Proc Natl Acad Sci USA 78:2737–2741
41. Williams DG, Stocks MR, Smith PR & Maini RN (1986) Murine lupus monoclonal antibodies define five epitopes on two different Sm polypeptides. Immunology 58:495-500
42. Duran N, Bach M, Puigdomenech P & Palau J (1984) Characterization of antigenic polypeptides of the RNP, Sm and SS-B nuclear antigens from calf thymus. Mol Immunol 21:731–739
43. Ishaq M & Ali R (1983) Purification and identification of antigenic polypeptides of Sm and RNP antigens of goat livers. Biochem Biophys Res Commun 114:564–570
44. White SD, Rosychuk RAW & Schur PH (1992) Investigation of antibodies to extractable nuclear antigens in dogs. Am J Vet Res 53:1019–1021
45. Billings PB & Hoch SO (1983) Isolation of intact Sm and RNP polypeptides from rabbit thymus. J Immunol 131:347–351
46. Zeller R, Nyffenegger T & Robertis EM (1983) Nucleocytoplasmic distribution of snRNPs and stockpiled snRNA-binding proteins during oogenesis and early development in Xenopus laevis. Cell 32:425–434
47. Thomas JD, Conrad RC & Blumenthal T (1988) The C. elegans *trans*-spliced leader RNA is bound to Sm and has a trimethylguanosine cap. Cell 54:533–539
48. Paterson T, Beggs JD, Finnegan DJ & Lührmann R (1991) Polypeptide components of *Drosophila* small nuclear ribonucleoprotein particles. Nucleic Acids Res 19:5877–5882
49. Francoeur AM, Gritzmacher CA, Peebles CL, Reese RT & Tan EM (1985) Synthesis of small nuclear ribonucleoprotein particles by the malarial parasite *Plasmodium falciparum*. Proc Natl Acad Sci USA 82:3635–3639
50. Tollervey D & Mattaj IW (1987) Fungal small nuclear ribonucleoproteins share properties with plant and vertebrate U-snRNPs. EMBO J 6:469–476
51. Pálfi Z, Bach M, Solymosy F & Lührmann R (1989) Purification of the major UsnRNPs from broad bean nuclear extracts and characterization of their protein constituents. Nucleic Acids Res 17:1445–1458
52. McNeilage LJ & Whittingham S (1984) Use of the Bio-Rad Stain to identify gel purified RNA Components of small nuclear ribonucleoprotein antigens. J Immunol Methods 66:253–260
53. Anderson J & Zieve GW (1991) Assembly and intracellular transport of snRNP particles. BioEssays 13:57–64
54. Tan EM (1989) Antinuclear antibodies: diagnostic markers for autoimmune diseases and probes for cell biology. Adv Immunol 44:93–151
55. Rokeach LA, Jannatipour M & Hoch SO (1990) Heterologous expression and epitope mapping of a human small nuclear ribonucleoprotein associated Sm-B'/B autoantigen. J Immunol 144:1015–1022
56. Williams DG, Sharpe NG, Wallace G & Latchman DS (1990) A repeated proline-rich sequence in Sm B/B' and N is a dominant epitope recognized by human and murine autoantibodies. J Autoimmun 3:715–725
57. Fischer U & Lührmann R (1990) An essential signaling role for the m_3G cap in the transport of U1 snRNP to the nucleus. Science 249:786–790
58. Fischer U, Sumpter V, Sekine M, Satoh T & Lührmann R (1993) Nucleo-cytoplasmic transport of U snRNPs: definition of a nuclear location signal in the Sm core domain that

binds a transport receptor independently of the m₃G cap. EMBO J 12:573–583

59. Guthrie C (1991) Messenger RNA splicing in yeast: clues to why the spliceosome is a ribonucleoprotein. Science 253:157–163

60. Schmauss C, McAllister G, Ohosone Y, Hardin JA & Lerner MR (1989) A comparison of snRNP-associated Sm-autoantigens: human N, rat N, and human B/B'. Nucleic Acids Res 17:1733–1743

61. Renz M, Heim C, Bräunling O, Czichos A, Wieland C & Seelig HP (1989) Expression of the major human ribonucleoprotein (RNP) autoantigens in *Escherichia coli* and their use in an EIA for screening sera from patients with autoimmune diseases. Clin Chem 35:1861–1863

62. Sharpe NG, Williams DG, Howarth DN, Coles B & Latchman DS (1989) Isolation of cDNA clones encoding the human Sm B/B' auto-immune antigen and specifically reacting with human anti-Sm auto-immune sera. FEBS Lett 250:585–590

63. Elkon KB, Hines JJ, Chu J-L & Parnassa A (1990) Epitope mapping of recombinant HeLa SmB and B' peptides obtained by the polymerase chain reaction. J Immunol 145:636–643

64. Ohosone Y, Mimori T, Griffith A, Akizuki M, Homma M, Craft J & Hardin JA (1989) Molecular cloning of cDNA encoding Sm autoantigen: derivation of a cDNA for a B polypeptide of the U series of small nuclear ribonucleoprotein particles. Proc Natl Acad Sci USA 86:4249–4253

65. Ohosone Y, Mimori T, Griffith A, Akizuki M, Homma M, Craft J & Hardin JA (1989) Molecular cloning of cDNA encoding Sm autoantigen: derivation of a cDNA for a B polypeptide of the U series of small nuclear ribonucleoprotein particles (correction). Proc Natl Acad Sci USA 86:8982

66. Gerrelli D, Grimaldi K, Horn D, Mahadeva U, Sharpe N & Latchman DS (1993) The cardiac form of the tissue-specific SmN protein is identical to the brain and embryonic forms of the protein. J Mol Cell Cardiol 25:321–329

67. Griffith AJ, Schmauss C & Craft J (1992) The murine gene encoding the highly conserved Sm B protein contains a nonfunctional alternative 3' splice site. Gene 114:195–201

68. Brunet C, Quan T & Craft J (1993) Comparison of the *Drosophila melanogaster*, human and murine Sm *B* cDNAs: evolutionary conservation. Gene 124:269–273

69. Rokeach LA, Haselby JA & Hoch SO (1988) Molecular cloning of a cDNA encoding the human Sm-D autoantigen. Proc Natl Acad Sci USA 85:4832–4836

70. Mitsuda T, Eisenberg RA & Cohen PL (1992) The murine Sm-D autoantigen: multiple genes, genetic polymorphism, evolutionary conservation and lack of intervening sequences in the coding region. J Immunol 5:277–287

71. Rymond BC, Rokeach LA & Hoch SO (1993) Human snRNP polypeptide D1 promotes pre-mRNA splicing in yeast and defines nonessential yeast Smd1p sequences. Nucleic Acids Res 21:3501–3505

72. Fautsch MP, Thompson MA, Holicky EL, Schultz PJ, Hallett JB & Wieben ED (1992) Conservation of coding and transcriptional control sequences within the snRNP E protein gene. Genomics 14:883–890

73. Etzerodt M, Vignali R, Ciliberto G, Scherly D, Mattaj IW & Philipson L (1988) Structure and expression of a *Xenopus* gene encoding an snRNP protein (U1 70K). EMBO J. 7:4311–4321

74. Bräunling O (1990) Inaugural-Dissertation. PhD Dissertation, University of Heidelberg, Germany

75. James JA & Harley JB (1992) Linear epitope mapping of an Sm B/B' polypeptide. J Immunol 148:2074–2079

76. Ohosone Y, Mimori T, Fujii T, Akizuki M, Matsuoka Y, Irimajiri S, Hardin JA, Craft J & Homma M (1992) Autoantigenic epitopes on the B polypeptide of Sm small nuclear RNP particles. Arthritis Rheum 35:960–966

77. Huntriss JD, Latchman DS & Williams DG (1993) Lupus autoantibodies discriminate between the highly homologous Sm polypeptides B/B' and SmN by binding an epitope restricted to B/B'. Clin exp Immunol 92:263–267

78. Sabbatini A, Bombardieri S & Migliorini P (1993) Autoantibodies from patients with systemic lupus erythematosus bind a shared sequence of SmD and Epstein-Barr virus-encoded nuclear antigen EBNA I. Eur J Immunol 23:1146–1152

79. Muller S, Barakat S, Watts R, Joubaud P & Isenberg D (1990) Longitudinal analysis of antibodies to histones, Sm-D peptides and ubiquitin in the serum of patients with systemic lupus erythematosus, rheumatoid arthritis and tuberculosis. Clin Exp Rheumatol 8:445–453

Manual of Biological Markers of Disease **B2.5**: 1–11, 1994.

Ribosomal RNP

KEITH B. ELKON

The Hospital for Special Surgery, Cornell University Medical Center, New York, NY 10021, U.S.A.

Abbreviations

A. salina: *Artemia salina*; *T. cruzi*: *Trypanosoma cruzi;* r: ribosomal

Key words: ribosome; SLE; synthetic peptide

1. The antigens

The major antigens on the ribosomal RNP are the P (phospho-) proteins, P0, P1 and P2 and 28S rRNA. Autoantibodies against the ribosomal proteins, S10 [1] and L12 [2] have also been detected. Since these occur at lower frequency, the focus of this section will be on the P proteins and 28S rRNA.

2. Synonyms

The A (acidic) proteins. The homologous P1/P2 proteins in lower organisms are: L7/L12, *E. coli*; L44', L44, L45 yeast; eL12, eL12', *A. salina*; and rpA1, *D. melanogaster*.

3. Antibody detection

3.1 Immunofluorescence

Anti-P antibodies can be detected by the indirect immunofluorescence technique [3]. The antibodies produce a typical fine dense staining of the cytoplasm (Fig. 1). For technical details, see this Manual, chapter A2.

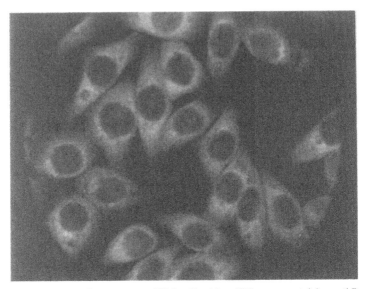

Fig. 1. Indirect immunofluorescence on HEp2 cells with a SLE serum containing anti-P autoantibodies. Note the dense granular cytoplasmic staining. Since the patient did not have antinuclear antibodies, the nucleus is unstained.

3.2 Counterimmunoelectrophoresis

Anti-P antibodies can be detected by immunodiffusion and counterimmuno-electrophoresis techniques [3] although the sensitivity of detection is rather low.

For technical details of these techniques, see this Manual, chapter A3.

3.3 Immunoblotting

Anti-P antibodies can be detected very well by the immunoblotting technique [4–6]. An example is given in Figure 2A. A typical profile of an anti-ribosomal P-protein antibody specificity is composed of 3 bands representing P0 (37 kDa), P1 (19 kDa) and P2 (17 kDa). For technical details, see this Manual, chapter A4.

3.4 Immunoprecipitation of labeled cell extracts

Anti-P antibodies can be detected by immunoprecipitation of ^{35}S methionine labeled cell extracts [5, 7]. For technical details, see this Manual,

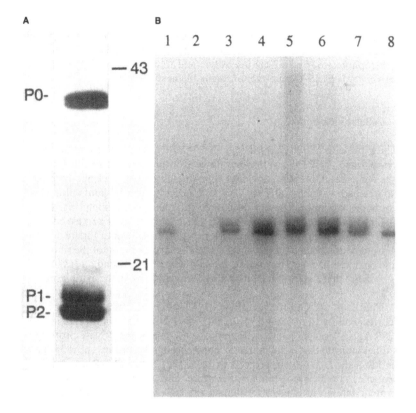

Fig. 2. A. Immunoblot of HeLa ribosomes showing binding to P0 (38 kd), P1 (19 kd) and P2 (17 kd) using a SLE anti-P serum [3, 4].
B. Immunoprecipitation of 28S rRNA by SLE sera.
[32]P-labeled 28S rRNA was transcribed *in vitro* and incubated with Sepharose-protein-A bound IgG obtained from 7 patients with anti-28S rRNA antibodies (lanes 1 and 3–8) or a normal control (lane 2). The RNA was isolated by phenol/chloroform extraction, resolved on a 7M urea/10% polyacrylamide gel and developed by autoradiography [12].

chapter A6. Anti-rRNA antibodies are detected by immunoprecipitation of RNA transcribed *in vitro* (Fig. 2B) [11, 12] (see also this Manual, chapter A7).

3.5 Elisa

The test with the greatest sensitivity and specificity for anti-P is ELISA using a synthetic peptide (NH2-KKEEKKEESEEEDEDMGFGLFD-COOH) corresponding to the C-terminus of *A. salina* or human P2 [8–10]. For technical details, see this Manual, chapter A5.

The relative sensitivities and specificities of the above methods are tabulated in the clinical section of ribosomal RNP (chapter C-2.4). It should be noted from this Table that immunofluorescence and counterimmuno-electrophoresis are relatively insensitive and that the test with the greatest sensitivity is ELISA. Immunoblotting is the next best test.

4. Cellular localization

The P proteins are 3 of approximately 80 proteins that make up the largest cytoplasmic RNP, the ribosome. Since ribosomes are assembled in the nucleoli, the P proteins are also components of the nucleoli. Immuno-fluorescence with high titer anti-P sera (especially those without antinuclear antibodies) therefore shows cytoplasmic and nucleolar staining (Fig. 1).

Unlike all other ribosomal proteins, which are present as a single copy on the ribosome, there are 5 ribosomal P proteins. As shown in Figure 3A, the P proteins are located on the stalk of the 60S (large) subunit of the ribosome and comprise a single copy of P0, and homodimers of P1 and P2 [13]. There is evidence that the P1 and P2 may also be exchangable in the cytoplasm (see [5]).

The 28S rRNA fragment, nt 1944–2002 of human 28S rRNA (corresponding to nt 1863–1921 of rat 28S rRNA) that is antigenic in SLE (28S Ag), is situated in close proximity to the P proteins. RNase protection studies indicate that the pentameric P protein complex protects nt 1838–1936 and that the L12 protein protects nt 1859–1921 of rat 28S rRNA [14]. The sequence, predicted secondary structure and relationship between the protein and RNA antigens for human 28S rRNA are shown in Figure 3B.

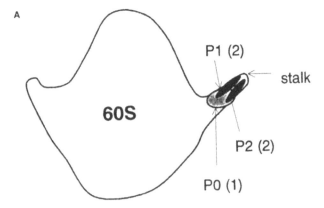

Fig. 3A. Structure of the ribosomal P proteins.
A pentavalent complex comprising P1$_2$•P2$_2$•P0 is situated on the stalk of the 60S (large) subunit of the ribosome.

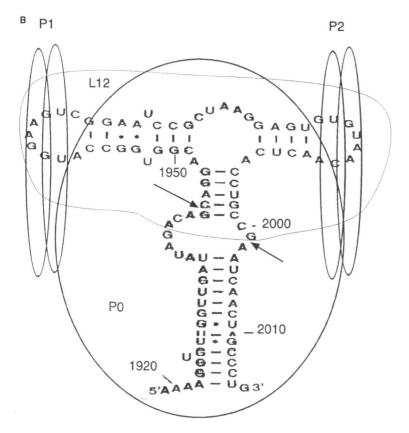

Fig. 3B. Proposed secondary structure of the human 28S Ag (the antigenic part of 28S rRNA situated between the arrows) and a schematic demonstration of the regions of RNA protected by L12 and the P proteins.
The exact location of each protein is hypothetical. The results are based on RNase protection studies performed with rat 28S rRNA [14].

5. Function

The exact functions of the individual P proteins on the ribosome are not fully understood. However, studies performed on the homologous proteins in prokaryotes, lower eukaryotes and mammalian P proteins indicate that the P proteins and the 28S Ag form a functional GTPase domain on the 60S stalk (see [10] and [12] for detailed discussion and references) (Fig. 4). P1/P2 are thought to be required for all 3 phases of protein synthesis *viz.* initiation, translocation, and termination [11, 15, 16]. Selective elution of P1 and P2

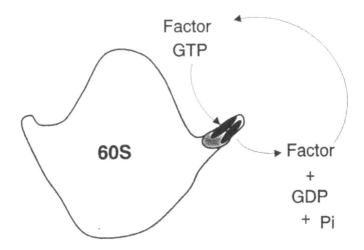

Fig. 4. Function of the P protein 28S Ag complex.
The P protein/RNA complex comprises a functional GTPase domain to which factor-GTP complexes bind. The interaction results in catalysis of the appropriate step in the cycle (initiation, elongation or release). GTP is hydrolyzed and the factor is released.

from ribosomes results in the total loss of GTPase activity and protein synthesis [15, 16]. Similarly, monoclonal [17] and polyclonal [18] anti-P sera inhibit GTPase activity and protein synthesis in general. Since the discovery of self-splicing RNA [19], considerable effort has been applied to investigation of possible catalytic functions of rRNA [20]. The remarkably high degree of conservation of rRNA in general, and the GTPase domain in particular (97% sequence identity between human and yeast), are compatible with an important role for rRNA in either directly catalyzing enzymatic reactions or providing the conformation necessary for these reactions to occur.

6. Presence in other species

SLE anti-P sera have been shown to bind to the P protein homologues from mammals, invertebrates, plants and unicellular eukaryotes [4, 5]. It should be noted that the P0 proteins from *T. cruzi* and Archebacteria are unusual and have several non-conservative amino acid substitutions in their C-termini [21, 22]. Anti-P sera bind weakly to these proteins [21]. *T. cruzi* and yeast have three, rather than two, P1/P2 homologues [23, 24]. Anti-P autoantibodies do not bind to the *E. coli* P protein homologues, L10, L7/L12 [3, 4].

P0

```
         10        20        30        40        50        60
          |         |         |         |         |         |
ATGCCCAGGGAAGACAGGGCGACCTGGAAGTCCAACTACTTCCTTAAGATCATCCAACTA
 M  P  R  E  D  R  A  T  W  K  S  N  Y  F  L  K  I  I  Q  L


         70        80        90       100       110       120
          |         |         |         |         |         |
TTGGATGATTATCCGAAATGTTTCATTGTGGGAGCAGACAATGTGGGCTCCAAGCAGATG
 L  D  D  Y  P  K  C  F  I  V  G  A  D  N  V  G  S  K  Q  M


        130       140       150       160       170       180
          |         |         |         |         |         |
CAGCAGATCCGCATGTCCCTTCGCGGGAAGGCTGTGGTGCTGATGGGCAAGAACACCATG
 Q  Q  I  R  M  S  L  R  G  K  A  V  V  L  M  G  K  N  T  M


        190       200       210       220       230       240
          |         |         |         |         |         |
ATGCGCAAGGCCATCCGAGGGCACCTGGAAAACAACCCAGCTCTGGAGAAACTGCTGCCT
 M  R  K  A  I  R  G  H  L  E  N  N  P  A  L  E  K  L  L  P


        250       260       270       280       290       300
          |         |         |         |         |         |
CATATCCGGGGGAATGTGGGCTTTGTGTTCACCAAGGAGGACCTCACTGAGATCAGGGAC
 H  I  R  G  N  V  G  F  V  F  T  K  E  D  L  T  E  I  R  D


        310       320       330       340       350       360
          |         |         |         |         |         |
ATGTTGCTGGCCAATAAGGTGCCAGCTGCTGCCCGTGCTGGTGCCATTGCCCCATGTGAA
 M  L  L  A  N  K  V  P  A  A  A  R  A  G  A  I  A  P  C  E


        370       380       390       400       410       420
          |         |         |         |         |         |
GTCACTGTGCCAGCCCAGAACACTGGTCTCGGGCCCGAGAAGACCTCCTTTTTCCAGGCT
 V  T  V  P  A  Q  N  T  G  L  G  P  E  K  T  S  F  F  Q  A


        430       440       450       460       470       480
          |         |         |         |         |         |
TTAGGTATCACCACTAAAATCTCCAGGGGCACCATTGAAATCCTGAGTGATGTGCAGCTG
 L  G  I  T  T  K  I  S  R  G  T  I  E  I  L  S  D  V  Q  L


        490       500       510       520       530       540
          |         |         |         |         |         |
ATCAAGACTGGAGACAAAGTGGGAGCCAGCGAAGCCACGCTGCTGAACATGCTCAACATC
    I  K  T  G  D  K  V  G  A  S  E  A  T  L  L  N  M  L  N  I


        550       560       570       580       590       600
          |         |         |         |         |         |
TCCCCCCTTCTCCTTTGGGCTGGTCATCCAGCAGGTGTTCGACAATGGCAGCATCTACAAC
 S  P  F  S  F  G  L  V  I  Q  Q  V  F  D  N  G  S  I  Y  N


        610       620       630       640       650       660
          |         |         |         |         |         |
CCTGAAGTGCTTGATATCACAGAGGAAACTCTGCATTCTCGCTTCCTGGAGGGTGTCCGC
 P  E  V  L  D  I  T  E  E  T  L  H  S  R  F  L  E  G  V  R


        670       680       690       700       710       720
          |         |         |         |         |         |
AATGTTGCCAGTGTCTGTCTGCAGATTGGCTACCCAACTGTTGCATCAGTACCCCATTCT
 N  V  A  S  V  C  L  Q  I  G  Y  P  T  V  A  S  V  P  H  S


        730       740       750       760       770       780
          |         |         |         |         |         |
ATCATCAACGGGTACAAACGAGTCCTGGCCTTGTCTGTGGAGACGGATTACACCTTCCCA
 I  I  N  G  Y  K  R  V  L  A  L  S  V  E  T  D  Y  T  F  P


        790       800       810       820       830       840
          |         |         |         |         |         |
CTTGCTGAAAAGGTCAAGGCCTTCTTGGCTGATCCATCTGCCTTTGTGGCTGCTGCCCCT
 L  A  E  K  V  K  A  F  L  A  D  P  S  A  F  V  A  A  A  P


        850       860       870       880       890       900
          |         |         |         |         |         |
GTGGCTGCTGCCACCACAGCTGCTCCTGCTGCTGCTGCAGCCCCAGCTAAGGTTGAAGCC
 V  A  A  A  T  T  A  A  P  A  A  A  A  A  P  A  K  V  E  A


        910       920       930       940       950
          |         |         |         |         |
AAGGAAGAGTCGGAGGAGTCGGACGAGGATATGGGATTTGGTCTCTTTGAC
 K  E  E  S  E  E  S  D  E  D  M  G  F  G  L  F  D
```

Fig. 5. Nucleotide and derived amino acid sequences of human P0, P1 and P2.
Only the coding regions are shown [25].

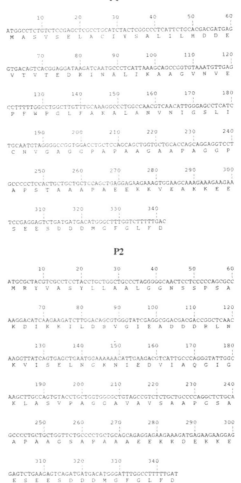

P1

```
        10         20         30         40         50         60
ATGGCCTCTGTCTCCGAGCTCGCCTGCATCTACTCGGCCCTCATTCTGCACGACGATGAG
 M   A   S   V   S   E   L   A   C   I   Y   S   A   L   I   L   H   D   D   E

        70         80         90        100        110        120
GTGACAGTCACGGAGGATAAGATCAATGCCCTCATTAAAGCAGCCGGTGTAAATGTTGAG
 V   T   V   T   E   D   K   I   N   A   L   I   K   A   A   G   V   N   V   E

       130        140        150        160        170        180
CCTTTTTGGCCTGGCTTGTTTGCAAAGGCCCTGCCCAACGTCAACATTGGGAGCCTCATC
 P   F   W   P   G   L   F   A   K   A   L   A   N   V   N   I   G   S   L   I

       190        200        210        220        230        240
TGCAATGTAGGGGCCGGTGGACCTGCTCCAGCAGCTGGTGCTGCACCAGCAGGAGGTCCT
 C   N   V   G   A   G   G   P   A   P   A   A   G   A   A   P   A   G   G   P

       250        260        270        280        290        300
GCCCCCTCCACTGCTGCTGCTCCAGCTGAGGAGAAGAAAGTGGAAGCAAAGAAAGAAGAA
 A   P   S   T   A   A   A   P   A   E   E   K   K   V   E   A   K   K   E   E

       310        320        330        340
TCCGAGGAGTCTGATGATGACATGGGCTTTGGTCTTTTTGAC
 S   E   E   S   D   D   D   M   G   F   G   L   F   D
```

P2

```
        10         20         30         40         50         60
ATGCGCTACGTCGCCTCCTACCTGCTGGCCTGCCCTAGGGGGCAACTCCTCCCCCAGCGCC
 M   R   Y   V   A   S   Y   L   L   A   A   L   G   G   N   S   S   P   S   A

        70         80         90        100        110        120
AAGGACATCAAGAAGATCTTGGACAGCGTGGGTATCGAGGCGGACGACGACCGGCTCAAC
 K   D   I   K   K   I   L   D   S   V   G   I   E   A   D   D   D   R   L   N

       130        140        150        160        170        180
AAGGTTATCAGTGAGCTGAATGGAAAAAACATTGAAGACGTCATTGCCCAGGGTATTGGC
 K   V   I   S   E   L   N   G   K   N   I   E   D   V   I   A   Q   G   I   G

       190        200        210        220        230        240
AAGCTTGCCAGTGTACCTGCTGGTGGGGGCGTGTAGCCGTCTCTGCTGCCCCAGGCTCTGCA
 K   L   A   S   V   P   A   G   G   A   V   A   V   S   A   A   P   G   S   A

       250        260        270        280        290        300
GCCCCTGCTGCTGGTTCTGCCCCTGCTGCAGCAGAGGAGAAGAAAGATGAGAAGAAGGAG
 A   P   A   A   G   S   A   P   A   A   A   E   E   K   K   D   E   K   K   E

       310        320        330        340
GAGTCTGAAGAGTCAGATGATGACATGGGATTTGGCCTTTTTGAT
 E   S   E   E   S   D   D   D   M   G   F   G   L   F   D
```

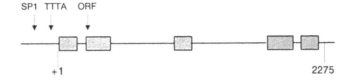

SP1   TTTA   ORF

+1

2275

AMAN-B2.5/8

7. CDNA and derived amino acid sequences

The cDNA sequences of human P0, P1 and P2 [25] are shown in Figure 5. The cDNA sequences for other species can be found in [21, 24–26] and references therein. Structure/function mapping of the proteins are not yet known.

8. Genomic structure

The genomic organization of the human P proteins are not available. However, it is clear from the human cDNAs and studies performed in yeast [24], that each P protein is encoded by a single copy gene. The structure of the gene encoding the rat P2 protein [26] is shown in Figure 6.

9. Relation to other nucleic acids and proteins

These are discussed above and illustrated in Figure 3.

10. Biochemical characteristics

The P proteins are distinctive in their overall net negative charge, high content of alanine residues (19% in P1) and predicted secondary structure ($\sim$70% helical). The C-terminal 17 amino acids of all 3 P proteins are virtually identical (with the exceptions mentioned above) and are highly conserved between species [21–25].

The P proteins are also unusual, but not unique, as ribosomal proteins that are phosphorylated. The physiological role of phosphorylation is unclear but may affect binding to the ribosome. Recent studies suggest that the C-terminal serine residue is phosphorylated by casein kinase II [27].

As shown in Figure 2, the apparent molecular masses of the human P0, P1 and P2 proteins are 38, 19 and 17 kd respectively [4, 5]. The proteins migrate more slowly in SDS-PAGE than would be predicted from their molecular masses. The aberrant migration may be due to the unusual amino acid composition, phosphorylation or other post-translational modification.

←

Fig. 6. Genomic structure of the rat P2 gene.

The gene [26] comprises 5 exons and 4 introns. The open reading frame (ORF) begins at the second nucleotide of exon 2. The TATA box is replaced by a TTTA transcription initiation site at position −30. A presumptive Sp1 (a transcription factor) binding site is shown at position −308. Map not drawn to scale.

The rat P2 gene, like most other mammalian ribosomal protein genes identified to date, share a number of interesting properties. These include the presence of a large number of pseudogenes (there are 8 rat P2 pseudogenes but only a single copy of the authentic P2 gene) and the presence of TTTA, rather than a TATA box, at the RNA polymerase II binding site [26].

Whereas almost all ribosomal proteins are basic, the P proteins are unusual in being acidic. The observed pIs range from 3 to 5 depending on the species and phosphorylation status [21–25].

11. B-cell epitopes

The major epitope on all 3 P proteins is located within the C-terminal 22 amino acids [7–10]. The striking immunodominance of this epitope was shown by the ability of a 22-mer synthetic peptide to completely absorb autoantibody binding to all 3 P proteins on an immunoblot [7]. This observation indicates that SLE sera recognize a linear/continuous epitope on the P proteins although immunoprecipitation studies provide evidence for recognition of conformational/discontinuous epitope(s) as well [7]. The C-termini of the P proteins are the most hydrophilic regions of the proteins and, therefore, predicted to be exposed and mobile. These properties are important for antigenicity as discussed elsewhere [27].

Detailed epitope mapping revealed that the shortest epitope recognized by all anti-P sera tested was 11 residues in length (Fig. 7) but that some sera bound to peptides as short as 7 residues [28]. It is interesting to note that the phosphorylation site of the P proteins is located within the C-terminus [27] (Fig. 7).

Fig. 7. The linear P protein epitope.
The amino acid sequence of the C-terminal 35 residues of human P2 is shown. The characteristic biochemical features of the protein in relation to the antigenicity are indicated. The asterisk at serine 105 indicates the site phosphorylated by casein kinase II [28].

12. T-cell epitopes

There is no published data on T-cell epitopes.

Acknowledgements

The help of Drs. N. Brot, H. Weissbach, W. Danho (Roche Institute of Molecular Biology, NJ), E. Bonfa and P. Hasler are gratefully acknowledged. This work was supported, in part, by the NIH, Bethesda, MD, USA.

References

1. Bonfa E, Parnassa AP, Rhoads DD, Roufa DJ, Wool IG & Elkon KB (1989) Arthritis Rheum 32:1252–1261
2. Sato T, Uchiumi T, Kominami R & Arakawa M (1990) Biochem Biophys Res Comm 172:496–502
3. Bonfa E & Elkon KB (1986) Arthritis Rheum 29:981–985
4. Elkon KB, Parnassa AP & Foster CL (1985) J Exp Med 162:459–471
5. Francoeur A-M, Peebles CL, Heckman KJ, Lee JC & Tan EM (1985) J Immunol 135:2378–2384
6. Nojima Y, Minota S, Yamada A, Takaku F, Aotsuka S & Yokohari R (1993) Ann Rheum Dis 51:1053–1055
7. Elkon KB, Skelly S, Parnassa AP, Moller W, Weissbach H & Brot N (1986) Proc Natl Acad Sci USA 83:7419–7423
8. Bonfa E, Golombek SJ, Kaufman LD, Skelly S, Weissbach H, Brot N & Elkon KB (1987) N Engl J Med 317:265–271
9. Schneebaum AB, Singleton JD, West SG, Blodgett JK, Allen LG, Cheronis JC & Kotzin BL (1991) Am J Med 90:54–62.
10. Teh LS, Hay EM, Amos N, Black D, Huddy A, Creed F, Bernstein RM, Holt PJL & Williams BD (1992) Br J Rheumatol 32:287–290
11. Uchiumi T, Traut RR, Elkon KB & Kominami R (1991) J Biol Chem 266:2054–2062
12. Chu JL, Brot N, Weissbach H & Elkon KB (1991) J Exp Med 174:507–514
13. Uchiumi T, Wahba AJ & Traut RR (1987) Proc Natl Acad Sci USA 84:5580–5584
14. Uchiumi T & Kominami R (1992) J Biol Chem 267:19179–19185
15. Van Agthoven AJ, Maasen JA & Moller W (1977) Biochem Biophys Res Comm 77:989–998
16. MacConnell WP & Kaplan NO (1982) J Biol Chem 257:5359–5366
17. Uchiumi T, Traut RR & Kominami R (1990) J Biol Chem 265:89–95
18. Stacey DW, Skelly S, Watson T, Elkon K, Weissbach H & Brot N (1988) Arch Biochem Biophys 267:398–403
19. Cech TR (1990) Annu Rev Biochem 59:543–558
20. Dahlberg AE (1989) Cell 57:525–529
21. Skeiky YAW, Benson DR, Parsons M, Elkon KB & Reed SG (1992) J Exp Med 176:201–211
22. Hansen TS, Andreasen PH, Dreisig H, Hojrup P, Nielsen H, Engberg J & Kristiansen K (1991) Gene 105:143–150
23. Vazquez MP, Schijman AG, Panebra A & Levin MJ (1992) Nucleic Acids Res 20:2893
24. Remacha M, Saenz-Robles MT, Vilella MD & Ballesta JPG (1988) J Biol Chem 263:9094–9101
25. Rich BE & Steitz JA (1987) Mol Cell Biol 7:4065–4074
26. Chan Y-L & Wool IG (1991) Nucleic Acids Res 19:4895–4900
27. Elkon KB, Bonfa E, Llovet R, Danho W, Weissbach H & Brot N (1988) Proc Natl Acad Sci USA 85:5186–5189
28. Hasler P, Brot N, Weissbach H, Parnassa AP & Elkon KB (1991) J Biol Chem 266:13,815–13,820

Manual of Biological Markers of Disease **B2.6**: 1–22, 1994.
© 1994 *Kluwer Academic Publishers. Printed in the Netherlands.*

Ku and related antigens

WESTLEY H. REEVES, MINORU SATOH, JINGSONG WANG and
AJAY K. AJMANI
*Departments of Medicine and Microbiology and Immunology, Thurston Arthritis Research
Center and UNC Lineberger Comprehensive Cancer Center, University of North Carolina,
Chapel Hill, NC 27599–7280, U.S.A.*

Abbreviations

NFIV: nuclear factor IV; DNA-PK: DNA-dependent protein kinase (350 kDa catalytic subunit); ELISA: enzyme linked immunosorbant assay; IF: immunofluorescence; IB: immunoblotting; CIE: counterimmunoelectrophoresis; ID: double immunodiffusion; IP: immunoprecipitation

Key words: Antinuclear antibodies, Ku antigen, systemic lupus erythematosus, DNA-dependent protein kinase

1. The antigen

Ku (p70/p80).

2. Synonyms

The Ku antigen was first described by Mimori *et. al.* [1] as a nonhistone nuclear antigen producing a precipitin line with sera from certain patients with scleroderma-polymyositis overlap syndrome. As has been common practice, the designation Ku was derived from the initial letters of a patient's name. A number of other nuclear antigens, including nuclear factor IV (NFIV), PSE1, TREF, Ku-2, EBP80, E1BF, and YPF1, are similar or identical to Ku (Table 1). Other synonyms for Ku include p70/p80 [2], 87–70 kDa protein complex [3, 4], and Ki_{86}, Ki_{66} [5]. Exchange of reference sera has established that all of these antigens are the same as Ku. *Note: There has been some confusion in the literature regarding the relationship of the Ku and Ki autoantigens [5, 6]. Although these specificities are frequently present in the same serum, the Ki antigen is a 29.5 kDa protein that appears to be immunologically and biochemically unrealted to Ku, but is identical to the SL antigen [7].*

AMAN-B2.6/1

Table 1. Factors with immunological or biochemical similarity to Ku

Factor	Ref.	Subunits	Proposed function	Similarity to Ku
NFIV[1]	[45]	72K, 84K	Binds adenovirus DNA replication origin, then translocates to internal sites	DNA sequence (p80); amino acid sequence (p70); reacts with anti-Ku antibodies
PSE1[2]	[29]	73K, 83K	Binds to proximal sequence element of U1 snRNA promoter; may stimulate transcription *in vitro*	Amino acid sequence (p70 and p80); reacts with anti-Ku antibodies
TREF[2]	[28, 29]	62K, 82K	Binds to transferrin receptor promoter	Amino acid sequence; reacts with anti-Ku antibodies
EBP-80[2]	[31]	70K, 85K	Methylation of intra-cisternal A-particle long terminal repeats	DNA binding properties; amino acid sequence (p70, positions 39–46, 207–218; p80 positions 906–919, 1984–1989); reacts with anti-Ku antibodies
E1BF[2]	[32]	72K, 85K	Regulation of rDNA transcription	Reacts with anti-Ku antibodies
YPF1[3]	[33, 34]	69K, 85K	Developmental regulation of yolk protein gene expression	DNA sequence homology, antibodies to 85K cross-react with Ku; DNA binding properties
Ku-2[4]	[30]	72K, 83K	B-cell specific factor that binds octamer motif (ATTTGCAT); may stimulate transcription *in vitro*	Amino acid sequence (p70, 7 of 9 positions; p80, 17 of 19 positions); reacts with anti-Ku antibodies
IEF 2505 5705 6707 6706[5]	[37]	64K, 81.2K, 81.2K, 81.9K	Primate-specific proliferation	Co-migration with NFIV in 2D gel database

[1] NFIV is a heterodimer of 72 and 84 kDa protein subunits that binds to the termini of adenovirus type 2 DNA [27]. The nucleotide sequence of cDNAs encoding the larger subunit of NFIV is identical to that of p80 Ku, and the amino acid sequence of a tryptic peptide derived from the small subunit matches amino acids 557–570 of p70 [45].

[2] The putative transcription factors PSE1, TREF, EBP-80, and E1BF are all immunologically cross-reactive with and display amino acid sequence similarity with Ku. PSE1 binds to the proximal and distal sequence elements of the U1 small nuclear RNA gene [29] and is immunologically cross-reactive with, and shares amino acid sequence with, TREF1 (82 kDa) and TREF2 (62 kDa) [28]. EBP-80 is a transcription factor that binds to a regulatory sequence

3. Antibody detection by if, ib, cie, id, or ip

3.1 Immunofluorescence

Autoantibodies to Ku react well by immunofluorescence using methanol or acetone-fixed cells (Fig. 1) as well as cells fixed with low concentrations of paraformaldehyde, but antigenicity is disrupted partially by high concentrations of aldehyde fixatives or extremes of pH (e.g. methanol-acetic acid fixation). The nuclear immunofluorescence pattern displayed by numerous murine anti-Ku mAbs is diffuse and/or speckled with intense homogeneous nucleolar staining [2, 4]. A similar pattern is displayed by human autoimmune sera, but surprisingly nucleolar immunofluorescence is not prominent [1]. The explanation for the strong nucleolar staining displayed by anti-Ku mAbs and the lack of nucleolar staining by many autoimmune sera is not yet known, but may be related to cell cycle-specific association of Ku with the nucleolus [8, 9].

---

in intracisternal A-particle long terminal repeats (LTRs) [73–75] and displays similar DNA binding properties to Ku [31]. EBP-80 is thought to be involved in the methylation response of intracisternal A-particle LTRs and to enhance methylation-sensitive transcription from the LTR [31, 75]. Enhancer 1 binding factor (E1BF), a rat nuclear factor, interacts with ribosomal gene core promoter and upstream enhancer elements, and stimulates RNA polymerase I directed transcription [32].

[3] Recent data suggest that the *Drosophila* yolk protein factor I (YPF1) is likely to be related to Ku (D. Jacoby and Pieter C. Wensink, personal communication). cDNA clones for the β subunit (69 kDa) reveal 23% identity with the human p70 sequence over a 495 amino acid overlap, suggesting that this subunit may be a *Drosophila* p70 homolog. The alpha subunit (85 kDa) is immunologically cross-reactive with Ku, but cDNA clones have not yet been obtained. YPF1, a heterodimer of the two subunits, displays high affinity binding to a 31 bp sequence within the coding region of the yolk protein 1 (yp1) gene as well as weaker "nonspecific" binding. DNA termini are required for both specific and nonspecific binding. YPF1 activity is required for normal steady state levels of yp1 RNA *in vivo*, and is detected in late stages of *Drosophila* oogenesis and embryogenesis, but not in larvae or adult males [33, 34].

[4] Ku-2, a factor with immunological cross-reactivity and amino acid sequence homology (but not identity) with Ku, was purified from Daudi cell lysates [30]. Based on differences in the amino acid sequences obtained by Edman degradation, it was proposed that Ku-2 is a B-cell homologue of Ku [30], however this has not been confirmed by DNA sequence analysis.

[5] A group of primate-specific, proliferation-sensitive nuclear proteins of 64–82 kDa identified by isoelectric focusing have been shown to be identical to Ku [36, 37]. IEFs 5705, 6707, and 6706 (also designated 8z30 and 8z31) appear to be identical to p80, whereas IEF 2505 appears to be the same as p70.

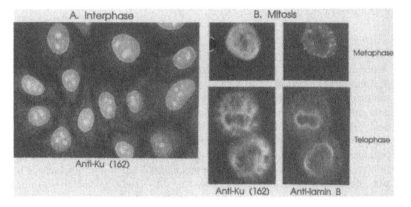

Fig. 1. Immunofluorescence of Ku antigen.
A. Interphase cells. HEp-2 cells were fixed with methanol (10 min. at −20 °C) and stained with
ascitic fluid from monoclonal antibody 162 (anti-p70/p80 dimer) at a dilution of 1:50 followed
by 1:50 FITC-conjugated goat anti-mouse IgG antibodies (Tago, Burlingame, CA). Cells were
viewed using a MRC-600 Confocal Laser Scanning System (Bio-Rad Microscience Division,
Cambridge, MA).
B. Mitotic cells. HEp-2 cells were fixed as in panel A, then stained with murine monoclonal
antibody 162 (1:50) plus human anti-lamin B antiserum (patient AB, [17]). Second antibodies
were rhodamine-conjugated goat anti-mouse IgG and FITC-conjugated goat anti-human IgG,
respectively. Cells were viewed with a MRC-600 Confocal Laser Scanning System using
appropriate filters for rhodamine or FITC fluorescence. Top, metaphase cell showing anti-Ku
(left) and lamin B (right) fluorescence patterns. Bottom, telophase cell showing anti-Ku (left)
and lamin B (right) fluorescence patterns.

3.2 Immunoblotting

Immunoblotting is less useful for screening sera for autoantibodies to Ku than
immunoprecipitation or immunodiffusion [10]. Sera that are reactive by
immunoblotting generally contain high titer antibodies to Ku (titer $\geq 10^{-6}$).
Nevertheless, immunoblotting may be useful in some cases for distinguishing
anti-p70 from anti-p80 autoantibodies [5]. Many high titer anti-Ku sera contain antibodies reactive with both p70 and p80 in immunoblots, but the relative
strength of the reaction with p70 and p80 can be variable, with some sera
showing significantly stronger reactivity with one or the other subunit [5, 11].

3.3 Counterimmunoelectrophoresis

The use of counterimmunoelectrophoresis (CIE) for detecting autoantibodies
to Ku has not been reported. Although CIE is likely to be useful, its sensitivity
and specificity compared with immunoprecipitation have not been
determined.

AMAN-B2.6/4

3.4 Immunodiffusion

The initial report of anti-Ku autoantibodies employed double immuno-diffusion (ID) [1], and this procedure remains a good choice for identifying these antibodies. Although most high titer anti-Ku sera contain precipitating antibodies, ID is relatively insensitive compared with immunoprecipitation and ELISA. Thus, the usefulness of ID is limited, in the Ku system, by its inability to detect low titer Ku positive sera. It is also not certain whether autoantibodies to DNA-PK (p350) or other Ku-associated antigens can be detected by this technique.

3.5 Immunoprecipitation

Autoantibodies to Ku in SLE and other autoimmune disorders recognize primarily conformational epitopes that do not survive immunoblotting [10]. For this reason, immunoprecipitation is the most sensitive and specific means of identifying autoantibodies to Ku (Fig. 2). Both murine monoclonal antibodies (Fig. 2A) and human autoimmune sera (Fig. 2B) characteristically immunoprecipitate prominent 70 and ~86 kDa proteins (p70 and p80, respectively) from extracts of ^{35}S-methionine labeled human cell lines such as K562 (human erythroleukemia) or HeLa (human cervical carcinoma) [2, 5, 12]. If the immunoprecipitates are washed stringently with 0.5 M or greater NaCl or strong detergents, only p70 and p80 appear on autoradiographs of the immunoprecipitated proteins (Fig. 2A,B), whereas if the immuno-precipitates are washed less stringently, additional proteins that form a complex with Ku are immunoprecipitated specifically by both human sera containing anti-Ku antibodies and by murine monoclonal antibodies [13, 14]. The most prominent of these is the 350 kDa subunit of DNA-dependent protein kinase (DNA-PK, see below), but other less strongly labeled proteins are also co-immunoprecipitated specifically. Since autoantibodies to these additional proteins are present in certain sera [14], immunoprecipitation with stringent washing is necessary to identify anti-Ku antibodies definitively.

3.6 ELISA

An antigen capture ELISA based on murine anti-Ku mAbs that correlates well with immunoprecipitation and employs stringent washing conditions to dissociate p350 DNA-PK and other proteins as well as DNA has been developed [14]. This assay is a simple alternative to immunoprecipitation for screening sera for autoantibodies to the native Ku antigen.

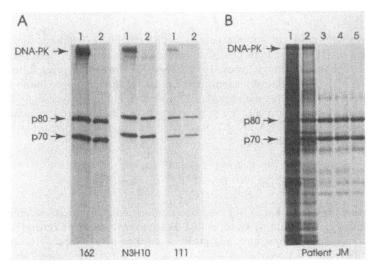

Fig. 2. Immunoprecipitation of Ku antigen.

A. Immunoprecipitation with murine monoclonal antibodies. Human K562 cells were labeled with ^{35}S-methionine and cysteine ("Translabel", ICN, Costa Mesa, CA). Cell lysate was cleared by centrifugation and immunoprecipitated using protein A sepharose beads coated with murine monoclonal antibodies specific for the p70/p80 dimer (mAb 162, [18]), p70 (mAb N3H10, [29, 38]), or p80 (mAb 111, [2, 38]). The beads were washed with 0.15 M NaCl, 50 mM Tris pH 7.5, 2 mM EDTA, 0.3% Nonidet P-40 (lane 1) or mixed micelle buffer (MMB = 150 mM NaCl / 50 mM Tris pH 7.5 / 2 mM EDTA / 0.25 M sucrose / 0.5% SDS / 2.5% Triton X-100, lane 2) and analyzed by SDS-PAGE and autoradiography. Note presence of the 350 kDa DNA-PK protein in lanes 1, but not lanes 2. Binding of the p70 and p80 Ku proteins to the beads was not affected by stringent washing with MMB. For further details on the procedure, see this Manual, Chapter A6.

B. Immunoprecipitation with human autoimmune serum. K562 cell lysate was immuno-precipitated using serum from patient JM as in panel A. Immunoprecipitates were washed with 50 mM Tris pH 7.5, 2 mM EDTA, 0.3% Nonidet P-40 (NP-40) containing 0, 0.15, 0.5, or 1.5 M NaCl (lanes 1–4, respectively) or with MMB (lane 5). As in panel A, the p350 subunit of DNA-PK was dissociated from Ku by washing with MMB (lane 5). The interaction between Ku and p350 was stable at 0 and 0.15 M NaCl in the presence of 0.3% NP-40 (lanes 1–2), but not at 0.5 or 1.5 M NaCl (lanes 3–4).

4. Cellular localization

Immunofluorescence with a variety of monoclonal antibodies suggests that Ku is localized primarily or exclusively to the nucleus and nucleolus [2, 4, 15]. Confocal microscopy confirms that Ku has a granular nucleoplasmic and homogeneous nucleolar distribution in interphase cells and suggests a higher concentration of Ku at the nuclear periphery (Fig. 1A). In mitosis, Ku co-localizes with the condensing chromosomes of early prophase cells, but appears to dissociate from the chromosomes during late prophase, remaining dispersed in the cytoplasm until cell division is complete [12, 16]. A recent study suggests that Ku may remain associated with the periphery of chromosomes throughout mitosis [15]. However, double immunofluorescent staining with a murine anti-Ku monoclonal antibody (162) specific for the p70/p80 heterodimer plus human antiserum specific for lamin B [17] is more consistent with the dispersal of Ku throughout the cytoplasm in mitosis (Fig. 1B). Examination of double-stained cells by confocal microscopy suggests that the cytoplasmic dispersal of Ku begins even before disassembly of the nuclear lamina, and that reassociation of Ku with the chromosomes occurs only after the nuclear lamina is repolymerized (Fig. 1B and M. Ajmani, et al., submitted).

Recent studies suggest that both p70 and p80 carry nuclear localization signal sequences, and that dimerization is not required for their transport to the nucleus [18]. However, based on hydrophobicity analysis of the p70 amino acid sequence, it has also been suggested that Ku might be membrane-associated [19]. More recently, antisera to p70 peptides were found to stain the surface of HeLa cells, as did polyclonal antisera to p70 and p80 [20, 21]. However, there is no direct proof that either subunit is an integral membrane protein, and the significance of these observations remains controversial.

5. Cellular function

Despite intensive study, the cellular role of Ku remains poorly understood. There is general agreement that Ku is a DNA-binding factor and is associated with the 350 kDa subunit of DNA-dependent protein kinase (DNA-PK), but the consequences of DNA binding and p350 catalytic activity are unclear. Based on its binding to DNA termini [22] and nicks or gaps [23], a role for Ku in DNA repair or transposition has been suggested [22]. More recently, a role as a modulator of checkpoint mechanisms activated by DNA damage has been proposed [24].

5.1 Possible role in transcription

A role in transcription has been proposed based on the association of Ku with DNase I hypersensitive sites and its concentration in certain transcriptionally active nuclear domains such as the nucleolus [3], the association of Ku in insects with transcriptionally active chromosome puffs [25], and the structure of the p70 subunit [26]. A number of studies have suggested that binding of Ku to specific DNA sequences can enhance transcription [27–35]. However, a consensus sequence for Ku binding has not emerged, and it is difficult at present to be certain whether the enhancement of transcription by Ku is mediated through binding to a particular sequence or class of sequences, or if it is a general property related to the "nonspecific" binding of Ku to DNA. The phosphorylation of several transcription factors as well as RNA polymerase II *in vitro* by DNA-PK (see below) provides additional indirect evidence that Ku is involved in transcriptional activation. Thus, at the present time, there is increasing evidence that Ku is involved in some aspect of transcriptional regulation. However, an effect on DNA synthesis, repair, or translocation has not been ruled out, and the possibilities are not mutually exclusive.

5.2 Cell cycle regulation of Ku

The expression of Ku mRNA and its nuclear localization appears to be cell cycle-dependent. p70 and p80 gene expression is activated in late G1 during the G_0 to G1 transition of PHA-stimulated human lymphocytes, although levels of the proteins do not change markedly due to their long half-life [8]. The intranuclear localization of Ku during the cell cycle is less clear. In one report, Ku was localized to both the nucleoplasm and nucleolus during early G1, but only to the nucleoplasm during late G1, S, and G2 [8]. However, others have suggested that Ku is primarily nucleoplasmic at the G1/S boundary, with subsequent accumulation within the nucleolus during S phase, reaching a maximum in late S or G2 [9].

6. Presence in other species

The Ku antigen is unusual among the common autoantigens in being expressed at high levels primarily in cells of primate origin [5, 36–38]. Two dimensional gels suggested the complete absence of spots corresponding to p70 and p80 in non-primate cells, leading to the suggestion that Ku might be a primate-specific antigen [36]. However, more recent studies have shown that the Ku antigen is present at low levels in a variety of non-primate cells, including bovine, rodent, and insect cells (Fig. 3, [25, 38, 39], and D. Jacoby and P. Wensink, personal communication). The cDNAs for murine p70 and p80 have been cloned and sequenced, revealing a high degree of amino acid sequence homology with the human homologs [39, 40]. Interestingly, mRNAs for p70 [39] and p80 (our unpublished observations) are present at comparable levels in murine and human cells, despite low levels of Ku protein in murine cells (Fig. 3). Although most murine cell lines and tissues, including spleen, liver, heart and kidney, contain low but detectable levels of Ku protein, L929 cells contain no detectable Ku [38]. The extremely low level, or complete absence, of Ku protein (but not mRNA) in L cells appears to reflect, in part, the rapid degradation of newly synthesized Ku in this cell line [18].

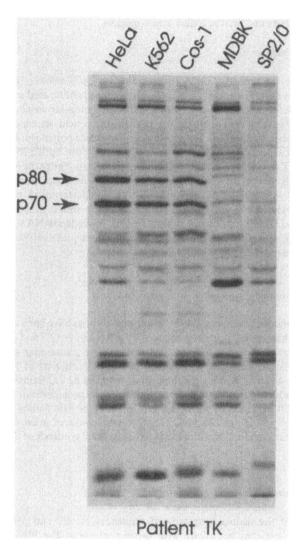

Fig. 3. Ku antigen in human, green monkey, bovine, and murine cell lines.

Human (HeLa and K562), African green monkey (Cos-1), bovine (MDBK), and murine (SP2/0) cells were metabolically labeled with ^{35}S-methionine and cysteine as in Figure 1. Radioactivity in the different cell extracts was determined by scintillation counting, and equal dpm were immunoprecipitated using serum from patient TK, which was shown previously to contain autoantibodies to conserved antigenic determinants of Ku. p70 and p80 are relatively abundant in human and green monkey cell lines, less abundant in bovine (MDBK) cells, and still less abundant in murine (SP2/0) cells.

AMAN-B2.6/9

7. The cDNAs and derived amino acid sequences

7.1 p70 sequence

Nucleotide sequences of cDNAs encoding the p70 protein have been reported by several groups [26, 41, 42]. The deduced amino acid sequence of p70 consists of 609 amino acids with a predicted molecular weight of 69,851 daltons (Fig. 4A), and matches the partial amino acid sequences of p70 chymotryptic peptides [26]. Several apparent sequence polymorphisms in the p70 cDNA sequence have been described, including nucleotide 266 (T or C) and 1778 (T or G) [26, 42], although most of these do not alter the amino acid sequence. Based on the deduced sequence at the N-terminus, it is likely that the initial methionine is removed by an amino-terminal methionine aminopeptidase, but this has not been verified by amino acid sequencing because the N-terminus of p70 is blocked [26]. A single mRNA of ~2.4 kb hybridizing with p70 cDNAs is identified by northern blot analysis in human cells and tissues [19, 26].

7.2 p80 sequence

cDNAs obtained from human liver and spleen libraries have been sequenced, revealing that the p80 protein is encoded by mRNAs of ~2.7 and 3.4 kd [41, 43] that may be derived from a common gene containing alternative polyadenylation sites [43]. The human p80 protein consists of 732 amino acids (Fig. 4B) with a predicted molecular weight of 82,713 daltons [41, 43–45]. The deduced amino acid sequence of p80 shows no significant homology with the p70 sequence, or with other sequences in the PIR database. The N-terminus of p80 is not blocked, and N-terminal protein microsequencing suggests that the initial methionine is removed after synthesis of the protein [41].

7.3 Structural features of p70 and p80

Analysis of the deduced amino acid sequences of p70 and p80 has not revealed significant homology with other proteins in the PIR database. However, several structural features with potential functional significance are apparent (Fig. 4C).

A.

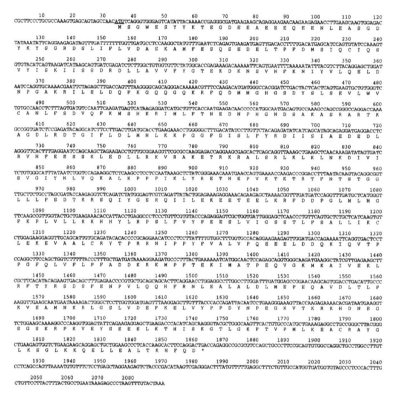

Fig. 4A. Sequences of human p70 and p80.

p70 nucleotide and amino acid sequences.

Sequence data is taken from Reeves and Sthoeger [26] (Genbank accession number J04611). Nucleotide sequence (2090 bp) was determined from cDNAs obtained from a HEpG2 cell library and encodes a 609 amino acid protein of predicted molecular mass 69,851. The murine p70 cDNA and deduced amino acid sequences have also been reported [39]. Murine and human p70 are 83% identical at the amino acid level, whereas the human and *Drosophila* p70 sequences display ~23% amino acid sequence identity.

B.

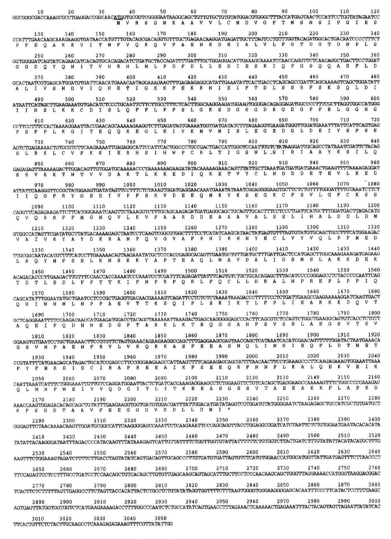

Fig. 4B. p80 nucleotide and amino acid sequences.

Sequence data is taken from Yaneva *et al.* [41] (Genbank accession number J04977). Nucleotide sequence (3052 bp) was determined from cDNAs obtained from human fetal liver, HEp-2, and placental libraries and encodes a 732 amino acid protein of predicted molecular mass 81,914. Murine p80 cDNA clones have also been obtained [38, 40]. Like p70, the murine p80 protein displays strong amino acid sequence homology with human p80.

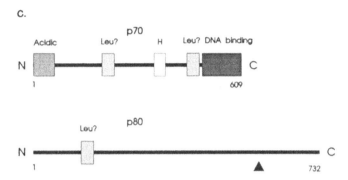

Fig. 4C. Structural features of p70 and p80 proteins.

Positions of structural/functional domains of p70 and p80 are indicated.

Structural features of p70. The amino terminus contains a 61 amino acid domain containing 31% Glu + Asp residues that is also rich in Ser and Thr residues [26]. This domain is similar in size and charge to the acidic transcriptional activator domains of certain transcription factors [64, 65], which may stimulate transcription by enhancing the association of the general transcription factors TFIIB and TFIID with preinitiation complexes [66, 67]. Since net negative charge is essential, but not sufficient, for transcriptional activation [68], it will be necessary to obtain direct experimental evidence that the acidic domain of p70 enhances transcription or binding of general transcription factors before concluding that it functions as a transcriptional activator. The DNA binding domain has been mapped to amino acids 536–609, and is comprised of a sequence similar in structure to the helix-turn-helix motif of lambda cro protein and an adjacent basic domain [53]. Both regions are required for DNA binding. The proposed helix-turn-helix region is a critical component of a major p70 autoepitope (see Fig. 5). Mutations that eliminate DNA binding also eliminate autoantibody recognition, and vice versa.

Two regions having some similarity to leucine zipper motifs [26] are present (Leu?, amino acids 215–242, and 483–504). One of these sequences (amino acids 483–504) is adjacent to the minimal DNA binding domain, as might be expected from the model of Landschultz *et. al.* [69, 70], but there is as yet no direct experimental evidence that this region mediates dimerization.

The existence of a hydrophobic transmembrane spanning domain (H, amino acids 374–390) has been suggested based on sequence analysis and fluorescence data [19–21]. However, direct proof that p70 is an intrinsic membrane protein is not available.

Structural features of p80. The structural features of p80 are less well defined than those of p70. A leucine zipper-like structure (Leu?, amino acids 138–165) has been identified, but like those of p70, its role in p70-p80 dimerization remains to be established. Site-specific, DNA binding-dependent, proteolytic cleavage of p80, producing fragments of 72 and 18 kDa has been reported [71]. Since cleavage greatly reduces reactivity of p80 with human autoimmune sera, it was suggested that the cleavage site may be located in the C-terminal region. The likely position of this site near the C-terminal end is indicated by a triangle.

8. The genes

The genomic organization of the p70 and p80 genes has not yet been determined. Southern blot analysis of human leukocyte genomic DNA digested with BamHI, EcoRI, HindIII, or PstI reveals a reproducible pattern of 6–14 fragments hybridizing with p70 probes, ranging in size from 1.35 to

25.1 kb, for all enzymes tested [42]. Further analysis with additional cDNA probes was consistent with the existence of multiple functional genes and/or pseudogenes. The genomic locus for p70 has been assigned to chromosome 22 [46, 47], although it is not clear that all copies of the gene reside on that chromosome.

Southern blot analysis of genomic DNA with p80 probes also reveals multiple fragments with EcoRI and HindIII, but in contrast to p70, these fragments appear to be generated from a single gene with multiple introns [43].

9. Relation with nucleic acids

An association of Ku with DNA was first suggested by immunoprecipitation data [2] and the release of Ku from nuclei by deoxyribonuclease I treatment [2, 4]. Further evidence for an association with chromatin came from the observation that Ku is associated with condensing chromosomes in early prophase [16].

9.1 DNA binding may be dependent on strand separation

Ku binds to the termini of double-stranded DNA *in vitro*, displaying little affinity for closed circular DNA, single-stranded DNA, DNA-RNA heteroduplexes, or tRNA [22]. In DNase I footprinting assays, both 3' and 5' end-labeled double-stranded DNA termini are protected [22], and binding to DNA termini can be visualized with the electron microscope [27]. Binding to the ends of double-stranded DNA is independent of whether the ends are blunt, or 5'- or 3'-recessed. However, preferential binding to A-T-rich DNA termini has been reported by several groups [27, 31, 45]. More recently, it has been shown that Ku binds to nicks or gaps in double-stranded DNA [30] as well as single- to double-strand transitions [31]. Based on these observations, it has been suggested that transient strand opening at the ends of linear duplex DNA provides an opportunity for Ku to bind [31].

Several studies have suggested that Ku may interact preferentially with particular DNA sequences based on gel shift assays and nuclease or chemical protection data [27–30, 35]. However, interpretation of these assays is complicated by the use of linear target DNA fragments [48, 49], and the preferential binding of Ku to A-T-rich DNA, which is common in upstream regions of eukaryotic genes [31]. Moreover, strand separation might be induced by the binding of other transcription factors to the target DNA. The available data are most consistent with a model in which Ku binds to regions of DNA strand separation and/or single-stranded nicks or gaps rather than a specific sequence.

9.2 Tracking of ku along duplex DNA

An unusual characteristic of the p70/p80 Ku heterodimer is its ability, once bound to DNA, to translocate along duplex DNA like a train on a railroad track, until it is blocked by another factor bound to the DNA [27, 50]. If linear duplex DNA is recircularized once Ku has moved to internal sites, the bound Ku becomes highly resistant to displacement by NaCl [50]. Whereas Ku can be eluted by 0.35 M NaCl from linear DNA, it is resistant to dissociation at up to 2 M NaCl when bound to closed circular DNA. Interestingly, certain bacteriophage T4 accessory factors display a similar property of moving along duplex DNA, and tracking along the DNA appears to be essential for transcriptional activation by these proteins [51]. However, it is not known whether Ku performs an analogous role.

9.3 DNA binding domain

The contribution of p70 and p80 to DNA binding is somewhat controversial. Southwestern blot analysis suggests that p70 can bind DNA whereas p80 cannot [22, 52, 53]. However, more recent studies utilizing *in vitro* translated p70 and p80 suggested that the p70/p80 dimer binds DNA efficiently, whereas p70 and p80 have little DNA binding activity by themselves [54]. Further studies using full-length p70 expressed in baculovirus [52] or p70 fragments expressed in E. coli [53] suggest that p70 binds efficiently to DNA after a prolonged renaturation step. The minimal DNA binding site of p70 has been localized to the carboxy-terminal 74 amino acids of the p70 subunit, which carry a sequence related to the helix-turn-helix of the lambda cro protein, and an adjacent highly basic region, both of which are necessary for DNA binding [53]. The monomeric p70 protein has been shown recently to associate with chromatin *in vivo* in cells infected with a recombinant vaccinia virus directing the synthesis of p70, providing further evidence that p70/p80 dimerization is not required for DNA binding [18]. Although DNA binding by p70 has been demonstrated, it is not yet clear if p70 can translocate along the DNA like the p70/p80 dimer.

10. Relation with other proteins

There is recent evidence that Ku is associated with other nuclear proteins including the p350 subunit of DNA-activated protein kinase (DNA-PK), and possibly other factors as well.

10.1 Association with p350

Recent studies have shown that the 350 kDa DNA-PK protein [55–57] co-immunoprecipitates with Ku [13, 58]. Moreover, p350 has little kinase activity unless bound to Ku, and cannot be cross-linked to DNA unless it is physically associated with Ku. These data strongly imply that Ku is the DNA binding component of DNA-PK, and that Ku is responsible for directing the kinase to its substrates [13, 58].

The 350 kDa DNA-PK component (p350) had been characterized extensively before it was realized that it binds to Ku [55–57, 59]. DNA-PK is an abundant nuclear serine/threonine protein kinase that phosphorylates a wide variety of transcription factors *in vitro*, including Sp1, CTF/NF-1, Oct-1, Oct-2, c-Myc, serum response factor (SRF), p53, c-Jun, c-Fos, the E2 protein of bovine papilloma virus, and the chicken progesterone receptor [59, 60], as well as hsp90, SV40 large T antigen, RNA polymerase II, and Ku [55, 58, 59]. The activity of highly purified DNA-PK is increased over 700-fold by any linear double-stranded DNA greater than 12 basepairs, whereas closed circular DNA is a poor activator [55, 56, 59]. Analysis of the sites phosphorylated by DNA-PK *in vitro* has revealed at least two classes of substrates. One motif consists of either serine or threonine followed by glutamine, although not every S-Q or T-Q sequence is a good substrate. Certain serines or threonines can be phosphorylated in the absence of an adjacent glutamine, notably in the case of RNA polymerase II, but the precise motif or motifs recognized by the kinase have not been identified.

10.2 Association with other factors

Recent studies suggest that, in addition to p350, other nuclear factors may be associated with Ku. A monoclonal antibody to one such factor has been obtained (M. Satoh *et al.*, manuscript in preparation). This antibody, designated F5, recognizes a chromosomal protein or proteins distinct from Ku and p350, and co-immunoprecipitates Ku efficiently under conditions preserving protein-protein interactions between Ku and p350. The function of this factor is currently under investigation, but in view of its chromosomal localization and binding to Ku, it might be speculated that it serves as a "receptor" for the Ku/p350 complex.

11. Biochemical characteristics

Initial biochemical characterization of the Ku autoantigen revealed that it is an acidic nuclear protein antigen that is sensitive to trypsin, mild heating, and pH > 10 or <5 [1]. Size exclusion chromatography suggests an apparent molecular weight of ~300,000 [1], and density gradient centrifugation demonstrates a mobility of $\sim 10S$, with larger forms in the range 10–20S representing the p70/p80 heterodimer associated with DNA fragments [2, 3]. The 70 and 80 kDa subunits are not covalently linked, and can be dissociated by high concentrations of NaCl plus detergents ([18] and M. Satoh, et al. submitted).

Electron microscopic examination of Ku reveals an ellipsoidal shape with a length of 16 nm and a width of 8 nm, corresponding to a M_r of $\sim 160,000$ [27]. In view of these data and the constant 1:1 ratio of p70 to p80 after electrophoresis and staining of purified Ku, it is likely that most Ku exists as a 1:1 heterodimer of p70 and p80, rather than a tetramer or higher order oligomer.

Analysis by isoelectric focusing reveals a pI of 5.6–6.0 for p80 and 6.5–6.77 for p70 [4, 36, 37]. The p80 subunit consists of at least three charge variants of pI 5.74, 5.67, and 5.60, respectively [37]. The relative amounts of the two major forms (pI 5.74 and 5.67) is cell cycle-dependent, with the more acidic form being present at 2.3-fold higher levels than the more basic form in quiescent MRC-5 cells. In proliferating cells, this ratio is altered to ~ 1.2-fold, and in SV40-transformed MRC-5 cells, the ratio is 0.6. Although the alteration in charge might be explained by post-translational modification, evidence of phosphorylation of p80 could not be obtained [37]. However, *in vivo* phosphorylation of both p70 and p80 on serine residues has been described by other investigators [3], and Ku is a target for phosphorylation by p350 DNA-PK *in vitro* [55]. It is not yet certain that phosphorylation is responsible for the proliferation-dependent alterations in the pI of p80.

12. B-cell epitopes

Immunoblotting and competition studies have shown that human anti-Ku autoantibodies bind to a variety of different antigenic determinants, some on p70, others on p80, and still others that are dependent on the quaternary structure of the p70/p80 dimer [2, 5, 18, 38, 61]. These autoepitopes have been defined more precisely using fragments of p70 and p80 expressed as fusion proteins in bacteria [10, 11, 39, 62, 63]. At least four B-cell epitopes of p70 have been described (Fig. 5A).

Like p70, one or more major autoepitope(s) lies near the C-terminus of p80 [10, 11, 62] (Fig. 5B). There is some evidence for a population of autoantibodies that cross-reacts with autoepitope(s) of both p70 and p80 [5], although this has not been confirmed by others. In view of the lack of significant sequence homology between p70 and p80, this might reflect antibodies to a conformational epitope.

Several of the autoepitopes defined by fusion proteins appear to refold at least partially under the conditions of immunoblotting, but other conformational epitopes of Ku do not reassemble after denaturation [10]. A majority of anti-Ku positive sera have little or no reactivity with p70 or p80

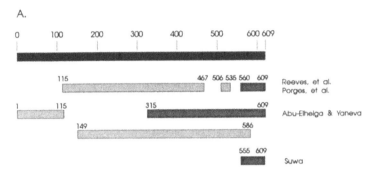

Fig. 5A. Autoepitopes of p70 and p80.
Autoepitopes of p70.
A diagram of the full-length p70 protein and amino acid number is shown at the top. Fragments of the p70 protein expressed as bacterial fusion proteins carrying antigenic determinants recognized by human autoantibodies are shown below. Major epitopes are indicated by dark shading and minor epitopes by light shading. Data from Reeves *et al.* [10], Porges *et al.* [39], Abu-Elheiga and Yaneva [63], and Suwa [11]. A major conformational autoepitope of p70 recognized by human autoimmune sera lies on amino acids 560–609 [10, 11, 39]. Mutational analysis indicates that three distinct segments contribute to the antigenicity of this region [10, 39]. Deletion of amino acids 560–571 or 601–609 eliminates antigenicity, and mutations in the vicinity of amino acids 585–595 reduce antigenicity substantially [39, 72]. However, in another study, amino acids 315–609 appeared to be required for antigenicity, and there was no reactivity of 10 human autoimmune sera with fusion proteins carrying amino acids 551–609 or 514–609 [63]. The explanation for this apparent discrepancy is not clear, but might be related to the use of TrpE fusion proteins in some studies [10, 11, 39] and β-galactosidase fusion proteins in the other [63]. Alternatively, it might reflect differences in the patient populations examined. Minor autoepitopes of p70 have been defined on amino acids 1–115, 506–535, and in the vicinity of amino acids 115–467 [10, 63].

on immunoblots, while reacting strongly with the native antigen. At present, reactivity with these additional conformational epitopes is best defined by competition ELISA based on murine monoclonal antibodies specific for the native Ku antigen. Although only some anti-Ku sera are reactive with p70 or p80 fusion proteins on immunoblots, nearly all of the sera compete with the binding of monoclonal antibody 162, 111, or both [10]. It is possible that recombinant Ku antigens expressed in "native" form in eukaryotic cells will permit these additional conformational determinants to be defined more precisely.

Although lupus autoantibodies tend to bind to epitopes that are highly conserved in evolution, the major epitope of p70 located on amino acids 560–609 is an exception to this rule. Human autoantibodies to this region react strongly with human p70, but only weakly with a recombinant murine p70 fusion protein consisting of amino acids 560–610 [39]. However, other autoepitopes appear to be well conserved in evolution ([38] and D. Jacoby and P.C. Wensink, personal communication).

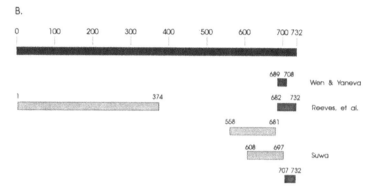

Fig. 5B. Autoepitopes of p80.

A diagram of the full-length p80 protein and amino acid number is shown at the top. Fragments of the p80 protein expressed as bacterial fusion proteins carrying antigenic determinants recognized by human autoantibodies are shown below. Major epitopes are indicated by dark shading and minor epitopes by light shading. Data from Wen and Yaneva [62], Reeves *et al.* [10], and Suwa [11]. Deletion analysis suggested that two autoimmune sera were reactive with an epitope dependent on amino acids 689–732, although it was not determined that this region was sufficient for reactivity [62]. Other studies suggest that the minimal epitope may be carried by amino acids 707–732 [11]. Additional minor autoepitopes of p70 have been localized to amino acids 1–374, and in the vicinity of amino acids 558–681 [10, 11].

13. T-cell epitopes

Not known.

14. Monoclonal antibodies

Several murine monoclonal antibodies specific for Ku are available. These show reactivity with p70, p80, or the p70/p80 heterodimer (Table 2).

Acknowledgements

We thank Dr. Mark Knuth for providing mAb N3H10, Dr. Bruce Batten (Bio-Rad) for assistance with confocal microscopy, and Jenifer J. Langdon for expert technical assistance. We are also grateful to Drs. Doug Jacoby and Pieter C. Wensink for sharing unpublished data on the Ku antigen in Drosophila. This work was supported by grants AR40391, P50-AR42573, P60-AR30701, T32-AR7416, and RR00046 from the United States Public Health Service. Dr. Wang is the recipient of an Arthritis Foundation Postdoctoral Fellowship.

Table 2. Some murine monoclonal antibodies specific for Ku

mAb	Specificity	Subclass	Reference	Reactivity[1]
162	p70/p80 dimer	IgG2a	[2, 18]	IF, IP
111	p80 (610–705)	IgG1	[2, 38]	IF, IP, IB
1177	p80	IgG1	[2]	IF, IP
RZ2	p80 (608–708)	n.a.	[4]	IF, IP, IB
D6C8	p80	n.a.	[62]	IB
2D9	p80	n.a.	[62]	IB
YE8	p80	n.a.	[62]	IB
S10B1	p80 (8–221)	IgG1	[29, 38]	IF, IP, IB
N9C1	p80 (1–374)	IgG1	[29, 38]	IF, IB
RN3	p80	n.a.	[76]	IP
1C4C10	p80	n.a.	[36]	IF, IB
LL1	p80	n.a.	[9]	IF, IB
S5C11	p70	IgG1	[29, 38]	IF, IP, IB
N3H10	p70 (506–541)	IgG2b	[29, 38]	IF, IP, IB
H6	p70	IgM	[15]	IF, IB

[1] Abbreviations – IF: immunofluorescence; IP: immunoprecipitation; IB: immunoblotting; n.a.: not available

References

1. Mimori T, Akizuki M, Yamagata H, Inada S, Yoshida S & Homma M (1981) J Clin Invest 68:611–620
2. Reeves WH (1985) J Exp Med 161:18–39
3. Yaneva M & Busch H (1986) Biochemistry 25:5057–5063
4. Yaneva M, Ochs R, McRorie DK, Zweig S & Busch H (1985) Biochim Biophys Acta 841:22–29
5. Francoeur AM, Peebles CL, Gompper PT & Tan EM (1986) J Immunol 136:1648–1653
6. Fritzler MJ (1992) In: Lahita RG (Ed) Systemic Lupus Erythematosus (pp 273–291). Churchill Livingstone, New York
7. Nikaido T, Shimada K, Shibata M, Hata M, Sakamotos M, Takasaki Y & Sato C (1990) Clin Exp Immunol 79:209–214
8. Yaneva M & Jhiang S (1991) Biochim Biophys Acta 1090:181–187
9. Li LL & Yeh NH (1992) Exp Cell Res 199:262–268
10. Reeves WH, Pierani A, Chou CH, Ng T, Nicastri C, Roeder RG & Sthoeger ZM (1991) J Immunol 146:2678–2686
11. Suwa A (1990) Keio Igaku 67:865–879
12. Mimori T, Hardin JA & Steitz JA (1986) J Biol Chem 261:2274–2278
13. Gottlieb TM & Jackson SP (1993) Cell 72:131–142
14. Satoh M, Langdon J & Reeves WH (1993) Clin Immunol Newsletter 13:23–31
15. Higashiura M, Shimizu Y, Tanimoto M, Morita T & Yagura T (1992) Exp Cell Res 201:444–451
16. Reeves WH (1987) J Rheumatol 14 (Suppl 13):97–105
17. Reeves WH, Chaudhary N, Salerno A & Blobel G (1987) J Exp Med 165:750–762
18. Wang J, Pierani A, Schmitt J, Chou CH, Satoh M, Stunnenberg HG, Roeder RG & Reeves WH (1993) J Cell Sci, in press
19. Chan JYC, Lerman MI, Prabhakar BS, Isozaki O, Santisteban P, Kuppers RC, Oates EL, Notkins AL & Kohn LD (1989) J Biol Chem 264:3651–3654
20. Prabhakar BS, Allaway GP, Srinivasappa J & Notkins AL (1990) J Clin Invest 86:1301–1305
21. Dalziel RG, Mendelson SC & Quinn JP (1992) Autoimmunity 13:265–267
22. Mimori T & Hardin JA (1986) J Biol Chem 261:10375–10379
23. Blier PR, Griffith AJ, Craft J & Hardin JA (1993) J Biol Chem 268:7594–7601
24. Anderson CW (1993) Trends Biochem Sci 18:433–437
25. Amabis JM, Amabis DC, Kaburaki J & Stollar BD (1990) Chromosoma 99:102–110
26. Reeves WH & Sthoeger ZM (1989) J Biol Chem 264:5047–5052
27. de Vries E, van Driel W, Bergsma WG, Arnberg AC & van der Vliet PC (1989) J Mol Biol 208:65–78
28. Roberts MR, Miskimins WK & Ruddle FH (1989) Cell Regulation 1:151–164
29. Knuth MW, Gunderson SI, Thompson NE, Strasheim LA & Burgess RR (1990) J Biol Chem 265:17911–17920
30. May G, Sutton C & Gould H (1991) J Biol Chem 266:3052–3059
31. Falzon M, Fewell JW & Kuff EL (1993) J Biol Chem 268:10546–10552
32. Hoff CM & Jacob ST (1993) Biochem Biophys Res Commun 190:747–753
33. Mitsis PG & Wensink PC (1989) J Biol Chem 264:5195–5202
34. Mitsis PG & Wensink PC (1989) J Biol Chem 264:5188–5194
35. Messier H, Fuller T, Mangal S, Brickner H, Igarashi S, Gaikwad J, Fotedar R & Fotedar A (1993) Proc Natl Acad Sci USA 90:2685–2689
36. Celis JE, Madsen P, Nielsen S, Ratz GP, Lauridsen JB & Celis A (1987) Exp Cell Res 168:389–401
37. Stuiver MH, Celis JE & van der Vliet PC (1991) FEBS 282:189–192

38. Wang J, Chou CH, Blankson J, Satoh M, Knuth MW, Eisenberg RA, Pisetsky DS & Reeves WH (1993) Mol Biol Rep 17:15–28
39. Porges A, Ng T & Reeves WH (1990) J Immunol 145:4222–4228
40. Falzon M & Kuff EL (1992) Nucl Acids Res 20:3784
41. Yaneva M, Wen J, Ayala A & Cook R (1989) J Biol Chem 264:13407–13411
42. Griffith AJ, Craft J, Evans J, Mimori T & Hardin JA (1992) Mol Biol Rep 16:91–97
43. Mimori T, Ohosone Y, Hama N, Suwa A, Akizuki M, Homma M, Griffith AJ & Hardin JA (1990) Proc Natl Acad Sci USA 87:1777–1781
44. Fletcher CF (1988) Purification and characterization of OTF-1, a transcription factor regulating cell cycle expression of a human histone H2b gene: demonstration of its functional identity with NF-III, a factor required for efficient initiation of adenovirus DNA replication. PhD Thesis, Rockefeller University, New York, NY
45. Stuiver MH, Coenjaerts FEJ & van der Vliet PC (1990) J Exp Med 172:1049–1054
46. Mitchell AL, Bale AE, Chan J, Kohn L, Gonzalez F & McBride OW (1989) Cytogenet Cell Genet 51:1045
47. Emanuel BS & Seizinger BR (1990) Cytogenet Cell Genet 55:245–253
48. Quinn JP & Farina AR (1991) FEBS Lett 286:225–228
49. Zhang WW & Yaneva M (1992) Biochem Biophys Res Commun 186:574–579
50. Paillard S & Strauss F (1991) Nucl Acids Res 19:5619–5624
51. Herendeen DR, Kassavetis GA & Geiduschek EP (1992) Science (Wash. DC) 256:1298–1303
52. Allaway GP, Vivino AA, Kohn LD, Notkins AL & Prabhakar BS (1989) Biochem Biophys Res Commun 168:747–755
53. Chou CH, Wang J, Knuth MW & Reeves WH (1992) J Exp Med 175:1677–1684
54. Griffith AJ, Blier PR, Mimori T & Hardin JA (1992) J Biol Chem 267:331–338
55. Lees-Miller SP, Chen YR & Anderson CW (1990) Mol Cell Biol 10:6472–6481
56. Carter T, Vancurova I, Sun I, Lou W & DeLeon S (1990) Mol Cell Biol 10:6460–6471
57. Jackson SP, MacDonald JJ, Lees-Miller S & Tjian R (1990) Cell 63:155–165
58. Dvir A, Peterson SR, Knuth MW, Lu H & Dynan WS (1992) Proc Natl Acad Sci USA 89:11920–11924
59. Anderson CW & Lees-Miller SP (1992) CRC Crit Rev Eukaryotic Gene Exp 2:283–314
60. Bannister AJ, Gottlieb TM, Kouzarides T & Jackson SP (1993) Nucl Acids Res 21:1289–1295
61. Reeves WH, Sthoeger ZM & Lahita RG (1989) J Clin Invest 84:562–567
62. Wen J & Yaneva M (1990) Mol Immunol 27:973–980
63. Abu-Elheiga L & Yaneva M (1992) Clin Immunol Immunopathol 64:145–152
64. Ma J & Ptashne M (1987) Cell 48:847–853
65. Hope IA & Struhl K (1986) Cell 46:885–894
66. Stringer KF, Ingles CJ & Greenblatt J (1990) Nature (Lond.) 345:783–786
67. Lin YS & Green M (1991) Cell 64:971–981
68. Cress WD & Treizenberg SJ (1991) Science (Wash. DC) 251:87–90
69. Landschulz WH, Johnson PF & McKnight SL (1988) Science (Wash. DC) 240:1759–1764
70. Landschulz WH, Johnson PF & McKnight SL (1989) Science (Wash. DC) 243:1681–1688
71. Paillard S & Strauss F (1993) Proteins: structure, function, and genetics 15:330–337
72. Chou CH, Satoh M, Wang J & Reeves WH (1992) Mol Biol Rep 16:191–198
73. Falzon M & Kuff EL (1989) J Biol Chem 264:21915–21922
74. Falzon M & Kuff EL (1990) J Biol Chem 265:13084–13090
75. Falzon M & Kuff EL (1991) Mol Cell Biol 11:117–125
76. Verheijen R, Kuijpers H, Van Venrooij W & Ramaekers F (1988) The 80 kilodalton component of the Ku autoantigen is associated with the nuclear matrix. PhD Thesis, University of Nijmegen, The Netherlands

Manual of Biological Markers of Disease **B2.7**: 1–12, 1994.
© 1994 *Kluwer Academic Publishers. Printed in the Netherlands.*

PCNA

YOSHINAO MURO and ENG M. TAN

W. M. Keck Autoimmune Disease Center, Department of Molecular and Experimental Medicine, The Scripps Research Institute, 10666 North Torrey Pines Road, La Jolla, CA 92037, U.S.A.

Abbreviations

PCNA: proliferating cell nuclear antigen; ELISA: enzyme-linked immuno-sorbent assay; UV: ultraviolet light; IVP: *in vitro* protein product; FP: fusion protein

Key words: autoantibody, autoantigen, PCNA, cell cycle, DNA polymerase δ, DNA replication, DNA repair, conformation-dependent epitopes

1. The antigen

Proliferating cell nuclear antigen (PCNA). In 1978, a new autoantibody specificity was identified by sera from certain patients with systemic lupus erythematosus. The antigen, which was highly expressed in actively dividing cells or in lymphocytes stimulated to proliferate with phytohemagglutinin or concanavalin A, was called proliferating cell nuclear antigen (PCNA) [1].

2. Synonyms

2.1 Cyclin

In 1980, Bravo and Celis [2] identified in two-dimensional gel electrophoresis a protein which was synthesized in greater amounts during the S phase of proliferating and transformed cells. This protein, designated cyclin [3], was subsequently shown to be identical to PCNA [4]. The name PCNA is preferable to avoid confusion because totally distinct proteins, which are involved in cell cycle regulation [5], are also called "cyclins".

2.2 DNA polymerase δ auxiliary protein

In 1986, a protein, which is required by DNA polymerase δ for its catalytic activity in DNA synthesis with templates having low primer/template ratios, was isolated from calf thymus [6]. It was reported from two laboratories that the auxiliary protein of DNA polymerase δ is structurally and immunologically identical to PCNA [7, 8].

3. Antibody detection

3.1 Immunofluorescence

In immunofluorescence studies using anti-PCNA antibodies, some cells are weakly positive whereas others are strongly positive and show a wide variety of nuclear staining morphology (see Section 4) when nonsynchronized but rapidly dividing tissue culture cells are examined (Fig. 1). PCNA can be induced in resting lymphocytes by mitogen stimulation [1, 9, 10]. It should be noted that it is possible to detect anti-PCNA antibodies by immuno-fluorescence only when autoimmune sera have only anti-PCNA antibody or much higher titer of anti-PCNA than other co-existing autoantibodies.

3.2 Double immunodiffusion

The double immunodiffusion (see this book Chapter A3) can be used to demonstrate and identify the precipitin reaction between PCNA antigen and anti-PCNA antibody [1, 9, 10].

3.3 Counterimmunoelectrophoresis

Counterimmunoelectrophoresis (see this book Chapter A3) is also used to detect anti-PCNA antibodies. This method has the advantage of requiring smaller amounts of serum than double immunodiffusion to yield a precipitin line and higher sensitivity [10].

3.4 Immunoblotting

Some anti-PCNA antibodies react with 36 kDa polypeptide in immunoblotting analysis using cell extracts [10–12]. However, some fail to react to 36 kDa cellular protein (our unpublished results) as well as recombinant bacterial fusion protein (see Section 11) [13]. This is related to the observation that human autoantibodies often recognize conformation-

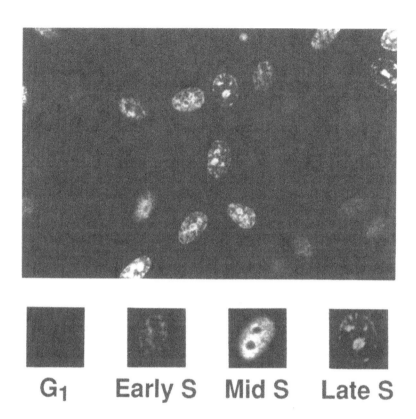

Fig. 1. Immunolabelling of PCNA and typical PCNA staining patterns from G_1 to late S-phase of HEp-2 cells. PCNA is seen in nucleoplasm with relative absence of nucleolar localization in some cells and prominence of nucleolar localization in other cells. There is also localization at the nuclear membrane in some cells. Some of these areas of PCNA localization have been shown to be associated with DNA synthesis at these sites.

dependent epitopes which might be denatured in this procedure. For technical details, see this Manual, chapter A4.

3.5 *Immunoprecipitation*

Anti-PCNA antibodies can be detected by immunoprecipitation using partially purified PCNA from rabbit thymus [10], ^{35}S-labeled cellular protein [14], or *in vitro* protein products [13]. For procedure, see this book, Chapter A6.

3.6 Enzyme-linked immunosorbent assay (ELISA)

Purified PCNA can be used as an antigen to develop ELISA systems (see this book Chapter A5) for rapid and specific detection of anti-PCNA antibodies [11].

4. Cellular localization

Immunofluorescence studies using anti-PCNA antibodies reveal dramatic changes in the nuclear distribution of this protein during the S phase of the cell cycle (Fig. 1) [9, 15, 16]. G_1, G_2 and mitotic cells react weakly with the antibody, while S phase cells show variable nuclear staining patterns both in terms of intensity and distribution of the antigen. Early in S phase, PCNA staining is distributed throughout the nucleoplasm with the exception of the nucleoli. Strong nuclear speckled staining is observed as the cells progress through S phase. Late in S phase, PCNA staining redistributes to reveal punctate pattern with foci throughout the nucleus, the nucleolus, and near the nuclear membrane at the time of maximum DNA synthesis, and again punctate pattern at the S/G_2 phase.

5. Cellular function

PCNA had been suspected of being involved in DNA replication because the distribution of PCNA during the cell cycle revealed dramatic changes and anti-PCNA staining pattern co-localized with areas of radiolabeled thymidine or bromodeoxyuridine uptake in immunofluorescence [16, 17]. Prelich *et al.* showed that PCNA was required for complete replication of simian virus 40 (SV40) *in vitro* [18] and that, in the presence of PCNA, the synthesis of both leading and lagging strands occurred, whereas in the absence of PCNA, only early replication intermediates and short Okazaki-like fragments were synthesized [19]. Exposure of cells to anti-sense oligodeoxynucleotides to PCNA resulted in the complete cessation of DNA synthesis [20]. Human anti-PCNA serum was inhibitory to oligonucleotide synthesis mediated by DNA polymerase δ in an *in vitro* system [21], to induction of DNA synthesis in isolated nuclei [22], and to pUC9 DNA replication in living eggs of *Xenopus leavis* [23].

DNA repair also involves PCNA as shown by increased PCNA staining in non-S phase transformed human amnion cells irradiated with ultraviolet light (UV) under conditions known to induce excision DNA repair synthesis [24], and by the appearance of PCNA immunofluorescence in quiescent human fibroblasts immediately after UV irradiation [25]. PCNA was shown to be required for DNA excision repair in a cell-free system consisting of UV-irradiated plasmid DNA and human cell extracts [26].

6. Presence in other species

PCNA is a highly conserved protein. Sequence data are now available for genes or cDNAs encoding PCNAs of the human [27, 28, 29], rat [30], mouse [31], *Xenopus* [32], *Plasmodium falciparum* [33], *Drosophila* [34], rice [35], soybean [36], *Saccharomyces cerevisiae* [37], and *Schizosaccharomyces pombe* [38]. There are also virus-encoded proteins partially homologous to mammalian PCNA. They include herpes simplex virus type I DNA binding protein [30], baculovirus ETL protein [39], and bacteriophage T4 gene 45 protein [40]. Although these proteins are associated with DNA polymerase function, the homology with mammalian PCNA in these proteins is less than 50%. The epitopes on PCNA recognized by human autoantibody have been observed in a wide range of species including human, rabbit, mouse [1], amphibian [23], ciliated protozoa [41], and soybean [33].

7. The human cDNA and derived amino acid sequence

A cDNA clone of human PCNA has been isolated by Almendral *et al.* [27]. The deduced protein sequence contains 261 amino acid residues with a predicted molecular weight of 29,261 and a high acidic-to-basic residues ratio (41 to 24). Putative DNA-binding domain, which has the potential to form an α-helix-turn-α-helix, was shown in the region of amino acids 61–80. Jaskulski *et al.* [28] isolated from an Okayama-Berg library a PCNA cDNA that was 41 nucleotides longer at the 5′ end than the one described by Almendral *et al.* [27], and was subsequently shown to be a full-length cDNA for human PCNA (Fig. 2) [29].

8. The gene

The gene for human PCNA has also been cloned, completely sequenced, and found to be a unique copy gene [29]. The gene (Fig. 3) was sequenced from a BamHI restriction site located 2.8 kilobases upstream from the preferred site for initiation of transcription (CAP site). The gene consists of six exons separated by five introns and spans 4961 basepairs from the CAP site to the poly(A) signal. In the 5′-flanking sequence, there is a region with promoter activity, as well as other structural elements common to other promoters. Extensive sequence similarities among introns and between introns and exons are interesting features. The human PCNA gene was shown to be localized to chromosome 20, region 20pter → 20q13 [42].

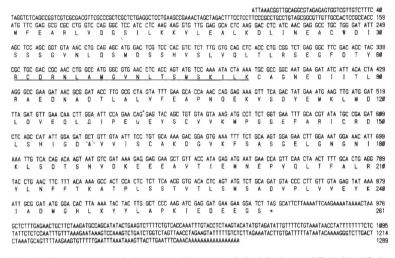

Fig. 2. cDNA sequence and the amino acid sequence derived for human PCNA [27, 28]. A homologous region to the α-helix-turn-α-helix is underlined.

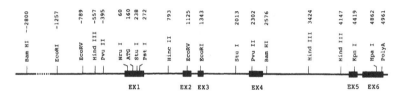

Fig. 3. Physical map of the human PCNA gene. The exons are shown as boxes. For details, see [29].

9. Relation with nucleic acids or with other proteins

The mode of interaction between PCNA, polymerase δ, and DNA is not well understood. The domain of PCNA for binding to polymerase δ has not been defined, and DNA binding property although implicated by the deduced amino acid sequence of PCNA (see Section 7) has not been shown experimentally. In the proposed models of eukaryotic DNA replication [43–45], PCNA is thought to be complexed with the replication factor RF-C (also known as activator 1), which consists of five discrete subunits of 145, 40, 38, 37, and 36.5 kDa protein [46], and DNA polymerase δ or DNA polymerase ε [47].

Among D type cyclins, which are associated with oncogenesis and cell cycle progression [5], D1 and D3 cyclins and their attendant CDKs can be co-immunoprecipitated with PCNA [14].

10. Biochemical characteristics

The antigenicity of PCNA was sensitive to trypsin but resistant to ribonuclease and deoxyribonuclease, suggesting that the antigenic determinant resided in protein and that the antigenicity was not dependent on association with DNA or RNA [1, 10]. On SDS polyacrylamide gels, PCNA was shown to be a single protein of 36 kDa molecular weight [10, 11]. Isoelectric focusing showed that the pI of PCNA protein was 4.8 [10]. Purification studies [6, 10, 48] have shown that PCNA acts as a dimer or multimer, however a conclusive determination has not been made.

It has been shown that there are at least two populations of PCNA in living cells [17]. One is the active form in DNA replication which is matrix bound and less soluble in physiological solvents and the other is the inactive form which is not matrix bound and more readily extracted. The latter cannot be detected immunocytologically after methanol fixation. Immunoreactivities of anti-PCNA antibodies depend on fixation/permeation procedures in immunofluorescence [17, 49] and in flow cytometric analysis [50, 51].

11. B-cell epitopes

Studies of the antigenicity of PCNA fragments generated by limited protease digestion of purified rabbit PCNA demonstrated differences between the epitopes recognized by human anti-PCNA antibodies and murine monoclonal antibodies [12]. Antigenic domains reacting with lupus sera were shown to be located in both the NH_2- and COOH-terminal halves of PCNA, whereas two murine monoclonal antibodies [52] reacted with a different domain located between the two lupus domains. Fine epitope mapping was performed by Huff *et al.* [13], using immunoprecipitation of *in vitro* translated PCNA and immunoblotting of fusion proteins. Overlapping 15-mer synthetic peptides that spanned the entire sequence of the protein were tested for reactivity with two murine monoclonals raised against purified PCNA and a rabbit anti-synthetic peptide serum. These experimentally induced antibodies reacted with expected synthetic peptides, whereas none of 14 human lupus sera containing anti-PCNA antibodies were reactive. Thus, the epitopes recognized by human antibodies were not present in continuous sequence 15-mer peptides of linearized PCNA.

Epitope mapping using deletion mutants of PCNA revealed heterogeneity of human autoantibodies to PCNA. All 14 human sera immunoprecipitated *in vitro* translated full-length protein, but had different reactivities with truncated products. In immunoblotting, nine human sera failed to react with any fusion proteins, whereas five others were reactive in different patterns. In general, the lupus sera could be divided into four subgroups as shown in Table 1. The epitopes recognized by the 14 human lupus sera could be

Table 1. Immunoreactivity of *in vitro* protein products (IVPs) and fusion proteins (FPs) of PCNA

Protein product	amino acid sequence	Reactivity with antibodies														
		SLE I					SLE II				SLE III			SLE IVA SLE IVB	
		AI	YK	MI	KU	JO	CR	OK	CA	EB	PT	AK	YO	NE	FL
IVP – Immunoprecipitation															
IVP 1	1–261	+	+	+	+	+	+	+	+	+	+	+	+	+	+
IVP 2	1–211	−	+	+	+	+	+	−	+	−	−	−	−	+	−
IVP 3	1–150	−	−	−	−	−	+	−	−	−	−	−	−	+	−
IVP 4	1–128	−	−	−	−	−	+	+	+	+	−	−	−	−	+
IVP 5	1– 85	−	−	−	−	−	−	+	+	+	−	−	−	−	+
IVP 6	1– 67	−	−	−	−	−	−	−	−	−	−	−	−	−	−
FP-Immunoblotting															
FP 1	4–261	−	−	−	−	−	−	−	−	−	+	+	+	+	+
FP 2	38–261	−	−	−	−	−	−	−	−	−	+	+	+	+	+
FP 3	67–261	−	−	−	−	−	−	−	−	−	+	+	+	+	+
FP 4	99–261	−	−	−	−	−	−	−	−	−	+	+	+	+	+
FP 8	4–150	−	−	−	−	−	−	−	−	−	+	+	+	+	+
FP 9	4–128	−	−	−	−	−	−	−	−	−	−	+	+	+	+
FP 11	4– 68	−	−	−	−	−	−	−	−	−	−	−	−	−	−
FP 13	67–150	−	−	−	−	−	−	−	−	−	−	+	−	−	+

mapped as illustrated in Figure 4. Lupus group II sera were reactive only by immunoprecipitation, and the regions necessary for reactivity could be localized to the stippled and hatched areas. Lupus group III sera were nonreactive in immunoprecipitation with truncated translation products but were reactive with fusion proteins, and the regions necessary for reactivity were mapped to the shaded areas. The minor groups IVA and IVB incorporated properties of both groups II and III. These observations strongly suggest that human autoantibodies to PCNA are directed to highly conformation-dependent epitopes.

Tsai *et al.* [53] showed that the epitopes recognized by human autoantibodies also involve certain sulfhydryl groups. Thimerosal, a mercury-containing sulfhydryl blocking compound, markedly reduced or abolished the reactivity of this autoantibody-defined PCNA epitope in indirect immunofluorescence, immunodiffusion, and immunoprecipitation.

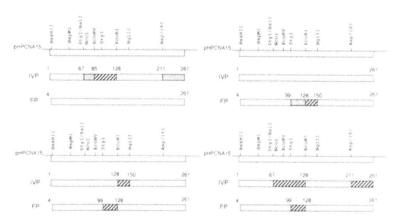

Fig. 4. Epitope mapping of reactive regions in PCNA deduced by subtraction analysis of data derived from immunoprecipitation and immunoblotting [13]. The top bar in each panel shows cDNA restriction enzyme map of PCNA. SLE group II sera (top left) were reactive only by immunoprecipitation, and the regions necessary for reactivity could be localized to the stippled areas by some sera and to the hatched area by other sera. SLE group III sera (top right) were nonreactive with truncated translation products but were reactive with fusion proteins, and the regions necessary for reactivity were mapped to the shaded areas. The reactivities of group IVA are shown in bottom left of the figure, and those of group IVB, in bottom right of the figure. IVP: *in vitro* protein product; FP: fusion protein.

12. T-cell epitopes

So far, there have been no published reports of T-cell epitopes on PCNA.

References

1. Miyachi K, Fritzler MJ & Tan EM (1978) J Immunol 121:2228–2234
2. Bravo R & Celis JE (1980) J Cell Biol 84:795–802
3. Bravo R, Fey SJ, Bellatin J, Mose Larsen P, Arevalo J & Celis JE (1981) Exp Cell Res 136:311–319
4. Mathews MB, Bernstein RM, Franza BR & Garrel JI (1984) Nature 309:374–376
5. Sherr CJ (1993) Cell 73:1059–1065
6. Tan C-K, Castillo C, So AG & Downey KM (1986) J Biol Chem 261:12310–12316
7. Bravo R, Frank R, Blundell PA & Macdonard-Bravo H (1987) Nature 326:515–517
8. Prelich G, Tan C-K, Kostura M, Mathews MB, So AG, Downey KM & Stillman B (1987) Nature 326:517–520
9. Takasaki Y, Deng J-S & Tan EM (1981) J Exp Med 154:1899–1909
10. Takasaki Y, Fishwild D & Tan EM (1984) J Exp Med 159:981–992
11. Ogata K, Ogata Y, Nakamura RM & Tan EM (1985) J Immunol 135:2623–2627
12. Ogata K, Ogata Y, Takasaki Y & Tan EM (1987) J Immunol 139:2942–2946
13. Huff JP, Roos G, Peebles CL, Houghten R, Sullivan KF & Tan EM (1990) J Exp Med 172:419–429
14. Xiong Y, Zhang H & Beach D (1992) Cell 71:505–514
15. Celis JE & Celis A (1985) Proc Natl Acad Sci USA 82:3262–3266
16. Bravo R & Macdonald-Bravo H (1985) EMBO J 4:655–661
17. Bravo R & Macdonald-Bravo H (1987) J Cell Biol 105:1549–1554
18. Prelich G, Kostura M, Marshak DR, Mathews MB & Stillman B (1987) Nature 326:471–475
19. Prelich G & Stillman B (1988) Cell 53:117–126
20. Jaskulski D, deRiel JK, Mercer WE, Calabretta B & Baserga R (1988) Science 240:1544–1546
21. Tan C-K, Sullivan K, Li X, Tan EM, Downey KM & So AG (1987) Nucleic Acids Res 15:9299–9308
22. Wong RL, Katz ME, Ogata K, Tan EM & Cohen S (1987) Cell Immunol 110:443–448
23. Zuber M, Tan EM & Ryoji M (1989) Mol Cell Biol 9:57–66
24. Celis JE & Madsen P (1986) FEBS Lett 209:277–283
25. Toschi L & Bravo R (1988) J Cell Biol 107:1623–1628
26. Shivji MKK, Kenny MK & Wood RD (1992) Cell 69:367–374
27. Almendral JM, Huebsch D, Blundell PA, Macdonald-Bravo H & Bravo R (1987) Proc Natl Acad Sci USA 84:1575–1579
28. Jaskulski D, Gatti G, Travali S, Calabretta B & Baserga R (1988) J Biol Chem 263:10175–10179
29. Travali S, Ku D-H, Rizzo MG, Ottavio L, Baserga R & Calabretta B (1989) J Biol Chem 264:7466–7472
30. Matsumoto K, Moriuchi T, Koji T & Nakane PK (1987) EMBO J 6:637–642
31. Yamaguchi M, Hayashi Y, Hirose F, Matsuoka S, Moriuchi T, Shiroishi T, Moriwaki K & Matsukage A (1991) Nucleic Acids Res 19:2403–2410
32. Leibovici M, Gusse M, Bravo R & Mechali M (1990) Dev Biol 141:183–192
33. Kilbey BJ, Fraser I, McAleese S, Goman M & Ridley RG (1993) Nucleic Acids Res 21:239–243
34. Yamaguchi M, Nishida Y, Moriuchi T, Hirose F, Hui C-C, Suzuki Y & Matsukage A (1990) Mol Cell Biol 10:872–879
35. Suzuka I, Daidoji H, Matsuoka M, Kadowaki K, Takasaki Y, Nakane PK & Moriuchi T (1989) Proc Natl Acad Sci USA 86:3189–3193
36. Suzuka I, Hata S, Matsuoka M, Kosugi S & Hashimoto J (1991) Eur J Biochem 195:571–575
37. Bauer GA & Burgers PMJ (1990) Nucleic Acids Res 18:261–265
38. Wassem NH, Labib K, Nurse P & Lane DP (1992) EMBO J 11:5111–5120
39. O'Rielly DR, Crawford AM & Miller LK (1989) Nature 337:606

40. Tsurimoto T & Stillman B (1990) Proc Natl Acad Sci USA 87:1023–1027
41. Olins DE, Olins AL, Cacheiro LH & Tan EM (1989) J Cell Biol 109:1399–1410
42. Ku D-H, Travali S, Calabretta B, Huebner K & Baserga R (1989) Somat Cell Molec Gen 15:297–307
43. Tsurimoto T & Stillman B (1989) EMBO J 8:3883–3889
44. Burgers PMJ (1991) J Biol Chem 266:22698–22706
45. Diffley JFX (1992) Trends Cell Biol 2:298–303
46. Lee S-H, Kwong AD, Pan Z-Q & Hurwitz J (1991) J Biol Chem 266:594–602
47. Lee S-H, Pan Z-Q, Kwong AD, Burgers PMJ & Hurwitz J (1991) J Biol Chem 266:22707–22717
48. Nichols AF & Sancar A (1992) Nucleic Acids Res 20:2441–2446
49. Van Dierendonck JH, Wijsman JH, Keijzer R, Van de Velde CJH & Cornelisse CJ (1991) Am J Pathol 138:1165–1172
50. Landberg G & Roos G (1991) Cancer Res 51:4570–4574
51. Landberg G & Roos G (1991) Acta Oncol 30:917–921
52. Ogata K, Kurki P, Celis JE, Nakamura RM & Tan EM (1987) Exp Cell Res 168:475–486
53. Tsai W-M, Roos G, Hugli TE & Tan EM (1992) J Immunol 149:2227–2233

Manual of Biological Markers of Disease **B2.8**: 1–14, 1994.

Phospholipids

E. NIGEL HARRIS

Department of Medicine, Division of Rheumatology, School of Medicine, University of Louisville, Louisville, Kentucky 40292, U.S.A.

Abbreviations

APS: Anti-phospholipid syndrome; ELISA: Enzyme-linked immunosorbent assay; SLE: systemic lupus erythematosus

Key words: anticardiolipin antibodies, antiphospholipid antibodies, lupus anticoagulant, Antiphospholipid Syndrome, anti-β_2 glycoprotein 1

1. The antigen

Phospholipids are antigens for a heterogeneous group of antibodies [1]. Different diseases may induce phospholipid binding antibodies with different specificities, but a single disease may also induce antibodies with different phospholipid binding specificities (Fig. 1).

2. Synonyms

Cardiolipin is the most frequently utilized antigen for detection of antiphospholipid antibodies. However, other negatively charged phospholipids, such as phosphatidyl serine and phosphatidic acid can be used in place of cardiolipin [2–5], particularly for detection of sera from patients with the Antiphospholipid Syndrome. Because of this cross-reactivity, the general term "antiphospholipid antibody" is preferred [1] as opposed to "anticardiolipin" or "anti-phosphatidylserine" antibody.

3. Antibody detection

Antiphospholipid antibodies are detected by a variety of techniques, but different tests may detect subgroups with differing specificities (Fig. 1).

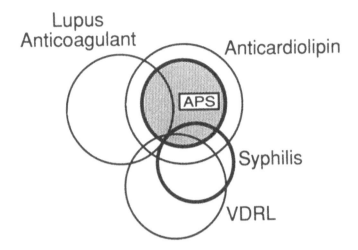

Fig. 1. Venn Diagram to show hypothetical grouping of antiphospholipid antibodies. There are 5 circles depicted in the diagram, the outer 3 circles representing the tests by which antiphospholipid antibodies are detected – VDRL, anticardiolipin, and lupus anticoagulant. The 2 inner circles represent 2 major but separate disorders in which these tests may be positive. In patients with the Antiphospholipid Syndrome and related autoimmune disorders such as Systemic Lupus Erythematosus (SLE), one, two, or all three tests may be positive. In Syphilis, the VDRL is always positive, the anticardiolipin test sometimes positive, and lupus anticoagulant test is almost always negative. It is very likely that antibodies in SLE and Syphilis have different specificities [11–13].

3.1 Agglutination and complement fixation tests

Either the VDRL test – an agglutination assay, or the Wasserman test – a complement fixation assay, can be used to detect antiphospholipid antibodies induced in syphilis [1]. The antigen used in these tests is a mixture of cardiolipin, phosphatidyl choline and cholesterol (the "VDRL" antigen) [6–10]. Phosphatidylcholine and cholesterol (in the VDRL antigen) are probably not bound but may alter the conformation of cardiolipin in such a way as to favour its binding by antiphospholipid antibodies of syphilis [2, 11–13]. Although antiphospholipid antibodies induced in the Antiphospholipid Syndrome or Systemic Lupus can produce positive VDRL or Wasserman tests [1, 14–17], titers are often low positive [16] suggesting that antibodies from patients with autoimmune disorders are not specific for cardiolipin in the "VDRL configuration".

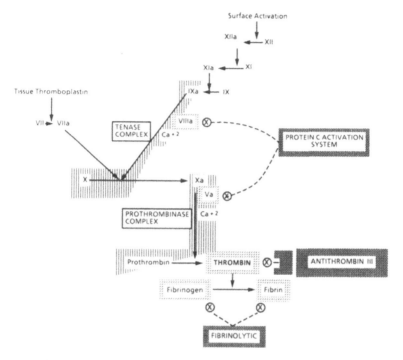

Fig. 2. A summary of the clotting cascade.

Antiphospholipid antibodies may act at a number of steps in the cascade, and these actions may explain their effects *in vitro* (inhibition of coagulation) and *in vivo* (promotion of coagulation and thrombosis). The clotting cascade results from the activation and interaction of proteases circulating in plasma. Fibrin (clot) formation is the end result of these interactions. There are 2 pathways by which the cascade might proceed, the "extrinsic pathway" (activation of the coagulation protein designated Factor VII initiates this pathway) or the "intrinsic pathway" initiated by surface activation of Factor XII). Once the pathways are initiated, a series of enzymatic reactions take place in an expanding fashion ending with formation of fibrin. Some of these reactions require phospholipids as catalysts. For example, conversion of factor X to its activated form, factor Xa, requires formation of a complex of proteins involving factors IXa, VIIIa, and X in the presence of calcium and phospholipids – together these are called the "tenase complex". Activation of prothrombin (factor II) to thrombin (factor IIa) requires formation of the "prothrombinase complex" (see Fig. 3). *In vitro*, antiphospholipid antibodies bind either the phospholipid catalyst [24–27] or a phospholipid-coagulation protein complex [28–30] and so delay fibrin formation – it is this inhibition of clotting and prolongation of clotting time that is referred to as "lupus anticoagulant" activity of the antibodies.

In vivo, once clotting is initiated, other systems are activated (e.g. the fibrinolytic system, protein C activation) to counteract clotting. Some of the latter systems, such as protein C activation also require phospholipid catalysts. These catalysts are provided by membranes of endothelial cells and platelets which contain phospholipids. Antiphospholipid antibodies can inhibit protein C activation, perhaps by binding the phospholipid membrane catalyst alone [31–35] or by binding the protein C – phospholipid complex [29], and, in this way, promote thrombosis.

3.2 Solid phase immunoassays

Antiphospholipid antibodies present in the Antiphospholipid Syndrome are best detected by solid phase immunoassays which use cardiolipin [18–20] (or other negatively charged phospholipids [4, 21–23]) as antigen – the "anticardiolipin test" [18]. Patients with syphilis can give positive anticardiolipin tests but this occurs only with some syphilis sera, and there is no correlation with VDRL titers [11].

3.3 The lupus anticoagulant test

Antiphospholipid antibodies in some patients with the Antiphospholipid Syndrome can be detected by an ability to prolong certain clotting tests. Prolongation is secondary to antibody inhibition of the "prothrombinase complex" and delay in thrombus (fibrin) formation (Figs. 2, 3). One or more of the following clotting tests may be prolonged – the partial thromboplastin time (PTT) [36–39], Kaolin Clotting time (KCT) [40, 41], Russell Viper Venom Time (RVVT) [42, 43] or the Tissue Thromboplastin Inhibition Time (TTIT) [44]. The ability to inhibit clotting is a "functional effect" of these antibodies, as opposed to "binding activity" measured by ELISA assays. Thus, the lupus anticoagulant and anticardiolipin tests detect antiphospholipid antibodies with potentially different characteristics accounting for an estimated 20% discordance between the two tests.

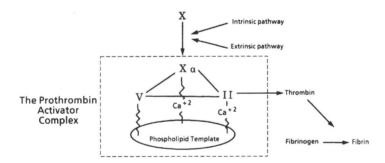

Fig. 3. A vital step in the coagulation cascade is thrombin generation, since thrombin plays a dominant role in fibrin (clot) formation and protein C activation (which counteracts clotting) – see legend of Figure 2 for more details. Prothrombin (factor II) is cleaved by activated Factor X in the presence of factor V. This reaction requires Calcium and is catalyzed by phospholipids – the phospholipid, phosphatidylserine, appears very important for catalysis. Antiphospholipid antibodies bind phosphatidylserine and in this manner they inhibit formation of the "prothrombinase" complex (Factor Xa, II, V in the presence of calcium and phospholipid as shown diagrammatically). Disruption of the complex delays thrombin generation and clot formation. It is this action of antiphospholipid antibodies that explains their ability to prolong clotting *in vitro*.

Antiphospholipid antibodies in syphilis do not exhibit lupus anticoagulant activity [45, 46].

4. Cellular localization

Phospholipids are major components of membranes of all living cells and of organelles within these cells [47, 48]. Phospholipids are usually arranged in these membrane structures as bilayers, with polar head groups facing outwards and inwards, towards the aequeous extra- and intracellular spaces, and the hydrophobic regions facing each other (Fig. 4). Phosphatidyl serine, a target antigen for some antiphospholipid antibodies [3–5, 24–27], is located primarily on the inner leaflet of the bilayer of mammalian cell membranes [48]. Phosphatidyl choline, phosphatidyl ethanolamine, and sphigomyelin

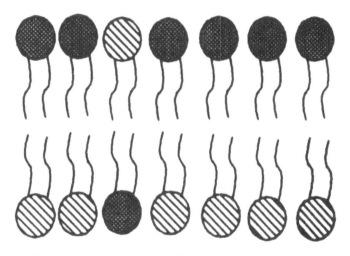

Fig. 4A. Arrangement of phospholipids in bilayer. One-dimensional view.

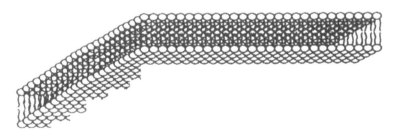

Fig. 4B. Three-dimensional view.

are other major phospholipids of mammalian cell membranes, but none of these three phospholipids are bound significantly by antiphospholipid antibodies [1–3, 10, 11].

Cardiolipin is located in bacterial membranes, mitochondria, and chloroplasts [49, 50]. Cardiolipin in mitochondria is almost exclusively located in the inner leaflet of the membrane. Cytochrome c is present in close proximity to cardiolipin within the mitochondrial membrane, and this arrangement appears to affect cytochrome c function [51, 52].

5. Cellular function

Several functions have been demonstrated.

5.1 Membrane barrier

The phospholipid bilayer of cellular and organelle membranes serve as a barrier primarily separating intracellular or intra-organellar constituents from the extracellular milieu [47, 48].

5.2 Functional effects on membrane proteins

A number of proteins are located within the lipid bilayer of cell membranes, and it is likely that the function of these proteins may be dependent on specific phospholipids located in close proximity [48–52]. The best example is cytochrome c, whose function within mitochondria appears dependent on cardiolipin located nearby in the membrane [51, 52].

5.3 Formation of membrane channels

By assuming a hexagonal (H_{II}) configuration (Fig. 5), membrane phospholipids may form "channels" within the cell membrane so allowing passage of small molecules such as calcium and other cations [49, 50].

5.4 Catalytic role in the clotting cascade

Of particular relevance to the Antiphospholipid Syndrome is the catalytic role of platelet and endothelial cell membrane phospholipids in the coagulation system (Fig. 2) [48].

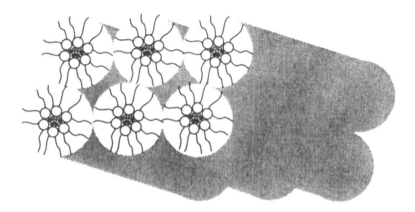

Fig. 5. "Three"-dimensional view of phospholipids arranged in tubular hexagonal structures. Formation of these tubular structures in the cell membrane can result in formation of channels wⁱich enable passage of small molecules.

6. Biochemical characteristics

Phospholipids are diacyl phosphoglycerides made up of a phosphate head group attached to a glyceride (Fig. 6A). Cardiolipin is made up of 2 molecules of phosphatidic acid joined by a glycerol molecule attached to the two phosphate groups (Fig. 6B).

Because phospholipids have both polar and hydrophobic regions, they are defined as being amphipathic. Such molecules aggregate spontaneously into highly organized structures in aequeous solution. These structures include *liposomes* (Fig. 7), *micelles* (Fig. 7), *hexagonal* (H_{II}) [53–55] (Fig. 5), cubic, inverted micellar, and other intermediate configurations [12]. Each configuration depends on the type of phospholipid, mixture of phospholipids, proteins with which the phospholipids are associated, the presence or absence of cations, temperature, and other physiologic conditions [12, 13, 47].

7. B-cell epitopes

Individual phospholipid molecules are relatively small (M.W. 700 to 1000). It is likely that antiphospholipid antibodies do not bind single phospholipids but large aggregates of these molecules in aequeous solution. Costello and Green estimated that 11,000 to 16,000 molecules of cardiolipin are bound per molecule of anticardiolipin antibody [56]. Thus, the macromolecular *configuration* of phospholipids in aequeous solution is probably as important as the *chemical structure* and *net charge* of individual molecules in determining antiphospholipid binding.

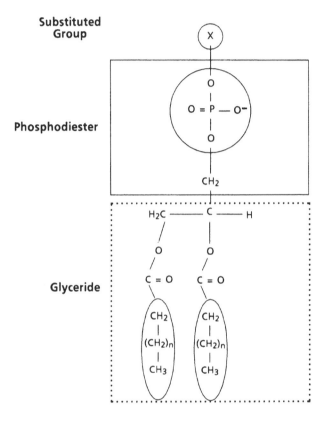

PHOSPHOLIPID

Fig. 6A. Basic structures of phospholipid and cardiolipin.
Phospholipids are made up of a phosphate group attached to a glyceride moiety by a phosphodiester bond. The (polar) phosphate group may have various attached molecules, such as serine, choline, or ethanolamine [47]. These attached molecules may influence antibody access to the phosphate (see text). The glyceride portion of the molecule is also important since the antibody will not bind phosphorylated molecules (e.g. DNA [1–3]) other than phospholipids [1], nor even phospholipids with glycerides containing fatty acid carbon chains with less than 14 carbons [10].

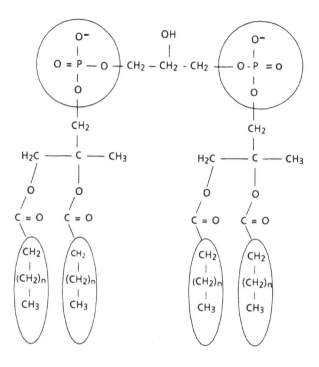

CARDIOLIPIN

Fig. 6B. Cardiolipin is made up of 2 phosphatidic acid groups joined by a central glycerol moiety. Anticardiolipin antibodies are believed to bind the 2 phosphate groups [7–9].

7.1 Chemical structure and charge

Basic requirements for antiphospholipid antibody binding to phospholipids are the presence of a phosphate head group with a net negative charge attached to a glyceride moiety containing fatty acids of 16 or more carbon atoms (Fig. 6) [1–3, 5–13]. Antibodies will not bind the phosphate head group if it is attached to a positively charged molecule, such as choline or sphigosine, possibly because the positively charged group blocks antibody access to the phosphate [4].

7.2 Configuration

Phospholipids aggregate into highly organized structures in aqueous solution (Figs. 5, 7), which influence antibody binding. Some workers report

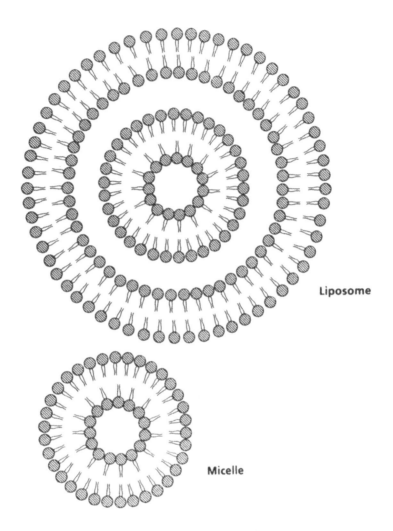

Liposome

Micelle

Fig. 7. Pictoral model of liposomes consisting of concentric bilayers of phospholipids separated by aequeous solution. Micelles are composed of a single sphere of phospholipids arranged as a bilayer enclosing aequeous solution in the center.

that antibodies with lupus anticoagulant activity bind egg phosphatidylethanolamine (EPE) in hexagonal (H_{II}) configuration, but not palmitoylethanolamine which forms bilayers [53, 54]. Cardiolipin binding by antiphospholipid antibodies of syphilis may be altered in the presence of Ca^{+2} or Mg^{+2} ions, possibly because cardiolipin molecules assume a hexagonal configuration in the presence of these cations [55, 56].

Antiphospholipid antibodies in patients with the Antiphospholipid Syndrome bind cardiolipin better than the VDRL (cardiolipin:phosphatidyl choline:cholesterol) antigen, but antibodies of syphilis do the reverse [11] – these differences may be explained by differences between cardiolipin presented alone compared to cardiolipin in the VDRL antigen [11–13].

7.3 β_2 glycoprotein 1

Recently, several groups have demonstrated that a plasma protein, β_2 glycoprotein 1 (β_2GP1) *enhances* binding of antiphospholipid to cardiolipin in standard ELISA tests [57–61] – the report by some groups [62–64] that anticardiolipin antibodies are specific for β_2GP1 *alone* has not been confirmed by others [57, 58, 60, 61, 65–67]. The degree of β_2GP1 enhancement of anticardiolipin binding is dependent on the antibody preparation [58, 60, 67, 68], the phospholipid antigen [69], and whether the antigen is presented in suspension or in solid phase immunoassays [60, 69, 70]. Enhanced binding to β_2GP1-cardiolipin complexes may be secondary to recognition of epitopes on both β_2GP1 and cardiolipin [57], or exposure of new epitopes on cardiolipin [71], or β_2GP1 [61, 65] (Fig. 8). Some investigators have suggested that antibodies with lupus anticoagulant activity may be specific for β_2GP1-phospholipid (23,57), prothrombin-phospholipid [28] or protein C-phospholipid [29] complexes. The latter reports require further documentation in light of much data showing lupus anticoagulant specificity for phospholipids alone [2, 5, 24–27].

Given the interest presently in anti-phospholipid specificity, issues about the epitopes bound by these antibodies will doubtless be resolved in the next few years.

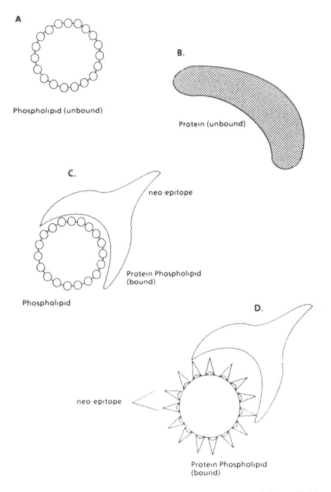

Fig. 8. Alternative interpretations of the heterogeneity of antiphospholipid specificities. The original view was of an antibody specific for phospholipids alone (A) [1, 2, 5, 24–27]. An alternative view is that these antibodies are specific for β_2GP1 alone (B) [62–64], or a β_2GP1-phospholipid [53] complex. It has been proposed that when β_2GP1 and cardiolipin bind each other, there may be alteration either in the β_2GP1 [61, 65] or cardiolipin [71] structures (or both) resulting in formation of "neo" (new) epitopes. Formation of neo-epitopes on the β_2GP1-cardiolipin complex may result in enhanced binding by antiphospholipid antibodies [57, 58, 65, 68].

References

1. Harris EN, Gharavi AE & Hughes GRV (1985) Clinics in Rheum Dis 11:591–609
2. Harris EN, Gharavi AE, Tincani A, Chan JKH, Englert H *et al.* (1985) J Clin Lab Immunol 17:155–162
3. Harris EN, Gharavi AE, Loizou S, Derue G, Chan JKH *et al.* (1985) J Clin Lab Immunol 16:1–6
4. Gharavi AE, Harris EN, Asherson RA & Hughes GRV (1987) Ann Rheum Dis 46:1–6
5. Pengo V, Heine MJ, Thiagarajan P & Shapiro SS (1987) Blood 70:69–76
6. Pangborn MC (1941) Proc Soc Exp Biol Med 48:484–486
7. DeHaas GH & Van Deenan LLM (1965) Nature 206:935–937
8. Faure M, Coulon-Morelec M (1963) Ann Inst Pasteur (Paris) 104:246–251
9. Inoue K & Nojima S (1969) Chem Phys Lipids 3:70–77
10. Levy RA, Gharavi AE, Sammaritano LR *et al.* (1990) J Clin Immunol 10:141–45
11. Harris EN, Gharavi AE, Wasley GD & Hughes GRV (1988) J Infect Dis 157:23–31
12. Janoff AD & Rauch J (1986) Chem Phys Lipids 40:315–322
13. Alving CR (1986) Chem Phys Lipids 40:303–314
14. Moore JE & Mohr CF (1952) J Am Med Assoc 150:467–473
15. Moore JE & Lutz WB (1955) J Chronic Dis 1:297–316
16. Fiumara NS (1963) NEJM 268:402–406
17. Harvey AM & Shulman LE (1966) Med Clin N Am 50:1271–1279
18. Harris EN, Gharavi AE, Boey ML *et al.* (1983) Lancet ii:1211–1214
19. Harris EN, Gharavi AE, Patel SP & Hughes GRV (1987) Clin Exp Immunol 68:215–222
20. Harris EN (1990) British J of Haematol 74:1–9
21. Triplett DA, Brandt JT, Musgrave KA & Orr CA (1988) JAMA 259:550–554
22. Rote NS, Dostal-Johnson D & Branch WD (1990) Am J Obstet Gynecol 163:575–584
23. McNeil HP, Chesterman CN & Krilis SA (1991) Adv Immunol 49:193–280
24. Shapiro SS & Thiagarajan P (1982) Progress in Haemostas. Thromb 6:263–285
25. Brandt JT (1991) Thromb Haemost 66:453–458
26. Galli M, Beguin S, Lindout T *et al.* (1989) Br J Haematol 72:549–555
27. Goldsmith GH, Pierangeli SS, Branch DW *et al.* (submitted)
28. Bevers EM, Galli M, Barbui T *et al.* (1991) Thromb Haemost 66:629–632
29. Oosting JD, Derksen RHWM, Bobbink IW *et al.* (1993) Blood 81:2618–2625
30. Galli M, Comfurius P, Barbui T, Zwaal RFA & Bevers EM (1992) Thromb Haemostas 68:297–300
31. Freyssinet JM, Wiesel ML, Gaudry J *et al.* (1986) Thromb Haemostas 55:309–314
32. Cariou R, Tobelin G, Bellucci S, Soria J *et al.* (1988) Thromb Haemostas 60:54–59
33. Marciniak E & Romond EH (1989) Blood 74:2426–2432
34. Malia RG, Kitchen S, Greaves M & Preston FE (1990) Br J Haematol 76:101–107
35. Esmon C'I (1989) J Biol Chem 264:4743–4746
36. Exner T & Triplett D (1991) In: Harris EN, Exner T, Hughes GRV & Asherson RA (Eds) Phospholipid Binding Antibodies (pp 141–158). CRC Press, Boca Raton
37. Proctor RR & Rapaport SI (1958) Am J Clin Pathol 11:406–410
38. Alving BM, Baldwin PE, Richards RL *et al.* (1985) Thromb Haemostas 54:709–713
39. Rosove MH, Ismail M, Koziol BJ *et al.* (1986) Blood 68:472–478
40. Margolis J (1958) J Clin Pathol 11:406–411
41. Exner T, Rickard KA & Kronenberg H (1978) Br J Haematol 40:143–150
42. Exner T, Rickard KA & Kronenberg H (1975) Pathology 7:319–326
43. Thiagarajan P, Pengo V & Shapiro SS (1986) Blood 68:869–877
44. Schleider MA, Nachman RL, Jaffe EA *et al.* (1976) Blood 48:499–503
45. Johannson EA & Lassus A (1974) Ann Clin Res 6:108–8
46. Pierangeli SS, Krnic S, Goldsmith G, Campbell A & Harris EN (1994) Infect Immun (In press)

47. Ching-hsien H (1991) In: Harris EN, Exner T, Hughes GRV & Asherson RA (Eds) Phospholipid Binding Antibodies (pp 3–30). CRC Press, Boca Raton
48. Zwaal RFA, Bevers EA & Rosing J (1991) In: Harris EN, Exner T, Hughes GRV & Asherson RA (Eds) Phospholipid Binding Antibodies (pp 31–56). CRC Press, Boca Raton
49. Rand RP & Sengupta S (1972) Biochim Biophys Acta 255:484–492
50. DeKruijf B, Verkleij AJ, Leunissen-Bijvelt J et al. (1982) Biochim Biophys Acta 163:1–12
51. Gray GM & MacFarlane MG (1958) Biochem J 70:409–425
52. Fry M & Green DE (1980) Biochim Biophys Res Commun 93:1238–1246
53. Rauch J, Tannenbaum M, Tannenbaum H et al. (1986) J Biol Chem 261:9672–9678
54. Rauch J, Tannenbaum M & Janoff AS (1989) Thromb Haemostas 62:892–896
55. Green FA & Costello PB (1987) Biochim Biophys Acta 896:47–51
56. Costello PB & Green FA (1988) Infect Immun 56:1738–1742
57. McNeil HP, Simpson RJ, Chesterman CN & Krilis SA (1990) Proc Natl Acad Sci 87:4120–4124
58. Sammaritano LR, Lockshin MD & Gharavi AE (1992) Lupus 1:51–56
59. Vermylen J & Arnaut J (1992) J Lab Clin Med 120:10–11
60. Pierangeli SS, Harris EN, Davis & DeLorenzo G (1992) Br J Haematol 82:565–570
61. Jones JV, James H, Tan MJ & Mansour M (1992) J Rheumatol 48:484–486
62. Galli M, Comfurius P, Maasen C et al. (1990) Lancet 336:1544–1547
63. Arvieux J (1992) Br J Haematol 81:568–573
64. Viard JP, Amoura Z & Bach JF (1992) Am J Med 93:181–186
65. Jones JV (1992) J Rheumatol 19:1774–1777
66. Gharavi AE, Harris EN, Sammaritano LR et al. (1993) J Lab Clin Med (In press)
67. Harris EN, Pierangeli SS & Birch D (1993) Am J Clin Pathol (In press)
68. Matsuura E, Igarashi V, Fujimoto M et al. (1992) J Immunol 148:3885–3891
69. Pierangeli SS, Davis SA & Harris EN (1992) Arthritis Rheum 35(Suppl):S207
70. Pierangeli SS, Harris EN, Gharavi AE et al. (1993) Br J Haematol (In press)
71. Harris EN & Pierangeli S (1990) Lancet 336:1550

Manual of Biological Markers of Disease **B3.1**: 1–20, 1994.

Autoantigens contained in the U1 small nuclear ribonucleoprotein complex

JACQUELINE M.T. KLEIN GUNNEWIEK and
WALTHER J. VAN VENROOIJ
Department of Biochemistry, University of Nijmegen, PO Box 9101, 6500 HB Nijmegen, The Netherlands

Abbreviations

snRNP: small nuclear ribonucleoprotein

Key words: U1 snRNP, autoantigen, ribonucleoprotein, autoimmunity

1. The antigen

The U1 snRNP particle contains the following antigenic components:
– the specific proteins U1-A, U1-70K and U1-C, of which the U1-70K and the U1-A polypeptides are the major targets of human autoantibodies (reviewed in [1])
– U1 snRNA [2, 3].
The U1 snRNP complex also contains Sm proteins, but these are not antigenic in sera containing antibodies to U1 snRNP only.

2. Synonyms

Nuclear RNP (nRNP), U1 RNP.

3. Antibody detection

3.1 Immunofluorescence

Anti-U1 snRNP antibodies produce a very typical, coarse speckled pattern, as has been discussed in this Manual, chapter A2. Figure 1 gives an example. It should be noted that the immunofluorescence technique cannot distinguish between anti-U1 snRNP and anti-Sm antibodies.

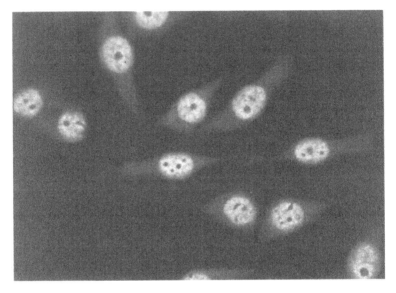

Fig. 1. Immunofluorescence pattern produced by human anti-U1 snRNP antibodies.
A human serum containing antibodies to U1-70K, U1-A, U1-C and U1 snRNA was used in immunofluorescence to demonstrate the nuclear location of U1 snRNP complexes in HEp-2 cells. The typical pattern of intermediate-sized speckles, generally associated with large speckles in the nucleoplasmic area of the cell can be seen. For further details, see this Manual, chapter A2.

3.2 Counterimmunoelectrophoresis and immunodiffusion

Autoantibodies directed to the protein components of the U1 snRNP complex can be detected by these techniques in a very specific manner, provided that a reference serum is available (for technical details, see this Manual, chapter A3). Both these techniques are able to distinguish between anti-U1 snRNP and anti-Sm antibodies.

3.3 Immunoblotting

Anti-U1 snRNP antibodies can specifically be detected using the immunoblotting technique (for technical details, see this Manual, chapter A4). The antibodies are in most cases directed to the U1-70K protein, decorating a very typical triplet pattern [4], and the U1-A protein. In about 60% of the cases, a somewhat broader smear of the U1-C protein is seen as well. Antibodies to the U1 snRNA molecule are not detectable by immunoblotting.

In principle, immunoblotting can distinguish between anti-U1 snRNP and anti-Sm activities because of the fact that decoration of the Sm-D band, mostly in combination with Sm-B'/B, is typical for anti-Sm activity [4].

3.4 RNA immunoprecipitation

RNA immunoprecipitation is by far the most sensitive technique to detect low titred anti-U1 snRNP and other anti-RNP activities (for technical details, see this Manual, chapter A7). Although this is not a routine technique, it is quite useful to distinguish between anti-U1 snRNP and anti-Sm activities.

Anti-U1 snRNA activities can be detected by this technique when radiolabeled total cellular RNA or *in vitro* made U1 snRNA is used as the substrate [3, 5].

3.5 Protein immunoprecipitation

Precipitation of ^{35}S-labeled proteins present in the RNP complex is a sensitive technique to detect anti-snRNP antibodies (see this Manual, chapter A6 for technical details). However, this technique is not useful to distinguish between anti-U1 snRNP and anti-Sm activities because in both cases the whole labeled U1 snRNP complex will be precipitated.

3.6 ELISA

An overview of the ELISA assays that can be used to detect anti-U1 snRNP antibodies has been given by Charles and Maini (part A of this Manual, chapter 5). Recombinant antigens can be used as substrate to distinguish between anti-U1-70K, U1-A and U1-C antibodies.

For the detection of anti-U1 snRNA antibodies one can use a dot-blot assay as published by Hoet *et al.* [5].

4. Cellular localization

U1 snRNP complexes are primarily localized in the nucleoplasm. Recent studies have shown that part of the snRNP complexes are concentrated in the so-called *coiled bodies* [6], the rest being localized in (filaments interconnecting) interchromatin granules (see Figs. 2A and 2B).

The assembly of snRNP complexes occurs primarily in the cytoplasm. After export from the nucleus, the newly synthesized U snRNA is further processed by 3'-end trimming, binding of Sm proteins and hypermethylation

A

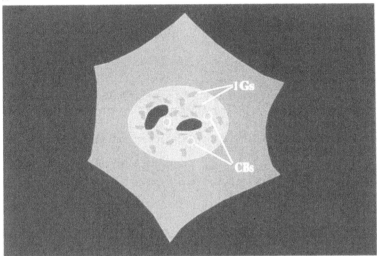

Fig. 2A. Cellular localization of U1 snRNP particles.
Schematic drawing of U1 snRNP location in the coiled bodies and interchromatin granules of the nucleoplasm. CBs: coiled bodies, spherical subnuclear domains which vary in size from 0.1–1 μm. They consist of both granules and coiled threads with a diameter of approximately 50 nm. Coiled bodies are ubiquitous and show striking evolutionary conservation [6]. They contain the autoantigen p80 coilin (see the chapter of M. Carmo-Fonseca and co-workers in this Manual, part B-8.2). IGs: interchromatin granules and perichromatin fibrils which interconnect these granules to form a three-dimensional network in the nucleus.

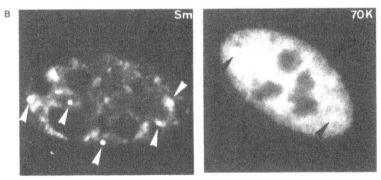

Fig. 2B. Confocal image of HeLa cells labeled with monoclonal antibodies directed to the Sm complex and to the U1-70K protein.
The images illustrate the typical widespread staining of the nucleoplasm and the additional concentration of U1 snRNPs and Sm antigens in coiled bodies (arrowheads). Note that U1 snRNP, as illustrated by the anti-U1-70K antibody, is always less prominent in the coiled bodies than the U2, U4, U5 or U6 snRNPs, decorated by the anti-Sm antibody. For further details, see [6]. Courtesy of A. Lamond.

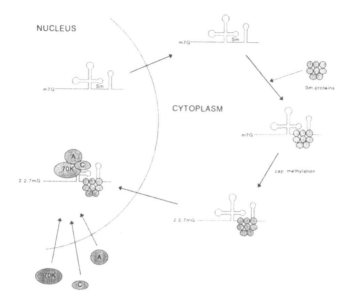

NUCLEUS

CYTOPLASM

cap methylation

Fig. 3. A model of U1 snRNP assembly and transport.
Upon transcription in the nucleus, the U1 snRNA is exported to the cytoplasm where it assembles with at least a subset of the Sm proteins. It also undergoes further processing including cap trimethylation and 3'-end trimming. Subsequently the core U1 snRNP complex returns to the nucleus [7, 8]. In *Xenopus laevis* oocytes the migration back to the nucleus is dependent on the presence of both the hypermethylated cap and the common snRNP proteins bound to the U1 snRNA [9, 10]. It is not precisely known where the association with the specific proteins U1-A, U1-70K and U1-C takes place but the binding of these proteins does not seem to be required for nuclear migration [7, 11]. m7G: monomethyl (m^7GpppG) cap structure; 2,2,7mG: trimethyl ($m^{2,2,7}$GpppG) cap structure; Sm: Sm-binding site, where the common proteins bind U1 snRNA. This site (also referred to as "domain A") consists of a single-stranded region, PuA (U)$_n$GPu with $n > 3–6$, flanked by double-stranded stems [12]. It has been shown that the Sm-G protein directly interacts with the AAU stretch within the 5'-terminal half of the Sm-site [13]. Adapted from [14].

of the 5' cap structure. The core snRNP particle is then re-imported into the nucleus (see Fig. 3).

5. Cellular function

Almost all precursor messenger RNAs (pre-mRNAs) contain coding sequences (exons) interrupted by non-coding sequences (introns). In a process called "splicing" the introns are removed and the exons are ligated in order to obtain functional mRNA. Splicing takes place in a large complex (the spliceosome), which contains snRNPs along with numerous other

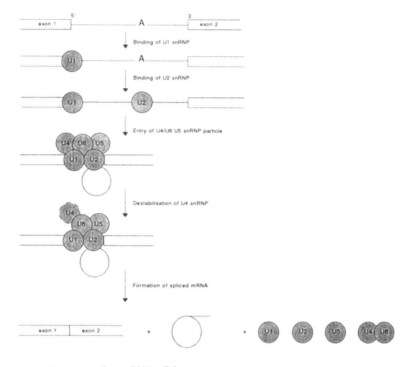

Fig. 4. Mechanism of pre-mRNA splicing.
The U1 snRNP binds the 5' splice site via basepairing between the 5' end of U1 snRNA and the 5' splice site of the pre-mRNA (for a review, see [17]). The exact role of the U1 snRNP-specific proteins A, 70K and C in this process is not yet known, but the binding of U1 snRNP to the pre-mRNA 5' splice site is dependent on the presence of these proteins [reviewed in 15, 16]. Heinrichs *et al.* suggested that U1-C is required for efficient U1 snRNP/5' splice site complex formation [18]. Furthermore, U1–70K is strongly phosphorylated in higher organisms. It has been suggested that the state of phosphorylation is critical for its participation in the splicing reaction [19–21]. The pre-mRNA is shown with exons as boxes and the intron as a line. 5': 5' splice site; 3': 3' splice site; A: branch point. Adapted from [17].

proteins (reviewed in [15, 16]). The U1 RNP complex is primarily active in the first step of splicing (Fig. 4).

6. Presence in other species

Both U1 snRNA and its specific proteins are found in many other eucaryotic cells [22–24] including plants and yeast [22]. Sequencing of U1 RNAs isolated from various cells and organisms has revealed extensive conservation of the RNA sequence and, in particular, of its secondary structure [15, 22].

7. The human cDNAs and derived amino acid sequences

The typical domains present in the U1-70K, U1-A and U1-C polypeptides are shown in Figure 5. The cDNA sequences and derived amino acid sequences are given in Figure 6A, B and C.

7.1 U1-70K

U1-70K contains a so-called RNP-80 motif [26, 31] also referred to as RNA-binding region [28] or RNA-recognition motif [25] (see Figs. 5 and 6A). This motif is common to many other RNA-binding proteins [15, 25] and contains the highly conserved RNP-1 or RNP consensus sequence in addition to a less conserved sequence referred to as RNP-2 [29]. The RNP-80 motif of U1-70K is involved in the binding of U1-70K to the first stemloop of U1 snRNA. Furthermore, U1-70K has two highly charged arginine/serine-rich regions and a small region with sequence similarities to p30^{gag} of type-C retroviruses [30].

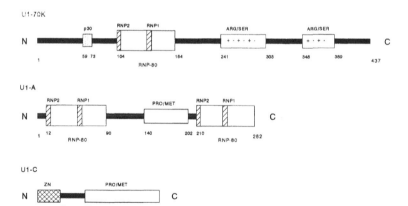

Fig. 5. Structural features of U1-A, U1-70K and U1-C polypeptides.
The RNA-binding domain of U1-70K (111 amino acids) encompasses the RNP-80 motif plus 11 N-terminal and 19 C-terminal flanking amino acids [25]. The region of U1-A required to bind to U1 snRNA consists of the N-terminal RNP-80 motif together with maximally 11 N-terminal and 10 C-terminal flanking amino acids [26]. U1-C however does not bind directly to the U1 snRNA but needs additional interactions with the U1-70K and the Sm-B'/B proteins for binding the U1 snRNP particle [27]. RNP-80: RNP-80 motif or RNA-binding region or RNA-recognition motif; RNP1 and RNP2: most highly conserved sequences within the RNP-80 motif; PRO/MET: proline/methionine-rich region; ARG/SER: arginine/serine-rich region; p30: p30^{gag} sequence similarity; Zn: zinc-finger-like region. Numbers refer to the amino acids of U1-A, U1-70K and U1-C. N is the amino-terminus, C the carboxy-terminus.

A U1-70K

```
          10        20        30        40        50        60        70        80        90
GGCTGAGCAGCGGCCTGGTGCGCTCGCTTAGCGGGCGACGGAATCAGACGGACGTGGACGCCCCGGAGTGGAAGCCGAAGCAGGAGTTG
```

```
         100       110       120       130       140       150       160       170       180
TTGTTGCTGAGGGGCTGCCGCAGCCGCCGCGAGCCTCCGGACAGACGCCAGAGCGAGGAGGGCGCTACGCGACTTGGCAAGATGACCCAG
                                                                                  M  T  Q
```

```
         190       200       210       220       230       240       250       260       270
TTCCTGCCGCCCAACCTTCTGGCCCTCTTTGCCCCCCGTGACCCTATTCCATACCTGCCACCCCTGGAGAAACTGCCACATGAAAAACAC
 F  L  P  P  N  L  L  A  L  F  A  P  R  D  P  I  P  Y  L  P  P  L  E  K  L  P  H  E  K  H
```

```
         280       290       300       310       320       330       340       350       360
CACAATCAACCTTATTGTGGCATTGCGCCGTACATTCGAGAGTTTGAGGACCCTCGAGATGCCCCTCCTCCAACTCGTGCTGAAACCCGA
 H  N  Q  P  Y  C  G  I  A  P  Y  I  R  E  F  E  D  P  R  D  A  P  P  P  T  R  A  E  T  R
```

```
         370       380       390       400       410       420       430       440       450
GAGGAGCGGCATGGAGAGGAAAAGACGGGAAAAGATTGAGCGGCGACAGCAAGAAGTGGAGACAGAGCTTAAAATGTGGGACCCTCACAAT
 E  E  R  M  E  R  K  R  R  E  K  I  E  R  R  Q  Q  E  V  E  T  E  L  K  M  W  D  P  H  N
```

```
         460       470       480       490       500       510       520       530       540
GATCCCAATGCTCAGGGGGATGCCTTCAAGACTCTCTTCGTGGCGAGAGTGAATTATGACACAACAGAATCCAAGCTCCGGAGAGAGTTT
 D  P  N  A  Q  G  D  A  F  K  T  L  F  V  A  R  V  N  Y  D  T  T  E  S  K  L  R  R  E  F
```

```
         550       560       570       580       590       600       610       620       630
GAGGTGTACGGACCTATCAAAAGAATACACATGGTCTACAGTAAGCGGTCAGGAAAGCCCCGTGGCTATGCCTTCATCGAGTACGAACAC
 E  V  Y  G  P  I  K  R  I  H  M  V  Y  S  K  R  S  G  K  P  R  G  Y  A  F  I  E  Y  E  H
```

```
         640       650       660       670       680       690       700       710       720
GAGCGAGACATGCACTCCGCTTACAAACACGCAGATGGCAAGAAGATTGATGGCAGGAGGGTCCTTGTGGACGTGGAGAGGGGCCGAACC
 E  R  D  M  H  S  A  Y  K  H  A  D  G  K  K  I  D  G  R  R  V  L  V  D  V  E  R  G  R  T
```

```
         730       740       750       760       770       780       790       800       810
GTGAAGGGCTGGAGGCCCCGGCGGCTAGGAGGAGGCCTCGGTGGTACCAGAAGAGGAGGGGCTGATGTGAACATCCGGCATTCAGGCCGC
 V  K  G  W  R  P  R  R  L  G  G  G  L  G  G  T  R  R  G  G  A  D  V  N  I  R  H  S  G  R
```

```
         820       830       840       850       860       870       880       890       900
GATGACACCTCCCGCTACGATGAGAGGCCCGGCCCCTCCCCGCTTCCGCACAGGGACCCGGGACCGGGACCGTGAGCGGGAGCGCAGAGAG
 D  D  T  S  R  Y  D  E  R  P  G  P  S  P  L  P  H  R  D  R  D  R  D  R  E  R  E  R  R  E
```

```
         910       920       930       940       950       960       970       980       990
CGGAGCCGGGAGCGAGACAAGGAGCGAGAACGGCGACGCTCCCGCTCCCGGGACCGGCGGAGGCGCTCACGGAGTCGCGACAAGGAGGAG
 R  S  R  E  R  D  K  E  R  E  R  R  R  S  R  S  R  D  R  R  R  S  R  S  R  D  K  E  E
```

```
        1000      1010      1020      1030      1040      1050      1060      1070      1080
CGGAGGAGCGCTCCAGGGAGCGGAGCAAGGACAAGGACCGGGACCGGAAGCGGCGAAGCAGCCGGAGTCGGGAGCGGGCCCGGCGGGAGCGG
 R  R  R  S  R  S  R  E  R  S  K  D  K  D  R  D  R  K  R  R  S  S  R  S  R  E  R  A  R  R  E
```

```
        1090      1100      1110      1120      1130      1140      1150      1160      1170
GAGCGCAAGGAGGAGCTGCGTGGTGGCGGTGGCGACATGGCGGAGCCCTCCGAGGCGGGTGACGGCGCCCCTGATGATGGGCCTCCAGGG
 E  R  K  E  E  L  R  G  G  G  D  M  A  E  P  S  E  A  G  D  A  P  P  D  D  G  P  P  G
```

```
        1180      1190      1200      1210      1220      1230      1240      1250      1260
GAGCTCGGGCCTGACGGCCCTGACGGTCCAGAGGAAAAGGGCCGGGATCGTGACCGGGACGCGACGGCGGAGCCACCGGAGCGAGCGCGAG
 E  L  G  P  D  G  P  D  G  P  E  E  K  G  R  D  R  D  R  E  R  R  R  S  H  R  S  E  R  E
```

```
        1270      1280      1290      1300      1310      1320      1330      1340      1350
CGGCGCGGGACCGGGATCGTGACCGTGACCGTGACCGCGAGCACAAACGGGGGGAGCGGGGCAGTGAGCGGGGCAGGGATGAGGCCCGA
 R  R  R  D  R  D  R  D  R  D  R  D  R  E  H  K  R  G  E  R  G  S  E  R  G  R  D  E  A  R
```

```
        1360      1370      1380      1390      1400      1410      1420      1430      1440
GGTGGGGGCGGTGGCCAGGACAACGGGCTGGAGGGTCTGGGCAACGACAGCCGAGACATGTACATGGAGTCTGAGGGCGGCGACGGCTAC
 G  G  G  G  G  Q  D  N  G  L  E  G  L  G  N  D  S  R  D  M  Y  M  E  S  E  G  G  D  G  Y
```

```
        1450      1460      1470      1480      1490      1500      1510      1520      1530
CTGGCTCCGGAGAATGGGTATTTGATGGAGGCTGCGCCGGAGTGAAGAGGTCGTCCTCTCCATCTGCTGTGTTTGGACGCGTTCCTGCCC
 L  A  P  E  N  G  Y  L  M  E  A  A  P  E
```

```
        1540      1550      1560      1570      1580      1590      1600      1610      1620
AGCCCCTTGCTGTCATCCCCTCCCCCAACCTTGGCCACTTGAGTTTGTCCTCCAAGGGTAGGTGTCTCATTTGTTCTGGCCCCTTGGATT
```

```
        1630      1640      1650      1660
TAAAAATAAAATTAATTTCCTGTTGAAAAAAAAAAAAAAAAAA
```

Fig. 6. The human cDNAs and the derived amino acid sequences of U1-70K, U1-A and U1-C. A. The first published cDNA for U1-70K [31] encoded a protein of 637 amino acids. Later it was found that the sequences upstream of amino acid 178 are not required to encode the full-length U1-70K protein [25, 32]. The calculated molecular weight of the U1-70K protein is 52 kDa [25, 32]. Accession number GenBank/EMBL X06815.

B

U1-A

```
          10        20        30        40        50        60        70        80        90
GAATTCCTGACTTCCTTTTCGGAGGAAGATCCTTGAGCAGCCGACGTTGGGACAAAGGATTTGGAGAAACCCAGGGCTAAAGTCACGTTT


         100       110       120       130       140       150       160       170       180
TTCCTCCTTTAAGACTTACCTCAACACTTCACTCCATGGCAGTTCCCGAGACCCGCCCTAACCACACTATTTATATCAACAACCTCAATG
                                          M  A  V  P  E  T  R  P  N  H  T  I  Y  I  N  N  L  N


         190       200       210       220       230       240       250       260       270
AGAAGATCAAGAAGGATGAGCTAAAAAAGTCCCTGTACGCCATCTTCTCCCAGTTTGGCCAGATCCTGGATATCCTGGTATCACGGAGCC
E  K  I  K  K  D  E  L  K  K  S  L  Y  A  I  F  S  Q  F  G  Q  I  L  D  I  L  V  S  R  S


         280       290       300       310       320       330       340       350       360
TGAAGATGAGGGGCCAGGCCTTTGTCATCTTCAAGGAGGTCAGCAGCGGCCACCAACGCCCTGCGCTCCATGCAGGGTTTCCCTTTCTATG
L  K  M  R  G  Q  A  F  V  I  F  K  E  V  S  S  A  T  N  A  L  R  S  M  Q  G  F  P  F  Y


         370       380       390       400       410       420       430       440       450
ACAAACCTATGCGTATCCAGTATGCCAAGACCGACTCAGATATCATTGCCAAGATGAAAGGCACCTTCGTGGAGCGGGACCGCAAGCGGG
D  K  P  M  R  I  Q  Y  A  K  T  D  S  D  I  I  A  K  M  K  G  T  F  V  E  R  D  R  K  R


         460       470       480       490       500       510       520       530       540
AGAAGAGGAAGCCCAAGAGCCAGGAGACCCCGGCCCACCAAGAAGGCTGTGCAAGGCGGGGGAGCCACCCCCGTGGTGGGGGCTGTCCAGG
E  K  R  K  P  K  S  Q  E  T  P  A  T  K  K  A  V  Q  G  G  G  A  T  P  V  V  G  A  V  Q


         550       560       570       580       590       600       610       620       630
GGCCTGTCCCGGGCATGCCGCCGATGACTCAGGCGCCCGCATTATGCACCACATGCCGGGCCAGCCGCCCTACATGCCGCCCCCTGGTA
G  P  V  P  G  M  P  P  M  T  Q  A  P  R  I  M  H  H  M  P  G  Q  P  P  Y  M  P  P  P  G


         640       650       660       670       680       690       700       710       720
TGATCCCCCCGCCAGGCCTTGCACCTGGCCAGATCCCACCAGGGGCCATGCCCCCGCAGCAGCTTATGCCAGGACAGATGCCCCCTGCCC
M  I  P  P  P  G  L  A  P  G  Q  I  P  P  G  A  M  P  P  Q  Q  L  M  P  G  Q  M  P  P  A


         730       740       750       760       770       780       790       800       810
AGCCTCTTTCTGAGAATCCACCGAATCACATCTTGTTCCTCACCAACCTGCCAGAGGAGACCAACGAGCTCATGCTGTCCATGCTTTTCA
Q  P  L  S  E  N  P  P  N  H  I  L  F  L  T  N  L  P  E  E  T  N  E  L  M  L  S  M  L  F


         820       830       840       850       860       870       880       890       900
ATCAGTTCCCTGGCTTCAAGGAGGTCCGTCTGGTACCCGGGCGGCATGACATCGCCTTCGTGGAGTTTGACAATGAGGTACAGGCAGGGG
N  Q  F  P  G  F  K  E  V  R  L  V  P  G  R  H  D  I  A  F  V  E  F  D  N  E  V  Q  A  G


         910       920       930       940       950       960       970       980       990
CAGCTCGCGATGCCCTGCAGGGCTTTAAGATCACGCAGAACAACGCCATGAAGATCTCCTTTGCCAAGAAGTAGCACCTTTTCCCCCCAT
A  A  R  D  A  L  Q  G  F  K  I  T  Q  N  N  A  M  K  I  S  F  A  K  K


        1000      1010      1020      1030      1040      1050      1060      1070      1080
GCCTGCCCCTTCCCCTGTTCTGGGGGCCACCCCTTTCCCCCTTGGCTCAGCCCCCTGAAGGTAAGTCCCCCCTTGGGGGCCTTCTTGGAGC


        1090      1100      1110      1120      1130      1140      1150      1160      1170
CGTGTGTGAGTGAGTGGTCGCCACACAGCATTGTACCCAGAGTCTGTCCCCAGACATTGCACCTGGCGCTGTTAGGCCGGAATTAAAGTG


        1180      1190      1200
GCTTTTTGAGGTTTGGTTTTTCACAAAAAAAAGGAATTC
```

Fig. 6B. The cDNA of U1-A (1.2 kb) codes for a polypeptide of 282 amino acids with a calculated molecular weight of 31,243 daltons [33]. Accession number GenBank/EMBL X06347.

7.2 U1-A

U1-A contains two RNP-80 motifs [26, 33]. The N-terminal RNP-80 motif is more conserved than the C-terminal motif [33] and is involved in the binding of the second stemloop of U1 RNA [26].

7.3 U1-C

U1-C lacks an RNP-80 motif but contains a Zn-finger-like region [34] (Figs. 5 and 6C). It is known that, in the U1 snRNP complex, this region of the U1-C protein interacts with the U1-70K and Sm-B'/B proteins and that these interactions are essential for the binding of U1-C to the core U1 snRNP complex [27].

C

U1-C

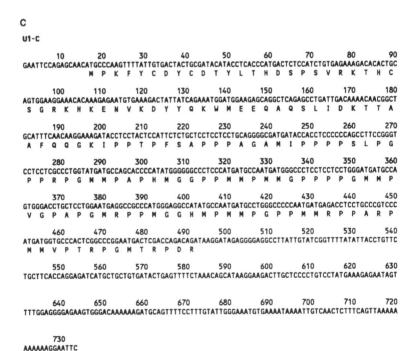

Fig. 6C. The cDNA of U1-C (733 bp) codes for a polypeptide of 159 amino acids with a calculated molecular weight of 17,373 daltons [23]. Accession number GenBank/EMBL X12517.

7.4 U1 snRNA

U1-70K and U1-A respectively bind stemloop I and stemloop II of U1 snRNA. The Sm proteins bind the U1 snRNA at the single-stranded Sm-binding site [12]. The U1-C protein is associated with the complex via protein-protein interactions with the U1-70K and Sm polypeptides (see Fig. 7).

8. The genes

8.1 U1-70K

U1-70K is encoded by a single gene located on chromosome 19, with no detectable pseudogenes [32]. cDNA cloning studies suggest that various U1-

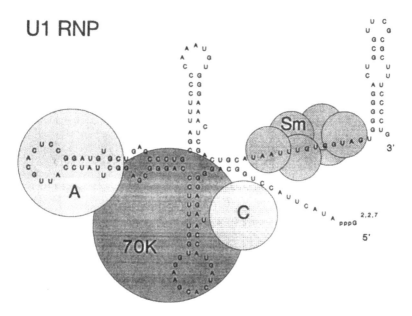

Fig. 7. Secondary structure of U1 snRNA showing binding domains for U1 snRNP proteins. The binding site of U1-A is localized to stemloop II (nt. 48 to 91) [26, 35]. It has been established that the N-terminal RNA-binding motif of the U1-A protein and the loop-sequence AUUGCA of the U1 snRNA are essential for this interaction [26, 36–38]. The U1-70K protein binding site is localized to stemloop I (nt. 18 to 48) [39]. Deletion and mutation analyses of U1 snRNA further showed that 8 of the 10 bases in the loop are required for binding [40]. The binding of the U1-C protein needs the presence of the U1-70K protein and the Sm proteins [27]. The amino-terminal 45 amino acids, including the CC-HH type zinc finger, are sufficient for binding to U1 snRNPs lacking protein C [27, 34].

70K mRNAs may be derived from a common U1-70K pre-mRNA by alternative splicing processes [31, 32]. In *Xenopus laevis* it has been found that the coding region for U1-70K is interrupted by several introns and that the *Xenopus laevis* U1-70K is encoded by at least two genes [41].

8.2 U1-A

U1-A is also encoded by a single-copy gene and its locus has been mapped to the q-arm of chromosome 19 (19q13.1). The gene is about 14–16 kilobases (kb) long and consists of six exons (see Fig. 8 and [24]).

Fig. 8. Schematic representation of the U1-A gene.
Open boxes (I to VI) represent the exons, the dashed lines the introns. The numbers below the boxes refer to the corresponding amino acids. Exon sequences encoding the N-terminal and the C-terminal RNP-80 motifs evolved differentially after a probable duplication within the gene [24].

8.3 U1-C

The gene coding for U1-C has not been identified yet.

8.4 U1 snRNA

The genes for U1 snRNA constitute a collection of multigene families (in mammals it has about 30 members). Each of these gene families arose by amplification of an ancient progenitor DNA sequence containing only one or a few snRNA genes. The genes are all mapped to the short arm of chromosome 1 (1p36) [42].

9. Relation with nucleic acids and other proteins

9.1 Nucleic acids

The 5′ end of U1 snRNA is known to interact with the 5′ splice site of the pre-mRNA in the commitment complex of the splicing reaction. This is only one of the many RNA-RNA interactions occurring in the spliceosome [43].

9.2 Proteins

The U1 snRNP proteins have an intimate interaction with each other and with some of the Sm proteins. For example, the U1-C protein is known to interact with the U1-70K and Sm-B'/B and the U1-70K with the U1-C and Sm-D2 proteins [27]. A cartoon of some of the interactions in the U1 snRNP complex is shown in Figure 7.

10. Biochemical characteristics

U1-70K and U1-C have an electrophoretic mobility that is different from what could be expected from their predicted molecular weights. This aberrant behaviour is due to the post-translational modifications of the two proteins [19, 44] and the highly charged C-terminal part of U1-70K [25, 32]. The U1-70K gives a typical triple-banded pattern on Western blots [4]. Furthermore, characterization by two-dimensional electrophoresis shows that the U1-70K is found in at least 13 isoelectric variants. The final charge of the U1-70K variants is determined both by phosphorylation and by other, as yet unidentified, post-translational modifications [19].

11. B-cell epitopes

Extensive mapping studies have been performed with synthetic peptides and recombinant proteins of U1-A, U1-70K and U1-C (see Fig. 9 and [1] for a review). These studies revealed that the anti-U1 RNP autoimmune response is polyspecific, polyclonal and remarkably heterogeneous in different patients [1].

The main B-cell epitopes on U1 snRNA are shown in Figure 10. They are located in the stem of stemloop II and in the loop of stemloop IV [57, 58].

12. T-cell epitopes

Autoantigen-reactive T cells to U1-A have been described and the epitope has been mapped [55]. All patients' sera recognized the 83-amino acid sequence located in the C-terminus of U1-A (see Fig. 9B). Nothing is known yet about T-cell epitopes on U1 snRNA, U1-70K or U1-C.

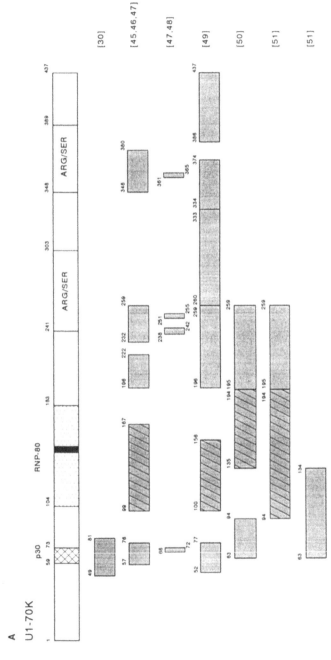

B

U1-A

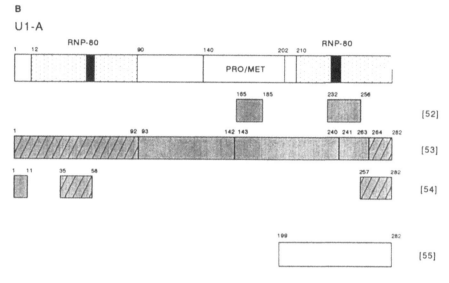

Fig. 9B.

←
Fig. 9. Schematic representation of the autoreactive domains on U1-70K (Fig. 9A; aa positions according to [25]), U1-A (Fig. 9B; aa positions according to [33]) and U1-C (Fig. 9C; aa positions according to [23]).

The major B-cell epitope of both U1-70K and U1-A co-localize with the RNA-binding domain [1, 46–48, 50–54]. On U1-70K, this domain has multiple, most probably, discontinuous epitopes and is recognized in a patient-specific manner. Furthermore, a variety of other B-cell epitopes are spread over almost the entire U1-70K protein with exception of the extreme N-terminus. A subset of patients' sera also recognizes the domain with p30^{gag} sequence similarity [30].

For U1-A it has been suggested that the sera recognize discontinuous epitopes spread all over the protein, with the N- and the C-terminal sequences contributing to the autoantigenic determinants [1, 52–54].

There is only one report describing B-cell epitopes on U1-C [56]. All anti-U1-C positive sera appeared to recognize the region between amino acid residues 102 and 125. Interestingly, this region contains an amino acid sequence similar to that of the herpes simplex virus type 1 ICP4 protein.

The T-cell epitope on U1-A [55] is indicated as an open box in Fig. 9B.

RNP-80: the RNP-80 motif with the highly conserved RNP-1; p30: p30^{gag} sequence similarity; PRO/MET: proline/methionine-rich; ARG/SER: arginine/serine-rich; ZN: zinc-finger-like sequence.

AMAN-B3.1/15

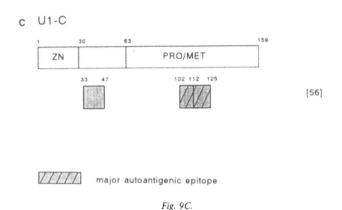

c **U1-C**

Fig. 9C.

13. Monoclonal antibodies

The following mouse monoclonal antibodies directed to U1 snRNP components have been published.

13.1 Anti-U1-70K

MoAb 2.73 [59]. This antibody is an IgG2a(x) that precipitates U1 snRNP particles only. The B-cell epitope is located in the Arg/Ser domain, a reason why this antibody might cross-react with other Arg/Ser proteins.

13.2 Anti-U1-A

MoAbs 9A9 and 11A1 [60]. Both antibodies are of the IgG1 class but precipitate both U1 snRNP and U2 snRNP because they recognize the U2-B'' protein as well.

Acknowledgements

We thank Dr. Angus Lamond (EMBL, Heidelberg) for providing Figure 2 and Dr. Celia van Gelder for providing Figure 10. This work was supported in part by "Het Nationaal Reumafonds" of the Netherlands and by the Netherlands Foundation for Chemical Research (SON) with financial aid from the Netherlands Organization for Scientific Research (NWO).

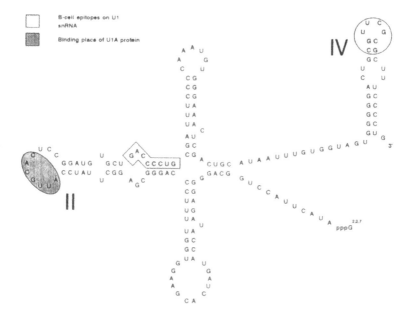

Fig. 10. B-cell epitopes on U1 snRNA.

The main epitopes are located in the stem structure of stemloop II and in the upper part of the stem and the loop of stemloop IV [57, 58]. The structure and conformation of both stemloops seem to be critical for antibody recognition [57].

References

1. Guldner HH (1992) Mol Biol Rep 16:155–164
2. Deutscher SL & Keene JD (1988) Proc Natl Acad Sci USA 85:3299–3304
3. van Venrooij WJ, Hoet R, Castrop J, Hageman B, Mattaj IW & van de Putte LB (1990) J Clin Invest 86:2154–2160
4. van Venrooij W (1987) J Rheumatol 14 (suppl. 13):78–82
5. Hoet RMA, Koornneef I, de Rooij DJ, van de Putte LB & van Venrooij WJ (1992) Arthitis Rheum 35:1202–1210
6. Lamond AI & Carmo-Fonseca M (1993) Trends Cell Biol 3:198–204
7. Mattaj IW, Boelens W, Izaurralde E, Jarmolowski A & Kambach C (1993) Mol Biol Rep 18:79–83
8. Zieve GW, Sauterer RA & Feeney RJ (1988) J Mol Biol 199:259–267
9. Feeney RJ & Zieve GW (1990) J Cell Biol 110:871–881
10. Kambach C & Mattaj IW (1992) J Cell Biol 118:11–21
11. Hamm J, Darzynkiewicz E, Tahara SM & Mattaj IW (1990) Cell 62:569–577
12. Branlant C, Krol A, Ebel JP, Lazar E, Haendler B & Jacob M (1982) EMBO J 1:1259–1265
13. Heinrichs V, Hackl W & Lührmann R (1992) J Mol Biol 227:15–28
14. Nigg EA, Baeuerle PA & Lührmann R (1991) Cell 66:15–22
15. Lührmann R, Kastner B & Bach M (1990) Biochim Biophys Acta 1087:265–292
16. Mattaj IW, Tollervey D & Séraphin B (1993) FASEB J 7:47–53
17. Green M (1991) Annu Rev Cell Biol 7:559–599
18. Heinrichs V, Bach M, Winkelmann G & Lührmann R (1990) Science 247:69–72
19. Woppmann A, Patschinsky T, Bringmann P, Godt F & Lührmann R (1990) Nucl Acids Res 18:4427–4438
20. Mermoud JE, Cohen P & Lamond AI (1992) Nucl Acids Res 20:5263–5269
21. Tazi J, Kornstädt U, Rossi F, Jeanteur P, Cathala G, Brunel C & Lührmann R (1993) Nature 363:283–286
22. Reddy R & Busch H (1988) In: Birnstiel ML (Ed) Structure and Function of the Major and Minor Small Nuclear Ribonucleoprotein Particles (pp 1–37). Springer-Verlag, New York
23. Sillekens PTG, Beijer RP, Habets WJ & van Venrooij WJ (1988) Nucl Acids Res 16:8307–8321
24. Nelissen RLH, Sillekens PTG, Beijer RP, van Kessel AHM & van Venrooij WJ (1991) Gene 102:189–196
25. Query CC, Bentley RC & Keene JD (1989) Cell 57:89–101
26. Scherly D, Boelens W, van Venrooij WJ, Dathan NA, Hamm J & Mattaj IW (1989) EMBO J 8:4163–4170
27. Nelissen R, Will C, van Venrooij WJ & Lührmann R Submitted
28. Dreyfuss G, Swanson MS & Piñol-Roma S (1988) Trends Biochem Sci 13:86–91
29. Kenan DJ, Query CC & Keene JD (1991) Trends Biochem Sci 16:214–220
30. Query CC & Keene JD (1987) Cell 51:211–220
31. Theissen H, Etzerodt M, Reuter R, Schneider C, Lottspeich F, Argos P, Lührmann R & Philipson L (1986) EMBO J 5:3209–3217
32. Spritz RA, Strunk K, Surowy CS, Hoch SO, Barton DE & Franke U (1987) Nucl Acids Res 15:10373–10391
33. Sillekens PTG, Beijer RP, Habets WJ & van Venrooij WJ (1987) EMBO J 6:3841–3848
34. Nelissen RLH, Heinrichs V, Habets WJ, Simons F, Lührmann R & van Venrooij WJ (1991) Nucl Acids Res 19: 449–454
35. Lutz-Freyermuth C & Keene JD (1989) Mol Cell Biol 9:2975–2982
36. Scherly D, Boelens W, Dathan NA, van Venrooij WJ & Mattaj IW (1990) Nature 345:502–506
37. Bentley RC & Keene JD (1991) Mol Cell Biol 11:1829–1839
38. Tsai RE, Harper DS & Keene JD (1991) Nucl Acids Res 19:4931–4936

39. Query CC, Bentley RC & Keene JD (1989) Mol Cell Biol 9:4872–4881
40. Surowy CS, van Santen VL, Scheib-Wixted SM & Spritz RA (1989) Mol Cell Biol 9:4179–4186
41. Etzerodt MR, Vignali R, Ciliberto G, Scherly D, Mattaj IW & Philipson L (1988) EMBO J 7:4311–4321
42. Dahlberg JA & Lund E (1988) In: Birnstiel ML (Ed) Structure and Function of the Major and Minor Small Nuclear Ribonucleoprotein Particles (pp 38–70). Springer-Verlag, New York
43. Lamond AI (1993) Current Biol 3:62–64
44. Fischer DE, Connor GE, Reeves WH, Wisniewolski R & Blobel G (1985) Cell 42:751–758
45. Guldner HH, Netter HJ, Szostecki C, Lakomek HJ & Will H (1988) J Immunol 141:469–475
46. Netter HJ, Guldner HH, Szostecki C & Will H (1990) Scand J Immunol 32:163–176
47. Guldner HH, Netter HJ, Szostecki C, Jaeger E & Will H (1990) J Exp Med 171: 819–829
48. Netter HJ, Will H, Szostecki C & Guldner HH (1991) J Autoimmunity 4: 651–663
49. Cram DS, Fisicaro N, Coppel RL, Whittingham S & Harrison LC (1990) J Immunol 145: 630–635
50. Nyman U, Lundberg I, Hedfors E & Petterson I (1990) Clin Exp Immunol 81:52–58
51. Tadeka Y, Nyman U, Winkler A, Wise KS, Hoch SO, Petterson I, Anderson SK, Wang RJ, Wang GS & Sharp GC (1991) Clin Immunol Immunopathol 61:55–68
52. Habets WJ, Sillekens PTG, Hoet MH, McAllister G, Lerner MR & van Venrooij WJ (1989) Proc Natl Acad Sci USA 86:4674–4678
53. Habets WJ, Hoet MH & van Venrooij (1990) Arthritis Rheum 33:834–841
54. Barakat S, Briand JP, Abuaf N, van Regenmortel MHV & Muller S (1991) Clin Exp Immunol 86:71–78
55. Okubo M, Yamamoto K, Kato T, Matsuura N, Nishimaki T, Kasukawa R, Ito K, Mizushima Y & Nishioka K (1993) J Immunol 151:1108–1115
56. Misaki Y, Yamamoto K, Yanagi K, Miura H, Ichijo H, Kato T, Mato T, Welling-Wester S, Nishioka K & Koji I (1993) Eur J Immunol 23:1064–1071
57. Hoet RM, de Weerd P, Klein Gunnewiek J, Koornneef I & van Venrooij WJ (1992) J Clin Invest 90:1753–1762
58. Tsai DE & Keene JD (1993) J Immunol 150:1137–1145
59. Billings PB, Allan RW, Jensen FC & Hoch SO (1982) J Immunol 128:1176–1180
60. Habets WJ, Hoet MH, de Jong BAW, van der Kemp A & van Venrooij WJ (1989) J Immunol 143:2560–2566

Manual of Biological Markers of Disease **B4.1**: 1–18, 1994.

The SS-A/Ro antigen

EDWARD K.L. CHAN and JILL P. BUYON[1]
From the W. M. Keck Autoimmune Disease Center, and DNA Core Laboratory for Structural Analysis, Department of Molecular and Experimental Medicine, The Scripps Research Institute, 10666 North Torrey Pines Road, La Jolla, CA 92037, U.S.A.; [1] Department of Medicine, Division of Rheumatology, New York University Medical Center and the Department of Rheumatic Diseases and Molecular Medicine at the Hospital for Joint Diseases, 301 East 17th Street, New York, New York 10003, U.S.A.

Abbreviations

ANA: autoantibody to nuclear antigen; CIE: counterimmunoelectrophoresis; hY RNA: human Y RNA; IP: immunoprecipitation; RBC: red blood cells; Ro52: 52-kD SS-A/Ro protein; Ro60: 60-kD SS-A/Ro protein; SS: Sjögren's syndrome; SLE: systemic lupus erythematosus

Key words: Antinuclear antibodies, Nuclear antigens, Sjögren's syndrome

1. The antigens: SS-A/Ro

In 1961 Anderson *et al.* [1] described two precipitating autoantibody specificities known as SjD and SjT from patients with Sjögren's syndrome. SjD antigen was reported to be insensitive to trypsin or heat treatment while SjT antigen could be destroyed by the same treatment. In 1969 Clark *et al.* [2] reported a novel antibody specificity known as Ro in 40% of patients with systemic lupus erythematosus (SLE) and the Ro antigen was described as a protease-, RNase-, and DNase-resistant cytoplasmic antigen observed in various tissue extracts. Aware of the work of Anderson *et al.* [1], Alspaugh and Tan [3] in 1975 described precipitin lines for two distinct autoantibody specificities known as SS-A and SS-B occurring predominantly in patients with Sjögren's syndrome (SS). Both SS-A and SS-B were characterized as trypsin-sensitive but RNase- and DNase-resistant nuclear antigens. It was not until 1979 that SS-A and Ro were shown to be identical as reported by Alspaugh and Maddison [4] and since then this antigen system has been referred to as SS-A/Ro or Ro/SS-A. However, in the area of molecular and cell biology, this system is often referred to as Ro ribonucleoprotein (Ro RNP) probably because of the report of Lerner *et al.* [5] showing the association of Ro antigens with small RNAs. It is clear now that human sera containing anti-SS-A/Ro antibodies recognize two proteins of 52 and 60kD and the latter is bound to Ro RNAs known as hY RNAs. Several recent reviews on the SS-A/Ro autoantibody-antigen system have been published [6–10]. Another protein, calreticulin, was initially reported to be the 60-kD SS-A/Ro protein [11, 12]. However more recent work revealed that calreticulin is an unrelated autoantigen [13].

2. Synonyms

SS-A, Ro, SjD.

3. Biochemical characteristics

The SS-A/Ro complex in human cells contains one of at least 4 hY RNAs associated with at least two proteins of 52 kD (Ro52) and 60 kD (Ro60) [14, 15]. hY RNAs are small RNAs known as hY1, hY3, hY4, and hY5 RNA ranging from 84–112 nucleotides and these RNAs are present in about 1–5 × 10^5 copies per cell [16, 17]. Within the past two years, Pruijn, Slobbe and van Venrooij *et al.* [18, 19] have described several important aspects of the Ro RNP complex leading to the model shown in Figure 1. The hY-RNA bound

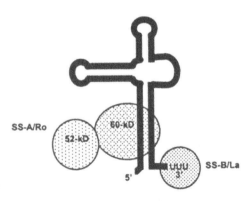

SS-A/Ro hY-RNA Complex

Fig. 1. A schematic diagram for the putative structure of SS-A/Ro hY-RNA particle. The associations of the Ro60 to the 5′ and 3′ stem of hY-RNAs and SS-B/La protein to the 3′ oligo U tail have been reported by several investigators [17–19, 22]. The presence of the Ro52 as shown here and its interaction with the 60-kD protein in the hY-RNP particle has also been reported by Slobbe *et al.* [19]. Furthermore, these investigators concluded that the region aa276–318 of the 60-kD protein was important for protein-protein interaction with the 52-kD protein. This model is simplified from [19].

directly to the 60-kD SS-A/Ro via the lower stem of the RNA formed by basepairing their 5′ and 3′ ends. The 60-kD protein belongs to a family of RNA-binding proteins and can bind to hY-RNA alone [18–20]. SS-B/La is known to bind to the 3′ oligo uridylate residues of hY RNAs [21, 22]. The association of Ro52 with the RNA-protein complex is more controversial and has been proposed to occur indirectly via the interaction with Ro60 [19].

4. Antibody detection

Many clinical laboratories still use immunodiffusion or counterimmuno-electrophoresis (CIE) for the detection of SS-A/Ro. For research laboratories, immunofluorescence, immunoblotting, immunoprecipitation (IP), and and ELISA are used for the identification of SS-A/Ro antibodies.

4.1 Immunofluorescence

For general technique refer to this Manual chapter A2. SS-A/Ro nuclear staining has been detected in many cultured cell lines provided that there is an adequate fixation procedure [23]. Acetone was found to be the most useful for stabilizing the SS-A/Ro antigen in KB or HEp-2 cells [23]. Ethanol or methanol alone were less efficient. HEp-2 cells are the most commonly used cell substrate in diagnostic ANA tests done currently (Fig. 2). It is important to note that some commercial HEp-2 cell slides are not suitable for the detection of SS-A/Ro and variations in different batches from the same manufacturer have also been observed. This is explained largely by the fact that SS-A/Ro is relatively soluble in many different fixation methods employed by manufacturers. Each batch of slides should be screened for nuclear staining using established prototype SS-A/Ro sera. This is especially important regarding the description of ANA negative SS-A/Ro sera in the literature.

4.2 Counterimmunoelectrophoresis/immunodiffusion

For the general technique refer to this Manual chapter A3. Earlier work leading to the discoveries of SS-A/Ro antibodies was based on immunodiffusion using extracts from human thyroid, spleen, and calf thymus [2, 3]. The other commonly used antigen sources are bovine spleen extracts, extracts from guinea pig organs [24], and pig spleen extracts [25]. Extracts of Wil-2 cells as well as human spleen extracts are probably the best for the detection of SS-A/Ro antibodies. The choice depends on the availablity of human spleen and the relatively high cost of tissue culture for Wil-2 cells. From the comparison of five methods Manoussakis *et al.* [26]

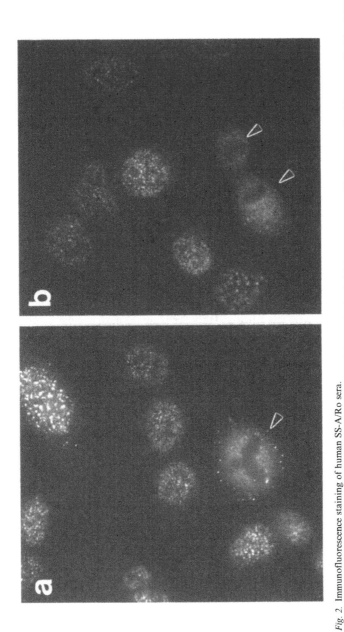

Fig. 2. Immunofluorescence staining of human SS-A/Ro sera.

HEp-2 cells were fixed in an ice-cold mixture of acetone and methanol (3:1) and stained with human prototype SS-A/Ro serum Ew (a) or serum Ge (b) at 1:100 dilution. Nuclear speckles of many sizes are observed in most interphase cells with relatively weak staining also detected in the cytoplasm. Many other SS-A/Ro sera also show similar staining including the speckles observed in mitotic cells (a, arrowhead) and newly divided cells (b, arrowheads). Nucleolar staining is generally not detected.

AMAN-B4.1/4

concluded that CIE using calf thymus extract as antigen was the most reliable method for the detection of SS-A/Ro autoantibodies. Similarly, Meilof *et al.* [27] also found that CIE is a highly sensitive assay for the detection of SS-A/Ro antibodies. Immunodiffusion and CIE generally favor high titered sera and cannot differentiate the response to Ro52 and Ro60. Note that Ro52 has not been detected in bovine substrate.

4.3 Immunoblotting

Extracts from many cell lines such as MOLT-4 and HeLa have been used successfully for the detection of SS-A/Ro antigens [14, 15]. The technique of Western blotting for ribonucleoproteins has been described in detail [28] (also see this Manual chapter A4). It is important to use the appropriate ratio of acrylamide to bisacrylamide (172:1) to obtain optimal separation of the 52-kD protein from the 48-kD SS-B/La protein coincident in the majority of sera [29, 30]. In all cell lines tested, many SS-A/Ro sera showed reactivities with Ro52 and Ro60 but some SS-A/Ro sera showed little or no reaction (Fig. 3A).

Rader *et al.* have described two Ro proteins of 60 kD and 54 kD in extracts of human red blood cells (RBC). The 60-kD and 54-kD RBC Ro proteins are immunologically related to the 60-kD and 52-kD SS-A/Ro proteins detected in lymphocytes respectively [15]. Manoussakis *et al.* [26] reported recently that in their analysis of RBC Ro preparation using the method of Rader *et al.* [15], they were unable to confirm the specific reactivity of the 60-kD RBC Ro protein since a significant percent of reactive sera did not immuno-precipitate Ro RNAs. Manoussakis *et al.* [26] suggested that there might be a second unrelated antigen co-migrating with the 60-kD protein in the RBC extract. It was also noted that the gel migration of the 54-kD protein of RBC and the 52-kD protein of lymphocytes was identical in SDS polyacrylamide [26].

4.4 RNA immunoprecipitation

For the general technique refer to this Manual chapter A7. The association of hY RNAs and Ro60 were initially characterized using IP of extracts from [^{32}P]-orthophosphate labeled cultured cells [17]. Five hY RNAs were identified as hY1–5 RNA and it was shown later that hY2 RNA derived from hY1 RNA (Fig. 3C). An alternative method using silver staining for the detection of RNA from unlabeled cell extracts has been described [31]. IP of hY RNAs is considered a definitive criterion for the identification of SS-A/Ro antibodies.

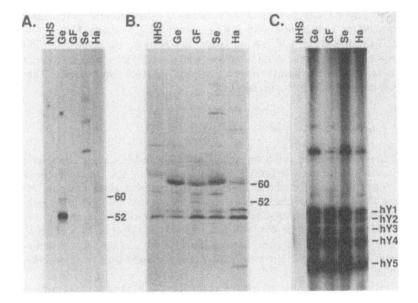

Fig. 3. Analysis of SS-A/Ro sera using Western blotting and immunoprecipitation.

Four anti-SS-A/Ro sera demonstrated by immunodiffusion were selected to illustrate the interesting features of this autoantibody system.

A. The reactivities of anti-SS-A/Ro sera with MOLT-4 cell extracts in Western blotting using standard SDS denaturing gel [14, 29]. As described in our earlier report [14], serum Ge recognizes both 52-kD and 60-kD proteins, a reactivity observed in approximately 50% of anti-SS-A/Ro sera. Sera GF, Se, and Ha showed very weak or no reactivity with the 52-kD and/or 60-kD protein in Western blotting further supporting the dependence of the anti-60-kD antibody on protein conformation.

B. The results of a standard IP assay using extracts from [35S]-methionine labeled HeLa cells that were labeled in an overnight culture. All four SS-A/Ro sera show a common 60-kD band consistent with the original report of Wolin and Steitz [17]. Although the reactivity to the 52-kD protein is not apparent in panel B, the 52-kD protein can be readily detected in Western blotting using a color development method (data not shown). Therefore the apparent reason for the failure to detect the 52-kD protein in panel B was that the HeLa 52-kD protein was not readily labeled under commonly used experimental conditions.

C. The results of RNA IP using extracts from [32P]-phosphate labeled HeLa cells. All four sera immunoprecipitated the hY RNAs as would be predicted by the results in immunodiffusion demonstrating that these four sera contained SS-A/Ro autoantibodies. All four sera were shown to recognize *in vitro* full-length translation products of the 52-kD and 60-kD proteins derived from their established cDNAs respectively (data not shown).

4.5 Protein immunoprecipitation

For the general technique refer to this Manual chapter A6. The reactivity with Ro60 can be readily identified in IP (Fig. 3B). Recent studies [22, 32, 33] have shown that native Ro RNP particles are recognized by many SS-A/Ro sera that were undetectable in Western blotting analysis (Fig. 3). IP using extracts from [^{35}S]-methionine labeled cells is not suitable for the detection of autoantibody to Ro52 protein since most sera are apparently negative in this assay. This may be due to the inability of Ro52 to be labeled under normal culture conditions because of its slow turnover rate. This has led to the misconception that autoantibodies to the 52-kD antigen were directed solely to denatured epitopes accounting for the preferential reactivity on Western blotting [34]. When Ro52 is labeled in an *in vitro* translation system using a cDNA template, anti-52-kD SS-A/Ro sera demonstrated by Western blot do react with the 52-kD antigen [35].

4.6 Elisa

For the general technique refer to this Manual chapter A5. The method of Yamagata *et al.* [36] for the affinity purification of calf thymus or human spleen SS-A/Ro that is suitable for ELISA has been used by several investigators before the description of the recombinant Ro52. Pig spleen was also reported as a useful source to isolate SS-A/Ro antigen suitable for ELISA [37]. With the recent identification of the 52-kD protein [14] and the cloning of both Ro52 and Ro60 proteins, purified recombinant proteins may become the preferred substrates for highly specific ELISA [35, 38].

5. Cellular localization

The cellular distribution of SS-A/Ro has been a controversial topic since the early description of these autoantigens. Ro antigen was isolated from a cytoplasmic fraction [2] while SS-A was identified as a nuclear autoantigen [3, 39]. The report that the Ro antigen was associated with cytoplasmic hY RNAs suggested that Ro RNPs are cytoplasmic RNP particles [5, 16, 17, 40]. The report of Harmon *et al.* [23] showed that specific SS-A antibody gave predominantly discrete nuclear speckles (Fig. 2). Ro52 has transcription factor-like protein motifs [35, 41] and this is consistent with its described location within the nucleus [14]. Furthermore, Mamula *et al.* showed that rabbit anti-Ro60 antibody gave a speckled nuclear pattern with HEp-2 cells and the staining was completely inhibited with purified 60-kD protein [42]. However, recent work reported by O'Brien *et al.* [43] examining the Ro RNPs in mouse and *Xenopus* have concluded that these RNPs resided primarily in the cytoplasm. Using cell enucleation for cell fractionation, Peek *et al.* [44] have described that about 30–40% of both SS-A/Ro and SS-B/La proteins are present in the nucleus while the remainder is in the cytoplasm; however, consistent with earlier findings, these investigators have noted that the Y RNAs are exclusively in the cytoplasm. Further examination of the cellular location will be important to provide insights into the cellular function of the Ro RNPs and to further resolve the discrepancy in the literature.

6. Cellular function

The function of Ro RNPs has eluded identificiation. Based on its similiarity with other closely related proteins which function in transcriptional regulation, it would be predicted that Ro52 has a similar role [35]. On the other hand, the 60-kD protein and hY RNAs distributed in the nucleus and cytoplasm might be carrying out other cellular function(s) yet to be determined. Although Ro52 and Ro60 have been proposed by some to reside in a single complex [14], discrete functions may exist for each. Recently the expression of different subsets of hY RNAs and Ro proteins in human RBC and platelets has been described [15, 45, 46]. It is interesting that hY1 and hY4 RNAs were detected in RBC [45] while hY3 and hY4 RNAs were found in platelets [46]. A specific form of the 52-kD Ro in human platelets and other Ro proteins in RBC have also been described [15, 46]. However, the role of hY RNAs and proteins in lymphocytes, RBC, and platelets remains to be determined.

7. Presence in other species

Using monospecific anti-SS-A/Ro sera, Harmon *et al.* [23] showed that the SS-A/Ro antigen was present in significant quantities as detected by immunofluorescence in cells of human, monkey, dog, and guinea pig, but absent to low amounts were detected in cells of mouse, rat, rabbit, hamster, and chicken. Calf thymus 60-kD SS-A/Ro was purified and partial amino acid sequences were reported [20]. Pig spleen has been used as an excellent source of antigen and the 60-kD protein has been identified [25, 30, 37]. Both 52-kD and 60-kD proteins have been detected in various organs of the guinea pig [24]. Recently *Xenopus* 60-kD SS-A/Ro was also cloned and comparison of the amino acid sequences between Xenopus and human 60-kD (60α) showed 78% identity [43]. In addition to hY RNAs, four bovine Y RNAs, three rabbit Y RNAs, and two duck Y RNAs were described and all the tested species possess an RNA which co-migrates with hY1 RNA [42]. Four hY-like RNAs were also observed in the guinea pig [24]. Four distinct *Xenopus* Y RNAs (xY) have also been recently described [43]. Three of the four xY RNAs are related to the human hY3, hY4, and hY5 RNAs. The fourth xY RNA, xYα, is not a homologue of any human Y RNAs [43]. All four xY RNAs also possess a conserved stem, as in hY RNAs, that is critical for binding to the 60-kD protein. Slobbe *et al.* [30] and Pruijn *et al.* [47] have performed an extensive survey of the conservation of the SS-A/Ro proteins and Y RNAs. In summary Ro60 and hY-like RNAs have been detected in many species while the equivalent of Ro52 has only been described in human, monkey [30], and guinea pig [24].

8. The cDNAs and derived amino acid sequences

The cloning of cDNAs for Ro52 has allowed the definitive characterization of this component [35, 41]. Figure 4 shows the features of the full-length Ro52. The full-length cDNAs for Ro60 have been reported by Deutscher *et al.* [20] and Ben-Chetrit *et al.* [48]. Comparison of the two deduced protein sequences showed that the only difference is in the C-terminal residues and the two forms are provisionally named 60α and 60β (Fig. 5). The 60α and 60β can be distinguished in an IP assay using antibodies specific for their respective C-termini [49]. Note that human anti-SS-A/Ro sera do not recognize the type-specific C-terminal peptides and therefore are unable to differentiate the two forms of Ro60.

A

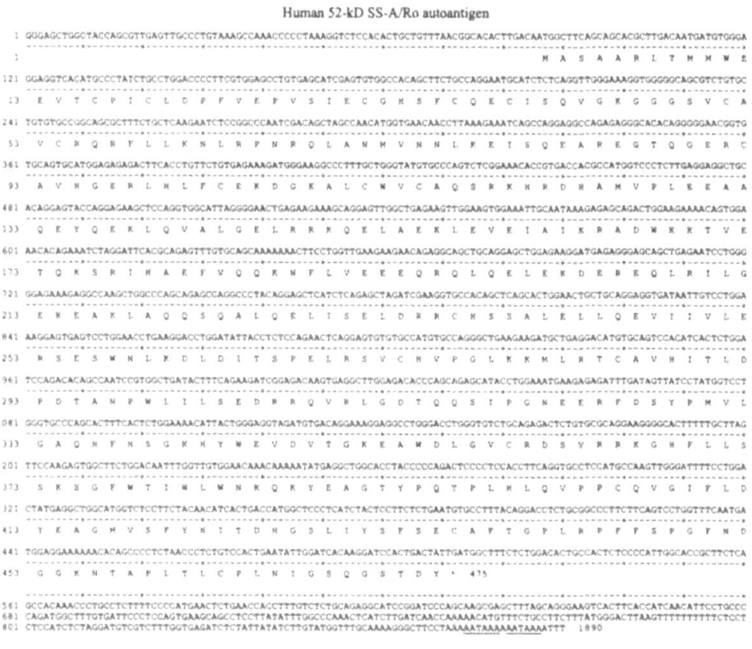

Human 52-kD SS-A/Ro autoantigen

Fig. 4A. Features and motifs of the 52-kD SS-A/Ro protein.
A. The complete cDNA and protein sequence for Ro52 (GenBank/EMBL accession numbers M35041). The full-length protein consists of 475 amino acid residues with a deduced molecular mass of 54,000 [35,41]; the deduced pI is 6.35 while the experimentally observed pI is 6.5 [35].

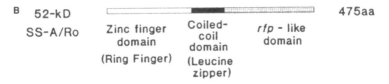

52-kD   475aa

SS-A/Ro   Zinc finger   Coiled-   *rfp* - like

                 domain   coil   domain

               (Ring Finger)   domain

                            (Leucine

                            zipper)

Fig. 4B. The analysis of Ro52 protein showed three domains: the N-terminal zinc finger domain, the central coiled-coil domain, and C-terminal *rfp*-like domain. The N-terminal zinc finger domain is a member of the newly described "Ring" finger proteins [65] while the C-terminal *rfp*-like domain belongs to a second protein domain subfamily named after the *rfp* protein [66]. Note that *rfp* has been reported to be a DNA-binding protein associated with the nuclear matrix [55]. The central coiled-coil domain contains a leucine zipper motif that is often responsible for the formation of protein dimers.

A

Human 60-kD SS-A/Ro autoantigen

```
   1 ATTTTGCCTTTTTGTTAGGTTTCCTAAAGACAAAAAAAATGGAGGAATCTGTAAACCAAATGCAGCCACTGAATGAGAAGCAGATAGCCAATTCTCAGGATGGATATGTATGGCAAGTC
     ---------+---------+---------+---------+---------+---------+---------+---------+---------+---------+---------+---------+
   1                                        M  E  E  S  V  N  Q  M  Q  P  L  N  E  K  Q  I  A  N  S  Q  D  G  Y  V  W  Q  V

 121 ACTGACATGAATCGACTACACCGGTTCTTATGTTTCGGTTCTGAAGGTGGGACTTATTATATCAAAGAACAGAAGTTGGGCCTTGAAAATGCTGAAGCTTTAATTAGATTGATTGAAGAT
     ---------+---------+---------+---------+---------+---------+---------+---------+---------+---------+---------+---------+
  28  T  D  M  N  R  L  H  R  F  L  C  F  G  S  E  G  G  T  Y  Y  I  K  E  Q  K  L  G  L  E  N  A  E  A  L  I  R  L  I  E  D

 241 GGCAGAGGATGTGAAGTGATACAAGAAATAAAGTCATTTAGTCAAGAAGGCAGAACCACAAAGCAAGAGCCTATGCTCTTTGCACTTGCCATTTGTTCCCAGTGCTCCGACATAAGCACA
     ---------+---------+---------+---------+---------+---------+---------+---------+---------+---------+---------+---------+
  68  G  R  G  C  E  V  I  Q  E  I  K  S  F  S  Q  E  G  R  T  T  K  Q  E  P  M  L  F  A  L  A  I  C  S  Q  C  S  D  I  S  T

 361 AAACAAGCAGCATTTAAAGCTGTTTCTGAAGTTTGTCGCATTCCTACCCATCTCTTTACTTTTATCCAGTTTAAGAAAGATCTGAAGGAAAGCATGAAATGTGGCATGTGGGGTCGTGCC
     ---------+---------+---------+---------+---------+---------+---------+---------+---------+---------+---------+---------+
 108  K  Q  A  A  F  K  A  V  S  E  V  C  R  I  P  T  H  L  F  T  F  I  Q  F  K  K  D  L  K  E  S  M  K  C  G  M  W  G  R  A

 481 CTCCCGGAAGGCTATAGCGGACTGGTACAATGAGAAAGGTGGCATGGCCCTTGCTCTGGCAGTTACAAAATATAAACAGAGAAATGGCTGGTCTCACAAAGATCTATTAAGATTGTCACAT
     ---------+---------+---------+---------+---------+---------+---------+---------+---------+---------+---------+---------+
 148  L  R  K  A  I  A  D  W  Y  N  E  K  G  G  M  A  L  A  L  A  V  T  K  Y  K  Q  R  N  G  W  S  H  K  D  L  L  R  L  S  H

 601 CTTAAAACCTTCCAGTGAAGGACTTGCAATTGTGACCAAATATATTACAAAGGGCTGGAAAGAAGTTCATGAATTGTATAAAGAAAAAGCACTCTCTGTGGAGACTGAAAAATTATTAAAG
     ---------+---------+---------+---------+---------+---------+---------+---------+---------+---------+---------+---------+
 188  L  K  P  S  S  E  G  L  A  I  V  T  K  Y  I  T  K  G  W  K  E  V  H  E  L  Y  K  E  K  A  L  S  V  E  T  E  K  L  L  K

 721 TATCTGGAGGCTGTAGAGAAAGTGAAGCGCACAAGAGATGAGCTAGAAGTCATTCATCTAATAGAAGAACATAGATTAGTTAGAGAACATCTTTTAACAAATCACTTAAAGTCTAAAGAG
     ---------+---------+---------+---------+---------+---------+---------+---------+---------+---------+---------+---------+
 228  Y  L  E  A  V  E  K  V  K  R  T  R  D  E  L  E  V  I  H  L  I  E  E  H  R  L  V  R  E  H  L  L  T  N  H  L  K  S  K  E

 841 GTATGGAAGGCTTTGTTACAAGAAATGCCGCTTACTGCATTACTAAGGAATCTAGGAAAGATGACTGCTAATTCAGTACTTGAACCAGGAAATTCAGAAGTATCTTTAGTATGTGAAAAA
     ---------+---------+---------+---------+---------+---------+---------+---------+---------+---------+---------+---------+
 268  V  W  K  A  L  L  Q  E  M  P  L  T  A  L  L  R  N  L  G  K  M  T  A  N  S  V  L  E  P  G  N  S  E  V  S  L  V  C  E  K

 961 CTGTGTAATGAAAAACTATTAAAAAAGGCTCGTATACATCCATTTCATATTTTGATCGCATTAGAAACTTACAAGACAGGTCATGGTCTCAGAGGGAAACTGAAGTGGCGCCCTGATGAA
     ---------+---------+---------+---------+---------+---------+---------+---------+---------+---------+---------+---------+
 308  L  C  N  E  K  L  L  K  K  A  R  I  H  P  F  H  I  L  I  A  L  E  T  Y  K  T  G  H  G  L  R  G  K  L  K  W  R  P  D  E

1081 GAAATTTTGAAAGCATTGGATGCTGCTTTTTATAAAACATTTAAGACAGTTGAACCAACTGGAAAACGTTTCTTACTAGCTGTTGATGTCAGTGCTTCTATGAACCAAAGAGTTTTGGGT
     ---------+---------+---------+---------+---------+---------+---------+---------+---------+---------+---------+---------+
 348  E  I  L  K  A  L  D  A  A  F  Y  K  T  F  K  T  V  E  P  T  G  K  R  F  L  L  A  V  D  V  S  A  S  M  N  Q  R  V  L  G

1201 AGTATACTCAACGCTAGTACAGTTGCTGCAGCAATGTGCATGGTTGTCACACGAACAGAAAAAGATTCTTATGTAGTTGCTTTTTCCGATGAAATGGTACCATGTCCAGTGACTACAGAT
     ---------+---------+---------+---------+---------+---------+---------+---------+---------+---------+---------+---------+
 388  S  I  L  N  A  S  T  V  A  A  A  M  C  M  V  V  T  R  T  E  K  D  S  Y  V  V  A  F  S  D  E  M  V  P  C  P  V  T  T  D

1321 ATGACCTTACAACAGGTTTTAATGGCTATGAGTCAGATCCCAGCAGGTGGAACTGATTGCTCTCTTCCAATGATCTGGGCTCAGAAGACAAACACACCTGCTGATGTCTTCATTGTATTC
     ---------+---------+---------+---------+---------+---------+---------+---------+---------+---------+---------+---------+
 428  M  T  L  Q  Q  V  L  M  A  M  S  Q  I  P  A  G  G  T  D  C  S  L  P  M  I  W  A  Q  K  T  N  T  P  A  D  V  F  I  V  F

1441 ACTGATAATGAGACCTTTGCTGGAGGTGTCCATCCTGCTATTGCTCTGAGGGAGTATCGAAAGAAAATGGATATTCCAGCTAAATTGATTGTTTGTGGAATGACATCAAATGGTTTCACC
     ---------+---------+---------+---------+---------+---------+---------+---------+---------+---------+---------+---------+
 468  T  D  N  E  T  F  A  G  G  V  H  P  A  I  A  L  R  E  Y  R  K  K  M  D  I  P  A  K  L  I  V  C  G  M  T  S  N  G  F  T

1561 ATTGCAGACCCAGATGATAGA|GCATGTTGGATATGTGCGGCTTTGATACTGGAGCTCTGGATGTAATTCGAAATTTCACATTAGATATGATTTAACCATAAGCAGCAGCCAGCACGATCCAGA
     ---------+---------+---------+---------+---------+---------+---------+---------+---------+---------+---------+---------+
 508  I  A  D  P  D  D  R |G  M  L  D  M  C  G  F  D  T  G  A  L  D  V  I  R  N  F  T  L  D  M  I  *  538

1681 GATCCATTGCCATCAGTGATCTCACTAAAAAAATATACAGCTACTTCCCAGCTAATCTCCACCCAATGAATGATGATGGTATAGTATGTGCATAATGGAAAGTTACCTTACTGAAAAAAAA
1801 AAAAAAAAAAAAAAAAA  1817
```

```
1561 ATTGCAGACCCAGATGATAGAG|CCTTGCAAAATACCCTACTAAATAAATCATTTTAGACATGGAGTGCAGGTGGACACTGTGTGAACTGTTTTTGGTCAGTTATTGTAGAAATTGATAGA
     ---------+---------+---------+---------+---------+---------+---------+---------+---------+---------+---------+---------+
 508  I  A  D  P  D  D  R |A  L  Q  N  T  L  L  N  K  S  F  *  525

1681 TGTACCAAATAAACTCTATGCACATTAAAAAAAAAAAAAAAA
```

Fig. 5A. Features and motifs of the 60-kD SS-A/Ro proteins.
Full-length cDNA sequences for Ro60 have been described in reports by Deutscher *et al.* [20] and Ben-Chetrit *et al.* [48] which showed identical sequences except the 5′ and 3′ nucleotide

B

60-kD
SS-A/Ro

RNA binding
domain

Zinc finger
motif

GMLDMCGFDTGA
LDVIRNFTLDMI
60α
538aa

ALQNTLLNKSF
60β
525aa

Fig. 5B. The RNA binding domain is essentially conserved between human and Xenopus 60-kD sequences [43] and may account for the binding of the 60-kD protein to hY-RNAs [17, 18, 20]. The function of the zinc finger domain is less obvious since the second of the two Cys residues in the putative zinc finger domain in the human sequence is changed to a Thr residue in the *Xenopus* sequence [43].

9. The genes

The genes for Ro52 and Ro60 have not been published yet. However, genomic clones for both genes are being characterized in one of our laboratories (EKLC). Using cDNA probes, Frank *et al.* have shown that the human Ro52 gene is located on chromosome 11 [50]. Human Ro60 gene has also been localized to Ig31 by fluorescence in situ hybridization [50a].

10. Relation with nucleic acids

The interaction of hY RNAs with SS-A/Ro is described in Paragraph 3. Note that in RBC, hY1 and hY4 RNAs have been described and are apparently associated with an RBC specific form of the 60-kD protein [15, 45].

11. Relation with other proteins

Ro60 is a member of the RNA binding protein family with RNA binding consensus sequences [51, 52]; however, the similarity of Ro60 to other

sequences. Analysis of their respective amino acid sequences revealed that C-terminal amino acid residues were different; however, differences in the 5' nucleotide sequence had no effect on the protein sequence. The two forms of Ro60 are provisionally termed 60α and 60β corresponding to the cDNA sequences reported by Deutscher *et al.* [20] and Ben-Chetrit *et al.* [48] respectively.
A. The complete cDNA and protein sequences for 60α (GenBank/EMBL accession numbers J04137) are shown and the boxed sequences represent the unique sequences in 60α. Separate 60β 3' nucleotide and amino acid sequences (GenBank/EMBL accession numbers M25077) are also shown with the unique sequence for 60β indicated (box). Ro60α is the predominant type expressed in all human cells tested; it has 538 amino acids, a deduced molecular mass of 60,670, and pI 7.97. In contrast, 60β has only 525 amino acids, a deduced molecular mass of 59,269, and pI 8.5, and it was originally described in MOLT-4 cells. Both 60α and 60β contain an RNA binding domain and a zinc finger domain.

members of the protein family is very low. There is no known protein with a high degree of similarity to Ro60. It has been reported that the nucleocapsid protein N of vesicular stomatitis virus showed sequence in common with short peptides of Ro60 [53] but cross-reactive antibodies between the N protein and Ro60 have not been described.

Ro52 belongs to the newly defined *rfp* protein family with its members containing a N-terminal zinc-finger domain (RING finger), a center coiled-coil domain (leucine zipper), and a C-terminal *rfp*-like domain. As much as 50% amino acid sequence identity has been described among other *rfp* proteins which are often found in developmental stages requiring rapid cell divisions [8]. It is interesting that many *rfp* proteins are DNA-binding proteins and are localized to the nucleus [54, 55]. Although the C-terminal domain of Ro52 is highly homologous to the corresponding domain in the *rfp* protein, rabbit antibodies directed against *rfp* do not immunoprecipitate human recombinant Ro52 [35]. While human anti-Ro52 sera also do not immunoprecipitate *rfp* [35], the interpretation is complicated by the recent observation that the C-terminal domain of Ro52 is not recognized by most human anti-Ro52 sera [56, 57].

When immunoprecipitation was performed using radiolabeled Ro52 derived from *in vitro* translation, it was noted that many normal human sera appeared to be weakly reactive, a problem that has not been observed in similar IP with labeled Ro60 or SS-B/La. This "stickiness" of Ro52 protein was observed in IP alone and not in ELISA or Western blotting; therefore, it is not likely to represent true immunoreactivity. Since the recombinant Ro52 was produced in a relatively "native" condition, this sticky feature of Ro52 may be of some potential biological significance. We have mapped the sticky domain to the C-terminal region of Ro52 [56]. A similar finding has been described by Bozic *et al.* [57] who attributed the "stickiness" to the binding to the C-terminal tail of Ro52 via hydrophobic protein-protein intereactions.

12. B-cell epitopes

It has been reported that antibodies to SS-A/Ro react preferentially with the human antigen [58]. While it has been appreciated that epitopes recognized by autoantibodies to Ro60 can be highly conformational [32, 33, 59], there is a general misconception that autoantibodies to Ro52 can only recognize denatured or linear epitopes. "Native" recombinant Ro52 synthesized in an *in vitro* translation system is clearly recognized in an IP assay [56, 57]. Figure 6 summarizes the results on the reactivities of Ro52 fragments with human anti-SS-A/Ro sera. Interestingly, Bozic *et al.* [57] have observed that most anti-SS-A/Ro sera from SLE patients recognized predominantly the region aa216–292 while anti-SS-A/Ro antibodies from SS patients recognized multiple epitopes located between aa55–292. While a similar study using

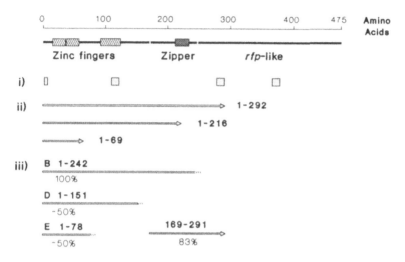

Fig. 6. B-cell epitopes of 52-kD SS-A/Ro protein.

i) Using 39 overlapping synthetic peptides spanning the entire sequence of Ro52 in ELISA, Ricchiuti *et al.* [67] reported that four peptides (aa2–11, aa107–126, aa277–292, and aa365–382) were recognized by many anti-SS-A/Ro sera. Some patients with SS associated with SLE also showed reactivities to these Ro52 synthetic peptides although they were not normally considered positive for anti-SS-A/Ro antibodies in other detection methods such as CIE, immunoblot, and recombinant Ro52 protein ELISA.

ii) Bozic et al. [57] have shown that recombinant Ro52 fragments from *in vitro* translation are recognized by SS-A/Ro sera. Fragment aa1–292 was the smallest C-terminal deletion fragment that was still reactive with all anti-SS-A/Ro sera to the same extent compared to full-length Ro52. Many anti-SS-A/Ro sera showed weaker or no reactivities when tested with the shorter fragment aa1–216. Fragment aa1–69 was the smallest C-terminal deletion fragment that was still recognized by some anti-SS-A/Ro sera.

iii) Based on the reactivity by IP of fragments B, D, and E generated from partial SaV8 protease digestion of full-length *in vitro* translated Ro52 – and other approximately equivalent recombinant Ro52 fragments, Buyon *et al.* [56] showed that the reactivities of human SS-A/Ro sera in IP could be separated into two broad types of approximately equal number. Type I sera recognized fragment B but did not recognize the smaller fragments D or E while Type II sera recognized fragments B, D and E. While most Type II responders had high titer anti-Ro52 antibodies, several Type I responders also had equally high anti-Ro52 antibody titers. The separation of SS-A/Ro sera into these two types was not directly correlated with the titer of anti-SS-A/Ro antibodies and did not segregate with clinical disease subset. These data suggest that there may be at least two B-cell epitopes: a "universal" epitope within the region aa169–242 recognized by both Type I and Type II sera and a second "restricted" epitope in the N-terminal zinc finger region (fragment E) recognized by Type II sera alone. This interpretation is supported by the observation that the majority of both Type I and II sera recognized fragment aa169–291 alone in this IP assay [56].

comparable recombinant fragments noted the same trend, the results did not reach statistical significance [56]. Both studies suggested that there are probably at least two or more B-cell epitopes recognized by human autoantibodies to Ro52. These data support that the autoimmune response to the 52-kD protein is polyclonal and most likely via an antigen-driven mechanism as described for many other autoantibody responses in systemic rheumatic diseases [60–62].

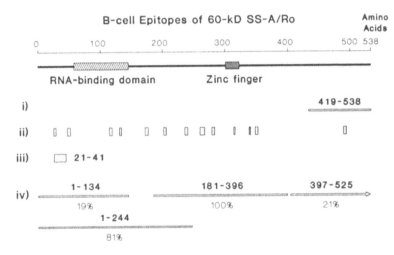

Fig. 7. B-cell epitopes of 60-kD SS-A/Ro protein.

i) Scofield *et al.* [68] reported that the C-terminal 13-kD fragment derived from SaV8 protease digestion of the 60-kD protein (60α) was recognized by 28 of 45 of SS-A/Ro sera (62%).

ii) Scofield and Harley [53] described the reactivities of human SS-A/Ro with overlapping synthetic octapeptides spanning the entire Ro60 protein. Many short regions were recognized by human sera and, furthermore, it was concluded that several of these short regions showed amino acid sequence similarity with the nucleocapsid N protein of vesicular stomatitis virus.

iii) Barakat *et al.* [69] described the reactivities of human anti-SS-A/Ro sera with five selected synthetic peptides: aa1–23, aa21–41, aa304–324, aa495–518, and aa524–538. Only oligopeptide aa21–41 showed considerable reactivity especially when this peptide was conjugated to bovine serum albumin.

iv) Using recombinant fusion proteins of Ro60β, Wahren *et al.* [70] showed that all SS-A/Ro sera recognized a universal epitope in the central region 181–396. Sera recognizing other fragments such as aa1–244, aa1–134, and aa397–525 were 81%, 19%, and 21% respectively. Furthermore, Wahren *et al.* [70] reported that 20-mer overlapping synthetic oligopeptides corresponding to the central region showed four patterns of reactivities among SS-A/Ro sera based on the types of reactive peptide(s) recognized. However, when 10-mer overlapping peptides for the same central region were used in ELISA, Wahren *et al.* [70] found that, with the exception of one patient, the reactivities of anti-SS-A/Ro sera in ELISA were equivalent to those obtained with control sera. Their data suggested that 10-mer peptides were not as useful as the 20-mers for the detection of autoantibody reactivity. The discrepancy among these studies especially with regard to the reactivities with short synthetic peptides and their relationship with antibodies to conformational epitopes of SS-A/Ro [32, 33] must be addressed in further studies.

Many SS-A/Ro sera have been reported to be nonreactive in Western blotting because they recognize conformational epitopes on the 60-kD protein [33] (see also Fig. 3). Most or all anti-SS-A/Ro sera recognize the 60-kD protein in IP assay but not or less efficiently in Western blot. Boire and Craft [32] have also described two sera with antibodies to a conformational epitope specific for the hY5–60-kD protein complex. Over the past two years, several epitope mapping studies have been published. Figure 7 summarizes the B-cell epitopes of the 60-kD SS-A/Ro protein reported by several investigators. In summary, like Ro52, there are at least two or more B-cell epitopes for Ro60.

13. T-cell epitopes

Unknown.

14. Monoclonal antibodies

Bachmann and co-workers [63, 64] have decribed a murine monoclonal antibody to Ro60. One of the authors has produced 5 murine IgG monoclonal antibodies to recombinant human Ro60 (antibody A6-A10, Chan, unpublished). Monoclonal antibody to Ro52 has not been published yet.

Acknowledgements

This is publication 8243-MEM from The Scripps Research Institute. EKLC is grateful for the support of Dr. Eng M. Tan over the past ten years as an inspiring mentor and a stimulating collaborator. EKLC also acknowledges the generous comments and support from the members of W.M. Keck Autoimmune Disease Center. This work is supported by NIH Grant AR41803 and AR42455 and also in part by SLE Foundation Inc., New York Chapter to JPB.

References

1. Anderson JR, Gray KG, Beck JS & Kinnear WF (1961) Lancet 2:456-460
2. Clark G, Reichlin M & Tomasi TB (1969) J Immunol 102:117-122
3. Alspaugh MA & Tan EM (1975) J Clin Invest 55:1067-1073
4. Alspaugh MA & Maddison P (1979) Arthritis Rheum 22:796-798
5. Lerner MR, Boyle JA, Hardin JA & Steitz JA (1981) Science 211:400-402
6. Slobbe RL, Pruijn GJM & van Venrooij WJ (1991) Ann Med Interne (Paris) 142:592-600
7. Ben-Chetrit E (1993) Br J Rheumatol 32:396-402
8. Chan EKL & Andrade LEC (1992) Rheum Dis Clin North Am 18:551-570
9. Reichlin M (1992) In: Lahita R (Ed) Systemic Lupus Erythematosus. Antibodies to Cytoplasmic Antigens (pp 237-246). Churchill Livingstone, New York
10. van Venrooij WJ, Slobbe RL & Pruijn GJM (1993) Mol Biol Reports 18:113-119
11. McCauliffe DP, Lux FA, Lieu TS, Sanz I, Hanke J, Newkirk MM, Bachinski LL, Itoh Y, Siciliano MJ, Reichlin M, Sontheimer RD & Capra JD (1990) J Clin Invest 85:1379-1391
12. McCauliffe DP, Zappi E, Lieu TS, Michalak M, Sontheimer RD & Capra JD (1990) J Clin Invest 86:332-335
13. Rokeach LA, Haselby JA, Meilof JF, Smeenk RJT, Unnasch TR, Greene BM & Hoch SO (1991) J Immunol 147:3031-3039
14. Ben-Chetrit E, Chan EKL, Sullivan KF & Tan EM (1988) J Exp Med 167:1560-1571
15. Rader MD, O'Brien C, Liu YS, Harley JB & Reichlin M (1989) J Clin Invest 83:1293-1298
16. Hendrick JP, Wolin SL, Rinke J, Lerner MR & Steitz JA (1981) Mol Cell Biol 1:1138-1149
17. Wolin SL & Steitz JA (1984) Proc Natl Acad Sci USA 81:1996-2000
18. Pruijn GJM, Slobbe RL & van Venrooij WJ (1991) Nucleic Acids Res 19:5173-5180
19. Slobbe RL, Pluk W, van Venrooij WJ & Pruijn GJM (1992) J Mol Biol 227:361-366
20. Deutscher SL, Harley JB & Keene JD (1988) Proc Natl Acad Sci USA 85:9479-9483
21. Mamula MJ, Silverman ED, Laxer RM, Bentur L, Isacovics B & Hardin JA (1989) J Immunol 143:2923-2928
22. Boire G & Craft J (1990) J Clin Invest 85:1182-1190
23. Harmon CE, Deng JS, Peebles CL & Tan EM (1984) Arthritis Rheum 27:166-173
24. Itoh Y, Kriet JD & Reichlin M (1990) Arthritis Rheum 33:1815-1821
25. Takano S, Matsushima H, Hiwatashi T & Miyachi K (1989) Jpn J Rheumatol 2:67-78
26. Manoussakis MN, Kistis KG, Liu X, Aidinis V, Guialis A & Moutsopoulos HM (1993) Br J Rheumatol 32:449-455
27. Meilof JF, Bantjes I, De Jong J, Van Dam AP & Smeenk RJ (1990) J Immunol Methods 133:215-226
28. Chan EKL & Pollard KM (1992) In: Rose NR, Conway de Macario E, Fahey JL, Friedman H & Penn GM (Eds) Manual of Clinical Laboratory Immunology. Antibodies to Ribonucleoprotein Particles by Immunoblotting (pp 755-761). American Society of Microbiology, Washington, DC
29. Buyon JP, Slade SG, Chan EKL, Tan EM & Winchester R (1990) J Immunol Methods 129:207-210
30. Slobbe RL, Pruijn GJM, Damen WG, van der Kemp JW & van Venrooij WJ (1991) Clin Exp Immunol 86:99-105
31. Forman MS, Nakamura M, Mimori T, Gelpi C & Hardin JA (1985) Arthritis Rheum 28:1356-1361
32. Boire G & Craft J (1989) J Clin Invest 84:270-279
33. Boire G, Lopez-Longo F-J, Lapointe S & Menard HA (1991) Arthritis Rheum 34:722-730
34. Tsuzaka K, Fujii T, Akizuki M, Mimori T, Tojo T, Fujii H, Tsukatani Y, Kubo A & Homma M (1994) Arthritis Rheum 37:88-92
35. Chan EKL, Hamel JC, Buyon JP & Tan EM (1991) J Clin Invest 87:68-76
36. Yamagata H, Harley JB & Reichlin M (1984) J Clin Invest 74:625-633
37. Kumagai T (1991) Jpn J Rheumatol 3:137-150

38. Julkunen H, Kurki P, Kaaja R, Heikkila R, Immonen I, Chan EKL, Wallgren E & Friman C (1993) Arthritis Rheum 36:1588–1598

39. Alspaugh MA, Talal N & Tan EM (1976) Arthritis Rheum 19:216–222

40. Wolin SL & Steitz JA (1983) Cell 32:735–744

41. Itoh K, Itoh Y & Frank MB (1991) J Clin Invest 87:177–186

42. Mamula MJ, OBrien CA, Harley JB & Hardin JA (1989) Clin Immunol Immunopathol 52:435–446

43. O'Brien CA, Margelot K & Wolin SL (1993) Proc Natl Acad Sci USA 90:7250–7254

44. Peek R, Pruijn GJM, van der Kemp AJW & van Venrooij WJ (1993) J Cell Sci 106:929–935

45. O'Brien CA & Harley JB (1990) EMBO J 9:3683–3689

46. Itoh Y & Reichlin M (1991) Arthritis Rheum 34:888–893

47. Pruijn GJM, Wingens PAETM, Peters ALM, Thijssen JPH & van Venrooij WJ (1993) Biochim Biophys Acta 1216:395–401

48. Ben-Chetrit E, Gandy BJ, Tan EM & Sullivan KF (1989) J Clin Invest 83:1284–1292

49. Chan EKL (1994) In: Homma M (Ed) Proceedings of the 4th International Symposium on Sjögren's Syndrome. Basis for Heterogeneity of SS-A/Ro Autoantigens. Kugler Publications, Amsterdam

50. Frank MB, Itoh K, Fujisaku A, Pontarotti P, Mattei MG & Neas BR (1993) Am J Hum Genet 52:183–191

50a. Chan EKL, Tan EM, Ward DC & Matera AG (1994) Genomics, in press.

51. Adam SA, Nakagawa T, Swanson MS, Woodruff TK & Dreyfuss G (1986) Mol Cell Biol 6:2932–2943

52. Kenan DJ, Query CC & Keene JD (1991) Trends Biochem Sci 16:214–220

53. Scofield RH & Harley JB (1991) Proc Natl Acad Sci USA 88:3343–3347

54. Bellini M, Lacroix J-C & Gall JG (1993) EMBO J 12:107–114

55. Isomura T, Tamiya-Koizumi K, Suzuki M, Yoshida S, Taniguchi M, Matsuyama M, Ishigaki T, Sakuma S & Takahashi M (1992) Nucleic Acids Res 20:5305–5310

56. Buyon JP, Slade SG, Reveille JD, Hamel JC & Chan EKL (1994) J Immunol 152:3675–3684

57. Bozic B, Pruijn GJM, Rozman B & van Venrooij WJ (1993) Clin Exp Immunol 94:227–235

58. Reichlin M & Reichlin MW (1989) J Autoimmun 2:359–365

59. Itoh Y, Itoh K, Frank MB & Reichlin M (1992) Autoimmunity 14:89–95

60. Tan EM & Chan EKL (1993) Clinical Investigator 71:327–330

61. Tan EM (1989) Adv Immunol 44:93–151

62. Tan EM, Chan EKL, Sullivan KF & Rubin RL (1988) Clin Immunol Immunopathol 47:121–141

63. Mayet WJ, Bachmann M, Pfeifer K, Schroder HC, Muller WE, Gudat W, Korting GW & Meyer zum Buschenfelde KH (1988) Eur J Clin Invest 18:465–471

64. Bachmann M, Mayet WJ, Schroder HC, Pfeifer K, Meyer zum Buschenfelde KH & Muller WE (1986) Proc Natl Acad Sci USA 83:7770–7774

65. Reddy BA, Etkin LD & Freemont PS (1992) Trends Biochem Sci 17:344–345

66. Takahashi M, Inaguma Y, Hiai H & Hirose F (1988) Mol Cell Biol 8:1853–1856

67. Ricchiuti V, Briand JP, Meyer O, Isenberg D, Pruijn GJM & Muller S (1994) Clin Exp Immunol 95:397–407

68. Scofield RH, Dickey WD, Jackson KW, James JA & Harley JB (1993) J Clin Immunol 11:378–388

69. Barakat S, Meyer O, Torterotot F, Youinou P, Briand JP, Kahn MF & Muller S (1992) Clin Exp Immunol 89:38–45

70. Wahren M, Ruden U, Andersson B, Ringertz NR & Pettersson I (1992) J Autoimmun 5:319–332

Manual of Biological Markers of Disease **B4.2**: 1–14, 1994.

The La (SS-B) antigen

GER J.M. PRUIJN

Department of Biochemistry, University of Nijmegen, PO Box 9101, NL-6500 HB Nijmegen, The Netherlands

Abbreviations

aa: amino acids; bLa: bovine La (SS-B); cDNA: complementary DNA; CIE: counterimmunoelectrophoresis; ELISA: enzyme-linked immunosorbant assay; hLa: human La (SS-B); IB: immunoblotting; ID: immunodiffusion; IF: immunofluorescence; IP: immunoprecipitation; kb: kilo basepairs; mLa: mouse La (SS-B); NLS: nuclear localization signal; NP-40: Nonidet P-40; PKR: double-stranded RNA-activated protein kinase; rabLa: rabbit La (SS-B); rLa: rat La (SS-B); RNP: ribonucleoprotein particle; rRNA: ribosomal RNA; SDS: sodium dodecyl sulphate; SLE: systemic lupus erythematosus; snRNA: small nuclear RNA; tRNA: transfer RNA; xLaA: *Xenopus laevis* La (SS-B), variant A; xLaB: *Xenopus laevis* La (SS-B), variant B

Key words: Autoantigen; Ribonucleoprotein; La (SS-B); Sjögren's syndrome; systemic lupus erythematosus

1. The antigen

A 46.7 kDa protein named La or SS-B. During the sixties and seventies the La autoantigen was identified by several researchers, using antisera from SLE and Sjögren's syndrome patients, and given several names. Later, interlaboratory exchange of sera and antigen extracts showed that the previously identified La, SS-B, Ha and SjT antigens were antigenically identical [1].

2. Synonyms

Ha [2], La [3], SjT [4], SS-B [5]. The most frequently used names are La and SS-B.

3. Biochemical characteristics

In cells La (SS-B) is a highly phosphorylated protein migrating at about 50 kDa in SDS-PAGE. Phosphorylated residues (serines and/or threonines) are present exclusively in the carboxy-terminal part of the protein. At least 8 isoelectric forms (pI 6 to 7) can be distinguished by two-dimensional gel electrophoresis [6–8].

All six methionines of human La (SS-B) are located in the amino-terminal half of the molecule.

The amino-terminal amino acid of La (SS-B) is probably blocked since amino-terminal amino acid sequencing could not be successfully performed [9].

Based upon structure predictions, the La (SS-B) protein includes a central alpha helical domain, containing a protease sensitive region that separates the phosphorylated carboxy-terminal domain from the amino-terminal methionine-containing domain, which contains the structures involved in RNA-binding. The frequently found degradation products of La (SS-B) of 29 kDa and 23 kDa (formerly designated domain X and domain Y, respectively [10]) result from proteolytic cleavage in the central alpha helical domain separating the methionine-containing amino-terminal part of the protein from the phosphorylated carboxy-terminal part [10].

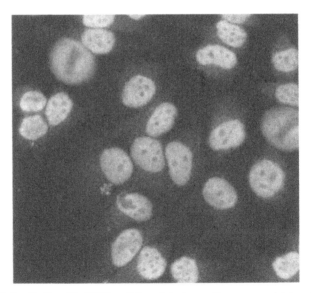

Fig. 1. A. Localization of La (SS-B) on HeLa cells by immunofluorescence using a monoclonal anti-La (SS-B) antibody SW5 (for experimental procedures see this Manual, Chapter A2). La (SS-B) is predominantly seen in the nucleoplasm.

4. Antibody detection

Representative IF, IB and IP patterns for anti-La (SS-B) antibody containing patient sera are shown in Figure 1A, 1B and 1C, respectively.

4.1 Immunofluorescence

La (SS-B) nuclear staining has been detected in many cultured cell lines. Although many different nuclear staining patterns were reported, in general a fine speckled (granular) pattern is observed with fine grainy staining in a uniform distribution, sometimes very dense so that an almost homogeneous pattern is obtained (Fig. 1A; see also this Manual, Chapter A2). Even nucleolar staining has been reported [11].

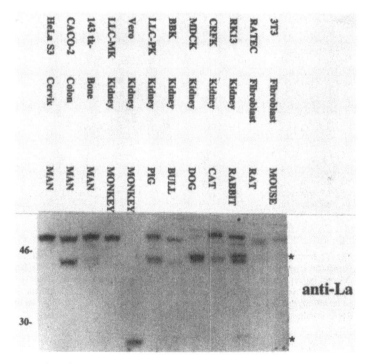

Fig. 1B. Detection of La (SS-B) in cell extracts from cell lines originating from different mammals (as indicated) by immunoblotting with an affinity-purified anti-La (SS-B) patient antibody (for experimental procedures see this Manual, Chapter A4). The position of molecular weight markers ($Mr \times 10^{-3}$) is indicated on the left. Asterisks mark prominent degradation products (43 and 29 kDa) of La (SS-B).

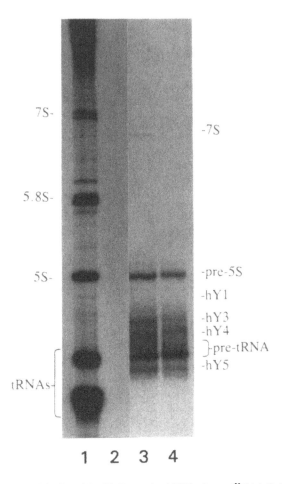

Fig. 1C. Immunoprecipitation of La (SS-B)-associated RNAs from a [^{32}P]-labelled HeLa cell extract (for experimental procedures see this Manual, Chapter A7). Lane 1 contains RNAs from the total extract. Lane 2 contains material precipitated by a normal human serum. Lanes 3 and 4 contain RNAs precipitated by an anti-La (SS-B) reference serum (CDC, Atlanta, USA) and by an affinity-purified anti-La (SS-B) patient antibody, respectively. The position of various RNAs is indicated.

4.2 *Immunodiffusion and counterimmunoelectrophoresis*

Although anti-La (SS-B) antibodies can be efficiently detected by ID and CIE (this Manual, Chapter A3), a distinct subpopulation of patient sera containing merely precipitin-negative anti-La (SS-B) antibodies has been described [12]. These antibodies, which, as a consequence, are undetectable by ID and CIE, are characterized by restricted epitope recognition [13].

4.3 Immunoblotting

IB provides an efficient and sensitive method to detect anti-La (SS-B) antibodies (Fig. 1B; see also this Manual, Chapter A4). Both nuclear and cytoplasmic extracts from many cell lines have been used successfully for the detection of these antibodies [14, 15]. The fact that La (SS-B) is rather sensitive to proteolysis and that La (SS-B) co-migrates with one of the Ro (SS-A) proteins (Ro52) in conventional SDS-polyacrylamide gel systems may complicate the interpretation of IB data. La (SS-B) from murine cells migrates somewhat faster (approx. 47 kDa) in SDS-polyacrylamide gels than La (SS-B) from human and bovine cells (approx. 50 kDa).

4.4 Immunoprecipitation

The first indications for La (SS-B) being a polypeptide of approx. 50 kDa came from IP experiments using extracts from [^{35}S]methionine-labeled cultured cells [16]. When [^{32}P]-labeled cell extracts are used for IP with anti-La (SS-B) antibodies, a characteristic set of small RNAs is precipitated (Fig. 1C), representing all newly synthesized RNA polymerase III transcripts (see par. 10) [reviewed in 17, 18]. The latter method provides a specific and sensitive method to confirm the presence of anti-La (SS-B) antibodies in patient sera (this Manual, Chapter A7).

4.5 ELISA

Recently developed ELISAs with bacterially expressed recombinant La (SS-B) offer highly sensitive, specific and, in addition, fast and non-laborious methods to detect anti-La (SS-B) antibodies [19–21] (for details, see this Manual, Chapter A5).

5. Cellular localization

Mammalian cells contain ~2×10^7 copies of the La (SS-B) protein. When analyzed by IF these molecules are localized predominantly in the nucleus (fine speckled pattern) (see Fig. 1A). By conventional cell-fractionation procedures (homogenization in a hypotonic buffer; cell-lysis in a buffer containing 0.5% NP-40) 20–50% of La (SS-B) is found in the cytoplasmic fraction [14]. Recent cell-enucleation experiments indicated that indeed a substantial amount of La (SS-B) may reside in the cytoplasm [22]. In several reports a redistribution (cytoplasmic accumulation as detected by IF) of La (SS-B) in cells that are stressed by various methods (virus infection; UV irradiation) has been described [23–28].

6. Cellular function

La (SS-B) has been shown to be required for efficient and correct termination of RNA polymerase III transcription [29, 30]. Furthermore, La (SS-B) may function as an ATP-dependent helicase able to melt RNA-DNA hybrids [31]. The latter activity may be related to its role in the RNA polymerase III transcription process. Although not proven yet, several observations suggest that La (SS-B) may be involved in various other processes as well, like maturation and/or nuclear export of RNA polymerase III products and some aspect(s) of translation [24, 32].

7. Presence in other species

Proteins immunoreactive with human and mouse monoclonal anti-La (SS-B) antibodies have been detected in all mammalian species examined (see Fig. 1B) and in amphibians and protozoa as well [15, 33, 34]. See also par. 8.

8. cDNA and amino acid sequence

The cDNA and derived amino acid sequence of human La (SS-B) are shown in Figure 2A [9, 35, 36]. The calculated molecular mass of La (SS-B) is 46.7 kDa. In the primary sequence several protein structural motifs can be discerned (Fig. 2B).

cDNAs encoding bovine, murine, rat, rabbit, *Xenopus laevis* and *Drosophila melanogaster* La (SS-B) have been isolated and characterized [9, 37–40]. At the amino acid level the homology with human La (SS-B) varies from about 60% (*Xenopus laevis*) to about 95% (bovine). Species-specific inserts are present in the carboxy-terminal parts of murine and *Xenopus laevis* La (SS-B). At the position of this insert in *Xenopus laevis* La (SS-B) some additional amino acids are also present in human, bovine and rabbit La (SS-B) (Fig. 2C). In contrast to La (SS-B) from *Xenopus laevis*, which migrates somewhat slower than the human antigen in SDS-polyacrylamide gels, most likely due to the insert of 22 amino acids (with respect to the human protein),

---

→

Fig. 2A. cDNA sequence and predicted amino acid sequence of human La (SS-B) (GenBank/EMBL data bases, accession nos. J04205, X13697).

```
          10        20        30        40        50        60        70        80        90
GCGCTTTAGGCTGGCCGGCGGCGCTGGGAGGTGGAGTCGTTGCTGTTGCTGTTTGTGAGCCTGTGCGGCGGCTTCTGTGGGCCGGAACCT

         100       110       120       130       140       150       160       170       180
TAAAGATAGCCGCAATGGCTGAAAATGGTGATAATGAAAAGATGGCTGCCCTGGAGGCCAAAATCTGTCATCAAATTGAGTATTATTTTG
                   M  A  E  N  G  D  N  E  K  M  A  A  L  E  A  K  I  C  H  Q  I  E  Y  Y  F    25

         190       200       210       220       230       240       250       260       270
GCGACTTCAATTTGCCACGGGACAAGTTTCTAAAGGAACAGATAAAACTGGATGAAGGCTGGGTACCTTTGGAGATAATGATAAAATTCA
 G  D  F  N  L  P  R  D  K  F  L  K  E  Q  I  K  L  D  E  G  W  V  P  L  E  I  M  I  K  F    55

         280       290       300       310       320       330       340       350       360
ACAGGTTGAACCGTCTAACAACAGACTTTAATGTAATTGTGGAAGCATTGAGCAAATCCAAGGCAGAACTCATGGAAATCAGTGAAGATA
 N  R  L  N  R  L  T  T  D  F  N  V  I  V  E  A  L  S  K  S  K  A  E  L  M  E  I  S  E  D    85

         370       380       390       400       410       420       430       440       450
AAACTAAAATCAGAAGGTCTCCAAGCAAACCCCTACCTGAAGTGACTGATGAGTATAAAAATGATGTAAAAACAGATCTGTTTATATTA
 K  T  K  I  R  R  S  P  S  K  P  L  P  E  V  T  D  E  Y  K  N  D  V  K  N  R  S  V  Y  I    115

         460       470       480       490       500       510       520       530       540
AAGGCTTCCCAACTGATGACAACTCTTGATGACATAAAAGAATGGTTAGAAGATAAAGGTCAAGTACTAAATATTCAGATGAGAAGAACAT
 K  G  F  P  T  D  A  T  L  D  D  I  K  E  W  L  E  D  K  G  Q  V  L  N  I  Q  M  R  R  T    145

         550       560       570       580       590       600       610       620       630
TGCATAAAGCATTTAAGGGATCAATTTTTGTTGTGTTTGATAGCATTGAATCTGCTAAGAAATTTGTAGAGACCCCTGGCCAGAAGTACA
 L  H  K  A  F  K  G  S  I  F  V  V  F  D  S  I  E  S  A  K  K  F  V  E  T  P  G  Q  K  Y    175

         640       650       660       670       680       690       700       710       720
AAGAAACAGACCTGCTAATACTTTTCAAGGACGATTACTTTGCCAAAAAAAATGAAGAAGAAAACAAAATAAAGTGGAAGCTAAATTAA
 K  E  T  D  L  L  I  L  F  K  D  D  Y  F  A  K  K  N  E  E  R  K  Q  N  K  V  E  A  K  L    205

         730       740       750       760       770       780       790       800       810
GAGCTAAACAGGAGCAAGAAGCAAAACAAAAGTTAGAAGAAGATGCTGAAATGAAATCTCTAGAAGAAAAGATTGGATGCTTGCTGAAAT
 R  A  K  Q  E  Q  E  A  K  Q  K  L  E  E  D  A  E  M  K  S  L  E  E  K  I  G  C  L  L  K    235

         820       830       840       850       860       870       880       890       900
TTTCGGGTGATTTAGATGATCAGACCTGTAGAGAGATTTACACATACTTTTCTCAAATCATGGTGAAATAAAATGGATAGACTTCGTCA
 F  S  G  D  L  D  D  Q  T  C  R  E  D  L  H  I  L  F  S  N  H  G  E  I  K  W  I  D  F  V    265

         910       920       930       940       950       960       970       980       990
GAGGAGCAAAAGAGGGGATAATTCTATTTAAAGAAAAAGCCAAGGAAGCATTGGGTAAAGCCAAAGATGCAAATAAGTGGTAACCTACAAT
 R  G  A  K  E  G  I  I  L  F  K  E  K  A  K  E  A  L  G  K  A  K  D  A  N  N  G  N  L  Q    295

        1000      1010      1020      1030      1040      1050      1060      1070      1080
TAAGGAACAAAGAAGTGACTTGGGAAGTACTAGAAGGAGAGGTGGAAAAAGAAGCACTGAAGAAAATAATAGAAGACCAACAAGAATCCC
 L  R  N  K  E  V  T  W  E  V  L  E  G  E  V  E  K  E  A  L  K  K  I  I  E  D  Q  Q  E  S    325

        1090      1100      1110      1120      1130      1140      1150      1160      1170
TAAACAAATGGAAGTCAAAAGGTCGTAGATTTAAAGGAAAAGGAAAGGGTAATAAAGCTGCCCAGCCTGGGTCTGGTAAAGGAAAAGTAA
 L  N  K  W  K  S  K  G  R  R  F  K  G  K  G  K  G  N  K  A  A  Q  P  G  S  G  K  G  K  V    355

        1180      1190      1200      1210      1220      1230      1240      1250      1260
AGTTTCAGGGCAAGAAAACGAAATTTGCTAGTGATGATGAACATGATGAACATGATGAAAATGGTGCAACTGGACCTGTGAAAAGAGCAA
 Q  F  Q  G  K  K  T  K  F  A  S  D  D  E  H  D  E  H  D  E  N  G  A  T  G  P  V  K  R  A    385

        1270      1280      1290      1300      1310      1320      1330      1340      1350
GAGAAGAAACAGACAAAGAAGAACCTGCATCCAAACAACAGAAAACAGAAAATGGTGCTGGAGACCAGTAGTTTAGTAAACCAATTTTTT
 R  E  E  T  D  K  E  E  P  A  S  K  Q  Q  K  T  E  N  G  A  G  D  Q    408

        1360      1370      1380      1390      1400      1410      1420      1430      1440
ATTCATTTTAAATAGGTTTTAAACGACTTTTGTTTGCGGGGCTTTTAAAAGGAAAACCGAATTAGGTCCACTTCAATGTCCACCTGTGAG

        1450      1460      1470      1480      1490      1500      1510      1520      1530
AAAGGAAAAATTTTTTTGTTGTTTAACTTGTCTTTTTGTTATGCAAATGAGATTTCTTTGAATGTATTGTTCTGTTTGTGTTATTTCAGA

        1540      1550      1560      1570      1580      1590      1600      1610      1620
TGATTCAAATATCAAAAGGAAGATTCTTCCATTAAATTGCCTTTGTAATATGAGAATGTATTAGTACAAACTAACTAATAAAATATATAC

        1630      1640      1650
TATATGAAAAGAGCAAAAAAAAAAAAAAAA
```

AMAN-B4.2/7

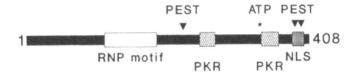

Fig. 2B. Schematic structure of the La (SS-B) protein. Several structural features can be discerned. First, an RNP motif (RNA binding domain, RNA recognition motif), which is found in many RNA binding proteins (reviewed by Mattaj [53]), is present in the amino-terminal half of the molecule. Indeed, evidence has been obtained that this domain in conjunction with the amino-terminal region of the protein flanking the RNP motif is involved in RNA binding. Second, a putative ATP-binding site has been identified (sequence GXXXXGK) [37]. No experimental evidence is available yet for this evolutionarily conserved motif to be a functional ATP-binding site. Third, three so-called PEST regions, sequences rich in the amino acids P, E, S and T, were identified, which are known to be susceptible to proteolytic degradation [9]. La (SS-B) is known to be fairly sensitive to endogenous proteases in cell extracts, leading to prominent degradation products of 43 kDa, 29 kDa and 23 kDa. Fourth, two sequence elements that are homologous to regions of the double-stranded RNA-activated protein kinase PKR in the carboxy-terminal half of La (SS-B) have been described [45]. Finally, a nuclear localization signal (NLS), structurally resembling the canonical bipartite nuclear localization signal, has been identified and mapped near the carboxy-terminus of the protein [54].

La (SS-B) from murine cells migrates faster than human La (SS-B) in spite of 7 additional amino acids in the complete polypeptide. In case of *Xenopus laevis* cDNAs encoding two highly related La (SS-B) proteins have been described [39].

9. Gene

The human La (SS-B) gene, which spans about 9.5 kb from the transcription initiation site to the polyadenylation site, is located on chromosome 2 and comprises 11 exons (Fig. 3). The structure of the promoter, in particular the absence of a TATA box and the presence of a G/C-rich region, is characteristic for a "housekeeping gene" [35].

10. Relation with nucleic acids

The La (SS-B) protein binds to all newly synthesized RNA polymerase III products, including (precursors of) tRNA, 5S rRNA, 7S RNAs, U6 snRNA, and Y RNAs, as well as some virally encoded RNAs like adenovirus VA RNAs and Epstein-Barr virus EBER RNAs. Additionally, binding to U1 snRNA (transcribed by RNA polymerase II), leader RNAs of vesicular stomatitis virus and rabies virus (transcribed by virus-specific polymerases) and poliovirus RNA [32] have been reported [reviewed in 17, 18]. The 3'-oligouridine stretch, common to all RNA polymerase III transcribed RNAs, constitutes the primary La (SS-B) binding region of the RNA [36, 41, 42].

```
hLa   1  MAENGDNEKM AALEAKICHQ IEYYFGDFNL PRDKFLKEQI KLDEGWVPLE IMIKFNRLNR
bLa   1  MAENGDNEKM AALEAKICHQ IEYYFGDFNL PRDKFLKEQI KLDEGWVPLE IMIKFNRLNR
mLa   1  MAENGDNEKM TALEAKICHQ IEYYFGDFNL PRDKFLKEQI KLDEGWVPLE TMIKFNRLNR
rLa   1  MAENGDNEKM AALEAKICHQ IEYYFGDFNL PRDKFLKEQI KLDEGWVPLE TMIKFNRLNR
xLaA  1  MAENGDKEQK LDSDTKICEQ IEYYFGDHNL PRDKFLKQQI LLDDGWVPLE TMIKFNRLSK
xLaB  1  MAENGDKEQ- LDLDTKICEQ IEYYFGDHNL PRDKFLKQQV LLDNGWVPLE TMIKFNRLSK


hLa      LTTDFNVIVE ALSKSKAELM EISEDKTKIR RSPSKPLPEV TDEYKNDVKN RSVYIKGFPT
bLa      LTTDFNVIVE ALSKSEAELM EISEDKTKIR RSPSKPLPEV TDEYKNDVKN RSVYIKGFPT
mLa      LTTDFNVIVQ ALSKSKAKLM EVSADKTKIR RSPSRPLPEV TDEYKNDVKN RSVYIKGFPT
rLa      LTTDFNVIVQ ALSKSKANLM EVSADKTKIR RSPSRPLPEV TDEYKNDVKN RSVYIKGFPT
xLaA     LTTDFNTILQ ALKKSKTELL EINEEKCKIR RSPAKPLPEL NDEYKNSLKH KSVYIKGFPT
xLaB     LTTDFNIILQ ALKKSKTELL EINEEKCKIR RSPAKPLPEL NEDYKNSFKH RSVYIKGFPT


hLa      DATLDDIKEW LEDKGQVLNI QMRRTLHKAF KGSIFVVFDS IESAKKFVET PGQKYKETDL
bLa      DAALDDIKEW LEDKGQVLNI QMRRTLHKAF KGSIFAVFDS IESAKKFVET PGQKYKDTDL
mLa      DATLDDIKEW LDDKGQILNI QMRRTLHKTF KGSIFAVFDS IQSAKKFVEI PGQKYKDTNL
rLa      DATLDDIKEW LDDKGQILNI QMRRTLHKTF KGSIFAVFDS IQSAKKFVDT PGQKYKDTNL
xLaA     SAILDDVKEW LKDKGPIENI QMRRTLQREF KGSIFIIFNT DDDAKKFLEN RNLKYKDNDM
xLaB     ITNLDEIKEW LNDKGPIENI QMRRTLQREF KGSVFLVFNT EDGAKKFLED KNLKYKDNDM


hLa      LILFKDDYFA KKNEERKQNK VEAKLRAKQE QEAKQKLEED AEMKSLEEKI GCLLKFSGDL
bLa      LILFKEDYFT KKNEERKQNK MEAKLRAKQE QEEKQKLAEN AEMKSLEEKI GCLLKFSGDL
mLa      LILFKEDYFA KKNEERKQSK VEAKLKAKQE HEGRHK-PGS TETRALEGKM GCLLKFSGDL
rLa      LILFKEDYFA KKNEERKQSK VEAKLKAKQE HEGRHK-PGS TETRALEGKM GCLLKFSGDL
xLaA     TVLSRSEYHA KKNEERKLNK SEEKAKSKQV KKEAQKQAED AERKLVEERV GSLLKFSGDL
xLaB     IILSREEYFA KKNEERKLNK SEEKAKSKQE KEEAQKQAED AERKLMEERV GCLLKFSGDL


hLa      DDQTCREDLH ILFSNHGEIK WIDFVRGAKE GIILFKEKAK EALGKAKDAN NGNLQLRNKE
bLa      DDQTCREDLH TLFSNHGEIK WIHFVRGAKE GIILFKEKAK EALDKAKEAN NGNLQLRNKE
mLa      DDQTCREDLH PLFSNHGEIK WVDFARGAKE GIILFKEKAK EALEKARNAN NGNLLLRNKK
rLa      DDQTCREDFH PLFSNHGEIK WIDFVRGAKE GIILFKEKAK DALEKARSAN NGNLLLRNKN
xLaA     DNMTSREDLH ALFQTHGDIE WIDFSRGAKE GIVLFKMNAK EALDKAKAAN SDNLKLKGKD
xLaB     DNMTSREDLH ALFQTHGEIE WIDFSRGAKE GIVLFKMNAK EALDKAKAAN NDNLKLKGKN


hLa      VTWEVLEGEV EKEALKKIIE DQQESLNKWK SKG------- ---------R RFKGKGKGNK
bLa      VTWEVLEGDV EKEALKKIIE DQQESLNKWK SKG------- ---------R RFKGKGKGNK
mLa      VTWKVLEGHA EKEALKKITD DQQESLNKWK SKGGHAGGRF KGSHVFTAAR RFKGKGKGNR
rLa      VTWKVLEGHA EKDAMKKITD DQQESLNKWK SKGGHAARRF KGSHVFTAAR RFKGRGKGNR
xLaA     VKWELIEGDT EKEALKKILE GKQESFNKRK GRDG------ ---------R KFKGKGRGGK
xLaB     VKWELIEGDA EKEALKKIME GKQESFNKRK GRDG------ ---------R KFKGKGRGGK


hLa      AAQPGSGKGK VQFQGKKTKF ASDDEHDEHD EN-------- ---------- ----GATGPV
bLa      AAQAGSAKGK VQFQGKKTKF DSDDERDE-- -N-------- ---------- ----GASRAV
mLa      PGYAGAPKGR GQFHGRRTRF DDDDRRR--- ---------- ---------- -------GPM
rLa      PAYAGAPKGR GQFQGRRTRF DDDDHRR--- ---------- ---------- -------GPV
rabLa                GK VEFQGKKTKF DSDDERNE-- -N-------- ---------- ----GAAGPV
xLaA     -GNDSSSRRK TQFQGKKKTF DSSDDEDDME ESESPQKASV KAEESAGTKN GAAAAPGSPK
xLaB     -GNDSSPRKK IQFQGKKKTF DSSDDEDDME ESESPQKVTI KAKETAGPKN GASAAPGSPK


hLa      KRAREETDKE EPASKQQKTE NGAGDQ   408
bLa      KRAREETDKE -PPSKQQKTE NGAGDQ   404
mLa      KRGRDGRDRE EPASKHKKRE NGARDK   415
rLa      KRGIDGRDRE EPASKHKKRE NGARDK   415
rabLa    KRAREETDKE EPASKQQKTE NGAGDQ     ?
xLaA     KRSLDDKAED GPAVKQSKTE VG--DQ   428
xLaB     KRALDDKAED GPAVKQSETE VG--DQ   427
```

Fig. 2C. Sequence alignment of predicted La (SS-B) amino acid sequences from various species (hLa: human; bLa: bovine; mLa: mouse; rLa: rat; rabLa: rabbit (only carboxy-terminal part is known); xLaA: *Xenopus laevis* variant A; xLaB: *Xenopus laevis* variant B; GenBank/EMBL data bases, accession nos. J04205, X13697, X13698, L00993, X67859, L08230, X68817, X68818). Sequence comparison and alignment were performed by Clustal V [55]. Structural features are indicated as follows: RNP motif: shading and double underlining and overlining; PEST regions: broken double underlining and overlining; PKR homologies: shading and single underlining and overlining; ATP-binding site: stars; NLS: shading.

The expression of the human La (SS-B) protein in murine cells by introducing the human La (SS-B) gene into cultured murine cell lines revealed that the human molecule is functionally highly conserved across species, despite 23% non-identity within the primary structure of human and murine La (SS-B) [56].

AMAN-B4.2/9

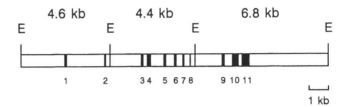

Fig. 3. Schematic structure of the human La (SS-B) gene.
Exons are numbered 1 through 11. The relative positions of EcoRI restriction enzyme recognition sites (E) are indicated. Fragment lengths are indicated in thousands of basepairs (kb) [35]. The two La (SS-B) variants observed in *Xenopus laevis* are probably each encoded by a single gene [39].

Since this sequence motif in most cases is lost upon maturation of the transcripts, the La (SS-B) protein binds to the precursor RNAs only transiently.

The RNP motif (amino acids 112–187) and, in addition, the amino-terminal 111 amino acids of the La (SS-B) protein (see Fig. 2) are required for its specific interaction with RNA [36, 43].

11. Relation with other proteins

Although the La (SS-B) protein is assembled into larger complexes like Ro RNPs in normal cells [44], EBER RNPs in Epstein-Barr virus infected or transformed cells [45] and poliovirus mRNPs in poliovirus infected cells [32], at present it is unknown whether the La (SS-B) protein is able to interact directly with other proteins.

12. B-cell epitopes

Cloning of cDNAs encoding the human La (SS-B) protein by several groups has allowed the generation of subclones to study the B-cell epitope topography. A summary of all these studies is presented in Figure 4 (see also [46]).

13. T-cell epitopes

Unknown.

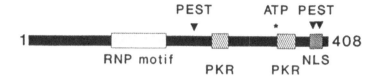

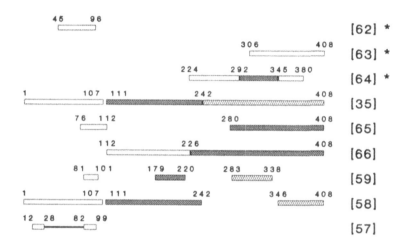

Fig. 4. B-cell epitope mapping data for human La (SS-B).

Subfragments of La (SS-B) reported to be reactive with patient anti-La (SS-B) antisera in references listed on the right-hand side are marked by bars with terminal amino acids numbered. Patterns within bars are only meant to distinguish between epitope regions. These regions may contain linear as well as discontinuous epitopes. Asterisks indicate that in these studies only part of the La polypeptide was analyzed. The epitope identified by McNeilage *et al.* [57] is discontinuous.

Taken together, the data suggest that there are at least three major B-cell epitopes. One of these, the amino-terminal aa 1–107 peptide, that contains at least one discontinuous epitope, may be immunodominant. McNeilage *et al.* [58] reported that in the very early stage of the anti-La (SS-B) response only the amino-terminal 107 aa are recognized and, with time, the response broadens to include other epitopes. Interestingly, this region contains a striking homology (aa 88–101) with an amino acid sequence in the gag protein of feline sarcoma virus [59, 60]. St. Clair *et al.* [61] obtained evidence that antibody responses to four different La fragments in individual patients vary in parallel over time, suggesting that the anti-La (SS-B) response to B-cell epitopes might be coordinated.

Gordon *et al.* [13] have demonstrated that the lack of precipitin formation observed for a substantial number of anti-La (SS-B) positive sera may be explained by the restricted epitope recognition of antibodies in these sera.

14. Monoclonal antibodies

Anti-La (SS-B) monoclonal antibodies (MoAbs) from both mouse and human have been generated.

Table 1.

MoAb	Subclass	Remarks	Ref.
Mouse:			
SW1	IgG2b	precipitates La RNPs	[47]
SW3	IgG2b	precipitates some La RNPs	[47]
SW5	IgG2b	precipitates La RNPs	[47]
La1	IgG1$\varkappa$	specific for bovine La	[48]
La1B5	IgG		[49]
A1	IgG1$\varkappa$	precipitates La RNPs	[50]
A2	IgG1$\varkappa$	precipitates La RNPs	[50]
A3	IgG1$\varkappa$	precipitates La RNPs; cross-reactive with other proteins	[50]
A4	IgG1$\varkappa$	cross-reactive with other proteins	[50]
A5	IgG1$\varkappa$		[50]
La11G7	IgM		[31]
La4B6	IgG1		[51]
Human:			
8G3	IgM$\varkappa$	precipitates La RNPs	[52]
9A5	IgG1$\varkappa$		[52]

Acknowledgements

The author thanks Frank Simons for providing results before publication and Ron Peek for Figure 1A. This work was supported in part by the Netherlands Foundation for Chemical Research (SON) with financial aid from the Netherlands Organization for Scientific Research (NWO). The work of Dr. G. Pruijn has been made possible by a fellowship of the Royal Netherlands Academy of Arts and Sciences.

References

1. Alspaugh MA & Maddison PJ (1979) Arthritis Rheum 22:796–798
2. Akizuki M, Powers R & Holman HR (1977) J Clin Invest 59:264–271
3. Clark G, Reichlin M & Tomasi TB (1969) J Immunol 102:117–122
4. Anderson JR, Gray KG, Beck JS & Kinnear WF (1961) Lancet 2:456–460
5. Alspaugh MA & Tan EM (1975) J Clin Invest 55:1067–1073
6. Pizer LI, Deng J-S, Stenberg RM & Tan EM (1983) Mol Cell Biol 3:1235–1245
7. Francoeur AM, Chan EKL, Garrels JI & Mathews MB (1985) Mol Cell Biol 5:586–590
8. Pfeifle J, Anderer FA & Franke M (1987) Biochim Biophys Acta 928:217–226
9. Chan EKL, Sullivan KF & Tan EM (1989) Nucleic Acids Res 17:2233–2244
10. Chan EKL, Francoeur AM & Tan EM (1986) J Immunol 136:3744–3749
11. Deng JS, Takasaki Y & Tan EM (1981) J Cell Biol 91:654–660
12. Meilof JF, Bantjes I, de Jong J, Van Dam AP & Smeenk RJ (1990) J Immunol Methods 133:215–226
13. Gordon T, Mavrangelos C & McCluskey J (1992) Arthritis Rheum 35:663–666
14. Habets WJ, Den Brok JH, Boerbooms AMT, Van de Putte LBA & Van Venrooij WJ (1983) EMBO J 2:1625–1631
15. Hoch SO & Billings PB (1984) J Immunol 133:1397–1403
16. Francoeur AM & Mathews MB (1982) Proc Natl Acad Sci USA 79:6772–6776
17. Pruijn GJM, Slobbe RL & Van Venrooij WJ (1990) Mol Biol Rep 14:43–48
18. Slobbe RL, Pruijn GJM & Van Venrooij WJ (1991) Ann Med Interne (Paris) 142: 592–600
19. St. Clair EW, Pisetsky DS, Reich CF, Chambers JC & Keene JD (1988) Arthritis Rheum 31:506–514
20. Owe Young RA, Horn S, Edmonds JP & Sturgess AD (1992) J Autoimmun 5:351–361
21. Veldhoven CH, Meilof JF, Huisman JG & Smeenk RJ (1992) J Immunol Methods 151:177–189
22. Peek R, Pruijn GJM, Van der Kemp AJW & Van Venrooij WJ (1993) J Cell Sci 106: 929–935
23. Bachmann M, Falke D, Schröder HC & Müller WEG (1989) J Gen Virol 70:881–891
24. Bachmann M, Pfeifer K, Schröder HC & Müller WEG (1989) Mol Cell Biochem 85:103–114
25. Baboonian C, Venables PJW, Booth J, Williams DG & Maini RN (1989) Clin Exp Immunol 78:454–459
26. Bachmann M, Chang S, Slor H, Kukulies J & Müller WEG (1990) Exp Cell Res 191:171–180
27. Furukawa F, Kashihara-Sawami M, Lyons MB & Norris DA (1990) J Invest Dermatol 94:77–85
28. Golan TD, Elkon KB, Gharavi AE & Krueger JG (1992) J Clin Invest 90:1067–1076
29. Gottlieb E & Steitz (1989) EMBO J 8:841–850
30. Gottlieb E & Steitz (1989) EMBO J 8:851–861
31. Bachmann M, Pfeifer K, Schröder HC & Müller WEG (1990) Cell 60:85–93
32. Meerovitch K, Svitkin YV, Lee HS, Lejbkowicz F, Kenan DJ, Chan EKL, Agol VI, Keene JD & Sonenberg N (1993) J Virol 67:3798–3807
33. Slobbe RL, Pruijn GJM, Damen WGM, van der Kemp JWCM & Van Venrooij WJ (1991) Clin Exp Immunol 86:99–105 [published erratum appears in Clin Exp Immunol 87:336 (1992)]
34. Francoeur AM, Gritzmacher CA, Peebles CL, Reese RT & Tan EM (1985) Proc Natl Acad Sci USA 82:3635–3639
35. Chambers JC, Kenan D, Martin BJ & Keene JD (1988) J Biol Chem 263:18043–18051
36. Pruijn GJM, Slobbe RL & Van Venrooij WJ (1991) Nucleic Acids Res 19:5173–5180
37. Topfer F, Gordon T & McCluskey J (1993) J Immunol 150:3091–3100
38. Semsei I, Tröster H, Bartsch H, Schwemmle M, Igloi GL & Bachmann M (1993) Gene 126:265–268
39. Scherly D, Stutz F, Lin-Marq N & Clarkson SG (1993) J Mol Biol 231:196–204

40. Wolin SL, O'Brien CA, Margelot K, Becker A & Yoo CJ (1993) In: Filipowicz W, Mattaj I & Van Venrooij WJ (Eds) Structure and Function of Eukaryotic RNPs (p 74).
41. Mathews MB & Francoeur AM (1984) Mol Cell Biol 4:1134–1140
42. Stefano JE (1984) Cell 36:145–154
43. Kenan DJ, Query CC & Keene JD (1991) Trends Biochem Sci 16:214–220
44. Van Venrooij WJ, Slobbe RL & Pruijn GJM (1993) Mol Biol Rep 18:113–119
45. Clemens MJ (1993) Mol Biol Rep 17:81–92
46. Whittingham S (1992) Mol Biol Rep 16:175–181
47. Smith PR, Williams DG, Venables PJW & Maini RN (1985) J Immunol Meth 77:63–76
48. Harley JB, Rosario MO, Yamagata H, Fox OF & Koren E (1985) J Clin Invest 76:801–806
49. Bachmann M, Mayet WJ, Schröder HC, Pfeifer K, Meyer zum Büschenfelde KH & Müller WEG (1986) Proc Natl Acad Sci USA 83:7770–7774
50. Chan EKL & Tan EM (1987) J Exp Med 166:1627–1640
51. Tröster H, Bartsch H, Metzger TE, Klein R, Pollak G, Semsei I, Pfeifer K, Schwemmle M, Pruijn GJM, Van Venrooij WJ & Bachmann M (1994) submitted
52. Mamula MJ, Silverman ED, Laxer RM, Benthur L, Isacovics B & Hardin JA (1989) J Immunol 143:2923–2928
53. Mattaj IW (1993) Cell 73:837–840
54. Simons FHM, Van Venrooij WJ & Pruijn GJM (1994) manuscript in preparation
55. Higgins DG & Sharp PM (1988) Gene 73:237–244
56. Keech CL, Gordon TP, Reynolds P & McCluskey J (1993) J Autoimmun 6: 543–555
57. McNeilage LJ, Umapathysivam K, Macmillan E, Guidolin A, Whittingham S & Gordon T (1992) J Clin Invest 89:1652–1656
58. McNeilage LJ, Macmillan EM & Whittingham SF (1990) J Immunol 145:3829–3835
59. Kohsaka H, Yamamoto K, Fujii H, Miura H, Miyasaka N, Nishioka K & Miyamoto T (1990) J Clin Invest 85:1566–1574
60. Horsfall AC (1992) Mol Biol Rep 16:139–147
61. St. Clair EW, Burch JA, Ward MM, Keene JD & Pisetsky DS (1990) J Clin Invest 85:515–521
62. Chambers JC & Keene JD (1985) Proc Natl Acad Sci USA 82:2115–2119
63. Sturgess AD, Peterson MG, McNeilage LJ, Whittingham S & Coppel RL (1988) J Immunol 140:3212–3218
64. Rauh AJG, Hornig H & Lührmann R (1988) Eur J Immunol 18:2049–2057
65. Chan EKL, Sullivan KF, Fox RI & Tan EM (1989) J Autoimmun 2:321–327
66. Bini P, Chu JL, Okolo C & Elkon K (1990) J Clin Invest 85:325–333

Manual of Biological Markers of Disease **B5.1**: 1–16, 1994.

Scleroderma-associated antigens

Introduction

As much as 95 to 98% of scleroderma patients have been shown to produce autoantibodies to nuclear, nucleolar or mitochondrial antigens [1–3]. These serologic markers have been shown to be useful in diagnosis and identifying disease subsets [3–5]. A number of autoantigens in scleroderma have been described. The most important are: DNA topoisomerase I (110 kDa) [3, see also this Manual, chapter B-5.1], the centromere proteins CENP-A (17 kDa), CENP-B (80 kDa) and CENP-C (140 kDa) [3, 5, see also this Manual chapter B-5.2], the small nucleolar RNA (snoRNA) associated protein fibrillarin (34 kDa) [3, 5, see also this Manual, chapter B-5.3], the 210 kDa protein component of the RNA polymerase I complex [3, 5], a 40 kDa protein associated with the 7–2 RNP complex [3, 6], nucleolin (110 kDa) [7], a 70 kDa protein of the mitochondrial M2 complex [3, 8], and a nucleolus-organizing region (NOR) associated 90 kDa protein (NOR-90) [3, 9], recently identified as transcription factor hUBF (human upstream binding factor) [10].

In addition, autoantibodies to the polymyositis-scleroderma (PM-Scl) associated 80 kDa and 110 kDa antigens [3, 11, 12, see also this Manual, chapter B-5.4] and to the 86 kDa and 70 kDa subunits of the Ku complex [13–15, see also this Manual, chapter B-2.6] (= nuclear factor IV [16]) can be detected in sera from certain patients with a polymyositis-scleroderma overlap syndrome [13, 17].

In the following chapters we will review the molecular characteristics of the scleroderma-associated antigens which have been studied most extensively.

References

1. Bernstein RM, Steigerwald JC & Tan EM (1982) Clin Exp Immunol 48:43–51
2. Steen VD, Powell DL & Medsger TA Jr (1988) Arthritis Rheum 31:196–203
3. Reimer G (1990) In: LeRoy EC (ed) Rheumatic disease clinics of North America Vol 16(1), pp 169–183. WB Saunders Company, Philadelphia
4. Tan EM (1982) Adv Immunol 33:167–240
5. Tan EM (1989) Adv Immunol 44:93–152
6. Kipnis RJ, Craft J & Hardin JA (1990) Arthritis Rheum 33:1431–1437
7. Minota S, Jarjour WN, Suziki N, Nojima Y, Roubey RAS, Mimura T, Yamada A, Hosoya T, Takaku F & Winfield JB (1991) J Immunol 146:2249–2252
8. Freqeau DR, Leung PSC, Coppel RL, McNeilage LJ, Medsger TA Jr & Gershwin ME (1988) Arthritis Rheum 31:386–392
9. Rodriquez-Sanchez JL, Gelpi C, Juarez C & Hardin JA (1987) J Immunol 139:2579–2584
10. Chan EKL, Imai H, Hamel JC & Tan EM (1991) J Exp Med 174:1239–1244

11. Alderuccio F, Chan EKL & Tan EM (1991) J Exp Med 173:941–952
12. Reimer G, Scheer U, Peters J-M & Tan EM (1986) J Immunol 137:3802–3808
13. Mimori T, Hardin JA & Steitz JA (1986) J Biol Chem 261:2274–2278
14. Reeves WH & Sthoeger ZM (1989) J Biol Chem 264:5047–5052
15. Yaneva M, Wen J, Ayala A & Cook R (1989) J Biol Chem 264:13407–13411
16. Stuiver MH, Coenjaerts FEJ & van der Vliet PC (1990) J Exp Med 172:1049–1054
17. Mimori T, Akizuki M, Yamagata H & Inada S (1981) J Clin Invest 68:611–620

DNA topoisomerase I

R. VERHEIJEN

Department of Biochemistry, University of Nijmegen, P.O. Box 9101, 6500 HB Nijmegen, The Netherlands

Abbreviations

aa: amino acids; CIE: Counterimmuno electrophoresis; ELISA: Enzyme-linked Immunosorbent Assay; nt: nucleotides; PKC: Protein Kinase C; topo I: DNA topoisomerase I

Key words: Autoantigen, autoantibody, scleroderma, DNA topoisomerase I

1. The antigen

DNA topoisomerase I (EC 5.99.1.2.).

2. Synonyms

Topoisomerase I, topo I or type I topoisomerase (110kDa), Scleroderma 70 (Scl-70) antigen (70 kDa) [1, 2], Scl-86 antigen (86 kDa) [3].
The Scl-70 antigen is a degradation product of the 100 kDa protein.

3. Antibody detection

3.1 Immunofluorescence

In immunofluorescence, anti-topo I antibodies can be identified by a fine speckled staining of the nucleoplasm which often has the appearance of a homogeneous pattern. When using human autoimmune sera, nucleoli are mostly excluded from labeling [4–6; see also this Manual, chapter A2 and Fig. 1]. In contrast, studies in which rabbit polyclonal antibodies or mouse monoclonal antibodies to topo I are used, a fluorescent staining of both nucleoplasm and nucleoli can be seen [7–9]. During mitosis the antigen remains associated with the condensed chromosomal material (Fig. 1).

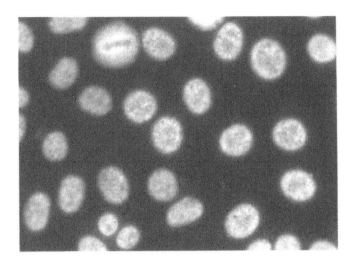

Fig. 1. Immunofluorescence localization of the topo I antigen in HEp-2 (human epidermoid larynx carcinoma) cells in interphase and mitosis, as detected by a human autoimmune serum positive for anti-topo I antibodies. In interphase cells a fine speckled staining of the nucleoplasm is seen that has the appearance of a homogeneous pattern. The nucleoli are excluded from labeling. In mitotic cells topo I remains associated with the condensed chromatin. For technical details see this Manual, chapter A2.

3.2 Counterimmunoelectrophoresis (CIE) and immunodiffusion

Both CIE [10] and immunodiffusion [11] can be used to detect anti-topo I antibodies. For technical details see this Manual, chapter A3.

3.3 Immunoblotting

In an immunoblotting analysis using a nuclear cell extract of human origin as antigen source, anti-topo I antibody will decorate the 110 kDa native topo I protein. The distinct immunoreactive 70 kDa degradation product, known as the Scl-70 antigen, is often seen as well (see this Manual, chapter A4).

3.4 Immunoprecipitation

Anti-topo I antibodies can be detected by immunoprecipitation of topo I from a [^{35}S]-methionine labeled cell extract, followed by visualization of the precipitated protein on an autoradiogram [12; see also this Manual, chapter A6].

AMAN-B5.1/4

3.5 ELISA

Both purified native topo I [10, 11, 13, 14] and recombinant topo I [15] have been applied as antigen in an ELISA for detecting anti-topo I antibodies. For technical details see this Manual, chapter A5.

4. Cellular localization

As concluded from its cellular function (see section 5) topo I is localized in both the nucleoplasm and the nucleoli of the interphase cell. When using human autoimmune sera in immunofluorescence studies, however, nucleoli are mostly excluded from labeling [4–6]. It is possible that we are dealing here with a fixation artifact since, in studies in which rabbit polyclonal antibodies or mouse monoclonal antibodies to topo I are used, a fluorescent staining of both nucleoplasm and nucleoli can be seen [7–9].

During mitosis the antigen remains associated with the condensed chromosomal material. In normal eukaryotic cells the concentration of topo I does not show significant fluctuations across the cell cycle [16].

5. Cellular function

In the eukaryotic cell, the topological states of DNA are modulated by both type I and type II DNA topoisomerase. The type I enzyme interconverts

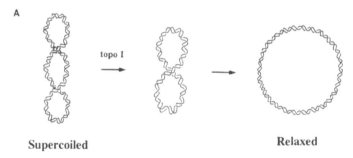

Supercoiled   **Relaxed**

Fig. 2A. Relaxation of supercoiled DNA by topo I.
Eukaryotic type I topoisomerases can relax negatively as well as positively supercoiled DNA, changing the linking number in steps of one. The reaction does not require ATP.
In the reaction intermediate the DNA is covalently linked to the enzyme. This transient intermediate involves a phosphodiester bond between a tyrosine-OH group on the enzyme and the 3′ terminus of the broken DNA strand [17]. Although topo I is not necessary for the viability of eukaryotic cells [35, 36], it does appear to play important roles in chromatin organization [35], mitosis [6], DNA replication [37, 38], recombination [39, 40], and transcription of both rRNA genes [31. 41. 42] and protein encoding genes [43. 44].

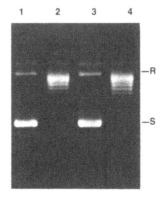

Fig. 2B. Gel electrophoresis patterns showing the relaxation of supercoiled plasmid DNA by topo I.

Purified HeLa topo I was used to demonstrate the inhibition of topo I activity by anti-topo I antibodies in a patients' serum. The assay was performed as described by Meesters *et al.* [19]. The incubation time was chosen such that the DNA was not fully relaxed, giving rise to the typical DNA relaxation ladder. The positions of relaxed (R) and supercoiled (S) forms of the plasmid DNA are indicated.

Lane 1: no enzyme and no serum added, showing the supercoiled DNA; lane 2: DNA and enzyme without serum, showing the partially relaxed DNA; lane 3: DNA with enzyme and an anti-topo I positive serum that inhibits topo I activity; lane 4: DNA with enzyme and an anti-topo I positive serum that does not inhibit topo I activity.

different topological forms of DNA by creating a transient single-stranded nick in the DNA backbone, passing the unbroken strand of the DNA through the nick, and resealing the original scission [17, 18] (Fig. 2A).

Topo I activity can be detected by an *in vitro* relaxation assay of supercoiled DNA [19]. Such an assay can be used in a topo I isolation procedure to follow the enrichment of enzymatic activities as well as to study the capability of anti-topo I antibodies to inhibit DNA relaxation (Fig. 2B).

6. Presence of topo I in other species

Topo I is a housekeeping protein present in relatively high amounts in the nucleus of all eukaryotic cells. It shows a significant degree of conservation among different species. As an example, Figure 3 shows a comparison of polypeptide sequences between the human and yeast enzyme.

Fig. 3. Schematic diagram of regions of similarity between the topo I amino acid sequences of man [20] and *Saccharomyces cerevisiae* [36]. Thin lines represent divergent regions, whereas well-conserved elements are indicated by heavy bars. These latter domains share 54% identical residues and display an amino acid sequence similarity of 69% at optimal alignment [20].

7. Human cDNA and predicted amino acid sequence

Cloning of the cDNA encoding the full-length human topo I was first described by D'Arpa *et al.* [20]. The cDNA sequence as well as the deduced amino acid sequence is shown in Figure 4.

8. The topo I gene

Human topo I is transcribed from a single-copy gene that encodes a 4.1 kb mRNA [20]. The complete gene has been localized on chromosome 20 (band q11.2–13.1), whereas two truncated pseudogene sequences have been identified on band q23–24 of chromosome 1 and band q11.2–13.1 of chromosome 22, respectively [21].

The molecular organization of the human topo I gene has been described by Kunze *et al.* [22]. The coding sequence is split into 21 exons (see Fig. 4) distributed over at least 85 kb of genomic DNA. The sizes of the 20 introns vary between 0.2 and at least 30 kb and all contain the sequence elements known to be required for pre-mRNA splicing. Several of the intron sequences separate exons encoding parts of the enzyme that are highly conserved between man and yeast, suggesting that at least some of the exons code for structurally or functionally important domains of the enzyme.

The combined coding sequence of the gene as published by Kunze *et al.*

$\rightarrow$

Fig. 4. Nucleotide sequence of the human topoisomerase I cDNA and the deduced amino acid sequence as published by D'Arpa *et al.* [20] (EMBL accession no. J03250). The 3645-base sequence begins with a 5' untranslated region of 211 nucleotides, followed by a 765-amino acid (2295 nt) open reading frame, and terminates with 1139 nucleotides of 3' untranslated sequence. The encoded polypeptide has a calculated molecular weight of 90,753 Da, which is smaller than the apparent molecular weight of 100-110 kDa on SDS-polyacrylamide gels.

The coding sequence of the topo I gene is split into 21 exons. With exception of exon 1 (280 nt) the first nucleotide of each exon is indicated by an arrowhead.

The presumed active site tyrosine 723 [27, 28] is shaded.

```
EcoRI  10        20        30        40        50        60        70        80        90
GAATTCGGGCGCCGCCCGCCCGGCAGTCAGGCAGCGTCGCCGCCGTGGTAGCAGCCTCAGCCGTTTCTGGAGTCTCGGGCCCACAGTCACC

        100       110       120       130       140       150       160       170       180
GCCGCTTACCTGCGCCTCCTCGAGCCTCCGGAGTCCCCGTCCGCCCGCACAGGCCGGTTCGCCGTCTGCGTCTCCCCCACGCCGCCTCGC
                                                                           2                 3
        190       200       210       220       230       240    ▼  250       260
CTGCCGCCGCCGCTCGTCCCTCCGGGCCGACATGAGTGGGGACCACCTCCACAACGATTCCCAGATCGAAGCGGATTTCCGATTGAATGAT
                                         M  S  G  D  H  L  H  N  D  S  Q  I  E  A  D  F  R  L  N  D    20

        280       290       300       310       320       330       340       350       360
TCTCATAAACACAAAGATAAACACAAAGATCGAGAACACCGGCACAAGAACACAAGAAGGAGAAGGACCGGGAAAAGTCCAAGCATAGC
 S  H  K  H  K  D  K  H  K  D  R  E  H  R  H  K  E  H  K  K  E  K  D  R  E  K  S  K  H  S    50
  4
   ▼370       380       390       400       410       420       430       440       450
AACAGTGAACATAAAGATTCTGAAAAGAAACACAAAGAGAAGGAGAAGACCAAACACAAAGATGGAAGCTCAGAAAAGCATAAAGACAAA
 N  S  E  H  K  D  S  E  K  K  H  K  E  K  E  K  T  K  H  K  D  G  S  S  E  K  H  K  D  K    80
                                       5
        460       470       480     490▼      500       510       520       530       540
CATAAAGACAGAGACAAGGAAAAACGAAAAGAGGAAAAAGGTTCGAGCCCTCTGGGGATGCAAAAATAAAGAAGGAGAAGGAAAATGGCTTC
 H  K  D  R  D  K  E  K  R  K  E  E  K  V  R  A  S  G  D  A  K  I  K  K  E  K  E  N  G  F    110
  6
   ▼550       560       570       580       590       600       610       620       630
TCTAGTCCACCACAAATTAAAGATGAACCTGAAGATGATGGCTATTTTGTTCCTCCTAAAGAGGATATAAAGCCATTAAAGAGACCTCGA
 S  S  P  P  Q  I  K  D  E  P  E  D  D  G  Y  F  V  P  P  K  E  D  I  K  P  L  K  R  P  R    140
  7                                                                                         8
        640    ▼  650       660       670       680       690       700       710
GATGAGGATGATGTTGATTATAAACCTAAGAAAATTAAAACAGAAGATACCAAGAAGGAGAAGAAAAGAAAACTAGAAGAAGAAGAGGAT
 D  E  D  D  V  D  Y  K  P  K  K  I  K  T  E  D  T  K  K  E  K  K  R  K  L  E  E  E  E  D    170

        730       740       750       760       770       780       790       800       810
GGTAAATTGAAAAAACCCAAGAATAAAGATAAAGATAAAAAAGTTCCTGAGCCAGATAACAAGAAAAAGAAGCCGAAGAAGAAGAGGAA
 G  K  L  K  K  P  K  N  K  D  K  D  K  K  V  P  E  P  D  N  K  K  K  K  P  K  K  E  E  E    200
  9
        820    ▼  830       840       850       860       870       880       890       900
CAGAAGTGGAAATGGTGGGAAGAAGAGCGCTATCCTGAAGGCATCAAGTGGAAATTCCTAGAACATAAAGGTCCAGTATTTGCCCCACCA
 Q  K  W  K  W  W  E  E  E  R  Y  P  E  G  I  K  W  K  F  L  E  H  K  G  P  V  F  A  P  P    230
                                                       10
        910       920       930     940▼      950       960       970       980       990
TATGAGCCTCTTCCAGAGAATGTCAAGTTTTATTATGATGGTAAAGTCATGAAGCTGAGCCCCAAAGCAGAGGAAGTAGCTACGTTCTTT
 Y  E  P  L  P  E  N  V  K  F  Y  Y  D  G  K  V  M  K  L  S  P  K  A  E  E  V  A  T  F  F    260
                                                                           11
        1000      1010      1020      1030      1040      1050      1060   ▼  1070      1080
GCAAAAATGCTCGACCATGAATATACTACCAAGGAAATATTTAGGAAAAATTTCTTTAAAGACTGGAGAAAGGAAATGACTAATGAAGAG
 A  K  M  L  D  H  E  Y  T  T  K  E  I  F  R  K  N  F  F  K  D  W  R  K  E  M  T  N  E  E    290

        1090      1100      1110      1120      1130      1140      1150      1160      1170
AAGAATATTATCACCAACCTAAGCAAATGTGATTTTACCCAGATGAGCCAGTATTTCAAAGCCCAGACGGAAGCTCGGAAACAGATGAGC
 K  N  I  I  T  N  L  S  K  C  D  F  T  Q  M  S  Q  Y  F  K  A  Q  T  E  A  R  K  Q  M  S    320
  12
        1180      ▼          1200      1210      1220      1230      1240      1250      1260
AAGGAAGAGAAACTGAAAATCAAAGAGGAGAATGAAAAATTACTGAAAGAATATGGATTCTGTATTATGGATAACCACAAAGAGAGGATT
 K  E  E  K  L  K  I  K  E  E  N  E  K  L  L  K  E  Y  G  F  C  I  M  D  N  H  K  E  R  I    350
        1270      1280      1290      1300      1310      1320      1330      1340      1350
GCTAACTTCAAGATAGAGCCTCCTGGACTTTTCCGTGGCCGCGGCAACCACCCCCAAGATGGGCATGCTGAAGAGACGAATCATGCCCGAG
 A  N  F  K  I  E  P  P  G  L  F  R  G  R  G  N  H  P  K  M  G  M  L  K  R  R  I  M  P  E    380
  13
        1360      1370    ▼  1380      1390      1400      1410      1420      1430      1440
GATATAATCATCAACTGTAGCAAAGATGCCAAGGTTCCTTCTCCTCCTCCAGGACATAAGTGGAAAGAAGTCCGGCATGATAACAAGGTT
 D  I  I  I  N  C  S  K  D  A  K  V  P  S  P  P  P  G  H  K  W  K  E  V  R  H  D  N  K  V    410
                                                                                   14
        1450      1460      1470      1480      1490      1500      1510     ▼        1530
ACTTGGCTGGTTTCCTGGACAGAGAACATCCAAGGTTCCATTAAATACATCATGCTTAACCCTAGTTCACGAATCAAGGGTGAGAAGGAC
 T  W  L  V  S  W  T  E  N  I  Q  G  S  I  K  Y  I  M  L  N  P  S  S  R  I  K  G  E  K  D    440

        1540      1550      1560      1570      1580      1590      1600      1610      1620
TGGCAGAAATACGAGACTGCTCGGCGGCTGAAAAAATGTGTGGACAAGATCCGGAACCAGTATCGAGAAGACTGGAAGTCCAAAGAGATG
 W  Q  K  Y  E  T  A  R  R  L  K  K  C  V  D  K  I  R  N  Q  Y  R  E  D  W  K  S  K  E  M    470
                                               15
        1630      1640      1650      1660   ▼  1670      1680      1690      1700      1710
AAAGTCCGGCAGAGAGCTGTAGCCCTGTACTTCATCGACAAGCTTGCTCTGAGAGCAGGCAATGAAAAGGAGGAAGGAGAAACAGCCGGAC
 K  V  R  Q  R  A  V  A  L  Y  F  I  D  K  L  A  L  R  A  G  N  E  K  E  E  G  E  T  A  D    500

        1720      1730      1740      1750      1760      1770      1780      1790      1800
ACTGTGGGCTGCTGCTCACTTCGTGTGGAGCACATCAATCTACACCCAGAGTTGGATGGTCAGGAATATGTGGTAGAGTTTGACTTCCTC
 T  V  G  C  C  S  L  R  V  E  H  I  N  L  H  P  E  L  D  G  Q  Q  E  Y  V  V  E  F  D  F  L    530
```

AMAN-B5.1/8

```
                                            16
      1810      1820      1830      1840     ▼    1860      1870      1880      1890
GGGAAGGACTCCATCAGATACTATAACAAGGTCCCTGTTGAGAAACGAGTTTTTAAGAACCTACAACTATTTATGGAGAACAAGCAGCCC
 G  K  D  S  I  R  Y  Y  N  K  V  P  V  E  K  R  V  F  K  N  L  Q  L  F  M  E  N  K  Q  P   560
            17
      1900      1910     ▼    1930      1940      1950      1960      1970      1980
GAGGATGATCTTTTTGATAGACTCAATACTGGTATTCTGAATAAGCATCTTCAGGATCTCATGGAGGGCTTGACAGCCAAGGTATTCCGT
 E  D  D  L  F  D  R  L  N  T  G  I  L  N  K  H  L  Q  D  L  M  E  G  L  T  A  K  V  F  R   590
                                          18
      1990      2000      2010      2020     ▼    2040      2050      2060      2070
ACGTACAATGCCTCCATCACGCTACAGCAGCAGCTAAAAGAACTGACAGCCCCGGATGAGAACATCCCAGCGAAGATCCTTTCTTATAAC
 T  Y  N  A  S  I  T  L  Q  Q  Q  L  K  E  L  T  A  P  D  E  N  I  P  A  K  I  L  S  Y  N   620

      2080      2090      2100      2110      2120      2130      2140      2150      2160
CGTGCCAATCGAGCTGTTGCAATTCTTTGTAACCATCAGAGGGCACCACCAAAAACTTTTGAGAAGTCTATGATGAACTTGCAAACTAAG
 R  A  N  R  A  V  A  I  L  C  N  H  Q  R  A  P  P  K  T  F  E  K  S  M  M  N  L  Q  T  K   650
 19
     ▼    2170      2180      2190      2200      2210      2220      2230      2240      2250
ATTGATGCCAAGAAGGAACAGCTAGCAGATGCCCGGAGACCTGAAAAGTGCTAAGGCTGATGCCAAGGTCATGAAGGATGCAAAGACG
 I  D  A  K  K  E  Q  L  A  D  A  R  R  D  L  K  S  A  K  A  D  A  K  V  M  K  D  A  K  T   680
     ▼    2270      2280      2290      2300      2310      2320      2330      2340
AAGAAGGTAGTAGAGTCAAAGAAGAAGGCTGTTCAGAGACTGGAGGAACAGTTGATGAAGCTGGAAGTTCAAGCCACAGACCGAGAGGAA
 K  K  V  V  E  S  K  K  K  A  V  Q  R  L  E  E  Q  L  M  K  L  E  V  Q  A  T  D  R  E  E   710
                                                        21
      2350      2360      2370      2380      2390      2400     ▼    2420      2430
AATAAACAGATTGCCCTGGGAACCTCCAAACTCAATTATCTGGACCCTAGGATCACAGTGGCTTGGTGCAAGAAGTGGGGTGTCCCAATT
 N  K  Q  I  A  L  G  T  S  K  L  N  Y  L  D  P  R  I  T  V  A  W  C  K  K  W  G  V  P  I   740

      2440      2450      2460      2470      2480      2490      2500      2510      2520
GAGAAGATTTACAACAAAACCCAGCGGGAGAAGTTTGCCTGGGCCATTGACATGGCTGATGAAGACTATGAGTTTTAGCCAGTCTCAAGA
 E  K  I  Y  N  K  T  Q  R  E  K  F  A  W  A  I  D  M  A  D  E  D  Y  E  F  *            765

      2530      2540      2550      2560      2570      2580      2590      2600      2610
GGCAGAGTTCTGTGAAGAGGAACAGTGTGGTTTGGGAAAGATGGATAAACTGAGCCTCACTTGCCCTCGTGCCTGGGGGAGAGAGGCAGC

      2620      2630      2640      2650      2660      2670      2680      2690      2700
AAGTCTTAACAAACCAACATCTTTGCGAAAAGATAAACCTGGAGATATTATAAGGGAGAGCTGAGCCAGTTGTCCTATGGACAACTTATT

      2710      2720      2730      2740      2750      2760      2770      2780      2790
TAAAAAATATTTCAGATATCAAAATTCTAGCTGTATGATTTGTTTTGAATTTTGTTTTTTATTTTCAAGAGGGCAAGTGGATGGGAATTTGT

      2800      2810      2820      2830      2840      2850      2860      2870      2880
CAGCGTTCTACCAGGCAAATTCACTGTTTCACTGAAATGTTTGGATTCTCTTAGCTACTGTATGCAAAGTCCGATTATATTGGTGCGTTT

      2890      2900      2910      2920      2930      2940      2950      2960      2970
TTACAGTTAGGGTTTTGCAATAACTTCTATATTTTAATAGAAATAAATTCCTAAACTCCCTTCCCTCTCTCCCATTTCAGGAATTTAAAA

      2980      2990      3000      3010      3020      3030      3040      3050      3060
TTAAGTAGAACAAAAAACCCAGCGCACCTGTTAGAGTCGTCACTCTCTATTGTCATGGGGATCAATTTTCATTAAACTTGAAGCAGTCGT

      3070      3080      3090      3100      3110      3120      3130      3140      3150
GGCTTTGGCAGTGTTTTGGTTCAGACACCTGTTCACAGAAAAAGCATGATGGGAAAATATTTCCTGACTTGAGTGTTCCTTTTTAAATGT

      3160      3170      3180      3190      3200      3210      3220      3230      3240
GAATTTTTTTTTTTTTTAATTATTTTAAAATATTTAAACCTTTTTCTTGATCTTAAAGATCGTGTAGATTGGGGTTGGGGAGGGATGAAG

      3250      3260      3270      3280      3290      3300      3310      3320      3330
GGCGAGTGAATCTAAGGATAATGAAATAATCAGTGACTGAAACCATTTCCCATCATCCTTTGTTCTGAGCATTCGCTGTACCCTTTAAG

      3340      3350      3360      3370      3380      3390      3400      3410      3420
ATATCCATCTTTTTCTTTTTAACCCTAATCTTTCACTTGAAAGATTTTATTGTATAAAAAGTTTCACAGGTCAATAAACTTAGAGGAAAA

      3430      3440      3450      3460      3470      3480      3490      3500      3510
TGAGTATTTGGTCCAAAAAAAGGAAAAATAATCAAGATTTTAGGGCTTTTATTTTTTCTTTTGTAATTGTGTAAAAAATGGAAAAAAACA

      3520      3530      3540      3550      3560      3570      3580      3590      3600
TAAAAAGCAGAATTTTAATGTGAAGACATTTTTTGCTATAATCATTAGTTTTAGAGGCATTGTTAGTTTTAGTGTGTGTGCAGAGTCCATT

      3610      3620      3630      3640
TCCCACATCTTTCCTCAAGTATCTTCTATTTTTTATCATGAATTC
                                        EcoRI
```

Fig. 4. Continued.

[22] is almost identical to the published topo I cDNA sequence by D'Arpa *et al.* [20], except for some minor differences:

a) The first nine nucleotides of the cDNA are not found in the genomic sequence. Most probably, these nucleotides are derived from the oligonucleotide linker used for the construction of the cDNA library.

b) In the genomic sequence a cytosine residue was found instead of a thymine residue at position 645 of the cDNA. This results in a Val → Ala exchange at position 145 of the protein sequence.

c) Two additional nucleotides were recognized in the 3′ untranslated region: an adenine residue at the cDNA position 3160 and a cytosine residue at position 3163. These latter changes have been noted before [23, 24].

9. Interaction with nucleic acids

In a study by Zechiedrich and Osheroff [25] it was shown that both type I and type II topoisomerases recognize supercoiled DNA (at least in part) by interacting preferentially at points of helix-helix juxtaposition. The authors concluded that the preference of these enzymes for crossovers may represent a general mechanism by which proteins can recognize the topological state of nucleic acids.

10. Interaction with other proteins

So far, there have been no published reports on the interaction of topo I with other proteins.

11. Biochemical characteristics of topo I

Human topo I (765 amino acids) has a calculated molecular weight of 90,753 Da, and an apparent molecular weight of 100–110 kDa on SDS-polyacrylamide gels. The entire protein contains 45% charged residues (200 basic (Arg + Lys + His) and 142 acidic (Asp + Glu)). The calculated isoelectric point is 10.09. The N-terminal fragment of 196 residues, which has a low similarity with the yeast topo I sequence, has a remarkable 67% content of charged residues (80 basic and 52 acidic).

In contrast to prokaryotic topo I, the eukaryotic type I topoisomerase does not require Mg^{2+} for its *in vitro* relaxation activity. Furthermore, there is no energy cofactor required for this activity [26].

The active site tyrosine of topo I

The active site tyrosine for topo I from both *Saccharomyces cerevisiae* and *Schizosaccharomyces pombe* has been mapped [27, 28]. Based on amino acid sequence similarities between topo I of man and yeast, tyrosine 723 of the human enzyme has been inferred as the active site tyrosine.

Inhibitors of topo I

A specific inhibitor of eukaryotic topo I is the alkaloid camptothecin. This anti-tumor drug inhibits the enzyme activity by stabilizing the covalent enzyme-DNA intermediate [29]. Inhibition by camptothecin has been shown to reduce the rate of transcription by both RNA polymerase I and II [30, 31].

Recently, it has been reported that physiologic concentrations of zinc ($\geq$ 80 μM) potently inactivate topo I [32].

Phosphorylation of topo I

It has been demonstrated that dephosphorylation of topo I by alkaline phosphatase abolishes its DNA relaxation activity as well as its sensitivity towards camptothecin. Topo I can be reactivated by incubation with the serine kinase Protein Kinase C (PKC) [33, 34]. After phosphorylation of such inactivated topo I by PKC, the reactivated enzyme behaves like native topo I with respect to processivity of DNA relaxation and sequence selectivity of DNA cleavage in the presence of camptothecin. These results show that active topo I is a phosphoprotein and suggest a possible regulatory role of PKC on topo I activity and on its sensitivity to camptothecin [33].

12. B-cell epitopes on topo I

The distribution of the epitope regions on topo I as reported by several research groups is shown in Figure 5 [see also 34].

13. T-cell epitopes on topo I

So far, no reports have been published on T-cell epitopes on topo I.

14. Mouse monoclonal antibodies to topo I

The published monoclonal antibodies directed to DNA topoisomerase I are tabulated in Table 1.

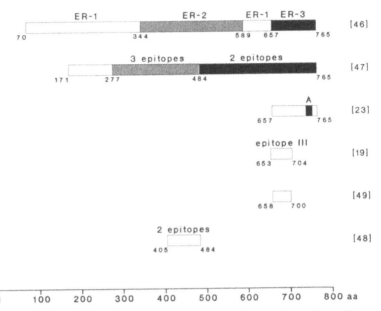

Fig. 5. Schematic representation of autoimmune epitope regions on the topo I recombinant protein. This figure is modified after [45]. References are indicated between brackets.

Based on the reaction patterns of recombinant topo I fragments with anti-topo I positive sera, three autoimmune epitope regions (ERs) were defined, i.e. ER-1 (aa 70–344 and/or 589–637), ER-2 (aa 344–589) and ER-3 (aa 657–765) [46]. In a more precise epitope mapping study by Meesters *et al.* [19] amino acids 653–704, designated as epitope III, emerged as a major epitope of topo I. This latter observation was confirmed in a recent study of Kuwana and co-workers [49].

An epitope region consisting of amino acids 675–765 was reported by Maul *et al.* [23]. In addition, it was shown that some patients' sera cross-reacted with a synthetic topo I peptide (Peptide A: aa 737–753) having a 6 amino acid sequence match with the mammalian retroviral p30^{gag} protein.

D'Arpa *et al.* [47] reported the finding of 6 independent epitopes distributed over the entire topo I molecule. Residues 277–484 were shown to harbor at least 3 epitopes, whereas the C-terminal end of the protein (aa 485–765) contained at least 2 epitopes. On fragment 171–484 a sixth epitope was detected that was not found on either subfragment 171–276 or 277–484.

A major epitope region consisting of amino acids 405–485 and containing at least two different epitopes was reported by Piccinini *et al.* [48]. A possible explanation for the discrepancy between these results and the results obtained by Meesters *et al.* [19] is the fact that Piccinini *et al.* [48] used fusion proteins with a short fusion segment. It is our experience that the antigenicity of especially the C-terminal part of topo I is highly influenced by the length of its fusion segment. In immunoblotting experiments using recombinant proteins with short fusion segments of ~10 amino acids, the antigenicity of segment ER-3 of topo I (data not published) was diminished dramatically. For this reason, Piccinini *et al.* [48] could have missed the C-terminally located epitopes on topo I. The finding that synthetic peptide A in the study of Maul *et al.* [23] was

Table 1. Mouse monoclonal antibodies directed to DNA topoisomerase I

Name	Subclass	Reference	Reactivity
8C2	IgM (*K*)	9	WB, IP, IF*
6B5	IgG2a (*K*)	9	WB, IP, IF*
A10	IgG1	8	ELISA, WB, IF*
B7	IgG2	8	ELISA, WB, IF*
E11	IgM	8	ELISA, IF*

* Staining of both the nucleoplasm and nucleoli.
WB: Western blot; IP: immunoprecipitation; IF: immunofluorescence.

Acknowledgements

I am grateful to Dr. W.C. Earnshaw for helpful comments and for critically reading the manuscript. The research of the author was supported in part by the Netherlands Foundation for Chemical Research (SON) with financial aid from the Netherlands Organization for Scientific Research (NWO).

---

immunopositive in the ELISA may be explained by the fact that the sensitivity of this detection method is much higher than that of immunoblotting.

Another consideration which plays an important role in epitope mapping of topo I is the choice of secondary antibodies in detecting the antigen-antibody complexes. Both Hildebrandt *et al.* [11] and Verheijen *et al.* [15] reported a high frequency of IgA anti-topo I antibodies in patients' sera. Furthermore, it was demonstrated that there can be considerable differences between the titers of IgG and IgA antibodies to each of the three epitope regions ER-1, ER-2 and ER-3. Thus, the IgG and IgA anti-topo I antibodies in a serum may recognize different epitopes on the antigen.

References

1. Douvas AS, Achten M & Tan EM (1979) J Biol Chem 254:10514–10522
2. Kumar V, Kowalewski C, Koelle M, Qutaishat S, Chorzelski TP, Beutner EH, Jarzabek-Chorzelska M, Kolacinska Z & Jablonska S (1988) J Rheumatol 15:1499–1502
3. van Venrooij WJ, Stapel SO, Houben H, Habets WJ, Kallenberg CGM, Penner E & van de Putte LB (1985) J Clin Invest 75:1053–1060
4. Tan EM (1982) Adv Immunol 33:167–240
5. Tan EM (1989) Adv Immunol 44:93–152
6. Maul GG, French BT, van Venrooij WJ & Jiminez SA (1986) Proc Natl Acad Sci USA 83:5145–5149
7. Muller MT, Pfund WP, Mehta VB & Trask DK (1985) EMBO J 4:1237–1243
8. Oddou P, Schmidt U, Knippers R & Richter A (1988) Eur J Biochem 177:523–529
9. Negri C, Chiesa R, Cerino A, Bestagno M, Sala C, Zini N, Maraldi NM & Astaldi Ricotti GCB (1992) Exp Cell Res 200:452–459
10. Ghirardello A, Ruffatti A, Calligaro A, Del Ross T & Gambari PF (1992) Clin Rheum 11:126
11. Hildebrandt S, Weiner E, Senécal J-L, Noell S, Daniels L, Earnshaw WC & Rothfield NF (1990) Arthritis Rheum 33:724–727
12. Kipnis RJ, Craft J, Hardin JA (1990) Arthritis Rheum 33:1431–1437
13. Juarez C, Vila JL, Gelpi C, Agusti M, Amengual MJ, Martinez MA & Rodriquez JL (1988) Arthritis Rheum 31:108–115
14. Hildebrandt S, Noell GS, Vazquez-Abad D, Earnshaw WC, Zanetti M & Rothfield NF (1991) Autoimmunity 10:41–48
15. Verheijen R, de Jong BAW & van Venrooij WJ (1992) Clin Exp Immunol 89:456–460
16. Heck MMS, Hittelman WN & Earnshaw WC (1988) Proc Natl Acad Sci USA 85:1086–1090
17. Wang JC (1985) Ann Rev Biochem 54:665–697
18. Osheroff N (1989) Pharmacol Ther 41:223–241
19. Meesters TM, Hoet M, van den Hoogen F, Verheijen R, Richter A, Habets WJ & van Venrooij WJ (1992) Mol Biol Rep 16:117–123
20. D'Arpa P, Machlin PS, Ratrie H, Rothfield NF, Cleveland DW & Earnshaw WC (1988) Proc Natl Acad Sci USA 85:2543–2547
21. Kunze N, Yang GC, Jiang ZY, Hameister H, Adolph S, Wiedorn K-H, Richter A & Knippers R (1989) Hum Genet 84:6–10
22. Kunze N, Yang G, Dölberg M, Sundarp R, Knippers R & Richter A (1991) J Biol Chem 266:9610–9616
23. Maul GG, Jimenez SA, Riggs E & Ziemnicka-Kotula (1989) Proc Natl Acad Sci USA 86:8492–8496
24. Zhou BS, Bastow KF & Cheng Y (1989) Cancer Res 49:3922–3927
25. Zechiedrich EL & Osheroff N (1990) EMBO J 9:4555–4562
26. Barrett JF, Sutcliffe JA & Gootz TD (1990) Antimicrobial Agents and Chemotherapy 34:1–7
27. Eng W-K, Pandit SD & Sternglanz R (1989) J Biol Chem 264:13373–13376
28. Lynn RM, Bjornsti M-A, Caron PR & Wang JC (1989) Proc Natl Acad Sci USA 86:3559–3563
29. Hsiang Y-H, Hertzberg R, Hecht S & Liu LF (1985) J Biol Chem 260:14873–14878
30. Gilmour DS & Elgin SC (1987) Mol Cell Biol 7:141–148
31. Zhang H, Wang JC & Liu LF (1988) Proc Natl Acad Sci USA 85:1060–1064
32. Douvas A, Lambie PB, Turman MA, Nitahara KS & Hammond L (1991) Biochem Biophys Res Commun 178:414–421
33. Pommier Y, Kerrigan D, Hartman KD & Glazer RI (1990) J Biol Chem 265:9418–9422
34. Samuels DS, Shimizu Y & Shimizu N (1989) FEBS Lett 259:57–60
35. Uemura T & Yanagida M (1984) EMBO J 3:1737–1744

36. Thrash C, Bankier AT, Barrell BG & Sternglanz R (1985) Proc Natl Acad Sci USA 82:4374–4378
37. Snapka RM (1986) Mol Cell Biol 6:4221–4227
38. Yang L, Wold MS, Li JJ, Kelly TJ & Liu LF (1987) Proc Natl Acad Sci USA 84:950–954
39. Bullock P, Champoux JJ & Botchan M (1985) Science 230:954–958
40. McCroubrey WK Jr & Champoux JJ (1986) J Biol Chem 261:5130–5137
41. Christiansen K, Boven BJ & Westergaard O (1987) J Mol Biol 193:517–525
42. Ness JP, Koller T & Thoma F (1988) J Mol Biol 200:127–139
43. Gilmour DS, Pflugfelder G, Wang JC & Lis JT (1986) Cell 44:401–407
44. Stewart AF & Schütz G (1987) Cell 50:1109–1117
45. Verheijen R (1992) Mol Biol Rep 16:183–189
46. Verheijen R, van den Hoogen F, Beijer R, Richter A, Penner E, Habets WJ & van Venrooij WJ (1990) Clin Exp Immunol 80:38–43
47. D'Arpa P, White-Cooper H, Cleveland DW, Rothfield NF & Earnshaw WC (1990) Arthritis Rheum 33:1501–1511
48. Piccinini G, Cardellini E, Reimer G, Arnett FC & Durban E (1991) Mol Immunol 28:333–339
49. Kuwana M, Kaburaki J, Mimori T, Tojo T & Homma M (1993) Arthritis Rheum 36:1406–1413

Manual of Biological Markers of Disease **B5.2**: 1–17, 1994.
© 1994 *Kluwer Academic Publishers. Printed in the Netherlands.*

Centromere proteins

RON VERHEIJEN

Department of Biochemistry, University of Nijmegen, P.O. Box 9101, 6500 HB Nijmegen, The Netherlands

Abbreviations

CENP: Centromere proteins

Key words: Autoantibody, autoantigen, centromere proteins, kinetochore, scleroderma

1. The antigens

The antigens are protein components present in the centromere complex and, therefore, are referred to as CENtromere Proteins (CENP). The major antigens are:
CENP-A (17 kDa)
CENP-B (80 kDa)
CENP-C (140 kDa)
Human autoimmune sera that recognized antigens localized uniquely at the centromere were first described by Moroi *et al.* [1]. The use of these autoimmune sera has led to the identification of three human centromere-associated proteins, i.e. CENP-A, CENP-B and CENP-C [2, 3]. CENP-B has been referred to as the major centromere antigen, since antibodies to it were found to be present at high titers in all anti-centromere positive patient sera, while the titers of antibodies to CENP-A and CENP-C were often lower [4].

2. Synonyms

CENP-B: major centromere antigen [4].

3. Antibody detection

3.1 Immunofluorescence

In immunofluorescence, anti-centromere antibodies can be identified by a very characteristic staining pattern consisting of distinct speckles (40 to 60 per nucleus) distributed uniformly throughout the entire interphase nucleus [5, 6; see also this Manual chapter A2]. In mitotic cells these speckles are found in the condensed chromosomal material (Fig. 1).

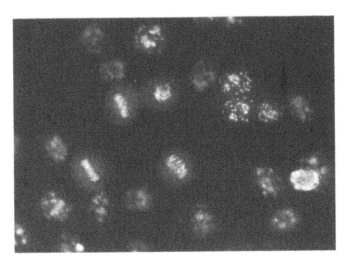

Fig. 1. Immunofluorescence localization of the centromere proteins in HEp-2 (human epidermoid larynx carcinoma) cells in interphase and mitosis, as detected by a human autoimmune serum positive for anti-centromere antibodies. In interphase cells, discrete speckles are seen which are distributed uniformly throughout the entire nucleoplasm. Mitotic cells show these speckles in the condensed chromatin. For technical details see this Manual chapter A2.

3.2 Counterimmunoelectrophoresis (CIE) and immunodiffusion

Both CIE and immunodiffusion are unsuitable methods for detecting antibodies directed to centromere proteins.

3.3 Immunoblotting

Anti-centromere antibodies can be identified in an immunoblotting assay using purified chromosomes [7, 8], a nuclear cell extract [9; see also this Manual chapter A4], or recombinant CENP-B protein [10–12] as antigen source.

3.4 Immunoprecipitation

This technique has not yet been applied for the detection of anti-centromere antibodies.

3.5 ELISA

Recombinant CENP-B has been applied as antigen in an ELISA [7, 8, 12] and radioimmunoassay (RIA) [11] for detecting anti-centromere antibodies. It is expected that commercial tests using recombinant CENP-B will be available soon.

4. Cellular localization

CENP-A, CENP-B and CENP-C are protein constituents of the centromere [reviewed in 13–15] (Fig. 2). Recently, interphase centromeres have been shown to be localized within nucleoli as well [16].

5. Cellular function

5.1 CENP-A

Based on similar biochemical characteristics [17] and amino acid sequence homology with core histone H3 [18], CENP-A has been implicated as a distinctive, centromere-specific histone. Although nothing is known about the distribution of CENP-A within the centromere, this suggests a direct role for CENP-A in centromeric chromatin packaging and function.

5.2 CENP-B

CENP-B is a DNA-binding protein which, via its N-terminal region, specifically interacts with a 17-bp sequence (CENP-B box: 5′ CTTCGTTGGAAACGGGA 3′) in α satellite DNA [19–22]. *In vitro* studies have shown that CENP-B participates in a dimeric structure composed of a CENP-B dimer and two molecules of α satellite DNA [20]. The dimerization activity of CENP-B is localized in the C-terminal region.

Immunoelectron microscopy has demonstrated that CENP-B is dispersed throughout the central domain beneath the kinetochore of the centromere [23], suggesting a role in the formation of the centromeric heterochromatin. The amount of CENP-B present at different centromeres varies because of differences in the sequence composition of the α satellite DNA [24].

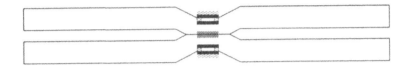

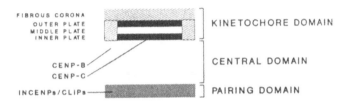

FIBROUS CORONA
OUTER PLATE
MIDDLE PLATE
INNER PLATE

KINETOCHORE DOMAIN

CENP-B
CENP-C

CENTRAL DOMAIN

INCENPs/CLIPs

PAIRING DOMAIN

Fig. 2. Diagrammatic representation of the mammalian centromere showing its domain organization as well as the localization of the CENP antigens.

Based on electron microscopic observations the kinetochore can be divided into three disk-shaped layers: an inner electron-dense layer overlying the central domain, a middle electron-translucent layer, and an outer electron-dense layer. CENP-C has been identified as a constituent of the kinetochore inner plate [25]. When microtubules are absent, a fourth layer, known as the fibrous corona, is observed as well.

The central domain consists largely of centromeric heterochromatin. This chromatin is composed primarily of various families of repetitive DNA elements (satellite DNAs) and their attendant proteins. CENP-B is most probably an α satellite binding protein and is distributed predominantly (> 95%) throughout the heterochromatin of the central domain [23].

The pairing domain represents the site of interaction between sister chromatids at metaphase. So far, two classes of proteins have been assigned to this domain, i.e. the INCENPs (INner CENtromere Proteins) and the CLIPs (Chromatid LInking Proteins) [15].

The precise localization of CENP-A within the centromere has not yet been established.

5.3 CENP-C

CENP-C has been identified as a component of the inner kinetochore plate [25]. Preliminary experiments have shown that CENP-C binds to DNA *in vitro* [25]. Recent evidence suggests that CENP-C is required for the formation of a stable and correctly sized kinetochore [Tomkiel *et al.*, submitted].

6. Presence of CENPs in other species

Several immunoblotting studies using human autoimmune sera have suggested that CENP-A, CENP-B and CENP-C are widely conserved among mammals [11, 15, 26–28]. However, since the precise antigenic relationship between these centromere proteins is still rather ambiguous, the use of such polyclonal antibodies does not rule out possible cross-reactivities. Nevertheless, comparison of the human and mouse CENP-B cDNAs reveals a 96% homology within the coding region [29].

7. cDNA and predicted amino acid sequence

7.1 CENP-A

A cDNA encoding the human CENP-A has not yet been described. The only sequence information is derived from a partial sequence analysis of purified bovine CENP-A by Palmer *et al.* [18], who have shown that selected regions of histone H3 and CENP-A are homologous.

7.2 CENP-B

Cloning of the cDNA encoding ~99% of human CENP-B was first described by Earnshaw *et al.* [10] (Fig. 3). This cDNA lacks 15 nucleotides at the 5' site encoding the first 5 N-terminal amino acids of CENP-B, i.e. M, G, P, K, R [29, 30, see also Fig. 5].

7.3 CENP-C

Cloning of the cDNA encoding the full-length human CENP-C was first described by Saitoh *et al.* [25]. The cDNA sequence as well as the deduced amino acid sequence is shown in Figure 4.

8. CENP genes

8.1 CENP-A

The molecular organization of the CENP-A gene has not yet been described.

8.2 CENP-B

Human CENP-B is transcribed from a single-copy gene that encodes a 2.9 kb mRNA [10].

The molecular organizations of both the human and mouse CENP-B genes have been described by Sullivan and Glass [29]. Mammalian CENP-B

```
            10        20        30        40        50        60        70        80        90
CGACAGCTGACGTTCCGGGAGAAGTCACGGATCATCCAGGAGGTGGAGGAGAATCCGGACCTGCGCAAGGGCGAGATCGCGCGGCGCTTC
 R  Q  L  T  F  R  E  K  S  R  I  I  Q  E  V  E  E  N  P  D  L  R  K  G  E  I  A  R  R  F        30

           100       110       120       130       140       150       160       170       180
AACATCCCGCCGTCCACGCTGAGCACGATCCTGAAGAACAAGCGCGCCATCCTGGCGTCGGAGCGCAAGTACGGGGTGGCCTCCACCTGC
 N  I  P  P  S  T  L  S  T  I  L  K  N  K  R  A  I  L  A  S  E  R  K  Y  G  V  A  S  T  C        60

           190       200       210       220       230       240       250       260       270
CGCAAGACCAACAAGCTGTCTCCCTACGACAAGCTCGAGGGCTTGCTCATCGCCTGGTTCCAGCAGATCCGCGCCGCCGGCCTGCCGGTC
 R  K  T  N  K  L  S  P  Y  D  K  L  E  G  L  L  I  A  W  F  Q  Q  I  R  A  A  G  L  P  V        90

           280       290       300       310       320       330       340       350       360
AAGGGCATCATCCTCAAGGAGAAGGCGCTGCGCATAGCCGAGGAGCTGGGCATGGACGACTTCACCGCCTCCAACGGCTGGCTGGACCGC
 K  G  I  I  L  K  E  K  A  L  R  I  A  E  E  L  G  M  D  D  F  T  A  S  N  G  W  L  D  R       120

           370       380       390       400       410       420       430       440       450
TTCCGCCGGCGCCACGGCGTGGTGTCCTGCAGCGGCGGTGGCCCGCCCGCGCGCGAAACGCTGCCCCCCGCACCCCGGCGGCGGCCTGCC
 F  R  R  R  H  G  V  V  S  C  S  G  V  A  R  A  R  A  R  A  R  N  A  A  P  R  T  P  A  A  P  A  150

           460       470       480       490       500       510       520       530       540
AGTCCGGCCGCGGTGCCCTCGGAGGGCAGTGGCGGGAGCACTACTGGTTGGCGCGCTCGGGAGGAGCAGCCGCCGTCGGTGGCCGAGGGC
 S  P  A  A  V  P  S  E  G  S  G  G  S  T  T  G  W  R  A  R  E  E  Q  P  P  S  V  A  E  G       180

           550       560       570       580       590       600       610       620       630
TACGCCTCGCAGGACGTGTTCAGCGCCACCGAGACCAGTCTATGGTACGACTTCCTGCCCGACCAGGCCGCGGGGCTGTGCGGAGGCGAC
 Y  A  S  Q  D  V  F  S  A  T  E  T  S  L  W  Y  D  F  L  P  D  Q  A  A  G  L  C  G  G  D       210

           640       650       660       670       680       690       700       710       720
GGACGGCCGCGTCAAGCCACCCAGCGCCTGAGCGTCCTGCTCATGCGCCAATGCCGACGGCAGCGAGAAGCTGCCCCCGCTGGTGGCCGGC
 G  R  P  R  Q  A  T  Q  R  L  S  V  L  L  C  A  N  A  D  G  S  E  K  L  P  P  L  V  A  G       240

           730       740       750       760       770       780       790       800       810
AAGTCGGCCAAGCCCCGCGGCAGGCCAAGCCGGCCTGCCCTGCGACTACACCGCCAACTCCAAGGGTGGTGTCACCACCCAGGCCCTGGCC
 K  S  A  K  P  R  A  G  Q  A  G  L  P  C  D  Y  T  A  N  S  K  G  G  V  T  T  Q  A  L  A       270

           820       830       840       850       860       870       880       890       900
AAGTACTTGAAGGCCTTGGACACCCGAATGGCTGCAGAGTCTCGCCGGGTCCTGCTGTTGGCCGGCCGCTTGGCTGCCCAGTCCTTGGAC
 K  Y  L  K  A  L  D  T  R  M  A  A  E  S  R  R  V  L  L  L  A  G  R  L  A  A  Q  S  L  D       300

           910       920       930       940       950       960       970       980       990
ACCTCGGGCCTGCGGCATGTGCAGCTGGCCTTCTTCCCTCCCGGCACCGTGCATCCGCTGGAGAGGGGAGTGGTCCAGCAGGTGAAGGGC
 T  S  G  L  R  H  V  Q  L  A  F  F  P  P  G  T  V  H  P  L  E  R  G  V  V  Q  Q  V  K  G       330

          1000      1010      1020      1030      1040      1050      1060      1070      1080
CACTACCGCCAGGCCATGCTGCTCAAGGCCATGGCCGCGCTAGAGGGCCAGGATCCCTCAGGCCTGCAGCTGGGTCTCACGGAGGCCCTG
 H  Y  R  Q  A  M  L  L  K  A  M  A  A  L  E  G  Q  D  P  S  G  L  Q  L  G  L  T  E  A  L       360

          1090      1100      1110      1120      1130      1140      1150      1160      1170
CACTTTGTGGCTGCCGCCTGGCAGGCAGTGGAGCCTTCGGACATAGCCGCCTGCTTTCGTGAGGCTGGCTTTGGGGGTGGCCCTAATGCC
 H  F  V  A  A  A  W  Q  A  V  E  P  S  D  I  A  A  C  F  R  E  A  G  F  G  G  G  P  N  A       390

          1180      1190      1200      1210      1220      1230      1240      1250      1260
ACCATCACCACTTCCCCTCAAGAGTGAGGGAGAGGAAGAGGAGGAGGAGGAAGAAGAAGAGGAGGAGGAAGAGGGTGAAGGAGAGGAAGAG
 T  I  T  T  S  L  K  S  E  G  E  E  E  E  E  E  E  E  E  E  E  E  E  E  E  E  G  E  G  E  E  E  420

          1270      1280      1290      1300      1310      1320      1330      1340      1350
GAGGAGGAAGGGGAGGAGGAGGAGGAGGAAGGGGGGAAGGAGAGGAATTGGGGGAGGAAGAGGAGGTGGAGGAGGAGGGTGATGTTGAT
 E  E  E  G  E  E  E  E  E  E  G  G  E  G  G  E  E  L  G  E  E  E  E  V  E  E  E  G  D  V  D    450

          1360      1370      1380      1390      1400      1410      1420      1430      1440
AGTGATGAAGAAGAGGAGGAAGATGAGGAGAGCTCCTCGGAGGGCTTGGAGGCTGAGGACTGGGCCCAGGGAGTAGTGGAGGCCGGTGGC
 S  D  E  E  E  E  E  D  E  E  S  S  S  E  G  L  E  A  E  D  W  A  Q  G  V  V  E  A  G  G       480

          1450      1460      1470      1480      1490      1500      1510      1520      1530
AGCTTCGGGGCTTATGGTGCCCAGGAGGAAGCCCAGTGCCCTACTCTGCATTTCCTGGAAGGTGGGGAGGACTCTGATTCAGACAGTGAG
 S  F  G  A  Y  G  A  Q  E  E  A  Q  C  P  T  L  H  F  L  E  G  G  E  D  S  D  S  D  S  E       510

          1540      1550      1560      1570      1580      1590      1600      1610      1620
GAAGAGGACGATGAGGAAGAGGATGATGAAGATGAAGACGACGATGATGATGAGGAGGATGGTGATGAGGTGCCTGTACCCAGCTTTGGG
 E  E  D  D  E  E  E  D  D  E  D  E  D  E  D  D  D  D  D  D  E  E  D  G  D  E  V  P  V  P  S  F  G  540

          1630      1640      1650      1660      1670      1680      1690      1700      1710
GAGGCCATGGCCTTACTTTGCCATGGTCAAGAGGTACCTGACCTCCTTCCCCATTGATGACCGCGTGCAGAGCCACATCCTCCACTTGGAA
 E  A  M  A  Y  F  A  M  V  K  R  Y  L  T  S  F  P  I  D  D  R  V  Q  S  H  I  L  H  L  E       570
```

AMAN-B5.2/6

```
        1720      1730      1740      1750      1760      1770      1780      1790      1800
CACGATCTGGTTCATGTGACCATGAAGAACCACGCCAGGCAGGCGGGACTTCTAGGTCTTGGACATCAAAGCTGAGTCACTGGACCTAGC
  H  D  L  V  H  V  T  M  K  N  H  A  R  Q  A  G  L  L  G  L  G  H  Q  S  *                          594

        1810      1820      1830      1840      1850      1860      1870      1880      1890
TGTGCCCCCAACCTAGATTGGCAGCACCACCCCAGGGCAGAGGACTCTCTGGGCACCCGCTGTGCATGGAGCCAGAGTGCAGAGCCCCAG

        1900      1910      1920      1930      1940      1950      1960      1970      1980
ATCCTTTAGTAATGCTTCCCCTGGTCCTGCAACAGGCCCGGTCACCTCGGCCGGGCCCGGGGCTGAGGTCAGCCTCACTGCCTGCTTATT

        1990      2000      2010      2020      2030      2040      2050      2060      2070
GCCTCTTTCTCAGAATCCTCTTTCCTCCCCATTTGGCCCTGGGCTCAGGGGACCAGGTGGGGCGGGTGGGGAGCTGTCCGGTGCTACCAC

        2080      2090      2100      2110      2120      2130      2140      2150      2160
ACCGTGCCCTCAGTGGACTAACCACAGCAGCAGCCAGGGATGGGCCCTGGAGGTTCCCGGCCGGAGAGTGCCTCTCCCCTCTGCCATCCA

        2170      2180      2190      2200      2210      2220      2230      2240      2250
CGTCAGGTCTTTGGTGGGGGGACCCCAAAGCCATTCTGGGAAGGGCTCCAGAAGAAGGTCCAGCCTAGGCCCCCTGCAAGGCTGGCAGCC

        2260      2270      2280      2290      2300      2310      2320      2330      2340
CCCACCCCCCACCCCCCAGGCCGCCTGAGAAGCACAGTTAACTCACTGCGGGCTCCTGAGCCTGCTTCTGCCTGCTTTCCACCTCCCCAG

        2350      2360      2370      2380      2390      2400      2410      2420      2430
TCCCTTTCTCTGGCCCTGTCCATGTGACTTTGGCCCTTGGTTTTCTTTCCAGATTGGAGGTTTCCAAGAGGCCCCCCACCGTGGAAGTAA

        2440      2450      2460      2470      2480      2490      2500      2510      2520
CCAAGGGCGCTTCCTTGTGGGCAGCTGCAGGCCCCATGCCTCTCCTCCCTCTCTCTGGCAGGGCCCATCCTGGGCAGAGGGGCCTGGGGC

        2530      2540      2550      2560      2570      2580      2590      2600      2610
TGGGCCCAGAGTCCAGCCGTCCAGNTGCTCCTTTCCCAGTTTGATTTCAATAAATCTGTCCACTCCCCTTTTGTGGGGGTGAACGTTTTA

ACAGCCA₂₆
```

Fig. 3. Continued.

is encoded by an intronless open reading frame that specifies a 599-amino acid protein sharing 96% sequence similarity between mouse and human. This high degree of homology is also found in the flanking 5' and 3' untranslated regions of the mRNA detected in both species. The sequence of the human CENP-B gene is shown in Figure 5.

8.3 *CENP-C*

Human CENP-C is transcribed from a single-copy gene that encodes a 3.6 kb mRNA [25]. The molecular organization of the CENP-C gene has not yet been described.

←

Fig. 3. Nucleotide sequence of the human CENP-B cDNA and the deduced amino acid sequence as published by Earnshaw *et al.* [10] (EMBL accession no. X05299). The 2642-base sequence begins with a 594 amino acid (1782 nt) open reading frame, followed by a 3' untranslated region of 834 nucleotides that terminates at a short poly-A sequence (26 nt). The most striking feature of the primary sequence of CENP-B is the presence of two highly acidic domains, together comprising 93 residues that consist for more than 80% of glutamic or aspartic acid. These acidic domains are shaded.

```
              10        20        30        40        50        60        70        80        90
CGGATCGCAGCTCTCGCGGCAGTCGCCTGAGACTTAAGGTTATTGCTTGGCCGCGGCCTGGTATTCCGGCGATTCGTTTCTTGCTCGGCT

              100       110       120       130       140       150       160       170       180
TCCTGGAGCTGTGGTCCGTGTGGGCTTCCACCTCAGACAGTTGCGCTGGCTCAGCGGGGCCGGAACATGGCTGCGTCCGGTCTGGATCAT          8
                                                                 M  A  A  S  G  L  D  H

              190       200       210       220       230       240       250       260       270
CTCAAAAATGGCTACAGAAGAAGATTTTGTCGACCTTCCAGGGCACGTGACATTAACACAGAGCAAGGCCAGAATGTTCTGGAAATCTTA          38
 L  K  N  G  Y  R  R  R  F  C  R  P  S  R  A  R  D  I  N  T  E  Q  G  G  Q  N  V  L  E  I  L

              280       290       300       310       320       330       340       350       360
CAAGACTGTTTTGAAGAAAAAGTCTTGCCAATGATTTTAGTACAAATTCTACAAAATCAGTGCCTAATTCAACACGCAAATAAAAGAC          68
 Q  D  C  F  E  E  K  S  L  A  N  D  F  S  T  N  S  T  K  S  V  P  N  S  T  R  K  I  K  D

              370       380       390       400       410       420       430       440       450
ACTTGTATTCAGTCACCAAGCAAAGAGTGCCAGAAATCACATCCAAAGTCAGTTCCAGTTTCTTCAAAGAAGAAAGAAGCCTCTCTACAG          98
 T  C  I  Q  S  P  S  K  E  C  Q  K  S  H  P  K  S  V  P  V  S  S  K  K  K  E  A  S  L  Q

              460       470       480       490       500       510       520       530       540
TTTGTTGTAGAACCAAGTGAAGCCACAAACAGATCAGTTCAGGCCCATGAAGTTCATCAGAAAATTCTGGCAACTGATGTTAGTTCCAAA          128
 F  V  V  E  P  S  E  A  T  N  R  S  V  Q  A  H  E  V  H  Q  K  I  L  A  T  D  V  S  S  K

              550       560       570       580       590       600       610       620       630
AATACACCTGACTCGAAAAAAATATCAAGTAGAAACATAAATGATCATCACAGTGAAGCTGATGAAGAATTTTACTTATCCGTTGGCTCA          158
 N  T  P  D  S  K  K  I  S  S  R  N  I  N  D  H  H  S  E  A  D  E  E  F  Y  L  S  V  G  S

              640       650       660       670       680       690       700       710       720
CCTTCTGTTCTTTTGGATGCAAAAACATCTGTATCACAAAATGTTATTCCATCTAGTGCCAAAAAGAGAGAGACTTACACTTTTGAAAAT          188
 P  S  V  L  L  D  A  K  T  S  V  S  Q  N  V  I  P  S  S  A  K  K  R  E  T  Y  T  F  E  N

              730       740       750       760       770       780       790       800       810
TCAGTAAATATGCTGCCTTCAAGTACAGAGGTTTCAGTTAAAACCAAAAAAAGGTTAAACTTTGATGATAAAGTTATGTTAAAGAAAATA          218
 S  V  N  M  L  P  S  S  T  E  V  S  V  K  T  K  K  R  L  N  F  D  D  K  V  M  L  K  K  I

              820       830       840       850       860       870       880       890       900
GAAATAGATAATAAAGTATCAGATGAAGAGGATAAAACATCGGAAGGACAAGAAAGAAAACCATCCAGGATCATCTCAGAATAGAATACGA          248
 E  I  D  N  K  V  S  D  E  E  D  K  T  S  E  G  Q  E  R  K  P  S  G  S  S  Q  N  R  I  R

              910       920       930       940       950       960       970       980       990
GATTCAGAATATGAAATTCAACGACAAGCTAAAAAAAGTTTTTCAACATTGTTTTTAGAAACAGTAAAACGAAAAAGTGAATCCAGTCCC          278
 D  S  E  Y  E  I  Q  R  Q  A  K  K  S  F  S  T  L  F  L  E  T  V  K  R  K  S  E  S  S  P

              1000      1010      1020      1030      1040      1050      1060      1070      1080
ATTGTTAGGCATGCGGCAACTGCTCCACCTCATTCGTGTCCTCCCGATGATACGAAGTTGATAGAGGATGAATTTATAATTGATGAGTCG          308
 I  V  R  H  A  A  T  A  P  P  H  S  C  P  P  D  D  T  K  L  I  E  D  E  F  I  I  D  E  S

              1090      1100      1110      1120      1130      1140      1150      1160      1170
GATCAAAGTTTTGCCAGTAGATCTTGGATTACAATACCAAGAAAGGCAGGGTCTCTGAAACAACGCACAATATCCCCGGCTGAGAGCACT          338
 D  Q  S  F  A  S  R  S  W  I  T  I  P  R  K  A  G  S  L  K  Q  R  T  I  S  P  A  E  S  T

              1180      1190      1200      1210      1220      1230      1240      1250      1260
GCACTCTTTCAAGGTAGAAAGTCAAGAGAAAAGCATCATAATATATTACCTAAGACTTTGGCAAATGACAAACATTCCCATAAACCTCAC          368
 A  L  F  Q  G  R  K  S  R  E  K  H  H  N  I  L  P  K  T  L  A  N  D  K  H  S  H  K  P  H
 ▲
              1270      1280      1290      1300      1310      1320      1330      1340      1350
CCAGTAGAGACATCTCAGCCCTCTCTGATAAAACAGTACTGGATACAAGTTATGCTTTGATAGATGAAACAGTAAATAATTATAGATCTACA          398
 P  V  E  T  S  Q  P  S  D  K  T  V  L  D  T  S  Y  A  L  I  D  E  T  V  N  N  Y  R  S  T

              1360      1370      1380      1390      1400      1410      1420      1430      1440
AAATATGAAATGTATTCCAAGAATGCAGAAAAAACCATCTAGAAGCAAAAGGACTATAAAACAAAAACAGAGAAGAAAATTCATGGCTAAA          428
 K  Y  E  M  Y  S  K  N  A  E  K  P  S  R  S  K  R  T  I  K  Q  K  Q  R  R  K  F  M  A  K

              1450      1460      1470      1480      1490      1500      1510      1520      1530
CCAGCTGAAGAACAGCTTGATGTGGGACAGTCTAAAGATGAAAACATACATACATCACATATTACCCAAGACGAATTTCAAAGAAATTCA          458
 P  A  E  E  Q  L  D  V  G  Q  S  K  D  E  N  I  H  T  S  H  I  T  Q  D  E  F  Q  R  N  S

              1540      1550      1560      1570      1580      1590      1600      1610      1620
GACAGAAATATGGAAGAGCATGAAGAGATGGGAAATGATTGTGTTTCCAAAAAACAGATGCCACCTGTGGGAAGCAAGAAAAGTAGCACT          488
 D  R  N  M  E  E  H  E  E  M  G  N  D  C  V  S  K  Q  M  P  P  V  G  S  K  K  S  S  T

              1630      1640      1650      1660      1670      1680      1690      1700      1710
AGAAAAGATAAGGAAGAATCTAAAAAGAAGCGCTTTTCCAGTGAGTCCAAGAACAAACTTGTACCTGAAGAAGTGACTTCAACTGTCACG          518
 R  K  D  K  E  E  S  K  K  K  R  F  S  S  E  S  K  N  K  L  V  P  E  E  V  T  S  T  V  T
```

AMAN-B5.2/8

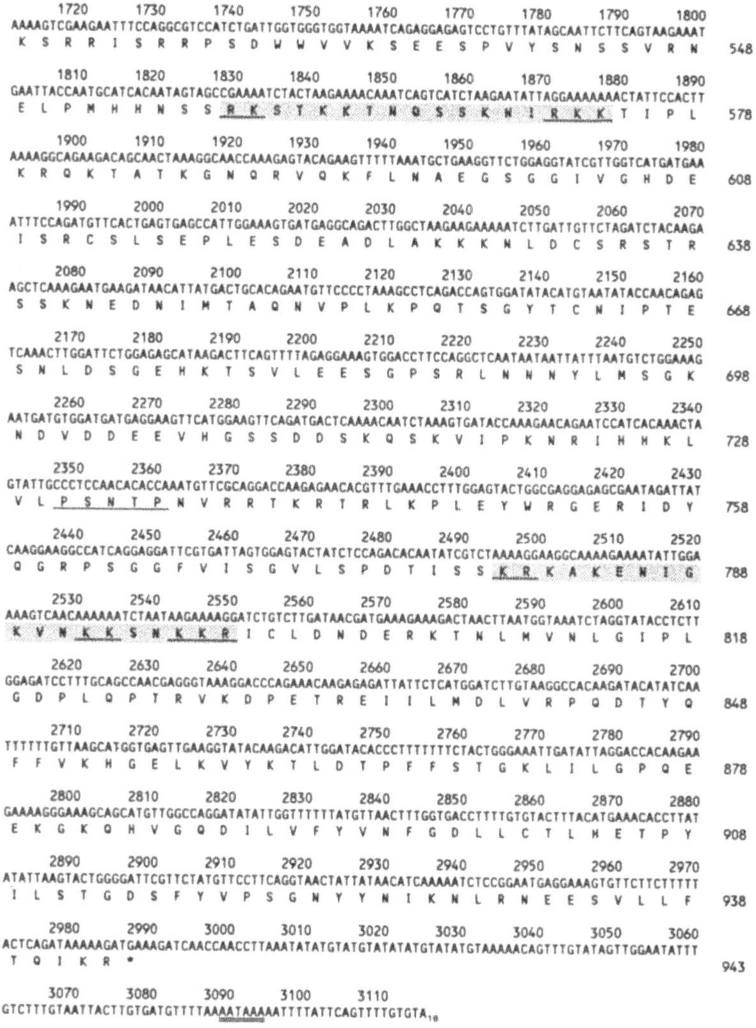

```
         1720      1730      1740      1750      1760      1770      1780      1790      1800
AAAAGTCGAAGAATTTCCAGGCGTCCATCTGATTGGTGGGTGGTAAAATCAGAGGAGAGTCCTGTTTATAGCAATTCTTCAGTAAGAAAT
  K  S  R  R  I  S  R  R  P  S  D  W  W  V  K  S  E  E  S  P  V  Y  S  N  S  S  V  R  N      548

         1810      1820      1830      1840      1850      1860      1870      1880      1890
GAATTACCAATGCATCACAATAGTAGCCGAAAATCTACTAAGAAAACAAATCAGTCATCTAAGAATATTAGGAAAAAAACTATTCCACTT
  E  L  P  M  H  H  N  S  S  R K S T K K T N Q S S K H I R K K  T  I  P  L                    578

         1900      1910      1920      1930      1940      1950      1960      1970      1980
AAAAGGCAGAAGACAGCAACTAAAGGCAACCAAAGAGTACAGAAGTGTTTTAAATGCTGAAGGTTCTGGAGGTATCGTTGGTCATGATGAA
  K  R  Q  K  T  A  T  K  G  N  Q  R  V  Q  K  F  L  N  A  E  G  S  G  G  I  V  G  H  D  E    608

         1990      2000      2010      2020      2030      2040      2050      2060      2070
ATTTCCAGATGTTCACTGAGTGAGCCATTGGAAAGTGATGAGGCAGACTTGGCTAAGAAGAAAAATCTTGATTGTTCTAGATCTACAAGA
  I  S  R  C  S  L  S  E  P  L  E  S  D  E  A  D  L  A  K  K  K  N  L  D  C  S  R  S  T  R    638

         2080      2090      2100      2110      2120      2130      2140      2150      2160
AGCTCAAAGAATGAAGATAACATTATGACTGCACAGAATGTTCCCCTAAAGCCTCAGACCAGTGGATATACATGTAATATACCAACAGAG
  S  S  K  N  E  D  N  I  M  T  A  Q  N  V  P  L  K  P  Q  T  S  G  Y  T  C  N  I  P  T  E    668

         2170      2180      2190      2200      2210      2220      2230      2240      2250
TCAAACTTGGATTCTGGAGAGCATAAGACTTCAGTTTTAGAGGAAAGTGGACCTTCCAGGCTCAATAATAATTATTTAATGTCTGGAAAG
  S  N  L  D  S  G  E  H  K  T  S  V  L  E  E  S  G  P  S  R  L  N  N  N  Y  L  M  S  G  K    698

         2260      2270      2280      2290      2300      2310      2320      2330      2340
AATGATGTGGATGATGAGGAAGTTCATGGAAGTTCAGATGACTCAAAACAATCTAAAGTGATACCAAAGAACAGAATCCATCACAAACTA
  N  D  V  D  D  E  E  V  H  G  S  S  D  D  S  K  Q  S  K  V  I  P  K  N  R  I  H  H  K  L    728

         2350      2360      2370      2380      2390      2400      2410      2420      2430
GTATTGCCCTCCAACACACCAAATGTTCGCAGGACCAAGAGAACACGTTTGAAACCTTTGGAGTACTGGCGAGGAGAGCGAATAGATTAT
  V  L  P  S  N  T  P  N  V  R  R  T  K  R  T  R  L  K  P  L  E  Y  W  R  G  E  R  I  D  Y    758

         2440      2450      2460      2470      2480      2490      2500      2510      2520
CAAGGAAGGCCATCAGGAGGATTCGTGATTAGTGGAGTACTATCTCCAGACACAATATCGTCTAAAAGGAAGGCAAAAGAAAATATTGGA
  Q  G  R  P  S  G  G  F  V  I  S  G  V  L  S  P  D  T  I  S  S  K R K A K E N I G             788

         2530      2540      2550      2560      2570      2580      2590      2600      2610
AAAGTCAACAAAAAATCTAATAAGAAAAGGATCTGTCTTGATAACGATGAAAGAAAGACTAACTTAATGGTAAATCTAGGTATACCTCTT
  K V N K K S N K K R  I  C  L  D  N  D  E  R  K  T  N  L  M  V  N  L  G  I  P  L               818

         2620      2630      2640      2650      2660      2670      2680      2690      2700
GGAGATCCTTTGCAGCCAACGAGGGTAAAGGACCCAGAAACAAGAGAGATTATTCTCATGGATCTTGTAAGGCCACAAGATACATATCAA
  G  D  P  L  Q  P  T  R  V  K  D  P  E  T  R  E  I  I  L  M  D  L  V  R  P  Q  D  T  Y  Q    848

         2710      2720      2730      2740      2750      2760      2770      2780      2790
TTTTTTGTTAAGCATGGTGAGTTGAAGGTATACAAGACATTGGATACACCCTTTTTTTCTACTGGGAAATTGATATTAGGACCACAAGAA
  F  F  V  K  H  G  E  L  K  V  Y  K  T  L  D  T  P  F  F  S  T  G  K  L  I  L  G  P  Q  E    878

         2800      2810      2820      2830      2840      2850      2860      2870      2880
GAAAAGGGAAAGCAGCATGTTGGCCAGGATATATTGGTTTTTTATGTTAACTTTGGTGACCTTTTGTGTACTTTACATGAAACACCTTAT
  E  K  G  K  Q  H  V  G  Q  D  I  L  V  F  Y  V  N  F  G  D  L  L  C  T  L  H  E  T  P  Y    908

         2890      2900      2910      2920      2930      2940      2950      2960      2970
ATATTAAGTACTGGGGATTCGTTCTATGTTCCTTCAGGTAACTATTATAACATCAAAAATCTCCGGAATGAGGAAAGTGTTCTTCTTTTT
  I  L  S  T  G  D  S  F  Y  V  P  S  G  N  Y  Y  N  I  K  N  L  R  N  E  E  S  V  L  L  F    938

         2980      2990      3000      3010      3020      3030      3040      3050      3060
ACTCAGATAAAAAGATGAAAGATCAACCAACCTTAAATATATGTATGTATATATGTATATGTAAAAACAGTTTGTATAGTTGGAATATTT
  T  Q  I  K  R  *                                                                            943

         3070      3080      3090      3100      3110
GTCTTTGTAATTACTTGTGATGTTTTAAAATAAAAATTTTATTCAGTTTTGTGTA_18
```

Fig. 4. Continued.

←

Fig. 4. Nucleotide sequence of the human CENP-C cDNA and the deduced amino acid sequence as published by Saitoh et al. [25] (EMBL accession no. M95724). The 3132-base sequence begins with a 5′ untranslated region of 156 nucleotides, followed by a 943-amino acid (2829 nt) open reading frame, and terminates with a 3′ untranslated region of 129 nucleotides and a short poly-A sequence (18 nt).

Potential phosphorylation sites for the kinetochore-associated kinase p34^{CDC2} [32, 33] and

9. Interaction with other proteins and nucleic acids

See section "Cellular function".

10. Biochemical characteristics of the CENPs

10.1 CENP-a

On SDS-polyacrylamide gels CENP-A has an apparent molecular weight of 17 kDa. It co-purifies with core histones and with nucleosome core particles [17, 18]. The isoelectric point is higher than 9.0.

10.2 CENP-b

Human CENP-B (599 amino acids) has a calculated molecular weight of 65,171 Da, and an apparent molecular weight of 80 kDa on SDS-polyacrylamide gels. The entire protein contains 33% charged residues (74 basic (Arg + Lys + His) and 126 acidic (Asp + Glu)). The calculated isoelectric point is 4.3.

CENP-B can be divided into four domains which may correspond to physical domains of the folded protein (Fig. 6).

→

Fig. 5. Sequence of the human CENP-B gene as published by Sullivan and Glass [29] (EMBL accession no. X55039). The 3717-base sequence begins with a 5' untranslated region of 935 nucleotides, followed by a 599 amino acid (1797 nt) open reading frame, and terminates with 985 nucleotides of 3' untranslated sequence. The polyadenylation signal AATAAA is double underlined. Corrections to the nucleotide sequence result in three amino acid alterations compared to the cDNA sequence of Earnshaw *et al.* [10], namely at positions 583 (M to R), 592 (L to V) and 593 (L to R). These corrections are indicated by arrowheads. Current nucleotide alterations are detailed in the EMBL submission.

No introns are present in the 3' untranslated region. Determination of the precise site of transcription initiation has been hindered because of the extremely high G + C content of the immediately flanking 5' untranslated sequence (> 90%). However, based on the observed size of the transcript and the homology of the 5' flanking sequence with the mouse CENP-B gene, it is presumed to be within 150 bp of the initiator codon [29].

Fig. 4. Continued

microtubule-associated protein (MAP) kinase [34] are amino acids 75 and 732, respectively (underlined). Sequences that fit the consensus for a nucleoplasmin-like nuclear localization signal [35] are shaded with conserved critical residues underlined. The polyadenylation signal AATAAA is double underlined.

By comparison of several human CENP-C cDNA sequences obtained from different libraries, three sites of polymorphism were found [25]. These polymorphisms all resulted in amino acid variations, i.e. at position 341 (F or L) and 361 (D or G) or at position 389 (D or V). These sites are indicated by arrowheads.

```
           10        20        30        40        50        60        70        80        90
TAGATTTACAACTGTAATGCTGATTTTAAAAATAACTTAAGTGTGTAACAATAGGTTACATACGCTAGTGTTCATTCATAGCTGGAATGC

          100       110       120       130       140       150       160       170       180
TACACAATCATTAAAAAGCGTGTTGGCAATGATACTTACCGATATGGCAAAATTCTCACTACTTGATGCCAGGTAAAACAGCAGGAGATG

          190       200       210       220       230       240       250       260       270
CCAGATATACAGTCTATACAGTCTGATGCCAGTTTTATATAATAAATTTTTAAAAAGAAAACGACATGCATCTAGAGAAAAGACTGGAAG

          280       290       300       310       320       330       340       350       360
GAAATGTATCAAATGTTACCTGTGCATCTCTCCAGGTGACGTGAAAAGCGTTTTTCAACTTCTTGAATTTTTCAGGTTCCGCAAAAAAA

          370       380       390       400       410       420       430       440       450
AAAGTGCGTTATTTTTACCTTAATTAAAAAAAAAAAAAAACAAACAAAAACCCTGGTTCTTATTCTTTTTGAGCAGCAAGTCGCTGTCCGC

          460       470       480       490       500       510       520       530       540
CGGGCGCAGTAGCGATGCACCGCGTCCCCGCCCGCGTCGCTGCCGGGTCGCCGGAGGCCGCAGCACGGGCAGGTCCAGCAGGCCGCAGCG

          550       560       570       580       590       600       610       620       630
CCCCCCCGCCGGTATGTGCCCGGCGGCGGGAAGGGGGTGGGCGCGGAGGGCGGGGAGGAGGCGCCGGGCGTCCCCGCGCTTCCCGCGAGA

          640       650       660       670       680       690       700       710       720
TCCCGCTCCGCCCCGCTCGCCGCGCGTCCCAGCTCCCCGCGCCGGCCACTTCCTGCCTTCCGCGCCCGCGCCGCCCGCCCTCGTCTGCGC

          730       740       750       760       770       780       790       800       810
CCGTCGCCTGCCGCCCGCCGCCCGGGACGCGGCCCGCCGGCGTCCCGGAGGTGCCCGGCCCGGCCGGGTCGTCGCCCCGCCGCCGCGCC

          820       830       840       850       860       870       880       890       900
GCGAGCCGCTTTGTCTCGGGCGGGGCGCGCGGGAGAGGCCGCCAGGTGCCCCCCGCCACGGGCCCGGGCCCCCGCCCCGGGGCGGCGGCG

          910       920       930       940       950       960       970       980       990
GCGGCGCGCGGGCCCCGGGGCGGGGGGGCGCGCCGGGATGGGCCCCAAGAGGCGACAGCTGACGTTCCGGGAGAAGTCACGGATCATCCAGG
                                        M  G  P  K  R  R  Q  L  T  F  R  E  K  S  R  I  I  Q     18

         1000      1010      1020      1030      1040      1050      1060      1070      1080
AGGTGGAGGAGAATCCGGACCTGCGCAAGGGCGAGATCGCGCGGCGCTTCAACATCCCGCCGTCCACGCTGAGCACGATCCTGAAGAACA
 E  V  E  E  N  P  D  L  R  K  G  E  I  A  R  R  F  N  I  P  P  S  T  L  S  T  I  L  K  N      48

         1090      1100      1110      1120      1130      1140      1150      1160      1170
AGCGCGCCATCCTGGCGTCGGAGCGCAAGTACGGGGTGGCCTCCACCTGCCGCAAGACCAACAAGCTGTCTCCCTACGACAAGCTCGAGG
 K  R  A  I  L  A  S  E  R  K  Y  G  V  A  S  T  C  R  K  T  N  K  L  S  P  Y  D  K  L  E      78

         1180      1190      1200      1210      1220      1230      1240      1250      1260
GCTTGCTCATCGCGTCGGTTCCAGCAGATCCGCGCCGCCGGCCTGCCGGTCAAGGGCATCATCCTCAAGGAGAAGGCGCTGCGCATAGCCG
 G  L  L  I  A  W  F  Q  Q  I  R  A  A  G  L  P  V  K  G  I  I  L  K  E  K  A  L  R  I  A      108

         1270      1280      1290      1300      1310      1320      1330      1340      1350
AGGAGCTGGGCATGGACGACTTCACCGCCTCCAACGGCTGGCTGGACCGCTTCCGCCGGCGGCCACGGCGTGGTGTCCTGCAGCGGCGTGG
 E  E  L  G  M  D  D  F  T  A  S  N  G  W  L  D  R  F  R  R  R  H  G  V  V  S  C  S  G  V      138

         1360      1370      1380      1390      1400      1410      1420      1430      1440
CCCGCGCCCGCGCGCGAAACGCTGCCCCCCGCACCCCGGCGGCGGCCTGCCAGTCCGGCCGCGGTGCCCTCGGAGGGCAGTGGCGGGAGCA
 A  R  A  R  A  R  N  A  A  P  R  T  P  A  A  P  A  S  P  A  A  V  P  S  E  G  S  G  G  S      168

         1450      1460      1470      1480      1490      1500      1510      1520      1530
CTACTGGTTGGCGCGCTCGGGAGGAGCAGCCGCCGCCGTCGGTGGCCGAGGGCTACGCCTCGCAGGACGTGTTCAGCGCCACCGAGACCAGTC
 T  T  G  W  R  A  R  E  E  Q  P  P  S  V  A  E  G  Y  A  S  Q  D  V  F  S  A  T  E  T  S      198

         1540      1550      1560      1570      1580      1590      1600      1610      1620
TATGGTACGACTTCCTGCCCGACCAGGCCGCGGGGCTGTGCGGAGGCGACGGACAGGCCGCGTCAAGCCCACCCAGCGCCTGAGCGTCCTGC
 L  W  Y  D  F  L  P  D  Q  A  A  G  L  C  G  G  D  G  R  P  R  Q  A  T  Q  R  L  S  V  L      228

         1630      1640      1650      1660      1670      1680      1690      1700      1710
TATGCGCCAATGCCGACGGCAGCGAGAAGCTGCCCCCGCTGGTGGCCGGCAAGTCGGCCAAGCCCCGCGCAGGCCAAGCCGGCCTGCCCT
 L  C  A  N  A  D  G  S  E  K  L  P  P  L  V  A  G  K  S  A  K  P  R  A  G  Q  A  G  L  P      258

         1720      1730      1740      1750      1760      1770      1780      1790      1800
GCGACTACACCGCCAACTCCAAGGGTGGTGTCACCACCCAGGCCCTGGCCAAGTACTTGAAGGCCTTGGACACCCGAATGGCTGCAGAGT
 C  D  Y  T  A  N  S  K  G  G  V  T  T  Q  A  L  A  K  Y  L  K  A  L  D  T  R  M  A  A  E      288

         1810      1820      1830      1840      1850      1860      1870      1880      1890
CTCGCCGGGTCCTGCTGTTGGCCGGCCGCTTGGCTGCCCAGTCCTTGGACACCTCGGGCCTGCGGCATGTGCAGCTGGCCTTCTTCCCTC
 S  R  R  V  L  L  L  A  G  R  L  A  A  Q  S  L  D  T  S  G  L  R  H  V  Q  L  A  F  F  P      318
```

AMAN-B5.2/11

```
        1900      1910      1920      1930      1940      1950      1960      1970      1980
CCGGCACCGTGCATCCGCTGGAGAGGGGAGTGGTCCAGCAGGTGAAGGGCCACTACCGCCAGGCCATGCTGCTCAAGGCCATGGCCGCGC
 P  G  T  V  H  P  L  E  R  G  V  V  Q  Q  V  K  G  H  Y  R  Q  A  M  L  L  K  A  M  A  A        348

        1990      2000      2010      2020      2030      2040      2050      2060      2070
TAGAGGGCCAGGATCCCTCAGGCCTGCAGCTGGGTCTCACGGAGGCCCTGCACTTTGTGGCTGCCGCCTGGCAGGCAGTGGAGCCTTCGG
 L  E  G  Q  D  P  S  G  L  Q  L  G  L  T  E  A  L  H  F  V  A  A  A  W  Q  A  V  E  P  S        378

        2080      2090      2100      2110      2120      2130      2140      2150      2160
ACATAGCCGCCTGCTTTCGTGAGGCTGGCTTTGGGGGGTGGCCCTAATGCCACCATCACCACTTCCCTCAAGAGTGAGGGAGGAGGAAGAGG
 D  I  A  A  C  F  R  E  A  G  F  G  G  G  P  N  A  T  I  T  T  S  L  K  S  E  G  E  E  E       408

        2170      2180      2190      2200      2210      2220      2230      2240      2250
AGGAGGAGGAGGAAGAAGAGGAGGAGGAAGAGGGTGAAGGAGAGGAAGAGGAGGAGGAAGGGGAGGAGGAGGAGGAGGAAGGGGGGAAG
 E  E  E  E  E  E  E  E  E  E  G  E  G  E  E  E  E  E  G  E  E  E  E  E  G  G  E               438

        2260      2270      2280      2290      2300      2310      2320      2330      2340
GAGAGGAATTGGGGGAGGAAGAGGAGGTGGAGGAGGAGGGTGATGTTGATAGTGATGAAGAAGAGGAAGATGAGGAGAGCTCCTCGG
 G  E  E  L  G  E  E  E  E  V  E  E  E  G  D  V  D  S  D  E  E  E  E  E  D  E  E  S  S         468

        2350      2360      2370      2380      2390      2400      2410      2420      2430
AGGGCTTGGAGGCTGAGGACTGGGCCCAGGGAGTAGTGGAGGCCGGTGGCAGCTTCGGGGGCTTATGGTGCCCAGGAGGAAGCCCAGTGCC
 E  G  L  E  A  E  D  W  A  Q  G  V  V  E  A  G  G  S  F  G  A  Y  G  A  Q  E  E  A  Q  C      498

        2440      2450      2460      2470      2480      2490      2500      2510      2520
CTACTCTGCATTTCCTGGAAGGTGGGGAGGACTCTGATTCAGACAGTGAGGAAGAGGACGATGAGGAAGAGGATGATGAAGATGAAGACG
 P  T  L  H  F  L  E  G  G  E  D  S  D  S  D  S  E  E  E  D  D  E  E  E  D  D  E  D  E  D       528

        2530      2540      2550      2560      2570      2580      2590      2600      2610
ACGATGATGATGAGGAGGATGGTGATGAGGTGCCTGTACCCAGCTTTGGGGAGGCCATGGCTTACTTTGCCATGGTCAAGAGGTACCTGA
 D  D  D  D  E  E  D  G  D  E  V  P  V  P  S  F  G  E  A  M  A  Y  F  A  M  V  K  R  Y  L       558

        2620      2630      2640      2650      2660      2670      2680      2690      2700
CCTCCTTCCCCATTGATGACCGCGTGCAGAGCCACATCCTCCACTTGGAACACGATCTGGTTCATGTGACCAGGAAGAACCACGCCAGGC
 T  S  F  P  I  D  D  R  V  Q  S  H  I  L  H  L  E  H  D  L  V  H  V  T  R  K  N  H  A  R       588
                                                                            ▲
        2710      2720      2730      2740      2750      2760      2770      2780      2790
AGGCGGGAGTTCGAGGTCTTGGACATCAAAGCTGAGTCACTGGACCTAGCTGTGCCCCCAACCTAGATTGGCAGCACCACCCCAGGGCAG
 Q  A  G  V  R  G  L  G  H  Q  S  *                                                           599
   ▲  ▲
        2800      2810      2820      2830      2840      2850      2860      2870      2880
AGGACTCTCTGGGCACCCGCTGTGCATGGAGCCAGAGTGCAGAGCCCCAGATCCTTTAGTAATGCTTCCCCTGGTCCTGCAACAGGCCCG

        2890      2900      2910      2920      2930      2940      2950      2960      2970
GTCACCTCGGCCGGGCCCGGGGCTGAGGTCAGCCTCACTGCCTGCTTATTGCCTCTTTCTCAGAATCCTCTTTCCTCCCCATTTGGCCCT

        2980      2990      3000      3010      3020      3030      3040      3050      3060
GGGCTCAGGGGACCAGGTGGGGCGGGTGGGGAGCTGTCCGGTGCTACCACACCGTGCCCTCAGTGGACTAACCACAGCAGCAGCCAGGGA

        3070      3080      3090      3100      3110      3120      3130      3140      3150
TGGGCCCTGGAGGTTCCCGGCCGGAGAGTGCCTCTCCCCTCTGCCATCCACGTCAGGTCTTTGGTGGGGGGACCCCAAAGCCATTCTGGG

        3160      3170      3180      3190      3200      3210      3220      3230      3240
AAGGGCTCCAGAAGAAGGTCCAGCCTAGGCCCCCTGCAAGGCTGGCAGCCCCCACCCCCACCCCCCAGGCCGCCTTGAGAAGCACAGTTT

        3250      3260      3270      3280      3290      3300      3310      3320      3330
AACTCACTGCGGGCTCCTGAGCCTGCTTCTGCCTGCTTTCCACCTCCCCAGTCCCTTTCTCTGGCCCTGTCCATGTGACTTTGGCCCTTG

        3340      3350      3360      3370      3380      3390      3400      3410      3420
GTTTTCTTTCCAGATTGGAGGTTTCCAAGAGGCCCCCCACCGTGGAAGTAACCAAGGGCGCTTCCTTGTGGGCAGCTGCAGGCCCCATGC

        3430      3440      3450      3460      3470      3480      3490      3500      3510
CTCTCCTCCCTCTCTCTGGCAGGGCCCATCCTGGGCAGAGGGGCCTGGGGCTGGGCCCAGAGTCCAGCCGTCCAGCTGCTCCTTTCCCAG

        3520      3530      3540      3550      3560      3570      3580      3590      3600
TTTGATTTCAATAAATCTGTCCACTCCCCTTTTGTGGGGGTGAACGTTTTAACAGCCAAGGGTGCATCCTTCATGGTCTGGGCTTGCGTC

        3610      3620      3630      3640      3650      3660      3670      3680      3690
TGTCTTGGGGGACTTATTCGTCCTGGCTCTCTTTGGTCCTTGCTCTGGTGGGACATGGAGGCAAGTGTTGAGAGGGTTGCCCTGACCGGA

        3700      3710
AGAGGGGCAGGAGGAGACCTCAAGCTT
```

Fig. 5. Continued.

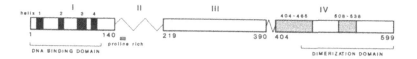

Fig. 6. Proposed domain structure of human CENP-B (compiled from data of [20, 21, 29]). The N-terminal domain I (amino acids 1–140) consists of a relatively basic domain of four predicted α helices, i.e. helix 1 (aa 10–22), helix 2 (aa 46–56), helix 3 (aa 77–93), and helix 4 (aa 100–111) [20]. Rather long loops intervene between helix 1 and helix 2 and between helix 2 and 3. Helix 3 is linked to helix 4 with a 6 aa stretch of β sheet structure. It was suggested that a tertiary structure composed of the four α helices in domain I is necessary for DNA binding activity, since this binding was completely abolished by deletion of helix 1 (10–22) or helix 4 (100–111) [20]. Helix 4 is amphipatic and might be important for the tertiary structure of the DNA binding domain of CENP-B.

Domain I is separated from the central domain III (amino acids 219–390) by an extended segment with a high predicted flexibility and turn content, interspersed with short β sheets (domain II: residues 141–218). This region contains a proline rich sequence (amino acids 148–157) identified by Joly *et al.* [36] that is similar to the protease-sensitive hinge region of the microtubule-associated protein-2 (MAP2). Domain III contains a series of alternating α helices and β sheets, connected by a long hydrophobic helix at the C-terminal site of the domain. This is separated from domain IV (amino acids 404–599) by a short flexible segment. The C-terminal domain contains two highly acidic segments (amino acids 404–465 and 508–538), together comprising 93 residues that account for more than 80% of glutamic or aspartic acid.

In vitro studies by Yoda *et al.* [20] have shown that CENP-B participates in a dimeric structure composed of a CENP-B dimer and two molecules of α satellite DNA. The model that was postulated for this stable complex, designated as complex A, consists of a DNA binding domain and a dimerization domain at independent locations in the N- and C-terminal regions, respectively. Indicated are the approximate lengths of the fragments necessary for DNA binding and dimerization.

10.3 CENP-C

Human CENP-C (943 amino acids) has a calculated molecular weight of 106,925 Da, and an apparent molecular weight of 140 kDa on SDS-polyacrylamide gels. The entire protein contains 34% charged residues (189 basic and 128 acidic). The calculated isoelectric point is 10.1.

11. B-cell epitopes on CENP-B

The distribution of the autoimmune epitope regions on CENP-B as reported by several research groups is shown in Figure 7 (see also [31]).

12. T-cell epitopes on the CENP antigens

So far, no reports have been published on T-cell epitopes on CENP-A, CENP-B or CENP-C.

B-CELL EPITOPES ON CENP-B (599 aa)

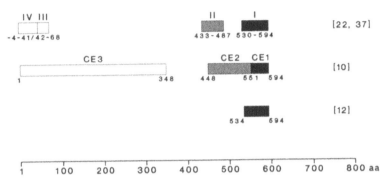

Fig. 7. Schematic representation of autoimmune epitope regions on the CENP-B recombinant protein. References are indicated between brackets.

Earnshaw *et al.* [10] identified three epitope regions on the recombinant CENP-B protein. The two segments consisting of amino acids 1–348 and 448–551 were found to form epitope regions CE3 and CE2, respectively. The C-terminal segment of 43 residues itself was not antigenic, but in combination with the 104 residues of CE2 a third epitope (CE1) was generated. Verheijen *et al.* [12] demonstrated that the last 60 C-terminal amino acids of CENP-B (534–594) constitute an important autoantigenic domain.

In a study of Sugimoto *et al.* [22, 37] four CENP-B epitopes emerged, i.e. epitope I (aa 530–594), epitope II (aa 433–487), epitope III (aa 42–68), and epitope IV (aa −4–41). Epitope I was designated the major epitope since it was recognized by all anti-centromere positive sera tested. In contrast, the other three epitopes were not recognized by all sera and can therefore be regarded as minor epitopes.

13. Mouse monoclonal antibodies to the CENP antigens

The published monoclonal antibodies directed to these centromeric antigens are tabulated in Table 1.

Table 1. Mouse monoclonal antibodies directed to the CENP antigens

Name	Subclass	Reference	Reactivity[a]
1H2	IgG2a	10	IF, WB
2D7	IgG1	10	IF, WB
IG11	IgG1	10	IF, WB

[a] Directed against the C-terminal 146 amino acids of CENP-B.
WB: Western blot; IF: immunofluorescence.

Acknowledgements

I am grateful to Dr. W.C. Earnshaw for helpful comments and for critically reading the manuscript. The research of the author was supported in part by the Netherlands Foundation for Chemical Research (SON) with financial aid from the Netherlands Organization for Scientific Research (NWO).

References

1. Moroi Y, Peebles C, Fritzler MJ, Steigerwald J & Tan EM (1980) Proc Natl Acad Sci USA 77:1627–1631
2. Earnshaw WC & Rothfield N (1985) Chromosoma 91:313–321
3. Valdivia MM & Brinkley BR (1985) J Cell Biol 101:1124–1134
4. Earnshaw W, Bordwell B, Marino C & Rothfield N (1986) J Clin Invest 77:426–430
5. Tan EM (1982) Adv Immunol 33:167–240
6. Tan EM (1989) Adv Immunol 44:93–152
7. Rothfield N, Whitaker D, Bordwell B, Weiner E, Senécal J-L & Earnshaw W (1987) Arthritis Rheum 30:1416–1419
8. Hildebrandt S, Weiner E, Senécal J-L, Noell S, Daniels L, Earnshaw WC & Rothfield NF (1990) Arthritis Rheum 33:724–727
9. Muro Y, Sugimoto K, Okazaki T & Ohashi M (1990) J Rheum 17:1042–1047
10. Earnshaw WC, Sullivan KF, Machlin PS, Cooke CA, Kaiser DA, Pollard TD, Rothfield NF & Cleveland DW (1987) J Cell Biol 104:817–829
11. Earnshaw WC, Machlin PS, Bordwell BJ, Rothfield NF & Cleveland DW (1987) Proc Natl Acad Sci USA 84:4979–4983
12. Verheijen R, de Jong BAW, Oberyé EHH & van Venrooij WJ (1992) Mol Biol Rep 16:49–59
13. Pluta AF, Cooke CA & Earnshaw EC (1990) Trends Biochem 15:181–185
14. Willard HF (1990) Trends Genet 6:410–416
15. Rattner JB (1991) BioEssays 13:51–56
16. Ochs RL & Press RI (1992) Exp Cell Res 200:339–350
17. Palmer DK, O'Day K, Wener MH, Andrews BS & Margolis RL (1987) J Cell Biol 104:805–815
18. Palmer DK, O'Day K, Trong HL, Charbonneau H & Margolis RL (1991) Proc Natl Acad Sci USA 88:3734–3738
19. Masumoto H, Masukata H, Muro Y, Nozaki N & Okazaki T (1989) J Cell Biol 109:1963–1973
20. Yoda K, Kitagawa K, Masumoto H, Muro Y & Okazaki T (1992) J Cell Biol 119:1413–1427
21. Pluta AF, Saitoh N, Goldberg I & Earnhaw WC (1992) J Cell Biol 116:1081–1093
22. Sugimoto K, Muro Y & Himeno M (1992) J Biochem 111:478–483
23. Cooke CA, Bernat RL & Earnshaw WC (1990) J Cell Biol 110:1475–1488
24. Willard HF & Waye JS (1987) Trends Genet 3:192–198
25. Saitoh H, Tomkiel J, Cooke CA, Ratrie H, Maurer M, Rothfield NF & Earnshaw WC (1992) Cell 70:115–125
26. Kingwell B & Rattner JB (1987) Chromosoma 95:403–407
27. Kremer L, Del Mazo J & Avila J (1988) Eur J Cell Biol 46:196–199
28. Palmer DK, O'Day K & Margolis RL (1990) Chromosoma 100:32–36
29. Sullivan KF & Glass CA (1991) Chromosoma 100:360–370
30. Sugimoto K & Himeno M (1991) Agric Biol Chem. 55:2687–2692
31. Verheijen R (1992) Mol Biol Rep 16:183–189
32. Rattner JB, Lew J & Wang JH (1990) Cell Motility and the Cytoskeleton 17:227–236
33. Moreno S & Nurse P (1990) Cell 61:549–551
34. Sturgill TW & Wu J (1991) Biochim Biophys Acta 1092:350–357
35. Robbins J, Dilworth SM, Laskey RA & Dingwall C (1991) Cell 64:615–623
36. Joly JC, Flynn G & Purich DL (1989) J Cell Biol 109:2289–2294
37. Sugimoto K, Migita H, Hagishita Y, Yata H & Himeno M (1992) Cell Structure and Function 17:129–138

Manual of Biological Markers of Disease **B5.3**: 1–10, 1994.

Fibrillarin

RON VERHEIJEN

Department of Biochemistry, University of Nijmegen, P.O. Box 9101, 6500 HB Nijmegen, The Netherlands

Abbreviations

DFC: dense fibrillar component; FC: fibrillar center; sno RNA: small nucleolar RNA

Key words: Autoantibody, autoantigen, fibrillarin, U3 small nuclear ribonucleoprotein, nucleolus

1. The antigen

Fibrillarin, a nucleolar protein of 34 kDa.

2. Synonyms

Human autoimmune sera that recognized a 34 kDa nucleolar protein were first described by Lischwe *et al.* [1]. Because of its specific immunolocalization within the fibrillar component of nucleoli, the protein was named fibrillarin [2]. Fibrillarin appeared to be the homologue of the B36 nucleolar protein characterized in the slime mold *Physarum polycephalum* [3]. In yeast, fibrillarin is referred to as "nucleolar protein-1" (NOP1) [4].

3. Antibody detection

3.1 Immunofluorescence

In immunofluorescence, anti-fibrillarin antibodies can be identified by a clumpy appearance of the interphase nucleoli (see this Manual, chapter A2 and Fig. 1). However, since this staining pattern very much resembles the staining patterns of other anti-nucleolar antibody activities for inexperienced investigators, confirmation of the presence of anti-fibrillarin antibodies by

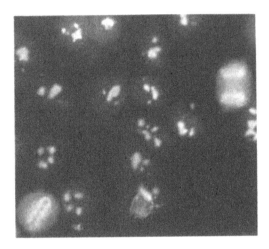

Fig. 1. Immunofluorescence localization of fibrillarin in HEp-2 (human epidermoid larynx carcinoma) cells in interphase and mitosis as detected by a human autoimmune serum positive for anti-fibrillarin antibodies. For technical details see this Manual chapter A2.

other techniques is recommended. In mitotic cells the antigen remains associated with the condensed chromosomal material (Fig. 1).

3.2 Immunodiffusion

Anti-fibrillarin antibody activity in a serum can be detected by the Ouchterlony double immunodiffusion technique using extracted rabbit thymus acetone powder as antigen source [5] (for technical details see also this Manual, chapter A3).

3.3 Immunoblotting

Using nuclear or nucleolar cell extracts as antigen source, anti-fibrillarin antibodies can also be detected by immunoblotting. For technical details see this Manual, chapter A4.

3.4 Immunoprecipitation

Immunoprecipitation from radioactively labeled cell extracts with patients' sera is one of the most commonly used methods to detect antibodies to fibrillarin. For detecting precipitated fibrillarin itself, such cell extracts are usually labeled with [^{35}S]-methionine or [^{3}H]-leucine [5, 6] (for technical details see this Manual, chapter A6).

Analysis of precipitated RNAs, with which fibrillarin is associated, requires cell extracts that have been labeled with [^{32}P]-orthophosphate [5] (for technical details see this Manual, chapter A7). Alternatively, when using unlabeled cell extracts, the recovered RNA can either be labeled at its 3′ terminus with [^{32}P]pCp [7] or identified by Northern blot analysis using antisense riboprobes [8].

3.5 ELISA

An ELISA for detecting anti-fibrillarin antibodies has not yet been described.

4. Cellular localization

Using immunoelectron microscopic techniques, the highest concentration of fibrillarin is found within the dense fibrillar component (DFC) of the nucleolus and the interface between DFC and the fibrillar center (FC), that is, at the sites were transcription of the ribosomal DNA is thought to take place [9]. Fibrillarin is also found in the nuclear body known as the coiled body [10]. See also this Manual, chapter B-8.2, for more detailed information on coiled bodies.

5. Cellular function

Fibrillarin has been shown to function in ribosomal RNA (rRNA) processing [11, reviewed in 12–14]. Mechanisms such as pre-rRNA processing, pre-rRNA modification and ribosome assembly are dependent on the presence of fibrillarin containing ribonucleoproteins.

6. Presence of fibrillarin in other species

Fibrillarin from man [15], mouse [16], *Xenopus laevis* [17], and *Saccharomyces cerevisiae* [4] show a striking degree of evolutionary conservation of primary structure. The human and amphibian sequences contain 81% identical and 90% homologous residues. The identity between the human and the yeast sequence is 67%, whereas the human and mouse sequences share more than 94% identical residues.

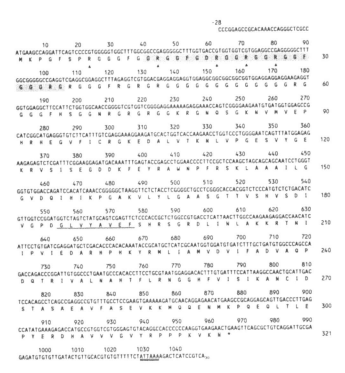

Fig. 2. Nucleotide sequence of the human fibrillarin cDNA and the deduced amino acid sequence.

The sequence given is the one as published by Aris and Blobel [15] (EMBL accession no. M59849). The 1104-base sequence starts with a 5′ untranslated region of 28 nucleotides, followed by a 321-amino acid (963 nt) open reading frame that terminates with a 3′ untranslated region of 83 nucleotides and a short poly-A sequence (30 nt). The encoded polypeptide has a calculated molecular weight of 33,818 Da. The complete protein contains 27% charged residues (59 basic (Arg + Lys + His) and 29 acidic (Asp + Glu)).

The N-terminal domain of fibrillarin is rich in glycine and arginine residues, which accounts for the abundance of glycine (20.2 mol%) and arginine (4.2 mol%) in the protein. The arginine residues in this domain are in similar relative positions as arginines known to be modified by methylation. This suggests that these arginines (arrowheads) may be N^G,N^G-dimethylated [15, see also 24].

The polyadenylation signal ATTAAA is double-underlined.

The cluster of RGG repeats with characteristic spacing (shaded) is also known as the RGG-box [27] or GAR domain [28]. This motif is found in several RNA-binding proteins such as the vertebrate heterogeneous nuclear ribonucleoprotein (hnRNP) proteins A1, hnRNP C, nucleolin and the yeast nucleolar proteins GAR1, SSB1, NSR1, and NOP3 [27, 29]. Recently, it was shown that the RGG-box of nucleolin destabilizes the structure of RNA duplexes in a non-sequence-dependent manner by altering base stacking, possibly to allow access of sequence-specific binding components [30].

7. Human cDNA and predicted amino acid sequence

Cloning of the cDNA encoding the full-length human fibrillarin was first described by Aris and Blobel [15]. The cDNA sequence as well as the deduced amino acid sequence is shown in Figure 2.

8. The fibrillarin gene

The molecular organization of the human fibrillarin gene has not yet been described.

9. Interaction with nucleic acids

Anti-fibrillarin antibodies precipitate a number of small nucleolar RNAs (snoRNAs), the most abundant one being U3 RNA (Table 1). U3, U8 and U13 RNA are transcribed from independent genes, whereas U14, U15, U16, U20, U21 and probably RNA Y as well are encoded within introns of mRNA-coding genes. All snoRNAs which are precipitable by anti-fibrillarin antibodies share two or three short, conserved single-stranded regions (boxes C and/or C' and box D). Box C (consensus 5' UGAUGAUYG 3') was shown to be important for fibrillarin binding to U3 RNA [18, 19, see also Fig. 3]. It is not known whether fibrillarin binds directly to the snoRNAs or indirectly via (an)other snoRNP protein(s). The fact that the presumed octameric RNP consensus motif in fibrillarin (Fig. 2) does not conform to the

Table 1. Small nucleolar RNAs (snoRNAs) associated with fibrillarin

RNA	Size (nt)	Species Copies/cell	TMG cap[a]	Gene	Conserved boxes[b]	Reference
U3	217	Human 2.10^5	+	U3	A, B, C, C', D	7, 12, 22, 31–34
U8	136	Human 4.10^4	+	U8	C, D	7, 12, 35, 36
U13	105	Human 1.10^4	+	U13	C, D	7, 12
U14	86–88	Human ?	−	hsc70	C, D	12, 37–39
U15[c]	146–148	Human ?	−	S3	C', D	12, 43
U16	106	Human ?	−	L1	C', D	12, 40
U20	~80	Human ?	−	Nucleolin	C, D	12
U21	?	Chicken ?	−	L5	C, D	12
Y	125	Human ?	−	?	C', D	7, 12, 43

Data taken from [12].
[a] 2,2,7-trimethylguanosine (m_3G) cap structure at the 5' end.
[b] Distinguishing between the C and C' boxes in RNAs other than U3 is somewhat arbitrary [43].
[c] U15 was formerly known as RNA X [7, 18, 43].

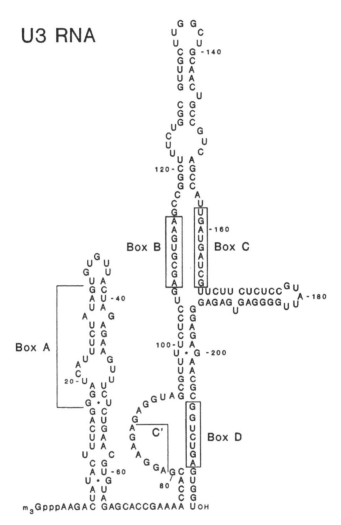

Fig. 3. Primary and possible secondary structure of human U3 RNA [7, 43]. The conserved sequence elements (boxes A, B, C, C' and D) are indicated. Boxes A and B are unique to U3 RNA, whereas boxes C (or C') and D occur in other RNAs as well (see also Table 1).

consensus motif in other RNA-binding proteins [20, 21] suggests an indirect binding.

10. Interaction with other proteins

So far, no reports have been published on the interaction of fibrillarin with other snoRNP proteins or with other proteins participating in ribosome synthesis. However, it has been described that fibrillarin exists in the U3 snoRNP particle with at least 5 other proteins of molecular weights of 12.5, 14, 31, 59 and 74 kDa [22].

11. Biochemical characteristics of fibrillarin

On SDS-polyacrylamide gels fibrillarin has an apparent molecular weight of 34–37 kDa. The isoelectric point has been determined to be approximately 10.7 [23]. Actually, another isoelectric form of fibrillarin with pI: 9.5 seems to exist as well [23]. Fibrillarin contains $\sim$4 mol% N^G,N^G-dimethylarginine and is the most heavily methylated protein identified to date [1, 24].

12. B-cell epitopes on fibrillarin

Reimer *et al.* [25] have shown that a monoclonal antibody derived from an autoimmune (New Zealand black $\times$ New Zealand white) F1 mouse reacted with an epitope on fibrillarin that could be blocked by human scleroderma anti-U3 RNP antibodies *in situ*.

A more recently published observation demonstrated that mercury chloride, when given orally to mice from susceptible strains, induced the production of antibodies to fibrillarin. Since these induced murine autoantibodies as well as scleroderma-specific human anti-U3 RNP antibodies reacted with nucleoli from a wide variety of species in immunofluorescence, it can be concluded that both types of antibodies recognize evolutionarily highly conserved epitopes on fibrillarin [26]. However, since the primary structure of fibrillarin from distantly related species shows a high degree of evolutionary conservation over the entire protein [15, 16], the human and murine autoantibodies to fibrillarin do not necessarily have to recognize the same epitopes.

13. T-cell epitopes on fibrillarin

So far, no reports have been published on T-cell epitopes on fibrillarin.

14. Mouse monoclonal antibodies to fibrillarin

An overview of published monoclonal antibodies directed to fibrillarin is given in Table 2. A monoclonal anti-fibrillarin autoantibody has been described from the (NZB × NZW)F1 autoimmune strain [24]. Monoclonal anti-fibrillarin autoantibodies have also been obtained from HgCl$_2$-treated mice [41]. A group of nine monoclonals has been raised against the fibrillarin of the slime mold *Physarum polycephalum*. Eight of these monoclonals recognize epitopes in the carboxy-terminal half of the molecule, while the remaining clone appears to recognize the amino-terminal GAR domain [42].

Table 2. Mouse monoclonal antibodies directed to fibrillarin

Name	Subclass	Reference	Reactivity
72B9	IgG2a	24, 31	IF, IP
17C12	IgG	41	IF, IP, WB*
PIG12	–	42	WB (amino terminal)
P2B11	–	42	WB (carboxy terminal)

* K.M. Pollard, personal communication
WB: Western blot; IP: Immunoprecipitation; IF: immunofluorescence

Acknowledgements

I am grateful to Dr. K.M. Pollard for helpful comments and for critically reading the manuscript. The research of the author was supported in part by the Netherlands Foundation for Chemical Research (SON) with financial aid from the Netherlands Organization for Scientific Research (NWO).

References

1. Lischwe MA, Ochs RL, Reddy R, Cook RG, Yeoman LC, Tan EM, Reichlin M & Busch H (1985) J Biol Chem 260:14304–14310
2. Ochs RL, Lischwe MA, Spohn WH & Busch H (1985) Biol Cell 54:123–134
3. Christensen ME, Moloo J, Swischuk JL & Schelling ME (1986) Exp Cell Res 166:77–93
4. Schimmang T, Tollervey D, Kern H, Frank R & Hurt EC (1989) EMBO J 8:4015–4024
5. Okano Y, Steen VD & Medsger TA Jr (1992) Arthritis Rheum 35:95–100
6. Kipnis RJ, Craft J & Hardin JA (1990) Arthritis Rheum 33:1431–1437
7. Tyc K & Steitz JA (1989) EMBO J 8:3113–3119
8. Verheijen R, Wiik A, de Jong BAW, Høier-Madsen M, Ullman S, Halberg P & van Venrooij WJ (1994) J Immunol Methods 169:173–182
9. Puvion-Dutilleul F, Mazan S, Nicoloso M & Christensen ME (1991) Eur J Cell Biol 56:178–186
10. Raska I, Andrade LEC, Ochs RL, Chan EKL, Chang CM, Roos G & Tan EM (1991) Exp Cell Res 195:27–37
11. Tollervey D, Lehtonen H, Jansen R, Kern H & Hurt EC (1993) Cell 72:443–457
12. Filipowicz W & Kiss T (1993) Mol Biol Rep 18:149–156
13. Fournier MJ & Maxwell ES (1993) Trends Biochem 18:131–135
14. Mattaj IW, Tollervey D & Séraphin B (1993) FASEB J 7:47–53
15. Aris JP & Blobel G (1991) Proc Natl Acad Sci USA 88:931–935
16. Turley SJ, Tan EM & Pollard KM (1993) Biochim Biophys Acta 1216:119–122
17. Lapeyre B, Mariottini P, Mathieu C, Ferrer P, Amaldi F, Amalric F & Caizergues-Ferrer M (1990) Mol Cell Biol 10:430–434
18. Baserga SJ, Yang XW & Steitz JA (1991) EMBO J 10:2645–2651
19. Baserga SJ, Gilmore-Hebert M & Yang XW (1992) Genes Dev 6: 1120–1130
20. Query CC, Bentley RC & Keene JD (1989) Cell 57:89–101
21. Bandziulus RJ, Swanson MS & Dreyfuss G (1989) Genes Dev 3:431–437
22. Parker KA & Steitz JA (1987) Mol Cell Biol 7:2899–2913
23. Celis JE, Rasmusssen HH, Madsen P, *et al.* (1992) Electrophoresis 13:893–959
24. Najbauer J, Johnson BA, Young AL & Aswad DW (1993) J Biol Chem 268:10501–10509
25. Reimer G, Pollard KM, Penning CA, Ochs RL, Lischwe MA, Busch H & Tan EM (1987) Arthritis Rheum 30:793–800
26. Reuter R, Tessars G, Vohr H-W, Gleichmann E & Lührmann R (1989) Proc Natl Acad Sci USA 86:237–241
27. Kiledjian M & Dreyfuss G (1992) EMBO J 11:2655–2664
28. Girard J-P, Lehtonen H, Caizergues-Ferrer M, Amalric F, Tollervey D & Lapeyre B (1992) EMBO J 11:673–682
29. Russell ID & Tollervey D (1992) J Cell Biol 119:737–747
30. Ghisolfi L, Joseph G, Amalric F & Erard M (1992) J Biol Chem 267:2955–2959
31. Reddy R & Busch H (1988) In: Birnstiel ML (ed) Structure and Function of Small Nuclear Ribonucleoprotein Particles, pp 1–37. Springer Verlag, Berlin
32. Kass S, Tyc K, Steitz JA & Sollner-Webb B (1990) Cell 60:897–908
33. Savino R & Gerbi SA (1990) EMBO J 8: 2299–2308
34. Tyc K & Steitz JA (1992) Nucleic Acids Res 20:5375–5382
35. Reddy R, Henning D & Busch H (1985) J Biol Chem 260:10930–10935
36. Peculis BA & Steitz JA (1993) Cell 73:1233–1245
37. Shanab GM & Maxwell ES (1991) Nucleic Acids Res 19:4891–4894
38. Liu J & Maxwell ES (1990) Nucleic Acids Res 18:6565–6571
39. Leverette R, Andrews M & Maxwell ES (1992) Cell 71:1215–1221
40. Fragapane P, Prislei S, Michienzi A, Caffarelli E & Bozzoni I (1993) EMBO J 12:2921–2929
41. Wu Z, Murphy C, Wu H, Tsvetkov A & Gall JG (1993) Cold Spring Harbor Symposia on Quantitative Biol LVIII (in press)

42. Christensen ME & Banker N (1992) Cell Biol Int Rep 16:1119–1131
43. Tycowski KT, Shu MD & Steitz JA (1993) Genes Dev 7:1176–1190

Manual of Biological Markers of Disease **B5.4**: 1–17, 1994.
© 1994 *Kluwer Academic Publishers. Printed in the Netherlands.*

The PM/Scl antigens

FRIEDLINDE A. BAUTZ AND MARTIN BLÜTHNER*
*Institute of Molecular Genetics, University of Heidelberg, Im Neuenheimer Feld 230, D-69120
Heidelberg, Germany; * Present Address: W.M. Keck Autoimmune Disease Center, Scripps
Clinic and Research Foundation, 10666 North Torrey Pines Rd., LaJolla, Ca 92037, U.S.A.*

Abbreviations

CTNE: Calf thymus nuclear extract; PM/Scl: Polymyositis-Scleroderma;
ELISA: Enzyme-linked immunosorbant assay; UTR: Untranslated region

Key words: autoantibodies, autoantigens, nucleolus, granular component,
multiprotein particle, PM/Scl

1. The antigen

The polymyositis/scleroderma, i.e. PM/Scl antigen is a complex of about 11
proteins of which two polypeptides, PM/Scl-100 and PM/Scl-75, are the
major antigenic targets [1, 5, 9–11].

2. Synonyms

PM-1 detected by double immunodiffusion using crude calf thymus nuclear
extract (CTNE) [2], was shown to be identical to the PM/Scl antigen [4].

3. Antibody detection

3.1 Immunofluorescence

Detection of PM/Scl antibodies can be done by indirect immunofluorescence
on human HeLa and Hep 2 cells [3–7, 12]. In all instances a strong nucleolar
fluorescence often accompanied by weak nucleoplasmic staining has been
reported (Fig. 1).

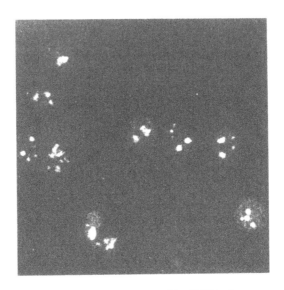

Fig. 1. Indirect immunofluorescence on Hep2 cells with a PM/Scl reference serum.
A strong nucleolar staining of the nucleolus can be observed, occasionally accompanied by a weak staining of the nucleoplasm. For technical details, see this Manual chapter A2.

3.2 Counterimmunoelectrophoresis

The PM/Scl autoantibody activity can be detected by counterimmuno-electrophoresis as discussed in this Manual chapter A3 by C. Bunn and T. Kveder.

3.3. Immunoblotting

In their attempt to identify the antigenic polypeptide(s) of the PM/Scl complex, Reimer *et al.*, using whole HeLa cell proteins as well as isolated nucleolar proteins in immunoblots, found only one protein of approximately 80,000 kD to be reactive with some PM/Scl sera [5]. From these experiments they concluded that the identity of the antigenic component of the PM/Scl complex had not been unequivocally ascertained. In immunoblotting experiments, with the use of highly purified nucleolar extracts from HeLa S3 cells, several investigators identified two polypeptides of 100,000 and 75,000 kD as the main antigenic components of the PM/Scl complex, whereas with some sera an additional reactivity with a nucleolar protein of 80,000 kD was observed [7, 9–12] (Fig. 2).

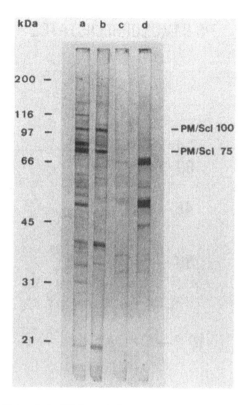

Fig. 2. Western blot analysis of HeLa nucleolar extracts.
Proteins were separated on 12.5% SDS polyacrylamide gels and probed on nitrocellulose with two typical PM/Scl sera (lanes a and b). The major antigens PM/Scl-100 and PM/Scl-75 are indicated. The PM/Scl serum used in lane a shows an additional immunoreactivity with a 80 kD antigen. In lanes c and d a normal human serum and a PBC serum (primary billiary cirrhosis), respectively, are used as controls. No cross-reactivity between the two major immunoreactive peptides PM/Scl 100 and PM/Scl 75 could be observed with affinity-purified antibodies, leading to the conclusion that they represent two distinct protein entities (Haidmann, diploma thesis). This finding was recently confirmed by sequence-comparision of the cloned c-DNAs of the 100,000 and 75,000 kD autoantigens of the PM/Scl complex by Ge *et al.*, Blüthner & Bautz and Alderuccio *et al.* [9–11]. For details of the Western blot procedure, see this Manual chapter A4: Protein Blotting by R. Verhijen, M. Salden and W.J. van Venrooij.

3.4 Immunoprecipitation

On the molecular level characterization of the PM/Scl antigen was first carried out by Reimer *et al.* in 1986 by the use of the immunoprecipitation technique [5]. Protein-A facilitated immunoprecipitation with 6 PM/Scl sera using $[^{35}S]$ methionine-labeled crude HeLa cell extract, followed by SDS-

³⁵S-IMMUNOPRECIPITATES

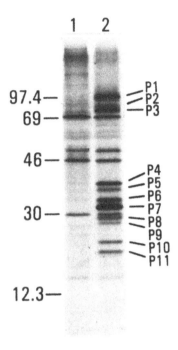

Fig. 3. PM/Scl immunoprecipitation pattern.
Autoradiogram of [³⁵S] methionine-labeled HeLa cell proteins immunoprecipitated by anti-PM/Scl antibodies and resolved in 17.5% SDS-polyacrylamide gel. For details of the procedure see [5] and this Manual part A, section 6 by E.M. Tan and C.L. Peebles. Lane 1 shows proteins precipitated by normal control serum. Lane 2 demonstrates the radiolabeled proteins from HeLa cells that are selectively brought down by anti-PM/Scl antibodies and are named P1–P11. Data kindly provided by Dr. G. Reimer.

PAGE analysis showed that all sera with PM/Scl specificity (characterized by double immunodiffusion) precipitated at least 11 polypeptides (P1-P11) with kD values of: 110, (100), 90, 80, (75), 39, 37, 33, 30, 27, 26, 22, 20 (Fig. 3). Immunoprecipitation of (³²P) orthophosphate labeled HeLa cells with PM/Scl sera showed that only the 80,000 (P3) and 20,000 kD (P11) polypeptides were phosphorylated. Upon urea-polyacrylamide gel electrophoresis of the immunoprecipitation reaction RNA was not found to be part of the PM/Scl complex. These data, albeit with minor alterations as to the protein components (kD's) of the PM/Scl complex, were subsequently confirmed by other investigatiors [7, 8, 12].

From the combined data obtained with the cellular and recombinant antigens we can assume that antibody to the 100-kD protein and, to a lesser degree, to the 75-kD protein of the PM/Scl complex represent the most common antibody specificity found so far for the disease [7, 9–11].

3.5 Elisa

No data on the detection of PM/Scl antibody by ELISA are available so far.

4. Cellular localization

4.1 Light microscopy

The main cellular localization of the PM/Scl antigen resides in the nucleolus, although weak nucleoplasmic staining can be observed occasionally by indirect immunofluorescence [3–7] (see Fig. 1). Reimer *et al.*, 1986 found a homogenous nucleolar staining in interphase cells of HeLa and rat kangaroo PTK2 cells with anti-PM/Scl sera. In metaphase cells, diffuse staining was observed in the area around the condensed chromosomes, whereas in anaphase cells, anti-PM/Scl staining was diffusely distributed around the chromosomes and in the perichromosomal cell plasm [5] (Fig. 4).

4.2 Electron microscopy

To study the subnucleolar location of the PM/Scl antigen electron microscopic immunocytochemistry on hepatocytes of regenerating rat liver was performed with the immunogold label technique [5]. It was found that the gold-antibody-antigen complexes were selectively enriched over the granular component and are seen at an even higher density in the periphery of the nucleoli. Immunogold was virtually absent in the dense fibrillar component or fibrillar centers (Fig. 5). Although gold-antibody complexes were also observed in the nucleoplasm, a finding which is supported by indirect immunofluorescence observations, one is led to conclude that the nucleolus is the main localization of the PM/Scl antigen.

5. Cellular function

Although the 100 kD and 75 kD PM/Scl antigens have been cloned and sequenced nothing is known so far about the structure and function of the PM/Scl complex. No identity on the DNA or protein level with known eukaryotic proteins in several data bases was found for the PM/Scl antigens.

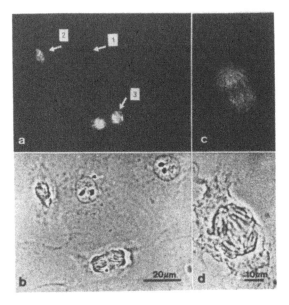

Fig. 4. Cell cycle-dependent localization of PM/Scl antigens.
Immunolocalization of anti-PM/Scl antibodies in rat-kangaroo (PtK 2) cells during different phases of the cell cycle as visualized by immunofluorescence microscopy (a and c) and phase-contrast (b and d). In interphase cells nucleolar staining by anti-PM/Scl antibodies can be observed (arrow 1). During metaphase the nucleolar staining relocates as the chromosomes condense giving rise to diffuse staining in the area of the condensed chromosomes (arrow 2). In late anaphase cells PM/Scl staining is diffusely distributed on the surface of the chromosomes and in the perichromosomal cell plasm (arrow 3). Individual chromosomes can be distinguished in the phase contrast optics (d) as elongated lighter structures surrounded by dark borders. Data kindly provided by Dr. G. Reimer.

The available data, such as their localization in the granular region of the nucleolus [5], their weak presence in the nucleoplasm and the apparent migration of nucleolar staining to the nucleoplasm after actinomycin D treatment [5, 7] suggest a possible association with ribosomal biogenesis [5, 9].

Furthermore, database comparision of the cDNA-derived amino acid sequence revealed close homology of the peptide stretch LHAKNIIRPQLK (amino acid position 175–186) of the PM/Scl-100 antigen to a potential consensus sequence, LHAKNIIHRDLK, of serine/threonine kinases, which are known to be involved in regulatory processes [10].

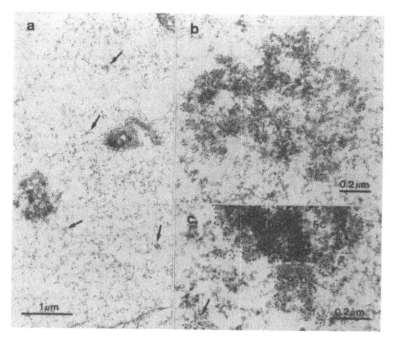

Fig. 5. Sub-nucleolar localization of PM/Scl antigens.
Immunolocalization of PM/Scl in sub-nucleolar compartments in the hepatocytes of regenerating rat liver was shown by the electron microscopic immunogold labeling technique.
a. Sections of two nucleoli and surrounding heterochromatin at low magnification.
b and c. Two sections of nucleoli at high magnification. The 5 mm gold particles (small black dots) indicating gold-antibody-antigen complexes are selectively enriched over the granular component (GC) and are distributed at higher density in the periphery of these nucleoli. The dense fibrillar component (DFC) and the fibrillar centers (FC) are free of gold particles.
Note that gold-antibody-antigen complexes are also present in clusters throughout the heterochromatin structures as indicated by arrows in (a) and (c). Data kindly provided by Dr. G. Reimer.

6. Presence in other species

The presence of the PM/Scl antigens as observed by indirect immuno-fluorescence has been reported for Hep2, HeLa, KB cells, T-cell lymphoma Molt 4 cells, rat RV cells, rat kangaroo PTK2 cells, fish FHM cells, turtle TH-1 cells, duck (Peking ducks), embryonic cells and rat liver cryostat sections [5]. These data suggest that epitopes recognized by anti-PM/Scl antibodies are well conserved in evolution.

-37   <u>ACAAGCTCTCGCGAGACGAGCCGTGCAGGCTGAAAA</u>

```
      ATGGCGCCACCCAGTACCCGGGAGCCCAGGGTCCTGTCGGCGACCAGCGCAACCAAATCCGACGGAGAGATGGTGCTGCCAGGCTTCCCGGACGCCGACAGCTTTGTGAAGTTTGCTCTTGGGTCCGTGGTGGCA
   1  M  A  P  P  S  T  R  E  P  R  V  L  S  A  T  S  A  T  K  S  Q  G  E  M  V  L  P  G  F  P  D  A  D  S  F  V  K  F  A  L  G  S  V  V  A

 136  GTCACCAAGGCATCTGGGGGCCTACCACAGTTTGGCGATGAGTATGATTTTTACCGAAGTTTTCCTGGCTTCCAAGCATTTTGCGAAACACAGGGAGACAGGTTGCTTCAGTGCATGAGCAGAGTAATGCAGTAC
  46  V  T  K  A  S  G  G  L  P  Q  F  G  D  E  Y  D  F  Y  R  S  F  P  G  F  Q  A  F  C  E  T  Q  G  D  R  L  L  Q  C  M  S  R  V  M  Q  Y

 271  CATGGGTGTCGCAGCAACATTAAGGATCGAAGTAAAGTGACTGAGCTGGAAGACAAGTTTGATTTACTAGTTGATGCCAATGATGTAATTCTGGAGAGAGTGGGTATTTTACTGGATGAAGCCTCAGGTGTAAAC
  91  H  G  C  R  S  N  I  K  D  R  S  K  V  T  E  L  E  D  K  F  D  L  L  V  D  A  N  D  V  I  L  E  R  V  G  I  L  L  D  E  A  S  G  V  N

 406  AAGAATCAACAGCCTGTCCTCCCTGCCGGCTTGCAGGTCCCCAAACGGGTAGTGTCCAGCTGGAACCGTAAGGCAGCAGAATATGGCAAAAAGCAAAATCTGAAACTTTCCGGCTGCTTCATGCAAAAAATATC
 136  K  N  Q  Q  P  V  L  P  A  G  L  Q  V  P  K  T  V  V  S  S  W  N  R  K  A  A  E  Y  G  K  K  A  K  S  E  T  F  R  L  L  H  A  K  R  I

 541  ATCCGACCTCAGCTCAAGTTTCGAGAGAAGATTGACAATTCCAACACACATTTCTTCCTAAAATCTTCATCAAACCCAATGCTCAGAAACCTCTCCCTCAAGCTCTCTCTAAGGAAAGGCGGGAACGCCCACAG
 181  I  R  P  Q  L  K  F  R  E  K  I  D  N  S  N  T  P  F  L  P  K  I  F  I  K  P  N  A  Q  K  P  L  P  Q  A  L  S  K  E  R  R  E  R  P  Q

 676  GATCGTCCTGAGGACTTGACGTCCCCCCTGCACTGGCTGATTTCATCCATCAGCAGAGAACCCAGCAGGTTGACAAGACATGTTTGCACATCCTTATCAATATGAACTAAATCACTTTACCCCAGCAGATGCA
 226  D  R  P  E  D  L  D  V  P  P  A  L  A  D  F  I  H  Q  Q  R  T  Q  Q  V  E  Q  D  M  F  A  H  P  Y  Q  Y  E  L  N  H  F  T  P  A  D  A

 811  GTGCTTCAAAAGCCACAACCCCAGTTATACAGACCTATAGAAGAGACACCATGCCATTTCATATCCTCCCTGGATGAACTCGTGGAACTCAACGAAAAGCTCTTGAATTGTCAGGAATTTGCAGTTGACTTGGAG
 271  V  L  Q  K  P  Q  P  Q  L  Y  R  P  I  E  E  T  P  C  H  F  I  S  S  L  D  E  L  V  E  L  N  E  K  L  L  N  C  Q  E  F  A  V  D  L  E

 946  CACCACTCTTACAGGAGCTTCCTGGGACTGACCTGCCTGATGCAAATTTCTACTCGGACGGAAGACTTCATCATTGACACCCTCGAGCTTGGAAGTGACATGTACATTCTCAATGAGAGCCTCAGCACAGACCAGCC
 316  H  H  S  Y  R  S  F  L  G  L  T  C  L  M  Q  I  S  T  R  T  E  D  F  I  I  Q  T  L  E  L  R  S  D  M  Y  I  L  N  E  S  L  T  D  P  A

1081  ATCGTTAAGGTCTTTCATGGTGCTGATTCAGACATAGAATGGCTACAGAAAGACTTTGGGTTGTATGTAGTAAACATGTTTGATACTCATCAGGCAGCACGCCTTCTTAACCTGGGCAGGCACTCACTCGATCAT
 361  I  V  K  V  F  H  G  A  D  Q  S  D  I  E  W  L  Q  K  D  F  G  L  Y  V  V  N  M  F  D  T  H  Q  A  A  R  L  L  N  L  G  R  H  S  L  D  H

1216  CTCCTGAAACTCTACTGCAACGTGGACTCAAACAAGCAAATATCAGCTGGCTGATTGGAGAATAGGCCCTCTGCCCGAGGAGATGCTCAGCTACGCCCGGGATGACACCCATTACCTGCTATATATCTATGACAAA
 406  L  L  K  L  Y  C  N  V  D  S  N  K  Q  Y  Q  L  A  D  W  R  I  R  P  L  P  E  E  M  L  S  Y  A  R  D  D  T  H  Y  L  L  Y  I  Y  D  K

1351  ATGAGGCTGGAGATGTGGGAGCGCGGCAACGGGCAGCCCGTGCAGCTCCAGCAGGTGGTGGCAACGGAGCAGGGACATCTGCCTCAAGAAATTCATCAAACCTATCTTCAGGATGAGTCCTACCTTGAACTCTAT
 451  M  R  L  E  M  W  E  R  G  N  G  Q  P  V  Q  L  Q  V  V  W  Q  R  S  R  D  I  C  L  K  K  F  I  K  P  I  F  I  D  E  S  Y  L  E  L  Y

1481  AGGAAGCAGAAGAAGCACCTTAACACACAGCAATTGACAGCCTTTCAGCTGCTGTTTGCCTGGAGGGATAAAACAGCTCGCAGGGAAGATGAAAGTTACGGATATGTACTGCCAAACCACATGATGCTGAAAATA
 496  R  K  Q  K  K  H  L  N  T  Q  Q  L  T  A  F  Q  L  L  F  A  W  R  D  K  T  A  R  R  E  D  E  S  Y  G  Y  V  L  P  N  H  M  M  L  K  I

1621  GCTGAAGAACTGCCTAAGGAACCTCAGGGCATCATAGCTTGCTGCAACCCAGTACCGCCCCTTGTGCGGCAGCAGATCAACGAAATGCACCTTTTAATCCAGCAGGCCCGAGAGATGCCCCTGCTCAAGTCTGAA
 541  A  E  E  L  P  K  E  P  Q  G  I  I  A  C  C  N  P  V  P  P  L  V  R  Q  Q  I  N  E  M  H  L  L  I  Q  Q  A  R  E  M  P  L  L  K  S  E

1756  GTTGCAGCCGGAGTGAAGAAGAGCGGACCGCTGCCCAGTGCTGAGAGATTGGAGAATGTTCTCTTTGGACCTCACGACTGCTCCCATGCCCTCCGGATGGCTATCCAATCATCCCAACCAGTGGATCTGTGCCA
 586  V  A  A  G  V  K  K  S  G  P  L  P  S  A  E  R  L  E  N  V  L  F  G  P  H  D  C  S  H  A  P  P  D  G  Y  P  I  I  P  T  S  G  S  V  P

1891  GTTCAGAAGCAGGCGAGCCTCTTCCCTGATGAAAAGAAGATAATGCTTGCTGGGTACCACATGCCTGATTGCCACAGCTGTCATCACGTTATTTAATGAACCTAGTGCTGAAGACAGTAAAAGGGTCCATTGACA
 631  V  Q  K  Q  A  S  L  F  P  D  E  K  E  D  N  L  L  G  T  T  C  L  I  A  T  A  V  I  T  L  F  N  E  P  S  A  E  D  S  K  K  G  P  L  T

2026  GTTGCACAGAAAAAAGCCCAGAACATCATGGAGTCCTTTGAAAATCCATTTAGGATGTTTCTGCCCTCACTGGGACACCGTGCTCCCGTCTCTCAGGCAGGGAAGTTCGATCCATCAACCAAATCTATGAAATC
 676  V  A  Q  K  K  A  Q  H  I  M  E  S  F  E  N  P  F  R  M  F  L  P  S  L  G  H  R  A  P  V  S  Q  A  A  K  F  D  P  S  T  K  I  Y  E  I

2161  AGCAACCGTTGGAAGCTGGCCCAGGTACAAGTACAAAAGACTCTAAAGAAGCTGTCAAGAAGAAGGCAGCTGAGCAAACAGCTGCCCGAGGACAGGCAAAGGAGGCGTGCAAAGTGCAGCAGAACAGGCCATC
 721  S  N  R  W  K  L  A  Q  V  Q  V  Q  K  D  S  K  E  A  V  K  K  K  A  A  E  Q  T  A  A  R  E  Q  A  K  E  A  C  K  A  A  A  E  Q  A  I

2296  TCCGTCCGACAGCAGGTCGTGCTAGAAAATGCTGCAAAGAAGAGAGAGCGAGCAACAAGCGACCCAAGGACCACAGAACAGAAACAAGAGAAGAAACGACTCAAAAATTTCCAAGAAGCCAAAGGACCCAGAGCCA
 766  S  V  R  Q  Q  V  V  L  E  N  A  A  K  K  R  E  R  A  T  S  Q  P  R  T  T  E  Q  K  Q  E  K  K  R  L  K  I  S  K  K  P  K  D  P  E

2431  CCAGAAAAAGAGTTTACGCCTTACGACTACAGCCAGTCAGACTTCAAGGCTTTTGCTGGAAACAGCAAATCCAAAGTTTCTTCTCAGTTTGATCCAAATAAACAGACCCCGTCTGGCAAGAAATGCATTGCAGCC
 811  P  E  K  E  F  T  P  Y  D  Y  S  Q  S  D  F  K  A  F  A  G  N  S  K  S  K  V  S  S  Q  F  D  P  N  K  Q  T  P  S  G  K  K  C  I  A  A

2566  AAAAAAATTAAACAGTCGGTGGGAAACAAAAGCATGTCCTTTCCAACTGGAAAGTCAGACAGAGGCTTCAGGTACAACTGGCCACAGAGA<u>TAG</u>TCTGGAAGACACGTGGCGCCTGTGGACCGGAAGCACCAAAT
 856  K  K  I  K  Q  S  V  G  N  K  S  M  S  F  P  T  G  K  S  D  R  G  F  R  Y  N  W  P  Q  R

2701  GCTGGTGCTGCTTTTGTACATACATATTTTTAAACCATT<u>AAAA</u>TTCTTCCTGAAAAAAAAAAAAAAAAAAAAAAAAAAAAAAAAAAAAAAA
```

AMAN-B5.4/8

7. CDNAs and derived amino acid sequences

Two proteins of 100 kD and 75 kD termed PM/Scl-100 and PM/Scl-75 are known so far to be the major autoantigenic targets within the PM/Scl-complex [9–11]. The cDNAs for these two proteins have been cloned and analyzed recently [9–11].

7.1 PM/Scl-100

The cDNA coding for the PM/Scl-100 antigen has been cloned and sequenced independently from HeLa and placenta λ gt11 libraries by Ge *et al.* and Blüthner and Bautz respectively [11, 10] (Fig. 6). Since these two cDNA sequences differ in one interesting aspect, we shall be referring to them as PM/Scl-100a (Ge *et al.*) and PM/Scl-100b (Blüthner and Bautz).

The PM/Scl-100a cDNA consists of 2718 bp with 2580 bp in the coding region and encodes 860 amino acids with a deduced molecular mass of 98.088 kD [11] (Fig. 7).

The PM/Scl-100b cDNA with an overall length of 2791 bp excluding the polyA-tail, has been isolated using affinity-purified anti-PM/Scl-100 antibodies. The 2655-bp long coding region encodes 885 amino acids with a deduced molecular mass of 100.8 kD and a deduced pI-value of 8.57 [10] (Fig. 7).

7.2 PM/Scl-75

The cDNA coding for the PM/Scl-75 antigen has been cloned and sequenced from a MOLT-4 Lambda gt 11 library by Alderuccio *et al.* [9]. The 1562-bp long cDNA sequence consists of a 419 bp 5′-UTR, a 1065-bp coding region, and a 75-bp 3′-UTR (Fig. 8). Northern blotting experiments identified a corresponding mRNA of about 1.6 kb. The PM/Scl-75 cDNA encodes 355 amino acids with a deduced molecular weight of 39.19 kD, which is strikingly different from the 75-kD value obtained in immunoblots.

This aberrant migration of the PM/Scl-75 protein in SDS-PAGE was analyzed in detail using truncated and *in vitro* translated cDNA fragments of

←

Fig. 6. cDNA sequence of PM/Scl-100b clone H2/8 and deduced amino acid sequence of the encoded nucleolar PM/Scl-100b antigen [10].

Complete cDNA sequence of PM/Scl-100b clone H2/8. The deduced amino acid sequence is displayed below the DNA sequence.The start and termination codons and the potential polyadenylation signal (ATTAAA) are underlined. Double-underlined sequences show the additional cDNA sequence obtained by reverse transcription followed by tailing, PCR, and direct sequencing of the resulting single-stranded PCR product. The sequence data are available from the EMBL Data bank under accession number X66113.

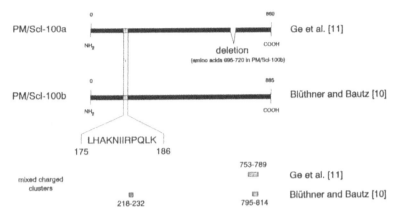

deduced amino acid sequences

Fig. 7. Analysis of the cDNA-derived amino acid sequences of the PM/Scl-100a and PM/Scl-100b antigens.

The cDNA-derived amino acid sequences of the PM/Scl-100a and PM/Scl-100b antigens consist of 860 or 885 amino acid residues, respectively. Comparision of the two cDNA-sequences (PM/Scl-100a and PM/Scl-100b) [10, 11] shows that they exactly match each other except that 75 nucleotides (positions 2082 to 2156 of the PM/Scl-100b cDNA) are deleted in the PM/Scl-100a cDNA sequence. This in-frame deletion results in an internal 25 amino acid loss in the PM/Scl-100a protein sequence, which corresponds to amino acids 695 to 720 of PM/Scl-100b [11]. Note the resulting different numbering of nucleotides and amino acids respectively starting from amino acid 695 (nucleotide position 2082). The amino acids LHAKNIIRPQLK (amino acids 175–186) showing close homology to a potential consensus sequence of serine/threonine kinases, LHAKNIIHRDLK, are indicated [10].

On further analysis of the deduced amino acid sequence of the PM/Scl-100a and PM/Scl-100b antigens, regions of so-called "mixed charged clusters" were reported (PM/Scl-100a: amino acid position 753–789; PM/Scl-100b: amino acid positions 218–232 and 795–814) which are displayed below. Such mixed clusters of charged amino acids have been reported to be a common feature of nuclear antigens, although their significance in the context of the autoimmune response remains unclear [13].

→

Fig. 8. cDNA sequence and deduced amino acid sequence of the encoded nucleolar PM/Scl-75 [9].

Complete nucleotide and deduced amino acid sequence of PM/Scl-75 polypeptide. Underlined are the first upstream in frame stop codon (TGA) at nucleotide −69, the putative nuclear or nucleolar localization signal KRRKKKRA starting at amino acid 346, and the polyadenylation signal ATTAAA at nucleotide 1099. These sequence data are available from EMBL/Genbank/DDBJ under accession number M58460.

```
-419 CGACCGGCACGTTCACCCCATCCCTCAGGCTTTATTTATTTTTTTTCGACACGGTTCTTTTCAAGG
     CTCCAGTCACCGCAGCAGTTGTCCATGCTGTAGTTTCCACTTTCCTGTATGGGCGGGCTGGTTAGG
     ATTCCACTTTCCCCCAAGTGCTTAGCCCAGGGCCAGACAAAAAGTAGTTGCTTAAGAAATACTTGT
     TGAAGGAATAAATTAATGAATGAATTTGTGCTTACAGCGGCTGGATGGCAGACAAACCTATGATTA
     TAGGAACATCACGATCTCATTTGGAACAGATTACGGATGCTGCATTGTGGAACTTGGAAAAACAAG
     AGTTCTTGGACAGGTTTCCTGTGAACTTGTGTCTCCAAAACTCAATCGGGCAACAGAAGGTATTCT
     TTTTTTAACCTTGAACTCTCTCAG
   1 ATGGCCCCTCCAGCTTTCGAACCTGGCAGGCAGTCAGATCTCTTGGTGAAGTTGAATCGACTCATG 66
   1 M  A  A  P  A  F  E  P  G  R  Q  S  D  L  L  V  K  L  N  R  L  M  22

  67 GAAAGATGTCTAAGAAATTCGAAGTGTATAGACACTGAGTCTCTCTGTGTTGTTGCTGGTGAAAAG 132
  23 E  R  C  L  R  N  S  K  C  I  D  T  E  S  L  C  V  V  A  G  E  K  44

 133 GTTTGGCAAATACGTGTAGACCTACATTTATTAAATCATGATGGAAATATTATTGATGCTGCCAGC 198
  45 V  W  Q  I  R  V  D  L  H  L  L  N  H  D  G  N  I  I  D  A  A  S  66

 199 ATTGCTGCAATCGTGGCCTTATGTCATTTCCGAAGACCTGATGTCTCTGTCCAAGGAGATGAAGTA 264
  67 I  A  A  I  V  A  L  C  H  F  R  R  P  D  V  S  V  Q  G  D  E  V  88

 265 ACACTGTATACACCTGAAGAGCGTGATCCTGTACCATTAAGTATCCACCACATGCCCATTTGTGTC 330
  89 T  L  Y  T  P  E  E  R  D  P  V  P  L  S  I  H  H  M  P  I  C  V  110

 331 AGTTTTGCCTTTTTCCAGCAAGGAACATATTTATTGGTGGATCCCAATGAACGAGAAGAACGTGTG 396
 111 S  F  A  F  F  Q  Q  G  T  Y  L  L  V  D  P  N  E  R  E  E  R  V  132

 397 ATGGATGGCTTGCTGGTGATTGCCATGAACAAACATCGAGAGATTTGTACTATCCAGTCCAGTGGT 462
 133 M  D  G  L  L  V  I  A  M  N  K  H  R  E  I  C  T  I  Q  S  S  G  154

 463 GGGATAATGCTACTAAAAGATCAAGTTCTGAGATGCAGTAAAATCGCTGGGGTGAAAGTAGCAGAA 528
 155 G  I  M  L  L  K  D  Q  V  L  R  C  S  K  I  A  G  V  K  V  A  E  176

 529 ATTACAGAGCTAATATTGAAAGCTTTGGAGAATGACCAAAAAGTAAGGAAAGAAGGTGGAAAGTTT 594
 177 I  T  E  L  I  L  K  A  L  E  N  D  Q  K  V  R  K  E  G  G  K  F  198

 595 GGTTTTGCAGAGTCTATAGCCAAATCAAAGGATCACAGCATTTAAAATGGAAAAGGCCCCTATTGAT 660
 199 G  F  A  E  S  I  A  N  Q  R  I  T  A  F  K  M  E  K  A  P  I  D  220

 661 ACCTCGGATGTAGAAGAAAAAGCAGAAGAAATCATTGCTGAAGCAGAACCTCCTTCAGAAGTTGTT 726
 221 T  S  D  V  E  E  K  A  E  E  I  I  A  E  A  E  P  P  S  E  V  V  242

 727 TCTACACCTGTGCTATGGACTCCTGGAACTGCCCAAATTGGAGAGGGAGTAGAAAACTCCTGGGGT 792
 243 S  T  P  V  L  W  T  P  G  T  A  Q  I  G  E  G  V  E  N  S  W  G  264

 793 GATCTTGAAGACTCTGAGAAGGAAGATGATGAAGGCGGTGGTGATCAAGCTATCATTCTTGATGGT 858
 265 D  L  E  D  S  E  K  E  D  D  E  G  G  G  D  Q  A  I  I  L  D  G  286

 859 ATAAAAATGGACACTGGAGTAGAAGTCTCTGATATTGGAAGCCAAGATGCTCCCATAATACTCTCA 924
 287 I  K  M  D  T  G  V  E  V  S  D  I  G  S  Q  D  A  P  I  I  L  S  308

 925 ,GATAGTGAAGAAGAAGAAATGATCATTTTGGAACCAGACAAGAATCCAAAGAAAATAAGAACACAG 990
 309 D  S  E  E  E  M  I  I  L  E  P  D  K  N  P  K  K  I  R  T  Q  330

 991 ACCACCAGTGCAAAACAAGAAAAAGCACCAAGTAAAAAGCCAGTGAAAAGAAGAAAAAAGAAGAGA 1056
 331 T  T  S  A  K  Q  E  K  A  P  S  K  K  P  V  K  R  R  K  K  K  R  352

1056 GCTGCCAATTAA 1068
 353 A  A  N  *  355
     GCTAACAGTTGTATATCTGTATATATAACTATTAAAAGGGATATTTATTCCATTAAAAAAAAAAAA
     AAAAAAAAA 1143
```

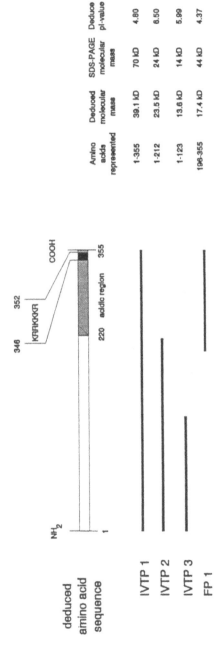

	Amino acids represented	Deduced molecular mass	SDS-PAGE molecular mass	Deduce pI-value
IVTP 1	1-355	39.1 kD	70 kD	4.80
IVTP 2	1-212	23.5 kD	24 kD	6.50
IVTP 3	1-123	13.6 kD	14 kD	5.99
FP 1	196-355	17.4 kD	44 kD	4.37

Fig. 9. Analysis of the cDNA derived PM/Scl-75 amino acid sequence.
The 1065 bp open reading frame of PM/Scl-75 codes for 355 amino acids with a deduced molecular mass of 39.1 kD in contrast to the observed molecular mass of 70 k–75 kD [9]. This discrepancy is due to a highly acidic region from amino acids 220 to 355 as indicated. Different truncated *in vitro* translation products (IVTP 1–3) and a fusion protein of the C-terminus (FP 1) were analysed in SDS-PAGE in order to determine the region on the deduced amino acid sequence responsible for the aberrant migration of the PM/Scl-75 antigen. The main reason for the aberrant migration of the PM/Scl-75 antigen in SDS-PAGE seems to be the negative net charge of the protein in this region. A stretch of 101 amino acid residues beginning at position 220 is composed almost exclusively of acidic amino acids. It is not clear how this acidic region is involved in the migration of the PM/Scl-75 autoantigen in SDS-PAGE but regions of high negative charge are thought to interfere with SDS binding and thus influence migration in SDS gels [9].
The corresponding amino acids, the deduced molecular masses, the observed molecular masses and the deduced pI-values are displayed together with the relative positions of the fragments analysed. An area of basic residues beginning at position 346 (KRRKKKR) of PM/Scl-75, that may encode a nuclear/nucleolar localization signal resembles the nuclear localization signal KKKRK of SV-40 large T antigen and the molecular signal RKKRRQRRRA found in the HIV tat protein.

the protein (Fig. 9). It could be shown that the C-terminal region of the protein (amino acids 196–355) seems to be responsible for this phenomenon.

8. The genes

Until now, nothing has been known about the chromosomal localization and the genomic organization of the PM/Scl-100 and PM/Scl-75 genes.

9. Relation with nucleic acids

As described before (section 4), the PM/Scl antigens are part of a multiprotein complex localized in the nucleolus. The nucleolus itself consists of a high proportion of nucleic acids, since it is the subcellular compartment where rRNA processing and ribosome-biogenesis and –maturation takes place. Furthermore the nucleolus is attached to DNA at the NORs. Therefore, it seems feasible that nucleic acids are somehow involved in the architecture of the PM/Scl particle. However, Wolfe et al. reported that when calf thymus nuclear extracts are pretreated with RNAse or DNAse, no change in the precipitin pattern could be observed in immunodiffusion [2]. This observation finds some support in the studies published by Reimer et al. where they investigated the influence of RNAseA and DNAseI in indirect immunofluorescence. They found that neither RNAse nor DNAse pretreatment of HeLa cells altered the nucleolar staining pattern. In addition no co-precipitation of RNA could be detected by polyacrylamide/urea gel analysis in immunoprecipitation assays [5, 8, 12].

In contrast Gelpi et al. report a partial sensitivity of the PM/Scl particle to Dnase I digestion [7]. They found that preincubation of HeLa cell extracts with Dnase I prior to immunoprecipitation resulted in an apparent depletion especially of the high molecular weight proteins (110, 100, 90, 80 and 20 kD) of the PM/Scl complex.

10. Relation with other proteins

Since the major nucleolar PM/Scl antigens (PM/Scl-100 and PM/Scl-75) are part of a multiprotein complex (section 3) the relationship of the non-immunogenic proteins of this particle with the antigenic proteins has to be a subject for further investigations.

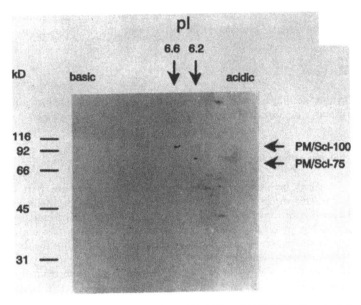

Fig. 10. Determination of the pI-values of the PM/Scl-100 and PM/Scl-75 antigens by 2-D gel electrophoresis.
HeLa nucleolar extracts were analyzed by isoelectric focussing followed by SDS-polyacrylamide gel electrophoresis and Western blot. For the PM/Scl-100 antigen a pI-value of 6.6 and for the PM/Scl-75 antigen a pI-value of 6.2 was determined (arrows) [Haidmann, Diploma thesis].

11. Biochemical characteristics

Sucrose gradient analysis and subsequent immunoblotting of the sucrose gradient fractions with a PM/Scl reference serum revealed a sedimentation coefficient of about 20 S for the PM/Scl particle (Haidmann, diploma thesis, data not shown). In the course of these studies pI-values of 6.6 and 6.2 for the two major antigens, PM/Scl-100 and PM/Scl-75 were determined by isoelectric focussing followed by SDS-PAGE and western blot (Fig. 10). These pI-values differ from the values calculated from the cDNA deduced amino acid sequences (see section 7) which are 8.57 and 4.8 respectively. Further experiments have to be performed to elucidate these discrepancies.

12. B-cell epitopes

The distribution of epitopes on the recombinant PM/Scl-100b antigen has been studied in some detail by Blüthner & Bautz and is presented in Figure 11 [10].

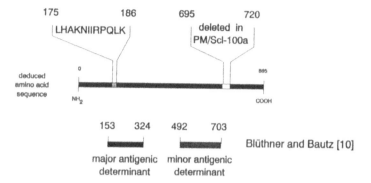

Fig. 11. Epitope distribution on the PM/Scl-100b antigen.
Defined cDNA fragments of the PM/Scl-100b were expressed and their immunoreactivities with several PM/Scl autoantisera were analysed in Western blots. All fifteen sera tested reacted with a peptide representing the amino acids 153 to 324 suggesting this part contains a major antigenic determinant. Four out of these fifteen sera additionally reacted with a peptide fragment representing the amino acids 492 to 703 of the PM/Scl-100b antigen leading to the conclusion that within this region a minor antigenic determinant is localized. It is interesting that the major antigenic determinant covers the amino acids LHAKNIIRPQLK which show close homology to the potential consensus sequence LHAKNIIHRDLK of serine/threonine kinases (see Fig. 7). Furthermore, the minor antigenic determinant partially overlaps with the region deleted in the sequence of PM/Scl-100a (see Fig. 7).

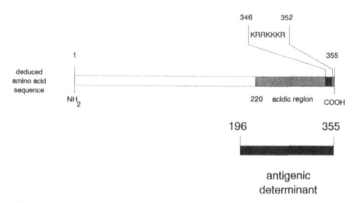

Fig. 12. Localization of an antigenic determinant on the PM/Scl-75 sequence.
The initially isolated cDNA clone represents the amino acids 196 to 355 of PM/Scl-75. The corresponding fusion protein was able to completely remove the anti-PM/Scl-75 activity from a PM/Scl autoantiserum leading to the conclusion that within this region an antigenic determinant must be localized [9]. In addition this region represents the highly acidic region responsible for the aberrant migration of the PM/Scl-75 antigen in SDS-PAGE and contains similarities to the nuclear localization signal of the SV-40 large T antigen and the HIV tat protein (see Fig. 9).

A peptide fragment of 138 amino acids (residues 218–355) containing at least one antigenic determinant shown in Figure 12 could be identified for the PM/Scl-75 antigen by Alderuccio *et al.* [9].

In summary, the antigenic determinants within the nucleolar PM/Scl-100 and the PM/Scl-75 autoantigens seem to be identical or in close proximity to potential functional domains of the proteins. These findings are in support of a hypothesis forewarded by Tan which suggests that immunogenic regions of autoantigens are often located within functional protein domains [14].

13. T-cell epitopes

Nothing is known so far about T-cell epitopes for the PM/Scl autoantigens.

References

1. Reichlin M & Mattioli M (1976) Description of a serological reaction characteristic of polymyositis. Clin Immunol Immunopathol 5:12–20
2. Wolfe JF, Adelstein E & Sharp GC (1977) Antinuclear antibody with distinct specificity for polymyositis. J Clin Invest 59:176–178
3. Bernstein RM, Bunn CC, Hughes GRV, Francoeur AM & Mathews MB (1984) Cellular protein and RNA antigens in autoimmune diseases. Mol Biol Med 2:105–120
4. Targoff IN & Reichlin M (1985) Nucleolar localization of the PM-Scl antigen. Arthritis Rheum 28:226–230
5. Reimer G, Scheer U, Peters JM & Tan EM (1986) Immunolocalization and partial characterization of a nucleolar autoantigen (PM-Scl) associated with polymyositis/scleroderma overlap syndromes. J Immunol 137:3802–3808
6. Reimer G, Steen VD, Penning CA, Medsger TA & Tan EM (1988) Correlates between autoantibodies to nucleolar antigens and clinical features in patients with systemic sclerosis (scleroderma). Arthritis Rheum 31:525–532
7. Gelpi C, Alguero A, Angeles Martinez M, Vidal S, Juarez C & Rodriguez-Sanchez JL (1990) Identification of protein components reactive with anti-PM/Scl autoantibodies. Clin Exp Immunol 81:59–64
8. Kipins RJ, Craft J & Hardin JA (1990) The analysis of antinuclear and antinucleolar autoantibodies of scleroderma by radioimmunoprecipitation assays. Arthritis Rheum 33:1431–1437
9. Alderuccio F, Chan EKL & Tan EM (1991) Molecular characterization of an autoantigen of PM-Scl in the Polymyositis/Scleroderma overlap syndrome: A unique and complete human cDNA encoding an apparent 75-kD acidic protein of the nucleolar complex. J Exp Med 173:941–952
10. Blüthner M & Bautz FA (1992) Cloning and characterization of the cDNA coding for a polymyositis-scleroderma overlap syndrome-related nucleolar 100-kD protein. J Exp Med 176:973–980
11. Ge Q, Frank MB, O'Brien C & Targoff IN (1992) Cloning of a complementary DNA coding for the 100-kD antigenic protein of the PM-Scl autoantigen. J Clin Invest 90:559–570
12. Bautz FA, Haidmann L & Genth E (1988) Nucleolar autoantibodies in patients with polymyositis-scleroderma overlap syndrome. Z Rheumatol 47:314
13. Brendel V, Dohlman J, Blaisdel BE & Karlin S (1991) Very long charge runs in systemic lupus erythematosus-associated autoantigens. Proc Natl Acad Sci USA 88:1536–1539
14. Tan EM (1989) Antinuclear antibodies: Diagnostic markers for autoimmune diseases and probes for cell biology. Adv Immunol 44:93–151

Manual of Biological Markers of Disease **B6.1**: 1–18, 1994.

Myositis-associated antigens. Aminoacyl-tRNA synthetases Jo-1, PL-7, PL-12, EJ, and OJ

PAUL PLOTZ[1] and IRA TARGOFF[2]

[1] *Arthritis and Rheumatism Branch, National Institute of Arthritis and Musculoskeletal and Skin Diseases, National Institutes of Health, Bethesda, MD 20892, U.S.A.;* [2] *Oklahoma Medical Research Foundation, 825 NE 13th Street, Oklahoma City, OK 73104, U.S.A.*

Abbreviations

Jo-1 = HRS = histidyl-tRNA synthetase; PL-7 = TRS = threonyl-tRNA synthetase; PL-12 = ARS = alanyl-tRNA synthetase; EJ = GRS = glycyl-tRNA synthetase; OJ = IRS = isoleucyl-tRNA synthetase

Key words: polymyositis, dermatomyositis, autoantibody, aminoacyl-tRNA synthetase, autoimmunity

1. The antigens

1.1 Jo-1 (histidyl-tRNA synthetase)

In 1980, Nishikai and Reichlin recognized a precipitating system by Ouchterlony immunodiffusion, labeled Jo-1 after the prototype patient, that was apparently specific for myositis [1]. In 1983, using ^{32}P-labeled HeLa cells or mouse Friend erythroleukemia cells as a source of antigen, Rosa *et al.* recognized that a tRNA^{his} was immunoprecipitated by anti-Jo-1-containing sera from myositis patients [2]. Shortly thereafter, Mathews and Bernstein showed that a protein of M_r 50,000 was immunoprecipitated from ^{35}S-labeled HeLa cell extracts by anti-Jo-1-containing sera, and that the enzyme which joins histidine to tRNA^{his}, histidyl-tRNA synthetase (HRS), was inhibited by these sera [3].

1.2 PL-7 (threonyl-tRNA synthetase)

Mathews *et al.* identified the antigen for the PL-7 precipitating system as another aminoacyl-tRNA synthetase, threonyl-tRNA synthetase [4].

1.3 PL-12 (alanyl-tRNA synthetase)

Bunn *et al.* determined that anti-PL-12 sera contained autoantibodies reacting with alanyl-tRNA synthetase, and associated autoantibodies to tRNA[ala] [5].

1.4 EJ (glycyl-tRNA synthetase)

In 1990, Targoff recognized that the EJ antigen is glycyl-tRNA synthetase [6].

1.5 OJ (isoleucyl-tRNA synthetase)

In 1990, Targoff showed that the OJ antigen is isoleucyl-tRNA synthetase [6]. Isoleucyl-tRNA synthetase is a component of a multi-enzyme complex of aminoacyl-tRNA synthetases containing synthetase activity for 9 amino acids, and sera that contain anti-OJ autoantibodies immunoprecipitate the entire complex. Some anti-OJ sera also contain antibodies to certain other components of the complex, including lysyl-tRNA synthetase, a 160 kD synthetase, and probably leucyl-tRNA synthetase. Demonstration of immunoprecipitation of specific tRNAs and of proteins of compatible size, and specific inhibition of aminoacylation activity, have been the principal methods used to identify the target antigens, supplemented in some cases with analysis of the tRNAs.

2. Synonyms

Jo-1 = histidyl-tRNA synthetase (HRS)
PL-12 = alanyl-tRNA synthetase (ARS)
PL-7 = threonyl-tRNA synthetase (TRS)
EJ   = glycyl-tRNA synthetase (GRS)
OJ   = isoleucyl-tRNA synthetase (IRS)

Aminoacyl-tRNA synthetases may also be referred to as aminoacyl-tRNA ligases.

3. Antibody detection

3.1 Immunofluorescence

By immunofluorescence with monospecific sera, antibodies directed against aminoacyl-tRNA synthetases often show a diffuse cytoplasmic fluorescence (Fig. 1). For technical details, see this Manual chapter A2.

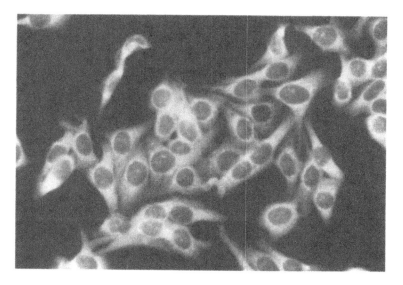

Fig. 1. Cytoplasmic staining of human epithelial cells (Hep 2) with a serum containing antibodies directed at histidyl-tRNA synthetase (Jo-1). Original magnification 1:400.

3.2 Counter-immunoelectrophoresis and immunodiffusion

Double immunodiffusion against calf thymus extract was originally used to identify and define anti-Jo-1 autoantibody. This is a reliable method when an extract rich in Jo-1 or purified Jo-1 antigen are used as antigen sources [7]. Specificity should be confirmed in all cases with standard serum regardless of the antigen preparation used (see this Manual chapter A3). It is rare to encounter patient sera that are positive by very sensitive methods such as ELISA, but negative by immunodiffusion against calf thymus or liver extract [7]; most are from treated patients or those in remission. Most anti-PL-7 sera are also detectable by immunodiffusion, but false negatives occur [8]. Anti-PL-12 was originally defined as a precipitin by counter-immuno-electrophoresis, but immunodiffusion of anti-PL-12 sera against calf thymus extract is usually negative. No precipitin line corresponding to anti-EJ or anti-OJ has been identified.

3.3 Immunoblotting

Most anti-Jo-1 sera are positive by Western blot, with occasional exceptions [7], as are most anti-EJ sera, but confirmation of specificity would be required. Most anti-PL-7 sera are negative by Western blot [11], and anti-PL-

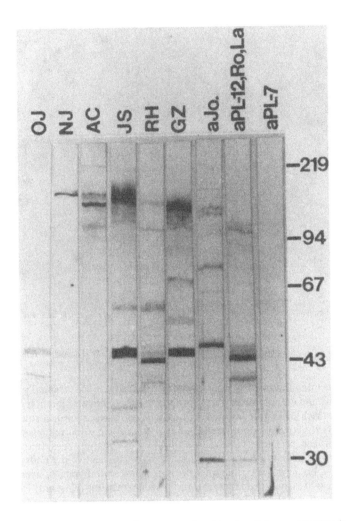

Fig. 2. SDS-PAGE of whole HeLa cell extract, transferred to nitrocellulose, and developed with anti-synthetase sera. Most anti-Jo-1 sera react strongly by immunoblot with a 50–52 kD protein; occasional weak sera may be negative. Other bands seen with anti-Jo-1 serum are due to co-existent antibodies. As is common with anti-PL-7 sera, no reaction is seen by immunoblot with the 80 kD threonyl-tRNA synthetase protein. Weak binding to a 110 kD protein is visible with the anti-PL-12 serum; reaction with other anti-PL-12 sera is variable. The strong reaction at 44 kD with this anti-PL-12 serum represents co-existent anti-La/SSB. Reaction of anti-OJ by immunoblot is also variable. Binding to the probable primary antigen, 150 kD isoleucyl-tRNA synthetase, is unusual, but seen here with serum JS. Sera NJ and AC react with the 160–170 kD synthetase component of the multi-enzyme synthetase complex. None of the bands seen with sera GZ, RH, or OJ represent staining of components of the synthetase complex. Sera JS, OJ, and GZ react with a 48 kD protein; GZ reacts with an unidentified 140 kD protein; and RH

12 is inconsistent. Most anti-OJ sera do not react with IRS, the primary antigen, by Western blot, but lysyl-tRNA synthetase and the 160 kD synthetase are stained by those anti-OJ sera that react with them [10] (Fig. 2).

3.4 Immunoprecipitation of trna

For research purposes, the best method for detection of each of these antibodies currently available in terms of combined sensitivity and specificity is protein A-assisted immunoprecipitation from cultured cells (see this Manual chapter A7). Using ^{32}P-labeled cells, tRNAs can be detected with all sera thus far described, and the tRNA patterns by PAGE that are characteristic of each anti-synthetase can be distinguished from the others [6] (Fig. 3). Immunoprecipitation for nucleic acid analysis using unlabeled cells developed by silver staining is capable of detecting tRNA in almost all sera that are positive using ^{32}P-labeled cells, but is quantitatively less sensitive [6, 9].

3.5 Immunoprecipitation of proteins

By immunoprecipitation for protein analysis using ^{35}S-methionine-labeled HeLa cells, antibodies to GRS, TRS, and ARS give very intense single protein bands of 75, 80 and 110 kD respectively, while the 52 kD Jo-1 band is of variable intensity. Anti-OJ autoantibodies precipitate the entire multi-enzyme complex of synthetases, and the resulting characteristic protein band pattern is specific [6, 10] (Fig. 4). The single protein bands of other anti-synthetases require confirmation.

3.6 Elisa

ELISA is highly sensitive for detection of anti-Jo-1, whether using immunoaffinity purified [7], biochemically purified [12], or recombinant antigen [13]. Specificity is dependent on the purity of the antigen (see also this Manual chapter A5). ELISA has also been used for anti-PL-7 [8].

---

reacts with anti-Ro/SSA and anti-La/SSB. Not shown is anti-EJ serum (anti-glycyl-tRNA synthetase), which reacts with a 75 kD protein.

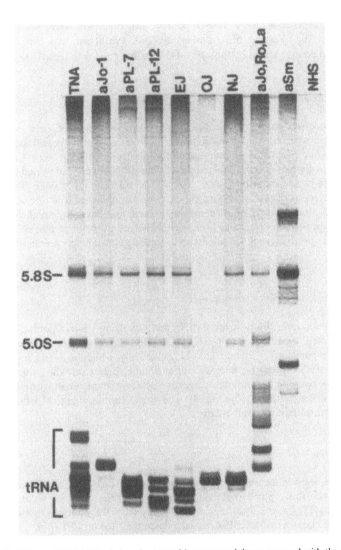

Fig. 3. 8M urea-10% PAGE of phenol-extracted immunoprecipitates prepared with the anti-synthetase sera indicated, developed with silver stain. Sera OJ and NJ both have anti-OJ antibodies. tRNAs are immunoprecipitated by all patient autoantibodies aminoacyl-tRNA synthetases. The tRNAs are highly restricted compared to total tRNA, but may include multiple forms of tRNA for the individual amino acid. The same set of tRNAs is immunoprecipitated by all sera with a particular antibody, and the pattern of these tRNAs on PAGE can specifically identify the particular anti-synthetase. TNA = total nucleic acid, NHS = normal human serum. (Reproduced from [6] with permission.)

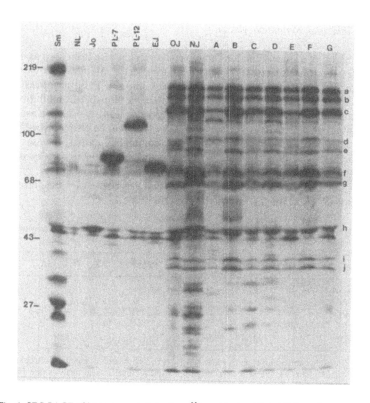

Fig. 4. SDS-PAGE of immunoprecipitates from ^{35}S-methionine-labeled HeLa cell extract. An abundant protein is immunoprecipitated by antibodies to 75 kD glycyl-tRNA synthetase (EJ), by antibodies to 80 kD threonyl-tRNA synthetase (anti-PL-7), and by antibodies 110 kD alanyl-tRNA synthetase (anti-PL-12). The weaker band of about 50 kD that is precipitated by anti-Jo-1 antibodies co-migrates with an artifact, but it can be distinguished by its intensity. Sera OJ, NJ, and A through G all have anti-OJ and immunoprecipitate the same series of proteins constituting the multi-enzyme complex of synthetases. The components of the complex are indicated by the letter at the right, and include synthetases for: glutamic acid-proline vs. glutamine (160 kD) (a), isoleucine (150 kD) (b), leucine (c), methionine (d), glutamine vs. glutamic acid (e), lysine (f), arginine (g), and aspartic acid (h); and non-synthetase components (i and j). A third, smaller, non-synthetase component is reported but not visible in this gel. Components a, b, and c are consistently more intense than other components. Other evidence indicates that the primary antigen is component b, isoleucyl-tRNA synthetase, but antibodies have been recognized by immunoblot that react with components a (160 kD synthetase) and f. Antibodies to component c (leucyl-tRNA synthetase) may also be present. (Reproduced from [10] with permission.)

AMAN-B6.1/7

3.7 Inhibition of aminoacylation

Inhibition of antigen function (aminoacylation) has also been used as a test to detect anti-synthetases [7, 14]. Quantitative serum inhibitory activity was correlated with ELISA activity of anti-Jo-1 sera [7]. False negatives may occur but are uncommon [10, 14]. There is variable non-specific inhibition by normal human serum of most synthetase assays [4, 7, 14], which decreases the quantitative sensitivity of this method, since it cannot be distinguished from low-level specific inhibition. Most sera containing anti-synthetases can nevertheless be detected because the specific inhibition is usually much greater (> 50% inhibition is often considered significant) (Fig. 5).

An overview of the various techniques that can be used for the detection of anti-tRNA synthetase activity is given in Table 1.

Table 1. Comparison of different techniques used for the detection of anti-tRNA synthetase activities

Name	DID	CIE	IPPna	IPPpro	WB	ELISA	AAI
Anti-Jo-1	+	+	+	+	+	+	+
Anti-PL-7	+	+	+	+	−	+	+
Anti-PL-12	−	+	+	+	+/−	nd	+
Anti-EJ	nd	nd	+	+	+	nd	+
Anti-OJ	−	nd	+	+	+/−*	nd	+

DID = double immunodiffusion (see Manual, Part A); CIE = counter immunoelectrophoresis (see Manual, Part A); IPPna = immunoprecipitation for nucleic acid analysis (see Fig. 3); IPPpro = immunoprecipitation for protein analysis (see Fig. 4); WB = Western blot (see Fig. 2); ELISA = Enzyme linked immunosorbent assay (see Manual, Part A); AAI = Aminoacylation inhibition (testing for specific inhibition of charging with the respective amino acid) (see Fig. 5); nd = not determined.
[a]Most anti-OJ sera do not react with isoleucyl-tRNA synthetase by WB, but some react with other components of the multienzyme complex by WB.

4. Cellular localization

Although early studies were unclear, it is now evident that all of these autoantigens are localized in the cytoplasm, the venue of protein synthesis. The enzymes are present in cytoplasmic extracts and sera stain cytoplasm [4, 6, 14–17] (see Fig. 1). Experiments showing apparent nuclear staining are likely to be the result of using patient sera which contain other autoantigen specificities.

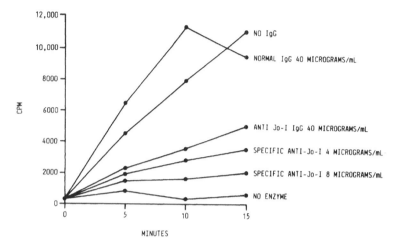

Fig. 5. Patient antibody was tested for ability to inhibit synthetase function by performing the histidine aminoacylation reaction with and without pre-incubation of enzyme with antibody. The reaction mixture included 2 μl ^{3}H-histidine, cold histidine to 16 μM final concentration, 4 A_{260} units of whole calf liver tRNA, 5mM ATP, 0.5 mM CTP, 0.5 mM GTP, 10 mM MgSO$_4$, 1 mM EDTA, 1 mM DTT, 20 mM NaCl, in Tris buffer at a final pH of 7.5. Partially purified calf liver extract as enzyme source was pre-incubated with an equal volume of normal or anti-Jo-1 IgG for 2 h at 4 °C. The reaction was started by addition of 6 μl of enzyme-antibody mixture to the reaction mixture at room temperature. Aliquots of reaction mixture were taken at intervals and placed on filters pre-treated with 10% trichloroacetic acid (TCA). Filters were washed with TCA, then with ethanol/ether mixture, then dried and counted. tRNA and histidine-tRNA are precipitated by TCA, but free histidine is washed away. HeLa cell extract has also been commonly used as enzyme source. The enzyme must be diluted until it is limiting in the reaction so that inhibition of enzyme activity will lead to decreased product. Serum may be used instead of purified IgG, but may have non-specific inhibitory activity and must be compared to a series of normals. Anti-Jo-1 IgG was DEAE-purified IgG fraction from anti-Jo-1-positive serum. Specific anti-Jo-1 was affinity-purified from immune precipitates of the same serum. (Reproduced from [7] with permission.)

5. Cellular function

All of these enzymes join an amino acid to its cognate tRNA, a two-step reaction in which AMP esterified to the amino acid is a transient intermediate (Fig. 6). The reaction actually esterifies the carboxyl group of the amino acid to a hydroxyl on the ribose of the 3′-terminal adenine of the tRNA. The resulting aminoacyl-tRNA is the substrate for a reaction joining the amino acid to the carboxyl terminus of a polypeptide chain being synthesized on the ribosome. Other activities have been suggested for aminoacyl-tRNA synthetases [18], and it would not be a surprise to uncover other functions of

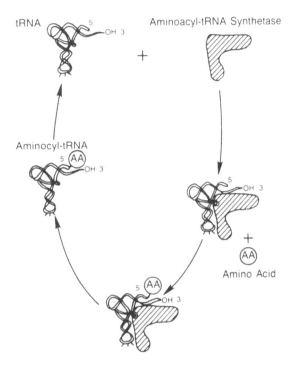

tRNA   Aminoacyl-tRNA Synthetase

+

Aminocyl-tRNA

+

Amino Acid

Fig. 6. A schematic drawing of the biochemical action of aminoacyl-tRNA synthetases as described in the test. (Figure kindly provided by Frederick Miller.)

the synthetase antigens. Essentially nothing is yet known of the aminoacyl-tRNA synthetases which are transported into the mitochondria.

6. Presence in other species

All synthetases are obligatorily present in all prokaryotes and eukaryotes, and there is considerable homology between synthetases for the same amino acid among all species. The homology is strongest in the regions which interact with the amino acid, with the ATP, and with the 3'-terminus of the tRNA. Two families of synthetases have been distinguished based on bacterial sequences with wholly distinct evolutionary origins. Within families, consensus sequences have been identified, and there are similarities of structure and organization which have been preserved in eukaryotes through humans, for those enzymes that have been studied. This complex and fascinating subject is beyond the scope of this book, but has been well-reviewed recently [19]. Of note, however, is that autoantibodies and animal

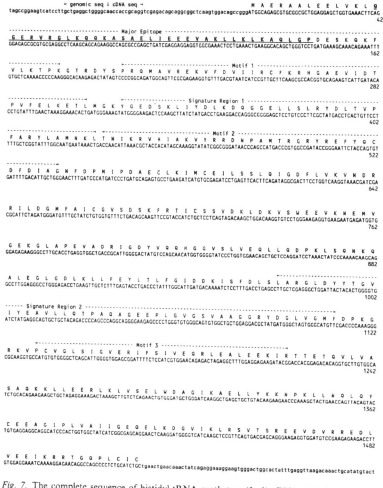

```
                              ← genomic seq ↓ cDNA seq →                    M  A  E  R  A  A  L  E  E  L  V  K  L  g
tagccggaagtcatccttgctgaggctggggcaaccaccgcaggtcgagacagcaggcggctcaagtggacagccgggATGGCAGAGCGTGCGGCGCTGGAGGAGCTGGTGAAACTTCAG
                                                                                                                    42

---------------------------------------------------- Major Epitope -----------------------------------------------------
G  E  R  V  R  G  L  K  Q  Q  K  A  S  A  E  L  I  E  E  E  V  A  K  L  L  L  K  L  K  A  Q  L  G  P  D  E  S  K  Q  K  F
GGAGAGCGCGTGCGAGGCCTCAAGCAGCAGAAGGCCAGCGCCGAGCTGATCGAGGAGGAGGTGGCGAAACTCCTGAAACTGAAGGCACAGCTGGGTCCTGATGAAAGCAAACAGAAATTT
                                                                                                                    162

V  L  K  T  P  K  G  T  R  D  Y  S  P  R  Q  M  A  V  R  E  K  V  F  D  V  I  I  R  C  F  K  R  H  G  A  E  V  I  D  T
GTGCTCAAAACCCCCAAGGGCACAAGAGACTATAGTCCCCGGCAGATGGCAGTTCGCGAGAAGGTGTTTGACGTAATCATCCGTTGCTTCAAGCGCCACGGTGCAGAAGTCATTGATACA
                                                                                                                    282

---------------------------------                    ------------------------------ Signature Region 1 ------------------
P  V  F  E  L  K  E  T  L  M  G  K  Y  G  E  D  S  K  L  I  Y  D  L  K  D  Q  G  G  E  L  L  S  L  R  Y  D  L  T  V  P
CCTGTATTTGAACTAAAGGAAACACTGATGGGAAAGTATGGGGAAGACTCCAAGCTTATCTATGACCTGAAGGACCAGGGCGGGGAGCTCCTGTCCCTTCGCTATGACCTCACTGTTCCT
                                                                                                                    402

------------------                   -------------------------------------------------- Motif 2 ----------------------
F  A  R  Y  L  A  M  N  K  L  T  N  I  K  R  V  H  I  A  K  V  Y  R  R  D  N  P  A  M  T  R  G  R  Y  R  E  F  Y  Q  C
TTTGCTCGGTATTTGGCAATGAATAAACTGACCAACATTAAACGCTACCACATAGCAAAGGTATATCGGCGGGATAACCCAGCCATGACCCGTGGCCGATACCGGGAATTCTACCAGTGT
                                                                                                                    522

------------------
D  F  D  I  A  G  N  F  D  P  M  I  P  D  A  E  C  L  K  I  M  C  E  I  L  S  S  L  Q  I  G  D  F  L  V  K  V  N  D  R
GATTTTGACATTGCTGGGAACTTTGATCCCATGATCCCTGATGCAGAGTGCCTGAAGATCATGTGCGAGATCCTGAGTTCACTTCAGATAGGCGACTTCCTGGTCAAGGTAAACGATCGA
                                                                                                                    642

R  I  L  D  G  M  F  A  I  C  G  V  S  D  S  K  F  R  T  I  C  S  S  V  D  K  L  D  K  V  S  W  E  E  V  K  N  E  M  V
CGGCATTCTAGATGGGATGTTTGCTATCTGTGGTGTTTCTGACAGCAAGTTCCGTACCATCTGCTCCTCAGTAGACAAGCTGGACAAGGTGTCCTGGGAAGAGGTGAAGAATGAGATGGTG
                                                                                                                    762

G  E  K  G  L  A  P  E  V  A  D  R  I  G  D  Y  V  Q  Q  H  G  G  V  S  L  V  E  Q  L  L  Q  D  P  K  L  S  Q  N  K  Q
GGAGAGAAGGGCCTTGCACCTGAGGTGGCTGACCGCATTGGGGACTATGTCCAGCAACATGGTGGGGTATCCCTGGTGGAACAGCTGCTCCAGGATCCTAAACTATCCCAAAACAAGCAG
                                                                                                                    882

A  L  E  G  L  G  D  L  K  L  L  F  E  Y  L  T  L  F  G  I  D  D  K  I  S  F  D  L  S  L  A  R  G  L  D  Y  Y  T  G  V
GCCTTGGAGGGCCTGGGAGACCTGAAGTTGCTCTTTGAGTACCTGACCCTATTTGGCATTGATGACAAAATCTCCTTTGACCTGAGCCTTGCTCGAGGGCTGGATTACTACACTGGGGTG
                                                                                                                    1002

------ Signature Region 2 ----------------------------------------------------------------------------------------
I  Y  E  A  V  L  L  Q  T  P  A  Q  A  G  E  E  P  L  G  V  G  S  V  A  A  G  G  R  Y  D  G  L  V  G  M  F  D  P  K  G
ATCTATGAGGCAGTGCTGCTACAGACCCCAGCCCAGGCAGGGGAAGAGCCCCTGGGTGTGGGCAGTGTGGCTGCTGGAGGACGCTATGATGGGCTAGTGGGCATGTTCGACCCCAAAGGG
                                                                                                                    1122

-------------------------------- Motif 3 --------------------------------
R  K  V  P  C  V  G  L  S  I  G  V  E  R  I  F  S  I  V  E  Q  R  L  E  A  L  E  E  K  I  R  T  T  E  T  Q  V  L  V  A
CGCAAGGTGCCATGTGTGGGGCTCAGCATTGGGGTGGAGCGGATTTTCTCCATCGTGGAACAGAGACTAGAGGCTTTGGAGGAGAAGATACGGACCACGGAGACACAGGTGCTTGTGGCA
                                                                                                                    1242

S  A  Q  K  K  L  L  E  E  R  L  K  L  V  S  E  L  W  D  A  G  I  K  A  E  L  L  Y  K  K  N  P  K  L  L  N  Q  L  Q  Y
TCTGCACAGAAGAAGCTGCTAGAGGAAAGACTAAAGCTTGTCTCAGAACTGTGGGATGCTGGGATCAAGGCTGAGCTGCTGTACAAGAAGAACCCAAAGCTACTGAACCAGTTACAGTAC
                                                                                                                    1362

C  E  E  A  G  I  P  L  V  A  I  I  G  E  Q  E  L  K  D  G  V  I  K  L  R  S  V  T  S  R  E  E  V  D  V  R  R  E  D  L
TGTGAGGAGGCAGGCATCCCACTGGTGGCTATCATCGGCGAGCAGGAACTCAAGGATGGGGTCATCAAGCTCCGTTCAGTGACGAGCAGGGAAGAGGTGGATGTCCGAAGAGAAGACCTT
                                                                                                                    1482

V  E  E  I  K  R  R  T  G  Q  P  L  C  I  C
GTGGAGGAAATCAAAAGGAGAACAGGCCAGCCCCTCTGCATCTGCtgaactgaacaaactatcagaggaaaggaagtgggactggcactatttgaggttaagacaaactgcatatgtact
```

Fig. 7. The complete sequence of histidyl-tRNA synthetase (Jo-1) cDNA with the complete amino acid sequence. The 5′ and 3′ untranslated regions are noted in lowercase. The Class II motifs, the histidyl-signature regions, and the major epitope are noted above the sequence (for details, see [24]).

antisera do not cross-react between different aminoacyl-tRNA synthetases.

Of the aminoacyl-tRNA synthetases which are targeted by myositis sera, IRS belongs to Class I, and ARS, GRS, HRS, and TRS belong to Class II. In general, the enzymatic core of the molecules is best preserved in evolution, and there tend to be additions at both the amino- and carboxy-termini of the enzymes. The function of these additions is not yet known. The species range of recognition by the naturally occurring autoantibodies has been incompletely studied for most of the specificities. Human anti-Jo-1 sera recognize the human enzyme best, and recognize the enzyme only from higher eukaryotes [20, 21].

7. cDNA

Although Tsui and colleagues published sequences for the cDNA of both the human and hamster histidyl-tRNA synthetases [22, 23], the published sequences contain many errors. Raben and her colleagues have published a corrected human sequence (EMBL accession – Z11518) (Fig. 7) and pointed out the likely errors in the hamster sequence [24]. The 5′ untranslated sequence of the mRNA was also published along with several hundred bases of the gene upstream of the transcription initiation site. Raben and her colleagues identified all the motifs characteristic of Class II synthetases and further noted the presence of "signature regions" characteristic of histidyl-tRNA synthetases which are present in the E. coli and yeast enzymes (Fig. 8). They also pointed out similar signature regions in the related seryl- and threonyl-tRNA synthetases [25, 26].

Cruzen and Arfin have determined the sequence of the cDNA for human threonyl-tRNA synthetase [25]. Although sequences for alanyl-, glycyl-, and isoleucyl-tRNA synthetases from prokaryotes and lower eukaryotes are available [27, 28], sequences from humans or other higher eukaryotes have not yet been published.

Fig. 8. A schematic drawing of histidyl-tRNA synthetase (Jo-1). The major epitope is found in the amino terminal 60 amino acids, a region that includes one of the two parts of the molecule that are predicted to assume a coiled-coil configuration.

8. Gene

Information regarding the structure of the genes themselves is unavailable for any of the human enzymes. A structure has been published for the hamster HRS gene but has not been verified and contains some sequencing errors [22].

The chromosomal location of 10 aminoacyl-tRNA synthetase genes has been determined for humans, and four are on chromosome 5, including those for the two primary antigenic synthetases that have been mapped (HRS and TRS), along with arginyl- and leucyl-tRNA synthetases [29, 30]. The 160 kD synthetase (which may include both prolyl- and glutamyl- activities, and which reacts with some anti-OJ sera) was mapped to chromosome 1 [31]. The five other mapped synthetases are each on separate chromosomes.

9. Relation with nucleic acids

All the synthetases interact with their cognate tRNAs and probably with other tRNAs as well. It is not yet entirely clear exactly what permits the astonishing specificity of the aminoacylation reaction since the synthetases within each family resemble one another, as do the tRNAs. The recognition of the anticodon and other discriminating bases at the 3′ end of the tRNA or other areas presumably provide much of the specificity, since modification of individual bases can sometimes convert the specificity of a tRNA from one amino acid to another [32–35]. This "second genetic code" by which synthetases recognize tRNAs is currently being deciphered. The modified bases characteristic of tRNAs may help keep inappropriate tRNAs from binding strongly enough to interfere with the proper function of the enzymes.

The exact way in which the tRNAs bind to these enzymes is not known with certainty, although the general features should resemble those of the other class II synthetases whose crystal structure has been published [19]. Of great interest is the translational regulation of the synthesis of both E coli and yeast TRS. In the 5′ untranslated portion of the messenger RNA is a region which assumes a tertiary structure homologous to the substrate tRNA^{thr}. The protein product of the gene binds to this region and feedback inhibits translation. The primary structure of the same region of the human cDNA does not show a similar potential binding site, but biochemical studies of translation have not been done [25].

10. Relation with other proteins

Sequence homology is strong with HRS from all species, especially in the regions identified as signature regions [22, 24] (see Fig. 7). There is also recognizable homology to all class II synthetases through motifs characteristic of that class, of which HRS possesses all three [24]. The homology in these regions is quite strong among the class IIa synthetases – histidyl-, threonyl-, seryl-, and prolyl-. The presumptive first two exons of human HRS contain a 32 amino acid region with strong homology to regions found in higher eukaryotic glutamyl-prolyl-tRNA synthetase [36, 38], glycyl-tRNA synthetase [27], and tryptophanyl-tRNA synthetase [37]. This region is predicted to assume a coiled-coil configuration. Regions of this configuration are especially likely to be found in proteins which are the targets of human disease-related autoantibodies. Furthermore, this region is present in the hamster HRS, which is recognized by anti-Jo-1 autoantibodies, but not in the yeast HRS, which is not [22, 24].

11. Autoantibodies to trna

When myositis sera are used to precipitate synthetases, tRNA generally co-precipitates with the protein. It is clear that in most cases, antibodies reacting directly with the tRNA are not present in these sera, and the sera will not precipitate tRNA from de-proteinized extracts [2, 6]. However, most anti-PL-12 sera contain two specificities: one directed at the protein, and the other directed at the tRNA [5] (Fig. 9). We and others (Bunn, personal communication) have observed a similar phenomenon with several sera containing antibodies to histidyl- or glycyl-tRNA synthetases as well. However, direct reaction with the enzyme proteins has been demonstrated for most anti-synthetases. As noted, most anti-Jo-1 sera react by Western blot with isolated protein [7], as do anti-EJ sera, and ELISAs using purified Jo-1 and PL-7 have been described. Recombinant HRS produced in baculovirus immunoprecipitates perfectly well although it has no detectable tRNA bound to it [13]. Presumably, the tRNA is co-precipitated because of its affinity for the enzyme. It has not yet been determined whether populations of antibodies exist in anti-synthetase sera that require the enzyme-tRNA complex, or react better with the enzyme when tRNA is bound to it. However, it is of note that while anti-OJ autoantibodies immunoprecipitate all the proteins in the multienzyme complex, they appear to immunoprecipitate a more restricted set of tRNAs, comprised mainly of tRNA^{ile} [10]. This suggests that the anti-synthetase facilitates precipitation of the tRNA. Animal antisera raised against synthetases often do not precipitate the tRNA.

12. Biochemical characteristics

HRS probably is a dimer within the cell, with identical 50–52 kD subunits, each with a bound tRNA^{his}. These enzymes generally have high affinity for the cognate tRNAs. It is thought that motif I is concerned with dimerization in class II synthetases, but there is no direct proof of this. The subunit molecular weight by PAGE of TRS is 80 kD [4], of ARS 110 kD [5], of GRS 75 kD [6], and of IRS 150 kD [6] (see Fig. 4). IRS is the second largest component of the multi-enzyme complex. The largest is a 160–170 kD protein that was originally identified as glutaminyl-tRNA synthetase based on homologies with bacterial sequences [36]. However, recent studies in

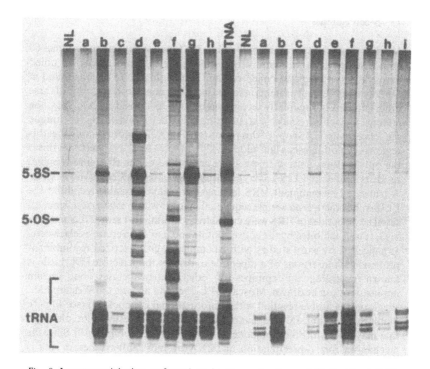

Fig. 9. Immunoprecipitation performed as in Figure 4 using anti-PL-12 (alanyl-tRNA synthetase) sera. In the left half of the figure, whole Hela cell extract was used as antigen; in the right half of the figure, the antigen was Hela cell extract that had been de-proteinized by incubation with proteinase K followed by phenol extraction and ethanol precipitation. Most anti-PL-12 sera also contain autoantibodies to tRNA^{ala}. The set of tRNAs precipitated seems similar with both extracts, but most sera precipitate less tRNA with de-proteinized extract (the de-proteinized extract was derived from twice as much original extract). As expected, co-existent anti-Ro/SSA, anti-La/SSB, anti-U1RNP, and anti-Sm did not immunoprecipitate their RNAs from de-proteinized extract. The immunoprecipitate of left lane a was lost. A small proportion of sera with other anti-synthetases also have anti-tRNA antibodies directed at tRNAs for the cognate synthetase, but most do not show such antibodies. (Reproduced from [14] with permission.)

Drosophila suggest that the largest component of the eukaryotic complex is a multi-functional synthetase with synthetase activity for both glutamic acid and proline [38]. About 1/4 of anti-OJ sera react with this 160 kD synthetase by Western blot, but do not inhibit any of the 3 synthetases [10]. Leucyl-tRNA synthetase, the third largest at 140 kD, is specifically inhibited by some anti-OJ sera, probably indicating autoantibodies reacting with it. Occasional anti-OJ sera reacted with lysyl-tRNA synthetase, a 75 kD protein that is the 6th largest complex component.

13. B-cell epitopes

Most or all of the epitope activity of HRS is found in the amino-terminal 60 amino acids which presumptively comprise the first two exons. As noted above, this is a predicted coiled-coil and is homologous to regions found in other synthetases and proteins involved in translation. However, none of these other proteins is apparently recognized by anti-Jo-1 positive sera. Thus, the epitope probably resides in a structural feature(s) related to the unique sequences in the HRS copy of this region (see Fig. 8). It is worth noting that this region is evolutionarily a late addition [24]. In some sera, it is likely that this is the only epitope, since a free peptide comprised of the 60 amino-terminal residues inhibits an ELISA assay essentially as well as pure recombinant HRS. A truncated recombinant HRS lacking this region is unable to inhibit the ELISA, but there is no simple sequential epitope within this region, since all serial hexapeptides of HRS were unreactive with anti-Jo-1 sera [20]. Some sera are less well-inhibited by the active peptide and may have epitopes elsewhere, as suggested by early studies with incompletely characterized recombinant proteins with insertions of a dipeptide to disrupt the structure [39]. Little is known regarding the epitopes of other antisynthetases, but certain implications can be drawn. Most anti-PL-7 sera do not react in Western blot, suggesting exclusive reaction with conformational epitopes [11] (see Fig. 2). Similarly, most anti-OJ sera do not react with IRS in Western blot, although they may react with other components of the synthetase complex [10]. Since the mechanism for co-precipitation of tRNA by most anti-synthetase sera is believed to be through affinity of tRNA for the synthetase, this implies that the tRNA binding site is specifically spared, whereas reaction with this site may be the explanation for lack of tRNA-precipitation by some animal antisera. The anti-tRNA^{ala} antibodies in anti-PL-12 sera may be absorbed out of anti-ARS antibodies by tRNA-Sepharose, and the remaining anti-PL-12 sera were no longer able to immunoprecipitate tRNA, suggesting that anti-ARS antibodies may bind to the tRNA binding site [5].

The anti-tRNA^{ala} antibodies of anti-PL-12 sera react with a select set of tRNA^{ala}, all of which had the inosine-guanine-cytosine anticodon, with none having other alanine anticodons [40]. The epitope recognized by PL-12 anti-tRNA^{ala} was localized to a 9-base region that included the anticodon, probably explaining the exclusivity of this reaction [41]. As noted, a few patients have antibodies to tRNA^{his} as well as to the protein, and while these appear to be specific for tRNA^{his}, the epitopes have not been identified (S. Cochran, E. Chabner and P. Plotz, unpublished observations; C. Bunn, personal communication).

14. T-cell epitopes

Unknown.

References

1. Nishikai M & Reichlin M (1980) Arthritis Rheum 23:881–888
2. Rosa MD, Hendrick JP Jr, Lerner MR, Steitz JA & Reichlin M (1983) Nucleic Acids Res 11:853–870
3. Mathews MB & Bernstein RM (1983) Nature 304:177–179
4. Mathews MB, Reichlin M, Hughes GRV & Bernstein RM (1984) J Exp Med 160:420–434
5. Bunn CC, Bernstein RM & Mathews MB (1986) J Exp Med 163:1281–1291
6. Targoff IN (1990) J Immunol 144:1737–1743
7. Targoff IN & Reichlin M (1987) J Immunol 138:2874–2882
8. Targoff IN, Arnett FC & Reichlin M (1988) Arthritis Rheum 31:515–524
9. Forman MS, Nakamura M, Mimori T, Gelpi C & Hardin JA (1985) Arthritis Rheum 28:1356–1361
10. Targoff IN, Trieu EP & Miller FW (1993) J Clin Invest 91:2556–2564
11. Dang CV, Tan EM & Traugh JA (1988) FASEB J 2:2376–2379
12. Biswas T, Miller FW, Takagaki Y & Plotz PH (1987) J Immunol Methods 98:243–248
13. Raben N, Nichols RC, Jain A, Amin J, Hyde C, Leff RL & Plotz PH (1992) Arthritis Rheum 35:S170 (Abstract)
14. Targoff IN & Arnett FC (1990) Am J Med 88:241–251
15. Shi MH, Tsui FWL & Rubin LA (1991) J Rheumatol 18:252–258
16. Dang CV, LaDuca FM & Bell WR (1986) Exper Cell Res 164:261–266
17. Targoff IN & Reichlin M (1985) Arthritis Rheum 28:S74 (Abstract)
18. Dang CV & Dang CV (1986) Mol Cell Biochem 71:107–120
19. Carter CW (1993) Ann Rev Biochem 62:715–748
20. Miller FW, Waite KA, Biswas T & Plotz PH (1990) Proc Natl Acad Sci USA 87:9933–9937
21. Roberts MM & Targoff IN (1989) Arthritis Rheum 32:S22 (Abstract)
22. Tsui FWL & Siminovitch L (1987) Nucleic Acids Res 15:3349–3366
23. Tsui FWL, Andrulis IL, Murialdo H & Siminovitch L (1985) Mol Cell Biol 5:2381
24. Raben N, Borriello F, Amin J, Horwitz R, Fraser DD & Plotz PH (1992) Nucleic Acids Res 20:1075–1081
25. Cruzen ME & Arfin SM (1991) J Biol Chem 266:9919–9923
26. Cusack S, Berthet-Colominas C, Hartlein M, Nassar N & Leberman R (1990) Nature 347:249–255
27. Nada S, Chang PK & Dignam JD (1993) J Biol Chem 268:7660–7667
28. Mirande M (1991) Prog Nucleic Acid Res Mol Biol 40:95–142
29. Gerken SC, Wasmuth JJ & Arfin SM (1986) Somatic Cell Mol Genet 12:519–522
30. Cruzen ME, Bengtsson U, McMahon J, Wasmuth JJ & Arfin SM (1993) Genomics 15:692–693
31. Kunze N, Bittler E, Fett R, Schray B, Hameister H, Wiedorn KH & Knippers R (1990) Hum Genet 85:527–530
32. McClain WH, Foss K, Jenkins RA & Schneider J (1991) Proc Natl Acad Sci USA 88:9272–9276
33. McClain WH & Foss K (1988) Science 240:793–796
34. McClain WH & Foss K (1988) Science 241:1804–1807
35. Sampson JR, DiRenzo AB, Behlen LS & Uhlenbeck OC (1989) Science 243:1363–1366
36. Fett R & Knippers R (1991) J Biol Chem 266:1448–1455
37. Garret M, Pajot B, Trezeguet V, Labouesse J, Merle M, Gandar J-C, Benedetto J-P, Sallafranque M-L, Alterio J, Gueguen M, Sarger C, Labouesse B & Bonnet J (1991) Biochemistry 30:7809–7817
38. Cerini C, Kerjan P, Astier M, Gratecos D, Mirande M & Semeriva M (1991) EMBO J 10:4267–4277
39. Ramsden DA, Chen J, Miller FW, Misener V, Bernstein RM, Siminovitch KA & Tsui FWL (1989) J Immunol 143:2267–2272

40. Bunn CC & Mathews MB (1987) Mol Biol Med 4:21–36
41. Bunn CC & Mathews MB (1987) Science 238:1116–1119

Manual of Biological Markers of Disease **B7.1**: 1–9, 1994.
© 1994 *Kluwer Academic Publishers. Printed in the Netherlands.*

ANCA antigens: Proteinase 3

JÖRGEN WIESLANDER and ALLAN WIIK
Department of Autoimmunology, Statens Seruminstitut, Artillerivej 5, DK-2300 Copenhagen, Denmark

Abbreviations

ANCA: anti-neutrophil cytoplasmic antibodies; cANCA: cytoplasmic ANCA; pANCA: perinuclear ANCA; AGP7: azurophil granule protein 7; bp: base pairs; IIF: indirect immunofluorescence; MPO: myeloperoxidase; PR-3: proteinase 3; GS-ANA: granulocyte specific antinuclear antibody; ELISA: enzyme linked immunosorbent assay; SDS-PAGE: sodium dodecyl sulphate polyacrylamide gel electrophoresis

Key words: Autoantigens, Autoantibodies, Vasculitis, Proteinase 3, Myeloperoxidase

1. The antigen

The antigen for the cANCA reactivity is the serine protease Proteinase 3 (PR-3). In 1988 several groups showed that the antigen was a protein with a molecular weight of 29 kDa [1–3] and in 1989 three groups showed by N-terminal sequencing that the antigen was PR-3 [4–6].

2. Synonyms

PR-3 was already described in 1973 by Ohlsson & Olsson [7, 8] under the name of neutrophil collagenase. This group also showed that the enzyme was a serine protease [9]. A third neutral protease was already suspected to be located in the neutrophil granules [10] and therefore neutrophil collagenase got the name "proteinase 4". After sequencing, however, it became clear that this protein was identical to PR-3 [11]. During the last few years other groups have isolated the same protein under different names like AGP7, p29 [12, 13] or myeloblastin [14, 15].

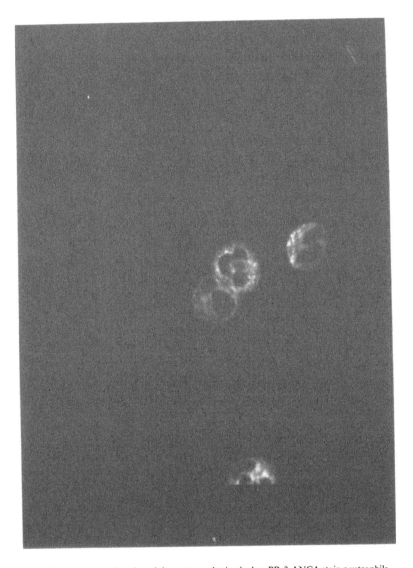

Fig. 1. The typical cytoplasmic staining pattern obtained when PR-3-ANCA stain neutrophils. The cells are fixed in ethanol for 5 min, incubated with serum from a patient with Wegeners granulomatosus and bound antibodies detected by FITC-labeled anti-human IgG as described in this Manual, chapter A9 [16].

3. Proteinase 3-ANCA detection

3.1 Immunofluorescence

The autoantibody is detected by IIF on ethanol-fixed human leukocytes. Figure 1 shows the typical granular cytoplasmic staining pattern which then can be observed. For technical details, see this Manual chapter A9 [16].

3.2 Counterimmunoelectrophoresis or immunodiffusion

These techniques have not been used to detect this type of antibody.

3.3 Immunoblotting

Immunoblotting can be used to detect anti-PR-3 antibodies [1]. Care, however, has to be taken to prevent denaturation of the antigen. Therefore boiling in SDS should be avoided.

3.4 Immunoprecipitation

Immunoprecipitation from ^{35}S-labeled cell extracts can be used to detect anti-PR-3 antibodies [1]. For technical details, see this Manual, chapter A6.

3.5 ELISA

ELISA procedures to detect antibody directed to PR-3 have been extensively discussed in this Manual, chapter A9 [16].

4. Cellular localization

PR-3 is present in the azurophilic granules of human neutrophilic granulocytes [17, 18] together with myeloperoxidase (MPO) and elastase (Fig. 2). It is also found in the granules of monocytes.

5. Cellular function

PR-3 destroys bacteria in the phagocytic vacuoles with the help of oxygen radicals, elastase and cathepsin G. In inflammatory conditions the oxygen radicals and enzymes are also released extracellularly and can degrade

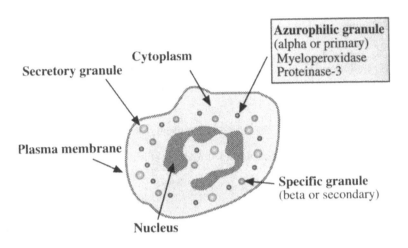

Fig. 2. The neutrophil granulocyte with its granules. PR-3 is present in the azurophilic granules as are MPO, elastase, cathepsin G and others.

collagens, proteoglycans as well as other connective tissue proteins. It can cause emphysema when administered into the bronchial system of hamsters [19]. PR-3 also has a non-proteolytic antimicrobial activity against bacteria and fungi [13].

6. Presence in other species

PR-3 is only present in higher primates like chimpanzees and humans. It is *not* present in mouse, rat, rabbit or cows [20].

7. The cDNA and derived amino acid sequence

The human cDNA is 1.3 kb. The derived amino acid sequence (Fig. 3) is highly homologous to that of other serine proteinases [21]. There is 54% homology with elastase and 35% with cathepsin G. The PR-3 polypeptide consists of 228 amino acids and has a molecular weight of 25 kDa. Two potential asparagine glycosylation sites are present. There are 8 cysteine residues and, thus, there are potentially four disulphide bonds, just like in elastase.

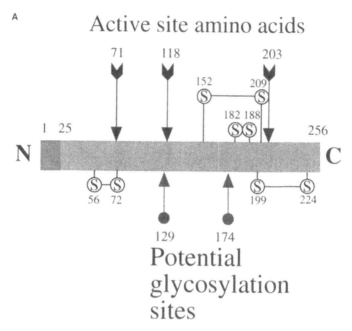

Fig. 3A. Schematic structure of PR-3. Major features are outlined. The sequence starts with a signal peptide from amino acids 1–25. Active site amino acids His[71], Asp[118] and Ser[203] are indicated by arrows above. The two potential glycosylation sites 129 and 174 are indicated by arrows below. The four potential disulphide bonds are indicated with S and drawn to be similar to other serine proteases. Amino acids are indicated by their number in the sequence.

B

```
              10        20        30        40        50
         ┴┴┴┴┴┴┴┴┴┴┴┴┴┴┴┴┴┴┴┴┴┴┴┴┴┴┴┴┴┴┴┴┴┴┴┴┴┴┴┴┴┴┴┴┴┴┴┴
MAHRPPSPALASVLLALLLSGAARAAEIVGGHEAQPHSRPYMASLQMRGN 50
PGSHFCGGTLIHPSFVLTAAHCLRDIPQRLVNVVLGAHNVRTQEPTQQHF 100
SVAQVFLNNYDAENKLNDILLIQLSSPANLSASVTSVQLPQQDQPVPHGT 150
QCLAMGWGRVGAHDPPAQVLQELNVTVVTFFCRPHNICTFVPRRKAGICF 200
GDSGGPLICDGIIQGIDSFVIWGCATRLFPDFFTRVALYVDWIRSTLRRV 250
             260       270       280       290       300
         ┴┴┴┴┴┴┴┴┴┴┴┴┴┴┴┴┴┴┴┴┴┴┴┴┴┴┴┴┴┴┴┴┴┴┴┴┴┴┴┴┴┴┴┴┴┴┴┴
EAKGRP 256
```

Fig. 3B. The amino acid sequence of PR-3, derived from the cDNA sequence.

8. The gene

The human gene is localized to chromosome 19p13.3 [22] and contains 6.5 kilobase pairs. It consists of five exons and four introns (Fig. 4) [23]. The catalytic amino acids His^{71}, Asp^{118} and Ser^{203} are located on different exons, His^{71} on exon 2, Asp^{118} on exon 3 and Ser^{203} on exon 5. In all enzymatically active serine proteases these three catalytic amino acids are conserved [24]. The organization with five exons and four introns is the same as in other serine proteinases, i.e., cathepsin G and elastase.

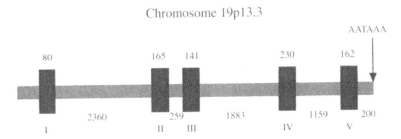

Fig. 4. The human gene for PR-3. The gene is localized to chromosome 19p13.3. The five exons (dark boxes) are numbered I–V with their sizes in bp on top. Introns are in the lighter colour with their sizes in bp below. Totally there are 6980 bp. AATAAA indicates the polyadenylation signal.

9. Relation with nucleic acids

Not known.

10. Relation with other proteins

Not known.

11. Biochemical characteristics

PR-3 has an apparent molecular weight of 29 kDa with a calculated isoelectric point of 7.9. Three bands are stained after separation by SDS-PAGE (Fig. 5). They are probably caused by post-translational modifications, e.g. glycosylations [1]. The substrate specificity of the enzyme is very similar to elastase and cathepsin G. PR-3 prefers, however, small aliphatic amino acids in the substrate pocket (alanine, serine and valine) [25–27].

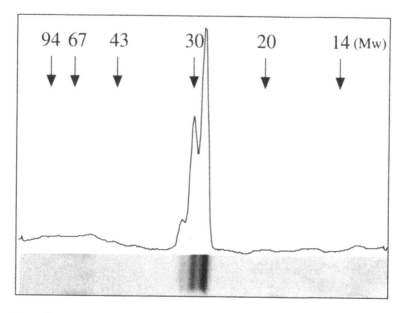

Fig. 5. SDS-PAGE analysis of PR-3 on a 10–16% gradient gel. Three bands with molecular weights of 29, 30.5 and 32 kDa are seen. A scan of the gel is also shown. Standard molecular weights (Mw) are shown in kDa on top.

12. B-cell epitopes

PR-3-ANCA predominantly react with conformational epitopes on the molecule [28]. These have not yet been characterized.

13. T-cell epitopes

Not known.

References

1. Goldschmeding R, van der Schoot CE, ten Bokkel Huinink D, Hack CE, van den Ende ME, Kallenberg CG & von dem Borne AE (1989) J Clin Invest 84:1577–1587
2. Ludemann J, Utecht B & Gross WL (1990) J Exp Med 171:357–362
3. Wieslander J, Rasmussen N & Bygren P (1989) APMIS 97 Suppl 6:42
4. Niles JL, McCluskey RT, Ahmad MF & Arnaout MA (1989) Blood 74:1888–1893
5. Jenne DE, Tschopp J, Ludemann J, Utecht B & Gross WL (1990) Nature 346: 520–521
6. Goldschmeding R, Dolman KM, van den Ende ME, van der Meer Gerritsen CH, Sonnenberg A & von dem Borne AE (1990) APMIS Suppl 19:26–27
7. Ohlsson K & Olsson I (1973) Eur J Biochem 36:473–481
8. Ohlsson K & Olsson I (1977) Scand J Haematol 19:145–152
9. Ohlsson K (1980) In: Wolley DE & Evanson JM (eds) Collagenase in Normal and Pathological Connective Tissue, pp 209–222. John Wiley & Sons, Chichester, Great Britain
10. Baggiolini M, Bretz U, Dewald B & Feigenson ME (1978) Agents and Actions 8:3–10
11. Ohlsson K, Linder C & Rosengren M (1990) Biol Chem Hoppe-Seyler 371:549–555
12. Wilde CG, Snable JL, Griffith JE & Scott RW (1990) J Biol Chem 265:2038–2041
13. Gabay JE, Scott RW, Campanelli D, Griffith J, Wilde C, Marra MN, Seeger M & Nathan CF (1989) Proc Natl Acad Sci USA 86:5610–5614
14. Bories D, Raynal M, Solomon DH, Darzynklewicz Z & Cayre Y (1989) Cell 59:959–968
15. Labbaye C, Musette P & Cayre YE (1991) Proc Natl Acad Sci USA 88:9253–9256
16. Wiik A, Rasmussen N & Wieslander J (1993) In: Van Venrooij WJ & Maini RN (eds) Manual of Biological Markers of Disease, pp A9, 1–14. Kluwer Academic Publishers, Dordrecht, The Netherlands
17. Calafat J, Goldschmeding R, Ringeling PL, Janssen H & van der Schoot CE (1990) Blood 75:242–250
18. Rasmussen N, Borregaard N & Wiik A (1987) Lancet i:1488
19. Kao RC, Wehner NG, Skubitz KM, Gray BH & Hoidal JR (1988) J Clin Invest 82:1963–1973
20. Rasmussen N, Borregaard N, Wieslander J & Worsaa A (1989) APMIS 97, Suppl 6:40
21. Campanelli D, Melchior M, Fu Y, Nakata M, Shuman H, Nathan C & Gabay JE (1990) J Exp Med 172:1709–1715
22. Sturrock AB, Espinosa R, Hoidal JR & Lebeau MM (1993) Cytogenetics and Cell Genetics 64:33–34
23. Sturrock AB, Franklin KF, Rao G, Marshall BC, Rebentisch MB, Lemons RS & Hoidal JR (1992) J Biol Chem 267:21193–21199
24. Neurath H (1984) Science 224:350–357
25. Rao NV, Wehner NG, Marshall BC, Gray WR, Gray BH & Hoidal JR (1991) J Biol Chem 266:9540–9548
26. Kam CM, Kerrigan JE, Dolman KM, Goldschmeding R, von dem Borne AE & Powers JC (1992) FEBS Lett 297:119–123
27. Brubaker MJ, Groutas WC, Hoidal JR & Rao NV (1992) Biochem Biophys Res Commun 188:1318–1324
28. Bini P, Gabay JE, Teitel A, Melchior M, Zhou JL & Elkon KB (1992) J Immunol 149:1409–1415

Manual of Biological Markers of Disease **B7.2**: 1–9, 1994.

ANCA antigens: Myeloperoxidase

JÖRGEN WIESLANDER and ALLAN WIIK
Department of Autoimmunology, Statens Seruminstitut, Artillerivej 5, DK-2300 Copenhagen, Denmark

Abbreviations

ANCA: anti-neutrophil cytoplasmic antibodies; cANCA: cytoplasmic ANCA; pANCA: perinuclear ANCA; AGP7: azurophil granule protein 7; bp: base pairs; IIF: indirect immunofluorescence; MPO: myeloperoxidase; PR-3: proteinase 3; GS-ANA: granulocyte specific antinuclear antibody; ELISA: enzyme linked immunosorbent assay; SDS-PAGE: sodium dodecyl sulphate polyacrylamide gel electrophoresis

Key words: Autoantigens, Autoantibodies, Vasculitis, Proteinase 3, Myeloperoxidase

1. The antigen

The main antigen for the pANCA reactivity is the enzyme myeloperoxidase (MPO). This was first shown by Falk and Jennette in 1988 [1].

2. Synonyms

MPO was first isolated by Agner in 1941, who gave it the name "verdoperoxidase" [2] because of its green colour and ability to catalyze peroxidative reactions. It is MPO that gives pus its greenish tone.

3. Myeloperoxidase–ANCA detection

3.1 Immunofluorescence

The autoantibody is detected by IIF on ethanol fixed granulocytes. Figure 1 shows the typical perinuclear staining pattern which then can be observed. For technical details, see this Manual chapter A9 [3].

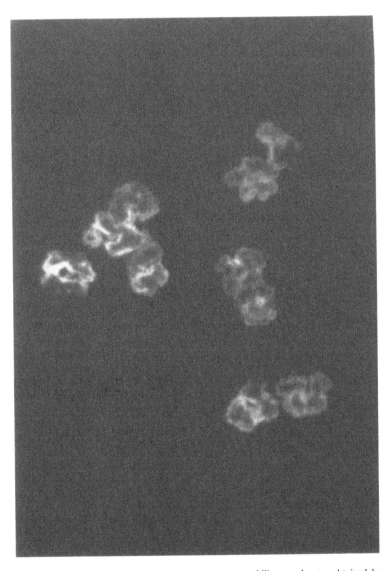

Fig. 1. The typical perinuclear staining pattern on neutrophilic granulocytes obtained by antibodies to MPO. The cells are fixed in ethanol for 5 min, incubated with serum from a patient with MPO-ANCA and bound antibodies were detected by FITC-labeled anti-human IgG as described in this Manual, chapter A9 [3].

3.2 Counterimmunoelectrophoresis or immunodiffusion

These techniques have not been used to detect this type of antibody.

3.3 Immunoblotting

Immunoblotting can also be used to detect anti-MPO antibodies. Care, however, has to be taken to prevent denaturation of the antigen. Therefore boiling in SDS should be avoided.

3.4 Immunoprecipitation

Immunoprecipitation has not been used for the detection of anti-MPO antibodies.

3.5 ELISA

ELISA procedures to detect antibody directed to MPO have been extensively discussed in this Manual, chapter A9 [3].

4. Cellular localization

MPO is present in and is a marker enzyme of the azurophilic granules of neutrophils [4]. The perinuclear distribution by IIF is due to an artifact occurring during ethanol fixation [1, 3, 5]. Ethanol disrupts the granule membranes and the positively charged MPO molecule is attracted by the negatively charged nucleus as shown in Figure 2. This artifactual distribution is not seen if the cells are fixed in formaldehyde.

5. Cellular function

MPO catalyses the hydrogen peroxide mediated peroxidation of chloride ion to hypochlorite [6].

$$H_2O_2 + Cl^- + H_3O^+ \rightarrow HOCl + 2H_2O$$

The hypochlorous acid is effective in killing phagocytized bacteria and viruses. Furthermore hypochlorous acid and its metabolites can inactivate proteinase inhibitors in the blood and tissues and thus have an indirect role in the tissue degradation of inflammatory disorders [7]. Through its ability to

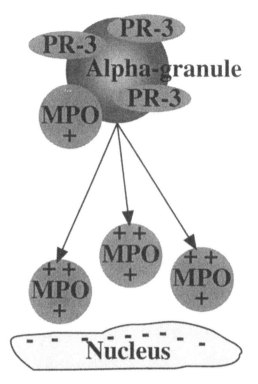

Fig. 2. Upon fixation in ethanol the granule membranes are opened and most of the positively charged MPO migrates to the negatively charged nucleus. PR-3 = proteinase 3; MPO = myeloperoxidase.

induce highly reactive free oxygen radical production, MPO may also be a key molecule for direct tissue peroxidation and possibly local autoantigen formation.

6. Presence in other species

MPO is present in most mammalian species.

7. The cDNA and derived amino acid sequence

The full-length human cDNA is 2.2 kb. The amino acid sequence deduced from cDNA shows the presence of two polypeptides, the α- and the β-chain, 113 and 467 amino acids long, respectively, and five potential sites for asparagine-linked glycosylation [8] (Fig. 3). MPO is synthezised as a 80 kDa precursor that is glycosylated and finally processed to the 59 kDa α-chain and the 13.5 kDa β-chain [9]. A recombinant MPO protein has been expressed in an enzymatically active form in Chinese hamster ovary cells [10]. This protein is recognized by the MPO-ANCA sera.

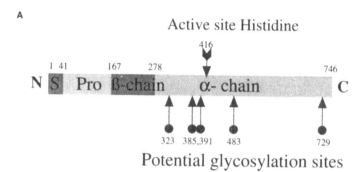

A

Active site Histidine

Potential glycosylation sites

Fig. 3A. Schematic structure of MPO. Major features are outlined. The active site histidine[416] is indicated with an arrow on top. The five potential glycosylation sites are indicated with arrows below. The β-chain starts at amino acid 167 and the α-chain at amino acid 278. S and Pro indicate signal and propeptides, respectively.

B

```
              10        20        30        40        50
MGVPFFSSLRCMVDLGPCWAGGLTAEMKLLLALAGLLAILATPQPSEGAA  50
PAVLGEVDTSLVLSSMEEAKQLVDKAYKERRESIKQRLRSGSASPMELLS 100
YFKQPVAATRTAVRAADYLHVALDLLERKLRSLWRRPFNVTDVLTPAQLN 150
VLSKSSGCAYQDVGVTCPEQDKYRTITGMCNNRRSPTLGASNRAFVRWLP 200
AEYEDGFSLPYGWTPGVKRNGFPVALARAVSNEIVRFPTDQLTPDQERSL 250
              260       270       280       290       300
MFMQWGQLLDHDLDFTPEPAARASFVTGVNCETSCVQQPPCFPLKIPPND 300
PRIKNQADCIPFFRSCPACPGSNITIRNQINALTSFVDASMVYGSEEPLA 350
RNLRNMSNQLGLLAVNQRFQDNGRALLPFDNLHDDPCLLTNRSARIPCFL 400
AGDTRSSEMPELTSMHTLLLREHNRLATELKSLNPRWDGERLYQEARKIV 450
GAMVQIITYRDYLPLVLGPTAMRKYLPTYRSYNDSVDPRIANVFTNAFRY 500
              510       520       530       540       550
GHTLIQPFMFRLDNRYQPMEPNPRVPLSRVFFASWRVVLEGGIDPILRGL 550
MATPAKLNRQNQIAVDEIRERLFEQVMRIGLDLPALNMQRSRDHGLPGYN 600
AWRRFCGLPQPETVGQLGTVLRNLKLARKLMEQYGTPNNIDIWMGGVSEP 650
LKRKGRVGPLLACIIGTQFRKLRDGDRFWWENEGVFSMQQRQALAQISLP 700
RIICDNTGITTVSKNNIFMSNSYPRDFVNCSTLPALNLASWREAS 745
```

Fig. 3B. The amino acid sequence of MPO.

8. The gene

The human gene for MPO is located on the long arm of chromosome 17 (17q22,23)[11]. It is about 14 kb long and contains 11 introns and 12 exons [12].

9. Relation with nucleic acids

In one report it is shown that MPO can bind to DNA and perhaps may play a role in protecting DNA from oxygen radicals produced during cell maturation [13].

10. Relation with other proteins

Not known.

11. Biochemical characteristics

MPO makes up almost 5% of the total protein of a neutrophilic granulocyte. It is a covalently linked dimer with a molecular weight of about 140 kDa that can be cleaved into two halves (Fig. 4). The halves are called hemi-MPO and are enzymatically active [14]. The isoelectric point is > 11. On SDS-PAGE additional MPO bands can be seen, and recently it has been shown that MPO has a methionine containing site that is easily cleaved by boiling [15] (Fig. 5). The protein has been crystallized and the 3Å resolution structure has been solved [16]. The central region consists of five α-helices and surrounding this are polypeptides whose secondary structure is predominantly α-helical with very little β-sheet structure.

12. B-cell epitopes

The MPO-ANCA react with conformational epitopes on the molecule, but the exact nature of these epitopes is not known [17].

13. T-cell epitopes

Not known.

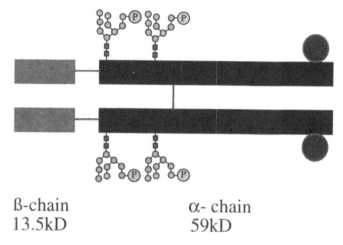

ß-chain
13.5kD

α- chain
59kD

Fig. 4. The peptide subunits of MPO. MPO is a covalently linked dimer with a molecular weight of 140 kDa. It is composed of two halves, each with one α-chain of 59 kDa and one β-chain of 13.5 kDa. Each half of the molecule contains one heme group. The heme group has an intense band of UV absorption around 680 nm that causes the green colour. Two of the glycosylation sites are shown here, but up to four can actually be found in the protein. The glycosylation is characterized by high mannose type oligosaccharide chains.

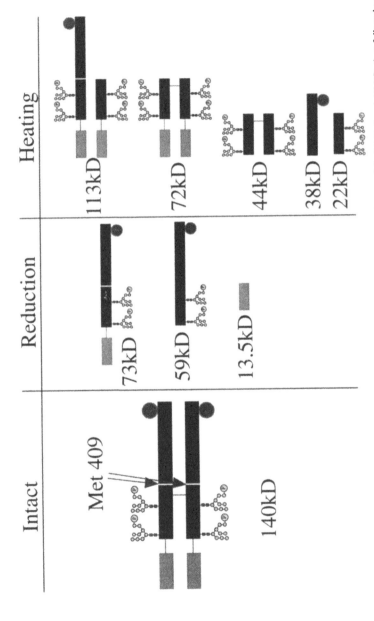

Fig. 5. The fragments of MPO. The complete molecule has a molecular weight of 140 kDa. Upon reduction of disulphide bonds 73 kDa (not fully reduced), 59 kDa and 14 kDa fragments are obtained. By heating without reduction, cleavage at the sensitive methionine can give rise to fragments of 22, 38, 44, 72 and 113 kDa. Met[409] indicates the heat labile methionine at position 409.

References

1. Falk RJ & Jennette JC (1988) N Engl J Med 318:1651–1657
2. Agner K (1941) Acta Chem Scand 2 Suppl 8:1–62
3. Wiik A, Rasmussen N & Wieslander J (1993) In: Van Venrooij WJ & Maini RN (eds) Manual of Biological Markers of Disease, pp A9, 1–14. Kluwer Academic Publishers, Dordrecht, The Netherlands
4. Bretz U & Baggiolini M (1974) J Cell Biol 63:251–269
5. Pryzwansky KB, Martin LE & Spitznagel JK (1978) J Reticuloendothel Soc 24:295–310
6. Harrison JE & Schultz J (1976) J Biol Chem 251:1371–1374
7. Weiss SJ (1989) N Engl J Med 320:365–376
8. Johnson KR, Nauseef WM, Care A, Wheelock MJ, Shane S, Hudson S, Koeffler HP, Selsted M, Millwe C & Rovera G (1987) Nucleic Acids Res 15:2013–2028
9. Nauseef WM, Olsson I & Arnljots K (1988) Eur J Haematol 40:97–110
10. Moguilevsky N, Garcia-Quintana L, Jacquet A, Tournay C, Fabry L, Pierard L & Bollen A (1991) Eur J Biochem 197:605–614
11. Chang KS, Schroeder W & Siciliano MJ (1987) Leukemia 1:458–462
12. Johnson KR, Gemberlein I, Hudson S, Shane S & Rovera G (1989) Nucleic Acids Res 17:7985–7986
13. Murao S, Stevens FJ, Ito A & Huberman E (1988) Proc Natl Acad Sci USA 85:1232–1236
14. Taylor KL, Guzman GS, Pohl J & Kinkade JMJ (1990) J Biol Chem 265:15938–15946
15. Taylor KL, Pohl J & Kinkade JMJ (1992) J Biol Chem 267:25282–25288
16. Zeng J & Fenna RE (1992) J Mol Biol 226:185–207
17. Falk RJ, Becker M, Terrell R & Jennette JC (1992) Clin Exp Immunol 89:274–278

Manual of Biological Markers of Disease **B8.1**: 1–14, 1994.

Mitochondrial autoantigens

PATRICK S.C. LEUNG[1], IAN MACKAY[2] and M. ERIC GERSHWIN[1]

[1] *Division of Rheumatology, Allergy and Clinical Immunology, University of California at Davis, TB 192, School of Medicine, Davis, CA 95616, U.S.A.;* [2] *Centre for Molecular Biology and Medicine, Monash University, Clayton, Victoria, 3168, Australia*

Abbreviations

AMA: Anti-mitochondrial antibodies; BCKD: Branched chain α-keto acids dehydrogenase; BCOADC: Branched chain 2-oxo-acid dehydrogenase complex; KGDC: α-ketoglutarate acid dehydrogenase complex; OGDC: 2-oxo-glutarate acid dehydrogenase complex; PBC: Primary Biliary Cirrhosis; PDC: Pyruvate dehydrogenase complex; PBS: Phosphate buffered saline; PDH: Pyruvate dehydrogenase

Key words: primary biliary cirrhosis, anti-mitochondrial antibodies, 2-oxo-acids dehydrogenase complex, pyruvate dehydrogenase complex, cDNA, lipoyl domain, T cells

1. The antigens

Subunits of the 2-oxo-acids dehydrogenase complexes. These include:
a) Dihydrolipoamide acetyltransferase (EC 2.2.1.12); the E2 subunit of pyruvate dehydrogenase [1, 2]
b) The E2 subunit of branched chain-keto acids dehydrogenase complex, (branched chain 2-oxo-acid dehydrogenase complex) [2]
c) The E2 subunit of α-ketoglutarate acid dehydrogenase (2-oxo-glutarate acid dehydrogenase complex) [3]
d) Protein X [4]
e) PDC-E1α and PDC-E1β [5]

2. Synonyms

a) PDC-E2 or PDH-E2
b) BCKD-E2 or BCOADC-E2
c) KGDC-E2 or OGDC-E2
d) Protein X
e) PDC-E1α and PDC-E1β

These antigens have also collectively been designated as M2, being a cluster of mitochondrial inner membrane proteins [6].

3. Autoantibody detection

Autoantibodies to mitochondrial antigens can be detected by the following methods.

3.1 Indirect immunofluorescence

A typical cytoplasmic fluorescent pattern showing large irregular granules often in a filamentous pattern (Fig. 1, mitochondrial pattern) [7–9]. For technical details, see this Manual, chapter A2 [9].

3.2 Immunoblotting

Anti-mitochondrial antibodies (AMA) can be detected by immunoblotting using, for example, a cytoplasmic extract of cultured cells. The antigen PDC-E2 of molecular weight 70–74 kDa is recognized by most sera from patients with PBC (Fig. 2) [1, 10, 11]. For technical details, see this Manual, chapter A4 [11].

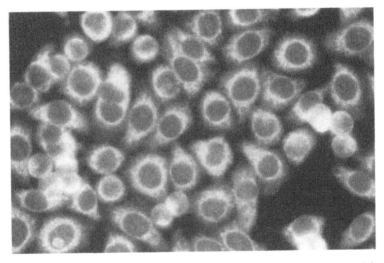

Fig. 1. Immunofluorescence of HEp-2 cells. Note the typical anti-mitochondrial staining pattern of reactivity with serum antibodies from patients with PBC.

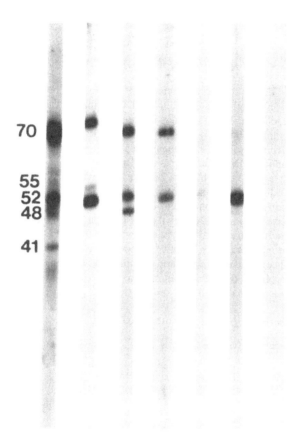

Fig. 2. Immunoblot of PBC sera with beef heart mitochondria. Beef heart mitochondrial extracts were resolved in 10% SDS-PAGE, transferred onto nitrocellulose filters and probed with PBC sera at 1:1,000 dilution. The immunoreactive reactive bands are 70kD PDC-E2, 55kD Protein X, 52KD BCOADC-E2, 48kD OGDC-E2 and 41kD PDC-E1α (lane 1). Serum from lane 2 reacted with the 70 kD, 55 kD and 52 kD protein bands, serum from lane 3 reacted with the 70 kD, 52 kD and 48 kD bands, serum from lane 4 reacted with the 70 and 52 kD protein bands, serum from lane 5 reacted with the 55 kD band only and sera from lane 6 reacted with the 52KD band only. Normal sera do not react (lane 7).

3.3 Elisa

AMA can be detected by ELISA techniques provided that the purified antigen is available [12, 13]. For technical details, see this Manual, part A5 [14].

3.4 Inhibition of catalytic activity of the enzyme

This assay to detect antibody directed to PDC [15] can be carried out in microtiter plates. Briefly, 100 μl of diluted PDC (Sigma, St. Louis, MO, USA) is added to each well. The reaction is initiated by adding 100 μl of reaction mixture containing 50 mM KH_2PO_4, pH 8.0, 1 mM $MgCl_2$, 2.5 mM NAD, 0.2 mM cocarboxylase, 0.12 mM coenzyme A, 2.5 mM β-mercaptoethanol, 0.25% BSA and 1.2 mM sodium pyruvate. Thereafter, absorbance at 340 nm is monitored at 0, 3, 10, 12 and 20 min. The reaction is validated by establishing a linear relationship between the amount of enzyme used and rate of reaction with 8 doubling dilutions of PDC.

To detect inhibition of PDC activity by PBC sera, 100 μl of serial diluted PBC sera are added to a microtiter plate, to which 100 μl of PDC (0.025 U/ml) is added. The reaction is then initiated by adding 100 μl of reaction mixture and absorbance at 340 nm is measured at 2-min intervals for 10 min. Control is set up using 100 μl of PBS instead of sera.

Rate of reaction is defined as:

(Final absorbance − Initial absorbance) / Time

Percentage Inhibition is defined as:

[(Control Rate − Test Rate) / Control Rate] × 100

4. Cellular localization

These antigens are located in the inner mitochondria membrane of eucaryotic cells (Fig. 3).

5. Cellular function

The primary function of the 2-oxo-acid dehydrogenase complexes is the channeling of electrons via oxidative decarboxylase reactions (Table 1). The 2-oxo-acid dehydrogenase complexes are similar in that each of the enzyme complexes consist of three nuclear coded subunits, E1, E2 and E3. Both PDC and BCOADC, have E1 subunits, namely E1α and E1β [16].

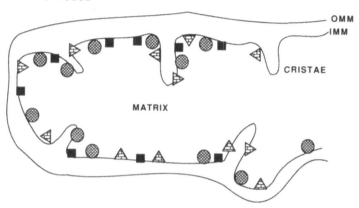

Fig. 3. Diagrammatic representation of intracellular localization of the major mitochondrial autoantigens in PBC. IMM: Inner mitochondrial membrane; OMM: Outer mitochondrial membrane; cross-hatched circles: pyruvate dehydrogenase complex; triangles: branched-chain 2-oxo-acid dehydrogenase complex; ■ : 2-oxo-glutarate dehydrogenase complex.

Table 1. Biochemical and functional characteristics of 2-oxo-acid dehydrogenase complexes in PBC

Enzymes	Molecular weight	Function	Autoantigen in PBC	% of AMA
PDC				
E1α decarboxylase	41	Decarboxylates pyruvate with TPP as a cofactor	+	66
E1β decarboxylase	36	Decarboxylates pyruvate with TPP as a cofactor	+	2
E2 acetyltransferase	74	Transfers acetyl group from E1 to CoA	+	>95
E3 lipoamide dehydrogenase	55	Regenerates disulfide of E2 by oxidation of lipoic acid	−	−
Protein X	56	Unknown	+	>95
BCOADC				
E1 α decarboxylase	46	Decarboxylates α-keto-acids derived from	−	−
E1β decarboxylase	38	leucine, isoleucine, and valine with TPP as a cofactor	−	−
E2 acyltransferase	52	Transfers acyl group from E1 to CoA	+	71
E3 lipoamide dehydrogenase	55	Regenerates disulfide for E2 by oxidation of lipoid acid	−	−

Table 1. Continued

Enzymes	Molecular weight	Function	Autoantigen in PBC	% of AMA
OGDC				
E1 decarboxylase	113	Decarboxylates α-ketogluterate with TPP as a cofactor	−	−
E2 succinyl transferase	48	Transfers succinyl group from E2 to CoA	+	72
E3 lipoamide dehydrogenase	55	Regenerates disulfide from E2 by oxidation of lipic acid	−	−

The PDC-E1 subunit acts to decarboxylase pyruvate, releasing CO_2 and an acetyl group. The E2 subunit, dihydrolipoamide acetyltransferase, transfers the acetyl group via a lipoic acid cofactor to coenzyme A (Co A) which enters the tricarboxylic acid cycle. The E3 subunit, a flavin adenine dinucleotide (FAD) containing dihydrolipoamide, reoxidizes lipoic acid, forming reduced nicotine amide adenine dinucleotide (NADH). Similarly, OGDC decarboxylases 2-oxo-glutarate to succinyl CoA, and BCOADC catalyses the catabolism of branched chain amino acids, leucine, valine and isoleucine [8, 17].

6. Presence in other species

2-oxo-acid dehydrogenase complexes are ubiquitous and present in all species, including yeast and bacteria. However, the AMA activity against mitochondrial preparations in mammalian species is much stronger than its immunoreactivity against 2-oxo-acid acid dehydrogenases in yeast and bacteria [18].

7. cDNA and derived amino acid sequences

The cDNA of PDC-E2 have been cloned and sequenced in rat [19], human [20, 21] (Fig. 4) and mouse (our unpublished data). The cDNA of BCOADC-E2 has been cloned and sequenced in human and bovine species [22]; a striking similarity is the conserved lipoic acid binding domain. The PDC E1α subunit human cDNA has been cloned and sequenced [23–25].

→

Fig. 4. Nucleotide sequence and deduced amino acid sequence of human PDC-E2. Note the lipoyl domains which are underlin d.

```
      R   V   T   S   R   S   G   P   A   P   A   R   R   N   S   V   T   T   G   Y
CGAGTGACCTCGCGATCTGGCCCGGCTCCCGCTCGTCGCAACAGCGTGACTACAGGGTAT
          10                  30                  50
      G   G   V   R   A   L   C   G   W   T   P   S   S   G   A   T   P   R   N   R
GGCGGGGTCCGGGCACTGTGCGGCTGGACCCCCAGTTCTGGGGCCACGCCGCGGAACCGC
          70                  90                  110
      L   L   L   Q   L   L   G   S   P   G   R   R   Y   Y   S   L   P   P   H   Q
TTACTGCTGCAGCTTTTGGGGTCGCCCGGCCGCCGCTATTACAGTCTTCCCCCGCATCAG
          130                 150                 170
      K   V   P   L   P   S   L   S   P   T   M   Q   A   G   T   I   A   R   W   K
AAGGTTCCATTGCCTTCTCTTTCCCCCACAATGCAGGCAGGCACCATAGCCCGTTGGAAA
          190                 210                 230
      K   K   E   G   D   K   I   N   E   G   D   L   I   A̲   E̲   V̲   E̲   T̲   D̲   K̲
AAAAAAGAGGGGGACAAAATCAATGAAGGTGACCTAATTGCAGAGGTTGAAACTGATAAA
          250                 270                 290
      A̲   T̲   V̲   G̲   F̲   E   S   L   E   E   C   Y   M   A   K   I   L   V   A   E
GCCACTGTTGGATTTGAGAGCCTGGAGGAGTGTTATATGGCAAAGATACTTGTTGCTGAA
          310                 330                 350
      G   T   R   D   V   P   I   G   A   I   I   C   I   T   V   G   K   P   E   D
GGTACCAGGGATGTTCCCATCGGAGCGATCATCTGTATCACAGTTGGCAAGCCTGAGGAT
          370                 390                 410
      I   E   A   F   K   N   Y   T   L   D   S   S   A   A   P   T   P   Q   A   A
ATTGAGGCCTTTAAAAATTATACACTGGATTCCTCAGCAGCACCTACCCCACAAGCGGCC
          430                 450                 470
      P   A   P   T   P   A   A   T   A   S   P   P   T   P   S   A   Q   A   P   G
CCAGCACCAACCCCTGCTGCCACTGCTTCGCCACCTACACCTTCTGCTCAGGCTCCTGGT
          490                 510                 530
      S   S   Y   P   P   H   M   Q   V   L   L   P   A   L   S   P   T   M   T   M
AGCTCATATCCCCCTCACATGCAGGTACTTCTTCCTGCCCTCTCTCCCACCATGACCATG
          550                 570                 590
      G   T   V   Q   R   W   E   K   K   V   G   E   K   L   S   E   G   D   L   L
GGCACAGTTCAGAGATGGGAAAAAAAAGTGGGTGAGAAGCTAAGTGAAGGAGACTTACTG
          610                 630                 650
      A̲   E̲   I̲   E̲   T̲   D̲   K̲   A̲   T̲   I̲   G̲   F̲   E   V   Q   E   E   G   Y   L
GCAGAGATAGAAACTGACAAAGCCACTATAGGTTTTGAAGTACAGGAAGAAGGTTATCTG
          670                 690                 710
      A   K   I   L   V   P   E   G   T   R   D   V   P   L   G   T   P   L   C   I
GCAAAAATCCTGGTCCCTGAAGGCACAAGAGATGTCCCTCTAGGAACCCCACTCTGTATC
          730                 750                 770
      I   V   E   K   E   A   D   I   S   A   F   A   D   Y   R   P   T   E   V   T
ATTGTAGAAAAAGAGGCAGATATATCAGCATTTGCTGACTATAGGCCAACCGAAGTAACA
          790                 810                 830
      D   L   K   P   Q   V   P   P   P   T   P   P   P   V   A   A   V   P   P   T
GATTTAAAACCACAAGTGCCACCACCTACCCCACCCCCGGTGGCCGCTGTTCCTCCAACT
          850                 870                 890
      P   Q   P   L   A   P   T   P   S   A   P   C   P   A   T   P   A   G   P   K
```

AMAN-B8.1/7

```
CCCCAGCCTTTAGCTCCTACACCTTCAGCACCCTGCCCAGCTACTCCTGCTGGACCAAAG
          910                 930                 950
G   R   V   F   V   S   P   L   A   K   K   L   A   V   E   K   G   I   D   L

GGAAGGGTGTTTGTTAGCCCTCTTGCAAAGAAGTTGGCAGTAGAGAAAGGGATTGATCTT
          970                 990                1010
T   Q   V   K   G   T   G   P   D   G   R   I   T   K   K   D   I   D   S   F

ACACAAGTAAAAGGGACAGGACCAGATGGTAGAATCACCAAGAAGGATATCGACTCTTTT
          1030                1050                1070
V   P   S   K   V   A   P   A   P   A   A   V   V   P   P   T   G   P   G   M

GTGCCTAGTAAAGTTGCTCCTGCTCCGGCAGCTGTTGTGCCTCCCACAGGTCCTGGAATG
          1090                1110                1130
A   P   V   P   T   G   V   F   T   D   I   P   I   S   N   I   R   R   V   I

GCACCAGTTCCTACAGGTGTCTTCACAGATATCCCAATCAGCAACATTCGTCGGGTTATT
          1150                1170                1190
A   Q   R   L   M   Q   S   K   Q   T   I   P   H   Y   Y   L   S   I   D   V

GCACAGCGATTAATGCAATCAAAGCAAACCATACCTCATTATTACCTTTCTATCGATGTA
          1210                1230                1250
N   M   G   E   V   L   L   V   R   K   E   L   N   K   I   L   E   G   R   S

AATATGGGAGAAGTTTTGTTGGTACCGAAAGAACTTAATAAGATATTAGAAGGGAGAAGC
          1270                1290                1310
K   I   S   V   N   D   F   I   I   K   A   S   A   L   A   C   L   K   V   P

AAAATTTCTGTCAATGACTTCATCATAAAAGCTTCAGCTTTGGCATGTTTAAAAGTTCCC
          1330                1350                1370
E   A   N   S   S   W   M   D   T   V   I   R   Q   N   H   V   V   D   V   S

GAAGCAAATTCTTCTTGGATGGACACAGTTATAAGACAAAATCATGTTGTTGATGTCAGT
          1390                1410                1430
V   A   V   S   T   P   A   G   L   I   T   P   I   V   F   N   A   H   I   K

GTTGCGGTCAGTACTCCTGCAGGACTCATCACACCTATTGTGTTTAATGCACATATAAAA
          1450                1470                1490
G   V   E   T   I   A   N   D   V   V   S   L   A   T   K   A   R   E   G   K

GGAGTGGAAACCATTGCTAATGATGTTGTTTCTTTAGCAACCAAAGCAAGAGAGGGTAAA
          1510                1530                1550
L   Q   P   H   E   F   Q   G   G   T   F   T   I   S   N   L   G   M   F   G

CTACAGCCACATGAATTCCAGGGTGGCACTTTTACGATCTCCAATTTAGGAATGTTTGGA
          1570                1590                1610
I   K   N   F   S   A   I   I   N   P   P   Q   A   C   I   L   A   I   G   A

ATTAAGAATTTCTCTGCTATTATTAACCCACCTCAAGCATGTATTTTGGCAATTGGTGCT
          1630                1650                1670
S   E   D   K   L   V   P   A   D   N   E   K   G   F   D   V   A   S   M   M

TCAGAGGATAAACTGGTCCCTGCAGATAATGAAAAAGGGTTTGATGTGGCTAGCATGATG
          1690                1710                1730
S   V   T   L   S   C   D   H   R   V   V   D   G   A   V   G   A   Q   W   L

TCTGTTACACTCAGTTGTGATCACCGGGTGGTGGATGGAGCAGTTGGAGCCCAGTGGCTT
          1750                1770                1790
A   E   F   R   K   Y   L   E   K   P   I   T   M   L   L   *
GCTGAGTTTAGAAAGTACCTTGAAAAACCTATCACTATGTTGTTGTAACTAACTCAAGAA
          1810                1830                1850
```

AMAN-B8.1/8

8. The gene

The gene of human PDC-E2 has been localized on human chromosome 11q band 23.1 [26]. Gene sequences have not been published.

9. Relation with nucleic acids

None of the 2-oxo-acid dehydrogenase complexes is known to bind to nucleic acids.

10. Relation with other proteins

Mammalian PDC-E2 is known to be associated with a polypeptide, protein X [6], which contains at least one lipoyl moiety that can be radioactively acetylated and appears to be necessary for the catalytic function of PDC.

11. Biochemical characteristics

Each of the 2-oxo-acid dehydrogenase complexes plays a key role in metabolism. PDC oxidatively decarboxylates pyruvate to acetyl CoA, OGDC oxidatively decarboxylates 2-oxo-glutarate to succinyl CoA and BCOADC oxidatively decarboxylates several branched chain amino acids derived 2-oxo-acids to acyl CoAs [14] (Table 1).

Each enzyme complex consists of multiple copies of three enzymes E1, E2 and E3. Using PDC as an example, E1 requires thiamine pyrophosphate as its cofactor; E2 is a dihydrolipoamide acetyltransferase which has a covalently attached lipoic acid as its cofactor; E3 is a flavin-adenine dinucleotide containing dihydrolipoamide dehydrogenase. Structurally, the subunits are arranged in a strategic position with respect to their function. E2 forms a central symmetrical core to which multiple copies of E1, E3 and protein X subunits are attached [6]. Studies by limited proteolysis, electron microscopy and nuclear magnetic resonance spectroscopy demonstrate that the lipoic acid regions of the E2 protrude from the E2 core and extend between E1 and E3 subunits [27, 28]. It is likely that this particular conformation allows mobility and greatly extends the working radius of a swinging arm for active enzyme substrate interaction [29] (Fig. 5).

←

Fig. 4. Continued.

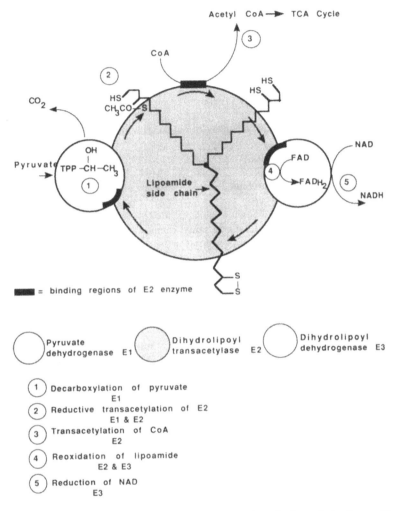

Fig. 5. Diagrammatic representation of the biochemical mechanisms of the 2-oxo-acid dehydrogenase complexes, illustrated by PDC as an example. Note the central location of the E2 core to which E1 and E3 are attached and the strategic location of the lipoic acid side chain which extends the working radius of a swinging arm for reaction.

12. B-cell epitopes

The B-cell epitope of human PDC-E2 has been determined and found to include the inner lipoyl domain (Fig. 6).

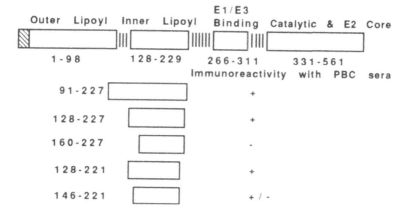

Fig. 6. Structural domains and B cell autoepitope of PDC-E2.

The relative positions of the structural domains are shown with the position of the amino acid residues. III indicates the hinge region between the structural domains. The immunodominant B-cell epitope is located in the inner lipoyl domain. +/− corresponds to positive and negative immunoreactivity of the constructs with PBC sera respectively [30].

Lipoic acid is covalently bound to the E amino group of the lysine residue. Several studies have indicated that lipoic acid binding site is the immunodominant epitope. Using synthetic peptides, it has been shown that a 20 amino acid peptide that corresponds to the lipoate-binding region of rat PDC-E2 can absorb autoantibodies reactivity against the recombinant rat PDC-E2 [1]. Using recombinant peptides corresponding to different domains of the human PDC-E2, it has been shown that PBC sera recognize specifically recombinant peptides corresponding to the outer lipoyl domain and inner lipoyl domain of human PDC-E2 [30]. Moreover, most of the reactivity is confined to amino acid residues 128–227 within the inner lipoyl domain. A recombinant peptide containing amino acid residues 160–227 was not reactive. From the C-terminus, deletion of amino acid residues beyond amino acid residue 221 abolished reactivity with PBC sera. Recombinant peptide containing amino acid residues 146–221 is required for detectable binding and amino acid residues 128–227 are required for strong binding. The data strongly suggest a conformational epitope between amino acid residues 128 and 227 in PDC-E2 [28].

The B-cell epitopes of BCOADC-E2, OGDC-E2, Protein X, PDC-E1α, PDC-E1β have not been reported.

13. T-cell epitopes

CD3+, CD4+, CD8- T lymphocytes are found in the portal tracts in late stage PBC. These T-cells infiltrates predominantly express TCR-$\alpha\beta$ chain, only a small percentage of CD3+ T lymphocytes have $\gamma\delta$ TCR on their surface [31–33]. Published data on T-cell epitopes are not available.

14. Monoclonal antibodies

A number of mouse monoclonal antibodies directed to PDC-E2, the major autoantigen of PBC have been described [34].

References

1. Van de Water J, Gershwin ME, Leung P & Coppel RL (1988) The autoepitope of the 74 kD mitochondrial autoantigen of primary biliary cirrhosis corresponds to the functional site of dihydrolipoamide acetyltransferase. J Exp Med 167:1791–1799
2. Yeaman S, Danner DJ, Mutimer DJ, Fussey SPM, James OFW & Bassendine MF (1988) Primary biliary cirrhosis: Identification of two major M2 mitochondrial autoantigens. Lancet i:1067–1070
3. Fussey SPM, Guest JR, James OFW, Bassendine MF & Yeaman SJ (1988) Identification and analysis of the major M2 autoantigens in primary biliary cirrhosis. Proc Natl Acad Sci USA 85:8654–8658
4. Surh CD, Roche TE, Danner DJ, Ansakri A, Coppel RL, Prindiville T, Dickson ER & Gershwin ME (1989) Antimitochondrial autoantibodies in primary biliary cirrhosis recognize cross-reactive epitope on protein X and dihydrolipoamide acetyltransferase of pyruvate dehydrogenase complex. Hepatology 10:127–133
5. Fregeau DR, Roche TE, Davis PA, Coppel RL & Gershwin ME (1990) Inhibition of pyruvate dehydrogenase complex activity by autoantibodies specific for E1α, a non-lipoic acid containing mitochondrial enzyme. J Immunol 144:1671–1676
6. Mackay IR & Gershwin ME (1989) Molecular basis of mitochondrial autoreactivity in primary biliary cirrhosis. Immunol Today 10:315–318
7. Walker JG, Doniach D, Roitt IM & Sherlock S (1986) Serological tests in the diagnosis of primary biliary cirrhosis. Lancet 1:827–831
8. Van de Water J, Surh CD, Leung PSC, Krams SM, Fregeau D, Davis P, Coppel RL, Mackay IR & Gershwin ME (1989) Molecular definitions, autoepitopes and enzyme activities of the mitochondrial autoantigens of primary biliary cirrhosis. Semin Liver Dis 9:132–137
9. Humbel RL (1993) Detection of antinuclear antibodies by immunofluorescence. Manual Biol Markers Dis A2:1–16
10. Lindenborn-Fotinos J, Baum H & Berg PA (1985) Mitochondrial antibodies in primary biliary cirrhosis: Species and nonspecies specific determinants of M2 antigen. Hepatology 5:763–769
11. Verheijen R, Salden M & Van Venrooij WJ (1993) Protein blotting. Manual Biol Markers Dis A4:1–25
12. Van de Water J, Cooper A, Surh CD, Coppel R, Danner D, Ansari A, Dickson R & Gershwin ME (1989) Detection of autoantibodies to the recombinant 74 kD and 52 kD mitochondrial autoantigens of primary biliary cirrhosis. N Engl J Med 320:1377–1380
13. Leung PSC, Iwayama T, Prindiville T, Chuang DT, Ansari AA, Wynn RM, Dickson R, Coppel R & Gershwin ME (1992) Use of designer recombinant mitochondrial antigens in the diagnosis of primary biliary cirrhosis. Hepatology 15:367–372
14. Charles PJ & Maini RN (1993) Enzyme-linked immunosorbant assay in the rheumatological laboratory. Manual Biol Markers Dis A5:1–23
15. Teoh KL, Rowley MJ & Mackay IR (1991) An automated microassay for enzyme inhibitory effects of M2 antibodies in primary biliary cirrhosis. Liver 11:287–291
16. Yeaman SJ (1986) The mammalian 2-oxo-acid dehydrogenases. A complex family. Trends Biochem Sci 11:293–396
17. Gershwin ME & Mackay IR (1991) Primary biliary cirrhosis: Paradigm or paradox for autoimmunity. Gastroenterology 100:822–833
18. Gershwin ME, Rowley R, Davis PA, Leung P, Coppel R & Mackay IR (1992) Molecular biology of the 2-oxo-acid dehydrogenase complexes and anti-mitochondrial antibodies. In: Progress in Liver Diseases, 10:47–61
19. Gershwin ME, Mackay IR, Sturgess A & Coppel RL (1987) Identification of a cDNA encoding the 70 kDa mitochondrial antigen recognized in the human autoimmune disease primary biliary cirrhosis. J Immunol 138:3525–3536

20. Coppel RL, McNeilage J, Surh CD, Van de Water J, Spithill TW, Whittingham S, Gershwin ME (1988) Primary structure of the human M2 mitochondrial autoantigen of primary biliary cirrhosis; dihydrolipoamide acetyltransferase. Proc Natl Acad Sci USA 85:7317–7321

21. Thekkumkara TJ, Jesse BW, Ho L, Raaesfky C, Pepin RA, Javed AA, Pons G & Patel MS (1987) Isolation of a cDNA clone for the dihydrolipoamide acetyltransferase component of the human liver pyruvate dehydrogenase complex. Biochem Biophys Res Comm 2:903–907

22. Lau KS, Griffin TA, Hu CWC & Chuang DT (1988) Conservation of primary structure in the lipoyl-bearing and dihydrolipoyl dehydrogenase binding domain of mammalian branched chain α-keto acid dehydrogenase complex: Molecular cloning of human and bovine transacylase (E2) cDNAs. Biochemistry 27:1972–1981

23. Dahl HMM, Hunt SM, Hutchison WH & Brown GK (1987) The human pyruvate dehydrogenase complex. Isolation of cDNA clones for the E1α subunit, sequence analysis and characterization of the mRNA. J Biol Chem 262:7398–7403

24. Koike K, Ohita S, Urata Y, Kagawa Y & Koike M (1988) Cloning and sequencing of cDNAs encoding α and β subunits of human pyruvate dehydrogenase. Proc Natl Acad Sci USA 85:41–45

25. DeMeirler I, Mackay N, Marie A, Lam HW & Robinson BH (1988) Isolation for a full length complementary DNA encoding for human E1α subunit of the pyruvate dehydrogenase complex. J Biol Chem 263:1991–1995

26. Leung PSC, Watanabe Y, Munoz S, Patel M, Korenberg JR, Hara P, Coppel R & Gershwin ME (1993) Chromosome localization and RFLP analysis of PDC-E2: the major autoantigen of primary biliary cirrhosis. Autoimmunity 14:335–340

27. Bleile DM, Munk P, Oliver RM & Reed LJ (1979) Subunit structure of dihydrolipoamide transacetylase component of pyruvate dehydrogenase complex from Escherichia coli. Proc Natl Acad Sci USA 76:4385–4389

28. Perham RN, Duckworth HW & Roberts GCK (1981) Mobility of polypeptide chain in the pyruvate dehydrogenase complex revealed by proton NMR. Nature 292:474–477

29. Leung PSC & Gershwin ME (1990) The molecular structure of autoantigens. Curr Opin Immunol 2:567–575

30. Surh CD, Coppel R & Gershwin ME (1990) Structural requirement for autoreactivity on human pyruvate dehydrogenase-E2, the major autoantigen of primary biliary cirrhosis. J Immunol 144:3367–3374

31. Van de Water J, Ansari AA, Surh CD, Coppel R, Roche T, Bonkovsky H, Kaplan M & Gershwin ME (1991) Evidence for the targeting by 2-oxo-dehydrogenase enzymes in the T cell response of primary biliary cirrhosis. J Immunol 146:89–94

32. Moebius U, Manns M, Hess G, Kober G, Meyerzum KH & Meuer SC (1990) T cell receptor gene rearrangements of T lymphocytes infiltrating the liver in chronic active hepatitis B and primary biliary cirrhosis (PBC) oligoclonality of PBC-derived T cell clones. Eur J Immunol 20:889–896

33. Krams SM, Van de Water J, Coppel RL, Esquivel C, Roberts J, Ansari A & Gershwin ME (1990) Analysis of hepatic T lymphocyte and immunoglobulin deposits in patients with primary biliary cirrhosis. Hepatology 12:306–313

34. Surh CD, Ahmed-Ansari A & Gershwin ME (1990) Comparative epitope mapping of murine monoclonal and human autoantibodies to human PDH-E2, the major mitochondrial autoantigen of primary biliary cirrhosis. J Immunol 144:2647–2652

Manual of Biological Markers of Disease **B8.2**: 1–11, 1994.
© 1994 *Kluwer Academic Publishers. Printed in the Netherlands.*

P80 coilin

MARIA CARMO-FONSECA[1], KERSTIN BOHMANN[2],
MARIA TERESA CARVALHO[1] and ANGUS I. LAMOND[2]
[1] *Institute of Histology and Embryology, Faculty of Medicine, University of Lisbon, Portugal;*
[2] *European Molecular Biology Laboratory, Heidelberg, Germany*

Abbreviations

RNP: ribonucleoprotein; rRNA: ribosomal RNA; snRNP: small nucleolar
RNP; snoRNA: small nucleolar RNA

Key words: coilin; coiled body; splicing; snRNP

1. The antigen

p80 coilin is a nuclear autoantigen with a very recent history. It was first
identified as an approximately 80 kDa protein recognized by several human
autoimmune sera that specifically labelled a discrete nuclear structure called
the coiled body [1, 2].

2. Synonyms

There are no synonyms for this antigen.

3. Antibody detection

At present, the activity of anti-coilin antibodies has been detected by
immunoblotting, immunoprecipitation, immunofluorescence and immuno-
electron microscopy.

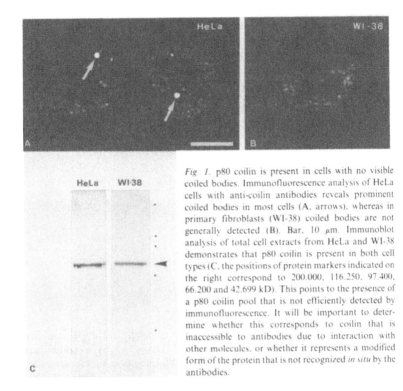

Fig. 1. p80 coilin is present in cells with no visible coiled bodies. Immunofluorescence analysis of HeLa cells with anti-coilin antibodies reveals prominent coiled bodies in most cells (A, arrows), whereas in primary fibroblasts (WI-38) coiled bodies are not generally detected (B). Bar, 10 μm. Immunoblot analysis of total cell extracts from HeLa and WI-38 demonstrates that p80 coilin is present in both cell types (C, the positions of protein markers indicated on the right correspond to 200.000, 116.250, 97.400, 66.200 and 42.699 kD). This points to the presence of a p80 coilin pool that is not efficiently detected by immunofluorescence. It will be important to determine whether this corresponds to coilin that is inaccessible to antibodies due to interaction with other molecules, or whether it represents a modified form of the protein that is not recognized *in situ* by the antibodies.

3.1 Immunofluorescence

Indirect immunofluorescence using anti-coilin antibodies reveals bright foci in the nucleoplasm of cells both in culture and in tissues (Figs. 1A and 2). These bright foci correspond to coiled bodies (see next section). Immunofluorescence microscopy shows that coiled bodies are usually present in most cells.

3.2 Immunoblotting

Western blotting analysis using either human autoimmune sera or rabbit antibodies raised against recombinant coilin reveal a cellular protein with an electrophoretic mobility of ~80 kDa (Fig. 1C) [1].

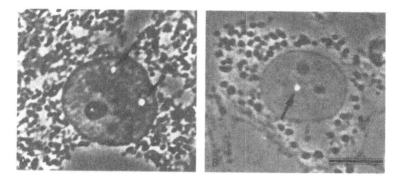

Fig. 2. Immunofluorescence localization of p80 coilin in normal tissue. Cryo-sections of rat liver were immunolabelled with affinity purified polyclonal antibodies raised against recombinant p80 coilin and observed with a confocal laser scanning microscope. (A) and (B) represent a superimposition of phase contrast and fluorescence images. Coiled bodies are indicated by arrows. Bar, 10 μm.

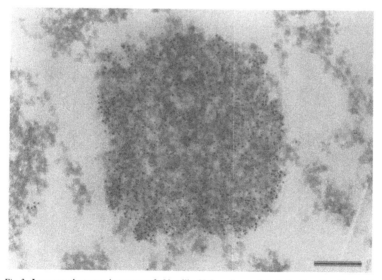

Fig. 3. Immuno-electron microscopy of p80 coilin. HeLa cells were labelled with anti-coilin rabbit polyclonal antibodies using a pre-embedding technique. The coiled body is composed of dense fibrillar threads or coils which are intensely labelled by immuno-gold particles (bar, 0.2 μm). Coiled bodies are more or less spherical aggregates, with 0.5–1 μm in diameter, composed of coiled threads which appear to correspond to bundles of twisted tiny fibrils. The term "coiled bodies" was first used by Monneron and Bernhard [4]. Subsequent studies have shown that the coiled body represents the ultrastructural counterpart of the "accessory body" originally described by Ramón y Cajal in 1903 [5, 6]. Coiled bodies have been described in a variety of cell types, including rat liver and pancreas, HeLa cells [4], neurons [5–9], plant cells [10] and rat germinal cells [11]. Possibly, coiled bodies are present in the great majority, if not all, eukaryotic cells.

3.3 Immunoprecipitation

Both human autoimmune sera and rabbit antibodies precipitate p80 coilin [3].

3.4 Immuno-electron microscopy

In the electron microscope anticoilin antibodies specifically label discrete nucleoplasmic structures called coiled bodies (Fig. 3) [1, 2].

4. Cellular localization

p80 coilin is localised in coiled bodies. However, immunoblot analysis demonstrates the presence of coilin even in cells that appear devoid of coiled bodies in the fluorescence microscope (Fig. 1) [3].

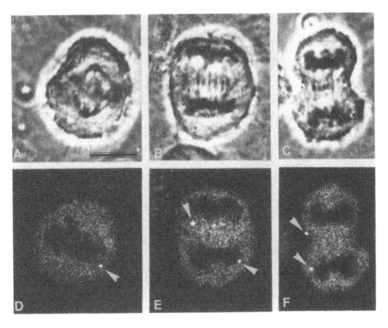

Fig. 4. Distribution of p80 coilin in mitotic cells. In metaphase (A & D), anaphase (B & E) and early telophase (C & F) cells anti-coilin antibodies diffusely stain the cytoplasm, excluding the chromosomes and, in addition, label small remnants of coiled bodies (arrows). The mitotic remnants of coiled bodies are smaller and more numerous than the typical interphase coiled bodies [24]. Bar, 10 μm.

During mitosis coilin is predominantly diffuse in the cell and, in addition, is present in small remnants of coiled bodies that persist scattered in the cytoplasm of metaphase, anaphase and early telophase cells (Fig. 4). Additionally, when cell extracts are separated into soluble and particulate fractions, interphase coilin is mostly present in the particulate fraction, whereas mitotic coilin is predominantly detected in the soluble fraction [16]. These results suggest that the coiled body breaks down during mitosis. Possibly, the disassembly of the coiled body is triggered by the mitotic phosphorylation of p80 coilin (see par. 11).

5. Cellular function

The cellular function of p80 coilin is not yet known. However, since this protein is present in coiled bodies, we may speculate that coilin and coiled bodies are functionally related.

Both immunofluorescence and immuno-electron microscopy demonstrate that coiled bodies contain splicing snRNPs (for a recent review see [12]). The association of splicing snRNPs with coiled bodies is dynamic and dependent upon the metabolic state of the cell. When transcription is shut off by drugs [13] or at late stages of erythroid differentiation [14], snRNPs no longer co-localize with p80 coilin. These findings argue against coiled bodies being storage sites where inactive snRNPs accumulate. In contrast, the number of coiled bodies increases when gene expression is stimulated [15, 16], suggesting that coiled bodies may be engaged in some form of RNA metabolism connected with splicing snRNPs and gene expression.

Coiled bodies are frequently observed in association with the periphery of the nucleolus and have thus been implicated in the metabolism of ribosomal RNA (reviewed in [17]). Fibrillarin, a protein involved in pre-rRNA processing in the nucleolus, co-localizes with p80 coilin in coiled bodies [18, 3] suggesting that the two nuclear organelles may share a common structure or function. At present no other nucleolar components have been detected in the coiled body, namely nucleolin, nucleolar protein B23, RNA polymerase I, 5S RNP [2], 5S rRNA and 28S rRNA [3]. The absence of rRNA from the coiled body strongly argues against this organelle having a role in the biogenesis of ribosomes. Very recent data raise an alternative hypothesis for a nucleolar-related function of the coiled body. It was shown that several nucleolar UsnoRNAs, including the U14, U15, U16 and U17 snoRNAs, are encoded within introns of pre-mRNAs [19–22]. Therefore, the coiled body could be involved in the specific processing and/or transport of these snoRNAs from the nucleoplasm to the nucleolus.

A

p80-Coilin

```
     CCCCGTGGTATCTCTCGGCTTCCGTTGAGCACCAAGCAAGATGGCAGCTTCCGAGACGGT
1    ---------+---------+---------+---------+---------+---------+
                                              M  A  A  S  E  T  V

     TAGGCTACGGCTTCAATTTGATTACCCGCCGCCAGCTACCCCGCACTGTACGGCCTTCTG
61   ---------+---------+---------+---------+---------+---------+
      R  L  R  L  Q  F  D  Y  P  P  P  A  T  P  H  C  T  A  F  W

     GCTTCTGGTCGACTTGAACAGATGCCGAGTCGTCACAGATCTCATTAGTCTCATCCGCCA
121  ---------+---------+---------+---------+---------+---------+
      L  L  V  D  L  N  R  C  R  V  V  T  D  L  I  S  L  I  R  Q

     GCGCTTCGGCTTCAGTTCTGGGGCCTTCCTAGGCCTCTACCTGGAGGGGGGGCTCTTGCC
181  ---------+---------+---------+---------+---------+---------+
      R  F  G  F  S  S  G  A  F  L  G  L  Y  L  E  G  G  L  L  P

     CCCCGCCGAGAGCGCGCGCCTTGTGAGAGACAACGACTGCCTCAGAGTTAAATTAGAAGA
241  ---------+---------+---------+---------+---------+---------+
      P  A  E  S  A  R  L  V  R  D  N  D  C  L  R  V  K  L  E  E

     GAGAGGAGTTGCTGAGAATTCTGTAGTCATCAGTAATGGTGACATTAATTTATCTCTTAG
301  ---------+---------+---------+---------+---------+---------+
      R  G  V  A  E  N  S  V  V  I  S  N  G  D  I  N  L  S  L  R

     AAAAGCAAAGAAGCGGGCATTTCAGTTAGAGGAGGGTGAAGAAACTGAACCAGATTGCAA
361  ---------+---------+---------+---------+---------+---------+
      K  A  K  K  R  A  F  Q  L  E  E  G  E  E  T  E  P  D  C  K

     ATATTCAAAGAAGCATTGGAAGAGTCGAGAGAACAATAACAATAATGAGAAGGTCTTGGA
421  ---------+---------+---------+---------+---------+---------+
      Y  S  K  K  H  W  K  S  R  E  N  N  N  N  E  K  V  L  D

     TCTGGAACCAAAAGCTGTCACAGATCAGACTGTCAGCAAAAAAAACAAGAGAAAAAATAA
481  ---------+---------+---------+---------+---------+---------+
      L  E  P  K  A  V  T  D  Q  T  V  S  K  K  N  K  R  K  N  K

     AGCAACCTGTGGCACAGTGGGTGATGATAACGAAGAGGCCAAAAGAAAATCACCAAAGAA
541  ---------+---------+---------+---------+---------+---------+
      A  T  C  G  T  V  G  D  D  N  E  E  A  K  R  K  S  P  K  K

     AAAGGAGAAATGTGAATATAAAAAAAAAGGCTAAGAATCCCAAGTCTCCGAAAGTACAGGC
601  ---------+---------+---------+---------+---------+---------+
      K  E  K  C  E  Y  K  K  K  A  K  N  P  K  S  P  K  V  Q  A

     AGTGAAAGACTGGGCCAATCAGAGATGTAGTTCTCCAAAAGGTTCTGCTAGAAACAGCCT
661  ---------+---------+---------+---------+---------+---------+
      V  K  D  W  A  N  Q  R  C  S  S  P  K  G  S  A  R  N  S  L

     TGTTAAAGCCAAAAGGAAAGGTAGTGTAAGCGTTTGCTCAAAAGAGAGTCCCAGTTCCTC
721  ---------+---------+---------+---------+---------+---------+
      V  K  A  K  R  K  G  S  V  S  V  C  S  K  E  S  P  S  S  S

     CTCGGAGTCTGAGTCTTGTGATGAATCTATCAGTGATGGTCCCAGCAAAGTCACTTTGGA
781  ---------+---------+---------+---------+---------+---------+
      S  E  S  E  S  C  D  E  S  I  S  D  G  P  S  K  V  T  L  E

     GGCCAGAAATTCCTCAGAGAAATTACCAACTGAGTTATCAAAGGAAGAACCCTCTACCAA
841  ---------+---------+---------+---------+---------+---------+
      A  R  N  S  S  E  K  L  P  T  E  L  S  K  E  E  P  S  T  K

     AAATACAACTGCAGACAAACTGGCTATAAAACTTGGCTTTAGCCTTACCCCCAGCAAGGG
901  ---------+---------+---------+---------+---------+---------+
      N  T  T  A  D  K  L  A  I  K  L  G  F  S  L  T  P  S  K  G

     CAAGACCTCTGGAACAACATCTTCCAGTTCAGACTCTAGTGCAGAGTCAGACGACCAATG
961  ---------+---------+---------+---------+---------+---------+
      K  T  S  G  T  T  S  S  S  S  D  S  S  A  E  S  D  D  Q  C

     CTTGATGTCATCGAGCACCCCGGAGTGTGTCTGCGGGTTTCTTAAAGACAGTAGGCCTTTT
1021 ---------+---------+---------+---------+---------+---------+
      L  M  S  S  T  P  E  C  A  A  G  F  L  K  T  V  G  L  F

     TGCAGGAAGAGGTCGTCCAGGCCCAGGGCTGTCATCACAGACTGCAGGTGCTGCTGGATG
1081 ---------+---------+---------+---------+---------+---------+
      A  G  R  G  R  P  G  P  G  L  S  S  Q  T  A  G  A  A  G  W

     GAGGCGTTCTGGCTCAAATGGTGGTGGACAGGCTCCTGGTGCTTCTCCCAGTGTGTCTCT
1141 ---------+---------+---------+---------+---------+---------+
      R  R  S  G  S  N  G  G  G  Q  A  P  G  A  S  P  S  V  S  L

     CCCCTGCTAGTTTAGGAAGAGGATGGGGTAGAGAAGAGAACCTTTTTTCTTGGAAGGGAGC
1201 ---------+---------+---------+---------+---------+---------+
      P  A  S  L  G  R  G  W  G  R  E  E  N  L  F  S  W  K  G  A
```

AMAN-B8.2/6

```
     TAAGGGACGGGGCATGCGGGGGAGAGGTCGAGGACGAGGGCATCCTGTTTCCTGTGTTGT
1261 ---------+---------+---------+---------+---------+---------+
     K  G  R  G  M  R  G  R  G  R  G  R  G  H  P  V  S  C  V  V

     AAATAGAAGCACTGACAACCAGAGGCAACAGCAATTAAATGACGTGGTAAAAAATTCATC
1321 ---------+---------+---------+---------+---------+---------+
     N  R  S  T  D  N  Q  R  Q  Q  Q  L  N  D  V  V  K  N  S  S

     TACTATTATCCAGAATCCAGTAGAGACACCCAAGAAGGACTATAGTCTGTTACCACTGTT
1381 ---------+---------+---------+---------+---------+---------+
     T  I  I  Q  N  P  V  E  T  P  K  K  D  Y  S  L  L  P  L  L

     AGCAGCTGCCCCTCAAGTTGGAGAAAAGATTGCATTTAAGCTTTTGGAGCTAACATCCAG
1441 ---------+---------+---------+---------+---------+---------+
     A  A  A  P  Q  V  G  E  K  I  A  F  K  L  L  E  L  T  S  S

     TTACTCTCCTGATGTCTCTGACTACAAGGAAGGAAGAATATTAAGCCACAATCCAGAGAC
1501 ---------+---------+---------+---------+---------+---------+
     Y  S  P  D  V  S  D  Y  K  E  G  R  I  L  S  H  N  P  E  T

     CCAGCAAGTAGATATAGAAATTCTTTCATCCTTACCTGCCTTGAGAGAACCTGGGAAATT
1561 ---------+---------+---------+---------+---------+---------+
     Q  Q  V  D  I  E  I  L  S  S  L  P  A  L  R  E  P  G  K  F

     TGATTTAGTTTATCACAATGAAAATGGAGCCGAGGTAGTGGAGTACGCTGTGACACAGGA
1621 ---------+---------+---------+---------+---------+---------+
     D  L  V  Y  H  N  E  N  G  A  E  V  V  E  Y  A  V  T  Q  E

     GAGCAAGATCACTGTATTTTGGAAAGAGTTGATTGACCCAAGACTGATTATTGAATCTCC
1681 ---------+---------+---------+---------+---------+---------+
     S  K  I  T  V  F  W  K  E  L  I  D  P  R  L  I  I  E  S  P

     AAGTAACACATCAAGTACAGAACCTGCCTGAGTATGACCTCTCCACCTTATAGTTTATGA
1741 ---------+---------+---------+---------+---------+---------+
     S  N  T  S  S  T  E  P  A  *

     ATGTCTTGTTTGTGAAAGTGACTATAACCCAAACTTTTTTTTTTTTTTTAAAGAGGATTTG
1801 ---------+---------+---------+---------+---------+---------+

     GAAGTTGTATGGATTTTTTTGTTATCTTCACTTTACTGCATAGGAAACAATCTACCTCAT
1861 ---------+---------+---------+---------+---------+---------+

     CATTTAAAATGACATGGGTGTCGGTTTTGTAGATCTTTGGTTTTTTTGTCAGGTTTAATT
1921 ---------+---------+---------+---------+---------+---------+

     TCAGTTAACAAAATGTAAAACATGACATTCCCTGCAGATATTGTTGTATACCAGTATGGT
1981 ---------+---------+---------+---------+---------+---------+

     TTCTTCTCTTTCTTTAAATGTTTTTGGCCATCAAGTAGCAGTCGTCAGTAGGAGTTTATA
2041 ---------+---------+---------+---------+---------+---------+

     ATACCAAGAATGTGCTGCGTATCTTGTCTCAATAAGTTTTAAGTAACATTTAAAAATATT
2101 ---------+---------+---------+---------+---------+---------+

     AAAGCATGTTATTTGACCTAATTTTTTAGCATTTGAGTTGTTCCATTAAATGGAGCATCT
2161 ---------+---------+---------+---------+---------+---------+

     TGTAAATTTCAAGTATTTTATACTTGCAATTGTTAAGAGTTAACAGGTAGTTGGATTTGT
2221 ---------+---------+---------+---------+---------+---------+

     CGCAGACAATGAGTTAAGGAATCCTTTCACGTTTTTCCCAACTTTAAAATTAAGGATTCT
2281 ---------+---------+---------+---------+---------+---------+

     CAGGTCCCTGTGTAGAGCAGTGAAAATAAGATGTGCGTATGTGTGTGTATGCCTGGAGAT
2341 ---------+---------+---------+---------+---------+---------+

     TGGTGTTTCACTTCAGTGAGAGGATTGGCTGTGAGCTTCAGACCAGAAATGTGTCATCTT
2401 ---------+---------+---------+---------+---------+---------+

     GCCAGCCCCTGGCTGAGTGTGCTGGAGTGAGGATCTTGAACAGAAACTTCCTTTTCTGTT
2461 ---------+---------+---------+---------+---------+---------+

     ATTATTCACTACGAAGCTAAAATGGCCAAATATATACCGTGAAAATTGGTTTCATTTAAC
2521 ---------+---------+---------+---------+---------+---------+

     AAAAGATCAGATCCCTCTTTCAGCTGTACACATTTTTAAATAAAATCATATTGAACT
2581 ---------+---------+---------+---------+---------+-------
```

←

Fig. 5A. Nucleotide and amino acid sequence of p80 coilin.
The cDNA and deduced amino acid sequence is shown for p80 coilin. The cDNA nucleotide sequence is compiled from the separate sequences determined by Andrade *et al.* (1991, EMBL/Genbank/DDBJ accession number M58411) and by Bohmann *et al.* (unpublished). In the region of overlap between these independently determined sequences there are 11 differences,

B

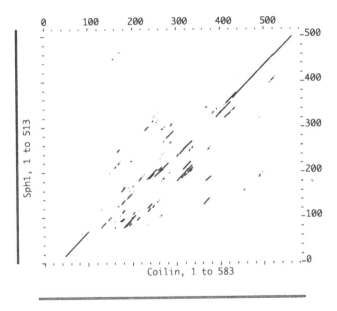

Fig. 5B. Dot plot comparison of the human p80 coilin and Xenopus SPH-1 protein sequences. Stringency used was 15.0. In this type of plot regions of identity between the two proteins appear as an unbroken diagonal line.

6. Presence in other species

At present only one coilin-related protein has been identified. It is a *Xenopus laevis* protein named SPH-1 [23]. The SPH-1 protein is present in the sphere organelle, a distinct nuclear structure of amphibian oocytes that may be related to mammalian coiled bodies. Overall there is 38.3% identity between coilin and SPH-1 (Fig. 5). It will be important to clarify whether SPH-1 is the amphibian homologue of human p80 coilin or whether it represents another member of a coilin-related protein family.

---

none of which alter the deduced amino acid sequence of p80 coilin. We show here the cDNA sequence according to Bohmann *et al.*, since it matches exactly with the corresponding genomic clone (Bohmann *et al.*, unpublished).

7. The cdna and derived amino acid sequence of p80 coilin

The sequence of a partial cDNA for p80 coilin has been reported by Andrade *et al.* [1]. We have recently obtained the full cDNA and genomic coding sequence of coilin (Bohmann *et al.*, unpublished data). The deduced amino acid sequence is shown in Figure 5. It corresponds to a protein of 576 amino acids with a predicted molecular mass of 62.57 kD and an estimated pI of 9.7. No consensus sequence motif was identified that gives a clue to the function of the protein.

8. The gene

The genomic organization of p80 coilin is not yet known.

9. Relation with nucleic acids

At present there is no evidence for a direct interaction of coilin with nucleic acids.

10. Relation with other proteins

Interactions of coilin with other proteins have not yet been established.

11. Biochemical characteristics

In vivo labelling of HeLa cells with [^{32}P] orthophosphate demonstrates that coilin is a phosphoprotein predominantly phosphorylated on serine residues both in interphase and in mitosis [3]. Comparison of tryptic peptides from *in vivo* labelled coilin extracted from interphase and mitotic cells demonstrates that p80 coilin is phosphorylated on at least two additional sites during mitosis [3], but the precise identification of these sites remains to be characterized. It will also be important to identify which kinase(s) are involved in the mitotic phosphorylation of coilin.

12. B-cell epitopes

B-cell epitopes are not yet known.

13. T-cell epitopes

T-cell epitopes are not yet known.

References

1. Andrade LEC, Chan EKL, Raska I, Peebles CL, Roos G & Tan EM (1991) J Exp Med 173:1407–1419
2. Raska I, Andrade LEC, Ochs RL, Chan EKL, Chang C-M, Roos G & Tan EM (1991) Exp Cell Res 195:27–37
3. Carmo-Fonseca M, Ferreira J & Lamond AI (1993) J Cell Biol 120:841–852
4. Monneron A & Bernhard W (1969) J Ultrastruct Res 27:266–288
5. Hardin JH, Spicer SS & Greene (1969) Anat Rec 164:403–432
6. Lafarga M, Hervas JP, Santa-Cruz MC, Villegas J & Crespo D (1983) Anat Embryol 166:19–30
7. Kinderman NB & LaVelle A (1976) J Neurocytol 5:545–550
8. Hervás JP, Villegas J, Crespo D & Lafarga M (1980) Am J Anat 159:447–454
9. Seite R, Pebusque MJ & Vio-Cigna M (1982) Biol Cell 46:97–100
10. Moreno Diaz de la Espina S, Sanchez-Pina MA & Risueño MC (1982) Cell Biol Int Rep 6:601–607
11. Schultz M (1990) Am J Anat 189:11–23
12. Lamond AI & Carmo-Fonseca M (1993) Trends Cell Biol 3:198–204
13. Carmo-Fonseca M, Pepperkok R, Carvalho MT & Lamond AI (1992) J Cell Biol 117:1–14
14. Antoniou M, Carmo-Fonseca M, Ferreira J & Lamond AI (1993) J Cell Biol 123:1055–1068
15. Lafarga M, Andres MA, Berciano MT & Maquiera E (1991) J Comp Neurol 308:329–339
16. Andrade LEC, Tan EM & Chan EKL (1993) Proc Natl Acad Sci USA 90:1947–1951
17. Brasch L & Ochs RL (1992) Exp Cell Res 202:211–223
18. Raska I, Ochs RL, Andrade LEC, Chan EKL, Burlingame R, Peebles C, Gruol D & Tan EM (1990) J Struct Biol 104:120–127
19. Leverette R, Andrews MT & Maxwell ES (1992) Cell 71:1215–1221
20. Kiss T & Filipowicz W (1993) EMBO J 12:2913–2920
21. Fragapane P, Prislei S, Michienzi A, Caffarelli E & Bozzoni I (1993) EMBO J 12:2921–2928
22. Tycowski KT, Shu MD & Steitz JA (1993) Genes & Dev 7:1176–1190
23. Tuma RS, Stolk JA & Roth MB (1993) J Cell Biol 122:767–773
24. Ferreira J, Carmo-Fonseca M & Lamond AI (1994) J Cell Biol (in press)